Pressure

1 pascal $= 1 \ \mathrm{N/m^2} = 1 \ \mathrm{kg/(m \cdot s^2)} = 1.4504 \times 10^{-4}$ lbf/in^2

1 lbf/in^2 = 6894.76 Pa = 0.0680 atm = 2.036 in. Hg

1 atm = 14.696 lbf/in.^{2} = 1.01325×10^5 Pa = 101.325 kPa = 760 mm Hg

1 bar = 10^5 Pa = 0.987 atm = 14.504 lbf/in^2

1 dyne/cm^2 = 0.1 Pa = 10^{-6} bar = 145.04×10^{-7} lbf/in^2

1 in. Hg = 3386.4 Pa = 0.491 lbf/in^2

1 in. H_2O = 249.1 Pa = 0.0361 lbf/in^2

Energy

1 joule = 1 N·m = 1 kg·m^2/s^2 = 9.479×10^{-4} Btu = 0.7376 ft·lbf
1 kJ = 1000 joule = 0.9479 Btu
1 Btu= 1055 joule =1.055 kJ = 778.169 ft·lbf
1 ft·lbf = 1.3558 joule
1 kW·h = 3600 kJ = 1.341 hp·h = 3412 Btu

Specific Energy

1 kJ/kg = 1000 m^2/s^2 = 0.4299 Btu/lbm

1 Btu/lbm = 2.326 kJ/kg = 2326 m^2/s^2

Power

1 watt = 1 joule/s = 1 kg·m^2/s^3 =3.412 Btu/h = 1.3405×10^{-3} hp
1 kW = 1000 watt = 3412 Btu/h = 737.6 ft·lbf/s = 1.3405 hp
1 Btu/h = 0.293 watt = 0.216 ft·lbf/s = 3.93×10^{-4} hp
1 hp = 550 ft·lbf/s = 33,000 ft·lbf/min = 2545 Btu/h = 746 watt = 0.746 kJ/s
1 ton of refrigeration or air conditioning = 12,000 Btu/h = 200 Btu/min = 4.716 hp

Specific Heat

1 Btu/(lbm·°F) = 1 Btu/(lbm·R)

1 kJ/(kg·K) = 0.2388 Btu/(lbm·R) = 185.8 ft·lbf/(lbm·R)

1 Btu/(lbm·R) = 778.16 ft·lbf/(lbm·R) = 4.186 kJ/(kg·K)

Temprerature

$T(°\mathrm{F}) = \frac{9}{5}T(°\mathrm{C}) + 32 = T(\mathrm{R}) - 459.67$

$T\ (°\mathrm{C}) = \frac{5}{9}[T\ (°\mathrm{F}) - 32] = T\ (\mathrm{K}) - 273.15$

$T\ (\mathrm{R}) = \frac{9}{5}T\ (\mathrm{K}) = (1.8)T\ (\mathrm{K}) = T\ (°\mathrm{F}) + 459.67$

$T(\mathrm{K}) = \frac{5}{9}T(\mathrm{R}) = T(\mathrm{R})/1.8 = T(°\mathrm{C}) + 273.15$

Electrical Units

1 watt = J/s = 1 V·A = 621 lumens at 5500 Angstroms
1 kilowatt hour (kW·h) = 3412 Btu = 1.341 horsepower hour (hp·h)
1 volt = 1 joule/coulomb = 1 watt/amp = 1 ohm·amp
1 amp = 1 coulomb/second

Modern Engineering Thermodynamics

Modern Engineering Thermodynamics

Robert T. Balmer

ELSEVIER

AMSTERDAM • BOSTON • HEIDELBERG • LONDON
NEW YORK • OXFORD • PARIS • SAN DIEGO
SAN FRANCISCO • SINGAPORE • SYDNEY • TOKYO
Academic Press is an imprint of Elsevier

Academic Press is an imprint of Elsevier
30 Corporate Drive, Suite 400, Burlington, MA 01803, USA
The Boulevard, Langford Lane, Kidlington, Oxford, OX5 1GB, UK

Library of Congress Cataloging-in-Publication Data
Balmer, Robert T.
Modern engineering thermodynamics / Robert T. Balmer
p. cm.
ISBN 978-0-12-374996-3
1. Thermodynamics. I. Title.
TJ265.B196 2010
621.402'1–dc22 2010034092

British Library Cataloguing-in-Publication Data
A catalogue record for this book is available from the British Library.

For information on all Academic Press publications, visit our website: *www.elsevierdirect.com*

Typeset by: diacriTech, India

Printed in the United States of America
10 11 12 13 6 5 4 3 2 1

Dedication

WHAT IS AN ENGINEER AND WHAT DO ENGINEERS DO?

The answer is in the word itself. An *er* word ending means "the practice of." For example, a farm*er* farms, a bak*er* bakes, a sing*er* sings, a driv*er* drives, and so forth. But what does an engine*er* do? Do they *engine*? Yes they do! The word *engine* comes from the Latin *ingenerare,* meaning "to create."

About 2000 years ago, the Latin word *ingenium* was used to describe the design of a new machine. Soon after, the word *ingen* was being used to describe all machines. In English, "ingen" was spelled "engine" and people who designed creative things were known as "engine-ers". In French, German, and Spanish today, the word for engineer is *ingenieur*.

So What Is an Engineer?

An engineer is a *creative* and *ingenious* person.

What Does an Engineer Do?

Engineers create ingenious solutions to society's problems.

This Book Is Dedicated to All the Future Engineers of the World.

Contents

Preface

TEXT OBJECTIVES

This textbook has two main objectives. The first is to provide students with a clear presentation of the fundamental principles of basic and applied engineering thermodynamics. The second is to help students develop skills as engineering problem solvers by nurturing the development of their confidence with basic engineering principles through the use of numerous solved example problems. Problem-solving skills are not necessarily learned simply by routinely solving more and more problems. The understanding of proven problem-solving strategies and techniques greatly accelerates the development of problem-solving skills. Throughout the text, learning assessment exercises are included that have proven to be effective in helping students to understand and develop confidence in their ability to solve engineering thermodynamics problems.

To meet these objectives, explanations are occasionally more detailed than those found in other texts, because common learning difficulties encountered by students have been anticipated. If students can understand the text by simply reading it, then the instructor has more flexibility in selecting lecture material. For example, an instructor might choose to develop a few salient points from the reading and then work a few interesting example problems, rather than present a complete derivation of all the assigned reading material.

CULTURAL INFRASTRUCTURE

What engineers do has an enormous impact on society and the world. Understanding how the great challenges of engineering were met in the past can help students understand the importance of the theory and practice of modern engineering principles. This text presents the historical background, the current uses, and the future importance of the thermodynamic topics treated. By understanding where ideas come from, how they were developed, and what external forces shaped the resulting technology, students will better understand their role as engineers of the future.

Engineering is an exciting and rewarding career. However, students occasionally become disenchanted with their engineering course work because they are unable to see the connection between what they are studying and what an engineer really does. To combat this problem, the thermodynamic concepts in this text are presented in a straightforward logical manner, and then applied to real-world engineering situations that are both timely and interesting.

TEXT COVERAGE

This text was designed for use in a standard two-semester engineering thermodynamics course sequence. The first part of the text (Chapters 1–10) contains material suitable for a Basic Thermodynamics course that can be taken by engineers from all majors. The second part of the text was designed for an Applied Thermodynamics course in a mechanical engineering program. Chapters 17, 18, and 19 present several unique topics (biothermodynamics, statistical thermodynamics, and coupled phenomena) for those wishing to glimpse the future of the subject.

TEXT FEATURES

1. **Style**. To make the subject as understandable as possible, the writing is somewhat conversational and the importance of the subject is evidenced in the enthusiasm of the presentation. The composition of the engineering student body has been changing in recent years, and it is no longer assumed that the students are all men and that they inherently understand how technologies (e.g., engines) operate. Consequently, the operation of basic technologies is explained in the text along with the relevant thermodynamic material.
2. **Significant figures**. One of the unique features of this text is the treatment of significant figures. Professors often lament about the number of figures provided by students on their homework and examinations. The rules for determining the correct number of significant figures are introduced in Chapter 1 and are followed consistently throughout the text. An example from Chapter 1 follows.

EXAMPLE 1.6

The inside diameter of a circular water pipe is measured with a ruler to two significant figures and is found to be 2.5 inches. Determine the cross-sectional area of the pipe to the correct number of significant figures.

Solution

The cross-sectional area of a circle is $A = \pi D^2/4$, so $A_{pipe} = \pi(2.5 \text{ inches})^2/4 = 4.9087 \text{ in}^2$, which must be rounded to 4.9 in^2, since the least accurate value in this calculation is the pipe diameter (2.5 inches), which has only two significant figures.

3. **Chapter overviews**. Each chapter begins with an overview of the material contained in the chapter.
4. **Problem-solving strategy**. A proven technique for solving thermodynamic problems is discussed early in the text and followed throughout in the solved examples. The technique follows these steps:

SUMMARY OF THE THERMODYNAMIC PROBLEM-SOLVING TECHNIQUE

Begin by carefully reading the problem statement completely through.

Step 1. Make a sketch of the system and put a dashed line around the system boundary.
Step 2. Identify the unknown(s) and write them on your system sketch.
Step 3. Identify the type of system (closed or open) you have.
Step 4. Identify the process that connects the states or stations.
Step 5. Write down the basic thermodynamic equations and any useful auxiliary equations.
Step 6. Algebraically solve for the unknown(s).
Step 7. Calculate the value(s) of the unknown(s).
Step 8. Check all algebra, calculations, and units.

Sketch → Unknowns → System → Process → Equations → Solve → Calculate → Check

5. **Solved example problems**. Over 200 solved example problems are provided in the text. These examples were carefully designed to illustrate the preceding text material. A sample from Chapter 5 follows.

EXAMPLE P.1

Read the problem statement. An incandescent lightbulb is a simple electrical device. Using the energy rate balance on a lightbulb, determine the heat transfer rate of a 100. W incandescent lightbulb.

Solution

Step 1. **Identify and sketch the system** (see Figure P.1 on the following page).
Step 2. **Identify the unknowns.** The unknown is $\dot{Q}$.
Step 3. **Identify the type of system.** It is a closed system.
Step 4. **Identify the process connecting the system states.** The bulb does not change its thermodynamic state, so its properties remain constant. The process path (after the bulb has warmed to its operating temperature) is U = constant.

Step 5. **Write down the basic equations.** The only basic equation thus far available for a closed system rate process is Eq. (4.21), the general closed system energy rate balance equation:

$$\dot{Q} - \dot{W} = \frac{d}{dt}(mu) + \frac{d}{dt}\left(\frac{mV^2}{2g_c}\right) + \frac{d}{dt}\left(\frac{mgZ}{g_c}\right) = \dot{U} + \dot{KE} + \dot{PE}$$

Assume $\dot{KE} = \dot{PE} = 0$, and since $U = \text{constant}$, $\dot{U} = 0$. This reduces the governing energy rate balance equation for this problem to $\dot{Q} - \dot{W} = 0$.
Write any relevant auxiliary equations. The only relevant auxiliary equation needed here is that the lightbulb has an electrical work *input* of 100. W, so that $\dot{W} = -100.$ W.
Step 6. **Algebraically solve for the unknown(s):** $\dot{Q} = \dot{W}$.
Step 7. **Calculate the value(s) of the unknowns:** $\dot{Q} = \dot{W} = -100.$ W (the minus sign tells us that the heat is leaving the system).
Step 8. **A check of the algebra, calculations, and units** shows that they are correct.

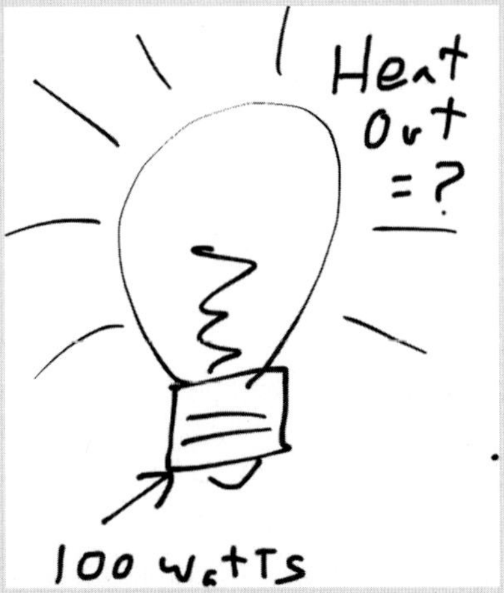

FIGURE P.1
Example P.1.

6. **Example problem exercises with answers.** Immediately following each solved example, several exercises are provided that are variations on the theme of the solved example. The answers to the exercises are also provided so that the student can build confidence in problem solving. For example, the exercises for the preceding example problem might look something like this:
   a. What would be the heat transfer rate if the lightbulb in the previous example is replaced by a 20.0 W fluorescent lightbulb? **Answer**: $\dot{Q} = \dot{W} = -20.0$ W.
   b. How would the lightbulb in the previous example behave if it were put into a small, sealed, rigid, insulated box? **Answer**: Since the box is insulated, the heat transfer rate would be zero.
   c. How would the internal energy of the incandescent lightbulb change if it were put into a small, sealed, rigid, insulated box? **Answer**: Then, since $\dot{Q} = 0$, $\dot{U} = -\dot{W} = 100.$ W, and the internal energy increases until the bulb overheats and fails.
7. **Unit systems.** Engineers today need to understand two types of units systems: classical Engineering English units and modern metric SI units. Both are used in this text, with SI units used in many of the example and homework problems.
8. **Critical Thinking boxes.** At various points in the chapters, special "Critical Thinking" boxes are introduced to challenge the students' understanding of the material. The example that follows is from Chapter 3.

CRITICAL THINKING

If we chose the *color* of a system as a thermodynamic property, would it be an extensive or intensive property?

9. **Question-and-answer boxes.** Students' questions are anticipated at various points throughout the text and are answered in a simple, direct manner. This example is from Chapter 4.

WHAT ARE *HEAT* AND *WORK*?

Heat is usually defined as energy transport to or from a system due to a temperature difference between the system and its surroundings. This can occur by only three modes: conduction, convection, and radiation.

Work is more difficult to define. It is often defined as a force moving through a distance, but this is only one type of work and there are many other work modes as well. Since the only energy transport modes for moving energy across a system's boundary are heat, mass flow, and work, the simplest definition of work is that it is as any energy transport mode that is neither heat nor mass flow.[1]

[1] *Work can also be defined using the concept of a "generalized" force moving through a "generalized" displacement.*

10. **Case studies in applied thermodynamics**. Scattered throughout the text are numerous case studies describing actual engineering applications of specific thermodynamic concepts. Typical case studies include the following topics:

    Supercritical wastewater treatment; The "drinking bird"; Heat pipes; Vortex tubes; A hypervelocity gun; GE 90 aircraft engine; Stirling engines; Stanley steamer automobile; Forensic analysis.

11. **Historical vignettes**. The text also contains numerous short stories describing human side of the development of important thermodynamic concepts and technologies. The following example is from Chapter 14.

IS IT DANGEROUS TO STUFF A CHICKEN WITH SNOW?

The great British philosopher and statesman Sir Francis Bacon (1561–1626) was keenly interested in the possibility of using snow to preserve meat. In March 1626, he stopped in the country on a trip to London and purchased a chicken. He had the chicken killed and cleaned on the spot, then he packed it with snow and took it with him to London. Unfortunately, the experiment caused his death only a few weeks later. The 65-year-old statesman apparently caught a chill while stuffing the chicken with snow and came down with terminal bronchitis. Refrigeration was clearly not something to be taken lightly.

12. **Chapter summaries**. Each chapter ends with a summary (including relevant equations) that reviews the important concepts covered in the chapter.
13. **End-of-chapter problems:**
    - **Homework problems**. At the end of each chapter, an extensive set of problems is provided that is suitable for homework assignments or solved classroom examples. The homework problems include traditional ones that have only one unique answer, as well as modern computer problems, design problems, and writing to learn problems that allow students more latitude.
    - The **computer problems** allow students to use spreadsheets and equation solvers in modern programming languages to address more complex problems requiring a range of solutions.
    - The **design problems** provide an opportunity for students to carry out a preliminary design requiring the use of the material presented in the chapter.
    - **Writing to learn** problems have a dual function. They allow students to enhance their understanding of the subject by expressing themselves verbally in short, written essays about topics presented in the chapter, and they also develop students' writing and communication skills.
    - **Create and solve** problems are designed to help students learn how to formulate solvable thermodynamics problems from engineering data. Engineering education tends to focus only on the process of solving problems. It ignores teaching the process of formulating solvable problems. However, working engineers are never given a well-phrased problem statement to solve. Instead, they need to react to situational information and organize it into a structure that can then be solved using the methods learned in college.
14. **Appendices**. There are two appendices in this text. *Appendix A* provides a list of unit conversions. Since thermodynamics is laced with a variety of technical terms, some having Greek or Latin origin, *Appendix B* provides a brief introduction to the etymology of these terms, in the belief that understanding the meaning of the words themselves enhances the learning of the subject matter.
15. *"Thermodynamic Tables to accompany Modern Engineering Thermodynamics"* is included with new copies of this text. This booklet contains *Appendices C* and *D*, tables, and charts essential for solving the text's thermodynamics problems.

The United States uses more than 10^{17} (100 quadrillion) Btu of energy every year. But the really surprising fact is that 45% of this energy ends up as waste heat dumped into the lakes, rivers, and atmosphere. Our energy conversion technologies today are inefficient because we still rely on the burning of fossil fuels. As the 21st century progresses and more and more countries strive to improve their standard of living, we will need to do a better job of providing nonthermal energy sources. We can and will develop new energy-conversion technologies through a detailed understanding and use of the principles of thermodynamics.

Acknowledgments

I wish to acknowledge help, suggestions, and advice from the following University of Wisconsin–Milwaukee student reviewers: Thomas Jobs, Christopher Zainer, Janice Fitzgerald, Karen Ali, David Hlavac, Paul Bartelt, Margaret Mikolajczak, Lisa Lee Winders, Andrew Hensch, Steven Wietrzny, and Brian Polly.

I also wish to acknowledge the assistance of Professors John Reisel and Kevin Renken at UW–Milwaukee, and Professor John L. Krohn at Arkansas Tech University, as well as the following reviewers for their valuable comments and suggestions:

Jorge L. Alvarado
Texas A&M University

Steven J. Brown
The Catholic University of America

Lorenzo Cremaschi
Oklahoma State University

Gregory W. Davis
Kettering University

Shoeleh Di Julio
California State University–Northridge

Haifa El-Sadi
Concordia University

Sebastien Feve
Iowa State University

Steven H. Frankel
Purdue University

Sathya Gangadharan
Embry-Riddle Aeronautical University

Rama S.R. Gorla
Cleveland State University

Pei-feng Hsu
Florida Institute of Technology

Matthew R. Jones
Brigham Young University

Y. Sungtaek Ju
University of California Los Angeles

Tarik Kaya
Carleton University

Michael Keidar
George Washington University

Joseph F. Kmec
Purdue University

Charles W. Knisely
Bucknell University

Kevin H. Macfarlan
John Brown University

Nathan McNeill
Purdue University

Daniel B. Olsen
Colorado State University

Patrick A. Tebbe
Minnesota State University, Mankato

Finally, I acknowledge the love and support of my wife, Mary Anne, for allowing me to spend endless hours in the dark, cold, spider-infested basement penning this tome.

The graphic illustrations in this book were produced by Ted Balmer at March Twenty Productions (http://www.marchtwenty.com)

Resources That Accompany This Book

A companion website containing interactive activities designed to test students' knowledge of thermodynamic concepts can be found at: http://booksite.academicpress.com/balmer.

For instructors, a solutions manual and PowerPoint slides are available by registering at: www.textbooks.elsevier.com.

Thermodynamic Tables to accompany Modern Engineering Thermodynamics

A separate booklet containing thermodynamic tables and charts useful for solving thermodynamics problems is included with new copies of this text. The booklet (ISBN: 9780123850386) can also be purchased separately through www.elsevierdirect.com or through any bookstore or online retailer.

Elsevier Online Testing

Elsevier Online Testing is a testing and assessment feature that is also available for use with this book. It allows instructors to create online tests and assignments that automatically assess students' responses and performance, providing them with immediate feedback. Elsevier's online testing includes a selection of algorithmic questions, giving instructors the ability to create virtually unlimited variations of the same problem. Contact your local sales representative or email textbooks@elsevier.com for additional information, or visit http://booksite.academicpress.com/balmer.

List of Symbols

A	Availability and area
a	Specific availability, A/m
B	Magnetic induction
COP	Coefficient of performance
CR	Compression ratio
c	Specific heat of a liquid or solid
c_p	Constant pressure specific heat
c_v	Constant volume specific heat
E	Energy and electric field strength
e	Specific energy, E/m
F	Force
G	Gibbs function
g	Acceleration of gravity
g_c	$32.174\ \mathrm{lbm \cdot ft/l(lbf \cdot s^2)}$
H	Total enthalpy and magnetic field strength
h	Specific enthalpy, H/m, and convective heat transfer coefficient
I	Irreversibility
$\dot{I}$	Irreversibility rate
i	Specific irreversibility, I/m, and electric current
KE	Kinetic energy
ke	Specific kinetic energy, KE/m
K_e	Chemical equilibrium constant
k_t	Thermal conductivity
k	specific heat ratio, c_p/c_v
L	Length
L_{ik}	Phenomenological coefficients
L_{qr}	Coupling coefficient
L_{sr}	Coupling coefficient
m	Mass
$\dot{m}$	Mass flow rate
M	Molecular mass
n	Number of moles
P	Polarization
PE	Potential energy
pe	Specific potential energy, PE/m
PR	Pressure ratio
p	Pressure
p_{sat}	Saturation pressure
p_i	Partial pressure of species i
p_r	Reduced pressure
q	Heat transfer per unit area, Q/A
Q	Heat transfer
$\dot{Q}$	Heat transfer rate
R	Individual gas constant and electrical resistance
$\mathfrak{R}$	Universal gas constant
S	Total entropy
s	Specific entropy, S/m
S_p	Entropy production
$\dot{S}_P$	Entropy production rate
t	Time
T	Temperature and torque
T_{sat}	Saturation temperature
T_r	Reduced temperature
U	Total internal energy
u	Specific internal energy, U/m
V	Velocity
∀	Volume
v	Specific volume, ∀/m
w	Mass fraction
W	Work
$\dot{W}$	Work rate (power)
x	Quality
X	Generalized thermodynamic force
y	Degree of chemical dissociation
Z	Height and compressibility factor

GREEK LETTERS

β	Isobaric coefficient of volume expansion
ε	Emissivity and second law efficiency
ε_0	Electric permittivity of vacuum
μ_0	Magnetic permeability
ϕ	Voltage
Γ	Explosive energy per unit volume
η	Efficiency
η_m	Mechanical efficiency
η_s	Isentropic efficiency
η_T	Thermal efficiency
κ	Isothermal coefficient of compressibility
θ	Angular displacement
μ	Viscosity
μ_i	Gibbs chemical potential of species i
μ_J	Joule-Thomson coefficient
ν	Kinematic viscosity
ν_i	Stoichiometric coefficient
ρ	Density
σ	Stefan-Boltzmann constant and entropy production rate density
χ_e	Electric susceptibility

Prologue

PARIS FRANCE, 10:35 AM, AUGUST 24, 1832

The nurse closed the door quietly behind her as she left his hospital room. She knew her patient was very sick, because for the past two days, he had been irritable and lethargic and now he was complaining of a fever and muscle cramps. His eyes looked sunken and he was constantly thirsty; yesterday, he vomited for hours. Sadi Carnot was only 36 years old, but that day he would die of cholera.

Sadi Carnot was born June 1, 1796, in the Luxembourg Palace in Paris. His father, Lazare Carnot, was one of the most powerful men in France and would eventually become Napoleon Bonaparte's war minister. He named his son Sadi simply because he greatly admired a medieval Persian poet called Sa'di of Shiraz.

At the age of 18, Sadi graduated from the École Polytechnique military academy and went on to a military engineering school. Sadi's friends saw him as reserved, but he became lively and excited when their discussions turned to science and technology.

After the defeat of Napoleon at Waterloo in October 1815, Sadi's father was exiled to Germany and Sadi's military career stagnated. Unhappy at his lack of promotion and his superiors' refusal to give him work that allowed him to use his engineering training, he took a half-time leave to attend courses at various institutions in Paris. He was fascinated by technology and began to study the theory of gases.

After the war with Britain, France began importing advanced British steam engines, and Sadi realized just how far French designs had fallen behind. He became preoccupied with the operation of steam engines; in 1824, he published his studies in a small book entitled *Reflections on the Motive Power of Fire*. At the time, his book was largely ignored, but today it represents the beginning of the field we call *thermodynamics*.

Because Sadi Carnot died of infectious cholera, all his clothes and writings were buried with him. Who knows what thermodynamic secrets still lie hidden in his grave?

CHAPTER 1

The Beginning

CONTENTS

1.1 WHAT IS THERMODYNAMICS?

Thermodynamics is the study of *energy* and the ways in which it can be used to improve the lives of people around the world. The efficient use of natural and renewable energy sources is one of the most important technical, political, and environmental issues of the 21st century.

In mechanics courses, we study the concept of force and how it can be made to do useful things. In thermodynamics, we carry out a parallel study of energy and all its technological implications. The objects studied in mechanics are called *bodies*, and we analyze them through the use of *free body* diagrams. The objects studied in thermodynamics are called *systems*, and the free body diagrams of mechanics are replaced by *system* diagrams in thermodynamics.

THERMO—WHAT?

The word *thermodynamics* comes from the Greek words θερμη (*therme*, meaning "heat") and δυναμις (*dynamis*, meaning "power"). Thermodynamics is the study of the various processes that change energy from one form into another (such as converting heat into work) and uses variables such as temperature, volume, and pressure.

Energy is one of the most useful concepts ever developed.[1] Energy can be possessed by an object or a system, such as a coiled spring or a chemical fuel, and it may be transmitted through empty space as electromagnetic radiation. The energy contained in a system is often only partially available for use. This, called the *available energy* of the system, is treated in detail later in this book.

One of the basic laws of thermodynamics is that *energy is conserved*. This law is so important that it is called the *first law* of thermodynamics. It states that energy can be changed from one form to another, but it cannot be created or destroyed (that is, energy is "conserved"). Some of the more common forms of energy are: gravitational, kinetic, thermal, elastic, chemical, electrical, magnetic, and nuclear. Our ability to efficiently convert energy from one form into a more useful form has provided much of the technology we have today.

1.2 WHY IS THERMODYNAMICS IMPORTANT TODAY?

The people of the world consume 1.06 cubic miles of oil each year as an energy source for a wide variety of uses such as the engines shown in Figures 1.1 and 1.2.[2] Coal, gas, and nuclear energy provide additional energy, equivalent to another 1.57 mi^3 of oil, making our total use of *exhaustible energy* sources equal to 2.63 mi^3 of oil every year. We also use *renewable energy* from solar, biomass, wind (see Figure 1.3), and hydroelectric, in amounts that are equivalent to an additional 0.37 mi^3 of oil each year. This amounts to a total worldwide

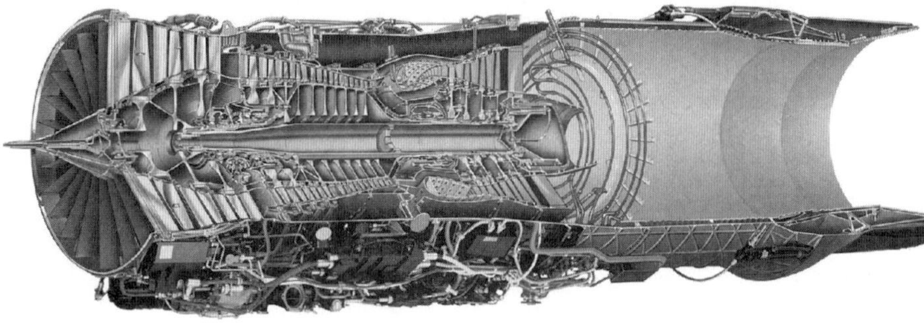

FIGURE 1.1
A cutaway of the Pratt & Whitney F-100 gas turbine engine.

FIGURE 1.2
Corvette engine.

---

[1] The word *energy* is the modern form of the ancient Greek term *energeia*, which literally means "in work" (*en* = in and *ergon* = work).
[2] One cubic mile of oil is equal to 1.1 trillion gallons and contains 160 quadrillion (160 × 10^{15}) kilojoules of energy.

FIGURE 1.3
Sustainable wind energy technology.

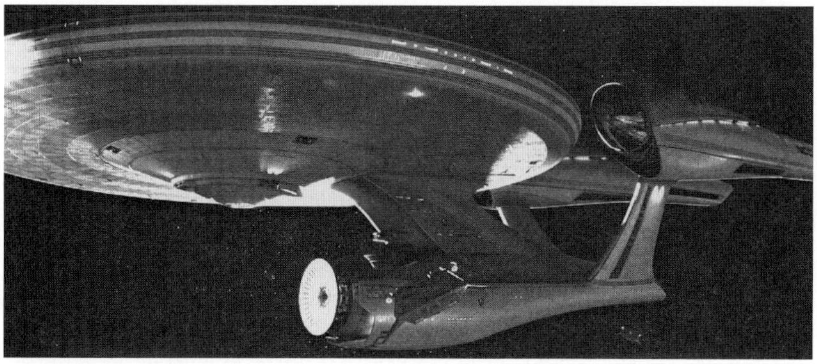

FIGURE 1.4
The *Starship Enterprise* in *Star Trek*. *(Photo credit: Industrial Light & Magic, Copyright © 2008 by PARAMOUNT PICTURES. All Rights Reserved.)*

energy use equivalent to 3.00 mi^3 of oil each year. If the world energy demand continues at its present rate to create the technologies of the future (e.g., the Starships of Figure 1.4), we will need an energy supply equivalent to consuming an astounding 270 mi^3 of oil by 2050 (90 times more that we currently use). Where is all that energy going to come from? How are we going to use energy more efficiently so that we do not need to use so much? We address these and other questions in the study of thermodynamics.

The study of energy is of fundamental importance to all fields of engineering. Energy, like momentum, is a unique subject and has a direct impact on virtually all technologies. In fact, things simply do not "work" without a flow of energy through them. In this text, we show how the subject touches all engineering fields through worked example problems and relevant homework problems at the end of the chapters.

HOW IS THERMODYNAMICS USED IN ENGINEERING?

- *Mechanical engineers* study the flow of energy in systems such as automotive engines (Figure 1.2), turbines, heat exchangers, bearings, gearboxes, air conditioners, refrigerators, nozzles, and diffusers.
- *Electrical engineers* deal with electronic cooling problems, increasing the energy efficiency of large-scale electrical power generation, and the development of new electrical energy conversion technologies such as fuel cells.
- *Civil engineers* deal with energy utilization in construction methods, solid waste disposal, geothermal power generation, transportation systems, and environmental impact analysis.
- *Materials engineers* develop new energy-efficient metallurgical compounds, create high-temperature materials for engines, and utilize the unique properties of nanotechnology.
- *Industrial engineers* minimize energy consumption and waste in manufacturing processes, develop new energy management methods, and improve safety conditions in the workplace.
- *Aerospace engineers* develop energy management systems for air and space vehicles, space stations, and planetary habitation (Figure 1.4).
- *Biomedical engineers* develop better energy conversion systems for the health care industry, design new diagnostic and treatment tools, and study the energy flows in living systems.

All engineering fields utilize the conversion and use of energy to improve the human condition.

1.3 GETTING ANSWERS: A BASIC PROBLEM SOLVING TECHNIQUE

Unlike mechanics, which deals with a relatively small range of applications, thermodynamics is truly global and can be applied to virtually any subject, technology, or object conceivable. You no longer can thumb through a book looking for the right equation to apply to your problem. You need a method or technique that guides you through the process of solving a problem in a prescribed way.

In Chapter 4, we provide a more detailed technique for thermodynamics problem solving, but for the present, here are seven basic problem solving steps you should know and understand.

1. **Read**. Always begin by carefully reading the problem statement and try to visualize the "thing" about which the problem is written (a car, engine, rocket, etc.). The "thing" about which the problem is written is called the *system* in thermodynamics. This may seem simple, but it is key to understanding exactly what you are analyzing.
2. **Sketch**. Now draw a simple sketch of the system you visualized and add as much of the numerical information given in the problem statement as possible to the sketch. *If you do not know what the "thing" in the problem statement looks like, just draw a blob and call it the system.* You will not be able to remember all the numbers given in the problem statement, so write them in an appropriate spot on your sketch, so that they are easy to find when you need them.
3. **Need**. Write down exactly what you need to determine—what does the problem ask you to find?
4. **Know**. Make a list of the names, numerical values, and units of everything else given in the problem statement. For example, Initial velocity = 35 meters per second, mass = 5.5 kilograms.
5. **How**. Because of the nature of thermodynamics, there are more equations than you are accustomed to working with. To be able to sort them all out, you need to get in the habit of listing the relevant equations and assumptions that you "might" be able to use to solve for the unknowns in the problem. Write down *all* of them.

A BASIC PROBLEM SOLVING TECHNIQUE

1. Carefully **read** the problem statement and visualize what you are analyzing.
2. Draw a **sketch** of the object you visualized in step 1.
3. Now write down what you **need** to find, that is, make a list of the unknown(s).
4. List everything else you **know** about the problem (i.e., all the remaining information given in the problem statement).
5. Make a list of relevant equations to see **how** to solve the problem.
6. **Solve** these equations algebraically for the unknown(s).
7. **Calculate** the value(s) of the unknown(s), and check the **units** in each calculation.

Read → Sketch → Need → Know → How → Solve → Calculate

6. **Solve**. Next, you need to algebraically solve the equations listed in step 5 for the unknowns. Because the number of variables in this subject can be large, the unknowns you need to determine may be inside one of your equations, and you need to solve for it algebraically.
7. **Calculate**. Finally, after you have successfully completed the first six steps, you compute the values of the unknowns, being careful to *check the units* in all your calculations for consistency.

This technique requires discipline and patience on your part. However, if you follow these basic steps, you will be able solve the thermodynamics problems in the first three chapters of this textbook. The following example illustrates this problem solving technique.

EXAMPLE 1.1

A new racecar with a JX-750 free-piston engine is traveling on a straight level test track at a velocity of 85.0 miles per hour. The driver accelerates at a constant rate for 5.00 seconds, at which point the car's velocity has increased to 120. miph. Determine the acceleration of the car as it went from 85.0 to 120. mph.[3]

Solution

1. *Read* the problem statement carefully. Sometimes you may be given miscellaneous information that is not needed in the solution. For example, we do not need to know what kind of engine is used in the car, but we do need to know that the car has a constant acceleration for the 5.00 s.
2. Draw a *sketch* of the problem, like the one in Figure 1.5. Transfer all the numerical information given in the problem statement onto your sketch so you need not search for it later.

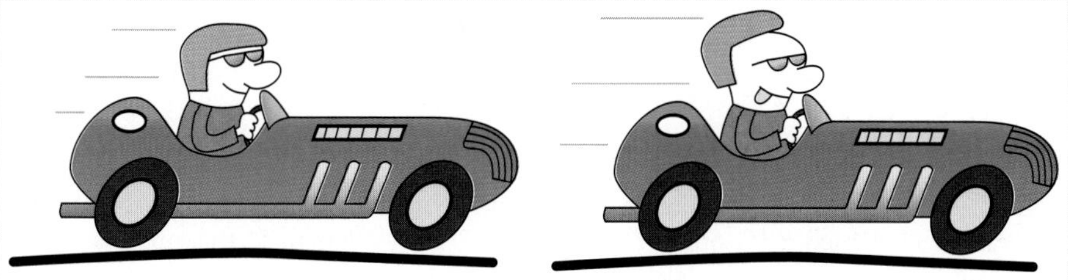

FIGURE 1.5
Example 1.1, solution step 2.

3. What are we supposed to find? We *need* the acceleration of the car.
4. We *know* the following things: The initial velocity = 85.0 mph, the final velocity = 120. mph, and the car accelerates for $t = 5.00$ s.
5. *How* are we going to find the car's acceleration? In this case, the basic physics equation that defines acceleration is $a = dx^2/dt^2 = dV/dt$, and if the acceleration a is constant, then we can integrate this equation to get $V_{final} = V_{initial} + at$. Note that the acceleration must be constant to use this equation. Aha, that is why the acceleration was specified as constant in the problem statement. No additional equations are needed to solve this problem.
6. Now we can *solve* for the unknown acceleration, a:

$$a = \frac{V_{final} - V_{initial}}{t}$$

7. Now all we have to do is to insert the given numerical values and *calculate* the solution:

$$a = \frac{\left(120 - 85\,\dfrac{\text{miles}}{\text{hour}}\right)}{5\text{ seconds}} = 7\,\frac{\text{miles}}{\text{hour/seconds}}$$

Now check the *units*. Miles per hour times seconds makes no sense. Let us convert the car's velocity from miles per hour to feet per second before we calculate the acceleration[4]:

$$V_{initial} = \left(85\,\frac{\text{miles}}{\text{hour}}\right)\left(\frac{5280\text{ feet/mile}}{3600\text{ seconds/hour}}\right) = 125\,\frac{\text{feet}}{\text{second}}$$

and

$$V_{final} = \left(120.\,\frac{\text{miles}}{\text{hour}}\right)\left(\frac{5280\text{ feet/mile}}{3600\text{ seconds/hour}}\right) = 176\,\frac{\text{feet}}{\text{second}}$$

(*Continued*)

EXAMPLE 1.1 *(Continued)*

Then, the acceleration becomes

$$a = \frac{176 - 125 \dfrac{\text{feet}}{\text{second}}}{5\text{ seconds}} = 10.3 \frac{\text{feet}}{\text{second}^2} = 10.3\text{ ft/s}^2$$

Remember, the answer is not correct if the units are not correct.

Following most of the Example problems in this text are a few Exercises, complete with answers, that are based on the Example. These exercises are designed to allow you to build your problem solving skills and develop self-confidence. The exercises are to be solved by following the solution structure of the preceding example problem. Here are typical exercise problems based on Example 1.1.

Exercises

1. Determine the acceleration of the race car in Example 1.1 if its final velocity is 130. mph instead of 120. mph.
   Answer: a = 13.2 feet/second^2.
2. If the racecar in Example 1.1 has a constant acceleration of 10.0 ft/s^2, determine its velocity after 6.00 s.
   Answer: V = 126 mph.
3. A dragster travels a straight level $\frac{1}{4}$ mile drag strip in 6.00 s from a standing start (i.e., $X_{\text{initial}} = V_{\text{initial}} = 0$). Determine the average constant acceleration of the dragster. Hint: The basic physics equation you need here is $X_{\text{final}} = V_{\text{initial}} \times t + (\frac{1}{2})at^2$.
   Answer: a = 73.3 ft/s^2.

[3] *You may be wondering why there are decimal points and extra zeros added to some of these numbers. This is because we are indicating the number of significant figures represented by these values. The subject of significant figures is covered later in this chapter.*
[4] *For future reference, there are "exactly" 5280 feet in one mile and "exactly" 3600 seconds in one hour.*

1.4 UNITS AND DIMENSIONS

In thermodynamics, you determine the energy of a system in its many forms and master the mechanisms by which the energy can be converted from one form to another. A key element in this process is the use of a consistent set of dimensions and units. A calculated engineering quantity always has two parts, the numerical value and the associated units. The result of any analysis must be correct in both categories: *It must have the correct numerical value and it must have the correct units.*

Engineering students should understand the origins of and relationships among the several units systems currently in use within the profession. Earlier measurements were carried out with elementary and often inconsistently defined units. In the material that follows, the development of measurement and units systems is presented in some detail. The most important part of this material is that covering modern units systems.

1.5 HOW DO WE MEASURE THINGS?

Metrology is the study of measurement, the source of reproducible quantification in science and engineering. It deals with the dimensions, units, and numbers necessary to make meaningful measurements and calculations. It does not deal with the technology of measurement, so it is not concerned with how measurements are actually made.

We call each measurable characteristic of a quantity a *dimension* of that quantity. If the quantity exists in the material world, then it automatically has three spatial dimensions (length, width, and height), all of which are called *length* (L) dimensions. If the quantity changes in time, then it also has a temporal dimension called *time* (t). Some dimensions are not unique because they are made up of other dimensions. For example, an area (A) is a measurable characteristic of an object and therefore one of its dimensions. However, the area dimension is the same as the length dimension squared ($A = L^2$). On the other hand, we could say that the length dimension is the same as the square root of the area dimension.

Even though there seems to be a lack of distinguishing characteristics that allow one dimension to be recognized as more fundamental than some other dimension, we easily recognize an apparent utilitarian hierarchy within a set of similar dimensions. We therefore choose to call some dimensions *fundamental* and all other dimensions related to the chosen fundamental dimensions *secondary* or *derived*. *It is important to understand that not all systems of dimensional analysis have the same set of fundamental dimensions.*

Units provide us with a numerical scale whereby we can carry out a measurement of a quantity. They are established quite arbitrarily and are codified by civil law or cultural custom. How the dimension of length ends up being measured in units of feet or meters has nothing to do with any physical law. It is solely dependent on the creativity and ingenuity of people. Therefore, whereas the basic concepts of dimensions are grounded in the fundamental logic of descriptive analysis, the basic ideas behind the units systems are often grounded in the roots of past civilizations and cultures.

ANCIENT UNITS SYSTEMS

Intuition tells us that civilization should have evolved using the decimal system. People have ten fingers and ten toes, so the base 10 (decimal) number system would seem to be the most logical system to be adopted by prehistoric people. However, archaeological evidence has shown that the pre-Egyptian Sumerians used a base 60 (sexagesimal) number system, and ancient Egyptians and early American Indians used a base 5 number system. A base 12 (duodecimal) number system was developed and used extensively during the Roman Empire. Today, mixed remains of these ancient number systems are deeply rooted in our culture.

A fundamental element of a successful mercantile trade is that the basic units of commerce have easily understood subdivisions. Normally, the larger the base number of a particular number system, the more integer divisors it has. For example, 10 has only three divisors (1, 2, and 5), but 12 has five integer divisors (1, 2, 3, 4, and 6) and therefore makes a considerably better fractional base. On the other hand, 60 has an advantage over 100 as a number base because the former it has 11 integer divisors whereas 100 has only 8.

The measurements of length and time were undoubtedly the first to be of concern to prehistoric people. Perhaps the measurement of time came first, because people had to know the relationship of night to day and understand the passing of the seasons of the year. The most striking aspect of our current measure of time is that it is a mixture of three numerical bases; decimal (base 10) for counting days of the year, duodecimal (base 12) for dividing day and night into equal parts (hours), and sexagesimal (base 60) for dividing hours and minutes into equal parts.

Nearly all early scales of length were initially based on the dimensions of parts of the adult human body because people needed to carry their measurement scales with them (see Figure 1.6). Early units were usually related to each other in a binary (base 2) system. For example, some of the early length units were: half-hand = 2 fingers; hand = 2 half-hands; span = 2 hands; forearm (cubit) = 2 spans; fathom = 2 forearms, and so forth. Measurements of area and volume followed using such units as handful = 2 mouthfuls, jack = 2 handfuls, gill = 2 jacks, cup = 2 gills, and so forth.

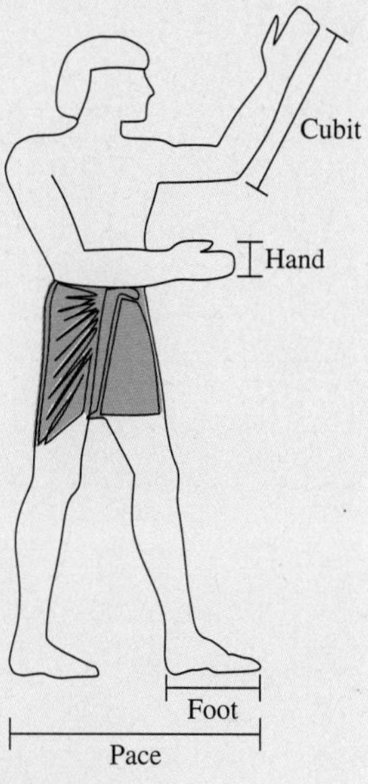

FIGURE 1.6
Egyptian man with measurements.

Weight was probably the third fundamental measure to be established, with the development of such units as the grain (i.e., the weight of a single grain of barley), the stone, and the talent (the maximum weight that could be comfortably carried continuously by an adult man).

CRITICAL THINKING

Where are Roman numerals still commonly used today? How would technology be different if we used Roman numerals for engineering calculations today?

NURSERY RHYMES AND UNITS

Many of the Mother Goose nursery rhymes were not originally written for children but in reality were British political poems or songs. For example, in 17th century England, the treasury of King Charles I (1625–1640) ran low, so he imposed a tax on the ancient unit of volume used for measuring honey and hard liquor, the jack (1 jack = 2 handfuls). The response of the people was to avoid the tax by consuming drink measured in units other than the jack. Eventually, the jack unit became so unpopular with the people that it was no longer used for anything. One of the few existing uses of the jack unit

(Continued)

NURSERY RHYMES AND UNITS *Continued*

today is in the term *jackpot*. Coincidentally, the next larger unit size, the gill (1 gill = 2 jacks), also fell into disuse. The political meaning of the following popular Mother Goose rhyme should now become clear (Figure 1.7):

> **Jack and Gill went up a hill to fetch a pail of water.**
> **Jack fell down and broke his crown and Gill came tumbling after.**

The Jack and Gill in this rhyme are not really a little boy and girl, they are the old units of volume measure. *Jack fell down* refers to the fall of the jack from popular usage as a result of the tax imposed by the *crown*, Charles I. The phrase *and Gill came tumbling after* refers to the subsequent decline in the use to the gill unit of volume measure. The "real" jack and gill of this rhyme are shown in Figure 1.8.

FIGURE 1.7
Jack and Jill.

FIGURE 1.8
The real jack and gill.

CRITICAL THINKING

What other Mother Goose rhymes or children's songs are not what they seem?

1.6 TEMPERATURE UNITS

The development of a temperature unit of measure came late in the history of science. The problem with early temperature scales is that all of them were empirical, and their readings often depended on the material (usually a liquid or a gas) used to indicate the temperature change. In a liquid-in-glass thermometer, the difference between the coefficient of thermal expansion of the liquid and the glass causes the liquid to change height when the temperature changes. If the coefficient of thermal expansion depends in some way on temperature, then an accurate thermometer cannot be made simply by defining two fixed (calibration) points and subdividing the difference between these two points into a uniform number of degrees. Unfortunately, the coefficients of thermal expansion of all liquids depend to some extent on temperature; consequently, the two-fixed-point method of defining a temperature scale is inherently prone to this type of measurement error.

In 1848, William Thomson (1824–1907), later to become Lord Kelvin, developed a thermodynamic absolute temperature scale that was independent of the measuring material. He was further able to show that his thermodynamic absolute temperature scale was identical to the ideal gas absolute temperature scale developed earlier, and therefore an ideal gas thermometer could be calibrated to measure thermodynamic absolute temperatures. Thereafter, the absolute Celsius temperature scale was named the *Kelvin scale* in his honor. Because it was a real thermodynamic absolute temperature scale, it could be constructed from a single fixed calibration point once the degree size had been chosen. The triple point of water (0.01°C or 273.16 K) was selected as the fixed point.

THE DEVELOPMENT OF THERMOMETERS

Thermometry is the technology of temperature measurement. Although people have always been able to experience the physiological sensations of hot and cold, the quantification and accurate measurement of these concepts did not occur until the 17th century. Ancient physicians judged the wellness of their patients by sensing fevers and chills with a touch of the hand (as we often do today). The Roman physician Galen (ca. 129–199) ascribed the fundamental differences in the health or "temperament" of a person to the proportions in which the four "humors" (phlegm, black bile, yellow bile, and blood) were mixed within the body.[5] Thus, both the term for wellness (temperament) and that for body heat (temperature) were derived from the same Latin root *temperamentum*, meaning "a correct mixture of things."

Until the late 17th century, thermometers were graduated with arbitrary scales. However, it soon became clear that some form of temperature standardization was necessary, and by the early 18th century, 30 to 40 temperature scales were in use. These scales were usually based on the use of two fixed calibration points (standard temperatures) with the distance between them divided into arbitrarily chosen equally spaced degrees.

The 100 division (i.e., base 10 or decimal) Celsius temperature scale became very popular during the 18th and 19th centuries and was commonly known as the *centigrade* (from the Latin *centum* for "100" and *gradus* for "step") scale until 1948, when Celsius's name was formally attached to it and the term *centigrade* was officially dropped.

[5] *It was thought that illness occurred when these four humors were not in balance, and that their balance could be restored by draining off one of them (i.e., by "bleeding" the patient).*

Table 1.1 Early Temperature Scales

Inventor and Date	Fixed Points
Isaac Newton (1701)	Freezing water (0°N) and human body temperature (12°N)
Daniel Fahrenheit (1724)[a]	*Old*: Freezing saltwater mixture (0°F) and human body temperature (96°F) *New*: Freezing water (32°F) and boiling water (212°F)
René Reaumur (1730)	Freezing water (0°Re) and boiling water (80°Re)
Anders Celsius (1742)[b]	Freezing water (0°C) and boiling water (100°C)

[a] *The modern Fahrenheit scale uses the freezing point of water (32°F) and the boiling point of water (212°F) as its fixed points. This change to more stable fixed points resulted in changing the average body temperature reading from 96°F on the old Fahrenheit scale to 98.6°F on the new Fahrenheit scale.*

[b] *Initially, Celsius chose the freezing point of water to be 100° and the boiling point of water to be 0°, but this scale was soon inverted to its present form.*

The difference between the boiling and freezing points of water at atmospheric pressure then became 100 K or, alternatively, 100°C, making the Kelvin and Celsius degree size the same.

Soon thereafter, an absolute temperature scale based on the Fahrenheit scale was developed, named after the Scottish engineer William Rankine (1820–1872).

Some early temperature scales with fixed calibration points are shown in Table 1.1. Note that both the Newton and the Fahrenheit scales are duodecimal (i.e., base 12).

EXAMPLE 1.2

Convert 55 degrees on the modern Fahrenheit scale (Figure 1.9) into (a) degrees Newton, (b) degrees Reaumer, and (c) Kelvin.

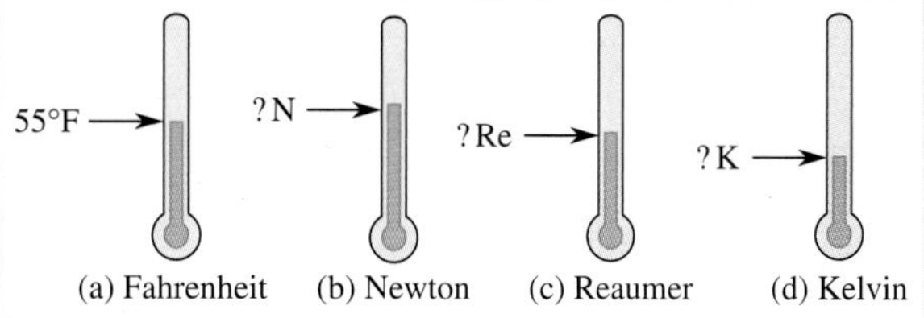

FIGURE 1.9
Example 1.2.

(Continued)

EXAMPLE 1.2 *(Continued)*

Solution

(a) From Table 1.1, we find that both 0°N and 32°F correspond to the freezing point of water, and body heat (temperature) corresponds to 12°N and 98.6°F (on the modern Fahrenheit scale) on these scales. Since both these scales are linear temperature scales, we can construct a simple proportional relation between the two scales as

$$\frac{98.6-55}{98.6-32}=\frac{12-x}{12-0}$$

where x is the temperature on the Newton scale that corresponds to 55°F. Solving for x gives

$$x=12\left(1-\frac{98.6-55}{98.6-32}\right)=4.14°\text{N}$$

(b) Since the Reaumur scale is also a linear scale with 0°Re and 80°Re corresponding to 32°F and 212°F, respectively, we can establish the following proportion for the Reaumur temperature y that corresponds to 55°F:

$$\frac{212-55}{212-32}=\frac{80-y}{80-0}$$

from which we can solve for

$$y=80\left(1-\frac{212-55}{212-32}\right)=10.2°\text{Re}$$

(c) Here we have 273.15 K and 373.15 K corresponding to 32°F and 212°F, respectively. The proportionality between these scales is then

$$\frac{212-55}{212-32}=\frac{373.15-z}{373.15-273.15}$$

from which we can compute the Kelvin temperature z that corresponds to 55°F as

$$z=373.15-(373.15-273.15)\left(\frac{212-55}{212-32}\right)=285.9\text{ K}$$

Notice that we do not use the degree symbol (°) with either the Kelvin or the Rankine absolute temperature scale symbols. The reason for this is by international agreement as explained later in this chapter.

Exercises

4. Convert 20.0°C into Kelvin and Rankine. **Answer**: 293.2 K and 527.7 R.
5. Convert 30°C into degrees Newton and degrees Reaumur. **Answer**: 9.7°N and 24°Re.
6. Convert 500. K into Rankine, degrees Celsius, and degrees Fahrenheit. **Answer**: 900 R, 226.9°C, and 440.3°F.

1.7 CLASSICAL MECHANICAL AND ELECTRICAL UNITS SYSTEMS

The establishment of a stable system of units requires the identification of certain measures that must be taken as absolutely fundamental and indefinable. For example, one cannot define length, time, or mass in terms of more fundamental dimensions. They all seem to be fundamental quantities. Since we have so many quantities that can be taken as fundamental, we have no single unique system of units. Instead, there are many equivalent units systems, built on different fundamental dimensions. However, all the existing units systems today have one thing in common—they have all been developed from the same set of fundamental equations of physics, equations more or less arbitrarily chosen for this task.

It turns out that all the equations of physics are mere proportionalities into which one must always introduce a "constant of proportionality" to obtain an equality. These proportionality constants are intimately related to the system of units used in producing the numerical calculations. Consequently, three basic decisions must be made in establishing a consistent system of units:

1. The choice of the *fundamental quantities* on which the system of units is to be based.
2. The choice of the *fundamental equations* that serve to define the secondary quantities of the system of units.
3. The choice of the magnitude *and* dimensions of the inherent *constants of proportionality* that appear in the fundamental equations.

With this degree of flexibility, it is easy to see why such a large number of measurement units systems have evolved throughout history.

Table 1.2 Five Units Systems in Use Today

System Name	Type	*F*	*M*	*L*	*t*	$g_c = 1/k_1$
MKS (SI)	MLt	newton (N)	kilogram (kg)	meter (m)	second (s)	1 (dimensionless)
CGS	MLt	dyne (d)	gram (g)	centimeter (cm)	second (s)	1 (dimensionless)
Absolute English	MLt	poundal (pd)	pound mass (lbm)	foot (ft)	second (s)	1 (dimensionless)
Technical English	FLt	pound force (lbf)	slug (sg)	foot (ft)	second (s)	1 (dimensionless)
Engineering English	FMLt	pound force (lbf)	pound mass (lbm)	foot (ft)	second (s)	$32.174\ \text{lbm}\cdot\text{ft/lbf}\cdot\text{s}^2$

The classical mechanical units system uses Newton's second law as the fundamental equation. This law is a proportionality defined as

$$F = k_1 ma \tag{1.1}$$

The wide variety of choices available for the fundamental quantities that can be used in this system has produced a large number of units systems. Over a period of time, three systems, based on different sets of fundamental quantities, have become popular:

- MLt system, which considers mass (M), length (L), and time (t) as independent fundamental quantities.
- FLt system, which considers force (F), length (L), and time (t) as independent fundamental quantities.
- FLMt system, which considers all four as independent fundamental quantities.

Table 1.2 shows the various popular mechanical units systems that have evolved along these lines. Also listed are the names arbitrarily given to the various derived units and the value and units of the constant of proportionality, k_1, which appears in Newton's second law, Eq. (1.1).

In Table 1.2, the four units in boldface type have the following definitions:

$$1\ \textit{newton} = 1\ \text{kg}\cdot\text{m/s}^2 \tag{1.2}$$

$$1\ \textit{dyne} = 1\ \text{g}\cdot\text{cm/s}^2 \tag{1.3}$$

$$1\ \textit{poundal} = 1\ \text{lbm}\cdot\text{ft/s}^2 \tag{1.4}$$

$$1\ \textit{slug} = 1\ \text{lbf}\cdot\text{s}^2/\text{ft} \tag{1.5}$$

These definitions are arrived at from Newton's second law using the fact that k_1 has been arbitrarily chosen to be unity and dimensionless in each of these units systems.

Because of the form of k_1 in the Engineering English system, engineering texts have evolved a rather strange and unfortunate convention regarding its use. It is common to let $g_c = 1/k_1$, where g_c in the Engineering English units system is simply

$$\text{Engineering English units: } g_c = \frac{1}{k_1} = 32.174\frac{\text{lbm}\cdot\text{ft}}{\text{lbf}\cdot\text{s}^2} \tag{1.6}$$

and in all the other units systems described in Table 1.2, it is

$$\text{All other units systems: } g_c = \frac{1}{k_1} = 1\ (\text{dimensionless}) \tag{1.7}$$

This symbolism was originally chosen apparently because the *value* (but not the *dimensions*) of g_c happens to be the same as that of standard gravity in the Engineering English units system. However, this symbolism is awkward because it tends to make you think that g_c is the same as local gravity, *which it definitely is not*. Like k_1, g_c is nothing more than a proportionality constant with dimensions of $ML/(Ft^2)$. Because the use of g_c is so widespread today and it is important that you are able to recognize the meaning of g_c when you see it elsewhere, it is used in all the relevant equations in this text. For example, we now write Newton's second law as

$$F = \frac{ma}{g_c} \tag{1.8}$$

Until the mid-20th century, most English speaking countries used the Engineering English units system. But, because of world trade pressures and the worldwide acceptance of the SI system, most engineering thermodynamics texts today (including this one) present example and homework problems in both the old Engineering English and the new SI units systems.

The dimensions of energy are the same as the dimensions of work, which are *force* × *distance*, and the dimensions of power are the same as the dimensions of work divided by time, or *force* × *distance* ÷ *time*. The corresponding units and their secondary names (when they exist) are shown in Table 1.3.

WHICH WEIGHS MORE—A POUND OF FEATHERS OR A POUND OF GOLD?

The avoirdupois (from the French meaning "to have weight") pound contains 7000 barleycorns and is divided into 16 ounces. It was used primarily for weighing ordinary commodities, such as wood, bricks, feathers, and so forth. The troy pound was named after the French city Troyes and was used to weigh only precious metals (gold, silver, etc.), gems, and drugs. The English troy pound contains only 5760 barleycorns and is subdivided into 12 ounces, as was the original Roman pound. The English word *ounce* is also derived from the Latin word *uncia,* meaning "the twelfth part of."

Consequently, the avoirdupois pound is considerably larger (by a factor of 7000/5760 = 1.215) than the troy pound and the coexistence of both pound units produced considerable confusion over the years. So a pound of feathers actually does weigh more than a pound of gold, because the weight of the feathers is measured with the avoirdupois pound, whereas the weight of the gold is measured with the troy pound. Today, all engineering calculations done in an English units system are done with the 16 ounce, 7000 grain, *avoirdupois* pound.

Table 1.3 Units of Energy and Power

System Name	Energy	Power
MKS (SI)	$N \cdot m = kg \cdot m^2/s^2$ = joule (J)	$N \cdot m/s = kg \cdot m^2/s^3$ = J/s = watt (W)
CGS	$dyn \cdot cm = g \cdot cm^2/s^2$ = erg	$dyn \cdot cm/s = g \cdot cm^2/s^3$ = erg/s
Absolute English	foot·poundal (ft·pdl)	ft·pdl/s
Technical English	ft·lbf	ft·lbf/s
Engineering English	ft·lbf (1 Btu = 778.17 ft·lbf)	ft·lbf/s (1 hp = 550 ft·lbf/s)

Note: 1 dyn = 10^{-5} N and 1 erg = 10^{-7} J.

EXAMPLE 1.3

In Table 1.2, the Technical English units system uses force (F), length (L), and time (t) as the fundamental dimensions. Then, the mass unit "slug" was defined such that k_1 and g_c came out to be unity (1) and dimensionless. Define a new units system in which the force, mass, and time dimensions are taken to be fundamental with units of lbf, lbm, and s, and the length unit is defined such that k_1 is unity (1) and dimensionless. Call this new length unit the *chunk* and find its conversion factor into the Engineering English and SI units systems (Figure 1.10).

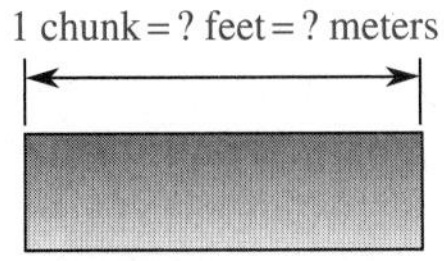

FIGURE 1.10
Example 1.3.

Solution

From Eq. (1.1), we see that the length unit must be defined via Newton's second law, $F = k_1 ma$. Since we want k_1 to be unity and dimensionless, we set

$$k_1 = \frac{F}{ma} = 1\,(\text{dimensionless})$$

In our new system, we arbitrarily require 1 lbf to be the force calculated from Newton's second law when 1 lbm is accelerated at a rate of 1 chunk/s^2. Then, from the preceding k_1 equation, we get

$$\frac{1\,\text{lbf}}{1\,\text{lbm}\,(1\,\text{chunk/s}^2)} = 1\,(\text{dimensionless})$$

so that

$$1\,\text{chunk} = 1\,\frac{\text{lbf}\cdot\text{s}^2}{\text{lbm}}$$

In the Engineering English units system, 1 lbf accelerates 1 lbm at a rate of

$$a = \frac{F}{m}(g_c) = \frac{1\,\text{lbf}}{1\,\text{lbm}}\left(32.174\frac{\text{lbm}\cdot\text{ft}}{\text{lbf}\cdot\text{s}^2}\right) = 32.174\frac{\text{ft}}{\text{s}^2}$$

Since the lbf, lbm, and s have the same meaning in both the new system and the traditional Engineering English units system, it follows that

$$1\,\frac{\text{chunk}}{\text{s}^2} = 32.174\,\frac{\text{ft}}{\text{s}^2}$$

and that

$$1\text{ chunk} = 32.174\text{ ft} = (32.174\text{ ft})\left(\frac{1\text{m}}{3.281\text{ ft}}\right) = 9.806\text{ m}$$

Exercises

7. Determine the weight at standard gravity of an object whose mass is 1.0 slug. **Answer**: Since force and weight are the same, Eq. (1.8) gives $F = W = mg/g_c$. From Table 1.2, we find that, in the Absolute English units system, $g_c = 1$ (dimensionless). So the weight of 1.0 slug is $W = (1.0\text{ slug})(32.174\text{ ft/s}^2)/1 = 32.174\text{ slug (ft/s}^2)$. But, from Eq. (1.8), we see that 1.0 slug = 1.0 lbf·s²/ft, so the weight of 1 slug is then $W = 32.174\text{ (lbf}\cdot\text{s}^2\text{/ft)(ft/s}^2) = 32.174\text{ lbf}$.
8. Determine the mass of an object whose weight at standard gravity is 1 poundal. **Answer**: Using the same technique as in Exercise 7, show that the mass of 1 poundal is $m = Fg/g_c = Wg/g_c = (1\text{ poundal})(1)/32.174\text{ ft/s}^2 = 0.03108\text{ pdl}\cdot\text{s}^2\text{/ft} = 0.03108\text{ lbm}$.
9. W. H. Snedegar whimsically suggested the following new names for some of the SI units[6]:

1 far = 1 meter (m);	1 jog = 1 m/s;	1 pant = 1 m/s^2
1 shove = 1 newton (N);	1 grunt = 1 joule (J);	1 varoom = 1 watt (W)
1 lump = 1 kilogram (kg);	1 gasp = 1 pascal (Pa);	1 flab = 1 $kg\cdot m^2$

and so forth. Of course the Snedegar units would use the same unit prefixes as SI (see Table 1.5 later). For example, a km would be a kilofar, a kJ would be a kilogrunt, a MPa would be a megagasp, and an incremental length (incremental far) would probably be called a *near*. In this system the fundamental mass, length, and time (M, L, t) units are the lump, far, and second. All other Snedegar units are secondary, being defined by some basic equation. For example, the secondary unit for velocity, the jog, is defined from the definition of the dimensions of velocity as length per unit time (L/t), or 1 jog = 1 far/s. This can, however, produce some problems in usage. In mechanics, the units of microstrain would be microfar/far. Since a microfar is closer to a near than a far, microstrain units would probably become a near/far. Such logistical inconsistency often adds confusion to an otherwise well-defined system of units.

Determine the relation between the primary and secondary Snedegar units for (a) force, (b) momentum (ML/t), (c) acceleration, (d) work, (e) power, and (f) stress (F/L^2). **Answers**: (a) 1 shove = 1 lump·far/s²; (b) 1 lump·jog = 1 lump·far/s²; (c) 1 pant = 1 far/s²; (d) 1 grunt = 1 shove·far = lump·far²/s²; (e) 1 varoom = 1 grunt/s = shove·far/s = lump·far²/s³; (f) 1 gasp = shove/far² = 1 lump/far·s².

[6] *Snedegar, W. H., "Letter to the Editor," 1983. Am. J. Phys. 51, 684.*

EXAMPLE 1.4

Time passes. You graduate from college and go on to become a famous NASA design engineer. You have sole responsibility for the design and launch of the famous Bubble-II space telescope system. The telescope weighs exactly 25,000 lbf on the surface of the Earth and is to be installed in an asynchronous Earth orbit with an orbital velocity of exactly 5000 mph (Figure 1.11).

a. What is the value of g_c (in lbm·ft/lbf·s²) in this orbit?
b. How much will the telescope weigh (in lbf) in Earth orbit where the local acceleration of gravity is only 2.50 ft/s².

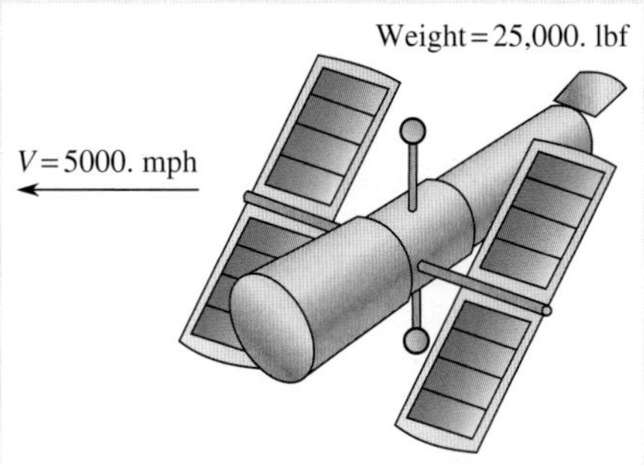

FIGURE 1.11
Example 1.4.

(*Continued*)

EXAMPLE 1.4 *(Continued)*

Solution

a. From the text, we see that g_c is always a constant. It does not depend on the local acceleration of gravity. From Eq. (1.6) and Table 1.2, we find that in the Engineering English units system $g_c = 32.174\ \text{lbm}\cdot\text{ft/lbf}\cdot\text{s}^2$.

b. Since weight is force due to gravity, we have $W = F = mg/g_c$, and the mass can be computed from $m = Wg_c/g$, or

$$m = \frac{(25{,}000\ \text{lbf})\left(32.174\,\frac{\text{lbm}\cdot\text{ft}}{\text{lbf}\cdot\text{s}^2}\right)}{32.174\,\frac{\text{ft}}{\text{s}^2}} = 25{,}000\ \text{lbm}$$

Then the weight in Earth orbit is

$$W_{\text{orbit}} = \frac{mg_{\text{orbit}}}{g_c} = \frac{(25{,}000\ \text{lbm})(2.50\ \text{ft/s}^2)}{32.174\,\frac{\text{lbm}\cdot\text{ft}}{\text{lbf}\cdot\text{s}^2}} = 1940\ \text{lbf}$$

Exercises

10. Suppose you use the SI units system in Example 1.4. What is the value of g_c in the orbit? **Answer**: 1.0 and dimensionless (see Table 1.2).
11. Suppose the orbit in Example 1.4 changes so that the local acceleration of gravity is decreased from 2.50 ft/s^2 to 1.75 ft/s^2. Determine the new weight of the telescope in orbit. **Answer**: $W_{\text{orbit}} = 1360$ lbf.
12. If the telescope in Example 1.4 weighs 112 kN on the surface of the Earth, how much does it weigh on the surface of the Moon, where the local gravity is only 1.60 m/s^2? **Answer**: $W_{\text{moon}} = 18.3$ kN.

1.8 CHEMICAL UNITS

A good deal of energy conversion technology comes from converting the chemical energy of fuels into thermal energy. Therefore, we need to be aware of the nature of units used in chemical reactions.

A chemical reaction equation is essentially a molecular mass balance equation. For example, the equation $A + B = C$ tells us that one molecule of A reacts with one molecule of B to yield one molecule of C. Since the molecular mass of substance A, M_A, contains the same number of molecules (6.022×10^{23}, Avogadro's constant) as the molecular masses M_B and M_C of substances B and C, the coefficients in their chemical reaction equation are also equal to the number of molecular masses involved in the reaction as well as the number of molecules.

Chemists find it convenient to use a mass unit that is proportional to the molecular masses of the substances involved in a reaction. Since chemists use only small amounts of chemicals in laboratory experiments, the centimeter-gram-second (CGS) units system has proven to be ideal for their work. Therefore, chemists defined their molecular mass unit as *the amount of any chemical substance that has a mass in grams numerically equal to the molecular mass of the substance* and gave it the name *mole*.

However, the chemists' mole unit is problematic, in that most of the other physical sciences do not use the CGS units system and the actual size of the molar mass unit depends on the size of the mass unit in the units system being used. Strictly speaking, the molar mass unit used by chemists should be called a *gram mole*, because the word *mole* by itself does not convey the type of mass unit used in the units system. Consequently, we call the molar mass of a substance in the SI system a *kilogram mole*; in the Absolute and Engineering English systems it is a *pound mole*; and in the Technical English system it is a *slug mole*. In this text, we abbreviate *gram mole* as gmole, *kilogram mole* as kgmole, and *pound mole* as lbmole. Clearly, these are all different amounts of mass, since 1 gmole ≠1 kgmole ≠ 1 lbmole ≠ 1 slug mole. For example, 1 pound mole of water would have a mass of 18 lbm, whereas 1 gram mole would have a mass of only 18 g (0.04 lbm), so that there is an enormous difference in the molar masses of a substance depending on the units system being used.

Since the molar amount n of a substance having a mass m is given by

$$n = \frac{m}{M} \tag{1.9}$$

where M is the molecular mass[7] of the substance, it is clear that the molecular mass must have units of mass/mass-mole. Therefore, we can write the molecular mass of water as

$$M_{H_2O} = 18\,\text{g/gmole} = 18\,\text{lbm/lbmole} = 18\,\text{kg/kgmole} = \ldots$$

The numerical value of the molecular mass is constant, but it has units that must be taken into account whenever it is used in an equation.

EXAMPLE 1.5

A cylindrical drinking glass, 0.07 m in diameter and 0.15 m high, is three-quarters full of water (Figure 1.12). Determine the number of kilogram moles of water in the glass. The density of liquid water is exactly 1000 kg/m^3.

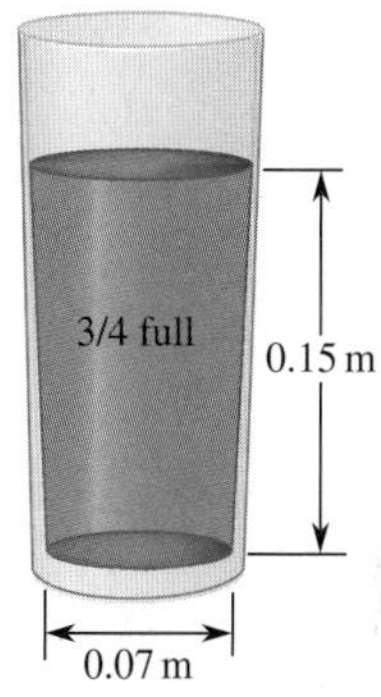

FIGURE 1.12
Example 1.5.

Solution

The mass of water in the glass is equal to the volume of water present multiplied by the density of water, or

$$m = (\pi R^2 L) \times \rho = \pi(0.035\,\text{m})^2(0.75 \times 0.15\,\text{m})(1000\,\text{kg/m}^3) = 0.433\,\text{kg}$$

The molecular mass of water is 18 kg/kgmole, and Eq. (1.9) gives the number of moles present as

$$n = \frac{m}{M} = \frac{0.433\,\text{kg}}{18\dfrac{\text{kg}}{\text{kgmole}}} = 0.024\,\text{kgmole}$$

Exercises

13. Determine the number of lbmole in a cubic foot of air whose mass is 0.075 lbm. The molecular mass of air is 28.97 lbm/lbmole. **Answer:** n = 0.00259 lbmole.

14. How many kilograms are contained in 1 kgmole of a polymer with a molecular mass of 2.5×10^6 kg/kgmole? **Answer:** $m = 2.5 \times 10^6$ kg.

15. Exactly 2 kgmole of xenon has a mass of 262.6 kg. What is the molecular mass of xenon? **Answer:** M = 131.3 kg/kgmole.

1.9 MODERN UNITS SYSTEMS

The units systems commonly used in thermodynamics today are the traditional Engineering English system and the metric SI system. Table 1.4 lists various common derived secondary units of the SI system, and Table 1.5 shows the approved SI prefixes, along with their names and symbols.

You need to understand the difference between the units of absolute pressure and gauge pressure. In the Engineering English units system, we add the letter *a* or *g* to the psi (pounds per square inch) pressure units to make this distinction. Thus, atmospheric pressure can be written as 14.7 psia or as 0 psig. In the SI units system, we add the *word* that applies (and *not* the letter *a* or *g*) immediately after the unit name or symbol. For example, atmospheric pressure in the SI system is 101,325 Pa absolute or 0 Pa gauge. When the words *absolute* or *gauge* do not appear on a pressure unit, assume it is absolute pressure.

In 1967, the degree symbol (°) was officially dropped from the absolute temperature unit, and the notational scheme was introduced wherein *all unit names were to be written without capitalization* (unless, of course, they

HOW DO I KNOW WHETHER IT IS ABSOLUTE OR GAUGE PRESSURE?

When the clarifying term *absolute* or *gauge* is not present in a pressure unit in the textbook, assume that pressure unit is *absolute*. For example, the pressure 15.2 kPa is interpreted to mean 15.2 kPa absolute.

---

[7] Most texts call M the molecular *weight*, probably out of historical tradition. However, M clearly has units of *mass*, not weight, and therefore is more appropriately named *molecular mass*.

Table 1.4 Some Common Derived SI Units

Dimension	Name	Symbol	Formula	Expression in Terms of SI Fundamental Units
Frequency	hertz	Hz	1/s	s^{-1}
Force	newton	N	$kg \cdot m/s^2$	$m \cdot kg \cdot s^{-2}$
Energy	joule	J	$N \cdot m$	$m^2 \cdot kg \cdot s^{-2}$
Power	watt	W	J/s	$m^2 \cdot kg \cdot s^{-3}$
Electric charge	coulomb	C	$A \cdot s$	$A \cdot s$
Electric potential	volt	V	W/A	$m^2 \cdot kg \cdot s^{-3} \cdot A^{-1}$
Electric resistance	ohm	Ω	V/A	$m^2 \cdot kg \cdot s^{-3} \cdot A^{-2}$
Electric capacitance	farad	F	C/V	$m^{-2} \cdot kg^{-1} \cdot s^4 \cdot A^2$
Magnetic flux	weber	Wb	$V \cdot s$	$m^2 \cdot kg \cdot s^{-2} \cdot A^{-1}$
Pressure or stress	pascal	Pa	N/m^2	$m^{-1} \cdot kg \cdot s^{-2}$
Conductance	siemens	S	A/V	$m^{-2} \cdot kg^{-1} \cdot s^3 \cdot A^2$
Magnetic flux density	tesla	T	Wb/m^2	$kg \cdot s^{-2} \cdot A^{-1}$
Inductance	henry	H	Wb/A	$m^2 \cdot kg \cdot s^{-2} \cdot A^{-2}$
Luminous flux	lumen	lm	$cd \cdot sr$	$cd \cdot sr$
Illuminance	lux	lx	lm/m^2	$m^{-2} \cdot cd \cdot sr$

Source: Adapted from the American Society for Testing and Materials, 1980. Standard for Metric Practice, ASTM 380-79. Copyright ASTM. Reprinted with permission. ASTM International, 100 Barr Harbor Drive, PO Box C700, West Conshohocken, PA 19428.

Table 1.5 SI Unit Prefixes

Multiples	Prefixes	Symbols
10^{18}	exa	E
10^{15}	peta	P
10^{12}	tera	T
10^{9}	giga	G
10^{6}	mega	M
10^{3}	kilo	k
10^{2}	hecto	h
10^{0}	—	—
10^{1}	deka	da
10^{-1}	deci	d
10^{-2}	centi	c
10^{-3}	milli	m
10^{-6}	micro	μ
10^{-9}	nano	n
10^{-12}	pico	p
10^{-15}	femto	f
10^{-18}	atto	a

Source: Adapted with permission from the American Society for Testing and Materials, 1980. Standard for Metric Practice, ASTM 380-79. ASTM International, 100 Barr Harbor Drive, PO Box C700, West Conshohocken, PA 19428.

appear at the beginning of a sentence) regardless of whether they were derived from proper names or not. Therefore the name of the SI absolute temperature unit was reduced from *degree Kelvin* to simply *kelvin* even though the unit was named after Lord Kelvin. However, when the name of a unit is to be abbreviated, it was decided that *the name abbreviation was to be capitalized if the unit was derived from a proper name*. Therefore, the kelvin absolute temperature unit is abbreviated as K (not °K, k, or °k). Similarly, the SI unit of force, the newton, named after Sir Isaac Newton (1642–1727), is abbreviated N. The following list illustrates a variety of units from the SI and other systems, all of which were derived from proper names:

> ampere (A), becquerel (Bq), celsius (°C), coulomb (C), farad (F), fahrenheit (°F), gauss (G), gray (Gy), henry (H), hertz (Hz), joule (J), kelvin (K), newton (N), ohm (Ω), pascal (Pa), poiseuille (P), rankine (R), siemens (S), stoke (St), tesla (T), volt (V), watt (W), weber (Wb).

CRITICAL THINKING

Suppose someone wanted to name a new unit of measure after you. What name would you choose and how would it be abbreviated so that your unit would not be confused with other existing unit abbreviations?

Note that we still use the degree symbol (°) with the celsius and fahrenheit temperature units. This is due partly to tradition and partly to distinguish their abbreviations from those of the coulomb and farad. In this text, we also drop the degree symbol on the rankine absolute temperature unit, even though it is not part of the SI system. This is done simply to be consistent with the SI notation scheme and because the rankine abbreviation, R, does not conflict with that of any other popular unit. Note that abbreviations use two letters *only* when necessary to prevent them from being confused with other established unit abbreviations or to express prefixes (e.g., kg for kilogram).[8]

All other units whose names were *not* derived from the names of historically important people are both written and abbreviated with lowercase letters; for example, meter (m), kilogram (kg), and second (s). Obvious violations of this rule occur when any unit name appears at the beginning of a sentence or when its abbreviation is part of a capitalized title, such as in the MKSA System of Units.

Also, a unit abbreviation is *never* pluralized, whereas the unit's name may be pluralized. For example, kilograms is abbreviated as kg and *not* kgs, and newtons as N and *not* Ns. Finally, unit name abbreviations are *never written with a terminal period* unless they appear at the end of a sentence. For example, the correct abbreviation of seconds is s, not sec. or secs.

1.10 SIGNIFICANT FIGURES

Using the proper number of significant figures in calculations is an important part of carrying out credible engineering work. Two types of numbers are used in engineering calculations: *exact values*, such as an integer number used in counting (e.g., 5 ingots of steel) or numbers fixed by definition (e.g., 3600 seconds = 1 hour); and *inexact values*, such as numbers produced by physical measurements (e.g., the diameter of a pipe or the velocity or height of an object).

Every physical measurement is inexact to some degree. The number of *significant figures* used to record a measurement is used as an indication of the accuracy of the measurement itself.

For example, if you measure the diameter of a shaft with a ruler that could be read to two significant figures, the result might be 3.5 inches, but if it were measured with a micrometer that could be read to four significant figures it might be 3.512 inches. So, when you are given a value for some variable as, say, the number 4, you see that it is measured with a precision of only one significant figure. But if the value you are given is 4.0, you see that it is measured with two significant figures, and 4.00 indicates it is measured with three significant figures.

"EXACT" NUMBERS HAVE AN INFINITE NUMBER OF SIGNIFICANT FIGURES

Exact numbers, such as the number of people in a room, have an infinite number of significant figures. Exact numbers are not measurements made with instruments. For example, there are defined numbers, such as 1 foot = 12 inches, so there are "exactly" 12 inches in 1 foot. If a number is "exact," it does not affect the accuracy of a calculation. Some other examples are 100 years in a century, 2 molecules of hydrogen react with 1 molecule of oxygen to form 1 molecule of water, 500 sheets of paper in 1 ream, 60 seconds in 1 minute, and 1000 grams in 1 kilogram.

WHAT IS A "SIGNIFICANT FIGURE"?

A significant figure is any one of the digits 1, 2, 3, 4, 5, 6, 7, 8, and 9. Zero is also a significant figure except when used simply to fix the decimal point or to fill the places of unknown or discarded digits.

---

[8] Non-SI units systems do not generally follow this simple rule. For example, the English length unit, foot, could be abbreviated f rather than ft. However, the latter abbreviation is well established within society and changing it at this time would only cause confusion.

A number reported as 0.000452 has only three significant figures (4, 5, and 2), since the leading zeros are used simply to fix the decimal point. But the number 7305 has four significant figures. The number 2300 may have two, three, or four significant figures. To convey which ending zeros of a number are significant, it should be written as 2.3×10^3 if it has only two significant figures, 2.30×10^3 if it has three, and 2.300×10^3 if it has four. Remember that the identification of the number of significant figures associated with a measurement comes only through a detailed knowledge of how the measurement is carried out.

Computations often deal with numbers having unequal numbers of significant figures. A number of rules have been developed for various computations. The rule for addition and subtraction of figures follows. Next comes the rule for multiplication and division of figures. The operation of rounding values up or down also follows specific rules.

Do you need to maintain the correct number of significant figures in all the steps of a calculation? No, just keep one or two more digits in intermediate results than you need in your final answer. These rules are summarized in Table 1.6.

RULE FOR ADDITION AND SUBTRACTION

The sum or difference of two numbers should contain no more significant figures *farther to the right of the decimal point* than occur in the *least* accurate number used in the operation. For example, $114.2 + 1.31 = 115.51$, which must be rounded to 115.5, since the least precise number in this operation is 114.2 (having only one place to the right of the decimal point). Similarly, $114.2 - 1.31 = 112.89$, which must now be rounded to 112.9.

This rule is vitally important when subtracting two numbers of similar magnitudes, since their difference may be much less significant than the two numbers that were subtracted. For example, $114.212 - 114.0 = 0.212$, which must be rounded to 0.2 since 114.0 has only one significant figure to the right of the decimal point. In this case, the result has only one significant figure even though the "measured" numbers each had four or more significant figures.

RULE FOR MULTIPLICATION AND DIVISION

The product or quotient should contain no more significant figures than are contained in the term with the *least* number of significant figures used in the operation. For example, $114.2 \times 1.31 = 149.602$, which must be rounded to 150, since the term 1.31 contains only three significant figures. Also, $114.2/1.31 = 87.1756$, which must be rounded to 87.2 for the same reason.

RULES FOR ROUNDING

1. When the discarded value is *less than 5*, the next remaining value should not be changed. For example, if we round 114.2 to three significant figures it becomes 114; if we rounded it to two significant figures it becomes 110; and rounding it to one significant figure produces 100.
2. When the discarded value is *greater than 5* (or is 5 followed by at least one digit other than 0), the next remaining value should be increased by 1. For example, 117.879 rounded to five significant figures is 117.88; rounded to four significant figures, it becomes 117.9; and rounding it to three significant figures produces 118.
3. When the discarded value is *exactly equal to 5* followed only by zeros, then the next remaining value should be *rounded up* if it is an odd number, but remain *unchanged* if it is an even number. For example, 1.55 rounds to two significant figures as 1.6, and 1.65 also rounds to two significant figures as 1.6.

Table 1.6 Significant Figures

Written Form of a Number	Number of Significant Figures Represented by These Numbers
3 or 0.1 or 0.01 or 0.001 or 3×10^{-5} or 5×10^4	One significant figure
3.1 or 50. or 0.010 or 0.00036 or 7.0×10^3	Two significant figures
3.14 or 500. or 0.0155 or 0.00106 or 7.51×10^4	Three significant figures
3.142 or 1,000. or 0.1050 or 0.0004570 or 3.540×10^8	Four significant figures
3.1416 or 10,000. or 0.0030078 or 1.2500×10^4	Five significant figures
3.14159 or 100,000. or 186,285	Six significant figures

WHAT ABOUT INTERMEDIATE CALCULATIONS?

When doing multi-step calculations, *keep one or two more digits in intermediate results* than needed in your final answer. If you round-off all your intermediate answers to the correct number of significant figures, you discard the information contained in the next digit, and the last digit in your final answer might be incorrect. For example, the calculation $12 \times 12 \times 1.5$ has an answer with two significant figures. But you should use the intermediate results without rounding because $12 \times 12 = 144$, and $144 \times 1.5 = 216 \rightarrow 220$. But, if you round 144 to 140, you obtain $140 \times 1.5 = 210$, which is pretty far off. It is best to wait until the end of a calculation to round to the correct number of significant figures.

Never round in the middle of a multi-step calculation, round only the final answer.

NOW TEST YOURSELF

a. The number 106.750 has ___ significant figures.
b The number 0.0003507 has ___ significant figures.
c. The number 3.7×10^4 has ___ significant figures.
d. The number 2.7182818 has ___ significant figures.

(Answers: a. six, b. four, c. two, d. eight)

In a textbook, it is often awkward to write each value with the proper number of significant figures. However, the examples and problems in this textbook have a specific number of significant figures indicated in the measured values. For example, a mass that has been measured to three significant figures is given as, say, 10.0 kg, and a temperature measured to three significant figures is given as, say, 200.°C (note the decimal point). This is followed throughout the remainder of this textbook.

EXAMPLE 1.6

The inside diameter of a circular water pipe is measured to two significant figures with a ruler measure and found to be 2.5 inches (Figure 1.13). Determine the cross-sectional area of the pipe to the correct number of significant figures.

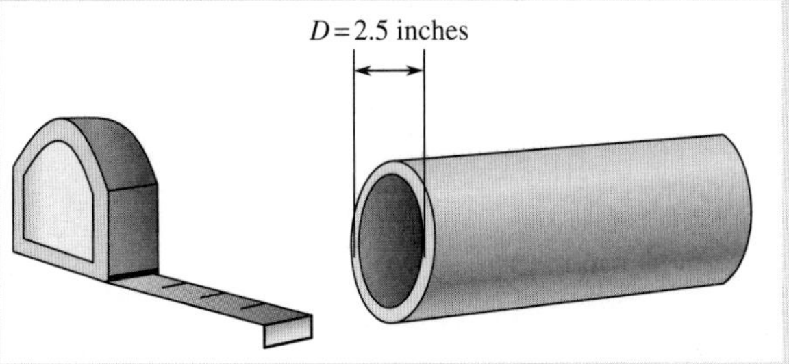

FIGURE 1.13
Example 1.6.

Solution
From elementary algebra, the cross-sectional area of a circle is $A = \pi D^2/4$, so $A_{pipe} = \pi(2.5 \text{ inches})^2/4 = 4.9087 \text{ in}^2$, which must be rounded to 4.9 in^2, since the least accurate value in this calculation is the pipe diameter, with only two significant figures.

Exercises
16. Determine the cross-sectional area of a circular metal rod measured to two significant figures with a tape measure and found to be 0.025 m. **Answer**: $A_{bar} = 0.00049 \text{ m}^2$ to two significant figures.

(*Continued*)

EXAMPLE 1.6 *(Continued)*

17. A cubical box is measured to three significant figures with a ruler and found to be 1.21 ft on one side, 1.22 ft on another side, and 1.20 ft on the third side. Determine the volume of the box to the proper number of significant figures. **Answer**: 1.77 ft^3.
18. A shaft is measured with a micrometer and found to have a diameter of 1.735 inches (to four significant figures). Determine the circumference of the shaft to the proper number of significant figures. **Answer**: 5.451 inches.

1.11 POTENTIAL AND KINETIC ENERGIES

In classical physics, the term *potential energy* usually refers to *gravitational* potential energy and represents the work done against the local gravitational force in changing the position of an object. It depends on the mass m of the object and its height Z above a reference level, written as

$$\text{Potential energy} = \text{PE} = k_1 mgZ = \frac{mgZ}{g_c} \tag{1.10}$$

where k_1 is defined by Eqs. (1.6) and (1.7) as $1/g_c$ (see Table 1.2).

Kinetic energy represents the work associated with changing the motion of an object and can occur in two forms: translational and rotational. The *total* kinetic energy of an object is the sum of both forms of its kinetic energy. The translational kinetic energy of an object is the kinetic energy resulting from a translation velocity V, written as

$$\text{Translational kinetic energy} = (\text{KE})_{\text{trans}} = k_1 \frac{mV^2}{2} = \frac{mV^2}{2g_c} \tag{1.11}$$

The rotational kinetic energy of an object is the kinetic energy resulting from a rotation about some axis with an angular velocity ω, written as

$$\text{Rotational kinetic energy} = (\text{KE})_{\text{rot}} = k_1 \frac{I\omega^2}{2} = \frac{I\omega^2}{2g_c} \tag{1.12}$$

where I is the mass moment of inertia of the object about the axis of rotation.

The mass moment of inertia of an object is the integral of a mass element dm located at a radial distance r from the axis of rotation:

$$I = \int r^2 dm \tag{1.13}$$

Table 1.7 provides equations for the mass moment of inertia of various common geometrical shapes with a total mass m.

Table 1.7 Mass Moments of Inertia of Various Common Shapes

Slender circular rod $I_x = mL^2/2$	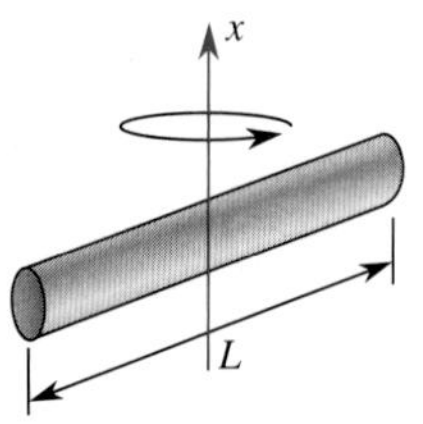	Thin rectangular plate $I_x = m(a^2 + b^2)/12$ $I_y = ma^2/12$ $I_z = mb^2/12$	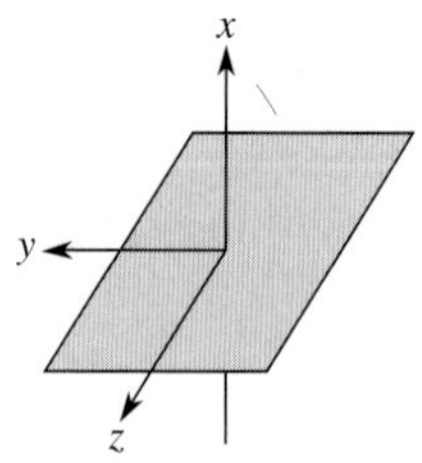
Solid rectangular prism $I_x = m(a^2 + b^2)/12$ $I_y = m(a^2 + L^2)/12$ $I_z = m(b^2 + L^2)/12$	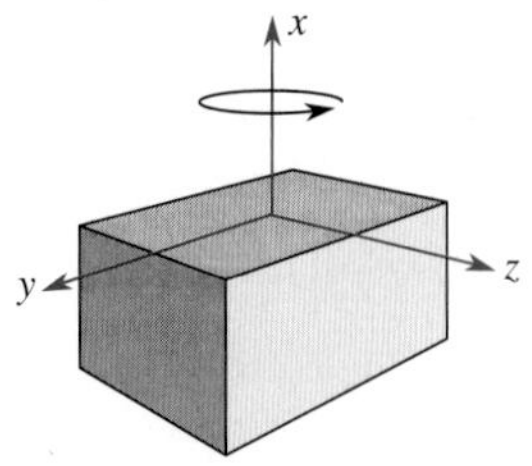	Thin disk $I_x = mR^2/2$ $I_y = I_z = mR^2/4$	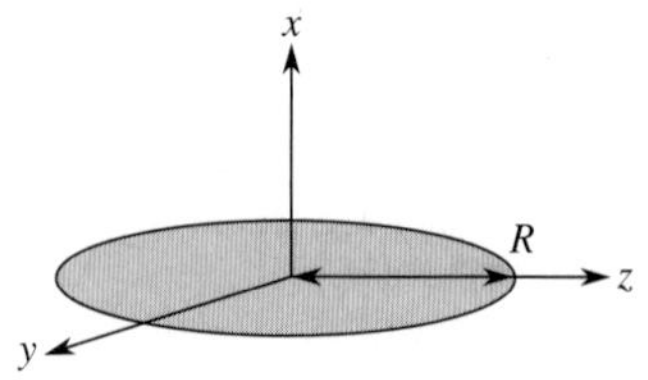

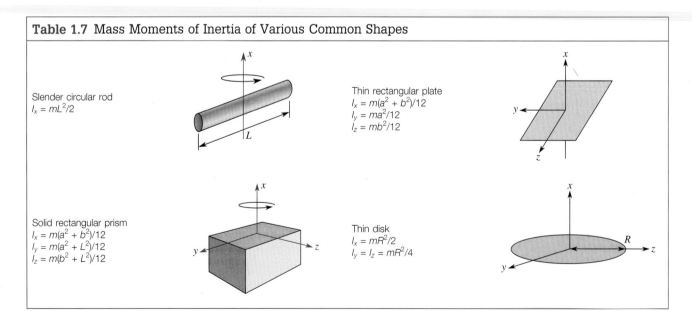

Table 1.7 *continued*

Solid circular cylinder $I_x = mR^2/2$ $I_y = I_z = m(3R^2 + L^2)/12$	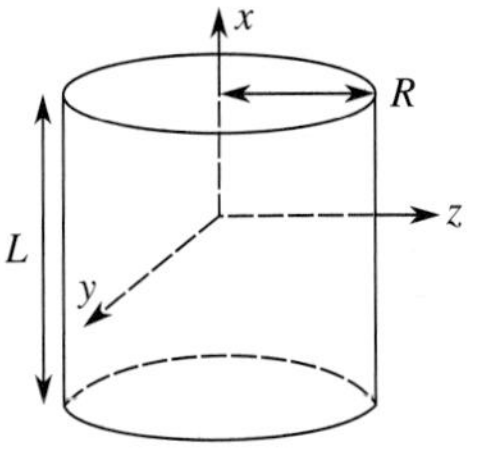	Thin circular cylinder $I_x = mR^2$	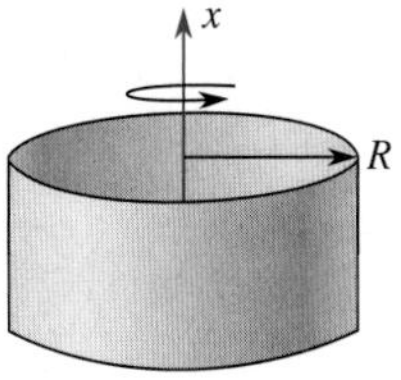
Hollow circular cylinder $I_x = m(R_1^2 + R_2^2)/2$	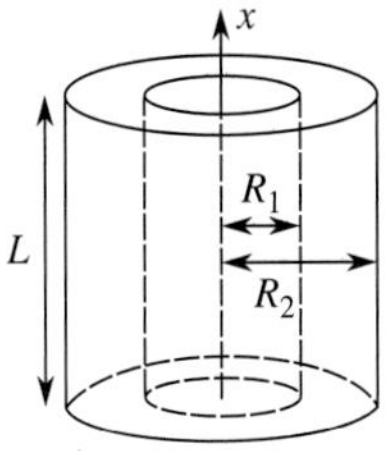	Solid circular cone $I_x = 3mR^2/10$ $I_y = I_z = 3m(R^2/4 + L^2)/5$	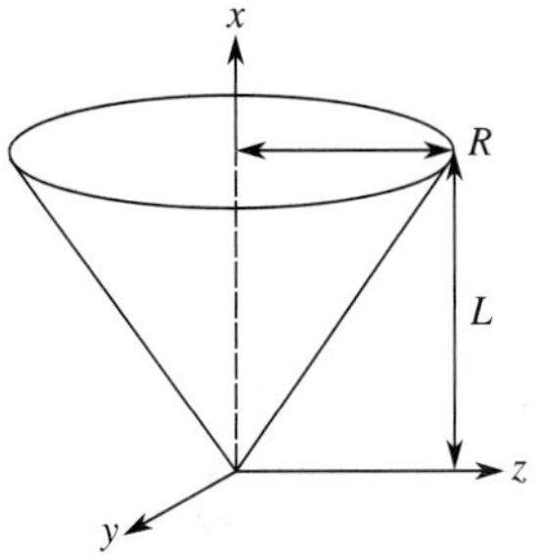
Solid sphere $I_x = I_y = I_z = 2mR^2/5$	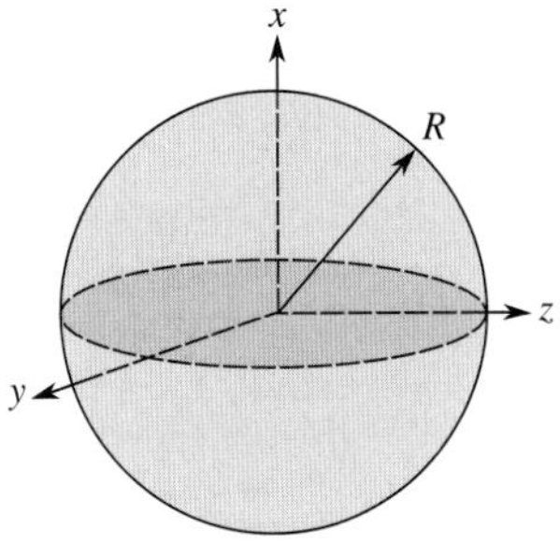	Hollow sphere $I_x = I_y = I_z = 2mR^2/3$	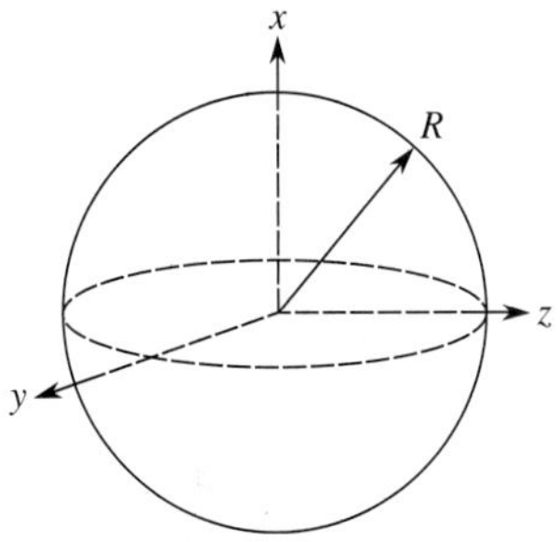

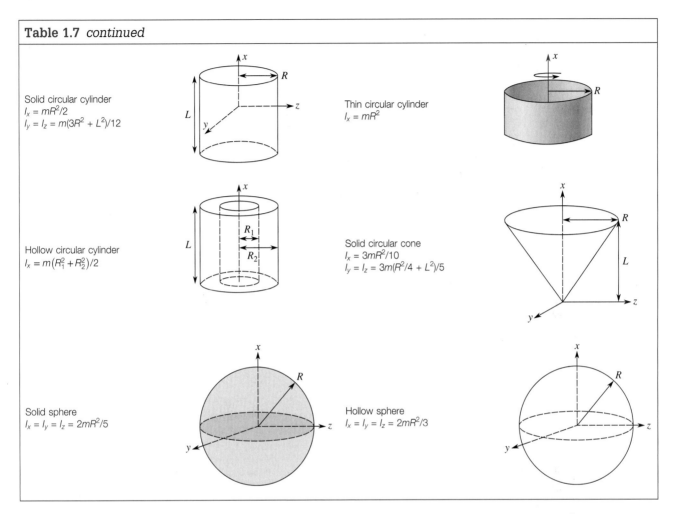

In the equations that follow in this chapter and throughout the text, the phrase *kinetic energy* can mean either translational or rotational. You must be alert to the conditions present in the problem to choose the correct form. Later in the text, general equations for the first law of thermodynamics are developed that include kinetic energy terms. Usually only the translational kinetic energy expression is written out in these equations, but be alert to the fact that this could change to rotational kinetic energy or a combination of both translational and rotational kinetic energy in any problem.

The following examples illustrate the calculation of kinetic and potential energies in both SI and Engineering English units systems.

EXAMPLE 1.7

Determine the potential energy of an automobile weighing 2000. lbf when it is 8.00 ft off the floor on a hoist in a repair shop (Figure 1.14). Express the result in both SI and Engineering English units.

Solution

The formula for the potential energy (PE) of an object of mass m at a distance Z above the reference height is given by Eq. (1.10) as

$$\text{PE} = k_1 mgZ = \frac{mgZ}{g_c}$$

We first calculate the automobile's mass from its weight using Newton's second law, $m = Fg_c/g = \text{Weight } g_c/g$. In the SI units system (see Table 1.2), we find that $g_c = 1$ (dimensionless), so

$$m = \frac{Fg_c}{g} = \frac{(2000.\ \text{lbf})\left(\dfrac{1\ \text{N}}{0.2248\ \text{lbf}}\right)(1)}{9.81\ \text{m/s}^2} = 907\ \frac{\text{N}\cdot\text{s}^2}{\text{m}} = 906.9\ \text{kg}$$

(Continued)

EXAMPLE 1.7 (*Continued*)

Now,

$$Z = (8.00\,\text{ft})\left(\frac{1\,\text{m}}{3.281\,\text{ft}}\right) = 2.438\,\text{m}$$

Therefore,

$$\text{PE} = \frac{mgZ}{g_c} = \frac{(907\,\text{kg})(9.81\,\text{m/s}^2)(2.438\,\text{m})}{1} = 21{,}700\,\text{kg}\cdot\text{m}^2/\text{s}^2$$
$$= 21{,}700\,\text{N}\cdot\text{m} = 21{,}700\,\text{J} = 21.7\,\text{kJ}$$

In the Engineering English units system, we have

$$m = \frac{Fg_c}{g} = \frac{(2000.\,\text{lbf})\left(32.174\,\frac{\text{lbm}\cdot\text{ft}}{\text{lbf}\cdot\text{s}^2}\right)}{32.174\,\text{ft/s}^2} = 2000.\,\text{lbm}$$

Here, $Z = 8.00$ ft, and from Table 1.2, we find that in the Engineering English units system $g_c = 32.174\ \text{lbm}\cdot\text{ft}/(\text{lbf}\cdot\text{s}^2)$. Therefore,

$$\text{PE} = \frac{mgZ}{g_c} = \frac{(2000.\,\text{lbm})(32.174\,\text{ft/s}^2)(8.00\,\text{ft})}{32.174\,\frac{\text{lbm}\cdot\text{ft}}{\text{lbf}\cdot\text{s}^2}} = 16{,}000\,\text{ft}\cdot\text{lbf}$$

$$= (16{,}000\,\text{ftl} \cong \text{bf})\left(\frac{1\,\text{Btu}}{778.17\,\text{ft}\cdot\text{lbf}}\right) = 20.6\,\text{Btu}$$

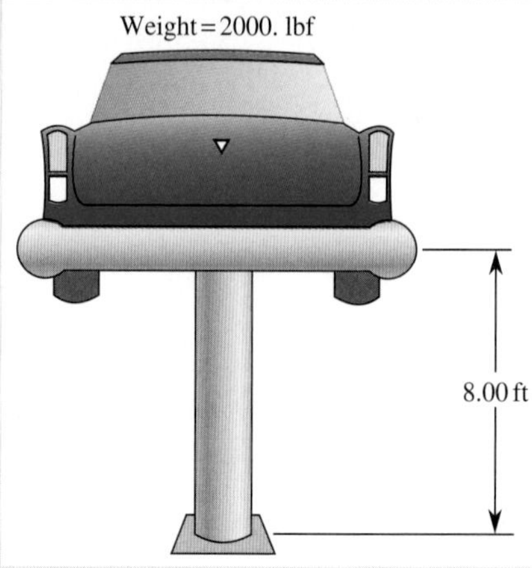

FIGURE 1.14
Example 1.7.

Exercises

19. If the automobile in Example 1.7 weighs 4000. lbf instead of 2000. lbf, determine its potential energy in SI and Engineering English units. Assume all other variables (i.e., the height) remain unchanged). **Answer**: PE = 43.4 kJ = 41.1 Btu.
20. Determine the potential energy in both SI and Engineering English units of a 3.00 lbf textbook sitting on a table 28.0 inches above the floor. **Answer**: PE = 9.49 J = 7.00 ft·lbf.
21. Determine the potential energy in SI and Engineering English units of a 4000. kg moose standing on the top of a house 5.00 m above the ground. **Answer**: PE = 196 kJ = 186 Btu.

EXAMPLE 1.8

Determine the translational kinetic energy of a bullet having a mass of 10.0 grams traveling at a velocity of 3000. ft/s in both SI and Engineering English units (Figure 1.15).

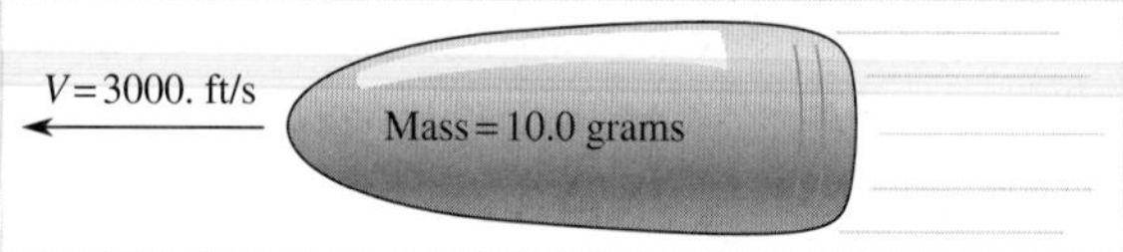

FIGURE 1.15
Example 1.8.

Solution

The formula for the kinetic energy (KE) of an object with mass m traveling at velocity V is given by Eq. (1.11) as

$$\text{KE} = \frac{1}{2}k_1 mV^2 = \frac{mV^2}{2g_c}$$

In the SI units system, $m = 10.0\ g = 0.0100$ kg, and

$$V = \left(3000.\,\frac{\text{ft}}{\text{s}}\right)\left(\frac{1\,\text{m}}{3.281\,\text{ft}}\right) = 914.4\,\text{m/s}$$

From Table 1.2, we find that $g_c = 1$ (dimensionless). Therefore,

$$\text{KE} = \frac{(0.0100\,\text{kg})(914.4\,\text{m/s})^2}{2(1)} = 4180\,\frac{\text{kg}\cdot\text{m}^2}{\text{s}^2} = 4180\,\text{N}\cdot\text{m} = 4180\,\text{J} = 4.18\,\text{kJ}$$

where 1 J = 1 N·m = 1 kg·m^2/s^2. In the Engineering English units system, V = 3000. ft/s, and m = 10.0 g = 0.0100 kg = (0.0100 kg)(2.205 lbm/kg) = 0.02205 lbm. From Table 1.2, we find that g_c = 32.174 lbm·ft/(lbf·s^2). Therefore,

$$\text{KE} = \frac{(0.02205\,\text{lbm})(3000.\,\text{ft/s})^2}{2\left(32.174\dfrac{\text{lbm}\cdot\text{ft}}{\text{lbf}\cdot\text{s}^2}\right)} = 3084\,\text{ft}\cdot\text{lbf}$$

$$= (3084\,\text{ft}\cdot\text{lbf})\left(\frac{1\,\text{Btu}}{778.17\,\text{ft}\cdot\text{lbf}}\right) = 3.96\,\text{Btu}$$

Exercises

22. If the bullet in Example 1.8 has a mass of 16.0 g instead of 10.0 g, determine its translational kinetic energy in both SI and Engineering English units. Assume all the other variables remain unchanged. **Answer**: KE = 6.69 kJ = 6.34 Btu.
23. Suppose the bullet in Example 1.8 has a mass of 8.00 g and travels at a velocity of 1000. m/s. Determine its translational kinetic energy in both SI and Engineering English units. **Answer**: KE = 4.00 kJ = 3.79 Btu.
24. Determine the translational kinetic energy of a baseball having a mass of 5.00 ounces thrown with a velocity of 90.0 mph. (Recall that 1 lbm = 16 oz.) **Answer**: KE = 115 J = 84.6 ft·lbf.

EXAMPLE 1.9

Determine the rotational kinetic energy in the armature of an electric motor rotating at 1800. rpm. The mass of the armature is 10.0 lbm, and its diameter is 4.00 inches (Figure 1.16).

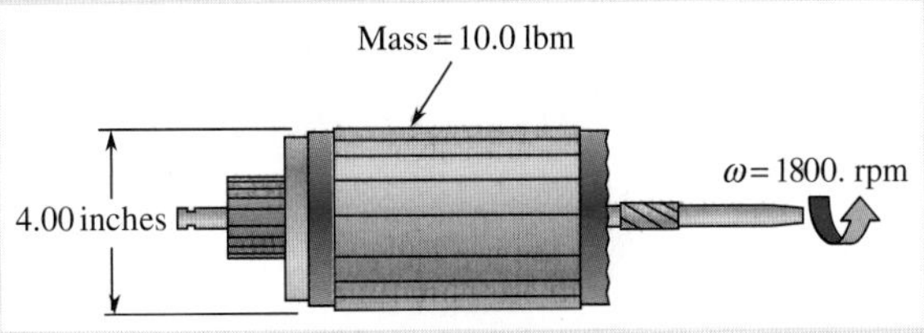

FIGURE 1.16
Example 1.9.

Solution

We approximate the armature as a solid cylinder rotating about its axis. Next, from Table 1.7, we find the equation for the mass moment of inertia of a solid cylinder, then calculate the mass moment of inertia of the armature to be

$$I = \frac{mR^2}{2} = \frac{(10.0\,\text{lbm})\left[(2.00\,\text{in})\left(\dfrac{1\,\text{ft}}{12\,\text{in}}\right)\right]^2}{2} = 0.139\,\text{lbm}\cdot\text{ft}^2$$

Equation (1.12) gives the rotational kinetic energy of the armature as

$$(\text{KE})_{\text{rot}} = \frac{I\omega^2}{2g_c} = \frac{(0.139\,\text{lbm}\cdot\text{ft}^2)\left[\left(1800.\dfrac{\text{rev}}{\text{min}}\right)\left(2\pi\dfrac{\text{rad}}{\text{rev}}\right)\left(\dfrac{1\,\text{min}}{60\,\text{s}}\right)\right]^2}{2\left(32.174\dfrac{\text{lbm}\cdot\text{ft}}{\text{lbf}\cdot\text{s}^2}\right)} = 76.8\,\text{ft}\cdot\text{lbf}$$

Exercises

25. If the armature in Example 1.9 rotates at 2000. rpm instead if 1800. rpm, determine its new rotational kinetic energy. Assume all the other variables remain unchanged. **Answer**: $(\text{KE})_{\text{rot}}$ = 94.8 ft·lbf.
26. If the armature diameter in Example 1.9 is increased from 4.00 inches to 12.0 inches, determine its new rotational kinetic energy. Assume all the other variables remain unchanged. **Answer**: $(\text{KE})_{\text{rot}}$ = 690. ft·lbf.
27. Determine the rotational kinetic energy of the Earth as it rotates on its axis once every 24.0 h. The mass and radius of the Earth are 5.976×10^{24} kg and 6.37×10^9 m. **Answer**: $(\text{KE})_{\text{rot}} = 2.56 \times 10^{35}$ J.

THERMODYNAMIC CASE STUDIES

Case study 1.1. The Anatomy of an Accident

A testing company is commissioned to build a facility to spin test the impellers for large centrifugal compressors. A test impeller is spun in a vacuum inside a thick-walled *spin chamber* until it bursts. The impeller is driven by a small air turbine with a shaft that enters through the top of the spin chamber, and the walls of the spin chamber are lined with thick lead bricks to absorb the rotational kinetic energy of the pieces of the impeller when it bursts (Figure 1.17).

During the initial test run, a 600. lbm, 30.0-inch diameter stainless steel impeller is to be spun until it bursts as part of the acceptance test of the facility. At about 1 AM, the rotor reaches 14,000 rpm and bursts. The people conducting the test are located in a room adjacent to and one floor above the test chamber room. The burst makes a single "thud" noise, typical of bursting rotors, but when the operators go to the test room they find the corridor and stairwell full of dust and debris. Some of the lead bricks from inside the spin chamber are found in the hallway, and one brick penetrated the test room wall and ended up in the kitchen of a neighborhood house. The 3000 lbm spin chamber cover had been blown up through the ceiling and fell back down (Figure 1.18). The entry door into the test room was blown into an adjacent parking lot, and the test room had extensive damage from flying lead bricks and pieces of the impeller penetrating the walls.

The cover was bolted to the chamber with 24 1-inch diameter bolts that could resist a total force of 1.90×10^6 lbf. It is concluded that the accident was caused by the impact of the impeller fragments extruding the lead bricks vertically in the spin chamber, ultimately exerting a force on the cover of about 2.10×10^6 lbf. The spin chamber contained all the radial burst forces, and only 5% of the rotational kinetic energy escaped the chamber via the lead extruded against the cover and forcing it off.

The moment of inertia of the impeller is measured and found to be 542 $\text{lbm} \cdot \text{ft}^2$, and the rotational kinetic energy of the impeller at the point where it burst is

$$\text{Rotational KE} = \frac{1}{2}(I\omega^2) = 7.80 \times 10^6 \text{ft} \cdot \text{lbf} = 10{,}000\,\text{Btu} = 10.6\,\text{MJ}$$

which is equivalent to the explosive power of about 7.20 lb of TNT (or about 60 hand grenades).

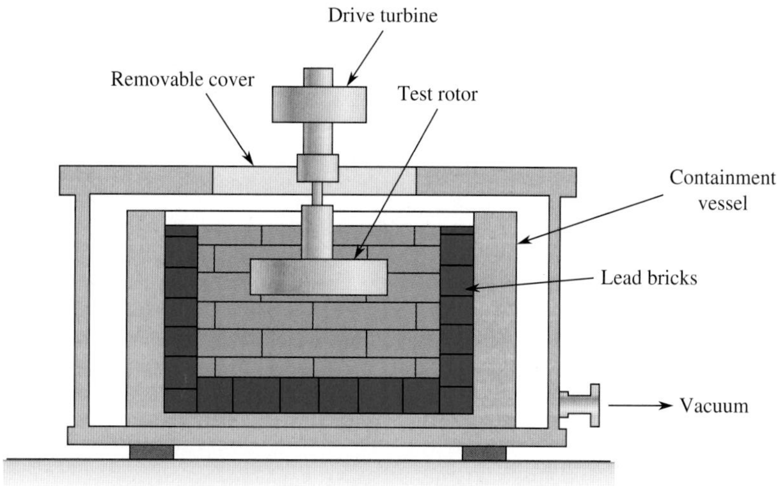

FIGURE 1.17
Case study 1 illustration.

FIGURE 1.18
Case study 1, damage to centrifuge.

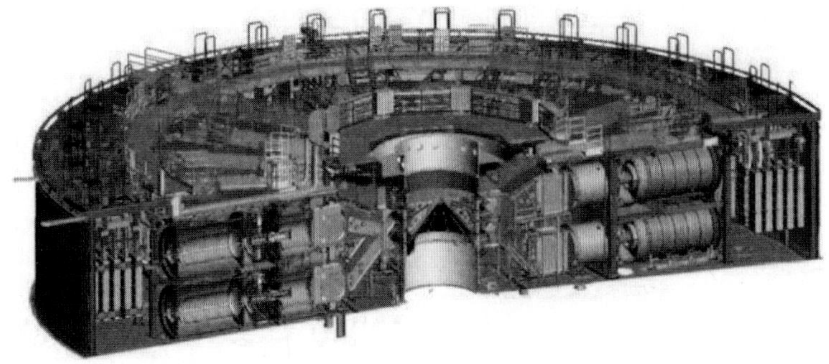

FIGURE 1.19
Case study 1.2, Z accelerator cross-section.

Case study 1.2. Sandia's hypervelocity gun

The high-velocity impact of even a small particle having a mass of only 1 g can have a disastrous effect on a spacecraft. To develop shields against such an eventuality, engineers at the Sandia National Laboratories developed a high-velocity launcher (gun) that allows the testing of materials and equipment here on Earth.

Sandia's hypervelocity launcher, known as the *Z accelerator,* is capable of accelerating dime-sized projectiles a few centimeters to gain information that can be used to simulate the effect of meteoroid impact on spacecraft (Figure 1.19). The propulsion technique uses the Z machine's 20 million amps to produce a huge magnetic field that expands in approximately 200 nanoseconds. The smooth acceleration produced by the expanding magnetic field produces a smooth projectile acceleration rather than that produced by shock of an explosion. When accelerated to a velocity of 20 km/sec, an aluminum projectile is liquefied but not vaporized.

Hypervelocity impact testing is also an accurate method of determining a material's "equation of state," which predicts how a material will react when the pressure and temperature are changed by specific amounts.

The energy required to launch a small projectile to 20 km/s is about 15 times the energy required to melt and vaporize the projectile. Therefore, the energy must be imparted in a well-controlled manner to prevent this from happening. This is achieved by using a variable density assembly to impact a stationary projectile to propel it to very high velocity without melting or fracturing.

The kinetic energy contained in a 1.00 g projectile launched at a velocity of 20.0 km/s is

$$(\mathrm{KE})_{\text{launch}} = (mV^2)/2 = (1.00\times10^{-3}\,\mathrm{kg})(20.0\times10^3\,\mathrm{m/s})^2/2$$
$$= 200.\times10^3\,\mathrm{kg(m/s)^2} = 200.\times10^3\,\mathrm{N\cdot m} = 200.\times10^3\,\mathrm{J} = 200.\,\mathrm{kJ}$$

which is about the same kinetic energy as contained in a 1000 kg (2200 lb) automobile traveling at 20 m/s (45 mph). The impact of a 1.00 g object traveling at 20.0 km/s is spread over a very small area, and the material damage produced is enormous.

SUMMARY

At the beginning of this chapter we saw the significance of understanding basic thermodynamics in a well-rounded engineering education. A working *definition of thermodynamics* is presented and the *value of thermodynamics* to all engineering fields discussed. A *basic problem solving* technique is presented that is used throughout the text and expanded on in later chapters.

Engineers must have a sound understanding of how units systems are constructed and how the various popular units systems relate to each other, because engineering units are not trivial. An accurate computation depends as much on correct units management as it does on correct numerical calculation. In this chapter, the concepts of *units, dimensions,* and *metrology* are also discussed. We see that *ancient units of measurement* evolved from a growing need to expand and quantify the elements of commerce and are undeniably woven into the history of civilizations. The historical evolution of these units often involved the binary doubling of size between successive units. It is pointed out that *temperature units* came into use quite recently, and they have their origin in the common medical practice of sensing fever in the human body.

By the turn of the 20th century, *classical mechanical and electrical units systems* had been developed and were in common use by engineers. Other units, such as *chemical units,* are also often used in engineering analysis.

The development of *modern unit systems* began in 1870 and is still going on. The United States is currently in the process of converting all its commerce and technology into the SI system. Since it is not known exactly how long this will take, textbooks such as this one present material in both the traditional Engineering English units system and the SI units system so that you, the next generation of engineers, will be able to work with both systems when necessary. Finally, we saw how to apply the basic units systems to the calculation of the *potential* and *kinetic energy* of systems.

Some of the more important equations developed in this chapter follow.

1. The equations for the conversion of temperature units:

$$T(°\text{F}) = \frac{9}{5} \times T(°\text{C}) + 32 = T(\text{R}) - 459.67$$

$$T(°\text{C}) = \frac{5}{9} \times [T(°\text{F}) - 32] = T(\text{K}) - 273.15$$

$$T(\text{R}) = \frac{9}{5} \times T(\text{K}) = 1.8 \times T(\text{K}) = T(°\text{F}) + 459.67$$

$$T(\text{K}) = \frac{5}{9} \times T(\text{R}) = \frac{T(R)}{1.8} = T(°\text{C}) + 273.15$$

2. By Newton's second law, $F = ma/g_c$, and the dimensional constant g_c

$$g_c = 32.174 \frac{\text{lbm} \cdot \text{ft}}{\text{lbf} \cdot \text{s}^2} \text{ for the Engineering English units system}$$

$$g_c = 1.0 \text{ (dimensionless) for the SI units system}$$

3. The relation between mass (m) and moles (n) of a chemical substance with a molecular mass M:

$$n = \frac{m}{M}$$

4. The definitions of potential and kinetic energies:

$$\text{Potential Energy} = \text{PE} = \frac{mgZ}{g_c}$$

$$\text{Translational Kinetic Energy} = (\text{KE})_{\text{trans}} = \frac{mV^2}{2g_C}$$

$$\text{Rotational Kinetic Energy} = (\text{KE})_{\text{rot}} = \frac{I\omega^2}{2g_C}$$

Some of the important technical terms introduced in this chapter are given in the glossary shown in Table 1.8. Many of these terms are used throughout the remainder of the text without further explanation.

Table 1.8 Glossary of Technical Terms Introduced in Chapter 1

Technical Term	Meaning
metrology	The study of measurement
dimension	A measurable characteristic
duodecimal	A base 12 number system
sexagesimal	A base 60 number system
Newton's second law	$F = ma/g_c$
newton	1 newton = 1 kg·m/s^2
dyne	1 dyne = 1 g·cm/s^2
poundal	1 poundal = 1 lbm·ft/s^2
slug	1 slug = 1 lbf·s^2/ft
g_c	The dimensional proportionality constant in Newton's second law. In the Engineering English units system, g_c = 32.174 lbm·ft/(lbf·s^2), and in the SI units system g_c = 1 and is dimensionless.
gmole	The amount of any chemical substance that has a mass in grams numerically equal to the molecular mass of the substance. This is called simply a *mole* in chemistry textbooks.
kgmole	The amount of any chemical substance that has a mass in kilograms numerically equal to the molecular mass of the substance.
lbmole	The amount of any chemical substance that has a mass in lbm (pounds mass) numerically equal to the molecular mass of the substance.
SI	*Le Systéme International d'Unités* (French)
Psia	lbf/in^2 *absolute* (pressure)
Psig	lbf/in^2 *gauge* (pressure)

Problems (* indicates problems in SI units)

1.* Using the problem solving technique described at the beginning of the chapter, work out your answer to the following question: A fresh egg is released by an ancient pterodactyl flying horizontally at 10. m/s at an altitude of 1500 km (Figure 1.20). If it takes 15 s for the egg to hatch, will it hatch before it hits the ground?

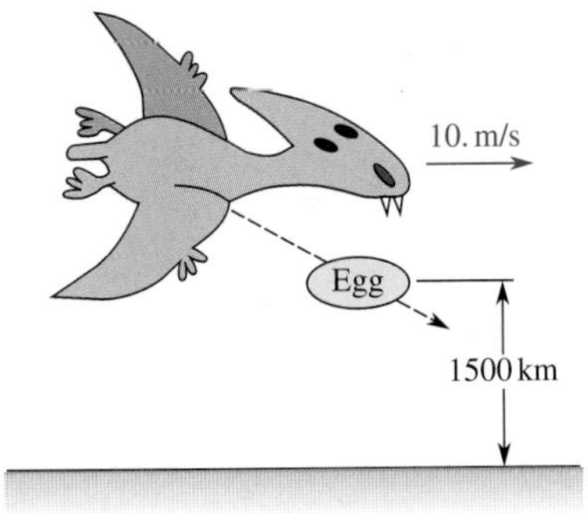

FIGURE 1.20
Problem 1.

2. During the construction of the Eads bridge across the Mississippi river at St. Louis in 1873, Theodore Cooper, a young assistant civil engineer, slipped on a loose board and fell 90. ft into the river (Figure 1.21). He later reported that, during the fall, he rolled himself into a ball and rapidly calculated the velocity with which he would hit the water. After a deep plunge, he came to the surface still clutching his pencil and was rescued by a nearby boat. Neglecting air resistance, determine (a) how long it took him to fall 90. ft, and (b) the velocity with which he hit the water.
3. If 1 gallon has a volume of 0.1337 ft^3, then how many mouthfuls of water are required to fill the moat of a castle that is 1.0 pole deep, 1.0 fathom wide, and 1.0 furlong long (Figure 1.22)? Note: 1.0 pole = 12 cubits = 18 feet, and 1.0 fathom = 4.0 cubits = 6.0 feet.
4. The *gauge* of shotguns is universally expressed as the number of spheres of the diameter of the bore of the gun that can be cast from 1 lb of lead (Figure 1.23). This standardization procedure came from the English Gun Barrel Proof Act of 1868. Taking the density of the lead as 705 lbm/ft^3, develop a formula relating the diameter of the gun barrel to the gauge of the gun. Compute the barrel diameters for 20. gauge, 12. gauge, and 10. gauge shotguns. (Note that the *caliber* of a gun is not the same as its gauge. The caliber of a weapon is just the diameter of the bore expressed in inches multiplied by 100. For example, a .38 caliber pistol has a bore diameter of 0.38 inch).
5. If lead is measured in avoirdupois ounces and silver is measured in troy ounces, which weighs more (a) an ounce of lead or an ounce of silver, and (b) a pound of lead or a pound of silver (Figure 1.24)?

FIGURE 1.22
Problem 3.

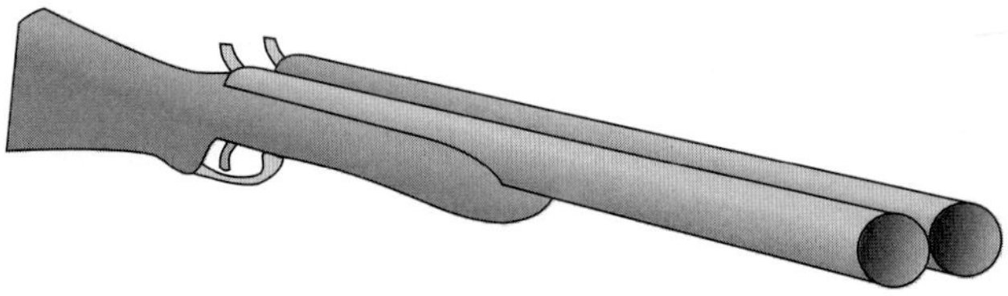

FIGURE 1.23
Problem 4.

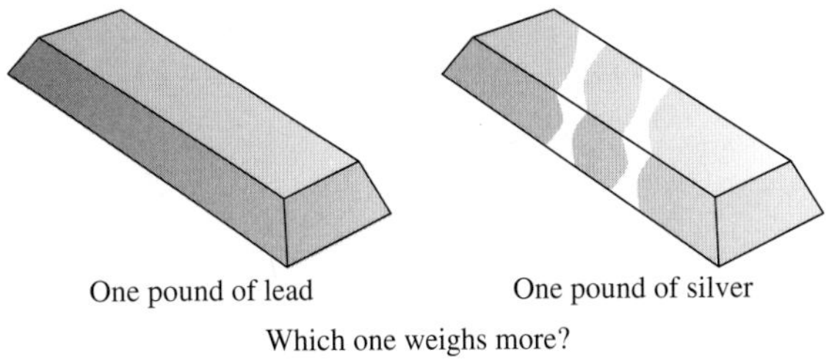

FIGURE 1.24
Problem 5.

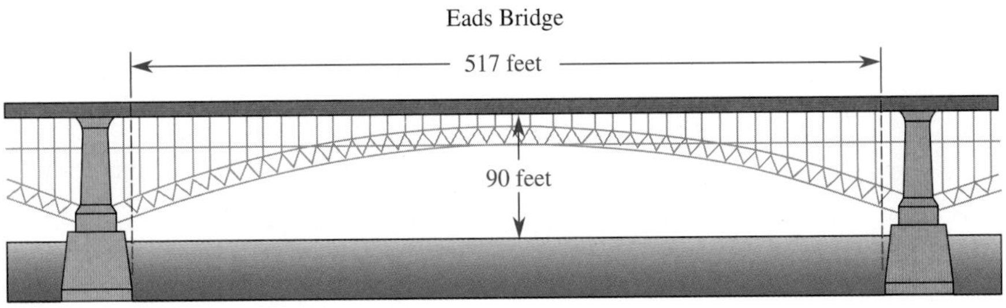

FIGURE 1.21
Problem 2.

6. Most people believe that the size of a shoe is the length of the shoe in inches (Figure 1.25). Curiously, this is not the case. Shoe sizes became standardized between 1850 and 1900 as factory-made shoes became popular. The size standardization consisted of defining the smallest shoe size at some fixed insole length, then increasing the insole length by a fixed amount for each size increment. The child's size 0 shoe was specified to have a $3\frac{11}{12}$ inch long insole and the adult size 1 shoe (there is no size 0 adult shoe) was specified to have an $8\frac{7}{12}$ inch long insole. It was also decided that each full size increment would represent an increase in insole length by one barleycorn ($\frac{1}{3}$ inch), and an increase in girth (the internal circumference of the shoe and the ball of the foot) by $\frac{1}{4}$ inch. Letters were chosen to denote shoe width increments, and the difference in girth between these width increments (for example, between a C and a D width shoe) is also $\frac{1}{4}$ inch.
   a. Determine the equations that relate the adult and child's size directly to insole length, and compute the insole length of an adult size 10 shoe.
   b. Compute the size of a child's shoe that has the same insole length as a size 1 adult shoe and explain why children's shoes are not available in size 14 or larger.

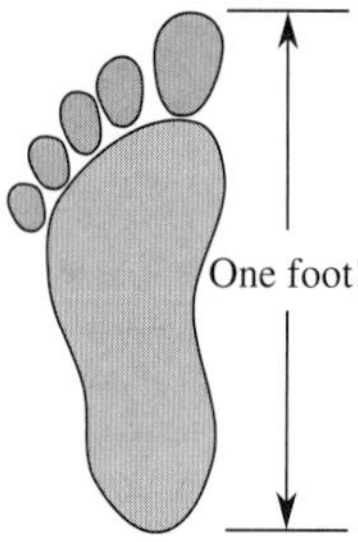

FIGURE 1.25
Problem 6.

7. The classification of carpenters' nails is based on a unit system that is at least six centuries old. The *penny* system of nail sizing, usually designated by *d*, the letter that was also the symbol for the monetary penny or pence (i.e., 3d = 3 penny nail), originated in medieval England.[9] At that time, nails were sold by the hundred, and originally a hundred 3 penny nails cost 3 pennies. From this practice came the classification of nail sizes according to the price per hundred (Figure 1.26). This system had the disadvantage that inflation in the monetary system caused the size of the nails to change. By the end of the 15th century, the classification became standardized according to Table 1.9[10]; and from this point on, the size of the nails no longer corresponded to their actual cost. Estimate the percent of monetary inflation since the 15th century if a pound of 6d nails currently costs \$1.00.

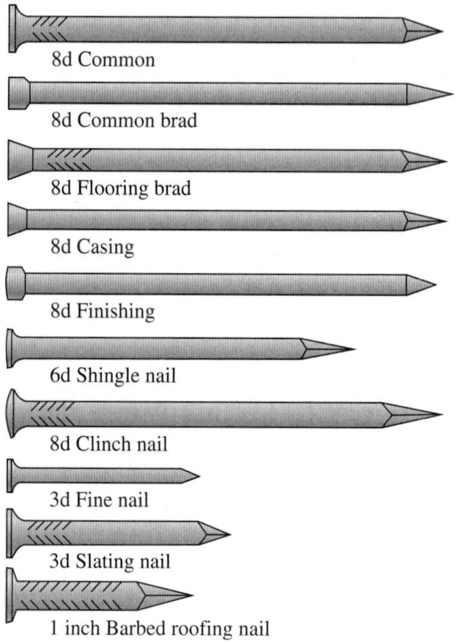

FIGURE 1.26
Problem 7.

8. By 1724, Gabriel Daniel Fahrenheit (1686–1736) had established his well-known temperature scale. This scale was based on two fixed points: the freezing point of a water and ammonium chloride solution (called 0°F) and the temperature of the human body (called 96°F). Later adjustments to the scale shifted the body temperature to 98.6 EF. What advantages did the number 96 have over, say, 100 as an upper end to this scale in 1724?
9. Determine the units of thermal conductivity, k_t, as defined by the following equation: $\dot{Q} = -k_t A(dT/dx)$, where $\dot{Q}$ is the heat transfer rate in watts, A is the cross-sectional area in m^2, T is the absolute temperature in K, and x is the distance in m.
10. Determine the units of viscosity, μ, in the following equation: $\tau = \mu(du/dy)$, where τ is a shear stress, u a velocity, and y is a distance in (a) the Engineering English system, and (b) the SI system.
11. Develop a unit conversion factor to convert the specific heat of a substance in calories/(g·K) into Btu/(lbm·R).
12. The specific internal energy of a system is 411.7 J/kg. Express this value in the following units: (a) ft·lbf/lbm, (b) kcal/kg, and (c) kW·h/lbm.
13. Determine the weight at standard gravity of 10.0 lbm in (a) lbf, (b) poundals, (c) dynes, and (d) newtons.

Table 1.9 Nail Sizes

	Nail Size															
	2d	3d	4d	5d	6d	7d	8d	9d	10d	12d	16d	20d	30d	40d	50d	60d
Length (inches)	1.00	1.25	1.50	1.75	2.00	2.25	2.50	2.75	3.00	3.25	3.50	4.00	4.50	5.00	5.50	6.00
Number per pound	845	540	290	250	165	150	100	90	65	60	45	30	20	17	13	10

[9] The *d* is from *denarius*, the name of an old Roman silver coin.
[10] Nails less than 2 penny in size are called *tacks* or *brads*, and nails larger than 60 penny are called *spikes*.

14. Determine the mass of an object whose weight on the Moon, where the local acceleration of gravity is 5.3 ft/s^2, is 10.0 poundals in (a) lbm, (b) slugs, (c) g, and (d) kg.
15. Determine the acceleration of gravity at the location where 3.0 slugs of mass weigh 50.0 N.
16. How much does 10.0 lbm weigh on a planet where g = 322 ft/s^2?
17. Determine the value of g_c at a location where a body with a mass of 270. lbm weighs 195 lbf.
18. Develop a mechanical units system in which the mass is the *stone*, the length is the *angstrom* (0.1 nm), and the time is the *century*. (a) Define your own force unit and choose the magnitude of k_1. (b) Discuss the problems that would be encountered in converting between your system and the SI system.
19. Develop a mechanical units system in which force (F), mass (M), length (L), and time (t) are independent quantities, using the kgf (kilogram force) for F, and kgm (kilogram mass) for M, the meter for L, the second for t, and 9.81 kgm·m/(kgf·s^2) as g_c. Note the similarities between this units system (which is used by some European engineers today) and the Engineering English units system.
20. Develop an FLt mechanical units system in which g_c = 1 and the force is the pound force (lbf), the length is the foot, and time is the second. Define the mass in this system to be the pound mass (lbm) and determine the conversion between the primary units (lbf, ft, and s) and the secondary mass (lbm) unit at standard gravity. Note that this is not the same FLMt system used in the Engineering English units system shown in Table 1.2. Explain the differences and similarities between these two systems.
21. Determine the mass of 18 lbm of water in (a) lbmoles, (b) gmoles, (c) kgmoles, and (d) slug moles.
22. How many kgmoles of nitroglycerine $C_3H_5(NO_3)_3$ are contained in 1.00 kg?
23. How many lbmoles of TNT (trinitrotoluene) $C_7H_5(NO_2)_3$ are contained in a 1.00 lbm stick?
24. How many lbm are contained in 1.00 lbmole of glucose $C_6H_{12}O_6$?
25. Determine the mass in lbm of 1.00 lbmole of Illinois coal having a molecular structure of $C_{100}H_{85}S_{2.1}N_{1.5}O_{9.5}$.
26. What will 3.0 kgmoles of CO_2 weigh at standard gravity in (a) N, and (b) lbf?
27. Determine the molecular mass of a substance for which 5.0 gmoles weighs 10. × 10^3 dynes at standard gravity.
28. Create an absolute temperature scale based on Reaumur's relative temperature scale defined in Table 1.1, and name it after you. Determine the boiling point of water in your new scale, and the conversion factors between your scale and the Kelvin and Rankine scales.
29. Create an absolute temperature scale based on Newton's relative temperature scale defined in Table 1.1, and name it after you. Determine the boiling point of water in your new scale, and the conversion factors between your scale and the Kelvin and Rankine scales.
30. Both the numerical value and dimensions of the universal gas constant $\Re$ in the ideal gas formula $p\forall = n\Re T$ depend on whether the temperature T is in Kelvin or Rankine absolute temperature units. In 1964, at Washington University, St. Louis, Missouri, Professor John C. Georgian recognized that, if the universal gas constant were set equal to unity and made dimensionless, then the ideal gas equation of state could be used to define an absolute temperature unit in terms of the traditional mass, length, and time dimensions from the result: $T = p\forall/n$, where p is the absolute pressure, $\forall$ is the total volume, $n = m/M$ is the number of moles, and m and M are mass and molecular mass, respectively.
    a. Using $T = p\forall/n$ (i.e., set $\Re = 1$), determine the equivalent Georgian temperature unit in terms of the standard SI units (m, kg, s). Call this new temperature unit the *georgian*, G.
    b. Find the conversion factor between G and the SI units system absolute temperature scale unit, K.
    c. Find the conversion factor between G and the Engineering English units system absolute temperature scale unit, R.
    d. Determine the triple and boiling points of water in G (Figure 1.27).

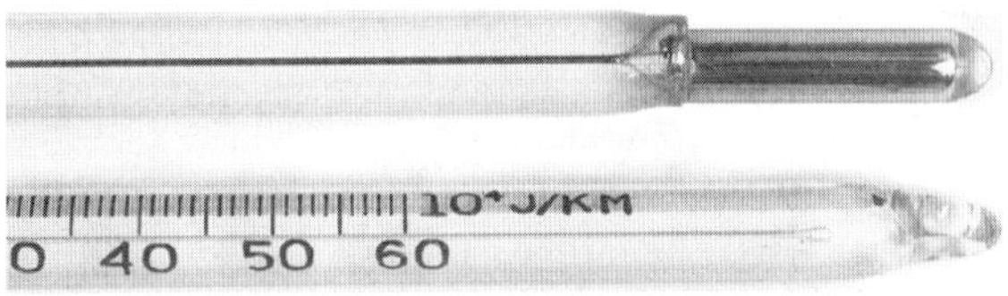

FIGURE 1.27
Problem 30, part d.

31. Show that $(4\pi\varepsilon_0)(c^2 \times 10^{-7})$ = 1 C^2/(kg·m), where c is the velocity of light = 2.998 × 10^8 m/s, and C is the charge in coulomb (1C = 1 A·s).
32. Show that $\varepsilon_0\mu_0 c^2$ = 1.0, where ε_0 is the electric permittivity of a vacuum = 8.8542 × 10^{-12} A^2·s^4/(kg·m^3), and μ_0 = the magnetic permeability of a vacuum = 4π × 10^{-7} kg·m/(A·s)^2.
33. If the pressure inside an automobile tire is 32.0 psig in Engineering English units, what is its pressure in SI units?
34. If your mass is 183 lbm in the Engineering English units system, what is your weight in (a) the Engineering English units system and (b) the SI units system?
35. If you weigh 165 lbf in the Engineering English units system, determine your mass in the following units systems: (a) Engineering English, (b) SI, and (c) Technical English.
36.* The potential of a typical storm cloud can be as high as 10^9 volts. When lightning is produced, a typical lightning strike can produces an electric current of 20,000 amps (Figure 1.28).
    a. Determine the power contained in a lightning strike (in kW).

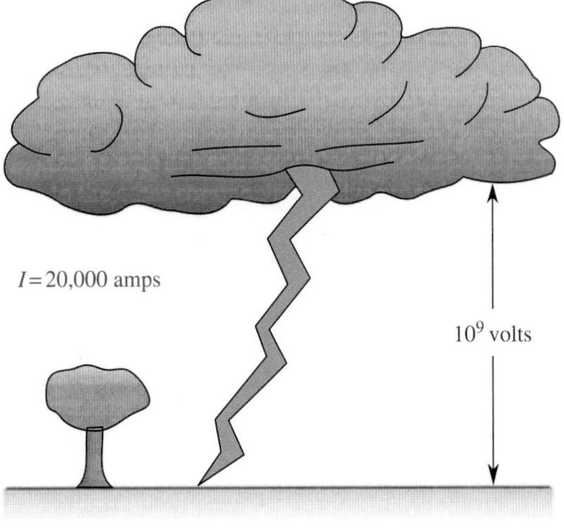

FIGURE 1.28
Problem 36.

b. If the earth is covered with 2000 lightning storms each producing 100 strikes per second, determine the total lightning electrical power available (in kW).

37. If a person reports the dimensions of a room from measurements with a tape measure as 12 feet, $6\frac{1}{4}$ inches by 14 feet, $3\frac{1}{2}$ inches, how many significant figures are being used?

38. If you measure time in hours with an accuracy of five significant figures then convert it into seconds, to how many significant figures should you report the answer in seconds?

39.* An engineer reports a value of 1.3695 m/s for the velocity of conveyer system.

a. How accurate (i.e., to how many significant figures) is the velocity measurement?

b. If this velocity is calculated from a measurement of a distance traveled divided by the time required, how accurately (i.e., to how many significant figures) must the distance and time be measured?

40. How accurate (i.e., how many significant figures can you measure) is (a) a bathroom scale graduated in quarter pound increments, (b) a yardstick graduated in eighth-inch increments, (c) a 6-inch machinist's pocket rule graduated in $\frac{1}{16}$th inch increments, (d) a 1-inch micrometer graduated in $\frac{1}{1000}$th inch increments, and (e) an analog stopwatch graduated in one hundredths of a second?

41. If you are reporting a distance of less than ten miles traveled in your car from reading the odometer, how many significant figures do you use?

42. If you are calculating the potential energy of an object for which you know its mass to three significant figures, its height to two significant figures, and the local gravity to four significant figures, how many significant figures do you use in your final answer?

43. If you are calculating the kinetic energy of an object for which you measure the distance it travels with an instrument having an accuracy of four significant figures, the time of travel with a stopwatch accurate to three significant figures, and a mass measured to an accuracy of three significant figures, how many significant figures do you use in your answer?

44.* Determine the potential energy of 1.00 kg of water at a height of 1.00 m above the ground at standard gravity in (a) the Engineering English units system and (b) the SI units system.

45. Compute the kinetic and potential energies of an airplane weighing 5.00 tons flying at a height of 30.0×10^3 ft at 500. mph (Figure 1.29). Give your answer in both SI and Engineering English units. Assume standard gravity.

46. Assume the binding energy per molecule for liquid water at 212.°F is about 7.00×10^{-20} ft·lbf/molecule. Then, assuming all the binding energy is converted into mass, determine the percent gain in mass when 1.00 lbm (10^{25} molecules) of liquid water vaporizes. Note: $E = mc^2/g_c$, where, in the Engineering English units system, $g_c = 32.174$ lbm·ft/(lbf·s^2) and $c = 9.84 \times 10^8$ ft/s.

47. The engine horsepower required to overcome rolling and air resistance for a passenger vehicle is given by the *dimensional* formula

$$\text{Horsepower} = \left[\left(53.0\frac{\text{hp}\cdot\text{h}}{\text{lbf}\cdot\text{mi}}\right)(WV) + \left(6.8\frac{\text{hp}\cdot\text{h}^3}{\text{ft}^2\cdot\text{mi}^3}\right)C_D A V^3\right] \times 10^{-6}$$

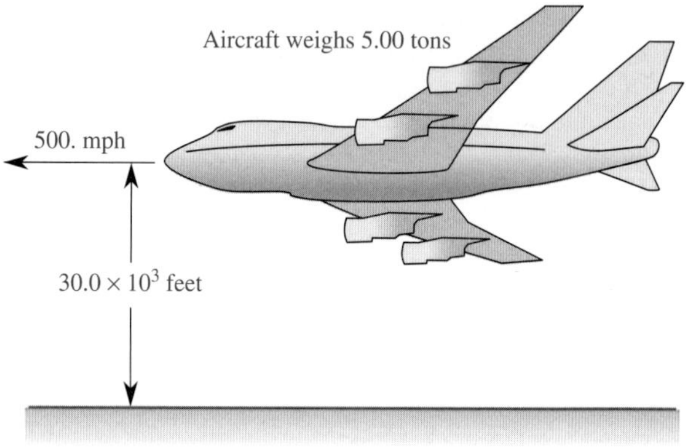

FIGURE 1.29
Problem 45.

where W is the vehicle weight in pounds force, V is the vehicle's road speed in miles per hour, A is the vehicle's frontal area in square feet, and C_D is a dimensionless drag coefficient. Convert this formula into a dimensional formula that uses only the four base units (i.e., eliminate all derived units such as horsepower and mile) of

a. The Engineering English system (lbf, lbm, ft, s).

b. The SI system (N, kg, m, s).

48.* You are suddenly transported through time and space to an unknown planet, where you find yourself face to face with a hungry giant quadroplex creature. Your finely honed survival skills as a successful, but mild-mannered, engineering student lead you conclude that the beast has a mass of 1.00×10^4 kg (gads!). In response to its unwanted affection, you quickly pick up a stone and throw it vertically with a carefully calibrated launch velocity of 10.0 m/s (Figure 1.30). Your well-trained eye determines that the stone flies to a height of 20.0 m before it begins to drop. Knowing that the initial kinetic energy and the final potential energy of the stone must be equal, determine (before the creature reaches you)

a. The value of g_c on this planet (in kg·m/N·s^2).

b. The value of the local acceleration of gravity (in m/s^2).

c. The local weight (in newtons) of the approaching giant bulbous creature.

FIGURE 1.30
Problem 48.

49. Using the CGS units system, determine the kinetic energy of an automobile weighing 1.60 billion dynes traveling at 3000. cm/s.
50. Using the CGS units system, determine the potential energy of a truck weighing 27.0 billion dynes at a height of 30.0×10^3 cm at standard gravity.
51. Using the Absolute English units system, determine the weight of an object whose kinetic energy is 306.2 ft·poundal, when it is traveling at a velocity of 10.0 ft/s.
52. Using the Absolute English units system, determine the kinetic energy of an object traveling at 15.3 ft/s and weighing 40.0 poundal at standard gravity.
53. Using the Absolute English units system, determine the potential energy of an object weighing 200. lbm·ft/s^2 at a height of 3000. ft at standard gravity.
54. Using the Technical English units system, determine the mass of an object having a potential energy of 705 ft·lbf when it is at a height of 25.0 ft at standard gravity.
55. Using the Technical English units system, determine the kinetic energy of a 197 slug mass traveling at a velocity of 33.5 ft/s.
56.* Micrometeoroids have space station impact velocities of 19.0 km/s. Determine the impact kinetic energy in SI and Engineering English units of a 1.00 g micrometeoroid traveling at this velocity.
57. A 2000. lbm meteoroid has a velocity of 23.0×10^3 mph (Figure 1.31). Determine the kinetic energy of the meteoroid in Engineering English and SI units.

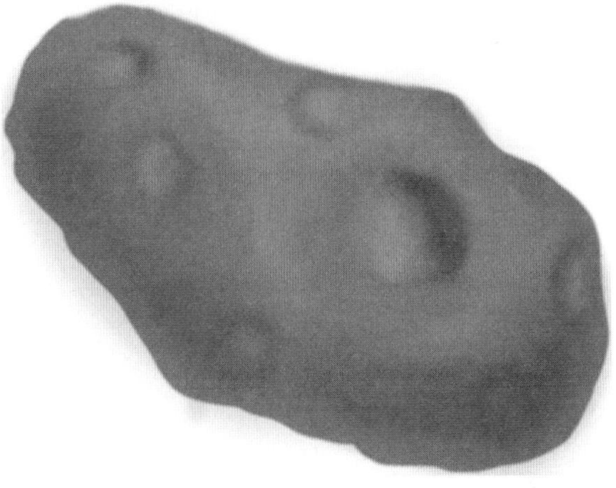

FIGURE 1.31
Problem 57.

58.* The Sandia National Laboratory hypervelocity two-stage light gas gun achieved muzzle velocities of 12.0 km/s with 0.500 g flat plate projectiles. Determine the muzzle kinetic energy of the projectiles in SI and Engineering English units.
59.* A thin disk with a diameter of 1.00 m and weighing 8.00 kg is spun about its axis at 30.0×10^3 revolutions per minute. Determine its rotational kinetic energy.
60.* The armature of a large electric motor can be thought of as being composed of a slender solid circular rod (the motor drive shaft) inside a large hollow circular cylinder (the armature windings). If the motor shaft has a mass of 150. kg and a diameter of 0.250 m, and the armature windings have a mass of 600. kg and an outside diameter of 1.50 m (the inside diameter is the same as the outside diameter of the shaft), determine the rotational kinetic energy stored in the motor when it is rotating at 3600. rpm.
61. A 0.3125 lbm baseball 2.866 inches in diameter is thrown with a velocity of 80.0 mph and it simultaneously spins about its axis at 5.00 rad per second (Figure 1.32). Determine the total (translational plus rotational) kinetic energy of the ball.

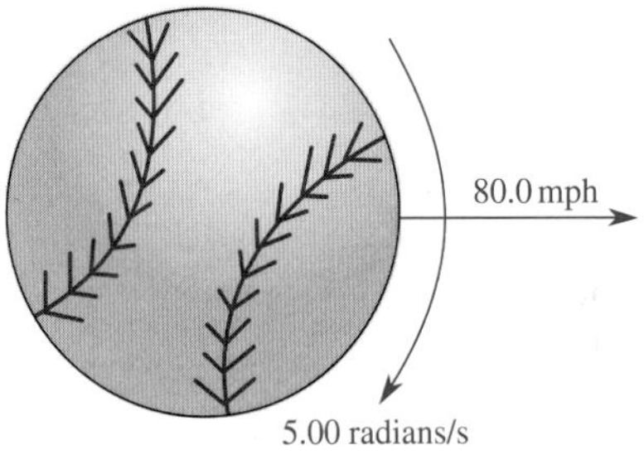

FIGURE 1.32
Problem 61.

62. A 12 lbm bowling ball 8.59 inches in diameter is given a spin of 1.0 revolution per second while traveling down the lane at 17 ft per second (Figure 1.33). Determine the total translational plus rotational kinetic energy of the ball.

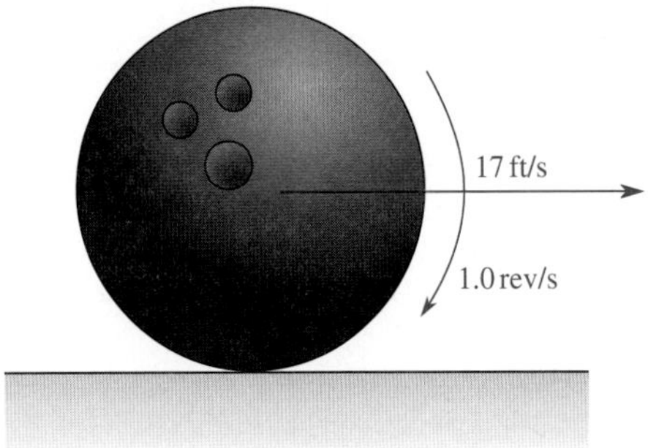

FIGURE 1.33
Problem 62.

63.* The NASA space shuttle's main engine high-pressure turbopump rotors are cryogenically spin tested in a vacuum at 40.0×10^3 rpm and −195°C. The rotor can be modeled as a thin disk 0.600 m in diameter with an effective mass of 8.50 kg. Determine the rotational kinetic energy of the rotor about its axis of rotation when it is running at the test speed.
64. A great deal of research has been carried out at the Oak Ridge National Laboratory in Tennessee on flywheel energy-storage systems. The lab developed a 27.0-inch diameter composite flywheel that can run at 40.0×10^3 rpm on

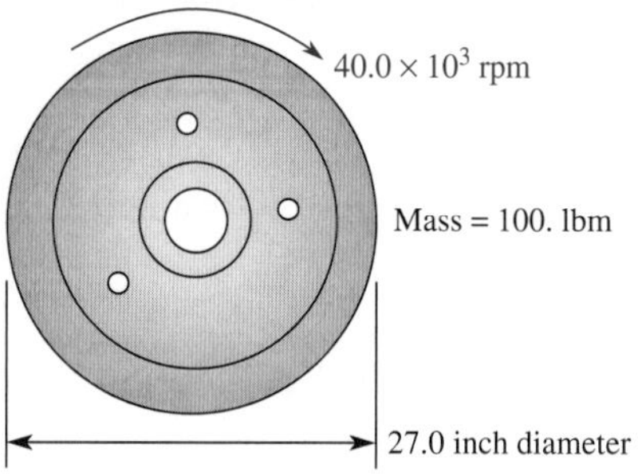

FIGURE 1.34
Problem 64.

magnetic bearings (Figure 1.34). If this flywheel can be represented as a thin disk with a mass of 100. lbm, determine

a. The rotational kinetic energy stored in the flywheel.

b. The energy storage capacity of this flywheel in W · h/lbm.

65. Determine the energy-storage capacity in W·h/lbm of the impeller in the Anatomy of an Accident case study presented in this chapter.

66.* A manufacturer of a thin disk energy-storage flywheels claims to have a flywheel made of a composite material mounted on magnetic bearings. If the flywheel turns at $200. \times 10^3$ rpm and has an energy-storage capacity of 50.0 W·h/kg, determine the diameter of the flywheel.

CHAPTER 2

Thermodynamic Concepts

CONTENTS

2.1 INTRODUCTION

Some students have difficulty with thermodynamics because it is such a broad subject. Engineering courses like statics, dynamics, and materials focus on just a few topics. Thermodynamics, on the other hand, deals with many issues that are common to a variety of engineering systems. A thermodynamic analysis can span the gamut from a huge power plant to the smallest microscopic system. It can often be applied in a fairly simple way to extremely complex systems (like biological systems) to provide profound results.

One of the most powerful aspects of thermodynamics is its "black box" approach to system analysis. It is not necessary to know what takes place *inside* the box, it is necessary only to watch the box's boundaries and see what, and how much, crosses them. This is the essence of the *balance concept*, discussed later in this chapter. But we begin by introducing some basic thermodynamic definitions.

2.2 THE LANGUAGE OF THERMODYNAMICS[1]

This section deals with a series of definitions and technical terms fundamental to understanding the language of thermodynamics. Some of these terms are already in our everyday vocabulary as a result of the broad-based use of thermodynamics concepts in everyday life. It was popular among the 19th century scientists to coin technical terms using Greek or Latin words instead of English. Consequently, many of the key terms (words) in thermodynamics are really Greek or Latin words, which, in the 21st century, are probably foreign to you. But when these terms are translated into English, you will find that their English meaning is identical to their thermodynamic use. For example, the English translation of the term *isothermal* is simply *constant temperature*, which is the physical meaning of what the term *isothermal* is meant to imply. Consequently, when Greek or Latin terms are introduced in this text, their equivalent English translations also are given at that point. Appendix B at the end of this book gives a more comprehensive analysis of the Greek and Latin origins of scientific and engineering terms. Though this may seem like a small point to you at this stage, your understanding and ease with this subject are greatly enhanced if you pay particular attention to the English meanings of these otherwise meaningless technical terms.

The name *thermodynamics* itself is an example of a Greek technical term. Basically, it means the process of converting heat (*thermo*) into mechanical power (*dynamics*). Modern thermodynamics deals with more than just thermal energy. It is more appropriately defined today as in the following box titled "Thermodynamics."

There are four basic laws of thermodynamics: the zeroth, first, second, and third laws. Like all of the other basic laws of physics, each of these laws is a generalization of observed events in the real world, and their "discovery" was the result of an individual's perception of how nature functions. Curiously, the order in which the thermodynamic laws are named does not correspond to the order of their discovery. The zeroth law is attributed to Fowler and Guggenheim in 1939; the first law to Joule, Mayer, and Colding in about 1845; the second law to Carnot in 1824; and the third law to Nerst in 1907. The first and second laws are the most pragmatic and consequently the most important to engineers. A thermodynamic analysis involves applying the laws of thermodynamics to a *thermodynamic system*.

A thermodynamic system often is referred to as just a *system*. Its *boundary* is defined simply as its surface.

The system and its boundary are always chosen by the analyst (i.e., you); they are almost never specified in a problem statement. It should be clear that, if different systems are used to analyze the same quantity, they should produce the same results in each case. A system does not have to be fixed in space. It can move, deform, and increase or decrease in size with time. Basically, there are three types of systems: isolated, closed, and open.

Figure 2.1 illustrates each of these types of systems. In Figure 2.1a, a pan of water is in a mass and energy impervious insulated box, thus forming an isolated system. In Figure 2.1b, we have a closed system, wherein the contents of the pan are closed by an airtight lid, but heat energy enters the pan from the burner. In Figure 2.1c, water (mass) enters the pan by crossing the system boundary, so here the pan is an open system.

Notice that whether a system is open or closed depends on how the analyst views the system. Figure 2.1c could be made into a closed system if the system boundary is extended to include the faucet, all the water pipe going back

THERMODYNAMICS

Thermodynamics deals with the laws that govern the transformation of energy from one form to another.

THERMODYNAMIC SYSTEM

The thermodynamic system is a volume of space containing the item chosen for thermodynamic analysis.

THERMODYNAMIC SYSTEM BOUNDARY

The surface of a thermodynamic system forms its boundary.

[1] Feel free to turn on your babble fish here, but do not put it in your ear just yet.

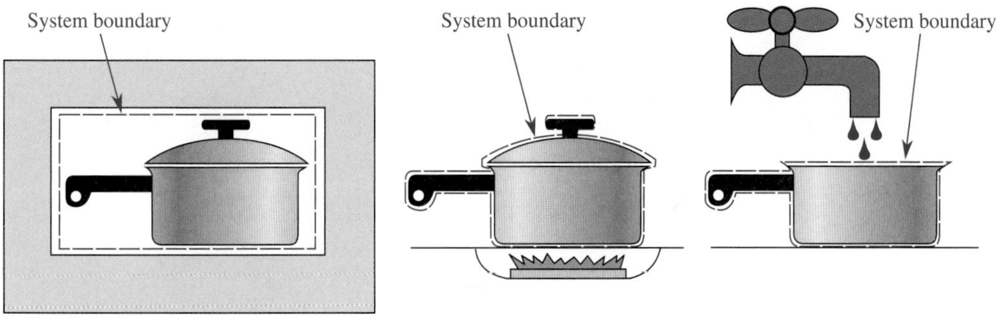
System boundary
System boundary
System boundary
(a) Isolated System: Neither mass nor energy can cross the system boundary.
(b) Closed System: Mass cannot cross the system boundary, but energy can.
(c) Open System: Both mass and energy can cross the system boundary.

FIGURE 2.1
The three types of thermodynamic systems.

TYPES OF THERMODYNAMIC SYSTEMS

Isolated system. Any system in which neither mass nor energy crosses the system boundary.
Closed system. Any system in which mass does not cross the system boundary, but energy may cross the system boundary.
Open system. Any system in which both mass and energy may cross the system boundary.

CRITICAL THINKING

If we select your body as a thermodynamic system, is it an open or closed system? What happens to you if we force you to be a closed system?

to the water treatment plant, and the water supply for the plant. But such a system would be too large to analyze properly, since we must be able to find all the energy that crosses its boundary, at any point along the boundary. Therefore, it is much easier to view Figure 2.1c as an open system with a small, well-defined system boundary.

The choices of the proper system, along with the proper form of the thermodynamic laws, always are decisions that you, the analyst, must make whenever beginning to solve a thermodynamics problem. Making a sketch of the system that shows the system boundary is a useful aid in making these decisions. The system sketch in thermodynamics is equivalent to the free body diagram sketch in mechanics. Its value cannot be overstated.

2.3 PHASES OF MATTER

The physical *phase* of a substance is defined by the molecular structure of the substance. For example, water can be described chemically as H_2O, but it may exist in a number of molecular configurations. At low temperatures, water takes on a rigid crystalline molecular structure, ice, but at higher temperatures its molecular structure becomes amorphous as it becomes a liquid and random as it becomes a vapor. We can easily identify three common structural phases of matter: solid, liquid, and vapor (or gas). But, whereas only one liquid phase or one vapor phase may be possible, many different solid molecular configurations of a substance may exist.

The term *homogeneous* can be used to describe either physical or chemical uniformity. Here we use the term *pure substance* to describe substances that are chemically uniform (Figure 2.2), and reserve the term *homogeneous* to describe substances that are physically uniform (i.e., have a single physical phase). Hence, we define a pure substance as anything that contains the same uniform chemical composition in all its physical phases. For example, a mixture of water vapor and liquid water is a pure substance. On the other hand, air is not really a pure substance, because when it is cooled sufficiently, some of its components condense into their liquid state, thus changing the composition of the remaining gases.

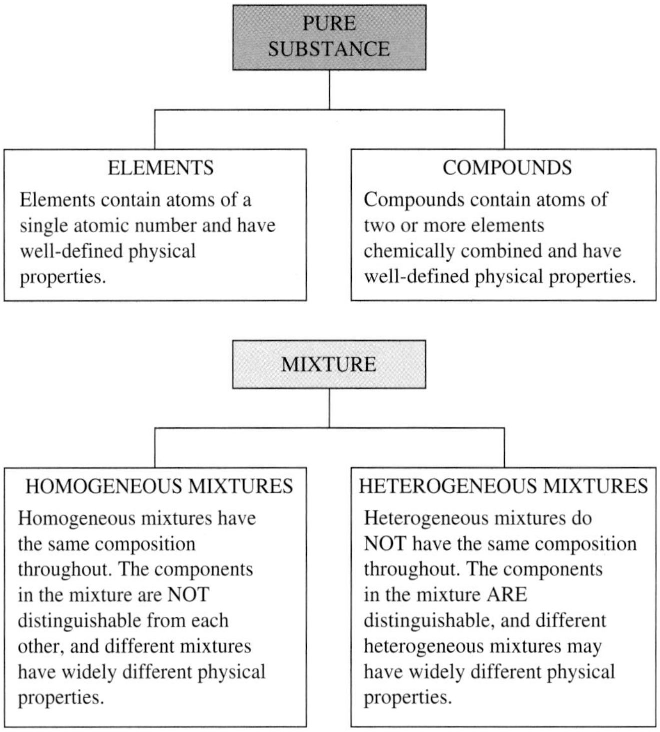

FIGURE 2.2
Pure substances and mixtures.

CRITICAL THINKING

From the previous discussion as to whether or not air is a pure substance, do you think that a more practical definition of a *pure substance* should require only that the substance not change chemical composition under the conditions of a given engineering process? For example, if air at atmospheric pressure is being heated from say 20°C to 200°C, then it would not change chemical composition and therefore could be considered as a pure substance. If, on the other hand, this air is being cooled from 20°C to −200°C, then it could not be considered a pure substance because the oxygen gas condenses into a liquid at −183°C and this would change the chemical composition of the remaining gas.

We further define a homogeneous substance as anything that contains a single physical phase. Air at normal atmospheric conditions is a homogeneous substance, but being a mixture of various gases, it is not a pure substance. A mixture of liquid water and ice on the other hand is not a homogeneous substance, but it is a pure substance. A pure substance that is also a homogeneous substance is called a *simple substance*. Liquid water is an example of a simple substance.

Pure substances must be chemically uniform but need not consist of a single chemical species. For example, a homogeneous mixture of uniform chemical composition can often be treated as if it were a single phase of a pure substance. Air in its gaseous state is usually treated as a pure substance, even though it does not satisfy the general definition of a pure substance.

2.4 SYSTEM STATES AND THERMODYNAMIC PROPERTIES

The thermodynamic *state* of a system can be either an *equilibrium* state or a *nonequilibrium* state. A thermodynamic equilibrium state is defined by the values of its thermodynamic properties. A nonequilibrium state is much more difficult to define and generally requires the existence of a condition called *local* thermodynamic equilibrium, which exists when thermodynamic equilibrium occurs locally within a series of small volumes that make up the system. Conversely, a thermodynamic *property* is any characteristic of a system whose numerical value depends only on the (local) thermodynamic equilibrium state of the system and is independent of how that state was attained. Mass, volume, temperature, pressure, color, viscosity, magnetization, and so forth are all possible properties.

The list of possible properties is quite long. Fortunately, not all properties are independent of each other. In fact, a homogeneous system contains relatively few independent properties. The formula relating the dependent and independent properties of a system is called a thermodynamic *equation of state*. Once the values of the independent properties are known for a particular state, this formula can be used to calculate the values of all the dependent properties at that state. The ideal gas formula, $pv = RT$, is an example of such an equation of state for a simple system. In classical thermodynamics, there are two types of properties, *intensive* and *extensive*.

Most extensive properties can be converted into intensive properties by dividing the extensive property by the system mass (or the number of moles) in the system. Intensive properties created in this way are called *specific* properties. For example, the total volume of a system divided by the total mass of the system is the intensive property called *specific volume*, and the total volume divided by the total number of moles of the system is the intensive property called *molar specific volume*. To be able to tell the difference between extensive and intensive properties in the formulae of this book, we adopt the notational scheme explained in the boxes.

Exceptions to this extensive property notation are uppercase T for temperature (which is an intensive property), lowercase m for mass (which is an extensive property), and lowercase n for the number of moles (also an extensive property). The letters T, m, and n for temperature, mass, and moles are the traditional symbols for these quantities, and the symbols t, M, and N are the traditional symbols for time, molecular mass, and Avogadro's number, respectively.

INTENSIVE PROPERTY

An *intensive property* is any thermodynamic property of a homogenous system that is *independent* of mass. Examples are the pressure, temperature, and density.

EXTENSIVE PROPERTY

An *extensive property* is any thermodynamic property of a homogenous system that *depends* on mass. Examples are the mass, volume, and energy.

EXTENSIVE PROPERTY NOTATION

Extensive properties are symbolized by uppercase (capital) letters. For example, Ꝟ, E, and U are the symbols for volume,[2] energy, and internal energy.

[2] *In this text, Ꝟ represents total volume, and V represents the magnitude of velocity.*

INTENSIVE PROPERTY NOTATION

Intensive mass-based properties are symbolized by lowercase letters, and intensive mole-based properties are symbolized by lowercase letters with overbars. For example, $v = $ Ꝟ$/m$,, $e = E/m$,, $u = U/m$ are the symbols for mass-based *specific* volume, *specific* energy, and *specific* internal energy. Similarly, $\overline{v}, \overline{e}$, and $\overline{u}$ are the symbols for *molar specific* volume, *molar specific* energy, and *molar specific* internal energy.

CRITICAL THINKING

If we chose the *color* of a system as a thermodynamic property, would it be an extensive or intensive property?

Table 2.1 Mass-Based and Mole-Based Specific Quantities

Mass-Based Specific Quantities	Mole-Based Specific Quantities
$v = \forall/m$	$\overline{v} = \forall/n$
$e = E/m$	$\overline{e} = E/n$
$\mathrm{ke} = \mathrm{KE}/m = V^2/2g_c$	$\overline{\mathrm{ke}} = \mathrm{KE}/n = (m/n)(V^2/2g_c)$
$\mathrm{pe} = \mathrm{PE}/m = gZ/g_c$	$\overline{\mathrm{pe}} = \mathrm{PE}/n = (m/n)(gZ/g_c)$

HOW DO I DETERMINE THE STATE?

The state of a pure substance subjected to only one work mode is determined by the values of *any pair of independent intensive properties*. If the pure substance is also homogeneous, then all its intensive properties are independent and any two of them fix the state.

Exceptions to this intensive property notation are again temperature T (an intensive property), mass m (an extensive property), and the number of moles n (another extensive property), as explained previously. Pressure, p, is a *natural* intensive property that is not obtained by dividing something by the system mass.

The uppercase-lowercase notational scheme is also used for other thermodynamic quantities, such as kinetic energy, potential energy, work, and heat, that are not thermodynamic properties. Total (mass dependent) values of these quantities are given the uppercase symbols KE, PE, W, and Q, respectively. If we divide these quantities by the system mass m, we get their *specific* (or per unit mass) forms, which are given the following lowercase symbols: $\mathrm{ke} = \mathrm{KE}/m$, $\mathrm{pe} = \mathrm{PE}/m$, $w = W/m$, and $q = Q/m$. If we divide by the number of moles n in the system, we get the *specific molar* values of these quantities, which are symbolized by lowercase letters with an overbar: $\overline{\mathrm{ke}} = \mathrm{KE}/n$, $\overline{\mathrm{pe}} = \mathrm{PE}/n$, $\overline{w} = W/n$, and $\overline{q} = Q/n$. These are summarized in Table 2.1.

Later, we discuss a general principle that provides an easy way to determine the number of independent properties in any system. In the meantime, you need to know that, for a pure substance (anything with a uniform chemical composition in all its physical phases) subjected to only one work mode[3] (type of work), only two independent intensive properties are required to determine its thermodynamic state.

A pure substance can be in any physical state—solid, liquid, vapor—or any combination of these states. Liquid water with ice cubes in a glass is a pure substance system if the system boundary is drawn so that it does not include the glass itself. If the system boundary is drawn outside the glass, then the system no longer contains a pure substance (it contains water and glass). This illustrates the importance of carefully considering exactly what the system is to be and where its boundaries are to be drawn.

2.5 THERMODYNAMIC EQUILIBRIUM

An equilibrium situation implies a condition of balance between opposing factions. There are many different kinds of equilibria. A *mechanical* equilibrium exists when all the mechanical forces within a system are balanced so that there is no acceleration (the study of mechanical equilibrium is called *statics*). A *thermal* equilibrium exists within a system if there is a uniform temperature throughout the system. An *electrostatic* equilibrium exists within a system when there is a balance of charge throughout the system. A *phase* equilibrium exists within a system when no phase transformations (such as vaporization or melting) occur within the system. A system is said to be in *chemical* equilibrium when no chemical reactions occur within the system. Since the subject matter of thermodynamics contains all these types of phenomena, we lump all these definitions together to define thermodynamic equilibrium.

Classical equilibrium thermodynamics is based on the analysis of equilibrium states and therefore is analogous to *statics* in mechanics. Since dynamic energy systems contain nonequilibrium thermodynamic states, they cannot be analyzed by the methods of classical thermodynamics. Hence, the term thermo*dynamics* appears to be a misnomer. Some authors have proposed that classical thermodynamics could be more accurately titled thermo*statics*, to keep it consistent with the titles used in mechanics. However, the origin of the term *thermodynamics* is more closely aligned with the concept of converting heat (the *thermo* part) into work (the *dynamics* part). Consequently, the *dynamics* in thermodynamics should be thought of as the dynamics of the various processes of converting heat into work (or power).

[3] A work mode may be mechanical, electrical, magnetic, etc., but only one may be present in this instance. More complex systems with multiple work modes are discussed in "The State Postulate" section of Chapter 4.

WHAT IS THERMODYNAMIC EQUILIBRIUM?

A system is said to be in thermodynamic equilibrium if it does not spontaneously change its state after it has been isolated.

2.6 THERMODYNAMIC PROCESSES

Engineering thermodynamics is primarily concerned with systems that undergo thermodynamic processes. A system subjected to a thermodynamic process normally experiences a change in its thermodynamic state. Consequently, we define a thermodynamic process as in the following box.

Neither the initial, final, nor any intermediate states need be in thermodynamic equilibrium during a thermodynamic process. A process can change a system from one nonequilibrium state to another nonequilibrium state via a path of nonequilibrium states. Figure 2.3 illustrates several process paths that change a system from an initial state *A* to a final state *B*.

If a process path closes back on itself so that it is repeated periodically in time, then the thermodynamic process is called a *thermodynamic cycle*. Figure 2.4 illustrates the definition of a thermodynamic cycle.

There is a difference between a thermodynamic cycle and a mechanical cycle. In a mechanical cycle, all the mechanical components begin and end in the same geometrical configuration. For example, the engine of an automobile goes through a mechanical cycle once per two crankshaft rotations for a four-stroke engine, but it does not go through a thermodynamic cycle. For an automobile engine to go through a thermodynamic cycle the engine's exhaust would have to be converted back into air and fuel (the initial state).

Understanding the process path is *extremely* important in thermodynamic analysis because it often determines the final state of the system. In most thermodynamic textbook problem statements, the process path is only vaguely alluded to or else is hidden in one or more of the technical terms used. Therefore, in addition to deciding on the type of system to use in the analysis of a problem and preparing a system sketch, you must also determine the type of process that is occurring.

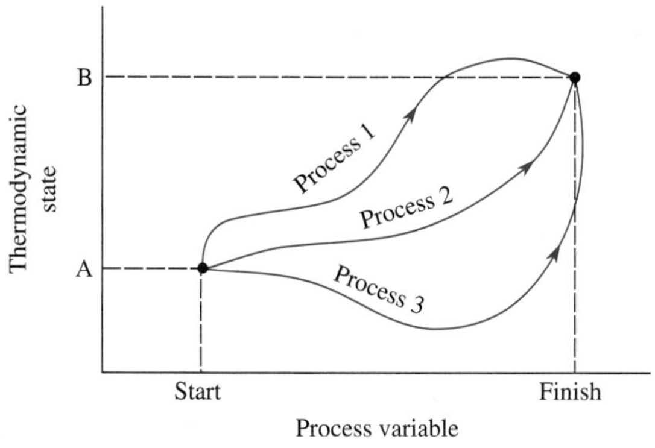

FIGURE 2.3
Three process paths that change the state of the system from *A* to *B*.

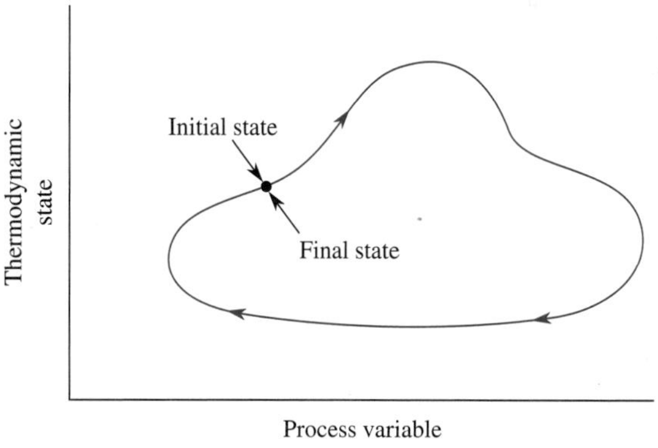

FIGURE 2.4
A thermodynamic cycle.

WHAT IS A THERMODYNAMIC PROCESS?

A thermodynamic process is the succession of thermodynamic states that a system passes through as it goes from an initial state to a final state.

WHAT IS A THERMODYNAMIC CYCLE?

A system process is said to go through a thermodynamic cycle when the final state of the process is the same as the initial state of the process.

2.7 PRESSURE AND TEMPERATURE SCALES

Because of the historical manner in which the concepts of pressure and temperature evolved, we are forced to deal with two scales for each. We have a *relative* and an *absolute* scale for both temperature and pressure measurement. Some formulae allow the use of either scale in calculations, but other formulae require the use of only absolute scales in calculations. Therefore, it is very important to know which scales are being used when you are given values for temperature and pressure.

As we saw in Chapter 1, there are two common absolute temperature scales, Rankine (R) and Kelvin (K). They are related as follows:[4]

$$T(\mathrm{R}) = \frac{9}{5}T(\mathrm{K}) \tag{2.1}$$

Each of these absolute scales has a relative scale, the common English Fahrenheit (°F) scale and the European Celsius (°C) scale.[5] These two relative scales are related to each other by

$$T(^\circ\mathrm{F}) = \frac{9}{5}T\,(^\circ\mathrm{C}) + 32 \tag{2.2}$$

and the respective absolute and relative scales are related by

$$T(\mathrm{R}) = T(^\circ\mathrm{F}) + 459.67 \tag{2.3}$$

and

$$T(\mathrm{K}) = T(^\circ\mathrm{C}) + 273.15 \tag{2.4}$$

Pressure can be viewed as a compressive stress. Thus, absolute zero pressure corresponds to a level of zero stress. However, even though we generally do not encounter negative absolute pressures in thermodynamics, any finite tensile stress in a fluid or a solid is equivalent to its being subjected to a negative absolute pressure. There is no lack of consistency here; this is merely a standard sign convention for stress.

Because most gauges manufactured to measure pressure were designed to read zero at local atmospheric pressure, their readings constitute a relative pressure scale, called *gauge pressure.*

To distinguish between gauge and absolute pressure values in our writing, we append the letter *g* or *a* to the English units of the term. Therefore, the English pressure units psia and psig are to be read "pounds per square inch absolute" and "pounds per square inch gauge," respectively. SI pressure units should carry the identifying words *absolute* or *gauge* (e.g., 3.75 MPa-absolute or 3.75 MPa-gauge). This is a clumsy indicator, and since thermodynamic tables are always given in absolute pressure units and thermodynamic equations work with absolute pressure units, SI pressures are generally assumed to be in absolute units even when not so specified.

Unless otherwise specified in a problem statement, the local atmospheric pressure should always be taken to be the standard atmospheric pressure, which is 14.696 psia (or 14.7 psia) or 101,325 Pa (or 101.3 kPa). Figure 2.5 illustrates the meanings of relative and absolute temperature and pressure.

If you are given a formula with a quantity such as p or T in it, how do you know which scale to use? The following boxed rule of thumb titled "How Do I Know When I *Have* to Use Absolute Pressure or Temperature?" provides the answer.

In the ideal gas equation of state,

$$p\maltese = mRT \tag{2.5}$$

both the quantities p and T stand alone, so that the values substituted for them must always be in an absolute scale (psia and R or Pa-absolute and K). On the other hand, if a formula contains the difference in a quantity not raised to a power, such as $p_2 - p_1$ (or Δp), or $T_2 - T_1$ (or ΔT), then the values assigned to that quantity may be in either absolute or relative scale units. For example, if we have an ideal gas in a closed system of constant volume $\maltese$, then when the gas is in state 1, we can write

$$p_1\maltese = mRT_1 \tag{2.6}$$

[4] Recall from Chapter 1 that, since 1967, we no longer use the degree prefix on the absolute temperature scales but retain it on the relative scales. Hence, we write 100 R for a temperature of 100 rankine, not 100°R.

[5] The Celsius scale was also commonly called the *centigrade scale*. However, the centigrade—from the Latin for 100 (*centi*) divisions (*grade*)—scale was developed by the Swedish astronomer Anders Celsius in about 1742; and in 1948, the centigrade scale was officially renamed the *Celsius scale*.

WHO DEVELOPED THE IDEAL GAS EQUATION OF STATE?

By 1662, the English chemist Robert Boyle (1627–1691) had conducted experiments establishing that the pressure of a gas varies inversely with the volume when the temperature is held constant. In the early 1800s, the French physicists Jacques Charles (1746–1823) and Joseph Gay-Lussac (1778–1850) independently determined that the volume $\forall$ of a gas increases linearly with temperature when the pressure is held constant. The Charles/Gay-Lussac relation can be written as

$$\frac{\forall}{\forall_0} = 1 + \alpha T$$

where T is in °C and $\forall_0$ is the volume of the gas at 0°C. The empirical constant α is the coefficient of thermal expansion of the gas and was found to have the *same value for all gases* as the pressure approached zero.

$$\alpha = 0.003661°\text{C}^{-1} = \frac{1}{273.15}°\text{C}^{-1}$$

Since α is the same for all gases at low pressure, the Charles/Gay-Lussac equation provides a single calibration point (at $\forall_0$) temperature scale that is independent of the type of gas used, plus it defines the "size" of the degree ($\alpha\forall_0$) on the scale.

By 1820, the Boyle and Charles/Gay-Lussac results had been combined to produce the "ideal" gas equation of state:

$$p\forall = mR\left[T(\text{in }°\text{C}) + \frac{1}{\alpha}\right] = mR[T(\text{in }°\text{C}) + 273.15]$$

and by then it was generally accepted that T (in °C) + 273.15 corresponded to some sort of ideal gas absolute temperature scale. However, the problem remained that this scale still appeared to depend on the thermometric measuring material (an ideal gas) and therefore did not constitute a genuine "thermodynamic" absolute temperature scale.

Since the concept of an absolute temperature scale was not firmly established until 1848 by Lord Kelvin, it is remarkable that the ideal gas equation of state, which depends on the use of an absolute temperature scale, was in use a full 30 years earlier. However, historically, we find that empirical equations often precede theoretical explanations.

HOW IS GAUGE PRESSURE RELATED TO ABSOLUTE PRESSURE?

Absolute pressure = Gauge pressure + Local atmospheric pressure

HOW DO I KNOW IF A GIVEN SI PRESSURE IS ABSOLUTE OR GAUGE?

When an SI pressure appears in a textbook without such an identifier (e.g., 3.75 MPa), assume that it is an absolute pressure (i.e., 3.75 MPa-absolute). Gauge pressures should always be identified as "gauge" to avoid confusion.

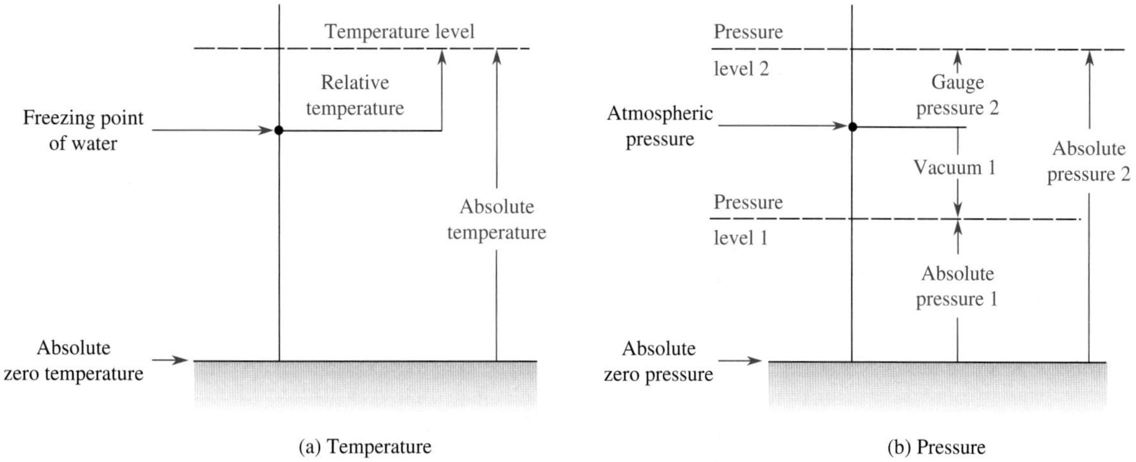

FIGURE 2.5
Relative and absolute temperatures and pressures.

HOW DO I KNOW WHEN I *HAVE* TO USE ABSOLUTE PRESSURE OR TEMPERATURE?

If pressure or temperature stands by itself or is raised to some power in an equation, then the values assigned to it *must be in an absolute scale of units.*

and when it is changed to state 2, we can write

$$p_2 \forall = mRT_2 \tag{2.7}$$

Now, if we subtract Eq. (2.6) from (2.7), we get

$$(p_2 - p_1)\forall = mR(T_2 - T_1) \tag{2.8}$$

Absolute pressure and temperature scales must be used in making calculations with Eqs. (2.6) and (2.7), but either absolute or relative scales may be used in Eq. (2.8). Relative scales can be used whenever the additive term that converts a relative scale to an absolute scale cancels out within the formula, as it does when a simple difference is taken.

Using relative scale values where absolute scale values should be used clearly leads to enormous calculational errors. If in doubt, use values in the absolute scale units.

2.8 THE ZEROTH LAW OF THERMODYNAMICS

As previously mentioned, the zeroth law was one of the last thermodynamic laws to be developed. It was introduced by R. H. Fowler and E. A. Guggenheim in 1939.[6]

This may seem trivial at first reading, but consider: If man A loves woman C and man B loves woman C, does it follow that man A loves man B? One of the major values of the zeroth law is that it forms the theoretical basis for temperature measurement technology. Consider the mercury in glass thermometer shown in Figure 2.6. The zeroth law tells us that if the glass is at the same temperature as (i.e., is in thermal equilibrium with) the surrounding fluid, and if the mercury is at the same temperature as the glass, then the mercury is at the same temperature as the surrounding fluid. Thus, the thermometer can be graduated to show the mercury temperature, and this temperature is automatically (via the zeroth law) equal to the temperature of its surroundings.

ZEROTH LAW OF THERMODYNAMICS

Consider three thermodynamic systems, A, B, and C. If system A is in thermal equilibrium with (i.e., is the same temperature as) system C and system B is in thermal equilibrium with system C, then system A is in thermal equilibrium with system B.

CRITICAL THINKING

The text describes a love triangle that would not necessarily satisfy the zeroth law of thermodynamics. Can you think of other human characteristics (e.g., hate) that might not satisfy this law? In the zeroth law, thermal equilibrium is the same as temperature equilibrium, so that if $T_A = T_C$ and $T_B = T_C$, then the zeroth law requires that $T_A = T_B$. Can we create yet another thermodynamic law based on requiring a different physical property of systems A, B, and C to be in another (say, mechanical rather than thermal) type of equilibrium? What thermodynamic value would this new "law" have?

[6] Fowler, R. H., Guggenheim, E. A., 1939. *Statistical Thermodynamics*. Cambridge University Press, Cambridge, MA.

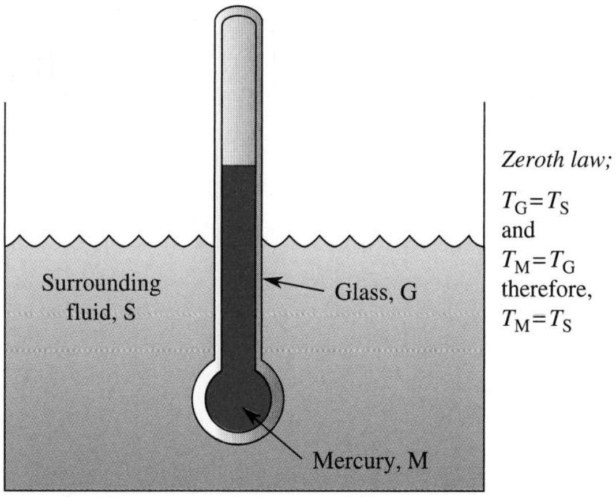

FIGURE 2.6
The zeroth law of thermodynamics applied to a mercury in a glass thermometer.

2.9 THE CONTINUUM HYPOTHESIS

While we recognize today the existence of the atomic nature of matter, we have not found an effective way to apply the basic laws of physics to large aggregates of atomic particles except by a statistical averaging technique. This is because the number of molecules in even a cubic centimeter of gas at standard atmospheric pressure and temperature is so large (about 10^{20}) that we cannot simultaneously solve all the equations of motion for each molecule. The statistical averaging process taken over large numbers of molecules produces a *continuum model* for matter, and the *continuum hypothesis* simply states that large systems made up of many discrete molecules or atoms may be treated as though they were made up of a *continuous* (i.e., nonmolecular) material.

The continuum approach to thermodynamics works well so long as the dimensions of the systems being analyzed are much larger than the dimensions of the molecules themselves and so long as the time interval over which a process takes place is very much greater than the average time between molecular collisions. The continuum thermodynamics breaks down when these conditions are violated, such as in the rarefied gas of outer space. When continuum thermodynamics breaks down, another type of thermodynamics, called *statistical thermodynamics*, can be used to solve problems.

The vast majority of engineering problems can be solved with continuum concepts, and they are the main focus of this text. Two other technical terms are used to express these ideas, *microscopic* and *macroscopic*.

When we deal with differential quantities in continuum analysis, such as dx/dt, we do not infer that the differentials shrink down to molecular dimensions and thus invalidate the continuum concept. Also, when we speak of evaluating thermodynamic properties at a *point* in a continuum system, we extrapolate the continuum concept in a mathematical sense only. The resulting mathematical functions and relations developed in macroscopic system analysis are not valid in, and cannot be accurately applied to, microscopic systems.

MICROSCOPIC SYSTEM ANALYSIS

Microscopic system analysis is the analysis of systems at the atomic level. This is the domain of statistical thermodynamics.

MACROSCOPIC SYSTEM ANALYSIS

Macroscopic system analysis is the analysis of systems at the continuum level (i.e., molecular dimensions and time scales do not enter into the analysis). This is the domain of classical and nonequilibrium thermodynamics.

2.10 THE BALANCE CONCEPT

The balance concept is one of the most important and, oddly enough, most underrated concepts in physical science today. It is basically nothing more than a simple accounting procedure. Consider some quantity X possessed by an arbitrary system. Then the balance of X over the system during a macroscopic time interval δt is

$$\begin{Bmatrix}\text{The gain in } X\\ \text{by the system}\\ \text{during time } \delta t\end{Bmatrix} = \begin{Bmatrix}\text{The amount of } X\\ \text{transported into the}\\ \text{system during time } \delta t\end{Bmatrix} - \begin{Bmatrix}\text{The net amount of } X\\ \text{leaving the system}\\ \text{during time } \delta t\end{Bmatrix} + \begin{Bmatrix}\text{The amount of } X\\ \text{produced by the system}\\ \text{during time } \delta t\end{Bmatrix} - \begin{Bmatrix}\text{The amount of } X\\ \text{destroyed by the system}\\ \text{during time } \delta t\end{Bmatrix} \tag{2.9}$$

By using the word *net* to signify the difference between like terms, Eq. (2.9) can be simplified to

$$\begin{Bmatrix}\text{The net gain in}\\ X \text{ by the system}\\ \text{during time } \delta t\end{Bmatrix} = \begin{Bmatrix}\text{The net amount of } X\\ \text{transported into the}\\ \text{system during time } \delta t\end{Bmatrix} + \begin{Bmatrix}\text{The net amount of } X\\ \text{produced by the system}\\ \text{during time } \delta t\end{Bmatrix} \tag{2.10}$$

In symbol form, Eq. (2.10) can be further simplified to

$$X_{\text{Gain}} = X_{\text{Transport}} + X_{\text{Production}} \text{ or } X_G = X_T + X_P \tag{2.11}$$

where the subscripts G, T, and P refer to net gain, net transport, and net production, respectively. In equilibrium systems, Eq. (2.11) is sufficient. But in nonequilibrium systems, X_G, X_T, and X_P may be functions of time. In systems in which X_G, X_T, and X_P change continuously in time, Eq. (2.11) can be differentiated with respect to time to give a *rate* balance equation of X as

$$\dot{X}_G = \dot{X}_T + \dot{X}_P \tag{2.12}$$

where $\dot{X}_G = dX_G/dt$, $\dot{X}_T = dX_T/dt$, and $\dot{X}_P = dX_P/dt$. Equations (2.11) and (2.12) provide a full and general account of the behavior of any property X of a system, and they are valid for any coordinate system.

EXAMPLE 2.1

The Rosalyn Computer Chip Manufacturing Company ships 120,000 chips per day to its customers and receives 100,000 chips per day from its suppliers (Figure 2.7). It manufactures 30,000 of its own chips per day, of which 3,000 are rejected as defective and are destroyed. Determine the change in chip inventory at the end of each day.

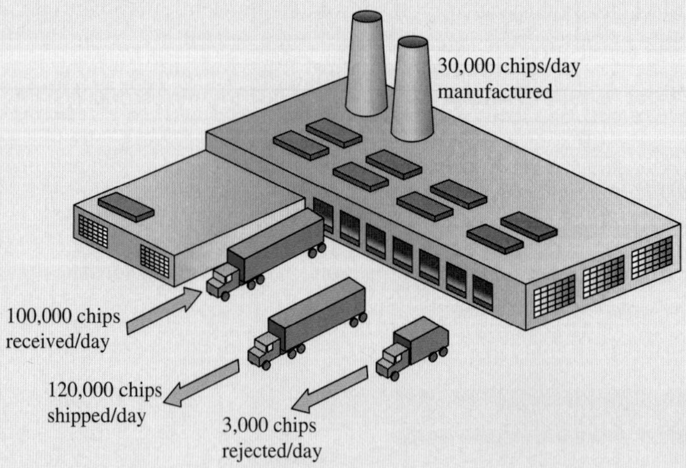

FIGURE 2.7
Example 2.1.

Solution

For each day of operation, Eq. (2.11) gives the net gain in chips as

$$X_G = X_T + X_P$$

where the net transport of chips into the facility is

$$X_T = 100{,}000 \text{ chips from suppliers} - 120{,}000 \text{ chips to customers} = -20{,}000 \text{ chips/day}$$

and the net production of chips is

$$X_P = 30{,}000 \text{ chips manufactured} - 3{,}000 \text{ chips rejected and destroyed} = 27{,}000 \text{ chips/day}$$

so the net gain in computer chips at the end of each day is

$$X_G = X_T + X_P = -20{,}000 + 27{,}000 = 7{,}000 \text{ chips/day}$$

So the chip inventory increases by 7,000 chips per day.

EXAMPLE 2.2

In 1798, the famous social scientist and economist Thomas Robert Malthus (1766–1834) discovered that, if relatively small groups of animals are left undisturbed (Figure 2.8), their population often grows such that the sum of their net birthrate and their net immigration rate into the population is directly proportional to the instantaneous value of the population. This, since known as Malthus's law of population growth, has been successfully applied to numerous types of populations, such as humans and bacteria. Write a rate balance equation for this type of population growth rate and determine how the instantaneous population varies with time.

FIGURE 2.8
Example 2.2.

Solution

Equation (2.12) gives the general rate balance: $\dot{X}_G = \dot{X}_T + \dot{X}_P$. Let N be the instantaneous population. Then, from the problem statement, we have

$$\dot{X}_G = \frac{dN}{dt}$$

and according to Malthus's law, the net birth and immigration rates are

$$\dot{X}_T + \dot{X}_P = \alpha N$$

where α is a constant of proportionality. Then, the complete Malthus population rate balance equation becomes

$$\frac{dN}{dt} = \alpha N$$

(Continued)

EXAMPLE 2.2 *(Continued)*

Since this is a simple first-order ordinary differential equation, we can separate the variables to obtain

$$\frac{dN}{N} = \alpha dt$$

Then defining N_0 as the population at time $t = 0$, this equation can be integrated as

$$\int_{N_0}^{N} \frac{dN}{N} = \int_{t=0}^{t=t} \alpha dt$$

to give

$$\ln\left(\frac{N}{N_0}\right) = \alpha t$$

And inverting the logarithm gives

$$e^{\ln(N/N_0)} = \frac{N}{N_0} = e^{\alpha t}$$

or

$$N = N_0 e^{\alpha t}.$$

Thus the population increases or decreases exponentially depending on the sign of α.

Exercises

1. Develop a balance equation for the number of hamburgers in your room. **Answer**: Net hamburgers in the room = Net hamburgers brought into the room + Net hamburgers made inside the room.
2. The growth rate discussed in Example 2.2 is often called a *geometric* growth rate. Malthus argued that the food supply of a population often grew only at a constant, or arithmetic, rate, $dF/dt = \beta$, where F is the size of the food supply at time t and β is a constant. Write a rate balance equation for the food supply F using this growth rate and solve it for F as a function of time t. **Answer**:

$$\left(\frac{dF}{dt}\right)_{\text{system}} = \dot{F}_G = \beta = \dot{F}_T + \dot{F}_P$$

   and solving for F gives $F = \beta t + F_0$, where F_0 is the size of the food supply at time $t = 0$.
3. In Example 2.2, the constant α is called the growth rate when it is greater than zero. Determine a general expression for the time t_D required for a population to double. **Answer**: $t_D = [\ln(2)]/\alpha$.

2.11 THE CONSERVATION CONCEPT

In classical physics, a quantity is said to be *conserved* if it can be neither created nor destroyed. The basic laws of physics would not produce unique balance equations if it were not for this concept. Whereas a balance equation can be written for any conceivable quantity, conserved quantities can be discovered only by human research and observation. The outstanding characteristic of conserved quantities is that their net production is always zero, and therefore their balance equations reduce to these simpler forms:

$$\text{When } X \text{ is conserved, } X_{\text{Production}} = \dot{X}_{\text{Production}} = 0, \text{ and } \begin{cases} X_{\text{Gain}} = X_{\text{Transport}} & (2.13) \\ \dot{X}_{\text{Gain}} = \dot{X}_{\text{Transport}} & (2.14) \end{cases}$$

This may not seem like much of a reduction at first, but it is a very significant simplification of the general balance equations. It means that we need not worry about property production or destruction mechanisms and how to calculate their effects. Equations (2.13) and (2.14) turn out to be very effective working equations for engineering design and analysis purposes.

Thus far, scientists have empirically discovered four major entities that are conserved: mass (in nonnuclear reactions), momentum (both linear and angular), energy (total), and electrical charge. These yield the four basic laws of physics: the conservation of mass, the conservation of momentum, the conservation of energy, and the conservation of charge. The conservation of energy is also called the *first law of thermodynamics*.

If we let E be the total energy of a system, then its conservation is written as

$$E_{\text{Production}} = E_P = 0 \qquad (2.15)$$

or

$$\dot{E}_P = 0 \tag{2.16}$$

and its resulting balance (or *conservation law*) equation is

$$E_G = E_T \tag{2.17}$$

or

$$\dot{E}_G = \dot{E}_T \tag{2.18}$$

Equations (2.15) through (2.18) are elementary forms of the first law of thermodynamics. They are elementary because, to be useful for calculation purposes, the *terms* E_T and $\dot{E}_T$, representing the system's energy transport, must be expanded into a sum of terms that accounts for all the energy transport mechanisms. This is taken up in detail in Chapter 4.

EXAMPLE 2.3

Develop an accurate verbal descriptive form of the energy balance equation that incorporates the conservation of energy principle for a system consisting of a cannon firing a projectile (Figure 2.9).

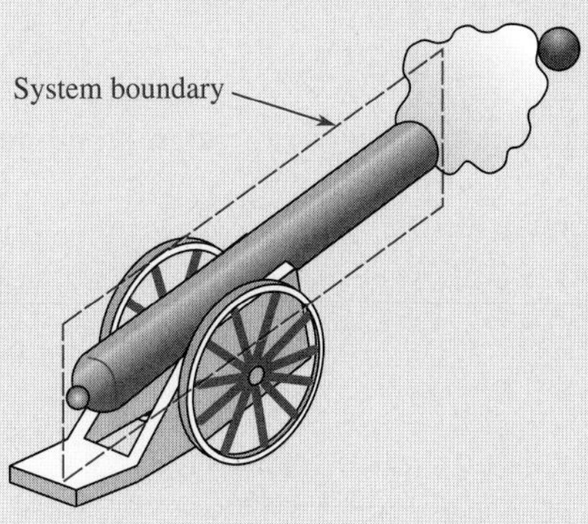

FIGURE 2.9
Example 2.3.

Solution

The complete literal descriptive energy balance equation is obtained from Eq. (2.10) as

$$\left\{\begin{array}{l}\text{Net energy of the projectile}\\ \text{and gases transported}\\ \text{into the system}\end{array}\right\} + \left\{\begin{array}{l}\text{Net production of}\\ \text{energy inside the}\\ \text{system}\end{array}\right\} = \left\{\begin{array}{l}\text{Net change in the}\\ \text{energy of the system}\end{array}\right\}$$

Since energy is a *conserved* quantity, the net production of energy inside the cannon must be zero. Now, the *net* transport of any quantity into a system is just the difference between the input and the output of that quantity; and since there is no transport of energy into this system, then the *net* transport of energy is just the negative of the energy output. So the resulting literal descriptive form of the combined conservation of energy equation and energy balance equation for this system is

$$\left\{\begin{array}{l}\text{Net energy of the projectile}\\ \text{and gases transported}\\ \text{into the system}\end{array}\right\} = -\left\{\begin{array}{l}\text{Energy of the projectile}\\ \text{and gases transported}\\ \text{out of the system}\end{array}\right\} = \left\{\begin{array}{l}\text{Net change in the}\\ \text{energy of the system}\end{array}\right\}$$

Exercises

4. Develop a conservation of momentum balance for a cannon firing a projectile. **Answer**: The conservation of momentum balance equation is

$$\left\{\begin{array}{l}\text{Net momentum of the projectile}\\ \text{and gases transported into the}\\ \text{system}\end{array}\right\} = -\left\{\begin{array}{l}\text{Momentum of the projectile}\\ \text{and gases transported out of}\\ \text{the system}\end{array}\right\} = \left\{\begin{array}{l}\text{Net change in}\\ \text{the momentum of}\\ \text{the system}\end{array}\right\}$$

(*Continued*)

EXAMPLE 2.3 (*Continued*)

5. Develop a balance equation for the conservation of electric charge in a system. **Answer**: The conservation of charge balance equation is

$$\left\{\begin{array}{l}\text{Net electric charge}\\ \text{transported into the system}\end{array}\right\} = \left\{\begin{array}{l}\text{Net change in electric}\\ \text{charge in (or on) the system}\end{array}\right\}$$

From the resulting conservation of energy and momentum balance equations developed in Example 2.3 and its Exercise 1, we can investigate the technology of the ballistic pendulum shown in Figure 2.10. The ballistic pendulum was developed in 1740 by the English mathematician and engineer Benjamin Robins (1707–1751) and operates on the principle that the deflection of the pendulum after impact is directly proportional to the projectile's impact velocity. Since the projectile is imbedded in the pendulum after impact, we choose to view this as a closed system consisting of the projectile and the pendulum. Since the system is closed, there can be no mass transport across the system's boundaries; and because momentum transport requires the mass to cross the system boundary, there is also no momentum transport in this system. Then, the conservation of momentum equation for a closed system reduces to {Net change in momentum of the system} = 0. Therefore, the initial momentum of the projectile/pendulum system must equal to the final momentum of this system, or

$$\left[m_{\text{projectile}} V_{\text{projectile}}\right]_{\text{initial}} = \left[\left(m_{\text{projectile}} + m_{\text{pendulum}}\right) V_{\text{pendulum}}\right]_{\text{final}}$$

After impact, the pendulum/projectile system swings through an angle θ, raising the center of gravity by an amount $h = R(1 - \cos\theta)$. Since the initial kinetic energy and the final potential energy of the pendulum/projectile system must be equal, we can write

$$\left(m_{\text{projectile}} + m_{\text{pendulum}}\right)\left(\frac{V^2_{\text{pendulum}}}{2g_c}\right) = \left(m_{\text{projectile}} + m_{\text{pendulum}}\right)\left(\frac{gh}{g_c}\right)$$

or

$$V_{\text{pendulum}} = [2gh]^{1/2} = [2gR(1-\cos\theta)]^{1/2}$$

Then, by combining these equations, we can determine the impact velocity of the projectile as

$$V_{\text{projectile}} = \left(\frac{m_{\text{projectile}} + m_{\text{pendulum}}}{m_{\text{projectile}}}\right) V_{\text{pendulum}} = \left(1 + \frac{m_{\text{pendulum}}}{m_{\text{projectile}}}\right)[2gR(1-\cos\theta)]^{1/2} \quad (2.19)$$

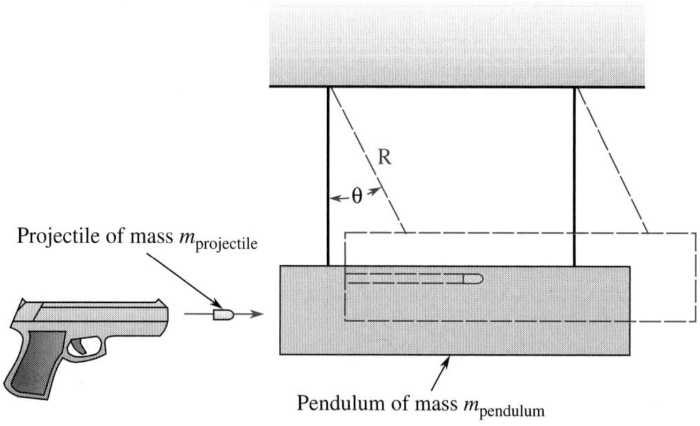

FIGURE 2.10
The operation of a ballistic pendulum.

Therefore, by knowing the masses of the pendulum and the projectile and measuring R and θ, we can easily calculate the impact velocity of the projectile. If the projectile is fired point blank into the pendulum, then the impact velocity is essentially the muzzle velocity of the projectile. These concepts are illustrated in Example 2.4.

WHAT HAPPENS ON IMPACT?

You might wonder why we do not set the initial kinetic energy of the projectile equal to the final potential energy of the pendulum/projectile system. Even though energy is conserved during the impact, this is not a perfectly elastic impact and other forms of energy are involved in addition to kinetic and potential. Some of the projectile's initial kinetic energy is converted into heat through the friction and deformation that occur during the impact. Figure 2.11 illustrates what happens when a .45 caliber bullet is fired into a $1\frac{1}{4}$ inch thick laminated Lexan plastic block. All the kinetic energy of the bullet has been absorbed by the plastic.

FIGURE 2.11
Bullet in a Lexan block.

EXAMPLE 2.4

Determine the muzzle velocity of a weapon fired point blank into a ballistic pendulum causing the pendulum to deflect 15°. The mass of the pendulum is 5.0 kg, the mass of the projectile is 0.01 kg, and the length of the pendulum support cable is 1.5 m (Figure 2.12).

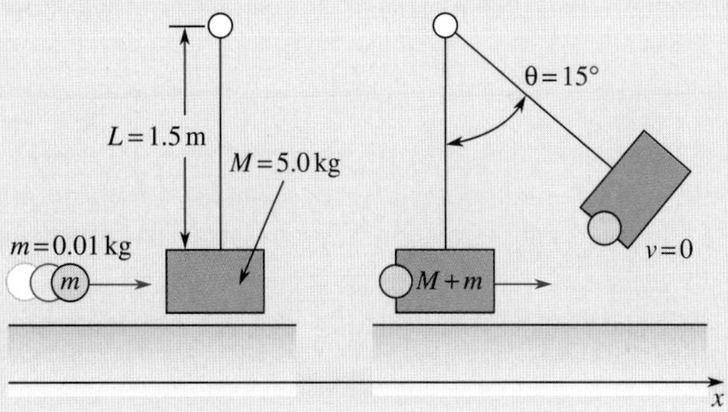

FIGURE 2.12
Example 2.4.

(Continued)

EXAMPLE 2.4 *(Continued)*

Solution

From Eq. (2.19), we compute the muzzle velocity as

$$V_{\text{projectile}} = \left(1 + \frac{m_{\text{pendulum}}}{m_{\text{projectile}}}\right)[2gR(1 - \cos\theta)]^{1/2} = \left(1 + \frac{5.0\ \text{kg}}{0.01\ \text{kg}}\right)[2(9.81\ \text{m/s}^2)(1.5\ \text{m})(1 - \cos 15°)]^{1/2}$$
$$= 5.0 \times 10^2\ \text{m/s}$$

Exercises

6. To bring down large game requires at least 2500. ft·lbf of impact energy (the impact energy of a high-speed projectile is its kinetic energy on impact). Determine the necessary impact velocity for the following projectiles (recall that there are 7000 grains in 1 lbm): (a) a 200. grain bullet, (b) 300. grain bullet, and (c) a 500. grain bullet. **Answers**: (a) 2370 ft/s, (b) 1940. ft/s, (c) 1500 ft/s.
7. Determine the displacement angle produced when a 0.3125 lbm baseball traveling at 90.0 miles per hour is "caught" by a ballistic pendulum having a 3.00 ft support cable and a mass of 180. lbm. **Answer**: 1.33°.

2.12 CONSERVATION OF MASS

An important application of the balance equation is to one of the basic conserved physical quantities, mass. Since mass is conserved in all nonnuclear reactions, its net production in any system is zero. Therefore, Eq. (2.10) tells us that the mass balance equation has the form

$$\left\{\begin{array}{c}\text{Net gain in mass by the}\\ \text{system during time } dt\end{array}\right\} = \left\{\begin{array}{c}\text{Net mass transported into}\\ \text{the system during time } dt\end{array}\right\} \tag{2.20}$$

This statement can be cast in mathematical form via Eq. (2.11) as $m_G = m_T$ (since $m_P = 0$), and Eq. (2.12) provides the rate form of this balance equation as $\dot{m}_G = \dot{m}_T$ (since $\dot{m}_P = 0$). In more precise mathematical language, the mass balance (MB) measured over some time interval δt can be written as

$$\text{Mass balance (MB) over time } \delta t = (\delta m)_{\text{system}} = \sum m_{\text{in}} - \sum m_{\text{out}} \tag{2.21}$$

and the mass *rate* balance (MRB) becomes

$$\text{Mass } rate \text{ balance (MRB)} = \left(\frac{dm}{dt}\right)_{\text{system}} = \sum \dot{m}_{\text{in}} - \sum \dot{m}_{\text{out}} \tag{2.22}$$

One of the most common uses of the conservation of mass balance equation is in chemistry. Chemical reaction equations are simply mass balances. Though reaction equations are not usually written as equalities, the left-hand side is the total mass of *reactants* used and the right-hand side is the total mass of *products* produced by the reaction. Since mass is conserved in chemical reactions, these two masses must be equal. For example, the reaction indicated by the equation $A + B = C + D$ means that the mass of A plus the mass of B is the same as the mass of C plus the mass of D. These equations are valid either for individual molecules or groups of molecules that bear the same reaction relation as the individual molecules. These molecular groups, called *moles*, form the macroscopic basis for chemistry and chemical engineering. This concept is illustrated in the following example.

EXAMPLE 2.5

The total mass of a system is conserved in a chemical reaction, but the mass of any particular chemical species is not necessarily conserved. Show that the following chemical reaction is just a closed system mass balance for the chemical species involved:

$$n_a A + n_b B \rightarrow n_c C + n_d D$$

where A, B, C, and D are the chemical species, and n_a, n_b, n_c, and n_d are their molar amounts.

Solution

When a chemical reaction occurs in a *closed system*, the mass transport term vanishes and then a mass balance for a chemical species X becomes "*the mass of X gained in the reaction must equal the mass of X produced by the reaction and this must equal the change in the mass of X due to the reaction"*: $m_{GX} = m_{PX} = \delta m_X$. Then, for a chemical reaction in a closed system, the mass balance for each of the chemical species present can be written as

$$m_{GA} = m_{PA} = \delta m_A$$

$$m_{GB} = m_{PB} = \delta m_B$$

$$m_{GC} = m_{PC} = \delta m_C$$

$$m_{GD} = m_{PD} = \delta m_D$$

If we add these equations together, we get

$$\delta m_A + \delta m_B + \delta m_C + \delta m_D = m_{PA} + m_{PB} + m_{PC} + m_{PD}$$

Now, since total mass must be conserved, it follows that

$$\sum m_P = m_{PA} + m_{PB} + m_{PC} + m_{PD} = 0$$

then the previous equation can be written as

$$\delta m_A + \delta m_B = -\delta m_C - \delta m_D$$

If we now convert this equation into a molar equation by dividing each mass term by its corresponding species molecular mass, then this equation becomes the stoichiometric chemical reaction equation $n_a A + n_b B \rightarrow n_c C + n_d D$, where

$$n_a = \frac{\delta m_A}{M_A}; \quad n_c = -\frac{\delta m_C}{M_C}$$

$$n_b = \frac{\delta m_B}{M_B}; \quad n_d = -\frac{\delta m_D}{M_D}$$

and M_A, M_B, M_C, and M_D are the molecular masses of chemical species *A*, *B*, *C*, and *D*. Consequently, all stoichiometric chemical reaction equations are just molar mass balances that utilize the conservation of mass law.

SUMMARY[7]

In this chapter, we provide the *definitions* of many of the concepts necessary in the development of thermodynamics. First, we review the *phases of matter* and define *thermodynamic equilibrium*. Then we see that one of the fundamental elements underlying modern technology is the *thermodynamic processes* that systems undergo. Next we reexamine and expand the *pressure and temperature scales* information given in Chapter 1. The *zeroth law of thermodynamics* is found to provide an axiomatic way to develop temperature measurement technology, and the *continuum hypothesis* is found to be a useful and valid analysis tool so long as the system's dimensions are very large compared to those of the molecules it contains. The *balance* and *conservation concepts* round out the chapter with an important discussion of the form of all the equations used to represent the basic thermodynamic laws in this book.

Some of the more important equations introduced in this chapter are as follows.

1. The equations for the conversion of temperature units, Eqs. (2.1)–(2.4):

$$T(°F) = \frac{9}{5}T(°C) + 32 = T(R) - 459.67$$

$$T(°C) = \frac{5}{9}[T(°F) - 32] = T(K) - 273.15$$

$$T(R) = \frac{9}{5}T(K) = 1.8T(K) = T(°F) + 459.67$$

$$T(K) = \frac{5}{9}T(R) = \frac{T(R)}{1.8} = T(°C) + 273.15$$

2. The equation for absolute to gauge pressure conversion:

$$\text{Absolute pressure} = \text{Gauge pressure} + \text{Local atmospheric pressure}$$

3. The general balance equations, (2.11) and (2.12), are

$$X_G = X_T + X_P \text{ and } \dot{X}_G = \dot{X}_T + \dot{X}_P$$

[7] Well, what do you think of this course so far? Isn't this a fascinating subject? I know, I know, it is a little vague right now, but it gets better.

4. The ballistic equation, (2.19), is

$$V_{\text{projectile}} = \left(\frac{m_{\text{projectile}} + m_{\text{pendulum}}}{m_{\text{projectile}}}\right) V_{\text{pendulum}} = \left(1 + \frac{m_{\text{pendulum}}}{m_{\text{projectile}}}\right)[2gR(1 - \cos\theta)]^{1/2}$$

5. The conservation mass, or mass balance (MB), Eq. (2.21), is

$$\text{Mass balance(MB) over time } \delta t = (\delta m)_{\text{system}} = \sum m_{\text{in}} - \sum m_{\text{out}}$$

and the mass rate balance (MRB), Eq. (2.22), is

$$\text{Mass rate balance(MRB)} = \left(\frac{dm}{dt}\right)_{\text{system}} = \sum \dot{m}_{\text{in}} - \sum \dot{m}_{\text{out}}$$

Important technical terms introduced in this chapter are given in Table 2.2.

Table 2.2 Glossary of Technical Terms Introduced in Chapter 2

thermodynamics	The science and technology that deal with the laws that govern the transformation of energy from one form to another
thermodynamic system	A volume containing the item chosen for thermodynamic analysis
system boundary	The surface of a thermodynamic system
isolated system	Any system in which neither mass nor energy crosses the system boundary
closed system	Any system in which mass does not cross the system boundary, but energy may cross the system boundary
open system	Any system in which both mass and energy may cross the system boundary
physical phase	A molecular configuration of matter, categorized as either solid, liquid, or vapor (or gas)
pure substance	A substance containing a uniform chemical composition in all its physical states
homogeneous system	A system containing only a single physical phase of a substance
simple substance	A homogeneous pure substance
thermodynamic state	The condition of a thermodynamic system as specified by the values of its independent thermodynamic properties
thermodynamic property	Any characteristic of a thermodynamic system that depends on the system's thermodynamic state and is independent of how that state is achieved
thermodynamic equation of state	A formula relating the dependent and independent properties of a system
intensive property	Any property of a homogeneous system that is independent of the system mass
extensive property	Any property of a homogeneous system that depends on the mass of the system
mechanical equilibrium	A situation where all the mechanical forces within a system are balanced so that there is no acceleration of the system
thermal equilibrium	A situation where there are no variations in temperature throughout the system
phase equilibrium	A situation where no phase transformations occur within the system
chemical equilibrium	A situation where no chemical reactions occur within the system
thermodynamic equilibrium	A situation where a system does not have the capacity to spontaneously change its state after it has been isolated
nonequilibrium thermodynamics	The study of systems that are not in thermodynamic equilibrium
thermodynamic processes	The path of thermodynamic states that a system passes through as it goes from an initial state to a final state
thermodynamic cycle	A situation where the final thermodynamic state of a process is identical with the initial thermodynamic state of the process
standard atmospheric pressure	14.696 psia, 29.92 inches of mercury, 101.325 kPa absolute
absolute pressure	Gauge pressure plus the local atmospheric pressure
gauge pressure	Absolute pressure minus the local atmospheric pressure
absolute zero temperature	−273.15°C or −459.67°F
zeroth law of thermodynamics	If system *A* is in thermal equilibrium with (i.e., is the same temperature as) system *C*, and system *B* is in thermal equilibrium with system *C*, then system *A* is in thermal equilibrium with system *B*
the continuum hypothesis	Large systems made up of many discrete molecules or atoms may be treated as though they were made up of a continuous material
microscopic system analysis	The analysis of a system at the atomic level.
statistical thermodynamics	The study of atomic level (i.e., microscopic) systems
macroscopic system analysis	The analysis of a system at the continuum level
the balance equation	An equation that accounts for all the changes in some quantity within a system
the conservation concept	If a quantity is neither produced nor destroyed, then it is said to be conserved

Problems (* indicates problems in SI units)

1. Define the following terms: (a) thermostatics, (b) open system, (c) extensive property, (d) equilibrium, and (e) zeroth law.
2. Define the following terms: (a) thermodynamics, (b) closed system, (c) intensive property, (d) macroscopic analysis, and (e) isolated system.
3.* A mixture of 1.0 kg of oxygen and 2.0 kg of hydrogen at atmospheric temperature and pressure are placed in a closed container. Explain (a) whether or not this mixture is a pure substance, and (b) whether or not it is a homogenous substance.
4. Dry ice (solid CO_2) and CO_2 vapor are in a sealed rigid container. Does the CO_2 in this system constitute (a) a pure substance, (b) a homogenous substance, (c) a simple substance?
5. Which of the following are extensive properties? (a) Temperature, (b) volume, (c) density, (d) work, (e) mass.
6. Are the following extensive or intensive properties? (a) Total energy, (b) temperature, (c) pressure, (d) mass.
7. Identify whether the following properties are intensive or extensive: (a) Specific energy, (b) total energy, (c) temperature, (d) molar mass.
8. Explain how color could be a thermodynamic property and indicate whether it would be an intensive or extensive property.
9. Let us say we have 2.00 ft^3 of a liquid/vapor mixture of motor oil at 70.0°F and 14.7 psia that weighs 97.28. lbf where g = 32.0 ft/s^2. List the values of three intensive and two extensive properties of the oil.
10.* There once was a man who had 1.00 m^3 of air at 20.0°EC and 0.100 MPa with a specific volume of 0.840 m^3/kg. List the values of three intensive and two extensive properties of his air (Figure 2.13).

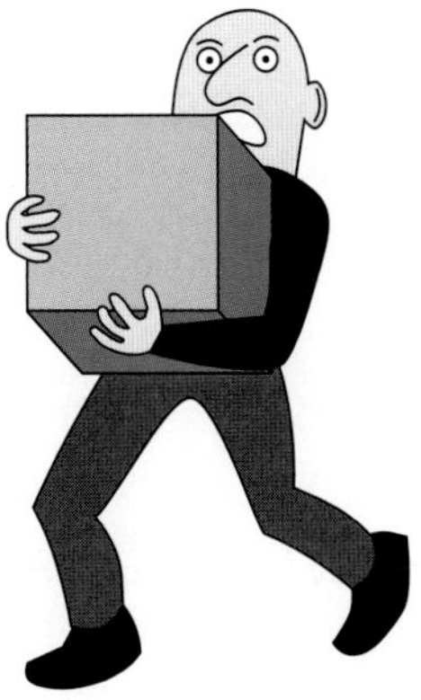

Intensive properties

1. ____________________

2. ____________________

3. ____________________

Extensive properties

1. ____________________

2. ____________________

FIGURE 2.13
Problem 10.

11. A woman once collected 2000. lbm of seawater at 50.0°F and 14.7 psia that occupied a volume of 56.0 ft^3. List the values of three intensive and two extensive properties of her seawater.
12. How many independent property values are required to fix the state of a pure substance subject to only one work mode?
13. Determine whether the following statements are true or false:
    a. The mass of a closed system is constant and its boundaries are not movable.
    b. An open system is defined as a system that can exchange only heat and work with its surroundings.
    c. An isolated system is completely uninfluenced by the surroundings.
    d. A thermodynamic property is a quantity that depends on the state of the system and is independent of the process path by which the system arrived at the given state.
    e. When a system in a given state goes through a number of processes and its final temperature is the same as the initial temperature, the system has undergone a thermodynamic cycle.
14. Identify the proper type of system (isolated, closed, or open) to be used in the analysis of each of the following and explain the reasons for your choice: (a) the Universe, (b) a kitchen refrigerator, (c) an electrical generator, (d) a hydraulic pump, (e) a living human being.
15. Determine the proper type of system (isolated, closed, or open) to be used in analyzing each of the following items and explain the reasons for your choice: (a) a bicycle, (b) a personal computer, (c) a stereo system, and (d) a lawn water sprinkler.
16. Establish the correct type of system (isolated, closed, or open) to be used in analyzing each of the following items and explain the reasons for your choice: (a) the solar system, (b) a carbonated drink dispensing machine, (c) an insulated box of fruit buried deep in the ground, and (d) a lightbulb.
17. Which of the basic types of systems (isolated, closed, or open) should be used to analyze each of the following items and explain the reasons for your choice: (a) a flying insect, (b) a black hole, (c) Niagara Falls, and (d) the great soaring whipple bird of Mars.
18.* As chief engineer for Thermodynamic Analysts Inc., your job is to determine the proper type of system (isolated, closed, or open) to be used in analyzing each of the following items submitted to your company and to explain the reasons for your choice: (a) a water pistol, (b) a flashlight battery, (c) a rubber boot, and (d) 2 kg chocolate fudge candy in a thermally insulated box buried deep in the Andes mountains.
19. An alien spacecraft has abducted you from a NASA space station and demands to know the type of system (isolated, closed, or open) NASA uses to analyze each of the following items, plus an explanation of the reasons for the choice: (a) the space station itself, (b) your pocket computer, (c) a can of tuna fish, and (d) an old Madonna movie.
20. An automobile internal combustion engine by itself is an open system because it draws in air and exhausts combustion products. How could one construct system boundaries around an operating internal combustion engine to cause it to become a closed system?
21.* Determine the specific volume of liquid water at 20.0°C that has a density of 998.0 kg/m^3.
22. Determine the mass of ammonia having a specific volume of 35.07 ft^3/lbm in a container measuring exactly 3 ft by exactly 2 ft by exactly 1 ft.
23.* Determine the volume of liquid mercury with a density of 3.87 kg/m^3 having a mass of 2.00 kg.
24. A system consists of a mixture of 2.00 ft^3 of a liquid having a density of 50.0 lbm/ft^3 and 4.00 lbm of a second liquid having a specific volume of 0.0400 ft^3/lbm. The specific volume of the mixture in the system in ft^3/lbm is (a) 0.208, (b) 2.08, (c) 0.022, (d) 2.048, or (e) none of these.

25. The equilibrium state of carbon at atmospheric pressure and temperature is graphite. If diamond is the equilibrium form of carbon only at very high pressures and temperatures, then why does diamond exist at atmospheric pressure and temperature?
26. Water in equilibrium at 70.0°F and 14.7 psia is in the liquid phase. If solid ice is in an equilibrium phase only at 32.0°F or lower at atmospheric pressure, why does solid ice still exist when you take it from the freezer and put it on the table at room temperature?
27. Sketch the following process paths on p–v coordinates starting from state (p_1, v_1).
   a. A constant pressure (isobaric) expansion from (p_1, v_1) to (p_2, v_2), where $v_2 = 2v_1$.
   b. A constant volume (isochoric) compression from (p_2, v_2) to (p_3, v_3), where $p_3 = 2p_2$.
   c. A process described by $p = p_1 + k(v - v_1)$, where k is a constant, from (p_3, v_3) back to (p_1, v_1) again.
28. Sketch the following thermodynamic cycle on $p - \mathcal{V}$ coordinates for a substance obeying the ideal gas equation of state, $p\mathcal{V} = mRT$.
   a. An isothermal compression (i.e., decreasing volume) from $(p_1, \mathcal{V}_1)$ to $(p_2, \mathcal{V}_2)$.
   b. An isobaric (i.e., constant pressure) expansion (i.e., increasing volume) from $(p_2, \mathcal{V}_2)$ to $(p_3, \mathcal{V}_3)$.
   c. An isothermal expansion from $(p_3, \mathcal{V}_3)$ to $(p_4, \mathcal{V}_4)$.
   d. An isochoric (i.e., constant volume) depressurization from $(p_4, \mathcal{V}_4)$ to $(p_5, \mathcal{V}_5)$.
   e. An isobaric compression from $(p_5, \mathcal{V}_5)$ back to $(p_1, \mathcal{V}_1)$.
29. A new thermodynamic cycle for an ideal gas is described by the following processes:
   a. An isothermal compression from $(p_1, \mathcal{V}_1)$ to $(p_2, \mathcal{V}_2)$.
   b. An isochoric compression from $(p_2, \mathcal{V}_2)$ to $(p_3, \mathcal{V}_3)$.
   c. An isobaric expansion from $(p_3, \mathcal{V}_3)$ to $(p_4, \mathcal{V}_4)$.
   d. An isothermal expansion from $(p_4, \mathcal{V}_4)$ to $(p_5, \mathcal{V}_5)$.
   e. An isochoric decompression from $(p_5, \mathcal{V}_5)$ back to $(p_1, \mathcal{V}_1)$.

   Sketch this cycle on pressure-volume coordinates.
30. Convert (a) 20.0°C into R, (b) 1.00°C into °F, (c) 56.0°F into °C, (d) 253°C into K, and (e) 1892°F into R.
31. Convert (a) 32.0°F into °C, (b) 500. R into °F, (c) 373 K into °C, (d) 20.0°C into R, and (e) −155°F into K.
32. Convert (a) 12.0°C into °F, (b) 6500. K into °C, (c) 1500. R into °F, (d) 120.°F into K, and (e) −135°C into K.
33. Convert (a) 8900. R into K, (b) −50.0°C into °F, (c) 3.00 K into °C, (d) 220.°C into R, and (e) 1.00 × 10^6°F into °C.
34. Convert the following temperatures into kelvin:(a) 70.0°F, (b) 70.0°C, (c) 70.0 R, and (d) 70.0° Reaumur. The Reaumur temperature scale was developed in 1730 by the French scientist René Antoine Ferchault de Réaumur (1683–1757). The freezing and boiling points of water at atmospheric pressure are defined to be 0° and 80.0° Reaumur, respectively.
35. Many historians believe that Gabriel Daniel Fahrenheit (1686–1736) had established his well-known temperature scale by 1724. It was based on three easily measured fixed points: the freezing temperature of a mixture of water and ammonium chloride (0.00°F), the freezing point of pure water (32.0°F), and the temperature of the human body (96.0°F). Later this scale was changed to read 212°F at the boiling point of pure water, which moved the body temperature from 96.0 to 98.6°F. Using the *original* Fahrenheit scale (freezing point of water = 32.0°F and body temperature = 96.0°F), determine
   a. The temperature of the boiling point of pure water.
   b. The conversion formula between the original Fahrenheit and the modem Celsius temperature scales.
36.* Convert the following pressures into the proper SI units:
   a. 14.7 psia.
   b. 5.00 atmospheres absolute.
   c. 1.00 × 10^5 dynes/cm^2 absolute.
   d. 30.0 lbf/ft^2 gauge.
   e. 12.4 poundals/ft^2 absolute.
37. Convert the following pressures into psia:
   a. 1000. N/m^2 gauge.
   b. 3.00 MPa-absolute.
   c. 11.0 Pa-gauge.
   d. 20.3 kN/m^2 absolute.
   e. 556 GPa-absolute.
38. Will the continuum hypothesis hold for the following thermodynamic states (and why)?
   a. Air at 20.0°C and atmospheric pressure.
   b. Liquid water at 70.0°F and 14.7 psia.
   c. Steam at 1.00 psia and 100.°F.
   d. Steam at 1.00 MPa absolute and 100.C.
   e. Air at 1.00 μN/m^2 absolute and 10.0 K.
39.* Convert the following pressures into MPa-absolute:
   a. 100. psig.
   b. 2,000. kPa-absolute.
   c. 14.7 psia.
   d. 1.00 Pa gauge.
   e. 500. N/m^2 absolute.
40. Convert the following pressures into lbf/ft^2 absolute:
   a. 14.7 psia.
   b. 100. lbf/ft^2 gauge.
   c. 0.200 MPa-absolute.
   d. 1200. kPa-gauge.
   e. 1500. psig.
41.* Convert the following pressures into N/m^2 absolute:
   a. 0.100 MPa-absolute.
   b. 14.7 psia.
   c. 25.0 psig.
   d. 100. Pa-absolute.
   e. 100. Pa-gauge.
42. In the late 18th century, it was commonly believed that heat was some kind of colorless, odorless, weightless fluid. Today we know that heat is not a fluid (it is primarily an energy transport due to a temperature difference), but we still have many old phrases and terms in our everyday and technical language that imply that heat is a fluid (e.g., heat "pours" out of a hot stove or heat always "flows" down a temperature gradient and so on). Discuss whether or not heat can be generated or absorbed, and using the balance concept, discuss whether or not it is a conserved quantity.
43. In Example 2.2, the Malthus population law was evaluated and found to produce an exponential growth or decay in the size of the population. A more sophisticated population model includes the effects of birth and death rates that vary linearly with the instantaneous size of the population as

$$\text{Birthrate} = \alpha_1 - \beta_1 N$$

and

$$\text{Death rate} = \alpha_2 + \beta_2 N$$

Ignoring the effects of immigration into the population, determine the rate balance equation for this population and show that it can be solved to produce the following population function:

$$N = \frac{\alpha/\beta}{(1 + e^{-\alpha t})}$$

where $\alpha = \alpha_1 - \alpha_2$ and $\beta = \beta_1 + \beta_2$. Note that this model predicts a limiting population size of α/β as $t \rightarrow \infty$.

44. Is the amount of gold reserves held by a nation a conserved quantity? Explain what happens if more currency is put into circulation while the currency base (e.g., gold reserves) is held constant.
45. Use the balance concept to explain the changes in the wealth of a nation. In particular, describe methods by which a nation can add or lose wealth by transport across its boundaries, and show how it can produce or destroy wealth within its boundaries.
46. Are the natural resources of a nation conserved in a thermodynamic sense? If not, explain what would have to be done to cause them to be conserved. Give a specific example where this is currently being done.
47. Using Eq. (2.9), write a balance equation for the total potential energy of a system during a time interval δt. Is this potential energy conserved? How can potential energy be produced or destroyed within a system?
48. From Eq. (2.9), develop a balance equation for the gain in kinetic energy of a system during a time interval δt. Is kinetic energy a conserved quantity? How can kinetic energy be produced or destroyed within a system?
49. Create a balance equation for the change in the number of chairs in a classroom during a time interval δt using Eq. (2.9). Are classroom chairs a conserved quantity? Describe how classroom chairs can be transported into and out of the classroom and produced or destroyed within the classroom.
50. Using Eq. (2.9), prepare a balance equation for the gain in the number of dollar bills in your pocket during a week. Are these dollar bills a conserved quantity? How can dollar bills be transported, created, or destroyed in your pocket?
51. Equation (2.10) provides a *net* balance equation for any quantity. Use this equation to construct a net balance equation for the change in the number of automobiles contained within the city limits of Detroit, Michigan, during a time interval of 1 year. What are some possible mechanisms for transport, production, and destruction of automobiles?
52. Reproduction and death are production and destruction mechanisms for humans. Using Eq. (2.10), develop a net balance equation to predict the net gain in people in any given family group over a time interval of 10 years.
53. Take a typical library as your system. Using Eq. (2.10), develop a net balance equation for the net gain in books in the library over the period of 6 months. Specify explicit transport, production, and destruction mechanisms for the books.
54.* Determine the muzzle velocity of a 10.0 g bullet that impacts a 5.00 kg ballistic pendulum 0.200 m below the fulcrum of the pendulum and produces a 25.0° deflection.
55.* You are designing a ballistic pendulum for a baseball throwing contest. The baseballs have a mass of 0.142 kg and the pendulum is a 10.0 kg horizontal hollow cylinder closed on the far end and suspended from the ceiling. Determine the length of the suspension cords if the deflection of the cylinder is not to exceed 20.0° when the baseball impact velocity is 40.0 m/s (89.5 mph).
56. The muzzle velocity of a Daisy Red Ryder BB gun is 260 ft/s and the mass of a BB is 7.5×10^{-4} lbm. Determine the angle of deflection of a 0.50 lbm ballistic pendulum suspended 2.0 ft from its support when impacted by the BB.
57. In 1945, the United States Army developed the largest artillery weapon ever constructed. Called the *Little David* (Figure 2.14), it has a bore of 900. mm and fires a 2.00 ton shell with a muzzle velocity of 200. ft/s that produces a crater 38.0 ft wide and 20.0 ft deep. As chief engineer of Army Ordnance, you are to design a ballistic pendulum for this gun. If the distance from the fulcrum of the pendulum to the point of impact is 10.0 ft, how much mass would the pendulum have to contain if deflection of its suspension cord is not to exceed 20.0°?

FIGURE 2.14
Problem 57.

58.* The Sandia National Laboratory hypervelocity two-stage light gas gun produces muzzle velocities of 12 km/s with 5.0 g projectiles. A ballistic pendulum is to be designed for this gun. It is to be suspended on cords 1.0 m long to produce a 15° deflection when impacted by the gun's projectile. Determine the mass required for the pendulum.
59. Since mass is a conserved quantity, use Eq. (2.21) to develop a conservation of mass balance equation over a time interval of 1 year for a system consisting of the entire Earth (do not forget to draw a sketch of your system).
60. A person is trying to fill a 10.0 gallon bucket by hand with a dipper (Figure 2.15). One dipper of water is added each second, and the dipper holds 1.0 lbm of water. Unfortunately, the bucket has a hole in it and water leaks out at a rate of 0.50 lbm/s. How long does it take for the person to fill the bucket? (Note: Water weighs 8.3 lbf/gal at standard gravity.)

FIGURE 2.15
Problem 60.

61.* Say you consume about 1.5 kg of solid food and about 1.0 kg of liquid beverage per day. If you do not produce any waste material during this time, draw a sketch of this system (you) and use Eq. (2.21) to determine the increase in your mass at the end of the day.

62.* The combustion chamber on a jet engine has air entering at a rate of 2.0 kg/s while the fuel enters through a fuel injector at a rate of 0.07 kg/s. Draw a sketch of this system and use Eq. (2.22) to determine the mass flow rate of exhaust gases from the engine.

63.* As your automobile travels down the highway, it consumes fuel at a rate of 2.0×10^{-3} kg/s and it consumes 20. kg of air for every kg of fuel burned. Draw a sketch of this system and use Eq. (2.22) to determine the mass flow rate of exhaust gases out the tail pipe of your automobile.

64.* A rigid tank is being filled with high-pressure oxygen gas at a rate of 1.3 lbm/h from an external source. Taking the tank as your system, draw a sketch of the system being filled and use Eq. (2.22) to determine the rate of gain in mass of the tank.

65.* A chemical reaction vessel has chemical *A* entering at a rate of 0.51 kg/m through a 0.020 m diameter pipe, chemical *B* entering at a rate of 0.75 kg/m through a 0.050 m diameter pipe, and chemical *C* entering at a rate of 0.011 kg/m through a 0.015 m diameter pipe. The reaction products are drawn off through two pipes at the bottom of the vessel at a rate of 0.35 kg/m in a small 0.015 m diameter pipe and 0.67 kg/m in a large 0.085 m diameter pipe. Determine the net of accumulation of chemicals in the vessel.

66. A new water spray nozzle head has 15 holes. If the mass flow rate of water into the nozzle head is 0.13 lbm/s, determine the mass flow rate of water through each hole.

67. A creative young engineer designed a hydraulic reaction multiplexer that contains a constant mass. The unit has five inlet pipes, numbered Inlet 1 through Inlet 5, and eight outlet pipes, numbered Outlet 1 through Outlet 8. Each inlet pipe has *twice* the mass flow rate of the previous numbered pipe (i.e., Inlet 2 has twice the mass flow rate of Inlet 1 and so forth), and each outlet pipe has *half* the mass flow rate of the previous numbered pipe (Figure 2.16). If the mass flow rate in Inlet 1 is 10. lbm/s, determine the mass flow rates in all the remaining inlet and outlet pipes.

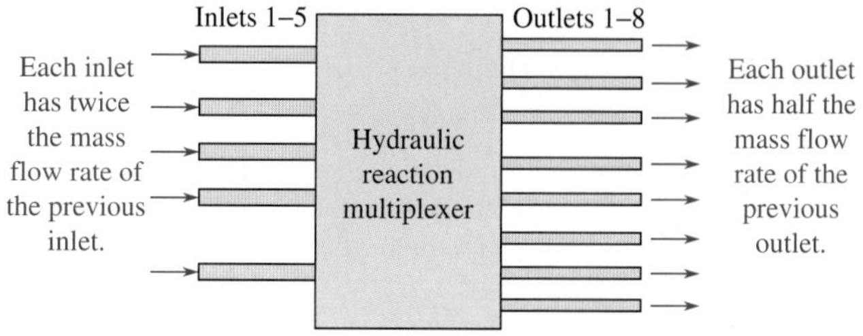

FIGURE 2.16
Problem 67.

68.* 2.0 kg of hydrogen (H_2) reacts with 16 kg of oxygen (O_2) to yield water (H_2O). Determine the chemical equation for this reaction on a kgmole basis, and find the amount of water formed in kg.

69. 12 lbm of carbon (C) reacts with 24 lbm of oxygen (O_2) to form 22 lbm of carbon dioxide (CO_2) plus an unknown amount of carbon monoxide (CO). Determine the amount of carbon monoxide formed in lbm, and find the chemical equation for this reaction on a lbmole basis.

Writing to Learn Problems

The following questions are designed to assist in the learning process through the development of writing skills. For these problems, you should develop a written answer containing an opening thesis sentence followed by the presentation of several supporting statements, ending with a concluding section that supports the thesis. Equations should be used only to supplement your written statements. Limit your response to about two double-spaced pages per question. You will need to find additional material in your library to complete these assignments.

70. Write a set of instructions to an engineering student friend defining a thermodynamic state and describing how to determine it from its thermodynamic properties. Illustrate your instructions with specific examples dealing with water.

71. Provide a detailed written explanation of a thermodynamic cycle. Give three specific examples of thermodynamic cycles. Chapter 9 contains numerous practical thermodynamic cycles from which you may choose.

72. Write a letter to a nontechnical friend in which you explain the zeroth law of thermodynamics. Define the law and create three nontechnical examples where it applies.

73. Write a short science fiction story based on the continuum hypothesis. First, describe the hypothesis as it is currently understood, then create an imaginary scenario where it does not work. Describe the consequences your new theory may have on world order.

74. Write a short science fiction story based on the balance concept. First, describe the concept as it is currently understood, then create an imaginary scenario in which a new, as yet undiscovered, term must be added to create a true balance. Describe the consequences your new theory has on physics today.

75. Write a 500 word article for your high school newspaper on the conservation of mass law. Is this a truly valid law or are there cases in which mass is not conserved? If it is not a truly valid law of physics, then why do we treat mass as conserved in most engineering applications?

CHAPTER 3

Thermodynamic Properties

CONTENTS

3.1 THE TREES AND THE FOREST

Thermodynamics is like a forest. The tall trees in the center of the forest are the laws of thermodynamics. They are surrounded by a thick underbrush of thorny bushes. These bushes are the thermodynamic properties. Some are known by their common names, such as *pressure*, *temperature*, and *volume*. Others have Latin and Greek names, like *energy*, *enthalpy*, *entropy*, and *exergy*. Before you can climb the tall trees of the thermodynamic laws to look out over your energetic future, you must find your way through the thermodynamic underbrush that surrounds them.

From the edge of the forest are barely visible paths that lead inward, but they intersect with other paths marked with obscure signs like "This Way to Adiabatic Heat Engines" and "This Way to the Isotropic Pumps." To survive in this wilderness, you need to understand the symbiotic relationships between the bushes and the trees—that is, the properties and the laws—and how to use them to solve engineering problems that enhance humankind. This chapter gives you the tools to clear the paths toward the thermodynamic laws of the next chapter.

3.2 WHY ARE THERMODYNAMIC PROPERTY VALUES IMPORTANT?

Since all of the basic laws of thermodynamics have terms containing key thermodynamic properties, we have to determine numerical values for these properties before the laws can be used to solve a thermodynamics problem. In other words, you cannot solve thermodynamic problems without accurate numerical values for the system's thermodynamic properties.

Thermodynamic property values can be determined from five sources:

1. Thermodynamic equations of state.
2. Thermodynamic tables.
3. Thermodynamic charts.
4. Direct experimental measurements.
5. The formulae of statistical thermodynamics.

This chapter deals with the first three sources. Source 4, the techniques of direct property measurement, are not discussed in this text, but the information given in many of the thermodynamic problem statements can be assumed to have come from such measurements. The last source, the formula of statistical thermodynamics, is covered in Chapter 18 of this textbook.

Property values are often given in thermodynamic problem statements in the *process path* designation. For example, if a system changes its state by an isothermal process at 250.°C, then we know that T_1 = 250.°C = T_2. Thus, the process path statement gives us the value of a thermodynamic property (temperature, in this case) in each of the two states. Process path statements that imply that some property is held constant during a change of state are quite common in thermodynamics.

3.3 FUN WITH MATHEMATICS

In the previous chapter, you were told that the values of any two thermodynamic properties are sufficient to fix the state of a homogeneous (single-phase) pure substance subjected to only one work mode. This means that each thermodynamic property of the a pure substance can be written as a function of any two independent thermodynamic properties. Thus, if x, y, and z are all intensive properties, we can write

$$f(x,y,z) = 0 \tag{3.1}$$

or

$$x = x(y,z)$$

$$y = y(x,z)$$

$$z = z(x,y)$$

Using the chain rule for differentiating the composite functions in the previous equations yields

$$dx = \left(\frac{\partial x}{\partial y}\right)_z dy + \left(\frac{\partial x}{\partial z}\right)_y dz$$

$$dy = \left(\frac{\partial y}{\partial x}\right)_z dx + \left(\frac{\partial y}{\partial z}\right)_x dz$$

$$dz = \left(\frac{\partial z}{\partial x}\right)_y dx + \left(\frac{\partial z}{\partial y}\right)_x dy$$

where the notation $(\partial x/\partial y)_z$ means the partial derivative of the function x with respect to the variable y while holding the variable z constant. Substituting the expression for dy into the expression for dx and rearranging gives

$$\left[1 - \left(\frac{\partial x}{\partial y}\right)_z \left(\frac{\partial y}{\partial x}\right)_z\right] dx = \left[\left(\frac{\partial x}{\partial y}\right)_z \left(\frac{\partial y}{\partial z}\right)_x + \left(\frac{\partial x}{\partial z}\right)_y\right] dz$$

Normally, the partial differential notation $(\partial x/\partial y)$ automatically implies that all the other variables of x are held constant while differentiation with respect to y is carried out. However, in thermodynamics, we always have a wide choice of variables with which to construct the function x, but when we change variables, we do not always change the functional notation. For example, we can write $x = x(y, z) = x(y, w) = x(y, q)$, where each of

CRITICAL THINKING

If you have a composite function of the form $f(x, y, z) = 0$, where $x = y + z$, $y = xz$, and $z = y/x$, then are the following chain rule differentials correct or not?

1. $dx = dy + dz$.
2. $dy = zdx + xdz$.
3. $dz = dy/x + dx/y$.

these three functions has a different form, even though they all yield x. Since these functions are *not* the same, if follows that their partial derivatives are *not* equal, or

$$\left(\frac{\partial x}{\partial y}\right)_z \neq \left(\frac{\partial x}{\partial y}\right)_w \neq \left(\frac{\partial x}{\partial y}\right)_q$$

That is why we *always* indicate which variables are held constant in partial differentiation. This also informs the reader as to which independent variables are being used in a functional relation.

Equation (3.1) tells us that two of the three variables are independent. If we choose the independent variables to be x and z, then the preceding equation, which relates x and z, is valid only if the coefficients of both dx and dz are equal to zero (otherwise, they would not be independent). Then we have

$$1 - \left(\frac{\partial x}{\partial y}\right)_z \left(\frac{\partial y}{\partial x}\right)_z = 0$$

or

$$\left(\frac{\partial x}{\partial y}\right)_z \left(\frac{\partial y}{\partial x}\right)_z = 1$$

so

$$\left(\frac{\partial x}{\partial y}\right)_z = \left[\left(\frac{\partial y}{\partial x}\right)_z\right]^{-1} \tag{3.2}$$

We also must have

$$\left(\frac{\partial x}{\partial y}\right)_z \left(\frac{\partial y}{\partial z}\right)_x + \left(\frac{\partial x}{\partial z}\right)_y = 0$$

or

$$\left(\frac{\partial x}{\partial y}\right)_z \left(\frac{\partial y}{\partial z}\right)_x = -\left(\frac{\partial x}{\partial z}\right)_y$$

Then, using the results of Eq. (3.2), we can write

$$\left(\frac{\partial x}{\partial y}\right)_z \left(\frac{\partial y}{\partial z}\right)_x \left(\frac{\partial z}{\partial x}\right)_y = -1 \tag{3.3}$$

EXAMPLE 3.1

Show that, if the pressure p of a substance is a function of its temperature T and its density ρ, we can write

$$\left(\frac{\partial p}{\partial T}\right)_p \left(\frac{\partial \rho}{\partial p}\right)_T = -\left(\frac{\partial \rho}{\partial T}\right)_p$$

Solution

From Eq. (3.3) with $x = p$, $y = T$, and $z = \rho$, we have

$$\left(\frac{\partial p}{\partial T}\right)_\rho \left(\frac{\partial T}{\partial \rho}\right)_p \left(\frac{\partial \rho}{\partial p}\right)_T = -1$$

(Continued)

EXAMPLE 3.1 (*Continued*)

Then, by multiplying this by $(\partial p/\partial T)_p$ and utilizing Eq. (3.2), we get the desired result:

$$\left(\frac{\partial p}{\partial T}\right)_\rho \left(\frac{\partial \rho}{\partial p}\right)_T = -\left(\frac{\partial \rho}{\partial T}\right)_p$$

Exercises

1. If $v = 1/\rho$ is the specific volume of the material in Example 3.1, show that the result in this example can be written as

$$\left(\frac{\partial p}{\partial T}\right)_v \left(\frac{\partial v}{\partial p}\right)_T = -\left(\frac{\partial v}{\partial T}\right)_p$$

2. If a, b, and c are three independent intensive thermodynamic properties, use Eqs. (3.2) and (3.3) to show that they can be related by

$$\left(\frac{\partial a}{\partial b}\right)_c = -\frac{\left(\frac{\partial a}{\partial c}\right)_b}{\left(\frac{\partial b}{\partial c}\right)_a}$$

3.4 SOME EXCITING NEW THERMODYNAMIC PROPERTIES

In Chapter 2, we introduced the specific volume v, an intensive property, as

$$v = \forall/m \tag{3.4}$$

where $\forall$ is the total volume[1] and m is the total mass of the system. We are now free to establish v as a function of any two other independent properties. For a single-phase (i.e., homogeneous) pure substance subjected to only one work mode, the pressure and temperature *are* independent properties, and for such a system, we can then write

$$v = v(p, T)$$

Differentiating this equation gives

$$dv = \left(\frac{\partial v}{\partial p}\right)_T dp + \left(\frac{\partial v}{\partial T}\right)_p dT$$

The coefficients of dp and dT in the previous equation reflect the dependence of volume on pressure and temperature, respectively. Because these terms have such important physical meaning, we introduce the following notation:

$$\frac{1}{v}\left(\frac{\partial v}{\partial T}\right)_p = \beta = \text{isobaric coefficient of volume expansion} \tag{3.5}$$

and

$$-\frac{1}{v}\left(\frac{\partial v}{\partial p}\right)_T = \kappa = \text{isothermal coefficient of compressibility} \tag{3.6}$$

where the thermodynamic term *isobaric* is from the Greek words *iso* meaning "constant" and *baric* meaning "weight" or "pressure"; the term is to be taken to mean *constant pressure* in this text. Therefore, we can write

$$dv = -v\kappa\, dp + v\beta\, dT$$

or

$$\frac{dv}{v} = \beta\, dT - \kappa\, dp \tag{3.7}$$

If κ and β are constant (or averaged) over small ranges of temperature and pressure, then Eq. (3.7) can be integrated to give

$$\ln\frac{v_2}{v_1} = \beta(T_2 - T_1) - \kappa(p_2 - p_1)$$

[1] Remember that, in this text, volume is represented by the symbol $\forall$.. The symbol V is reserved for the magnitude of velocity.

or

$$v_2 = v_1\{\exp[\beta(T_2 - T_1) - \kappa(p_2 - p_1)]\} \tag{3.8}$$

Thus, v is seen to have an exponential dependence on p and T when β and κ are constant. Table 3.1 gives values of β and κ for copper as a function of temperature, and Table 3.2 gives values of β and κ of various liquids at 20.°C (68°F).

Table 3.1 Values of β and κ for Copper as a Function of Temperature

	$\beta \times 10^6$		$\kappa \times 10^{11}$	
***T* (K)**	**R^{-1}**	**K^{-1}**	**ft^2/lbf**	**m^2/N**
100.	17.5	31.5	34.51	0.721
150.	22.8	41.0	35.08	0.733
200.	25.3	45.6	35.80	0.748
250.	26.7	48.0	36.47	0.762
300.	27.3	49.2	37.14	0.776
500.	30.1	54.2	40.06	0.837
800.	33.7	60.7	44.13	0.922
1200.	38.7	69.7	49.30	1.030

Source: Material drawn from Zemansky, M. W., 1957. Heat and Thermodynamics, fourth ed. McGraw-Hill, New York. Reprinted by permission of the publisher.

Table 3.2 Values of β and κ for Various Liquids at 20.°C (68°F)

	$\beta \times 10^6$		$\kappa \times 10^{11}$	
Substance	**R^{-1}**	**K^{-1}**	**ft^2/lbf**	**m^2/N**
Benzene	0.689	1.24	4550	95
Diethyl ether	0.922	1.66	8950	187
Ethyl alcohol	0.622	1.12	5310	111
Glycerin	0.281	0.505	1010	21
Heptane (*n*)	0.683	1.23	6890	144
Mercury	0.101	0.182	192	4.02
Water	0.115	0.207	2200	45.9

Source: Adapted by permission of the publisher from Zemansky, M. W., Abbott, M. M., Van Ness, H. C., 1975. Basic Engineering Thermodynamics, second ed. McGraw Hill, New York.

EXAMPLE 3.2

A 1.00 cm^3 copper block at 250. K is heated in the atmosphere to 800. K (Figure 3.1). Find the volume of the block at 800. K.

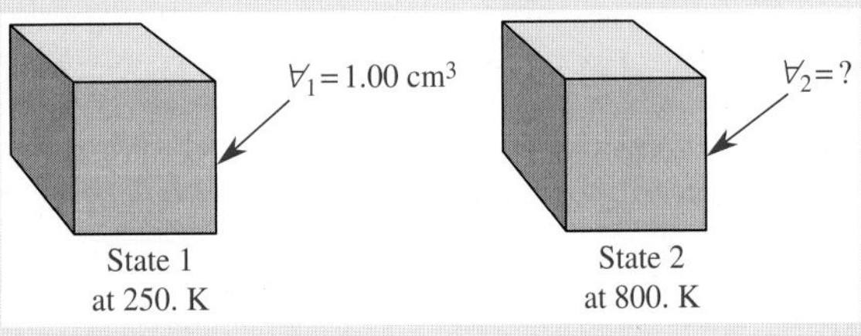

FIGURE 3.1
Example 3.2, problem.

Solution

Since the copper block changes state under atmospheric (constant) pressure, it undergoes an isobaric process. When p = constant, Eq. (3.8) reduces to

$$v_2 = v_1\{\exp[\beta(T_2 - T_1)]\}$$

(Continued)

EXAMPLE 3.2 *(Continued)*

and multiplying both sides of this equation by the mass m of the block gives the total volume $\forall$ as

$$\forall_2 = mv_2 = \forall_1\{\exp[\beta(T_2 - T_1)]\}$$

It can be seen from Table 3.1 that β for copper is not constant in the temperature range of 250. to 800. K. To come up with a reasonable value for an average β, we must see how β varies with temperature. Figure 3.2 shows the data for β vs. T for copper taken from Table 3.1.

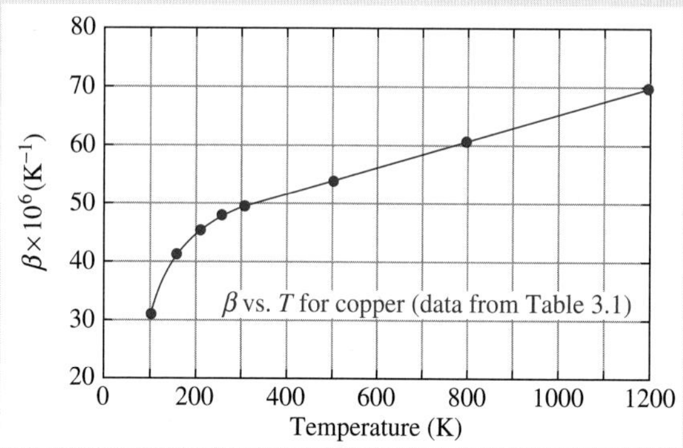

FIGURE 3.2
Example 3.2, solution.

This figure shows that β varies linearly with T in the range of 250. to 800. K. The average value of β in this temperature range is easily found to be

$$\beta_{avg} = \frac{60.7 \times 10^{-6} + 48.0 \times 10^{-6}}{2} = 54.4 \times 10^{-6}\,\text{K}^{-1}$$

Now we can calculate the final volume as

$$\begin{aligned}\forall_2 &= mv_2 = \forall_1\{\exp[\beta(T_2 - T_1)]\}\\ &= (1.00\ \text{cm}^3)\{\exp[(54.4 \times 10^{-6}\,\text{K}^{-1})(800. - 250.\ \text{K})]\}\\ &= 1.03\ \text{cm}^3\end{aligned}$$

Note that we could also fit a straight line to the β vs. T data between 250. and 800. K and come up with a formula of the form $\beta = C_1T + C_2$. Inserting this formula into Eq. (3.7) and integrating it (with $dp = 0$) yields a different (but equally valid) relation among v_1, v_2, T_1, and T_2. This is left as an exercise at the end of this chapter.

Exercises

3. Use Tables 3.1 and 3.2 to find values for the isobaric coefficient of volume expansion β and the isothermal coefficient of compressibility κ for (a) copper at 1200. K, (b) benzene at 68°F, and (c) mercury at 20.°C. **Answers**: (a) $\beta = 38.7 \times 10^{-6}$ R^{-1} = 69.7×10^{-6} K^{-1} and $\kappa = 49.30\ 10^{-11}$ ft^2/lbf = 1.030×10^{-11} m^2/N; (b) $\beta = 0.689 \times 10^{-6}$ R^{-1} = 1.240×10^{-6} K^{-1} and $\kappa = 4550 \times 10^{-11}$ ft^2/lbf = 95.0×10^{-11} m^2/N; (c) $\beta = 0.101 \times 10^{-6}$ R^{-1} = $0.182 \times 10^{-}$ K^{-1} and $\kappa = 192 \times 10^{-11}$ ft^2/lbf = 4.02×10^{-11} m^2/N.
4. Rework Example 3.2 for a 1.00 cm^3 block of solid platinum, whose average isobaric coefficient of volume expansion over the temperature range from 250. K to 800. K is 3.00×10^{-5} K^{-1}. **Answer**: $\forall$ = 1.017 cm^3.
5. Liquid water at 68°F is isothermally compressed from 14.7 psia to 3000. psia. Determine the percent change in the volume of the water, $[(v_1 - v_2)/v_1] \times 100$. **Answer**: 0.94%.

3.5 SYSTEM ENERGY

For historical reasons, the total energy of a system that has no magnetic, electric, surface, or other effects is divided into three parts. Classical physicists recognized two easily observable forms of energy: (1) the total kinetic energy KE = $mV^2/2g_c$, and (2) the total potential energy PE = mgZ/g_c. The remaining unobservable part of the total energy is simply called the *total internal energy* U. Thus, the total energy E of a system is written as

$$E = U + mV^2/2g_c + mgZ/g_c \tag{3.9}$$

or

$$E = U + \text{KE} + \text{PE} \tag{3.10}$$

WHO WAS AMALIE EMMY NOETHER?

PART 1

Emmy Noether was born on March 23, 1882, the first of four children. Her first name was Amalie, after her mother and paternal grandmother, but she began using her middle name at a young age. Emmy was taught to cook and clean—as were most girls of the time—and she took piano lessons. She pursued none of these activities with passion, although she loved to dance.

Emmy attended the Höhere Töchter Schule in Erlangen, Germany, from 1889 until 1897. She studied German, English, French, and arithmetic and was given piano lessons. In 1900, became a certificated teacher of English and French in Bavarian girls' schools.

We use the simpler Eq. (3.10) when writing general expressions and the more complete Eq. (3.9) when calculations are required. Total internal energy is an all-inclusive concept that includes chemical, nuclear, molecular, and other energies within the system. Since all mass has internal energy, the only systems that have zero internal energy are devoid of matter.

We define a system's *specific internal energy* u as

$$u = U/m \tag{3.11}$$

where m is the system mass. We can now write Eqs. (3.9) and (3.10) as

$$e = \frac{E}{m} = u + \frac{V^2}{2g_c} + \frac{gZ}{g_c} \tag{3.12}$$

and

$$e = u + \text{ke} + \text{pe} \tag{3.13}$$

where ke $= V^2/2g_c$ and pe $= gZ/g_c$.

The specific internal energy u is an intensive property, like pressure, temperature, and specific volume; so, it too can be written as a function of any other two independent properties. For a simple (i.e., homogeneous and pure) substance, the temperature and specific volume are independent thermodynamic properties. So we can write

$$u = u(T, v)$$

then

$$du = \left(\frac{\partial u}{\partial T}\right)_v dT + \left(\frac{\partial u}{\partial v}\right)_T dv \tag{3.14}$$

The first term in Eq. (3.14) describes the temperature dependence of u, and the coefficient of dT is written as c_v, where

$$\left(\frac{\partial u}{\partial T}\right)_v = c_v = \text{constant volume specific heat} \tag{3.15}$$

Then Eq. (3.14) becomes

$$du = c_v dT + \left(\frac{\partial u}{\partial v}\right)_T dv$$

Many of the equations of thermodynamics have groupings of similar terms. It is convenient to simplify the writing of these equations by assigning a single symbol and name to such a grouping. This is what was done in Eq. (3.9) in defining the total system energy as the sum of three other energy terms. Also, it should be quite clear that any function of a system's thermodynamic properties is also a thermodynamic property itself.

3.6 ENTHALPY

When we introduce the open system energy balance later in this text, we find that the properties u and pv are consistently grouped together. For simplicity, then, we combine these two properties into a new thermodynamic property called *enthalpy*, whose total and specific forms are defined as

$$H = U + p\forall = \text{total enthalphy} \tag{3.16}$$

and

$$h = H/m = u + pv = \text{specific enthalpy} \tag{3.17}$$

WHY IS IT CALLED *ENTHALPY*?

The quantity $u + pv$ has had many different names over the years. In the early years of thermodynamics, it was known at various times as the *heat function*, the *heat content*, and the *total heat*. The term *enthalpy* comes from the Greek ενθαλποσ, meaning "in warmth," and was introduced in 1922 by Professor Alfred W. Porter. He credited the coining of the name to the Dutch physicist Kamerlingh Onnes (1853–1926). This name was officially adopted by the American Society of Mechanical Engineers (ASME) in 1936.

EXAMPLE 3.3

The specific internal energy and specific volume of liquid water at a temperature of 20.0°C and a pressure of 20.0 MPa are 82.77 kJ/kg and 0.0009928 m³/kg, respectively (Figure 3.3). Determine the specific enthalpy of the water under these conditions.

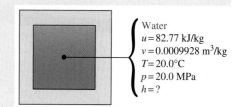

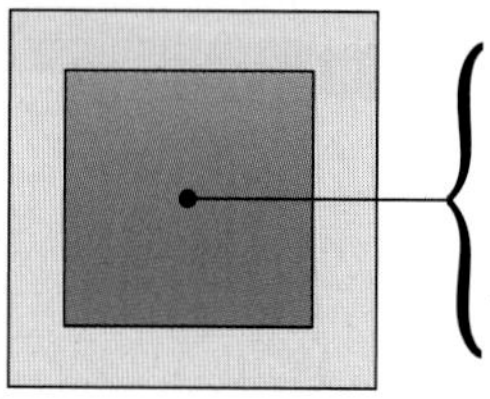

FIGURE 3.3
Example 3.3.

Solution

From Eq. (3.18), we have

$$h = u + pv = 82.77\,\text{kJ/kg} + (20.0\times 10^3\,\text{kN/m}^2)(0.0009928\,\text{m}^3/\text{kg})$$

$$= 82.77\,\text{kJ/kg} + 19.856\,\text{kN}\cdot\text{m/kg} = 103\,\text{kJ/kg}$$

where we use the fact that 1 kJ = 1 kN · m. Notice that the units of the specific internal energy u and the pv product must be exactly the same before these terms can be added. *Consequently, we converted the pressure into units of kN/m² here so that the pv product comes out in units of kN · m/kg = kJ/kg to match the units of u.*

Exercises

6. If the specific internal energy and specific volume of raw sewage at 150.0 psia and 500.0°F are 115.0 Btu/lbm and 0.01700 ft³/lbm, respectively, determine the specific enthalpy of this material under these conditions. **Answer**: h = 115.5 Btu/lbm. (Hint: Check the units carefully and note that u is in Btu/lbm whereas the pv product is in ft · lbf/lbm when the pressure is converted from lbf/in² into lbf/ft². Since the u and pv terms must be in the exact same units before they can be added, the pv product must be divided by 778.17 ft · lbf/Btu.)
7. If the specific enthalpy and specific volume of mercury vapor at 6.000 MPa and 719.7°C are 381.0 kJ/kg and 0.006930 m³/kg, respectively, determine the specific internal energy of this material under these conditions. **Answer**: u = 339.4 kJ/kg. (Hint: Check the units carefully and note that h is in kJ/kg or kN · m/kg, whereas the pv product is in MN · m/kg. Convert the pressure into kN/m² before or during the calculation so that h and the pv product have the same units.)
8. The specific internal energy and specific enthalpy of compressed liquid water at 5000. psia and 700.°F are 721.8 Btu/lbm and 746.6 Btu/lbm, respectively. Determine the specific volume of this material under these conditions. **Answer**: v = 0.0268 ft³/lbm. (Hint: See the hint for Exercise 6.)

Like specific internal energy, specific enthalpy can be a function of any two independent properties for a simple substance subjected to only one work mode. For such a simple substance, the temperature and pressure are independent, so we can write

$$h = h(T, p)$$

and

$$dh = \left(\frac{\partial h}{\partial T}\right)_p dT + \left(\frac{\partial h}{\partial p}\right)_T dp \tag{3.18}$$

The temperature dependence of h is important in classical physics, and the coefficient of dT is written as c_p, where

$$\left(\frac{\partial h}{\partial T}\right)_p = c_p = \text{constant pressure specific heat} \tag{3.19}$$

Then Eq. (3.18) becomes

$$dh = c_p\,dT + \left(\frac{\partial h}{\partial p}\right)_T dp \tag{3.20}$$

WHO WAS EMMY NOETHER?

PART 2

Emmy Noether never became a language teacher; instead she decided to attend the University of Erlangen to study mathematics. Unfortunately, at that time, women were not allowed to enroll because the faculty felt that allowing female students would "overthrow all academic order." She could only audit classes with the permission of each professor whose lectures she wished to attend. Nonetheless, on July 14, 1903, she passed the graduation exam.

During the winter of 1903–1904, she studied at the University of Göttingen, attending lectures by astronomer Karl Schwarzschild and mathematicians Hermann Minkowski, Otto Blumenthal, Felix Klein, and David Hilbert. By then, restrictions on women's rights in Erlangen were rescinded and she returned there. She officially reentered the university on October 24, 1904, and declared her intention to focus solely on mathematics. In 1907, she received a doctorate in mathematics.

Other thermodynamic properties, such as *entropy* and *availability*, are introduced later in this text when they are needed. It must be remembered, however, that not all thermodynamic properties are directly measurable. A pressure gauge and a thermometer give us numerical values for p and T, but there are no instruments that give us values of u and h directly. It takes much more sophisticated measurements to allow us to calculate accurate values for u and h. More complex mathematical relations between thermodynamic properties are developed after the reader is thoroughly familiar with the concept of entropy discussed in Chapter 7.

3.7 PHASE DIAGRAMS

A pure substance is composed of a single chemical compound, which may itself be composed of a variety of chemical elements. Water (H_2O), ammonia (NH_3), and carbon dioxide (CO_2) are all pure substances, but air is not because it is a mixture of N_2, O_2, H_2O, CO_2, and so forth. All substances can exist in one or more of the gaseous (or vapor), liquid, or solid physical states, and some solids can have a variety of molecular structures. In 1875, the American physicist Josiah Willard Gibbs (1839–1903) introduced the term *phase* to describe the different forms in which a pure substance can exist. We now speak of the gaseous, liquid, and solid phases of a pure substance, and we recognize that a pure substance may have a number of different solid phases.[2] Multiple solid phases are called *allotropic*, a term that comes from the Greek words *allos*, meaning "related to," and *trope* meaning "forms of the same substance." For example, graphite and diamond are allotropic forms of carbon.

A substance made up of only one physical phase is called *homogeneous*; if it is composed of two or more phases it is called *heterogeneous*. Coexistent phases are separated by an interface, called the *phase boundary*, of finite thickness across which the property values change uniformly. A system in which two phases coexist in equilibrium is called *saturated*.

The number of degrees of freedom within a heterogeneous mixture of pure substances is given by Gibbs's phase rule as

$$f = C - P + 2$$

where f is the number of degrees of freedom, C is the number of components (pure substances) in the mixture, and P is the number of phases. Also, f can be interpreted to be the number of intensive properties of the individual phases required to fix the state of the individual phases. For example, a homogeneous ($P = 1$) pure substance ($C = 1$) requires $f = 1 - 1 + 2 = 2$ intensive properties to fix its state. Similarly, a homogeneous ($P = 1$) mixture of two pure substances ($C = 2$) requires $f = 2 - 1 + 2 = 3$ intensive properties to fix its state, and so forth. The case of a two-phase ($P = 2$) pure substance ($C = 1$), however, is misleading, because $f = 1 - 2 + 2 = 1$, but this simply means that each

SATURATED WITH WHAT?

The term *saturated* comes from the 18th century, when heat was thought to be a fluid. At that time, it was thought that a substance could be saturated with heat, just like water can become saturated with salt or sugar. Today we recognize that heat is not a fluid, and therefore the use of the word *saturation* in reference to a thermodynamic phase change is really a misnomer. However, this term is now completely entrenched in modern thermodynamic literature and cannot be changed.

[2] Actually, matter can exist in a bewildering variety of phases beyond the common solid, liquid, and vapor forms. Ferromagnetic, antiferromagnetic, ferroelectric, superconducting, superfluid, nematic, smectic, and so on are all valid phases.

phase requires one intensive property to fix its state. Hence, two independent properties are required to fix the state of the complete two-phase system. To find the state of a mixture of two phases, we need to know how much of each phase is present, that is, the composition of the mixture. The phase composition in a liquid-vapor mixture is given by a new thermodynamic property called the *quality* of the mixture, which is defined shortly.

A phase diagram is made by plotting thermodynamic properties as coordinates. Figure 3.4 illustrates typical *p-T* and *p-v* phase diagrams for a substance that expands on freezing (such as water or antimony). When the *p-T* and *p-v* diagrams are combined to form a three-dimensional *p-v-T* surface, thermodynamic surfaces arise, as shown in Figure 3.5. Figures 3.6 and 3.7 show the similar plots for a substance that contracts on freezing (such as carbon dioxide and most other substances).

The expansion or contraction behavior of a substance on solidification can be deduced either from the increase or decrease in specific volume as the substance goes from a liquid to a solid, or from the slope of the fusion line on the *p-T* diagram. If the *p-T* fusion line has a negative slope, then the substance contracts on melting; if it has a positive slope, then it expands on melting.

The pure substance *p-T* phase diagram shown in Figures 3.4 and 3.5 is composed of three unique curves. The *fusion line* represents the region of two-phase solid-liquid equilibrium, the *vaporization line* represents the region

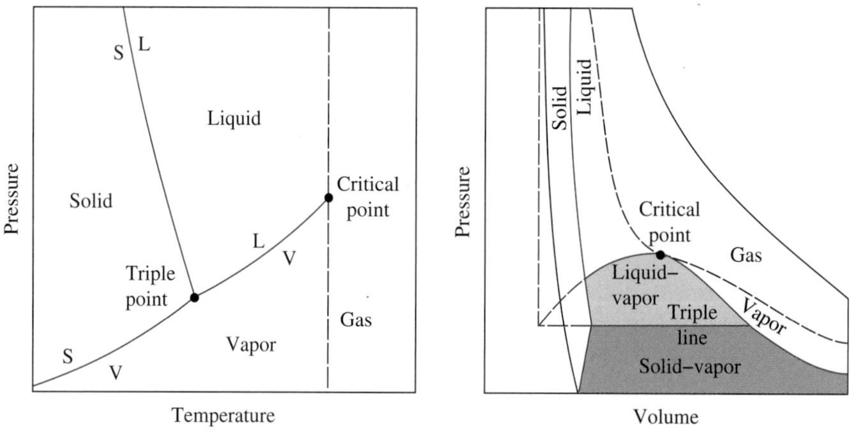

FIGURE 3.4

Pressure-temperature and pressure-volume diagrams for a substance that expands on freezing (for example, water).

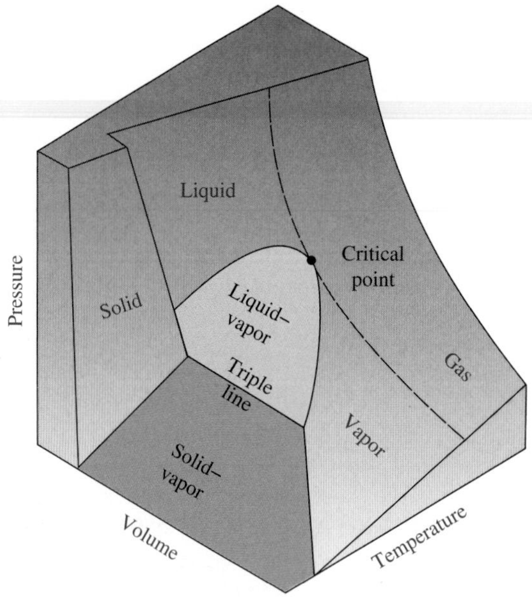

FIGURE 3.5

The *p-v-T* surface for a substance that expands on freezing (for example, water).

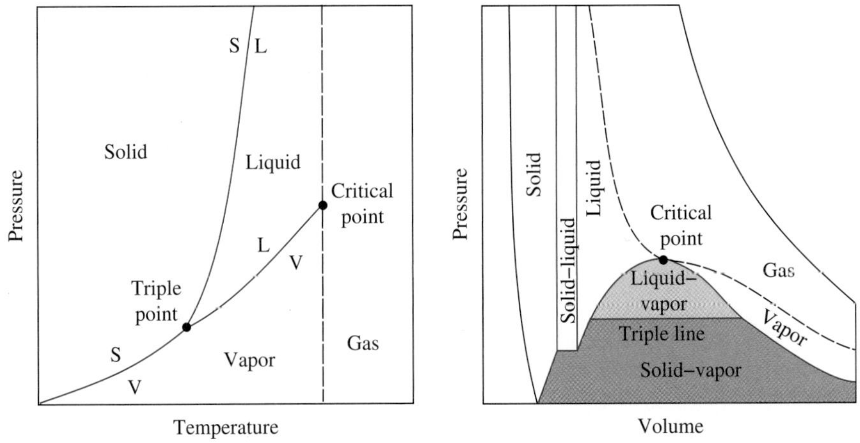

FIGURE 3.6

Pressure-temperature and pressure-volume diagrams for a substance that contracts on freezing (for example, carbon dioxide).

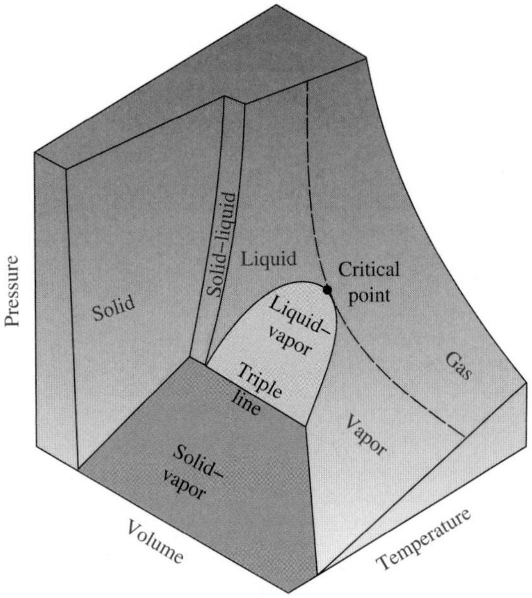

FIGURE 3.7

The p-v-T surface for a substance that contracts on freezing (for example, carbon dioxide).

of two-phase liquid-vapor equilibrium, and the *sublimation line* represents the region of two-phase solid-vapor equilibrium. These three lines intersect at one point, called the *triple point,* which is the only point where all three phases can be in equilibrium simultaneously. The triple point on the p-T diagram appears as a line on the p-v diagram, with the triple point simply being an end view of this line. Table 3.3 gives the property values at the solid-liquid-vapor triple point of various substances. At the triple point of a pure substance, $C = 1$, $p = 3$, and the number of degrees of freedom are $f = 1 - 3 + 2 = 0$; that is, there is no flexibility in the thermodynamic state and none of the properties can be varied and still keep the system at the triple point. The properties can be varied along the various two-phase boundary lines but not at the three-phase triple point. If a substance has more than one solid phase, then it also has more than one triple point.

When pressurized, most liquids freeze at a higher temperature because the pressure forces the molecules together. However, at pressures higher than 1 atmosphere, water remains liquid at a temperature below 0°C due to the strong hydrogen bonds in water. This is why ice melts under an ice skater's blades and lubricates her or his movement. The melting of ice under high pressures is also thought to contribute to the movement of glaciers.

When subjected to high pressures, water can form at least 15 solid phases. These phases differ by their crystalline structure, ordering, and density. In 2009, ice XV was found at extremely high pressures and −143°C. Figure 3.8

Table 3.3 Triple Point Data for Various Materials

Substance	*T* (R)	*T* (K)	*p* (psia)	*p* (kPa)
Ammonia (NH_3)	351.7	195.4	0.89	6.16
Carbon dioxide (CO_2)	389.9	216.6	75.98	523.8
Helium-4 (λ point)	3.9	2.17	0.74	5.11
Hydrogen (H_2)	24.9	13.84	1.03	7.13
Neon (Ne)	44.2	24.57	6.35	43.77
Nitrogen (N_2)	113.7	63.18	1.84	12.67
Oxygen (O_2)	97.8	54.36	0.02	0.15
Sulfur dioxide (SO_2)	355.9	197.7	0.02	0.17
Water (H_2O)	491.7	273.16	0.09	0.62

Source: Adapted by permission of the publisher from Zemansky, M. W., Abbott, M. M., Van Ness, H. C., 1975. Basic Engineering Thermodynamics, second ed. McGraw-Hill, New York.

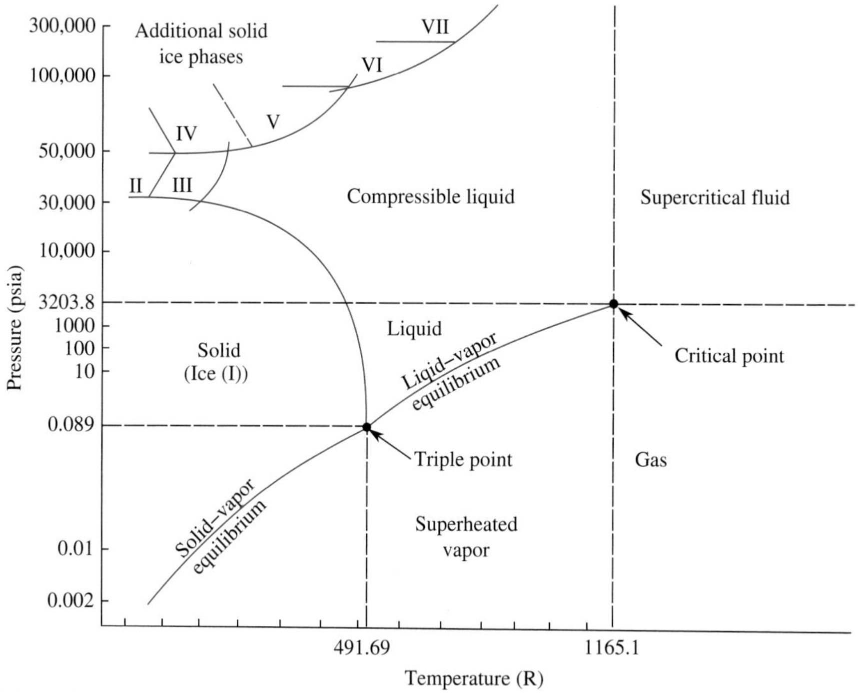

FIGURE 3.8
Phase diagram for water (not drawn to scale).

shows a more complete *p-T* phase diagram for water including 7 of its 15 known solid phases. Each intersection of three phase transition lines forms a new triple point.

The vaporization curve for all known substances has a peak at a curious point, known as the *critical point*. This is the state at which the densities of the liquid and the vapor phases become equal and, consequently, where the physical interface between the liquid and the vapor disappears. At or above the critical state, there is no longer any physical difference between a liquid and a vapor. Substances existing under these conditions are called *gases*. In this text, we use the term *gas* to describe the state of any substance whose temperature is greater than its critical state temperature. A substance in the vapor phase that does not meet the definition of a gas is called a *superheated vapor* (sometimes just *vapor*). These definitions are illustrated in Figures 3.8 and 3.10. Table 3.4 gives the critical state temperature, pressure, and specific volume for various common pure substances. A larger critical state data table is given in Table C.12 of *Thermodynamic Tables to accompany Modern Engineering Thermodynamics*.

Notice in Table 3.4 that, at 14.7 psia and 70.°F (530. R), ammonia is a *vapor* ($T_c > 530.$ R). Also, it should be clear from Figure 3.10 that, to liquefy any gas whose pressure is initially less than its critical pressure simply by increasing its pressure alone, the gas must first be made into a vapor by lowering its temperature below its critical temperature. Vapor-liquid condensation is shown by process *A-B* in Figure 3.10. Thus, for example, no matter how high the applied pressure, hydrogen cannot be liquefied unless its temperature is below 59.9 R (see Table 3.4).

WHAT IS A "PHASE"?

A material "phase" is a physically distinct region of space that is chemically uniform with homogenous physical properties. For example, imagine a system consisting of ice cubes and liquid water in drinking glass. The ice cubes are one phase (solid), the water is a second phase (liquid). The glass itself is a different material in a solid phase.

Material phases are different states of matter, such as solid, liquid, gas, or plasma. It is possible for a material to have more than one liquid or solid phase. For example, depending on the cooling process, metals can solidify into several distinct crystal phases.

The liquid to vapor phase transformation is called *vaporization* (Figure 3.9), and the vapor to liquid phase transformation is called *condensation*. Similarly, the solid to liquid phase transformation is known as *melting*, and the liquid to solid phase transformation is called *freezing* or *solidification*. Finally, the solid to vapor phase transformation is known as *sublimation*, and the vapor to solid phase transformation is *deposition* (or *frost* in the case of water).

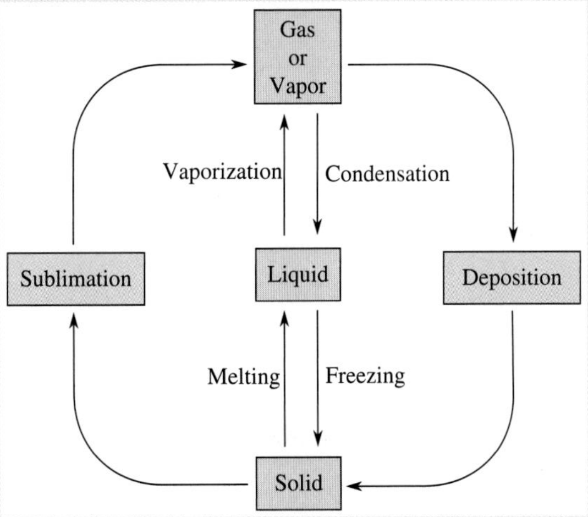

FIGURE 3.9
What is a phase?

FROM WHENCE COMETH THE GAS?

The term *gas* was coined by the Belgian chemist Jan Bapist Van Helmont (1577–1644), derived from the Greek word κεøσ, meaning "gaping void."

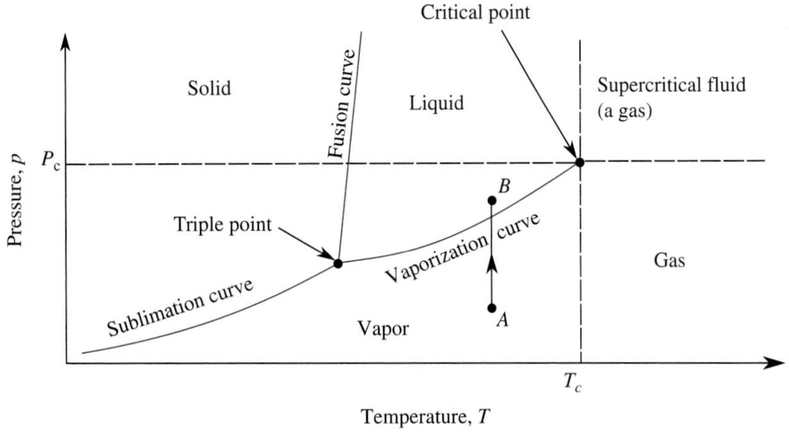

FIGURE 3.10
The definitions of a gas and a vapor.

Several thermodynamic properties have discontinuities at the critical state; β, κ, and c_p become infinite there. Near the critical state, a transparent substance becomes almost opaque due to light scattering caused by large fluctuations in local density. This phenomenon, called *critical opalescence*, is illustrated in Figure 3.11. Notice the appearance of the liquid-vapor interface in Figure 3.11b when the temperature becomes less than the critical temperature.

Table 3.4 Critical State Properties for Various Substances (see also Table C.12)

Substance	T_c (R)	T_c (K)	p_c (psia)	p_c (MPa)	v_c (ft^3/lbm)	v_c (m^3/kg)
Ammonia (NH_3)	729.9	405.5	1636	11.28	0.068	0.0043
Carbon dioxide (CO_2)	547.5	304.2	1071	7.39	0.034	0.0021
Carbon monoxide (CO)	240.0	133.0	507.0	3.50	0.053	0.0033
Helium (He)	9.5	5.3	33.2	0.23	0.231	0.0144
Hydrogen (H_2)	59.9	33.3	188.1	1.30	0.516	0.0322
Nitrogen (N_2)	227.1	126.2	491.68	3.39	0.051	0.0032
Oxygen (O_2)	278.6	154.8	736.9	5.08	0.039	0.0024
Sulfur dioxide (SO_2)	775.2	430.7	1143	7.88	0.030	0.0019
Water (H_2O)	1165.1	647.3	3203.8	22.09	0.051	0.0032

Source: Van Wylen, G. J., Sonntag, R. E., 1986. Fundamentals of Classical Thermodynamics, third ed. Wiley, New York. Reprinted by permission of John Wiley & Sons.

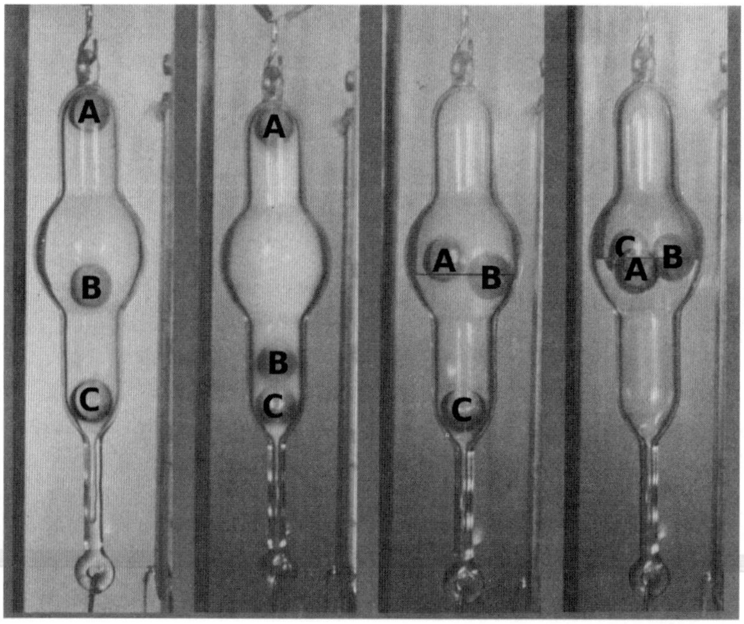

FIGURE 3.11

The glass bulb contains carbon dioxide near the critical density $\rho_{critical}$, and three balls with densities $\rho_A \lesssim \rho_{critical}$ $\rho_B = \rho_{critical}$, and $\rho_C \gtrsim \rho_{critical}$. In (a), the temperature is well above the critical temperature, leaving all the carbon dioxide in the gaseous state. In (b), the temperature is only slightly above the critical temperature and the carbon dioxide has become foggy. In (c), the temperature is slightly below the critical temperature and a meniscus has developed separating the gaseous and liquid states. In (d), the temperature is far below the critical temperature and the density of the liquid has increased to the point where all three balls now float on the surface of the liquid. *(Source: Reprinted with permission from Sengers J. V., Sengers, A. L., 1968. The critical region. Chem. Eng. News 48, 104.)*

HOW DO YOU MAKE A DIAMOND?

Diamond is the hardest naturally occurring material known, and it is also the most popular gemstone (Figure 3.12). For centuries, it has been one of the most desirable and mysterious materials available. In 1772, the French Chemist Antoine-Laurent Lavoisier (1743–1794) proved that diamond was just another crystalline form of carbon. He invested a considerable sum of money to purchase a small diamond, then he burned it in a controlled oxygen environment. When he analyzed the resulting combustion gas, he found it to be just carbon dioxide.

FIGURE 3.12
A real diamond.

From that time forward, many attempts have been made to make synthetic diamond from pure carbon (graphite). However, diamond was not successfully synthesized until 1955, at the General Electric Corporation in Schenectady, New York, when GE researchers compressed graphite to a pressure exceeding 1.5×10^6 psi (10. GPa) and 5000.°F (~3000.°C). Industrial and gemstone quality synthetic diamonds have been commercially available since 1960.

The pressure-temperature phase diagram for carbon is shown in Figure 3.13. At low pressure and temperature, the solid carbon phase is called *graphite*. At very high pressures and temperatures, a second solid carbon phase appears with a different atomic structure. This phase is the valuable gemstone *diamond* with which we are all familiar. However, the phase diagram clearly indicates that graphite, not diamond, is the equilibrium form of solid carbon at room temperature and pressure.

Since phase changes are rate processes that increase rapidly with increasing temperature, what is happening to all diamonds that are exposed to room temperature and pressure? What would happen to a diamond ring if you put it into an oven at 1000 or 2000°F? Can you suggest a practical way of making a synthetic diamond? Do you think that exposing graphite to explosive loading using dynamite under controlled conditions would work?

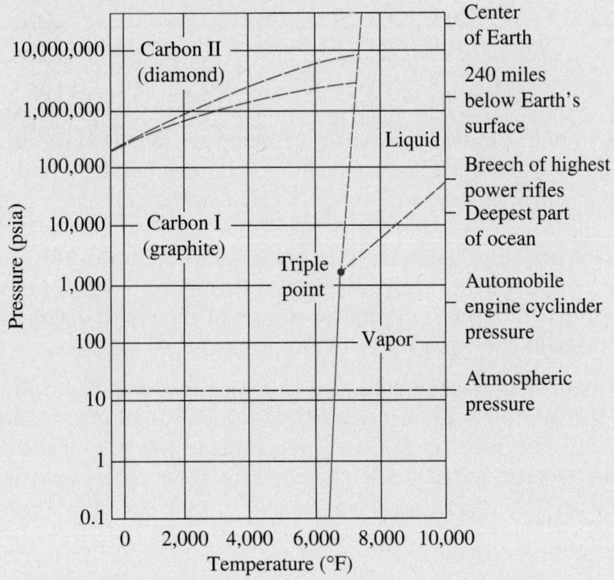

FIGURE 3.13
Phase diagram of a diamond.

CASE STUDY: A NEW SUPERCRITICAL WATER TREATMENT

A major engineering challenge today is the development of an effective waste disposal or destruction technique that produces no toxic waste or emission itself. Hazardous wastes have historically been discarded in ocean dumping, landfills, incineration, or long-term storage. However, ocean dumping is now illegal and landfill sites are becoming increasing difficult to manage due to concerns about groundwater contamination.

A new technology has emerged as an effective way to eliminate thousands of tons of organic wastes that are the by-products of modern society. Called *supercritical water oxidation,* the technique exploits the fact that, while most organic wastes are not soluble in water at normal temperatures, they are dissolved at high pressure and temperature (Figure 3.14). Once the organic wastes are dissolved, oxygen is added and an oxidation-reduction reaction occurs, much like a controlled combustion process. Operating at supercritical conditions results in a single-phase homogenous reaction environment that causes rapid oxidation of the organics, producing carbon dioxide, water, nitrogen, and small amounts of other compounds, such as ammonia and acids. Since the entire process is in a closed system, no harmful products are released into the environment.

This technique has also been found to be an effective disposal method for surplus military chemical wastes, such as nerve agents, mustard gas, rocket fuels, TNT, and other explosives. More than 99.9% of the explosives are destroyed in less than 30 s at 600.°C. Even radioactive wastes can be concentrated and stabilized by eliminating their organic components. The resulting radioactive components can then be encased in molten glass and stored deep underground.

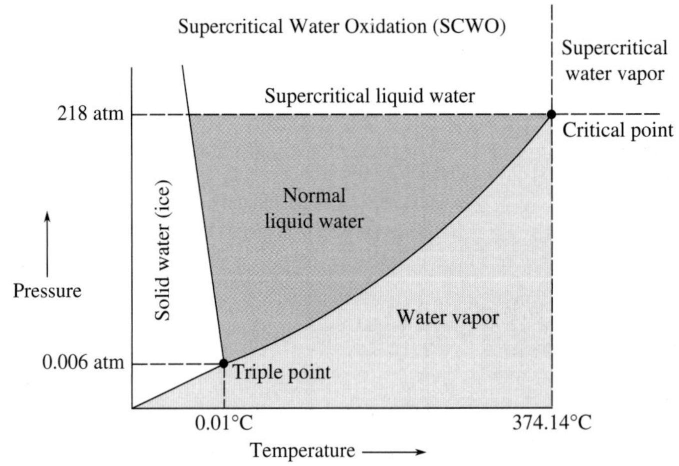

FIGURE 3.14
Supercritical water oxidation.

3.8 QUALITY

As mentioned earlier, in an equilibrium two-phase mixture, temperature and pressure cannot be varied independently; therefore, either one or the other can be taken as an independent thermodynamic property, but not both. Figure 3.15a shows the actual *p-v* diagram for water on log-log coordinates. Notice that, in the two-phase regions (liquid plus vapor and solid plus vapor), the isotherms (lines of constant temperature) are parallel to the isobars (lines of constant pressure), showing that pressure and temperature are not independent in this region. To determine other thermodynamic properties of a mixture of phases, we need to know the amount of each phase present. We do this with a *lever rule* applied to one of the phase diagram coordinates. Consider the simplified liquid-vapor *p-v* diagram shown in Figure 3.15b.

Substances whose states lie *on* the saturation curve are called *saturated.* Substances whose states lie *under* the saturation curve are called *wet.* Substances whose states are on the saturation curve but to the *left* of the critical state are called *saturated liquids,* and those on the saturation curve to the *right* of the critical state are called *saturated vapors.* Substances whose states are to the left of the saturation curve are called *compressed* or *subcooled liquids,* and those to the right are called *superheated vapors.*

To help identify the properties of a system, we adopt the convention of using an *f* subscript on the symbols of all thermodynamic properties of saturated liquids and a *g* subscript on the symbols of all thermodynamic properties of saturated vapors. Thermodynamic properties in the compressed (or subcooled) liquid region, the wet (or mixture) region, and the superheated vapor (or gas) region carry *no* subscripts. Consequently, the specific

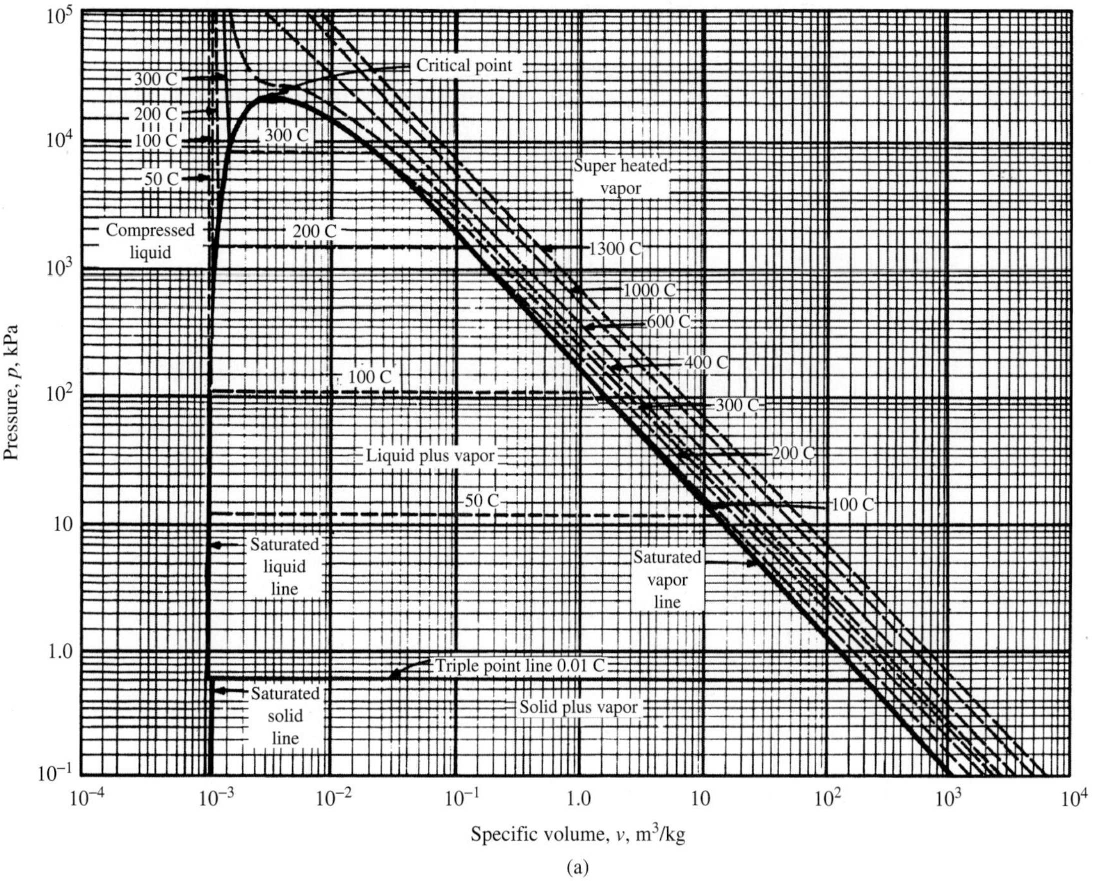

FIGURE 3.15a

p-v diagram notation. The actual *p-v* diagram for water plotted on log-log coordinates. (*Source: Wood, B. D., 1982. Applications of Thermodynamics, second ed. Addison-Wesley Publishing Co., Inc., Reading, MA. Reprinted with permission.*)

volume of saturated liquid is written as v_f and that of saturated vapor as v_g, and the associated specific internal energies, enthalpies, and masses are written as u_f, u_g, h_f, h_g, m_f, and m_g.

From Figure 3.16, we see that the *total* volume of a substance whose state is in the wet (liquid plus vapor) region is given by

$$\forall = mv = m_f v_f + m_g v_g$$

where m is the total mass given by

$$m = m_f + m_g$$

Dividing the equation for $\forall$ by m gives

$$\forall/m = v = m_f v_f/m + m_g v_g/m \quad (3.21)$$

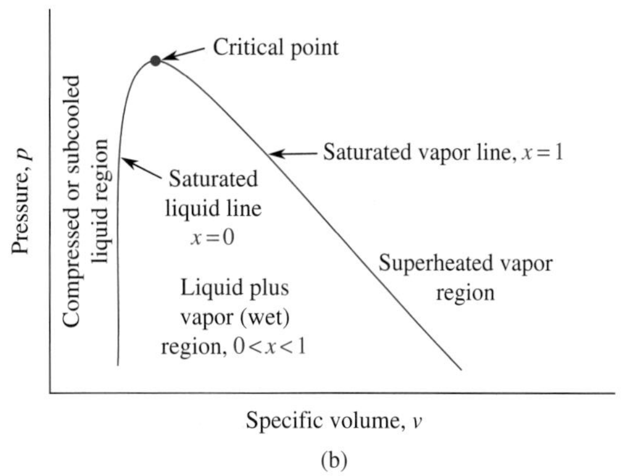

FIGURE 3.15b

Schematic *p-v* diagram for the liquid, mixture, and vapor regions.

WHY USE *F* AND *G* SUBSCRIPTS?

The use of *f* and *g* as subscripts here is out of tradition. They come from the first letters of the German words *flussig* (for liquid) and *gas*. More appropriate English subscripts might be *l* for liquid and *v* for vapor, but they are not used.

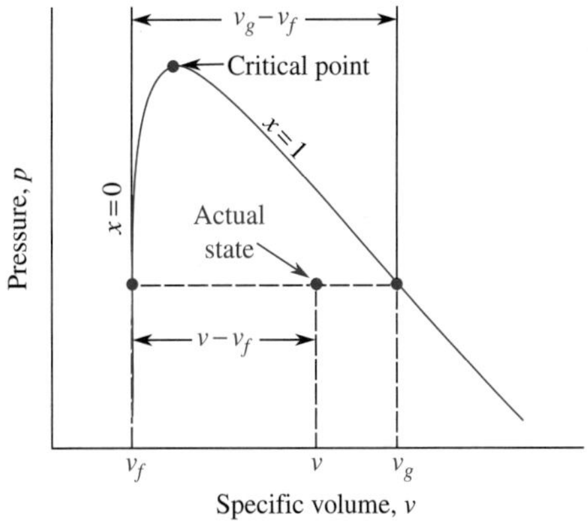

FIGURE 3.16
The lever rule for calculating quality.

Equation (3.21) is just a simple mass-based lever rule equation relating a mixture thermodynamic property (v) to the thermodynamic properties of the components of the mixture (v_f and v_g). We now define the *quality x* of a liquid plus vapor mixture as the relative amount of vapor present, or

$$\text{Quality} = x = \frac{\text{Mass of vapor}}{\text{Mass of vapor + Mass of liquid}} = \frac{\text{Mass of vapor}}{\text{Total mass}}$$

or

$$x = \frac{m_g}{m_f + m_g} = \frac{m_g}{m} \tag{3.22}$$

Therefore, Eq. (3.21) can be written as

$$v = (1 - x)v_f + xv_g \tag{3.23}$$

which can be rearranged as

$$v = v_f + x(v_g - v_f) = v_f + xv_{fg} \tag{3.24}$$

where we define the magnitude of the liquid to vapor property change as

$$v_{fg} = v_g - v_f \tag{3.25}$$

From Figure 3.16, we see that another definition of quality is

$$x = \frac{v - v_f}{v_{fg}} \tag{3.26}$$

It should be clear from the definition of quality that its *value* has the following bounds

$$\text{Saturated liquid}: \ x = 0$$

$$\text{Saturated vapor}: \ x = 1$$

$$\text{Wet (liquid plus vapor) region}: \ 0 < x < 1$$

WHY DO THEY CALL IT *QUALITY*?

In the 19th century, there were a lot of steam engines. Railroads, factories, ships, and so on all used steam engines as their source of power. The people responsible for keeping these engines running noticed that they worked better if the steam contained more vapor than liquid. So, a mixture that contained a lot of vapor and little liquid was said to be of high *quality*. They defined this *quality* to be the ratio of the mass of vapor to total mass of liquid plus vapor (Figure 3.17), or quality = $x = m_g/(m_f + m_g) = m_g/m$. On the other hand, the amount of liquid present in a mixture is called the *moisture* of the mixture, defined as moisture = $m_f/m = 1 - x$. Since these definitions apply only to mixtures of liquid plus vapor, they do not extend outside of the vapor dome. The condition where the quality $x = 1.0$ is reserved for saturated vapor and does not apply to superheated vapor. Similarly, the condition where quality $x = 0$ is reserved for saturated liquid and does not apply to compressed liquid. Note that the quality x of a liquid-vapor mixture *can never be less than 0 or greater than 1.0. That is, x always falls in the range* $0 < x < 1.0$.

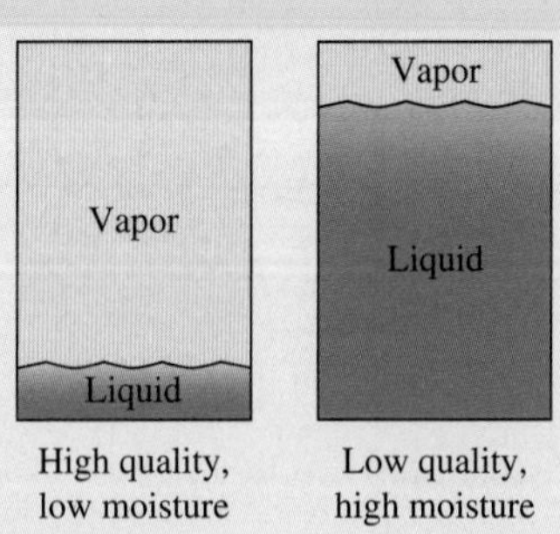

FIGURE 3.17
Why do they call it *quality*? In all other single-phase regions (compressed liquid, superheated vapor, gaseous), *x is not defined*, because these are single-phase homogeneous regions. Note that, since the numerical value of quality is restricted to lie in the range from 0 to 1, *no correct calculation can ever give a value of x less than 0 or greater than 1.0.*

Although Eq. (3.26) was developed using specific volume, an identical argument can be used to expand it to all other intensive properties (except pressure and temperature), resulting in equations of the form

$$x = \frac{v - v_f}{v_{fg}} = \frac{u - u_f}{u_{fg}} = \frac{h - h_f}{h_{fg}} \tag{3.27}$$

In addition, the term $m_f/m = 1 - x$ represents the relative amount of liquid present in the mixture, called the *moisture* of the mixture.

EXAMPLE 3.4

Saturated water at 14.696 psia and 212°F has the following properties:

$v_f = 0.01672\ \text{ft}^3/\text{lbm}$ $v_g = 26.80\ \text{ft}^3/\text{lbm}$
$u_f = 180.1\ \text{Btu/lbm}$ $u_g = 1077.6\ \text{Btu/lbm}$
$h_f = 180.1\ \text{Btu/lbm}$ $h_g = 1150.5\ \text{Btu/lbm}$

If 0.200 lbm of saturated water at 14.696 psia is put into a sealed rigid container whose total volume is 3.00 ft^3 (Figure 3.18), determine the following properties of the system:

a. The specific volume v.
b. The quality, x, and moisture, $1 - x$.
c. The specific internal energy u.
d. The specific enthalpy h.
e. The mass of water in the liquid and vapor phases, m_f and m_g.

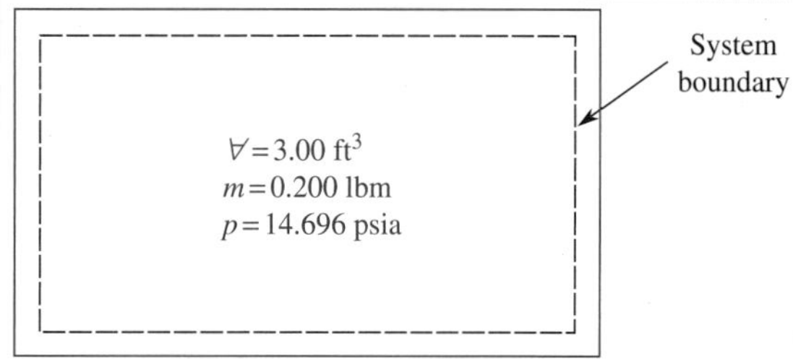

FIGURE 3.18
Example 3.4.

Solution
The system is a closed rigid container.

a. The specific volume can be calculated directly from its definition, Eq. (3.4). as

$$v = \forall/m = 3.00/0.200 = 15.0\,\text{ft}^3/\text{lbm}$$

b. The quality can be calculated from Eq. (3.26) or (3.27) and Eq. (3.25) as

$$x = \frac{v - v_f}{v_{fg}} = \frac{15.0 - 0.01672}{26.80 - 0.01672} = 0.559$$

or x = 55.9% vapor. Therefore, the amount of moisture present is $1 - x = 0.441$, or the mixture consists of 44.1% moisture.

c. The specific internal energy can be obtained by combining Eq. (3.27) with the definition $u_{fg} = u_g - u_f$ to give $u = u_f + xu_{fg} = u_f + x(u_g - u_f)$, or

$$u = 180.1 + (0.559)(1077.6 - 180.1) = 682\,\text{Btu/lbm}$$

d. The specific enthalpy can be obtained by combining Eq. (3.27) with the definition $h_{fg} = h_g - h_f$ to give $h = h_f + xh_{fg} = h_f + x(h_g - h_f)$, or

$$h = 180.1 + (0.559)(1150.5 - 180.1) = 722\,\text{Btu/lbm}$$

e. To obtain the mass of water in the liquid and vapor phases, we can use the original definition of quality given in Eq. (3.22) to get $m_g = xm = 0.559(0.2) = 0.112$ lbm of saturated water vapor and then $m_f = (1 - x)m = m - m_g = 0.088$ lbm of saturated liquid water.

EXAMPLE 3.5

What total mass of saturated water (liquid plus vapor) should be put into a 0.500 ft^3 sealed, rigid container at 14.696 psia so that the water passes exactly through the critical state when the container is heated? Also, determine the initial quality in the vessel.

Solution
Processes carried out in sealed rigid containers are constant volume (or *isochoric*) processes. Therefore, the process path on a p-v diagram is a vertical straight line, as shown in the p-v diagram of Figure 3.19. In this problem, we are given the final

(*Continued*)

EXAMPLE 3.5 (*Continued*)

state (the critical state), and we are asked to determine a thermodynamic property (the mass) at the initial state. In Table 3.4 or Table C.12a of *Thermodynamic Tables to accompany Modern Engineering Thermodynamics*, we find for water that p_c = 3203.8 psia, T_c = 1165.1R, and v_c = 0.05053 ft^3/lbm. Also, since both the volume and mass are constant here, $v_2 = v_1 = v_c$. This process can then be diagrammed as follows:

Initial state	Constant volume $\longrightarrow$ process	Final state
p_1 = 14.696 psia		$p_2 = p_c$
T_1 = 212°F (saturated)		$T_2 = T_c$
$v_1 = v_2$ (from the process path)		$v_2 = v_c$

Therefore,

$$m = \frac{V}{v_1} = \frac{V}{v_2} = \frac{V}{v_c} = \frac{0.500\ \text{ft}^3}{0.05053\ \text{ft}^3/\text{lbm}} = 9.90\ \text{lbm}$$

We can now find the quality in the initial state by using Eq. (3.26) and the data given in Example 3.2 as

$$x_1 = \frac{v - v_{f1}}{v_{fg1}} = \frac{v_c - v_{f1}}{v_{g1} - v_{f1}} = \frac{(0.05053 - 0.01672)\ \text{ft}^3/\text{lbm}}{(26.8 - 0.01672)\ \text{ft}^3/\text{lbm}} = 1.26 \times 10^{-3} = 0.126\%\ \text{vapor}$$

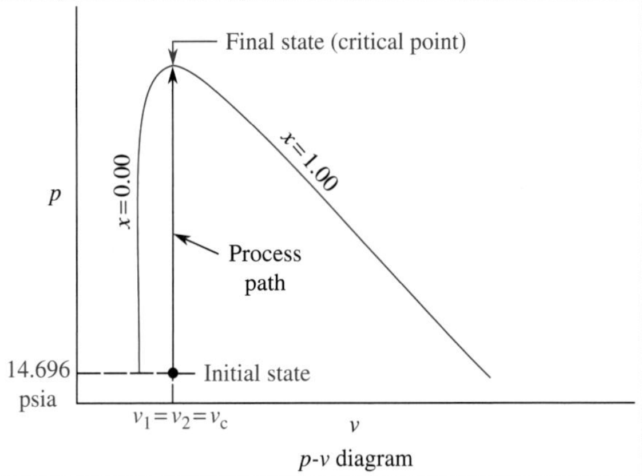

FIGURE 3.19

Example 3.5. Note that, since the quality is not defined at the critical state (quality is one of the properties that has a discontinuity there), no value for $x_2 = x_c$ can be given.

3.9 THERMODYNAMIC EQUATIONS OF STATE

In this section, we discuss some of the basic *p-v-T* equations of state for various substances. These equations can be easily typed into a computer spreadsheet, which greatly simplifies the calculations and allows for studying the effect of varying individual terms and plotting of the results. All engineering students today have access to a computer, and numerous computer programs can be found on the Internet that calculate the thermodynamic properties of water, refrigerants, and various other substances.

WHO WAS EMMY NOETHER?

PART 3

From 1908 to 1915, Emmy Noether taught at the University of Erlangen's Mathematical Institute without pay, occasionally substituting for her father when he was too ill to lecture. In the spring of 1915, she was invited to return to the University of Göttingen by David Hilbert and Felix Klein. Their effort to recruit her, however, was blocked by the faculty: Women, they insisted, should not become faculty. One faculty member protested: "What will our soldiers think when they return to the university and find that they are required to learn at the feet of a woman?" Hilbert responded with indignation, stating, "I do not see that the sex of the candidate is an argument against her admission as faculty.[3] After all, we are a university, not a bath house."

When World War I ended, significant social changes occurred in Germany, including more rights for women. In 1919, the University of Göttingen allowed Noether to proceed with her *habilitation* (eligibility for tenure), and she was given tenure in June 1919.

She had a very successful career in advanced mathematics, and one of her most important but unheralded discoveries is Noether's theorem, which proves a relationship between symmetries and conservation principles. This basic result was praised by Albert Einstein in a letter to David Hilbert, when he referred to Noether's "penetrating mathematical thinking." It was her work that led to formulations for several concepts of Einstein's general theory of relativity.

[3] *The actual term is* privatdozent, *which means an unsalaried university "private" lecturer or teacher paid directly by the students.*

CRITICAL THINKING

If an equation of state is an equation that relates thermodynamic properties of the system when it is in different thermodynamic states, then what can you say about a system that has an equation of state of the form pT = constant, where p is the absolute pressure and T is the corresponding absolute temperature of the system at any time?

Most materials have very complex thermodynamic equations of state, which are not given in textbooks. However, these complex equations are easily solved by computer programs like spreadsheets, Engineering Equation Solver,[4] Matlab, and so forth, which are readily available today. Thermodynamic property programs can also be found for PDAs and smart cell phones.[5]

While engineers in the 21st century use computer programs to find the thermodynamic property values they need, thermodynamic courses still rely on the use of printed property tables and charts in textbooks. However, students are encouraged to search out and use modern computer programs to solve problems or to verify that their table or chart solutions are correct.

In this section, we focus on the relatively simple equations of state of incompressible substances, ideal gases, and a few variations on the ideal gas equation. A more comprehensive discussion of the behavior of real gases is given in a subsequent chapter.

3.9.1 Incompressible Materials

The simplest equation of state is that for an incompressible material. One of its equations of state is merely v = constant. This equation can be used for either a solid or a liquid, but it cannot be used for a vapor or a gas. Incompressible substances also have one other state equation, which specifies that the specific internal energy of an incompressible material is a function of only one variable, temperature. Thus, the full set of state equations that characterize an incompressible material are

$$v = \text{constant} \tag{3.28}$$

and

$$u = u(T) \tag{3.29}$$

The constant volume specific heat of an incompressible material is given by Eq. (3.15) as

$$c_v = \left(\frac{\partial u}{\partial T}\right)_v = \frac{du}{dT} \quad \text{(since } u \text{ is independent of } v \text{ here)} \tag{3.30}$$

Because the specific enthalpy is defined as $h = u + pv$, Eq. (3.19) gives the constant pressure specific heat of an incompressible material as

$$c_p = \left(\frac{\partial h}{\partial T}\right)_p = \frac{du}{dT} + \left(\frac{\partial(pv)}{\partial T}\right)_p = \frac{du}{dT} + pv\beta = c_v = c \text{ (when } \beta = 0) \tag{3.31}$$

[4] F-Chart Software, Box 44042, Madison, WI 53744 (info@fchart.com).

[5] For example, see enggtools.com, processacesoftware.com, and appstorehq.com/engineeringtables.

since for incompressible materials v = constant and $\beta = 0$. However, note that, since β is very small for most liquids and solids (see Tables 3.1 and 3.2), these substances can be accurately modelled as incompressible materials. The subscripts p and v are meaningless for an incompressible material, and the simple phrase *specific heat,* represented by the symbol c with no subscript, is sufficient. Thus, for all incompressible substances, $c_p = c_v = c$. Consequently, for these materials, we can write

$$u_2 - u_1 = \int_{T_1}^{T_2} c dT \tag{3.32}$$

and if c is constant over the temperature range from T_1 to T_2, then Eq. (3.32) becomes

$$u_2 - u_1 = c(T_2 - T_1) \tag{3.33}$$

Also, since $v_2 = v_1 = v$ here,

$$h_2 - h_1 = c(T_2 - T_1) + v(p_2 - p_1) \tag{3.34}$$

Tables 3.5 and 3.6 list the specific heats of some materials whose liquid and solid phases accurately approximate incompressible substances.

Table 3.5 Specific Heats of Various Liquids at Atmospheric Pressure

	T		c	
Substance	**R**	**K**	**Btu/lbm · R**	**kJ/kg · K**
Benzene	520.	289	0.430	1.800
Butane (*n*)	492	273	0.550	2.303
Glycerin	510.	283	0.576	2.320
Mercury	510.	283	0.033	0.138
Propane	492	273	0.576	2.412
Water	492	273	1.007	4.186

Source: Some material drawn from Wark, K., Jr. 1988. Thermodynamics, fifth ed. McGraw-Hill, New York. Reprinted by permission of the publisher.

Table 3.6 Specific Heats of Various Solids at Atmospheric Pressure

	T		c	
Substance	**R**	**K**	**Btu/lbm · R**	**kJ/kg · K**
Aluminum	360.	200.	0.190	0.797
	540.	300.	0.215	0.902
	720.	400.	0.227	0.949
	900.	500.	0.238	0.997
Copper	540.	300.	0.092	0.386
	851	473	0.096	0.403
Graphite	527	293	0.170	0.712
Iron	527	293	0.107	0.448
Lead	540.	300.	0.031	0.129
	851	473	0.032	0.136
Rubber	527	293	0.439	1.84
Silver	527	293	0.056	0.233
Water (ice)	492	273	0.504	2.11
Wood	527	293	0.420	1.76

Source: Excerpted from Wark, K., Jr. 1988. Thermodynamics, fifth ed. McGraw-Hill, New York. Reprinted by permission of the publisher.

EXAMPLE 3.6

Determine the change in specific internal energy and specific enthalpy of an incompressible hardwood as it is heated from a temperature and pressure of 20.0°C, 0.100 MPa to a temperature and pressure of 100.°C and 1.00 MPa (Figure 3.20). Assume the wood has a constant density of 515 kg/m^3 and a constant specific heat over this temperature and pressure range.

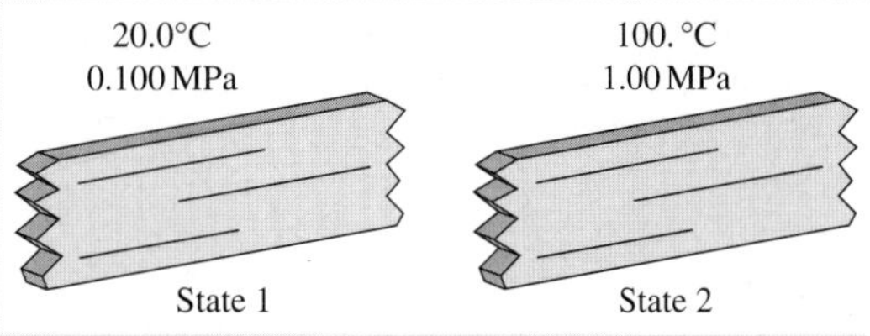

FIGURE 3.20
Example 3.6.

Solution

The changes in specific internal energy and specific enthalpy of a constant specific heat incompressible material are given by Eqs. (3.33) and (3.34), and the specific heat of wood is found in Table 3.6 to be $c = 1.76$ kJ/kg · K. Then, Eq. (3.33) gives

$$u_2 - u_1 = c(T_2 - T_1)$$
$$= (1.76\,\text{kJ/kg}\cdot\text{K})[(100 + 273.15) - (20.0 + 273.15)\,\text{K}] = 141\,\text{kJ/kg}$$

Notice that we could have used either °C or K for the temperature difference $T_2 - T_1$. This is because the Celsius and Kelvin degree sizes are exactly the same, only their zero points differ. From Eq. (3.34), we have $v = 1/\rho = 1/515 = 0.00194$ m^3/kg and

$$h_2 - h_1 = c(T_2 - T_1) + v(p_2 - p_1) = u_2 - u_1 + v(p_2 - p_1)$$
$$= 141 + (0.00194\,\text{m}^3/\text{kg})(1.00\times 10^3 - 100.\,\text{kN/m}^2) = 143\,\text{kJ/kg}$$

where we have converted the units of pressure into kN/m^2 so that the units of u and of the pv product match exactly.

Exercises

9. Determine the changes in specific internal energy and specific enthalpy of liquid propane as it is heated from −90.0°C, 0.100 MPa to −50.0°C, 1.00 MPa. Assume that liquid propane has a constant density of 615 kg/m^3 and a constant specific heat over this temperature and pressure range. **Answer**: $u_2 - u_1 = 96.5$ kJ/kg and $h_2 - h_1 = 97.9$ kJ/kg.
10. Determine the changes in specific internal energy and specific enthalpy of a block of iron as it is heated in an oven at atmospheric pressure from 70.0°F to 250.°F. Assume that iron has a density of 490. lbm/ft^3 and a constant specific heat over this temperature and pressure range. **Answer**: $u_2 - u_1 = 19.3$ Btu/lbm and $h_2 - h_1 = 19.3$ Btu/lbm.

3.9.2 Ideal Gases

The next simplest equation of state is that of an ideal gas. It is important because all gases approach ideal gas behavior at low pressure. Like an incompressible substance, an ideal gas is also defined by two state equations, both of which must be obeyed if a gas is to be called ideal. The first equation of state is the common ideal gas law, which has the following four equivalent forms:

$$p\forall = mRT \tag{3.35a}$$

$$pv = RT \tag{3.35b}$$

$$p\forall = n\Re T \tag{3.35c}$$

$$p\bar{v} = \Re T \tag{3.35d}$$

where $n = m/M$ is the number of moles, $\bar{v} = \forall/n$ is the molar specific volume, and $\Re$ is the universal gas constant whose value is

$$\Re = 1545.35\,\text{ft}\cdot\text{lbf/(lbmole}\cdot\text{R)} = 1.986\text{Btu/(lbmole}\cdot\text{R)}$$
$$= 8314\,\text{joule/(kgmole}\cdot\text{K)} = 8.314\,\text{kJ/(kgmole}\cdot\text{K)}$$

The second state equation used to define an ideal gas is that its specific internal energy is only a function of temperature, or

$$u = u(T) \tag{3.36}$$

ARE "GREENHOUSE" GASES ALSO "IDEAL" GASES?

Many of the gases found in the Earth's atmosphere behave as ideal gases, and a few are classified as "greenhouse gases." Some atmospheric gases trap the heat of sunlight that enters the Earth's atmosphere just like the glass of a greenhouse traps the heat of incoming sunlight. Many people now believe that increasing the atmospheric concentrations of these gases is producing a global warming that will reach 3–10°F by 2100.

Atmospheric carbon dioxide is a major greenhouse gas. Oceans and growing plants remove billions of tons of atmospheric CO_2 from the atmosphere every year, but since the 1700s, the burning of oil, coal, and gas and continued deforestation have increased the atmospheric CO_2 concentration by about 30%.

Carbon dioxide is used extensively in carbonated beverages. It gives the beverage its sparkle and tangy taste, and because it forms a weak acidic solution in water (carbonic acid), it inhibits the growth of mold and bacteria. Soft drinks are carbonated by chilling the water and cascading it in thin sheets in an enclosure containing pressurized CO_2 gas, then flavoring is added.

If the amount of CO_2 absorbed in water increases with increased surface area, then does the pressure in a soda can increase or decrease when you shake it? **Answer:** The pressure actually goes down a little as you shake it because more CO_2 is dissolved due to the increased surface area produced by the shaking. But when you open it after shaking, it squirts a lot of bubbles because there is now too much CO_2 in solution and it comes out rapidly, as the can is depressurized when you open it.

As in the case of an incompressible substance, Eq. (3.15) gives the constant volume specific heat of an ideal gas (since u does not depend on v) as

$$c_v = \left(\frac{\partial u}{\partial T}\right)_v = \frac{du}{dT} \tag{3.37}$$

and if c_v is constant over the temperature range from T_1 to T_2, then integration of Eq. (3.37) gives

$$u_2 - u_1 = c_v(T_2 - T_1) \tag{3.38}$$

Thus, for a *constant specific heat ideal gas*, Eq. (3.38) is valid for *any* process (not just a constant volume process), because the internal energy of an ideal gas does not depend on its volume. Note that, even for a constant pressure (isobaric) process, Eq. (3.38) is valid when a constant specific heat ideal gas is used.

Combining Eqs. (3.17) and (3.35b) gives the specific enthalpy of an ideal gas as

$$h = u + pv = u + RT \tag{3.39}$$

From Eqs. (3.19) and (3.39), we see that the constant pressure specific heat does not depend on the pressure, so

$$c_p = \left(\frac{\partial h}{\partial T}\right)_p = \frac{dh}{dT} = \frac{du}{dT} + R \tag{3.40}$$

And for an ideal gas, $du/dT = c_v$, so Eq. (3.40) becomes

$$c_p = c_v + R \tag{3.41}$$

If c_p is constant over the temperature range from T_1 to T_2, integration of Eq. (3.40) gives

$$h_2 - h_1 = c_p(T_2 - T_1) \tag{3.42}$$

Thus, for a *constant specific heat ideal gas*, Eq. (3.42) is valid for *any* process (not just a constant pressure process), because the enthalpy of an ideal gas does not depend on its pressure. Thus, even for an isochoric (constant volume) process, Eq. (3.42) is valid when a constant specific heat ideal gas is used. Values of c_p, c_v, and the gas constant R are given in Table 3.7 for a variety of common gases at low pressure that behave as ideal gases. A larger table can be found in Table C.13 of *Thermodynamic Tables to accompany Modern Engineering Thermodynamics*.

3.9.3 Variable Specific Heats

Note that, even though the values of c_p and c_v for an ideal gas do not depend on p and v, they may depend on temperature. We can improve the accuracy of an ideal gas calculation by utilizing the concept of variable specific heats. By integrating Eqns. (3.37) and (3.40), we obtain

$$u_2 - u_1 = \int_{T_1}^{T_2} c_v dT$$

Table 3.7 Properties of Various Gases at Low Pressure (also see Table C.13)

		R		c_p		c_v		
Substance	***M***	**Btu/lbm · R**	**kJ/kg · K**	**Btu/lbm · R**	**kJ/kg · K**	**Btu/lbm · R**	**kJ/kg · K**	$k = c_p/c_v$
Air	28.97	0.0685	0.286	0.240	1.004	0.172	0.718	1.40
Argon (Ar)	39.94	0.0497	0.208	0.125	0.523	0.075	0.315	1.67
Carbon dioxide (CO_2)	44.01	0.0451	0.189	0.202	0.845	0.157	0.656	1.29
Carbon monoxide (CO)	28.01	0.0709	0.297	0.249	1.042	0.178	0.745	1.40
Helium (He)	4.003	0.4961	2.077	1.24	5.200	0.744	3.123	1.67
Hydrogen (H_2)	2.016	0.9850	4.124	3.42	14.32	2.435	10.19	1.40
Methane (CH_4)	16.04	0.1238	0.518	0.532	2.227	0.408	1.709	1.30
Nitrogen (N_2)	28.02	0.0709	0.296	0.248	1.038	0.177	0.742	1.40
Oxygen (O_2)	32.00	0.0621	0.260	0.219	0.917	0.157	0.657	1.39

Source: Reprinted by permission of the publisher from Reynolds, W. C., Perkins, H. C., 1977. Engineering Thermodynamics, second ed. McGraw-Hill, New York.

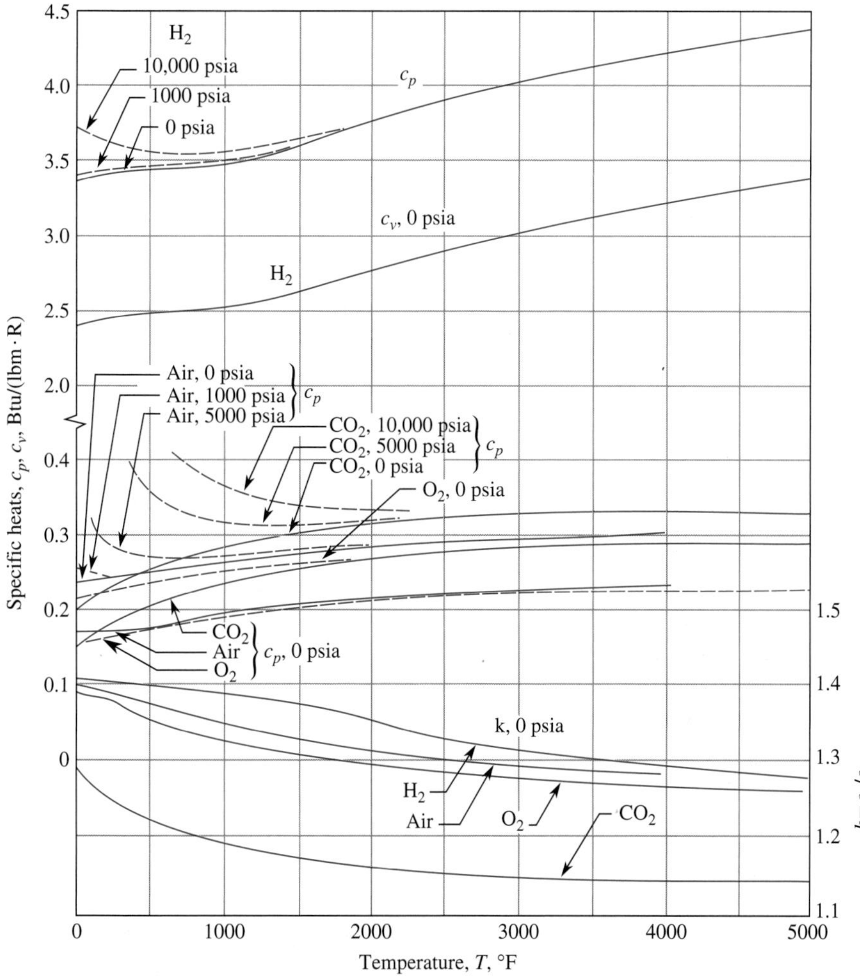

FIGURE 3.21

Specific heats of selected gases (data from the National Bureau of Standards). (*Source: Reprinted by permission of the publisher from Reynolds, W. C., Perkins, H. C., 1977. Engineering Dynamics. McGraw-Hill, New York.*)

and

$$h_2 - h_1 = \int_{T_1}^{T_2} c_p dT$$

Table C.16 in *Thermodynamic Tables to accompany Modern Engineering Thermodynamics* contains values for these integrals for air. (Note: The p_r and v_r columns are used for entropy values introduced later in this textbook).

Figure 3.21 illustrates the temperature and pressure dependence of c_p and c_v for various common gases. Note that the specific heat temperature dependence is fairly weak, and most ideal gases can be considered to have constant specific heats over temperature ranges of a few hundred degrees.

EXAMPLE 3.7

Determine the change in specific internal energy and specific enthalpy of air as it is cooled in a closed, rigid tank from a temperature and pressure of 240.°F, 150 psia to a temperature and pressure of 80.0°F and 14.7 psia (Figure 3.22). Assume the air behaves as (a) a constant specific heat ideal gas and (b) as a variable specific heat ideal gas.

Solution

a. The changes in specific internal energy and specific enthalpy of an ideal gas are given by Eqs. (3.38) and (3.42). The constant pressure and constant volume specific heats of air are found in Table 3.7 (or Table C.13) as c_p = 0.240 Btu/lbm · R and c_v = 0.172 Btu/lbm · R. Then Eq. (3.38) gives

$$u_2 - u_1 = c_v(T_2 - T_1) = (0.172\,\text{Btu/lbm}\cdot\text{R})[(80.0 + 459.67) - (240. + 459.67)\,\text{R}] = -27.5\,\text{Btu/lbm}$$

and from Eq. 3.42, we have

$$h_2 - h_1 = c_p(T_2 - T_1) = (0.240\,\text{Btu/lbm}\cdot\text{R})(80.0 - 240.°\text{F}) = -38.4\,\text{Btu/lbm}$$

Notice that you can use either fahrenheit or rankine values when computing the temperature difference $T_2 - T_1$ because the fahrenheit and rankine degree sizes are the same, only their zero points are different.

b. Values for u and h for variable specific heat air can be found in Table C.16. At T_1 = 240 + 459.67 = 700 R, h_1 = 167.56 Btu/lbm and u_1 = 119.58 Btu/lbm; and at T_2 = 80 + 459.67 = 540. R, h_1 = 129.06 Btu/lbm and u_2 = 92.04 Btu/lbm. Then the changes are

$$u_2 - u_1 = 92.04 - 119.58 = -27.5,$$

and

$$h_2 - h_1 = 129.06 - 167.56 = -38.5$$

Exercises

11. Determine the changes in specific internal energy and specific enthalpy as air is heated at constant pressure of 0.100 MPa from 300. K to 1500. K. Assume air behaves as (a) a constant specific heat ideal gas, and (b) as a variable specific heat ideal gas. **Answer:** (a) $u_2 - u_1$ = 862 kJ/kg and $h_2 - h_1$ = 1205 kJ/kg; (b) $u_2 - u_1$ = 991.38 kJ/kg and $h_2 - h_1$ = 1335.8 kJ/kg. The difference in the results of (a) and (b) is due to the large temperature difference between the two states.

12. Determine the changes in specific internal energy and specific enthalpy as methane is compressed at constant temperature of 20.0°C from 0.100 MPa to 10.0 MPa. Assume that methane behaves as a constant specific heat ideal gas. **Answer:** $u_2 - u_1 = h_2 - h_1$ = 0. The specific internal energy and specific enthalpy of an ideal gas depend only on temperature, so changing just the pressure on the gas does not alter the values of either u or h.

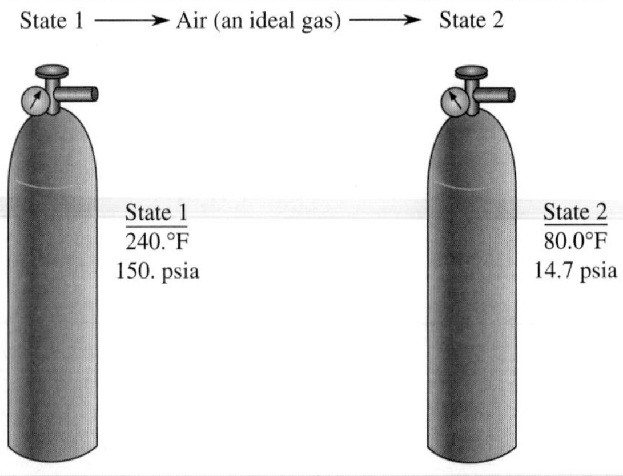

FIGURE 3.22
Example 3.7.

Normally, only low molecular mass real gases at high temperature or low pressure obey the ideal gas equation of state with good accuracy. For real gases with complex molecular structures or real gases approaching their saturated vapor region, more complex equations of state are required. The following equations have modifications to the ideal gas p-v-T equation that are intended to account for observed real gas behavior.

3.9.4 Real Gases

The Clausius equation of state accounts for the volume actually occupied by the gas molecules themselves. If we let b represent the specific volume of the molecules themselves, then the Clausius equation of state is

$$p(v-b) = RT \tag{3.43}$$

In 1873, van der Waals included a second correction factor to account for the forces of molecular attraction. These forces produce a net decrease in the observed pressure that is inversely proportional to v^2. The van der Waals equation of state has the form

$$\left(p+\frac{a}{v^2}\right)(v-b) = RT \tag{3.44}$$

The values of the molecular coefficients a and b in Eqs. (3.43) and (3.44) can be found in Table C.15 of *Thermodynamic Tables to accompany Modern Engineering Thermodynamics*.

Other important real gas equations of state that are commonly used in engineering analysis are the Dieterici equation,

$$p(v-b) = RT\exp[-a/(RTv)] \tag{3.45}$$

and the Berthelot equation,

$$p(v-b) = RT - \frac{a}{T}\left(\frac{v-b}{v^2}\right) \tag{3.46}$$

But perhaps the most useful, best known, and most accurate equations of state for real gases are those of Beattie and Bridgeman,

$$p = \left(\frac{1-\varepsilon}{v^2}\right)(v+B)RT - \frac{A}{v^2} \tag{3.47}$$

where

$$A = A_0(1-a/v),\ B = B_0(1-b/v),\ \text{and}\ \varepsilon = \frac{c}{vT^3}$$

and Redlich and Kwong,

$$p(v-b) = RT - \frac{a}{v\sqrt{T}}\left(\frac{v-b}{v+b}\right) \tag{3.48}$$

where A_0, B_0, a, b, and c are constants, whose values for various gases can be found in Table C.15.

A more general form for a real gas equation of state is a power series expansion such as

$$pv = RT + \frac{A}{v} + \frac{B}{v^2} + \frac{C}{v^3} + \ldots \tag{3.49}$$

where $A, B, C, \ldots$ are all empirically determined functions of temperature. These equations are called *virial expansions*, and the temperature dependent coefficients $A, B, C, \ldots$ are called the *virial coefficients*.

EXAMPLE 3.8

When an artillery cannon using a nitrocellulose propellant is fired, a maximum temperature of 2830°C is measured in the breech behind the moving projectile. The density of the propellant gases at this temperature is 200. kg/m^3, and the molecular mass of the propellant gases is 23.26 kg/kgmole. The volume occupied by the molecules of the propellant gases is $b = 0.960 \times 10^{-3}$ m^3/kg (Figure 3.23). Determine the maximum pressure in the breech as the cannon fires.

FIGURE 3.23
Example 3.8.

(Continued)

EXAMPLE 3.8 *(Continued)*

Solution

Since the temperature is very high, we can ignore the intermolecular forces in the propellant gases and use the Clausius equation of state (this equation is known as the Noble-Abel equation in ballistics literature): $p(v - b) = RT$, where $R = \Re/M$ and $\Re = 8314.3\ \text{N}\cdot\text{m}/(\text{kgmole}\cdot\text{K})$ is the universal gas constant. Then,

$$p_{\max} = \frac{\Re T_{\max}}{M(v-b)}$$

where $v = 1/\rho = 1/(200.\ \text{kg/m}^3) = 5.00 \times 10^{-3}\ \text{m}^3/\text{kg}$; and

$$p_{\max} = \frac{\Re T_{\max}}{M(v-b)} = \frac{[8314.3\ \text{Ngm/(kgmolegK)}](2830 + 273.15\ \text{K})}{(23.26\ \text{kg/kgmole})(5.00 \times 10^{-3} - 0.960 \times 10^{-3}\ \text{m}^3/\text{kg})}$$

$$= 2.7456 \times 10^8\ \text{N/m}^2 = (2.7456 \times 10^8\ \text{N/m}^2)\left(\frac{1\ \text{lbf/in}^2}{6894.76\ \text{N/m}^2}\right)$$

$$= 39{,}800\ \text{lbf/in}^2\ \text{absolute} = 39{,}800\ \text{psia}$$

Exercises

13. Determine the breech temperature in Example 3.8 if the breech pressure is 60.0×10^3 psia and all the remaining variables are as given in the example. **Answer**: $T_{\text{breech}} = 4400°\text{C}$.
14. Use the van der Waals equation of state to determine the pressure of water vapor at 100°C when the specific volume is 57.79 m³/kg. **Answer**: $p = 2.98$ kPa. (Hint: The values of a and b for water vapor can be found in Table C.15.)

WHO WAS EMMY NOETHER?

PART 4

When Adolf Hitler became chancellor of Germany in January 1933, one of the first actions of his administration was to remove all Jews from government positions (including university professors). In April 1933, Noether received a notice that her right to teach at the University of Göttingen had been withdrawn.

She joined the ranks of dozens of newly unemployed German professors who were searching for positions outside of Germany. Albert Einstein and Hermann Weyl were subsequently moved to the Institute for Advanced Study in Princeton, and late in 1933, Emmy Noether accepted a position at Bryn Mawr College, which is located ten miles west of Philadelphia, Pennsylvania.

In 1934, Noether began lecturing at the Institute for Advanced Study in Princeton (then an all-male university), but she felt that she was not welcome at the "men's university, where nothing female is admitted."

Emmy Noether once said, "If one proves the equality of two numbers a and b by showing first that a is less than or equal to b, and then a is greater than or equal to b, it is unfair, one should instead show that they are really equal by disclosing the inner ground for their equality."

On April 14, 1934, Emmy Noether died suddenly after an operation for a pelvic tumor. Her body was cremated and her ashes interred under the walkway around the cloisters of the M. Carey Thomas Library at Bryn Mawr College.

3.10 THERMODYNAMIC TABLES

Thermodynamic tables are generated from complex equations of state, which in turn were developed from accurate experimental data. These tables are quick and easy to use, but they are not available for all materials of engineering interest. Tables C.1 through C.13 in the *Thermodynamic Tables to accompany Modern Engineering Thermodynamics* give the thermodynamic properties of a variety of substances. Basically, only three types of tables are given there: pressure and temperature entry saturation tables, superheated vapor tables, and compressed or subcooled liquid tables. The saturation tables contain properties only along the saturation curve ($x = 0$ and $x = 1$) and no property values of liquid-vapor mixtures. These mixture properties must be calculated from the saturation values and the quality using Eq. (3.27). The superheated vapor and compressed liquid tables provide values throughout their regions of definition. Figure 3.24 illustrates the range of applicability of these tables.

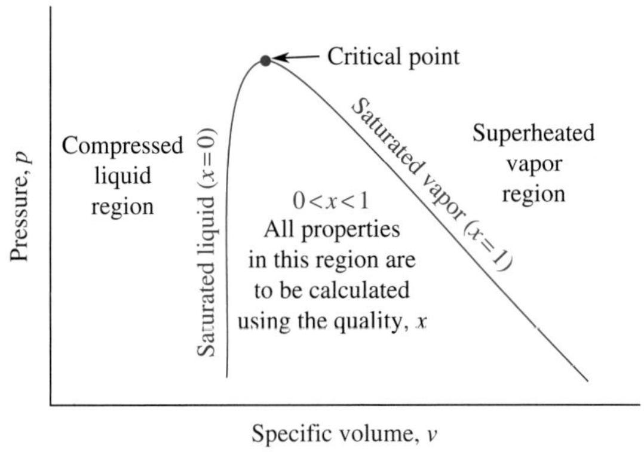

FIGURE 3.24
Regions of application of thermodynamic tables.

When thermodynamic data are given in problem statements, you normally are not told whether the state of the system is compressed, saturated, or superheated. To decide which table to use, you must be able to deduce the state of the system from the information given. This can be done by comparing the given properties with the saturation properties at the same temperature or pressure. For example, suppose you are given water at 500°F and 1000. psia. How can you tell if it is a compressed liquid, saturated liquid, a mixture of liquid plus vapor (i.e., wet), a saturated vapor, or a superheated vapor? The answer is obtained from the saturation data in Table C.1a or C.2a of *Thermodynamic Tables to accompany Modern Engineering Thermodynamics*. These tables tell you that, at 500°F, the saturation pressure is 680.8 psia, and at 1000. psia, the saturation temperature is 544.61°F. First of all, we could use the saturation pressure of 680.8 psia as a guide and note that the actual state (500°F, 1000. psia) is at a pressure greater than that required to produce a saturated liquid at 500°F, consequently the water must be in a compressed liquid state. Alternatively, we could use the saturation temperature of 544.61°F as a guide and note that the actual state has a temperature (500.°F) that is less than that required for a saturated liquid at 1000. psia (544.61°F), so again the water must be in a compressed (or subcooled) liquid state. Consequently, we obtain all other desired property information from Table C.4a, the compressed water table.

Similarly, in metric units, if you have water at 1.00 MPa and 200°C, a check of the saturation data in Table C.1b reveals that, at 200.°C, the saturation pressure is 1.554 MPa, which is greater than the actual pressure of 1.00 MPa. Therefore, the actual state of the water must be in the superheated vapor region. A check of Table C.3b reveals that this state can be easily found in the table.

How do you decide which table to use when you are given properties other than pressure and temperature? You use the same basic technique. For example, suppose you are given 3.00 lbm of water in a 15.0 ft^3 closed, rigid container at 14.696 psia. The specific volume of the system, then, is $v = 15.0/3.00 = 5.00$ ft^3/lbm. A check of Table C.2a reveals that, at 14.696 psia, $v_f = 0.01672$ ft^3/lbm, and $v_g = 26.80$ ft^3/lbm. Since the actual specific volume (5.00 ft^3/lbm) falls between these two values ($v_f < v < v_g$), the state of the water must be in the liquid plus vapor (wet) region, and it therefore has a quality of $x = (5.00 - 0.01672)/(26.8 - 0.01672) = 0.186$, or 18.6%. To get more familiar with these tables, it is recommended that you verify the states given in Table 3.8 for water.

Table 3.8 The States of Water Fixed by Various Combinations of Property Pairs

Pair of Independent Properties	State (Correct Table to Use)
T = 500.°F, p = 1000. psia	Compressed or subcooled liquid (C.4a)
p = 1.00 MPa, T = 200.°C	Superheated vapor (C.3b)
T = 170.°F, x = 1.0	Saturated vapor (C.1a)
p = 14.696 psia, v = 5.00 ft^3/lbm	Liquid-vapor mixture (C.2a)
u = 500. Btu/lbm, p = 100. psia	Liquid-vapor mixture (C.2a)
h = 1192.6 Btu/lbm, T = 300.°F	Superheated vapor (C.3a)
p = 0.100 MPa, h = 200. kJ/kg	Compressed or subcooled liquid (C.4b)
T = 100.°C, v = 8.585 m^3/kg	Superheated vapor (C.3b)
v = 0.10 m^3/kg, x = 1.0	Saturated vapor (C.1b or C.2b)
h = 3157.7 kJ/kg, u = 2875.2 kJ/kg	Superheated vapor (C.3b)

3.11 HOW DO YOU DETERMINE THE "THERMODYNAMIC STATE"?

First, remember that you need the values of only two independent intensive thermodynamic properties to fix the state of a homogeneous material, and the problem statement always provides these values. Usually, you are given the values of pressure, temperature, or specific volume in the problem statement. Sometimes specific internal energy or specific enthalpy also is one of the values given.

Second, choose one of the two given values and look up the corresponding saturation values (f and g) of the other given property from the saturation tables in *Thermodynamic Tables to accompany Modern Engineering Thermodynamics*, Tables C.1 and C.2, for your material and compare them with the given value. You can then determine the state of your material by following the rules in Table 3.9.

For example, if you are given water at a pressure of 1.0 MPa and a temperature of 200.°C, then you could choose p_{given} = 1.0 MPa and look up T_{sat} at that value of p_{given}. From Table C.1 at p_{given} = 1.0 MPa, you find that T_{sat} = 179.9°C. Since your value of T_{given} = 200.°C > T_{sat} = 179.9°C, the water must be in a superheated vapor state. Similarly, if you choose T_{given} = 200.°C instead of p_{given}, then you would look up p_{sat} at that value of T_{given}. From Table C.1 at T_{given} = 200.°C, you find that p_{sat} = 1.554 MPa. Since your value of p_{given} = 1.0 MPa < p_{sat} = 1.554 MPa, you again conclude that the water must be in a superheated vapor state.

While it is true that any pair of independent properties fix the state of a simple substance subjected to only one work mode, you must be able to deduce the system's thermodynamic state (compressed liquid, saturated liquid or vapor, liquid-vapor mixture, or superheated vapor) from the data given in a problem statement to know in which table to find the other properties required in the analysis. It is important to remember that thermodynamic states are unique and a given pair of independent properties fix the state at only one point in the tables. It is therefore essential to understand how to determine which table to use in the solution of a thermodynamics problem.

In Example 3.9, we introduce notation of the form $v(X°F, Y \text{ psia})$, which represents the value of the specific volume evaluated at X°F and Y psia. For example, v(100.°F, 50. psia) means the value of the specific volume at 100.°F and 50. psia. This is a convenient way of recording the pair of independent intensive properties used to determine the value of v. The same notation is used with the intensive properties u and h.

Table 3.9 How to Find the Thermodynamic "State"

Properties Given in the Problem Statement	Choose	Look Up in Appropriate Table	Then You Have a Compressed Liquid *If*	Mixture of Liquid and Vapor *If*	Superheated Vapor *If*
p_{given}, T_{given}	p_{given}	T_{sat} at p_{given}	$T_{given} < T_{sat}$	$T_{given} = T_{sat}$	$T_{given} > T_{sat}$
p_{given}, T_{given}	T_{given}	p_{sat} at T_{given}	$p_{given} > p_{sat}$	$p_{given} = p_{sat}$	$p_{given} < p_{sat}$
p_{given}, v_{given}	p_{given}	v_f and v_g	$v_{given} < v_f$	$v_f < v_{given} < v_g$	$v_{given} > v_g$
p_{given}, u_{given}	p_{given}	u_f and u_g	$u_{given} < u_f$	$u_f < u_{given} < u_g$	$u_{given} > u_g$
p_{given}, h_{given}	p_{given}	h_f and h_g	$h_{given} < h_f$	$h_f < h_{given} < h_g$	$h_{given} > h_g$
T_{given}, v_{given}	T_{given}	v_f and v_g	$v_{given} < v_f$	$v_f < v_{given} < v_g$	$v_{given} > v_g$
T_{given}, u_{given}	T_{given}	u_f and u_g	$u_{given} < u_f$	$u_f < u_{given} < u_g$	$u_{given} > u_g$
T_{given}, h_{given}	T_{given}	h_f and h_g	$h_{given} < h_f$	$h_f < h_{given} < h_g$	$h_{given} > h_g$

EXAMPLE 3.9

Find the specific volume and specific enthalpy of Refrigerant-134a at 100.°F and 95.0 psia (Figure 3.25).

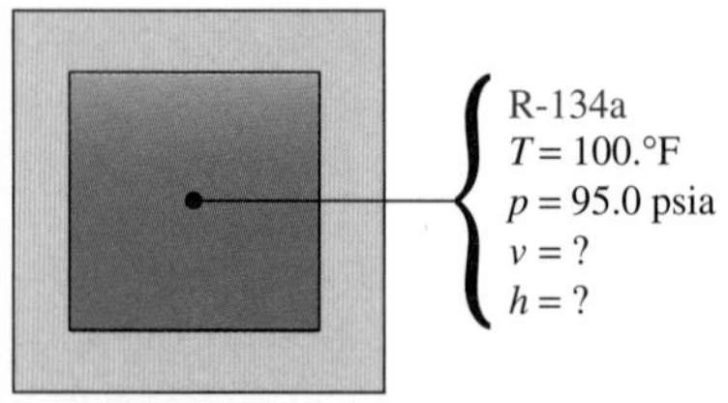

FIGURE 3.25
Example 3.9.

Solution

A check of Table C.7a of *Thermodynamic Tables to accompany Modern Engineering Thermodynamics* reveals that the saturation pressure of Refrigerant-134a at 100.°F is 138.83 psia. Since our actual pressure is less than the saturation pressure, we must have superheated vapor. A check of Table C.8a reveals that 100.°F and 95.0 psia is indeed in the superheated region. However, 95.0 psia is not a direct entry into this table, so we must use linear interpolation to find the needed values. This is how a linear interpolation for v is carried out:

$$\frac{v(100.°\text{F}, 95.0\text{ psia}) - v(100.°\text{F}, 90\text{ psia})}{95.0\text{ psia} - 90.0\text{ psia}} = \frac{v(100.°\text{F}, 100.\text{ psia}) - v(100.°\text{F}, 90.0\text{ psia})}{100.\text{ psia} - 90.0\text{ psia}}$$

or

$$\begin{aligned} v(100.°\text{F}, 95.0\text{ psia}) &= v(100.°\text{F}, 90\text{ psia}) \\ &+ \left(\frac{95.0\text{ psia} - 90.0\text{ psia}}{100.\text{ psia} - 90.0\text{ psia}}\right)[v(100.°\text{F}, 100.\text{ psia}) - v(100.°\text{F}, 90.0\text{ psia})] \\ &= 0.5751 + 0.500(0.5086 - 0.5751) = 0.54185\text{ ft}^3/\text{lbm} \end{aligned}$$

And interpolating for the specific enthalpy h gives

$$\frac{h(100.°\text{F}, 95.0\text{ psia}) - h(100.°\text{F}, 90.0\text{ psia})}{95.0\text{ psia} - 90.0\text{ psia}} = \frac{h(100.°\text{F}, 100.\text{ psia}) - h(100.°\text{F}, 90.0\text{ psia})}{100.\text{ psia} - 90.0\text{ psia}}$$

or

$$\begin{aligned} h(100°\text{F}, 95\text{ psia}) &= h(100.°\text{F}, 90\text{ psia}) \\ &+ \left(\frac{95.0\text{ psia} - 90.0\text{ psia}}{100.\text{ psia} - 90.0\text{ psia}}\right)[h(100.°\text{F}, 100.\text{ psia}) - h(100.°\text{F}, 90.0\text{ psia})] \\ &= 118.39 + 0.500(117.73 - 118.39) = 118.06\text{ Btu/lbm} \end{aligned}$$

Exercises

15. Use Table C.1b to find the values of p_{sat}, v_f, v_g, u_f, u_g, h_f, and h_g for saturated water at 100.°C. **Answers:** p_{sat} = 0.1013 MPa, v_f = 0.001044 m³/kg, v_g = 1.673 m³/kg, u_f = 418.9 kJ/kg, u_g = 2506.5 kJ/kg, h_f = 419.0 kJ/kg, and h_g = 2676.0 kJ/kg.
16. Use Table C.3a to find the values of v, u, and h for superheated water vapor at 2000. psia and 1000.°F. **Answers:** v = 0.3945 ft³/lbm, u = 1328.1 Btu/lbm, and h = 1474.1 Btu/lbm.
17. Use Table C.4b to find the values of v, u, and h for compressed liquid water at 30.0 MPa and 200.°C. **Answers:** v = 0.0011302 m³/kg, u = 831.4 kJ/kg, and h = 865.3 kJ/kg.
18. Use Table C.5a to find the values of T_{sat}, p_{sat}, v_g, and h_g for saturated ammonia when v_f = 0.02446 ft³/lbm and h_f = 53.8 Btu/lbm. **Answers:** T_{sat} = 10.0°F, p_{sat} = 38.51 psia, v_g = 7.304 ft³/lbm, and h_g = 614.9 Btu/lbm.
19. Use Table C.8b to find the values of T and h for superheated Refrigerant-134a when the pressure is 0.500 MPa and the specific volume is 0.06524 m³/kg. **Answers:** T = 140.°C and h = 382.42 kJ/kg.
6. Use Table C.11a to find the values of T, v, and h for saturated mercury at 1.00 psia and a quality of 50.0%. **Answers:** $T = T_{sat}$ = 457.72°F, v = 24.211 ft³/lbm, and h = 77.321 Btu/lbm.

3.12 THERMODYNAMIC CHARTS

Experimental data, equations of state, and statistical thermodynamics results can be combined into very accurate thermodynamic phase diagrams, called *thermodynamic charts.* These two-dimensional property diagrams can be constructed with various useful thermodynamic properties as coordinates. For example, Figure 3.26 shows a specific volume vs. specific internal energy chart for water. This chart also includes lines of constant pressure, temperature, and quality. Thus, given a pair of independent properties, such as p and T (or p and x in the wet region), the u and v values can be immediately read from the coordinate axes. Notice that, in the wet region, where $0 < x < 1$, the constant temperature and constant pressure lines lie on top of each other, since p and T are not independent in this region.

A series of similar charts for a variety of substances can be found in the charts portion of *Thermodynamics Table to accompany Modern Engineering Thermodymanics*. It must be emphasized, however, that since the physical size of these charts is very small, the values taken from them are not as accurate as those taken from a table for the same substance, even if interpolation must be used within the table. Therefore, small charts like these are used only when appropriate tables are not available or a state is to be fixed without using either pressure or temperature. For example, given values for u and v for water, it would be much easier to find the other thermodynamic properties at that state using Figure 3.26 than to do a double interpolation within the water tables (however, the accuracy still is not as good as using the tables).

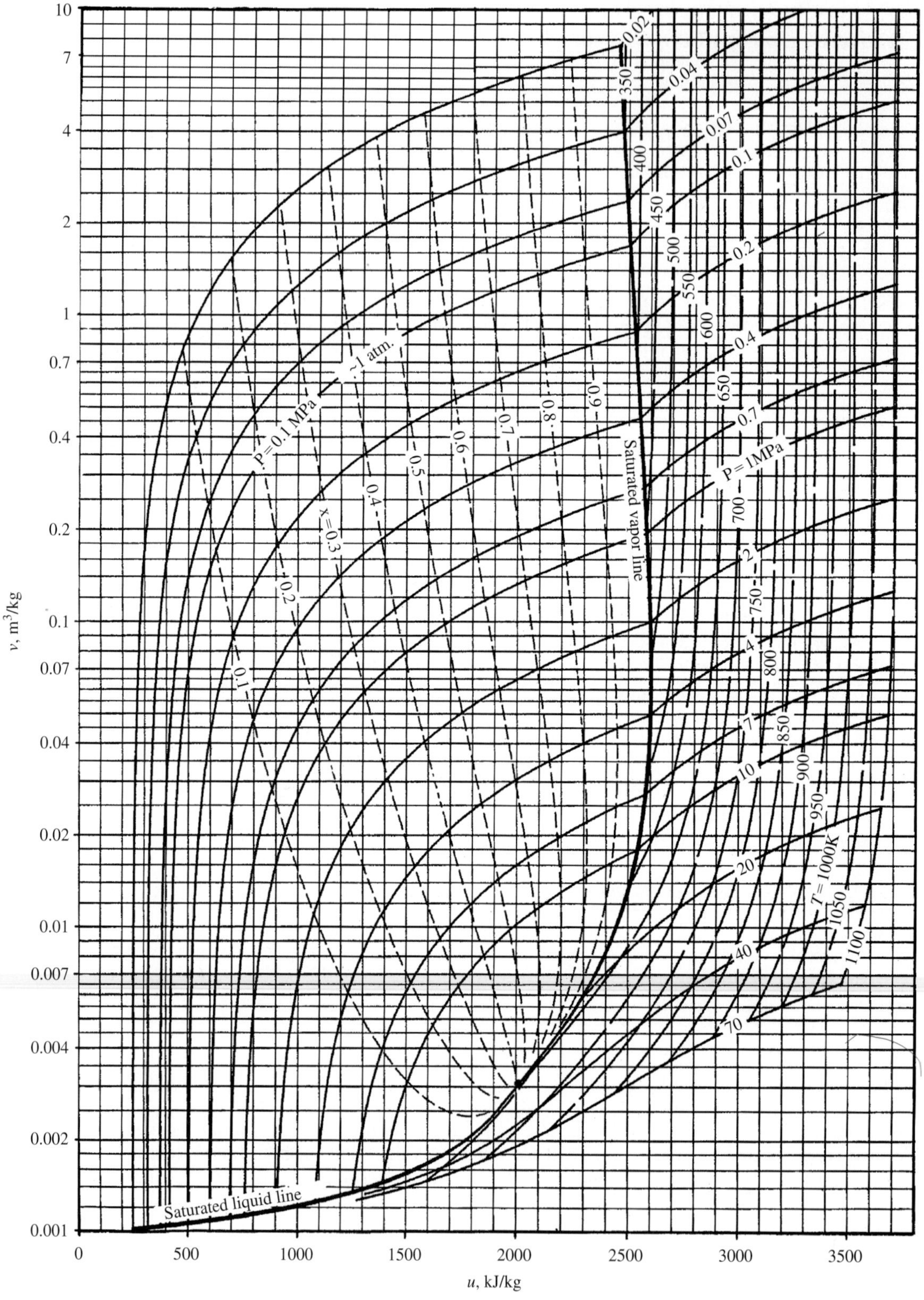

FIGURE 3.26

Thermodynamic properties of steam (H_2O). (*Source: Reprinted by permission of the publisher from Reynolds, W. C., Perkins, H. C., 1977. Engineering Thermodynamics, second ed. McGraw-Hill, New York.*)

WHO WAS EMMY NOETHER?

PART 5

To the Editor of *The New York Times*, May 5, 1935:

Within the past few days a distinguished mathematician, Professor Emmy Noether (Figure 3.27), formerly connected with the University of Göttingen and for the past two years at Bryn Mawr College, died in her fifty-third year. In the judgment of the most competent living mathematicians, Fräulein Noether was the most significant creative mathematical genius thus far produced since the higher education of women began. In the realm of algebra, in which the most gifted mathematicians have been busy for centuries, she discovered methods which have proved of enormous importance in the development of the present-day younger generation of mathematicians. Pure mathematics is, in its way, the poetry of logical ideas. One seeks the most general ideas of operation which will bring together in simple, logical and unified form the largest possible circle of formal relationships. In this effort toward logical beauty spiritual formulas are discovered necessary for the deeper penetration into the laws of nature.

Born in a Jewish family distinguished for the love of learning, Emmy Noether, who, in spite of the efforts of the great Göttingen mathematician, Hilbert, never reached the academic standing due her in her own country, none the less surrounded herself with a group of students and investigators at Göttingen, who have already become distinguished as teachers and investigators. Her unselfish, significant work over a period of many years was rewarded by the new rulers of Germany with a dismissal, which cost her the means of maintaining her simple life and the opportunity to carry on her mathematical studies. Farsighted friends of science in this country were fortunately able to make such arrangements at Bryn Mawr College and at Princeton that she found in America up to the day of her death not only colleagues who esteemed her friendship but grateful pupils whose enthusiasm made her last years the happiest and perhaps the most fruitful of her entire career.

Albert Einstein
Princeton University, May 1, 1935

FIGURE 3.27
Emmy Noether.

3.13 THERMODYNAMIC PROPERTY SOFTWARE

Very few 21st century engineers use tables and charts. Numerous computer programs can provide the numerical values of properties. Some of the more common are listed in Table 3.10.

Since these programs do not all use the same property equations, they do not give exactly the same numerical results. However, any differences are insignificant. In the examples and problems in this textbook, a variety of sources (tables and computer programs) have been used, so you may expect to see differences between the thermodynamic property values used here and values you find from other sources.

Table 3.10 Thermodynamic Property Software

Program Name	Source	Comments
Mini-NIST Reference Fluid Thermodynamic and Transport Properties (**REFPROP**)[a]	U.S. National Institute of Standards and Technology	This free program contains properties for water, CO_2, N_2, CH_4, R134a, propane, and dodecane
EES (Engineering Equation Solver)	F-Chart Software	This is an excellent program for solving thermodynamics problems
MathCAD (Functions are available for the thermodynamic properties of various materials[b])	Parametric Technology Corporation (PTC)	MathCAD can perform calculations with automatic unit conversion and checking.
CATT (Computer-Aided Thermodynamic Tables)	John Wiley and Sons	Incorporates color phase diagrams showing calculated points
Microsoft Excel (Spreadsheets available for thermodynamic properties of various materials)	Microsoft Corporation	Numerous Excel spreadsheets for thermodynamic properties are available on the Internet

[a] *This free program can be found at www.boulder.nist.gov/div838/theory/refprop/MINIREF/MINIREF.HTM.*

[b] *For example, see www.icee.usm.edu/ICEE/conferences/Conference%20Files/ASEE2006/P2006072MCC.pdf.*

Table 3.11 Glossary of Technical Terms Introduced in Chapter 3

Isobaric process	Constant pressure process
Isochoric process	Constant volume process
Internal energy	Total energy minus kinetic and potential energy
Enthalpy	Internal energy plus the product of pressure and volume
Constant volume specific heat (c_v)	The variation in specific internal energy with respect to temperature while holding volume constant
Constant pressure specific heat (c_p)	The variation in specific enthalpy with respect to temperature while holding pressure constant
Allotropic	Different solid forms of the same substance
Triple point	The point where the solid, liquid, and vapor phases coexist in thermal equilibrium.
Vaporization	The transformation of a liquid into a vapor
Condensation	The transformation of a vapor into a liquid or a solid
Melting	The transformation of a solid into a liquid (synonymous with fusion)
Solidification	The transformation of a liquid into a solid (synonymous with freezing)
Sublimation	The transformation of a solid into a vapor
Saturation	A condition that exists when two or more phases coexist in equilibrium
Critical state	The peak of the vaporization curve
Gas	The state of any substance whose temperature is greater than that at the critical state
Quality	The ratio of the mass of vapor present to the total mass present
Moisture	The ratio of the mass of liquid present to the total mass present (1.0 minus the quality)
Wet vapor	A substance whose state is under the saturation dome
Phase	The physical state (or molecular configuration) of matter

SUMMARY

In this chapter, three of the five main techniques used in obtaining values for thermodynamic properties are discussed. Equations of state, thermodynamic tables, and thermodynamic charts are valuable tools needed in the thermodynamic analyses that occur in the following chapters.

This chapter also introduces many new technical thermodynamic terms, most of which are listed in the glossary in Table 3.11. The reader is urged to learn the definitions of these terms. They are freely used in the remaining chapters under the assumption that their meaning is fully understood by the reader.

Here are some of the more important equations introduced in this chapter. Be careful not to try to use them blindly without understanding their limitations.

1. General property relations, Eq. (3.2):

$$\left(\frac{\partial x}{\partial y}\right)_z = \left[\left(\frac{\partial y}{\partial x}\right)_z\right]^{-1}$$

and Eq. (3.3):

$$\left(\frac{\partial x}{\partial y}\right)_z \left(\frac{\partial y}{\partial z}\right)_x \left(\frac{\partial z}{\partial x}\right)_y = -1$$

2. The definitions of two new physical properties, Eqs. (3.5) and (3.6)

$$\beta = \frac{1}{v}\left(\frac{\partial v}{\partial T}\right)_p = \text{isobaric coefficient of volume expansion}$$

and

$$\kappa = -\frac{1}{v}\left(\frac{\partial v}{\partial p}\right)_T = \text{isothermal coefficient of compressibility}$$

3. The definitions of the total and specific energy of a system from Eqs. (3.9) and (3.12):

$$E = U + \frac{mV^2}{2g_c} + \frac{mgZ}{g_c}$$

and

$$e = E/m = u + \frac{V^2}{2g_c} + \frac{gZ}{g_c}$$

4. The definitions of the constant volume and constant pressure specific heats from Eqs. (3.15) and (3.18):

$$c_v = \left(\frac{\partial u}{\partial T}\right)_v = \text{constant volume specific heat}$$

and

$$c_p = \left(\frac{\partial h}{\partial T}\right)_p = \text{constant pressure specific heat}$$

5. The general definition of enthalpy, Eq. (3.17):

$$h = u + pv = \text{specific enthalpy}$$

6. The general definition of *quality* from Eq. (3.22):

$$x = \frac{m_g}{m_g + m_f} = \frac{m_g}{m} = \text{quality}$$

7. The definition of the *specific volume* of a mixture of liquid and vapor using quality, Eqs. (3.23) and (3.24):

$$v = (1 - x)v_f + xv_g = v_f + xv_{fg}$$

where $v_{fg} = v_g - v_f$.

8. A more general definition of *quality* using other specific properties from Eq. (3.27):

$$x = \frac{v - v_f}{v_{fg}} = \frac{u - u_f}{u_{fg}} = \frac{h - h_f}{h_{fg}}$$

9. For incompressible materials, we have from Eqs. (3.28), (3.33), and (3.34),

$$v_{\text{incompressible material}} = (\forall/m)_{\text{incmopressible material}} = \text{constant}$$

and

$$(u_2 - u_1)_{\text{incompressible material}} = c(T_2 - T_1)$$

so that

$$(h_2 - h_1)_{\text{incompressible material}} = c(T_2 - T_1) + v(p_2 - p_1)$$

10. For constant specific heat ideal gases, we have, from Eqs. (3.35),

$$p\forall = mRT = n\Re T$$

or

$$pv = RT$$

or

$$p\bar{v} = \Re T$$

from Eq. (3.38),

$$(u_2 - u_1)_{\text{ideal gas}} = c_v(T_2 - T_1),$$

and, from Eq. (3.42),

$$(h_2 - h_1)_{\text{ideal gas}} = c_p(T_2 - T_1)$$

11. For ideal gases, only the following relation, from Eq. (3.41), also holds:

$$c_p = c_v + R$$

Problems (* indicates problems in SI units)

1. If p, v, and T are all intensive independent properties.
   a. Show that the following relation is always valid:
   $$\left(\frac{\partial p}{\partial v}\right)_T = -\left(\frac{\partial p}{\partial T}\right)_v \left(\frac{\partial T}{\partial v}\right)_p$$
   b. Verify this relation for the equation of state of an ideal gas.
2. Show that the following equations are valid:
   a. $$c_v = -\left(\frac{\partial v}{\partial T}\right)_u \left(\frac{\partial u}{\partial v}\right)_T$$
   b. $$c_p = -\left(\frac{\partial p}{\partial T}\right)_h \left(\frac{\partial h}{\partial p}\right)_T$$
3. Show that
   $$\beta = -(1/v)\left(\frac{\partial v}{\partial p}\right)_T \left(\frac{\partial p}{\partial T}\right)_v$$
4. Show that
   $$\kappa = (1/v)\left(\frac{\partial v}{\partial T}\right)_p \left(\frac{\partial T}{\partial p}\right)_v$$
5. Show that
   $$\left(\frac{\partial p}{\partial T}\right)_v = \beta/\kappa$$
6.* A 0.200 m diameter sphere of solid copper is isothermally compressed from 0.100 to 1000. MPa at 500.°C. Determine the sphere's diameter after the compression process.
7. Assuming that the isothermal coefficient of compressibility is constant, determine the percent decrease in the volume of liquid water that undergoes an isothermal increase in pressure of $100. \times 10^3$ psia.
8. Assuming that the isobaric coefficient of volume expansion is constant, determine the percent increase in the volume of liquid water that is heated at constant pressure from 50.0 to 212°F.
9. Some historical researchers believe that Gabriel Daniel Fahrenheit (1686–1736) constructed his well-known temperature scale based on the isobaric coefficient of volume expansion of mercury rather than with fixed reference temperatures. It is thought that he may have defined his degree size to be the temperature change required to isobarically change the volume of mercury by 1/10,000th of its value at the zero point of his scale. Modern measurements established that the isobaric coefficient of volume expansion of mercury at 0.00°F is 1.015×10^{-4} R^{-1}. Consider an ordinary glass thermometer with a bore of radius r and a reservoir bulb of volume $\forall_0$, at the bottom. Ignoring the expansion of the glass itself and assuming that the isobaric coefficient of volume expansion is a constant, determine the relationship between the change in length of the mercury column (ΔL) and the bulb volume ($\forall_0$), the initial length of the mercury column (L_0), the isobaric coefficient of volume expansion (β), and the temperature change (ΔT). Is ΔL independent of L_0? If not, then the temperature interval divisions are not the same size along the length of the thermometer. Determine the percent difference in ΔL between $L_0 = 0$ and $L_0 = 30.0$ cm if $(\forall_0) = 0.500$ cm^3 and $r = 0.100$ mm.
10.* Assuming that all physical properties are constant, use Table 3.2 to find the percent change in the volume of liquid glycerin as it is heated from 20.0 to 150.°C while simultaneously being pressurized from 0.100 to 10.0 MPa.
11.* Use Table 3.2 to find the gauge pressure that would have to be exerted on liquid diethyl ether to prevent any change in its volume as it is heated from 0.00 to 50.0°C. Assume all the physical properties are constant for this process.
12. Use Table 3.2 to find the temperature to which mercury needs to be heated to prevent any change in its volume as it is pressurized at 70.0°F from 14.7 to 1000. psia. Assume all physical properties are constant for this process.
13. Using the relations p = constant and $\beta = C_1T + C_2$, integrate Eq. (3.23) to obtain the result
    $$v_2 = v_1 \exp\{[C_1(T_2 + T_1)/2 + C_2](T_2 - T_1)\}$$
    and show that this is the same as
    $$v_2 = v_1 \exp[\beta_{avg}(T_2 - T_1)]$$
    where $\beta_{avg} = (\beta_2 + \beta_1)/2$.
14. a. For an ideal gas, mathematically evaluate the partial derivative $(\partial u/\partial v)_T$.
    b. What is the partial derivative $(\partial h/\partial T)_p$ called for a real gas?
15. The enthalpy of a certain gas can be obtained from the following equation:
    $$h = (0.21)T + (1.2 \times 10^{-4})T^2 + (0.32)p + (3.6)p^2$$
    where h is in Btu/lbm, T is in R, and p is in psia. Determine the specific heat at constant pressure (c_p) for this gas when the temperature is 500. R and the pressure is 1 atm.
16. Using the property data given in the superheated steam tables, estimate the specific heat at constant pressure for steam at 400. psia and 1000.°F.
17.* Using the property data given in the superheated steam tables, estimate the specific heat at constant pressure for steam at 30.0 MPa and 700.°C.
18. Sketch (neatly) the common p-T and p-v diagrams for water and label
    a. The critical state.
    b. The triple point and triple point line.
    c. The solid, liquid, and vapor regions.
    d. Indicate the correct slope of the fusion line (i.e., either a positive or negative slope).
19. Are the following statements true or false?
    a. The specific volume of mercury is a function of temperature only.
    b. If ice is heated sufficiently, it always melts to form a liquid.
    c. If water is at a pressure lower than the critical pressure, it is always in the liquid phase.
    d. If a mixture of liquid ammonia and ammonia vapor is heated sufficiently in a rigid, sealed tube, the content of the tube always becomes a vapor.
20. Define the following terms: (a) internal energy, (b) saturation, (c) critical state, and (d) moisture.
21. Define the following terms: (a) isobaric, (b) isochoric, (c) enthalpy, (d) quality, and (e) triple point.
22. a. Is quality (x) a thermodynamic property? Explain.
    b. Mathematically define the specific heat at constant pressure.

c. For a saturated mixture of liquid and vapor, explain whether or not the pressure and temperature can be varied independently.

23. A vessel with a volume of 10.0 ft^3 contains 3.00 lbm of a mixture of liquid water and water vapor in equilibrium at a pressure of 100. psia (Figure 3.28). Determine
   a. The mass of liquid present.
   b. The mass of vapor present.

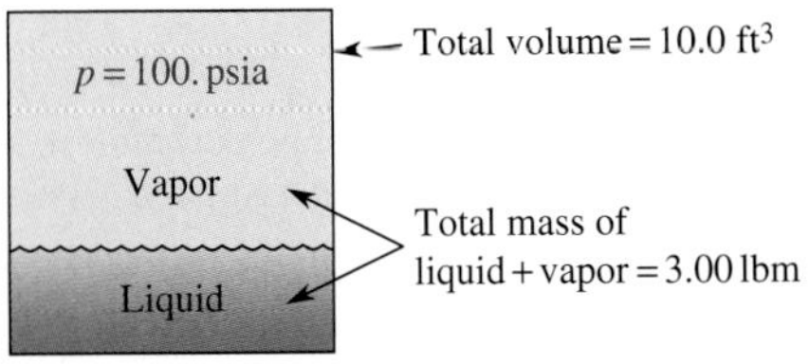

FIGURE 3.28
Problem 23.

24.* Determine the change in the specific internal energy of 3.00 kg of graphite as it is heated at atmospheric pressure from 20.0 to 200.°C. Assume a constant specific heat.

25. Determine the change in the enthalpy of 5.00 lbm of ice as it is heated from 22.0 to 32.0°F under constant atmospheric pressure. Assume a constant specific heat.

26.* Determine the change in the specific internal energy of solid aluminum as it is heated at atmospheric pressure from 300. to 500. K. Use an average specific heat over this temperature range.

27. Determine the change in the specific enthalpy of solid lead as it is heated from 14.7 psia, 80.0°F to 1000. psia, 200.°F. The density of lead is 710. lbm/ft^3. Assume a constant specific heat.

28. Determine the change in the specific internal energy of 7.00 lbm of methane gas as it is heated from 32.0 to 200.°F at atmospheric pressure. Assume ideal gas behavior.

29.* Determine the change in the specific enthalpy of carbon dioxide gas as it is heated at a constant pressure of 1 atm from 300. to 500. K. Assume ideal gas behavior.

30.* Argon gas is heated in a constant pressure process from 20.0 to 500.°C. Assuming ideal gas behavior, determine
   a. The ratio of the final to initial volumes.
   b. The change in specific internal energy.
   c. The change in specific enthalpy of the argon.

31. Helium gas is heated in a constant volume process from −200. to 500.°F. Assuming ideal gas behavior, determine
   a. The ratio of the final to initial pressures.
   b. The change in specific internal energy.
   c. The change in specific enthalpy of the helium.

32. Gaseous oxygen is heated in a constant temperature process until its volume is doubled. Assuming ideal gas behavior, determine
   a. The ratio of the final to initial pressures.
   b. The change in specific internal energy.
   c. The change in specific enthalpy of the oxygen.

33. Using Figure 3.21, estimate the average values for the constant pressure and constant volume specific heats for the following gases and processes:
   a. Carbon dioxide is heated at a constant pressure of 10,000 psia from 1000. to 2000.°F.
   b. Carbon dioxide gas is compressed isothermally at 1000.°F from 0 to 10,000. psia.
   c. Hydrogen gas is heated at a constant pressure of 0.00 psia from 0 to 5000.°F.
   d. Air is compressed from 0.00°F, 0.00 psia to 1000.°F, 5000. psia.

34. Determine the changes in specific internal energy and specific enthalpy as air is compressed from 0.00°F, 14.7 psia to 1000.°F, 5000. psia (Figure 3.29). Assume variable specific heat ideal gas behavior.

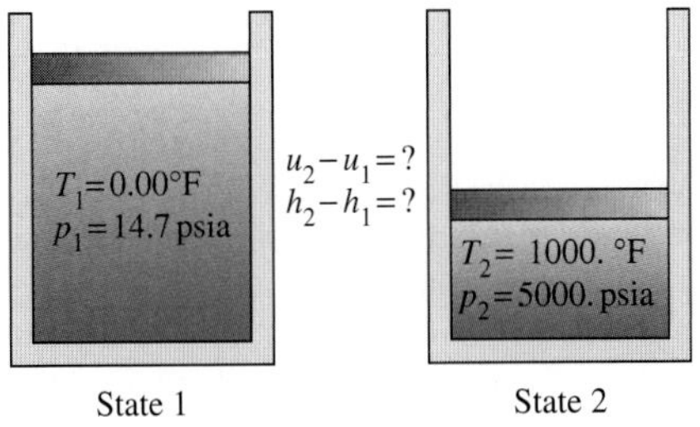

FIGURE 3.29
Problem 34.

35. Professor John L. Krohn at Arkansas Tech University invented a process whereby air is heated at constant volume from 60.0°F and v = 3.30 ft^3/lbm to a pressure of 180. psi. The air then expands adiabatically to atmospheric pressure and v = 14.6 ft^3/lbm. Assuming ideal gas behavior with variable specific heat, determine
   a. The temperature of the heated air (T_2) in °F.
   b. The heat transfer for the first process in Btu/lbm.
   c. The work for the second process in Btu/lbm,

36. Professor Krohn uses the constant pressure specific heat equation for water vapor given in by $c_p = A(B + CT + DT^2 + ET^3 + FT^4)$, where A = 0.1102 Btu/lbm · R, B = 4.070, $C = -0.000616$ R^{-1}, $D = 1.281 \times 10^{-6}$ R^{-2}, $E = -0.508 \times 10^{-9}$ R^{-3}, $F = 0.0769 \times 10^{-12}$ R^{-4}, and T is in Rankine (R). He wants you to estimate the change in enthalpy for water vapor from p_1 = 14.7 psi, T_1 = 250.°F to p_2 = 14.7 psi, T_2 = 500.°F, and compare this result to the change in enthalpy found in the superheated steam tables.

37.* In 1879, the French physicist Emile Amagat generated experimental data in a mine shaft at Verpilleux, France, for his research on the compressibility of gases. There, he used a vertical column of mercury 327 m high to measure the compressibility of nitrogen at a pressure of 430. atm. Assuming the temperature at the bottom of the mine shaft was 30.0°C, determine the specific volume of the nitrogen, assuming it is an ideal gas with constant specific heats.

38.* Calculate the specific volume of hydrogen (H_2) gas at a temperature of 20.0°C and a pressure of 11.0 MPa using
   a. The ideal gas equation of state.
   b. The Clausius equation of state (use the van der Waals value for b).

39. Determine the temperature of water vapor at 200. psia when it has a specific volume of 2.724 ft^3/lbm using
   a. The ideal gas equation of state.
   b. The van der Waals equation of state.
   c. The steam tables (Table C.3a).

   Then compute the percentage error of a and b with the actual value given in c.

40. Determine the temperature of carbon dioxide (CO_2) gas when it is at a pressure of 2500. psia and has a density of 32.0 lbm/ft^3. Assume constant specific heat ideal gas behavior.

41. Calculate the specific volume of propane at 1000. psia and 300.°F using
    a. The ideal gas equation of state.
    b. The Clausius equation of state (use the van der Waals value for *b*).

42. Determine the pressure exerted by 10.00 lbm of steam at a temperature of 1300.°F in a volume of 3.285 ft^3 using
    a. The steam tables.
    b. The ideal gas equation of state.
    c. The van der Waals equation of state.

43. a. Write down the van der Waals equation.
    b. Indicate which term corrects for the fact that the molecules occupy a finite volume.
    c. Indicate which term corrects for the fact that there are attractive forces between the molecules.
    d. How are the constants *a* and *b* in van der Waals equation determined, and are they the same for all gases?

44. For superheated Refrigerant-134a at 100. psia and 100.°F, determine the value of the specific volume
    a. From the superheated vapor table.
    b. Assuming it to be an ideal gas.
    c. From the van der Waals equation of state.

45. Estimate the temperature to which water at the bottom of a 500. ft deep lake would have to be heated before it would begin to boil (Figure 3.30). (Note: Hydrostatic pressure = γz, where γ = 62.4 lbf/ft^3 is the specific weight of water, and *z* is the depth below the free surface.)

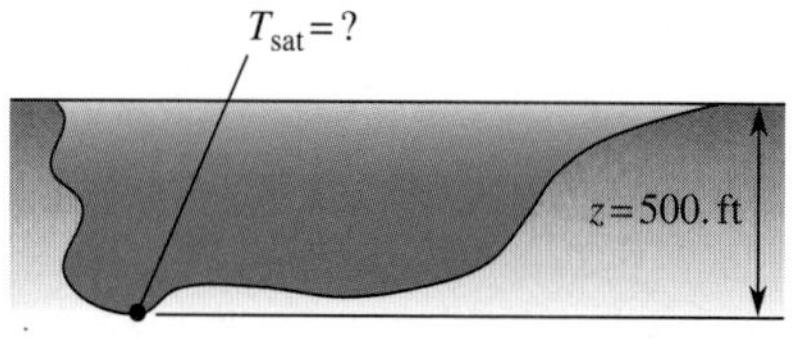

FIGURE 3.30
Problem 45.

46. One of the reasons for wearing a pressure suit in high altitude or space work is that, without it, the pressure in the body might become low enough to cause the blood to boil. Assume blood behaves essentially as pure water (which is its primary component) and that the body core temperature is 100.°F. Find the pressure at which this blood begins to "boil."

47. Refrigerant-134a contained in a tank at a pressure of 101.37 psia has a specific volume of 0.4682 ft^3/lbm. Using the proper thermodynamic table, determine the value of the enthalpy of the Refrigerant-134a under these conditions.

48.* The vapor produced when the pressure on saturated liquid water is suddenly reduced during a constant enthalpy process is called *flash steam,* because it occurs so quickly that part of the liquid appears to "flash" into vapor. Determine the final temperature and the percentage of flash steam (i.e., the quality) produced as the pressure on saturated liquid water at 2.00 MPa is suddenly reduced to 1.00 MPa in a constant enthalpy process (Figure 3.31)

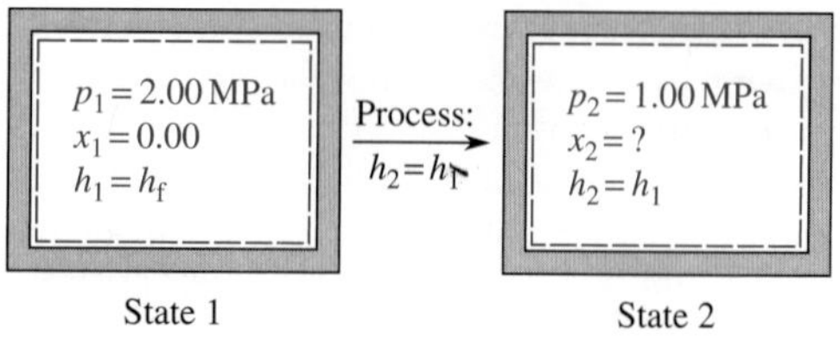

FIGURE 3.31
Problem 48.

49. What total mass of water must be put into a 1.00 ft^3 sealed, rigid container so that, when the container is heated, the contents pass through the saturated vapor curve exactly at the point where *p* = 2000. psia (Figure 3.32)

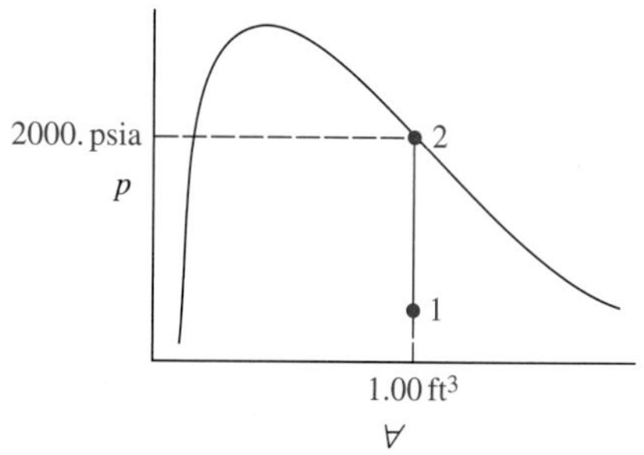

FIGURE 3.32
Problem 49.

50.* A rigid container contains 1.00 kg of water at the critical state. Determine the volume of the vapor present in the container after it has been cooled to 100.°C.

51. Suppose 0.667 lbm of water is put into a 1.00 ft^3 rigid container at 14.7 psia and 212°F and sealed. The container is then heated.
    a. At what temperature do the contents become a saturated vapor or saturated liquid?
    b. Which will it be—a saturated vapor or a saturated liquid?
    c. Sketch this process on a *p*-*v* diagram.

52. A closed rigid container contains water in an equilibrium mixture of liquid and vapor at 70.0°F. The mass of the liquid initially present is 10.236 times the mass of the vapor. The container is then heated until all the liquid becomes vapor. Determine
    a. The initial quality.
    b. The pressure in the container when the last bit of water becomes vapor.
    c. Sketch this process on a *p*-*v* diagram.

53.* It is desired to carry out an experiment that allows a visual observation of a material passing through the critical state. An empty, transparent, rigid, sealed container with a 2.00×10^{-6} m^3 internal volume is to be used.
    a. How many kilograms of solid CO_2 (dry ice) should be put into the container so that, when it is sealed and heated, its contents pass directly through the critical state?
    b. To what temperature (in °C) must the contents be heated to be at the critical state?
    c. What will be the pressure (in MPa) inside the container at the critical state?

54.* Using the tables for compressed liquid water (Tables C.4), determine the pressure increase required to raise the specific enthalpy of saturated liquid water at 50.0°C by 1.00 kJ/kg.

55. Determine the properties required in Table 3.12.

Table 3.12 Problem 55

Substance	Given	Find
a. Ammonia	T = 0.00°C x = 0.200	v = ?
b. Water	p = 400. psia h = 1000. Btu/lbm	x = ?
c. Water	T = 500.°F p = 400. psia	u = ?
d. Refrigerant-134a	p = 185.82 psia h = 51.47 Btu/lbm	v = ?

56.* The *Kara Maru* is a transport vessel that has inadvertently entered an enemy neutral zone. She carries 600 passengers and has radioed that she has blown her super-lumen drive system. The Star Command distress codes are taken from the tables of the thermodynamic properties of water, because the enemy has a very poor knowledge of this substance. Therefore, if there are any errors in the code, it may be a trap and not a real distress call. Is the transmission in Table 3.13 correct? If not, what are the errors? (Variations of less than 1% are not errors.)

Table 3.13 Problem 56

p (kPa)	T (°C)	v (m³/kg)	u (kJ/kg)	h (kJ/kg)	x (quality)
600.	600.	0.6696	2801.7	3700.7	
0.6113	0.0100	206.1	2375.3	2501.3	1.00
100.	99.6	0.001043	417.3	417.4	0.00

a. The standard reply to all such distress messages is to transmit the properties of water at 100% quality and 3000 kPa. What are those values?

p = 3000. (kPa)	u = ?
T = ?	h = ?
v = ?	

b. At this point, your reply is acknowledged with T = 200.°C, p = 1554.9 kPa, x = 0.23.
Is the acknowledgment from the *Kara Maru* or the enemy? Explain.

57. Using the tables in *Thermodynamic Tables to accompany Modern Engineering Thermodynamics*, fill in the missing properties in Table 3.14.

58. Using the tables and charts in *Thermodynamic Tables to accompany Modern Engineering Thermodynamics*, fill in the missing properties in Table 3.15.

59. Using the tables and charts in *Thermodynamic Tables to accompany Modern Engineering Thermodynamics* fill in the missing properties in Table 3.16.

60. Using the tables and charts in *Thermodynamic Tables to accompany Modern Engineering Thermodynamics*, fill in the missing properties in Table 3.17.

61. a. Using only the thermodynamic *tables* in *Thermodynamic Tables to accompany Modern Engineering Thermodynamics*, fill in the missing properties in Table 3.18.
b. Using only the thermodynamic *charts* in the tables book, fill in the missing properties in Table 3.19.

62. Using the tables and charts in *Thermodynamic Tables to accompany Modern Engineering Thermodynamics*, fill in the missing properties in Table 3.20.

63. Using the tables and charts in *Thermodynamic Tables to accompany Modern Engineering Thermodynamics*, fill in the missing properties in Table 3.21.

64. Using the tables and charts in *Thermodynamic Tables to accompany Modern Engineering Thermodynamics*, fill in the missing properties in Table 3.22.

65. Using the tables and charts in *Thermodynamic Tables to accompany Modern Engineering Thermodynamics*, fill in the missing properties in Table 3.23.

66. Using the tables and charts in *Thermodynamic Tables to accompany Modern Engineering Thermodynamics*, fill in the missing properties in Table 3.24.

67. Using the tables and charts in *Thermodynamic Tables to accompany Modern Engineering Thermodynamics*, fill in the missing properties in Table 3.25.

68. Using the tables and charts in *Thermodynamic Tables to accompany Modern Engineering Thermodynamics*, fill in the missing properties in Table 3.26.

Table 3.14 Problem 57

Material	T (°F)	p (psia)	u (Btu/lbm)	v (ft³/lbm)	ρ (lbm/ft³)	x
Water	?	60.0	?	?	?	1.00
Water	?	80.0	?	?	?	0.600
Ref.-134a	?	23.805	62.124		?	?

Table 3.15 Problem 58

Substance	p (psia)	T (°F)	v (ft³/lbm)	h (Btu/lbm)	u (Btu/lbm)
H_2O	300.	?	0.7811	?	?
H_2O	300.	600.	?	?	?
Ref.-134a	?	70.0	0.2526	?	?
Nitrogen	50.0	?	1.00	?	?

Table 3.16 Problem 59

Substance	*p* (psia)	*T* (°F)	*v* (ft³/lbm)	*h* (Btu/lbm)
Water	1000.	?	0.2326	?
Ref.-134a	85.788	?	?	111.33
Water	1000.	?	?	1505.9
Nitrogen	?	–160.	0.250	?

Table 3.17 Problem 60

Material	*p* (psia)	*T* (°F)	*v* (ft³/lbm)	*h* (Btu/lbm)	*u* (Btu/lbm)	*x* (if applicable)
Water	?	35.0	?	?	?	0.00
Water	1.00	?	?	?	?	1.00
Water	14.7	1000.	?	?	?	?
Ref.-134a	?	–40.0	?	?	?	0.500

Table 3.18 Problem 61a

Material	*p* (psia)	*T* (°F)	*v* (ft³/lbm)	*h* (Btu/lbm)	*u* (Btu/lbm)	*x* (if applicable)
Water	14.7	300.	?	?	?	?
Ref.-134a	23.805	?	?	?	?	1.00

Table 3.19 Problem 61b

Material	*T* (°F)	*p* (psia)	*h* (Btu/lbm)	*v* (ft³/lbm)	*x* (if applicable)
Carbon dioxide	0.00	?	?	?	0.20
Nitrogen	?	100.	?	1.000	?

Table 3.20 Problem 62

Material	*p* (psia)	*T* (°F)	*v* (ft³/lbm)	*x* (if applicable)
Water	5.00	300.	?	?
Water	100.	?	8.053	?
Water	1000.	544.8	0.100	?
Ref.-134a	?	0.00	?	0.00
Mercury	1.00	?	?	1.00

Table 3.21 Problem 63

Material	*p* (psia)	*T* (°F)	*v* (ft³/lbm)	*h* (Btu/lbm)	*x* (if applicable)
Water	1.20	?	?	?	0.00
Water	?	220.	?	?	1.00
Water	?	32.018	?	?	0.500
Water	8000.	2000.	?	?	?
Ref.-134a	21.203	0.000	0.01185	11.63	?

Table 3.22 Problem 64

Material	*p* (psia)	*T* (°F)	*v* (ft³/lbm)	*h* (Btu/lbm)	*x* (if applicable)
H_2O	600.	600.	?	?	?
H_2O	?	200.	?	?	1.00
H_2O	200.	1500.	?	?	?
H_2O	14.696	?	?	?	0.00
Ammonia	?	100.	?	?	0.00

Table 3.23 Problem 65

Material	p (psia)	T (°F)	v (ft³/lbm)	x (if applicable)
Water	?	300.	4.00	?
Water	300.	?	?	0.500
Water	1.00	1000.	?	?
Mercury	1.00	?	?	1.00
Ideal gas*	100.	?	5.00	?

* Use the ideal gas equation of state with R = 50 ft · lbf/(lbm · R.).

Table 3.24 Problem 66

Material	p (psia)	T (°F)	v (ft³/lbm)	h (Btu/lbm)	x (if applicable)
Water	40.0	?	?	?	0.00
Water	?	?	51.03	1240.5	?
Water	?	50.0	?	?	1.00
Ref.-134a	243.86	?	?	?	0.500
Ref.-134a	?	160.	?	?	1.00
Mercury	100.	?	?	?	1.00

Table 3.25 Problem 67

Material	T (°F)	p (psia)	h (Btu/lbm)	x (if applicable)
Ammonia	60.0	60.0	?	?
Ammonia	60.0	?	?	0.100
Mercury	?	60.0	38.44	?
Ref.-134a	60.0	?	?	1.00
Water	?	1.00	1336.1	?
Water	?	1.00	?	0.00

Table 3.26 Problem 68

Material	p (psia)	T (°F)	v (ft³/lbm)	x (if applicable)
H_2O	466.3	460.	?	0.00
H_2O	160.	363.6	?	1.00
H_2O	40.0	?	6.00	?
H_2O	1000.	1000.	?	?
Ammonia	?	105	1.00	?
Ammonia	100.	100.	?	?
Ref.-134a	?	200.	?	0.500
Ref.-134a	325	?	?	0.00
Mercury	1000.	?	?	1.00

Computer Problems

These problems are designed to be done on a personal computer using a spreadsheet or equation solver. The problems cannot be done easily without the use of a computer. They are meant to furnish an additional learning experience by providing new insights into the operation of complex thermodynamic systems and demonstrating the power of the personal computer in generating and manipulating thermodynamic properties. In these problems, log is the base 10 logarithm and ln is the base e (i.e., natural) logarithm.

69. In 1849, William Rankine proposed the following pressure-temperature relation for saturated water:

$$\log p_{sat} = 6.1007 - 2731.62/T_{sat} - 396{,}945/T_{sat}^2$$

where p_{sat} is in psia and T_{sat} is in R. Develop a computer program that returns values for p_{sat} in psia when T_{sat} is input in °F. Be sure to include proper units on all input and output values. Using the steam tables in Table C.1a of *Thermodynamic Tables to accompany Modern Engineering Thermodynamics*, plot the percent error in your calculated saturation pressure vs. input temperature utilizing data at 32.0, 100., 200., 300., 400., 500., 600., and 700.°F.

70. In 1905, Knoblauch, Linde, and Klebe proposed the following equation for the specific volume of superheated steam:

$$v = 0.5962T/p - (1 + 0.0014p)(150{,}300{,}000/T^3 - 0.0833)$$

where p is in psia, T is in R, and v is in ft^3/lbm. Develop a computer program that returns v when p and T are input. Allow the use of either SI or Engineering English units. Compare your results with steam table values at 0.10, 0.50, 1.00, 1.50, 2.00, 2.50, and 3.00 MPa along the 200.°C isotherm. Plot these results as a percent error in v vs. p for T = 200.°C.

71. Develop a computer program that calculates the pressure of superheated ammonia vapor from the Beattie-Bridgeman equation of state when the specific volume and temperature are input from the keyboard. Allow the use of either SI or Engineering English units. Using the superheated ammonia tables (Table C.6), determine the percent error between your calculated values of pressure and the correct values along the 100.°F isotherm. Plot this percent error vs. p utilizing actual pressure data of 10.0, 30.0, 50.0, 70.0, 90.0, 140., and 180. psia.

72. The pressure-temperature relation for saturated ammonia can be written as

$$\log p_{sat} = C_1 - C_2/T_{sat} - C_3\log(T_{sat}) - C_4T_{sat} + C_5T_{sat}^2$$

where

$$C_1 = 25.5743247$$

$$C_2 = 3295.1254$$

$$C_3 = 6.4012471$$

$$C_4 = 4.148279 \times 10^{-4}$$

$$C_5 = 1.4759945 \times 10^{-6}$$

In this equation p_{sat} is in psia and T_{sat} is in R. Develop a computer program that calculates in either SI or Engineering English units (your choice) p_{sat} in either psia or kPa when T_{sat} is entered in either °F or °C. Make sure the screen clearly indicates the proper units on the input information and all output values. Compare the resulting output values with a series of corresponding saturation values given in Table C.5 of *Thermodynamic Tables to accompany Modern Engineering Thermodynamics*.

73. The p-v-T relation for superheated mercury vapor is

$$pv = RT - (T/v)\exp(10.3338 - 312.095/T - 2.07951 \ln T)$$

where p is in N/m^2, v is in m^3/kg, T is in K, and R = 41.45 J/kg · K. Develop a computer program that outputs p, v, and T with their appropriate units when either (a) p and T are input or (b) v and T are input. Allow the user to work in either the SI or Engineering English units and to choose which type of input he or she wishes to use. For extra credit, create an isometric three-dimensional plot of a p-v-T surface using this equation of state.

CHAPTER 4

The First Law of Thermodynamics and Energy Transport Mechanisms

CONTENTS

4.1 INTRODUCCIÓN (INTRODUCTION)

In this chapter, we begin the formal study of the first law of thermodynamics. The theory is presented first, and in subsequent chapters, it is applied to a variety of closed and open systems of engineering interest. In Chapter 4, the first law of thermodynamics and its associated energy balance are developed along with a detailed discussion of the energy transport mechanisms of work and heat. To understand the usefulness of the first law of thermodynamics, we need to study the energy transport modes and investigate the energy conversion efficiency of common technologies.

In Chapter 5, the focus is on applying the theory presented in Chapter 4 to a series of steady state closed systems, such as sealed, rigid containers; electrical apparatuses; and piston-cylinder devices. Chapter 5 ends with a brief discussion of the behavior of unsteady state closed systems.

The first law of thermodynamics is expanded in Chapter 6 to cover open systems, and the conservation of mass law is introduced as a second independent basic equation. Then, appropriate applications are presented, dealing with a variety of common open system technologies of engineering interest, such as nozzles, diffusers, throttling devices, heat exchangers, and work-producing or work-absorbing machines. Chapter 6 ends with a brief discussion of the behavior of unsteady state open systems.

4.2 EMMY NOETHER AND THE CONSERVATION LAWS OF PHYSICS

Throughout the long history of physics and engineering, we believed that the conservation laws of momentum, energy, and electric charge were unique laws of nature that had to be discovered and verified by physical experiments. And, in fact, these laws were discovered in this way. They are the heart and soul of mechanics, thermodynamics, and electronics, because they deal with things (momentum, energy, charge) that cannot be created nor destroyed and therefore are "conserved." These conservation laws have broad application in engineering and physics and are considered to be the most fundamental laws in nature.

We have never been able explain where these laws came from because they seem to have no logical source. They seemed to be part of the mystery that is nature. However, almost 100 years ago, the mathematician Emmy Noether developed a theorem that uncovered their source,[1] yet few seem to know of its existence. Emmy Noether's theorem is fairly simple. It states that:

> For every *symmetry* exhibited by a system, there is a corresponding observable quantity that is *conserved.*

The meaning of the word *symmetry* here is probably not what you think it is. The symmetry that everybody thinks of is called *bilateral* symmetry, when two halves of a whole are each other's mirror images (bilateral symmetry is also called *mirror* symmetry). For example, a butterfly has bilateral symmetry. Emmy Noether was talking about symmetry with respect to a mathematical operation. We say that something has *mathematical* symmetry if, when you perform some mathematical operation on it, it does not change in any way. For example, everyone knows that the equations of physics remain the same under a translation of the coordinate system. This really says that there are no absolute positions in space. What matters is not where an object is in absolute terms, but where it is relative to other objects, that is, its coordinate differences.

The impact of Emmy Noether's studies on symmetry and the behavior of the physical world is nothing less than astounding. Virtually every theory, including relativity and quantum physics, is based on symmetry principles. To quote just one expert, Dr. Lee Smolin, of the Perimeter Institute for Theoretical Physics, "The connection between symmetries and conservation laws is one of the great discoveries of twentieth century physics. But very few non-experts will have heard either of it or its maker—Emily Noether, a great German mathematician. But it is as essential to twentieth century physics as famous ideas like the impossibility of exceeding the speed of light."[2]

Noether's theorem proving that symmetries imply conservation laws has been called the most important theorem in engineering and physics since the Pythagorean theorem. These symmetries define the limit of all possible conservation laws. Is it possible that, had Emmy Noether been a man, all the conservation laws of physics would be called Noether's laws?

---

[1] Noether, E., 1918. Invariante variationsprobleme. *Nachr. D. König. Gesellsch. D. Wiss. Zu Göttingen, Math-phys. Klasse 1918*, pp. 235–257. An English translation can be found at http://arxiv.org/PS_cache/physics/pdf/0503/0503066v1.pdf.

[2] Dr. Lee Smolin was born in New York City in 1955. He held faculty positions at Yale, Syracuse, and Penn State Universities, where he helped to found the Center for Gravitational Physics and Geometry. In September 2001, he moved to Canada to be a founding member of the Perimeter Institute for Theoretical Physics.

AN EXAMPLE OF MATHEMATICAL SYMMETRY

Here is a story about Carl Friedrich Gauss (1777–1855). When he was a young child, his teacher wanted to occupy him for a while, so he asked him to add up all the numbers from 1 to 100. That is, find $X = 1 + 2 + 3 + \ldots + 100$. To the teacher's surprise, Gauss returned a few minutes later and said that the sum was 5050.

Apparently Gauss noticed that the sum is the same regardless of whether the terms are added forward (from first to last) or backward (from last to first). In other words, $X = 1 + 2 + 3 + \ldots + 100 = 100 + 99 + 98 + \ldots + 1$. If we then add these two ways together, we get

$$\begin{aligned} X &= 1+2+3+\ldots+100 \\ \underline{X} &= \underline{100+99+98+\ldots+1} \\ 2X &= 101+101+\ldots+101 \end{aligned}$$

So $2X = 100 \times 101$ and $X = (100 \times 101)/2 = 5050$. Gauss had found a mathematical symmetry, and it tremendously simplified the problem. What is conserved here? It is the sum, X. It does not change no matter how you add the numbers.

Table 4.1 Relation of Conservation Laws to Mathematical Symmetry

Conservation Law	Mathematical Symmetry
Linear momentum	The laws of physics are the same regardless of where we are in space. This positional symmetry implies that linear momentum is conserved.
Angular momentum	The laws of physics are the same if we rotate about an axis. This rotational symmetry implies that angular momentum is conserved.
Energy	The laws of physics do not depend on what time it is. This temporal symmetry implies the conservation of energy.
Electric charge	The interactions of charged particles with an electromagnetic field remain the same if we multiply the fields by a complex number $e^{i\varphi}$. This implies the conservation of charge.

In summary, Emmy Noether's theorem shows us that (Table 4.1)

- Symmetry under translation produces the *conservation of linear momentum.*
- Symmetry under rotation produces the *conservation of angular momentum.*
- Symmetry in time produces the *conservation of energy.*
- Symmetry in magnetic fields produces the *conservation of charge.*

4.3 THE FIRST LAW OF THERMODYNAMICS

In this chapter, we focus our attention on the detailed structure of the first law of thermodynamics. To completely understand this law, we need to study a variety of work and heat energy transport modes and to investigate the basic elements of energy conversion efficiency. An effective general technique for solving thermodynamics problems is presented and illustrated. This technique is used in Chapters 5 and 6 and the remainder of the book.

The simplest, most direct statement of the first law of thermodynamics is that *energy is conserved.* That is, energy can be neither created nor destroyed. The condition of zero energy production was expressed mathematically in Eq. (2.15):

$$E_P = 0 \tag{2.15}$$

By differentiating this with respect to time, we obtain an equation for the condition of a zero energy production *rate*:

$$\frac{dE_P}{dt} = \dot{E}_p = 0 \tag{2.16}$$

Whereas Eqs. (2.15) and (2.16) are accurate and concise statements of the first law of thermodynamics, they are relatively useless by themselves, because they do not contain terms that can be used to calculate other variables. However, if these equations are substituted into the energy balance and energy rate balance equations, then the following equations result. For the energy balance,

$$E_G = E_T + E_P \text{ (as required by the first law)}$$

or

$$E_G = E_T \tag{4.1}$$

The energy rate balance is

$$\dot{E}_G = \dot{E}_T + \dot{E}_P \text{ (as required by the first law)}$$

or

$$\dot{E}_G = \dot{E}_T \tag{4.2}$$

From now on, we frequently use the phrases *energy balance* and *energy rate balance* in identifying the proper equation to use in an analysis. So, for simplicity, we introduce the following abbreviations:

EB = energy balance

and

ERB = energy rate balance

In Chapter 3, we introduce the components of the total system energy E as the internal energy U, the kinetic energy $mV^2/2g_c$, and the potential energy mgZ/g_c, or[3]

$$E = U + \frac{mV^2}{2g_c} + \frac{mgZ}{g_c} \tag{3.9}$$

In this equation, V is the magnitude of the velocity of the center of mass of the entire system, Z is the height of the center of mass above a ground (or zero) potential datum, and g_c is the dimensional proportionality factor (see Table 1.2 of Chapter 1). In Chapter 3, we also introduce the abbreviated form of this equation:

$$E = U + \text{KE} + \text{PE} \tag{3.10}$$

and similarly for the specific energy e,

$$e = \frac{E}{m} = u + \frac{V^2}{2g_c} + \frac{gZ}{g_c} \tag{3.12}$$

and

$$e = u + \text{ke} + \text{pe} \tag{3.13}$$

In these equations, we continue the practice introduced in Chapter 2 of using uppercase letters to denote *extensive* properties and lowercase letters to denote *intensive* (specific) properties. The energy concepts described in these equations are illustrated in Figure 4.1.

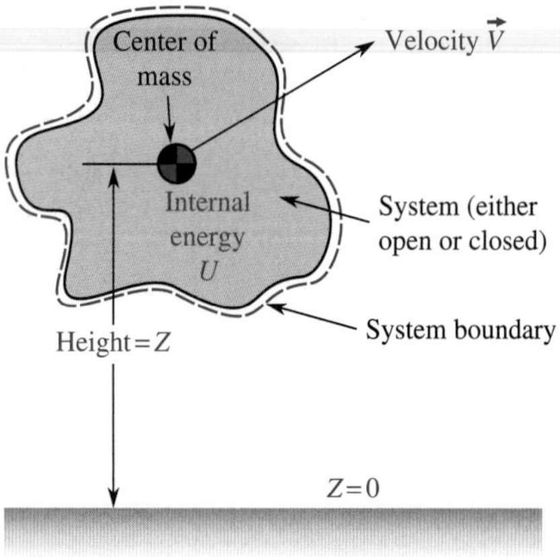

FIGURE 4.1
System energy components.

In equilibrium thermodynamics, the proper energy balance is given by Eq. (4.1), where the gain in energy E_G is to be interpreted as follows. The system is initially in some equilibrium state (call it state 1), and after the application of some "process," the system ends up in a different equilibrium state (call it state 2). If we now add a subscript to each symbol to denote the state at which the property is to be evaluated (E_1 is the total energy of the system in state 1 and so forth), then we can write the energy gain of the system as

$$E_G = \text{Final total energy} - \text{Initial total energy} \tag{4.3}$$

or

$$E_G = E_2 - E_1 \tag{4.4}$$

and extending this to Eq. (3.9), we obtain

$$E_G = U_2 - U_1 + \frac{m}{2g_c}(V_2^2 - V_1^2) + \left(\frac{mg}{g_c}\right)(Z_2 - Z_1) \tag{4.5}$$

or

$$E_G = m\left[u_2 - u_1 + \frac{V_2^2 - V_1^2}{2g_c} + \frac{g}{g_c}(Z_2 - Z_1)\right] \tag{4.6}$$

[3] In this text, we use the symbol V to represent the magnitude of the average velocity $|V|$, and the symbol $\forall$ to represent volume.

alternatively,

$$E_G = U_2 - U_1 + \mathrm{KE}_2 - \mathrm{KE}_1 + \mathrm{PE}_2 - \mathrm{PE}_1 \tag{4.7}$$

and

$$E_G = m(u_2 - u_1 + \mathrm{ke}_2 - \mathrm{ke}_1 + \mathrm{pe}_2 - \mathrm{pe}_1) \tag{4.8}$$

In most of the engineering situations we encounter, either the system is not moving at all or it is moving without any change in velocity or height. In these cases,

$$E_G = U_2 - U_1 = m(u_2 - u_1) = E_T$$

EXAMPLE 4.1

Figure 4.2 shows that 3.00 lbm of saturated water vapor at 10.0 psia is sealed in a rigid container aboard a spaceship traveling at 25,000. mph at an altitude of 200. mi. What energy transport is required to decelerate the water to zero velocity and bring it down to the surface of the Earth such that its final specific internal energy is 950.0 Btu/lbm? Neglect any change in the acceleration of gravity over this distance.

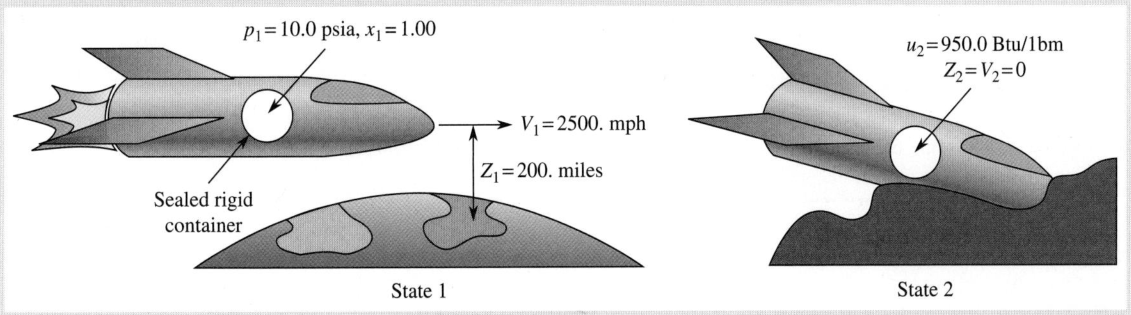

FIGURE 4.2
Example 4.1.

Solution

Let the system in this example be just the water in the container, then the process followed by the water is a constant volume process (the water is in a "rigid, sealed container"). Therefore, the problem statement can be outlined as follows:

State 1	$m = 3.0\,\text{lbm},\ \forall = \text{constant}$ →	State 2
$p_1 = 10.0\,\text{psia}$		$u_2 = 950.0\,\text{Btu/lbm}$
$x_1 = 1.00(\text{saturated vapor})$		$v_2 = v_1 = 38.42\,\text{ft}^3/\text{lbm}$
$v_1 = v_g(\text{at }10.0\,\text{psia}) = 38.42\,\text{ft}^3/\text{lbm}$		

Notice how the process path gives us the value of a property (v_2) in the final state. To determine the required energy transport, we use the energy balance Eq. (4.1), along with the definition of the energy gain term E_G from Eq. (4.5):

$$\text{EB: } E_G = E_T + E_{P_0} \text{ (as required by the first law)}$$

and, assuming g is constant during this process,

$$E_G = E_T = U_2 - U_1 + \frac{m}{2g_c}(V_2^2 - V_1^2) + \frac{mg}{g_c}(Z_2 - Z_1)$$

Here, $V_2 = Z_2 = 0$, so

$$E_T = U_2 - U_1 - \frac{m}{2g_c}V_1^2 - \frac{mg}{g_c}Z_1$$

Table C.2a in *Thermodynamic Tables to accompany Modern Engineering Thermodynamics* gives

$$u_1 = u_g(10.0\,\text{psia}) = 1072.2\,\text{Btu/lbm}$$

and the problem statement requires that $u_2 = 950.0$ Btu/lbm. Therefore,

$$U_1 = mu_1 = (3.00\,\text{lbm})(1072.2\,\text{Btu/lbm}) = 3216.6\,\text{Btu}$$

(Continued)

EXAMPLE 4.1 *(Continued)*

and

$$U_2 = mu_2 = (3.00\,\text{lbm})(950.0\,\text{Btu/lbm}) = 2850\,\text{Btu}$$

so

$$E_T = (2850 - 3216.6)\,\text{Btu} - \frac{3.00\,\text{lbm}}{2}\left[(25{,}000.\ \text{mile/h})\left(\frac{5280\ \text{ft/mile}}{3600\ \text{s/h}}\right)\right]^2$$
$$\times \frac{\dfrac{1\ \text{Btu}}{778.16\,\text{ft}\cdot\text{lbf}}}{32.174\dfrac{\text{lbm}\cdot\text{ft}}{\text{lbf}\cdot\text{s}^2}} - \frac{3.00\ \text{lbm}(32.174\ \text{ft/s}^2)}{\left(32.174\dfrac{\text{lbm}\cdot\text{ft}}{\text{lbf}\cdot\text{s}^2}\right)}(200.\ \text{miles})(5280\ \text{ft/mile})\left(\frac{1\ \text{Btu}}{778.16\,\text{ft}\cdot\text{lbf}}\right)$$
$$= -366.6 - 80{,}550 - 4071 = -85{,}000\,\text{Btu (to three significant figures)}$$

Therefore, 85,000 Btu of energy must be transferred *out of* the water (E_T is negative here) by some mechanism. This can be done, for example, by having the spaceship (and the water) do work on the atmosphere by aerodynamic drag as it lands.

Exercises

1. What would be the value of u_2 in Example 4.1 if E_T were zero? **Answer**: $u_2 = 29{,}300$ Btu/lbm. (What is the physical state of the water now?)
2. Which causes the larger change in E_G:
   a. A velocity increase from 0 to 1 ft/s or an increase in height from 0 to 1 ft?
   b. A velocity increase from 0 to 100 ft/s or a height increase from 0 to 100 ft?
   Answers: (a) height, (b) velocity.
3. Determine the value of E_T that must occur when you stop a 1300. kg automobile traveling at 100. km/h on a level road with no change in internal energy. **Answer**: $E_T = 502$ kJ.

In nonequilibrium systems, we use the energy rate balance equation with $\dot{E}_G$ defined as

$$\dot{E}_G = \frac{d}{dt}\left(U + \frac{m}{2g_c}V^2 + \frac{mg}{g_c}Z\right)_{\text{system}} = \dot{E}_T \tag{4.9}$$

Equation (4.9) can become quite complicated for open systems whose total mass is rapidly changing (such as with rockets), because it expands as follows (using $U = mu$):

$$\dot{E}_G = m\left[\dot{u} + \frac{V}{g_c}(\dot{V}) + \frac{g}{g_c}(\dot{Z})\right] + \left(u + \frac{V^2}{2g_c} + \frac{gZ}{g_c}\right)\dot{m} = \dot{E}_T \tag{4.10}$$

Notice that, in this equation, $\dot{V} = dV/dt$ is the magnitude of the instantaneous acceleration, and $\dot{Z}$ is the magnitude of the instantaneous vertical velocity.

The equilibrium thermodynamics energy balance and the nonequilibrium energy rate balance are fairly simple concepts; however, their implementation can be quite complex. Each of the gain, transport, and production terms may expand into many separate terms, all of which must be evaluated in an analysis. Next, we investigate the structure of the energy transport and energy transport rate terms.

4.4 ENERGY TRANSPORT MECHANISMS

There are three energy transport mechanisms, any or all of which may be operating in any given system: (1) heat, (2) work,[4] and (3) mass flow. These three mechanisms and their sign conventions are illustrated in Figure 4.3.

Note that the sign conventions for heat and work shown in Figure 4.3 are not the same. Heat transfer *into* a system is taken as positive, whereas work must be produced by or come *out of* a system to be positive. This is the conventional mechanical engineering sign convention and reflects the traditional view that heat coming out

[4] The types of work transports of energy included here are only those due to dissipative or nonconservative forces. For example, the work associated with gravitational or electrostatic forces is not considered a work mode because it is conservative (i.e., it is representable by the gradient of a scalar quantity) and is consequently nondissipative. Energy transports resulting from the actions of conservative forces have their own individual terms in the energy balance equation (such as mgZ/g_c for the gravitational potential energy).

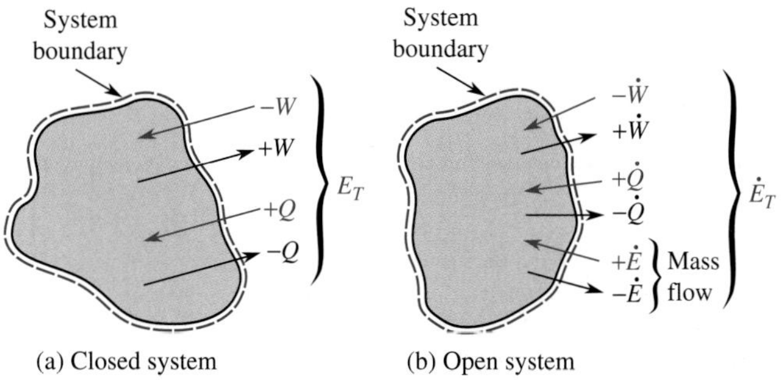

FIGURE 4.3
Energy transport mechanisms.

WHAT ARE *HEAT* AND *WORK* ANYWAY?

Heat is usually defined as energy transport to or from a system due to a temperature difference between the system and its surroundings. This can occur by only three modes: conduction, convection, and radiation.

Work is more difficult to define. It is often defined as a force moving through a distance, but this is only one type of work; there are many other work modes as well. Since the only energy transport modes for moving energy across a system's boundary are heat, mass flow, and work, the simplest definition of work is that it is any energy transport mode that is neither heat nor mass flow.[5]

[5] *Work can also be defined using the concept of a "generalized" force moving through a "generalized" displacement, see Table 4.2 later in this chapter.*

of a system is "lost" (i.e., negative), while work produced by a system (such as an engine) should be assigned a positive value.

By definition, a closed system has no mass crossing its system boundary, so it can experience only work and heat transport mechanisms. Also, since the gain, transport, and production terms in the balance equation are defined to be *net* values (see Eq. (2.10)), we define

1. The *net* heat transport of energy *into* a system = $\sum_i Q_i = Q$ and the *net* heat transport rate of energy *into* a system = $\sum_i \dot{Q}_i = \dot{Q}$.
2. The *net* work transport of energy *out* of a system = $\sum_i W_i = W$ and the *net* work transport rate of energy *out* of a system = $\sum_i \dot{W}_i = \dot{W}$.
3. The *net* mass transport of energy *into* the system = $\sum_i E_i = \sum E_{\text{mass flow}}$ and the *net* mass transport rate of energy *into* the system = $\sum_i \dot{E}_i = \sum \dot{E}_{\text{mass flow}}$.

Thus, for a *closed system*, the total energy transport becomes

$$E_T = Q - W \tag{4.11}$$

and the total energy transport rate is

$$\dot{E}_T = \dot{Q} - \dot{W} \tag{4.12}$$

For *open systems*, the same quantities are

$$E_T = Q - W + \sum E_{\substack{\text{mass}\\ \text{flow}}} \tag{4.13}$$

and

$$\dot{E}_T = \dot{Q} - \dot{W} + \sum \dot{E}_{\substack{\text{mass}\\ \text{flow}}} \tag{4.14}$$

In Eqs. (4.13) and (4.14), note that we write the summation signs on the net mass transport of energy terms, but for simplicity, we do not write the summation signs on the work or heat transport terms. This is because you often have open systems with more than one mass flow stream, but seldom do you have more than one type of work or heat transport present. However, you must always remember that W, $\dot{W}$, Q, and $\dot{Q}$ are also *net* terms and represent a summation of all the different types of work and heat transports of energy present. This is illustrated in the following example.

EXAMPLE 4.2

Determine the energy transport rate for the system shown in Figure 4.4.

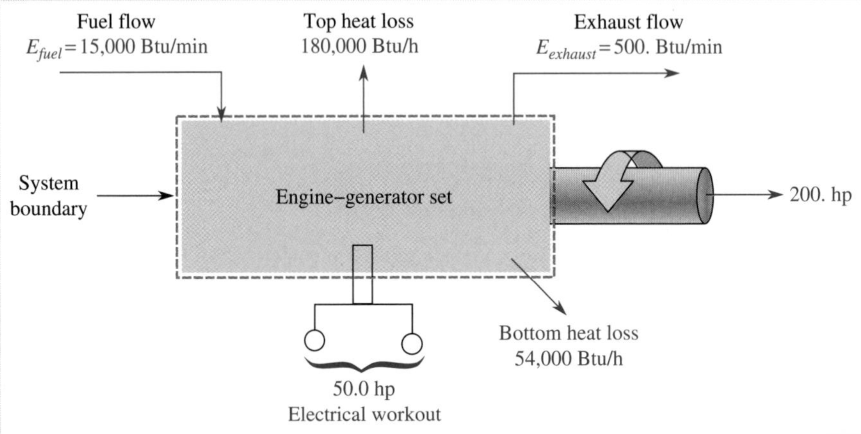

FIGURE 4.4
Example 4.2

Solution

From Eq. (4.14), the total energy transport rate is

$$\dot{E}_T = \dot{Q} - \dot{W} + \sum \dot{E}_{\substack{\text{mass}\\ \text{flow}}}$$

where

$$\begin{aligned}\dot{Q} &= \text{net heat transfer } \textit{into} \text{ the system}\\ &= -180. \times 10^3 \text{ Btu/h} - 54.0 \times 10^3 \text{ Btu/h} = -234 \times 10^3 \text{ Btu/h}\end{aligned}$$

and

$$\dot{W} = \text{net work rate } \textit{out of} \text{ the system} = 200. \text{ hp} + 50.0 \text{ hp} = 250. \text{ hp}$$

while

$$\begin{aligned}\sum \dot{E}_{\substack{\text{mass}\\ \text{flow}}} &= \text{net mass flow of energy } \textit{into} \text{ the system}\\ &= 15.0 \times 10^3 \text{Btu/min} - 500.\text{Btu/min} = 14.5 \times 10^3 \text{Btu/min}\end{aligned}$$

So

$$\dot{E}_T = (-234 \times 103 \text{ Btu/h})[1 \text{ h}/(60 \text{ min})] - (250. \text{ hp})[42.4 \text{ Btu/(hp·min)}] + 14.5 \times 103 \text{ Btu/min} = 0.00 \text{ Btu/min}$$

Exercises

4. Determine the energy transport rate that occurs in Example 4.2 when the work mode directions are reversed. **Answer:** $\dot{E}_T = 21.2 \times 10^3$ Btu/min.
5. Determine the net rate of energy gain of a closed system that receives heat at a rate of 4500. kJ/s and produces work at a rate of 1500. kJ/s. **Answer:** $\dot{E}_G = 3000.$ KJ/s.
6. An insulated open system has a net gain of 700. Btu of energy while producing 500. Btu of work. Determine the mass flow energy transport. **Answer:** $E_{\text{mass flow}} = 1.20 \times 10^3$ Btu.

The system of Example 4.2 has no *net* energy transport rate, even though it has six energy transport rates. Note that the energy rate balance (Eq. (4.2)) for this system is $\dot{E}_G = \dot{E}_T$; therefore, this system also has no net gain of energy. That is, the total energy E of this system is constant in time.

4.5 POINT AND PATH FUNCTIONS

A quantity, say y, that has a value at every point within its range is called a *point function*. Its derivative is written as dy, and its integral from state 1 to state 2 is

$$\int_1^2 dy = y_2 - y_1$$

Thus, the value of the integral depends only on the values of y at the end points of the integration path and is independent of the actual path taken between these end points. This is a fundamental characteristic of point functions. *All intensive and extensive thermodynamic properties are point functions.* Therefore, we can write

$$\int_1^2 dE = E_2 - E_1; \quad \int_1^2 du = u_2 - u_1; \quad \int_1^2 dm = m_2 - m_1$$

and so forth.

A quantity, say x, whose value depends on the *path* taken between two points within its range is called a *path function*. Since path functions do not differentiate or integrate in the same manner as point functions, we cannot use the same differential and integral notation for both path and point functions. Instead, we let $\bar{d}x$ denote the differential of the path function x, and we define its integral over the path from state 1 to state 2 as

$$\int_1^2 \bar{d}x = {}_1x_2 \quad \left[\text{Note: } \int_1^2 \bar{d}x \neq (x_2 - x_1)\right] \tag{4.15}$$

A path function does not have a value at a point. It has a value only for a path of points, and this value is directly determined by all the points on the path, not just its end points. For example, the area A under the curve of the point function $w = f(y)$ is a path function because

$$\bar{d}A = w\,dy = f(y)\,dy$$

and

$$\int_1^2 \bar{d}A = {}_1A_2 = \int_{Y1}^{Y2} f(y)\,dy = \text{area under } f(y) \text{ between the points } y_1 \text{ and } y_2$$

Clearly, if the path $f(y)$ is changed, then the area ${}_1A_2$ is also changed. Consequently, we say that ${}_1A_2$ is a path function.

We see in the next sections that *both the work and heat transports of energy are path functions*. Therefore, we write the differentials of these quantities as $\bar{d}W$ and $\bar{d}Q$, and their integrals as

$$\int_1^2 \bar{d}W = {}_1W_2 \tag{4.16}$$

and

$$\int_1^2 \bar{d}Q = {}_1Q_2 \tag{4.17}$$

Since the associated rate equations contain the time differential, we define *power* as the work rate, or

$$\dot{W} = \bar{d}W/dt \tag{4.18}$$

and, similarly, the heat transfer rate is

$$\dot{Q} = \bar{d}Q/dt \tag{4.19}$$

Each of the different types of work or heat transport of energy is called a *mode*. *A system that has no operating work modes is said to be aergonic*. Similarly, a system that changes its state without any work transport of energy having

NOTE!

Since work and heat are not thermodynamic properties and therefore not point functions, $\int_1^2 \bar{d}W \neq W_2 - W_1$ and $\int_1^2 \bar{d}W \neq \Delta W$. Similarly, $\int_1^2 \bar{d}Q \neq Q_2 - Q_1$, and $\int_1^2 \bar{d}Q \neq \Delta Q$. *Equations (4.16) and (4.17) are the only correct ways to write these path function integrals.*

WHAT IS *AERGONIC* ANYWAY?

The term aergonic comes from the Greek roots *a* meaning "not" and *ergon* meaning "work," and it should be interpreted to mean "no work has occurred." It is the analog of the word *adiabatic*, meaning no heat transfer has occurred, introduced later in this chapter.

Substituting Eqs. (4.8) and (4.11) into Eq. (4.1) and rearranging gives the general closed system energy balance equation for a system undergoing a process from state 1 to state 2 as

General closed system energy balance:

$$ {}_1Q_2 - {}_1W_2 = (E_2 - E_1)_{\text{system}} = m[(u_2 - u_1) + (V_2^2 - V_1^2)/(2g_c) + (Z_2 - Z_1)g/g_c]_{\text{system}} \tag{4.20} $$

and substituting Eq. (4.10) with m = constant and Eq. (4.12) into Eq. (4.2) gives the general closed system energy rate balance as

General closed system energy *rate* balance:

$$ \dot{Q} - \dot{W} = (dE/dt)_{\text{system}} = (m\dot{u} + mV\dot{V}/g_c + mg\,\dot{Z}/g_c)_{\text{system}} \tag{4.21} $$

Similarly, substituting Eqs. (4.9) and (4.14) into Eq. (4.2) gives the general open system energy rate balance as

General open system energy *rate* balance:

$$ \dot{Q} - \dot{W} + \sum_{\text{flow}} \dot{E}_{\text{mass}} = (d/dt)(mu + mV^2/2g_c + mZg/g_c)_{\text{system}} \tag{4.22} $$

where the mass of the system is no longer required to be constant.

occurred is said to have undergone an *aergonic process*. While there are only three modes of heat transport, there are many modes of work transport. In the following segments, four mechanical work modes and five nonmechanical work modes are studied in detail.

4.6 MECHANICAL WORK MODES OF ENERGY TRANSPORT

In mechanics, we recognize that work is done whenever a force moves through a distance. When this force is a mechanical force $\vec{F}$, we call this work mode *mechanical work* and define it as

$$ (\bar{d}W)_{\text{mechanical}} = (\vec{F}_{\text{applied by the system}})\cdot d\vec{x} - (\vec{F}_{\text{applied on the system}})\cdot d\vec{x} \tag{4.23} $$

CAN YOU ANSWER THIS QUESTION FROM 1936?

On page 66 of the October 1936 issue of *Modern Mechanix* is a discussion of the oddities of science that reads: "Modern science states that energy cannot be destroyed. Scientists are now wondering what happens to the energy contained in a compressed spring destroyed in acid." How would you answer this question more than 70 years later?

The person who wrote this in 1936 did not understand the concept of internal energy. Then, neglecting any changes in kinetic or potential energy, an energy balance on the system gives

$$ {}_1Q_2 - {}_1W_2 = (E_2 - E_1)_{\text{system}} = (U_2 - U_1)_{\text{system}} $$

where $U_1 = U_{\text{acid}} + U_{\text{spring}} = (m_{\text{acid}}u_{\text{acid}} + m_{\text{spring}}u_{\text{spring}})$. Now, $U_{\text{spring}} = F(\Delta X)$, the work done in compressing the spring. Finally, $U_2 = U_{\text{acid+spring}} = (m_{\text{acid}} + m_{\text{spring}})u_{\text{acid+spring}}$. If we make the reasonable assumption that the spring dissolved without any heat transfer (${}_1Q_2 = 0$) and aergonically (${}_1W_2 = 0$), then the energy balance equation gives $U_2 = U_1$, and solving it for the final specific internal energy of the acid-spring solution, we find that

$$ u_{\text{acid+spring}} = \frac{m_{\text{acid}}u_{\text{acid}} + m_{\text{spring}}u_{\text{spring}}}{m_{\text{acid}} + m_{\text{spring}}} $$

So the answer to the 1936 question is this: *The energy contained in the compressed spring ends up as part of the energy of the combined acid-spring solution.* That is, since the mechanical work that went into compressing the spring ended up as part of the spring's internal energy, when the spring was dissolved in the acid, the internal energy in the spring became part of the internal energy of the acid-spring solution.

Also, if we assume that the acid-spring solution is a simple incompressible liquid with an internal energy that depends only on temperature, then we can write $u_{\text{acid+spring}} = cT$, and we see that the energy contained in the compressed spring reappears as an increase in the temperature of the resulting acid-spring solution.

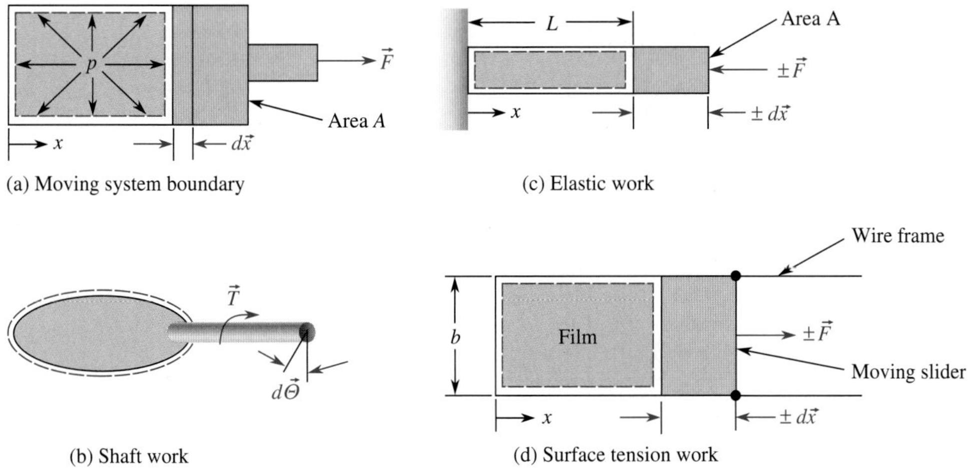

FIGURE 4.5
Four classical types of mechanical work.

or

$$({}_1W_2)_{\text{mechanical}} = \int_{x1}^{x2} (\vec{F}_{\text{applied by the system}})\cdot d\vec{x} - \int_{x1}^{x2} (\vec{F}_{\text{applied on the system}})\cdot d\vec{x} \tag{4.24}$$

Note that our sign convention requires that work done by the system be *positive*, while work done *on* the system be *negative*.

In thermodynamics, the four classical types of mechanical work (Figure 4.5) are

1. Moving system boundary work.
2. Rotating shaft work.
3. Elastic work.
4. Surface tension work.

These are very important work modes in engineering analysis and the following material provides a detailed discussion of their major characteristics.

4.6.1 Moving System Boundary Work

Whenever a system boundary moves such that the total volume of the system changes, moving system boundary work occurs. This is sometimes called *expansion* or *compression work,* and it has wide application in mechanical power technology. In this case, the force is applied by the system through the pressure p (see Figure 4.5a), so $\vec{F} = p\vec{A}$ and $\vec{F}\cdot d\vec{x} = p\vec{A}\cdot d\vec{x} = pd\mathcal{V}$, where p is the pressure acting on the system boundary, $\vec{A}$ is the area vector (defined to be normal to the system boundary and pointing outward), $d\vec{x}$ is the differential boundary movement, and $d\mathcal{V}$ is the differential volume $\vec{A}\cdot d\vec{x}$. Consequently,

$$(\bar{d}W)_{\substack{\text{moving}\\\text{boundary}}} = p\,d\mathcal{V} \tag{4.25}$$

and for moving boundary work,

Moving boundary work:

$$({}_1W_2)_{\substack{\text{moving}\\\text{boundary}}} = \int_1^2 p\,d\mathcal{V} \tag{4.26}$$

EXAMPLE 4.3

The sealed, rigid tank shown in Figure 4.6 contains air at 0.100 MPa and 20.0 °C. The tank is then heated until the pressure in the tank reaches 0.800 MPa. Determine the mechanical moving boundary work produced in this process.

(Continued)

EXAMPLE 4.3 (*Continued*)

Solution

Let the system be the material inside the tank. The process of heating the tank is one of constant volume (the tank is "rigid"). Therefore, since the system volume, $\mathcal{V}$, is constant, $d\mathcal{V} = 0$ and the moving boundary work is:

$$({}_1W_2)_{\substack{\text{moving}\\ \text{boundary}}} = \int_1^2 p\,d\mathcal{V} = 0$$

Therefore, no moving boundary work occurs during this process.

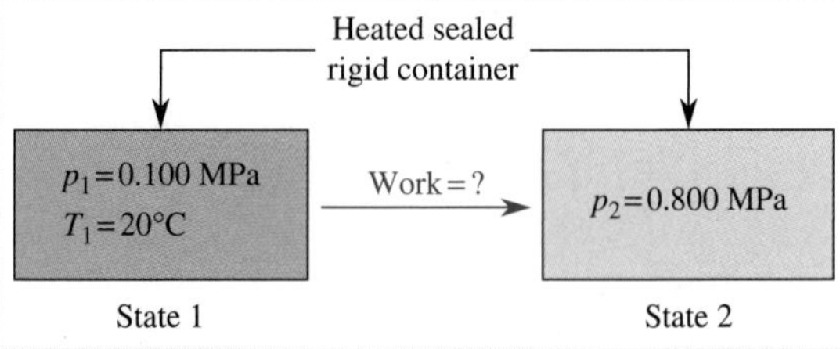

FIGURE 4.6
Example 4.3.

Since a "rigid" container cannot change its volume, its moving boundary work is always zero regardless of the process it undergoes.

EXAMPLE 4.4

The weather balloon in Figure 4.7 is inflated from a constant pressure, compressed gas source at 20.0 psia. Determine the moving system boundary work as the balloon expands from a diameter of 1.00 ft to 10.0 ft.

Solution

Assume the balloon is a sphere, then $\mathcal{V} = 4/3\,\pi R^3 = 1/6\,\pi D^3$. The process here is one of constant pressure, so $p = \text{constant}$, and

$$({}_1W_2)_{\substack{\text{moving}\\ \text{boundary}}} = \int_1^2 p\,d\mathcal{V} = p\int_1^2 d\mathcal{V} = p(\mathcal{V}_2 - \mathcal{V}_1)$$

$$= \left(20.0\,\frac{\text{lbf}}{\text{in}^2}\right)\left(\frac{144\ \text{in}^2}{\text{ft}^2}\right)\left(\frac{\pi}{6}\right)[(10.0^3 - 1.00^3)\ \text{ft}^3]$$

$$= 1.51 \times 10^6\ \text{ft}\cdot\text{lbf}$$

The work is positive because the balloon does work on the atmosphere as it expands and pushes the atmosphere out of the way.

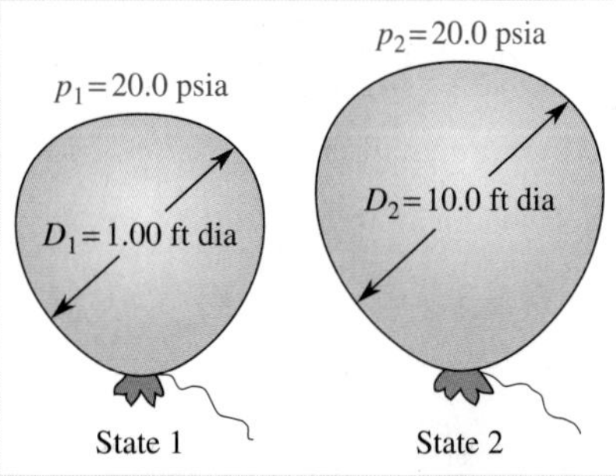

FIGURE 4.7
Example 4.4.

Exercises

7. In Example 4.3, is the moving boundary work always zero for a sealed, rigid container? Are any other work modes always zero for this type of system? Could a piston-cylinder apparatus be modeled as a sealed, rigid system? **Answers**: Yes, no, no. (It is sealed and the components, the piston and the cylinder, are rigid, but the piston can move, producing a change in the enclosed volume.)
8. Determine the moving boundary work for the balloon in Example 4.4 as it deflates from a diameter of 10. ft to a diameter of 5.0 ft at a constant pressure of 20. psia. What does the work on the balloon? **Answer**: $({}_1W_2)_{\text{moving boundary}} = -1.3 \times 10^6$ ft·lbf. The surrounding atmosphere does work on the balloon as it deflates, that is why the work is negative.
9. If the pressure inside a system depends on volume according to the relation $p = K_1 + K_2\mathcal{V} + K_3/\mathcal{V}$, where K_1, K_2, and K_3 are constants, determine the appropriate equation for the moving boundary work done as the volume changes from $\mathcal{V}_1$ to $\mathcal{V}_2$. **Answer**: $({}_1W_2)_{\text{moving boundary}} = K_1(\mathcal{V}_2 - \mathcal{V}_1) + K_2(\mathcal{V}_2^2 - \mathcal{V}_1^2)/2 + K_3 \ln(\mathcal{V}_2/\mathcal{V}_1)$.

To carry out the integration indicated in Eq. (4.26), the exact $p = p(\mathcal{V})$ pressure volume function must be known. This function is usually given in the process path specification of a problem statement. For example, in Example 4.3, the process is one of constant volume (the container is rigid), so $d\mathcal{V} = 0$; and in Example 4.4, the filling process is isobaric ($p = \text{constant}$), so the integral of Eq. (4.26) is very easy. In general, outside of these two cases, the integration of Eq. (4.26) is not trivial and must be determined with great care.

As an example of a nontrivial integration of Eq. (4.26), consider a process that obeys the relation

$$p\mathcal{V}^n = \text{constant} \tag{4.27}$$

or

$$p_1\mathcal{V}_1^n = p_2\mathcal{V}_2^n$$

where the exponent n is a constant. Such processes are called *polytropic processes.*[6] The moving system boundary work of any substance undergoing a polytropic process is

$$({}_1W_2)_{\substack{\text{polytropic}\\ \text{moving boundary}}} = \int_1^2 p\,d\mathcal{V} = \int_1^2 \frac{\text{constant}}{\mathcal{V}^n}\,d\mathcal{V}$$

For $n = 1$, this integral becomes

$$({}_1W_2)_{\substack{\text{polytropic }(n=1)\\ \text{moving boundary}}} = p_1\mathcal{V}_1 \ln\frac{\mathcal{V}_2}{\mathcal{V}_1} = p_2\mathcal{V}_2 \ln\frac{\mathcal{V}_2}{\mathcal{V}_1} \tag{4.28}$$

and for n $\neq$ 1, it becomes

$$({}_1W_2)_{\substack{\text{polytropic }(n\neq1),\\ \text{moving boundary}}} = \frac{p_2\mathcal{V}_2 - p_1\mathcal{V}_1}{1-n} \tag{4.29}$$

If the material undergoing a polytropic process is an ideal gas, then it must simultaneously satisfy both of the following equations:

1. The ideal gas equation of state, $p\mathcal{V} = mRT$.
2. The polytropic process equation, $p\mathcal{V}^n$ = constant.

Combining these two equations by eliminating the pressure p gives

$$mRT\mathcal{V}^{n-1} = \text{constant}$$

or, for a fixed mass system,

$$T_1\mathcal{V}_1^{n-1} = T_2\mathcal{V}_2^{n-1}$$

or

$$\frac{T_2}{T_1} = \left(\frac{\mathcal{V}_2}{\mathcal{V}_1}\right)^{1-n} = \left(\frac{v_2}{v_1}\right)^{1-n} \tag{4.30}$$

Similarly, eliminating $\mathcal{V}$ in these two equations (for a fixed mass system) gives the polytropic process equations for an ideal gas:

Polytropic process equations for an ideal gas

$$\frac{T_2}{T_1} = \left(\frac{p_2}{p_1}\right)^{(n-1)/n} = \left(\frac{v_2}{v_1}\right)^{1-n} \tag{4.31}$$

Finally, if we have an ideal gas undergoing a polytropic process with $n \neq 1$, then its moving system boundary work is given by Eq. (4.29), with $p_2\mathcal{V}_2 - p_1\mathcal{V}_1 = mR(T_2 - T_1)$ as the polytropic work equation for an ideal gas ($n \neq 1$):

Polytropic work equation for an ideal gas ($n \neq 1$)

$$({}_1W_2)_{\substack{\text{polytropic }(n\neq1)\\ \text{ideal gas}\\ \text{moving boundary}}} = \frac{mR}{1-n}(T_2 - T_1) \tag{4.32}$$

---

[6] The term *polytropic* comes from the Greek roots *poly* meaning "many" and *trope* meaning "turns" or "paths."

EXAMPLE 4.5

Figure 4.8 shows a new process in which 0.0100 kg of methane (an ideal gas) is compressed from a pressure of 0.100 MPa and a temperature of 20.0 °C to a pressure of 10.0 MPa in a polytropic process with $n = 1.35$. Determine the moving boundary work required.

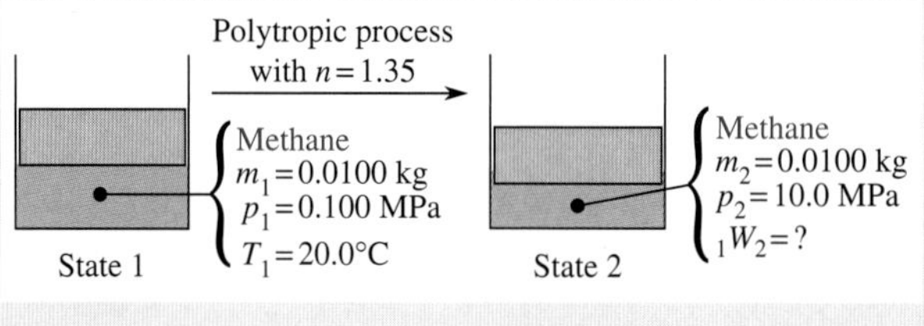

FIGURE 4.8
Example 4.5.

Solution

Since the methane behaves as an ideal gas and $n \neq 1$, we can find the work required from Eq. (4.32):

$$({}_1W_2)_{\substack{\text{polytropic}\,(n\neq 1)\\ \text{ideal gas}\\ \text{moving boundary}}} = \frac{mR}{1-n}(T_2 - T_1)$$

where the value of T_2 can be found from Eq. (4.31):

$$T_2 = T_1\left(\frac{p_2}{p_1}\right)^{(n-1)/n} = (20.0 + 273.15\text{ K})\left(\frac{10.0\text{ MPa}}{0.100\text{ MPa}}\right)^{(1.35-1)/1.35} = 967\text{ K} = 694\text{ °C}$$

Using Table C.13b of *Thermodynamic Tables to accompany Modern Engineering Thermodynamics* to find the value of the gas constant for methane, $R_{\text{methane}} = 0.518$ kJ/kg·K, Eq. (4.32) then gives

$$({}_1W_2)_{\substack{\text{polytropic}\,(n\neq 1)\\ \text{ideal gas}\\ \text{moving boundary}}} = \frac{(0.0100\text{ kg})(0.518\text{ kJ/kg·K})}{1-1.35}(967 - 293.15) = -9.98\text{ kJ}$$

The work comes out negative, because it is being done *on* the system.

Exercises

10. Determine the work required in Example 4.5 if the final pressure of the methane is 0.500 MPa. **Answer**: −2.25 kJ.
11. If the work required in Example 4.5 is −5.00 kJ, determine the final temperature and pressure of the methane. Answer: $T_2 = 631$ K, $p_2 = 1.92$ MPa.
12. If the gas used in Example 4.5 were air, determine the work required to compress it polytropically from 14.7 psia, 70.0°F to 150.°F with $n = 1.33$. **Answer**: ${}_1W_2 = -285.1$ ft·lbf

4.6.2 Rotating Shaft Work

Whenever a rotating shaft carrying a torque load crosses a system boundary, *rotating shaft work* is done. In this case (see Figure 4.5b),

$$(\bar{d}W)_{\substack{\text{rotating}\\ \text{shaft}}} = \vec{T}\cdot d\vec{\theta} \tag{4.33}$$

and, for rotating shaft work,

Rotating shaft work

$$({}_1W_2)_{\substack{\text{rotating}\\ \text{shaft}}} = \int_1^2 \vec{T}\cdot d\vec{\theta} \tag{4.34}$$

where $\vec{T}$ is the torque vector produced *by* the system on the shaft and $d\vec{\theta}$ is its angular displacement vector. These two vectors are in the direction of the shaft axis. Normally, thermodynamic problem statements do not require rotating shaft work to be calculated from Eq. (4.34). The rotating shaft work is usually openly given as part of the problem statement. For example, if you are analyzing an automobile internal combustion engine producing 150. ft·lbf of work at the crankshaft, you must be able to recognize that $({}_1W_2)_{\text{rotating shaft}} = 150.\ \text{ft·lbf}$.

WHEN IS SHAFT WORK NOT SHAFT WORK?

Suppose you have a system that contains a fluid, and this fluid is in contact with a mixing blade or an impeller driven by a shaft passing through the system boundary (see Figure 4.9). This would constitute an example of shaft work.

The shaft and the blade or impeller are inside the system and their physical and thermodynamic properties are part of the system's properties. You have a heterogenous system made up of the fluid and the solid shaft and blade. If the mass of the fluid is large enough and the size of the shaft and blade is small enough, then their impact on the system's properties can be neglected and the system can be considered to consist of the fluid alone. However, this is not always the case. Suppose now you exclude the shaft and the blade or impeller from the system by restricting the system to be only the fluid and redraw the system boundaries so that they pass along the surface of the shaft and blade (see Figure 4.10). Now, your system consists of a pure substance (the fluid), but what kind of work mode do you now have?

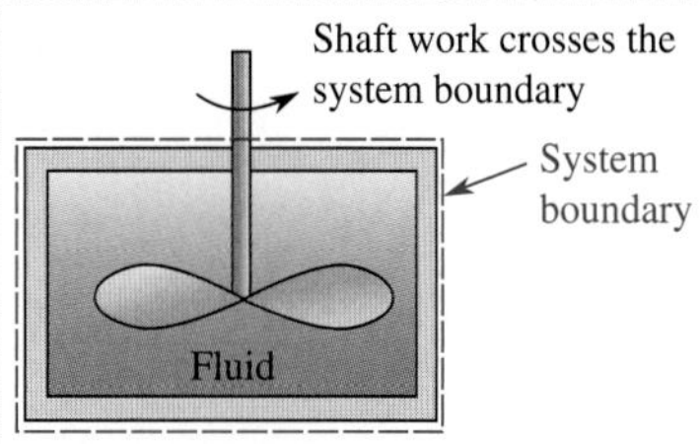

FIGURE 4.9
Shaft work in a system containing a fluid.

Since the only work modes we can analyze are "reversible," the fluid medium cannot possess viscosity (fluid friction), and consequently, there can be no shear forces on the blade. The only force a viscousless fluid can exert on the blades is a pressure force, p. As the blade moves, the system boundary must move accordingly to keep up with it, and the pressure force on the blade must also move. This is just the definition of the *moving boundary* work mode. Consequently, this type of shaft work is not really shaft work at all, it is really moving boundary work.

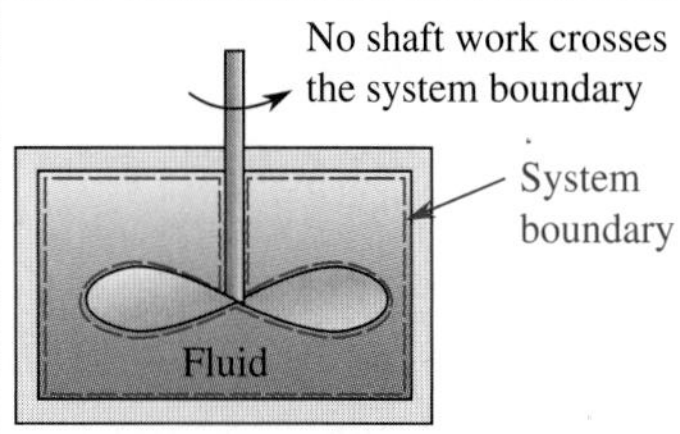

FIGURE 4.10
A new system boundary that omits the shaft and the blade.

Another example is the shaft work from an internal combustion engine. It is produced inside the engine by moving boundary piston-cylinder work, and in a frictionless reversible engine, these two work modes are equivalent. However, in a real engine, where friction and other losses are present, these two work modes are not equivalent (see Figure 4.11).

Not all shaft work can be viewed as moving boundary work. The shaft work from an electric motor or a mechanical gearbox is not equivalent to moving boundary work (see Figure 4.12).

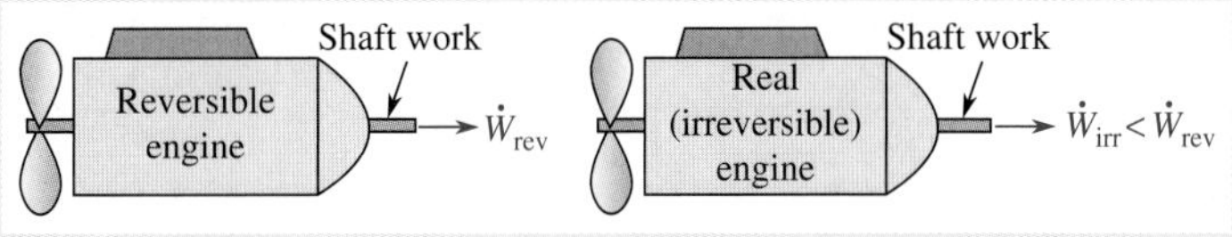

FIGURE 4.11
Reversible and irreversible work in an IC engine.

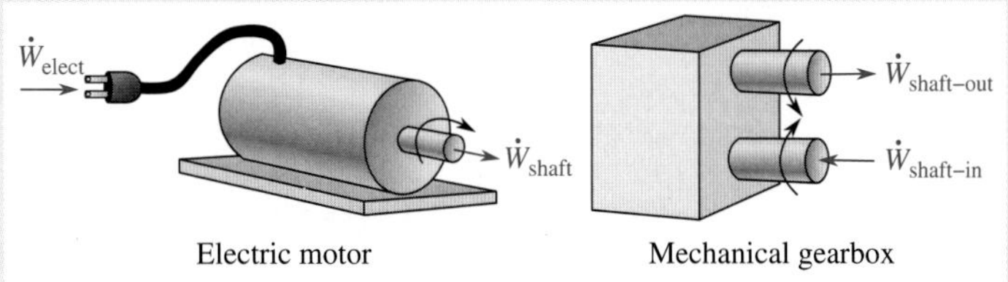

FIGURE 4.12
Shaft work from systems without internal moving boundaries.

4.6.3 Elastic Work

Whenever we compress or extend an elastic solid (like a spring), we perform elastic work. Consider a force $\pm\vec{F}$ applied on the end of an elastic rod (see Figure 4.5c). The normal stress σ in the rod is

$$\sigma = \pm\frac{|\vec{F}|}{A} \tag{4.35}$$

where $|\vec{F}|$ is the magnitude of the force and A is the cross-sectional area of the rod. Since the force $\vec{F}$ and its corresponding displacement $d\vec{x}$ are always in the same direction, the vector dot product $\vec{F}\cdot d\vec{x}$ always reduces to Fdx, where $F = |\vec{F}|$ and $dx = |d\vec{x}|$, and when the force is applied *on* the system from the surroundings rather than being produced *by* the system, the work is negative and its increment is

$$đW = -\vec{F}\cdot d\vec{x} = -Fdx = -\sigma A dx \tag{4.36}$$

The strain ε in the rod is defined as

$$d\varepsilon = \frac{dx}{L} = \frac{A\,dx}{AL} = \frac{A\,dx}{V\!\!\!\!-} = \frac{dV\!\!\!\!-}{V\!\!\!\!-} \tag{4.37}$$

where L is the length of the rod and AL is its volume $V\!\!\!\!-$. Then,

$$Adx = dV\!\!\!\!- = V\!\!\!\!-\,d\varepsilon \tag{4.38}$$

and Eq. (4.36) becomes

$$đW = -\sigma Adx = -\sigma V\!\!\!\!-\,d\varepsilon \tag{4.39}$$

Therefore, for elastic work,

Elastic work

$$({}_1W_2)_{\text{elastic}} = -\int_1^2 \sigma V\!\!\!\!-\,d\varepsilon \tag{4.40}$$

EXAMPLE 4.6

Determine an expression for the work involved in deforming a constant volume elastic solid that obeys Hooke's law of elasticity (see Figure 4.13).

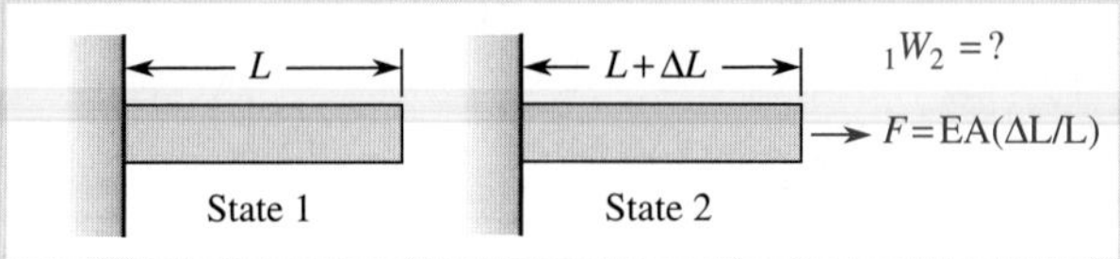

FIGURE 4.13
Example 4.6.

Solution

Here we have $V\!\!\!\!- =$ constant. Also, from strength of materials we can write Hooke's law as $\sigma = E\varepsilon$, where E is Young's modulus of elasticity. Then, Eq. (4.40) becomes

$$\begin{aligned}({}_1W_2)_{\text{elastic}} &= -\int_1^2 \sigma V\!\!\!\!-\,d\varepsilon = -\int_1^2 EV\!\!\!\!-\,\varepsilon\,d\varepsilon = -EV\!\!\!\!-\int_1^2 \varepsilon\,d\varepsilon \\ &= -EV\!\!\!\!-\left(\frac{\varepsilon_2^2-\varepsilon_1^2}{2}\right) = -\frac{V\!\!\!\!-}{2E}(\sigma_2^2-\sigma_1^2)\end{aligned}$$

Thus, if $\varepsilon_2^2 > \varepsilon_1^2$, then $({}_1W_2)_{\text{elastic}}$ is negative and work is being put into the system; and if $\varepsilon_2^2 < \varepsilon_1^2$, then $({}_1W_2)_{\text{elastic}}$ is positive and work is being produced by the system. Note that both tensile strains ($\varepsilon > 0$) and compressive strains ($\varepsilon < 0$) are possible here. But, the resulting work formula deals only with ε^2 and consequently gives the correct result regardless of the strain direction.

Exercises

13. What type of rigid system has zero elastic work regardless of the loading? **Answer**: A perfectly rigid system ($E = \infty$).
14. If the system analyzed in Example 4.6 was a rectangular steel bar, 1.0 inch square by 12 inches long, determine the elastic work required to stress it from 0.0 to 10. × 10^3 lbf/in^2. Use $E_{steel} = 30. \times 10^6$ lbf/in^2. **Answer**: $({}_1W_2)_{elastic} = -1.7$ ft·lbf.
15. Ten joules of elastic work is applied to a circular brass rod 0.0100 m in diameter and 1.00 m long. Determine the resulting stress and strain in the bar if it is initially unloaded. Use $E_{brass} = 1.05 \times 10^{11}$ Pa. **Answer**: $\sigma = 164$ MPa and $\varepsilon = 1.56 \times 10^{-3}$ m/m.

4.6.4 Surface Tension Work

Surface tension work is the two-dimensional analog of the elastic work just considered. Figure 4.5d shows a soap film on a wire loop. One side of the loop has a movable wire slider that can either compress or extend the film. As in the case of the elastic solid, the force and deflection are always in the same direction and the force is applied *to* the system, so we can modify Eq. (4.36) to read

$$\bar{d}W = -\vec{F}\cdot d\vec{x} = -F\,dx = -(2\sigma_s b)\,dx \tag{4.41}$$

where σ_s is the surface tension of the film, and b is the length of the moving part of the film. The factor of 2 appears because the film normally has two surfaces (top and bottom) in contact with air. Now, $2b\cdot dx = dA =$ change in the film's surface area, so Eq. (4.41) becomes

$$\bar{d}W = -\sigma_s dA \tag{4.42}$$

and, for the surface tension work,

Surface tension work

$$({}_1W_2)_{\substack{\text{surface}\\\text{tension}}} = -\int_1^2 \sigma_s\, dA \tag{4.43}$$

EXAMPLE 4.7

Determine the amount of surface tension work required to inflate the soap bubble shown in Figure 4.14 from a diameter of zero to 0.0500 m. The surface tension of the soap film can be taken to be a constant 0.0400 N/m.

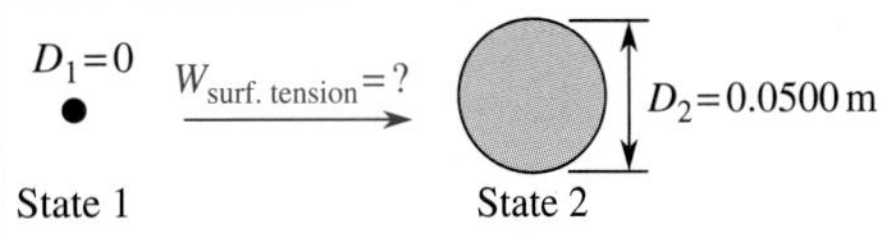

FIGURE 4.14
Example 4.7.

Solution

Here, σ_s = constant = 0.0400 N/m. Note that we are not calculating the surface area of the bubble here from its geometric elements, but wish only to find the change in area between states 1 and 2. Consequently, the area integral in this instance can be treated as a point function rather than as a path function. So Eq. (4.43) becomes

$$({}_1W_2)_{\substack{\text{surface}\\\text{tension}}} = -\sigma_s \int_1^2 dA = -\sigma_s(A_2 - A_1)$$

where $A_1 = 0$. Now, since a soap bubble has *two* surfaces (the outside and inside films),

$$A_2 = 2(4\pi R^2) = 2(4\pi)\left(\frac{0.0500\,\text{m}}{2}\right)^2 = 0.0157\,\text{m}^2$$

and

$$\begin{aligned}({}_1W_2)_{\substack{\text{surface}\\\text{tension}}} &= -(0.0400\,\text{N/m})(0.0157 - 0\,\text{m}^2)\\ &= -6.28 \times 10^{-4}\,\text{N}\cdot\text{m} = -6.28 \times 10^{-4}\,\text{J}\\ &= -(6.28 \times 10^{-4}\,\text{J})(1\,\text{Btu}/1055\,\text{J}) = -5.96 \times 10^{-7}\,\text{Btu}\end{aligned}$$

Table 4.2 Generalized Forces and Generalized Displacements

Work Mode	Generalized Force F	Generalized Displacement $d\chi$
Moving system boundary	p (pressure)	$d\forall$ (volume)
Shaft	T (torque)	$d\theta$ (angular displacement)
Elastic	$-\sigma$ (stress)	$\forall d\varepsilon$ (volume)
Surface tension	$-\sigma_s$ (surface tension)	dA (surface area)

Example 4.7 shows that it would take all of the surface tension energy stored in nearly 2 million 5 cm diameter soap bubbles to raise the temperature of one pound-mass of water by one degree Fahrenheit.

Notice that, in each of the four cases of classical mechanical work, the work differential $\bar{d}W$ was given by the product of what we can call a *generalized force* F and a *generalized displacement* $d\chi$; that is,

$$\bar{d}W = Fd\chi \tag{4.44}$$

where F and $d\chi$ for each of the four classical mechanical work modes are identified in Table 4.2. In Eq. (4.44), the scalar or dot product is implied if F and $d\chi$ are vectors.

The application of these work modes may change the thermodynamic state of the system and thus may produce a change in the system's thermodynamic properties. Finally, note that the generalized forces are all intensive properties, whereas the generalized displacements are all extensive properties.

We can generalize the work concept to nonmechanical systems by including any work mode given by Eq. (4.44) when the generalized force F is an intensive property *forcing function* and the generalized displacement $d\chi$ is an extensive property *response function*. We are now in a position to analyze the remaining work mode energy transport mechanisms.

4.7 NONMECHANICAL WORK MODES OF ENERGY TRANSPORT

Of the wide variety of nonmechanical work modes available, the following five are of significant engineering value:

1. Electrical current flow.
2. Electrical polarization.
3. Magnetic.
4. Chemical.
5. Mechanochemical.

Materials are electrically classified as conductors, nonconductors (dielectrics or insulators), and semiconductors. A pure *conductor* is a substance that has mobile charges (electrons) free to move in an applied electric field. They constitute the flow of electrical current. Pure *nonconductors* have no free electrons whatsoever, and a *semiconductor* is a material that behaves as a dielectric (nonconductor) at low temperatures but becomes conducting at higher temperatures.

As an electric field E is applied to a pure conductor, the free electrons migrate to the conductor's outer surface, where they create their own electric field, which opposes the applied field. As more and more electrons reach the outer surface, the electric field inside the object grows weaker and weaker, eventually vanishing altogether. At equilibrium, there is no electric field within a pure conductor.

A pure nonconductor has no free electrons with which to neutralize the applied electric field. The externally applied field therefore acts on the internal molecules, and normally nonpolar molecules become polar and develop *electric dipoles*. Some molecules are naturally polar in the absence of an electric field (e.g., water). The applied electric field rotates and aligns the newly created or naturally polar molecules. Complete alignment is normally prevented by molecular vibrations. But, when the applied field is strong enough to overcome the vibration randomizing effects and further increases in field strength have no effect on the material, the material is said to be *saturated* by the applied field. The process of electric dipole creation, rotation, and alignment in an applied electric field is known as dielectric *polarization*.

Therefore, two work modes arise from the application of an electric field to a material. The first is the work associated with the free electron (current) flow, and the second is the work associated with dielectric polarization. For a pure conductor, the polarization work is always zero; and for a pure nonconductor, the current flow work is always zero. We always treat these as separate work modes.

4.7.1 Electrical Current Flow Work

Electrical current flow work occurs whenever current-carrying wires (pure conductors) cross the system boundary. This is the most common type of nonmechanical work mode encountered in thermodynamic system analysis. The generalized force here is the intensive property *voltage* (the electric potential) ϕ, and the extensive property generalized displacement is the charge q.[7] Then, assuming the voltage is applied to the system,

$$(\bar{d}W)_{\substack{\text{electrical}\\\text{current}}} = -\phi dq$$

and

$$({}_1W_2)_{\substack{\text{electrical}\\\text{current}}} = -\int_1^2 \phi\, dq \tag{4.45}$$

Electrical current i is defined as

$$i = \frac{dq}{dt}$$

so $dq = i\, dt$, and

$$(\bar{d}W)_{\substack{\text{electrical}\\\text{current}}} = -\phi i\, dt \tag{4.46}$$

Then, electric current work is

Electrical current work

$$({}_1W_2)_{\substack{\text{electrical}\\\text{current}}} = \int_1^2 \phi i\, dt \tag{4.47}$$

From Ohm's law, the instantaneous voltage ϕ across a pure resistance R carrying an alternating current, described by $i = i_{max}\sin(2\pi ft)$, is

$$\phi = Ri = Ri_{max}\sin(2\pi ft)$$

where f is the frequency and $\phi_{max} = Ri_{max}$. Thus, Eq. (4.47) gives the electrical current work of n cycles of an alternating electrical current applied to a pure resistance from time 0 to time $t = n/f$ as

$$\begin{aligned}({}_1W_2)_{\substack{\text{electrical}\\\text{current}}} &= -\phi_{max} i_{max}\int_0^{t=n/f}\sin^2(2\pi ft)\, dt\\ &= -\phi_{max} i_{max}(t/2)\\ &= -\phi_e i_e t = -\phi_e^2(t/R) = -i_e^2 Rt\end{aligned} \tag{4.48}$$

where ϕ_e and i_e are the *effective* voltage and current defined by $\phi_e = \phi_{max}/\sqrt{2}$ and $i_e = i_{max}/\sqrt{2}$.

Electrical work can exist in either open or closed systems (we do not consider the flow of electrons across a system boundary to be a mass flow term). When the electron supply is going into a *finite* system, such as a battery or a capacitor, Eq. (4.45) or (4.47) is convenient to use. But, when an essentially *infinite* supply of voltage and current is used, it is more convenient to use the instantaneous rate at which electrical work is done, or the electrical *power*, defined as

$$(\dot{W})_{\substack{\text{electrical}\\\text{current}}} = \frac{\bar{d}W}{dt} = -\phi i \tag{4.49}$$

OHM'S LAW

This law was discovered experimentally by George Simon Ohm (1787–1854) in 1826. Basically, it states that, for a given conductor, the current is directly proportional to the potential difference, usually written as $\phi = Ri$, where R is the *electrical resistance* in units of ohms, where 1 ohm = 1 volt/ampere.

---

[7] The electrical potential ϕ and the electric field strength vector E are related by $E = -\nabla(\varphi)$, where $\nabla(\ )$ is the gradient operator.

The instantaneous electrical power $-\phi i$ of an alternating current circuit varies in time with the excitation frequency f. However, it is common to report the electrical power of an ac device as the instantaneous power averaged over one cycle of oscillation, or

$$\begin{aligned}(\dot{W})_{\substack{\text{electrical}\\ \text{(pure resistance)}}} &= -f\int_0^{1/f} \phi i\, dt = -f\phi_{\max} i_{\max}\int_0^{1/f} \sin^2(2\pi f t)\, dt \\ &= -\phi_{\max} i_{\max}/2 = -\phi_e i_e = -\phi_e^2/R = -i_e^2 R\end{aligned} \tag{4.50}$$

where ϕ_e and i_e are the effective voltage and current defined earlier.

EXAMPLE 4.8

Consider the 120. V, 144 Ω (ohm), alternating current incandescent lightbulb shown in Figure 4.15 to be a pure resistance. Determine

a. The electrical current work when the bulb is operated for 1.50 h.
b. Its electrical power consumption.

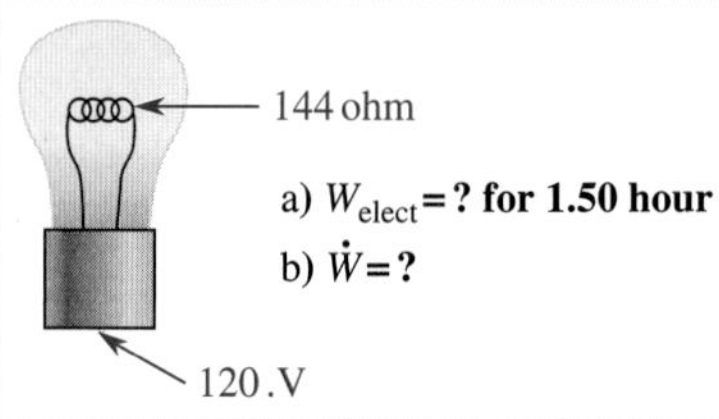

FIGURE 4.15
Example 4.8.

Solution

a. Since the voltage and current ratings of ac devices are always given in terms of their effective values, $\varphi_e = 120.$ V and, from Ohm's law, $i_e = \phi_e/R = 120./144 = 0.833$ A. Then, from Eq. (4.48),

$$\begin{aligned}({}_1W_2)_{\substack{\text{electrical}\\ \text{current}}} &= -\phi_e i_e t = -(120.\,\text{V})(0.833\,\text{A})(1.50\,\text{h}) \\ &= -150.\,\text{V·A·h} = -150.\,\text{W·h}\end{aligned}$$

b. From Eq. (4.50),

$$(\dot{W})_{\substack{\text{electrical}\\ \text{current}}} = -\phi_e i_e = -(120.\,\text{V})(0.833\,\text{A}) = -100.\,\text{V·A} = -100.\text{W}$$

The minus signs appear because electrical work and power go *into* the system.

Exercises

16. Determine the work and power consumption in Example 4.8 when the bulb is operated for 8.00 h instead of 1.50 h. Answer: $({}_1W_2)_{\text{electrical}} = -800.$ W·h, and $\dot{W}_{\text{electrical}} = -100.$ W.
17. Determine the effective current drawn by a 1.00 hp ac electric motor operating on a standard 120. V effective power line. Answer: $i_e = 6.22$ A.
18. Determine the electrical power dissipated by an 8-bit microprocessor computer chip that draws 90.0 mA at 5.00 V dc. Answer: $\dot{W}_{\text{electrical}} = -450.$ mW.

4.7.2 Electrical Polarization Work

The electric dipole formation, rotation, and alignment that occur when an electric field is applied to a nonconductor or a semiconductor constitutes an electric polarization work mode. The generalized force is the intensive property $\vec{E}$ (in V/m), the electric field strength vector, and the generalized displacement is the extensive property $\vec{P}$ (in A·s/m²), the polarization vector of the medium (defined to be the sum of the electric dipole rotation moments of all the molecules in the system). Then, assuming the electric field is applied *to* the system,

$$(đW)_{\substack{\text{electrical}\\ \text{polarization}}} = -\vec{E}\cdot d\vec{P} \tag{4.51}$$

and

$$({}_1W_2)_{\substack{\text{electrical}\\ \text{polarization}}} = -\int_1^2 \vec{E}\cdot d\vec{P} \tag{4.52}$$

Table 4.3 The Electric Susceptibility of Various Materials

Material	Temperature (°C/°F)	χ_e (dimensionless)
Air (14.7 psia)	20/68	5.36×10^{-4}
Plexiglass	27/81	2.40
Neoprene rubber	24/75	5.7
Glycerine	25/77	41.5
Water	25/77	77.5

Source: Reprinted by permission of the publisher from Zemansky, M. W., Abbott, M. M., Van Ness, H. C., 1975. Basic Engineering Thermodynamics, second ed. McGraw-Hill, New York.

Since the effect of the electric field is to orient the dipoles coincident with the field, then $\vec{E}$ and $\vec{P}$ are always parallel and point in the same direction. Therefore, if we let the magnitude of $\vec{E}$ be E and the magnitude of $\vec{P}$ be P, then Eqs. (4.51) and (4.52) reduce to

$$(\bar{d}W)_{\substack{\text{electrical}\\ \text{polarization}}} = -E\,dP \tag{4.53}$$

and

$$({}_1W_2)_{\substack{\text{electrical}\\ \text{polarization}}} = -\int_1^2 E\,dP \tag{4.54}$$

Many substances (particularly gases) correlate well with the following dielectric equation of state:

$$P = \varepsilon_0 \chi_e \forall E \tag{4.55}$$

where $\forall$ is the volume of the dielectric substance, ε_0 is the electric permittivity of vacuum (8.85419×10^{-12} N/V^2), and χ_e is the *electric susceptibility* (a dimensionless number) of the material. Table 4.3 gives values of χ_e for various materials.

EXAMPLE 4.9

The parallel plate capacitor shown in Figure 4.16 is charged to a potential difference of 120. V at 25.0°C. The plates are square with a side length of 0.100 m and are separated by 0.0100 m. If the gap between the plates is filled with water, determine the polarization work required in the charging of the capacitor.

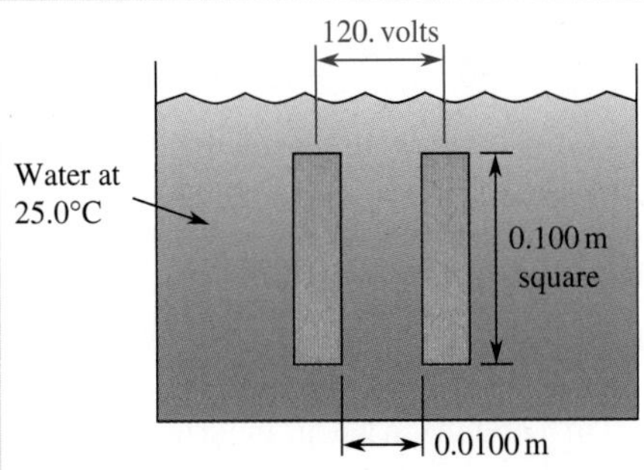

FIGURE 4.16
Example 4.9.

Solution

Here, we can use the dielectric equation of state, Eq. (4.55). Then, Eq. (4.54) becomes

$$({}_1W_2)_{\substack{\text{electric}\\ \text{polarization}}} = -\int_1^2 E\,dP = -\int_1^2 (\varepsilon_0\chi_e\forall E)\,dE = -\varepsilon_0\chi_e\forall(E_2^2 - E_1^2)/2$$

From the problem statement, we have

$$\forall = AL = (0.100\,\text{m})^2(0.0100\,\text{m}) = 1.00 \times 10^{-4}\,\text{m}^3$$

If we assume that the electrical potential ϕ varies linearly between the plates, then we can write

$$E = |-\nabla(\phi)| = (\text{voltage difference})/(\text{plate gap}) \text{ with } E_1 = 0 \text{ (uncharged plates)}$$

and

$$E_2 = \frac{120.\,\text{V}}{0.0100\,\text{m}} = 1.20 \times 10^4\,\text{V/m (charged plates)}$$

(Continued)

EXAMPLE 4.9 *(Continued)*

From Table 4.3, we find that, for water, $\chi_e = 77.5$. Then,

$$\begin{aligned}({}_1W_2)_{\substack{\text{electric}\\\text{polarization}}} &= -(8.85419\times10^{-12}\ \text{N/V}^2)(77.5)(1.00\times10^{-4}\text{m}^3)\times[(1.20\times10^4)^2 - 0^2\text{V}^2/\text{m}^2]/2\\ &= -4.94\times10^{-6}\ \text{N}\cdot\text{m} = -4.94\times10^{-6}\ \text{J}\end{aligned}$$

The work is negative since it went *into* the capacitor (the system).

Exercises

19. How much voltage would be required to store 1.00 MJ of electrical polarization work in the capacitor of Example 4.9? **Answer**: $V = 3.82\times10^7$ V.
20. Determine the electrical polarization work in Example 4.9 when the gap between the capacitor plates is filled with air at 20.0°C. **Answer**: $({}_1W_2)_{\text{polarization}} = -3.42\times10^{-11}$ J.
21. A capacitor is made from two concentric cylinders 0.100 m long. The diameter of the outer cylinder is 0.0200 m and the diameter of the inner cylinder is 0.0100 m. The gap between the cylinders is filled with glycerine at 25.0°C. Determine the electrical polarization work required to charge the capacitor when 120. V is applied. **Answer**: $({}_1W_2)_{\text{polarization}} = -1.04\times10^{-10}$ J.

The polarization work is a small fraction of the total energy required to charge an entire capacitor. The total work required to charge a capacitor is divided into two parts. The largest fraction goes into increasing the electric field strength $\vec{E}$ itself, and the remaining goes into the polarization of the material exposed to the electric field. Consequently, if the thermodynamic system you are analyzing is just the material between the plates of a capacitor, then the only polarization work is done on the material and Eq. (4.54) gives the correct electrical work mode value. On the other hand, if you are analyzing the entire capacitor (plates and dielectric), then Eq. (4.47) must be used to determine the correct electrical work mode value.

4.7.3 Magnetic Work

Materials are classified as either diamagnetic, paramagnetic, or ferromagnetic. Diamagnetic materials have no permanently established molecular magnetic dipoles. However, when they are placed in a magnetic field, their molecules develop magnetic dipoles whose magnetic field opposes the applied field (the Greek prefix *dia* means "to oppose"). Paramagnetic materials have naturally occurring molecular magnetic dipoles. When placed in a magnetic field, these dipoles tend to align themselves parallel to the field (the Greek prefix *para* means "beside"). Ferromagnetic materials retain some magnetism after the removal of a magnetic field. The thermodynamic state of these materials depends not only on the present values of their thermomagnetic properties, but also on their magnetic history. In this sense, ferromagnetic materials have a "memory" of their previous magnetic exposure.

As in the case of an electric field, the work associated with the initiation or destruction of a magnetic field consists of two parts. The first part is the work required to change the magnetic field itself (as though it existed within a vacuum), and the second part is the work required to change the magnetization of the material present inside the magnetic field.

For calculating the total work of magnetization, the generalized force is the intensive property $\vec{H}$ (in A/m^2), the magnetic field strength, and the generalized displacement is the extensive property $\forall\vec{B}$, the product of the system volume $\forall$ (in m^3) and the magnetic induction $\vec{B}$ (in tesla or V·s/m^2). Thus, assuming the magnetic field is applied to the system,

$$(\bar{d}W)_{\text{magnetic}} = -\vec{H}\cdot d(\forall\vec{B}) \tag{4.56}$$

and since $\vec{H}$ and $\vec{B}$ are always parallel and point in the same direction in magnetic materials, this reduces to

$$(\bar{d}W)_{\text{magnetic}} = -H\cdot d(\forall B) \tag{4.57}$$

where H is the magnitude of $\vec{H}$ and B is the magnitude of $\vec{B}$. The magnetic induction can be decomposed into two vectors as

$$\vec{B} = \mu_0\vec{H} + \mu_0\vec{M} \tag{4.58}$$

where $\vec{M}$ is the magnetization vector per unit volume of material exposed to the magnetic field (in a vacuum, $\vec{M}$ is equal to the null vector $\vec{0}$), and $\mu_0 = 4\pi\times10^{-7}$ V·s/(A·m) is a universal constant called the *magnetic permeability*. Inserting this information into Eq. (4.57) gives

$$(\bar{d}W)_{\substack{\text{magnetic}\\\text{(total)}}} = -\mu_0 H\,d(\forall H) - \mu_0 H\,d(\forall M) \tag{4.59}$$

Equation (4.59) is the differential of the total work associated with changing a material's magnetic field. The first term corresponds to the work required just to change the field itself (in a vacuum); and the second term corresponds to the work associated with the alignment of the molecular magnetic dipoles of the material present inside the magnetic field and represents the work of magnetization of the material exposed to the magnetic field. Hence, we can write

$$(\bar{d}W)_{\substack{\text{material}\\ \text{magnetization}}} = -\mu_0 H\, d(\forall M) \tag{4.60}$$

A simple and useful equation of state for a magnetic field is

$$M = \chi_m H \tag{4.61}$$

where χ_m is the *magnetic susceptibility* (a dimensionless number) of the material. The magnetic susceptibility is negative for diamagnetic materials and positive for paramagnetic materials (see Table 4.4). For a constant volume magnetization process, Eq. (4.61) can be used in Eq. (4.59) to give

$$(\bar{d}W)_{\substack{\text{magnetic}\\ \text{(total)}}} = -\mu_0 \forall (1+\chi_m) H\, dH$$

and assuming a constant volume and a constant magnetic susceptibility, this can be integrated to give the total magnetic work:

Total magnetic work

$$({}_1W_2)_{\substack{\text{magnetic}\\ \text{(total)}}} = -\mu_0 \forall (1+\chi_m)\left(\frac{H_2^2 - H_1^2}{2}\right) \tag{4.62}$$

where the increment to the total work due to the actual magnetization of the exposed material is just the actual magnetic work:

Actual magnetic work

$$({}_1W_2)_{\substack{\text{material}\\ \text{magnetization}}} = -\mu_0 \forall \chi_m \left(\frac{H_2^2 - H_1^2}{2}\right) \tag{4.63}$$

Table 4.5 summarizes the electrical and magnetic symbols used in this section.

Table 4.4 The Magnetic Susceptibility of Various Materials

Material	Temperature (°C/°F)	χ_m (dimensionless)
Mercury	18/26	-3.2×10^{-5}
Quartz	25/77	-1.65×10^{-5}
Ice	0/32	-0.805×10^{-5}
Nitrogen (14.7 psia)	20/68	-0.0005×10^{-5}
Oxygen (14.7 psia)	20/68	0.177×10^{-5}
Aluminum	18/64	2.21×10^{-5}
Platinum	18/64	29.7×10^{-5}

Source: Reprinted by permission of the publisher from Zemansky, M. W., Abbott, M. M., Van Ness, H. C., 1975. Basic Engineering Thermodynamics, second ed. McGraw-Hill, New York.

Table 4.5 Summary of Electrical and Magnetic Terms

Symbol	Name	SI Units
E	Electric field strength	V/m
P	Polarization	A·s/m^2
ε_0	Permittivity of free space	8.85419×10^{-12} N/V^2
χ_e	Electric susceptibility	Dimensionless
H	Magnetic field strength	A/m
B	Magnetic induction	Tesla or V·s/m^2
M	Magnetization	A/m
μ_0	Magnetic permeability	$4\pi \times 10^{-7}$ V·s/A·m

EXAMPLE 4.10

The magnetic susceptibility of the diamond in the gold engagement ring shown in Figure 4.17 is -2.20×10^{-5} at 20.0°C. Determine the (a) total magnetic and (b) material magnetic work required to change the magnetic field of a 1 carat diamond having a volume of 5.00×10^{-6} m^3 from 0.00 to 1.00×10^3 A/m.

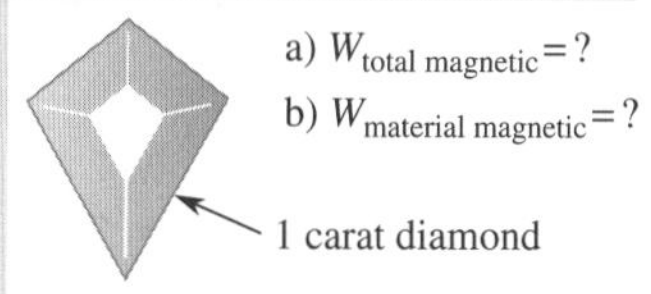

FIGURE 4.17
Example 4.10.

Solution

a. The total magnetic work required is given by Eq. (4.62) as

$$({}_1W_2)_{magnetic} = -\mu_0 V(1 + \chi_m)\left(\frac{H_2^2 - H_1^2}{2}\right)$$

where $\mu_0 = 4\pi \times 10^{-7}$ V·s/A·m and $\chi_m = -2.20 \times 10^{-5}$. Then,

$$({}_1W_2)_{magnetic} = -\left(4\pi \times 10^{-7} \frac{V \cdot s}{A \cdot m}\right)(5.00 \times 10^{-6}\, m^3)(1 - 2.20 \times 10^{-5})\left(\frac{1.00 \times 10^6 - 0\ A^2/m^2}{2}\right)$$
$$= -3.14 \times 10^{-6}\ J$$

b. The magnetic work required to change the magnetic field strength inside the diamond alone is given by Eq. (4.63) as

$$({}_1W_2)_{\substack{material\\magnetization}} = -\mu_0 V \chi_m \left(\frac{H_2^2 - H_1^2}{2}\right)$$

and, using the values from part a, we get

$$\left({}_2W^2\right)_{magnetic} = -\left(4\pi \times 10^{-7} \frac{V \cdot s}{A \cdot m}\right)(5.00 \times 10^{-6}\, m^3)(-2.20 \times 10^{-5})\left(\frac{1.00 \times 10^6 - 0\ A^2/m^2}{2}\right)$$
$$= 6.91 \times 10^{-11}\ J$$

Exercises

22. The magnetic susceptibility of gold is -3.60×10^{-5}. If the gold in the ring of Example 4.10 has a volume of 1.00×10^{-5} m^3, determine the total magnetic work required to change the magnetic field strength of the ring (the gold plus the diamond) from 0 to 1.00×10^3 A/m. **Answer:** $({}_1W_2)_{magnetic} = -9.42 \times 10^{-6}$ J.
23. The magnetic susceptibility of a ferromagnetic material such as iron varies with the applied magnetic field. However, if we assume it is constant over a small range of field strength at a value of 1800, then determine the (a) total work and (b) the material work required to magnetize a rectangular iron bar 0.500 inches square by 6.00 inches long from an initial magnetic field strength of zero to a magnetic field strength of 100. A/m. **Answer:** $({}_1W_2)_{total} = ({}_1W_2)_{iron} = -2.78 \times 10^{-4}$ J.

4.7.4 Chemical Work

Chemical work occurs whenever a specific chemical species is added to or removed from a system. Here, the generalized force is the intensive property μ_i, the Gibbs chemical potential of chemical species i, and the generalized displacement is the extensive property m_i, the mass of the chemical species added or removed.[8] Since any number of chemical species may be involved in a process, we write the chemical work as the sum over all k of the i species that are moved from the system to the surroundings as

$$(\bar{d}W)_{chemical} = -\sum_{i=1}^{k} \mu_i\, dm_i \tag{4.64}$$

and so

$$({}_1W_2)_{chemical} = -\int_1^2 \sum_{i=1}^{k} \mu_i\, dm_i \tag{4.65}$$

[8] In chemistry texts, the chemical potential is usually defined on a molar (i.e., per unit gram mole) basis. In this text, we define it as a standard intensive (per unit mass) property.

When the chemical potential is constant during the mass transfer from state 1 to state 2, Eq. (4.65) can be integrated to give the chemical work of adding chemical species:

Chemical work of adding chemical species

$$({}_1W_2)_{\substack{\text{chemical} \\ \mu_i = \text{constant}}} = -\sum_{i=1}^{k}\mu_i(m_2 - m_1)_i \tag{4.66}$$

Chemical work does not include the energy transports produced by chemical reactions, nor does it include the energy transported across the system boundary with the mass transport itself. Mass flow energy transport is considered later in this chapter, and the energy transports of chemical reactions are studied in detail in Chapter 9. The chemical work presented here essentially deals only with those energy transports involved in the mixing or separating of chemical species.

4.7.5 Mechanochemical Work

Mechanochemical work occurs whenever there is a direct energy conversion from chemical to mechanical energy. Animal muscles are examples of mechanochemical systems. Small mechanochemical engines have also been built using this work mode, and Figure 4.18 shows a small hydraulic pump driven by a mechanochemical contractile fiber. The "fuel" used in mechanochemical engines is not "burned," as in a standard heat engine. Often it is merely diluted and a small amount of chemical work is simultaneously extracted.

Mechanochemical work is calculated as basic mechanical work. The generalized force is the intensive property f, the force generated *by* or *within* the mechanochemical system, and the generalized displacement is the extensive property ℓ, the mechanical displacement of the system. Therefore,

$$(\bar{d}W)_{\text{mechanochemical}} = f\,d\ell \tag{4.67}$$

Generally, the mechanochemical force f is not constant during the contraction-expansion cycle, so the total mechanochemical work must be determined by a careful integration:

Mechanochemical work

$$({}_1W_2)_{\text{mechanochemical}} = \int_1^2 f\,d\ell \tag{4.68}$$

Note that, since the mechanochemical force comes from inside the system, a negative sign is not needed in Eqs. (4.67) and (4.68).

A system may be exposed to only one of these work modes of energy transport, or it may be exposed to several of them simultaneously. Since work is an additive quantity, to get the total (or net) work of a system that has more than one work mode present, we simply add all these work terms together:

Total differential work of all the work modes present

$$(\bar{d}W)_{\text{total}} = p\,d\forall + T\cdot d\theta - \sigma\,d\varepsilon - \sigma_s\,dA - \phi i\,dt - E\,dP - \mu_0 H\,d(\forall M) - \sum_{i=1}^{k}\mu_i\,dm_i + f d\ell + \cdots \tag{4.69}$$

It is generally the engineer's responsibility to determine the number and type of work modes present in any problem statement or real world situation. Often, the work modes of a problem are affected by how the system boundaries are drawn (recall that boundary definition is a prerogative of the problem solver). For example, if a system contains an electrical heater, then electrical current work is done on the system. However, if the boundary is drawn to exclude the heating element itself, then no electrical work occurs and the energy transport becomes a heat transport from the surface of the heating element into the system.

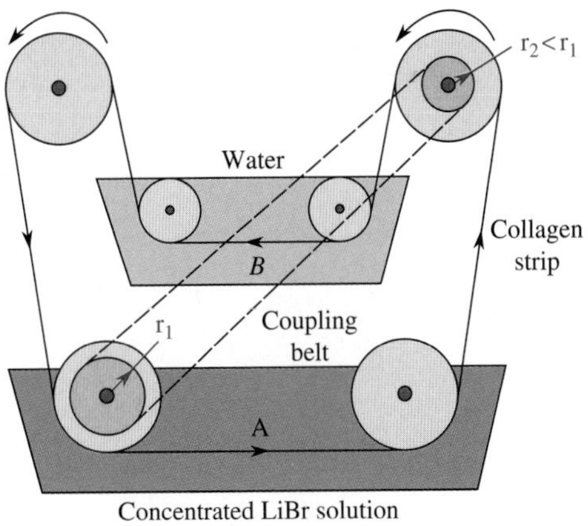

FIGURE 4.18
A simple mechanochemical Katchalsky engine.

4.8 POWER MODES OF ENERGY TRANSPORT

In thermodynamics, the time rate of change of a work mode, $\bar{d}W/dt$, is called *power*, and it represents the *power mode of an energy transport* $\dot{W}$. Dividing each of the previous nine differential work mode equations by the time differential dt produces an equation for the associated power mode. These results, summarized in Table 4.6, are useful in calculating the power (i.e., work rates) in problems in which continuous rate processes occur. While continuous rate processes can occur in both closed and open systems, they are more common in open systems.

4.9 WORK EFFICIENCY

Notice that, in all the work mode formulae given so far, no mention was made of the *efficiency* of the work transport of energy. This is because all the mechanical and nonmechanical work mode formulae discussed earlier were developed under the presumption of ideal circumstances, in which there were no friction losses or other inefficiencies within the system. Under these conditions the work process could ideally be reversed at any time, and all the work put into a system could be removed again simply by reversing the direction of the generalized force. Therefore, we call all the mechanical and nonmechanical work (or power) mode formulae developed previously *reversible* work (or power) formulae. Consequently—and this is very important—work or power calculations made with these formulae do not agree with the measurement of *actual* work that occurs in a real system. In real systems that absorb work, *more* actual work than that calculated from the previous formulae are required to produce the same effect on the system, and in real work producing systems, *less* actual work is produced than calculated from the previous formulae.

In the real world, nothing is reversible. Not one of the work modes discussed earlier can actually be carried out with 100% efficiency. Some are very close to being reversible (i.e., they have very high efficiencies) but none is completely reversible. This lack of reversibility in the real world is due to a phenomenon of nature that we describe with the second law of thermodynamics, which is discussed in detail in Chapter 7. Work modes with a low degree of reversibility (i.e., high irreversibility) are those carried out with systems far from thermodynamic equilibrium. Heat transfer, rapid chemical reactions (explosions), mechanical friction, and electrical resistance are all common sources of irreversibility in engineering systems.

Engineers use the concept of a work transport energy conversion efficiency to describe the difference between reversible and actual work. A general definition of the concept of an energy conversion efficiency is

$$\text{Energy conversion efficiency} = \eta_E = \frac{\text{Desired energy result}}{\text{Required energy input}} \tag{4.70}$$

Table 4.6 Power Modes of Energy Transport

Work Mode	Power Equation
Mechanical moving boundary	$(\dot{W})_{\substack{\text{moving}\\\text{boundary}}} = p\frac{d\mathcal{V}}{dt} = p\dot{\mathcal{V}}$
Mechanical rotating shaft	$(\dot{W})_{\substack{\text{rotating}\\\text{shaft}}} = T\left(\frac{d}{dt}\right) = T\omega$
Mechanical elastic	$(\dot{W})_{\text{elastic}} = -\sigma\mathcal{V}\left(\frac{d\varepsilon}{dt}\right) = -\sigma\mathcal{V}\dot{\varepsilon}$
Mechanical surface tension	$(\dot{W})_{\substack{\text{surface}\\\text{tension}}} = -\sigma_s\left(\frac{dA}{dt}\right) = -\sigma_s\dot{A}$
Electrical current	$(\dot{W})_{\substack{\text{electrical}\\\text{current}}} = -\phi i$
Electrical polarization	$(\dot{W})_{\substack{\text{electrical}\\\text{polarization}}} = -E\left(\frac{dP}{dt}\right) = -E\dot{P}$
Magnetic	$(\dot{W})_{\text{magnetic}} = -\mu_0\mathcal{V}(1+\chi_m)H\left(\frac{dH}{dt}\right) = -\mu_0\mathcal{V}(1+\chi_m)H\dot{H}$
Chemical	$(\dot{W})_{\text{chemical}} = -\sum\mu_i\left(\frac{dm_i}{dt}\right) = -\sum\mu_i\dot{m}_i$
Mechanochemical	$(\dot{W})_{\text{mechanochemical}} = f\left(\frac{d\ell}{dt}\right) = f\dot{\ell}$

In the case of work-absorbing systems, such as pumps or compressors, we can use an equation similar to Eq. (4.70) to define a *work transport energy conversion efficiency*, or *reversible efficiency*, η_W, as work efficiency for work-absorbing systems:

Work efficiency for work-absorbing systems

$$\eta_W(\%) = \frac{W_{rev}}{W_{act}} \times 100 = \frac{\dot{W}_{rev}}{\dot{W}_{act}} \times 100 \tag{4.71}$$

In the case of work-producing systems, such as engines or electrical generators, the reversible or work transport energy conversion efficiency becomes:

Work efficiency for work-producing systems

$$\eta_W(\%) = \frac{W_{act}}{W_{rev}} \times 100 = \frac{\dot{W}_{act}}{\dot{W}_{rev}} \times 100 \tag{4.72}$$

When these systems consist only of mechanical components, as, for example, in an internal combustion engine, the work transport energy conversion efficiency is simply called the *mechanical efficiency* and η_W is usually written as η_m.

Even though work transport energy conversion efficiencies are always less than 100%, not all energy conversion efficiencies are less than 100%. The value of the efficiency depends on the nature of the desired result in Eq. (4.70). An electrical resistance can convert electrical energy (the energy input) into heat (the desired result) with an energy conversion efficiency of 100%, but when this process is reversed, we find that the conversion of heat into work occurs with a much lower efficiency (a consequence of the second law of thermodynamics). On the other hand, refrigeration systems normally produce more "desired result" (cooling) than it actually costs in required energy input. Such systems normally have energy conversion efficiencies far in excess of 100%, not because they violate any law of physics, but simply because of the way their energy conversion efficiency is defined. Because it seems paradoxical to most people to speak of efficiencies in excess of 100%, we call such efficiencies *coefficients of performance* (COPs) instead. For example,

$$(\text{COP})_{\text{refrigerator}} = \frac{\text{Refrigerator cooling rate}}{\text{Refrigerator power input}}$$

EXAMPLE 4.11

The automobile engine shown in Figure 4.19 produces 150. hp on a test stand while consuming fuel with a heat content of 20.0×10^3 Btu/lbm at a rate of 1.10 lbm/min. A design engineer calculates the reversible power output from the engine as 223 hp. Determine

a. The energy conversion efficiency of the engine.
b. The work efficiency of the engine.

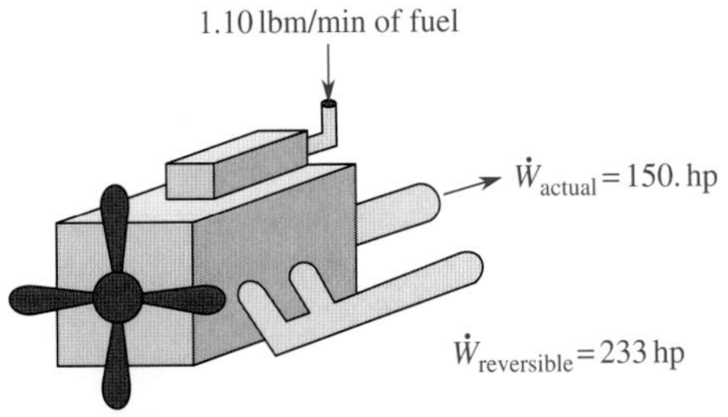

FIGURE 4.19
Example 4.11.

Solution

a. The energy conversion efficiency is given by Eq. (4.70) as

$$\eta_E = \frac{\text{Desired energy result}}{\text{Required energy input}}$$

The desired energy result here is the engine output power, 150. hp. The required energy input here is the energy coming from the fuel, 20.0×10^3 Btu/lbm $\times$ 1.10 lbm/min $\times$ 60 min/h = 1320×10^3 Btu/h $\times$ (1 hp)/(2545 Btu/h) = 519 hp. Then,

$$\eta_E = \frac{150.\text{ hp}}{519\text{ hp}} = 0.289 = 28.9\%$$

b. Since an engine is a work producing machine, Eq. (4.72) gives the work efficiency as

$$\eta_W = \frac{\dot{W}_{actual}}{\dot{W}_{reversible}} \times 100 = \frac{150.\text{ hp}}{223\text{ hp}} \times 100 = 67.3\%$$

(Continued)

EXAMPLE 4.11 *(Continued)*

Exercises

24. If the energy conversion efficiency in Example 4.11 were 15.5%, what would be the power output of the engine measured on the test stand for the same fuel flow rate? **Answer:** $\dot{W}_{actual} = 80.4$ hp.

25. An engineer designs a pump that requires 1.30 kW of reversible power to operate. A prototype pump is made and taken to the test laboratory. The actual power required to operate the prototype pump is measured at 1.50 kW. Determine the work (or mechanical) efficiency of this pump. **Answer:** $\eta_W = 86.7\%$.

26. A refrigeration system is powered by a 5.0 kW electric motor. It removes 18×10^3 J/s from the cold storage space. What is the coefficient of performance of this refrigeration system? **Answer:** COP = 3.6.

Because of the many irreversibilities that occur within a system, we cannot calculate actual work absorbed or produced from a theoretical formula. All efficiency values are determined from laboratory or field measurements on the actual work of real operating systems. When energy conversion efficiencies are to be taken into account in textbook problems, the efficiency values usually are provided within the problem statement. Experienced engineers often have a "feel" for what the efficiencies of certain devices should be, and they can use these efficiency estimations in their design calculations. Student engineers, however, are not presumed to be innately blessed with this knowledge.

The general form of Eq. (4.70) allows the creation of many different types of efficiencies. There are thermal, mechanical, volumetric, thermodynamic, and total efficiencies (to name just a few) in today's engineering literature. One should always be sure to understand the type of efficiency being used in any calculation.

4.10 THE LOCAL EQUILIBRIUM POSTULATE

Surprisingly, there is no adequate definition for the thermodynamic properties of a system that is not in an equilibrium state. Some extension of classical equilibrium thermodynamics is necessary for us to be able to analyze nonequilibrium (or irreversible) processes. We do this by subdividing a nonequilibrium system into many small but finite volume elements, each of which is larger than the local molecular mean free path, so that the continuum hypothesis holds. We then assume that each of these small volume elements is in *local* equilibrium. Thus, a nonequilibrium system can be broken down into a very large number of very small systems, each of which is at a different equilibrium state. This technique is similar to the continuum hypothesis, wherein continuum equations are used to describe the results of the motion of discrete molecules (see Chapter 2).

The differential time quantity dt used in nonequilibrium thermodynamic analysis cannot be allowed to go to zero as in normal calculus. We require that $dt > \sigma_s$, where σ is the time it takes for one of the volume elements of the subdivided nonequilibrium system just described to "relax" from its current nonequilibrium state to an appropriate equilibrium state. This is analogous to not allowing the physical size of the element to be less than its local molecular mean free path, as required by the continuum hypothesis. The error incurred by these postulates is really quite small, because they are the result of second-order variations of the thermodynamic variables from their equilibrium values. However, just as the continuum hypothesis can be violated by systems such as rarefied gases, the local equilibrium postulate can also be violated by highly nonequilibrium systems such as explosive chemical reactions. In the case of such violations, the analysis must be carried forward with techniques of statistical thermodynamics.

Because of the similarity between the local equilibrium postulate and the continuum hypothesis, it is clear that the local equilibrium postulate could as well be called the *continuum thermodynamics hypothesis*.

SIMPLE SYSTEM

Any two independent intensive property values are sufficient to determine (or "fix") the local equilibrium state of a simple system.

4.11 THE STATE POSTULATE

To carry out a reversible work mode calculation using the formulae given earlier, we must know the exact behavior of both the generalized force (an intensive property) and the generalized displacement (an extensive property) for each work mode. Systems with multiple work modes have a variety of property values that must be monitored during the work process to utilize the proper work mode formulae. Therefore, it seems reasonable to expect that a simple relation exists between the number of work modes present in any given system and the number of independent property values required to fix the state of that system. This is the purpose of the following *state postulate*:

The number of independent intensive thermodynamic property values required to fix the state of a closed system that is

1. Subject to the conditions of local equilibrium,
2. Exposed to n (nonchemical) work modes of energy transport, and
3. Composed of m pure substances is $n + m$.

Therefore, a pure substance ($m = 1$) subjected to only one work mode ($n = 1$) requires two ($n + m = 2$) independent property values to fix its state. Such systems are called *simple systems*, and any two independent intensive properties determine (or "fix") its state.

The compression or expansion of a pure gas or vapor is a simple system. The work mode is moving system boundary work, and any two independent intensive property values (p, v; p, T; v, T, etc.) fix its state. In fact, a simple system occurs when each of the nonchemical reversible work modes just discussed is individually applied to a pure substance. On the other hand, if two of them are simultaneously applied to a pure substance, then $n + m = 3$ and three independent intensive property values are required to fix the state of the system.

4.12 HEAT MODES OF ENERGY TRANSPORT

We now introduce the three basic modes of heat transport of energy. Since a good heat mode analysis is somewhat more complex than a work mode analysis and since its understanding is very important to a good engineering education, most mechanical engineering curricula include a separate heat transfer course on this subject. Consequently, this section is meant to be only an elementary introduction to this subject.

A system with no heat transfer is said to be *adiabatic*, and all well-insulated systems are considered to be adiabatic. A process that occurs with no heat transport of energy is called an *adiabatic process*.

In the late 18th century, heat was thought to be a colorless, odorless, and weightless fluid, then called *caloric*. By the middle of the 19th century, it had been determined that heat was in fact not a fluid but rather it represented energy in transit. Unfortunately, many of the early heat-fluid technical terms survived and are still in use today. This is why we speak of heat transfer and heat flow, as though heat were something physical, but it is not. Because these conventions are so deeply ingrained in our technical culture, we use the phrases *heat transfer, heat transport*, and the *heat transport of energy* interchangeably.

After it was determined that heat was not a fluid, late 19th century physicists defined heat transfer simply as energy transport due to a temperature difference. In this framework, temperature was the only intensive property driving force for the heat transport of energy.

Today, the simplest way to define heat transport of energy is as any energy transport that is neither a work mode nor a mass flow energy transport mode. More precisely, modern nonequilibrium thermodynamics defines heat transfer as just the transport of internal energy into or out of a system. With this definition, all other energy transport modes are automatically either work or mass flow modes.

The basic heat transfer formulae were developed empirically and, unlike the previous work mode formulae, give actual rather than reversible heat transport values. In fact, since heat transfer always occurs as a result of energy

WHAT DOES THE WORD *ADIABATIC* MEAN?

The term *adiabatic* was coined in 1859 by the Scottish engineer William John Macquorn Rankine (1820–1872). It comes from the Greek word, αδιαβατοσ, meaning "not to pass through." In thermodynamics, it means heat does not pass through the system boundary, or simply that there is no heat transfer. *Adiabatic* is the analog of the word *aergonic* (meaning "no work") introduced earlier in this chapter.

spontaneously moving *down* a potential gradient (such as from high to a low temperature) and the reverse cannot spontaneously occur, no heat transfer process can be reversed in any way whatsoever. Therefore, all finite heat transfer processes are irreversible.

4.13 HEAT TRANSFER MODES

Heat transfer is such a large and important mechanical engineering topic that most curricula have at least one required course in it. Heat transfer equations are always cast as heat transfer *rate* (i.e., $\dot{Q}$) equations. To determine the amount of heat energy transport that occurs as a system undergoes a process from one equilibrium state to another you must integrate $\dot{Q}$ over the time interval of the process, or ${}_1Q_2 = \int_1^2 \dot{Q}\,dt$. Normally, we choose processes in which $\dot{Q}$ is constant in time so that the integral becomes simply ${}_1Q_2 = \dot{Q}\,(t_2 - t_1) = \dot{Q}(\Delta t)$, where Δt is the time required for the process to occur.

Historically, the field has been divided into three heat transfer modes: conduction, convection, and radiation. These three modes are briefly described next.

4.13.1 Conduction

The basic equation of conduction heat transfer is Fourier's law:

$$\dot{Q}_{\text{cond}} = -k_t A\left(\frac{dT}{dx}\right) \tag{4.73}$$

where $\dot{Q}_{\text{cond}}$ is the conduction heat transfer rate, k_t is the thermal conductivity of the material, A is the cross-sectional area normal to the heat transfer direction, and dT/dx is the temperature gradient in the direction of heat transfer. The algebraic sign of this equation is such that a positive $\dot{Q}_{\text{cond}}$ always corresponds to heat transfer in the positive x direction, and a negative $\dot{Q}_{\text{cond}}$ always corresponds to heat transfer in a negative x direction. Since this is not the same sign convention adopted earlier in this text, the sign of the values calculated from Fourier's law may have to be altered to produce a positive when it enters a system and a negative when it leaves a system.

For steady conduction heat transfer through a plane wall (Figure 4.20), Fourier's law can be integrated to give

$$(\dot{Q}_{\text{cond}})_{\text{plane}} = -k_t A\left(\frac{T_2 - T_1}{x_2 - x_1}\right) \tag{4.74}$$

and for steady conduction heat transfer through a hollow cylinder of length L, Fourier's law can be integrated to give

$$(\dot{Q}_{\text{cond}})_{\text{cylinder}} = -2\pi L k_t \left[\frac{T_{\text{inside}} - T_{\text{outside}}}{\ln(r_{\text{inside}}/r_{\text{outside}})}\right] \tag{4.75}$$

Table 4.7 gives thermal conductivity values for various materials.

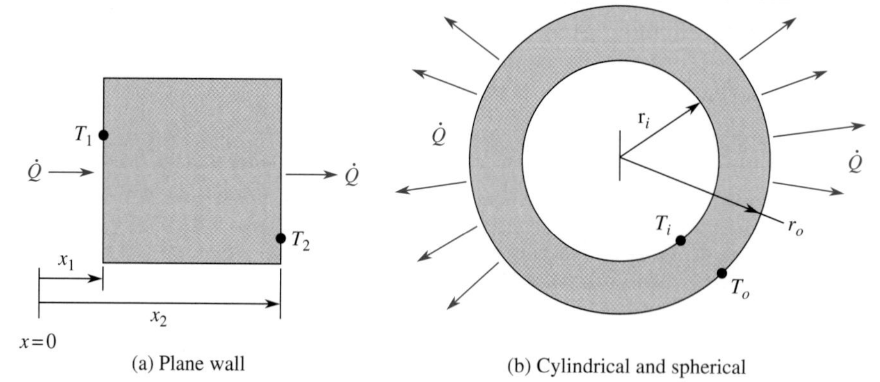

FIGURE 4.20
Thermal conduction notation in plane, cylindrical, and spherical coordinates.

Table 4.7 Thermal Conductivity of Various Materials

Material	Temperature (°C/°F)	Thermal Conductivity k_t Btu/(h·ft·R)	W/(m·K)
Air (14.7 psia)	27/81	0.015	0.026
Hydrogen (14.7 psia)	27/81	0.105	0.182
Saturated water vapor (14.7 psia)	100/212	0.014	0.024
Saturated liquid water (14.7 psia)	0/32	0.343	0.594
Engine oil	20/68	0.084	0.145
Mercury	20/68	5.02	8.69
Window glass	20/68	0.45	0.78
Glass wool	20/68	0.022	0.038
Aluminum (pure)	20/68	118.0	204.0
Copper (pure)	20/68	223.0	386.0
Carbon steel (1% carbon)	20/68	25.0	43.0

4.13.2 Convection

Convective heat transfer occurs whenever an object is either hotter or colder than the surrounding fluid. The basic equation of convection heat transfer is Newton's law of cooling:

$$\dot{Q}_{\text{conv}} = hA(T_\infty - T_s) \tag{4.76}$$

where $\dot{Q}_{\text{conv}}$ is the convection heat transfer rate, h is the convective heat transfer coefficient, A is the surface area of the object being cooled or heated, T_∞ is the bulk temperature of the surrounding fluid, and T_s is the surface temperature of the object. The algebraic sign of Newton's law of cooling has been chosen to be positive for $T_\infty > T_s$ (i.e., for heat transfer into the object). This corresponds to our thermodynamic sign convention for heat transfer when the object is the system. The convective heat transfer coefficient h is always a positive, empirically determined value. Table 4.8 lists typical heat transfer coefficients.

4.13.3 Radiation

All electromagnetic radiation is classified as radiation heat transfer. Infrared, ultraviolet, visible light, radio and television waves, X rays, and so on are all forms of radiation heat transfer. The radiation heat transfer between two objects situated in a nonabsorbing or emitting medium is given by the Stefan-Boltzmann law:

$$\dot{Q}_{\text{rad}} = F_{1-2}\varepsilon_1 A_1 \sigma(T_2^4 - T_1^4) \tag{4.77}$$

where $\dot{Q}_{\text{rad}}$ is the radiation heat transfer rate, F_{1-2} is called the *view factor* between objects 1 and 2 (it describes how well object 1 "sees" object 2), ε_1 is the dimensionless emissivity or absorptivity (the hotter object is said to *emit* energy while the colder object *absorbs* energy) of object 1, A_1 is the surface area of object 1, σ is the Stefan-Boltzmann constant (5.69×10^{-8} W/m$^2 \cdot$K^4 or 0.1714×10^{-8} Btu/h$\cdot$ft$^2 \cdot$R^4), and T_1 and T_2 are the surface temperatures of the objects. A *black* object is defined to be any object whose emissivity is $\varepsilon = 1.0$. Table 4.9 lists some typical emissivity values. Also, if object 1 is completely enclosed by object 2, then $F_{1-2} = 1.0$. For a completely enclosed black object, the Stefan-Boltzmann law reduces to

$$(\dot{Q}_{\text{rad}})_{\substack{\text{black}\\\text{enclosed}}} = A_1\sigma(T_2^4 - T_1^4) \tag{4.78}$$

Table 4.8 Typical Values of the Convective Heat Transfer Coefficient

Type of Convection	Convective Heat Transfer Coefficient h Btu/(h·ft^2·R)	W/(m^2·k)
Air, free convection	1–5	2.5–25
Air, forced convection	2–100	10–500
Liquids, forced convection	20–3000	100–15,000
Boiling water	500–5000	2500–25,000
Condensing water vapor	1000–20,000	5000–100,000

Table 4.9 Typical Emissivity Values for Various Materials

Material	Temperature (°C/°F)	Emissivity ε (dimensionless)
Aluminum	100/212	0.09
Iron (oxidized)	100/212	0.74
Iron (molten)	1650/3000	0.28
Concrete	21/70	0.88
Flat black paint	21/70	0.90
Flat white paint	21/70	0.88
Aluminum paint	21/70	0.39
Water	0–100/32–212	0.96

The sign convention in the Stefan-Boltzmann law has been chosen to be positive when $T_2 > T_1$; therefore, the "system" should be object 1 to achieve the correct thermodynamic sign convention. Also note that this equation contains the temperature raised to the fourth power. This means that *absolute* temperature units must always be used.

4.14 A THERMODYNAMIC PROBLEM SOLVING TECHNIQUE

The previous 11 example problems have been relatively straightforward, mainly illustrating the use of specific energy and work mode equations. However, most thermodynamics problems are not so straightforward, and now we are ready to introduce a comprehensive thermodynamic problem solving technique that allows you to set up and solve even the most complex thermodynamics problems.

Thermodynamic problem statements sometimes have the appearance of being stories full of technical jargon, liberally sprinkled with numbers. All too often, your first instinct on being faced with such a situation is to calculate something—anything—because the act of calculation brings about the euphoria of apparent progress toward a solution. However, this approach is quickly stalled by the inability to reach the final answer, followed by long frustrating periods of shoe shuffling and window staring until either enlightenment, discouragement, or sleep occurs. This is definitely the wrong problem solving technique. A good technique must have definite starting and ending points, and it must contain clear and logical steps that carry you toward a solution.

As a prelude to discussing the details of the problem solving technique, you should realize that the general structure of a thermodynamic word problem usually contains the following three features.

1. A thermodynamic problem statement is usually a small "story" that is too long to be completely and accurately memorized no matter how many times you read it. So simply reading the problem statement once is usually not enough; you must translate it into your own personal environment by adding a schematic drawing, writing down relevant assumptions, and beginning a structured solution.
2. To completely understand the problem statement, you must first "decode" it. That is, you must dissect and rearrange the problem statement until it fits into a familiar pattern. Any problem solving technique is, of course, based on the premise that the problem has a solution. Curiously, it is very easy to construct problem statements that are not solvable without the introduction of extraneous material (judiciously called *assumptions*).
3. Thermodynamic problem statements tend to be very wide ranging. They can be written about virtually any type of system and can deal with virtually any form of technology. To give the problem statements a pragmatic engineering flavor, they are usually written as tiny stories that are designed to reflect what you will encounter as a working engineer.

Unfortunately, many students facing thermodynamics for the first time are overwhelmed by these factors. How are you supposed to know anything about how a nuclear power plant operates, how the combustion chamber of a turbojet engine functions, or how a boiler feed pump works if you have never actually seen one in operation? The key is that you really do not have to know that much about how these things work to carry out a good thermodynamic analysis of them. But, you do have to understand how problem statements are written and how to analyze them correctly. This is the core of the problem solving technique.

In fact, it would be possible to write a computer program that could solve any thermodynamic word problem. What we are going to do is to show you how to solve thermodynamic problems by using a computerlike flowchart approach, as in Figure 4.21.

The technique is really very simple. First, you must learn to formulate a general starting point. Then you must learn to identify the key logical decisions that have to be made as the solution progresses. Finally, when all the

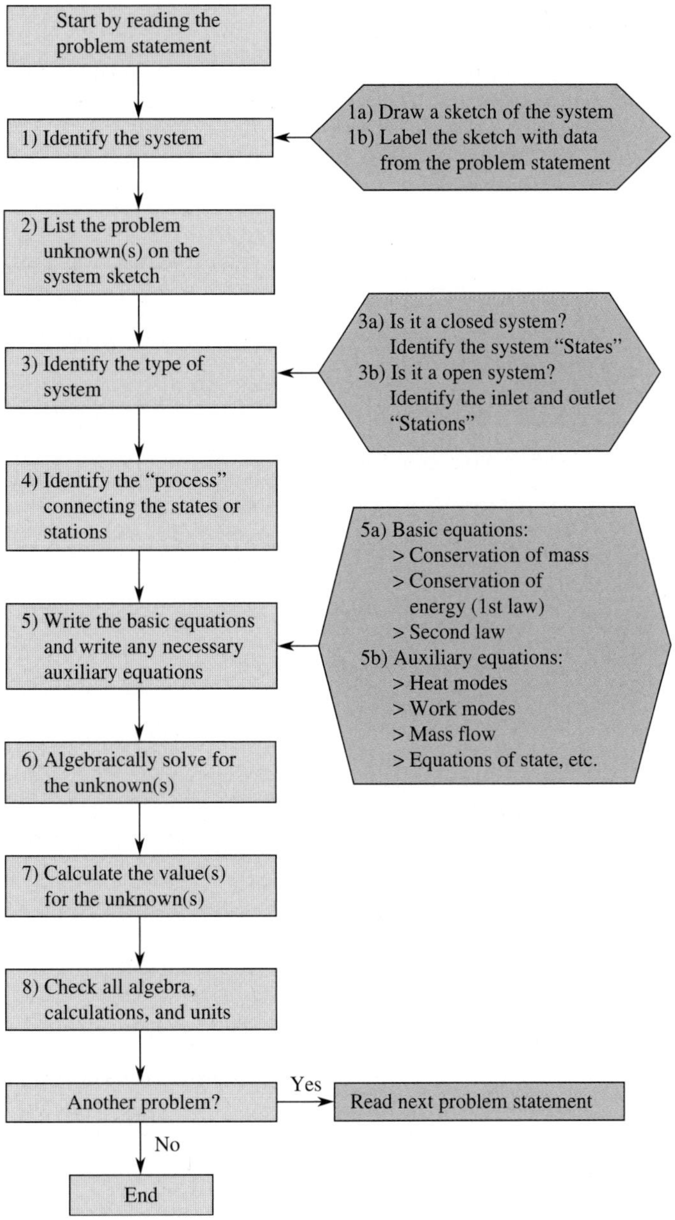

FIGURE 4.21
Flowchart for solving thermodynamic problems.

analysis and algebraic manipulations are complete, you make the necessary calculations (paying close attention to units and significant figures) to obtain the desired results.

The steps to be followed are shown in Figure 4.21, and each step is discussed in detail next.

Begin by carefully reading the problem statement completely through.

Step 1. **Make a sketch** of the system or device described in the problem statement and determine the material (air, steam, liquid water, etc.) with which you are working. Then, carefully define the part(s) you choose to analyze by inserting a dashed line to identify the system boundary.

Step 2. **Identify the problem's unknown(s)** by rereading the problem statement and picking out all the things you are supposed to determine. Write them on your system sketch.

Step 3. **Determine whether it is a closed system or an open system**. If your system is closed, identify as many of the state properties as you can. Most problems have only two states (initial and final), but some also have intermediate states with which you have to contend. To keep the numerical values and units of the state properties straight, list each one under a "state" heading.

WHAT IS THE SECRET TO SOLVING THERMODYNAMICS PROBLEMS?

The secret to solving thermodynamic problems is to *do the analysis first* and *do the calculations last*, not the other way around. The basic process for solving a thermodynamics problem is this:

Begin by carefully reading the problem statement completely through.

Step 1. Make a sketch of the system and put a dashed line around the system boundary.
Step 2. Identify the unknown(s) and write them on your system sketch.
Step 3. Identify the type of system (closed or open) you have.
Step 4. Identify the process that connects the states or stations.
Step 5. Write down the basic thermodynamic equations and any useful auxiliary equations.
Step 6. Algebraically solve for the unknown(s).
Step 7. Calculate the value(s) of the unknown(s).
Step 8. Check all algebra, calculations, and units.

The process is this:

Sketch → Unknowns → System → Process → Equations → Solve → Calculate → Check

For example, if you have a closed system that is initially at 14.7 psia with a specific volume of 0.500 ft^3/lbm and by some process it ends up at 200. psia at a quality of 90.0%, you should write this information on your work sheet in the following form (always be sure to include the units on these values):

State 1 $\xrightarrow{\text{Process path}}$	State 2
$p_1 = 14.7$ psia	$p_2 = 200.$ psia
$v_1 = 0.500$ ft^3/lbm	$x_2 = 0.900$

For example, if you have a flow stream entering the system at station 1 with a temperature of 300.°C and a pressure of 1.00 MPa, and a flow stream exiting the system at station 2 with a specific volume of 26.3 m^3/kg and a quality of 99.0%, you should write this information on your work sheet as (always be sure to include the *units* on these values):

Station 1 $\xrightarrow{\text{Process path}}$	Station 2
$p_1 = 1.00$ MPa	$v_2 = 26.3$ m^3/kg
$T_1 = 300.$°C	$x_2 = 0.990$

Here, too, we are trying to identify two independent property values at each station, because in simple systems, they fix the state of the material at that station.

Notice that, for "simple" thermodynamic systems, we always are looking for the values of two independent properties in each state. These two property values fix (i.e., determine) the state and we can then find the values of any of the other properties needed at that state.

Often a problem statement gives only one property value at a system state. In this case, the remaining independent property value at that state is usually given by the process path statement that indicates how that state was achieved (e.g., an isothermal process tells us that $T_2 = T_1$) or else it may be a problem unknown to be determined.

If it is an open system, we are interested in any changes that occur in the system bulk properties of the system plus all the properties of the entering and exiting flow streams. Flow stream properties are referred to as monitoring *station* properties, to clearly separate them from bulk system properties.

Step 4. Now **identify the process connecting the state or stations**. The process path statement is usually given in technical terms such as *a closed, rigid vessel*, meaning an isochoric (or constant volume) process will occur. Proper identification of the process path is very important, because it often provides numerical values for state properties (e.g., $v_2 = v_1$ for a closed, rigid vessel) or heat, work, or other thermodynamic quantities (e.g., an insulated or adiabatic system has ${}_1Q_2 = \dot{Q} = 0$, an aergonic system has ${}_1W_2 = \dot{W} = 0$, and so forth). When two independent property values are given in the problem statement for each state or station of the system, the process path is not necessary unless it provides values for heat, work, kinetic energy, or potential energy.

Step 5. **Write down all the basic equations**. Your work sheet should now have all the details of the problem on it and you should not have to look at the problem statement again. The actual solution to the problem is begun by automatically writing down (whether you think you need them or not) all the relevant basic equations. Thermodynamics has only three basic equations:

a. The *conservation of mass* (which is also called the *mass balance*).
b. The *first law of thermodynamics* (which is also called the *energy balance* or the *conservation of energy*).
c. The *second law of thermodynamics* (which is also called the *entropy balance*).

In closed systems, the conservation of mass is automatically satisfied and need not be written down. Also, since the entropy balance is not be introduced until Chapter 7, it does not enter into the solution of any problems until then. So, for solving the closed system problems of Chapter 5, there is really only one relevant basic equation: the first law of thermodynamics. In solving the open system problems of Chapter 6, there are two relevant basic equations: the conservation of mass and the first law of thermodynamics.

Write any necessary auxiliary equations. All the equations developed in this book that are not one of the three basic equations discussed previously are called *auxiliary equations*. For example, all equations of state (ideal gas and incompressible materials), all work mode equations (mechanical, electrical, etc.), all heat mode equations (conduction, convection, radiation), all property-defining equations (specific heats,

The easiest way to show the process path on your work sheet is to write the statement "Process: process name" on a connecting arrow between the state or station data sets. In the closed system example used in step 3, if the state change occurs in a closed, rigid vessel and we do not know the final quality, then we would write

State 1	Process: v = constant →	State 2
$p_1 = 14.7$ psia		$p_2 = 200.$ psia
$v_1 = 0.500\ \text{ft}^3/\text{lbm}$		$v_2 = v_1 = 0.500\ \text{ft}^3/\text{lbm}$

And, if the open system of step 3 is operated at a constant pressure (i.e., an isobaric process) and we do not know the final quality, then we would write

Station 1	Process: p = constant →	Station 2
$p_1 = 1.00$ MPa		$v_2 = 26.3\ \text{m}^3/\text{kg}$
$T_1 = 300.°\text{C}$		$p_2 = p_1 = 1.00$ MPa

Always write down the complete *general* form of the basic equations. Do not try to second-guess the problem by writing the shorter specialized forms of the basic equations that were developed for specific applications. Then, cross out all terms that vanish as a result of given constraints or process statements. For example, for a closed, adiabatic, stationary system, we write the energy balance as (see Eq. (4.20), where we have used the abbreviation KE = $mV^2/2g_c$ and PE = mgZ/g_c)

$$\underbrace{{}_1Q_2}_{=0\ \text{(adiabatic (insulated) system)}} - {}_1W_2 = m(u_2 - u_1) + \underbrace{\text{KE}_2 - \text{KE}_1 + \text{PE}_2 - \text{PE}_1}_{=0\ \text{(stationary - i.e., not moving)}}$$

Notice that we write why each crossed out term vanishes ("adiabatic" and "stationary" in this case). This makes the solution easier to follow and to check later if the correct answer was not obtained.

Unlike in some other engineering subjects, you will not be able to find all the algebraic manipulations already done for you in example problems within the text or by the instructor in class. There are simply too many possible variations on a problem theme to do this. Therefore, you have to carry out the mathematical manipulations suggested here to develop your own working formulae in almost every problem. This is a fact of thermodynamic problem solving.

By this point you should be able to see your way to the end of the problem, because the mechanism for finding each of the unknowns should now be clear. Determine the units on each value calculated and make sure that all values that are added together or subtracted from each other have the same units. Often one of the unknowns is needed to find another; for example, you may need to find ${}_1W_2$ from a work mode auxiliary equation to solve for ${}_1Q_2$ from the energy balance equation.

enthalpy, etc.), and all specialized equations (such as $KE = mV^2/2g_c$, etc.) are auxiliary equations. If the problem statement describes a mechanical, electrical, or other work mode, then write the equation for calculating the value of that work mode. Auxiliary equations ultimately provide numerical values for use in the basic mass, energy, and entropy balance equations.

Step 6. **Algebraically solve for the unknown(s)**. Do not calculate anything yet. By algebraically manipulating the basic and auxiliary equations you should be able to develop a separate equation for each unknown. Remember, you can solve for only as many unknowns as you have independent equations. All of the basic equations and most of the auxiliary equations are independent, so many times unknowns are determined directly from an auxiliary equation. For example, in the problem statements dealing with closed systems, we have only *one* applicable basic equation, the first law of thermodynamics (the energy balance). Therefore, if there is more than one unknown in these problem statements, then all but one of these unknowns must be determined directly from an appropriate auxiliary equation.

Step 7. **Calculate the value(s) of the unknown(s)**. Once all the algebra has been completed, then and only then should you begin to calculate numerical values.

Step 8. **Check all algebra, calculations, and units**. This is self-explanatory, but pay particular attention to checking the units. With the calculational accuracy of today's inexpensive electronic calculators and microcomputers, most of your errors occur as a result of poor units handling rather than from numerical manipulations.

These eight steps are illustrated in detail in the examples in the next chapter. They will lead you through even the most difficult thermodynamic problems. Once you become familiar with them, the solutions flow quite rapidly and naturally. It must be emphasized that these steps are not the only solution technique possible, but they have proven successful for many engineering students.

4.15 HOW TO WRITE A THERMODYNAMICS PROBLEM

A good test of your problem solving skills is to see whether or not you can write a thermodynamics problem that can be solved. The technique of writing your own thermodynamics problem is just the reverse of solving one. It is as simple as A, B, C.

A. First, you first decide (1) the type of system (closed or open) you want to use, (2) the equations you want to use in the solution (thermodynamic laws, equations of state, work mode equation, and so forth), and (3) the unknown(s) you want to find in the solution.

B. Next, you write a short story that provides physical motivation for the problem that contains all the numerical values necessary to find state properties and any geometry, height, or velocity information needed to solve for your chosen unknowns.

C. Finally, you solve your problem in a forward direction to see if you have specified all the necessary information for someone to produce an accurate solution.

This is easier than it sounds. First, let us look at the equations that can be used in a problem solution.

A. Select the working equations and unknowns

The problem unknowns can be any of the variables carried within the basic laws of thermodynamics and any of the related auxiliary equations introduced thus far. For simplicity, let us limit the discussion to a closed system analysis. The general closed system energy balance is

$$ {}_1Q_2 - {}_1W_2 = m\left[(u_2 - u_1) + \frac{V_2^2 - V_1^2}{2g_c} + \frac{g(Z_2 - Z_1)}{g_c}\right]_{\text{system}} $$

and the general closed system energy rate balance is

$$ \dot{Q} - \dot{W} = \frac{d}{dt}\left(mu + \frac{mV^2}{2g_c} + \frac{mgZ}{g_c}\right)_{\text{system}} $$

Any of the variables listed in these equations can be an unknown in a problem statement. In addition to the basic balance equations, we have numerous auxiliary equations, such as

- Equations of state for ideal gases, incompressible fluids, or other materials.
- Process path equations such as polytropic, isobaric, and the like.
- Various work mode equations for mechanical, electrical, and other work modes.

List all the *basic* (thermodynamics laws) and *auxiliary* equations you want the person who solves your problem to use in the solution. Then choose the variables you want to use as unknowns. Remember, you need as many independent equations in your list of equations as the number of unknowns you choose, so do not choose too many. Then, assign numerical values to all the remaining variables in the equations that are to be used to solve for the unknowns. Do not be too concerned about the actual values you pick at this point; if you choose the wrong values, it will show up in step C, and you can correct them later.

B. Write a short story that contains all the information needed to solve the problem

It would be helpful if we could categorize to some degree the wide variety of problem types or scenarios commonly encountered in thermodynamics. The first classification is by the *thermodynamic process* used in the problem scenario, the second classification is by the *engineering technology* used in the problem scenario, and the third is by *problem unknown*. Since the number of variations within these classifications is quite large, they are explained in detailed here.

Problem classification by thermodynamic process. A problem statement could involve more than one process or involve unknown processes. Therefore, the process for changing the state of a system could be the focal point of a problem statement. For example, we might want to find how the temperature changes during a constant pressure process. This would then be the central theme of the problem statement.

Problem classification by problem technology. The list of possible engineering technologies is much longer than the list of known processes. Actually, any device or technology can be analyzed thermodynamically. A series of "typical" technology based scenarios appear in engineering thermodynamics textbooks. For example, you might want to find the work required to compress a gas with a piston, the change in temperature across a nozzle, the power produced by a turbine, and so forth. Then, the problem statement focuses on these technologies, providing numerical values for all the variables except the problem unknowns.

Problem classification by problem unknown. These problems are usually the simplest, since they do not depend on a specific technology or process path. The unknowns are simply calculated directly from the thermodynamic laws (i.e., Q, W, KE, PE, etc.) or from an auxiliary equation (i.e., the ideal gas equation, etc.).

C. Solve the problem in the forward direction

Here you (or a friend) must actually solve the problem you wrote using the data you provided in the problem statement. You will usually find that you get stuck part way through the problem and have to go back and modify the problem statement. That is OK, do it quickly and go on with the solution. Sometimes, values you originally chose cannot be found easily in the tables or are unreasonable (for example, maybe you wanted a state to be a vapor but the values you originally specified for pressure and temperature are for a liquid). Using the tables in the tables book and your emerging solution, change the original values in your problem statement so that the problem solution moves along smoothly. Be careful to check the units on each calculation.

Now Let us Write a Thermodynamics Problem

Step A. We limit it to a closed system and use the energy balance as our primary equation:

$$ {}_1Q_2 - {}_1W_2 = m\left[(u_2 - u_1) + \frac{V_2^2 - V_1^2}{2g_c} + \frac{g(Z_2 - Z_1)}{g_c}\right]_{\text{system}} $$

Choose the material. Let the system contain an ideal gas. Then auxiliary equations $pv = RT$ and $u_2 - u_1 = c_v(T_2 - T_1)$ can be used.

Choose the unknowns. With two independent equations, we can have two unknowns. Let us choose ${}_1Q_2$ and p_2 as the unknowns. We can solve for ${}_1Q_2$ from the energy balance and solve for $p_2 = RT_2/v_2$. If we put the system in a rigid container, then ${}_1W_2 = m\int pdv = 0$, because for a sealed rigid container, v = constant, then $dv = 0$. Let us also add the condition that the process must be isothermal, then $T_2 = T_1$ and thus $u_2 - u_1 = c_v(T_2 - T_1) = 0$. Further, let us also require that $V_2 = V_1$, then the energy balance reduces to

$$ {}_1Q_2 - 0 = m\left[0 + 0 + \frac{g(Z_2 - Z_1)}{g_c}\right]_{\text{system}} $$

Now all we need to do is specify m, Z_1, and Z_2 and we can compute ${}_1Q_2$.

Step B. The next step is to write a scenario, or a short story, that uses these processes and values to create a thermodynamic problem. Let us try this:

There are 5.00 kg of hydrogen gas (an ideal gas) at 20.0°C and 0.300 MPa sealed inside a wooden barrel (a rigid container) at the top of Niagara Falls. The barrel is not insulated and is maintained at a constant temperature (i.e., isothermal) as it travels over the falls in contact with the water. Determine

a. The heat transfer from the barrel as it travels 50.0 m vertically between the top and bottom of the falls.
b. The final pressure inside the barrel at the bottom of the falls.

Note that the problem scenario does not have to be deadly serious, you can write problem statements around anything your imagination can conceive.

Step C. Now we must work the problem in the forward direction to see if all the necessary information has been provided, so let us try it.

Solution

The *problem solving technique* requires that we start by reading the problem statement carefully.

Step 1 ask us to draw a sketch of the system (the barrel going over the Niagara Falls, see Figure 4.22) and identify the material in the system, it is the hydrogen in the barrel.

Step 2 asks us to identify the unknowns. Even though we just wrote the problem statement, it is important to read it again, *carefully*, to check for errors and completeness. The problem statement should contain clarifying statements so that the reader need not make any unreasonable assumptions. For example, in our problem statement, we identified the hydrogen as an *ideal gas*, because it is not obvious to a beginning thermodynamics student which materials behave like an ideal gas and which do not. Also, while it may be obvious to you when you wrote the problem statement that a barrel is to be modeled as a *sealed, rigid container*, it is advisable to tell the reader this in clear terms, since the purpose of the problem should be to test the problem solving skills of the reader, not his or her ability to read your mind about how to interpret unfamiliar things. The unknowns here are clearly specified in items (a) and (b) at the end of the problem statement. They are find ${}_1Q_2$ and p_2.

Step 3 asks us to identify the system's type and its states. The system here is closed (because the barrel is *sealed*). We should be able to identify the system states from the information given in the problem statement:

State 1	State 2
$T_1 = 20.0°\text{C}$	$T_2 = T_1 = 20.0°\text{C}$
$p_1 = 0.300$ Mpa	?

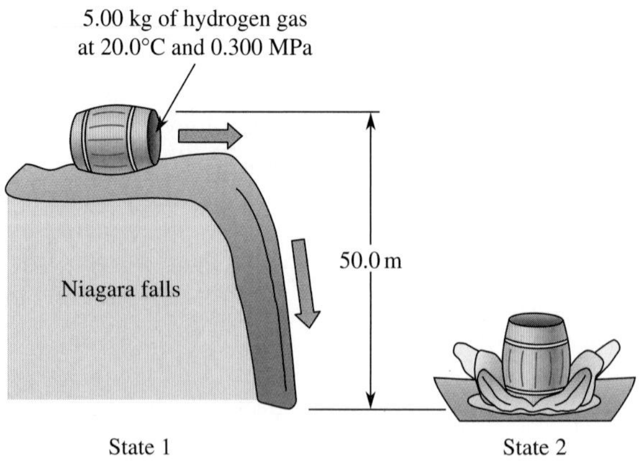

FIGURE 4.22
A barrel going over Niagara Falls.

We have now identified two properties in the first state but only one in the second. This is a common structure at this point in the solution. The missing property must come from somewhere else in the solution, either from the process path or from the working equations used in the solution.

In *step 4* we have to identify the process path taken by the system as it moves between states 1 and 2. Note that, since the system is at a *constant temperature* and is a *sealed, rigid container*, the process path here has both constant temperature and constant volume and mass, so it is also has a constant specific volume. Now, we can add the process path line to the state information and the missing second state property, so that it looks like this:

State 1	- - - - T = costant and v = constant - - - - >	State 2
$T_1 = 20.0°C$		$T_2 = T_1 = 20.0°C$
$p_1 = 0.300$ Mpa		$v_2 = v_1 = RT_1/p_1$

We now have two properties in each state and can continue with the solution. Note that we do not need to calculate the value of v_2 yet, since we are not sure we need it in the solution.

Step 5 is to write the basic equations. Since this is a closed system, the conservation of mass equation yields no useful information, as the mass of the system is constant. However, the conservation of energy (the *first law of thermodynamics*) is very useful here:

$$ {}_1Q_2 - {}_1W_2 = m\left[(u_2 - u_1) + \frac{V_2^2 - V_1^2}{2g_c} + \frac{g(Z_2 - Z_1)}{g_c}\right]_{\text{system}} $$

Next, we write all the relevant auxiliary equations. If you do not know whether an auxiliary equation is relevant or not, write it down anyway and decide later. Let us start with equations of state. If the material in the system were steam or refrigerant or anything for which there is a table in the tables book, we would use those tables rather than an equation of state. However, hydrogen was given in the problem statement as an *ideal gas*, so we can write its equations of state as

$$ pv = RT \quad \text{and} \quad u_2 - u_1 = c_v(T_2 - T_1) $$

Note that the first equation of state can be used as both $p_1v_1 = RT_1$ and $p_2v_2 = RT_2$.

Next, let us look at work mode equations. No rotating shafts or wires cross the system boundary nor has any reference been made in the problem statement to any electric or magnetic fields. Consequently, no shaft, electrical, polarization, magnetic, or other work mode is present. We also need to check for moving boundary work, $({}_1W_2)_{\text{moving boundary}} = m\int pdv$. Since the system is *closed* (the mass is constant) and *rigid* (so the volume is constant), the specific volume (total volume divided by mass) is constant. Then $dv = 0$ and there is no moving boundary or any other type of work. So, ${}_1W_2 = 0$.

At this point we should also identify any changes in kinetic or potential energy. Our problem statement specifies the change in height over the falls as 50.0 m, but it does not mention anything about velocity. The intent here is to have the initial and final velocities of the system be the same, but that might be too much to ask the reader to assume. Therefore, we should alter the problem statement by replacing the word *Determine* with the phrase *Assuming the initial and final velocities of the barrel are the same, determine*. Then, the problem statement reads as follows:

> Five kilograms of hydrogen gas (an ideal gas) at 20.0°C and 0.300 MPa are sealed inside a wooden barrel (a sealed, rigid container) at the top of Niagara Falls. The barrel is not insulated and is maintained at a constant temperature (i.e., isothermal) as it travels over the falls in contact with the water. Assuming the initial and final velocities of the barrel are the same, determine
>
> **a.** The heat transfer from the barrel as is travels 50.0 m vertically between the top and bottom of the falls.
> **b.** The final pressure inside the barrel at the bottom of the falls.

In *step 6* we are ready to algebraically solve for the unknowns. From the energy balance, we can solve for the heat transfer required in part (a) as

$$ {}_1Q_2 = m\left[(u_2 - u_1) + \frac{V_2^2 - V_1^2}{2g_c} + \frac{g(Z_2 - Z_1)}{g_c}\right]_{\text{system}} + {}_1W_2 $$

Now, we incorporate our earlier results that ${}_1W_2 = 0$ and $u_2 - u_1 = c_v(T_2 - T_1) = 0$, because $T_2 = T_1$ here (the process is also isothermal). Our latest rendition of the problem statement makes it clear that $V_2 = V_1$, and when these conditions are incorporated into the energy balance, we obtain our final equation for the heat transfer as

$$ {}_1Q_2 = m\left[0 + 0 + \frac{g(Z_2 - Z_1)}{g_c}\right]_{\text{system}} + 0 = \frac{mg}{g_c}(Z_2 - Z_1) $$

and, from the equation of state, we can determine the solution to part (b) as

$$ p_2 = RT_2/v_2 = RT_1/v_2 = RT_1/v_1 = p_1 $$

since $T_2 = T_1$ and $v_2 = v_1$.

Step 7 allows us to calculate the values of the unknowns:

$$ \text{(a) } {}_1Q_2 = \frac{(5.00\,\text{kg})(9.81\,\text{m/s}^2)}{1}(0 - 50.0\,\text{m}) = -2450\,\text{kg}\cdot\text{m}^2/\text{s}^2 $$
$$ = -2450\,\text{N}\cdot\text{m} = -2450\,\text{J} = -2.45\,\text{kJ} $$

and

$$ \text{(b) } p_2 = p_1 = 0.300\,\text{MPa} $$

The negative sign in the answer for part (a) tells us that the heat transfer is *out of* the system. Note that the answer in part (b) was not the result of a complex calculation. However, it did result from a rather complex analysis and, therefore, is not trivial. Also note that we did not need the value of v_2 in the solution of the problem, so it would have been a waste of time to have calculated it early in the solution.

In *step 8*, since the solution now seems to work well, the problem statement is complete and accurate. We should now check all the algebra, units, and calculations before creating and solving additional problems with similar or different scenarios.

Exercises for the problem solved in Steps 1–8

1. Rewrite this problem and make the barrel insulated but not isothermal. (Can it be both insulated and isothermal?) Resolve the problem with these new conditions. Is any additional information needed to find T_2 and p_2?
2. Write a thermodynamics problem about a computer chip. Look up the steady state voltage and current required by a typical computer chip in a handbook and supply these values in the problem statement. This is a closed system, and the chip cannot be insulated (otherwise, it would overheat). Use the energy rate balance in the formulation of your problem scenario.
3. Write a thermodynamics problem about an electrical generator. Use the closed system energy rate balance. Make the process steady state. You may have the generator insulated or uninsulated. Note that there are two work modes here, shaft work and electrical work.
4. Write a thermodynamics problem about an airplane. Make it a closed system and have it change altitude and speed. Choose an appropriate unknown and provide all the necessary values for the remaining variables.

SUMMARY

In this chapter, we discover that the first law of thermodynamics is simply the conservation of energy principle. Since energy is conserved in all actions, the change in a system's energy can be equated to the net transport of energy into the system. Only three possible energy transport mechanisms are available to us: (1) heat transport of energy (commonly called *heat transfer*), (2) work transport of energy (commonly called *work*), and (3) energy transported with a mass flowing across a system's boundary. This information produced the very powerful energy balance and energy rate balance equations.

The general closed system energy balance:

$$ {}_1Q_2 - {}_1W_2 = (E_2 - E_1)_{\text{system}} $$
$$ = m[(u_2 - u_1) + (V_2^2 - V_1^2)/(2g_c) + (Z_2 - Z_1)g/g_c]_{\text{system}} $$

The general closed system energy *rate* balance:

$$ \dot{Q} - \dot{W} = (dE/dt)_{\text{system}} = (m\dot{u} + mV\dot{V}/g_c + mg\dot{Z}/g_c)_{\text{system}} $$

The general open system energy *rate* balance:

$$\dot{Q} - \dot{W} + \sum \dot{E}_{\substack{\text{mass}\\\text{flow}}} = (d/dt)\left(mu + mV^2/2g_c + mZg/g_c\right)_{\text{system}}$$

Work modes of energy transport are not discussed in any course outside of thermodynamics and are very important for utilizing the full capacity of the first law of thermodynamics. We need to understand and master the work mode auxiliary equations, because they are often required in the solution of thermodynamic problems. Some of the important work mode auxiliary equations are given in Table 4.10. The associated power equations are given in Table 4.6 of the text.

The local equilibrium postulate allows us to deal with nonequilibrium states, and the state postulate defines the number of independent thermodynamic properties required to determine the local equilibrium state (two, for a simple system).

Heat transport of energy (heat transfer ${}_1Q_2$ and heat transfer rate $\dot{Q}$) is categorized into three modes: (1) conduction, (2) convection, and (3) radiation. Heat transfer is sufficiently important to mechanical engineers that most curricula have separate heat transfer courses. Consequently, the details of this subject are not emphasized in a thermodynamics course. The heat transfer rate modes are summarized in Table 4.11.

Generally, if you are asked to determine a heat transfer in a problem statement, you should calculate it from the first law energy balance rather than from one of the heat transfer mode auxiliary equations.

Table 4.10 Work Mode Auxiliary Equations

Work Mode	Equation
Moving boundary (general)	$({}_1W_2)_{\substack{\text{moving}\\\text{boundary}}} = \int_1^2 p\,d\forall$
Polytropic moving boundary ($n \neq 1$)	$({}_1W_2)_{\substack{\text{polytropic }(n\neq 1)\\\text{ideal gas}\\\text{moving boundary}}} = \frac{mR}{1-n}(T_2 - T_1)$
Rotating shaft	$({}_1W_2)_{\substack{\text{rotating}\\\text{shaft}}} = \int_1^2 \vec{\mathbf{T}} \cdot d$
Elastic	$({}_1W_2)_{\text{elastic}} = -\int_1^2 \sigma\forall\,d\varepsilon$
Surface tension	$({}_1W_2)_{\substack{\text{surface}\\\text{tension}}} = -\int_1^2 \sigma_s\,dA$
Electrical current	$({}_1W_2)_{\substack{\text{electrical}\\\text{current}}} = \int_1^2 \phi i\,dt$
Electrical polarization	$({}_1W_2)_{\substack{\text{electrical}\\\text{polarization}}} = -\int_1^2 E\,dP$
Magnetic	$({}_1W_2)_{magnetic} = -\mu_0\forall(1+\chi_m)\left(\frac{H_2^2 - H_1^2}{2}\right)$
Chemical	$({}_1W_2)_{\substack{\text{chemical}\\\mu_i=\text{constant}}} = -\sum_{i=1}^{k} \mu_i(m_2 - m_1)_i$
Mechanochemical	$({}_1W_2)_{\text{mechanochemical}} = \int_1^2 f\,d\ell$

Table 4.11 Heat Transfer Rate Modes

Heat Transfer Mode	Equation
Conduction	$\dot{Q}_{\text{cond}} = -k_t A\left(\frac{dT}{dx}\right)$
Convection	$\dot{Q}_{\text{conv}} = hA(T_\infty - T_s)$
Radiation	$\dot{Q}_{\text{rad}} = F_{1-2}\varepsilon_1 A_1 \sigma(T_2^4 - T_1^4)$

Finally, we study a special technique that maps the solution of any thermodynamic problem. If you follow the format given in Figure 4.21, you will breeze through the solution maze. But you must follow it religiously and take no shortcuts. As an extension of your problem solving skills, you are also shown how to write and solve your own thermodynamic problems. If you can do this successfully, you have mastered the subject.

Problems (* indicates problems in SI units)

1.* Determine the energy transport required to increase the temperature of 3.50 kg of air from 20.0 to 100.°C (Figure 4.23). Assume the air is stationary and behaves as an ideal gas with constant specific heats.

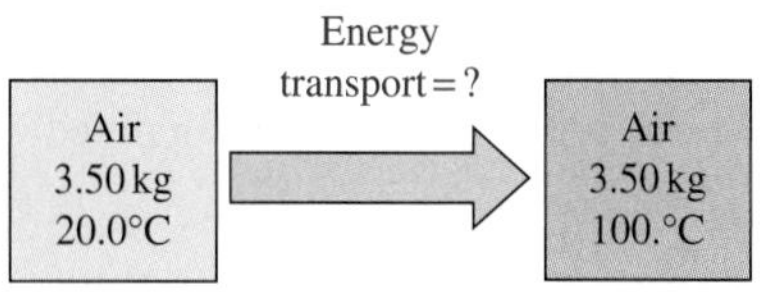

FIGURE 4.23
Problem 1.

2.* Determine the energy transport necessary to decrease the temperature of 15.0 kg of methane from 500. to 20.0°C. Assume the methane is stationary and behaves as an ideal gas with constant specific heats.

3. Determine the gain in energy of a stationary system of 5.00 lbm of argon whose temperature is increased from 70.0 to 1000.°F. Assume ideal gas behavior with constant specific heats.

4.* Determine the gain in energy of a stationary system of 11.0 kg of oxygen whose pressure is increased from 0.100 to 100. MPa isothermally. Assume ideal gas behavior with constant specific heats.

5. If 150. Btu are transported into a system via a work mode while 75.0 Btu are removed via heat transfer and mass flow modes (Figure 4.24), determine the net energy gain for this system.

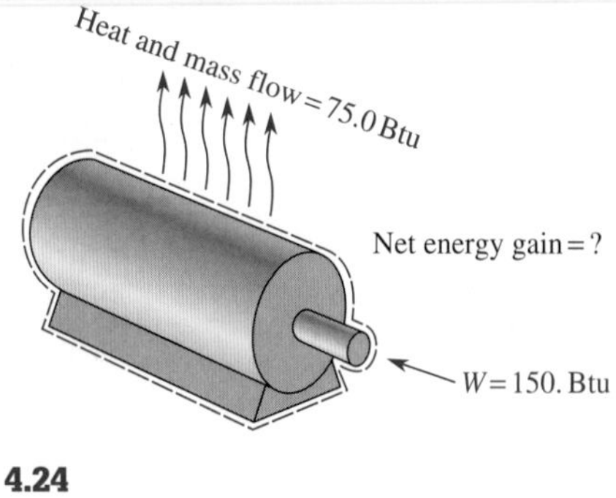

FIGURE 4.24
Problem 5.

6. A jet aircraft with a constant specific internal energy of 3500. Btu/lbm consumes fuel at a rate of 50.0 lbm/min while flying horizontally at an altitude of 30,000. ft with a constant velocity of 500. ft/s (Figure 4.25). Determine the net energy transport rate of the aircraft.

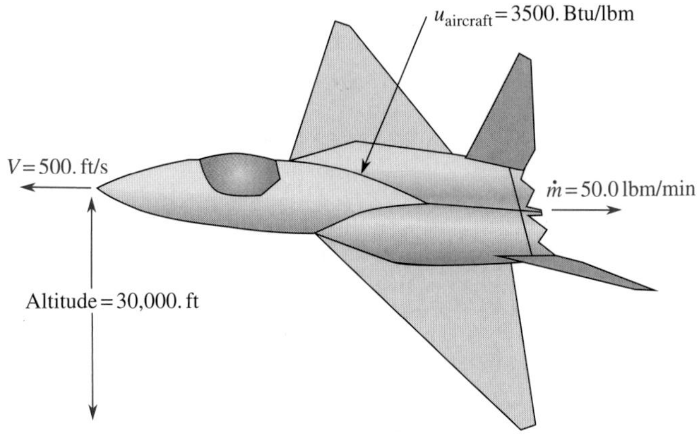

FIGURE 4.25
Problem 6.

7. An automobile transmission has 175 hp of power entering from the engine, 167 hp leaving to the wheels, while losing 5000. Btu/h to the surroundings as heat. What is the net energy transport rate of the transmission?

8. To keep the transmission in the previous problem from overheating, it was decided to cool it by circulating a coolant through its case. If the coolant enters the transmission with a mass flow energy rate of 10.0 Btu/s, what is its mass flow energy rate as it leaves the transmission?

9. Determine the heat transfer rate, in Btu/h, required to cool a 200. kW electric generator that is driven by a 300. hp diesel engine (Figure 4.26). Note: The generator runs cool if it has a zero net energy transport rate.

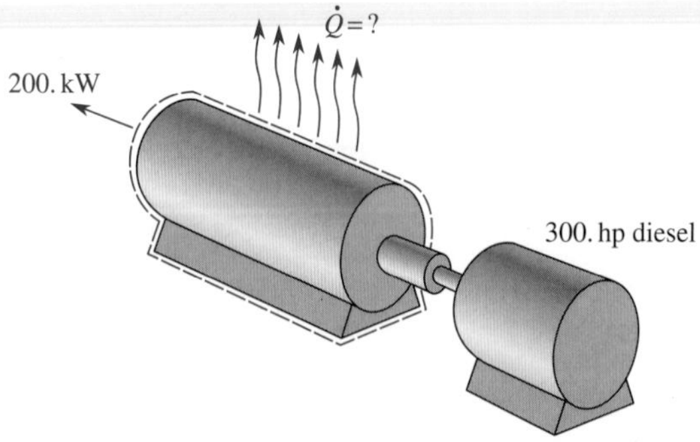

FIGURE 4.26
Problem 9.

10. In a stationary dynamometer test, an internal combustion automobile engine has a fuel energy input rate of 1.90 million Btu/h while producing 150. hp of output power. What other energy transport mechanisms are present and what are their magnitudes. Assume that the net energy transport rate is zero.

11.* Determine the heat transfer per kg necessary to raise the temperature of a closed rigid tank of saturated water vapor originality at 0.140 MPa to a temperature of 800.°C.

12. A closed rigid vessel of volume 5.00 ft^3 contains steam at 100. psia with 83.91% moisture. If 9490.4 Btu of heat are added to the steam, find the final pressure and quality (if wet) or temperature (if superheated).

13.* A closed rigid vessel having a volume of 0.566 m^3 is filled with steam at 0.800 MPa and 250.°C. Heat is transferred from the steam until it exists as saturated vapor. Calculate the amount of heat transferred during this process.

14. A sealed, rigid tank of 10.0 ft^3 capacity is initially filled with steam at 100. psia and 500.°F. The tank and its contents are then cooled to 260.°F. Find (a) the final quality in the container and the amounts of liquid water and water vapor (in lbm), and (b) the amount of heat transfer required (in Btu).

15.* A sealed rigid vessel contains 5.00 kg of water (liquid plus vapor) at 100.°C and a quality of 30.375%.

   a. What is the specific volume of the water?
   b. What is the mass of water in the vapor phase?
   c. What would be the saturation pressure and temperature of this water if it had the specific volume determined in part a and a quality of 100%?
   d. What heat transfer would be required to completely condense the saturated vapor of part c into a saturated liquid?

16. One pound of saturated liquid water at a pressure of 40.0 psia is contained in a rigid, closed, stationary tank. A paddle wheel does 3000. ft·lbf of the work on the system, while heat is transferred to or from the system. The final pressure of the system is 20.0 psia. Calculate the amount of heat transferred and indicate its direction.

17. Identify the following as either point or path functions:

   a. $u^2 + 3u - 5$.
   b. $T(h^2 - u^2) - 3(u - pv) + 4$.
   c. $\sin u^3 + \sin h^3$.
   d. $\int_1^2 \forall\, dp$, where $\forall = \forall(p)$.

18. Identify the following as either point or path functions:

   a. RT/v.
   b. $\int_1^2 p\, d\forall$, where $p = p(\forall)$.
   c. $h + pv$.
   d. $u + V^2/2g_c + gZ/g_c$.

19. Explain whether $u_2 - u_1 = \int_1^2 c_v\, dT$ is a point or a path function for a given system.

20. Explain whether $h_2 - h_1 = \int_1^2 c_p\, dT$ is a point or a path function for a given system.

21. Explain the meaning of the notation ${}_1Q_2$ and ${}_1W_2$. Why do we not write ${}_1E_2$, ${}_1u_2$, or ${}_1h_2$?

22.* Determine the moving boundary work transport of energy when 4.5 kg of water expands at constant pressure from saturated liquid to saturated vapor while at 20.0°C.

23. Determine the moving boundary work done by the atmosphere (14.7 psia) as a cube of ice 2.00 in on a side melts into a pool of liquid water (Figure 4.27). At 32.0°F, the density of ice is 57.2 lbm/ft^3 and that of liquid water is 62.4 lbm/ft^3.

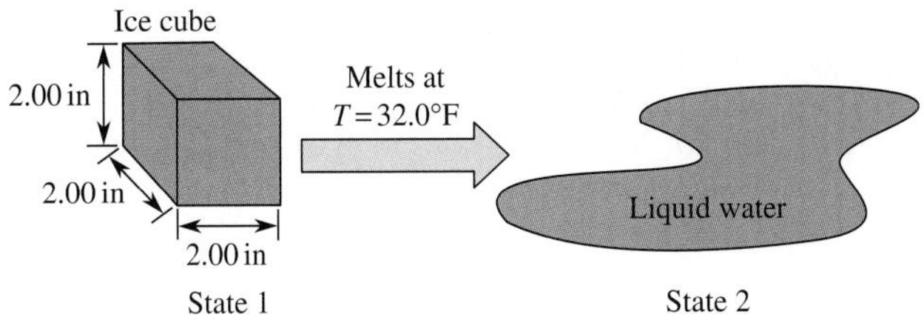

FIGURE 4.27
Problem 23.

24. Determine the moving boundary work done by a cube of solid CO_2 2.00 in on a side as it vaporizes at atmospheric pressure (14.7 psia) (Figure 4.28). The density of solid CO_2 is 97.561 lbm/ft^3 and that of CO_2 vapor is 0.174 lbm/ft^3.

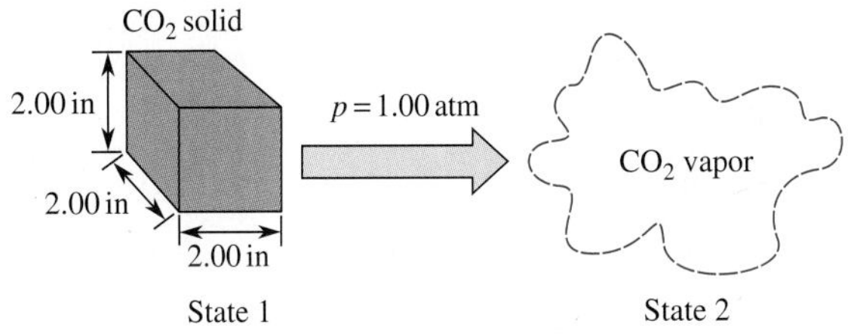

FIGURE 4.28
Problem 24.

25. A weather balloon is filled with helium at 50.0°F so that its volume is 500. ft^3. The balloon is left anchored in the sun and its temperature rises to 110.°F. How much moving boundary work is done by the balloon on the atmosphere as its volume increases due to the increase in temperature? Assume that helium is an ideal gas and the balloon skin is sufficiently thin that the pressure in the balloon remains approximately atmospheric.

26.* Suppose 2.00 m^3 of air (considered an ideal gas) is initially at a pressure of 101.3 kPa and a temperature of 20.0°C. The air is compressed at a constant temperature in a closed system to a pressure of 0.500 MPa (Figure 4.29). (a) How much work is done on the air to compress it? (b) How much energy is transferred as heat during the compression process?

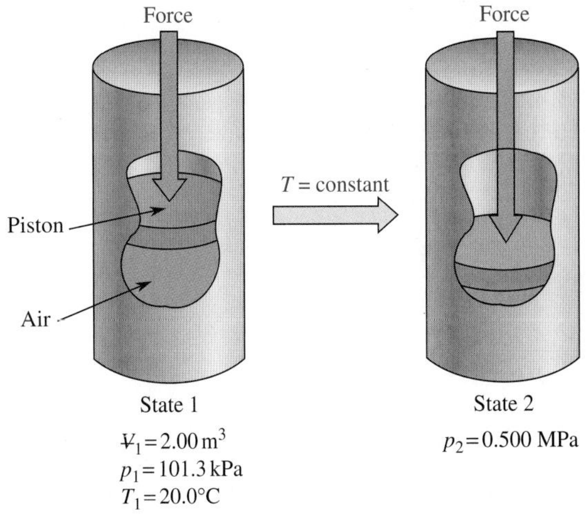

FIGURE 4.29
Problem 26.

27. Show that the first law of thermodynamics requires that, for an ideal gas with a constant specific heat ratio $c_p/c_v = k$ undergoing a polytropic process (i.e., pv^n = constant),
   a. n must be greater than k for $T_2 < T_1$ when there is heat transfer from the gas.
   b. n must be less than k for $T_2 < T_1$ when there is a heat transfer to the gas.

28. Find the moving boundary work done on a gas in compressing it from $V_1 = 10.0\,\text{ft}^3, p_1 = 10.0\,\text{psia}$ to $V_2 = 1.000\,\text{ft}^3$ according to the relation pV^3 = constant (Figure 4.30).

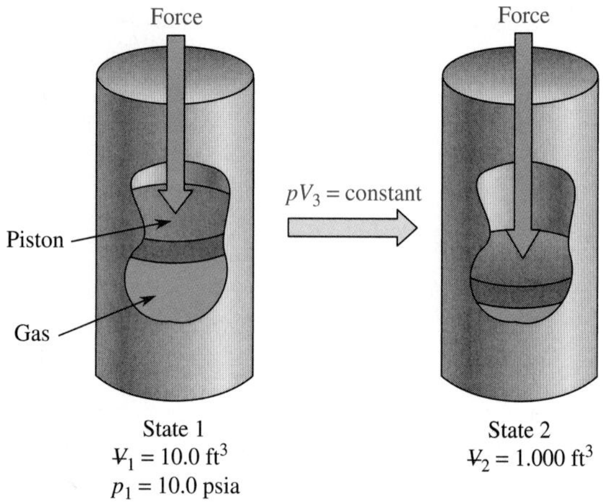

FIGURE 4.30
Problem 28.

29.* A brilliant young engineer claims to have invented an engine that runs on the following thermodynamic cycle:
   a. An isochoric pressurization from p_1 to $p_2 = p_1$.
   b. An isobaric expansion from V_2 to $V_3 = 2V_2$.
   c. An isochoric depressurization from p_3 to $p_4 = p_1$.
   d. An isobaric compression back to the initial state, p_1, V_1.
   Determine the net moving boundary work done during this cycle if p_1 = 25.0 kPa and $V_1 = 0.0300\,\text{m}^3$. Sketch this cycle on a $p-V$ diagram.

30.* A balloon filled with air at 0.100 MPa-absolute is heated in sunlight. As the balloon is heated, it expands according to the following pressure-volume relation:

$$p = 0.1 + 0.15V + 0.06V^2$$

where p is in MPa and V is in m³ (Figure 4.31). Determine the moving boundary work transport of energy as the balloon expands from 1.00 to 2.00 m³.

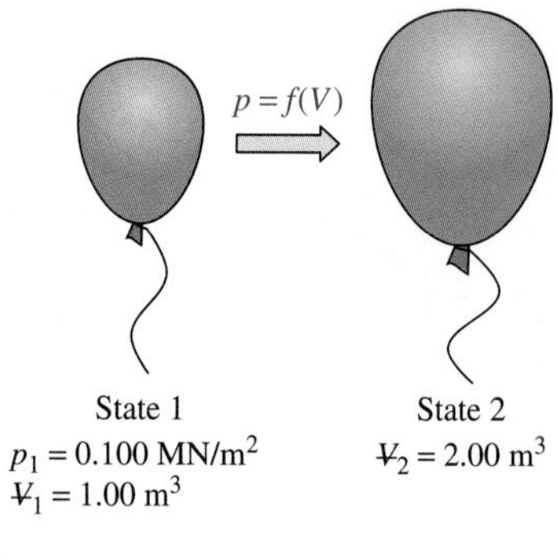

FIGURE 4.31
Problem 30.

31. One lbm of an ideal gas with molecular weight 6.44 lbm/lbmole is compressed in a closed system from 100. psia, 600. R to a final specific volume of 8.00 ft³/lbm. At all points during the compression, the pressure and specific volume are related by

$$p = 50 + 4v + 0.1v^2$$

where p is in psia and v is in ft³/lbm. Determine the moving boundary work required and the heat transfer during this compression if the gas has a constant volume specific heat of 0.200 Btu/(lbm·R).

32. Three lbm of a substance is made to undergo a reversible expansion process within a piston-cylinder device, starting from an initial pressure of 100 psia and an initial volume of 2.00 ft³. The final volume is 4.00 ft³. Determine the moving boundary work produced by this expansion for each of the following process paths. Note which process produces the maximum work and which produces the minimum.
   a. Pressure remains constant ($p = K$)
   b. Pressure times volume remains constant ($pV = K$).
   c. Pressure is proportional to volume ($p = KV$).
   d. Pressure is proportional to the square of volume ($p = KV^2$).
   e. Pressure is proportional to the square root of volume ($p = K\sqrt{V}$), where K is a constant in each case.

33.* The magnitude of the torque T on a shaft is given in N·m by

$$T = 6.3\cos\theta$$

where θ is the angular displacement. If the torque and displacement vectors are parallel, determine the work required to rotate the shaft through one complete revolution.

34. The magnitude of the torque vector normal to the axis of a shaft is given in ft·lbf by

$$\begin{aligned} T &= 21.7\sin\theta && \text{for } 0 < \theta \le \pi \\ &= 0 && \text{for } \pi < \theta \le 3\pi/2 \\ &= 50.4 && \text{for } 3\pi/2 < \theta \le 2\pi \end{aligned}$$

Determine the work done in one complete revolution of the shaft.

35. When the torque and angular displacement vectors are parallel, the torque displacement relation for the drive shaft of a 1909 American Underslung automobile is given by

$$T\theta^n = K$$

where K and n are constants. Determine a general formula for the shaft work when (a) n = 1.0, and (b) $n \neq 1.0$.

36. How much elastic work is done in uniaxially stretching an initially unstrained elastic steel bar (Young's modulus = 3.0 × 10⁷ psi = constant) whose volume (also a constant) is 5.00 in³ to a total strain of 0.00200 in/in?

37. When a rubber band is stretched, it exerts a restoring force (F) that is a function of its initial length (L) and displacement (x). For a certain rubber band this relation is

$$F = K\left[\frac{x}{L} + \left(\frac{x}{L}\right)^2\right],$$

where $K = 0.810$ lbf. Determine the elastic work (with the appropriate sign) required to stretch the rubber band from an initial length of 2.00 in to a final length of 3.00 in (Figure 4.32).

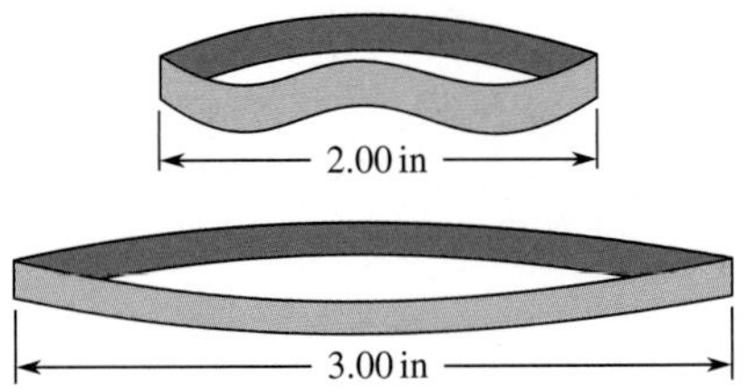

FIGURE 4.32
Problem 37.

38.* A 10.0 cm soap bubble is blown on the end of a large-diameter blowpipe. When the blowpipe end opposite the bubble is uncovered, the surface tension in the soap bubble causes it to collapse, thus sending its contents through the blowpipe and the atmosphere. Estimate the velocity of the air in the blowpipe as the bubble collapses. For the soap bubble, $\sigma_s = 0.0400$ N/m.

39. At 68.0°F the surface tension of acetic acid is 1.59×10^{-4} lbf/in. A film of acetic acid is maintained on the wire frame as shown in Figure 4.33. Determine the surface tension work done when the wire is moved 1.00 inch in the direction indicated.

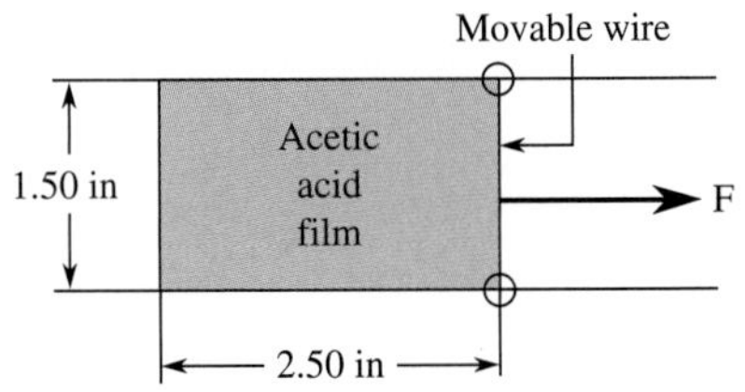

FIGURE 4.33
Problem 39.

40.* A 12.0 V automobile battery receives a constant charge from the engine's alternator. The voltage across the terminals is 12.5 V dc, and the current is 9.00 A. Determine the electrical work energy transport rate from the automobile's engine to the battery in both watts and horsepower.

41. A battery powered wheelchair uses a standard 12.0 V automotive lead-acid battery with a capacity of 20.0 A·h. Peukert's law for the discharge of lead-acid batteries is

$$\sigma i^{1.4} = K$$

where σ is the discharge time, i is the discharge current, and K is a constant that depends on the battery size. The capacity of the battery is given by capacity = $\sigma i = K i^{-0.4}$, and the average voltage during discharge is given by $\phi = 11.868 - 0.0618i$.

a. How much current is drawn from the battery if the torque on the drive shaft is 1.00 ft·lb when it is rotating at 1.00 rev/s?
b. How long will the wheelchair operate with this current drain before the battery is discharged?
c. Evaluate the constant K for this battery with this current drain.

42. Determine the electrical current power averaged over one period, T, for a sawtooth current waveform passing through a pure resistance R described by $i = i_{max}(t/T)$ for $0 < t < T$.

43. In an ac circuit in which a phase angle θ exists, the voltage and current are written as

$$\phi = \phi_{max}\cos(2\pi f t)$$
$$i = i_{max}\cos(2\pi f t - \theta)$$

Show that the electrical current power averaged over one period $(1/f)$ is

$$(\dot{W})_{\text{electrical avg.}} = -\left(\frac{1}{2}\right)(\phi_{max}) i_{max}\cos(\theta) = -\phi_e i_e \cos(\theta)$$

and thus the average power of any purely reactive $(\theta = \pi/2)$ circuit consisting entirely of ideal inductors and capacitors is zero. The term $\cos(\theta)$ is called the *power factor*, and the product $\phi_e i_e$ is called the *apparent* power. For a purely resistive circuit, $\theta = 0$ and the average power equals the apparent power.

44. Show that the polarization work required to charge a parallel plate capacitor is given by?A3B2 tptxb +2pt?>

$${}_1W_2 = -C\phi^2/2$$

where $C = \varepsilon_0 \chi_e A/d$ is the capacitance, ϕ is the voltage difference, A is the area of the plates, and d is their separation distance.

45. An electrical capacitor constructed of two parallel conducting plates of area A, separated by a distance d, has a capacitance C given by

$$C = \varepsilon_0 \chi_e A/d$$

where C is in faradays ($1\ \text{F} = 1\ \text{J/V}^2$). Determine the polarization work required to charge an initially discharged 10.0 μF parallel plate capacitor when the plates are separated by 5.00×10^{-3} m of Plexiglas and subjected to a potential difference of 300. V at 27.0°C.

46. A typical storm cloud at an altitude of 3000. ft has a cross-sectional area of 1.00×10^8 ft^2 and a surface potential relative to the earth of 1.00×10^8 V. Determine the amount of electrical energy stored in the cloud by calculating the polarization work required to charge the earth-cloud capacitor.

47.* A square aluminum bar 0.0300 m on a side and 1.00 m long is wrapped with a current-carrying wire (Figure 4.34). When the current in the wire is turned on, it exposes the aluminum core to a magnetic field strength of 456×10^3 A/m. Determine the total magnetic work that occurs when the current is turned on and determine what percentage of this work is associated with the alignment of the aluminum's molecular magnetic dipoles.

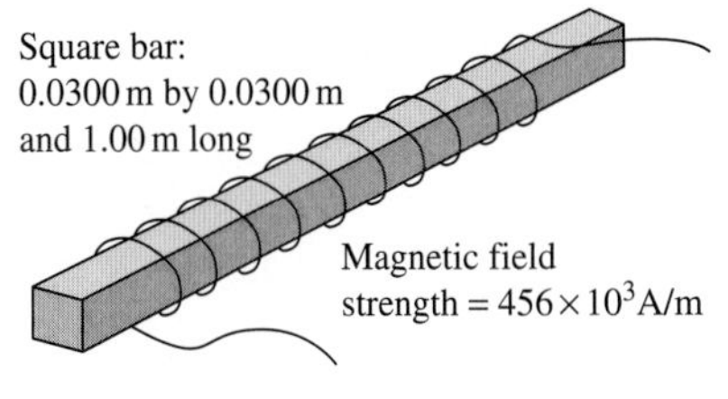

FIGURE 4.34
Problem 47.

48.* A quartz rod 0.0100 m in diameter and 0.100 m long is to be subjected to a magnetic intensity of 10,000. A/m. Determine the total magnetic work required for this process if the initial magnetic intensity of the rod is zero.

49.* A Curie substance has a magnetic susceptibility given by

$$\chi_m = C'/T$$

where C' is the Curie constant for the substance and T is its absolute temperature. Determine an expression for the work per unit volume for isothermal material magnetization of a constant volume Curie substance. Evaluate this for $C'' = 153$ K, $T = 300.$ K, $M_1 = 0$, $M_2 = 1000.$ A/m.

50.* The chemical potential of a professor's brain in a single species cranium is constant at −13.2 MJ/kg. Determine the chemical work required to remove 3.77 kg of this valuable substance from the cranium.

51. 2.00 lbm of chemical species A ($\mu_A = -5700.$ Btu/lbm) is removed from a system while 7.30 lbm of species B ($\mu_B = -3850$ Btu/lbm) and 11.1 lbm of species C ($\mu_C = 1050$ Btu/lbm) are added to the system. Determine the net chemical work involved. Assume constant chemical potentials.

52. If the total internal energy of an adiabatic, stationary, closed system is given by

$$U = -p\mathcal{V} + \sum \mu_i m_i - f\ell$$

Show that the following formula must hold:

$$-\mathcal{V}\,dp + \sum m_i\,d\mu_i - \ell\,df = 0$$

(Hint: Start from the differential form of the energy balance, $\bar{d}Q - \bar{d}W = dU$ and use Eq. (4.69)).

53. A simple mechanochemical engine operates on the thermodynamic cycle shown in Figure 4.35. The mechanochemical contractile work output ($fd\ell$) comes from a chemical work input (μdm) due to the aqueous dilution of a single chemical species ($i = 1$).

a. Show that the net chemical transport per cycle of this engine is given by

$$(W)_{\substack{\text{chemical}\\ \text{cycle net}}} = (\mu_1 - \mu_2)(\Delta m)$$

where $\Delta m = m_3 - m_2 = m_4 - m_1$.

b. Write an expression for the work transport energy efficiency of this engine.

54. A refrigeration cycle is chosen to maintain a freezer compartment at 10.0°F in a room that is at 90.0°F. If 200. Btu/min are extracted from the freezer compartment by heat transfer and the freezer is driven by a 1.00 hp electric motor, determine the dimensionless coefficient of performance (COP) of the unit, defined as the cooling rate divided by the input power.

55. An automobile engine produces 127 hp of actual output power. If the friction, heat transfer, and other losses consume 23.0 hp, determine the work transport energy efficiency of this engine.

56.* 60.0 kW enter a mechanical gearbox at its input shaft but only 55.0 kW exit at its output shaft. Determine its work transport energy efficiency.

57. Find the heat transport rate of energy from a circular pipe with a 2.00 inch outside diameter, 20.0 ft long, and a wall thickness of 0.150 in. The inside and outside surface temperatures of the pipe are 212 and 200.°F, respectively. The pipe is made of carbon steel.

58.* A wall is made up of carbon steel 1.00 cm thick. Determine the conduction heat transport rate per unit area through the wall when the outside temperature is 20.0°C and the inside temperature is −10.0°C.

59. A window consists of a 0.125 in glass pane. Determine the conduction heat transport rate per unit area of window pane when the inside and outside temperatures to be 70.0 and 0.0°F, respectively.

60.* Find the surface temperature of a bare 40.0 W fluorescent light tube, 3.60 cm in diameter and 1.22 m long in room air at 20.0°C. The convective heat transfer coefficient of the tube is 4.80 W/(m^2·K).

61.* An experiment has been conducted on a small cylindrical antenna 12.7 mm in diameter and 95.0 mm long. It was heated internally with a 40.0 W electric heater. During the experiment, it was put into a cross flow of air at 26.2°C and 10.0 m/s. Its surface temperature was measured and found to be 127.8°C. Determine the convective heat transfer coefficient for the antenna.

62. An automobile is parked outdoors on a cold evening, when the surrounding air temperature is 35.0°F. The convective heat transfer between the roof of the automobile and the surrounding

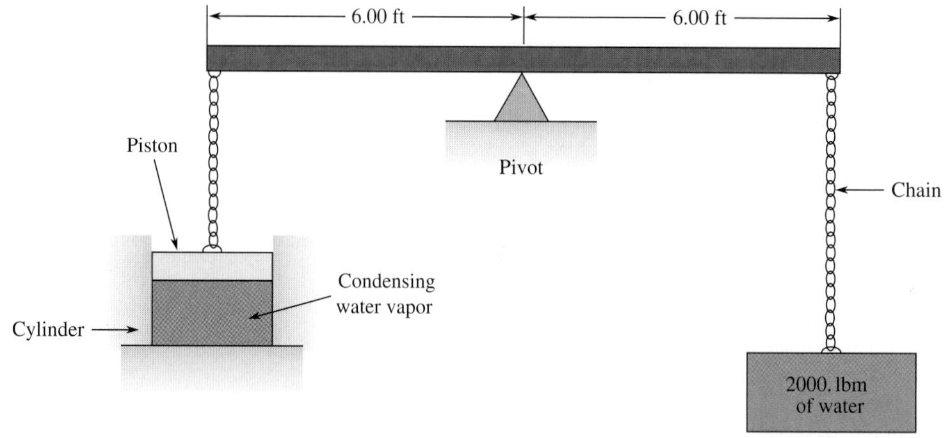

FIGURE 4.35
Problem 53.

Table 4.12 Problem 67

Process	Q_{cond}	Q_{conv}	Q_{rad}	W_{mech}	W_{elect}	W_{magn}	W_{chem}	$E_2 - E_1$
1–2	5	13	−34	45	2	−23	11	?
2–3	12.3	56.1	121.	0.0	85.0	0.0	?	211.0
3–1	1.1	−23.3	?	−44.8	89.9	−47.3	14.2	0.0

air is 1.50 Btu/(h·ft^2·R). The night sky is cloudless and forms a black body at a temperature of −30.0°F. By performing a convective-radiation balance on the roof, determine

a. The roof temperature.
b. Whether or not frost will form on the roof (and why).

63.* Determine the radiation heat transfer rate per unit area between an infant at 37.0°C in a crib and a nearby window at −10.0°C in the winter. The view factor between the infant and the window is 0.310.

64. Determine the radiation heat transfer rate per unit area from a nuclear fireball at 10,000.°F and a nearby building at 70.0°F covered with white paint. The view factor between the building and the fireball is 0.0100.

65. Define the following terms:
a. Adiabatic.
b. Mechanical work.
c. Reversible.
d. The state postulate.

66. Define the following terms:
a. Aergonic.
b. The local equilibrium postulate.
c. Enthalpy.
d. Work efficiency.

67.* A closed system undergoes a cycle made up of three processes. Fill in the missing data in Table 4.12. All the values are in kilojoules.

Computer Problems

The following computer assignments are designed to be carried out on a personal computer using a spreadsheet or equation solver. They are exercises that use some of the basic formulae of Chapter 4. They may be used as part of a weekly homework assignment.

68. Develop a program that calculates the work transport for an ideal gas undergoing a polytropic moving boundary process. Have the user input all necessary data from the keyboard by responding to properly worded screen prompts. Make sure that units are specified when requesting user data input. Output the polytropic work and all the input data (with corresponding units).

69. Develop a program that calculates the work transport for a Hookean elastic solid. Have the user input all necessary data from the keyboard by responding to properly worded screen prompts. Make sure that units are specified when requesting user data input. Output the elastic work and all the input data (with corresponding units).

70. Develop a program that determines the work transport in a constant volume magnetization process. Have the user input all necessary data from the keyboard by responding to properly worded screen prompts. Make sure that units are specified when requesting user data input. Output the magnetic work, the work of magnetization of the exposed material, and all the input data (with corresponding units).

71. Develop a program that determines the chemical work transport for a system with constant chemical potentials. Have the user input all the μ_i and the initial and final m_i from the keyboard by responding to properly worded screen prompts. Make sure that units are specified when requesting user data input. Output the chemical work (with corresponding units).

72. Develop a program that determines the total heat transfer rate from the sum of one or more of the three heat transport modes (conduction, convection, and radiation). Have the user select from a menu which heat transport mode or combination of heat transport modes he or she wishes to use. Then have the user input all necessary data from the keyboard by responding to properly worded screen prompts. Output the heat transport rate and all input data (with corresponding units).

Create and Solve Problems

Engineering education tends to focus only on the process of solving problems. It ignores teaching the process of formulating solvable problems. However, working engineers are never given a well-phrased problem statement to solve. Instead, they need to react to situational information and organize it into a structure that can be solved using the methods learned in college.

Also, if you see how problems are written (created), then you have a better chance of mastering the solution technique and of understanding how to structure information as a working engineer into solvable situations. These "Create and Solve" problems are designed to help you learn how to formulate solvable thermodynamics problems from engineering data. Since you provide the numerical values for some of the variables, these problems have no unique solutions. Their solutions depend on the assumptions you need to make and how you set them up to create a solvable problem.

73.* You are a design engineer working on a robotic system. The robot contains an imbedded circuit board that draws 30.0 mA at 5.00 V. Someone mentions that the circuit board might overheat during its 30 min. operating cycle. Write and solve a problem that provides (a) the heat generation rate of the circuit board, and (b) the temperature of the circuit board if it is insulated and operated for 30.0 minutes. Choose relevant values for the necessary variables. Hint: Your problem statement might read something like this:

> An insulated circuit board draws 30.0 mA at 5.00 V. Determine its heat generation rate and its temperature after 30.0 min of operation. The board has a mass of 1.00×10^{-3} kg and its specific heat is 0.500 kJ/kg·K.

Now you have to solve your problem to determine the answers to (a) and (b).

74. You are designing a new mechanical transmission large rock crusher used in the mining industry. The transmission is driven by a 300. hp engine but transmits only 290. hp to the rock crusher. You need to prevent the transmission from overheating, so how much cooling is needed to keep it at ambient temperature?

75.* Dave, your boss, wants you to estimate the amount of heat that has to be removed from an iron ingot to cool it from 900°C to 150°C. Make this request into a thermodynamic problem statement and solve it.

76.* You are a new engineer at a company that manufactures gas-filled shock absorbers for racing cars. The chief engineer wants to understand the relation between the gas pressure inside the shock absorber and the compression of the gas. The shock absorbers are essentially piston-cylinder devices that are initially filled with nitrogen gas at 0.345 MPa and 20.0°F. When the piston compresses the gas by 20%, the pressure increases to 0.414 MPa. You think this is a polytropic compression process. Write and solve a thermodynamics problem to determine the polytropic exponent for this process.

CHAPTER 5

First Law Closed System Applications

CONTENTS

5.1 INTRODUCTION

In this chapter, we present a series of detailed engineering analyses of the application of the first law of thermodynamics to closed systems. This material demonstrates good thermodynamic problem solving technique through a variety of worked examples. In the first three examples that follow, the numbered steps in the solution are the same as the steps shown in Figure 4.21. As we continue with the examples and the reader becomes more familiar with the technique, we condense the solutions by omitting the description of each solution step. In so doing, we also introduce some flexibility into the technique.

SUMMARY OF THE THERMODYNAMIC PROBLEM SOLVING TECHNIQUE

Begin by carefully reading the problem statement completely through.

Step 1. Make a sketch of the system and put a dashed line around the system boundary.
Step 2. Identify the unknown(s) and write them on your system sketch.
Step 3. Identify the type of system (closed or open) you have.
Step 4. Identify the process that connects the states or stations.
Step 5. Write down the basic thermodynamic equations and any useful auxiliary equations.
Step 6. Algebraically solve for the unknown(s).
Step 7. Calculate the value(s) of the unknown(s).
Step 8. Check all algebra, calculations, and units.

Sketch → Unknowns → System → Process → Equations → Solve → Calculate → Check

5.2 SEALED, RIGID CONTAINERS

One of the most innocuous technical incantations in basic thermodynamics is the phrase *sealed, rigid container* (or *tank* or *vessel*). This phrase is composed of the following three technical terms:

1. **Sealed** means the system is *closed.*
2. **Rigid** means the system has a *constant volume,* $\forall$ = constant and $d\forall = 0$. Therefore, there is no moving boundary mechanical work (i.e., $\int p\, d\forall = 0$).
3. **Container** (sometimes *tank* or *vessel*) means *the system boundary lies inside the enclosure,* because the material we want to analyze is inside the enclosure.

The following example illustrates a typical problem of this type.

EXAMPLE 5.1

Read the problem statement. A sealed, rigid container whose volume is 1.00 m^3 contains 2.00 kg of liquid water plus water vapor at 20.0°C. The container is heated until the temperature inside is 95.0°C.[1] Determine

a. The quality in the container when the contents are at 20.0°C.
b. The quality in the container when the contents are at 95.0°C.
c. The heat transport of energy required to raise the temperature of the contents from 20.0 to 95.0°C.

Solution

Step 1. Identify and sketch the system. Take the system to be the material *inside* the closed, rigid container as shown in Figure 5.1. Since we do not know the type or amount of material making up the container itself, the container cannot be part of the system. Also, since all the unknowns pertain to the container's contents, detailed knowledge of the container's construction is not relevant to the solution.

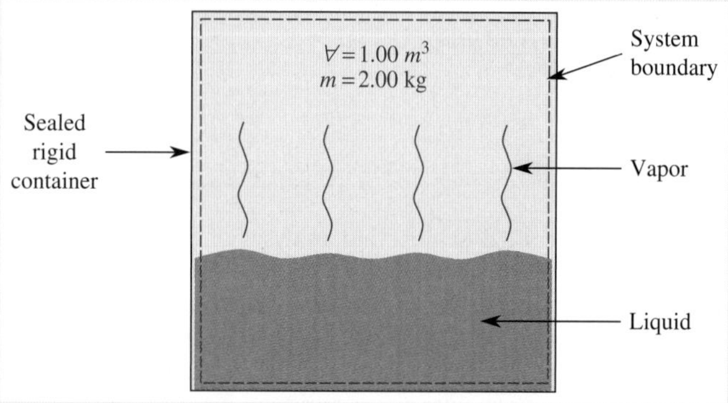

FIGURE 5.1
Example 5.1.

Step 2. Identify the unknowns. Here, there are three unknowns: x_1, x_2, and ${}_1Q_2$.

Step 3. It is a **closed system.** To fix the system's states, we note that we are given the initial and final temperatures of the water, but to find other properties (such as quality), we need the value of one more independent property in each state. Notice, however, that we are given both the total volume and the total mass and these do not change during the change of state. Therefore, we can calculate the system's specific volume in each state as

$$v_1 = v_2 = \forall/m = 1.00\,\text{m}^3/2.00\,\text{kg} = 0.500\,\text{m}^3/\text{kg}$$

Now we can write the states and process path as

State 1	Process: "*Rigid and Sealed*" *means* $v_1 = v_2$ constant →	State 2
$T_1 = 20.0°\text{C}$		$T_2 = 95.0°\text{C}$
$v_1 = 0.500\,\text{m}^3/\text{kg}$		$v_2 = v_1 = 0.500\,\text{m}^3/\text{kg}$

Step 4. Identify the process connecting the system states. The process here is one of constant volume (the container was specified as rigid), and the process path has already been indicated in the state property value listing.

Step 5. Write down the basic equations. The only basic equation we have thus far for closed systems is Eq. (4.20), the energy balance (EB) equation:

$${}_1Q_2 - {}_1W_2 = m[(u_2 - u_1) + \text{ke}_2 - \text{ke}_1 + \text{pe}_2 - \text{pe}_1]_{\text{system}}$$

In this case, nothing in the problem statement leads us to believe that the vessel undergoes any change in specific kinetic or potential energy during the heating process, so we assume that $ke_2 = ke_1$ and $pe_2 = pe_1$. Also, since the container is rigid, $\mathcal{V} = \text{constant}$ and $d\mathcal{V} = 0$, the mechanical moving boundary work is zero (i.e., $\int p\, d\mathcal{V} = 0$). Since no other work modes are suggested in the problem statement, we assume that ${}_1W_2 = 0$. Applying these results to the previous general energy balance yields the following simplified energy balance equation as the governing equation for this problem:

$$ {}_1Q_2 = m(u_2 - u_1) $$

Write any relevant auxiliary equations. Since we now know the values of two independent properties in each state (T_1, v_1, and T_2, v_2), we can find the values of any other properties in those states by use of thermodynamic tables, charts, or equations of state. In particular, the qualities can be determined from the saturation tables and the auxiliary equations

$$ x_1 = \frac{v_1 - v_{f1}}{v_{fg1}} = \frac{v_1 - v_f(20.0°\text{C})}{v_{fg}(20.0°\text{C})} $$

and

$$ x_2 = \frac{v_2 - v_{f2}}{v_{fg2}} = \frac{v_2 - v_f(95.0°\text{C})}{v_{fg}(95.0°\text{C})} $$

Since we are not given enough information to use the conduction, convection, or radiation heat transfer equations, we must find ${}_1Q_2$ from this energy balance. The values of u_1 and u_2 can be found by using the saturation tables and the following auxiliary equations:

$$ u_1 = u_{f1} + x_1 u_{fg1} = u_f(20.0°\text{C}) + x_1 u_{fg}(20.0°\text{C}) $$

and

$$ u_2 = u_{f2} + x_2 u_{fg2} = u_f(95.0°\text{C}) + x_2 u_{fg}(95.0°\text{C}) $$

Step 6. Algebraically solve for the unknown(s). At this point, we have algebraic equations for all the unknowns and we know where all the numbers in these equations are to be found.

Step 7. Calculate the value(s) of the unknowns. We are now ready to make the calculations. From Table C.1b of *Thermodynamic Tables to accompany Modern Engineering Thermodynamics*, we find that

a. At 20.0°C, $v_{f1} = 0.001002\ \text{m}^3/\text{kg}$, $v_{g1} = 57.79\ \text{m}^3/\text{kg}$, and $v_{fg1} = v_{g1} - v_{f1} = 57.789\ \text{m}^3/\text{kg}$. Also, $u_{f1} = 83.9\ \text{kJ/kg}$, $u_{g1} = 2402.9\ \text{kJ/kg}$, and $u_{fg1} = u_{g1} - u_{f1} = 2319.0\ \text{kJ/kg}$.

b. At 95.0°C, $v_{f2} = 0.00104\ \text{m}^3/\text{kg}$, $v_{g2} = 1.982\ \text{m}^3/\text{kg}$, and $v_{fg2} = v_{g2} - v_{f2} = 1.981\ \text{m}^3/\text{kg}$. Also, $u_{f2} = 397.9\ \text{kJ/kg}$, $u_{g2} = 2500.6\ \text{kJ/kg}$, and $u_{fg2} = u_{g2} - u_{f2} = 2102.7\ \text{kJ/kg}$.

So the unknowns can now be determined as

a. $x_1 = \dfrac{0.500 - 0.001002}{57.789} = 8.63 \times 10^{-3} = 0.863\%$.

b. $x_2 = \dfrac{0.500 - 0.00104}{1.981} = 0.252 = 25.2\%$.

c. $u_1 = 83.9 + (8.63 \times 10^{-3})(2319.0) = 103.9\ \text{kJ/kg}$ and $u_2 = 397.9 + (0.252)(2102.7) = 927.8\ \text{kJ/kg}$, so that ${}_1Q_2 = m(u_2 - u_1) = (2.0\ \text{kg})(927.8 - 103.9\ \text{kJ/kg}) = 1650\ \text{kJ}$.

Step 8. A check of the algebra, calculations, and units shows that they are correct.

[1] *We want three significant figures in the answer, so the data need to be given to three significant figures.*

DO THERMO COMPUTER PROGRAMS GIVE THESE SAME ANSWERS?

Suppose you use a computer program like EES or CATT2 or NIST to solve this problem. Would the answer be the same? Using EES you get (a) $x_1 = 0.864\%$, (b) $x_2 = 25.2\%$, and (c) ${}_1Q_2 = 1650\ \text{kJ}$; using CATT2 you get (a) $x_1 = 0.864\%$, (b) $x_2 = 25.2\%$, and (c) ${}_1Q_2 = 1650\ \text{kJ}$; and using NIST you get (a) $x_1 = 0.864\%$, (b) $x_2 = 25.2\%$, and (c) ${}_1Q_2 = 1650\ \text{kJ}$ (all to three significant figures). So the source of the thermodynamic properties has no significant effect on the answer.

5.3 ELECTRICAL DEVICES

In a vast number of closed and open systems the primary work mode is electrical. We call these systems *electrical devices* and recognize that one of the appropriate auxiliary equations to be used in their analysis is the electrical work or power mode equation introduced earlier. The following example illustrates a typical closed system electrical device problem.

EXAMPLE 5.2

Read the problem statement. An incandescent lightbulb is a simple electrical device. Using the energy rate balance on a lightbulb determine

a. The heat transfer rate of an illuminated 100. W incandescent lightbulb in a room.
b. The rate of change of its internal energy if this bulb were put into a small sealed insulated box.

Solution

Step 1. Identify and sketch the system. We imagine and sketch the system in each case as the entire lightbulb, the glass bulb plus its contents, rather than just the contents as in Example 5.1 (Figure 5.2).

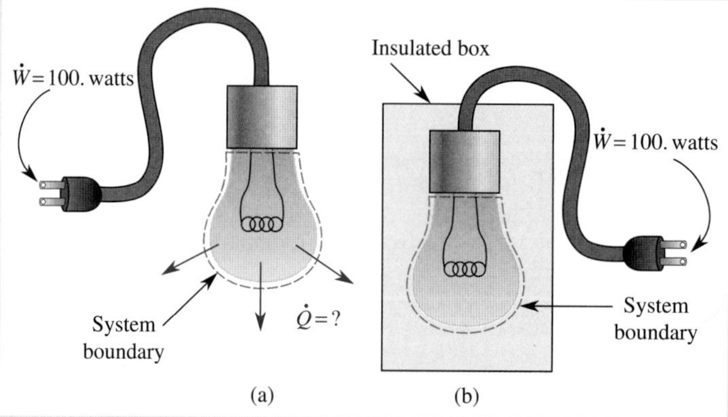

FIGURE 5.2
Example 5.2.

Step 2. Identify the unknowns. The unknowns are (a) $\dot{Q}$ and (b) $\dot{U}$.

Step 3. It is a closed system. This is a closed system for which we are not given specific thermodynamic properties in the problem statement. Presumably, they are not needed in the solution.

Step 4. Identify the process connecting the system states. The following processes occur:

- In part a, the bulb does not change its thermodynamic state, so its properties must remain constant. In particular, the process path (after the bulb has warmed to its operating temperature) is U = constant.
- In part b, the bulb is insulated, so it undergoes an adiabatic (i.e., $\dot{Q} = 0$) process.

Step 5. Write down the basic equations. The only basic equation thus far available for a closed system rate process is Eq. (4.21), the general closed system energy rate balance (ERB) equation:

$$\dot{Q} - \dot{W} = \frac{d}{dt}(mu) + \frac{d}{dt}\left(\frac{mV^2}{2g_c}\right) + \frac{d}{dt}\left(\frac{mgZ}{g_c}\right) = \dot{U} + \dot{KE} + \dot{PE}$$

Since both parts a and b imply that the lightbulb is to be stationary during analysis, we assume $\dot{KE} = \dot{PE} = 0$. This reduces the governing ERB equation for this problem to

$$\dot{Q} - \dot{W} = \dot{U} = \frac{d}{dt}(mu)$$

Write any relevant auxiliary equations. The only relevant auxiliary equation needed here is the recognition that the lightbulb has an electrical work *input* of 100. W, so that

$$\dot{W} = -100.\,\text{W}$$

Step 6. Algebraically solve for the unknown(s). Algebraically solving for the unknowns, we have, for part a)

$$\dot{Q} = \dot{U} + \dot{W}$$

and, for part b,

$$\dot{U} = \dot{Q} - \dot{W}$$

Step 7. Calculate the value(s) of the unknowns. Since we are assuming a constant bulb temperature in part a, U = constant and $\dot{U} = 0$. Then, our calculations give

a. $\dot{Q} = \dot{W} = -100.\ \text{W}$

(the minus sign tells us that heat is *leaving* the system). Thus, all the electrical work put into a lighting system ends up as heat. Architects use the lighting within a building to supply part of the heating requirements of the building. In the second part of this problem, the bulb is inside a small insulated box, so it cannot transport any heat energy through its boundaries. Therefore, $\dot{Q} = 0$ here (the bulb undergoes an adiabatic process), and the reduced energy rate balance yields

b. $\dot{U} = -\dot{W} = 100.\ \text{W}$

Step 8. A **check** of the algebra, calculations, and units shows that they are correct.

Note that the internal energy of the lightbulb must increase at a rate of 100. J/s. This means that its temperature must continually increase. Treating the bulb as a simple incompressible substance, we can write (see Eq. (3.33))

$$\dot{U} = mc\dot{T} = 100.\ \text{W}$$

where m is the mass of the bulb, c is its specific heat, and $\dot{T}$ is the time rate of change of its temperature. So long as $\dot{U}$ is constant and positive, the temperature of the bulb continually increases until the glass or the filament eventually melts.

5.4 POWER PLANTS

An electrical power generating facility is a very complex set of open and closed systems. However, if the entire facility is taken to be the system and the system boundaries are carefully chosen, then it can be modeled as a closed system. We call such systems *power plants*, and a simple thermodynamic analysis can provide important information about their operation, as the next example illustrates.

EXAMPLE 5.3

Read the problem statement. A basic vapor cycle power plant consists of the following four parts:

a. The boiler, where high-pressure vapor is produced.
b. The turbine, where energy is removed from the high-pressure vapor as shaft work.
c. The condenser, where the low-pressure vapor leaving the turbine is condensed into a liquid.
d. The boiler feed pump, which pumps the condensed liquid back into the high-pressure boiler for reheating.

In this power plant (to three significant figures), the boiler receives $950. \times 10^5$ kJ/h from the burning fuel and the condenser rejects $600. \times 10^5$ kJ/h to the environment. The boiler feed pump requires a 23.0 kW input, which it receives directly from the turbine. Assuming that the turbine, pump, and connecting pipes are all insulated, determine the net power of the turbine.

Solution

Step 1. Identify and sketch the system. Sketch the system as the entire power plant (Figure 5.3). If we choose only the turbine as the system, it would be an open system; and we do not wish to deal with open systems until Chapter 6.

Step 2. Identify the unknowns. The unknown here is $(\dot{W}_T)_{\text{net}}$.

Step 3. It is a **closed system.** This is closed system, and no specific information is given to identify the thermodynamic states of the system. Presumably, they are not needed in the solution.

Step 4. Identify the process connecting the system states. We assume a steady state process with no changes in kinetic or potential energy. Then, U, KE, and PE are all constants.

Step 5. Write down the basic equations. The only basic equation applicable here is the general closed system energy rate balance, Eq. (4.21):

$$\dot{Q} - \dot{W} = \dot{U}_0^{\searrow 0} + \dot{KE}^{\searrow 0} + \dot{PE}^{\searrow 0} = 0$$

which reduces to

$$\dot{W}_{\text{net}} = \dot{Q}_{\text{net}} = (\dot{W}_T)_{\text{net}}$$

Write any relevant auxiliary equations. No auxiliary equations are needed here.

(Continued)

EXAMPLE 5.3 (*Continued*)

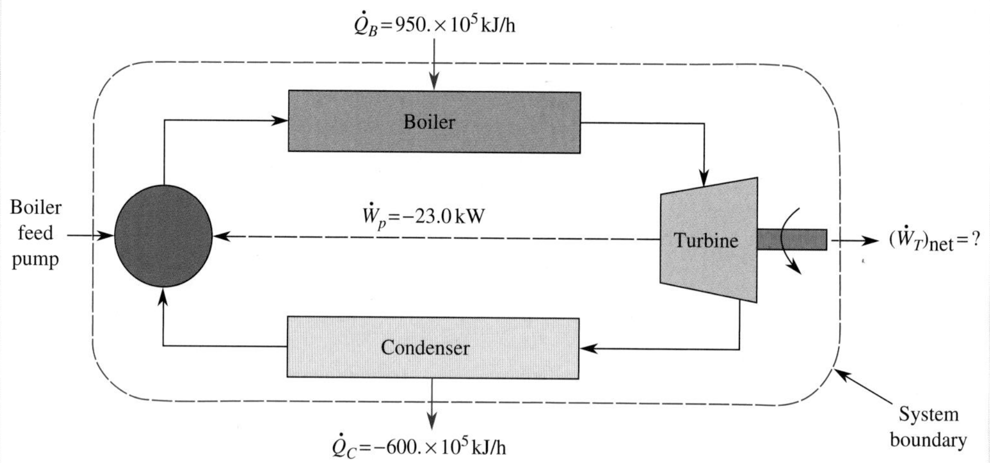

FIGURE 5.3
Example 5.3.

Step 6. Algebraically solve for the unknown(s). Algebraically solving for the net power of the turbine gives

$$(\dot{W}_T)_{net} = \dot{Q}_{net} = (\dot{Q}_{boiler} + \dot{Q}_{condenser})$$

Step 7. Calculate the value(s) of the unknowns. The calculations then give

$$\begin{aligned}(\dot{W}_T)_{net} &= [950. \times 10^5 + (-600. \times 10^5)] \text{ kJ/h} = 350. \times 10^5 \text{ kJ/h}\\ &= (350. \times 10^5 \text{ kJ/h})\left(\frac{1 \text{ h}}{3600 \text{ s}}\right) = 9720 \text{ kW} = 9.72 \text{ MW}\end{aligned}$$

Step 8. A **check** of the algebra, calculations, and units shows that they are correct.

The positive sign tells us that the net power is coming *out* of the turbine. Since the turbine must also power the boiler feed pump,

$$(\dot{W}_T)_{total} = (\dot{W}_T)_{net} + (\dot{W}_T)_{bioler\ feed\ pump} = 9720 + 23 = 9740 \text{ kW (to 3 significant figures)}$$

To simplify the example solutions from this point on, we omit the description of each step in the solution technique. The steps are all there, but now the solutions flow in a more continuous manner.

5.5 INCOMPRESSIBLE LIQUIDS

Perhaps the auxiliary equations most often used in thermodynamic analysis are equations of state. The two most common equations of state are those for ideal gases and incompressible liquids. Since most students are more familiar with ideal gases than they are with incompressible liquids, we chose the next example to illustrate the latter case. Note that this example could also be described as another illustration of the analysis of an electrical device.

EXAMPLE 5.4

A food blender has a cutting/mixing blade driven by a 0.250 horsepower (hp) electric motor. The machine is initially filled with 1.00 quart of water at 60.0°F, 14.7 psia. It is then turned on at full speed for 10.0 min. Assuming the entire machine is insulated and that the mixing takes place at constant pressure, determine the temperature of the water when the machine is turned off.

Solution

First, draw a sketch of the system (Figure 5.4).

The unknown is T_2. The system is closed, and the material is the 1.00 qt of liquid water.

The system states and processes are

State 1	Process	State 2
$p_1 = 14.7$ psia	Process: Constant pressure → mechanical mixing	$p_2 = p_1 = 14.7$ psia
$T_1 = 60.0°F$		

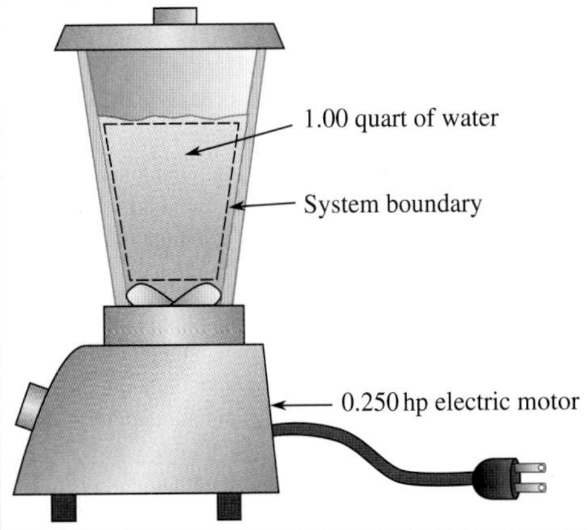

FIGURE 5.4
Example 5.4.

Note that, in this problem, we do not know the values of two independent properties in the second state, nor does the process path give us any information about an additional second-state property. This example illustrates how the energy balance itself can be used to find the value of a state property.

The basic energy balance (EB) equation for this system is

$$_1Q_2 - {_1W_2} = m(u_2 - u_1) + \text{KE}_2 - \text{KE}_1 + \text{PE}_2 - \text{PE}_1$$

Since we are given no information about the kinetic or potential energies of the system, we assume that they do not change during the process under analysis; that is,

$$\text{KE}_2 - \text{KE}_1 = \text{PE}_2 - \text{PE}_1 = 0$$

The auxiliary equations needed here are for the heat and work energy transport modes. They are $_1Q_2 = 0$ (insulated system), and in this case, the work mode is shaft work, but it can be calculated from the definition of power as $_1W_2 = \dot{W}(\Delta t)$, where Δt is the time interval of the process.

Since 14.7 psia is much greater than the saturation pressure at 60.0°F (which is 0.2563 psia), state 1 is seen to be a compressed liquid. We could find u_1 by interpolating the pressure between the saturation and compressed liquid tables at 60.0°F, then use the energy balance to find u_2. With u_2 and p_2, we could presumably find T_2 by again interpolating in the tables. Or else, we could treat the water as a simple incompressible material and use the auxiliary equation for specific heat (Eq. (3.33)), $u_2 - u_1 = c(T_2 - T_1)$. The latter approach is the simplest in this case, so we use it and take the specific heat of water to be 1.00 Btu/(lbm·R)

The mass of 1 qt of water at 60.0°F is given by $m = \rho \forall = \forall / v$, where

$$\forall = (1.00\,\text{qt})\left(\frac{1\,\text{gal}}{4\,\text{qt}}\right)(0.133\,68\,\text{ft}^3/\text{gal}) = 0.0334\,\text{ft}^3$$

and $v = v_f(60.0°\text{F}) = 0.01603\,\text{ft}^3/\text{lbm}$ (from Table C.1a of *Thermodynamic Tables to accompany Modern Engineering Thermodynamics*). Therefore,

$$m = \frac{0.0334\,\text{ft}^3}{0.01603\,\text{ft}^3/\text{lbm}} = 2.08\,\text{lbm}$$

Then, the energy balance gives

$$u_2 - u_1 = c(T_2 - T_1) = \frac{_1Q_2}{m} - \frac{_1W_2}{m}$$

and

$$T_2 = T_1 + \frac{_1Q_2}{mc} - \frac{_1W_2}{mc}$$

$$= 60.0°\text{F} + 0 - \frac{(-0.250\,\text{hp})(10.0\,\text{min})(1\,\text{h}/60\,\text{min})[2545\,\text{Btu}/(\text{hp·h})]}{(2.08\,\text{lbm})[1.00\,\text{Btu}/(\text{lbm·R})]} = 111°\text{F}$$

Exercises

1. Is the result in Example 5.4 of $T_2 = 111°$F an unreasonably high fluid temperature, since a blender of this type is never actually insulated (adiabatic)? In other words, what is the direction of the heat transfer in an uninsulated blender and what effect does this heat transfer have on the temperature T_2? **Answer**: An uninsulated blender has heat transfer from the mixing fluid to the surroundings, producing a lower value of T_2 than that calculated in Example 5.4.
2. What properties other than those used in the solution to Example 5.4 affect the designer's choice of the motor power for the blender? **Answer**: The fluid viscosity and the blender speed.
3. If the fluid temperature in Example 5.4 reaches only 85.0°F instead of 111°F after 10.0 min of operation, determine the power delivered by the motor. **Answer**: $\dot{W}_{\text{motor}} = 0.123$ hp.

5.6 IDEAL GASES

Ideal gas equations are usually quite familiar to engineering students. You see them in chemistry courses, fluid mechanics courses, and of course thermodynamics courses. They are perhaps the most used equations of state ever devised. The next example illustrates the use of the ideal gas equations in conjunction with solving a basic thermodynamics problem.

EXAMPLE 5.5

A new radiation heat transfer sensor consists of a small, closed, rigid, insulated 0.0400 m^3 box containing a 0.0100 m^3 rubber balloon. Initially, the box is evacuated but the balloon contains argon (an ideal gas) at 20.0°C and 0.0100 MPa. When the balloon receives 0.100 kJ of radiation energy through an uninsulated window in the box, it bursts. The resulting pressure change is sensed by a transducer and an alarm is sounded. Determine the pressure and temperature inside the box after the balloon bursts.

Solution

First, draw a sketch of the system (Figure 5.5).

The unknowns here are the final pressure and temperature inside the box, p_2 and T_2. The system is closed and consists of the argon in the balloon. Initially, it is at $T_1 = 20.0°\text{C}$ and $p_1 = 0.0100\,\text{MPa}$. Then, it undergoes a process in which ${}_1Q_2 = 0.100\,\text{kJ}$ of heat is added until the balloon bursts. After it bursts, the argon occupies the volume of the entire box, but its mass has not changed.

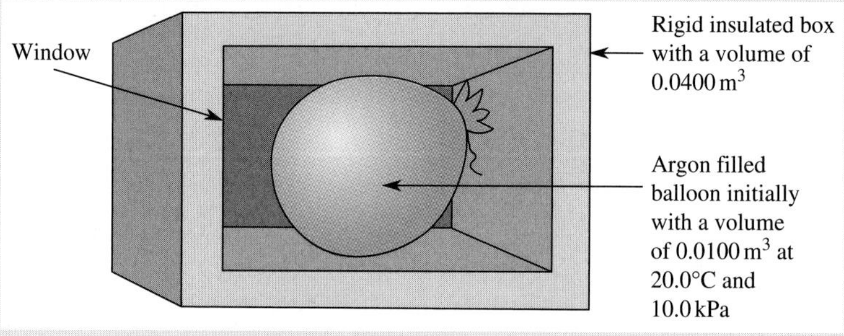

FIGURE 5.5
Example 5.5.

The basic conservation of energy is

$$ {}_1Q_2 - {}_1W_2 = m\left[(u_2 - u_1) + \frac{V_2^2 - V_1^2}{2g_c} + \frac{g}{g_c}(Z_2 - Z_1)\right]_{\text{argon}} $$

and since we are given no indication that the box is moving either before or after the heat transfer process, we set the changes in kinetic and potential energy equal to zero. Three of the relevant auxiliary equations for an ideal gas are (a) the equation of state, $p\mathcal{V} = mRT$, (b) the internal energy equation, Eq. (3.38), $u_2 - u_1 = c_v(T_2 - T_1)$, and (c) the work mode equation for moving boundary work, Eq. (4.26), ${}_1W_2 = \int_1^2 p\,d\mathcal{V}$. In this problem, the argon certainly changes volume as the balloon bursts, but the argon expands into a vacuum (the box was initially evacuated), so it expands against zero resistance. Therefore, the moving boundary work is zero. Putting these results into the basic energy balance equation gives

$$ {}_1Q_2 - 0 = mc_v(T_2 - T_1) + 0 + 0 $$

from which we can solve for T_2 as

$$ T_2 = T_1 + \frac{{}_1Q_2}{mc_v} $$

where m is the mass of argon present and ${}_1Q_2 = 0.100\,\text{kJ}$. Since this is a closed system, the mass can be determined from the ideal gas equation of state:

$$ m = \frac{p_1\mathcal{V}_1}{RT_1} = \frac{p_2\mathcal{V}_2}{RT_2} $$

and since we know that, for the argon, $p_1 = 0.0100\,\text{MPa} = 10.0\,\text{kPa} = 10.0\,\text{kN/m}^2$, $T_1 = 20.0°\text{C}$, and $\forall_1 = 0.0100\,\text{m}^3$, we can find the value of R from Table C.13b of *Thermodynamic Tables to accompany Modern Engineering Thermodynamics* as $R = 0.208\,\text{kJ/kg·K}$; then, determine the mass as

$$m = \frac{(10.0\,\text{kN/m}^2)(0.0100\,\text{m}^3)}{(0.208\,\text{kJ/kg·K})(20.0 + 273.15\,\text{K})} = 0.00164\,\text{kg}$$

This equation for T_2 can be solved to give

$$T_2 = 20.0°\text{C} + \frac{0.100\,\text{kJ}}{(0.00164\,\text{kg})(0.315\,\text{kJ/kg·K})} = 214°\text{C}$$

where we use the value of $c_v = 0.315\,\text{kJ/kg·K}$ for argon found in Table C.13b. Note that, if you express T_1 in °C, then T_2 is in °C, but if you express T_1 in K, then T_2 is in K also. Finally, we can compute the final pressure from the ideal gas equation of state as

$$p_2 = \frac{mRT_2}{\forall_2} = \frac{(0.00164\,\text{kg})(0.208\,\text{kJ/kg·K})(214 + 273.15\,\text{K})}{0.0400\,\text{m}^3} = 4.15\,\text{kPa}$$

where in the final state the argon fills the entire volume of the box.

Exercises

4. Suppose the balloon in Example 5.5 is designed to burst after absorbing 0.200 kJ of radiation heat transfer. What would the final temperature and pressure be inside the box after the balloon burst? **Answer**: $T_2 = 407°\text{C}$ and $p_2 = 5.8\,\text{kPa}$.
5. If air were substituted for the argon in Example 5.5 with no changes in the remaining parameters, what would the final temperature and pressure be inside the box after the balloon burst? **Answer**: $T_2 = 137°\text{C}$, $p_2 = 3.5\,\text{kPa}$.
6. Determine the radiation heat transfer in Example 5.5 that would produce a final pressure of 20.0 kPa in the box after the balloon bursts. What would be the corresponding temperature in the box? **Answers**: ${}_1Q_2 = 1.06\,\text{kJ}$, and $T_2 = 2070°\text{C}$.

5.7 PISTON-CYLINDER DEVICES

One of the oldest pieces of effective technology is the piston in a cylinder apparatus. It was used in early Roman pumps, and its use in the steam engine of the 18th century brought about the Industrial Revolution. It is still commonly used today in piston-type pumps and compressors and in a wide variety of internal and external combustion engines. The following example illustrates its use in a refrigeration process.

EXAMPLE 5.6

You have 0.100 lbm of Refrigerant-134a initially at 180.°F and 100. psia in a cylinder with a movable piston that undergoes the following two-part process. First, the refrigerant is expanded adiabatically to 30.0 psia and 120.°F, then it is compressed isobarically (i.e., at constant pressure) to half its initial volume.

Determine

a. The work transport of energy during the adiabatic expansion.
b. The heat transport of energy during the isobaric compression.
c. The final temperature at the end of the isobaric compression.

Solution (This is an example of a multiple-part solution process)

First, draw a sketch of the system (Figure 5.6).

The unknowns are (a) ${}_1W_2$, (b) ${}_2Q_3$, and (c) T_3. The system is closed, consisting of the R-134a in the cylinder.

Because this is a two-part process, three states are involved, as follows.

State 1	Process: Adiabatic expansion →	State 2	Process: Isobaric compression →	State 3
$p_1 = 100.\,\text{psia}$		$p_2 = 30.0\,\text{psia}$		$p_3 = p_2 = 30.0\,\text{psia}$
$T_1 = 180.°\text{F}$		$T_2 = 120.°\text{F}$		$v_3 = v_1/2$

The basic energy balance equations for these two processes are

$${}_1Q_2 - {}_1W_2 = m(u_2 - u_1) + \text{KE}_2 - \text{KE}_1 + \text{PE}_2 - \text{PE}_1$$

(*Continued*)

EXAMPLE 5.6 *(Continued)*

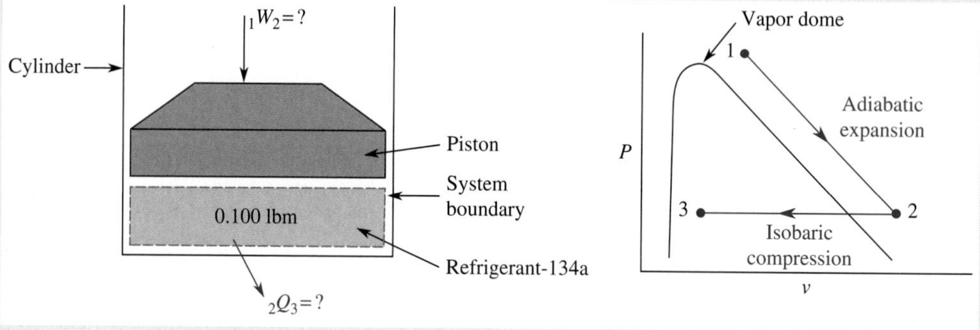

FIGURE 5.6
Example 5.6.

and

$$_2Q_3 - {}_2W_3 = m(u_3 - u_2) + \text{KE}_3 - \text{KE}_2 + \text{PE}_3 - \text{PE}_2$$

Since we are not given any potential or kinetic energy information, we assume that no changes occur in these variables. As auxiliary equations we have $_1Q_2 = 0$ (because the process from state 1 to state 2 is adiabatic); consequently, the resulting energy balance equation for the solution to part a) is

$$_1W_2 = -m(u_2 - u_1)$$

Also, since the process from state 2 to state 3 is isobaric, the work for this process is given by $_2W_3 = m\int_2^3 p\,dv = mp_3(v_3 - v_2)$ and the energy balance for the process from 2 to 3 gives the equation for the solution to part b as

$$_2Q_3 = m(u_3 - u_2) + mp_3(v_3 - v_2)$$

The solution to part c must be determined from the values of the independent properties p_3 and v_3 and the use of the R-134a tables. From Table C.7e of *Thermodynamic Tables to accompany Modern Engineering Thermodynamics*, we find that at $p_1 = 100.$ psia and $T_1 = 180.°\text{F}$, $v_1 = 0.6210\,\text{ft}^3/\text{lbm}$, $u_1 = 125.99\,\text{Btu/lbm}$. Similarly, at $p_2 = 30.0\,\text{psia}$ and $T_2 = 120°\text{F}$, we find from Table C.7e that $v_2 = 1.966\,\text{ft}^3/\text{lbm}$ and $u_2 = 115.47\,\text{Btu/lbm}$.

The final answers are

a.

$$\begin{aligned} _1W_2 &= -m(u_2 - u_1) \\ &= -(0.100\ \text{lbm})(115.47 - 125.99\ \text{Btu/lbm}) = 1.05\ \text{Btu} \end{aligned}$$

b. $_2Q_3 = m(u_3 - u_2) + mp_3(v_3 - v_2)$, where we already have numerical values for m, u_2, p_3, and v_2, but we also need values for v_3 and u_3. From the problem statement for the process from 2 to 3, we find that

$$v_3 = \frac{v_1}{2} = \frac{0.6210}{2} = 0.3105\,\text{ft}^3/\text{lbm}$$

From Table C.7b, we find that, since state 3 is at 30.0 psia, $v_f(30.0\,\text{psia}) < v_3 < v_g\,(30.0\ \text{psia})$, state 3 is a mixture of liquid plus vapor. It therefore has a quality given by

$$x_3 = \frac{v_3 - v_{f3}}{v_{fg3}}$$

Table C.7b at 30.0 psia gives $v_{f3} = 0.01209\,\text{ft}^3/\text{lbm}$, $v_{g3} = 1.5408\,\text{ft}^3/\text{lbm}$, $u_{f3} = 16.24\,\text{Btu/lbm}$, and $u_{g3} = 95.40\,\text{Btu/lbm}$. We can now calculate the quality at state 3 as

$$x_3 = \frac{v_3 - v_{f3}}{v_{g3} - v_{f3}} = \frac{0.3105\,\text{ft}^3/\text{lbm} - 0.01209\,\text{ft}^3/\text{lbm}}{1.5408\,\text{ft}^3/\text{lbm} - 0.01209\,\text{ft}^3/\text{lbm}} = 0.195 = 19.5\%$$

Finally, we get the other state 3 properties as

$$\begin{aligned} u_3 &= u_{f3} + x_3(u_{g3} - u_{f3}) \\ &= 16.24\,\text{Btu/lbm} + 0.1952(95.40 - 16.24\,\text{Btu/lbm}) = 31.7\,\text{Btu/lbm} \end{aligned}$$

then

$$
\begin{aligned}
{}_2Q_3 &= m(u_3 - u_2) + mp_3(v_3 - v_2) \\
&= (0.100\,\text{lbm})(31.7 - 115.47\,\text{Btu/lbm}) \\
&\quad + (0.100\,\text{lbm})(30.0\,\text{lbf/in.}^2)(144\,\text{in.}^2/\text{ft}^2)(0.3105 - 1.9662\,\text{ft}^3/\text{lbm})\left(\frac{1\,\text{Btu}}{778.17\,\text{ft}\cdot\text{lbf}}\right) \\
&= -9.31\,\text{Btu}
\end{aligned}
$$

c. Since state 3 is saturated (a mixture of liquid and vapor), T_3 must be equal to the saturation temperature at 30.0 psia, which, from Table C.7b, is $T_3 = 15.38°\text{F}$.

WHY AREN'T THE VALUES OF *u* AND *h* IN THE TABLES THE SAME AS THOSE YOU GET FROM THERMO COMPUTER PROGRAMS?

To develop a table or thermodynamic program of values of properties like internal energy and enthalpy, a zero reference point for these properties has to be chosen. This is called a *reference state*, and whoever develops the table or computer program is free to chose his or her own reference state. So the values of u and h for a particular material may not be the same from table to table or the same as those given by computer programs. However, it turns out that this is not important because the first law uses only the *differences*, $u_2 - u_1$ or $h_2 - h_1$, and when you subtract two values, the reference state used (whatever it was) cancels out. This is like calculating $T_2 - T_1$. You can use either °F or R (or °C or K) and get the same answer.

Using EES, the answers to Example 5.6 are (a) ${}_1W_2 = -1.02\,\text{Btu}$, (b) ${}_2Q_3 = -9.29\,\text{Btu}$, and (c) $T_3 = 15.37°\text{F}$. While these are not exactly the same as those given in Example 5.6, they differ by less than 3%, which is quite acceptable for this calculation.

Exercises

7. Find the work transport of energy in part a of Example 5.6 if the working fluid is air (an ideal gas) instead of Refrigerant-134a. **Answer:** $({}_1W_2)_a = -1.03\,\text{Btu}$.
8. Find the final temperature in part c of Example 5.6 if the working fluid is an ideal gas. **Answer:** $T_3 = 96\,\text{R}$.
9. Find the heat transport of energy in part b of Example 5.6 if we set $v_3 = v_g(30.0\,\text{psia})$ instead of $v_1/2$. **Answer:** $({}_2Q_3)_b = -2.20\,\text{Btu}$.

5.8 CLOSED SYSTEM UNSTEADY STATE PROCESSES

One of the most difficult thermodynamic processes to analyze is an unsteady state process. This is largely due to the fact that there are many more unknowns in these processes. In addition, they usually involve integrating the rate form of the basic equations, so some knowledge of the solution techniques for ordinary differential equations is essential before a complete thermodynamic analysis can be carried out. The following example illustrates this type of problem.

EXAMPLE 5.7

A microwave antenna for a space station consists of a 0.100 m diameter rigid, hollow, steel sphere of negligible wall thickness. During its fabrication the sphere undergoes a heat-treating operation in which it is initially filled with helium at 0.140 MPa and 200.°C, then it is plunged into cold water at 15.0°C for exactly 5.00 seconds. The convective heat transfer coefficient of the sphere in the water is $3.50\,\text{W/(m}^2\cdot\text{K)}$. Neglecting any changes in kinetic or potential energy and assuming the helium behaves as an ideal gas, determine

a. The final temperature of the helium.
b. The change in total internal energy of the helium.

Solution

First, draw a sketch of the system (Figure 5.7).

(*Continued*)

EXAMPLE 5.7 *(Continued)*

The unknowns are, after 5 seconds have passed, (a) T_2 = ? and (b) $U_2 - U_1$ = ? The system is closed, and the material is the helium gas in the sphere.

The basic equations here are the closed system energy balance (EB)

$$ {}_1Q_2 - {}_1W_2 = U_2 - U_1 + \underbrace{KE_2 - KE_1 + PE_2 - PE_1}_{\text{Neglect}} $$

and neglecting $\dot{KE}$ and $\dot{PE}$, the closed system energy rate balance (ERB) is

$$ \dot{Q} - \dot{W} = \dot{U} $$

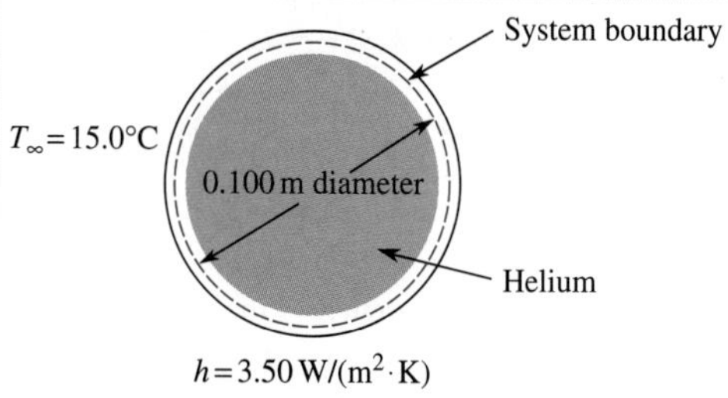

FIGURE 5.7
Example 5.7.

The auxiliary equations needed here are

1. The mechanical work mode ${}_1W_2 = \dot{W} = 0$ (a rigid hollow sphere).
2. The convective heat transfer mode $\dot{Q} = -hA(T_s - T_\infty)$ (the negative sign is necessary here because the helium loses heat).
3. Assuming the helium to be an ideal gas, the internal energy can be represented as $du = c_v(dT)$ and $\dot{U} = m\dot{u} = mc_v\dot{T}_s$. Putting these results into the energy rate balance equation gives (assuming T_∞ = constant)

$$ \dot{Q} = -hA(T_s - T_\infty) = \dot{U} = mc_v\dot{T}_s $$

or

$$ \dot{T}_s = \frac{dT_s}{dt} = \frac{d(T_s - T_\infty)}{dt} = -\frac{hA}{mc_v}(T_s - T_\infty) $$

This is a first-order ordinary differential equation with the initial condition $T_s = T_1$ when $t = 0$. Its solution is

$$ T_s = (T_1 - T_\infty)\, exp\left(-\frac{hAt}{mc_v}\right) + T_\infty $$

where $T_2 = T_s$ at $t = 5.00$ s. The remaining part of the solution is given by the energy balance equation as

$$ U_2 - U_1 = mc_v(T_2 - T_1) $$

The auxiliary equations and calculations are

$$ h = 3.50.\ \text{W}/(\text{m}^2\cdot\text{K}) $$
$$ T_\infty = 15.0°\text{F} $$
$$ \forall = \frac{\pi}{6}(D)^3 = \frac{\pi}{6}(0.100\ \text{m})^3 = 5.24\times10^{-4}\,\text{m}^3 $$
$$ A = \pi D^2 = \pi(0.100\ \text{m})^2 = 0.0314\ \text{m}^2 \text{ (the surface area of a sphere)} $$
$$ c_v = 3.123\ \text{kJ}/(\text{kg}\cdot\text{K}) \text{ (from Table C.13b)} $$

and from the ideal gas equation of state for helium, $m = P\forall/RT$, Table C.13b gives $R = 2.077$ kJ/(kg·K). Then,

$$ m = \frac{(140.\ \text{kN/m}^2)(5.24\times10^{-4}\,\text{m}^3)}{[2.077\ \text{kJ}/(\text{kg}\cdot\text{K})][(200.+273.15)\text{K}]} = 7.46\times10^{-5}\ \text{kg} $$

$$ \frac{hA}{mc_v} = \frac{[3.50\ \text{W}/(\text{m}^2\cdot\text{K})](0.0314\ \text{m}^2)}{(7.46\times10^{-5}\ \text{kg})[3.123\ \text{kJ}/(\text{kg}\cdot\text{K})]} = 0.472\ \text{s}^{-1} $$

and at $t = 5.00$ s,

$$ \begin{aligned} T_2 &= (T_1 - T_\infty)\exp(-hAt/mc_v) + T_\infty \\ &= [(200. - 15.0)°\text{C}]\exp[-(0.472\ \text{s}^{-1})(5.00\ \text{s})] + 15.0°\text{C} \\ &= 32.5°\text{C} \end{aligned} $$

Then,

$$ \begin{aligned} U_2 - U_1 &= mc_v(T_2 - T_1) \\ &= (7.46\times10^{-5}\ \text{kg})[3.123\ \text{kJ}/(\text{kg}\cdot\text{K})][(32.5 - 200.)\ \text{K}] \\ &= -0.039\ \text{kJ} \end{aligned} $$

Exercises

10. Determine the final temperature of the helium in Example 5.7 when the sphere is left in the cold water for 10.0 s rather than 5.00 s. **Answer**: $T_2 = 16.6°\text{C}$.
11. Determine the final temperature in the sphere in Example 5.7 after 5.00 s of immersion in the cold water if the sphere is filled with air instead of helium (everything else remains the same). **Answer**: $T_2 = 60°\text{C}$.
12. Describe in words how the solution to Example 5.7 changes if the material in the sphere is an incompressible liquid instead of an ideal gas. **Answer**: Change c_v to c and the rest of the solution is the same.

5.9 THE EXPLOSIVE ENERGY OF PRESSURE VESSELS

The explosion of a pressure vessel, such as a steam boiler, is an example of a very unsteady state process. But, since it is such an important topic from a safety point of view, it is treated as a separate subject here.

Not many engineers realize just how dangerous a high-pressure gas or vapor can be. The explosive energy of a pressure vessel is defined to be its capacity to do work adiabatically on its surroundings. Consider a pressure vessel whose initial state is just before the explosion and whose final state is immediately after all the debris has come to rest and thermodynamic equilibrium has been reestablished. The explosion process is considered to be adiabatic with no net change in system kinetic or potential energies. The explosive energy can be determined from the closed system energy balance as

$$\text{Explosive energy} = -{}_1W_2 = m(u_1 - u_2)$$

where we have introduced the minus sign because we want the work done by the system on the environment not that done on the system. The explosive energy per initial unit volume of the pressure vessel is defined to be Γ, where

$$\Gamma = m(u_1 - u_2)/\mathcal{V}_1 = (u_1 - u_2)/v_1 \tag{5.1}$$

If the pressure vessel contains an ideal gas with constant specific heats, then Eqs. (3.38) and (3.35b) can be used to give the explosive energy per unit volume as

$$(\Gamma)_{\text{ideal gas}} = c_v(T_1 - T_2)/(RT_1/p_1) = p_1(1 - T_2/T_1)/(k-1) \tag{5.2}$$

where $R/c_v = k - 1$. If, on the other hand, the pressure vessel contains an incompressible liquid, then Eq. (3.33) can be used to give

$$(\Gamma)_{\text{incompressible liquid}} = c(T_1 - T_2)/v = \rho c(T_1 - T_2) \tag{5.3}$$

A liquid that does not change phase during decompression undergoes the process very nearly isothermally, so $T_2 = T_1$ and its explosive energy is zero. Therefore, the explosive energy of high-pressure liquids is very slight in comparison with gases and vapors at the same pressure, and this is why liquids are often used to hydrostatically test pressure vessels to failure.

EXAMPLE 5.8

On March 10, 1905, a catastrophic boiler explosion occurred in a shoe factory in Brockton, Massachusetts, that killed 58 and injured 150 people (Figure 5.8).[2] This and similar explosions brought about the development of the ASME Boiler and Pressure Vessel Code in 1915. Suppose that the Brockton shoe factory had a 250. ft^3 boiler and right before the explosion it contained superheated steam at 600. psia and 800.°F. After the explosion, the steam quickly condensed into saturated liquid water at 70.0°F.

a. Determine the explosive energy per unit volume of superheated steam.
b. How many 1 lbm sticks of TNT would it take to equal the explosion of the boiler? The explosive energy per unit mass of TNT is 1400. Btu/lbm.

FIGURE 5.8
Example 5.8, Brockton Shoe Factory.

Solution

First, draw a sketch of the system.

(Continued)

EXAMPLE 5.8 *(Continued)*

The unknowns are explosive energy per unit volume of superheated steam and the number of 1 lbm sticks of TNT it would it take to equal the explosion of the boiler. The system is closed, and the material is steam.

a. From the superheated steam table, Table C.3a of *Thermodynamic Tables to accompany Modern Engineering Thermodynamics*, we find that at 600. psia and 800.°F, $u_1 = 1275.4$ Btu/lbm and $v_1 = 1.190\ \text{ft}^3/\text{lbm}$; from the saturated steam table, Table C.1a, we have $u_2 = u_f(70°\text{F}) = 38.1$ Btu/lbm. So Eq. (5.1) gives

$$\Gamma = (1275.4 - 38.1\ \text{Btu/lbm})/(1.190\text{ft}^3/\text{lbm}) = 1039.7\ \text{Btu/ft}^3$$

b. For a 250. ft^3 boiler, the explosive energy is then $(1039.7\ \text{Btu/ft}^3)(250.\ \text{ft}^3) = 2.60 \times 10^5$ Btu. Therefore, it would take $(2.60 \times 10^5\ \text{Btu})/(1400.\ \text{Btu/lbm}) = 186$ one-pound sticks of TNT to match the boiler explosion.

Exercises

13. Using saturated liquid water at 70.0°F as the postexplosion state, determine the explosive energy per unit volume of superheated steam at (a) 100. psia and 1000.°F, (b) 1000. psia and 1000.°F, (c) 80.0 MPa and 1000.°C. **Answer:** (a) $\Gamma = 154.0\ \text{Btu/ft}^3$, (b) $\Gamma = 1580\ \text{Btu/ft}^3$, (c) $\Gamma = 5.32 \times 10^5\ \text{kJ/m}^3$.

14. Determine the explosive energy of the boiler in Example 5.8 in lbm of TNT if it had a volume of 1500. ft^3. **Answer:** 1114 lbm TNT.

15. How could Eq. (5.2) lead to an incorrect conclusion regarding the explosive energy (and danger) of a compressed ideal gas? **Answer:** If the initial and final temperatures of the ideal gas before and after the explosion are taken to be the same, then $\Gamma = 0$ and you would conclude that a compressed ideal gas is not dangerous. However, the final state of the explosive process must be the state that occurs *immediately* after the debris from the explosion has come to rest. If we model the explosion as a reversible and adiabatic process, we see in Chapter 7 that $T_2 = T_1(p_2/p_1)^{(k-1)/k}$, where k is the specific heat ratio c_p/c_v, then Eq. (5.2) becomes

$$\Gamma_{\text{ideal gas reversible \& adiabatic}} = \frac{P_1}{k-1}\left[1 - \left(\frac{p_2}{p_1}\right)^{\frac{k-1}{k}}\right]$$

and the explosive danger of a compressed ideal gas becomes more apparent.

[2] *At about 8:00 AM there were around 400 employees at the R. B. Grover & Company shoe factory in Campello, when the boiler exploded, shot through the roof, and caused the building to collapse. The boiler traveled several hundred feet, damaging a number of buildings and coming to rest in the wall of a house. Thirty-six of the victims were never identified and were buried in a common grave, where a monument to the victims was later erected by the city.*

SUMMARY

In this chapter, we investigate a series of closed system examples and carry out a first law analysis of them using the energy balance or the energy rate balance. The primary purpose of these examples is to illustrate the material presented in Chapter 4.

The only new equations introduced in this chapter are those associated with the explosive energy per unit initial volume of pressure vessels, Γ, where in general

$$\Gamma = \frac{u_2 - u_1}{v_1} \tag{5.1}$$

and for ideal gases

$$\Gamma_{\text{ideal gas}} = \frac{p_1}{k-1}\left(1 - \frac{T_2}{T_1}\right) \tag{5.2}$$

or for incompressible liquids

$$\Gamma_{\text{incompressible liquid}} = \frac{c(T_1 - T_2)}{v} \tag{5.3}$$

Problems (* indicates problems in SI units)

1.* One kg of liquid water at 20.0°C is poured from a height of 10.0 m directly onto the floor. After a short time, the specific internal energy of the water on the floor has returned to its initial value before it was poured.

   a. What total heat transport of energy occurred during this process (ignore evaporation effects)?

   b. In which direction was this heat transport of energy, into or out of the water?

2. Determine the direction and amount of heat transfer required to raise the temperature of the contents of a rigid, sealed, subterranean silicon sphere containing 10.0 lbm of saturated water vapor from 280. to 1000.°F, 100. psia. Fill in the following table (with correct units) and show all calculations. Unknown: ${}_1Q_2$ = ?

State 1	Process = ? →	State 2
$x_1 = 1.00$		$p_2 = 100.$ psia
$T_1 = 208°F$		$T_2 = 1000.°F$
$p_1 = ?$		$x_2 = ?$
$u_1 = ?$		$u_2 = ?$

3.* Determine the direction and amount of heat transfer that occurs as 3.00 kg of superheated blood (essentially steam) expands isothermally from 800.°C, 80.0 MPa to 0.100 MPa doing 500. kJ of work in the process (Figure 5.9). Fill in the following table (with correct units) and show all calculations. Unknown: ${}_1Q_2$ = ?

State 1	Process = ? → ${}_1W_2 = 500$ kJ	State 2
$p_1 = 80.0$ MPa		$p_2 = 0.100$ MPa
$T_1 = 800.°C$		$T_2 = ?$
$v_1 = ?$		$v_2 = ?$
$u_1 = ?$		$u_2 = ?$

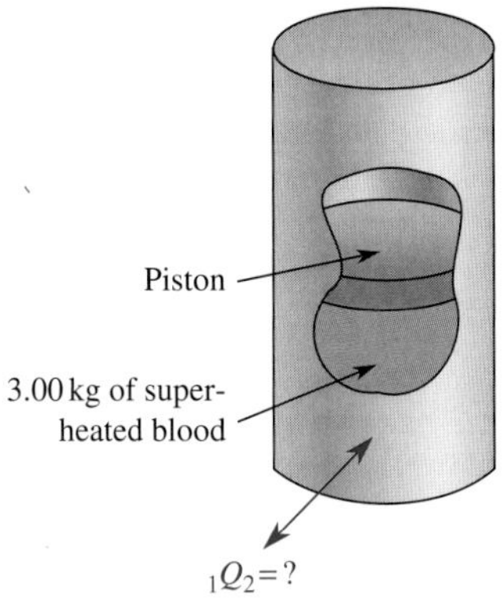

FIGURE 5.9
Problem 3.

4. A rigid, sealed 1936 Ford coupe contains an equilibrium water liquid-vapor mixture with a quality of 8.8333% at 3.00 psia. After 500. Btu of heat energy are added to the coupe, the contents becomes a saturated vapor. What is the total mass of water in the coupe? Fill in the following table (with correct units) and show all calculations. Unknown: m = ?

State 1	Process = ? → ${}_1Q_2 = 500$ Btu	State 2
$p_1 = 3.00$ psia		$p_2 = ?$
$x_1 = 0.088333$		$x_2 = 1.00$
$v_1 = ?$		$v_2 = ?$
$T_1 = ?$		$T_2 = ?$
$u_1 = ?$		$u_2 = ?$

5.* The makers of a new breakfast cereal have a process in which a rigid, sealed vessel contains 1.00 kg of saturated water vapor at 10.0 MPa. Energy is removed from the vessel as heat transfer until the pressure drops to 0.100 MPa. Determine the heat transfer, state its direction, fill in the following table (with correct units), and show all calculations. Unknown ${}_1Q_2$ = ?

State 1	Process = ? →	State 2
$x_1 = 1.0$		$T_2 = ?$
$P_1 = 10.0$ MPa		$P_2 = 0.100$ MPa
$v_1 = ?$		$v_2 = ?$
$T_1 = ?$		$x_2 = ?$
$u_1 = ?$		$u_2 = ?$

6. A rigid sealed fossilized goat's bladder contains water in a liquid-vapor equilibrium at 70.0°F. After 300. Btu of heat are added to the bladder, its contents are converted into a saturated vapor with a specific volume of 50.2 ft^3/lbm. What is the total mass of water in the bladder?

7.* A rigid, sealed 1939 Buick having an internal volume of 6.00 m^3 is filled with steam at 0.70 MPa and 300.°C. Heat is then transferred from the steam until it has a quality of 100.% while the contents of the Buick are stirred with a blade requiring 10.0 W·h of work input. Determine (a) the final pressure in the vessel and (b) the total heat transfer.

8.* Exactly 1.73 kg of water vapor is contained in a piston-cylinder assembly at a pressure of 1.00 MPa and temperature 600.°C. The vapor is isothermally compressed to 80.0 MPa. Determine the sum of the work and heat energy transports in this process.

9. One pound of Refrigerant-134a is put into a piston-cylinder assembly at an initial pressure and temperature of 200. psia and 200.°F. The R-134a is then slowly heated at constant pressure until the temperature reaches 300.°F. Determine the work done on or by the system and the heat transferred to or from the system.

10. The pressure in an isochoric automobile tire increases from 28.0 psia at 70.0°F to 35.0 psia on a trip during hot weather. Assume the air behaves as an ideal gas. (a) What is the air temperature inside the tire at the end of the trip. (b) How much heat is absorbed per unit mass of air in the tire during the trip?

11.* A small room 5.0 × 5.0 × 3.0 m high contains air at 20.0°C and 0.101 MPa. It is the camera stage of a television broadcasting studio and contains many bright lights for illumination. The room is closed, sealed, and insulated to isolate the performers from outside distractions. Assuming air is an ideal gas, find the temperature and pressure in the room 1.00 h after eight 1000. W lights are turned on. Assume there is no ventilation or air conditioning and ignore the effect of people in the room.

12. A room heating system uses steam radiators to heat the room air. A radiator that has a volume of 3.00 ft^3 is filled with saturated vapor at a pressure of 15.3 psig and the inlet and exit valves are closed. How much energy is transferred to the room air as heat at a time when the pressure in the radiator reaches 3.30 psig.

Assume the room air is at 14.7 psia and 70.0°F during the entire process.

13.* The human body under the stress of exercise can release 230. W as heat. Assume the human body to be a closed system and neglect any work or change in kinetic energy. Determine the rate of change in internal energy of the human body as a 68.0 kg person runs at a constant velocity up a staircase having a vertical height of 15.0 m in 60.0 s.

14.* Exactly 14.0 kg of herpes duplex virus scum is compressed from a volume of 4.50 to 1.50 m^3 in a process where p is in N/m^2 is given by $p = 60.0/\mathcal{V} + 30.0$, when $\mathcal{V}$ is in m^3. During the compression process the virus gives off 20.0 J of heat and turns a putrid yellow in color. Determine the change in the specific internal energy of the virus for this process.

15.* Exactly 3.70 kg of nitrogen gas at exactly 0°C and 0.100 MPa is put into a cylinder with a piston and compressed in a process defined by $p\mathcal{V}^2$ = constant. When the final pressure in the cylinder reaches 10.0 MPa, and assuming ideal gas behavior, determine (a) the amount of work done on the nitrogen by the piston and (b) the final temperature of the nitrogen.

16. Heat is transferred to 0.100 lbm of air contained in a frictionless piston-cylinder apparatus until its volume expands from an initial value of 1.00 ft^3 to a final value of 1.50 ft^3. Calculate the work transport of energy and the heat transfer when the system is the air in the cylinder. The initial temperature of the air is 70.0°F. Consider air to be an ideal gas.

17. Exactly 0.100 lbm of air (an ideal gas) initially at 50.0 psia and 100.°F in a cylinder with a movable piston undergoes the following two-part process. First, the air is expanded adiabatically to 30.0 psia and 24.0°F, then it is compressed isobarically (i.e., at constant pressure) to half its initial volume. Determine

a. The final temperature at the end of the isobaric compression.
b. The work produced during the adiabatic expansion.
c. The heat transfer during the isobaric compression.

18.* A 1000. kg battery powered adiabatic electric vehicle has a fully charged battery containing 20.0 MJ of stored energy. If it requires 12.0 kW of power to keep it moving at a constant velocity on a horizontal road, determine how long the vehicle will operate before its battery is fully discharged.

19. How many watt hours of electricity are needed to heat the contents of a sealed, rigid, insulated chamber pot containing 0.300 lbm of water from 50.0°F with a quality of 1.00% to a saturated vapor. The chamber pot has an internal electrical resistance heater with a power cord that plugs into a standard 110. V ac outlet.

20.* A small, sealed, rigid container holding 0.500 kg of water is heated in a microwave oven drawing 1600. W at 2460 MHz. The oven's timer is set for exactly 1 min. The initial thermodynamic state of the water is 20.0°C at 1.00 atm. After the 1 min heating period, determine (a) the water's work transport of energy, (b) the water's heat transport of energy, (c) the change in specific internal energy of the water, and (d) the final temperature and pressure assuming the liquid water to be an incompressible liquid with a specific heat of 4.50 kJ/(kg·K).

21.* 30.5 kg of H_2O contained in a 1.00 m^3 rigid tank are at an initial pressure of 10.0 MPa. The contents of the tank are cooled at constant volume until a final pressure of 2.00 MPa is reached. Determine the final temperature, the final value of the specific internal energy, and the process heat transfer.

22. A small rigid tank 1.00 ft^3 in volume contains saturated water vapor at 300.°F. An initially evacuated rigid container 3.4549 ft^3 in volume is then attached to the first tank and the interconnecting valve is opened. The combined system is then brought to equilibrium at 300.°F by an appropriate heat transport of energy. Determine the final pressure in the system and the required heat transfer.

23. A pressure vessel that has a volume of 0.200 ft^3 is filled with saturated liquid Refrigerant-22 at 70.0°F. An evacuated container 4.00 ft^3 in volume is attached to the vessel and the interconnecting valve is opened. The combined system is then brought to equilibrium at 70.0°F Calculate the heat transport of energy to (or from) the system.

24.* A mixture of hydrazine and cow manure happens to have the same thermodynamic properties as pure water. A secret process requires that this mixture be vaporized then injected into light bulbs. Determine the work and heat transport of energy that occurs when 1.30 kg of this mixture is isothermally converted from a saturated liquid to a saturated vapor at 40.0°C.

25. A lead bullet weighing 0.0200 lbf and traveling horizontally at 3000. ft/s is suddenly stopped by a perfectly rigid object that does not deform during the impact. Find the temperature rise of the bullet assuming the impact occurs so rapidly that the impact process can be considered to be adiabatic. For lead, use $\Delta u = 0.0130(\Delta T)$ in Btu/lbm, where T is in °F or R.

26. As a bullet travels down the barrel of a pistol, the pressure from the burning propellant behind it increases linearly with the volume $\mathcal{V}$ displaced by the bullet as $p = \mathcal{V} \times 10^3$ in psia, where $\mathcal{V}$ is in $in.^3$. The total volume of the barrel is $\pi R^2 L$, where R is the radius of the bore and L is its total length. Determine the velocity of the bullet at the end of the barrel if it travels horizontally and adiabatically down the barrel without changing its internal energy and with no friction.

Data

Barrel length = 6.00 in
Barrel diameter = 0.380 in
Bullet mass = 5.00 g

27. A rubber band weighing 1.00×10^{-3} lbf that obeys Hooke's law of elasticity is stretched horizontally and adiabatically from an initial length of 3.00 to 4.00 in.

a. Determine the change in total internal energy of the rubber band when it is stretched, if its elastic modulus is 1.00×10^3 lbf/in^2 and its cross-sectional area remains approximately constant at 7.80×10^{-3} in^2.
b. If the stretched rubber band is suddenly released horizontally and adiabatically, determine its final velocity neglecting air friction and any height change during its flight.

28. A thin glass sphere 0.0250 ft^3 in volume is completely filled with 1.00 lbm of saturated liquid nitrogen. The glass sphere is sealed inside a large rigid, evacuated, insulated container whose volume is 10.0 ft^3. What are the final pressure and quality (if any) inside the larger container if the glass sphere breaks.

29. 1.00 ft^3 of saturated liquid water at 14.7 psia is poured into an initially evacuated, rigid, insulated vessel whose volume is 100. ft^3. Inside the vessel is an electric heater that draws an effective 10.0 A at an effective 110. V. Once this heater is turned on, how long will it take the contents of the vessel to reach 40.0 psia?

30.* A rigid vessel having a volume of 3.00 m^3 initially contains steam at 0.400 MPa and a quality of 40.2%. If 23.79 MJ of heat is added to the steam, determine its final pressure and temperature.

31. A thermoelectric generator consists of a series of semiconductor elements heated on one side and cooled on the other. It is a type of *thermal engine*, except that the output is electrical rather than mechanical work. Electric direct current output is produced as a result of an input heat transport of energy. In a particular experiment, the steady state direct current is measured to be 0.500 A and the potential across the unit is 0.800 V. The heat input to the hot side is 5.50 W. Determine the heat transfer rate from the cold side and the energy conversion efficiency of this device.

32. A rigid, sealed pressure cooker has a volume of 0.700 ft^3 and contains 0.1279 lbm of water (liquid plus vapor) in equilibrium at 14.7 psia. The pressure cooker is then slowly heated until all the water inside becomes a vapor.
   a. What are the internal temperature and pressure when the last bit of liquid vaporizes.
   b. How much heat transfer is required (in Btu) to vaporize all the water.
   c. Sketch the process path on a p-v diagram for water.

33.* A pressure cooker whose volume is 0.300 m^3 contains 2.00 kg of water. It is placed on a heating element of an electric stove that continuously draws 220. V (effective) and 0.500 A. Assuming all the heat generated in the element goes into the pressure cooker, determine the rate of heat loss from the pressure cooker to the environment when it has reached steady state conditions (i.e., $(dE/dt)_{\text{system}} = 0$).

34. A teakettle initially contains 5.00 lbm of water (liquid plus vapor) and has a total volume of 0.500 ft^3. The atmospheric pressure (and thus the initial pressure in the teakettle) is 14.7 psia. The kettle has a "pop-off" valve that keeps the water vapor in the kettle until its pressure reaches 5.30 psig. At this internal pressure, the valve opens and allows the vapor to escape into the atmosphere in such a way as to maintain the internal pressure constant. The kettle is heated on a stove until all the remaining water inside becomes saturated vapor.
   a. Take the water that remains in the kettle at the final state as a system. Sketch the p-v diagram for this system for the process just described.
   b. List two intensive properties at each of the states shown in Table 5.1.
   c. Determine the mass of water in the kettle when it reaches the final state.
   d. Determine the work done by the escaping steam in pushing aside the atmosphere.

Table 5.1 Problem 34, Part b

Initial State	State when Valve Opens	Final State
1.	1.	1.
2.	2.	2.

35. A small electrically heated steam boiler with a total volume of 10.0 ft^3 can be considered to be a perfectly rigid, insulated vessel with three valves: an inlet valve, an exit valve, and a safety relief valve. During a test, the boiler operator closed both the inlet and exit valves while leaving the heater on. The safety relief valve is to stay closed until a pressure of 160. psia is reached. If there are 4.477 lbm of water in the boiler and the pressure is 100. psia at the time the valves are closed, how much energy will have been transferred to the water as heat when the safety relief valve first opens?

36. Helium contained in a cylinder fitted with a piston expands according to the relation $p\mathcal{V}^{1.5} = \text{constant}$. The initial volume of the helium is 2.00 ft^3, the initial pressure is 70.0 psia, and the initial temperature is 400. R. After expansion, the pressure is 30.0 psia. The specific heat of the helium is given by the relation $c_v = a + bT$, where $a = 0.400\,\text{Btu/(lbm}\cdot\text{R)}$ and $b = 1.00 \times 10^{-3}\,\text{Btu/(lbm}\cdot\text{R}^2)$. Determine the heat transfer and indicate its direction.

37.* A student weighs 1333 N and wishes to lose weight. The student climbs with a constant velocity to the top of a staircase with a vertical height of 250. m.
   a. Assuming the student is a closed adiabatic system (which is really not a very accurate assumption here), determine the change in total internal energy of the student.
   b. How much weight would the student lose if his total internal energy change were the result of the conversion of body fat, where 1.00 kg of body fat contains 32,300 kJ of energy?
   c. The student decides to take more drastic action and designs a machine that squashes him from an initial volume of 0.300 m^3 to a final volume of 0.100 m^3 according to the relation $p\mathcal{V}^{0.5} = \text{constant}$. If the student's initial internal pressure is 0.110 MPa, determine his final internal pressure and the work done in squashing the student.

38. A Newcomen steam engine, built in 1720, pumped water from a coal mine by condensing water vapor in a piston-cylinder device. If the piston had a cross-sectional area of 1.50 ft^2, determine
   a. The work done by the atmosphere (at 14.7 psia) on the piston in the cylinder when the water vapor volume is decreased by 6.00 ft^3.
   b. The work done in lifting the water from the mine for the same process as part a.

39.* Determine the surface temperature of an automobile engine, initially at 90.0°C, 4 h after it has stopped running on a winter day, when the air temperature is –30.0°C and the convective heat transfer coefficient is $h = 70.0\,\text{W/(m}^2\cdot\text{K)}$. Assume the engine to be approximately spherical in shape with the following physical properties: density $= 7750\,\text{kg/m}^3$, specific heat $= 0.4645\,\text{kJ/(kg}\cdot\text{K)}$, volume $= 0.500\,\text{m}^3$, and thermal conductivity $= 36.0\,\text{W/(m}\cdot\text{K)}$.

40. A lunar orbiting module is on its way back to Earth. At 200. miles above the surface of the Earth, the module's velocity is 2000. mi/h. At this point an astronaut seals a rigid insulated container holding saturated water vapor at 10.0 psia. You are a NASA engineering supervisor at Control Headquarters. Suddenly, two wild-eyed engineers run up to you with the following emergency:

ENGINEER A: "That sealed container aboard the lunar module may explode when it lands! Its bursting pressure is only 80.0 psia, and the internal energy of the water *must* increase due to the decrease in the potential and kinetic energies on landing."

ENGINEER B: "Engineer A is incorrect! That container is a sealed, rigid, insulated vessel, so it cannot do any work or have any heat transfer. Therefore, its internal energy cannot change on landing."

Write a brief paragraph stating (a) which engineer you support, (b) why (make this part very clear), and (c) what action (if any) you would take as engineering supervisor.

41.* When the pressure on saturated liquid water is suddenly reduced to a lower pressure in an adiabatic and aergonic process, the liquid's temperature must also be reduced to reach a new equilibrium state. Consequently, part of the initial liquid is very quickly converted into a saturated vapor at the lower pressure, and the resulting heat of vaporization cools the remaining liquid to the proper temperature. Vapor formed in this manner is called *flash steam*, because the liquid appears to "flash" into a vapor as the pressure is reduced. Determine the final temperature and the percent of flash steam produced as a closed system containing saturated liquid water suddenly bursts and the pressure drops from 1.00 to 0.100 MPa in an adiabatic and aergonic process.

42. In 1798, the American Benjamin Thompson (Count Rumford, 1753–1814) carried out a series of cannon-boring experiments in which he established that heat was not a material substance. (It was commonly believed at that time that heat was a colorless, odorless, weightless fluid called *caloric*.) In his third experiment, he noted that the "total quantity of ice-cold water which with the heat actually generated by friction, and accumulated in $2^h\ 30^m$, might have been heated 180°, or made to boil, = 26.58 lb." He also stated that "the machinery used in the experiment could easily be carried round by the force of one horse." Use this crude data of Rumford to estimate the mechanical equivalent of heat (i.e., the number of ft · lbf per Btu). Take the specific heat of liquid water to be 1.00 Btu/(lbm · R).

43. The mechanical equivalent of heat (i.e., the number of ft · lbf per Btu) was first established accurately by James Prescott Joule (1818–1889) in a long series of experiments carried out between 1849 and 1878. In one of his first experiments, the work done by falling weights caused the rotation of a paddle wheel immersed in water. The weights had a mass of 57.8 lbm and fell 105 ft. The resulting paddle wheel motion caused an increase in temperature of 0.563°F in 13.9 lbm of water in an insulated container. Using a specific heat of c = 1.00 Btu/(lbm · R), determine the mechanical equivalent of heat from these early data of Joule.

44. Determine the heat generated (in Btu/year) by the brakes of 100. million 3000. lbm automobiles that isothermally brake to a stop on a horizontal surface from 55.0 mph ten times per day. Convert your answer into equivalent barrels of crude oil per year then into quads per year, where one barrel of crude oil contains 5.80×10^6 Btu of energy and one quad is defined to be exactly 10^{15} Btu.

45.* An insulated vessel contains an unknown amount of ammonia. A 600. W electrical heater is put into the vessel and turned on for 30.0 min. The heater raises the temperature of the ammonia from 20.0 to 100.°C in a constant pressure process at 100. kPa. Determine the mass of ammonia in the vessel.

46. Using the general energy rate balance equation for a closed system, show that, under adiabatic, isothermal, and aergonic conditions, the acceleration of an object falling vertically downward in a vacuum is simply the local acceleration of gravity, g.

47.* Determine the difference in water temperature between the top and the bottom of a waterfall 35.0 m high. Choose as your system 1.00 kg of water at the top of the falls and follow its change of state as it moves to the bottom of the falls. Assume water to be an incompressible liquid and neglect any heat loss. Also assume a constant water velocity for this process.

48. In days of yore, a bow and arrows were an archer's best friend. Determine

a. The maximum velocity of a 0.400 lbm arrow shot horizontally from a bow in which 100. ft · lbf is required to draw back the arrow before releasing it.

b. The maximum height this arrow would reach if aimed vertically.

49.* A 5.00 cm diameter steel sphere initially at 20.0°C is to be heated by immersing it in boiling water at 100.°C with a convective heat transfer coefficient of 2000. W/(m^2·K). Determine the time required to raise the bulk temperature of the sphere to 90.0°C. The specific heat of the steel is 0.500 kJ/(kg·K) and its density is 7800. kg/m^3.

50.* An asteroid enters the Earth's atmosphere and descends vertically with a constant velocity of 100. m/s. Determine the rate of change of the asteroid's temperature at the point where its temperature exactly equals the surrounding air temperature. The specific heat of the asteroid is 0.300 kJ/(kg·K).

51.* 50,000. kg of saturated liquid water at 20.0°C is to be heated in a mass-energy conversion oven in which 1.00×10^{-6} kg of mass is converted into pure thermal energy ($Q = mc^2$). Assuming that the water is an incompressible liquid with a specific heat of 4.20 kJ/(kg · K), determine the final temperature of the water. The velocity of light is 2.998×10^8 m/s.

52. A hand grenade contains 1.90 ounces (0.120 lbm) of TNT. Determine the number of hand grenades it would take to produce an explosion equivalent to the Brockton shoe factory boiler explosion discussed in Example 5.8. The explosive energy of TNT is 1400. Btu/lbm.

53. Between 1897 and 1927, the Stanley brothers of Newton, Maine, manufactured steam-powered automobiles. They had a steam boiler 23.0 inches in diameter and 14.0 inches high that contained steam at 600. psia and 600.°F. Determine the explosive energy of these boilers and the number of 1.00 lbm sticks of TNT that would contain the equivalent amount of explosive energy. Assume the ambient temperature is 70.0°F.

54. The greatest steam explosion in history is thought to have occurred on August 27, 1883, when the volcano Krakatoa in Sunda Strait, Indonesia, erupted and its molten lava vaporized an estimated 1 mi^3 of seawater. The entire 2600. ft high mountain was disintegrated and a crater 1000. ft deep was produced. More than 36,000. people were killed, most by the 120. ft tidal wave created by the eruption. Assuming that the seawater is simply saturated liquid water at 60.0°F, determine the number of tons of TNT that would have the same explosive energy as this eruption. For reference, the total military production of explosives for both world wars was equivalent to 32.0 million tons of TNT. The explosive energy of TNT is 1400. Btu/lbm.

55. Show that, if a pressure vessel filled with a constant specific heat ideal gas ruptures and the gas follows a polytropic process during the subsequent depressurization, then the maximum explosive energy of this system can be written as

$$\Gamma_{max} = p_{initial}/(k-1)$$

56. Consider a gaseous star undergoing a gravitational collapse. Assume the star to be a closed system and composed of an ideal gas with constant specific heats. The collapse process is given by the relations

$$v/r^3 = \text{constant} \quad \text{and} \quad Tr = \text{constant}$$

where v, T, and r are the specific volume, temperature, and radius of the star.

a. Show that the collapse process is a polytropic process with $n = 4/3$.

b. Beginning with the per unit mass differential form of the first law, find an expression for the star's heat transfer as a function of its specific heats and temperature. Note that the star does $p - \forall$ work on itself as it collapses.

c. Using the expression found in part b, along with c_v = 0.200 Btu/(lbm · R) and k = 1.4, calculate the amount of heat transfer per unit mass of star as its temperature changes from 5000. to 10,000. R.

d. Explain the critical condition that exists when the specific heat ratio k takes on the value of 4/3.

Computer Problems

The following computer problems are designed to be completed using a spreadsheet or equation solver. They may be used as part of a weekly homework assignment.

57. Develop a computer program that performs an energy balance on a closed system containing an incompressible substance (either a liquid or a solid). Include the following input (in proper units): the heat and work transports of energy, the system volume, the initial temperature of the system, and the density and specific heat of the incompressible material contained in the system. Output to the screen the system mass and final temperature.

58. Develop a computer program that performs an energy balance on a closed system containing an ideal gas with constant specific heats. Include the following input (in proper units): the heat and work transports of energy, the system volume, the initial temperature and pressure of the system, and the constant volume specific heat and gas constant of the gas contained in the system. Output to the screen the system mass and the final pressure and temperature.

59. Repeat Problem 58, except allow the user to choose the system ideal gas from a screen menu, and omit the prompts for gas properties. Use the data in Table C.13 of *Thermodynamic Tables to accompany Modern Engineering Thermodynamics* for the properties of the gases in your menu.

60. Develop a computer program that generates data to allow you to plot on the computer the explosive energy contained in a 1000. ft^3 pressure vessel containing compressed air vs.

a. The vessel's initial temperature when the initial pressure is held constant at 100. psia.

b. The vessel's initial pressure when the initial temperature is held constant at 80.0°F.

c. Create a three-dimensional plot with the explosive energy on the vertical axis and the initial pressure and initial temperature on the horizontal axes.

Assume the final temperature and pressure of the air after the vessel has ruptured are 70.0°F and 14.7 psia in each case and that the air undergoes a polytropic decompression process with n = 1.25 when the vessel ruptures. Also assume that the air behaves as a constant specific heat ideal gas with k = 1.40.

61. A white dwarf is a spherical mass of gas in outer space. Its radial pressure gradient must always be in equilibrium with its own gravitational force field, or

$$\frac{dp}{dr} = -Gm\rho/r^2$$

where G is the gravitational constant, ρ is the density of the gas at radius r (i.e., $\rho = \rho(r)$), and m is the mass of gas inside a sphere of radius r,

$$m = 4\pi \int_0^r \rho r^2 dr$$

During its formation, the gas of a white dwarf obeys the polytropic equation

$$pv^{-5/3} = \alpha = \text{constant}$$

These relations can be combined to yield a differential equation for the density field $\rho = \rho(r)$ inside a white dwarf of the form

$$\frac{d^2\phi}{dx^2} + \frac{2}{x}\left(\frac{d\phi}{dx}\right) + \phi^{2/3} = 0$$

where $\phi = (\rho/\rho_0)^{2/3}$, $\rho_0 = \rho(r = 0)$, and $x = r/r^*$, where $r^* = 5\alpha/\left(8\pi G\rho_0^{1/3}\right)$.

a. Solve the preceding differential equation for $\phi(x)$ using a computer numerical solution with the boundary conditions $\phi = 1$ and $d\phi/dx = 0$ at $x = 0$ to show that $\phi(x) = 0$ at $x = 3.6537$.

b. Show that $\phi(x) = 0$ corresponds to the radius R of the white dwarf and a white dwarf therefore has a mass m given by $m = -45.91\rho_0(r^*)^3(d\phi/dx)|_{x=R}$.

CHAPTER 6

First Law Open System Applications

CONTENTS

6.1 INTRODUCTION

This chapter contains detailed solutions to a variety of classical open system thermodynamic problems. These solutions use the generalized problem solving procedure discussed in Chapters 4 and 5 (see Figure 4.21) and focus on illustrating the use of the conservation of mass law and the first law of thermodynamics. The availability of these two basic equations plus many auxiliary formulae means that there are usually more unknowns to be solved for in open system problems than in closed system problems.

SUMMARY OF THE THERMODYNAMIC PROBLEM SOLVING TECHNIQUE

Begin by carefully reading the problem statement completely through.

Step 1. Make a sketch of the system and put a dashed line around the system boundary.
Step 2. Identify the unknown(s) and write them on your system sketch.
Step 3. Identify the type of system (closed or open) you have.
Step 4. Identify the process that connects the states or stations.
Step 5. Write down the basic thermodynamic equations and any useful auxiliary equations.
Step 6. Algebraically solve for the unknown(s).
Step 7. Calculate the value(s) of the unknown(s).
Step 8. Check all algebra, calculations, and units.

Sketch → Unknowns → System → Process → Equations → Solve → Calculate → Check

Open system problems are written with their flow stream thermodynamic properties evaluated at inlet and outlet data monitoring stations. This is done in an attempt to simulate the way in which engineering data are provided from experimental or field measurements. The bulk properties inside an open system do not normally change from one equilibrium thermodynamic state to another during the process of interest, as do closed system bulk properties. In fact, the vast majority of open systems of engineering interest are not in any equilibrium state, since the thermodynamic properties of the material passing through them are continually changing inside the system between the inlet and the outlet flow streams. However, most open systems do reach a "steady state" nonequilibrium operating condition in which the total mass and energy they contain does not change with time. Any open or closed system can be indefinitely maintained in a steady nonequilibrium state if it has the proper energy or mass flows passing through it.

In addition, many thermodynamic properties are mathematically defined only for equilibrium conditions. If the steady state properties within a system do not exhibit large variations between two neighboring points, then we say that these points are in *local thermodynamic equilibrium*. The local equilibrium postulate introduced in an earlier chapter states that a small volume, large enough for the continuum hypothesis to hold, is in local equilibrium so long as its internal properties do not vary significantly within its borders. This means that the properties cannot change significantly in a distance on the order of the molecular mean free path at the point in question.[1] Most nonequilibrium processes of engineering interest obey this postulate. A few systems, such as those containing shock waves, do not. For example, if rapid explosions occur within a piston-cylinder apparatus (as in an internal combustion engine) or if the piston speed exceeds the speed of sound in the cylinder, then the gas in the cylinder is far from equilibrium and an accurate thermodynamic analysis becomes very difficult, from both a measurement and a theoretical point of view.

6.2 MASS FLOW ENERGY TRANSPORT

Mass flow energy transport occurs whenever mass crosses the system boundary. It consists of two parts. The first part is the total energy associated with the flow stream mass itself, and the second is the energy required to push the flow stream mass across the system boundary (this part is often called the *flow work*). Let an increment of flow stream mass dm be added to or removed from a system. The total energy dE_m associated with dm crossing the system boundary is given by

$$dE_m = (u + \text{ke} + \text{pe})\, dm$$

Figure 6.1 shows an incremental slug of mass with velocity $\mathbf{V}$ crossing a system boundary. The slug's volume is $d\mathcal{V} = A\, dL$, and its mass is $dm = \rho\, d\mathcal{V} = \rho A\, dL$, where ρ is the mass density of the slug. In the time increment dt,

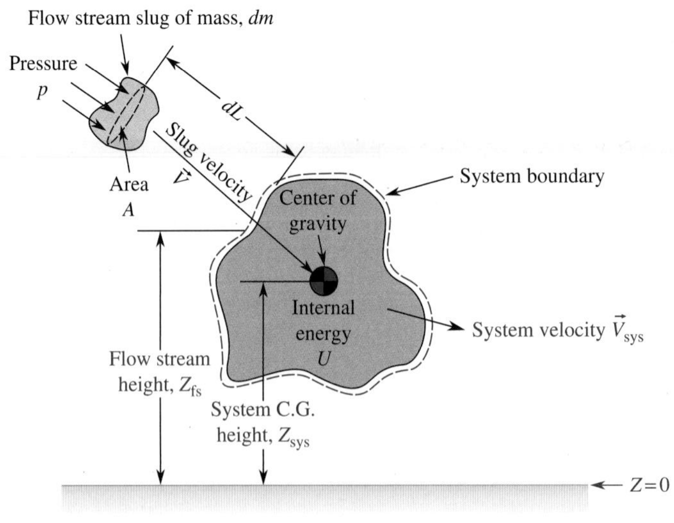

FIGURE 6.1
Open system flow stream and system energies.

[1] In air at standard temperature and pressure (STP), the molecular mean free path is approximately 8×10^{-8} m, or 3×10^{-6} in.

the slug moves a distance $|\vec{V}|dt = V\,dt$ and sweeps out an incremental volume $d\forall = AV\,dt$, which has an associated mass of $dm = \rho\,d\forall = \rho AV\,dt$. Dividing by dt gives

$$\left(\frac{dm}{dt}\right)_{\text{flow stream}} = (\dot{m})_{\text{flow stream}} = (\rho AV)_{\text{flow stream}} \tag{6.1}$$

This equation is a very convenient way to calculate the flow stream mass flow rate from the easily measured variables of density (ρ), cross-sectional area (A), and average fluid velocity ($|\vec{V}| = V$).

The incremental energy required to push the mass slug across the system boundary and into the system is the product of the force acting on it, $p \times A$, and the distance moved, dL. Consequently, the flow work energy increment is

$$\bar{d}W_{\substack{\text{mass}\\ \text{flow}}} = -pA\,dL = -p\,d\forall = -\frac{p}{\rho}dm = -pv\,dm$$

where $v = 1/\rho$ is the specific volume of the slug. The flow work is negative here because adding dm to the system represents work done *on* the system.

The total mass flow energy entering the system with this incremental mass is then

$$\begin{aligned} dE_{\substack{\text{mass}\\ \text{flow}}} &= dE_m - \bar{d}W_{\substack{\text{mass}\\ \text{flow}}} \\ &= (u + \text{ke} + \text{pe})\,dm + (pv)\,dm \\ &= (u + pv + \text{ke} + \text{pe})\,dm \end{aligned}$$

Because the sum of the terms u and pv always appears in this equation, it is convenient to combine them into a single term (as explained in Chapter 3) called *specific enthalpy*, $h = u + pv$.

In general, we have more than one mass flow stream in any given open system. To accurately account for all the mass flow energies, we sum them in two groups. One group accounts for all inlet flow streams and the other for all exiting flow streams. Therefore, we write

$$dE_{\substack{\text{mass}\\ \text{flow}}} = \sum_{\text{inlet}}(h + \text{ke} + \text{pe})\,dm - \sum_{\text{outlet}}(h + \text{ke} + \text{pe})\,dm$$

On dividing this equation through by dt, we obtain

$$\dot{E}_{\substack{\text{mass}\\ \text{flow}}} = \sum_{\text{inlet}} \dot{m}(h + \text{ke} + \text{pe}) - \sum_{\text{outlet}} \dot{m}(h + \text{ke} + \text{pe}) \tag{6.2}$$

and on integration of this equation, we obtain

$$\left(E_{\substack{\text{mass}\\ \text{flow}}}\right)_{1\,2} = \sum_{\text{inlet}} \int_1^2 \dot{m}(h + \text{ke} + \text{pe})\,dt - \sum_{\text{outlet}} \int_1^2 \dot{m}(h + \text{ke} + \text{pe})\,dt \tag{6.3}$$

where $\text{ke} = V^2/2g_c$ and $\text{pe} = gZ/g_c$ are the specific kinetic and potential energies of the flow streams at the point where they cross the system boundary. Note that these equations already contain the proper thermodynamic signs for input (+) and output (−) mass flow energy transport.

Each flow stream has its own average velocity V and height Z; in addition, the center of gravity of the entire system has unique and usually different V and Z values (see Figure 6.1). The student must be careful not to get these velocities and heights confused.

EXAMPLE 6.1

Determine the mass flow energy transport rate of steam at 100. psia, 500.°F leaving a system through a 6.00-inch inside diameter pipe at a velocity of 300. ft/s at a height of 15.0 ft above the floor (the zero height potential).

Solution

First, draw a sketch of the system (Figure 6.2).

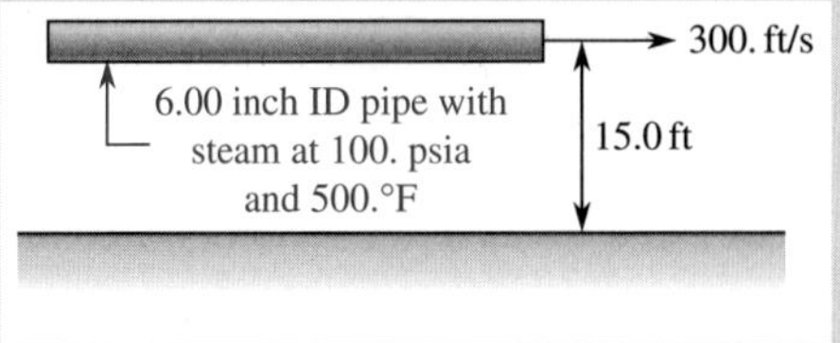

FIGURE 6.2
Example 6.1.

This is an open system, and the unknown is $\dot{E}_{\substack{\text{mass}\\\text{flow}}}$. The material is steam.

From the superheated steam table, Table C.3a in *Thermodynamic Tables to accompany Modern Engineering Thermodynamics*, we find that, at 100. psia and 500.°F,

$$v = 5.587\ \text{ft}^3/\text{lbm}$$

$$h = 1279.1\ \text{Btu/lbm}$$

Now, since $\rho = 1/v$, we can find $\dot{m}$ from Eq. (6.1):

$$\dot{m} = \rho A V = \frac{AV}{v} = \frac{\left[\pi\left(\frac{3}{12}\right)^2 \text{ft}^2\right](300.\ \text{ft/s})}{5.587\ \text{ft}^3/\text{lbm}} = 10.5\ \text{lbm/s}$$

Then,

$$\text{ke} = \frac{V^2}{2g_c} = \frac{(300.)^2\ \text{ft}^2/\text{s}^2}{2\left(32.174\dfrac{\text{lbm}\cdot\text{ft}}{\text{lbf}\cdot\text{s}^2}\right)} = 1398\frac{\text{ft}\cdot\text{lbf}}{\text{lbm}}$$

$$= \left(1398\frac{\text{ft}\cdot\text{lbf}}{\text{lbm}}\right)\left(\frac{1\ \text{Btu}}{778.16\ \text{ft}\cdot\text{lbf}}\right) = 1.80\frac{\text{Btu}}{\text{lbm}}$$

and

$$\text{pe} = \frac{gZ}{g_c} = \frac{(32.174\ \text{ft/s}^2)(15.0\ \text{ft})}{32.174\dfrac{\text{lbm}\cdot\text{ft}}{\text{lbf}\cdot\text{s}^2}} = 15.0\frac{\text{ft}\cdot\text{lbf}}{\text{lbm}}$$

$$= \left(15.0\frac{\text{ft}\cdot\text{lbf}}{\text{lbm}}\right)\left(\frac{1\ \text{Btu}}{778.16\ \text{ft}\cdot\text{lbf}}\right) = 0.019\ \text{Btu/lbm}$$

In this problem, we have only one flow stream, so

$$\dot{E}_{\substack{\text{mass}\\\text{flow}}} = -[\dot{m}(h + \text{ke} + \text{pe})]_{\text{out}}$$

$$= -(10.5\ \text{lbm/s})[(1279.1 + 1.80 + 0.019)\ \text{Btu/lbm}]$$

$$= -1.35 \times 10^4\ \text{Btu/s}$$

Exercises

1. Determine the percentage contribution to the mass flow energy transport rate in Example 6.1 of each of the following terms: (a) enthalpy, (b) kinetic energy, and (c) potential energy. **Answers**: (a) 99.86%, (b) 0.14%, and (c) 0.0015%.
2. Determine the percent error incurred in the answer to Example 6.1 if the kinetic and potential energy terms are neglected. **Answer**: Percent error = 0.142%.
3. Suppose the fluid leaving the system through the 6.00-inch pipe in Example 6.1 were saturated liquid water at 50.0°F, and determine the percentage error incurred in neglecting the kinetic and potential energies of the flow stream. **Answer**: Percent error = 9.14%.

Note that the ke and pe terms in Example 6.1 amount respectively to only 0.14% and 0.0015% of the total mass flow energy. This is because the specific enthalpy of steam (and most vapors) usually has a large numerical value, while the specific kinetic and potential flow energies for most engineering problems usually have much smaller values when converted into the same units.[2]

[2] *Actually, because of the form of the first law of thermodynamics, we normally compare values of the change in enthalpy, Δh, with the changes in ke, Δke, and pe, Δpe. Here, too, we find that Δh usually dominates.*

6.3 CONSERVATION OF ENERGY AND CONSERVATION OF MASS EQUATIONS FOR OPEN SYSTEMS

To obtain a general working formula for the first law of thermodynamics for open systems, we begin by constructing a general energy rate balance (ERB) equation for these systems. The general open system energy rate balance is given by Eq. (4.22) as

$$\dot{Q} - \dot{W} + \dot{E}_{\substack{\text{mass}\\\text{flow}}} = \dot{E}_G$$

where the rate of gain of total system energy $\dot{E}_G$ is given by Eq. (4.9) as

$$\dot{E}_G = \frac{d}{dt}\left(U + \frac{m}{2g_c}V^2 + mgZ/g_c\right)_{\text{system}}$$

and the mass flow energy transports are given by Eq. (6.2) as

$$\dot{E}_{\substack{\text{mass}\\\text{flow}}} = \sum_{\text{inlet}} \dot{m}(h + \text{ke} + \text{pe}) - \sum_{\text{outlet}} \dot{m}(h + \text{ke} + \text{pe})$$

where

$$(\text{ke})_{\text{inlet}} = \frac{V_{\text{inlet}}^2}{2g_c}$$

$$(\text{ke})_{\text{outlet}} = \frac{V_{\text{outlet}}^2}{2g_c}$$

and

$$(\text{pe})_{\text{inlet}} = \frac{gZ_{\text{inlet}}}{g_c}$$

$$(\text{pe})_{\text{outlet}} = \frac{gZ_{\text{outlet}}}{g_c}$$

are the specific kinetic and potential energies of each inlet and outlet flow stream. Combining these equations gives the general energy rate balance (ERB) for open systems:

General open system energy rate balance

$$\dot{Q} - \dot{W} + \sum_{\text{inlet}} \dot{m}(h + V^2/2g_c + gZ/g_c) - \sum_{\text{outlet}} \dot{m}(h + V^2/2g_c + gZ/g_c) = \frac{d}{dt}(U + mV^2/2g_c + mgZ/g_c)_{\text{system}} \tag{6.4}$$

It must be remembered that the $\dot{Q}$ and $\dot{W}$ terms in this equation are the *net* heat and work transport rate terms; that is,

$$\dot{Q} = \sum_{\substack{\text{all}\\\text{boundaries}}} \dot{Q} \quad \text{and} \quad \dot{W} = \sum_{\substack{\text{all}\\\text{boundaries}}} \dot{W} \tag{6.5}$$

where proper input and output signs are to be used in the summations. Also, the kinetic and potential energy terms on the right side of Eq. (6.4) are of the center of gravity of the entire system, whereas the kinetic and potential energy terms in the flow stream summation terms on the left side of this equation apply only to the point of entry or exit of the flow stream from the system (see Figure 6.1 for an illustration of this notation).

As a working equation, Eq. (6.4) is really too complex to remember or write down conveniently during the solution of each thermodynamic problem we face. Since most of our problems involve systems operating at steady state with a single inlet and a single outlet flow stream, we simplify Eq. (6.4) to fit this case. For a steady state process, the entire right side of Eq. (6.4) vanishes:

Steady state

$$\dot{E}_G = \frac{d}{dt}(U + mV^2/2g_c + mgZ/g_c)_{\text{system}} = 0 \tag{6.6}$$

Note that this does not necessarily mean that $\dot{U}$, $\dot{\text{KE}}$, and $\dot{\text{PE}}$ are all zero but only that their sum vanishes.

At this point, we introduce the conservation of mass law for open systems. This law can easily be cast into the form of a rate balance by using the general form of Eq. (2.14) as $\dot{m}_G = \dot{m}_T$, where the mass transport rate is simply given by

$$\dot{m}_T = \sum_{\text{inlet}} \dot{m} - \sum_{\text{outlet}} \dot{m} \tag{6.7}$$

Thus, the general mass rate balance for the rate of gain of mass $\dot{m}_G$ for an open system is simply

$$\dot{m}_G = \left(\frac{dm}{dt}\right)_{\text{system}} = \sum_{\text{inlet}} \dot{m} - \sum_{\text{outlet}} \dot{m} \tag{6.8}$$

Now, if a system is operating at steady state, then, by definition,

$$\left(\frac{dE}{dt}\right)_{\text{system}} = \left(\frac{dm}{dt}\right)_{\text{system}} = \dot{E}_G = \dot{m}_G = 0 \tag{6.9}$$

so that Eq. (6.8) gives the steady state mass rate balance as

$$\sum_{\text{inlet}} \dot{m} = \sum_{\text{outlet}} \dot{m}$$

The condition of equal mass inflows and outflows is called a *steady flow*:

$$\sum_{\text{inlet}} \dot{m} = \sum_{\text{oulet}} \dot{m} \tag{6.10}$$

It should be clear from this development that any steady state open system is also (by definition) a steady flow system. To keep this clearly in mind, we often write both statements, steady state and steady flow, explicitly, even though it is not really necessary to do so.

If the system has only one inlet and one outlet flow stream, then the summation signs can be dropped in Eqs. (6.4), (6.7), (6.8), and (6.10). The steady flow condition for a system with a single inlet and a single outlet flow stream then becomes

$$\dot{m}_{\text{inlet}} = \dot{m}_{\text{outlet}} = \dot{m} \tag{6.11}$$

Note that the inlet-outlet direction subscripts on the mass flow rate term can now be dropped because they are superfluous.

Substituting Eqs. (6.6) and (6.11) into Eq. (6.4), and abbreviating the terms inlet and outlet as simply *in* and *out* gives a simplified energy rate balance. We call the resulting formula the *modified energy rate balance* (MERB).

Thus, the open system modified energy rate balance applies only to systems that are

1. Steady state: ($\dot{E}_G = 0$).
2. Steady flow: ($\dot{m}_G = 0$).
3. Single inlet and single outlet: ($\dot{m}_{\text{inlet}} = \dot{m}_{\text{outlet}} = \dot{m}$).

and has the following form:

The open system modified energy rate balance (MERB)

$$\dot{Q} - \dot{W} + \dot{m}[h_{\text{in}} - h_{\text{out}} + (V_{\text{in}}^2 - V_{\text{out}}^2)/(2g_c) + (Z_{\text{in}} - Z_{\text{out}})(g/g_c)] = 0 \tag{6.12}$$

Integrating Eq. (6.12) over time gives the open system modified energy balance (MEB) as

The open system modified energy balance (MEB)

$$ {}_1Q_2 - {}_1W_2 + \int_1^2 \dot{m}[h_{\text{in}} - h_{\text{out}} + (V_{\text{in}}^2 - V_{\text{out}}^2)/(2g_c) + (Z_{\text{in}} - Z_{\text{out}})(g/g_c)]\,dt = 0 \tag{6.13} $$

However, the vast majority of open system problems are set up on a rate basis, so the modified energy rate balance is the equation most often used in open system analysis.

When the conditions of steady state (steady flow), single inlet, or single outlet do not exist in any particular problem, we must return to the more general energy rate balance of Eq. (6.4) as a starting point for the analysis. This is illustrated with the unsteady state examples presented later in this chapter.

6.4 FLOW STREAM SPECIFIC KINETIC AND POTENTIAL ENERGIES

Before we can begin analyzing thermodynamic problems, we must establish a criterion for when the specific kinetic and potential energy flow stream terms of Eqs. (6.4) and (6.12) are important and when they are not. To get some feeling for the importance of these terms, we look at how their magnitude varies over a wide range of velocities and heights. First, consider the specific kinetic energy term $V^2/2g_c$. If we work in the Engineering English units system, then V is normally in feet per second, and $g_c = 32.174\,\text{lbm}\cdot\text{ft}/(\text{lbf}\cdot\text{s}^2)$. Hence,

$$ \frac{V^2}{2g_c} = \frac{[V]^2\,\text{ft}^2/\text{s}^2}{2\times\left(32.174\dfrac{\text{lbm}\cdot\text{ft}}{\text{lbf}\cdot\text{s}^2}\right)} = \frac{[V]^2}{64.348}\,\frac{\text{ft}\cdot\text{lbf}}{\text{lbm}} $$

where the symbol $[V]$ stands for the numerical value of V in units of ft/s. The remaining term in the flow stream energy transport equation to which the specific kinetic and potential energy terms are to be added is the specific enthalpy h. In the Engineering English units system, the specific enthalpy has units of Btu/lbm. If we convert the specific kinetic energy into these units, we get

$$ \frac{V^2}{2g_c} = \left(\frac{[V]^2}{64.348}\,\frac{\text{ft}\cdot\text{lbf}}{\text{lbm}}\right)\left(\frac{1\,\text{Btu}}{778.16\,\text{ft}\cdot\text{lbf}}\right) = \frac{[V]^2}{50{,}070}\,\frac{\text{Btu}}{\text{lbm}} $$

In the SI units system, $g_c = 1.0$ and is dimensionless, so

$$ \frac{V^2}{2g_c} = \frac{[V]^2 m^2/s^2}{2(1)} = \frac{[V]^2}{2}\,\frac{\text{J}}{\text{kg}} = \frac{[V]^2}{2000}\,\frac{\text{kJ}}{\text{kg}} $$

where $1\,\text{m}^2/\text{s}^2 = 1\,\text{J/kg} = 10^{-3}\,\text{kJ/kg}$.

Table 6.1 gives values of the specific kinetic energy for various velocities using these equations. Since the specific enthalpy values for most substances fall roughly between 100 and 1000 Btu/lbm, Table 6.1 shows that the specific kinetic energy is very small when compared to these h values for velocities less than about 250 ft/s (76 m/s), or $V^2/2g_c$ is less than about 1.0 Btu/lbm (2.3 kJ/kg). Consequently, it is common to neglect the effect of a flow stream's specific kinetic energy when the flow stream velocity is less than about 250 ft/s (76 m/s). This is a relatively high velocity (~170 mi/h or 270 km/h), and most engineering applications do not have such rapid flow streams.

Table 6.1 The Effect of Velocity on Kinetic Energy

Velocity, V		Kinetic Energy, $V^2/2g_c$	
ft/s	m/s	Btu/lbm	kJ/kg
0	0	0	0
1	0.3	2×10^{-5}	4.7×10^{-5}
10.	3.0	2.0×10^{-3}	4.7×10^{-3}
100.	30.5	2.0×10^{-1}	4.70×10^{-1}
1000.	305	20.0	470.

Table 6.2 The Effect of Height on Potential Energy

Height, Z		Potential Energy, gZ/g_c	
ft	m	Btu/lbm	kJ/kg
0	0	0	0
1.0	0.30	1.3×10^{-3}	2.9×10^{-3}
10.	3.0	1.3×10^{-2}	2.9×10^{-2}
100.	30.5	1.3×10^{-1}	2.9×10^{-1}
1000.	305	1.3	2.9

There are, of course, exceptions to this rule of thumb. If $h_{in} - h_{out} \approx 0$, then the enthalpy term loses its dominance. In this case, a small specific kinetic energy term may be quite significant to the analysis. A nozzle or a diffuser is an example of such an exception.

Now consider the specific potential energy term, gZ/g_c. Taking $g = 32.174$ ft/s^2 and, $g_c = 32.174$ lbm $\cdot$ ft/(lbf $\cdot$ s^2), then

$$\frac{gZ}{g_c} = Z\left(\frac{g}{g_c}\right) = ([Z]\text{ ft})\times\left(\frac{32.174\text{ ft/s}^2}{32.174\dfrac{\text{lbm}\cdot\text{ft}}{\text{lbf}\cdot\text{s}^2}}\right) = [Z]\frac{\text{ft}\cdot\text{lbf}}{\text{lbm}}$$

where the symbol $[Z]$ stands for the numerical value of Z in units of feet. Again, converting to Btu/lbm gives

$$\frac{gZ}{g_c} = \left([Z]\frac{\text{ft}\cdot\text{lbf}}{\text{lbm}}\right)\times\left(\frac{1\text{ Btu}}{778.16\text{ ft}\cdot\text{lbf}}\right) = \frac{[Z]}{778.16}\frac{\text{Btu}}{\text{lbm}}$$

In the SI units system, $g_c = 1.0$ and is dimensionless, so

$$\frac{gZ}{g_c} = Z\left(\frac{g}{g_c}\right) = ([Z]\text{ m})\times\left(\frac{9.807\text{ m/s}^2}{1}\right) = [Z]\times 9.807\frac{\text{J}}{\text{kg}} = [Z]\frac{9.807}{1000}\frac{\text{kJ}}{\text{kg}}$$

where 1 m^2/s^2 = 1 J/kg = 10^{-3} kJ/kg.

Table 6.2 gives values of specific potential energy for various heights using these equations. Note that, for systems with normal engineering dimensions, say less than 1000 ft (305 m) high, the specific potential energy is very small. Consequently, it is common to neglect the effect of a flow stream's specific potential energy when the flow stream enters the system less than 1000 ft (305 m) above or below the potential energy baseline of the system, or $gZ/g_c < \sim 1.0$ Btu/lbm (2.3 kJ/kg).

There are also exceptions to this rule of thumb. Again, if $h_{in} - h_{out} \approx 0$, then flow stream height changes may be very important in the analysis. A hydroelectric power plant is an example of such an exception.

Deciding whether to neglect factors such as specific kinetic and potential energies is not always an easy task for the beginner. Self-confidence comes only with experience. However, one more rule of thumb applies to most textbook thermodynamic problems: *If values for velocity and height are not given in the problem statement and are not among the problem's unknowns, then you are supposed to neglect the kinetic and potential energy terms in your analysis.* This means that the person who wrote the problem knew that either $V_{in} \approx V_{out}$ and $Z_{in} \approx Z_{out}$ or that all the velocities and heights were relatively small. The only exception to this last rule of thumb is when you know the mass flow rate $\dot{m}$, the diameter D, or cross-sectional area A, and the fluid density ρ or specific volume v of a flow stream. With this information you can calculate the flow stream velocity using Eq. (6.1) as

$$V = \frac{\dot{m}}{\rho A} = \frac{\dot{m}v}{A} = \frac{4\dot{m}v}{\pi D^2} \tag{6.14}$$

If you can make this calculation for V, then you might as well use it in your energy rate balance equation, unless it is so small that you are certain it will not affect the results of your analysis.

6.5 NOZZLES AND DIFFUSERS

Nozzle is the generic name of any device whose primary function is to convert the pressure energy $\dot{m}pv$ of an inlet flow stream into the kinetic energy $\dot{m}V^2/2$ of an outlet flow stream. Thus, a nozzle is a very simple energy conversion device. Similarly, *diffuser* is the generic name of any device whose primary function is to convert the kinetic energy of an inlet flow stream into the pressure energy of an outlet flow stream. Note that nozzles and

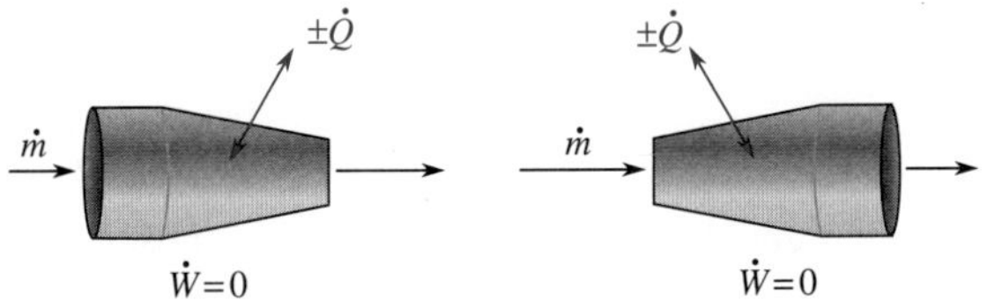

(a) Nozzle (b) Diffuser

FIGURE 6.3
Nozzles and diffusers.

diffusers perform opposite functions. In their simplest form, a nozzle is merely a converging duct and a diffuser is merely a diverging duct, as shown schematically in Figure 6.3.[3]

Most commercial nozzles and diffusers are well insulated (adiabatic). However, they need not be, and therefore may have either a heat loss or a heat gain. On the other hand, the simple mechanical nature of nozzles and diffusers prevents them from either performing or absorbing work. Therefore, they can generally be taken to be aergonic devices.

Since both nozzles and diffusers are clearly single inlet, single outlet devices, we can carry out an analysis of their steady state operation by using the modified energy rate balance of Eq. (6.12). Also, both nozzles and diffusers are either oriented horizontally, as shown in Figure 6.3, so that $Z_{in} = Z_{out}$ or have such small changes in height between the inlet and outlet that the enthalpy change dominates the specific potential energy change, as discussed previously. This allows us to neglect the change in flow stream specific potential energy in nozzle and diffuser analysis.

However, the flow stream specific kinetic energies are not necessarily negligible, because in both nozzles and diffusers, at least one of the flow streams normally has a high velocity. Consequently, we ignore the low-speed flow stream specific kinetic energy in each case and set $V_{in} \approx 0$ for the nozzle and $V_{out} \approx 0$ for the diffuser.

At this point, we have developed the following set of assumptions for these devices:

Nozzle	Diffuser
$\dot{W} = 0$	$\dot{W} = 0$
$Z_{in} - Z_{out} \approx 0$	$Z_{in} - Z_{out} \approx 0$
$V_{in} \approx 0$	$V_{out} \approx 0$

Applying these assumptions for nozzles to the modified energy rate balance of Eq. (6.12) gives the following results:

$$\dot{Q} - 0\,\dot{m}(h_{in} - h_{out} - V_{out}^2/2g_c + 0) = 0$$

or

$$V_{out}\Big|_{nozzle} = [2g_c(\dot{Q}/\dot{m} + h_{in} - h_{out})]^{1/2} \tag{6.15}$$

Notice that adding heat to the nozzle increases the outlet velocity, whereas removing heat decreases it. If the nozzle is insulated (adiabatic), then $\dot{Q} = 0$ and

$$V_{out}\Big|_{\substack{\text{adiabatic}\\ \text{nozzle}}} = \sqrt{2g_c(h_{in} - h_{out})} \tag{6.16}$$

Applying the previous assumptions for diffusers to the modified energy rate balance of Eq. (6.12) gives the following results:

$$\dot{Q} + \dot{m}(h_{in} - h_{out} + V_{in}^2/2g_c) = 0$$

or

$$h_{out}\Big|_{diffuser} = h_{in} + V_{in}^2/2g_c + \dot{Q}/\dot{m} \tag{6.17}$$

Thus, heat added to a diffuser increases the outlet specific enthalpy, whereas heat removal reduces it. For an insulated (adiabatic) diffuser, we have

$$h_{out}\Big|_{\substack{\text{adiabatic}\\ \text{diffuser}}} = h_{in} + V_{in}^2/2g_c \tag{6.18}$$

[3] This figure is accurate only for subsonic flow. When the flow becomes supersonic, the relative shapes of nozzles and diffusers are not the same as those shown here.

For an incompressible substance such as a liquid flowing through these systems, Eq. (3.34) gives the specific enthalpy change as

$$h_{in} - h_{out} = c(T_{in} - T_{out}) + v(p_{in} - p_{out}) \tag{6.19}$$

where c is the specific heat of the material and v is its specific volume. Combining Eq. (6.19) with Eqs. (6.15) and (6.16), we obtain

$$V_{out}\Big|_{\substack{\text{nozzle with}\\ \text{incompressible}\\ \text{fluid}}} = \left\{2g_c\left[\dot{Q}/\dot{m} + c(T_{in} - T_{out}) + v(p_{in} - p_{out})\right]\right\}^{1/2} \tag{6.20}$$

and

$$V_{out}\Big|_{\substack{\text{adiabatic}\\ \text{nozzle with}\\ \text{incompressible}\\ \text{fluid}}} = \sqrt{2g_c[c(T_{in} - T_{out}) + v(p_{in} - p_E)]} \tag{6.21}$$

Combining Eq. (6.19) with Eqs. (6.17) and (6.18), we can solve for the diffuser outlet pressure:

$$p_{out}\Big|_{\substack{\text{diffuser}\\ \text{incompressible}\\ \text{fluid}}} = p_{in} + (1/v)[c(T_{in} - T_{out}) + V_{in}^2/2g_c + \dot{Q}/\dot{m}]$$

and

$$p_{out}\Big|_{\substack{\text{adiabatic}\\ \text{diffuser}\\ \text{incompressible}\\ \text{fluid}}} = p_{in} + (1/v)[c(T_{in} - T_{out}) + V_{in}^2/2g_c]$$

For an ideal gas with constant specific heats (such as air at atmospheric pressure and temperature), Eq. (3.42) gives[4]

$$h_{in} - h_{out} = c_p(T_{in} - T_{out}) \tag{6.22}$$

where c_p is the constant pressure specific heat. Then, Eqs. (6.15) and (6.16) become

$$V_{out}\Big|_{\substack{\text{nozzle with}\\ \text{ideal gas}}} = \left\{2g_c\left[\dot{Q}/\dot{m} + c_p(T_{in} - T_{out})\right]\right\}^{1/2}$$

and

$$V_{out}\Big|_{\substack{\text{adiabatic}\\ \text{nozzle with}\\ \text{ideal gas}}} = \sqrt{2g_c c_p(T_{in} - T_{out})}$$

and Eqs. (6.17) and (6.18) become:

$$T_{out}\Big|_{\substack{\text{diffuser with}\\ \text{ideal gas}}} = T_{in} + (1/c_p)(V_{in}^2/2g_c + \dot{Q}/\dot{m})$$

and

$$T_{out}\Big|_{\substack{\text{adiabatic}\\ \text{diffuser with}\\ \text{ideal gas}}} = T_{in} + V_{in}^2/(2g_c c_p)$$

---

[4] Note that this formula can be used here even though this is not a constant pressure process, because the enthalpy of an ideal gas depends only on temperature and is therefore independent of pressure (see Chapter 3).

We are now ready to carry out a thermodynamic analysis of an open system. We begin with simple examples and work toward more difficult ones. The example problems are designed to illustrate the text material that immediately precedes the example. Therefore, the analysis section of the solution often is abbreviated, with appropriate reference made to the previous text material wherein the analysis has already been carried out. Do not be misguided by this.

Homework and examination problems usually do not simply mimic textbook examples. The purpose of these examples is *not* to give you a set of ready-made formulae into which you can plug numbers to solve specific problems. Their function is to teach you analysis techniques so that you have the ability and self-confidence to solve any thermodynamics problem whatsoever, whether you have seen one similar to it or not.

EXAMPLE 6.2

The nozzle on a lawn or garden hose has a 1.00 inch inlet inside diameter and an inlet pressure of 80.0 psig at 60.0°F. The mass flow rate of water through the nozzle is 0.800 lbm/s. Assuming the water flows through the nozzle isothermally, determine

a. The outlet velocity from the nozzle, $(V_{out})_a$.
b. The height to which the stream of water rises above the nozzle outlet when the nozzle is pointed straight up, $(Z_{out})_b$.

Solution

First, draw a sketch of the system (Figure 6.4).

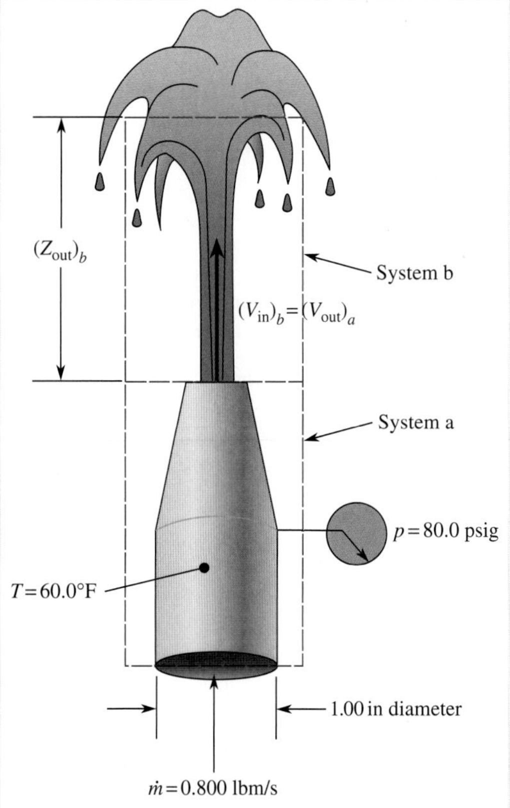

FIGURE 6.4
Example 6.2.

This is an open system, and the unknowns are $(V_{out})_a$ and $(Z_{out})_b$ (see Figure 6.4). The material is liquid water. We assume from our experience with garden hose nozzles that the system is a steady state, steady flow, single-inlet, single-outlet open system.

(Continued)

EXAMPLE 6.2 *(Continued)*

The material flowing through this system is water at 80.0 psig (94.7 psia) and 60.0°F. A check of the saturation tables for water shows us that the water is in a compressed liquid state and therefore can be considered to be an incompressible fluid. Also, since the amount of excess pressure here is relatively small (94.7 psia vs. the saturation pressure of 0.2563 psia at 60.0°F), we need not use the compressed water tables but can get sufficient accuracy using the saturated liquid tables at 60.0°F for all the inlet properties we may need.

The first law formulation that applies to this problem is the modified energy rate balance, Eq. (6.12). The auxiliary equations needed include the equation of state for an incompressible fluid (Eq. (6.19)) and the mass flow rate formula (Eq. (6.1)). For the case of the system shown in Figure 6.4, we have the standard nozzle configuration, except that we can calculate V_{in} in this problem from Eq. (6.14) as

$$V_{in} = \frac{4\dot{m}v}{\pi D^2}$$

where $v \approx v_f(60.0°F) = 0.01603\ \text{ft}^3/\text{lbm}$. Then, for system a,

$$(V_{in})_a = \frac{4(0.800\ \text{lbm/s})(0.01603\ \text{ft}^3/\text{lbm})}{\pi(1.00\ \text{in.})^2(1\ \text{ft}/12\ \text{in.})^2} = 2.35\ \text{ft/s}$$

Now, this $(V_{in})_a$ probably produces a negligible inlet kinetic energy, but since we have its value, we carry it along in the solution for the time being.

The solution to **part a** is obtained by solving Eq. (6.12), the modified energy rate balance, for $(V_{out})_a$, with the enthalpy values given by Eq. (6.19). This produces an equation similar to Eq. (6.20) except with a V_{in} term included. Since we have no information about any heat transfer to or from the nozzle, we assume that there is none. This is justifiable on the following basis: Since the nozzle is very small (hand size), the water is not inside of it long enough for any significant heat transfer of energy to occur. This is a common circumstance that often occurs in obviously uninsulated small systems with no significant heat transfer. If the residence time of the fluid in the system is very small, then in the absence of an extraordinarily large temperature difference between the environment and the system, the time is simply insufficient for any significant heat transfer to occur, regardless of whether the system is insulated or not. The modified energy rate balance for **system a** reduces to a form of Eq. (6.21):

$$(V_{out})_a = \{V_{in}^2 + 2g_c[c(T_{in} - T_{out}) + v(p_{in} - p_{out})]\}_a^{1/2}$$

The problem statement told us to assume the water flow through the nozzle is isothermal, so we set $T_{in} = T_{out}$. Actually, the water flow is not exactly isothermal, due to an increase in internal energy of the water from viscous effects, turbulence, and so forth. However, for a small nozzle, these effects are negligible. For an isothermal flow, we obtain

$$(V_{out})_a = [(V_{in}^2)_a + 2g_c v(p_{in} - p_{out})_a]^{1/2}$$

The data for **system a** are as follows:

$$\begin{aligned}(V_{in})_a &= 2.35\ \text{ft/s}\\ v &= v_f(60.0°F) = 0.01603\ \text{ft}^3/\text{lbm}\\ (p_{in})_a &= 80.0\ \text{psig} = 94.7\ \text{psia}\\ (p_{out})_a &= 0.00\ \text{psig} = 14.7\ \text{psia}\end{aligned}$$

Then

$$(V_{out})_a = \left\{(2.35\ \text{ft/s})^2 + 2\left(32.174\frac{\text{lbm}\cdot\text{ft}}{\text{lbf}\cdot\text{s}^2}\right)(0.01603\ \text{ft}^3/\text{lbm})[(80.0 - 0.00)\ \text{lbf/in}^2] \times (144\ \text{in}^2/\text{ft}^2)\right\}^{1/2} = 109\ \text{ft/s}$$

Notice that $(V_{in})_a$ was only about 2% of $(V_{out})_a$ and therefore could have been neglected in this case.

Part b of this example is basically a mechanics problem, but it can be easily solved using system b in Figure 6.4 and the modified energy rate balance. The following assumptions are now made for **system b**:

$$\begin{aligned}&\dot{Q} = \dot{W} = 0 && (Z_{out})_b = ?\\ &(V_{out})_a = (V_{in})_b = 109\ \text{ft/s} && (p_{in})_b = (p_{out})_b = 14.7\ \text{psia}\\ &(V_{out})_b \approx 0\ \text{ft/s} && (T_{in})_b = (T_{out})_b = 60.0°F\\ & && (Z_{in})_b = 0\end{aligned}$$

$(V_{out})_b$ is $\ll (V_{in})_b$ here because the water stream spreads into a large fan at the top of its trajectory and therefore exits system b through a large surface area. The conservation of mass law for an incompressible fluid requires that $\dot{m}$ = constant and so $(V_{out})_b = (V_{in})_b(A_{in}/A_{out})_b \ll (V_{in})_a$ here.

These assumptions imply a negligible aerodynamic drag on the water stream and negligible viscous dissipation within the stream itself. When these conditions are applied to the modified energy rate balance for **system b**, we obtain

$$\underset{0}{\dot{Q}} - \underset{0}{\dot{W}} + \dot{m}[h_{in} - h_{out} + (V_{in}^2 - 0)/2g_c + (0 - Z_{out})g/g_c]_b = 0$$

or

$$(Z_{out})_b = [(g_c/g)(h_{in} - h_{out}) + V_{in}^2/2g]_b$$

The change in specific enthalpy for this example is again given by Eq. (6.19), and under the assumptions listed for system b, it is clear that this change is zero. Then, our working modified energy rate balance reduces to

$$(Z_{out})_b = (V_{in}^2)_b/2g$$

or

$$(Z_{out})_b = \frac{(109\ \text{ft/s})^2}{2(32.174\ \text{ft/s}^2)} = 185\ \text{ft}$$

Note that our calculations for both $(V_{out})_a$ and $(Z_{out})_b$ in this example gave numbers somewhat higher than we would observe if we measured these values in an experiment. This is because we ignore the viscous dissipation effects in the water and surrounding air. Dissipation effects are considered in more detail in the next chapter, on the second law of thermodynamics.

Exercises

4. Determine the exit diameter of the garden hose nozzle used in Example 6.2. Assume the water is incompressible so that $v_{in} = v_{out}$. **Answer**: $(D_{out})_a = 0.147$ in.
5. Determine the height to which the stream of water rises when the garden hose in Example 6.2 is pointed straight up and the nozzle removed. **Answer**: $(Z_{out})_b = 1.03$ in.
6. If the exit diameter of the garden hose nozzle used in Example 6.2 is reduced to 0.100 in, determine the exit velocity and the height to which the stream of water rises when pointed straight up. **Answers**: $(V_{out})_a = 235$ ft/s, and $(Z_{out})_b = 858$ ft.

6.6 THROTTLING DEVICES

Throttling device is the generic name of any device or process that simply dissipates pressure energy $\dot{m}pv$ by irreversibly converting it into thermal energy. Unlike nozzles and diffusers, throttling devices provide no form of useful energy recovery. They merely convert pressure energy into thermal energy through dissipative viscous flow (usually turbulent) processes. In fact, any device that incurs a large irreversible pressure drop can be thought of as a throttling device. Figure 6.5 schematically illustrates a variety of common throttling devices.

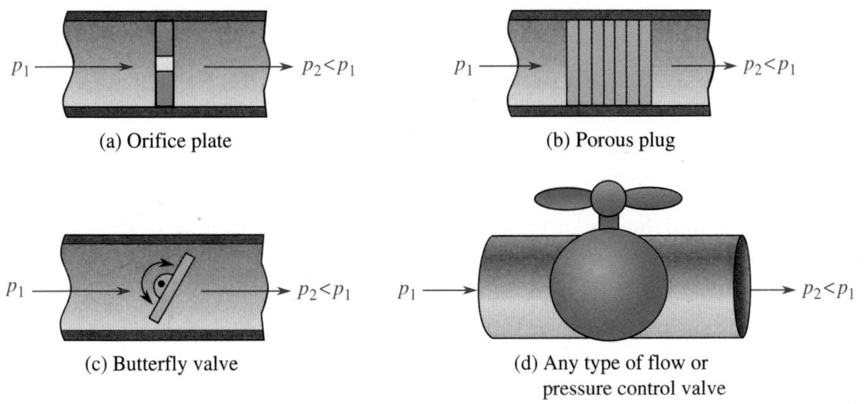

FIGURE 6.5
Some common throttling devices.

A throttling device may be thought of as any aergonic device whose primary purpose is to offer a resistance to flow. Throttles may or may not be insulated. But they are usually such small devices and have such high flow rates that the residence time of the fluid in them is too short for significant heat transport of energy to occur. Consequently, a throttling device is commonly taken to be adiabatic regardless of whether it is actually insulated or not.

The small physical size of most throttling devices also prevents them from having a significant change in specific potential energy between their inlet and outlet flow streams. However, a throttle need not have the same inlet and outlet flow velocities, and therefore, it may have a significant specific kinetic energy change across it.

Consequently, we define a throttling device with the following set of thermodynamic conditions:

Throttling Devices Have

$$\dot{Q} = 0$$
$$\dot{W} = 0$$
$$Z_{in} - Z_{out} \approx 0$$

Applying these conditions to the modified energy rate balance of Eq. (6.12) gives

$$0 - 0 + \dot{m}[h_{in} - h_{out} + (V_{in}^2 - V_{out}^2)/2g_c + 0] = 0$$

or

$$h_{out} = h_{in} + (V_{in}^2 - V_{out}^2)/2g_c \tag{6.23}$$

If $V_{in} = V_{out}$, as when the fluid is incompressible and the inlet and outlet areas of the throttle are equal (e.g., cases a–d in Figure 6.5), then Eq. (6.23) reduces to the simpler form

$$h_{out} = h_{in} \tag{6.24}$$

Such throttling devices are said to be *isenthalpic* (i.e., they have a constant enthalpy).

Even if the inlet and outlet velocities are clearly unequal in some problem, you may still be able to justify using the simpler Eq. (6.23) as the result of your analysis. The high-velocity flow stream of an unequal area throttling device is always limited by the speed of sound in the flowing medium.[5]

Consequently, if h is large, say on the order of 1000 Btu/lbm (2300 kJ/kg), then the specific kinetic energy of the flow stream can never be more than 2 or 3% of this value and may therefore be considered negligible. The rule of thumb discussed earlier in this chapter can be applied as follows: *If you are given a throttling device problem without adequate velocity information and where a velocity is not an unknown that you are required to find as part of the solution, then you should assume that the specific kinetic energy terms are either equal (and therefore cancel each other) or that they are negligible.*

For an incompressible fluid flowing through a throttling device, we can use Eq. (6.19) in Eq. (6.23) to produce

$$c(T_{in} - T_{out}) + v(p_{in} - p_{out}) + (V_{in}^2 - V_{out}^2)/2g_c = 0$$

and if we neglect the specific kinetic energy terms (or have $V_{in} = V_{out}$), then this equation can be rearranged to give

$$T_{out} = T_{in} + (v/c)(p_{in} - p_{out})$$

and since p_{in} is usually greater than p_{out}, this equation tells us that there is normally a temperature rise in an incompressible fluid flowing with a negligible specific kinetic energy change through a throttling device.

[5] Supersonic nozzles or diffusers usually have a flow stream velocity greater than the sonic velocity. But, with the rare exception of supersonic flow at the inlet to a throttling device, subsonic flow prevails throughout throttling devices.

For an ideal gas with constant specific heats, we can substitute Eq. (6.22) into Eq. (6.23) to obtain

$$T_{out} = T_{in} + (V_{in}^2 - V_{out}^2)/(2g_c c_p)$$

This equation tells us that, in the case of negligible change in specific kinetic energy, the throttling of an ideal gas is an isothermal process.

The actual throttling device outlet temperature for a pure substance is dependent on its Joule-Thomson coefficient μ_J, defined as

$$\mu_J = (\partial T/\partial p)_h \tag{6.25}$$

Since μ_J is defined completely in terms of intensive thermodynamic properties, it too is an intensive thermodynamic property. A throttling process that has a negligible change in specific kinetic energy is a process of constant h, so the Joule-Thomson coefficient for any pure substance can be approximated from data taken during such a throttling process as

$$\mu_J \approx (\Delta T/\Delta p)_{\substack{\text{throttling}\\ \text{process}}} \tag{6.26}$$

If we take $\Delta p = p_{out} - p_{in}$, then Δp normally is a negative number for such a process. Clearly, a positive value for μ_J means that the temperature drops during such a throttling process ($\Delta T = T_{out} - T_{in} < 0$) and a negative value for μ_J means that the temperature increases. For an isothermal throttling process (such as occurs with an ideal gas), $\mu_J = 0$.

A gaseous pure substance that has a positive Joule-Thomson coefficient could undergo a continuous decrease in temperature and eventually be liquified by a properly designed throttling process. This was the basis of a process introduced in 1895 by Karl von Linde (1842–1934) for the large-scale production of liquid air. The temperature at which $\mu_J = 0$ for a real pure substance is called its *inversion temperature* T_{inv}, and $\mu_J > 0$ for $T < T_{inv}$ and $\mu_J < 0$ for $T > T_{inv}$. Thus, the temperature of a real gas decreases in a throttling process if its inlet temperature is less than its inversion temperature. However, the temperature of a gas cannot be lowered via the Joule-Thomson effect if the gas inlet temperature exceeds its "maximum" inversion temperature (see Table 6.3).[6]

Figure 6.6 shows the variation in the Joule-Thomson coefficient with pressure and temperature for air and carbon dioxide.

Table 6.3 The Maximum Joule-Thomson Inversion Temperature for Various Common Gases

Substance	Maximum Inversion Temperature	
	K	R
Air	659	1186
Argon	780	1404
Carbon dioxide	1500	2700
Helium	40	72
Hydrogen	202	364
Neon	231	416
Nitrogen	621	1118
Oxygen	764	1375

Source: Reprinted by permission of the publisher from Zemansky, M. W., Abbott, M. M., Van Ness, H. C., 1975. Basic Engineering Thermodynamics, second ed. McGraw-Hill, New York.

---

[6] Since the condition $\mu_J = 0$ can occur at more than one temperature, a gas may have several inversion temperatures, the largest of which is its "maximum" inversion temperature.

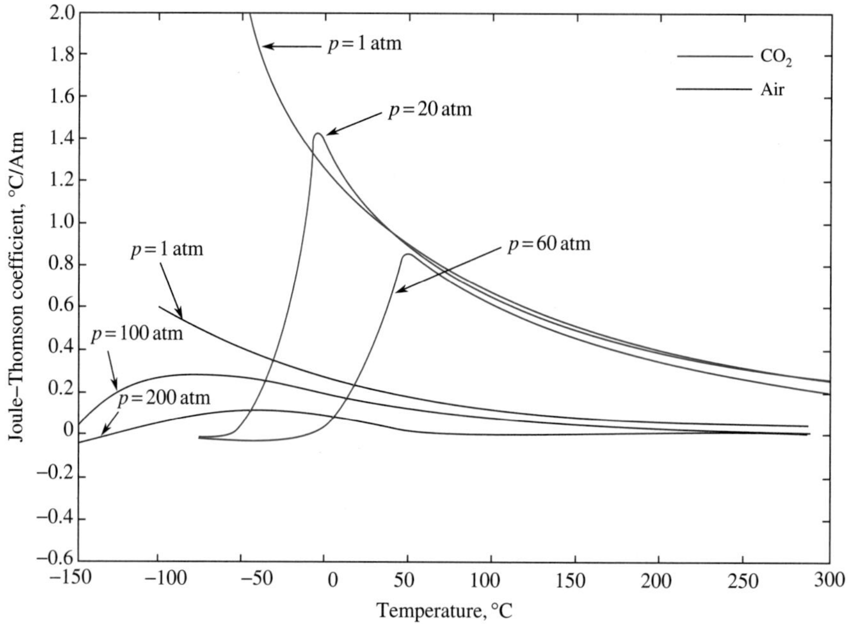

FIGURE 6.6
The variation in the Joule-Thomson coefficient of air and carbon dioxide with pressure and temperature.

6.7 THROTTLING CALORIMETER

A *throttling calorimeter* is a device that expands (i.e., throttles) a mixture of liquid plus vapor into the superheated vapor region. Under the vapor dome, temperature and pressure are not independent properties; therefore, their measurement alone cannot be used to fix the thermodynamic state of a substance. If, however, the thermodynamic state can be moved into a region where pressure and temperature are independent properties, then its state can be determined from a pressure gauge and a thermometer reading. This is the purpose of a throttling calorimeter, as illustrated in the following example.

EXAMPLE 6.3

Wet (i.e., a mixture of liquid plus vapor) steam flows in a pipe at 2.00 MPa. An insulated throttling calorimeter is attached to the pipe and a small portion of the steam is withdrawn and throttled to atmospheric pressure. The temperature and pressure of the throttled steam in the calorimeter are 150.°C and 0.100 MPa. Determine the quality of the wet steam in the pipe and estimate its Joule-Thomson coefficient μ_J.

Solution

First, draw a sketch of the system (Figure 6.7).

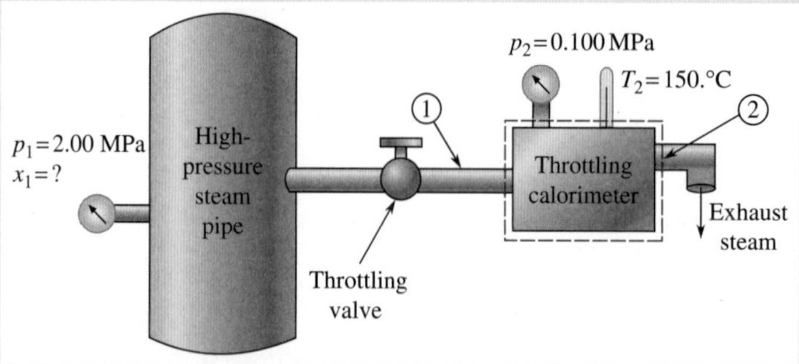

FIGURE 6.7
Example 6.3.

The unknowns are the quality of the wet steam in the pipe and its Joule-Thomson coefficient. The system is open and the material is wet steam.

A throttling calorimeter is clearly a steady state, steady flow, single-inlet, single-outlet open system. The material flowing is steam, and the unknown is the quality of the inlet flow stream ($x_{in} = x_1 = ?$) and the steam's Joule-Thomson coefficient μ_J.

The system and its properties follow:

$$\underline{\text{Station 1}}\text{ (inlet)} \rightarrow \frac{\text{Throttling}}{\text{process}} \rightarrow \underline{\text{Station 2}}\text{ (outlet)}$$

Station 1 (inlet)	Station 2 (outlet)
$p_1 = p_{sat} = 200.\ \text{MPa}$	$p_2 = 0.100\ \text{MPa}$
$x_1 = ?$	$T_2 = 150.°\text{C}$
	$h_2 = 2776.4\ \text{kJ/kg}$

This is an example of a problem where the process path plus three of the four state properties needed to fix the two states are given, and the unknown is a property in the undetermined state. This is a common problem format.

A quick check of the steam tables shows that the outlet state is in the superheated region, and therefore all the outlet properties are easily found from the superheated steam table. Since we are given no information on mass flow rates or velocities, we assume that the changes in flow stream specific kinetic and potential energies are negligible. The calorimeter was stated to be insulated, so it will have no heat transfer, and we acknowledge that this device can neither do work nor have work done on it. Under these conditions, the modified energy rate balance reduces to Eq. (6.24), or

$$h_{in} = h_1 = h_{out} = h_2$$

From the superheated steam table (Table C.3b of *Thermodynamic Tables to accompany Modern Engineering Thermodynamics*), we find that

$$h_2 = h(0.100\ \text{MPa}, 150.°\text{C}) = 2776.4\ \text{kJ/kg}$$

and this value is listed previously as part of the data for station 2. Therefore, from the modified energy rate balance, we have

$$h_1 = h(2.00\ \text{MPa}, ?) = 2776.4\ \text{kJ/kg}$$

and the pair of independent properties $p_1 = 2.00$ MPa and $h_1 = 2776.4$ kJ/kg now fix the inlet state. From the saturation tables for water (Table C.2b), we find that, at 2.00 MPa,

$$h_{f1} = h_f(2.00\ \text{MPa}) = 908.8\ \text{kJ/kg}$$
$$h_{fg1} = h_{fg}(2.00\ \text{MPa}) = 1890.7\ \text{kJ/kg}$$
$$h_{g1} = h_g(2.00\ \text{MPa}) = 2799.5\ \text{kJ/kg}$$

Since $h_{f1} < h_1 < h_{g1}$, we can now use the auxiliary formula for quality x to determine its value at station 1 as

$$x_1 = (h_1 - h_{f1})/h_{fg1} = (2776.4 - 908.8)/1890.7 = 0.9878$$

So $x_1 = 98.8\%$, which is the quality of the steam in the pipe.

A rough estimate for the Joule-Thomson coefficient for this process is given by Eq. (6.26) as

$$\mu_J \approx (\Delta T/\Delta p)_{\substack{\text{throttling}\\ \text{process}}}$$

where, from the saturation tables, $T_1 = T_{sat}(2.00\ \text{MPa}) = 212.4°\text{C}$.

$$\mu_J \approx (212.4 - 150.°\text{C})/(2.00 - 0.100\ \text{MPa}) = 32.8°\text{C/MPa}$$

Note that this is not a particularly accurate value, since μ_J is a point function and consequently the values of ΔT and Δp used in its calculation should really be much smaller than those used in the preceding calculation. This calculation does, however, provide a reasonable *average* value of μ_J for this throttling process.

Exercises

7. What would be the quality of the steam in the pipe in Example 6.3 if the pressure in the pipe were 3.00 MPa instead of 2.00 MPa and everything else remained the same? **Answer**: $x = 0.985 = 98.5\%$.
8. What would be the quality of the steam in the pipe in Example 6.3 if the temperature in the calorimeter was 100.°C instead of 150.°C and everything else remained the same? **Answer**: $x = 0.935 = 93.5\%$.

6.8 HEAT EXCHANGERS

Heat exchanger is the generic name of any device whose primary function is to promote a heat transport of energy from one fluid to another fluid. Most heat exchangers have two separate fluid flow paths, which do not mix the fluids but instead promote the transfer of heat from one fluid to another across a thermally conducting but otherwise impermeable barrier. These heat exchangers have four flow streams, two inlets and two outlets. Since the primary function of a heat exchanger is a heat transfer process, they are characteristically aergonic devices. Also, all of the heat transfer should take place *inside* the heat exchanger; therefore, most heat exchangers are normally adiabatic devices when the entire heat exchanger is insulated and taken to be the system. It is normal for an uninsulated heat exchanger to have a net heat transfer to or from its surroundings, but when environmental heat transfer values are not supplied in a problem statement and are not an unknown or otherwise determinable, you are to assume that the entire heat exchanger is an adiabatic system. Figure 6.8 illustrates a typical heat exchanger schematic and operating characteristics.

A heat exchanger can be considered to be a pair of steady state, steady flow, single-inlet, single-outlet systems that have equal but opposite heat transfer rates. It can also be analyzed as a steady state, steady flow, double-inlet, double-outlet system that has no (assuming it is insulated) external heat transfer. In both cases it is common to neglect any changes in flow stream specific kinetic or potential energy across the system. The general ERB (Eq. (6.4)) for the latter case reduces to

$$\dot{Q} + \dot{m}_1 h_1 + \dot{m}_3 h_3 - \dot{m}_2 h_2 - \dot{m}_4 h_4 = 0$$

The conservation of mass law requires that $\dot{m}_1 = \dot{m}_2 = \dot{m}_a$ and that $\dot{m}_3 = \dot{m}_4 = \dot{m}_b$, where the subscripts a and b refer to the two different fluids. Therefore, the ERB becomes

$$\dot{Q} + \dot{m}_a(h_1 - h_2) + \dot{m}_b(h_3 - h_4) = 0 \tag{6.27}$$

and if the heat exchanger is insulated, then $\dot{Q} = 0$ and this equation further reduces to

$$\dot{m}_a(h_1 - h_2) = \dot{m}_b(h_4 - h_3) \tag{6.28}$$

If both fluids a and b are incompressible (e.g., liquids), then Eq. (6.19) can be used to give

$$\dot{m}_a[c_a(T_1 - T_2) + v_a(p_1 - p_2)] = \dot{m}_b[c_b(T_4 - T_3) + v_b(p_4 - p_3)]$$

where c_a and c_b are the specific heats of fluids a and b, respectively. For liquids, not only are v_a and v_b small numbers, but the pressure drops $p_1 - p_2$ and $p_3 - p_4$ across the heat exchanger are also small. Therefore, it is common to ignore the pressure terms in the previous equation, giving the final incompressible fluid heat exchanger ERB as

ERB when both fluids are incompressible liquids

$$\dot{m}_a c_a(T_1 - T_2) = \dot{m}_b c_b(T_4 - T_3)$$

If both fluids are ideal gases with constant specific heats, then the use of Eq. (6.22) gives

ERB when both fluids are ideal gases

$$\dot{m}_a(c_p)_a(T_1 - T_2) = \dot{m}_b(c_p)_b(T_4 - T_3)$$

where $(c_p)_a$ and $(c_p)_b$ are the constant pressure specific heats of gases a and b, respectively.

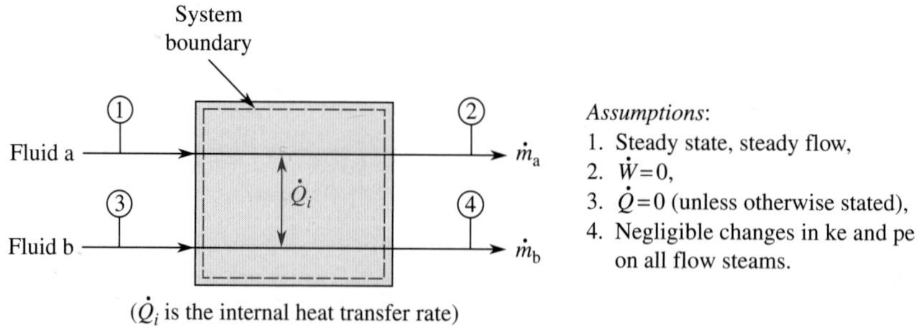

FIGURE 6.8
Typical heat exchanger operating characteristics.

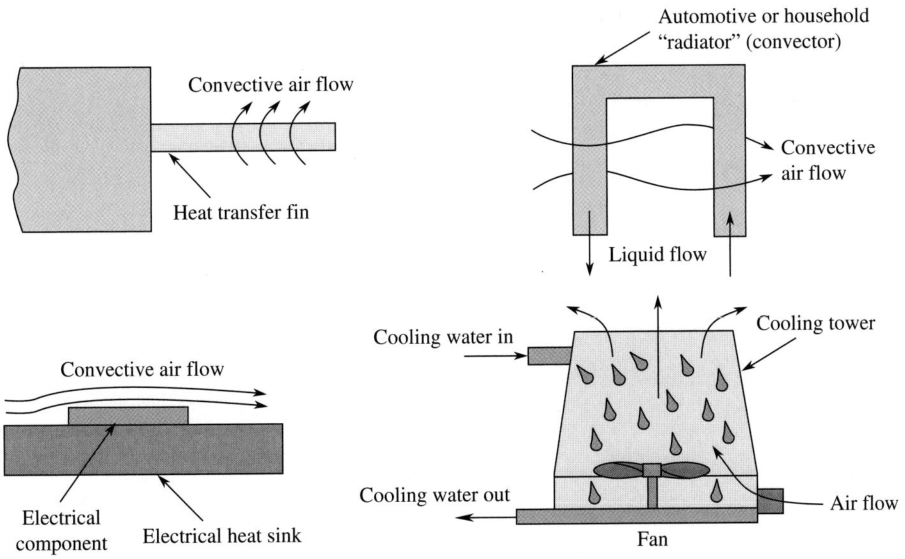

FIGURE 6.9
Heat exchangers where the environment is one of the heat transfer fluids.

If one fluid, say a, is an ideal gas (e.g., air) and the other fluid is an incompressible liquid with a negligible pressure drop across the device, then the insulated heat exchanger ERB becomes

ERB when one fluid is an ideal gas and the other fluid is an incompressible liquid

$$\dot{m}_a (c_p)_a (T_1 - T_2) = \dot{m}_b c_b (T_4 - T_3) \tag{6.29}$$

Several other combinations of flow stream fluid types are also possible.

If we now return to the two-system heat exchanger model and treat the two fluids as separate systems, we can apply the MERB individually to each of them to obtain

$$\dot{Q}_a + \dot{m}_a(h_1 - h_2) = \dot{Q}_b + \dot{m}_b(h_3 - h_4) = 0$$

or

$$\dot{Q}_a = \dot{m}_a(h_2 - h_1)$$

and

$$\dot{Q}_b = \dot{m}_b(h_4 - h_3)$$

Since the internal fluid to fluid heat transfer rate is $\dot{Q}_a = -\dot{Q}_b$, adding the previous two equations produces Eq. (6.28). Thus, both types of system analysis give the same results.

In some cases, the environment itself is one of the heat transfer fluids. Heat transfer fins, automobile radiators, electrical heat sinks, and so on are all designed to transfer heat to or from the environment. Some of these devices are single flow stream systems and must be analyzed using Eq. (6.29). These systems are illustrated in Figure 6.9.

EXAMPLE 6.4

A *condenser* is a heat exchanger designed to condense a vapor into a liquid. Determine the flow rate of cooling water taken from a local river required to condense 12.0 kg/min of water vapor at 1.00 MPa and 500.°C into a saturated liquid at 1.00 MPa. The river water can be considered to be an incompressible fluid with an inlet temperature of 15.0°C. The cooling water must be returned to the river and is restricted by environmental code requirements not to exceed 20.0°C.

(*Continued*)

EXAMPLE 6.4 *(Continued)*

Solution

First, draw a sketch of the system (Figure 6.10).

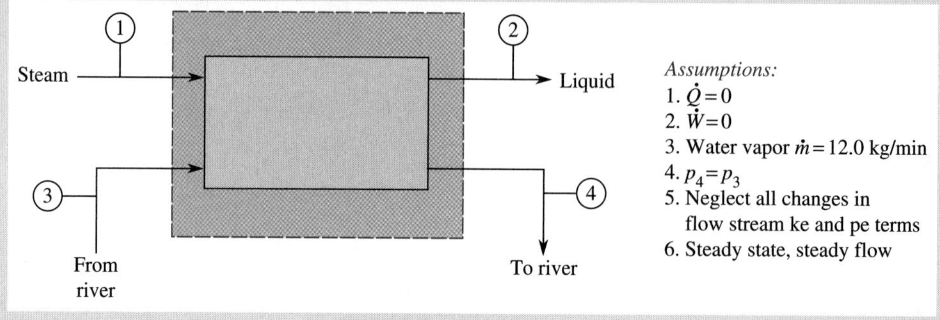

FIGURE 6.10
Example 6.4.

The system is open, and the unknown is the river water mass flow rate $\dot{m}_w$. We treat this as a steady state, steady flow, double-inlet, double-outlet open system problem. Since no information is given about heat loss or gain from the environmental surface of the condenser we assume that it is insulated and therefore adiabatic. Also, since no pressure loss information is given for the river cooling water flow stream, we assume that it is negligible.

This particular case is not covered by any of the equations developed in the previous discussion of heat exchangers. It is a combination of a pure substance (water vapor) and an incompressible fluid (river water). However, its ERB can be quickly arrived at by beginning with the general ERB for heat exchangers, Eq. (6.27), and adding the auxiliary formulae for the specific enthalpy change of an incompressible fluid, given in Eq. (6.19). This gives

$$\underbrace{\dot{Q}}_{0} + \dot{m}_s(h_1 - h_2) = \dot{m}_w(h_4 - h_3)$$

$$= \dot{m}_w[c_w(T_4 - T_3) + \underbrace{v_w(p_4 - p_3)}_{0}]$$

Then, solving for the river water mass flow rate $\dot{m}_w$, we get

$$\dot{m}_w = \dot{m}_s\{(h_1 - h_2)/[c_w(T_4 - T_3)]\}$$

This equation tells us what properties we need to calculate the unknown. The station data are in Table 6.4.

Table C.3b gives us $h_1 = 3478.4$ kJ/kg, and Table C.2b gives us $h_2 = h_f(1.0 \text{ MPa}) = 762.8$ kJ/kg. Both these values have been added to the station data list in Table 6.4. Also, Table 3.5 in Chapter 3 gives the specific heat of liquid water as $c = 4.2$ kJ/(kg·K). The condensate flow rate is given in the problem statement to be $\dot{m}_s = 12.0$ kg/min, so that the required river water flow rate is

$$\dot{m}_w = (12.0\,\text{kg/min})[(3478.4 - 762.8)\,\text{kJ/kg}]/\{[4.2\,\text{kJ/(kg·K)}](20 - 15\,\text{K})\}$$
$$= 1552\,\text{kg/min}$$

Table 6.4 Data for Example 6.4

Station 1	Station 2	Station 3	Station 4
$p_1 = 1.0$ MPa	$x_2 = 0.0$ (sat. liq.)	$T_3 = 15°C$	$T_4 = 20°C$
$T_1 = 500°C$	$p_2 = 1.0$ MPa		
$h_1 = 3478.4$ kJ/kg	$h_2 = 762.8$ kJ/kg		

Exercises

9. What cooling water flow rate would be required from the river in Example 6.4 if the steam flow rate is 500. kg/s instead of 12.0 kg/min and everything else remained the same? **Answer**: $\dot{m}_w = 3.88 \times 10^6$ kg/min.
10. Suppose air instead of river water is used to cool the steam in Example 6.4. Determine the mass flow rate of air required if the air has an inlet temperature of 20.0°C and an exit temperature of 30.0°C. **Answer**: $\dot{m}_{air} = 3250$ kg/min.
11. If the pumps that supply the river cooling water in Example 6.4 could deliver only 1000. kg/min, what would be the new temperature of the cooling water as it exits the heat exchanger? **Answer**: $(T_w)_{exit} = 22.8°C$.

6.9 SHAFT WORK MACHINES

Shaft work machines are devices whose primary function is to promote a work input or output through a rotating or reciprocating shaft. Common shaft work machines are hydraulic pumps, pneumatic compressors and fans, gas or hydraulic turbines, electric motors and generators, and external and internal combustion engines. Most shaft work machines are steady state, steady flow, single-inlet single-outlet devices (electric motors and generators are exceptions, since they have no flow streams). The work produced or absorbed by such devices can then be determined from the MERB of Eq. (6.12) as

$$\dot{W}_{\text{shaft}} = \dot{m}[h_{\text{in}} - h_{\text{out}} + (V_{\text{in}}^2 - V_{\text{out}}^2)/2g_c + (Z_{\text{in}} - Z_{\text{out}})g/g_c] + \dot{Q} \tag{6.30}$$

This equation shows that the effect of heat loss ($\dot{Q} < 0$) from a work-producing device ($\dot{W} > 0$) is to *reduce* the device's power output. Therefore, most work-producing systems (engines, turbines, etc.) are insulated to improve their efficiency. Similarly, heat loss from a work-absorbing device (such as a compressor) requires that more work be supplied to produce the same state change in the flow streams. Consequently, most of these devices are also insulated to increase their efficiency. Massive amounts of heat loss from these systems by external cooling usually indicate the need to lower their internal temperatures due to the existence of large internal irreversibilities. This is a consequence of the second law of thermodynamics and is discussed in the next chapter.

Most shaft work machines have negligible change in the specific kinetic and potential energies of their flow streams. Obvious exceptions are hydroelectric water turbines, in which the specific potential energy change of the water is the energy source for the turbine, and windmills, in which the specific kinetic energy change of the air is the energy source for the windmill. The resulting ERB for an insulated shaft work machine with negligible changes in specific kinetic and potential flow stream energies, operating in a steady state, steady flow, single-inlet, single-outlet manner is obtained from Eq. (6.30) as

$$\dot{W}_{\text{shaft}} = \dot{m}(h_{\text{in}} - h_{\text{out}}) \tag{6.31}$$

Figure 6.11 illustrates the graphical symbols used to represent several common shaft work machines.

If an incompressible fluid is used in a shaft work machine described by Eq. (6.31), then Eq. (6.19) can be used to describe the change in specific enthalpy as

$$\dot{W}_{\text{shaft}}\Big|_{\substack{\text{incomp.}\\\text{fluid}}} = \dot{m}[c(T_{\text{in}} - T_{\text{out}}) + v(p_{\text{in}} - p_{\text{out}})]$$

Normally, there is very little temperature change across such devices as hydraulic pumps, motors, and turbines, so that the previous equation reduces to

$$\dot{W}_{\text{shaft}}\Big|_{\substack{\text{isothermal}\\\text{incomp.}\\\text{fluid}}} = \dot{m}v(p_{\text{out}} - p_{\text{in}}) \tag{6.32}$$

Notice that, in this equation,

$$\dot{m}v = (\rho AV)v = \frac{AV}{v}(v) = AV$$

where AV is the volume flow rate.

If an ideal gas with constant specific heats is used in a shaft work machine described by Eq. (6.31), then Eq. (6.22) can be used to describe the change in specific enthalpy, and Eq. (6.31) becomes

$$\dot{W}_{\text{shaft}}\Big|_{\substack{\text{ideal}\\\text{gas}}} = \dot{m}c_p(T_{\text{in}} - T_{\text{out}})$$

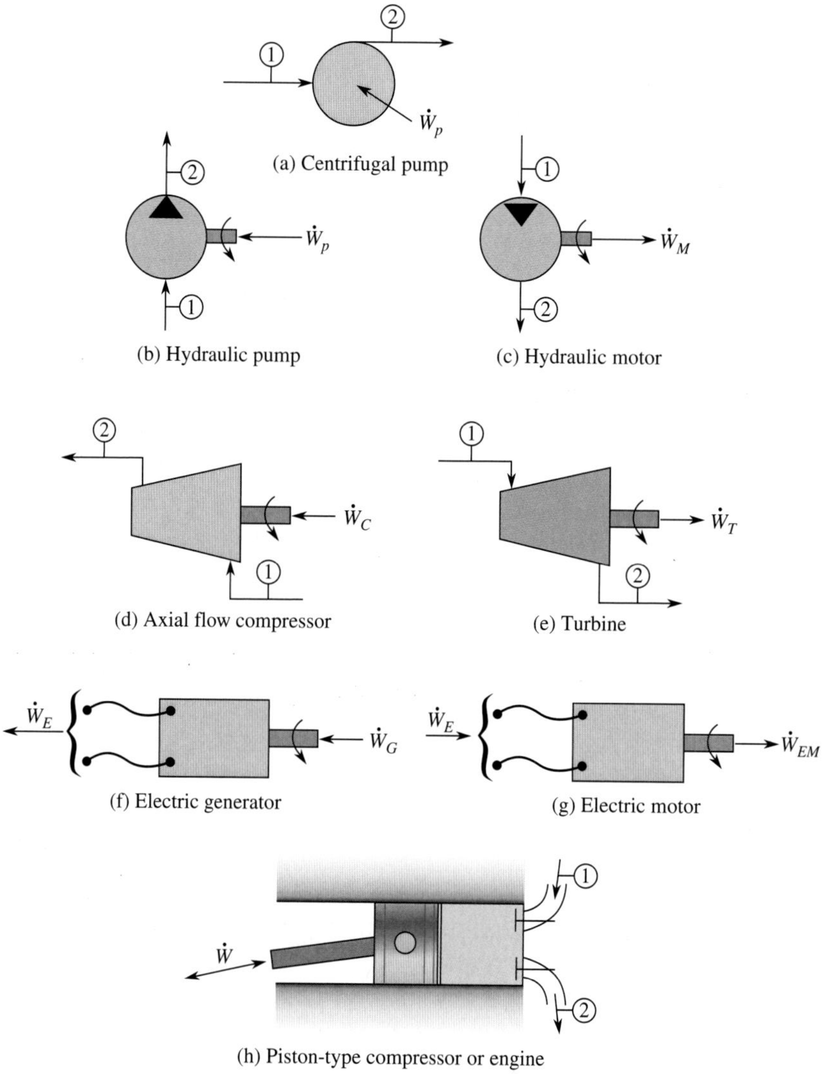

FIGURE 6.11
Graphical symbols for common shaft work machines.

EXAMPLE 6.5

Nearly every urban home has water supplied from a local water main. This water is used in washing and cooking, but its pressure could also be used as an energy supply. Suppose you installed a small hydraulic motor or turbine on the inlet water pipe of your house. Every time water is used in the house, the motor or turbine produces shaft work that could be used to run a small appliance or drive an electric generator and charge a battery. How much power could you realize in this way if you use an average of 20.0 gal of water over an eight (8.00) hour period, with an inlet water pressure of 85.0 psig and an exit water pressure of 10.0 psig?

Solution

First, draw a sketch of the system (Figure 6.12).

The system is open, and the unknown is $\dot{W}_{\text{shaft}}$. The material is liquid water, an incompressible fluid, and we ignore any changes in specific kinetic or potential energy of the flow stream plus any heat transfer that may occur. For steady state, steady flow, isothermal conditions, our modified energy rate balance (Eq. (6.12)) becomes Eq. (6.32), or

$$\dot{W}_{\text{shaft}} = \dot{m}v(p_{\text{out}} - p_{\text{in}})$$

where $\dot{m}v$ = 20.0 gal/8.00 h = 2.50 gal/h (on average).

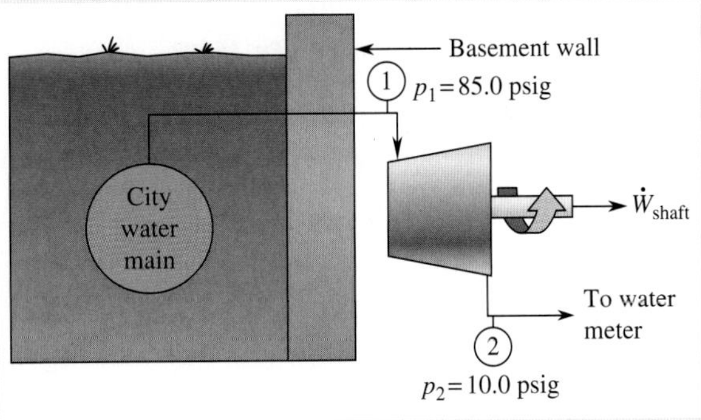

FIGURE 6.12
Example 6.5.

Now, (2.50 gal/h)(0.13368 ft^3/gal)(1 h/3600 s) = 9.283 × 10^{-5} ft^3/s. So,

$$\begin{aligned}\dot{W}_{shaft} &= (9.283 \times 10^{-5}\,\text{ft}^3/\text{s})[(85.0 - 10.0)\,\text{lbf/in}^2](144\,\text{in}^2/\text{ft}^2)\\ &= 1.00\,\text{ft}\cdot\text{lbf/s}.\\ &= (1.00\,\text{ft}\cdot\text{lbf/s})[1\,\text{hp}/(550\,\text{ft}\cdot\text{lbf/s})] = 1.82 \times 10^{-3}\,\text{hp}\\ &= (1.82 \times 10^{-3}\,\text{hp})(746\,\text{W/hp}) = 1.36\,\text{W}\end{aligned}$$

Therefore, the amount of power we get out of such a device would be extremely low and probably would not justify its initial expense.[7]

Suppose, now, we calculate an instantaneous power instead of the average power. For this calculation, we assume an instantaneous water flow of 5 gal/min. Then, the hydraulic power produced is

$$\begin{aligned}\dot{W}_{Shaft}\Big|_{instantaneous} &= (1.36\,\text{W})(5\,\text{gal/min})(60\,\text{min/h})[1/(2.50\,\text{gal/h})]\\ &= 163\,\text{W}\end{aligned}$$

Which is more reasonable. This is enough power to light two 75 W lightbulbs, but since this water flow rate does not occur continuously the bulbs would not be lit very often.

Exercises

12. What flow rate of water would be required in Example 6.5 to produce 1.00 hp of shaft output power?
Answer: $\dot{m}_w$ = 22.9 gal/min.
13. Determine the inlet water pressure required in Example 6.5 to produce a continuous shaft output power of 163 W with a water flow rate of 2.50 gal/h. **Answer**: $(p_{inlet})_w$ = 9.00 × 10^3 psig.
14. It has been suggested that, by installing a hydraulic turbine powered electric generator on the main water pipe supplying all the open flow devices in a house (sinks, toilets, washing machines, etc.), some of the water flow energy that is ordinarily wasted could be converted into useful electrical energy. Discuss the feasibility of this proposal in a short paragraph. **Hint**: Consider the economic costs of such a system and the required payback time. Estimate how much electrical power could be generated by analyzing the water flow in your home.

[7] *This conclusion might change if we were dealing with the domestic water flow into a large factory or multistory office or apartment building.*

EXAMPLE 6.6

Determine the quality of the steam at the outlet of an insulated steam turbine producing 2000. kJ of energy per kilogram of steam flowing through the turbine. The steam at the inlet of the turbine is at 2.00 MPa, 800.°C and the outlet pressure is 1.00 kPa. Neglect any changes in specific kinetic or potential energy of the flow stream, and assume a steady state operation.

Solution

First, draw a sketch of the system (Figure 6.13).

The system is open, and the unknown is the turbine's outlet steam quality x_2, as shown in Figure 6.13. The material is steam.

(*Continued*)

EXAMPLE 6.6 (Continued)

The MERB for the conditions described in the problem statement is Eq. (6.31):

$$\dot{W}_{\text{shaft}} = \dot{m}(h_1 - h_2)$$

We do not know $\dot{m}$ here, but we do know the energy produced per kilogram of steam flowing:

$$W/m = \dot{W}/\dot{m} = 2000.\ \text{kJ/kg}, \quad \text{so} \quad 2000.\ \text{kJ/kg} = h_1 - h_2$$

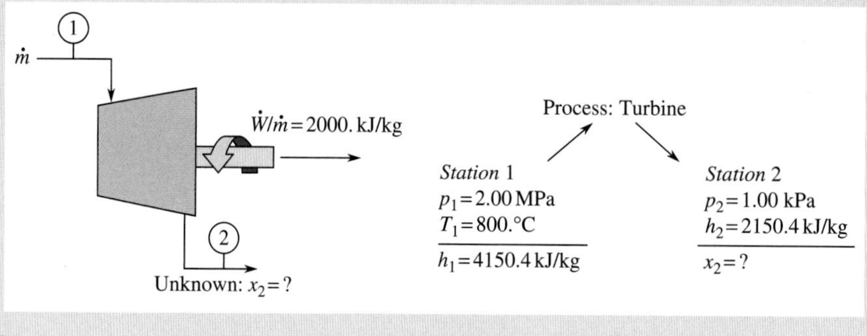

FIGURE 6.13
Example 6.6.

From Table C.3b of *Thermodynamic Tables to accompany Modern Engineering Thermodynamics*, we have $h_1 = 4150.4$ kJ/kg; and from Table C.2b, we have $h_{f2} = 29.30$ kJ/kg, $h_{fg2} = 2484.9$ kJ/kg, and $h_{g2} = 2514.2$ kJ/kg (all at 1.0 kPa). Therefore,

$$h_2 = h_1 - \dot{W}/\dot{m} = 4150.4 - 2000. = 2150.\ \text{kJ/kg}$$

These values of h_1 and h_2 have been added to our station data list in Figure 6.13. Hence, the values of $h_2 = 2150.$ kJ/kg and $p_2 = 1.00$ kPa are a pair of independent properties in the outlet state. Therefore, the outlet quality can be found from the auxiliary formula for quality:

$$x_2 = [h_2 - h_{f2}(1.00\ \text{kPa})]/h_{fg2}(1.00\ \text{kPa}) = (2150. - 29.30)/2484.9 = 0.854$$

$$= 85.4\%\ \text{vapor at the turbine's outlet}$$

Exercises

15. Determine the exit steam quality in Example 6.6 if the exit pressure is atmospheric pressure (0.101 MPa) instead of 1.00 kPa and everything else remains the same. **Answer**: $x_2 = 76.8\%$.
16. Determine the exit steam quality in Example 6.6 if the inlet steam temperature is 900.°C instead of 800.°C and everything else remains the same. **Answer**: $x_2 = 95.1\%$.
17. Suppose the shaft power produced by the turbine in Example 6.6 is 2000. kW and the steam mass flow rate is 2.00 kg/s. Determine the exit steam quality if everything else remains the same. **Answer**: Since $h_2 > h_g$ (1 kPa) here, the exit steam is superheated and quality is not a valid property.

6.10 OPEN SYSTEM UNSTEADY STATE PROCESSES

There are a wide variety of open system unsteady state processes in industry. Most are too complex to analyze easily, but one of the simpler cases involves the filling or emptying of a rigid tank or vessel.

Consider the tank-filling process illustrated in Figure 6.14. In this system, a rigid tank is connected through a valve to a high-pressure supply pipe. When the valve is opened, the rigid tank is filled from the supply pipe until the tank pressure is equal to that of the supply pipe. This is the filling process that we analyze.

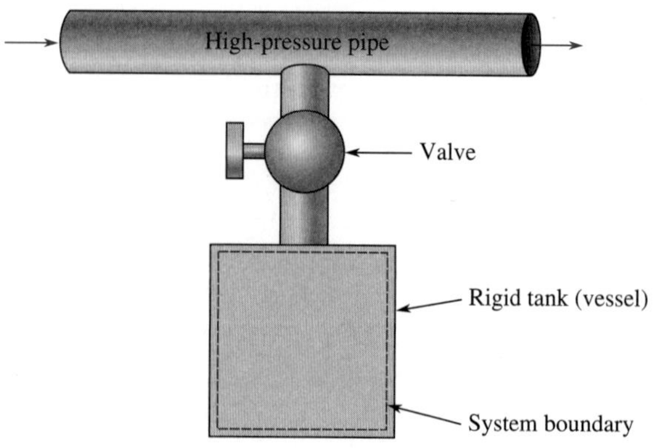

FIGURE 6.14
Filling a rigid vessel.

The filling process is neither steady state nor steady flow, since the mass of the system is continually changing.

Also, the system has a single-inlet flow stream but no outlet flow stream. To carry out the analysis, we note that the system does not contain a work mode, so $\dot{W} = 0$. Also, the system is not moving, so

$$[d(\mathrm{KE}+\mathrm{PE})/dt]_{\text{system}} = 0$$

Finally, we assume that the tank is filled slowly enough that we can ignore the inlet flow stream's specific kinetic and potential energy terms. Under these conditions, the generalized energy rate balance equation becomes

$$\dot{Q} - 0 + \dot{m}_{\text{in}}(h_{\text{in}} + 0 + 0) - 0 = (dU/dt + 0 + 0)_{\text{tank}}$$

or

$$\dot{Q} + \dot{m}_{\text{in}} h_{\text{in}} = [d(mu)/dt]_{\text{tank}}$$

Then, multiplying through by dt and integrating over the filling process gives

$${}_1Q_2 + \int_1^2 h_{\text{in}} dm_{\text{in}} = (m_2u_2 - m_1u_1)_{\text{tank}}$$

where state 1 is the initial state inside the tank and state 2 is the state inside the tank after it has been filled. When the tank is filled from a pipe of unlimited supply, as shown in Figure 6.14, h_{in} = constant and this equation becomes

$${}_1Q_2 + h_{\text{in}} m_{\text{in}} = (m_2u_2 - m_1u_1)_{\text{tank}}$$

where the conservation of mass law for the filling process gives

$$m_{\text{in}} = (m_2 - m_1)_{\text{tank}}$$

Combining the previous two equations, we obtain

$$u_2 = u_1(m_1/m_2) + {}_1Q_2/m_2 + h_{\text{in}}(1 - m_1/m_2) \tag{6.33}$$

where the numerical subscripts refer exclusively to states inside the tank from this point on.

By knowing p_2 from the filling process and calculating u_2 from Eq. (6.33), we have fixed the final thermodynamic state of the filled tank and can then find the value of any other property we desire, say its final temperature T_2. To illustrate this, let us assume the tank is insulated (${}_1Q_2 = 0$) and initially evacuated ($m_1 = 0$). Then, Eq. (6.33) becomes:

$$u_2 = h_{\text{in}} \tag{6.34}$$

If we now assume that the tank is filled with an incompressible fluid, then we can utilize Eq. (6.19) and write

$$u_2 = cT_2 = h_{\text{in}} = cT_{\text{in}} + vp_{\text{in}}$$

or

$$T_2\Big|_{\substack{\text{filling}\\ \text{incomp.}\\ \text{fluid}}} = T_{\text{in}} + (vp_{\text{in}})/c \tag{6.35}$$

and the compression process of pressurizing the tank would cause T_2 to be greater than T_{in} by an amount $(vp_{\text{in}})/c$.

EXAMPLE 6.7

A high-pressure water storage system is used to fill initially empty, rigid, insulated tanks with liquid water. The temperature of the water entering the tank is 20.0°C and the final pressure of the water in the tank is 50.0 MPa. Determine the final temperature of the water in the tank immediately after it has been filled.

Solution

First, draw a sketch of the system (Figure 6.15).

(Continued)

EXAMPLE 6.7 *(Continued)*

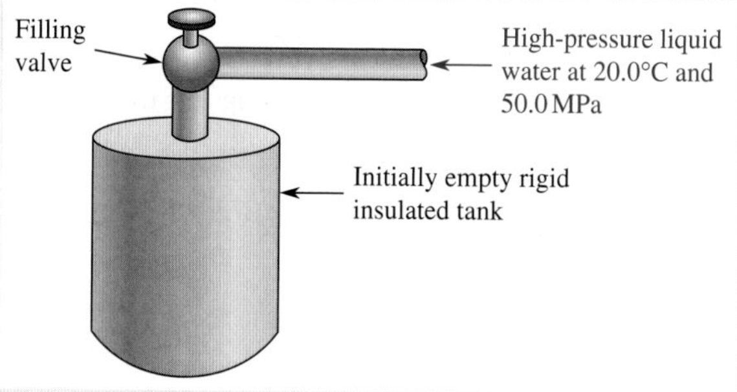

FIGURE 6.15
Example 6.7.

The unknown is the final temperature in the tank. The material is liquid water.

The final temperature in the tank immediately after it has been filled is given by Eq. (6.35) as

$$T_{\text{final filled}} = T_{\text{in}} + \frac{v p_{\text{in}}}{c}$$

In this problem T_{in} = 20.0°C, p_{in} = 50.0 MPa, and the specific heat of liquid water can be found in Table 3.5 as c = 4.216 kJ/kg·K. Even though the pressure range is quite high here (0 to 50.0 MPa), it is still a good assumption to take the specific volume of the water to be a constant at the value of v_f(20.0°C) = 0.001002 m^3/kg; the value of v(20.0°C, 50.0 MPa) from Table C.4b of *Thermodynamic Tables to accompany Modern Engineering Thermodynamics* is 0.0009804 m^3/kg, which is approximately the same as v_f(20.0°C). Then, Eq. (6.35) gives

$$T_{\text{final filled}} = 20.0°\text{C} + \frac{(0.001002\,\text{m}^3/\text{kg})(50.0 \times 10^3\,\text{kN/m}^2)}{4.216\,\text{kN}\cdot\text{m/kg}\cdot\text{K}} = 31.9°\text{C}$$

Exercises

18. An insulated, initially empty rigid container is filled with water from a water faucet in a house to a pressure of 0.700 MPa. The temperature of the water in the faucet and entering the container is 20.0°C. Determine the final temperature in the container immediately after it has been filled. **Answer**: T_{filled} = 20.2°C.
19. Determine the final temperature in the tank in Example 6.7 if it is filled with liquid mercury instead of water under the same conditions. Use v = 7.50 × 10^{-5} m^3/kg for mercury. **Answer**: T_{filled} = 20.9°C.

On the other hand, if we assume that the tank is filled with an ideal gas, then we can utilize Eq. (6.22) to get

$$u_2 = c_v T_2 = h_{\text{in}} = c_p T_{\text{in}}$$

or

$$T_2\Big|_{\substack{\text{filling}\\\text{ideal}\\\text{gas}}} = (c_p/c_v)T_{\text{in}} = kT_{\text{in}} \tag{6.36}$$

In the case of an ideal gas, the compression process generates a considerable amount of internal energy, as the following example illustrates.

EXAMPLE 6.8

A scuba diving air tank is filled from a 20.0°C, 1.40 MPa air supply. Neglecting the effect of any air initially in the tank and assuming the tank is insulated during the filling process, determine the final temperature of the air in the tank immediately after it is filled.

Solution

First, draw a sketch of the system (Figure 6.16).

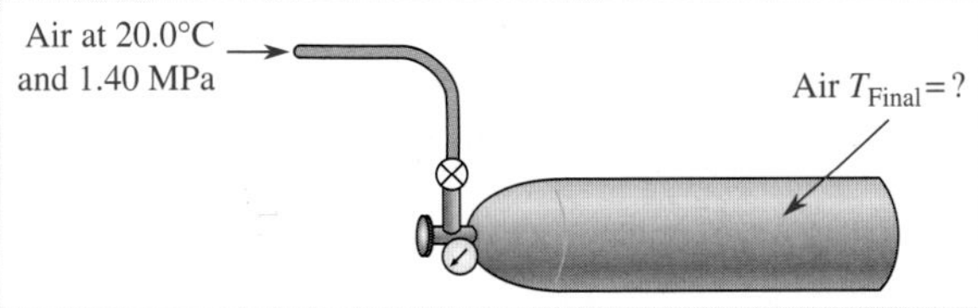

FIGURE 6.16
Example 6.8.

The unknown is the final temperature in the tank immediately after filling. The material is air.

The final temperature in the scuba tank immediately after it has been filled is given by Eq. (6.36) as

$$T_{\substack{\text{final}\\\text{filling}}} = kT_{\text{in}}$$

In this problem, T_{in} = 20.0°C and Table C.13b gives k = 1.40 for air. Then, Eq. (6.36) gives

$$T_{\substack{\text{final}\\\text{filling}}} = kT_{\text{in}} = 1.40(20.0 + 273.15) = 410\,\text{K} = 137°\text{C}$$

Therefore, bottled gas tanks can get quite hot during their filling process and should be cooled to minimize any rupture potential.

Exercises

20. If the scuba tank in Example 6.8 were filled with air at 70.0°F and 2000. psia, determine the final temperature, neglecting any air initially in the tank and assuming the tank is insulated during the filling process. **Answer**: T_{final} = 742 R = 282°F.

21. How do you account for the fact that the final temperature given by Eq. (6.36) is independent of the ideal gas pressure? **Answer**: Since the internal energy of an ideal gas is independent of pressure (see Chapter 3) and the property that defines the final state after the filling is complete is u_2, the final pressure cannot affect the final temperature when an ideal gas is used in the filling operation.

22. If the scuba tank in Example 6.8 is filled very quickly, it appears to be insulated whether it is actually insulated or not (why?) and the air inside gets quite hot. However, the tank is a heavy walled steel vessel with a large capacity to absorb the heat produced in filling the tank. If the tank itself has a mass of 10.0 kg, a specific heat of 0.503 kJ/kg · K, and initially is at 20.0°C, and it contains 0.500 kg of air with $(c_v)_{air}$ = 0.718 kJ/kg · K that is initially at 137°C, determine the final equilibrium temperature of the tank-air combined system. **Answer**: $T_{combined}$ = 27.8°C.

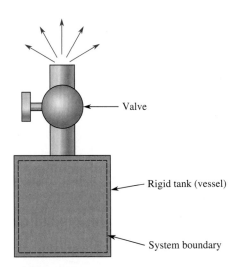

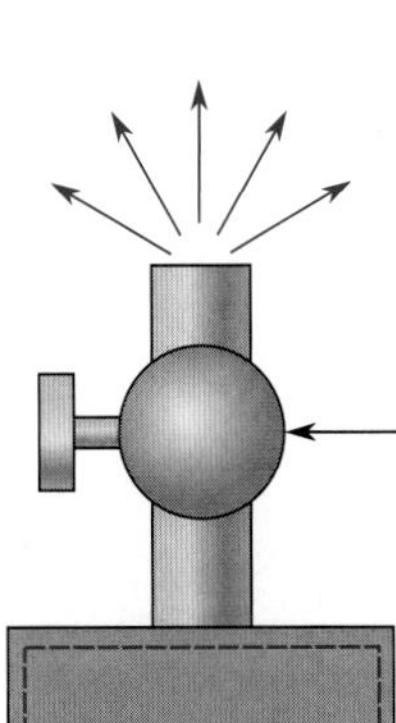

FIGURE 6.17
Emptying a rigid vessel.

The tank-emptying process is illustrated in Figure 6.17. Again we neglect flow stream specific kinetic and potential energies and require that the tank remain stationary and have no work transport of energy.

When the valve is opened, the initially pressurized rigid tank empties into the environment. The generalized energy rate balance for this process reduces to

$$\dot{Q} - 0 + 0 - \dot{m}_{\text{out}}(h_{\text{out}} + 0 + 0) = (dU/dt + 0 + 0)_{\text{tank}}$$

or

$$\dot{Q} - \dot{m}_{\text{out}} h_{\text{out}} = [d(mu)/dt]_{\text{tank}}$$

Again, multiplying through by dt and integrating gives

$$_1Q_2 - \int_1^2 h_{\text{out}} dm_{\text{out}} = (m_2u_2 - m_1u_1)_{\text{tank}} \tag{6.37}$$

where state 1 is the filled state and state 2 is the empty state (the reverse of the filling process). Unlike the filling process described earlier, the emptying process has no constant flow stream specific enthalpy ($h_{out} \neq$ constant); therefore, the integral in Eq. (6.37) cannot be evaluated. However, we can devise a way of dealing with this by defining an average specific enthalpy for the discharge flow stream as

$$h_{avg} = (1/m_{out}) \int_1^2 h_{out} dm_{out} \tag{6.38}$$

where, from the conservation of mass law, $m_{out} = (m_1 - m_2)_{tank}$. Approximating h_{avg} as

$$h_{avg} = [(h_1 + h_2)/2]_{tank}$$

Eq. (6.37) becomes

$${}_1Q_2 - h_1(m_1 - m_2)/2 - h_2(m_1 - m_2)/2 = m_2u_2 - m_1u_1$$

where the numerical subscripts refer exclusively to states inside the tank from this point on.

The effect of the emptying process can be more easily seen by simplifying the analysis somewhat. Let us stipulate that the tank is insulated (so ${}_1Q_2 = 0$) and that $m_1 \gg m_2 \approx 0$. Then, the previous equation can be rearranged to give

$$h_2 = 2u_1 - h_1 \tag{6.39}$$

In the case of emptying a pressurized incompressible fluid, we can write $u = cT$ and $h = cT + vp$, then Eq. (6.39) gives

$$T_2\Big|_{\substack{\text{emptying}\\\text{incomp.}\\\text{fluid}}} = T_1 - (v/c)(p_1 + p_2) \tag{6.40}$$

Equation (6.40) tells us that the expansion process that accompanies the emptying process of an incompressible fluid always causes the final temperature inside the tank to be less than the initial temperature inside the tank. In the case of emptying a pressurized ideal gas, we can write $u = c_vT$ and $h = c_pT$, then Eq. (6.39) gives

$$T_2\Big|_{\substack{\text{emptying}\\\text{ideal}\\\text{gas}}} = T_1[(2/k) - 1] \tag{6.41}$$

Since a continuous expansion process lowers the temperature of the remaining contents, this explains why pressurized cans of paint, deodorant, and the like become very cold when they are continuously discharged.

EXAMPLE 6.9

Immediately after the scuba tank in Example 6.8 is filled, the air is released into the atmosphere through an open valve on the top of the tank. Assuming that the emptying process is adiabatic and ignoring the mass of any air remaining in the tank when its pressure reaches atmospheric, determine the final temperature inside the tank immediately after the tank is empty.

Solution

First, draw a sketch of the system (Figure 6.18).

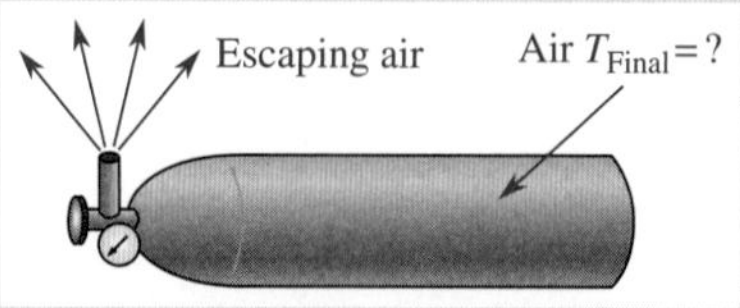

FIGURE 6.18
Example 6.9.

The unknown is the final temperature inside the tank immediately after the tank is emptied. The material is air.

The final temperature immediately after the emptying process is given by Eq. (6.41) as

$$T_{\substack{\text{final}\\\text{emptying}}} = T_{\text{initial}}\left[\frac{2}{k} - 1\right]$$

From Example 6.8, we find that the temperature in the tank immediately after it is filled is 137.3°C or 410.4 K. This is the initial temperature in the reverse emptying process. Again, from Table C.13b, we find for air that $k = 1.4$, so Eq. (6.41) gives

$$T_{\substack{\text{final}\\\text{emptying}}} = (410.\text{ K})\left[\frac{2}{1.40} - 1\right] = 176\text{ K} = -97.4°\text{C}$$

Since this is the reverse of the filling process, it should produce a final temperature of 20°C, the initial temperature in the filling process. The large error here is primarily due to the use of a simplified average enthalpy $[(h_1 + h_2)/2]_{\text{tank}}$ in place of a more accurate way of evaluating Eq. (6.38). Since we have no analytic information about h_{out}, we have no other choice if we are to obtain a solution. The error would not be so large if the temperatures were smaller.

Exercises

23. How could you correct the problem with the error introduced by using the simple average exit enthalpy shown in Example 6.9? **Answer**: Compute the average exit enthalpy over a series of small mass flow steps as the tank is emptied, or devise a correction factor for the offending equations.
24. A small, rigid, insulated tank is filled with helium at 70.0°F. A valve is opened on the tank, and it is completely emptied. Determine the temperature in the tank immediately after it is emptied. **Answer**: $T_{\text{final}} = -355$°F.
25. Immediately after the scuba tank in Example 6.9 is filled, the air is released into the atmosphere through an open valve on the top of the tank. Assuming that the emptying process is adiabatic and ignoring the effect of any air remaining in the tank when its pressure reaches atmospheric, use Eq. (6.40) to determine the final temperature inside the tank immediately after the tank is emptied. How does this temperature compare with the initial temperature used in Example 6.9? **Answer**: $T_{\text{final}} = 20.0$°C, the same as the initial temperature in Example 6.9.

EXAMPLE 6.10

An insulated, rigid tank on a spacecraft contains nitrogen at 2000. psig and 70.0°F. It is desired to discharge the tank isothermally to supply constant temperature nitrogen to the attitude control thrusters. This can be done if a portion of the discharged nitrogen is recycled back to the tank through a heater and compressor as shown in Figure 6.19. Assuming nitrogen to be an ideal gas and ignoring any changes in tank or flow stream kinetic and potential energies, determine

a. An expression for the ratio of recycled mass flow rate ($\dot{m}_R$) to discharge mass flow rate ($\dot{m}_D$) so that the temperature of the nitrogen in the tank (T_T) is constant in time.
b. The values of $\dot{m}_R/\dot{m}_D$ and $\dot{Q}_H$ for $T_R = 200.$°F, $T_T = 70.0$°F, $\dot{m}_R = 0.500$ lbm/s, and $\dot{W}_C = -3.00$ hp.

Solution

First, draw a sketch of the system (Figure 6.19).

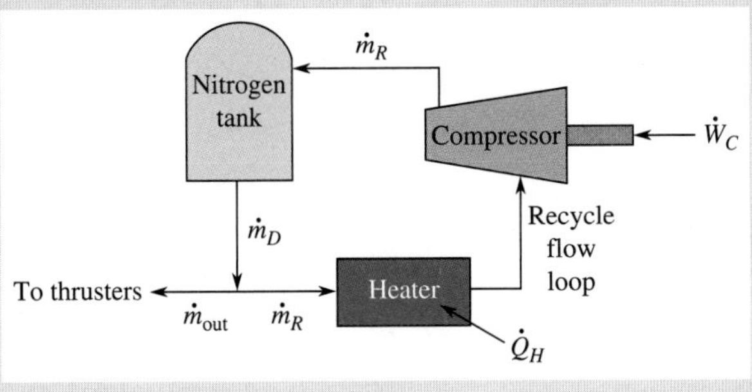

FIGURE 6.19
Example 6.10, schematic.

The unknown here is to find a formula for $\dot{m}_R/\dot{m}_D$ then to find its value for a specific set of conditions. Since $\dot{m}_R$ and $\dot{m}_D$ are mass flow rates into and out of the tank, let us first apply the general energy rate balance to the tank alone (Figure 6.20) and see what happens.

(Continued)

EXAMPLE 6.10 *(Continued)*

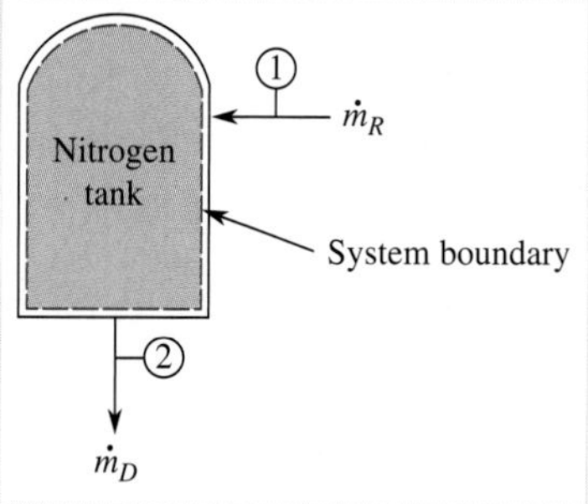

FIGURE 6.20
Example 6.10, tank alone.

Assumptions

1. $\dot{Q} = 0$ (tank is insulated).
2. $\dot{W} = 0$ (no work is done on or by the tank itself).
3. $T_T = T_2 = 70.0°\text{F}$ = constant (discharge is isothermal).
4. Neglect changes in KE and PE of the flow streams and the tank.
5. Treat N_2 as an ideal gas.

Then, the general energy rate balance becomes

$$\underbrace{\dot{Q}}_{0} - \underbrace{\dot{W}}_{0} + \dot{m}_R(h_1 + \underbrace{V_1^2/2g_c + gZ_1/g_c}_{0}) - \dot{m}_D(h_2 + \underbrace{V_2^2/2g_c + gZ_2/g_c}_{0})$$

$$= d(mu)_T/dt + \underbrace{d(\text{KE}+\text{PE})_T/dt}_{0}$$

or

$$\dot{m}_R h_1 - \dot{m}_D h_2 = d(mu)_T/dt = \dot{m}_T u_T + m_T \dot{u}_T$$

Now, $\dot{u}_T = du_T/dt = d(c_v T_T)/dt = 0$, since T_T = constant. The conservation of mass law gives

$$\dot{m}_T = \dot{m}_R - \dot{m}_D.$$

Then, the energy rate balance becomes

$$\dot{m}_R h_1 - \dot{m}_D h_2 = (\dot{m}_R - \dot{m}_D) u_T$$

or

$$\dot{m}_R/\dot{m}_D = (h_2 - u_T)/(h_1 - u_T)$$

Now,

$$h_2 = c_p T_2 = c_p T_T$$

and

$$u_T = c_v T_T$$

Then,

$$\dot{m}_R/\dot{m}_D = [(c_p - c_v)T_T]/(c_p T_1 - c_v T_T) = R/[c_p(T_1/T_T) - c_v]$$
$$= 1/[(c_p/R)(T_1/T_T) - c_v/R]$$

Since

$$c_p/R = c_p/(c_p - c_v) = k/(k-1)$$

and

$$c_v/R = c_v/(c_p - c_v) = 1/(k-1)$$

we have

$$\dot{m}_R/\dot{m}_D = (k-1)/[k(T_1/T_T)-1]$$

In part b, we are given that T_T = 70.0°F = 530. R and T_1 = 200.°F = 660. R.

Table C.13a gives us the specific heat ratio for nitrogen as k = 1.40. Therefore, our resultant equation gives

$$\dot{m}_R/\dot{m}_D = (1.40-1)/[1.40(660./530.)-1] = 0.538$$

Thus, 53.8% of the discharge mass must be recycled to keep the tank temperature constant.

To find the rate of recycle heat transfer required in part b, we must analyze the heater and compressor as a separate system (Figure 6.21).

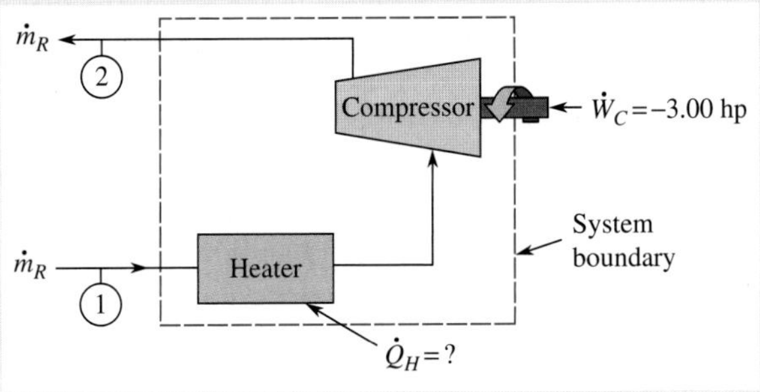

FIGURE 6.21
Example 6.10, heater and compressor as a separate system.

In general, $\dot{m}_R$ is not constant in time in this system. We assume that a feedback control system exists that automatically scales down $\dot{W}_C$ and $\dot{Q}_H$ as $\dot{m}_R$ goes to zero and the tank empties. We also assume that the internal, kinetic, and potential energies of the entire system are constant in time, so that we have a steady state, steady flow, single-inlet, single-outlet system. Under these conditions, the modified energy rate balance becomes

$$\dot{Q}_H - \dot{W}_C + \dot{m}_R(h_1 - h_2) = 0$$

where we again neglect any changes in flow stream specific kinetic and potential energies. Using the ideal gas relationship for specific enthalpy gives

$$\dot{Q}_H = \dot{m}_R c_p(T_2 - T_1) + \dot{W}_C$$

Table 13a gives c_p = 0.248 Btu/(lbm · R) for nitrogen. Then,

$$\begin{aligned}\dot{Q}_H &= (0.500\,\text{lbm/s})[0.248\,\text{Btu/(lbm·R)}](660. - 530.\,\text{R}) \\ &\quad + (-3.00\,\text{hp})[550\,\text{ft·lbf/(s·hp)}][1\,\text{Btu/(778.16 ft·lbf)}] \\ &= 14.0\,\text{Btu/s}\end{aligned}$$

SUMMARY

In this chapter, we investigate a series of open system examples and carry out a first law analysis of them using the energy balance or the energy rate balance. The primary purpose of this is to illustrate the material presented in Chapter 4. Several new equations are introduced in this chapter. Some of the more important equations are these.

1. The one-dimensional mass flow equation:

$$\left(\frac{dm}{dt}\right)_{\text{flow stream}} = (\dot{m})_{\text{flow stream}} = (\rho A V)_{\text{flow stream}} \qquad (6.1)$$

2. The general open system energy rate balance (ERB) equation:

$$\dot{Q} - \dot{W} + \sum_{\text{inlet}} \dot{m}(h + V^2/2g_c + gZ/g_c) - \sum_{\text{outlet}} \dot{m}(h + V^2/2g_c + gZ/g_c) = \frac{d}{dt}(U + mV^2/2g_c + mgZ/g_c)_{\text{system}} \quad (6.4)$$

3. The modified open system energy rate balance equation:

$$\dot{Q} - \dot{W} + \dot{m}[h_{\text{in}} - h_{\text{out}} + (V_{\text{in}}^2 - V_{\text{out}}^2)/(2g_c) + (Z_{\text{in}} - Z_{\text{out}})(g/g_c)] = 0 \quad (6.12)$$

Other equations are developed in this chapter that apply to specific geometries and boundary conditions. These equations are best studied in the context of the specific examples in which they were developed. The secret to mastering this material is to become competent at developing the equations that fit your specific problem by starting from the basic energy and auxiliary equations and applying the relevant boundary conditions (adiabatic, negligible kinetic or potential energies, and so forth). Consequently, not all the equations developed in this chapter are listed here so that you are encouraged to learn how to use the basic equations by studying the example problems and solutions presented.

The example problems discussed in this chapter are not meant to cover all the possible aspects of open system energy analysis. They were chosen to illustrate the problem solving technique, thermodynamic table usage, and how to make basic assumptions about process variables. You must learn how to successfully apply a generalized problem solution technique, such as the one used in this chapter and illustrated in the flowchart of Figure 4.21. More problem solving skills are gained by doing the problems at the end of this chapter.

Table 6.5 lists some of the new technical thermodynamic terms introduced in this chapter and earlier chapters. These terms are used without further explanation in the chapters that follow. It is recommended that the student learn their definitions before proceeding to the next chapter.

Table 6.5 Glossary of Technical Terms Introduced in Chapter 6 and Earlier Chapters

EB	The energy balance
ERB	The energy rate balance
${}_1Q_2$ and $\dot{Q}$	Heat transfer and heat transfer rate
${}_1W_2$ and $\dot{W}$	Work and work rate (power)
Aergonic	No work
Reversible work	No losses (i.e., no friction, heat transfer, etc.)
Work efficiency	A measure of the losses within a machine
Coefficient of performance (COP)	The name we give energy conversion efficiency when it is more than 100%
Conduction heat transfer	The heat transport of energy that obeys Fourier's law
Convection heat transfer	The heat transport of energy that obeys Newton's law of cooling
Radiation heat transfer	The heat transport of energy that obeys the Stefan-Boltzmann law
Adiabatic	No heat transfer
Steady state	A thermodynamic state that is constant in time
Steady flow	A state wherein the mass of an open system is constant in time
Flow stream	Where mass crosses a system boundary
Station	A data monitoring point on a flow stream
MEB	The modified energy balance
MERB	The modified energy rate balance
Pressure energy	Sometimes called the pressure *head*, $\dot{m}pv$
Nozzle	A device for converting pressure energy into kinetic energy
Diffuser	A device for converting kinetic energy into pressure energy
Heat exchanger	A device for promoting heat transfer from one fluid to another

Problems (* indicates problems in SI units)

1. Determine an expression for the time rate of change of the total internal energy of a submarine that is
   a. Not insulated.
   b. Being propelled by its propeller shaft.
   c. Taking on ballast water at only one opening in the submarine.
   d. Is diving and accelerating.
2. Write the complete energy rate balance (ERB) for an automobile accelerating up a hill and provide a physical interpretation for each term in the balance.
3. A new perfume is produced in the rigid, insulated reactor vessel shown in Figure 6.22. Determine the electrical power required to maintain the process in a steady state, steady flow condition. Neglect any changes in kinetic or potential energy.

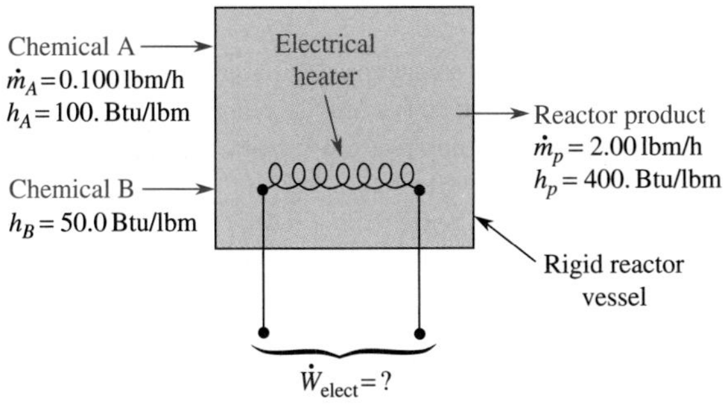

FIGURE 6.22
Problem 3.

4.* Determine the adiabatic change in temperature of a river that aergonically drops 1.00 m over a waterfall without a change in velocity. The specific heat of the river water is 4.20 kJ/(kg · K).
5. A small hydroelectric power plant discharges 200. ft^3/s of water. If the elevation difference ($Z_{in} - Z_{out}$) between the inlet and outlet is 15.0 ft and the temperature difference ($T_{in} - T_{out}$) is −0.0100°F, determine the mass flow energy transport rate. Assume the inlet and outlet velocities are identical: c = 1.00 Btu/(lbm · R) and ρ = 62.4 lbm/ft^3.
6. Air flows aergonically at a constant rate of 8.00 lbm/min down a horizontal duct so that its enthalpy remains constant. As the air flows down the duct, its velocity increases from 500. to 650. ft/s. Find the heat transfer rate and indicate whether it is to or from the system. Assume steady state operation.
7. A steady state air compressor takes in air at atmospheric pressure and discharges it at 100. psia. The inlet enthalpy is 120. Btu/lbm and the exit enthalpy is 176 Btu/lbm. Heat is transferred out of the compressor to cooling water at the rate of 1600. Btu/min. If the air flow rate through the compressor is 10.0 lbm/min, what horsepower must be supplied to the compressor? Neglect the kinetic and potential energies of the inlet and outlet flow streams.
8. Refrigerant-134a enters a constant area tube at 100.°F with a quality of 75.0%. Heat is transferred in a steady flow aergonic process until the R-134a leaves as a saturated liquid at exactly 0°F. Determine the heat transfer per lbm of R-134a flowing. Neglect any changes in kinetic and potential energies. Assume steady state operation.
9. An architect designed a 2.00 mile-high skyscraper. Steam is used for heating and is to be supplied to the top floor via a vertical pipe. The steam enters the pipe at the bottom as dry saturated vapor at 30.0 psia. At the top floor, the pressure is to be 16.0 psia, and the heat transfer from the steam as it flows up the pipe is to be 50.0 Btu/lbm. What is the quality of the steam at the top floor?
10.* How many watts of power could be recovered by decelerating 0.500 kg/s of air in a ventilating system from 10.0 m/s to 0.100 m/s before discharging it to the atmosphere?
11. Water initially at 300. psia and 500.°F is expanded isothermally and adiabatically to 14.7 psia in a horizontal steady flow process. In the absence of work modes, determine the change in kinetic energy per pound of water.
12. Refrigerant-134a expands in a steady flow diffuser from 300. psia, 180.°F to 35.0 psia in an isothermal process. During this process the heat transfer from the R-134a is 3.10 Btu/lbm. Assuming a negligible exit velocity, determine the inlet velocity to the diffuser.
13. A steam whistle is devised by attaching a simple converging nozzle to a steam line. At the inlet to the whistle, the pressure is 60.0 psia, the temperature is 600.°F, and the velocity is 10.0 ft/s. The steam expands and accelerates horizontally to the outlet, where the pressure and temperature are 14.7 psia and 500.°F. Determine the steam velocity at the whistle outlet. Assume the process is adiabatic, aergonic, and steady flow.
14.* Water vapor enters a diffuser at a pressure of 0.070 MPa, a temperature of 150.°C, and a velocity of 100 m/s. The inlet area of the diffuser is 0.100 m^2. By removing 288.2 kJ/kg in the form of heat across the duct walls, the velocity is reduced to 1.00 m/s and the pressure is increased to 0.200 MPa at the outlet. Determine the outlet area of the diffuser.
15. Air at 70.0°F, 30.0 psia, and a velocity of 3.00 ft/s enters an insulated steady state nozzle. The inlet area of the nozzle is 0.0500 m^2. The nozzle contains an operating 1500. W electrical heater. The air exits the nozzle at 14.7 psia and 300. ft/s. Determine the temperature of the air at the exit of the nozzle. Assume ideal gas behavior with constant specific heats and neglect any changes in flow stream potential energy.
16.* Air at 20.0°C, 0.500 MPa, and a velocity of 1.00 m/s enters an insulated nozzle. The inlet area of the nozzle is 0.0500 m^2. The nozzle contains an operating 500. W electrical resistance heater. The air exits the nozzle at atmospheric pressure and 100. m/s. Assuming ideal gas behavior with constant specific heats, determine the exit temperature.
17. Air at 70.0°F and 30.0 psia enters an insulated nozzle with a mass flow rate of 3.00 lbm/s. The nozzle contains an operating 1000. W electrical resistance heater. The air exits the nozzle at 14.7 psia. The inlet and exit areas of the nozzle are 0.500 and 0.100 ft^2, respectively. Determine the velocity and temperature of the air at the exit of the nozzle. Assume air to be an ideal gas.
18. The adiabatic, aergonic throttling calorimeter shown in Figure 6.23 is a device by which the quality of wet steam flowing in a pipe may be determined. Determine (a) the enthalpy of the steam in the pipe and (b) the quality of the steam in the pipe.

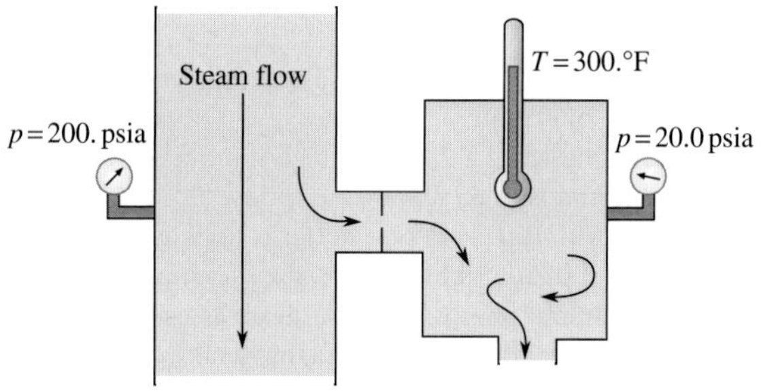

FIGURE 6.23
Problem 18.

19. Wet steam is throttled adiabatically and aergonically from 800. psia to 5.00 psia and 200.°F. If the inlet and exit velocities and heights are equal, what is the ratio of exit area to inlet area for this device?

20.* When the pressure on saturated liquid water is suddenly reduced in an adiabatic, aergonic, steady flow process, the exit state temperature must also be reduced to reach the new equilibrium state. Consequently, part of the initial liquid is very quickly converted into a saturated vapor at the lower pressure, with the vaporization energy (i.e., heat of vaporization) coming from the remaining liquid. In this way, the remaining liquid is cooled to the new (lower) equilibrium temperature. The resulting vapor is called *flash steam* because the liquid appears to "flash" into a vapor as it expands into the low-pressure region. Determine the exit temperature and percent of flash steam produced as saturated liquid water at 10.0 MPa is throttled through a partially open valve and discharged into the atmosphere adiabatically, aergonically, and with no change in kinetic or potential energy.

21. Refrigerant-22 is flowing steadily through a refrigerator throttling valve at the rate of 10.0 lbm/min. At the valve inlet, the R-22 is a saturated liquid at 80.0°F. At the valve outlet, the pressure is 31.162 psia. If the process can be considered aergonic, adiabatic, and with no change in kinetic or potential energy, find the quality at the valve outlet.

22.* The insulated vortex tube shown in Figure 6.24 contains no moving mechanical parts, yet it has the ability to separate the inlet air flow stream into hot and cold outlet air flow streams. Recorded test data are

Inlet pressure, $p_1 = 0.690$ MPa gauge
Inlet temperature, $T_1 = 20.0$°C
Hot side outlet temperature, $T_H = 82.0$°C
Hot side mass flow rate, $\dot{m}_H = 0.136$ kg/min
Cold side mass flow rate, $\dot{m}_C = 0.318$ kg/min

Calculate the cold side temperature, T_c.

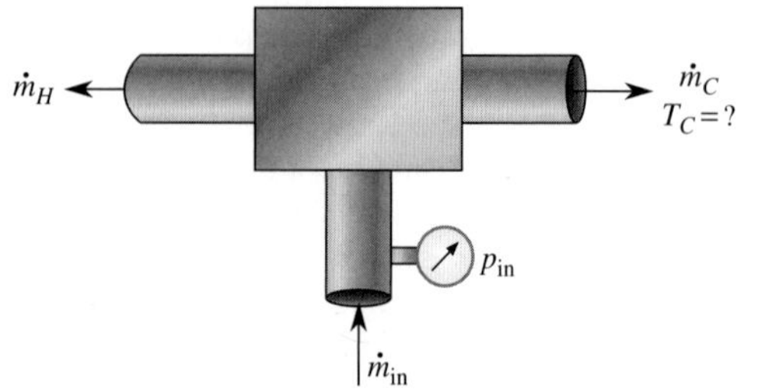

FIGURE 6.24
Problem 22.

23. Aerosol sprays are commonly used today for such things as hair sprays, shaving creams, deodorants, paints, and perfumes. At times, various inert gases have been used as the propellant medium for the active chemicals. Consider the design of a new deodorant that uses ammonia as the propellant medium. The spraying process is a simple throttling process (neglect kinetic and potential energy terms). If the can is at 80.0°F and 70.0 psia and it is spread into the atmosphere at 15.0 psia, then (a) at what temperature does the ammonia spray enter the atmosphere? (b) Draw this process on an *h-T* diagram and label all the relevant enthalpies and temperatures.

24.* Determine the inlet quality and the Joule-Thomson coefficient of wet steam that is throttled from 1.00 to 0.100 MPa and 150.°C.

25.* Estimate the Joule-Thomson temperature change that occurs as air is throttled from a pressure of 100. atm and 50.0°C to 1.00 atm.

26.* Estimate the Joule-Thomson temperature change as carbon dioxide is throttled from a pressure of 60.0 atm and exactly 0°C to 1.00 atm.

27. Refrigerant-22 enters a condenser at 30.0°F with a quality of 85.0% at a mass flow rate of 5.00 lbm/min. What is the smallest diameter tubing that can be used if the velocity of the refrigerant must not exceed 20.0 ft/s?

28.* How much electrical power (in kilowatts) is required to isothermally convert 10.0 kg/min of water from a saturated liquid to a saturated vapor at 100.°C in an electrically heated and completely insulated electric boiler?

29.* The hot and cold water faucets on a bathroom sink have water available at 80.0 and 15.0°C, respectively. When the faucets are opened, the sink drain is also open so that water leaves the sink as fast as it enters. Determine the ratio of hot water to cold water mass flow rates needed to produce a mixture temperature of 30.0°C in the sink.

30. The steady state, steady flow, adiabatic, aergonic feedwater heater shown in Figure 6.25 is used in an electric power plant. It mixes superheated steam with saturated liquid water to produce a low-quality outflow, in which 10.0 lbm/s of superheated steam at 80.0 psia and 500.°F is mixed with saturated liquid water at 80.0 psia. The outlet stream has a quality of 10.0% at 80.0 psia. What is the mass flow rate of the saturated liquid water flow stream?

FIGURE 6.25
Problem 30.

31.* An insulated aergonic condenser for a large power plant receives 3.00×10^6 kg/h of saturated water vapor at 6.00 kPa from a turbine and condenses it to saturated liquid at 6.00 kPa. Lake water is used to condense the steam and it is desired to maintain the inlet water temperature at 4.50°C and the outlet water temperature at 15.5°C. (a) What flow rate of lake water is required for an adiabatic, aergonic condenser and (b) what is the rate of heat transfer from the condensing steam to the lake water?

32.* A commercial slide projector contains a 500. W lightbulb. The bulb is to be air cooled. Determine the steady flow mass flow rate of air required if it enters the projector at 0.101 MPa and 22.0°C, and leaves the projector at 0.101 MPa and 50.0°C (neglect changes in kinetic and potential energies). Assume the air is an ideal gas.

33. Saturated liquid water at 70.0°F enters an aergonic device at a rate of 1.00 lbm/s. Heat is transferred to the water so that it exits the device as superheated steam at 100. psia and 600.°F. Determine the steady state heat transfer rate (ignore kinetic and potential energy effects).

34. Liquid water (ρ = 62.4 lbm/ft^3) enters one end of a 6.00 ft long, 1.00 in. diameter pipe with a uniform velocity. The entering pressure and velocity are 20.0 psia and 1.00 ft/s, respectively. Heat is added to the water as it flows down the pipe such that it exits the pipe as a saturated vapor at 14.7 psia. Determine the exit velocity.

35. Saturated liquid mercury at 100. psia enters an electrically heated 1.00 inch diameter horizontal pipe at a rate of 10.0 lbm/s with a negligibly small velocity. What is the steady flow heat transfer in Btu/s if the mercury exits the pipe at 80.0 psia as a saturated vapor with a velocity of 500. ft/s?

36.* A straight, horizontal, constant-diameter pipe contains an internal electrical heating coil (a resistance heater), and the outside of the pipe is insulated. Water enters the pipe as a saturated liquid at 0.500 kg/s and 0.200 MPa. How much electrical power must be dissipated in the electrical heater (in kilowatts) to produce saturated vapor at 1.00 MPa at the outlet?

37. An engineer wants to make a steady state, steady flow steam-cleaning jet by wrapping an electric heater around a water pipe. Water enters the pipe at 2.00 lbm/min as a slightly compressed liquid at 50.0°F and exits the pipe as a jet of saturated vapor at 14.7 psia. If the electric heater is plugged into a standard 110. V ac outlet, how much effective current does it draw? Ignore any kinetic or potential energy effects.

38.* The proposed solar collector installation shown in Figure 6.26 has a frontal area (exposed to the sun) of 50.0 m^2 and combines a thermoelectric generator with the air heating system of a building. The thermoelectric generator produces dc power with an efficiency of 5.00%. Solar radiation provides a net heat transfer rate to the absorber plate of 1000. W/m^2. The incoming air is at 18.0°C, and to obtain "good" operating efficiency, the exit air temperature must be maintained at 30.0°C. (a) How many watts of dc power are produced by the generator, and (b) what is the required air flow rate?

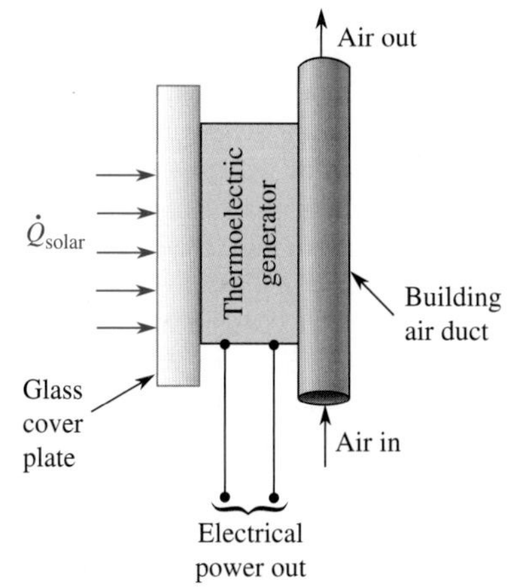

FIGURE 6.26
Problem 38.

39. Determine the final temperature and the power required to compress 10.0 ft^3/s of air from 14.7 psia and 80.0°F to a state where its specific volume is 2.84 ft^3/lbm in a steady state, steady flow process where $pv^{1.4}$ = constant. Assume ideal gas behavior.

40.* Find the power delivered by an adiabatic, isenthalpic turbine in which the mass flow rate is 2.00 kg/s and the flow enters at 1667 m/s and leaves at 404 m/s.

41. Liquid nitrogen can be made by a simple adiabatic expansion process through a turbine, in which 10.0 lbm/h N_2 enters the turbine at 500. R and 2000. psia and leaves the turbine at 1.00 atm as a liquid-vapor mixture. If the turbine produces work at a rate of 1500. Btu/h, what is the liquid nitrogen mass flow rate at the exit of the turbine? Neglect kinetic and potential energy effects.

42.* Determine the power required to compress a gas at a rate of 3.00 kg/s in a steady flow process from 0.100 MPa, 25.0°C to 0.200 MPa, 60.0°C. The specific enthalpy of the gas increases by 34.8 kJ/kg as it passes through the compressor, and the heat loss rate from the compressor is 16.0 kJ/s. Neglect any changes in flow stream kinetic and potential energies.

43.* Calculate the power required to compress air in a steady state, steady flow process with no change in elevation at a rate of 2.00 kg/s from 0.101 MPa, 40.0°C, 10.0 m/s to 0.300 MPa, 50.0°C, at 125 m/s. During this process, the enthalpy of the air increases by 40.15 kJ/kg, while 8.00 kJ/s of heat is lost to the environment.

44. Mercury enters the steady flow, steady state, adiabatic turbine of a starship warp drive system as a saturated vapor at 300. psia and exits the turbine with a quality of 75.0% at 1.00 psia. Determine

a. The mass flow rate of mercury required to produce 100. hp of turbine output power.

b. The inlet flow area if the inlet velocity is 1.00 ft/s.

45.* A simple air conditioner can be made by isothermally compressing air at atmospheric conditions of 0.101 MPa and 20.0°C to 0.700 MPa then adiabatically expanding it through a turbine back to its initial pressure. Determine the turbine outlet temperature if the turbine produces 750. W of power at an air flow rate of 0.100 kg/s. Assume ideal gas behavior.

46. The water pump on the engine of an automobile has a mass flow rate of 8.30 lbm/s. The water enters at 0.00 psig with a velocity of 1.00 ft/s and leaves at 10.0 psig with a velocity of 10.0 ft/s with no change in height or temperature. Assuming that the water is an incompressible liquid with a density of 62.4 lbm/ft^3 and the pump is adiabatic, determine the power (in horsepower) required to drive the pump.

47. A 20.0 hp aircraft engine is used to supply air at a rate of 0.982 lbm/s to support the ground effect vehicle shown in Figure 6.27. The vehicle has a support area of 50.0 ft^2. Estimate the maximum weight that this system can lift. Assume that the environmental temperature and pressure are 80.0°F and 14.7 psia, respectively, and the process path is pv^k = constant (where $k = c_p/c_v$).

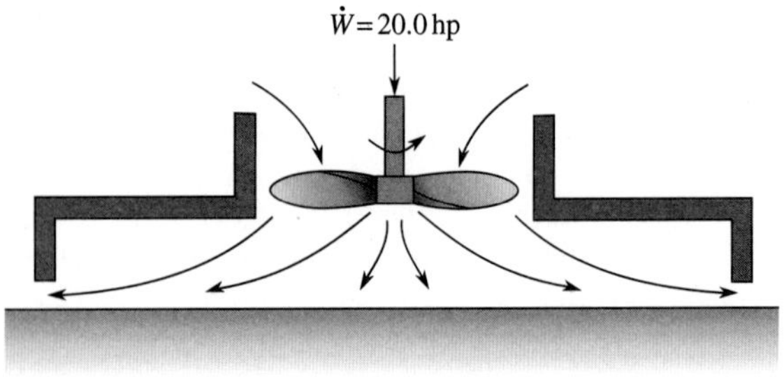

FIGURE 6.27
Problem 47.

48.* Determine the power required to drive a boiler feed pump that isothermally pumps 400. kg/s of saturated liquid water at 30.0°C to 8.00 MPa.

49. An adiabatic Refrigerant-22 turbine is mechanically coupled to an adiabatic steam compressor. Saturated R-22 vapor at 100.°F enters the turbine and exits as saturated vapor at −20.0°F. The steam enters the compressor as a saturated vapor at 14.7 psia and exits at 1000. psia, 1600.°F. If the steam mass flow rate is 5.00 lbm/s, find the R-22 mass flow rate.

50.* 3.00 kg/s of air is compressed in the steady flow, steady state, two-stage compressor shown schematically in Figure 6.28. Find the interstage temperature (T_2). Assume the air is an ideal gas with constant specific heats.

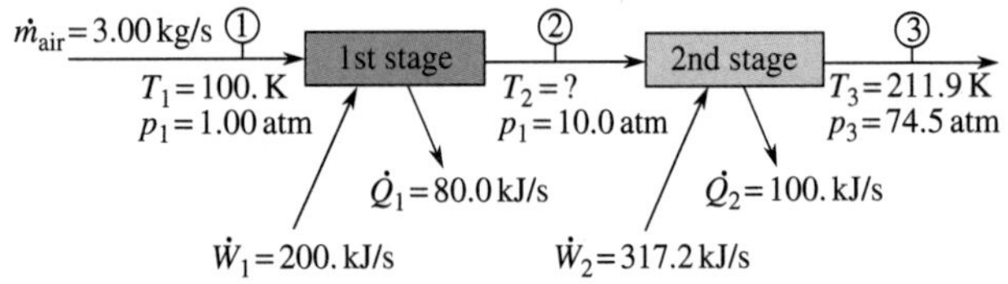

FIGURE 6.28
Problem 50.

51. In a steady flow process, a 1300. hp adiabatic steam turbine is supplied with 10.0 × 10^3 lbm of steam per hour. At the inlet to the turbine, the pressure of the steam is 500. psia and its velocity is 100. ft/s. The temperature of the steam leaving the turbine is 60.0°F, its quality 0.870, and its velocity is 700. ft/s. On leaving the turbine, the steam is condensed at constant pressure and exits the condenser as a saturated liquid at 60.0°F with negligible velocity. Find the temperature of the steam supplied to the turbine and the heat transfer rate in the condenser.

52.* Saturated liquid water at 70.0°C enters an aergonic boiler at station 1 in Figure 6.29. The boiler receives heat energy at a rate of 10.0 × 10^3 kJ/s. Superheated steam at 20.0 MPa and 800.°C leaves the boiler and enters an insulated turbine at station 2. The turbine exhausts to an aergonic condenser at a pressure of 200. kPa and a quality of 80.0% at station 3. The condenser cools the water to a saturated liquid at 100. kPa at station 4. Determine (a) the mass flow rate of water, (b) the power of the turbine, and (c) the heat transfer rate of the condenser.

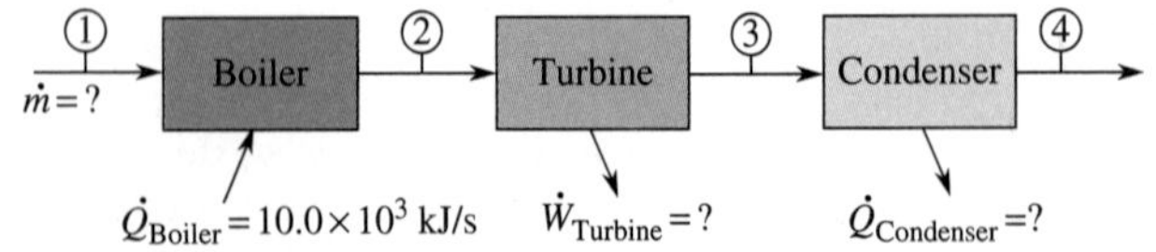

FIGURE 6.29
Problem 52.

53. 0.500 lbm/s of hydraulic oil (density = 55.6 lbm/ft^3 and specific heat = 0.520 Btu/(lbm · R)) is adiabatically pumped from 14.7 psia to 3014.7 psia with a 10.0 hp gear-type hydraulic pump. Determine the temperature rise in the oil as it passes through the pump.

54. An adiabatic air turbine is used to drive a compressor plus another device as shown in Figure 6.30. Assuming the working fluid (air) to be an ideal gas, find
   a. The mass flow rate of the air
   b. The power required to drive the compressor.

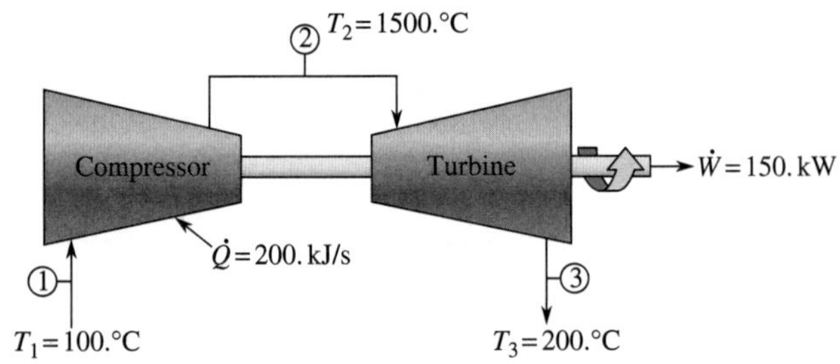

FIGURE 6.30
Problem 54.

55. It is proposed to construct a power plant on the shores of Lake Michigan. To preserve the essential qualities of the lake, a local environmental activist organized the community, which passed an ordinance requiring that condenser coolant obtained from the lake be returned to the lake at temperatures no warmer than 5.00°F above the temperature at which the water was withdrawn from the lake. The following are some of the design parameters of the proposed plant:
   a. Steam flow through condenser: 10.0 × 10^3 lbm/h.
   b. Inlet steam conditions: saturated vapor at 1.00 psia.
   c. Outlet condensate conditions: saturated liquid at 1.00 psia.
   d. External heat loss from condenser: equal to 8.00% of the energy extracted from the steam during condensation.
   e. Lake water has a specific heat of 1.00 Btu/(lbm · R)

   Find the required flow rate of coolant from Lake Michigan.

56. Determine the air velocity in the 0.250 in diameter neck of a balloon required to inflate an initially empty balloon to a diameter of 1.00 ft in 60.0 s. Assume the density of the air in the balloon remains constant during the inflation process.

57. Explain why the final temperature resulting from the adiabatic filling of a rigid vessel with an ideal gas is independent of the filling pressure.

58.* Incompressible hydraulic oil with a density of ρ = 880. kg/m^3 and specific heat of c = 2.10 kJ/(kg · K) is pumped from a reservoir at 35.0°C into a fully extended rigid hydraulic cylinder. Determine the temperature of the oil in the cylinder when its pressure reaches 35.0 MPa.

59. Incompressible liquid water—density = 62.4 lbm/ft^3 and specific heat = 1.00 Btu/(lbm · R)—at 70.0°F is pumped into a rigid insulated hollow bowling ball. Determine the temperature of the water in the bowling ball when its pressure reaches 100. × 10^3 psia.
60.* A 0.100 m^3 rigid tank is filled adiabatically to 20.0 MPa with helium. If the helium enters the tank at 20.0°C, determine the final temperature in the tank after it is filled. Assume ideal gas behavior with constant specific heats.
61.* Determine the heat transfer required to fill an initially empty rigid vessel isothermally with 15.0 kg of pure oxygen at 20.0°C. Assume ideal gas behavior with constant specific heats.
62. Determine the heat transfer required to cause a tank initially pressurized with air to discharge isenthalpically. The initial state inside the tank is p_1 = 1500. psia, T_1 = 100.°F, m_1 = 10.0 lbm and the final state is p_2 = 14.7 psia and m_2 = 0.098 lbm. Assume ideal gas behavior.
63.* A rigid tank with a volume of 0.500 m^3 contains superheated steam at 40.0 MPa and 500.°C. A valve on the tank is suddenly opened and steam is allowed to escape until the pressure in the tank is 1.00 MPa. While the steam is escaping, heat is simultaneously added to the tank in a manner that causes the specific enthalpy inside the tank to remain constant throughout the emptying process. Determine the total heat transfer required for this process.
64. Consider a rigid tank of volume $\forall$.
   a. Show that the heat transfer rate required to empty *or* fill the tank isenthalpically (note that you must show that this is true for both cases) is given by

$$\dot{Q}_{\substack{\text{isenthalpic}\\ \text{empty or fill}}} = -\forall(dp/dt)$$

   b. Then show that the total heat transfer required to carry out this isenthalpic process from state 1 to state 2 is given by

$$({}_1Q_2)_{\substack{\text{isenthalpic}\\ \text{empty or fill}}} = \forall(p_1 - p_2)$$

Computer Problems

The following computer problems were designed to be completed using a spreadsheet or equation solver. They may be used as part of a weekly homework assignment.

65. Develop a computer program that calculates the output velocity of an incompressible fluid flowing through an adiabatic nozzle. Input all the necessary variables with the proper units.
66. Develop a computer program that calculates the output pressure of an incompressible fluid from an adiabatic diffuser. Input the necessary variables with the proper units.
67. Develop a computer program that calculates the temperature of one of the four flow streams of a heat exchanger having two inlets and two outlets, when the mass flow rates and fluid properties of both of the flow stream fluids are known. Assume the fluids do not mix inside the heat exchanger and have one flow stream be an incompressible liquid and the other a constant specific heat ideal gas. Input the necessary variables with the proper units.
68. Develop a computer program that performs an energy rate balance on a gas turbine engine. Input the appropriate gas properties (in the proper units), the turbine's heat loss or gain rate, and the input mass flow rate, and inlet and exit temperatures. Output the turbine's output power. Assume the gas behaves as a constant specific heat ideal gas and neglect all kinetic and potential energy terms.
69. Develop a computer program that performs a steady state energy rate balance on an open system containing an ideal gas with constant specific heats. Input all but *one* (you choose which one) of the following quantities (in proper units): the heat and work energy transport rates, the mass flow rate, temperature, velocity and height of each flow stream entering and exiting the system, and the constant pressure specific heat of the gas contained in the system. Calculate the specific enthalpy of the gas using $h = c_pT$, where T is in absolute temperature units. One of these items is not supplied by the user and therefore becomes the unknown to be determined by the program. Output all the input variables plus the value of the unknown.
70. Repeat Problem 69, except use an incompressible liquid as the working fluid. Allow the user of your program to choose which variable is to be the unknown from a screen menu, then prompt for all the remaining variables. Use the conservation of mass law to determine or check the balance of the mass flows.

Create and Solve Problems

Engineering education tends to focus only on the process of solving problems. It ignores teaching the process of formulating solvable problems. However, working engineers are never given a well-phrased problem statement to solve. Instead, they need to react to situational information and organize it into a structure that can then be solved using the methods learned in college.

These "Create and Solve" problems are designed to help you learn how to formulate solvable thermodynamics problems from engineering data. Since you provide the numerical values for some of the variables, these problems do not have unique solutions. Their solutions depend on the assumptions you need to make and how you set them up to create a solvable problem.

71.* A canned pickle manufacturer wants to carry out the canning and sterilization process at 100. kPa. You have a 1.50. MPa steam line available at the sterilizer, containing steam with 5.00% moisture. You decide to drop the pressure from 250. to 100. kPa through a throttling orifice. Write and solve a thermodynamics problem that determines the 100. kPa (a) at exit temperature and (b) the exit quality.
72.* A design for a pump has been proposed involving the adiabatic, steady flow of liquid water through the pump. Saturated liquid water at 180.°C enters the pump and compressed water leaves at 2.20 MPa. Your section chief wants to know the work per unit mass of water flowing through the pump. Write and solve a thermodynamics problem that provides her with the answer.
73. Your boss comes to you with a new design for an insulated domestic electric water heater that is supposed to heat 1.00 gallons per minute of water from 50.0°F to 140.°F, but the outlet temperature reaches only 105°F. Your boss wants you to determine how much electrical power is needed to meet the design specifications. Write and solve a thermodynamics problem that provides him with the answer.

CHAPTER 7

Second Law of Thermodynamics and Entropy Transport and Production Mechanisms

CONTENTS

7.1 INTRODUCTION

In this chapter, we introduce the second law of thermodynamics and an important new thermodynamic property called *entropy*. The theory is presented first then applied to a variety of closed and open systems of engineering interest in the following chapters.

The details of the second law of thermodynamics and its associated entropy balance are presented, along with a detailed discussion of the entropy transport mechanisms associated with the energy transports of heat and work. Unlike mass, energy, and momentum, entropy is not conserved. Consequently, the mechanisms of entropy production must be well understood to produce an effective entropy balance equation.

In Chapter 8, the focus is on applying the theory presented in this chapter to the same steady state closed systems considered in Chapter 5. The second law of thermodynamics is expanded in Chapter 9 to cover open systems, as in Chapter 6. Then appropriate applications are presented, dealing with a variety of common open systems of engineering interest, such as nozzles, diffusers, throttling devices, heat exchangers, and mixing. Chapter 9 ends with a brief discussion of shaft work machines and unsteady state processes.

7.2 WHAT IS ENTROPY?

When we discussed the first law of thermodynamics in Chapter 4, it was fairly easy to apply the general balance equations to the energy concept and to invoke the conservation of energy principle to obtain a workable energy balance equation. *Energy* is a common English word, and it is also a well-accepted technical term. Everyone has a basic understanding of what the word means, though we would all have a difficult time defining it precisely. The same can be said for the words *force* and *momentum*. They are such familiar words that we easily accept mathematical formulae and logical arguments structured around them.

Most people are intrigued by seeing a movie run backward because it produces images of things never observed in the real world. What they do not realize is that they are seeing the effects of the second law of thermodynamics in action. The second law dictates the direction of the *arrow of time*. That is, things occur only in a certain way in the real world; and by applying the second law to an observation (like the screening of a movie), we can determine whether the event is running forward or backward in time. The second law of thermodynamics is what prohibits us from actually traveling backward in time. Curiously, it is the only law of nature that has such a restriction. All the other laws of mechanics and thermodynamics are valid regardless of whether time is moving forward or backward. Only the second law of thermodynamics is violated when time is reversed.

At this point we need to introduce a new thermodynamic property that is simply a measure of the amount of *molecular disorder* within a system. The name of this new property is *entropy*. The meaning of this particular name is explained later, but note that a system that has a high degree of molecular disorder (such as a high-temperature gas) has a very high entropy value and, conversely, a system that has a very low degree of molecular disorder (such as a crystalline solid) has a very low entropy value. This new property is very important because entropy is at the core of the second law of thermodynamics.

A system that has all its atoms arranged in some perfectly ordered manner has an entropy value of zero. This is the substance of the third law of thermodynamics. This law, introduced in 1906 by Walter H. Nernst (1864–1941), states that the entropy of a pure substance is a constant at absolute zero temperature. In 1911, Max Planck modified this law by setting the entropies of all pure substances equal to zero at absolute zero temperature. This had the effect of normalizing entropy values and thus creating a uniform absolute entropy scale for all substances. Therefore, we can write the following simple mathematical statement of the third law of thermodynamics:

The third law of thermodynamics: The entropy of a pure substance is zero at absolute zero temperature, or

$$\lim_{T\to 0}(\textbf{Entropy of a pure substance}) = 0 \tag{7.1}$$

With this simple entropy-disorder concept in mind, we would expect that the entropy of solid water (ice), with its highly ordered molecular structure, would be less than that of liquid water with its amorphous molecular structure, which in turn would be less than that of water vapor with its highly random molecular order. This is in fact true, for at the triple point of water (the only point where the solid, liquid, and vapor phases coexist in equilibrium), the values of the specific entropies of these phases are

Specific entropy of *ice* at the triple point = −1.221 kJ/(kg·K) = −0.292 Btu/(lbm·R)
Specific entropy of *liquid water* at the triple point = 0.0 kJ/(kg·K) = 0.0 Btu/(lbm·R)
Specific entropy of *water vapor* at the triple point = 9.157 kJ/(kg·K) = 2.187 Btu/(lbm·R)

So clearly the entropy of a solid < the entropy of a liquid < the entropy of a vapor or gas.

IS ENTROPY LIMITED TO MOLECULES?

Even though the physical concept of entropy is based on the behavior of molecular systems, order and disorder phenomena exist at all levels. For example, if entropy is a measure of disorder, then what is the entropy of your bedroom? If you have a messy room it is very disordered, so its entropy is high, but if you are a neat person and keep things picked up and put away, then it has a low entropy. Things seem to get messy easily, and to maintain cleanliness and order requires constant effort. This is a fundamental characteristic of the *second law of thermodynamics*. The natural progression of things is from ordered to disorganized and to keep it organized requires the input of energy. If entropy had been named *disorder*, perhaps it would not be so difficult to understand.

FOR WHAT IS THE THIRD LAW USED?

This law is used to define an absolute measurement scale for entropy, but it does not otherwise contribute to a thermodynamic analysis of an engineering system. Numerical values for specific entropy are listed in thermodynamic tables along with values for specific volume, specific internal energy, and specific enthalpy. For convenience, most thermodynamic tables are developed around a "relative" measurement scale, where the values of entropy and internal energy are arbitrarily set equal to zero at a point other than at absolute zero temperature. For example, in the steam tables, the specific internal energy and specific entropy of saturated liquid water are arbitrarily set equal to zero at the triple point of water (0.01°C, 0.6113 kPa or 32.018°F, 0.0887 psia). Thus, the specific internal energies and specific entropies of the less-disordered molecular states of water (like ice) have negative values on this relative scale.

THE TRUTH ABOUT ENTROPY

1. Entropy is a measure of the amount of *molecular disorder* within a system.
2. Entropy can only be *produced* (but not destroyed) within a system.
3. The entropy of a system can be *increased or decreased* by entropy transport across the system boundary.

Since it always takes an input of energy to create order within a system, it seems reasonable to postulate that a relation exists between the energy transports of a system and its order, or entropy value. Thus, we arrive at the three basic elements of the second law of thermodynamics:

We begin this chapter by assuming the existence of a disorder-measuring thermodynamic property that we call *entropy*. We use the symbol S to represent the total entropy (an extensive property), and use $s = S/m$ for the specific entropy (an intensive property).

7.3 THE SECOND LAW OF THERMODYNAMICS

We can use the general balance equation of Chapter 2 to analyze any concept whatsoever. Introducing the total entropy S into balance Eq. (2.11) provides the following total *entropy balance* (SB):

$$S_G = S_T + S_P \tag{7.2}$$

where S_G is the gain or loss of total entropy of the system due to the transport of total entropy S_T into or out of the system and the production or destruction of total entropy S_P by the system. A total *entropy rate balance* (SRB) is easily obtained from Eq. 7.2 by differentiating it with respect to time to give

$$\dot{S}_G = \dot{S}_T + \dot{S}_P \tag{7.3}$$

where the overdot indicates material time differentiation (i.e., $\dot{S} = dS/dt$).

Unlike energy, mass, and momentum, entropy is not conserved in any real process. Processes that have zero entropy production are called *reversible* and are characterized by the fact that they can occur equally well in either the forward or backward direction of time. The thing that makes entropy a unique concept worthy of a thermodynamic law of its own is that entropy is *never destroyed* in any real process. Now, it happens that some processes have very small amounts of entropy production, and it is a useful approximation for these processes to set their entropy production equal to zero. This can be stated in a very succinct mathematical form as

The second law of thermodynamics

$$\textbf{The entropy production, } S_P \geq 0 \tag{7.4a}$$

and

$$\textbf{The entropy production rate, } \dot{S}_P \geq 0 \tag{7.4b}$$

where the equality sign applies only to a reversible process. Equations (7.2) and (7.3), as modified by Eqs. (7.4a) and (7.4b), form the mathematical basis for a working form of the entropy balance and the entropy rate balance.

IS ENTROPY CONSERVED LIKE ENERGY AND MOMENTUM?

No, but it turns out that, if you "assume" the entropy production equal to zero, then the second law becomes a conservation law. This makes entropy problems easier to solve, because it eliminates the entropy production term in the entropy balance equation. But, since entropy is never actually conserved in any real process, processes that "assume" zero entropy production are only approximations of real world behavior.

We have a special word that we use to tell you when you are to assume the entropy production is zero. The word is *reversible*. When you see that word in a problem statement, you know that the entropy production term is assumed to be zero. However, remember that reversible processes are only approximations of real (irreversible) processes.

At this point we must develop the auxiliary formulae for the entropy transport and production terms before Eqs. (7.2) and (7.3) can be put to any practical use. Unfortunately, this is not an easy task. To understand the concepts of entropy transport and production, we must go back to the original 19th century classical ideas of Carnot, Clausius, and Thomson (Lord Kelvin), the early developers of this field. When this is completed, we bring the subject forward to a modern formulation.

7.4 CARNOT'S HEAT ENGINE AND THE SECOND LAW OF THERMODYNAMICS

The origins of the second law of thermodynamics lie in the work of a young 19th century French military engineer named Nicolas Leonard Sadi Carnot[1] (1796–1832). Sadi was the son of one of Napoleon's most successful generals, Lazare Carnot, and was educated at the famous Ecole Polytechnique in Paris. This institution was established in 1794 as an army engineering school and provided a rigorous program of study in chemistry, physics, and mathematics. Between 1794 and 1830, the Ecole Polytechnique had such famous instructors as Lagrange, Fourier, Laplace, Ampere, Cauchy, Coriolis, Poisson, Guy-Lussac, and Poiseuille.

After his formal education, Carnot chose a career as an army officer. At that time, Britain was a powerful military force, primarily as a result of the Industrial Revolution brought about by the British development of the steam engine. French technology was not developing as fast as Britain's, and in the 1820s, Carnot became convinced that France's inadequate utilization of steam power had made it militarily inferior. He began to study the fundamentals of steam engine technology, and in 1824, he published the results of his studies in a small book entitled *Reflections on the Motive Power of Fire* (the French word for *fire* was then a common term for what we call *heat* today).

Sadi Carnot was trained in the basic principles of hydraulics, pumps, and water wheels at the Ecole Polytechnique. It was clear to him that the power of a steam engine was released as the heat fluid (caloric) fell from the high temperature of the boiler to the lower temperature of the condenser, in much the same way that water falls through a water wheel to produce a mechanical shaft work output. He conjectured that

> According to established principles at the present time, we can compare with sufficient accuracy the motive power of heat to that of a waterfall. The motive power of a waterfall depends on its height and on the quantity of liquid; the motive power of heat depends also on the quantity of caloric used, and on what may be termed, on what in fact we will call, the *height of its fall*, that is to say, the difference of temperature of the bodies between which the exchange of caloric is made.

WHAT IS HEAT ANYWAY?

The essence of the concept of *heat* was a very actively debated scientific topic at the time. In 1789, the great French chemist Antoine Lavoisier (1743–1794) proposed the caloric theory of heat, in which heat was presumed to be a colorless, odorless, weightless fluid called *caloric* that could be poured from one object to another. When an object became full of caloric, it was said to be *saturated* with it. This was the origin of the terms *saturated liquid, saturated vapor,* and so on that we use in thermodynamics today. These terms were introduced into the scientific literature in the early 19th century, when the caloric theory of heat was popular, and they were never removed when it was later proven that heat was not a fluid. Today, they are simply misnomers. Now, when we use the word *heat* in a technical sense, we normally mean an *energy transport arising from a temperature difference*.

[1] Pronounced *car-no*. Many French words have a silent *t* ending. For example, Peugot, Tissot, Monet, ballet, chalet, Chevrolet, and Renault. Sadi was named after the medieval Persian poet Saadi Musharif ed Din, whose poems became popular in France in the late 18th century.

By the 1820s, a great deal of work had already been done on the efficiency of water wheels, and the water wheel–steam engine analogy must have seemed to Carnot like a good way to approach the problem of improving steam engine efficiency. Two important conclusions came from his work with this analogy.

First, he knew that no one could build a water wheel that would produce a continuous work output unless water actually entered and exited the wheel. And, if water with a certain kinetic and potential energy entered the wheel, then the same amount of water with a lower energy level must also exit the wheel. In other words, it is impossible to make a water wheel that converts all the energy of the inlet water into output shaft work. There must also be an outflow of water from the wheel, and this outflow must have some energy. These rather obvious statements are illustrated in Figure 7.1.

Now, if you extend this idea to a steam engine (or any type of heat engine) by replacing the word *water* by the hypothetical heat fluid *caloric,* then it is easy to conclude that when caloric at a certain energy level (temperature) enters a work-producing heat engine, it must also exit the engine at a lower energy level (temperature). This concept was later refined into the following form, known today as the Kelvin-Planck statement of the second law of thermodynamics: *You cannot make a continuously operating heat engine that converts all of its heat input directly into work output* (see Figure 7.2).

Second, Carnot observed that the maximum efficiency of a water wheel was independent of the type of liquid used and depended only on the inlet and outlet flow energies. This led him to the conclusion that

> The motive power of heat is independent of the agents employed to realize it; its quantity is fixed solely by the temperatures of the bodies between which is effected, finally, the transfer of caloric.

Or, the maximum efficiency of a steam engine (or any type of heat engine) is dependent only on the temperatures of the high- and low-temperature thermal reservoirs of the engine (the boiler and condenser temperatures in the case of a steam engine) and is independent of the working fluid of the engine (water in the case of a steam engine). Of course, to achieve the maximum possible efficiency, the water wheel and the heat engine must be completely *reversible*; that is, they cannot possess any mechanical friction or other losses of any kind.

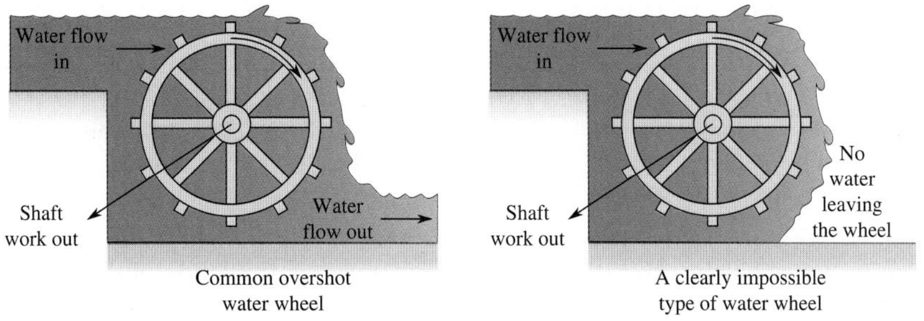

FIGURE 7.1
Water wheel operation.

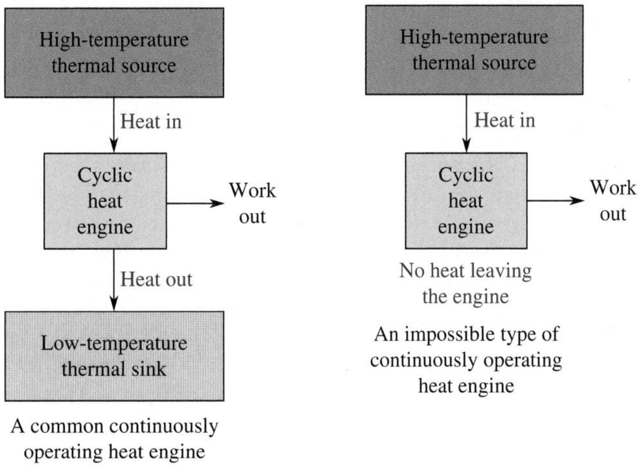

FIGURE 7.2
Heat engine operation.

CLAUSIUS STATEMENT OF THE SECOND LAW OF THERMODYNAMICS

It is impossible to build a continuously operating device that will cause heat energy to be transferred from a low-temperature reservoir to a high-temperature reservoir without the input of work energy (Figure 7.3).

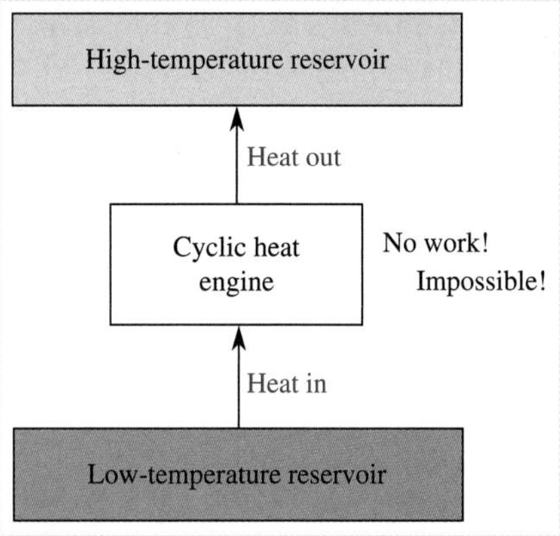

FIGURE 7.3
Clausius statement.

An early nonmathematical version of the second law of thermodynamics, expressed in words, is the Clausius statement[2] of the second law. It is easily understood using the water wheel analogy.

Another early nonmathematical verbalized version of the second law is by William Thomson (Lord Kelvin), later modified by Max Planck[3] (1858–1947).

Both the Clausius and the Kelvin-Planck "statements" of the second law are really just verbalized consequences of the second law. Its development by recourse to water wheel technology is clearly wrong, since a water wheel is not a heat engine, but the conclusions are nonetheless correct. They have historical explanatory value but cannot be used in engineering design. The power of the second law becomes apparent only when it has a mathematical formulation. This comes later in the chapter.

The significance of Carnot's conclusions was not recognized until the 1850s, when Rudolph Clausius (1822–1888) and William Thomson (Lord Kelvin, 1824–1907) worked out a clear formulation of the conservation of energy principle, which was then named the *first* law of thermodynamics. Carnot's *first* conclusion was then named the *second* law of thermodynamics by Clausius, who also expanded Carnot's energy transformation concepts into a new property he called *entropy*.

The classical Kelvin-Planck and Clausius statements of the second law also provide a classical means for defining the concept of reversibility. The Kelvin-Planck statement limits the efficiency of a heat engine to something less than 100%, but as yet, we have no idea how much less. To establish a more realistic efficiency limit, we need to define the simplest possible heat engine, one that operates with simple idealized processes. Such an engine would be frictionless and not have any losses. We call such an idealized engine *reversible* because the energy flow through it could be reversed without leaving any trace on the environment.

Since the first law of thermodynamics is a conservative law (i.e., energy cannot be produced or destroyed), it has no effect on the reversibility of a system. But the second law is not a conservative law, since entropy is produced in every real process. Therefore, the second law is what dictates whether or not a system and its surroundings can be returned to their original states, and in general, this is not possible if entropy is produced by the process. Therefore, a reversible process is really synonymous with a zero entropy production process.

Processes that are not reversible are called *irreversible*. Phenomena that cause processes to be irreversible are called process *irreversibilities*. Some typical process irreversibilities within a system are shown in the following table.

[2] The word *statement* is used here, because it is a verbalized rather than mathematical form of the second law.

[3] In the 1890s, Planck added the concept of "continuously operating" to Kelvin's 1850s verbalized version of the second law, so this statement now has both names associated with it.

mechanical friction	fluid viscosity	electrical resistance
shock waves	mixing	chemical reactions
heat transfer	plastic deformation	hysteresis

Thus, it is easy to see that all engineering processes of interest are really irreversible processes, and aside from a few mathematical formalities, we have very little need for the reversibility concept. However, we are not yet capable of analyzing all the irreversibilities that occur in the complex real world, so we indeed need to model our device as *reversible* during the design stage, then build and test a prototype of it to determine the effect of

KELVIN-PLANCK STATEMENT OF THE SECOND LAW OF THERMODYNAMICS[4]

You cannot make a continuously operating heat engine that converts all of its heat input directly into work output (Figure 7.4).

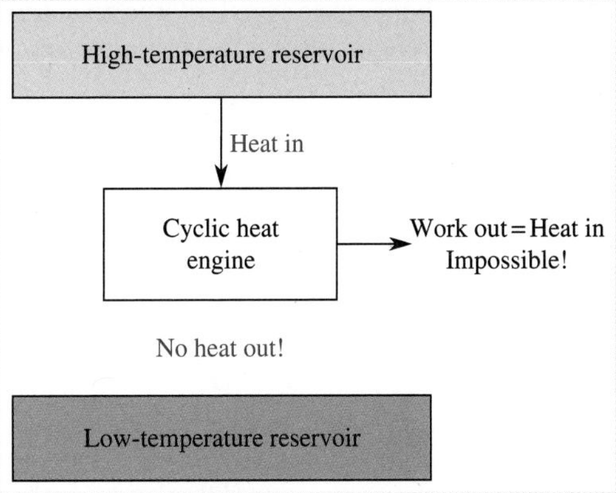

FIGURE 7.4
Kelvin-Planck statement.

[4] *The Kelvin-Planck statement has many different written forms. For example, it can be expressed as* "It is impossible to build a continuously operating device that produces a work output while absorbing heat from a single thermal reservoir," *and as* "No heat engine can be more than 100% efficient." *Another form of the Kelvin-Planck statement is* "It is impossible to build a continuously operating device that produces a work output while absorbing heat energy from a single thermal reservoir."

WHAT DOES THE WORD *ENTROPY* MEAN?

Rudolph Clausius was German, and the word he chose for Carnot's energy transformation concept was *verwandlungsinhalt*, meaning "transformation content." Fortunately, in 1865, he chose to rename this concept by choosing the Greek word *entropy*, meaning simply "to change, or transform." Later, there was an unsuccessful attempt to name an entropy unit the *clausius*, Cl, after him. It was defined as 1 Cl = 1 kcal/K = 4.186 J/K, but it was not universally accepted.

REVERSIBLE PROCESSES: CLASSICAL DEFINITION

A process is called *reversible* if, at any time, both the system and the environment can be returned to their original states.

REVERSIBLE PROCESS: MODERN DEFINITION

A *reversible process* is defined to be any process for which the entropy production or the entropy production rate for the process is zero (i.e., $S_P = \dot{S}_P = 0$).

the inherent irreversibilities. We then correct for using the reversibility assumption in the design through an appropriate efficiency calculation.

Though reversible processes do not actually exist in nature, they are conceptually necessary for creating performance limits for heat engine technology. William Thomson used Carnot's second conclusion regarding maximum (i.e., reversible) engine energy conversion efficiency to develop the concept of an absolute temperature scale.

7.5 THE ABSOLUTE TEMPERATURE SCALE

After studying the operation of steam engines for several years, Sadi Carnot concluded in 1824 that the efficiency of a heat engine depended *only* on the temperatures of the engine's hot and cold thermal reservoirs and *not* on the fluid used inside the engine. In 1848, Thomson used Carnot's conclusion to develop the concept of an absolute temperature scale. Soon afterward, an absolute temperature scale based on the size of the celsius degree (°C) became popular and was given his titled name *kelvin* (K) by his admirers.

By using Eq. (4.70), we can define the thermal energy conversion efficiency (also called the *thermal efficiency*) η_T of a continuously operating closed system heat engine with a net output work or power as

$$\eta_T = \frac{(W_{out})_{net}}{Q_{in}} = \frac{(\dot{W}_{out})_{net}}{\dot{Q}_{in}} \tag{7.5}$$

A closed system heat engine can operate continuously only if it operates in a thermodynamic cycle. A system that undergoes a thermodynamic cycle must end up at the same thermodynamic state at the end of the cycle as it started from at the beginning of the cycle. Because the total system energy E is a point function, the closed system first law of thermodynamics energy balance (EB) applied to a cyclic process yields

$$(Q - W)_{cycle} = (E_2 - E_1)_{cycle} = 0 \tag{7.6}$$

Now, from Figure 7.5, we see that the heat input to a cyclic heat engine is

$$(Q)_{cycle} = Q_{in} - |Q_{out}| \tag{7.7}$$

and

$$(W)_{cycle} = W_{out} \tag{7.8}$$

where $|Q_{out}|$ is the absolute value of this energy flow.

Note that we introduce the correct *sign* with the absolute value of the symbol in Eqs. (7.7) and (7.8) to indicate the proper flow direction (+ for heat in and work out, and – for heat out and work in). Normally we do not introduce the sign convention directly into the equations themselves. The usual custom is to attach the correct flow direction sign to the *number* and not the *symbol*. However, we change this notational scheme here to help you understand the operation of closed system heat engines. Later in this chapter, we revert to the conventional notation scheme for algebraic signs.

Combining Eqs. (7.5) through (7.8) and using the simplified notation shown in Figure 7.5 yields

$$\eta_T = \frac{(W_{out})_{net}}{Q_{in}} = \frac{(Q_{in} - |Q_{out}|)}{Q_{in}} = 1 - \frac{|Q_{out}|}{Q_{in}} = 1 - \frac{|Q_L|}{Q_H} \tag{7.9}$$

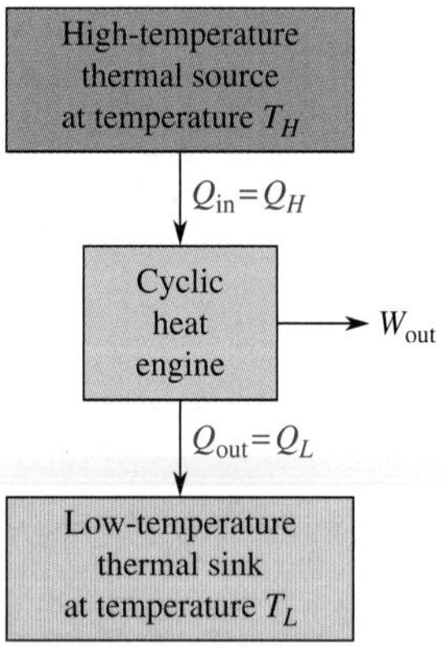

FIGURE 7.5
Schematic of a cyclic heat engine.

IS IT K OR °K?

In 1967, the International Bureau of Weights and Measures dropped the prefix *degree* from the SI absolute temperature scale. So we say "100 degrees celsius is equal to 373.15 kelvin" (or 100.00°C = 373.15 K). Notice that we do not capitalize the terms *Celsius* and *Kelvin*, even though they are proper names. Remember that, in Chapter 1, we discussed why (a) we do not capitalize the first letter of a unit whose name is derived from that of a person when the unit's name is written out and (b) the first letter is capitalized when the unit's name is abbreviated.

Also, in this book, we follow the same scheme of omitting the degree symbol on the Rankine absolute temperature scale, so 100.00°F = 559.67 R (and "559.67 R" is written out in lower case as "559.67 rankine").

If we now follow Carnot's lead and presume that the thermal efficiency of a reversible heat engine $(\eta_T)_{\text{rev}}$ depends only on the absolute temperatures of the thermal reservoirs, then we can write

$$(\eta_T)_{\text{rev}} = 1 - \left(\frac{|Q_{\text{out}}|}{Q_{\text{in}}}\right)_{\text{rev}} = 1 - \left(\frac{|Q_L|}{Q_H}\right)_{\text{rev}}$$

or

$$1 - (\eta_T)_{\text{rev}} = \left(\frac{|Q_L|}{Q_H}\right)_{\text{rev}} = f\left(\frac{T_L}{T_H}\right) \tag{7.10}$$

where $f(\,)$ is an unknown function that eventually is used to define the absolute temperature scale, and the subscripts L and H refer to the low- and high-temperature reservoirs, respectively.

Now consider the two reversible heat engines connected in series shown in Figure 7.6. The thermal efficiency of each of the individual reversible heat engines is determined from an analysis of systems A and B individually.

The thermal efficiency of the engine in system A is[5]

$$(\eta_T)_A = \frac{W_A}{Q_1} = 1 - \frac{|Q_2|}{Q_1} = 1 - f\left(\frac{T_2}{T_1}\right)$$

and that in system B is

$$(\eta_T)_B = \frac{W_B}{Q_2} = 1 - \frac{|Q_3|}{Q_2} = 1 - f\left(\frac{T_3}{T_2}\right)$$

Now, if we include both engines inside the system boundary, as in system C of Figure 7.6, then we have $W_C = W_A + W_B$, and utilizing the previous results, we can write

$$\begin{aligned}(\eta_T)_C &= \frac{W_C}{Q_1} = \frac{W_A + W_B}{Q_1} = \frac{(Q_1 - |Q_2|) + (|Q_2| - |Q_3|)}{Q_1}\\ &= 1 - \frac{|Q_2|}{Q_1} + \left(\frac{|Q_2| - |Q_3|}{|Q_2|}\right)\left(\frac{|Q_2|}{Q_1}\right)\\ &= 1 - f\left(\frac{T_2}{T_1}\right) + \left[1 - f\left(\frac{T_3}{T_2}\right)\right] f\left(\frac{T_2}{T_1}\right)\\ &= 1 - f\left(\frac{T_2}{T_1}\right) f\left(\frac{T_3}{T_2}\right)\end{aligned} \tag{7.11}$$

We can also compute the heat engine thermal efficiency of system C as

$$(\eta_T)_C = \frac{W_C}{Q_1} = \frac{Q_1 - |Q_3|}{Q_1} = 1 - \frac{|Q_3|}{Q_1} = 1 - f\left(\frac{T_3}{T_1}\right) \tag{7.12}$$

Comparing Eqs. (7.11) and (7.12), we conclude that the following functional relation must hold for the unknown temperature function, $f(\,)$:

$$f\left(\frac{T_3}{T_1}\right) \equiv f\left(\frac{T_2}{T_1}\right) f\left(\frac{T_3}{T_2}\right) \tag{7.13}$$

Many common functions do not satisfy this equation. For example,

$$\sin\frac{T_3}{T_1} \neq \left(\sin\frac{T_2}{T_1}\right)\sin\frac{T_3}{T_2}$$

$$\log\frac{T_3}{T_1} \neq \left(\log\frac{T_2}{T_1}\right)\log\frac{T_3}{T_2}$$

$$\exp\frac{T_3}{T_1} \neq \left(\exp\frac{T_2}{T_1}\right)\exp\frac{T_3}{T_2}$$

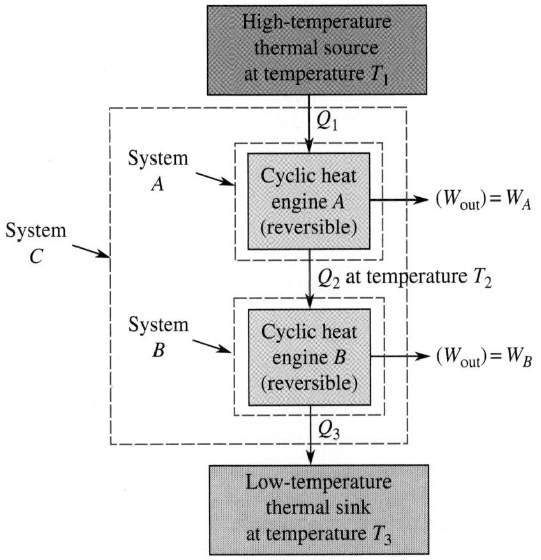

FIGURE 7.6
Two reversible heat engines connected in series.

[5] Since these engines are defined at the outset to be reversible, the rev subscript on the η_T, Q, and W terms in these equations has been dropped for simplicity. This subscript reappears in the equations at the end of this analysis.

WHAT IS A CARNOT ENGINE?

A Carnot engine is a *reversible heat engine* that operates between a high-temperature heat source at T_H and a low-temperature heat sink at T_L. The thermal efficiency of a Carnot engine is simply $(\eta_T)_{\text{Carnot}} = 1 - T_L/T_H$.

It turns out that no one has ever actually made a running Carnot engine (although Rudolph Diesel thought he did, but he invented the Diesel engine instead). The reversible constant temperature heat transfers T_H and T_L require the engine to have infinite heat transfer surface areas, and this is not practical.

The "concept" of a Carnot engine is important because it turns out that no other heat engine can have a thermal efficiency higher than that of a Carnot engine operating between the same two temperature limits. So the value of the Carnot engine is only as a benchmark with which to compare the thermal efficiencies of other actual operating heat engines. The Carnot cycle has become the universal standard by which the performance of other heat engine cycles can be measured.

and so forth. However, any simple power function of the form $f(T_3/T_1) = (T_3/T_1)^n$ does satisfy Eq. (7.13), since

$$\left(\frac{T_3}{T_1}\right)^n = \left(\frac{T_2}{T_1}\right)^n \left(\frac{T_3}{T_2}\right)^n$$

The simplest such power function is a linear one ($n = 1$), and this is what Thomson chose to establish his absolute temperature scale. Therefore, if we take

$$f\left(\frac{T_3}{T_1}\right) = \frac{T_3}{T_1} = \left(\frac{T_2}{T_1}\right)\left(\frac{T_3}{T_2}\right) = f\left(\frac{T_2}{T_1}\right) f\left(\frac{T_3}{T_2}\right) \tag{7.14}$$

then Eq. (7.10) becomes

$$\left(\frac{|Q_{\text{out}}|}{Q_{\text{in}}}\right)_{\text{rev}} = \left(\frac{|Q_L|}{Q_H}\right)_{\text{rev}} = \frac{T_L}{T_H} \tag{7.15}$$

It should be noted that Eq. (7.14) is not the only function that accurately defines an absolute temperature scale (but it is the simplest). Many other functions also work. However, they produce nonlinear temperature scales in which the size of the temperature unit is not constant but depends on the temperature level. This might be a useful technique to expand or condense a temperature scale in certain temperature regions, but the additional complexity associated with a nonlinear temperature scale makes it generally unsuitable for common usage.[6]

Now, clearly, the *maximum* possible thermal energy conversion efficiency of any real irreversible closed system cyclic heat engine is equal to the thermal energy conversion efficiency that the same heat engine would have if it were somehow made to run reversibly like a Carnot engine. Then, from Eq. (7.9),

$$(\eta_T)_{\max} = (\eta_T)_{\text{rev}} = (\eta_T)_{\text{Carnot}} = 1 - \frac{T_L}{T_H} \tag{7.16}$$

EXAMPLE 7.1

If a heat engine burns fuel for its thermal energy source and the combustion flame temperature is 4000.°F, determine the maximum possible thermal efficiency of this engine if it exhausts to the environment at 70.0°F.

Solution

First, draw a sketch of the system (Figure 7.7).

The unknown is the maximum possible thermal energy conversion efficiency of any heat engine. The "maximum" efficiency occurs when an engine operates reversibly (i.e., with no internal losses due to friction, etc.). Since all reversible engines must have the same thermal energy conversion efficiency when operated between the same high- and low-temperature reservoirs, we can apply the results of the reversible Carnot engine analysis to this problem. Equation (7.16) gives the maximum possible thermal efficiency as

$$(\eta_T)_{\max} = (\eta_T)_{\text{Carnot}} = 1 - \frac{T_L}{T_H} = 1 - \frac{(70.0 + 459.67)\text{ R}}{(4000. + 459.67)\text{ R}} = 0.881 = 88.1\%$$

[6] It has been suggested that, since many thermal phenomena are inherently nonlinear, the use of a nonlinear (e.g., logarithmic) temperature scale might have some engineering merit.

The results of Example 7.1 are highly unrealistic since no real heat engine can ever be reversible. The irreversibilities within modern heat engines limit their actual operating thermal energy conversion efficiency to around 30%.

Exercises

1. Rework Example 7.1 for a flame temperature of 2500.°C and an environmental temperature of 20.0°C. **Answer:** $(\eta_T)_{max} = 89.4\%$.
2. If the engine described in Example 7.1 has a maximum (reversible or Carnot) thermal efficiency of 60.0% when the environmental temperature is 70.0°F, determine the flame temperature of the combustion process. **Answer:** $T_{flame} = T_H = 1324$ R = 865°F.

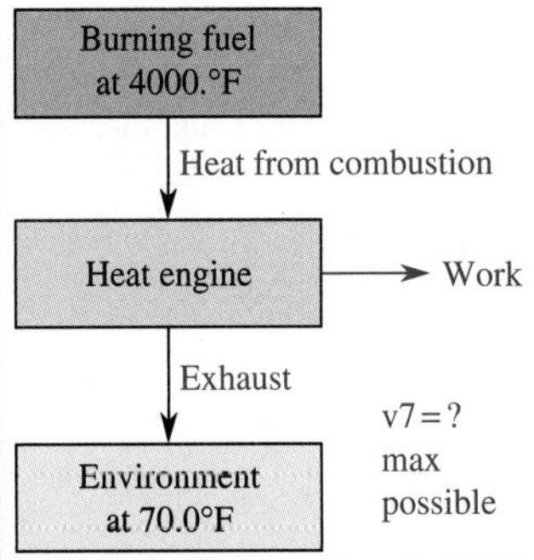

FIGURE 7.7
Example 7.1.

EXAMPLE 7.2

A coal-fired electrical power plant produces 5.00 MW of electrical power while exhausting 8.00 MW of thermal energy to a nearby river at 10.0°C. The power plant requires an input power of 100. kW to drive the boiler feed pump.

Determine

a. The actual thermal efficiency of the power plant.
b. The equivalent heat source temperature if the plant operated on a reversible Carnot cycle.

Solution

First, draw a sketch of the system (Figure 7.8).

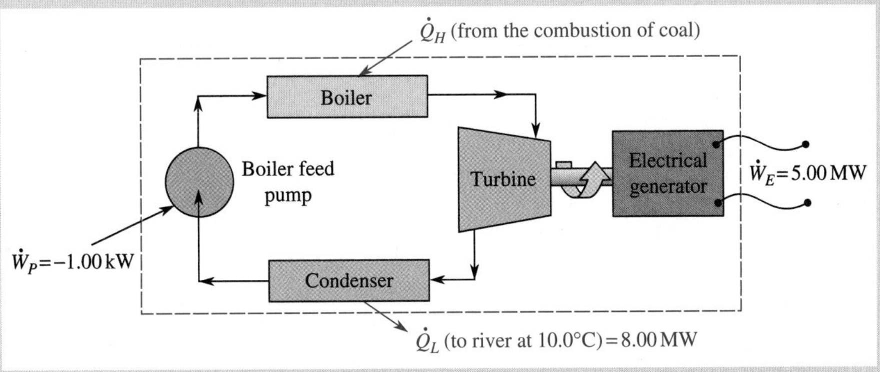

FIGURE 7.8
Example 7.2.

The unknowns are the actual thermal efficiency of the power plant and the equivalent heat source temperature if the plant operates on a reversible Carnot cycle. If we construct the system boundary as shown in the sketch, the power plant is a closed system.

a. The actual thermal efficiency of this system is given by Eq. (7.9) as

$$\eta_T = \frac{(\dot{W}_{out})_{net}}{\dot{Q}_{in}} = \frac{\dot{W}_E - |\dot{W}_P|}{\dot{Q}_H} = \frac{\dot{Q}_H - |\dot{Q}_L|}{\dot{Q}_H}$$

and the energy rate balance (ERB) for the steady state operation of this system is

$$\dot{Q}_H - |\dot{Q}_L| - (\dot{W}_E - |\dot{W}_P|) = 0$$

or

$$\begin{aligned}\dot{Q}_H &= |\dot{Q}_L| + (\dot{W}_E - |\dot{W}_P|)\\ &= |-8.00\text{ MW}| + (5.00\text{ MW} - |-0.100\text{ MW}|)\\ &= 12.9\text{ MW}\end{aligned}$$

(Continued)

EXAMPLE 7.2 (*Continued*)

and thus the actual thermal efficiency is

$$\eta_T = \frac{5.00\,\text{MW} - |-0.100\,\text{MW}|}{12.9\,\text{MW}} = \frac{12.9\,\text{MW} - 8.00\,\text{MW}}{12.9\,\text{MW}} = 0.380 = 38.0\%$$

b. From Eq. (7.16), we have

$$(\eta_T)_{\max} = (\eta_T)_{\text{Carnot}} = 1 - \frac{T_L}{T_H} = 0.380$$

so that

$$T_H = \frac{T_L}{1-0.380} = \frac{(10.0+273.15)\,\text{K}}{0.620}$$

$$= 457\,\text{K} = 184°\text{C}$$

The calculations of part a are perfectly valid for this power plant since they deal with *actual* input and output energy values. The answer to part b, however, is unrealistically lower than the actual coal flame temperature in the boiler due to the many irreversibilities that exist within a real power plant.

Exercises

3. If the combustion temperature of the power plant discussed in Example 7.2 were 2000.°C, determine the maximum (reversible or Carnot) thermal efficiency of the facility. **Answer:** $(\eta_T)_{\max} = 87.5\%$.
4. If the heat transfer to the boiler in Example 7.2 were 3.50×10^7 Btu/h, the heat transfer from the condenser were 2.10×10^7 Btu/h, and the power into the boiler feed pump were 1.50 hp, determine (a) the power output from the turbine/generator in MW and (b) the actual thermal efficiency of the power plant. **Answers:** (a) $W_{\text{act}} = 4.10\,\text{MW}$, (b) $\eta_T = 40.\%$.

7.6 HEAT ENGINES RUNNING BACKWARD

When a heat engine is run thermodynamically backward, it becomes a heat pump, a refrigerator, or an air conditioner, depending on your point of view. Figure 7.9 shows that, when a heat engine is thermodynamically reversed, the directions of all the energy flows are reversed. Thus, a work *input* W_{in} causes a thermal energy transfer Q_L *from* a low-temperature reservoir and a thermal energy transfer Q_H *to* a high-temperature reservoir. Consequently, the backward running heat engine appears to "pump" heat from a low-temperature reservoir to a high-temperature reservoir. However, since heat is really a thermal energy transport phenomenon and not a fluid, it is somewhat misleading to refer to it as being "pumped." Yet it is common practice in the heating, ventilating, and air conditioning (HVAC) industry to refer to these devices as *heat pumps* when they are used to provide a thermal energy transfer to a warm environment (e.g., a house) from a cold environment (e.g., the outside air).

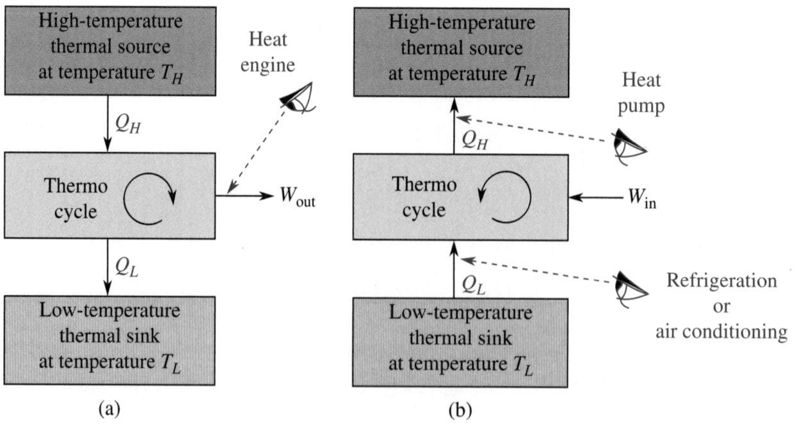

FIGURE 7.9
(a) Heat engine; (b) thermodynamically reversed heat engine (heat pump, refrigerator, or air conditioner).

IS THE EFFICIENCY OF HEAT PUMPS, AIR CONDITIONERS, AND REFRIGERATORS GREATER THAN 100%?

If you look closely at the thermal efficiency equations for heat pumps, air conditioners, and refrigerators, you see that their efficiency is going to be more than 100%, because under normal operating conditions, the numerator in their efficiency equation is usually greater than the denominator. Consequently, their energy conversion efficiency is usually greater than 100%.

How can that be—nothing should have an energy conversion efficiency greater then 100%. But, it is correct. This is simply due to the way in which the thermal efficiency formula (Eq. (4.70)) is structured:

$$\text{Energy conversion efficiency} = \eta_E = \frac{\text{Desired energy result}}{\text{Required energy input}} \tag{4.70}$$

This makes a heat pump much more attractive for domestic heating than, say, a purely resistive electrical heater. Electrical heaters convert all their input electrical energy directly into thermal energy and therefore have energy conversion efficiencies of 100%, whereas most heat pumps have energy conversion efficiencies far in excess of 100% for the same electrical energy input.

Since this could be a problem in public advertising, the industry uses the phrase *coefficient of performance* (COP) instead of *efficiency*. The COP is simply the pure efficiency number before it is converted into a percentage. For example, the COP of a heat pump with an energy conversion efficiency of 450% is 4.5.

$$\text{COP}_{\substack{\text{heat}\\ \text{pump}}} = \eta_{\substack{\text{heat}\\ \text{pump}}} = \frac{|Q_H|}{|W_{\text{in}}|} = \frac{|Q_H|}{|Q_H| - Q_L} = \frac{|\dot{Q}_H|}{|\dot{W}_{\text{in}}|} = \frac{|\dot{Q}_H|}{|\dot{Q}_H| - \dot{Q}_L} \tag{7.17}$$

The *desired energy result* in the operation of a heat pump is heat addition to an already warm environment. Therefore, its energy conversion efficiency can be determined from Eq. (4.70) and an energy balance on the device (see Figure 7.9b) as

$$\eta_{\substack{\text{heat}\\ \text{pump}}} = \frac{\text{Desired energy result}}{\text{Required energy input}} = \frac{|Q_H|}{|W_{\text{in}}|} = \frac{|Q_H|}{|Q_H| - Q_L} = \frac{|\dot{Q}_H|}{|\dot{W}_{\text{in}}|} = \frac{|\dot{Q}_H|}{|\dot{Q}_H| - \dot{Q}_L}$$

where, as in the previous section, we use the absolute values of certain terms to avoid improper or confusing algebraic signs.

If the heat pump is modeled as a backward running Carnot heat engine, then Eqs. (7.15) and (7.17) can be combined to yield the COP for a "reversible" (i.e., frictionless, etc.), or Carnot, heat pump as

$$\text{COP}_{\substack{\text{Carnot}\\ \text{heat pump}}} = \frac{T_H}{T_H - T_L} \tag{7.18}$$

If the removal of heat Q_L from a space is the desired result of a backward running heat engine, then the engine is called a *refrigerator* when food is stored in the cooled space and an *air conditioner* when people occupy the cooled space.

The energy conversion efficiency of a refrigerator or air conditioner can also be obtained from Eq. (4.70). As in the case of a heat pump, these efficiencies are also normally greater than 100% and they too are commonly represented with the pure number *coefficient of performance* label

$$\begin{aligned}\text{COP}_{\substack{\text{refrig. or}\\ \text{air cond.}}} = \eta_{\substack{\text{refrig. or}\\ \text{air cond.}}} &= \frac{\text{Desired energy result}}{\text{Required energy input}}\\ &= \frac{Q_L}{|W_{\text{in}}|} = \frac{Q_L}{|Q_H| - Q_L} = \frac{\dot{Q}_L}{|\dot{W}_{\text{in}}|} = \frac{\dot{Q}_L}{|\dot{Q}_H| - \dot{Q}_L}\end{aligned} \tag{7.19}$$

For a backward running Carnot (i.e., reversible) heat engine, Eqs. (7.15) and (7.19) can be combined to give the COP for a reversible refrigerator or air conditioner as

$$\text{COP}_{\substack{\text{Carnot}\\ \text{refrig. or}\\ \text{air cond.}}} = \frac{T_L}{T_H - T_L} \tag{7.20}$$

CRITICAL THINKING

Notice that, the *smaller* the difference between T_H and T_L, the *larger* the COP defined in Eqs. (7.18) and (7.20) becomes. To understand this on a physical basis, note that, since the numerator is a fixed value (Q_L or Q_H) in both equations and the denominator represents the net work input (W_{in}) to the device, it stands to reason that, the smaller is the work input for a given output, the better the efficiency (i.e., COP) is.

Comparing Eqs. (7.17) and (7.19), we see that

$$\text{COP}_{\text{heat pump}} = \text{COP}_{\substack{\text{refrig. or}\\ \text{air cond.}}} + 1$$

EXAMPLE 7.3

The temperature outside on a hot summer day is 95°F. You would like your room to be at 70.°F, so you go out shopping for an air conditioner. If you are going to buy a Carnot air conditioner, what should be its coefficient of performance?

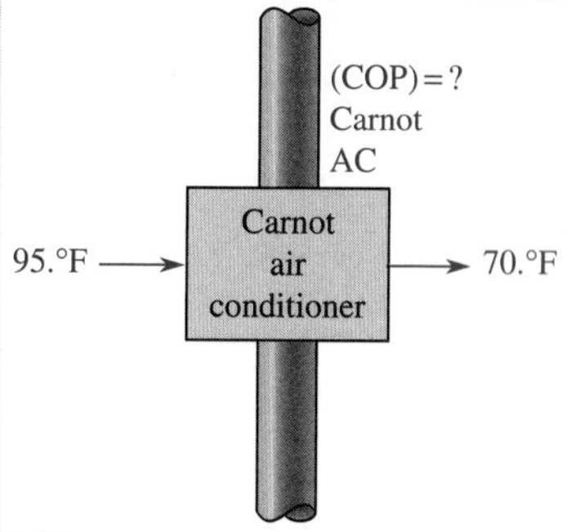

FIGURE 7.10
Example 7.3.

Solution

First, draw a sketch of the system (Figure 7.10).

The unknown is the coefficient of performance of an air conditioner operating between 70.°F and 95°F. Equation (7.20) gives the coefficient of performance for a Carnot air conditioner as

$$\text{COP}_{\substack{\text{Carnot}\\ \text{refrig. or}\\ \text{air cond.}}} = \frac{T_L}{T_H - T_L} = \frac{70. + 459.67}{(95 + 459.67) - (70. + 459.67)} = 21$$

Note that, in the denominator, the temperature *difference* in this equation can be in either relative or absolute temperature units because (95 + 459.67) − (70. + 459.67) = 95 − 70. = 25°F or R. The degree *size* is the same for both the absolute and the relative temperature scales.

The coefficient of performance of a real air conditioner is sometimes called its *energy efficiency rating* (EER) and usually ranges between 3 and 9. The value of 21 calculated in Example 7.3 is unreasonably high, because a Carnot air conditioner is reversible and has no friction or other internal losses. Therefore, it requires less input work than a real air conditioner.

Exercises

5. The refrigerator in your kitchen maintains a temperature difference of 2.00°C inside when the outside kitchen temperature is 22.0°C. If it is a Carnot refrigerator, what is its coefficient of performance? **Answer**: $(\text{COP})_{\text{Carnot ref.}} = 13.8$.
6. If the Carnot air conditioner purchased in Example 7.3 is inserted into the window backward during the winter and operated as a heat pump, determine its coefficient of performance as a heat pump. **Answer**: $(\text{COP})_{\text{Carnot HP}} = 22$.

7.7 CLAUSIUS'S DEFINITION OF ENTROPY

Rudolph Clausius extended Thomson's absolute temperature scale work by rearranging Eq. (7.15) to read

$$\frac{(Q_H)_{\text{rev}}}{T_H} = \frac{|Q_L|_{\text{rev}}}{T_L}$$

and since this applies only to a closed system undergoing a thermodynamic cycle, it can also be written as

$$\sum_{\text{cycle}} \left(\frac{Q}{T}\right) = \frac{(Q_H)_{\text{rev}}}{T_H} - \frac{|Q_L|_{\text{rev}}}{T_L} = 0$$

If we now take an arbitrary thermodynamic closed system cycle and overlay it with an infinite number of infinitesimal heat engine cycles, as shown in Figure 7.11, then we can extend the finite summation process of the previous equation into a cyclic integral. Also, since each of these infinite number of heat engines is now

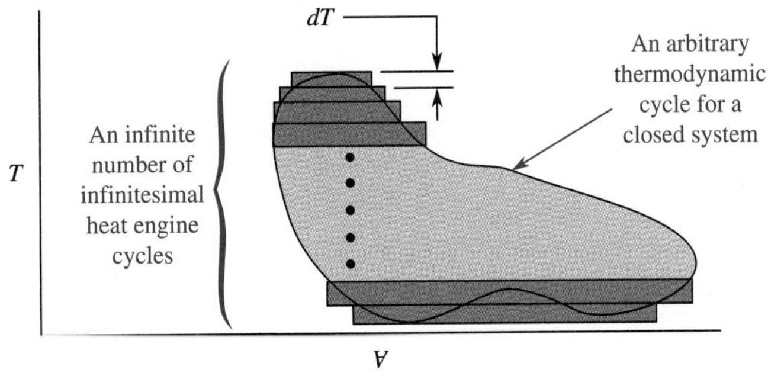

FIGURE 7.11
An infinite number of infinitesimal heat engine cycles approximating an arbitrary closed system thermodynamic cycle.

operating over an infinitely small temperature difference, $T_H \approx T_L = T$ and $(Q_H)_{\text{rev}} - |Q_L|_{\text{rev}} \approx (\bar{d}Q)_{\text{rev}}$. Then, in the limit, the previous equation becomes

$$\lim_{n\to\infty}\left[\sum_{n_{\text{cycles}}}\left(\frac{Q}{T}\right)\right] = \oint_{\text{cycle}}\left(\frac{\bar{d}Q}{T}\right)_{\text{rev}} = 0 \tag{7.21}$$

The temperature T in this equation is the *absolute* temperature at the point where the heat transfer $\bar{d}Q$ occurs.

Clausius then noted the remarkable result that, since, by definition,

$$\oint_{\text{cycle}} (\text{Any thermodynamic property differential}) = 0$$

the argument of the integral in Eq. (7.21) *must* define a thermodynamic property. That is,

$$\left(\frac{\bar{d}Q}{T}\right)_{\text{rev}} = \text{Differential of some thermodynamic property}$$

But, which property? The term $(\bar{d}Q)_{\text{rev}}$ by itself is a path function and thus cannot be a thermodynamic property differential. However, when $(\bar{d}Q)_{\text{rev}}$ is divided by T, a property differential results. Clausius realized that he had discovered a new thermodynamic property and he chose to name it *entropy*[7] and represent the total entropy of a system by the symbol S, where

$$dS = \left(\frac{\bar{d}Q}{T}\right)_{\text{rev}} \tag{7.22}$$

or

$$S_2 - S_1 = \int_1^2 \left(\frac{\bar{d}Q}{T}\right)_{\text{rev}} \tag{7.23}$$

or

$$s_2 - s_1 = \frac{S_2 - S_1}{m} = \frac{1}{m}\int_1^2 \left(\frac{\bar{d}Q}{T}\right)_{\text{rev}} \tag{7.24}$$

Be careful to note that Eqs. (7.22)–(7.24), which define entropy, are for a *closed system* of fixed mass m only. The effect of mass flow on system entropy is taken up in a separate section of this chapter.

The use of a relative temperature scale in a grouping of units can sometimes be confusing. For example, when a temperature unit appears in the denominator of a units grouping, it can be written either as °F or R (or °C or K in SI) because only the degree size there is important. Therefore, Eq. (7.24) indicates that the units of specific entropy can be written correctly in either of the following forms:

$$s\,[\text{in Btu/(lbm}\cdot{}^\circ\text{F)}] \equiv s\,[\text{in Btu/(lbm}\cdot\text{R)}]$$

---

[7] Here is a translation of how Clausius, in 1865, described why he chose the word *entropy* for the name of his new property. "We might call S the *transformational* content of the body, just as we termed the quantity U the *heat* and *work* content of the body. But since I believe it is better to borrow terms for important quantities from the ancient languages so that they may be adopted unchanged in all modern languages, I propose to call the quantity S the *entropy* of the body, from the Greek word ητροπή, meaning *a transformation*."

or

$$s\,[\text{in kJ/(kg}\cdot{}^\circ\text{C)}] \equiv s\,[\text{in kJ/(kg}\cdot\text{K)}]$$

This does not mean that the °F and R (or °C and K) scales are equal but only that their *degree sizes* are equal. Therefore, when you have units like Btu/(lbm·°F), you need not use any mathematical formula to convert °F to R in order to write this grouping as Btu/(lbm·R). This is a simple but often confusing point.

However, the temperature unit you choose to place in the denominator of a term's units grouping may depend on how the term is to be used in relation to other temperature terms in the equation. An example of where this occurs is in the use of specific heats. Equation (3.15) is

$$c_v = \left(\frac{\partial u}{\partial T}\right)_v \tag{3.15}$$

where c_v is the constant volume specific heat.

We discussed reversible processes briefly in Chapter 4 and noted that there are few reversible processes in the real world. In fact, every heat transport of energy through a finite temperature difference is irreversible. We are able to write Eqs. (7.21) through (7.24) as reversible heat transfers only because we created a very special situation, in which the heat transport of energy was assumed to take place through an infinitesimal temperature difference. But, in the real world, it would require an infinite amount of time to transport a finite amount of energy by this method. If we try to alter the results of Eq. (7.22) by considering only real irreversible heat transports of energy, we immediately realize that the amount of work done by the cyclic heat engines must be less than in the reversible case. Then, for an actual heat engine,

$$W_{\text{actual}} < W_{\text{reversible}}$$

and using the first law of thermodynamics, we conclude that, since the system total energy E is a point function and therefore independent of whether the process path is reversible or irreversible,

$$dE = \bar{d}Q_{\text{rev}} - \bar{d}W_{\text{rev}} = \bar{d}Q_{\text{act}} - \bar{d}W_{\text{act}}$$

WHEN DO WE USE °F (OR °C) AND WHEN DO WE USE R (OR K)?

Whether you use relative or absolute temperature units in an equation depends on whether the temperature appears in an equation as a difference or stands alone. For example, assuming c_v is a constant, integrating Eq. (3.15) gives $u_2 - u_1 = c_v(T_2 - T_1)$, and since the temperature appears as a difference here, we can use either °F or R (or °C or K) temperature units, because (T_1 in °F + 459.67 R) − (T_2 in °F + 459.67 R) = (T_1 in °F − T_2 in °F), as the conversion from °F to R cancels out. You can use either °F or R and you get the same answer in each case.

Also, the numerical value of c_v in Btu/(lbm·°F) has the same value in Btu/(lbm·R). For example, for air, c_v = 0.172 Btu/(lbm·°F) = 0.172 Btu/(lbm·R). That is because the temperature unit appears in the denominator as "per degree," and the Fahrenheit degree is the same size as a Rankine degree (only their zero point is different). Similarly, c_v = 0.718 kJ/(kg·K) = 0.718 kJ/(kg·°C) for the same reason.[8] This also applies to numerical values of c_p, entropy s, and the gas constant $R = \mathscr{R}/M$.

However, in Eqs. (7.18) and (7.20), the temperature stands alone in the numerator, but the denominator has a temperature difference. What do you do now?

$$\text{COP}_{\substack{\text{Carnot}\\\text{heat pump}}} = \frac{T_H}{T_H - T_L} \tag{7.18}$$

$$\text{COP}_{\substack{\text{Carnot}\\\text{refrig. or}\\\text{air cond.}}} = \frac{T_L}{T_H - T_L} \tag{7.20}$$

The rule is that, whenever you have an equation in which the temperature T *stands alone* (and not as a temperature difference), the temperature must *always be in an absolute unit* (R or K). So the numerators on Eqs. (7.18) and (7.20) must be in absolute temperature units (R or K), but since the denominator has a temperature difference, the temperatures here can be in either relative or absolute temperature units. If you are unsure whether to use absolute or relative temperature units in an equation, use absolute temperature units, since they always give the correct answer.

[8] *But be careful if you use a table with c_v or s in Btu/(lbm·°F) because you might be tempted to use T in °F to cancel the temperature unit. This would be incorrect. You have to understand that c_v or s in Btu/(lbm·°F) has the same numerical value as c_v or s in Btu/(lbm·R).*

WHAT ARE PERPETUAL MOTION MACHINES?

Devices that supposedly operate using processes that violate either the first or second laws of thermodynamics or that are required to be reversible represent various forms of perpetual motion machines. When the operation of a device depends on the violation of the first law of thermodynamics it is called a perpetual motion machine of the *first kind* (e.g., a heat engine that produces power but does not absorb heat from the environment). When the operation of a device depends on the violation of the second law, it is called a perpetual motion machine of the *second kind* (e.g., an adiabatic air compressor in which the air exits at a lower temperature than it entered), and when it requires a reversible process to operate it is called a perpetual motion machine of the *third kind* (e.g., a wheel on a shaft that, once started, continues to rotate indefinitely).

No perpetual motion machines operate as proposed and have for centuries been the source of frauds brought on the unsuspecting public by unscrupulous or naive inventors.

and dividing by the appropriate absolute temperature and rearranging gives

$$dS = \left(\frac{\bar{d}Q}{T}\right)_{\text{rev}} = \left(\frac{\bar{d}Q}{T}\right)_{\text{act}} + \frac{(\bar{d}W)_{\text{rev}} - (\bar{d}W)_{\text{act}}}{T} \tag{7.25}$$

For a work-producing heat engine, $(\bar{d}W)_{\text{act}} \le (\bar{d}W)_{\text{rev}}$ and both are positive work quantities, since they represent energy leaving the system; Eq. (7.25) can be rearranged to produce

$$dS > \left(\frac{\bar{d}Q}{T}\right)_{\text{act}} \tag{7.26}$$

Equation (7.26) is known as the *Clausius inequality*. It is Clausius's mathematical form of the second law of thermodynamics for a closed system. Dropping the subscript on the bracketed term and thus allowing it to represent either a reversible or actual process produces the following somewhat more general mathematical second law expression:

$$dS \ge \left(\frac{\bar{d}Q}{T}\right) \tag{7.27}$$

and

$$\oint_{\text{cycle}} \left(\frac{\bar{d}Q}{T}\right) \le 0 \tag{7.28}$$

where the equality sign is used for a reversible heat transport of energy.

7.8 NUMERICAL VALUES FOR ENTROPY

In Chapter 3, we discussed five methods for finding numerical values for properties: thermodynamic equations of state, thermodynamic tables, thermodynamic charts, direct experimental measurements, and the formulae of statistical thermodynamics. The same five methods can be used to find numerical values for the specific entropy. In this section, we focus on the use of thermodynamic equations of state, tables, and charts.

Energy and entropy are thermodynamic properties and therefore mathematical point functions. Consequently, the energy and entropy changes of a system depend only on the beginning and ending states of a process and not on the actual thermodynamic path taken by the process between these states. Therefore, for a closed system, we can write the differential energy and entropy balances as

$$(dE)_{\text{rev}} = (dE)_{\text{act}} = dE = (\bar{d}Q)_{\text{rev}} - (\bar{d}W)_{\text{rev}} = (\bar{d}Q)_{\text{act}} - (\bar{d}W)_{\text{act}}$$

and

$$(dS)_{\text{rev}} = (dS)_{\text{act}} = dS = \left(\frac{\bar{d}Q}{T}\right)_{\text{rev}}$$

Combining the "reversible" path parts of these two equations, we get

$$(\bar{d}Q)_{\text{rev}} = TdS = dE + (\bar{d}W)_{\text{rev}} \tag{7.29}$$

For a stationary differential closed system at a uniform temperature T containing a pure substance that is subjected to only a mechanical moving boundary work mode, Eq. (7.29) becomes

$$T\,dS = dU + p\,d\mathbb{V}$$

and on dividing through by the system mass m and the absolute temperature T,

$$ds = \frac{du}{T} + \frac{p}{T}dv \tag{7.30}$$

Since $u = h - pv$, this equation can also be written as

$$ds = \frac{dh}{T} - \frac{v}{T}dp \tag{7.31}$$

In Chapter 3, we define the constant volume and constant pressure specific heats for an incompressible substance as

$$c_v = c_p = c = \frac{du}{dT}$$

Since v = constant and $dv = 0$ for an incompressible material, then Eq. (7.30) becomes

$$(ds)_{\text{incomp.}} = c\left(\frac{dT}{T}\right)$$

or

$$(s_2 - s_1)_{\text{incomp.}} = \int_{T_1}^{T_2} c\left(\frac{dT}{T}\right) \tag{7.32}$$

If the specific heat c is constant over the temperature range from T_1 to T_2, then this equation can be integrated to give

$$(s_2 - s_1)_{\substack{\text{incompressible material}\\ \text{with a constant } c}} = c\ln(T_2/T_1) \tag{7.33}$$

In Chapter 3, we also define the constant volume and constant pressure specific heats for an ideal gas as

$$c_v = \frac{du}{dT} \tag{3.37}$$

and

$$c_p = \frac{dh}{dT} \tag{3.40}$$

Consequently, we can now write Eqs. (7.30) and (7.31) as

$$(ds)_{\substack{\text{ideal}\\ \text{gas}}} = c_v\left(\frac{dT}{T}\right) + \frac{p}{T}dv = c_p\left(\frac{dT}{T}\right) - \frac{v}{T}dp$$

Further, for an ideal gas, $p/T = R/v$ and $v/T = R/p$, so this equation can be integrated to give

$$(s_2 - s_1)_{\substack{\text{ideal}\\ \text{gas}}} = \int_1^2 c_v\left(\frac{dT}{T}\right) + R\,\ln\frac{v_2}{v_1} \tag{7.34}$$

$$= \int_1^2 c_p\left(\frac{dT}{T}\right) - R\,\ln\frac{p_2}{p_1} \tag{7.35}$$

and if the specific heats are constant over the temperature range from T_1 to T_2, then these equations become

$$(s_2 - s_1)_{\substack{\text{ideal gas}\\ \text{constant}\\ c_p \text{ and } c_v}} = c_v \ln\frac{T_2}{T_1} + R\,\ln\frac{v_2}{v_1} \tag{7.36}$$

$$= c_p \ln\frac{T_2}{T_1} - R\,\ln\frac{p_2}{p_1} \tag{7.37}$$

DO ALL ELASTIC MATERIALS HAVE ENTROPIC ELASTICITY?

No, most elastic solids do not have entropic elasticity, but the elasticity present in rubber and polymers is largely entropic. When work is done adiabatically on a material with entropic elasticity, the temperature of the material increases. You can demonstrate this by stretching a rubber band rapidly then immediately touching it to your lips (which are very sensitive to temperature). The rubber band is warmer than it was before it was stretched. Then, if you hold the stretched rubber band long enough for it to return to room temperature and suddenly release it and touch it to your lips, it is colder than it was before it was stretched.

When an elastic deformation produces a *decrease* in the specific entropy of a material, it is said to have *entropic elasticity*. In the case of an ideal gas, Eq. (7.36) shows that an isothermal ($T_2 = T_1$) compression ($v_2 < v_1$) produces a decrease in the specific entropy of the gas. Consequently, ideal gases have entropic elasticity.

Any process in which entropy remains constant is called an *isentropic* process. The term *isentropic* comes from the Greek words for "constant entropy."

If an ideal gas with constant specific heats undergoes an isentropic process, then $s_2 - s_1$ and Eqs. (7.36) and (7.37) give

$$\ln\frac{T_2}{T_1} = -\frac{R}{c_v}\ln\frac{v_2}{v_1} = \frac{R}{c_p}\ln\frac{p_2}{p_1}$$

or

For an isentropic process with an ideal gas,

$$\frac{T_2}{T_1} = \left(\frac{v_2}{v_1}\right)^{1-k} = \left(\frac{p_2}{p_1}\right)^{(k-1)/k} \tag{7.38}$$

and

$$p_1 v_1^k = p_2 v_2^k = \text{constant} \tag{7.39}$$

where

$$k = c_p/c_v \quad \text{and} \quad R = c_p - c_v \tag{7.40}$$

Consequently, from Eq. (4.27), we see that, in the case of an ideal gas with constant specific heats, an isentropic process is the same as a polytropic process with $n = k$.

These equations for the specific entropy of an incompressible substance and an ideal gas are the only such formulae to be introduced at this point. Specific entropy equations for more complex substances are introduced later in the text as they are needed.

EXAMPLE 7.4

An insulated apparatus contains 1.5 kg of saturated liquid water at 20.°C. Determine the change in specific entropy of the water as it is pressurized from 0.10 MPa to 10. MPa. Assume the liquid water is an incompressible material.

Solution

First, draw a sketch of the system (Figure 7.12).

The unknown is the change in specific entropy, $s_2 - s_1$, for the system. The material is liquid water.

An energy balance for this process gives

$${}_1Q_2 - {}_1W_2 = m(u_2 - u_1)$$

and since the apparatus is insulated ${}_1Q_2 = 0$. The only possible work mode here is moving boundary work, so ${}_1W_2 = \int p d\forall$.

But the water is to taken as incompressible, so $\forall$ = constant and $d\forall = 0$. Also, Eq. (3.33) gives the specific internal energy change of an incompressible material as $u_2 - u_1 = c(T_2 - T_1)$, where c is the specific heat of the material. Then, the energy balance equation gives for this process

$$0 - 0 = mc(T_2 - T_1)$$

(Continued)

EXAMPLE 7.4 *(Continued)*

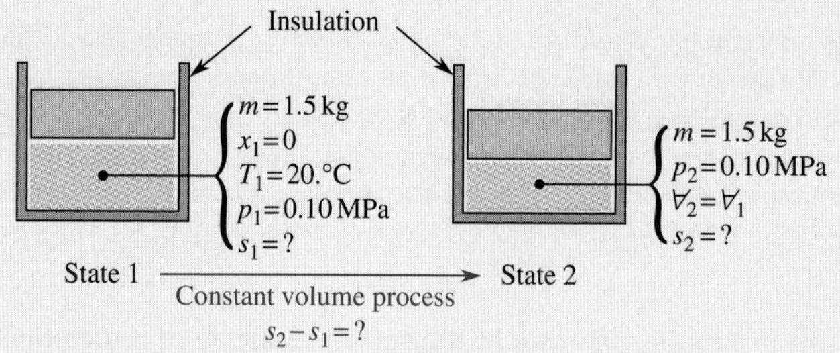

FIGURE 7.12
Example 7.4.

or, $T_2 = T_1$. Therefore, the process must also be isothermal, and Eq. (7.33) gives the specific entropy change as

$$s_2 - s_1 = c\ln(T_2/T_1) = c\ln(1) = 0$$

Consequently, the entropy of an incompressible material is not altered by changing its pressure.

EXAMPLE 7.5

An apparatus contains 0.035 kg of air (an ideal gas). The apparatus is used to compress the air isentropically (i.e., at constant entropy) from a pressure of 0.100 MPa to a pressure of 5.00 MPa. If the initial temperature of the air is 20.0°C, determine the final temperature and specific volume of the air.

Solution

First, draw a sketch of the system (Figure 7.13).

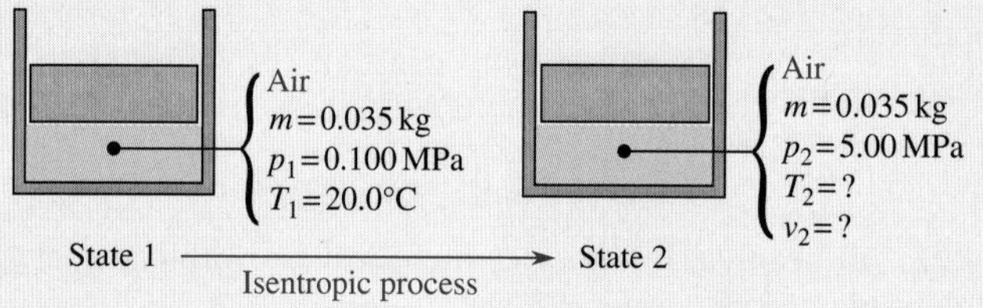

FIGURE 7.13
Example 7.5.

The unknowns are the final temperature, T_2, and specific volume, v_2, of the air in the system. Since $p_1 = 0.100$ MPa, $T_1 = 20.0$°C, and $p_2 = 5.00$ MPa, then Eq. (7.38) can be used to find T_2 and v_2 as follows. From *Thermodynamic Tables to accompany Modern Engineering Thermodynamics*, Table C.13b, we find for air that $k = 1.4$; and solving Eq. (7.38) for T_2 gives

$$T_2 = T_1\left(\frac{p_2}{p_1}\right)^{\frac{k-1}{k}} = (20.0 + 273.15\,\text{K})\left(\frac{5.00\,\text{MPa}}{0.100\,\text{MPa}}\right)^{\frac{0.4}{1.4}} = 896\,\text{K} = 623°\text{C}$$

Using the ideal gas equation of state and Table C.13b for the gas constant of air ($R_{air} = 0.286$ kJ/kg·K), we find the initial specific volume of the air to be

$$\begin{aligned} v_1 &= \text{m}RT_1/p_1 \\ &= (0.035\,\text{kg})(0.286\,\text{kJ/kg·K})(20.0 + 273.15\,\text{K})/(0.100 \times 10^3\,\text{kN/m}^2) \\ &= 0.02934\,\text{m}^3/\text{kg} \end{aligned}$$

Then, solving Eq. (7.38) for v_2 gives

$$v_2 = v_1\left(\frac{T_2}{T_1}\right)^{\frac{1}{1-k}} = (0.02934\,\text{m}^3/\text{kg})\left(\frac{623 + 273.15}{20.0 + 273.15}\right)^{-\left(\frac{1}{0.4}\right)} = 0.00180\,\text{m}^3/\text{kg}$$

Exercises

7. Determine the change in specific entropy as a 1.00 kg block of solid incompressible iron is heated from 20.0°C to 100.°C. (See Table 3.6 for specific heat values.) **Answer:** $(s_2 - s_1)_{iron} = 0.108$ kJ/kg·K.
8. Determine the change in specific entropy of air as it is heated from 20.0°C to 100.°C in a constant pressure (isobaric) process. Assume air behaves as an ideal gas. **Answer:** $(s_2 - s_1)_{isobaric\ air} = 0.242$ kJ/kg·K.
9. Determine the change in specific entropy of air as it is heated from 20.0°C to 100.°C in a constant volume (isochoric) process. Assume air behaves as an ideal gas. **Answer:** $(s_2 - s_1)_{isochoric\ air} = 0.173$ kJ/kg·K.

The tables and charts in *Thermodynamic Tables to accompany Modern Engineering Thermodynamics* list specific entropy along with the specific properties v, u, and h. Specific entropy values are obtained from these sources in the same way that any of the other specific properties are obtained. In particular, the quality x of a liquid-vapor mixture is computed using the same type of lever rule relation as was used with v, u, and h; that is,

$$x = \frac{v - v_f}{v_{fg}} = \frac{u - u_f}{u_{fg}} = \frac{h - h_f}{h_{fg}} = \frac{s - s_f}{s_{fg}} \tag{7.41}$$

EXAMPLE 7.6

Determine the change in total entropy of 3.00 lbm of steam at 100.°F and 80.0% quality when it is heated in an unknown process to 200. psia and 800.°F.

Solution

First, draw a sketch of the system (Figure 7.14).

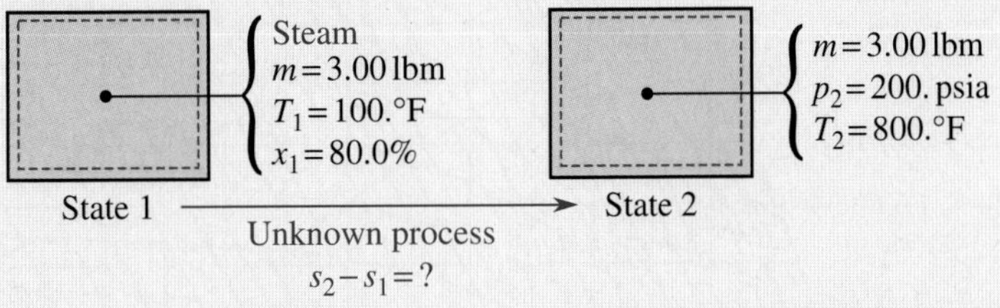

FIGURE 7.14
Example 7.6.

The unknown is the change in total entropy of the steam. Since we are given two independent properties in each state in this problem, we do not need to know how the heating process (i.e., the path) took place. We have a closed system consisting of 3.0 lbm of water, for which

State 1	Unknown process path →	State 2
$T_1 = 100.°\text{F}$		$p_2 = 200.\ \text{psia}$
$x_1 = 0.800$		$T_2 = 800.°\text{F}$
$s_1 = s_f(100.°\text{F}) + x_1 s_{fg}(100.°\text{F})$		$s_2 = 1.7662\ \text{Btu/lbm·R}$
$= 0.1296 + 0.800(1.8528)$		
$= 1.6118\ \text{Btu/lbm·R}$		

where the specific entropy values have been found in Tables C.1a and C.3a. Then,

$$\begin{aligned} S_2 - S_1 &= m(s_2 - s_1) \\ &= (3.00\ \text{lbm})[(1.7662 - 1.6118)\ \text{Btu/(lbm·R)}] \\ &= 0.463\ \text{Btu/R} \end{aligned}$$

Exercises

10. Determine the change in specific entropy of steam as it is cools from a saturated vapor (x = 100.%) at 100.°F to a saturated liquid (x = 0%) at 100.°F. **Answer:** $s_2 - s_1 = s_f(100.°\text{F}) - s_g(100.°\text{F}) = -s_{fg}(100.°\text{F}) = -1.8528$ Btu/lbm·R.

11. Saturated liquid ammonia at 0.00°C is heated in a constant pressure process until it has a quality of 50.0%. Determine the change in specific entropy of the ammonia. **Answer:** $s_2 - s_1 = 2.311$ kJ/kg·K.

12. Determine the magnitude of the change in specific entropy of a water liquid-vapor mixture at 100.°C as its quality decreases from 1.000 by (a) 1.00%, (b) 10.0%, (c) 50.0%. **Answers:** (a) $|s_2 - s_1| = 0.0605$ kJ/kg·K, (b) $|s_2 - s_1| =$ 0.605 kJ/kg·K, and (c) $|s_2 - s_1| = 3.024$ kJ/kg·K.

Figure 7.15 is an *h-s* plot for water. It is called a *Mollier diagram* after the German engineer Richard Mollier (1863–1935), who developed it in 1904. States 1 and 2 of Example 7.6 are shown on this chart to illustrate its use. Small charts like this are usually inaccurate for engineering problem solving. Professional engineers used much larger charts like this before thermodynamic property software became available.

At this point, we must expand the classical concepts presented thus far so that they fit into the more general balance equations introduced at the beginning of this chapter. We must now look for a set of general entropy transport mechanisms, valid for both open and closed systems, that are consistent with Eq. (7.28) when applied to closed systems.

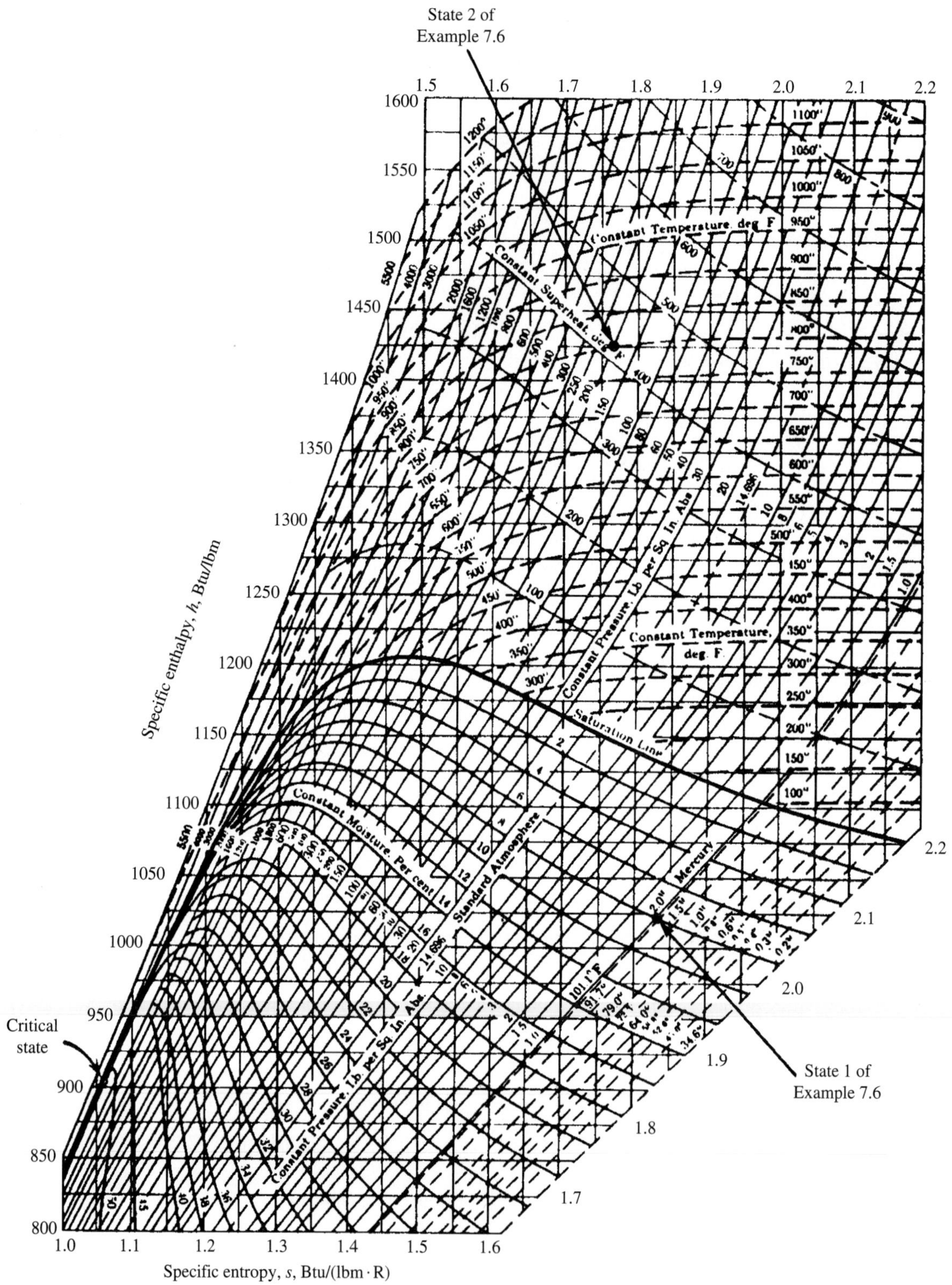

FIGURE 7.15

The Mollier diagram for water. (*Source: Keenan, J. H., Keyes, J., 1936. Thermodynamic Properties of Steam. Wiley, New York. Reprinted by permission of John Wiley & Sons.*)

7.9 ENTROPY TRANSPORT MECHANISMS

Several conceptual problems occur when Eq. (7.22) is used to define the entropy of a system. First of all, this equation is limited to closed systems, and at this point, we do not know how it must be altered to accommodate open systems. Second, it does not indicate how system entropy may be influenced by the work transport of energy; third, it deals only with hypothetical "reversible" processes. The third point is particularly bothersome, since all of our auxiliary formulae for heat transfer have been developed from an empirical basis and therefore always give the *actual* rather than the *reversible* heat transport of energy (see Chapter 4).

From the work of Carnot, Thomson, and Clausius discussed earlier in this chapter, it seems clear that energy and entropy are related in some way. In this chapter, we investigate the possibility that the energy transport mechanisms of heat transfer and work modes are also mechanisms for entropy transport. In Chapter 9, we expand this investigation to include mass flow transport of entropy.

First, we investigate heat and work transports of entropy by again restricting our analysis to closed systems. In Chapter 4, we note a modern definition of *heat transfer*: It is an energy transport mechanism that is neither a work mechanism nor a mass flow mechanism. It is often conveniently viewed as a nonwork, nonmass flow mechanism for transporting internal energy.

7.10 DIFFERENTIAL ENTROPY BALANCE

In a reversible process, the production of entropy is always zero, by definition. Therefore, if Eq. (7.22) is viewed as a differential entropy balance for a closed system undergoing a reversible heat transport of energy, then from a differential form of Eq. (7.2), it is clear that the heat transport of entropy is given by

$$(dS_G)_Q = (dS)_Q = (dS_T)_{\substack{\text{rev}\\Q}} + \underbrace{(dS_P)_{\substack{\text{rev}\\Q}}}_{0} = \left(\frac{\bar{d}Q}{T}\right)_{\text{rev}}$$

So, the differential entropy transport due to a hypothetical reversible heat transfer is simply

$$(dS_T)_{\substack{\text{rev}\\Q}} = \left(\frac{\bar{d}Q}{T}\right)_{\text{rev}} \tag{7.42}$$

Unfortunately, the two sides of Eq. (7.42) do not have the same differential form. The left side is the total differential of S_T, whereas the right side is a differential divided by an absolute temperature. To integrate this equation over the surface area of the system, we need to know the exact mathematical relationship between Q and the temperature of the boundary T_b at the point where this heat transfer occurs. For a reversible Carnot cycle, this relation is very simple, since a reversible heat transfer occurs only during an isothermal process; so T_b must be a constant. In this case, Eq. (7.42) can be integrated to give

$${}_1(S_T)_2\Big|_{\substack{\text{rev Q and}\\\text{isothermal}}} = \frac{{}_1Q_2}{T_b}\Big|_{\substack{\text{rev Q and}\\\text{isothermal}}} \tag{7.43}$$

However, Eq. (7.42) cannot be as easily integrated for common nonisothermal heat transfer processes. This problem can be solved by changing the form of Eq. (7.42) by introducing the following mathematical identity

$$\frac{\bar{d}Q}{T} = d\left(\frac{Q}{T}\right) + \frac{Q}{T^2}dT \tag{7.44}$$

Then, Eq. (7.42) becomes

$$(dS_T)_{\substack{\text{rev}\\Q}} = d\left(\frac{Q}{T}\right)_{\text{rev}} + \left(\frac{Q}{T^2}dT\right)_{\text{rev}}$$

Recall that heat transfer irreversibility is simply due to the heat transport of energy through a finite temperature difference, so that for all reversible heat transfers, we must have $dT = 0$, or

$$\left(\frac{Q}{T^2}dT\right)_{\text{rev}} = 0$$

then

$$(dS_T)_{\substack{\text{rev}\\Q}} = d\left(\frac{Q}{T}\right)_{\text{rev}} \tag{7.45}$$

which, when integrated, produces Eq. (7.43) again.

In an irreversible process, the total production of entropy is always positive, by virtue of the second law of thermodynamics. Therefore, Eq. (7.25) can be viewed as an entropy balance equation for a closed system undergoing irreversible heat and work transports of energy. Then,

$$\begin{aligned} dS &= dS_T + dS_P \\ &= \left(\frac{\bar{d}Q}{T}\right)_{\text{act}} + \frac{(\bar{d}W)_{\text{rev}} - (\bar{d}W)_{\text{act}}}{T} \end{aligned}$$

where $dS_P > 0$ by the second law of thermodynamics. To reconcile the difference between the actual work and the reversible work terms in this equation, we use the concept of a work transport energy efficiency, η_W, which was introduced in Chapter 4. For a work-absorbing system such as a pump, we have

Work-absorbing system

$$W_{\text{act}} = W_{\text{rev}}/\eta_W \tag{4.71}$$

and, for a work-producing system such as an engine, we have

Work-producing system

$$W_{\text{act}} = \eta_W W_{\text{rev}} \tag{4.72}$$

Let us define an irreversible work component, W_{irr}, which is *always* a positive number, as

$$W_{\text{irr}} = (1 - 1/\eta_W)W_{\text{rev}}, \text{ for a work-absorbing system}$$

and

$$W_{\text{irr}} = (1 - \eta_W)W_{\text{rev}}, \text{ for a work-producing system} \tag{7.46}$$

Note that η_W is a positive number between zero and unity ($0 < \eta_W < 1$) and that W_{rev} is negative for a work-absorbing system and positive for a work-producing system. Consequently, Eq. (7.46) produces positive values for W_{irr} for both work-absorbing *and* work-producing systems. Then, we can write for either a work-absorbing or a work-producing system that

$$W_{\text{rev}} - W_{\text{act}} = W_{\text{irr}}$$

and the entropy balance equation becomes[9]

$$dS = dS_T + dS_P = \left(\frac{\bar{d}Q}{T}\right)_{\text{act}} + \left(\frac{\bar{d}W}{T}\right)_{\text{irr}} \tag{7.47}$$

We now wish to identify the individual heat and work components of the entropy transport and production terms. For a closed system, we may decompose the transport and production terms as follows:

$$dS_T = (dS_T)_Q + (dS_T)_W$$

and

$$dS_P = (dS_P)_Q + (dS_P)_W$$

so that

$$dS = (dS_T)_Q + (dS_T)_W + (dS_P)_Q + (dS_P)_W \tag{7.48}$$

Substituting Eq. (7.44) into (7.47) gives

$$dS = d\left(\frac{Q}{T}\right)_{\text{act}} + \left(\frac{Q}{T^2}dT\right)_{\text{act}} + \left(\frac{\bar{d}W}{T}\right)_{\text{irr}} \tag{7.49}$$

[9] Note that W_{irr} is always positive for both work-absorbing and work-producing systems, so its sign is correct here.

and comparing Eqs. (7.48) and (7.49) allows us to identify the following terms:[10]

$$(dS_T)_Q = d\left(\frac{Q}{T}\right)_{\text{act}} \tag{7.50}$$

$$(dS_T)_W = 0 \tag{7.51}$$

$$(dS_P)_Q = \left(\frac{Q}{T^2}dT\right)_{\text{act}} \tag{7.52}$$

and

$$(dS_P)_W = \left(\frac{\bar{d}W}{T}\right)_{\text{irr}} \tag{7.53}$$

Integrating Eq. (7.49) and then differentiating it with respect to time produces an entropy rate balance equation for a system as

$$\begin{aligned} \dot{S} = \frac{dS}{dt} &= \frac{d}{dt}\int d\left(\frac{Q}{T}\right)_{\text{act}} + \frac{d}{dt}\int \left(\frac{Q}{T^2}dT\right)_{\text{act}} + \frac{d}{dt}\int \left(\frac{\bar{d}W}{T}\right)_{\text{irr}} \\ &= (\dot{S}_T)_Q + (\dot{S}_T)_W + (\dot{S}_P)_Q + (\dot{S}_P)_W \end{aligned} \tag{7.54}$$

so that

$$(\dot{S}_T)_Q = \frac{d}{dt}\int d\left(\frac{Q}{T}\right)_{\text{act}} \tag{7.55}$$

$$(\dot{S}_T)_W = 0 \tag{7.56}$$

$$(\dot{S}_P)_Q = \frac{d}{dt}\int \left(\frac{Q}{T^2}dT\right)_{\text{act}} \tag{7.57}$$

and

$$(\dot{S}_P)_W = \frac{d}{dt}\int \left(\frac{\bar{d}W}{T}\right)_{\text{irr}} \tag{7.58}$$

7.11 HEAT TRANSPORT OF ENTROPY

The integration of Eq. (7.50) gives the heat transport of entropy as

$$(S_T)_Q = \int_\Sigma d\left(\frac{Q}{T_b}\right)_{\text{act}} = \sum \left(\frac{Q}{T_b}\right)_\Sigma \tag{7.59}$$

where Σ is the surface area of the system and T_b is the local absolute temperature of the system boundary corresponding to the value of the local heat transfer at the boundary, Q. However, if Q and T_b vary continuously along the boundary, then Eq. (7.59) is not easy to evaluate. To produce a more useful version of this equation, let q be the heat transfer per unit area and let $\dot{q}$ be the heat transfer *rate* per unit area (i.e., the heat "flux"). Then, define the heat transport of entropy per unit area as $q/T_b = (dQ/dA)/T_b$ and define the heat transport *rate* of entropy as $\dot{q}/T = (d^2Q/dA\,dt)/T_b$ so that Eq. (7.59) becomes

$$(S_T)_Q = \int_\Sigma \left(\frac{q}{T_b}\right)_{\text{act}} dA = \int_\Sigma\int_\tau \left(\frac{\dot{q}}{T_b}\right)_{\text{act}} dA\,dt \tag{7.60}$$

[10] Note that we could also attempt to use the identity of Eq. (5.44) on the irreversible work term and decompose it into $\bar{d}W_{\text{irr}}/T = d(W/T)_{\text{irr}} + (W_{\text{irr}}/T^2)$, then we would be tempted to equate $d(S_T)_W = d(W/T)_{\text{irr}}$. This would be incorrect because the irreversible work always occurs *inside* the system boundary and therefore cannot be associated with a transport term that measures quantities crossing the system boundary.

HOW DOES HEAT TRANSPORT ENTROPY?

Take a glass of water, for example. If we heat it up by "transporting" heat into it, the water molecules move faster and become more disordered. If we add enough heat, the water boils and becomes steam, and the water molecules become even more disordered. If we cool the water by transporting heat out of it, the molecules slow down, and when it freezes they form a very ordered crystalline structure. However, if you cool down your bedroom it does not become more organized in the sense that all the stuff on the floor ends up in the closet. It only gets more organized at the molecular level as everything freezes.

where τ is the time over which the heat transport occurs. Differentiating this equation with respect to time gives the corresponding transport rate term as

$$(\dot{S}_T)_Q = \int_\Sigma \left(\frac{\dot{q}}{T_b}\right)_{act} dA \tag{7.61}$$

In the case where $\dot{q}$ and T_b are constant for time τ over the surface Σ of area A, Eq. (7.60) reduces to

Heat transport of entropy when $\dot{q}$ and T_b are constant:

$$(S_T)_Q = \left(\frac{\dot{q}}{T_b}\right)_{act}(\tau A) = \left(\frac{{}_1Q_2}{T_b}\right)_{act} \tag{7.60a}$$

and Eq. (7.61) reduces to

Heat transport rate of entropy when $\dot{q}$ and T_b are constant:

$$(\dot{S}_T)_Q = \left(\frac{\dot{Q}}{T_b}\right)_{act} \tag{7.61a}$$

Otherwise, the exact relations between $(\dot{q}/T_b)_{act}$, time, and surface area must be known before the integral in Eqs. (7.60) and (7.61) can be evaluated.

EXAMPLE 7.7

A dishwashing process in a restaurant has 3.00 kg/min of saturated liquid water heated in a steady flow process at 100.°C until it has a quality of 75.0%. Determine the heat transport rate of entropy for this process.

Solution

First, draw a sketch of the system (Figure 7.16).

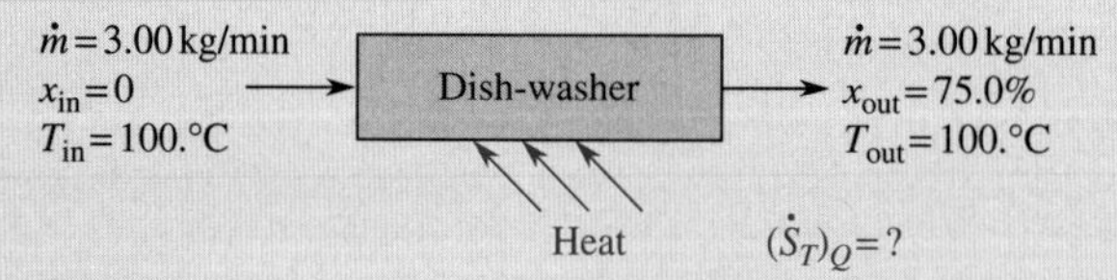

FIGURE 7.16
Example 7.7.

The unknown is the heat transport rate of entropy for this process, and the material is water. The heat transport rate of entropy is given by Eq. (7.61a) as

$$(\dot{S}_T)_Q = \left(\frac{\dot{Q}}{T_b}\right)_{act}$$

Neglecting any changes in flow stream kinetic and potential energy, an energy rate balance for this steady state, steady flow system is

$$\dot{Q} - \dot{W} + \dot{m}(h_1 - h_2 + 0 + 0) = 0$$

and since $\dot{W} = 0$ here (no work modes are present), we have

$$\begin{aligned}\dot{Q} &= \dot{m}(h_2 - h_1)\\ &= \dot{m}\{[h_f(100.°\text{C}) + x_2 h_{fg}(100.°\text{C})] - h_f(100.°\text{C})\}\\ &= \dot{m}x_2 h_{fg}(100.°\text{C}) = (3.00\,\text{kg/min})(0.750)(2257\,\text{kJ/kg}) = 5078\,\text{kJ/min}\end{aligned}$$

Then, Eq. (7.61a) gives

$$(\dot{S}_T)_Q = \left(\frac{5078\,\text{kJ/min}}{100. + 273.15\,\text{K}}\right) = 13.6\,\text{kJ/min}\cdot\text{K}$$

Exercises

13. Suppose the system in Example 7.7 is a closed, rigid vessel containing 3.00 kg of water, and the water is heated in the same manner from a saturated liquid at 100.°C to a liquid-vapor mixture at 100.°C with quality of 75.0%. Determine the heat transport of entropy for this process. **Answer:** $(S_T)_Q = 12.6$ kJ/K.
14. A new heat exchanger has been designed, where the local heat flux (heat "flux" is heat transfer per unit area) is directly proportional to the local surface temperature T_b of the heat transfer area A. Then, $\dot{q} = KT_b$ over the surface area A of the heat exchanger. Determine an expression for the heat transport rate of entropy for this system. **Answer:** $(\dot{S}_T)_Q = KA$.
15. An electric motor draws 800. W of electrical power and is 95.0% efficient. This means that 0.950(800.) = 760. W leaves the motor as mechanical shaft power and (1 − 0.950)(800.) = 40.0 W leaves as heat generated by the electrical and mechanical losses in the motor. Determine the heat transport rate of entropy of the motor if it has a uniform surface temperature of 30.0°C. **Answer:** $(\dot{S}_T)_Q = 0.132$ W/K.

7.12 WORK MODE TRANSPORT OF ENTROPY

Integration of Eq. (7.51) for all possible work modes clearly gives

Work mode transport of entropy

$$(S_T)_W = 0 \tag{7.62}$$

and, from Eq. (7.56), we also have

Work mode transport rate of entropy

$$(\dot{S}_T)_W = 0 \tag{7.63}$$

This produces the surprising result that none of the work modes discussed in Chapter 4 transports entropy into or out of a system. However, as is shown later, the irreversibilities of these work modes always contribute to the production of entropy within the system.

7.13 ENTROPY PRODUCTION MECHANISMS

One of the main problems with Eq. (7.22) is that it applies only to reversible processes. All of our auxiliary formulae for heat transfer have been developed on an empirical basis and therefore always give the *actual* or *irreversible* rather than the *reversible* heat transport of energy. Also, the process inefficiencies are mostly due to losses

WHY DO WORK MODES NOT TRANSPORT ENTROPY?

Let us suppose you have an electric ceiling fan in your room. When you turn it on, you have an electric work "mode" coming into the room. The fan blows your papers around and makes a mess (disorder). So why is this not a "work mode" transport of entropy (disorder)?

Well, "transports" have to be able to go both ways, in and out of the system, like heat transfer. If we run the fan backward, it does not put all your papers back where they were. So, in this case, your ceiling fan does not "transport" disorder (entropy) into your room, it "produces" disorder (entropy) inside your room. It is an entropy production process, not an entropy transport process. Work modes can only produce entropy; they cannot transport it across the system boundary like heat transfer.

produced by entropy production, so it is useful to have a tool to investigate these losses in the design and analysis process.

From the work of Carnot, Thomson, and Clausius discussed earlier, it seems clear that energy and entropy are related in some way. Therefore, in this section, we investigate the possibility that the energy transport mechanisms of heat, work, and mass flow are also mechanisms for entropy production. First, we investigate heat and work production of entropy by again restricting our analysis to closed systems. In Chapter 9, we expand this investigation to include mass flow entropy production.

7.14 HEAT TRANSFER PRODUCTION OF ENTROPY

To integrate Eq. (7.52) properly, we define a one-dimensional thermal entropy production rate per unit volume, σ_0, which is the rate of entropy production per unit volume due to the actual heat transfer, as

$$\frac{d^2(S_P)_Q}{dt\,d\mathcal{V}} = \frac{d(\dot{S}_P)_Q}{d\mathcal{V}} = \sigma_Q = -\left[\frac{\dot{q}}{T^2}\left(\frac{dT}{dx}\right)\right]_{\text{actual}} \tag{7.64}$$

where the minus sign appears because dT/dx always is negative when $\dot{q}$ is positive. The temperature T in this equation is the local absolute temperature *inside* the system boundary evaluated at the point where the local internal heat flux $\dot{q}$ occurs. It is generally not the same as the local system boundary temperature T_b except in the case of an isothermal system. Equation (7.64) can then be integrated to give

$$d(S_P)_Q = \frac{Q}{T^2}dT = -\int_\tau\left[\frac{\dot{q}}{T^2}\left(\frac{dT}{dx}\right)\right]_{\text{actual}} dt\,d\mathcal{V} = \int_\tau \sigma_Q dt\,d\mathcal{V}$$

and, with a second integration, we have

Heat transfer entropy production

$$(S_P)_Q = -\int_{\mathcal{V}}\int_\tau\left[\frac{\dot{q}}{T^2}\left(\frac{dT}{dx}\right)\right]_{\text{actual}} dt\,d\mathcal{V} = \int_{\mathcal{V}}\int_\tau \sigma_Q\,dt\,d\mathcal{V} \tag{7.65}$$

Differentiation of Eq. (7.65) yields the heat transfer entropy production rate term as

Heat transfer entropy production rate

$$(\dot{S}_P)_Q = -\int_{\mathcal{V}}\left[\frac{\dot{q}}{T^2}\left(\frac{dT}{dx}\right)\right]_{\text{actual}} d\mathcal{V} = \int_{\mathcal{V}} \sigma Q\,d\mathcal{V} \tag{7.66}$$

EXAMPLE 7.8

An electric motor has a volume of 2.50×10^{-3} m^3 and operates with a constant internal thermal entropy production rate per unit volume of $\sigma_Q = 53.7\,\text{W/K}\cdot\text{m}^3$. Determine the heat production of entropy inside this motor for the time period τ = 30.0 min of operation.

Solution

First, draw a sketch of the system (Figure 7.17).

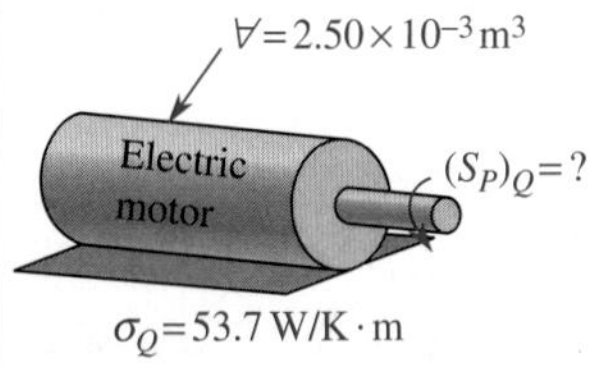

FIGURE 7.17
Example 7.8.

The unknown is the heat production of entropy inside this motor for the time period τ = 30.0 min of operation. The heat production of entropy for this system is given be Eq. (7.65) as

$$(S_P)_Q = -\int_{\mathcal{V}}\int_\tau\left[\frac{\dot{q}}{T^2}\left(\frac{dT}{dx}\right)\right]_{\text{actual}} dt\,d\mathcal{V} = \int_{\mathcal{V}}\int_\tau \sigma_Q\,dt\,d\mathcal{V}$$

Since σ_Q is a constant for τ = 30.0 min of operation, this equation reduces to

$$(S_P)_Q = \sigma_Q\tau\mathcal{V}$$

and we can calculate

$$(S_P)_Q = (53.7\,\text{W/K}\cdot\text{m}^3)(2.50\times10^{-3}\,\text{m}^3)(30.0\,\text{min})(60\,\text{sec/min}) = 242\,\text{J/K}$$

where we have used the fact that 1 W = 1 J/s.

EXAMPLE 7.9

Determine an equation for the steady state entropy production rate due to pure heat conduction in an insulated horizontal rod connecting a high-temperature (T_1) thermal reservoir with a low-temperature (T_2) thermal reservoir.

Solution

First, draw a sketch of the system (Figure 7.18).

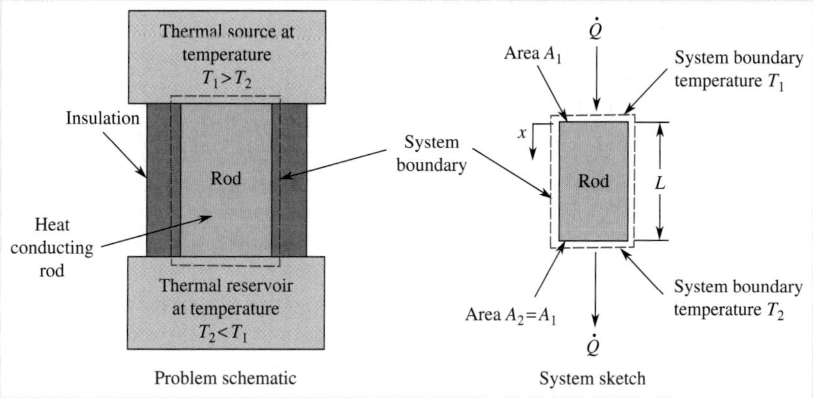

FIGURE 7.18
Example 7.9.

The unknown is an equation for the steady state entropy production rate for the system. The entropy production rate due to heat transfer is given by Eq. (7.66) as

$$(\dot{S}_P)_Q = -\int_{\forall}\left[\frac{\dot{q}}{T^2}\left(\frac{dT}{dx}\right)\right]_{\text{actual}} d\forall$$

For steady state conditions, $\dot{q}$ = constant across areas A_1 and A_2, but $\dot{q} = 0$ on the remaining surfaces of the system boundary. Since $A_1 = A_2 = A$ and $\dot{Q} = \dot{q}A$ is a constant, then using $d\forall = A\,dx$, we can write

$$(\dot{S}_P)_Q = -\int_0^L\left[\frac{\dot{q}}{T^2}\left(\frac{dT}{dx}\right)\right]A\,dx = -\dot{Q}\int_0^L\frac{1}{T^2}\left(\frac{dT}{dx}\right)dx = -\dot{Q}\int_{T_1}^{T_2}\frac{dT}{T^2}$$

$$= \dot{Q}\left(\frac{1}{T_2} - \frac{1}{T_1}\right) = \frac{\dot{Q}}{T_1T_2}(T_1 - T_2) > 0$$

For a pure one-dimensional, steady state conduction heat transfer, Fourier's law gives

$$\frac{dT}{dx} = -\frac{T_1 - T_2}{L} = -\frac{\dot{Q}}{k_tA} = \text{constant}$$

or

$$\dot{Q} = k_tA(T_1 - T_2)/L$$

Putting this information into the preceding entropy production rate formula gives

$$(\dot{S}_P)_Q = \frac{\dot{Q}}{T_1T_2}(T_1 - T_2) = \frac{k_tA}{T_1T_2L}(T_1 - T_2)^2 > 0$$

Notice that, in Example 7.9, the entropy production rate becomes zero only when $T_1 = T_2$ (i.e., when $\dot{Q} = 0$), so that a reversible conduction heat transfer is actually impossible. Also note that, the larger the temperature difference $T_1 - T_2$ through which a given heat transfer $\dot{Q}$ occurs, the larger the associated heat transfer entropy production rate becomes.

CASE STUDY: ENTROPY PRODUCTION IN HEAT PIPES

The application of basic thermodynamic principles recently produced a new heat transfer technology. In 1939. the German engineer E. Schmidt demonstrated that a hollow, sealed tube filled with a liquid-vapor mixture could transfer several thousand times more thermal energy than pure conduction in a solid copper rod with the same dimensions as the tube.

Heat applied to the liquid region at the lower end of the tube causes the quality of the mixture in the remainder of the tube to increase. The additional vapor thus produced rises inside the tube and condenses at the cooler end. This condensate then runs down the inside wall of the tube to replenish the liquid at the lower end. Figure 7.19a illustrates this process. A continuous circulation of vapor and condensate occurs when a steady state condition has been reached.

Since vaporization and condensation occur at the same temperature under constant pressure conditions, the entire inside volume of the tube reaches a constant temperature. Pure thermal conduction in a solid rod requires both a radial and an axial temperature gradient (see Figure 7.19c) to transport thermal energy. Schmidt's device, however, transports a great deal more thermal energy very efficiently with essentially no temperature gradient (see Figure 7.19d).

The name *heat pipe* was suggested for this device in 1963 by George M. Grover of the Los Alamos Scientific Laboratory. The first significant application of heat pipe technology occurred in the U.S. space program, then it spread into a wide variety of commercial areas including home furnaces and solar water heaters.

The steady state entropy production rate of a closed system heat pipe with isothermal input and output surfaces is obtained from the entropy rate balance as

$$(\dot{S}_P)_{\text{heat pipe}} = \sum\left(\frac{\dot{Q}}{T}\right)_{\text{net}} = \left(\frac{\dot{Q}}{T}\right)_{\text{out}} - \left(\frac{\dot{Q}}{T}\right)_{\text{in}} = \dot{Q}\left(\frac{T_{\text{in}} - T_{\text{out}}}{T_{\text{in}}T_{\text{out}}}\right) \quad (7.67)$$

If the heat pipe were truly isothermal throughout, then $T_{\text{in}} = T_{\text{out}}$ and the entropy production rate would be zero. But T_{in} is always slightly greater than T_{out} due to a small radial temperature gradient in the tube wall. However, these two temperatures are really very close to each other so that the entropy production rate of the heat pipe is actually quite small, making it a much more efficient heat transfer device than pure thermal conduction alone.

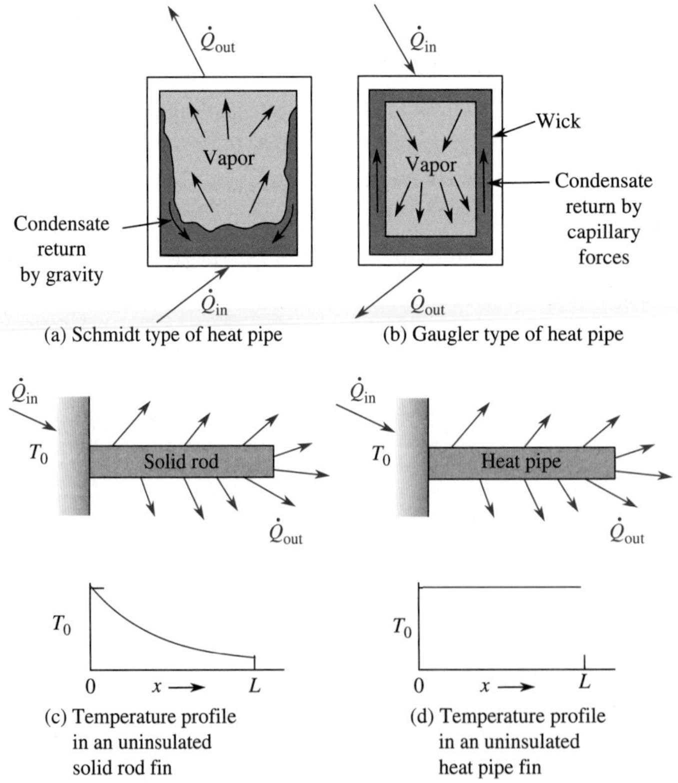

FIGURE 7.19
Heat pipe construction.

7.15 WORK MODE PRODUCTION OF ENTROPY

From Eq. (7.53), we have $(dS_P)_W = (\bar{d}W/T)_{\text{irr}}$, where T is the local absolute temperature inside the system boundary evaluated at the point where the work irreversibility (e.g., friction) occurs. Then.

$$(S_P)_W = \int \left(\frac{\bar{d}W}{T}\right)_{\text{irr}} \tag{7.68}$$

and, if the system is isothermal throughout at temperature T as it changes from state 1 to state 2, then

Isothermal work mode entropy production:

$$(S_P)_{W \atop T=\text{constant}} = {}_1\left(\frac{W_{\text{irr}}}{T}\right)_2 \tag{7.68a}$$

The corresponding entropy production rate equation is Eq. (7.58),

$$\left(\dot{S}_P\right)_W = \frac{d}{dt}\int \left(\frac{\bar{d}W}{T}\right)_{\text{irr}} \tag{7.58}$$

When the system local internal temperature T in Eq. (7.58) is independent of time, this equation simplifies to

Steady state work mode entropy production *rate*:

$$\left(\dot{S}_P\right)_W = \left(\frac{\dot{W}}{T}\right)_{\text{irr}} \tag{7.58a}$$

Therefore, only the work mode energy dissipated within the system contributes to the entropy production of the system. This dissipated energy has been defined to be the difference between the reversible work (as given by the work mode formula of Chapter 4) and the actual work (for which we have no specific formula except the empirically based efficiency Eq. (7.46)). If one has experimentally measured the work efficiency, then W_{irr} can be found from Eq. (7.46). However, to evaluate $(S_P)_W$ or $\left(\dot{S}_P\right)_W$ from Eq. (7.68) or (7.58), we need to know the mathematical functional relation between W_{irr} and the local absolute temperature T inside the system at all the points where the irreversibility occurs, in order to carry out the integration.

EXAMPLE 7.10

Determine the work mode entropy production when a measured 42.0×10^3 ft·lbf of work are used to compress 1.00 lbm of air (an ideal gas) from 14.7 psia to 50.0 psia isothermally at 70.0°F in a closed system.

Solution

First, draw a sketch of the system (Figure 7.20).

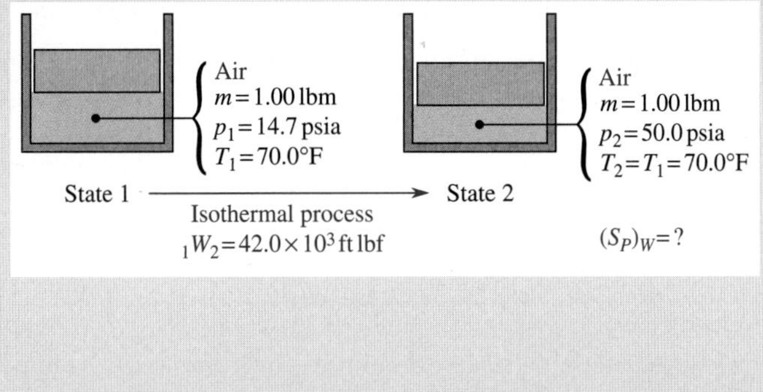

FIGURE 7.20
Example 7.10.

The unknown is the work mode entropy production for this system. Here, $W_{\text{act}} = 42.0 \times 10^3$ ft·lbf. For the isothermal compression of an ideal gas, Eq. (4.28) gives

$$\begin{aligned} W_{\text{rev}} &= \int_1^2 p\,d\forall = p_1\forall_1 \int_1^2 \frac{d\forall}{\forall} \\ &= p_1\forall_1 \ln\left(\forall_2/\forall_1\right) = p_1\forall_1 \ln(p_1/p_2) \end{aligned}$$

(Continued)

EXAMPLE 7.10 *(Continued)*

where

$$p_1 = 14.7\,\text{psia} = (14.7\,\text{lbf/in}^2)(144\,\text{in}^2/\text{ft}^2) = 2117\,\text{lbf/ft}^2$$

$$\forall_1 = \frac{mRT_1}{p_1} = \frac{(1.00\,\text{lbm})[53.34\,\text{ft}\cdot\text{lbf/(lbm}\cdot\text{R)}](70+459.67\,\text{R})}{2117\,\text{lbf/ft}^2} = 13.35\,\text{ft}^3$$

Then,

$$W_{\text{rev}} = (2117\,\text{lbf/ft}^2)(13.35\,\text{ft}^3)\ln\frac{14.7}{50.0} = -34{,}600\,\text{ft}\cdot\text{lbf}$$

Now, for an isothermal process, Eq. (7.68a) gives

$$(S_P)_W = \int_1^2 \frac{\bar{d}W_{\text{irr}}}{T} = \frac{1}{T}\int_1^2 \bar{d}W_{\text{irr}} = {}_1\!\left(\frac{W_{\text{irr}}}{T}\right)_2$$

Equation (7.59) gives us W_{irr} as

$$W_{\text{irr}} = W_{\text{rev}} - W_{\text{act}} = -34{,}600 - (-42{,}000) = 7{,}400\,\text{ft}\cdot\text{lbf}$$

Therefore,

$$(S_P)_W = \frac{7400\,\text{ft}\cdot\text{lbf}}{70+459.67\,\text{R}} = 13.97\,\text{ft}\cdot\text{lbf/R}$$

$$= (13.97\,\text{ft}\cdot\text{lbf/R})[1\,\text{Btu}/(778.16\,\text{ft}\cdot\text{lbf})] = 0.0179\,\text{Btu/R}$$

This example illustrates that you *must* know the relation between W_{irr} and T before Eq. (7.68) can be integrated. The simplest possible case occurs if T is a constant throughout the system volume, as in this example.

Exercises

16. Determine the work mode production of entropy in Example 7.10 if carbon dioxide gas (an ideal gas) is used instead of air. **Answer:** $(S_P)_W = 122.3$ ft·lbf/R.

17. If the losses in the electric motor in Example 7.7, Exercise 15, were interpreted as a work mode irreversibility with a work transport energy efficiency of $\eta_W = 95.0\%$, then Eq. (7.46) gives $W_{\text{irr}} = (1-\eta_W)W_{\text{rev}} = 0.050(800.\text{ W}) = 40.0$ W. Determine the work mode production rate of entropy for this system if it has a uniform internal temperature of 30.0°C. **Answer:** $(S_P)_W = 0.132$ W/K.

18. The actual work to compress a spring was measured and found to be 3.00 J. The reversible work required to compress the same spring was calculated to be 2.90 J. Determine the work mode production of entropy in the spring if the internal temperature of the spring is uniform at 20.0°C. **Answer:** $(S_P)_W = 3.41\times10^{-4}$ J/K.

Note that you must also know the exact relation between Q and T or T_b before Eqs. (7.60), (7.61), (7.65), and (7.66) can be integrated. These relations are often empirically derived auxiliary formulae known as *constitutive equations*. Fourier's law, Newton's law of cooling, and Planck's radiation law are three such constitutive equations, briefly discussed in Chapter 4, that relate heat transfer rate and temperature.

An alternative approach to evaluating the entropy production due to work mode irreversibilities is to attempt to identify the sources of the irreversibilities and to mathematically model them with appropriate equations. This is normally done by deriving relations for the work mode entropy production per unit time per unit volume σ_W. Since σ_W has a value at every point within the system, the total entropy production is determined by integrating σ_W over time and the system volume as

Work mode entropy production:

$$(S_P)_w = \int_\forall \int_\tau \sigma_w\,dt\,d\forall \tag{7.69}$$

and the entropy production rate is determined by integrating σ_W over the system volume as

Work mode entropy production *rate*:

$$(\dot{S}_P)_w = \int_\forall \sigma_w\,d\forall \tag{7.70}$$

These results are available for various work mode dissipation mechanisms, such as viscous dissipation within a Newtonian fluid, electrical energy resistive dissipation, diffusion of dissimilar chemicals, and so on. However, these mathematical models are normally quite complex, and therefore only the viscous dissipation and the electrical resistance mechanism models are introduced at this point. Dissipation resulting from diffusion is discussed in Chapter 9 in the section on mass flow production of entropy.

For the one-dimensional flow of a Newtonian fluid with viscosity μ, velocity distribution $V = V(x)$, and local internal absolute temperature T, we have

$$(\sigma_W)_{\text{vis}} = \frac{\mu}{T}\left(\frac{dV}{dx}\right)^2$$

Then, we can write

Entropy production due to fluid viscosity:

$$({}_{S_P})_{\substack{W\\\text{vis}}} = \int_\tau \int_{\forall} \frac{\mu}{T}\left(\frac{dV}{dx}\right)^2 dt\, d\forall \tag{7.71}$$

and

Entropy production *rate* due to fluid viscosity:

$$(\dot{S}_P)_{\substack{W\\\text{vis}}} = \int_{\forall} \frac{\mu}{T}\left(\frac{dV}{dx}\right)^2 d\forall \tag{7.72}$$

To carry out these integrations, we need to know in advance the velocity distribution function $V = V(x)$ and how μ and V depend on the local internal absolute temperature T.

EXAMPLE 7.11

Determine the entropy production rate per unit volume due to the flow of a lubricating oil ($\mu = 0.10\ \text{N}\cdot\text{s/m}^2$) at 30.°C in a sliding bearing. Assume the lubricant is isothermal throughout its volume.

Solution

First, draw a sketch of the system (Figure 7.21).

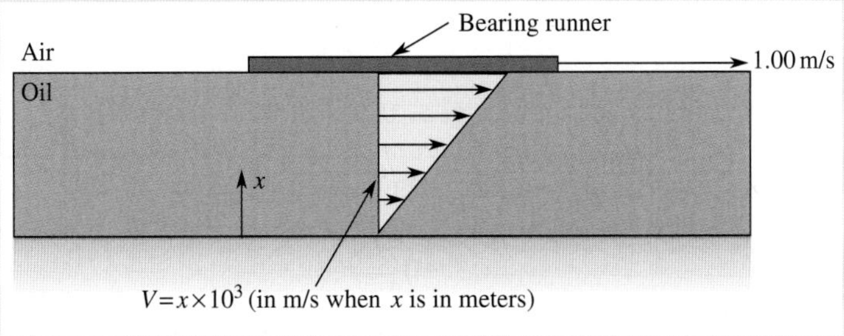

FIGURE 7.21
Example 7.11.

The unknown is the entropy production rate per unit volume due to the flow of the lubricating oil. In this case, $dV/dx = 1000\ \text{s}^{-1}$ = constant, and since μ, T, and dV/dx are constant here, Eq. (7.72) can be integrated to give

$$(\dot{S}_P)_{\substack{W\\\text{vis}}} = \frac{\mu}{T}\left(\frac{dV}{dx}\right)^2 \forall$$

and

$$\sigma_{W\text{-vis}} = \frac{(\dot{S}_P)_{W\text{-vis}}}{\forall} = \frac{\mu}{T}\left(\frac{dV}{dx}\right)^2 = \frac{(0.10\ \text{N}\cdot\text{s/m}^2)}{(30.+273.15\ \text{K})}(1000\ \text{s}^{-1})^2$$

$$= 330\frac{\text{N}}{\text{m}^2\cdot\text{s}\cdot\text{K}} = 0.33\ \text{kJ}/(\text{m}^3\cdot\text{s}\cdot\text{K})$$

Exercises

19. Determine the entropy production rate per unit volume in Example 7.11 when bearing grease ($\mu = 30.\ \text{N}\cdot\text{s/m}^2$) is used instead of lubricating oil. With the heavier lubricant, the bearing temperature is now 50.°C. **Answer**: $\sigma_{W\text{-vis}} = 93\ \text{kJ}/(\text{m}^3\cdot\text{s}\cdot\text{K})$.
20. Determine the work mode entropy production due to viscosity in a system where the fluid has a constant velocity. **Answer**: $(S_P)_{W\text{-vis}} = 0$.
21. If a fluid flowing in a pipe has a constant entropy production rate per unit volume due to viscosity of $0.100\ \text{Btu}/(\text{ft}^3\cdot\text{s}\cdot\text{R})$ and the volume of the fluid in the pipe is $0.475\ \text{ft}^3$, determine the work mode production rate of entropy for this system. **Answer**: $(\dot{S}_P)_{W\text{-vis}} = 0.0475\ \text{Btu/s}\cdot\text{R}$.

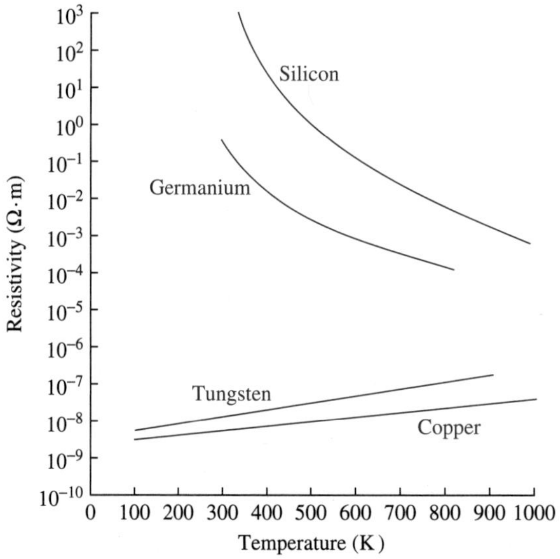

FIGURE 7.22
The variation of electrical resistivity with temperature. (*Source: Reprinted by permission of the author from Lenert, L. H., 1968. Semiconductor Physics, Devices, and Circuits. Charles E. Merrill Publishing, Columbus, OH.*)

Ohm's law is a simple mathematical model used for the resistive dissipation of electrical work mode energy. In this model, the entropy production rate per unit volume due to electrical resistance is

$$(\sigma_W)_{\text{elect}} = J_e^2 \rho_e / T$$

where $J_e = I/A$ is the electrical current per unit area (i.e., the electrical current flux), $\rho_e = R_e A/L$ is the electrical resistivity, R_e is the total electrical resistance of the conductor, L and A are the length and cross-sectional area of the conductor, and T is the local internal absolute temperature of the conductor. Then, Eq. (7.70) gives

$$(\dot{S}_P)_{\substack{W \\ \text{elect}}} = \int_{\mathcal{V}} \frac{J_e^2 \rho_e}{T} \, d\mathcal{V} \tag{7.73}$$

The electrical current flux J_e is often expressed by the one-dimensional Ohm's law as

$$J_e = -k_e \left(\frac{d\phi}{dx} \right)$$

where $k_e = 1/\rho_e$ is the electrical conductivity and ϕ is the electrical potential (i.e., voltage). The relation between ρ_e and T is shown in Figure 7.22 for various materials.

For the special case of an *isothermal system* with *uniform properties* and a *constant current density*, Eq. (7.73) reduces to this special case:

Entropy production rate due to electrical resistance:

$$(\dot{S}_P)_{\substack{W \\ \text{elect} \\ \text{(special)}}} = \frac{J_e^2 \rho_e \mathcal{V}}{T} = \left(\frac{(I/A)^2 \times (R_e A/L) \times LA}{T} \right) = \frac{I^2 R_e}{T} \tag{7.74}$$

EXAMPLE 7.12

A new high-temperature silicon computer chip 1.00×10^{-3} m by 5.00×10^{-3} m by 10.0×10^{-3} m long with uniform properties operates isothermally at 600. K and draws a constant electrical current of 0.10 A. Determine the entropy production rate of the chip.

Solution

First, draw a sketch of the system (Figure 7.23).

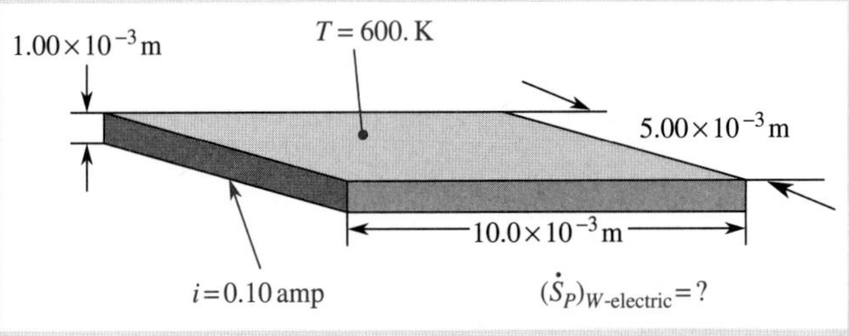

FIGURE 7.23
Example 7.12.

The unknown is the entropy production rate of the chip. Since the chip has uniform properties, is isothermal, and has a constant current, we can use Eq. (7.74) to determine its electrical work mode entropy production rate as:

$$(\dot{S}_P)_{\substack{W\\ \text{elect}}} = \frac{I^2 R_e}{T}$$

where the electrical resistance $R_e = \rho_e(L/A)$. The electrical *resistivity* can be found for silicon at 600. K from Figure 7.22 as $(\rho_e)_{\text{silicon}} = 0.10\ \Omega \cdot \text{m}$. Then, from the dimensions given for the chip, we have $A = 0.00100 \times 0.00500 = 5.00 \times 10^{-6}\ \text{m}^2$, $L = 0.0100$ m, and

$$R_e = (0.100\,\text{ohm}\cdot\text{m})(0.0100\,\text{m})/(5.00\times10^{-6}\,\text{m}^2) = 200.\,\Omega = 200.\,\text{W/A}^2$$

Then, Eq. (7.75) gives

$$(\dot{S}_P)_{\substack{W\\ \text{elect}}} = \frac{(0.10\,\text{A})^2(200\,\text{W/A}^2)}{600.\,\text{K}} = 0.0033\,\text{W/K}$$

Exercises

22. Determine the percent reduction in the entropy production rate in Example 7.12 if germanium (see Figure 7.22) is used in place of silicon in the chip and all the other parameters in the example remain the same. **Answer:** % reduction = 99%.

23. An incandescent lightbulb has a filament 0.13 mm in diameter and 0.076 m long. It is made of tungsten wire and operates at 900. K while carrying 0.90 A. Use Figure 7.22 to determine the entropy production rate due to the electrical work mode in the filament. **Answer:** $(\dot{S}_P)_{W\text{-elect}} = 0.0052\,\text{W/K}$.

24. A solenoid is made of copper wire 0.0500 mm in diameter and 200. m long. The solenoid operates at 300. K and carries a current of 0.500 A. Determine the entropy production rate inside the solenoid due to the electrical work mode. **Answer:** $(\dot{S}_P)_{W\text{-elect}} = 0.424\,\text{W/K}$.

7.16 PHASE CHANGE ENTROPY PRODUCTION

In any process, the change in entropy is independent of the actual process path used, because entropy is a state property (or a point function). Therefore, we may write $(\Delta S)_{\text{rev}} = (\Delta S)_{\text{act}}$ or $(\Delta s)_{\text{rev}} = (\Delta s)_{\text{act}}$ for any process whatsoever. The entropy change produced by an actual process can occasionally be found by assuming that the system has undergone a hypothetical reversible process that is easier to evaluate than the actual irreversible process. As an example of this technique, consider the entropy change and associated entropy production that occur in a phase change. For a reversible phase change carried out in a closed system, we have $(S_P)_{\text{rev}} = 0$, and the entropy balance for an isothermal system then gives

$$(S_2 - S_1)_{\text{rev}} = (S_2 - S_1)_{\text{act}} = (Q/T_b)_{\text{rev}} = (Q/T_b)_{\text{act}} + (S_P)_{\text{phase change}}$$

or

$$(S_P)_{\text{phase change}} = \frac{Q_{\text{rev}} - Q_{\text{act}}}{T_b} > 0$$

Thus, for an exothermic (heat liberating) phase change (e.g., a condensation or solidification process), the heat transfers are negative and it follows that $|Q_{\text{act}}| > |Q_{\text{rev}}|$ and for an endothermic phase change (e.g., a vaporization process), $Q_{rev} > Q_{act}$. The irreversibilities involved in a phase change process arise largely from the heat transfer required to produce the phase change and from the mechanical moving boundary work associated with any volume change between the phases. If the real system is truly isothermal, then heat transfer irreversibilities may be allocated to the system's surroundings. As for the work mode irreversibilities, for a reversible process, $(\bar{d}W)_{rev} = -mp\,dv$, and the differential energy balance then gives $(\bar{d}Q)_{rev} = m(du + p\,dv) = m(dh)$. However, for an actual irreversible process $(\bar{d}W)_{act} = f(\eta_W)(\bar{d}W)_{rev}$, where $f(\eta_W)$ is a function of the work transport energy efficiency, given in Eqs. (4.71) and (4.72), immediately preceding Eq. (7.46). Then, $(\bar{d}Q)_{act} = m[du + f(\eta_W)\,p\,dv] \neq m(dh)$. Consequently, if there is negligible change in system volume during the phase change or the moving boundary work is carried out very efficiently, then the work mode irreversibilities also are insignificant. Under these conditions, the actual phase change can be accurately approximated as a reversible process.

7.17 ENTROPY BALANCE AND ENTROPY RATE BALANCE EQUATIONS

The resulting entropy balance for a closed system of mass m is given by integrating Eq. (7.48) and substituting in Eqs. (7.60), (7.62), (7.65), and (7.69) to produce

General closed system entropy balance (SB)

$$\int_\tau \int_\Sigma \left(\frac{\dot{q}}{T_b}\right)_{act} dA\,dt + {}_1(S_P)_2 = (S_2 - S_1)_{system} = [m(s_2 - s_1)]_{system} \tag{7.75}$$

For the simplified case of isothermal boundaries, this equation reduces to

Isothermal boundary closed system entropy balance

$$\left(\frac{{}_1Q_2}{T_b}\right)_{act} + {}_1(S_P)_2 = m(s_2 - s_1) \tag{7.76}$$

where in each case

$${}_1(S_P)_2 = \int_\tau \int_{\forall} \left\{-\left[\frac{\dot{q}}{T^2}\left(\frac{dT}{dx}\right)\right] + \sigma_W\right\} d\forall\,dt$$

In Chapter 9, Eq. (7.75) is expanded into the following open system general entropy rate balance equations.

General open system entropy *rate* balance (SRB)

$$\int_\Sigma \left(\frac{\dot{q}}{T_b}\right)_{act} dA + \sum_{inlet} \dot{m}s - \sum_{outlet} \dot{m}s + \dot{S}_P = \dot{S}_{system} \tag{9.4}$$

and, for the simplified case of isothermal boundaries, this equation reduces to

Isothermal boundary open system entropy *rate* balance

$$\left(\frac{\dot{Q}}{T_b}\right)_{act} + \sum_{in} \dot{m}s - \sum_{out} \dot{m}s + \dot{S}_p = \dot{S}_{system} \tag{9.5}$$

There are two effective ways for calculating the entropy production or the entropy production rate for any process: the direct and the indirect methods.

- **Direct method** involves calculating the amount of entropy produced for a process from its defining equations. For example, for closed systems, the direct method of calculating the entropy production rate is

$$\dot{S}_P = \int_{\forall} (\sigma_Q + \sigma_W)\,d\forall$$

where

$$\sigma_Q = \left[\frac{\dot{q}}{T^2}\left(\frac{dT}{dx}\right)\right]_{actual} \quad \text{and} \quad \sigma_W = \sigma_{viscous} + \sigma_{electrical} + \cdots$$

- **Indirect method** involves calculating the amount of entropy production for a process from an entropy balance on the system. For example, for closed systems with isothermal boundaries, the indirect method of calculating the entropy production rate is

$$\dot{S}_P = \left(\frac{dS}{dt}\right)_{\text{system}} - \left(\frac{\dot{Q}}{T_b}\right)_{\text{act}}$$

Both the direct method and indirect methods give accurate answers for entropy production if applied correctly. Which method you choose to solve a particular problem depends entirely on the type of information given to you in the problem statement (usually only one of the two methods works for a given problem scenario). The examples presented in the next two chapters illustrate the use of both of these methods.

SUMMARY

In this chapter, we study the classical and the modern development of the second law of thermodynamics and the resulting entropy and entropy rate balances for closed systems. We find that entropy, unlike energy, is not conserved in any process and the second law of thermodynamics requires that entropy always be produced in a process. Kelvin used the classical results of Carnot to develop an absolute temperature scale and produced a definition of thermal efficiency based only on temperature. Numerical values for entropy can be computed from simple entropy equations of state for incompressible materials and ideal gases, but the tables in *Thermodynamic Tables to accompany Modern Engineering Thermodynamics* must be used for more complex materials, such as mixtures of liquid and vapor. Two important new process terms introduced in this chapter are *reversible process,* a process in which the entropy production inside the system is always zero ($S_P = \dot{S}_P = 0$), and *isentropic process,* a process in which the system's entropy is maintained constant ($s_2 - !s_1 = 0$).

Then we develop entropy balance (SB) and entropy rate balance (SRB) equations by using the same three transport modes (heat, work, and mass flow) as to develop the energy and energy rate balance equations in Chapter 4. However, since entropy is such an ambiguous concept, we have to be very careful how we define entropy transport and production modes. The resulting closed system constant boundary temperature entropy and entropy rate balance equations and other important equations introduced in this chapter follow.

1. The general closed system entropy balance equation for a system that has a constant temperature T_b on the boundaries where the heat transfer occurs is

$$\left(\frac{{}_1Q_2}{T_b}\right)_{\text{act}} + {}_1(S_P)_2 = m(s_2 - s_1) \tag{7.76}$$

2. The general closed system entropy *rate* balance equation for a system that has a constant temperature T_b on the boundaries where the heat transfer occurs is

$$\left(\frac{\dot{Q}}{T_b}\right)_{\text{act}} + \dot{S}_P = \left(\frac{dS}{dt}\right)_{\text{system}} \tag{7.78}$$

3. The *second law* of thermodynamics is simply

$$S_P \geq 0 \tag{7.4a}$$

or

$$\dot{S}_P \geq 0 \tag{7.4b}$$

4. The *third law* of thermodynamics is this: The entropy of a pure substance at absolute zero temperature is zero, or

$$\lim_{T=0} (\text{Entropy of a pure substance}) = 0 \tag{7.1}$$

5. The definition of the absolute temperature scale based on the heat transports of a (reversible) Carnot engine is

$$\left(\frac{|Q_{\text{out}}|}{Q_{\text{in}}}\right)_{\text{rev}} = \left(\frac{|Q_L|}{Q_H}\right)_{\text{rev}} = \frac{T_L}{T_H} \tag{7.15}$$

6. The thermal efficiency of a *Carnot* reversible heat engine and the *maximum* thermal efficiency of *any other* heat engine operating between the temperature limits of T_H and T_L is

$$(\eta_T)_{\text{max}} = (\eta_T)_{\text{rev}} = (\eta_T)_{\text{Carnot}} = 1 - \frac{T_L}{T_H} \tag{7.16}$$

7. The coefficient of performance (COP) of a Carnot engine operating in reverse as a heat pump, refrigerator, or air conditioner is

$$\text{COP}_{\substack{\text{Carnot}\\ \text{heat pump}}} = \frac{T_H}{T_H - T_L} \tag{7.18}$$

and

$$\text{COP}_{\substack{\text{Carnot}\\ \text{refrig. or}\\ \text{air cond.}}} = \frac{T_L}{T_H - T_L} \tag{7.20}$$

8. The change in specific entropy of an incompressible material (solid or liquid) with constant specific heat c is

$$(s_2 - s_1)_{\substack{\text{incompressible material}\\ \text{with a constant } c}} = c\ln(T_2/T_1) \tag{7.33}$$

and, for an ideal gas with constant specific heats c_p and c_v, it is

$$(s_2 - s_1)_{\substack{\text{ideal gas}\\ \text{constant}\\ c_p \text{ and } c_v}} = c_v \ln\frac{T_2}{T_1} + R\ln\frac{v_2}{v_1} \tag{7.36}$$

$$= c_p \ln\frac{T_2}{T_1} - R\ln\frac{p_2}{p_1} \tag{7.37}$$

9. When an *ideal gas* is used in an *isentropic* process between states 1 and 2, the following is valid:

$$\frac{T_2}{T_1} = \left(\frac{v_2}{v_1}\right)^{1-k} = \left(\frac{p_2}{p_1}\right)^{(k-1)/k} \tag{7.38}$$

10. The heat transport of entropy $(S_T)_Q$ and the heat transport rate of entropy $(\dot{S}_T)_Q$ across a system boundary at a constant temperature T_b are

$$(S_T)_Q = \left(\frac{\dot{q}}{T_b}\right)_{\text{act}}(\tau A) = \left(\frac{{}_1Q_2}{T_b}\right)_{\text{act}} \tag{7.60a}$$

and

$$(\dot{S}_T)_Q = \left(\frac{\dot{Q}}{T_b}\right)_{\text{act}} \tag{7.61a}$$

11. The work transport of entropy $(S_T)_W$ and the work transport rate of entropy $(\dot{S}_T)_W$ across a system boundary are both zero:

$$(S_T)_W = \text{constant} = 0 \tag{7.62}$$

and

$$(\dot{S}_T)_W = 0 \tag{7.63}$$

12. The entropy production $(S_P)_Q$ due to an actual heat transfer inside a system of volume $\forall$ during the time interval $0 - \tau$ is

$$(S_P)_Q = -\int_\forall\int_\tau\left[\frac{\dot{q}}{T^2}\left(\frac{dT}{dx}\right)\right]_{\text{act}} dt\,d\forall = \int_\forall\int_\tau \sigma_Q\,dt\,d\forall \tag{7.65}$$

and the entropy production rate $(\dot{S}_P)_Q$ due to an actual heat transfer inside a system of volume $\forall$ is

$$(\dot{S}_P)_Q = -\int_\forall\left[\frac{\dot{q}}{T^2}\left(\frac{dT}{dx}\right)\right]_{\text{act}} d\forall = \int_\forall \sigma_Q\,d\forall \tag{7.66}$$

13. The equations for the entropy production $(S_P)_W$ due to the presence of work modes by a system depends on the type of work mode present. For example, if the system is isothermal throughout at temperature T and an

amount of work W_{irr} is dissipated within the system due to irreversibilities inside the system, then the entropy produced by these irreversibilities is

$$\left(S_P\right)_{\substack{W \\ T=\text{constant}}} = {}_1\left(\frac{W_{irr}}{T}\right)_2 \tag{7.68a}$$

and the corresponding entropy production rate is

$$\left(\dot{S}_P\right)_W = \left(\frac{\dot{W}}{T}\right)_{irr} \tag{7.58a}$$

14. But if we know the irreversibilities are due to a velocity gradient dV/dx in a liquid with viscosity μ at a temperature T and that they occur in a system of volume $\forall$ during a time interval 0–τ, then we can calculate the entropy production and entropy production rates directly from

$$\left(S_P\right)_{\substack{W \\ \text{vis}}} = \int_\tau \int_\forall \frac{\mu}{T}\left(\frac{dV}{dx}\right)^2 dt\, d\forall \tag{7.71}$$

and

$$\left(\dot{S}_P\right)_{\substack{W \\ \text{vis}}} = \int_\forall \frac{\mu}{T}\left(\frac{dV}{dx}\right)^2 d\forall \tag{7.72}$$

Alternatively, if the irreversibilities are due to the flow of a constant electrical current I flowing in an isothermal system with uniform electrical resistance R_e at temperature T, then we have

$$\left(\dot{S}_P\right)_{\substack{W \\ \text{elect} \\ \text{(special)}}} = \frac{I^2 R_e}{T} \tag{7.74}$$

Problems (* indicates problems in SI units)

The first ten problems are designed to review some basic thermodynamic concepts of this and earlier chapters. They may have more than one correct answer.

1. A closed system becomes an open system when
   a. There is no heat transfer to energy.
   b. There is no work transfer of energy.
   c. There is no mass flow.
   d. There is no entropy production.
   e. There is no kinetic or potential energy.
   f. None of the above.
2. Which of the following are *intensive* properties: (a) pressure, (b) temperature, (c) volume, (d) mass, (e) quality, (f) power.
3. The entropy change of a closed system is zero for which of the following processes: (a) adiabatic, (b) isothermal, (c) isentropic, (d) isenthalpic, (e) aergonic, (f) reversible.
4.* An insulated, rigid container is divided into two compartments separated by a partition. One compartment contains air at 15°C and 0.101 MPa, the other compartment contains air at 40.°C and 0.101 MPa. When the dividing partition is removed, the total internal energy of the system (a) increases, (b) decreases, (c) does not change, (d) is converted into entropy, (e) is converted into temperature, (f) is converted into heat.
5. A rigid container contains air (an ideal gas), at 70.0°F and 14.7 psia. If the air is heated to 510.°F, its pressure (a) increases, (b) decreases, (c) does not change, (d) causes moving boundary work to occur, (e) causes polytropic work to occur, (f) is converted into thermal energy.
6. A constant velocity throttling process (a) is reversible, (b) is isothermal, (c) is isentropic, (d) is isenthalpic, (e) is aergonic, (f) does not exist in the real world.
7. Heat and work are both (a) intensive properties, (b) extensive properties, (c) process path dependent, (d) zero for an ideal gas, (e) zero for an adiabatic process, (f) zero for an aergonic process.
8. An ideal gas must satisfy (a) $pv^n = \text{constant}$, (b) $u = u(T)$, (c) $s = \text{constant}$, (d) $p_2/p_1 = v_1/v_2$, (e) $pv = nRT$, (f) $pv = mRT$.
9. In a steady flow *irreversible* process, the total entropy of a system
   a. Always increases.
   b. Always decreases.
   c. Always remains constant.
   d. Can increase, decrease, or remain constant.
   e. Cannot decrease.
   f. Cannot remain constant.
10. To determine the maximum possible work that a heat engine could produce, one must assume
    a. No entropy is produced by the engine.
    b. No heat energy is discharged to the environment.

c. No mechanical friction occurs anywhere in the engine.
d. No chemical reactions occur anywhere in the engine.
e. No irreversibilities occur anywhere within the engine.
f. No heat transfer occurs to or from the engine.

11. a. Write either the Clausius or the Kelvin-Planck word statements of the second law of thermodynamics.
b. Write an accurate mathematical equation for the second law.
c. Given any process, how can you determine whether it is physically possible?

12. An inventor claims to have developed an engine that operates on a cycle that consists of two reversible adiabatic processes and one reversible isothermal heat addition process (see Figure 7.24). Explain whether this engine violates either the first or second laws of thermodynamics. (Hint: Recall that the net work for the process is the area enclosed on the $p - \forall$ diagram.)

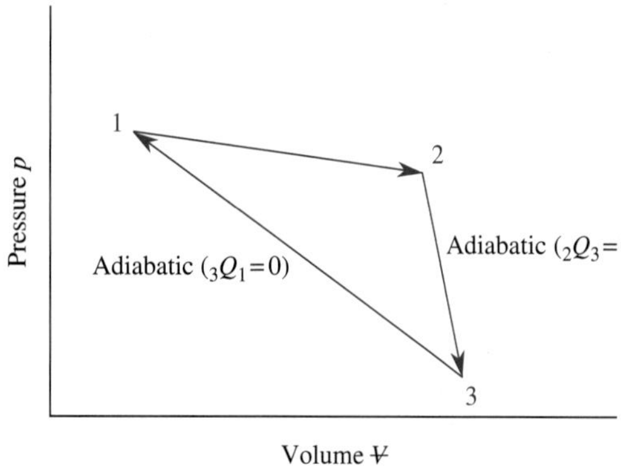

FIGURE 7.24
Problem 12.

13.* If the human body is modeled as a heat engine with its heat source at body temperature, what is its maximum (reversible or Carnot) efficiency when the ambient temperature is 20.0°C?

14.* An engine that operates on a reversible Carnot cycle transfers 4.00 kW of heat from a reservoir at 1000. K. Heat is then rejected to the atmosphere at 300. K. What is the thermal efficiency and the power output of this engine?

15. Determine whether each of the following functions could be used to define an absolute temperature scale:
a. $f(x) = \cos x$.
b. $f(x) = \tan x$.
c. $f(x) = x^4$.
d. $f(x) = 1 + x$.

16.* A closed system undergoes a cycle consisting of the following four processes:
Process 1. 10 kJ of heat are added to the system and 20 kJ of work are done by the system.
Process 2. The system energy increases by 30 kJ adiabatically.
Process 3. 10 kJ of work are done on the system while the system gains 50 kJ of energy.
Process 4. The system does 40 kJ of work while returning to its initial state.
a. Complete Table 7.1 (all values in kJ).
b. Find the thermal efficiency of this cycle.

Table 7.1 Problem 16

Process	$_iQ_j$	$_iW_j$	$E_i - E_j$
1			
2			
3			
4			
Totals			

17.* It is proposed to heat a house using a Carnot heat pump. The heat loss from the house is 50.0×10^3 J/s. The house is to be maintained at 25.0°C while the outside air is at −10.0°C. What coefficient of performance should the selected heat pump have, and what minimum horsepower of the motor is required to drive the heat pump?

18. A reversible Carnot refrigerator is to be used to remove 400. Btu/h from a region at −60.0°F and discharge heat to the atmosphere at 40.0°F. The reversible Carnot refrigerator is to be driven by a reversible heat engine operating between thermal reservoirs at 1040.°F and
40.0°F. How much heat must be supplied to the reversible Carnot heat engine from the 1040.°F reservoir?

19. A reversible Carnot refrigerator is used to maintain food in a refrigerator at 40.0°F by rejecting heat to the atmosphere at 80.0°F. The owner wishes to convert the refrigerator into a freezer at 0.00°F with the same atmospheric temperature of 80.0°F. What percent increase in reversible work input is required for the new freezer unit over the existing refrigerated unit for the same quantity of heat removed?

20.* What is the cooling capacity Q_L of a refrigerator with a coefficient of performance of 3.00 that is driven by a heat engine whose thermal efficiency is 25.0%? Both the engine and the refrigerator are reversible, and the engine receives 600. kW of heat energy from its high-temperature source.

21. A heat pump in a home is to serve as a heater in winter and an air conditioner in summer. This device transfers heat from its working fluid to air inside the house during the winter and to air outside the house during the summer. The design conditions (worst case) are as follows:

Winter	Summer
$T_{house} = 70.0°F$	$T_{house} = 70.0°F$
$T_{outside} = 20.0°F$	$T_{outside} = 100.°F$
$\dot{Q}_{house} = -50.0 \times 10^3$ Btu/h	$\dot{Q}_{house} = +30.0 \times 10^3$ Btu/h

Use the reversible Carnot cycle to determine the minimum power required to drive the heat pump.

22.* The air inside a garage is to be heated using a heat pump driven by a 500. W electric motor. The outside air is at a temperature of −20.0°C and provides the low-temperature heat source for the heat pump. The heat loss from the garage to the outside through the walls and roof is 12.5×10^3 kJ/h. If the heat pump operates on a reversible cycle, what is the highest temperature that can be maintained in the garage?

23. A heat pump, with the elements shown in Figure 7.25, is to be used to heat a home. On a given day, the evaporator receives heat at 30.0°F and the condenser rejects heat at 70.0°F. The required heat transfer rate from the condenser to the home is 50.0×10^3 Btu/h.

a. If the heat pump is operated reversibly, what is the rate of work transfer of energy to the compressor?

b. Express this compressor work rate in kilowatts and calculate the cost per 24.0 hour day if electricity is purchased at 10.0 cents/(kW·h).

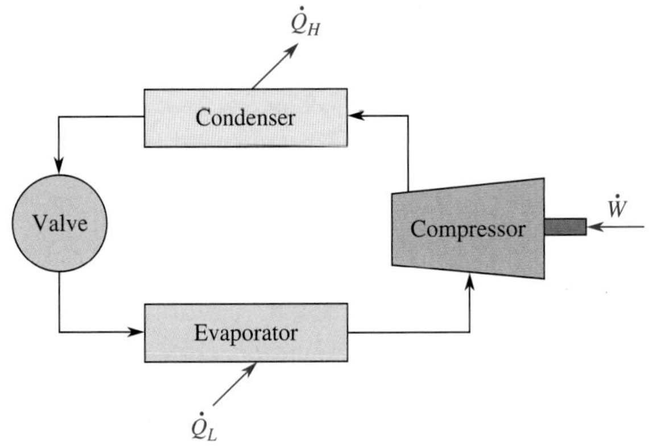

FIGURE 7.25
Problem 23.

24.* An automobile air conditioner removes 8000. kJ/h from the vehicle's interior. Determine the amount of engine horsepower required to drive the air conditioner if it has a coefficient of performance of 2.50.

25. A thermodynamic cycle using water is shown on the *T-s* diagram of Figure 7.26. Calculate the coefficient of performance for this cycle using the equation

$$\text{COP} = \frac{h_A - h_D}{(h_B - h_C) - (h_A - h_D)}$$

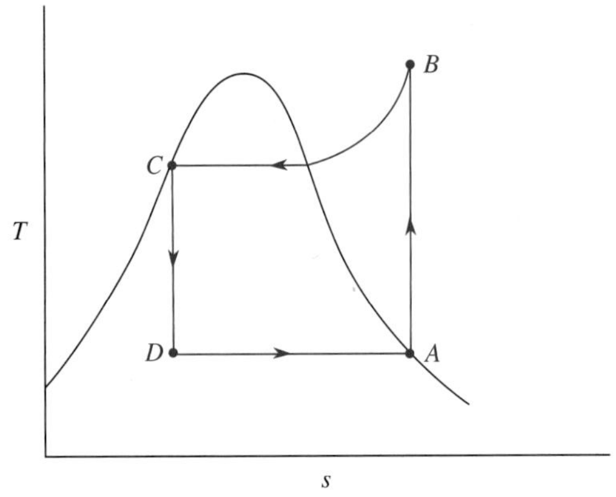

FIGURE 7.26
Problem 25.

State A	State B	State C	State D
$x_A = 1.00$	$p_B = 300.$ psia	$x_C = 0.00$	$p_D = 10.0$ psia
$P_A = 10.0$ psia	______	______	______

The processes are

$A \rightarrow B$ = isentropic compression,
$B \rightarrow C$ = isobaric expansion,
$C \rightarrow D$ = isentropic expansion,
$D \rightarrow A$ = isothermal compression.

26.* Determine the change in total entropy of 3.00 kg of incompressible liquid water with a specific heat of 4.20 kJ/(kg·K) as it is heated at atmospheric pressure from 20.0°C to its boiling point.

27. An ideal gas is compressed from 1.00 atm and 40.0°F to 3.00 atm and 540.°F. For this gas, $c_p = 0.280$ Btu/(lbm·R) and $c_v = 0.130$ Btu/(lbm·R). Calculate the change in specific entropy of this gas for this process.

28.* Determine the rate of heat transport of entropy through an isothermal boundary at 50.0°C when the heat transfer rate at the boundary is 350. kJ/min.

29.* Determine the heat transport of entropy into a pan of boiling water at 100.°C that has been sitting on a kitchen stove for 30.0 min. During this time, 0.300 kg of water is converted from a saturated liquid to a saturated vapor.

30.* If the heat entropy flux at a 2.70 m^2 boundary is given by $\dot{q}/T_b = 500. \times t + 243$ in W/(m^2·K), where *t* is in seconds, determine the total heat transport of entropy across this boundary during the time period from $t = 0.00$ to $t = 60.0$ s.

31.* If the thermal entropy production rate per unit volume is constant at 0.315 W/(m^3·K) throughout a 0.500 m^3 system, determine the system's rate of heat production of entropy.

32. Show that the one-dimensional thermal entropy production rate per unit volume for pure thermal conduction in a material with a constant thermal conductivity k_t is given by

$$\sigma_Q = (k_t/T^2)(dT/dx)^2.$$

33. Show that the one-dimensional temperature profile inside a system that has a constant thermal entropy production rate per unit volume σ_Q and a constant thermal conductivity k_t, restricted to pure conduction heat transfer, is given by $T = T_o \exp[(\sigma_Q/k_t)^{1/2}(x - x_o)]$, where T_o is the temperature at $x = x_o$.

34. If the temperature profile in Example 7.9 is given by

$$T = T_1\left[\exp\left(\frac{x}{L}\ \ln\ \frac{T_2}{T_1}\right)\right]$$

then

a. Show that for $\dot{Q}_{\text{in}} = \dot{Q}_{\text{out}}$ the rod cannot have a constant cross-sectional area and the areas of the ends of the rod are related by $A_2 = A_1\ (T_1/T_2)$.

b. Show that the entropy production rate under these circumstances is given by

$$(\dot{S}_P)_Q = \left(\frac{k_t A_1}{L}\right)\left(\frac{T_1}{T_2} - 1\right) \ln \frac{T_1}{T_2}$$

c. Show that the entropy production rate of part b is greater than that obtained with the linear temperature profile used in Example 7.9 when $A = A_1$.

35.* A stainless steel heat pipe, $k_t = 30.0$ W/(m·K), has an isothermal external surface temperature of 130.°C when its heat transfer rate is 1000. W. The surface area and wall thickness are 1.00×10^{-3} m^2 and 1.00×10^{-3} m, respectively. Determine the heat pipe's entropy production rate.

36.* If 0.101 grams of liquid plus vapor water at 0.0100 MPa-absolute are put into a heat pipe 1.00 m long with an inside diameter of 5.00×10^{-3} m, determine the temperature at which the heat pipe phenomena cease to operate due to the complete vaporization of the water.

37. Determine the work mode (a) entropy transport rate and (b) entropy production rate as 100. hp is continuously dissipated in a mechanical brake operating isothermally at 300.°F.

38.* Determine the work mode entropy production as a 300. kg steel block slides 10.0 m down a 60.0° incline. The coefficient of friction between the block and the incline is 0.100, and the bulk mean temperature of the sliding surface is 50.0°C.

39. A mechanical gearbox at a uniform temperature of 160.°F receives 150. hp at the input shaft and transmits 145 hp out the output shaft. Determine its work mode (a) entropy transport rate and (b) entropy production rate.

40.* Determine the work mode entropy production as 0.500 m^3 of air is compressed adiabatically from 200. kPa, 20.0°C to 0.100 m^3 in a piston-cylinder apparatus with a mechanical efficiency of 85.0%. Assume constant specific heat ideal gas behavior.

41. The velocity profile for the steady laminar flow of an incompressible Newtonian fluid in a horizontal circular pipe (Figure 7.27) is

$$V = V_{max}[1 - (x/R)^2]$$

where V_{max} is the centerline ($x = 0$) velocity, R is the pipe radius, and x is the radial coordinate measured from the centerline.

a. Determine the position in this flow where the entropy production rate per unit volume is a minimum.

b. Determine the position in this flow where the entropy production rate per unit volume is a maximum.

c. Comment on how you can minimize the total entropy production rate for this flow.

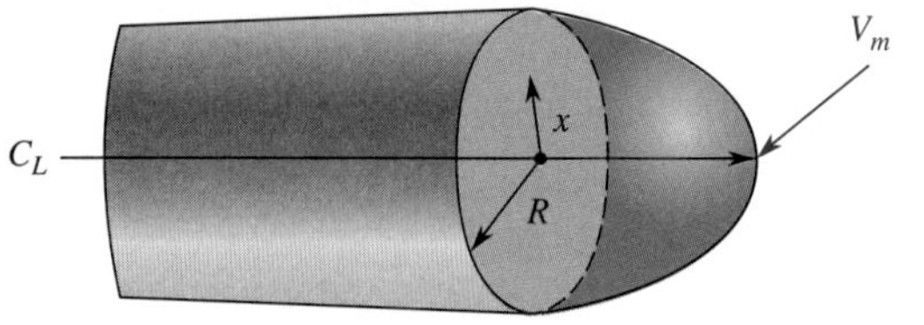

FIGURE 7.27
Problem 41.

42. The viscous work entropy production rate per unit volume in three-dimensional Cartesian coordinates is

$$(\sigma_W)_{vis} = \frac{\mu}{T}\left\{\frac{2}{3}\left[\left(\frac{\partial V_x}{\partial x} - \frac{\partial V_y}{\partial y}\right)^2 + \left(\frac{\partial V_y}{\partial y} - \frac{\partial V_z}{\partial z}\right)^2 + \left(\frac{\partial V_z}{\partial z} - \frac{\partial V_x}{\partial x}\right)^2\right] + \left(\frac{\partial V_x}{\partial y} + \frac{\partial V_y}{\partial x}\right)^2 + \left(\frac{\partial V_y}{\partial z} + \frac{\partial V_z}{\partial y}\right)^2 + \left(\frac{\partial V_z}{\partial x} + \frac{\partial V_x}{\partial z}\right)^2\right\}$$

Show that this can be written as

$$(\sigma_W)_{vis} = \frac{\mu}{T}\left\{-\frac{2}{3}\left(\frac{\partial V_x}{\partial x} + \frac{\partial V_y}{\partial y} + \frac{\partial V_z}{\partial z}\right)^2 + 2\left[\left(\frac{\partial V_x}{\partial x}\right)^2 + \left(\frac{\partial V_y}{\partial y}\right)^2 + \left(\frac{\partial V_z}{\partial z}\right)^2\right] + \left(\frac{\partial V_x}{\partial y} + \frac{\partial V_y}{\partial x}\right)^2 + \left(\frac{\partial V_y}{\partial z} + \frac{\partial V_z}{\partial y}\right)^2 + \left(\frac{\partial V_z}{\partial x} + \frac{\partial V_x}{\partial z}\right)^2\right\}$$

43. Show that the three-dimensional Cartesian coordinate viscous work entropy production rate per unit volume given in Problem 42 reduces to the following for two-dimensional incompressible flow:

$$(\sigma_W)_{vis} = \frac{\mu}{T}\left(\frac{\partial V_x}{\partial y} + \frac{\partial V_y}{\partial x}\right)^2$$

$$\left[\text{Hint: For incompressible fluids, } \frac{\partial V_x}{\partial x} + \frac{\partial V_y}{\partial y} + \frac{\partial V_z}{\partial z} = 0.\right]$$

44.* Determine the entropy production rate of a $10.0 \times 10^3\ \Omega$ electrical resistor that draws a constant 10.0 mA of current. The temperature of the resistor is constant throughout its volume at 35.0°C.

Computer Problems

The following computer programming assignments are designed to be carried out on any personal computer using a spreadsheet or equation solver. They are meant to be exercises in the manipulation of some of the basic formulae of this chapter. They may be used as part of a weekly homework assignment.

45. Develop a program that allows you to input *any* temperature (i.e., value plus unit symbol) in either the relative or absolute Engineering English or SI units system at the keyboard and converts this input into *all* the following temperatures and outputs them to the screen:
°C, °F, R, and K.

46. Develop a program that computes the change in specific internal energy, enthalpy, and entropy for an incompressible material. Input the initial and final temperatures and pressures, the specific heat, and the specific volume or density of the material. Output $u_1 - u_2$, $h_2 - h_1$, and $s_2 - s_1$ to the screen along with their proper units. Allow the choice of working in either the Engineering English or the SI units system.

47. Develop a program that computes the change in specific internal energy, enthalpy, and entropy for a constant specific heat ideal gas. Input the initial and final temperatures and pressures or specific volumes (allow the user the choice of which to input), the specific heats and gas constant. Output $u_2 - u_1$, $h_2 - h_1$, and $s_2 - s_1$ to the screen along with their proper units. Allow the choice of working in either the Engineering English or the SI units system.

48. Repeat Problem 47 except allow the user to choose a gas from a menu. Have all the specific heats and gas constants for the gases resident in your program.

49. Develop a program that outputs the heat production rate $(\dot{S}_P)_Q$ of entropy due to steady state, one-dimensional thermal conduction. Utilize Fourier's law of conduction and input the appropriate temperatures, thermal conductivity, cross-sectional area, and length in the proper units. Allow the choice of working in either the Engineering English or the SI units system.

50. Develop a program that outputs the work mode entropy production rate $(\dot{S}_P)_w$ due to the viscous dissipation in the steady one-dimensional flow of a Newtonian fluid in a circular pipe with the velocity profile given in Problem 41. Input the fluid's viscosity, density, and mass flow rate and the appropriate pipe dimensions in proper units. Allow the choice of working in either the Engineering English or the SI units system.

51. Develop a program that outputs the electrical work mode entropy production rate $(\dot{S}_P)_w$ due to resistive dissipation in an isothermal system with uniform properties and a constant current density. Input the appropriate variables in proper units. Allow the choice of working in either the Engineering English or the SI units system.

52. Since Figure 7.22 is a semi-logarithmic plot, the straight line for copper has an equation of the form $\rho_e = A(e^{BT})$. Estimate the coefficients A and B for copper from this figure and develop a program that outputs the electrical work mode entropy production rate $(\dot{S}_P)_w$ due to the temperature-dependent resistive dissipation in an isothermal copper wire. Input the appropriate variables in proper units. Allow the choice of working in either the Engineering English or the SI units system.

Create and Solve Problems

Engineering education tends to focus on the process of solving problems. It ignores teaching the process of formulating solvable problems. However, working engineers are never given a well-phrased problem statement to solve. Instead, they need to react to situational information and organize it into a structure that can be solved using the methods learned in college.

These "Create and Solve" problems are designed to help you learn how to formulate solvable thermodynamics problems from engineering data. Since you provide the numerical values for some of the variables, these problems have no unique solutions. Their solutions depend on the assumptions you need to make and how you set them up to create a solvable problem.

53.* You are an engineer at a company that manufactures domestic cookware. You are working on a new electric pot to be used to heat water to make spaghetti. The pot holds 2.5 kg of liquid water. You need to know the heat transfer rate and the electrical power required to heat the water from 15.0°C to 100.°C in 10.0 min. Tests have shown that 30% of the heat from the electrical heater is lost to the environment during the 10.0 min heating process. You also need to know the entropy production rate for this process (assume the average pot surface temperature is 55°C). Write and solve a thermodynamics problem that answers these questions.

54. You are an engineer at a diesel truck manufacturing company. Your work involves designing suitable braking systems for the trucks. To properly size the brakes, you need to know the work mode entropy transport rate and entropy production rate as 450. hp is absorbed in the brakes during an emergency stop. Tests show that the average brake temperature during this test is 350.°F. Write and solve a thermodynamics problem that provides the answers to these questions.

55. You have been transferred to the transmission section of the diesel truck manufacturing company. The transmission gearbox has a uniform surface temperature of 145°F when it receives 400. hp at the input shaft and transmits 375 hp out the output shaft. To improve this design of the transmission, you need to know the work mode entropy transport rate and entropy production rate of the existing transmission. Write and solve a thermodynamics problem that provides the answers to these questions.

56.* You are a newly employed engineer at a factory that produces scuba diving gear. The apparatus that fills the scuba tanks adiabatically compresses 1.50 m^3 of air from 100. kPa, 20.0°C to 0.067 m^3 in a piston-cylinder apparatus with a mechanical efficiency of 85.0%. You need to know the work mode entropy production for this process. Write and solve a thermodynamics problem that provides the answer you need.

CHAPTER 8

Second Law Closed System Applications

CONTENTS

8.1 INTRODUCTION

Chapters 8 and 9 provide closed and open system applications of the second law of thermodynamics in the same way that Chapters 5 and 6 dealt with closed and open system applications of the first law of thermodynamics. In this chapter, we present a series of applications of the closed system entropy balance and entropy production equations developed in Chapter 7. This material is organized into two major subdivisions: applications involving reversible processes and applications involving irreversible processes. Because the second law is seldom used alone, most of these examples also involve the application of the energy balance.

Our discussion of the applications involving reversible processes is similar to the way the second law is treated in many classical thermodynamics textbooks. Restricting consideration to reversible processes significantly simplifies the analysis, because the entropy production is zero and the second law is reduced to a simple conservation of entropy law. However, few real processes are truly reversible, so that any analysis that requires (or specifies) that a reversible process be assumed to get a solution always is somewhat in error. It should be remembered that a "reversible" process is just an idealization (a model) of some real irreversible process, much in the same way that we often model complex real gas behavior with the simple ideal gas equation of state. On the other hand, the study of reversible processes does provide an easy introduction to the use of the second law and the entropy balance equations, and they are accurate approximations to real processes in systems that have low entropy production values.

The section dealing with purely irreversible processes begins by expanding the closed system energy balance (first law) examples presented in Chapter 5 to include an entropy balance (second law) analysis. In Chapter 7, we introduce two methods for determining entropy production, the direct and the indirect methods. In the *direct method*, the amount of entropy produced for a process is calculated from its defining equations (e.g., Eqs. (7.65) and (7.66)); and in the *indirect method*, the amount of entropy production for a process is calculated from an entropy balance on the system.

The *indirect method* requires detailed temperature and heat flow information evaluated at the *boundary* of the system plus specific information about changes in system entropy, whereas the *direct method* requires detailed information about temperature, heat flow, and work mode irreversibilities spread throughout the *interior* of the system. The examples presented in this chapter illustrate the use of both methods. The following entropy production formulae correspond to these definitions.

8.1.1 Closed System Indirect Method

1. The entropy produced by a change of state is given by Eq. (7.75) as

$$ {}_1(S_P)_2 = m(s_2 - s_1) - \int_\tau \int_\Sigma \left(\frac{\dot{q}}{T_b}\right)_{\text{act}} dA\, dt \tag{8.1} $$

2. The entropy production *rate* of a closed system is given by differentiating Eq. (8.1) with respect to time, or

$$ \dot{S}_P = \dot{S} - \int_\Sigma \left(\frac{\dot{q}}{T_b}\right)_{\text{act}} dA \tag{8.2} $$

where T_b is the local system boundary temperature evaluated at the point where $\dot{q}$ crosses the system boundary Σ.

8.1.2 Closed System Direct Method

1. The entropy produced by a change of state is given by

$$ {}_1(S_P)_2 = \int_1^2 \left[\left(\frac{Q}{T^2}dT\right)_{\text{act}} + \left(\frac{\bar{d}W}{T}\right)_{\text{irr}}\right] = \int_\tau \int_{V\!\!\!\!-} \left[\left(-\frac{\dot{q}}{T^2}\frac{dT}{dx}\right) + \sigma_W\right] dV\!\!\!\!-\, dt \tag{8.3} $$

2. The entropy production *rate* of a closed system is given by

$$ \dot{S}_P = \int_{V\!\!\!\!-} \left\{\left[-\frac{\dot{q}}{T^2}\left(\frac{dT}{dx}\right)\right] + \sigma_W\right\} dV\!\!\!\!- \tag{8.4} $$

where $\sigma_W = (\sigma_W)_{\text{viscouw}} + (\sigma_W)_{\text{electrical}} + (\sigma_W)_{\text{diffusion}} + \cdots$ and T is the local temperature inside the system volume evaluated at the point where the heat transfer and work irreversibilities occur.

Both the *direct method* and the *indirect method* give accurate answers for entropy production if applied correctly. Which method you choose to solve a particular problem depends entirely on the type of information given to you in the problem statement (usually only one of the two methods works for a given problem scenario).

In the examples presented in this chapter, we are concerned mainly with determining the amount of entropy production for a process by calculating it from the closed system entropy balance (i.e., by using the *indirect method*).

Last, before we begin the example problems, keep in mind the basic reason why the evaluation of entropy production is important. The entropy production is a measure of the "losses" within the system, so the larger the entropy production, the more inefficient the system is at carrying out its function. We continually look for ways to minimize a system's entropy production and thus improve its overall operating efficiency.

8.2 SYSTEMS UNDERGOING REVERSIBLE PROCESSES

In Chapter 7, we define a *reversible* process as any process for which the entropy production or entropy production rate is zero. Thus, a system is said to be *reversible* if there are no losses due to friction, viscosity, heat transfer, diffusion, or the like anywhere within the system. Consequently, the entropy production of a system undergoing a reversible process is always zero, or

$$ \dot{S}_P = {}_1(S_P)_2 = 0 \quad \text{for all reversible processes} $$

and the closed system entropy balance (SB) given in Eq. (7.75) reduces to

$$ S_2 - S_1 = m(s_2 - s_1) = \int_\tau \int_\Sigma \left(\frac{\dot{q}}{T_b}\right)_{\text{rev}} dA\, dt \tag{8.5} $$

where $\dot{q}$ is the heat flux, T_b is the system boundary absolute temperature, A is the system boundary area, and τ is the time required to change from state 1 to state 2. The closed system entropy rate balance (SRB) is obtained from this equation by differentiating it with respect to time as[1]

$$ \frac{dS}{dt} = \dot{S} = m\dot{s} = \int_\Sigma \left(\frac{\dot{q}}{T_b}\right)_{\text{rev}} dA \tag{8.6} $$

---

[1] Remember that the system mass m is always constant in a closed system.

For the case of constant heat flux and isothermal system boundaries (i.e., $\dot{q}$ and T_b both constant), Eqs. (8.5) and (8.6) reduce to

Closed system reversible processes :

$$\text{Entropy balance (SB): } S_2 - S_1 = m(s_2 - s_1) = \left(\frac{{}_1Q_2}{T_b}\right)_{\text{rev}} \tag{8.7}$$

and

$$\text{Entropy rate balance (SRB): } \dot{S} = m\dot{s} = \left(\frac{\dot{Q}}{T_b}\right)_{\text{rev}} \tag{8.8}$$

where ${}_1Q_2$ and $\dot{Q}$ are the total heat transfer and total heat transfer rate, respectively, and T_b is again the absolute temperature of the system boundary (assumed isothermal here).

If the system boundary temperature is not constant, then the exact analytical dependence of $\dot{q}$ on system boundary temperature T_b, system boundary area A, and process time t must be known before the integrals in Eqs. (8.5) and (8.6) can be evaluated. This information must be provided in the problem statement (or determined by measurement or hypothesis in the case of real engineering situations).

EXAMPLE 8.1

Water in the amount of 2.00 kg undergoes a reversible isothermal expansion from a saturated liquid at 50.0°C to a superheated vapor at 50.0°C and 5.00 kPa. Determine the heat and work transports of energy for this process.

Solution

First, draw a sketch of the system (Figure 8.1).

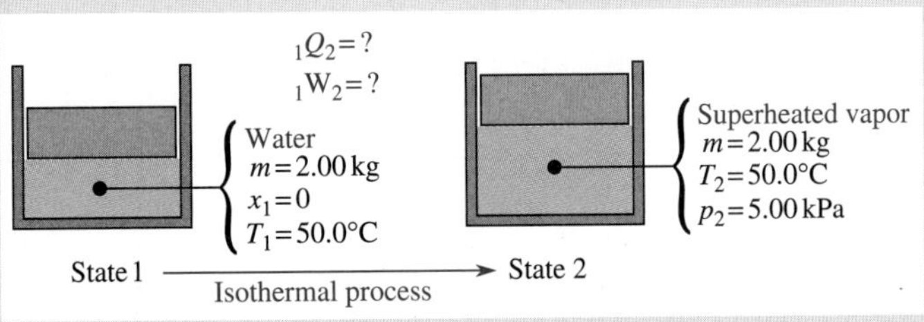

FIGURE 8.1
Example 8.1.

The unknowns here are ${}_1Q_2$ and ${}_1W_2$. The material is the water inside the cylinder (a closed system). The thermodynamic states are:

State 1	Reversible and Isothermal →	State 2
$T_1 = 50.0\,°\text{C}$		$T_2 = T_1 = 50.0\,°\text{C}$
$x_1 = 0$		$p_2 = 5.00\,\text{kPa}$

The basic equations here are the energy balance, EB,

$$ {}_1Q_2 - {}_1W_2 = m(u_2 - u_1) + \underbrace{\text{KE}_2 - \text{KE}_1 + \text{PE}_2 - \text{PE}_1}_{0 \text{ (assume the system is stationary)}} $$

and the entropy balance, SB,

$$m(s_2 - s_1) = \int_\tau \int_\Sigma \left(\frac{\dot{q}}{T_b}\right)_{\text{rev}} dA\,dt$$

Since this process is isothermal, the system boundary absolute temperature T_b is a constant and

$$\int_\tau \int_\Sigma \left(\frac{\dot{q}}{T_b}\right)_{\text{rev}} dA\,dt = \left(\frac{{}_1Q_2}{T_b}\right)_{\text{rev}}$$

(Continued)

EXAMPLE 8.1 (*Continued*)

and the SB reduces to

$$m(s_2 - s_1) = \left(\frac{{}_1Q_2}{T_b}\right)_{\text{rev}}$$

We know that the work mode involved in this system is of the moving boundary type $\left(\int p\,d\forall\right)$, but we do not know the $p-\forall$ relation for the process. Therefore, we cannot evaluate ${}_1W_2$ directly from its auxiliary equation. However, we can solve for ${}_1Q_2$ from the SB, then we can find ${}_1W_2$ from the EB:

$${}_1Q_2 = mT_b(s_2 - s_1)$$

and

$${}_1W_2 = m(u_1 - u_2) + {}_1Q_2$$

Here,

$$s_1 = s_f(50.0°\text{C}) = 0.7036\,\text{kJ/(kg·K)}$$
$$s_2 = s(50.0°\text{C and } 5.00\,\text{kPa}) = 8.4982\,\text{kJ/(kg·K)}$$

and

$$u_1 = u_f(50.0°\text{C}) = 209.3\,\text{kJ/kg}$$
$$u_2 = u(50.0°\text{C and } 5.00\,\text{kPA}) = 2444.7\,\text{kJ/kg}$$

so that

$${}_1Q_2 = (2.00\,\text{kg})(50 + 273.15\,\text{K})[8.4982 - 0.7036\,\text{kJ/(kg·K)}]$$
$$= 5040\,\text{kJ}$$

and

$${}_1W_2 = (2.00\,\text{kg})(209.3 - 2444.7\,\text{kJ/kg}) + 5040\,\text{kJ} = 569\,\text{kJ}$$

HOW DO YOU MAKE THE ASSUMPTIONS USED TO SOLVE EXAMPLE 8.1?

In Example 8.1, we had to deduce certain things based on information that was not explicitly given in the problem statement. For example, we knew that the mass was constant but the specific volume increased as the water went from a saturated liquid to a superheated vapor. Therefore, the total volume had to increase, and its expansion do moving boundary mechanical work. This is a typical engineering situation, which requires the use of commonsense assumptions that often come from practice and experience.

Exercises

1. Determine the heat transfer in Example 8.1 when the final state is a saturated (rather than a superheated) vapor at 50.0°C. **Answer**: ${}_1Q_2 = 4770$ kJ.
2. If the system in Example 8.1 is insulated so that no heat transfer occurs and the final expansion pressure is still 5.00 kPa, determine the final temperature and quality in the system. **Answer**: $T_2 = 32.6$°C and $x_2 = 2.92\%$.

EXAMPLE 8.2

The solar power plant shown in Figure 8.2 utilizes the thermal energy of the sun to drive a heat engine. Solar collectors with a constant surface temperature of 200.°F absorb 100. × 10^3 Btu/h of solar energy and deliver it to the heat engine. The heat engine rejects heat to a condenser in a river at 40.0°F. What is the maximum steady state electrical power (in kW) that can be produced by this power plant?

Solution

First, draw a sketch of the system.

Here, the unknown is $(\dot{W}_{electrical})_{max}$, and the system is the entire power plant as shown in Figure 8.2. You must now realize that a system like this produces maximum work output when the internal losses (friction etc.) are a minimum, and the absolute maximum occurs when the system is reversible. Therefore, what we wish to find here is $(\dot{W}_{electrical})_{rev}$.

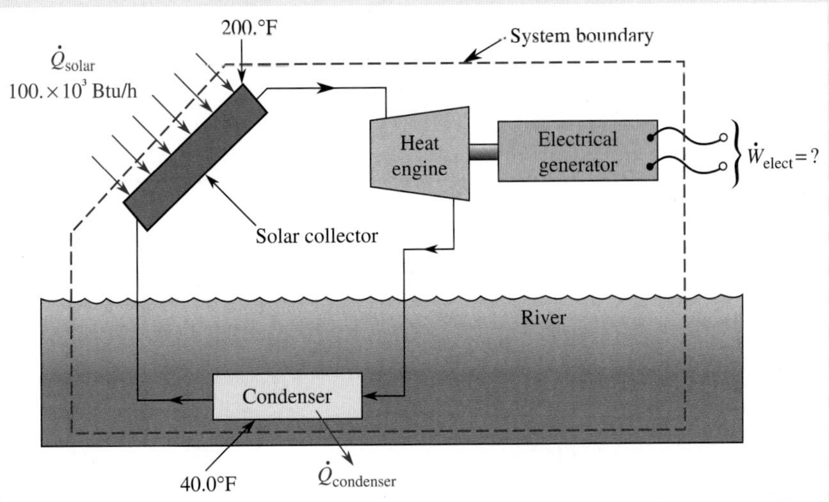

FIGURE 8.2
Example 8.2.

Since no clearly defined system states are given in the problem statement and all the given values are rate values, we recognize that this problem requires a *rate analysis*. The energy rate balance (ERB) equation for this system is

$$\dot{Q}_{net} - \dot{W}_{net} = \underbrace{\dot{U}}_{\substack{0 \\ \text{(steady state)}}} + \underbrace{\dot{KE} + \dot{PE}}_{\substack{0 \\ \text{(assume the system is stationary)}}}$$

or

$$(\dot{W}_{electrical})_{rev} = \dot{W}_{net} = \dot{Q}_{net} = (\dot{Q}_{solar} - |\dot{Q}|_{condenser})$$

Note that there are two heat transfer surfaces in this system (the solar collector and the condenser), each at a different isothermal temperature, and one work mode (electrical). The entropy rate balance (SRB) equation for this system is

$$\underbrace{\dot{S} = \dot{m}s}_{\substack{0 \\ \text{(steady state)}}} = \int_\Sigma \left(\frac{\dot{q}}{T_b}\right)_{rev} dA + \underbrace{\dot{S}_P}_{\substack{0 \\ \text{(reversible system)}}}$$

The system surface area Σ_{system} is the sum of the surface areas of the collector, condenser, and the remaining surfaces where no heat transfer occurs

$$\sum_{system} = \sum_{collector} + \sum_{condenser} + \sum_{\text{no heat transfer sufaces}}$$

Since the heat transfer surfaces are all isothermal, we can set

$$\int_\Sigma \left(\frac{\dot{q}}{T_b}\right) dA = \int_{\Sigma_{collector}} \left(\frac{\dot{q}}{T_b}\right) dA + \int_{\Sigma_{condenser}} \left(\frac{\dot{q}}{T_b}\right) dA = \frac{\dot{Q}_{solar}}{T_{collector}} - \left|\frac{\dot{Q}_{condenser}}{T_{river}}\right|$$

(*Continued*)

EXAMPLE 8.2 *(Continued)*

Therefore,

$$\left|\dot{Q}_{\text{condenser}}\right| = \dot{Q}_{\text{solar}}\left(\frac{T_{\text{river}}}{T_{\text{collector}}}\right)$$

and the ERB becomes

$$\left(\dot{W}_{\text{electrical}}\right)_{\text{rev}} = \dot{Q}_{\text{solar}}\left(1-\frac{T_{\text{river}}}{T_{\text{collector}}}\right) = (100.\times 10^3\ \text{Btu/h})\left(1-\frac{40.0+459.67}{200.+459.67}\right)$$

$$= (24{,}300\ \text{Btu/h})\left(\frac{1\ \text{kW}}{3412\ \text{Btu/h}}\right) = 7.11\ \text{kW}$$

The positive sign indicates that the electrical power is *out of* the system.

This example could also have been solved by using the Carnot heat engine efficiency (which is defined only for a reversible system) and the definition of the absolute temperature scale given in Eq. (7.15).

EXAMPLE 8.3

Determine an expression for the minimum isothermal system boundary temperature required by the second law of thermodynamics as an incompressible material is heated or cooled from a temperature T_1 to a temperature T_2 in a closed system with a constant heat flux.

Solution

First, draw a sketch of the system (Figure 8.3).

Here, we are asked to derive a formula (the unknown) for the minimum value of an isothermal boundary temperature, $(T_b)_{\text{min}}$ = ? The system is closed and made up exclusively of an incompressible material (solid or liquid). The energy balance equation for this system is

$${}_1Q_2 - {}_1W_2 = m(u_2-u_1) + \underbrace{\text{KE}_2-\text{KE}_1+\text{PE}_2-\text{PE}_1}_{0\ \text{(assume the system is stationary)}}$$

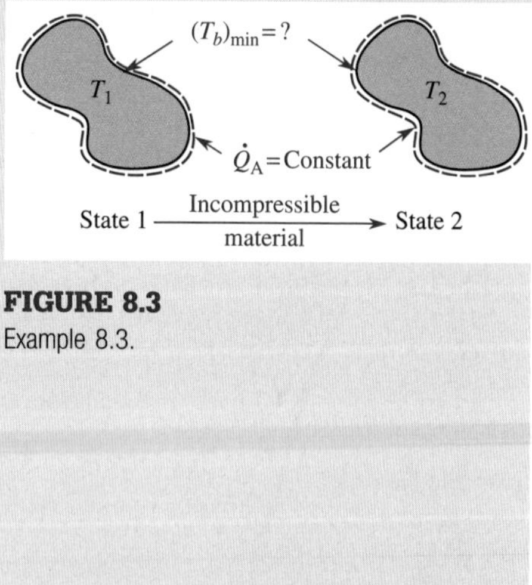

FIGURE 8.3
Example 8.3.

For an incompressible substance, $\forall$ = constant, so $d\forall = 0$, and ${}_1W_2 = -\int_1^2 p\,d\forall = 0$. Since no other work modes are mentioned in the problem statement, we assume that there are none. The resulting energy balance is then

$${}_1Q_2 = m(u_2-u_1)$$

The entropy balance equation for this system is

$$m(s_2-s_1) = \int_\tau\int_\Sigma\left(\frac{\dot{q}}{T_b}\right)dA\,dt + {}_1(S_P)_2$$

which, for a constant heat flux ($\dot{q}$) and isothermal boundaries, reduces to

$$m(s_2-s_1) = \frac{{}_1Q_2}{T_b} + {}_1(S_P)_2$$

Combining the energy balance and the entropy balance, we get

$$T_b = \frac{m(u_2-u_1)}{m(s_2-s_1) - {}_1(S_P)_2}$$

Now, since ${}_1(S_P)_2 \geq 0$, clearly T_b is at a *minimum* when ${}_1(S_P)_2 = 0$ (a reversible process). Therefore,

$$(T_b)_{\text{min}} = \frac{u_2-u_1}{s_2-s_1}$$

For an incompressible material, $u_2 - u_1 = c(T_2 - T_1)$ and $s_2 - s_1 = c\ \ln(T_2/T_1)$. Then,

$$(T_b)_{\min} = \frac{T_2 - T_1}{\ln(T_2/T_1)}$$

This is the required final formula.

If you look closely at the last equation in Example 8.3, you see that $(T_b)_{\min}$ always falls between T_1 and T_2. This means that, to change the temperature of an incompressible substance by a heat transfer process from an external source *without* producing entropy (i.e., reversibly), the system must be constructed with a well-defined boundary that is somehow maintained isothermal at the temperature given by this equation. This causes all the entropy generated by the heat transfer process to be produced outside the system.

EXAMPLE 8.4

Show that a reversible and adiabatic process carried out in a closed system results in the system having a constant entropy; that is, that a closed system reversible adiabatic process is also isentropic.

Solution

First, draw a sketch of the system (Figure 8.4).

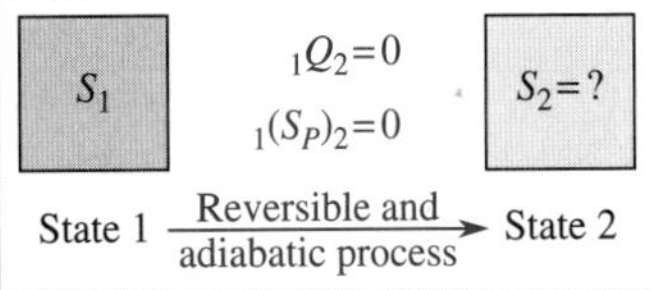

FIGURE 8.4
Example 8.4.

Here, we are to show that a reversible and adiabatic process carried out in a closed system results in the system having a constant entropy. We begin with just the entropy balance equation and note that

$$m(s_2 - s_1) = \underbrace{\int_\tau \int_\Sigma \left(\frac{\dot{q}}{T_b}\right) dA\, dt}_{0\ \text{(adiabatic)}} + \underbrace{{}_1(S_P)_2}_{0\ \text{(reversibile)}} = 0$$

and therefore $s_2 = s_1$ and we have a constant entropy or isentropic process. Similarly, the entropy rate balance equation gives

$$\dot{S} = \underbrace{\int_\Sigma \left(\frac{\dot{q}}{T_b}\right) dA}_{0\ \text{(adiabatic)}} + \underbrace{\dot{S}_P}_{0\ \text{(reversibile)}} = 0$$

and therefore S (and, of course, s) is again constant.

Exercises

4. If the size of the solar collectors in Example 8.2 is doubled and the collectors capture 200. × 10^3 Btu/h of solar energy, determine the new maximum steady state electrical power output of the power plant. **Answer**: $(\dot{W}_{\text{electrical}})_{\text{rev}} = 14.2$ kW.
5. Using the result of Example 8.3, determine the minimum isothermal boundary temperature as liquid water is heated from 50.0°F to 70.0°F at atmospheric pressure. **Answer**: $(T_b)_{\min} = 520$°F.
6. Example 8.4 shows that a reversible and adiabatic process carried out in a closed system is also an isentropic process. Use a similar argument to show that an isentropic process in a closed system is *not necessarily* reversible and adiabatic. **Answer**: $s_2 - s_1 = 0$ simply implies that $\dot{S}_P = -\int_\Sigma (\dot{q}/T_b)dA$, not that $\dot{S}_P = 0$ and $\dot{q} = 0$.

Isentropic processes are an important new category that we add to our list of *constant* property processes (isothermal, isobaric, isochoric, and isenthalpic).

ARE ALL ISENTROPIC PROCESSES REVERSIBLE AND ADIABATIC?

Example 8.4 shows that all reversible and adiabatic processes are isentropic but not all isentropic processes are necessarily reversible and adiabatic. It does not automatically go both ways. For example, heat loss from a system results in an entropy loss for that system; and if this entropy loss exactly balances the entropy production for the process the system is undergoing, then the process also is isentropic without being either reversible or adiabatic.

8.3 SYSTEMS UNDERGOING IRREVERSIBLE PROCESSES

We begin our treatment of irreversible processes by extending Examples 5.1 through 5.7 of Chapter 5 using the entropy balance or entropy rate balance equations in their analysis to obtain additional results from the problem. To do this effectively we have to add some information to these problem statements, in the way of additional unknowns or additional data. When additional wording has been added to these problem statements it appears in italic type so that you can clearly see what changes have been made. Also, the analysis in these examples contains less commentary, as they are designed to be straightforward applications.

EXAMPLE 8.5 A CONTINUATION OF EXAMPLE 5.1, WITH THE ADDED MATERIAL SHOWN IN *ITALIC TYPE*

A sealed, rigid container whose volume is 1.00 m^3 contains 2.00 kg of liquid water plus water vapor at 20.0°C. The container is heated until the temperature inside is 95.0°C. Determine

a. The quality in the container when the water is 20.0°C.
b. The quality in the container when the water is at 95.0°C.
c. The heat transport of energy required to raise the temperature of the contents from 20.0 to 95.0°C.
d. *The entropy production that occurs if the boundary of the container is maintained isothermal at 100.°C during the heat transfer process by condensing steam at atmospheric pressure on the outside of the tank.*

Solution

First, draw a sketch of the system (Figure 8.5).

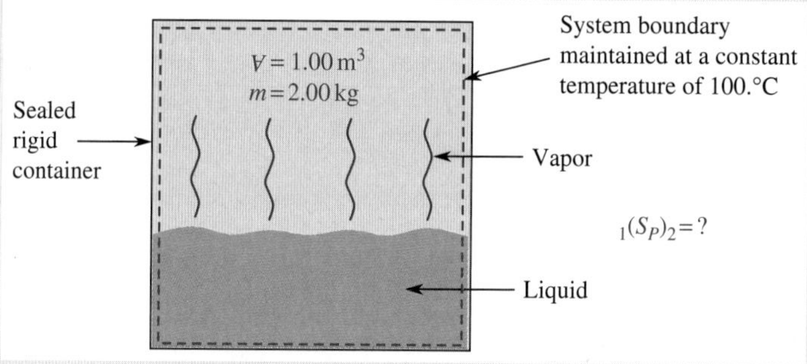

FIGURE 8.5
Example 8.5.

The unknowns are (a) x_1, (b) x_2, (c) ${}_1Q_2$ and (d) ${}_1(S_p)_2$. The system is the water and is closed. The thermodynamic states are

State 1	Isochoric $\xrightarrow{v_2 = v_1}$	State 2
$T_1 = 20.0\,°C$		$T_2 = 95.0\,°C$
$v_1 = 0.500\,m^3/kg$		$v_2 = v_1 = 0.500\,m^3/kg$

The answers to a, b, and c can be found in Example 5.1 as

a). $x_1 = 0.863\%$.
b). $x_2 = 25.2\%$.
c). ${}_1Q_2 = 1650\,kJ$.

The answer to part d is obtained from the entropy rate balance or the *indirect method* since we lack the detailed information about the interior of the system required to use the *direct method*. Equation (8.1) for an isothermal boundary becomes

$${}_1(S_P)_2 = m(s_2 - s_1) - \frac{{}_1Q_2}{T_b}$$

where

$$\begin{aligned} s_1 &= s_{f1} + x_1 s_{fg1} \\ &= 0.2965 + (0.00863)(8.3715) = 0.3687\,kJ/(kg \cdot K) \end{aligned}$$

and

$$\begin{aligned} s_2 &= s_{f2} + x_2 s_{fg2} \\ &= 1.2503 + (0.252)(6.1664) = 2.8042\,kJ/(kg \cdot K) \end{aligned}$$

Then,

$$ {}_1(S_P)_2 = (2.00\,\text{kg})[2.8042 - 0.3687\,\text{kJ/(kg}\cdot\text{K)}] - \frac{1650\,\text{kJ}}{100. + 273.15\,\text{K}} $$
$$ = 0.449\,\text{kJ/K} = 449\,\text{J/K} $$

Notice that the process described in this example also has a minimum isothermal system boundary temperature, like that described in Example 8.3. In this case, ${}_1(S_P)_2 = 0$, and the minimum boundary temperature is

$$ (T_b)_{\text{minimum}} = \frac{{}_1Q_2}{m(s_2 - s_1)} = 339\,\text{K} = 65.6°\text{C} $$

WHAT HAPPENS IF YOU TRY TO HEAT THE WATER IN EXAMPLE 8.5 WITH AN ISOTHERMAL BOUNDARY TEMPERATURE LESS THAN 65.2°C?

Any attempt to carry out the *heating* process in Example 8.5 with an isothermal boundary temperature lower than the minimum of 65.6°C fails because it violates the second law by requiring a negative entropy production. But, what would happen if you tried? You could add heat to the system with an isothermal boundary temperature of less than 65.6°C, but it would not reach 95.0°C.

However, *cooling* the tank from 95.0°C back to 20.0°C with an isothermal boundary reverses all the signs in the entropy production calculation so that ${}_2(S_P)_1 = {}_1(S_P)_2$. This cooling process requires an isothermal boundary temperature *less* than 65.6°C to satisfy the second law.

Exercises

7. Determine the entropy production that occurs in Example 8.5 if the surface temperature of the container is maintained at 80.0°C rather than 100.°C. **Answer:** ${}_1(S_P)_2 = 199$ J/K.
8. Determine the entropy production that occurs in Example 8.5 if the two states are fixed as $T_1 = 20.0°\text{C}$ and $x_1 = 0.00\%$, $T_2 = 95.0°\text{C}$ and $x_2 = 100.\%$. **Answer:** ${}_1(S_P)_2 = 9.83$ kJ/K.

EXAMPLE 8.6 A CONTINUATION OF EXAMPLE 5.2, WITH THE ADDED MATERIAL SHOW IN *ITALIC TYPE*

An incandescent lightbulb is a simple electrical device. Using the energy rate balance *and the entropy rate balance* on a lightbulb, determine

a. The heat transfer rate of an illuminated 100. W incandescent lightbulb in a room.
b. The rate of change of its internal energy if this bulb were put into a small, sealed, insulated box.
c. *The value of the entropy production rate for part a if the bulb has an isothermal surface temperature of 110.°C.*
d. *An expression for the entropy production rate as a function of time for part b.*

Solution

First, draw a sketch of the system (Figure 8.6).

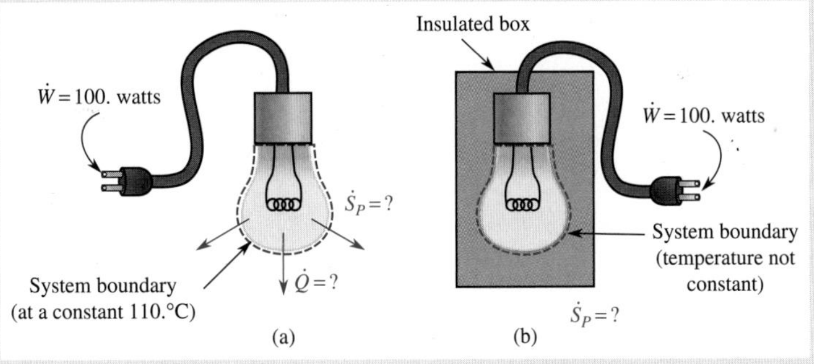

FIGURE 8.6
Example 8.6.

(*Continued*)

EXAMPLE 8.6 *(Continued)*

Here, the unknowns are (a) $\dot{Q}$, (b) $\dot{U}$, (c) $\dot{S}_P$, and (d) an expression for $\dot{S}_P$ as a function of time. The system is the lightbulb itself, and apparently, we do not need to know any specific system properties to solve this problem.

The answers for parts a and b can be found in Example 5.2 as

a). $\dot{Q} = -100.\text{ W}$.
b). $\dot{U} = 100.\text{ W}$.

The answer to part c can be obtained by the *indirect method* of the SRB. From Eq. (8.2), we have

$$\dot{S}_P = \dot{S} - \frac{d}{dt}\int_\Sigma \left(\frac{\bar{d}Q}{T_b}\right)_{\text{act}}$$

Since we are given that the surface temperature of the bulb is isothermal,

$$\frac{d}{dt}\int_\Sigma \left(\frac{\bar{d}Q}{T_b}\right) = \frac{1}{T_b}\frac{d}{dt}\int_\Sigma \bar{d}Q = \frac{\dot{Q}}{T_b}$$

and

$$\dot{S}_P = \dot{S} - \frac{\dot{Q}}{T_b}$$

In part a, we have steady state operation so $\dot{S} = 0$, then the answer to part c is

$$\dot{S}_P = -\frac{\dot{Q}}{T_b} = \frac{100.\text{ W}}{(110. + 273.15\text{ K})} = 0.261\text{ W/K}$$

The solution to part d is obtained by recognizing that. in part b. The bulb is insulated, so $\dot{Q} = 0$, then

$$\dot{S}_P = \dot{S}$$

The surface and internal temperatures of the bulb are not constant here. If we recognize that most of the mass of the bulb is made up of incompressible material (glass and tungsten wire), then we can write

$$s - s_{\text{ref}} = c\ln(T/T_{\text{ref}}) = c\ln(T) - c\ln(T_{\text{ref}})$$

where s_{ref} and T_{ref} are values chosen at some arbitrary reference state. Then,

$$\dot{S} = m\dot{s} = \frac{mc}{T}\left(\frac{dT}{dt}\right) = \frac{mc}{T}(\dot{T})$$

Similarly, we can write

$$m(u - u_{\text{ref}}) = c(T - T_{\text{ref}})$$

so that

$$\dot{U} = m\dot{u} = mc\left(\frac{dT}{dt}\right) = mc(\dot{T})$$

Therefore,

$$\dot{S} = \frac{\dot{U}}{T} = \dot{S}_P$$

The temperature in this equation is the mean temperature of the bulb and can be evaluated from the answer to part b, where we find that $\dot{U} = 100.\text{ W}$.

Therefore,

$$\dot{T} = \frac{dT}{dt} = \frac{\dot{U}}{mc} = \text{constant}$$

and integration of this equation gives

$$T = \frac{\dot{U}}{mc}t + T_0$$

where T_0 is the bulb temperature immediately before the insulation is applied. Therefore, the answer to part d is

$$\dot{S}_P = \frac{\dot{U}}{\dfrac{\dot{U}t}{mc} + T_0} = \frac{mc}{t + \dfrac{mcT_0}{\dot{U}}}$$

Since $\dot{U} = 100.\text{W}.$ and m, c, and T_0 are all constant measurable quantities, it is clear from this result that $\dot{S}_p$ slowly decays to zero as time t goes to infinity. However, the bulb temperature *increases* linearly with time, so the bulb overheats and burns out quickly.

Exercises

9. If the surface temperature of the bulb in Example 8.6 were 40.0°C (which is typical of a fluorescent light), determine the entropy production rate of the bulb. **Answer:** $\dot{S}_P = 0.319\,\text{W/K}$.

10. Take $\dot{U} = 100.$ watts, $T_0 = T_b = 110.$°C, and for a glass bulb take $m = 0.050$ kg and $c = 0.80$ kJ/kg·K. Then, using the last expression in Example 8.6 for the decay in entropy production rate for an insulated lightbulb, determine the entropy production rate at $t = 0$, 300., 600., and 1000. s. **Answers:** $\dot{S}_P = 0.26, 0.088, 0.053, \text{and } 0.035\,\text{W/K}$.

EXAMPLE 8.7 A CONTINUATION OF EXAMPLE 5.3, WITH THE ADDITION SHOWN IN *ITALIC TYPE*

A basic vapor cycle power plant consists of the following four parts:

a. The boiler, where high-pressure vapor is produced.
b. The turbine, where energy is removed from the high-pressure vapor as shaft work.
c. The condenser, where the low-pressure vapor leaving the turbine is condensed into a liquid.
d. The boiler feed pump, which pumps the condensed liquid back into the high-pressure boiler for reheating.

In such a power plant, the boiler receives 950. × 10^5 kJ/h from the burning fuel, and the condenser rejects 600. × 10^5 kJ/h to the environment. The boiler feed pump requires 23.0 kW input, which it receives directly from the turbine. Assuming that the turbine, pump, and connecting pipes are all insulated, determine the net power of the turbine *and the rate of entropy production of the plant if the boiler temperature is 500.°C and the condenser temperature is 10.0°C.*

Solution

First, draw a sketch of the system (Figure 8.7).

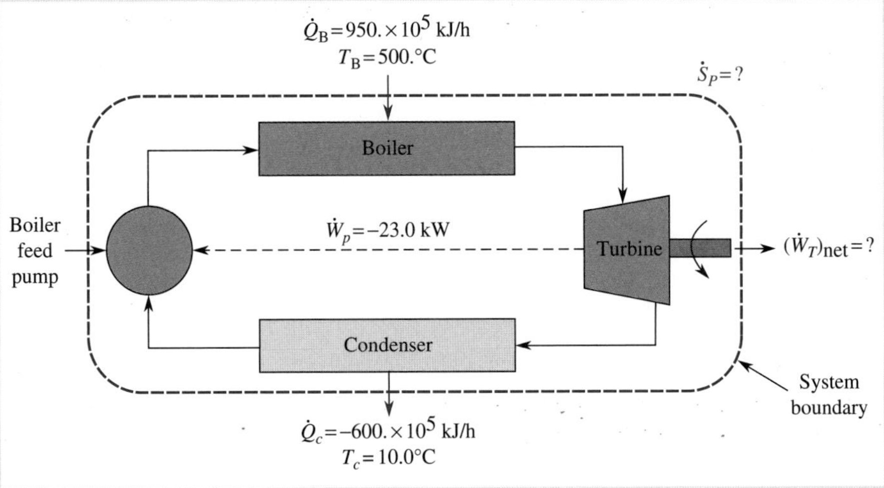

FIGURE 8.7
Example 8.7.

The unknowns here are $(\dot{W}_T)_{net}$ and $\dot{S}_P$ for the entire power plant. Therefore, the system is the entire power plant. Since we have few specific details on the internal operation of the plant, the thermodynamic properties within the power plant are apparently not needed in the solution.

The net turbine work output is determined in Example 5.3 to be

$$(\dot{W}_T)_{net} = 9720\,\text{kW}$$

The answer to the second part of this problem can be obtained by the *indirect method* from Eq. (8.2) as

$$\dot{S}_P = \dot{S} - \frac{d}{dt}\int_\Sigma \left(\frac{đQ}{T_b}\right)_{act}$$

(*Continued*)

EXAMPLE 8.7 *(Continued)*

The surface area of our system can be divided into three major parts: the boiler's surface area, the condenser's surface area, and all the remaining surface areas. Therefore, the surface area Σ_{system} of the system is composed of the boiler, the condenser, and everything else, or

$$\sum_{\text{system}} = \sum_{\text{boiler}} + \sum_{\text{condenser}} + \sum_{\text{everything else}}$$

Now, both the boiling and the condensing processes are isothermal phase changes; and since no heat transfer occurs at any other point in the system, we can write

$$\frac{d}{dt}\int_{\Sigma}\frac{\bar{d}Q}{T_b} = \frac{d}{dt}\left(\int_{\Sigma_{\text{boiler}}}\frac{\bar{d}Q}{T_b} + \int_{\Sigma_{\text{condenser}}}\frac{\bar{d}Q}{T_b} + \int_{\Sigma_{\text{remainder}}}\frac{\bar{d}Q}{T_b}\right)$$

where

$$\frac{d}{dt}\int_{\Sigma_{\text{boiler}}}\frac{\bar{d}Q}{T_b} = \frac{\dot{Q}_{\text{boiler}}}{T_{\text{boiler}}}$$

$$\frac{d}{dt}\int_{\Sigma_{\text{condenser}}}\frac{\bar{d}Q}{T_b} = \frac{\dot{Q}_{\text{condenser}}}{T_{\text{condenser}}}$$

and

$$\frac{d}{dt}\int_{\Sigma_{\text{remainder}}}\frac{\bar{d}Q}{T_b} = 0 \text{ (no heat transfer across the remaining surface area)}$$

Then, we have

$$\dot{S}_P = \dot{S} - \frac{\dot{Q}_{\text{boiler}}}{T_{\text{boiler}}} - \frac{\dot{Q}_{\text{condenser}}}{T_{\text{condenser}}}$$

and, for steady state operation ($\dot{S} = 0$), this reduces to

$$\dot{S}_p = -\left(\frac{\dot{Q}_{\text{boiler}}}{T_{\text{boiler}}} + \frac{\dot{Q}_{\text{condenser}}}{T_{\text{condenser}}}\right)$$

$$= -\left(\frac{950.\times 10^5}{500.+273.15} + \frac{-600.\times 10^5}{10.0+273.15}\right)\left(\frac{\text{kJ/h}}{\text{K}}\right) = 89.0\times 10^3 \text{ kJ/(h}\cdot\text{K)}$$

Note that the actual thermal efficiency of this power plant is given by Eq. (7.9) as

$$(\eta_T)_{\text{act}} = 1 - \frac{|\dot{Q}_{\text{out}}|}{\dot{Q}_{\text{in}}} = 1 - \frac{600.\times 10^5}{950.\times 10^5} = 0.368 = 36.8\%$$

whereas its theoretical reversible (Carnot) efficiency is given by Eq. (7.16) as

$$(\eta_T)_{\text{rev}} = 1 - \frac{T_{\text{condenser}}}{T_{\text{boiler}}} = 1 - \frac{10.0+273.15}{500.+273.15} = 0.634 = 63.4\%$$

Therefore, the actual efficiency is less than the theoretical maximum (reversible) efficiency, as it should be.

Exercises

11. Determine the condenser temperature that would cause the entropy production rate of the power plant in Example 8.7 to be 500. kJ/h·K. **Answer**: $T_{\text{condenser}}$ = 486 K = 213°C.
12. Determine the condenser temperature that would cause the entropy production rate of the power plant in Example 8.7 to equal zero. Why can't this occur? **Answer**: $T_{\text{condenser}}$ = 488 K = 215°C. This cannot occur because it would require the plant to be reversible throughout (i.e., it could have no internal friction, heat transfer, chemical reactions, or anything else that would naturally be irreversible).
13. If the heat transfer rate into the boiler is doubled in Example 8.7 and all the other parameters remain constant, determine the new entropy production rate of the power plant. Is this possible? **Answer**: $\dot{S}_P = -33{,}900$ kJ/h·K. This is not possible because it violates the second law of thermodynamics ($\dot{S}_P$ must be > 0).

EXAMPLE 8.8 A CONTINUATION OF EXAMPLE 5.4, WITH THE ADDITION SHOWN IN *ITALIC TYPE*

A food blender has a cutting-mixing blade driven by a 0.250 hp electric motor. The machine is initially filled with 1.00 qt of water at 60.0°F, 14.7 psia. It is turned on at full speed for 10.0 min. Assuming the entire machine is insulated and the mixing takes place at constant pressure, determine the temperature of the water *and the amount of entropy produced* when the machine is turned off.

Solution

First, draw a sketch of the system (Figure 8.8).

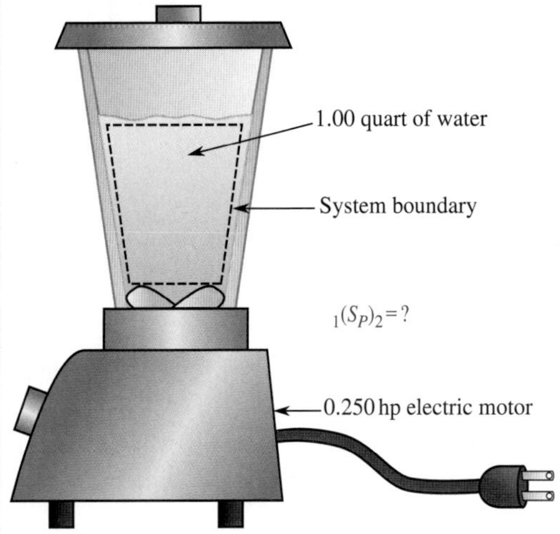

FIGURE 8.8
Example 8.8.

The unknowns here are T_2 and $_1(S_P)_2$. The system is the water in the blender, which we assume to be an incompressible material. The data for the water are as follows:

State 1	Isobaric mixing →	State 2
$p_1 = 14.7$ psia		$p_2 = p_1 = 14.7$ psia
$T_1 = 60.0°F$		

The solution to the first part of this problem is given in Example 5.4 as $m = 2.08$ lbm and $T_2 = 111°F$.

The solution to the second part is determined by the *indirect method.* Equation (8.1) gives

$$_1(S_P)_2 = m(s_2 - s_1) - \underbrace{\int_\Sigma \left(\frac{dQ}{T_b}\right)}_{0 \text{ (insulated system)}}$$

and, for an incompressible substance,

$$s_2 - s_1 = c \ln \frac{T_2}{T_1}$$

so that

$$_1(S_P)_2 = mc \ln \frac{T_2}{T_1}$$

or

$$\begin{aligned} _1(S_P)_2 &= (2.08\,\text{lbm})[1.0\,\text{Btu/(lbm}\cdot\text{R)}]\ln \frac{111 + 459.67}{60.0 + 459.67} \\ &= 0.195\,\text{Btu/R} \end{aligned}$$

If we wished to know the entropy production *rate* for this example, our analysis and results would be the same as that for part b of Example 8.6. The entropy production rate would not be constant in time, but would decrease as the water became hotter. Since it is stated that the blender is insulated, eventually enough mixing energy would be converted into internal (thermal) energy to cause the water to completely vaporize, whereupon the assumption of an incompressible fluid no longer applies.

EXAMPLE 8.9 A CONTINUATION OF EXAMPLE 5.5, WITH THE ADDITION SHOWN IN *ITALIC TYPE*

A new radiation heat transfer sensor consists of a small, closed, rigid, insulated 0.0400 m^3 box containing a 0.0100 m^3 rubber balloon. Initially, the box is evacuated but the balloon contains argon (an ideal gas) at 20.0°C and 0.0100 MPa. When the balloon receives 0.100 kJ of radiation energy through an uninsulated window in the box, it bursts. The resulting pressure change is sensed by a pressure transducer and an alarm is sounded. Determine the pressure and temperature inside the box after the balloon bursts *and the entropy produced during this process if the average surface temperature of the heat transfer window is 400.K.*

Solution

First, draw a sketch of the system (Figure 8.9).

(*Continued*)

EXAMPLE 8.9 (*Continued*)

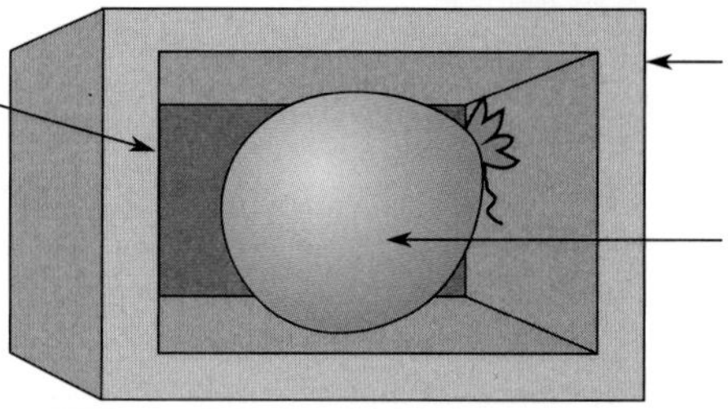

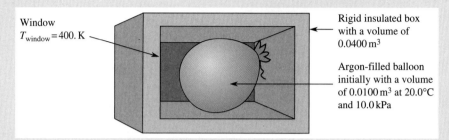

FIGURE 8.9
Example 8.9.

The unknown is the entropy produced during this process if the average surface temperature of the heat transfer window is 400. K. The material is argon gas.

In Example 5.5, we use an energy balance to find the final pressure and temperature as p_2 = 4.15 kPa and T_2 = 214°C = 487 K. Now, an entropy balance gives

$$\frac{{}_1Q_2}{T_b} + {}_1(S_P)_2 = m(s_2 - s_1)_{\text{argon}}$$

which can be solved for the entropy production as

$$_1(S_P)_2 = m(s_s - s_1) - \frac{{}_1Q_2}{T_b}$$

and, since argon is an ideal gas, we can write

$$s_2 - s_1 = c_p \ln(T_2/T_1) - R\ln(p_2/p_1)$$

From Table C.13b of *Thermodynamic Tables to accompany Modern Engineering Thermodynamics*, we find that for argon, c_p = 0.523 kJ/kg·K and R = 0.208 kJ/kg·K; and from Example 5.5, the mass of the argon is m = 0.00164 kg. Then, with $_1Q_2$ = 0.100 kJ and T_b = 400. K, we get

$$_1(S_P)_2 = (0.00164\text{ kg})\left[(0.523\text{ kJ/kg}\cdot\text{K})\ln\left(\frac{214+273.15\text{ K}}{20.0+273.15\text{ K}}\right) - (0.208\text{ kJ/kg}\cdot\text{K})\ln\left(\frac{4.15\text{ kPa}}{10.0\text{ kPa}}\right)\right]$$
$$-\frac{0.100\text{ kJ}}{400.\text{ K}} = 0.486\times10^{-3}\text{ kJ/K} = 0.486\text{ J/K}$$

In this example, the heat transfer occurs over only a portion of the system's surface (the uninsulated window). The rest of the enclosure is insulated to prevent the sensor from being influenced by anything except the heat source in front of the window. Though the surface temperature of the window probably changes during the heat transfer process, an adequate solution is obtained simply by using an *average* surface temperature during the process of T_b = 400. K.

Exercises

14. Suppose the balloon in Example 8.9 is designed to burst after absorbing 0.200 kJ of radiation heat transfer. Determine the entropy produced during this process if the average surface temperature of the heat transfer window is 500. K. **Answer:** $_1(S_P)_2$ = 0.336 J/K.

15. If air were substituted for the argon in Example 8.9 with no changes in the remaining parameters, what would be the entropy produced inside the box during this process? **Answer:** $_1(S_P)_2$ = 0.510 J/K.

16. Determine the entropy produced when the radiation heat transfer is increased to produce a final pressure of 20 kPa in the box after the balloon bursts. Assume that the average temperature of the heat transfer window is 1200 K for this process. **Answers:** $_1(S_P)_2$ = 0.664 J/K.

EXAMPLE 8.10 A CONTINUATION OF EXAMPLE 5.6, WITH THE ADDITION SHOWN IN *ITALIC TYPE*

Suppose 0.100 lbm of Refrigerant-134a initially at 180.°F and 100. psia in a cylinder with a movable piston undergoes the following two-part process. First, the refrigerant is expanded adiabatically to 30.0 psia and 120.°F, then it is isobarically compressed to half its initial volume. Determine

a. The work transport of energy during the adiabatic expansion.
b. The heat transport of energy during the isobaric compression.

c. The final temperature at the end of the isobaric compression.

d. *The total entropy production for both processes if the heat transport of energy and the boundary temperature are related by the formula* $dQ = KdT_b$, *where* $K = 5.00$ *Btu/R and* T_b *is in R.*

Solution

First, draw a sketch of the system (Figure 8.10).

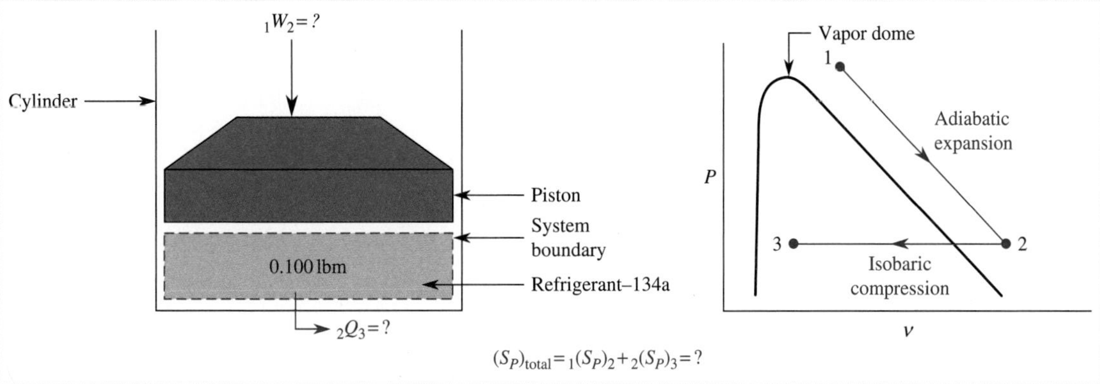

FIGURE 8.10
Example 8.10.

The unknowns here are (a) $_1W_2$, (b) $_2Q_3$, (c) T_3, and (d) $_1(S_P)_3$. The system is just the R-134a, and the state variables are as follows:

State 1	Adiabatic expansion →	State 2	Isobaric compression →	State 3
$p_1 = 100.$ psia		$p_2 = 30.0$ psia		$p_3 = p_2 = 30.0$ psia
$T_1 = 180.°F$		$T_2 = 120.°F$		$v_3 = v_1/2$
$v_1 = 0.6210\,ft^3/lbm$		$v_2 = 1.9662\,ft^3/lbm$		$v_3 = 0.3105\,ft^3/lbm$
$h_1 = 137.49$ Btu/lbm		$h_2 = 126.39$ Btu/lbm		$x_3 = 0.1952$
$s_1 = 0.2595$ Btu/(lbm·R)		$s_2 = 0.2635$ Btu/(lbm·R)		$s_3 = 0.07241$ Btu/(lbm·R)

The solutions to the first three parts of this problem are given in Example 5.6 as

a. $_1W_2 = 1.05$ Btu.

b. $_2Q_3 = -9.31$ Btu.

c. $T_3 = 15.4°F$.

The solution to part d can be determined by the *indirect method* (entropy balance) as follows. From Eq. (8.1), we have

$$_1(S_P)_2 = m(s_2 - s_1) - \underbrace{\int_\Sigma \left(\frac{\bar{d}Q}{T_b}\right)_{act}}_{0 \text{ (adiabatic process)}}$$

From Table C.7e, we find that

$$s_1 = 0.2595 \text{ Btu/(lbm·R)}$$

and

$$s_2 = 0.2635 \text{ Btu/(lbm·R)}$$

Since the process from 1 to 2 is adiabatic, $\bar{d}Q = 0.$ and

$$_1(S_P)_2 = (0.100\,\text{lbm})[0.2635 - 0.2595\,\text{Btu/(lbm·R)}] - 0 = 4.00 \times 10^{-4}\,\text{Btu/R}$$

(*Continued*)

EXAMPLE 8.10 (Continued)

Similarly,

$$_2(S_P)_3 = m(s_3 - s_2) - \int_\Sigma \frac{\bar{d}Q}{T_b}$$

In this process, we are given that $\bar{d}Q = KdT_b$, where $K = 5.00$ Btu/R. Therefore, $_2Q_3 = K(T_{b3} - T_{b2}) + C = -9.31$ Btu, where C is an integration constant. Then,

$$\int_\Sigma \frac{\bar{d}Q}{T_b} = \int_{T_{b2}}^{T_{b3}} K\left(\frac{dT_b}{T_b}\right) = K \ln \frac{T_{b3}}{T_{b2}}$$

Therefore,

$$_2(S_P)_3 = m(s_3 - s_2) - K \ln \frac{T_{b3}}{T_{b2}}$$

where $T_{b2} = T_2$ and $T_{b3} = T_3$.

Now, $s_3 = s_{f3} + x_3 s_{fg3}$, and from Table C.7b in *Thermodynamic Tables to accompany Modern Engineering Thermodynamics*, at 30.0 psia, we get

$$s_{f3} = 0.0364 \text{ Btu/(lbm}\cdot\text{R)}$$
$$s_{fg3} = 0.2209 \text{ Btu/(lbm}\cdot\text{R)}$$

Then,

$$s_3 = 0.0364 + (0.1952)(0.2209 - 0.0364) = 0.07241 \text{ Btu/(lbm}\cdot\text{R)}$$

and

$$_2(S_P)_3 = (0.100 \text{ lbm})[0.07241 - 0.2635 \text{ Btu/(lbm}\cdot\text{R)}] - (5.00 \text{ Btu/R})\ln \frac{15.38 + 459.67}{120. + 459.67}$$
$$= 0.976 \text{ Btu/R}$$

Finally, the entropy production for the entire process is given by

$$_1(S_P)_3 = {_1(S_P)_2} + {_1(S_P)_2} = 4.00 \times 10^{-4} + 0.976 = 0.976 \text{ Btu/R (to 3 significant figures)}$$

In this example, a special $Q = Q(T_b)$ relation is introduced, which is similar to that used to describe convection heat transfer processes. These relations, often called *thermal constitutive equations*, are mathematical models developed to describe specific heat transfer mechanisms. The following exercises illustrate this concept.

Exercises

17. Find $_2(S_P)_3$ in Example 8.10 when $T_{b2} = T_{b3} = T_b = 60.0°$C, and $_2Q_3 = -9.31$ Btu = constant. **Answer**: This is not possible since $_2(S_P)_3$ is negative for this process and violates the second law of thermodynamics.

18. Rework part d in Example 8.10 using the relation $dQ = K_2 T_b dT_b$, where $K_2 = 0.001$ Btu/R^2. Note K_2 is not the same as K in Example 8.10. Keep all other variables the same as in Example 8.10. You have to reevaluate the integral $\int dQ/T_b$ for this exercise. **Answer**: $_1(S_P)_3 = 0.0855$ Btu/R.

19. Resolve part d in Example 8.10 for radiation heat transfer where $dQ = K_3 T_b^3 dT_b$, where $K_3 = 6.30 \times 10^{-6}$. Note that K_4 is not the same as K in Example 8.10. Keep all other variables the same as in Example 8.10. You have to re-evaluate the integral $\int dQ/T_b$ for this exercise. **Answer**: $_1(S_P)_3 = 184$ Btu/R.

EXAMPLE 8.11 A CONTINUATION OF EXAMPLE 5.7, WITH THE ADDITION SHOWN IN *ITALIC TYPE*

A microwave antenna for a space station consists of a 0.100 m diameter rigid, hollow, steel sphere of negligible wall thickness. During its fabrication, the sphere undergoes a heat treating operation in which it is initially filled with helium at 0.140 MPa and 200.°C, then it is plunged into cold water at 15.0°C for exactly 5.00 s. The convective heat transfer coefficient of the sphere in the water is 3.50 W/(m$^2\cdot$K). Neglecting any changes in kinetic or potential energy and assuming the helium behaves as an ideal gas, determine

a. The final temperature of the helium.
b. The change in total internal energy of the helium.
c. *The total entropy production in the helium.*

Solution

First, draw a sketch of the system (Figure 8.11).

The unknowns here are, after 5 s have passed, (a) T_2, (b) $U_2 - U_1$, and (c) ${}_1(S_P)_2$; the material is helium gas.

The solutions to the first two parts of this problem can be found in Example 5.7 as

a. $T_2 = 32.5°C$.
b. $U_2 - U_1 = -0.039$ kJ.
The solution to part c can be found again by the *indirect method* (entropy balance) as follows. From Eq. (8.1), we have

$$ {}_1(S_P)_2 = m(s_2 - s_1) - \int_\Sigma \left(\frac{\bar{d}Q}{T}\right)_{act} $$

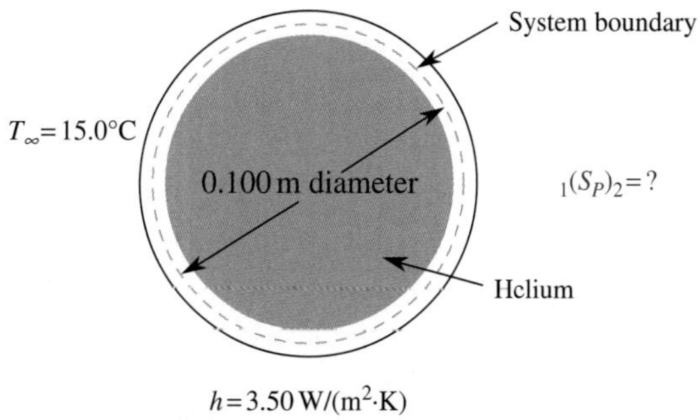

FIGURE 8.11
Example 8.11.

We can assume that helium behaves as an ideal gas, then since $v_2 = v_1$ = constant, we can use Eq. (7.36) to produce

$$ \begin{aligned} s_2 - s_1 &= c_v \ln\frac{T_2}{T_1} + R\ln\frac{v_2}{v_1} \\ &= [3.123\ \text{kJ/(kg·K)}]\ln\frac{32.5+273.15}{200.+273.15} + 0 \\ &= -1.37\ \text{kJ/(kg·K)} \end{aligned} $$

and, from Example 5.7, we have

$$ \dot{Q} = -hA(T_s - T_\infty) = \frac{\bar{d}Q}{dt} $$

so

$$ \bar{d}Q = -hA(T_s - T_\infty)\,dt $$

and

$$ \frac{\bar{d}Q}{T_b} = \frac{\bar{d}Q}{T_\infty} = -hA[(T_s - T_\infty)/T_\infty]\,dt $$

where we have set $T_b = T_\infty$ (i.e., we put the system boundary slightly outside the sphere itself). Also, in Example 5.7, we discover that

$$ T_s = T_\infty + (T_1 - T_\infty)\exp\left(-\frac{hAt}{mc_v}\right) $$

where $T_1 = T_s$ evaluated at $t = 0$. Then,

$$ \begin{aligned} \int_\Sigma \left(\frac{\bar{d}Q}{T_b}\right)_{act} &= -hA\int_0^{5s}\left(\frac{T_s - T_\infty}{T_\infty}\right)dt \\ &= mc_v\left(\frac{T_1 - T_\infty}{T_\infty}\right)\left[\exp\left(-\frac{hAt}{mc_v}\right)\Big|_0^{5s}\right] \end{aligned} $$

Now, from Example 5.7, we have the following numerical values:

$$ \begin{aligned} m &= 7.46\times10^{-5}\ \text{kg} \\ c_v &= 3.123\ \text{kJ/kg·RK} \\ T_1 &= 200.°C = 473.15\ K \\ T_\infty &= 15.0°C = 288.15\ K \end{aligned} $$

and $hA/(mc_v) = 0.472 s^{-1}$. Substituting these values into the preceding integration result gives

$$ \int_\Sigma \left(\frac{\bar{d}Q}{T_b}\right)_{act} = -1.35\times10^{-4}\ \text{kJ/K} $$

then

$$ \begin{aligned} {}_1(S_P)_2 &= (7.46\times10^{-5}\ \text{kg})[-1.365\ \text{kJ/(kg·K)}] - (-1.35\times10^{-4}\ \text{kJ/K}) \\ &= 3.32\times10^{-5}\ \text{kJ/K} = 0.0332\ \text{J/K} \end{aligned} $$

(Continued)

EXAMPLE 8.11 (*Continued*)

Exercises

20. Rework part c in Example 8.11 with an immersion time of 1.00 s rather than 5.00 s. Note that T_2 now becomes 130.5°C rather than 32.5°C. Keep the values of all the other variables the same as in the example. **Answer**: ${}_1(S_P)_2 = 0.0192$ J/K.
21. Determine ${}_1(S_P)_2$ in Example 8.11 when $T_4 = 30.0$°C rather than 15.0°C. With this value of T_4, the value of T_2 changes from 32.5°C to 136°C. Use this new value of T_2 but keep the values of all of the other variables the same as in the example. **Answer**: ${}_1(S_P)_2 = 0.0845$ J/K.
22. Plot ${}_1(S_P)_2$ vs. immersion time t over the range $0 \le t \le 30.0$ s for the process described in Example 8.11. Note that T_2 varies between $15.0°C \le T_2 \le 200.°C$ over this range of t. Do this by using commercial equation solver software or by choosing several values of t in the range $0 \le t \le 30.0$ s and carry out the calculations and plot the results with a spreadsheet. **Answer**: Your results should look like Figure 8.12.

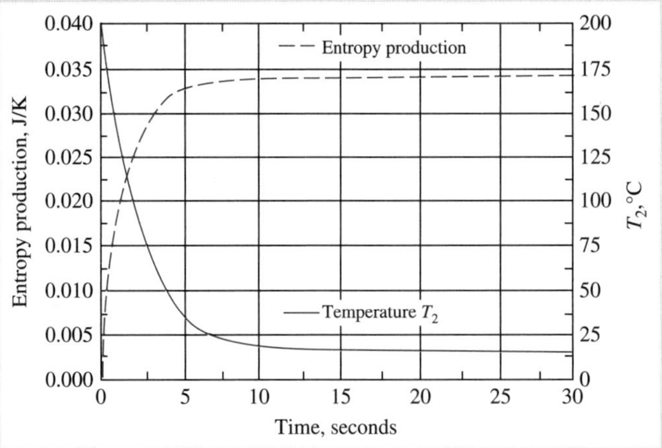

FIGURE 8.12
Example 8.11, Exercise 22.

The next three examples illustrate the use of the *direct method* of determining the entropy production. This method is usually more difficult than the entropy balance or *indirect method*, because it requires detailed information about *local* property values (i.e., properties at each point inside the system boundary) and the integration of Eq. (8.6) or (8.7) is often quite difficult. The formulae for the work mode entropy production per unit time per unit volume (σ_W) are given in Chapter 7 for viscous and electrical work mode irreversibilities.

EXAMPLE 8.12

The heat transfer rate from a very long fin of constant cross-section is given by

$$\dot{Q} = \sqrt{hPk_tA}\left(T_f - T_\infty\right)$$

where h is the convective heat transfer coefficient, P is the perimeter of the fin in a plane normal to its axis, k_t is the thermal conductivity of the fin, and A is the cross-sectional area of the fin (again in a plane normal to its axis). T_∞ is the temperature of the fin's surrounding (measured far from the fin itself) and T_f is the temperature of the foot (or base) of the fin. The temperature profile along the fin is given by

$$T(x) = T_\infty + \left(T_f - T_\infty\right)e^{-mx}$$

where

$$m = \left(\frac{hP}{k_tA}\right)^{1/2}$$

The fin is attached to an engine whose surface temperature is 95.0°C. Determine the entropy production rate for the fin if it is a very long square aluminum fin, 0.0100 m on a side, in air at 20.0°C. The thermal conductivity of aluminum is 204 W/(m·K) and the convective heat transfer coefficient of the fin is 3.50 W/(m²·K).

Solution

First, draw a sketch of the system (Figure 8.13).

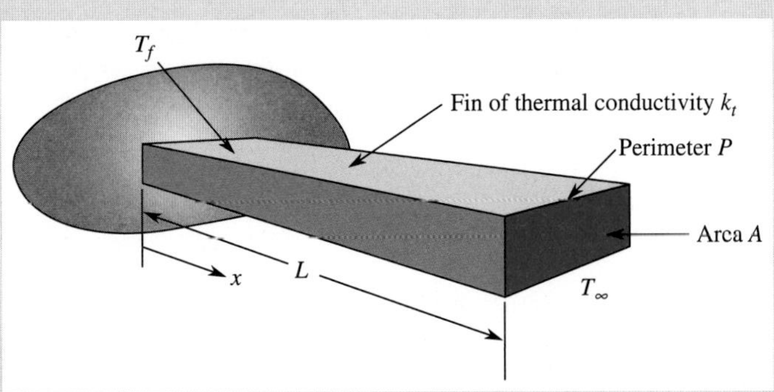

FIGURE 8.13

Example 8.12.

The unknown is the entropy production rate for the fin. The material is aluminum.

Since there is no work mode entropy production here, Eq. (7.66) is used to find the fin's entropy production rate by the *direct method* (note that we have insufficient information to use the more convenient *indirect method* or entropy balance here) as

$$\dot{S}_P = (\dot{S}_P)_Q = -\int_{\mathcal{V}} \left[\frac{\dot{q}}{T^2}\left(\frac{dT}{dx}\right)\right]_{\text{act}} d\mathcal{V}$$

Since this is a one-dimensional heat transfer problem, we can substitute $A\,dx$ for $d\mathcal{V}$. Then,

$$(\dot{S}_P)_Q = -\int_0^L \left[\frac{\dot{q}}{T^2}\left(\frac{dT}{dx}\right)\right] A\,dx = \int_0^\infty \left[\frac{k_t A}{T^2}\left(\frac{dT}{dx}\right)^2\right] dx$$

where we have used Fourier's law, $\dot{q} = -k_t(dT/dx)$, and have let $L \to \infty$ for a very long fin. We can differentiate the fin's temperature profile given previously to obtain

$$\frac{dT}{dx} = -m(T_f - T_\infty)e^{-mx}$$

then

$$k_t\left(\frac{A}{T^2}\right)\left(\frac{dT}{dx}\right)^2 = \frac{\sqrt{hPk_tA}(m)(T_f - T)^2 e^{-2mx}}{[T_\infty + (T_f - T_\infty)e^{-mx}]^2}$$

and

$$(S_P)_Q = m(T_f - T_\infty)^2\sqrt{hPk_tA}\int_0^\infty \frac{e^{-2mx}dx}{[T_\infty + (T_f - T_\infty)e^{-mx}]^2}$$

This expression can be integrated using a table of integrals and a change of variables (e.g., let $y = e^{-mx}$) to obtain

$$(\dot{S}_P)_Q = \sqrt{hPk_tA}\left(\ln\frac{T_f}{T_\infty} + \frac{T_\infty}{T_f} - 1\right)$$

In this problem, we have

$$h = 3.50\,\text{W}/(\text{m}^2\cdot\text{K}) \qquad A = 1.00\times10^{-4}\,\text{m}^2$$

$$P = 0.0400\,\text{m} \qquad T_\infty = 20.0°\text{C} = 293.15\,\text{K}$$

$$k_t = 204\,\text{W}/(\text{m}\cdot\text{K}) \qquad T_f = 95.0°\text{C} = 368\,\text{K}$$

Then,

$$\begin{aligned}\dot{S}_P &= (\dot{S}_P)_Q \\ &= \sqrt{[3.50\,\text{W}/(\text{m}^2\cdot\text{K})](0.0400\,\text{m})[204\,\text{W}/(\text{m}\cdot\text{K})](1.00\times10^{-4}\,\text{m}^2)}\left(\ln\frac{368}{293.15} + \frac{293.15}{368} - 1\right) \\ &= 0.00128\,\text{W/K}\end{aligned}$$

Note that the entropy production rate in this example is quite small.

WHY IS THE ENTROPY PRODUCTION RATE IN EXAMPLE 8.12 SO SMALL?

This is due to the fact that the fin temperature $T(x)$ is close to the environmental temperature T_∞ over most of the length of the fin. Entropy production due to heat transfer is minimized when the temperature difference producing the heat transfer is small. Check it out for yourself to see that $(\dot{S}_P)_Q$ becomes zero in the following equation when $T_f = T_\infty$ (remember, $\ln(1) = 0$).

$$(\dot{S}_P)_Q = \sqrt{hPk_tA}\left(\ln\frac{T_f}{T_\infty} + \frac{T_\infty}{T_f} - 1\right)$$

And the maximum entropy production rate occurs when the temperature difference is a maximum. This occurs in this example when $T_\infty = 0$ K or $T_f = \infty$. In this case,

$$(\dot{S}_P)_{max} = \lim_{T_\infty \to 0} \dot{S}_P = \lim_{T_f \to \infty} \dot{S}_P = \infty$$

EXAMPLE 8.13

The velocity profile in the steady isothermal laminar flow of an incompressible Newtonian fluid contained between concentric cylinders in which the inner cylinder is rotating and the outer cylinder is stationary is given by

$$V = \frac{\omega R_1^2}{R_2^2 - R_1^2}\left(\frac{R_2^2}{x} - x\right) \quad \text{for} \quad R_1 \le x \le R_2$$

where ω is the angular velocity of the inner cylinder, and x is measured radially outward. Determine the rate of entropy production due to laminar viscous losses for SAE-40 engine oil at 20.0°C in the gap between cylinders of radii 0.0500 and 0.0510 m when the inner cylinder is rotating at 1000. rev/min. The viscosity of the oil is 0.700 N·s/m^2 and the length of the cylinder is 0.100 m.

Solution

First, draw a sketch of the system (Figure 8.14).

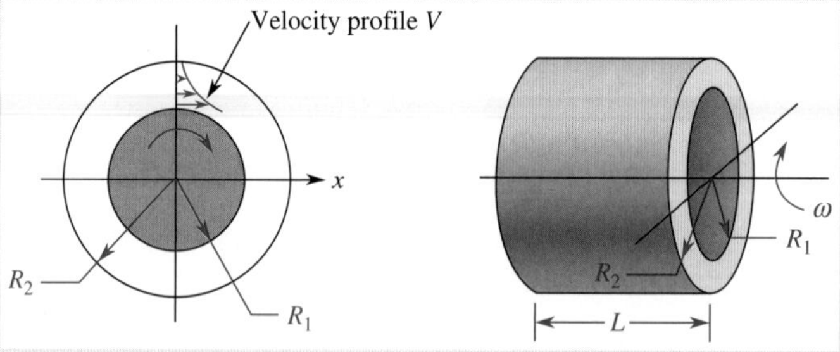

FIGURE 8.14
Example 8.14, system.

The unknown is the rate of entropy production due to laminar viscous losses. The material is SAE-40 engine oil at 20.0°C.

Here, we use Eq. (7.72) for the viscous dissipation of mechanical work. By differentiating the velocity formula just given, we get

$$\frac{dV}{dx} = -\frac{\omega R_1^2}{R_2^2 - R_1^2}\left(\frac{R_2^2}{x^2} + 1\right)$$

then

$$(\sigma_W)_{vis} = \frac{\mu}{T}\left(\frac{dV}{dx}\right)^2 = \frac{\omega^2 R_1^4 \mu}{(R_2^2 - R_1^2)^2 T}\left(\frac{R_2^2}{x^2} + 1\right)^2$$

Equation (7.69) gives the entropy production rate for viscous effects as

$$(\dot{S}_P)_{\substack{W\\vis}} = \int_{\forall} (\sigma_W)_{vis}\, d\forall$$

For the differential volume element $d\forall$, we use the volume of an annulus of thickness dx, or $d\forall = 2\pi Lx\,dx$ (Figure 8.15).

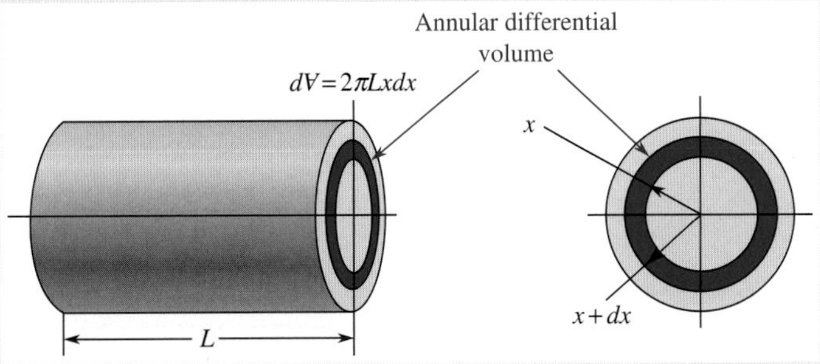

FIGURE 8.15
Example 8.13, annular differential volume.

Putting these expressions for $(\sigma_W)_{vis}$ and $d\forall$ into Eq. (7.70) and carrying out the integration gives

$$(S_P)_{\substack{W\\vis}} = \frac{2\pi L\omega^2 R_1^4 \mu}{(R_2^2 - R_1^2)^2 T}\left(2R_2^2 \ln\frac{R_2}{R_1} + \frac{R_2^4}{2R_1^2} - \frac{R_1^2}{2}\right)$$

where $\mu = 0.700\ \text{N}\cdot\text{s/m}^2$, $L = 0.100$ m, and

$$\omega = \left(1000.\frac{\text{rev}}{\text{min}}\right)\left(\frac{2\pi\,\text{rad}}{\text{rev}}\right)\left(\frac{1\,\text{min}}{60\,\text{s}}\right) = 104.7\,\text{rad/s}$$

$R_1 = 0.0500$ m, $R_2 = 0.0510$ m, and $T = 20.0°\text{C} = 293.15$ K.

Substituting these values into the preceding formula gives

$$(\dot{S}_P)_{\substack{W\\vis}} = 2.08\,\text{W/K}$$

In this example, we have a reasonably high entropy production rate. This is due to the very small gap between the cylinders and the high viscosity of the engine oil. Check for yourself to see that $(\dot{S}_P)_{W\text{-vis}} \to 0$ as R_2 becomes much larger than R_1, or as $\mu \to 0$. Conversely, check to see that $(\dot{S}_P)_{W\text{-vis}} \to \infty$ as $R_1 \to R_2$ or as $\mu \to \infty$. Some of these elements are explored in the following exercises.

Exercises

23. Determine the entropy production rate $\dot{S}_P$ in Example 8.13 if the oil is changed from SAE-40 motor oil with a viscosity of $\mu = 0.700\ \text{N}\cdot\text{s/m}^2$ to SAE-10 motor oil with $\mu = 0.150\ \text{N}\cdot\text{s/m}^2$ at 20.0°C. Keep the values of all the other variables the same as they are in Example 8.13. **Answer:** $(\dot{S}_P)_{W\text{-vis}} = 0.455\,\text{W/K}$.
24. Recalculate the entropy production rate $\dot{S}_P$ in Example 8.13 when the bearing gap is reduced from 1.00 mm to 0.500 mm by making $R_1 = 0.0500$ m and $R_2 = 0.0505$ m. Keep the values of all the other variables the same as they are in Example 8.13. **Answer:** $\dot{S}_P = 4.13\,\text{W/K}$.
25. Determine the entropy production rate $\dot{S}_P$ in Example 8.13 when the shaft rotation is increased to 8000. rpm. Keep the values of all the other variables the same as they are in Example 8.13. Plot the entropy

(*Continued*)

EXAMPLE 8.13 (*Continued*)

production rate $\dot{S}_P$ vs. shaft angular velocity ω as ω varies from 0 to 10,000. rpm (Figure 8.16).
Answer: $\dot{S}_P = 133\,\text{W/K}$.

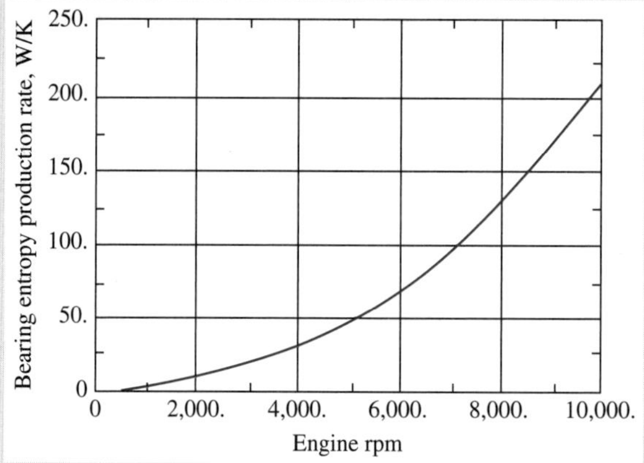

FIGURE 8.16
Example 8.13, Exercise 25.

EXAMPLE 8.14

An electrical circuit board contains a variety of digital logic elements. When operating, the board draws 10.0 mA at 5.00 V dc and it has a steady state surface temperature of 30.0°C. Estimate the entropy production rate of the circuit board.

Solution

First, draw a sketch of the system (Figure 8.17).

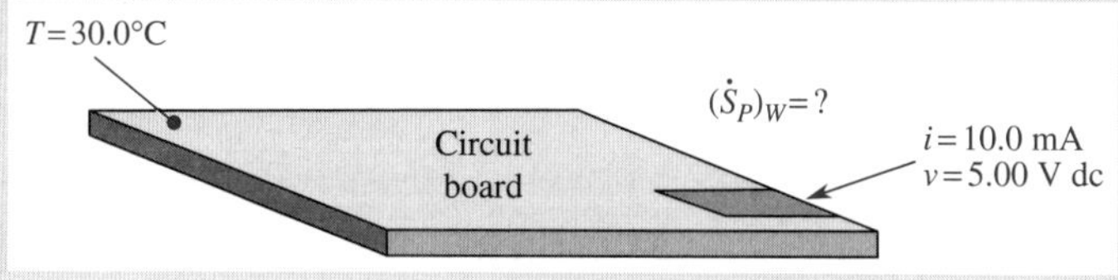

FIGURE 8.17
Example 8.14.

The unknown is entropy production rate of the circuit board. The system is the circuit board.

Since nearly all the electrical energy entering the circuit board is being converted into heat, we could calculate the entropy production rate of the board from its heat loss characteristics (convective heat transfer coefficient, surface temperature, environmental temperature, surface area, etc.); however, none of this information is supplied in the problem statement. We could also calculate the entropy production rate for electrical work mode dissipation directly from Eq. (7.73) if we knew how the current density J_e, electrical resistivity ρ_e, and local internal temperature T are distributed throughout the circuit board. But we do not know this information either. However, the problem statement asks for only an estimate of the entropy production rate, and we can obtain this from the special electrical work mode dissipation Eq. (7.74) if we lump all the components on the board into one uniform, isothermal, constant current density system. From Ohm's law, $\phi = IR_e$, we can write Eq. (7.74) as

$$(\dot{S}_P)_W = \frac{I^2 R_e}{T} = \frac{\phi I}{T} = \frac{(5.00\,\text{V})(10.0\times 10^{-3}\,\text{A})}{30.0 + 273.15\,\text{K}}$$
$$= 1.65\times 10^{-4}\,\text{W/K}$$

This is only a lumped parameter estimate of the entropy production rate for this system. The actual entropy production rate is somewhat larger due to the nonuniform distribution of entropy-producing electrical components within the system volume.

Exercises

26. Recompute the entropy production rate $\dot{S}_P$ in Example 8.14 when the steady state surface temperature of the board increases to 38.0°C. Keep the values of all the other variables the same as they are in Example 8.14. **Answer:** $(\dot{S}_P)_W = 1.61 \times 10^{-4}$ W/K.

27. Determine the entropy production rate $\dot{S}_P$ in Example 8.14 if the current is increased to 50.0 mA. Keep the values of all the other variables the same as they are in Example 8.14. **Answer:** $(\dot{S}_P)_W = 8.25 \times 10^{-4}$ W/K.

28. If the circuit voltage is increased from 5.00 V to 30.0 V in Example 8.14, determine the new entropy production rate. Keep the values of all the other variables the same as they are in Example 8.14. **Answer:** $(\dot{S}_P)_W = 9.90 \times 10^{-4}$ W/K.

8.4 DIFFUSIONAL MIXING

Here, we wish to use the *indirect method* to analyze the entropy production that results from a simple diffusion type of mixing process. Consider the rigid, insulated container shown in Figure 8.18. It contains the same substance on both sides of the partition but generally at different pressures, temperatures, and amounts. When the partition is removed, the material in the chambers mixes by diffusion, resulting in a final temperature T_2 and final pressure p_2. We are interested in the amount of entropy produced by this mixing process.

The energy balance equation for this system gives

$$\underbrace{{}_1Q_2}_{\substack{0\\(\text{adiabatic})}} - \underbrace{{}_1W_2}_{0\ (\text{adiabatic})} = U_2 - U_1 + \underbrace{\mathrm{KE}_2 - \mathrm{KE}_1 + \mathrm{PE}_2 - \mathrm{PE}_1}_{0\ (\text{stationary system})}$$

$$U_2 = m_2 u_2 = U_1 = m_a u_a + m_b u_b$$

or

$$u_2 = \frac{m_a u_a + m_b u_b}{m_a + m_b} = u_b + \gamma(u_a - u_b) \tag{8.9}$$

where

$$\gamma = \frac{m_a}{m_a + m_b}$$

and

$$1 - \gamma = \frac{m_b}{m_a + m_b}$$

The entropy balance equation gives

$${}_1(S_P)_2 = S_2 - S_1 - \underbrace{\frac{{}_1Q_2}{T_b}}_{\substack{0\\(\text{adiabatic})}}$$

or

Entropy production in two-component mixing:

$$\begin{aligned} {}_1(S_P)_2 &= m_2 s_2 - m_a s_a - m_b s_b p \\ &= (m_a + m_b)[s_2 - s_b + \gamma(s_b - s_a)] \end{aligned} \tag{8.10}$$

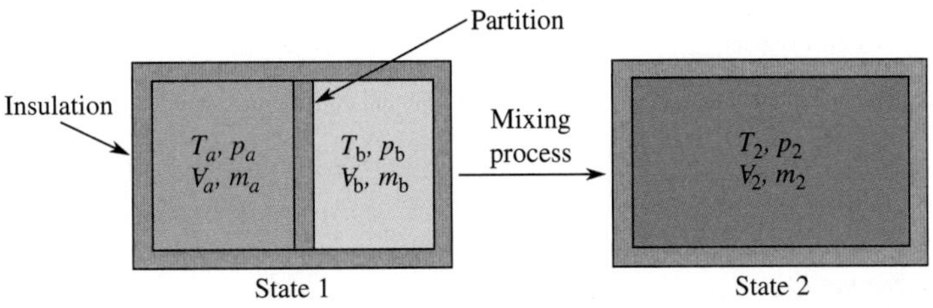

FIGURE 8.18
Mixing of single-species substances.

When the chambers contain identical ideal gases, Eqs. (3.38) and (8.9) give

$$u_2 - u_b = c_v(T_2 - T_b) = \gamma c_v(T_a - T_b)$$

or

$$T_2 = T_b + \gamma(T_a - T_b) \tag{8.11}$$

and Eqs. (7.37) and (8.10) give

$$_1(S_P)_2 = (m_a + m_b)\left[c_p \ln \frac{T_2}{T_b} - R \ln \frac{p_2}{p_b} + \gamma\left(c_p \ln \frac{T_b}{T_a} - R \ln \frac{p_b}{p_a}\right)\right]$$

or

$$_1(S_P)_2 = (m_a + m_b)\{c_p \ln [(T_2/T_b)(T_b/T_a)^\gamma] - R \ln [(p_2/p_b)(p_b/p_a)^\gamma]\}$$

and inserting Eq. (8.11) for T_2 gives

Entropy production when two identical ideal gases are mixed

$$_1(S_P)_2 = (m_a + m_b)\{c_p \ln [(1 + \gamma(T_a/T_b - 1))(T_b/T_a)^\gamma] - R \ln [(p_2/p_b)(p_b/p_a)^\gamma]\} \geq 0 \tag{8.12}$$

When the chambers contain identical incompressible liquids, Eqs. (3.33), (8.9), and (8.10) can be combined to yield

Entropy production when two identical incompressible liquids are mixed

$$_1(S_P)_2 = (m_a + m_b)c \ln [(1 + \gamma(T_a/T_b - 1))(T_b/T_a)^\gamma] \geq 0 \tag{8.13}$$

The actual mixing process need not have the same two-chamber geometry shown in Figure 8.18, as illustrated by the following example.

EXAMPLE 8.15

Determine the entropy produced when 3.00 g of cream at 10.0°C are added adiabatically and without stirring to 200. g of hot coffee at 80.0°C. Assume both the coffee and the cream have the properties of pure water.

Solution

First, draw a sketch of the system (Figure 8.19).

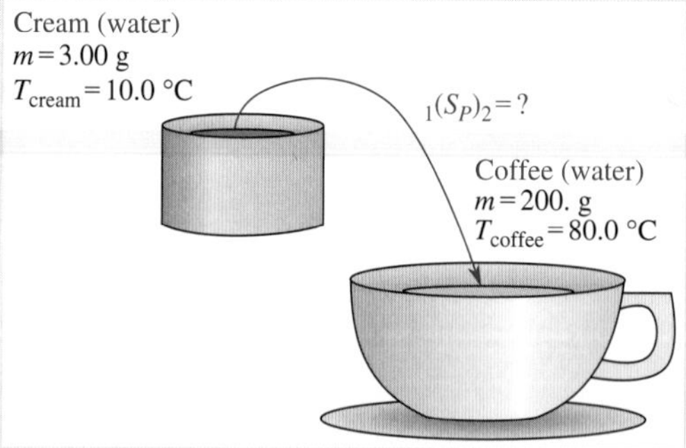

FIGURE 8.19
Example 8.15.

The unknown is the entropy produced by mixing, and the materials are coffee and cream, both modeled as liquid water.

Let a = cream and b = coffee. Then, γ = 3.00/203 = 0.0148. Assuming both the coffee and the cream are incompressible liquids with the specific heat of water, c = 4186 J/(kg·K), Eq. (8.13) gives

$$_1(S_P)_2 = (0.203\text{ kg})[4186\text{ J/(kg}\cdot\text{K)}] \ln\left\{\left[1 + 0.0148\left(\frac{10.0 + 273.15}{80.0 + 273.15} - 1\right)\right] \times \left(\frac{80.0 + 273.15}{10.0 + 273.15}\right)^{0.0148}\right\}$$

$$= 0.282\text{ J/K}$$

Note that this example does not include the entropy production due to the heat transfer required to cool the mixture down to a drinkable temperature.

Exercises

29. Recalculate the entropy production in Example 8.15 when the average specific heat of the mixture of coffee and cream of 3856 J/(kg·K) is used instead of the value for pure water. Keep the values of all the other variables the same as they are in Example 8.15. **Answer:** ${}_1(S_P)_2 = 0.259$ J/K.

30. Determine the entropy production in the mixing process described in Example 8.15 when the temperature of the cream is 20.0°C and the temperature of the coffee is 90.0°C. Keep the vales of all the other variables the same as they are in Example 8.15. **Answer:** ${}_1(S_P)_2 = 0.265$ J/K.

31. (a) Determine the entropy produced in the process described in Example 8.15 when the mass of the cream added is increased from 3.00 g to 10.0 g. Keep the values of all the other variables the same as they are in Example 8.15. (b) Plot ${}_1(S_P)_2/(m_a + m_b)$ vs. the mass fraction y over the range $0 \leq y \leq 1$ for the values of c, T_a, and T_b used in Example 8.15. (c) What value of y maximizes ${}_1(S_P)_2$ in part b? **Answers:** (a) ${}_1(S_P)_2 = 0.911$ J/K, (b) See Figure 8.20, (c) $[{}_1(S_P)_2]_{max}$ occurs at $y = 0.52$.

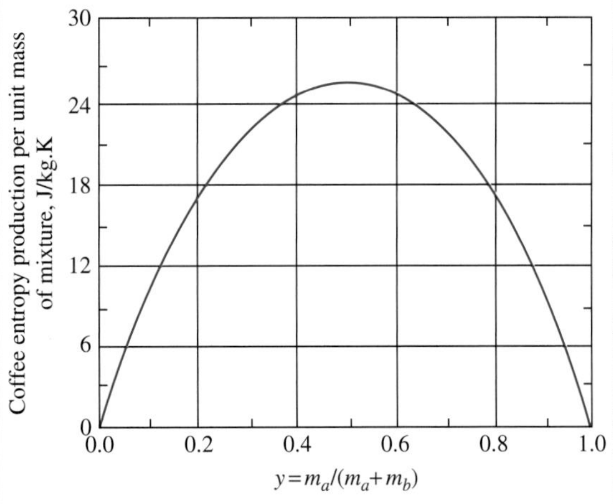

FIGURE 8.20
Example 8.15, Exercise 31, part b.

Equations (8.12) and (8.13) show that the amount of entropy production in this type of mixing process depends on the amounts mixed and their initial states. The larger the property differences between the initial states, the larger is entropy production on mixing. That is, the farther the initial states are from the final equilibrium state, the larger the associated entropy production is. This is a general characteristic of the second law.

SUMMARY

In this chapter, the energy and entropy balance and rate balance equations are used to investigate various closed system applications. The *indirect method* of calculating values for S_P and $\dot{S}_P$ occurs when they are calculated from the entropy balance or the entropy rate balance equations. But, when these values are calculated from the applicable entropy production rate density equations (σ_W) defined in Chapter 7, we called this the *direct method.*

Reversible processes are defined as processes that occur with no internal irreversibilities, such as friction, viscosity, heat transfer, or diffusion. Then, S_P and $\dot{S}_P = 0$, and the closed system entropy balance and entropy rate balance equations for a reversible process reduce to

$$\text{Entropy balance (SB): } S_2 - S_1 = m(s_2 - s_1) = \left(\frac{{}_1Q_2}{T_b}\right)_{\text{rev}} \tag{8.7}$$

and

$$\text{Entropy rate balance (SRB): } \dot{S} = m\dot{s} = \left(\frac{\dot{Q}}{T_b}\right)_{\text{rev}} \tag{8.8}$$

However, most systems of engineering interest do not undergo reversible processes, and the entropy produced by the irreversiblilties inherent in the system is of great value in the design process. Both the *direct method* and the *indirect method* can be used to find values for the entropy produced in any irreversible process. The *indirect method* is used in several examples developed in this chapter to determine the entropy produced by heating a fluid in a simple closed container, a lightbulb, a food blender, a typical electrical power plant, and so forth. The more complex *direct method* is then used to determine the entropy produced by convective heat transfer from a cooling fin, the viscosity of the lubricant liquid in an engine bearing, the electrical resistance of a circuit board, and the diffusional mixing of identical fluids.

Problems (* indicate problems in SI units)

1.* Air contained in a cylinder fitted with a piston initially at 2.00 MPa and 2000.°C expands to 0.200 MPa in an isentropic process. Assuming the air behaves as a constant specific heat ideal gas, determine the following:
   a. The final temperature.
   b. The change in specific internal energy.
   c. The change in specific enthalpy.
   d. The change in specific entropy.
   e. The work done per lbm of air during the expansion.

2. A constant specific heat ideal gas has a gas constant of 42.92 ft·lbf/(lbm·R) and a constant pressure specific heat of 0.200Btu/(lbm·R). Determine the heat transferred and the change of total entropy if 9.00 lbm of this gas is heated from 40.0°F to 340.°F in a rigid container.

3. A reversible Carnot heat engine has 1.00 lbm of air as the working fluid. Heat is received at 740.°F and rejected at 40.0°F. At the beginning of the heat addition process, the pressure is 100. psia and during this process the volume triples. Calculate the net cycle work per lbm of air. Assume the air behaves as a constant specific heat ideal gas.

4. A 1.000 ft^3 glass bottle is initially evacuated then has 1.000 g of water added that eventually comes to equilibrium at 70.00°F. The pressure in the bottle is increased by 11.1668 psia during a reversible adiabatic compression:
   a. What is the work done during compression?
   b. What is the entropy production for the compression?

5. 3.00 lbm of Refrigerant-134a (*not* an ideal gas) is compressed adiabatically in a closed piston-cylinder device from 5.00 psia, 220.°F to 200. psia, 340.°F.
   a. Determine the work for this process.
   b. Show whether or not this process violates the second law of thermodynamics.

6.* An engineer claims to be able to compress 0.100 kg of water vapor at 200.°C and 0.100 MPa in a piston-cylinder arrangement in an isothermal *and* adiabatic process. The engineer claims that the final volume is 6.10% of the initial volume. Determine
   a. The final temperature and pressure.
   b. The work required.
   c. Show whether the process is thermodynamically possible.

7. An inventive engineer claims to have designed a mechanochemical, single-stroke closed system that compresses 10.0 lbm of air isothermally from 14.7 psia, 100.°F to 200. psia while inputting 500. Btu of mechanochemical compression work. Assuming constant specific heat ideal gas behavior,
   a. What heat transfer is required for this process to occur?
   b. Does this process violate the second law of thermodynamics?

8.* Saturated liquid water at 8.58 MPa undergoes a reversible isothermal process in a cylinder until the pressure reaches 0.100 MPa. Calculate the heat transfer and work per kg of water for this process. Show the process on a *T-s* diagram. Neglect any changes in kinetic and potential energies.

9.* Air is compressed in a steady state reversible adiabatic process from 25.0°C and 0.150 MPa to 1.70 MPa. Determine the change of specific enthalpy in this process and find the density of the exit air. Assume the air behaves as an ideal gas with constant specific heats. Neglect any changes in kinetic and potential energies.

10.* One cubic meter of hydrogen (a constant specific heat ideal gas) expands from an initial pressure of 0.500 MPa to a final pressure of 0.100 MPa. The gas temperature before expansion is 27.0°C.
   a. Determine the final temperature if the process is isentropic.
   b. Determine the final temperature if the process is polytropic with $n = 1.30$.
   c. Calculate the heat transfer required for the polytropic case.

11. 2.00 lbm of saturated water vapor at 247.1 psia undergoes a reversible isothermal expansion until the pressure reaches 20.0 psia. Determine the heat transfer and the work done for this process. The system boundary temperature is the same as the process temperature.

12. Consider a fixed mass of a constant specific heat ideal gas in a piston-cylinder device undergoing a compression process for which $p\mathcal{V}^n$ = constant (a polytropic process). Show that the work done per unit mass of gas in such a process is given by $(p_2v_2 - p_1v_1)/(n-1)$ if $n \neq 1$. If the process is isentropic, show that this reduces to $c_v(T_2 - T_1)$.

13.* 0.130 kg of a constant specific heat ideal gas is compressed in a closed system from 1.00 atm and 40.0°C to 11.39 atm in an isothermal process. For this gas, $c_p = 523$ J/(kg·K), $c_v = 315$ J/(kg·K), and $R = 208$ J/(kg·K). For this process, determine
   a. The work required.
   b. The resulting heat transfer.
   c. The amount of entropy produced.
   d. Explain whether this process violates the second law of thermodynamics.

14. Show that a constant specific heat ideal gas undergoing a constant heat flux polytropic process (pv^n = constant with $n \neq 1$) has a limiting isothermal system boundary temperature corresponding to a reversible process given by

$$(T_b)_{rev} = \frac{T_2 - T_1}{\ln(T_2/T_1)}$$

(Hint: Recall that ${}_1W_2 = mR(T_2 - T_1)/(n-1)$ for such a polytropic process with an ideal gas.)

15. A 20.0 ft^3 tank contains air at 100. psia, 100.°F. A valve on the tank is opened and the pressure in the tank drops to 20.0 psia. If the air that remains in the tank is considered to be a closed system undergoing a reversible adiabatic process, calculate the final mass of air in the tank. Assume constant specific heat ideal gas behavior, and neglect any changes in kinetic and potential energies.
16. A pressure vessel contains ammonia at a pressure of 100. psia and a temperature of 100.°F. A valve at the top of the vessel is opened, allowing vapor to escape. Assume that, at any instant, the ammonia that remains in the vessel has undergone an isentropic process. When the ammonia remaining in the vessel becomes a saturated vapor, the valve is closed. The mass of ammonia in the vessel at this moment is 2.00 lbm. Find the mass of ammonia that escaped into the surroundings. Neglect any changes in kinetic and potential energies.
17.* A 2.00 m^3 tank contains air at 0.200 MPa and 35.0°C. A valve on the tank is opened and the pressure in the tank drops to 0.100 MPa. If the process is isentropic, calculate the final mass of air in the tank. Assume the air behaves as a constant specific heat ideal gas. Neglect any changes in kinetic and potential energies.
18. A 58.0 ft^3 tank contains air at 30.0 psia and 100.°F. A valve on the tank is opened and the pressure in the tank drops to 10.0 psia. If the air that remains in the tank has gone through an adiabatic polytropic process with $n = 1.33$, calculate the final mass of air in the tank and the entropy production that occurred in this mass. Assume the air behaves as a constant specific heat ideal gas, and neglect any changes in kinetic and potential energies.
19. An operating gearbox (transmission) has 200. hp at its input shaft while 190. hp are delivered to the output shaft. The gearbox has a steady state surface temperature of 140.°F. Determine the rate of entropy production by the gearbox.
20.* A gearbox (transmission) operating at steady state, receives 100. kW of power from an engine and delivers 97.0 kW to the output shaft. If the surface of the gearbox is at a uniform temperature of 50.0°C and the surrounding temperature is 20.0°C, what is the rate of entropy production?
21.* Determine the amount of entropy produced in the process described in Problem 1 at the end of Chapter 5, when both the specific internal energy and the specific entropy of the water have returned to their initial values.
22.* Determine the amount of entropy produced in the process described in Problem 3 at the end of Chapter 5. Assume that the system boundary temperature is the same as its bulk isothermal temperature.
23. Determine the amount of entropy produced in the process described in Problem 4 at the end of Chapter 5 if the 500. Btu heat transfer occurred across an isothermal system boundary at 250.°F.
24.* Determine the amount of entropy produced in the process described in Problem 8 at the end of Chapter 5 if the work transport is 90.0% of the magnitude of the heat transport. Assume that the system boundary temperature is the same as its bulk isothermal temperature.
25.* Determine the amount of entropy produced in the process described in Problem 13 at the end of Chapter 5. Assume the human body is a steady state closed system with an isothermal surface temperature of 36.0°C during the exercise process.
26. Determine the amount of entropy produced during the adiabatic expansion process described in Problem 17 at the end of Chapter 5. Discuss the difficulty encountered in determining the entropy production during the final isobaric compression process.
27. 1.00 lbm of saturated water vapor at 212°F is condensed in a closed, nonrigid system to saturated liquid at 212°F in a constant pressure process by a heat transfer across a system boundary with a constant temperature of 80.0°F. What is the total entropy production for this process?
28. A rigid container encloses 150. lbm of air at 15.0 psia and 500. R. We wish to increase the temperature to 540. R. Assuming constant specific heat ideal gas behavior,
    a. Determine the heat transfer to the air for this change of state.
    b. Determine the entropy production if this change of state is accomplished by using a constant system boundary temperature of 300.°F.
29. A sealed kitchen pressure cooker whose volume is 1.00 ft^3 contains 2.20 lbm of saturated water (liquid plus vapor) at 14.7 psia. The pressure cooker is then heated until its internal pressure reaches 20.0 psia. Determine
    a. The work done during the process.
    b. The heat transfer during the process.
    c. The entropy produced during the process if the inner surface of the pressure cooker is constant at 250.°F.
30. A closed, sealed, rigid container is filled with 0.05833 ft^3 of liquid water and 0.94167 ft^3 of water vapor in equilibrium at 1.00 psia.
    a. What is the quality in the vessel at this state?
    The vessel is then heated until its contents become a saturated vapor.
    b. What are the temperature and pressure in the vessel at this state?
    The heating process just described is done irreversibly.
    c. Determine the total entropy produced for this process if the surface temperature of the vessel is maintained constant at 300.°F.
31.* Determine the entropy produced as a 4.00 g, 80.0°C lead bullet traveling at 900. m/s impacts a perfectly rigid surface aergonically and adiabatically. The specific heat of lead at the mean temperature of the bullet is 167 J/(kg·K).
32.* Determine the minimum isothermal system boundary temperature required by the second law as a 1500. kg iron ingot is heated from 20.0°C to 1000.°C. Assume the ingot is incompressible.
33.* In the 21st century, the Earth will be terrorized by Zandar the Wombat, an asexual rebel engineer from the planet Q-dot. Earth's only hope for survival lies in your ability to determine the entropy production rate of Zandar. To do this, you cleverly trick Zandar into completely wrapping himself (herself?) with insulation and holding his breath. You then quickly measure his body temperature and find that it is increasing at a constant rate of 2.00°C per minute. Zandar weighs 981 N and has the thermodynamic properties of liquid water. Determine Zandar's entropy production rate when his body temperature reaches 50.0°C.
34.* The surface temperature of a 100. W incandescent lightbulb is 60.0°C. The surface temperature of a 20.0 W fluorescent tube producing the same amount of light as the 100. W incandescent lightbulb is 30.0°C. Determine the steady state entropy

production rate of each light source and comment on which is the most efficient.

35. Rework Example 8.11 by setting $T_b = T_s$. Using the formula for T_s given in the example, show that

$$\frac{\bar{d}Q}{T_b} = -\frac{hA\,dt}{1 + T_\infty e^{\alpha t}/(T_1 - T_\infty)}$$

where $\alpha = hA/(mc_v)$. Integrate this from $t = 0.00$ to 5.00 s and combine it with the $m(s_2 - s_1)$ result from the example to get $_1(S_P)_2$ under this condition. What is the significance of this result?

36. Integrate Eq. (7.52) to determine the entropy production and use Eq. (7.77) to find the total entropy change for an aergonic closed system in which the temperature increases from T_1 = 70.0°F to T_2 = 200.°F for the cases where the heat transfer varies with the system absolute temperature according to the relations
   a. $Q = K_1 T$ (convection).
   b. $Q = K_2 T^4$ (radiation), where K_1 = 3.00 Btu/R and $K_2 = 3.00 \times 10^{-4}$ Btu/R^4. The system boundary is maintained isothermal at 212°F.

37. Determine the entropy production rate due to conduction heat transfer inside a system having a volume of 3.00 ft^3 and a thermal conductivity of 105 Btu/(h·ft·R) that contains the following temperature profile:

$$T = 300. \times [\exp(x/3.00)]$$

where T is in R and x is in feet.

38. Using Eq. (7.64), show that the entropy production rate per unit volume due to heat transfer (σ_Q) is a constant if the temperature distribution due to conduction heat transfer is given by

$$T = C_1[\exp(C_2 x)]$$

where C_1 and C_2 are constants. (Hint: Use Fourier's law of heat conduction to eliminate the $\dot{q}$ term.)

39. The temperature distribution due to conduction heat transfer inside a flat plate with an internal heat generation (Figure 8.21) is given by

$$T = T_0 + (T_s - T_0)(x/L)^2$$

where T_s is the surface ($x = L$) and T_0 is the centerline ($x = 0$) temperature. Determine a formula for the entropy production rate ($\dot{S}_P$) for this system.

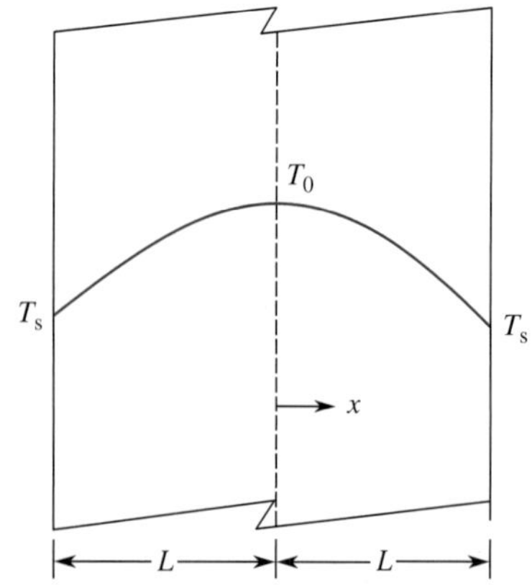

FIGURE 8.21
Problem 39.

40. Example 8.13 deals with the velocity profile in a liquid contained in the gap between two concentric cylinders of radii R_1 and $R_2 > R_1$ in which the inner cylinder is rotating with an angular velocity ω and the outer cylinder is stationary. If this gap is very small, the velocity profile can be approximated by the linear relation

$$V = R_1\omega(R_2 - x)/(R_2 - R_1)$$

where $R_2 >> (R_2 - R_1)$ and x is measured radially outward from the center of the inner cylinder. Using the data given in Example 8.13, determine the entropy production rate using this simpler velocity profile and compare your answer to that given in Example 8.13 for the more complex nonlinear velocity profile.

41. Example 8.13 deals with the entropy production rate in the flow between concentric rotating cylinders in which the outer cylinder was stationary and the inner cylinder rotates with a constant angular velocity ω. If, instead, we allow both cylinders to rotate in the same direction with constant angular velocities ω_2 at the outer cylinder and ω_1 at the inner one, then the velocity profile in the gap between the cylinders becomes

$$V = \left[(\omega_2 R_2^2 - \omega_1 R_1^2)x - (\omega_2 - \omega_1)(R_1^2 R_2^2)/x\right]/(R_2^2 - R_1^2)$$

Find the expression for the entropy production rate due to viscous effects in the fluid of viscosity μ contained between these rotating cylinders of radii R_1 and $R_2 > R_1$ and length L. Assume the fluid is maintained isothermal at temperature T.

42.* A *dipstick heater* is an electrical resistance heater plugged into a regular 110. V ac outlet and inserted into the dipstick tube of an automobile engine. Its purpose is to keep the engine oil warm during the winter when the car is not in use, thus allowing the engine to start more easily. Determine the entropy produced during an 8.00 h period by a 100. W steady state dipstick heater whose surface is isothermal at 90.0°C.

43.* The potential difference across the tungsten filament operating at 2400.°C in a cathode ray vacuum tube is 25.0×10^3 V. The filament is a small disk 2.00×10^{-3} m in diameter and 1.00×10^{-4} m thick having a resistivity of 6.00×10^{-4} Ω·m. Assuming all the voltage drop occurs uniformly across the thickness of the disk, determine its entropy production rate.

44. A current of 100. A is passed through a 6.00 ft long stainless steel wire 0.100 inch in diameter. The electrical resistivity of the wire is 197×10^{-5} Ω·in, and its thermal conductivity is 12.5 Btu/(h·ft·R). The outer surface temperature of the wire is maintained constant at 300.°F and the temperature profile inside the wire is given by

$$T = T_w + \rho_e J_e^2 (R^2 - x^2)/(4k_t)$$

where T_w is the wall temperature of the wire, R is its radius, and x is measured radially out from the center of the wire. Determine the total entropy production rate within the wire due to the flow of electricity through it. Assume all the physical properties are independent of temperature.

45. Determine the entropy produced when 3.00 lbm of carbon dioxide at 70.0°F and 30.0 psia are adiabatically mixed with 7.00 lbm of carbon dioxide at 100.°F and 15.0 psia. The final mixture pressure is 17.0 psia. Assume the carbon dioxide behaves as a constant specific heat ideal gas.

46. a. Determine a formula for the final pressure (p_2) that results when two volumes of the same constant specific heat ideal gas initially at p_a, T_a and p_b, T_b are mixed isentropically.
    b. Does this mixture pressure represent an upper or lower bound when this mixing is done adiabatically but not isentropically?
47. Determine the entropy produced as 5.0×10^{-3} lbm of human saliva at 98.6°F is adiabatically mixed with 3.00×10^{-3} lbm of human saliva at 103.2°F in a passionate and infectious kiss. The specific heat of the saliva is 0.950 Btu/(lbm·R).
48.* Determine the entropy produced as 10.0 kg of liquid water at 10.0°C is adiabatically mixed with 20.0 kg of liquid water at 80.0°C. The specific heat of the water is 4.20 kJ/(kg·K).
49. Here is the complete classical coffee and cream problem. Which of the following processes produces less entropy:
    a. Mixing cream with hot coffee and letting the mixture cool to the drinking temperature.
    b. Letting the coffee cool to a temperature such that, when the cream is added, the mixture will be at the drinking temperature? Do not ignore the cooling heat transfer entropy production.

Design Problems

The following are elementary, open-ended design problems. The objective is to carry out a preliminary thermal design as indicated. A detailed design with working drawings is not expected unless otherwise specified. These problems do not have specific answers, so each student's design is unique.

50.* Carry out a preliminary thermodynamic design of a system that heats 20.0 kg of liquid water from 20.0 to 80.0°C in 15.0 min at atmospheric pressure in a closed vessel. Use an electrical heating system and determine the electrical power and current requirements (assume standard line voltage values). Include a means of relieving any pressure buildup, and discuss safety considerations.
51. Carry out a preliminary thermodynamic design of a single piston-cylinder apparatus that produces 25.0 hp as it moves through a mechanical cycle in 10.0 s. The piston is to be drawn into the cylinder by condensing steam that enters the apparatus as a saturated vapor at 212°F. No work is done during the return stroke of the piston, during which time a fresh charge of steam is drawn into the apparatus. Continuous motion of this type can be accomplished through the use of a flywheel.
52.* Carry out a preliminary thermodynamic design of a system that mixes by only diffusive processes 5.00 kg of gaseous CO_2 with 10.0 kg of air in a maximum of 6.00 min in a closed, rigid vessel. Specify the vessel material, size, and internal geometry and discuss any relevant safety considerations.

Computer Problems

The following computer assignments are designed to be carried out on a personal computer using a spreadsheet or equation solver. They are meant to be exercises using some of the basic formulae of this chapter. They may be used as part of a weekly homework assignment.

53. Develop a program that determines the power output from a reversible solar power plant similar to that discussed in Example 8.2. Input all the relevant variables (in proper units), and output the net reversible electrical power produced. At your instructor's discretion, add screen graphics depicting a diagram of the power plant and the input and output variables. Allow the choice of working in either Engineering English or SI units.
54. Develop a program that performs an energy *and* entropy balance on a closed system with an isothermal boundary. The system contains an incompressible substance (either a liquid or a solid) that is undergoing an irreversible process. Input (in proper units) the heat and work transports of energy, the system volume, the initial internal temperature and the isothermal boundary temperature of the system, and the density and specific heat of the incompressible material contained in the system. Output to the screen the system mass, final temperature, and entropy production. Note that, if the entropy production becomes negative (an impossible physical situation), then the system boundary temperature was not properly specified. Check for this possibility and prompt the user for another boundary temperature if it occurs. Allow the choice of working in either Engineering English or SI units.
55. Develop a program that performs an energy *and* an entropy balance on a closed system with an isothermal boundary. The system contains an ideal gas with constant specific heats that is undergoing an irreversible process. Input (in proper units): the heat and work transports of energy, the system volume, the initial temperature and pressure of the system, and the constant volume specific heat and gas constant of the gas contained in the system. Output to the screen the system mass, the final pressure and temperature, and the entropy production for the process. Check to make sure the entropy production is positive and prompt the user for corrected input if it is not. Allow the choice of working in either Engineering English or SI units.
56. Repeat Problem 55, except allow the user to choose the system ideal gas from a screen menu and omit the prompts for gas properties. Use the data in Table C.13 of *Thermodynamic Tables to accompany Modern Engineering Thermodynamics* for the properties of the gases in your menu.
57.* Develop a program that allows you to plot the entropy production rate due to the heat transfer from the fin in Example 8.12 vs. the base temperature of the fin T_f. Allow the T_f to range from 20.0 to 200.°C. Keep all the remaining variables constant.
58.* The temperature profile for the fin discussed in Example 8.12 is for a "very long" (i.e., infinite) fin. A more accurate equation for a finite fin of length L is

$$T(x) = T_\infty + (T_f - T_\infty)\left\{\frac{\cosh[m(L-x)] + [h/(mk)]\,\sinh[m(L-x)]}{\cosh(mL) + [h/(mk)]\,\sinh(mL)}\right\}$$

where the remaining variables are defined in Example 8.12. Using this temperature profile, rework Example 8.12 and Problem 57 to produce a new plot of entropy production rate vs. fin base temperature (you may wish to use a numerical integration technique here). Which has the smaller entropy production rate, the infinite fin or the finite fin?

59.* In Example 8.13, the fluid in the gap between the cylinders was maintained isothermal. If, instead, the outer and inner cylinder surfaces are maintained isothermal at temperatures T_2 and T_1, respectively, and the gap between the cylinders is very small, then the fluid in the gap is not isothermal but has a temperature and velocity profile given by

$$T(x) = T_1 + (T_2 - T_1)[1 + \text{Br}/2)][1 - (x - R_1)/t](x - R_1)/t$$
$$V(x) = R_1\omega(R_2 - x)/t$$

where $t = R_2 - R_1$ and Br is the dimensionless Brinkman number,[2]

$$\text{Br} = \mu R_1^2\omega^2/[k_t(T_2 - T_1)]$$

Rework Example 8.13 using these temperature and velocity profiles with T_1 = 40.0°C, T_2 = 20.0°C, and k_t = 0.130 W/(m·K). Use the values given in Example 8.13 for the remaining variables. Since the resulting integrals involve a considerable amount of algebraic manipulation, you may wish to use a numerical integration technique.

[2] The Brinkman number is named after H. C. Brinkman, who solved the equations for the flow of a fluid with viscous heat generation in a circular tube in 1951. This dimensionless number is approximately the ratio of the viscous heat generation rate to the rate of conduction heat transfer due to the imposed temperature difference, $T_2 - T_1$. Note that there is a maximum in the temperature profile between the two cylinders if $|\text{Br}| > 2$.

CHAPTER 9

Second Law Open System Applications

CONTENTS

9.1 INTRODUCTION

This chapter is an extension of Chapter 6, except here we use both the first and second laws of thermodynamics to analyze open systems. As in Chapter 8, we have two types of system processes to consider, reversible and irreversible. A reversible process is easier to analyze, because its entropy production is always equal to zero. However, reversible process models are often unrealistic in actual engineering applications because they require that the system have no losses (i.e., no friction or no heat transfer through a finite temperature difference etc.). On the other hand, irreversible process models are very realistic, because they take into account all the losses within the system, but they are often very difficult to analyze because of the complex entropy production terms that must be evaluated. We are therefore faced with the choice of carrying out a quick but possibly inaccurate analysis based on hypothetical reversible processes or a more complex but accurate analysis based on real irreversible processes. The material presented in this chapter focuses primarily on the latter by utilizing the appropriate entropy production formulae developed in Chapter 7.

Since we have not yet considered the impact of flow streams on the general entropy balance, we must now introduce the mass flow transport and production of entropy characteristic of open systems.

9.2 MASS FLOW TRANSPORT OF ENTROPY

Mass flow transport of entropy occurs every time mass crosses the system boundary. Every element of mass *dm* is assumed to be in local equilibrium and hence has a well-defined specific entropy *s*. Therefore, *dm* transports an amount of entropy *sdm* when it crosses a system boundary, and we can set

$$(dS_T)_m = s\,dm$$

Modern Engineering Thermodynamics

and

$$(S_T)_m = \int s\,dm$$

This equation can also be written as

$$(S_T)_m = \int s\,dm = \int s\left(\frac{dm}{dt}\right)dt = \int \dot{m}s\,dt$$

where $\dot{m}$ is the mass flow rate crossing the system boundary. Differentiation of this equation with respect to time yields the mass transport rate term as

$$(\dot{S}_T)_m = \dot{m}s$$

Unlike heat and work transports of entropy, it is customary in mass flow transport to write a more explicit formula for the *net* mass flow entropy transport rate as

Entropy transport rate due to mass flow

$$\left[(\dot{S}_T)_m\right]_{\text{net}} = \sum_{\text{in}} \dot{m}s - \sum_{\text{out}} \dot{m}s \tag{9.1}$$

where the summations are over all inlet and outlet flow streams (the flow streams normally are pipes or ducts that convey mass into or out of the system).

9.3 MASS FLOW PRODUCTION OF ENTROPY

Mass flow entropy production is due to mass flow that occurs *inside* the system. No entropy production is due solely to mass crossing the system boundary, because it is an imaginary boundary of zero thickness. The two main sources for this type of internal mass flow entropy production are viscous dissipation and diffusion of dissimilar chemical species.

Viscous dissipation is really a work mode entropy production mechanism and has already been treated in Eqs. (7.71) and (7.72). Diffusion of dissimilar chemical species inside the system is an advanced topic not treated in this text. Consequently, if we neglect diffusion, there is no mass flow production of entropy inside a system, and

$$(\dot{S}_P)_m = 0$$

9.4 OPEN SYSTEM ENTROPY BALANCE EQUATIONS

The general entropy rate balance given by Eq. (7.3) equates the net rate of gain of total entropy by a system $\dot{S}_G$ to the sum of the net entropy transport rate into the system $\dot{S}_T$ plus the net entropy production rate within the system $\dot{S}_P$, or

$$\dot{S}_T + \dot{S}_P = \dot{S}_G \tag{7.3}$$

where

$$\dot{S}_G = \dot{S}_{\text{system}} = \left(\frac{dS}{dt}\right)_{\text{system}}$$

and the transport and production rates of entropy have been identified as caused by heat transfer, work, and mass flow, or

$$\dot{S}_T = (\dot{S}_T)_Q + (\dot{S}_T)_W + (\dot{S}_T)_m \tag{9.2}$$

and

$$\dot{S}_P = (\dot{S}_P)_Q + (\dot{S}_P)_W + (\dot{S}_P)_m \tag{9.3}$$

where $(\dot{S}_P)_m = 0$, since we are neglecting diffusion of dissimilar species inside the system. The individual entropy transport rate terms are given in Eqs. (7.61), (7.63), and (9.1) as

$$(\dot{S}_T)_Q = \int_\Sigma \left(\frac{\dot{q}}{T_b}\right)_{\text{act}} dA \tag{7.61}$$

$$(\dot{S}_T)_W = 0 \tag{7.63}$$

$$[(\dot{S}_T)_m]_{\text{net}} = \sum_{\text{in}} \dot{m}s - \sum_{\text{out}} \dot{m}s \tag{9.1}$$

Using the concept of entropy production rate per unit volume, σ, we can express the total entropy production rate of Eq. (9.3) (neglecting $(\dot{S}_P)_m$) as

$$\dot{S}_P = (\dot{S}_P)_Q + (\dot{S}_P)_W = \int_{\mathcal{V}} (\sigma_Q + \sigma_W) d\mathcal{V} = \int_{\mathcal{V}} \sigma d\mathcal{V}$$

where the total entropy production per unit volume is $\sigma = \sigma_Q + \sigma_W$. The relation for σ_Q is given by Eq. (7.64) as

$$\sigma_Q = -\left[\frac{\dot{q}}{T^2}\left(\frac{dT}{dx}\right)\right]_{\text{actual}} \tag{7.64}$$

and the equations for σ_W for viscous flow and electrical resistance are given by Eqs. (7.70) and (7.73) as

$$(\sigma_W)_{\text{vis}} = \frac{\mu}{T}\left(\frac{dV}{dx}\right)^2 \quad \text{and} \quad (\sigma_W)_{\text{elect}} = \frac{J_e^2 \rho_e}{T}$$

Putting Eqs. (7.61) and (9.1) into Eq. (7.3) produces a general open system entropy rate balance (SRB) equation as

General open system entropy rate balance (SRB)

$$\int_\Sigma \left(\frac{\dot{q}}{T_b}\right)_{\text{act}} dA + \sum_{\text{inlet}} \dot{m}s - \sum_{\text{outlet}} \dot{m}s + \dot{S}_P = \dot{S}_{\text{system}} \tag{9.4}$$

and, when the system boundaries are isothermal (i.e., T_b is constant), this equation reduces to

Isothermal boundary open system entropy rate balance

$$\left(\frac{\dot{Q}}{T_b}\right)_{\text{act}} + \sum_{\text{in}} \dot{m}s - \sum_{\text{out}} \dot{m}s + \dot{S}_p = \dot{S}_{\text{system}} \tag{9.5}$$

where $\dot{Q}/T_b = \int_\Sigma (\dot{q}/T_b) dA$. In this equation, $\dot{Q}$ is the *actual* heat transfer rate, and T_b is the temperature of the system boundary where $\dot{Q}$ occurs. The equation for the *direct method* for determining the entropy production rate is given by Eq. (9.6) as[1]

$$\dot{S}_P = \int_{\mathcal{V}} \sigma \, d\mathcal{V} > 0 \tag{9.6}$$

where σ is the total entropy production rate density (EPRD) for the system, given by

$$\sigma = \sigma_Q + (\sigma_W)_{\text{vis}} + (\sigma_W)_{\text{elect}} + \cdots$$

However, since the use of Eq. (9.6) for the direct determination of $\dot{S}_P$ is often quite difficult, the example problems presented in this chapter use the *indirect method* for determining the entropy production rate ($\dot{S}_P$) in the entropy rate balance, Eq. (9.5). When this is done, the entropy rate balance can be used to generate

---

[1] In this chapter, we assume that the processes of interest are truly irreversible, so we write $\dot{S}_P > 0$, rather than the less restrictive $\dot{S}_P \geq 0$.

values for only the entropy production rate, $\dot{S}_P$, and nothing else. This is the major disadvantage of the *indirect method,* since if we could determine $\dot{S}_P$ directly from Eq. (9.6), we could use the entropy rate balance to determine other important information about the system, such as $\dot{Q}$, T_b, $\dot{m}$, s_{in}, and s_{out}. Determining the value of $\dot{S}_P$ (by either method) provides us with a measure of how well or how poorly a system is operating. Large $\dot{S}_P$ values may indicate excessive losses within a particular system or a particularly inefficient mode of operation. Some of the examples in this chapter explore the possibility of reducing $\dot{S}_P$ values through alternative realistic processes that produce the same system operating goals. These analyses lead to methods of increasing the overall system efficiency (and thus reduce operating costs) by using processes that dissipate less useful energy as "losses."

In Chapter 6, we introduce the modified energy balance (MEB) and the modified energy rate balance (MERB), so that we need not continually deal with the complex general open system first law formulae. These formulae were designed to work only for steady state, steady flow, single-inlet, single-outlet systems, but the conditions fit most of the applications we were interested in analyzing. When one or more of these four conditions did not exist in a particular problem, we carried out the analysis by reverting to the complete, accurate, general energy rate balance equation.

The general entropy rate balance given in Eq. (9.5) is not as mathematically complex as the general energy rate balance given in Eq. (6.4), but we still find it convenient to develop similar modified entropy balance and modified entropy rate balance equations.

The MSB and MSRB formulae require the four conditions used in the MEB and MERB formulae plus one more condition. In the SRB, the steady state condition requires that

$$\textbf{Steady state: } \dot{S}_{\text{system}} = \left(\frac{dS}{dt}\right)_{\text{system}} = 0 \tag{9.7}$$

and the steady flow condition requires that

$$\textbf{Steady flow: } \sum_{\text{inlet}} \dot{m} = \sum_{\text{outlet}} \dot{m} \tag{9.8}$$

finally, the steady flow, single-inlet, single-outlet condition requires that

$$\textbf{Single-inlet, single-outlet: } \sum_{\text{inlet}} \dot{m} = \sum_{\text{outlet}} \dot{m} = \dot{m}$$

In addition to these four conditions, we add the fifth condition of *isothermal boundaries* at all points along the system boundary where heat transport of energy occurs. Under this condition, the entropy transport term due to the heat transport of energy becomes

$$\int_{\Sigma}\left(\frac{\dot{q}}{T_b}\right)_{\text{act}} dA = \sum_{\Sigma}\left(\frac{\dot{Q}}{T_b}\right)_{\text{act}} = \frac{\dot{Q}}{T_b} \tag{9.9}$$

In this equation, the simplified notation $\dot{Q}/T_b$ is used to describe the *net* (or total) value of the "actual" $\dot{Q}/T_b$ summed over the entire system boundary Σ. This simplification is also used in the EB and ERB equations, where Q, W, $\dot{Q}$, and $\dot{W}$ are used to represent their net (or total) values (e.g., see Eq. (6.5)). The *act* subscript and the summation sign has been dropped in the last term of Eq. (9.9) to simplify the notation, but it must always be considered to be present. Under these five restrictive conditions, the general open system entropy rate balance of Eq. (9.4) becomes the modified entropy rate balance (MSRB), defined as

Modified entropy rate balance

$$\frac{\dot{Q}}{T_b} + \dot{m}(s_{in} - s_{out}) + \dot{S}_P = 0 \tag{9.10}$$

Multiplying this equation through by dt and integrating over time from system state 1 to state 2 gives the open system modified entropy balance (MSB) equation as

Modified entropy rate balance

$$\frac{{}_1Q_2}{T_b} + \int_1^2 \dot{m}(s_{in} - s_{out})dt + {}_1(S_P)_2 = 0 \tag{9.11}$$

EXAMPLE 9.1

You are now an attorney in a patent office. An inventor wants to patent a new domestic hot water heater. The inventor claims to have a secret process that heats liquid water from 15.0°C to 50.0°C using only 100. W of electrical energy. Evaluate the inventor's claim and decide whether or not to issue a patent.

Solution

First, draw a sketch of the system (Figure 9.1).

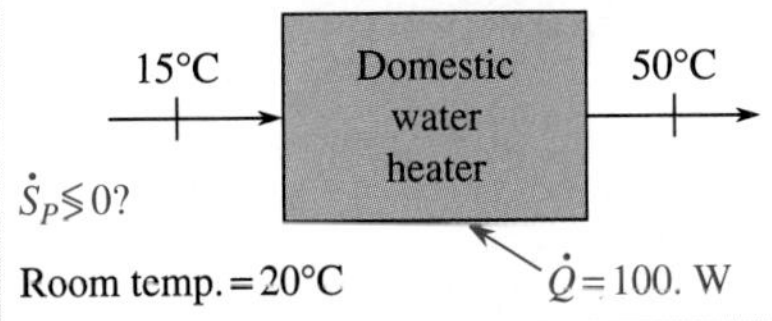

FIGURE 9.1
Example 9.1.

Since the water heater is a steady state open system with an isothermal boundary, we can provide a preliminary evaluation of the inventor's claim with Eq. (9.10). If this equation produces a positive entropy production rate, then the claim is at least possible. If it produces a negative entropy production rate, then we can say without equivocation that the claim is impossible to achieve.

Assumptions: (1) all the input electrical energy is converted into heating the water, (2) the system boundary temperature is the same as the ambient temperature, or 20°C.

Since the inventor does not provide us with the water flow rate, we can calculate it from the modified energy rate balance as $\dot{m} = \dot{Q}/(h_2 - h_1)$, where for an incompressible liquid, $h_2 - h_1 = c(T_2 - T_1) + v(p_2 - p_1)$. Neglecting the pressure drop across the water heater, we can now compute the expected water flow rate as

$$\dot{m} = \frac{\dot{Q}}{h_2 - h_1} = \frac{0.100\text{ kJ/s}}{(4.186\text{ kJ/kg}\cdot\text{K})[50 + 273.15 - (15 + 273.15)\text{ K}]} = 0.000686\text{ kg/s}$$

Not a very fast water flow rate. For liquid water, $s_{out} - s_{in} = c\ln(T_{out}/T_{in})$, so

$$s_{out} - s_{in} = c\ln(T_{out}/T_{in}) = (4.186\text{ kJ/kg}\cdot\text{K})\ln\left(\frac{50.0 + 273.15}{15.0 + 273.15}\right) = 0.480\text{ kJ/kg}\cdot\text{K}$$

Then, Eq. (9.10) gives

$$\dot{S}_P = \dot{m}(s_{out} - s_{in}) - \frac{\dot{Q}}{T_b} = (0.000686\text{ kg/s})(0.480\text{ kJ/kg}\cdot\text{K}) - \frac{0.100\text{ kJ/s}}{20.0 + 273.15\text{ K}} = -1.36 \times 10^{-5}\text{ kJ/s}\cdot\text{K}$$

Since the entropy production rate is negative, this water heater cannot possibly meet the claims of the inventor, so we should reject the patent application.

HOW CAN YOU HEAT THE WATER IN EXAMPLE 9.1 IF THE SYSTEM BOUNDARY TEMPERATURE IS ONLY 20°C?

Doesn't the boundary have to be higher than the fluid temperature for heat to go into the water? The boundary temperature in Eq. (9.10) can be the "average" boundary temperature, but if 20°C is the "average" boundary temperature, then 20°C = $(T_{max} + 15°C)/2$, and the maximum boundary temperature is T_{max} = 2(20) – 15 = 25°C, and that is not hot enough to heat the water to 50°C.

This is what happens when you make incorrect assumptions. The average boundary temperature cannot be 20°C, it has to be higher. Suppose the inventor now tells us that it is 150°C. Then, we get

$$\dot{S}_P = (0.000686\text{ kg/s})(0.480\text{ kJ/kg}\cdot\text{K}) - \frac{0.100\text{ kJ/s}}{150.0 + 273.15\text{ K}} = 9.12 \times 10^{-5}\text{ kJ/s}\cdot\text{K}$$

The patent application is now alright, since the entropy production rate is positive. But the water flow rate here is only 6.86×10^{-4} kg/s = 0.686 grams per second, or 42.2 grams per minute, or 2.47 kg per hour. Not a very effective water heater.

Exercises

1. Suppose the inventor in Example 9.1 corrected his patent claim and said the heater was 1000. W instead of 100. W and the heat transfer boundary temperature was 150°C instead of 20°C. What would be the water flow rate end entropy production rate under these conditions? **Answer:** $\dot{m}$ = 6.83 grams per second and $\dot{S}_P = 9.12 \times 10^{-4}$ kJ/s·K.
2. OK, so now the inventor says the heater is 10.0 kW, the heater boundary temperature is 150°C and the water flow rate is 0.500 kg/s. What are the water outlet temperature and the unit's entropy production rate? **Answer:** T_{out} = 19.8°C and $\dot{S}_P = 0.0108$ kJ/s·K.
3. Alright, now the inventor hires an engineer to determine the heat transfer rate and entropy production rate needed to heat 0.500 kg/s from 15°C to 50°C. You are the engineer, so what are the answers? Hint: Is $\dot{Q}$ (a) 20.9 kW, (b) 73.3 kW, or (c) 103. kW? Is $\dot{S}_P$ (a) 0.0220 kJ/s · K, (b) 0.137 kJ/s · K, or (c) 0.0669 kJ/s · K?

Note that the entropy production rate in Example 9.1 is actually positive if there is no heat transfer into the water ($\dot{Q} = 0$). This is because the water temperature could increase from 15°C to 50°C by internal viscous friction alone. However, that would require a really long pipe inside the water heater and a pretty big pressure drop (which we neglected in the solution).

9.5 NOZZLES, DIFFUSERS, AND THROTTLES

In Chapter 6, we see that nozzles, diffusers, and throttling devices are normally steady state, steady flow, single-inlet, single-outlet open systems with approximately constant surface temperature and they may or may not be adiabatic. Therefore, Eq. (9.10) can be applied to all three of these types of open systems. Solving Eq. (9.10) for $\dot{S}_P$ gives

1. For *adiabatic* nozzles, diffusers, and throttling devices:

$$\dot{S}_P = \dot{m}(s_{out} - s_{in}) > 0 \tag{9.12}$$

2. For *isothermal surface* nozzles, diffusers, and throttling devices:

$$\dot{S}_P = \dot{m}(s_{out} - s_{in}) - \frac{\dot{Q}}{T_b} > 0 \tag{9.13}$$

where the second law condition that $\dot{S}_P > 0$ has been added.

If the fluid flowing through these systems is incompressible and has a constant specific heat c, then from Chapter 7, we have

$$s_{out} - s_{in} = c \ln \frac{T_{out}}{T_{in}} \tag{7.33}$$

Equations (9.12) and (9.13) then become

Entropy production rate of an incompressible fluid in an adiabatic nozzle, diffuser, or throttle

$$\dot{S}_P \Big|_{\substack{\text{adiabatic} \\ \text{incompressible} \\ \text{fluid}}} = \dot{m} c \ln \frac{T_{out}}{T_{in}} > 0 \tag{9.14}$$

and

Entropy production rate of an incompressible fluid in a nozzle, diffuser, or throttle with heat transfer

$$\dot{S}_P \Big|_{\substack{\text{incompressible} \\ \text{fluid}}} = \dot{m} c \ln \frac{T_{out}}{T_{in}} - \frac{\dot{Q}}{T_b} > 0 \tag{9.15}$$

Equation (9.14) shows us that the outlet temperature must always be greater than the inlet temperature for an insulated (adiabatic) open system with an incompressible fluid. This is because all the dissipation due to the irreversibilities within the system simply goes into increasing the temperature of an incompressible fluid.

Nozzles, diffusers, and throttling devices are all physically small (i.e., $Z_{out} \approx Z_{in}$), aergonic systems, so their modified energy rate balance becomes

$$\dot{Q} = \dot{m}\left(h_{out} - h_{in} + \frac{V_{out}^2 - V_{in}^2}{2g_c}\right) \tag{9.16}$$

and, for constant specific heat incompressible fluids, we have

$$h_{out} - h_{in} = c(T_{out} - T_{in}) + v(p_{out} - p_{in}) \tag{6.19}$$

Combining Eqs. (9.15), (9.16), and (6.19) gives the combined nonadiabatic first and second law relation for an incompressible fluid with a constant specific heat as

Equation (9.15) with the heat transfer rate evaluated using the ERB

$$\dot{S}_P \Big|_{\substack{\text{incompressible} \\ \text{fluid}}} = \dot{m}\left[c \ln \frac{T_{out}}{T_{in}} - \frac{c(T_{out} - T_{in})}{T_b} - \frac{v(p_{out} - p_{in})}{T_b} - \frac{V_{out}^2 - V_{in}^2}{2g_c T_b}\right] \tag{9.17}$$

Notice that Eq. (9.17) is now written completely in terms of directly measurable physical quantities (m, c, T, v, p, and V).

Similarly, if the fluid flowing through these devices is an ideal gas with constant specific heats c_p and c_v, then from Chapter 7, we have

$$s_{out} - s_{in} = c_v \ln\frac{T_{out}}{T_{in}} + R\ln\frac{v_{out}}{v_{in}} \tag{7.36}$$

$$= c_p \ln\frac{T_{out}}{T_{in}} - R\ln\frac{p_{out}}{p_{in}} \tag{7.37}$$

and, from Chapter 6, we have

$$h_{out} - h_{in} = c_p(T_{out} - T_{in}) \tag{6.22}$$

Combining Eqs. (9.12) and (7.37) gives the adiabatic modified energy rate balance equation for an ideal gas with constant specific heats as

Entropy production rate of an ideal gas in an adiabatic nozzle, diffuser, or throttle

$$\dot{S}_P\Big|_{\substack{\text{adiabatic}\\ \text{ideal gas}}} = \dot{m}\left(c_p\ln\frac{T_{out}}{T_{in}} - R\ln\frac{p_{out}}{p_{in}}\right) > 0 \tag{9.18}$$

In the case of diffusers, $p_{out} > p_{in}$ and Eq. (9.18) requires that $T_{out} > T_{in}$, as in the case of incompressible fluid flow. However, for nozzles and throttling devices, $p_{out} < p_{in}$ and T_{out} can be either greater or less than T_{in}, depending on the value of the Joule-Thomson coefficient (see Eq. (6.25)). Combining Eqs. (9.13), (9.16), (7.37), and (6.22) gives the combined nonadiabatic first and second law relation for an ideal gas with constant specific heats as

Equation (9.18) with the heat transfer rate evaluated using the ERB

$$\dot{S}_P\Big|_{\text{ideal gas}} = \dot{m}\left(c_p\ln\frac{T_{out}}{T_{in}} - R\ln\frac{p_{out}}{p_{in}} - c_p\frac{T_{out} - T_{in}}{T_b} - \frac{V_{out}^2 - V_{in}^2}{2g_cT_b}\right) > 0 \tag{9.19}$$

Finally, if the fluid flowing through these devices is neither an incompressible fluid nor an ideal gas, then Eqs. (9.13) and (9.16) can still be combined and rearranged to give a combined nonadiabatic first and second law relation of the form

The entropy production rate for a general fluid in a nozzle, diffuser, or throttle

$$\dot{S}_P = -\frac{\dot{m}}{T_b}\left[(h_{out} - T_b s_{out}) - (h_{in} - T_b s_{in}) + \frac{V_{out}^2 - V_{in}^2}{2g_c}\right] > 0 \tag{9.20}$$

EXAMPLE 9.2

Determine the rate of entropy production as 0.2000 lbm/s of liquid water at 50.00°F, 95.00 psia flows through the nozzle on the end of a garden hose and exits at 14.70 psia. The inlet and outlet diameters of the nozzle are 1.000 and 0.2500 in., respectively. Assume that the flow through the nozzle is too fast to allow a significant heat transfer to occur.[2]

Solution

First, draw a sketch of the system (Figure 9.2).

The unknown is the rate of entropy production ($\dot{S}_P = ?$) for this system. The material is liquid water and the thermodynamic station conditions are:

Station 1	Nozzle process →	Station 2
$p_1 = 95.00$ psia		$p_2 = 14.70$ psia
$T_1 = 50.00$°F		

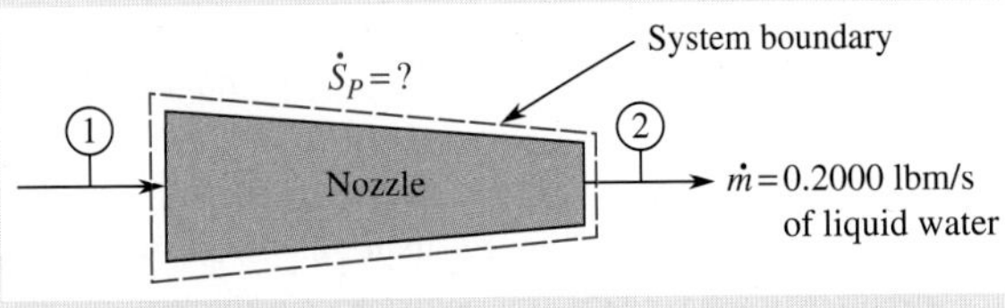

FIGURE 9.2
Example 9.2.

(Continued)

EXAMPLE 9.2 *(Continued)*

The modified energy rate balance equation for this system is

$$\underset{0}{\dot{Q}} - \underset{0}{\dot{W}} + \dot{m}\left[h_1 - h_2 + (V_1^2 - V_2^2)/2g_c + \underset{0}{\underbrace{g(Z_1 - Z_2)/g_c}}\right] = 0$$

Assuming liquid water is incompressible with a constant specific heat under these conditions and using the incompressible liquid auxiliary equation for enthalpy (Eq. (6.19)) allows the preceding modifies energy rate balance equation to be written as

$$h_2 - h_1 = c(T_2 - T_1) + v(p_2 - p_1) = (V_1^2 - V_2^2)/2g_c$$

or

$$T_2 = T_1 + \frac{v}{c}(p_1 - p_2) - \frac{(V_2^2 - V_1^2)}{2cg_c}$$

where $v = v_f$ (at 50.0°F) = 0.01602 ft^3/lbm (from Table C.1a of *Thermodynamic Tables to accompany Modern Engineering Thermodynamics*). From the data given in the problem statement, we can compute V_1 and V_2 as

$$V_1 = \frac{\dot{m}}{\rho A_1} = \frac{4\dot{m}v}{\pi D_1^2} = \frac{4(0.2000\text{ lbm/s})(0.01602\text{ ft}^3/\text{lbm})(144\text{ in.}^2/\text{ft}^2)}{\pi(1.000\text{ in.})^2} = 0.5874\text{ ft/s}$$

and

$$V_2 = \frac{4\dot{m}v}{\pi D_2^2} = V_1\left(\frac{D_1}{D_2}\right)^2 = (0.5874)\left(\frac{1.000}{0.2500}\right)^2 = 9.399\text{ ft/s}$$

Then,

$$T_2 = (50.00 + 459.67\text{ R}) + (0.01602\text{ ft}^3/\text{lbm})\left[\frac{(95.00 - 14.70\text{ lbf/in.}^2)(144\text{ in.}^2/\text{ft}^2)}{[1.0\text{ Btu/(lbm}\cdot\text{R)}](778.17\text{ ft}\cdot\text{lbf/Btu})}\right]$$
$$- \frac{(9.399)^2 - (0.5874)^2\text{ft}^2/\text{s}^2}{2[1.0\text{ Btu/(lbm}\cdot\text{R)}][32.174\text{ lbm}\cdot\text{ft/(lbf}\cdot\text{s}^2)](778.17\text{ ft}\cdot\text{lbf/Btu})}$$
$$= 509.9 - 0.0018\text{ R} = 509.9\text{ R}$$

and Eq. (9.14) gives[3]

$$\dot{S}_P = \dot{m}c\ln\frac{T_2}{T_1} = (0.2000\text{ lbm/s})[1.000\text{ Btu/(lbm}\cdot\text{R)}]\ln\frac{509.9}{509.7}$$
$$= [7.846\times10^{-5}\text{ Btu/(s}\cdot\text{R)}](778.17\text{ ft}\cdot\text{lbf/Btu}) = 0.0611\text{ ft}\cdot\text{lbf/(s}\cdot\text{R)}$$

Notice that the kinetic energy term in this example (0.0018 R) provides a negligible contribution to the exit temperature and the vast majority of the entropy production results from the pressure loss across the nozzle. The increase in velocity across the nozzle as converted pressure energy does decrease the entropy production rate slightly (but only by less than 1% in this case). Therefore, this nozzle is quite inefficient at converting pressure energy into kinetic energy. Its efficiency could be improved, however, by making the nozzle outlet diameter smaller, so that the outlet velocity is substantially increased.

Exercises

4. Determine the entropy production rate in Example 9.2 if the mass flow rate is increased from 0.2000 to 0.8000 lbm/s. Recalculate the values of V_1, V_2, and T_2, then keep the values of all the remaining variables the same as they are in Example 9.2. **Answer:** $\dot{S}_P = 0.288$ ft·lbf/(s·R).
5. Determine the entropy production rate in Example 9.2 if the water hose has been lying in the sun and the water temperature is 80.00°F rather than 50.00°F. Using $v = v_{sat}$ (80°F), recalculate V_1, V_2, and T_2, then keep the values of all the remaining variables the same as they are in Example 9.2. **Answer:** $\dot{S}_P = 0.068$ ft·lbf/(s·R).
6. The exit diameter on the nozzle in Example 9.2 is reduced to 0.1250 in. Determine the new entropy production rate for this system. Keep the values of all the variables except V_2 and T_2 the same as they are in Example 9.2. **Answer:** $\dot{S}_P = 0.064$ ft·lbf/(s·R).

---

[2] *To get a meaningful answer, this problem needs to be specified to four significant figures.*
[3] *Note that, without carrying four significant figures in this example, the logarithm is zero.*

EXAMPLE 9.3

Suppose 0.800 kg/s of argon flows at 93.0 m/s through an insulated diffuser from 97.0 kPa, 80.0°C to 101.3 kPa. Assuming the argon to be an ideal gas with constant specific heats, determine the rate of entropy production within the diffuser.

Solution

First, draw a sketch of the system (Figure 9.3).

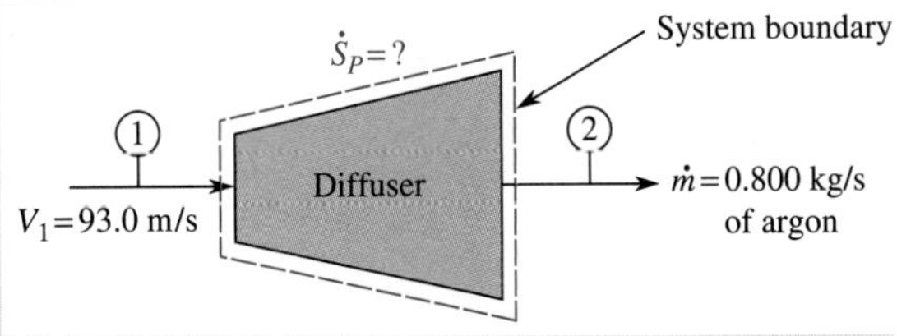

FIGURE 9.3
Example 9.3.

The unknown is $\dot{S}_P = ?$, and the material is argon gas. The thermodynamic station conditions are

Station 1	Diffuser process →	Station 2
$p_1 = 97.0$ kPa		$p_2 = 101.3$ kPa
$T_1 = 80.0$°C		

The modified energy rate balance equation for this system is

$$\dot{Q} - \dot{W} = 0 = \dot{m}\left(h_2 - h_1 + \frac{V_2^2 - V_1^2}{2g_c}\right)$$

Assuming $V_2 \approx 0$, and using the ideal gas auxiliary formula (Eq. (6.22)) with data for argon from Table C.13b, we find that

$$T_2 = T_1 + \frac{V_1^2}{2g_c c_p} = (80.0 + 273.15\text{ K}) + \frac{(93.0\text{ m/s})^2}{2(1)[523\text{ J/(kg·K)}]}$$
$$= 353.15 + 8.27 = 361.42\text{ K}$$

Then, Eq. (9.18) gives

$$\dot{S}_P = \dot{m}\left(c_p \ln\frac{T_2}{T_1} - R\ln\frac{p_2}{p_1}\right) = (0.800\text{ kg/s})\left[[523\text{ J/(kg·K)}]\ln\frac{361.42}{353.15} - [208\text{ J/(kg·K)}]\ln\left(\frac{101.3}{97.0}\right)\right]$$
$$= 2.47\text{ J/(s·K)} = 2.47\text{ W/K}$$

In this example, both the pressure and the temperature of the gas increase as it passes through the diffuser.

Exercises

7. Determine the entropy production rate in Example 9.3 when the mass flow rate of the argon is increased from 0.800 kg/s to 1.30 kg/s. Keep the values of all the other variables the same as they are in Example 9.3. **Answer:** $\dot{S}_P = 4.01$ W/K.
8. Determine the entropy production rate in Example 9.3 as the inlet velocity is increased from 93.0 m/s to 155 m/s. Keep the values of all the other variables the same as they are in Example 9.3. **Answer:** $\dot{S}_P = 19.1$ W/K.
9. Could the gas in Example 9.3 be changed from argon to air, keeping the values of all the other variables (except c_p and R) the same as they are in Example 9.3? Hint: Check the sign of $\dot{S}_P$. **Answer:** No. When the values of c_p and R for air are used along with the values of the other variables given in Example 9.3, Eq. (9.18) gives $\dot{S}_P = -0.187$. Since $\dot{S}_P < 0$ in this case, the process as described in Example 9.3 with air replacing argon cannot occur.

EXAMPLE 9.4

Suppose 0.100 lbm/s of Refrigerant-134a is throttled across the expansion valve in a refrigeration unit. The R-134a enters the valve as a saturated liquid at 100.°F and exits at 20.0°F with a quality of 53.0%. If the inlet and exit velocities are equal, determine

a. The entropy production rate inside the valve if the valve is *not* insulated and has an isothermal external surface temperature of 60.0°F.
b. The entropy production rate inside the valve if it *is* insulated and assuming it has the same inlet conditions and exit temperature as just stated.
c. The percent decrease in the entropy production rate of part a brought about by adding the insulation in part b.

Solution

First, draw a sketch of the system (Figure 9.4).

The unknowns are $(\dot{S}_P)_{\text{uninsulated}}$, $(\dot{S}_P)_{\text{insulated}}$, and the percent decrease in $\dot{S}_P$ due to the insulation. The material is R-134a. The thermodynamic station conditions are

Station 1	Throttling process →	Station 2
$x_1 = 0.00$ kPa		$x_2 = 0.530$ (only in part a)
$T_1 = 100.$°F		$T_2 = 20.0$°F

(Continued)

EXAMPLE 9.4 (*Continued*)

a. The MERB for this aergonic device with negligible change in kinetic and potential energy is

$$\dot{Q} - 0 = \dot{m}(h_2 - h_1 + 0)$$

and, from Table C.7a for R-134a, we find

$$h_1 = h_f(100°\text{F}) = 44.23 \text{ Btu/lbm}$$

$$s_1 = s_f(100°\text{F}) = 0.0898 \text{ Btu/(lbm}\cdot\text{R)}$$

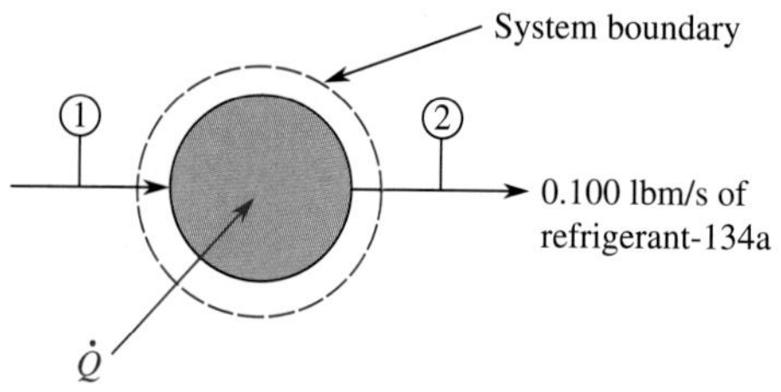

FIGURE 9.4
Example 9.4.

and

$$h_2 = h_f(20.0°\text{F}) + x_2 h_{fg}(20.0°\text{F})$$
$$= 17.74 + (0.530)(86.87) = 63.78 \text{ Btu/lbm}$$
$$s_2 = s_f(20.0°\text{F}) + x_2 s_{fg}(20.0°\text{F})$$
$$= 0.0393 + (0.530)(0.2206 - 0.0393) = 0.1353 \text{ Btu/(lbm}\cdot\text{R)}$$

Then, the MERB gives

$$\dot{Q} = (0.100 \text{ lbm/s})(63.78 - 44.23 \text{ Btu/lbm}) = 1.955 \text{ Btu/s}$$

and Eq. (9.13) gives

$$\dot{S}_P = \dot{m}(s_2 - s_1) - \frac{\dot{Q}}{T_b} = (0.100 \text{ lbm/s})[0.1353 - 0.0898 \text{ Btu/(lbm}\cdot\text{R)}]$$
$$- \left(\frac{1.955 \text{ Btu/s}}{60.0 + 459.67 \text{ R}}\right) = 7.88 \times 10^{-4} \text{ Btu/(s}\cdot\text{R)}$$
$$= [7.88 \times 10^{-4} \text{Btu/(s}\cdot\text{R)}](778.17 \text{ ft}\cdot\text{lbf/Btu}) = 0.613 \text{ ft}\cdot\text{lbf/(s}\cdot\text{R)}$$

b. The MERB for this device as an adiabatic, aergonic, negligible change in kinetic and potential energy system is

$$0 - 0 = \dot{m}(h_2 - h_1 + 0)$$

or

$$h_2 = h_1$$

Now, station 2 is fixed by the pair of properties T = 20.0°F and $h_2 = h_1$ = 44.23 Btu/lbm. Consequently, the quality at station 2 cannot be 53.0% but is instead

$$x_2 = \frac{h_2 - h_{f2}}{h_{fg2}} = \frac{44.23 - 17.74}{86.87} = 0.3049 = 30.5\%$$

then,

$$s_2 = 0.0393 + 0.3049(0.2206 - 0.0393) = 0.0946 \text{ Btu/(lbm}\cdot\text{R)}$$

Then, with $\dot{Q} = 0$, Eq. (9.13) gives

$$\dot{S}_P = \dot{m}(s_2 - s_1) - \frac{\dot{Q}}{T_b} = (0.100 \text{ lbm/s})[0.0946 - 0.0898 \text{ Btu/(lbm}\cdot\text{R)}] - 0$$
$$= 4.80 \times 10^{-4} \text{Btu/(s}\cdot\text{R)} = [4.80 \times 10^{-4} \text{Btu/(s}\cdot\text{R)}](778.17 \text{ ft}\cdot\text{lbf/Btu})$$
$$= 0.374 \text{ ft}\cdot\text{lbf/(s}\cdot\text{R)}$$

c. The percentage decrease in $\dot{S}_P$ brought about by adding the insulation is

$$\frac{7.88 \times 10^{-4} \text{ Btu/(s}\cdot\text{R)} - 4.80 \times 10^{-4} \text{ Btu/(s}\cdot\text{R)}}{7.88 \times 10^{-4} \text{ Btu/(s}\cdot\text{R)}} \times 100 = 39.1\%$$

Note that there is a substantial decrease in the entropy production rate of Example 9.4 due to simply insulating the valve. This results from the elimination of the entropy generated by the heat transfer present in the uninsulated valve.

Exercises

10. If the surface temperature of the valve in Example 9.4 is decreased from 60.0°F to 40.0°F, resolve part a to find the new entropy production rate of the valve. Keep the values of all the other variables the same as they are in Example 9.4. **Answer:** $\dot{S}_P = 6.35 \times 10^{-4}$ Btu/(s·R).

11. If the mass flow rate through the expansion valve discussed in Example 9.4 is increased from 0.100 lbm/s to 0.500 lbm/s, determine the new entropy production rate for parts a and b of the example. Keep the values of all the other variables the same as they are in Example 9.4. **Answer:** $(\dot{S}_P)_a = 0.0191$ Btu/(s·R), $(\dot{S}_P)_b = 2.35 \times 10^{-3}$ Btu(s·R).

12. Resolve part a of Example 9.4 if the R-134a exits the valve at 80.0% quality rather than 53.0%, determine the new entropy production rate for the system. Keep the values of all the other variables (except $\dot{Q}$ and s_{out}) the same as they are in Example 9.4. **Answer:** $(\dot{S}_P)_a = 1.16 \times 10^{-3}$ Btu/(s·R).

9.6 HEAT EXCHANGERS

Heat exchangers are discussed briefly in the energy balance examples of Chapter 6. This section expands the earlier material on this subject by introducing some of the basic concepts of heat exchanger design and analysis. Heat exchangers are normally classified as either *parallel flow, counterflow,* or *cross flow,* as shown in Figure 9.5. If both fluids flow in the same direction, it is said to be a parallel flow heat exchanger; if they flow in opposite directions, it is said to be a counterflow heat exchanger; and if they flow at right angles to each other, it is said to be a cross flow heat exchanger.

The two most common types of heat exchangers are *shell and tube* and *plate and tube.* The simplest type of shell and tube heat exchanger is the double-pipe system shown in Figures 9.1a and 9.1b. Figure 9.1c illustrates a simple plate and tube geometry. The efficiency of a shell and tube heat exchanger can be improved

THE WORLD'S LARGEST HORIZONTAL SHAFT HEAT EXCHANGER

Removing sulfur and nitrogen oxides from the combustion products of large industrial or electrical power plant furnaces is important for the preservation of the environment. If the catalytic reduction of nitrogen oxides in the furnace exhaust gas occurs after the desulfurization process, the exhaust gas must be reheated to a temperature of about 320°C. However, most of the thermal energy used to reheat the gas can be recovered using a heat exchanger so that a temperature differential of only about 30°C needs to be produced with an auxiliary heater.

In 1988, the largest horizontal shaft heat exchanger then existing was installed at the Heilbronn electrical power plant in Germany. It has a rotor diameter of 15.5 m, is about 46 m long, and weighs 870 tons. It is a counterflow heat exchanger that handles about 900,000 m^3/h of exhaust gas.

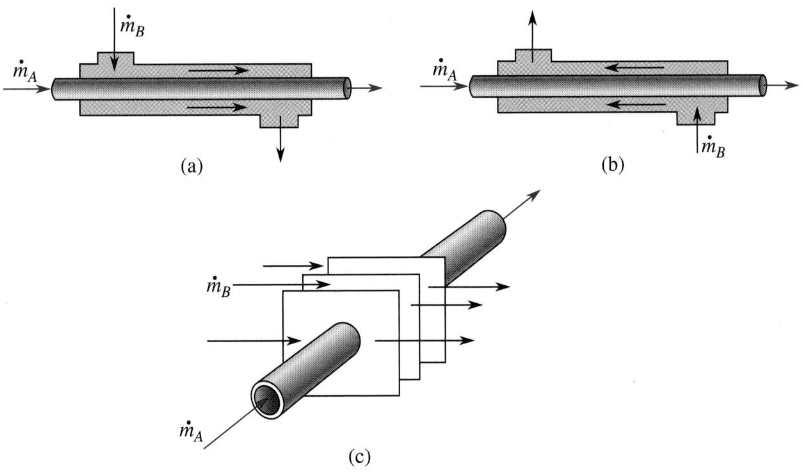

FIGURE 9.5
Single-tube, single-pass heat exchanger geometries: (a) parallel flow; (b) counterflow; (c) cross flow.

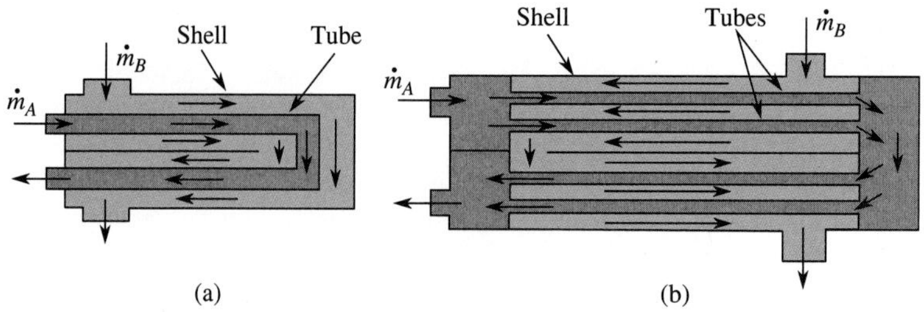

FIGURE 9.6
Multiple-tube, multiple-pass heat exchanger geometries: (a) single-tube, double-pass, parallel flow; (b) double-tube, double-pass, counterflow.

FIGURE 9.7
A cutaway of a single-pass shell and tube heat exchanger.

by using multiple tubes and multiple passes, as shown in Figure 9.6. Figure 9.7 illustrates a typical commercial multiple tube heat exchanger.

The temperature profiles inside single-tube, single-pass heat exchangers are shown in Figure 9.8. In the parallel flow arrangement, the outlet temperature of the cold flow stream can never exceed the outlet temperature of the hot flow stream. However, in the counterflow arrangement, this situation can occur, and consequently the required surface area to produce a given amount of heat transfer is less in the counterflow than the parallel flow configuration.

In heat exchanger design, the basic formula used to determine the internal heat transfer rate for a single-pass heat exchanger is

Internal heat transfer rate for a single-pass heat exchanger

$$\dot{Q}\Big|_{\substack{\text{heat}\\ \text{exchanger}\\ \text{(internal)}}} = UA(\Delta T)_{\text{LMTD}} \tag{9.21}$$

where U is the overall heat transfer coefficient (see Table 9.1), A is the total internal heat transfer area, and $(\Delta T)_{\text{LMTD}}$ is the *log mean temperature difference*, defined for a single-tube, single-pass heat exchanger as

Log mean temperature difference

$$(\Delta T)_{\text{LMTD}} = \frac{(T_H - T_C)|_{x=L} - (T_H - T_C)|_{x=0}}{\ln\left[(T_H - T_C)|_{x=L}/(T_H - T_C)|_{x=0}\right]} \tag{9.22}$$

Heat exchangers are normally two-fluid aergonic devices with dual inlets and dual outlets. If the entire heat exchanger is taken as the system, it is normally adiabatic (the main heat transfer takes place inside the heat exchanger not across its external boundary) and the modified energy rate balance equation for negligible change in kinetic and potential energy reduces to (see Eq. (6.28))

$$\dot{m}_H(h_{\text{in}} - h_{\text{out}})_H = \dot{m}_C(h_{\text{out}} - h_{\text{in}})_C \tag{9.23}$$

and the MSRB for this system yields

Heat exchanger entropy production rate

$$\dot{S}_P = \dot{m}_H(s_{\text{out}} - s_{\text{in}})_H + \dot{m}_C(s_{\text{out}} - s_{\text{in}})_C \tag{9.24}$$

If both fluids are incompressible liquids with constant specific heats, Eqs. (6.19) and (7.33) convert Eqs. (9.23) and (9.24) into

$$\dot{m}_H[c(T_{\text{in}} - T_{\text{out}}) + v(p_{\text{in}} - p_{\text{out}})]_H = \dot{m}_C[c(T_{\text{out}} - T_{\text{in}}) + v(p_{\text{out}} - p_{\text{in}})]_C$$

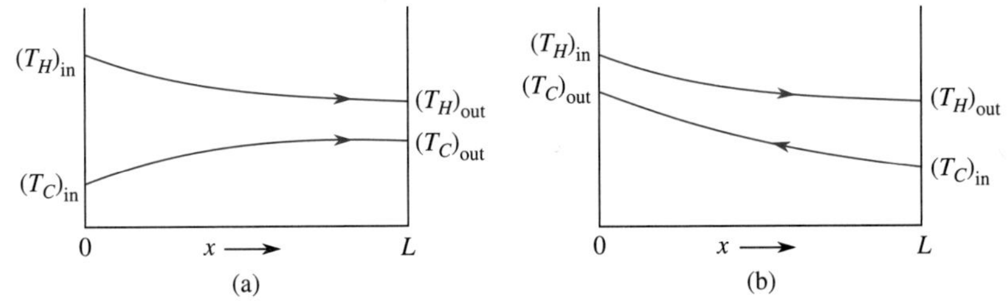

FIGURE 9.8
Temperature profiles inside single-tube, single-pass heat exchangers: (a) parallel flow; (b) counterflow.

TABLE 9.1 Typical Ranges for the Overall Heat Transfer Coefficient (U)

Fluids Used	Btu/(h · ft² · R)	W/(m² · K)
Water and		
Water	200–250	1140–1420
Gasoline	60–100	340–570
Fuel oil	15–25	85–140
Compressed air	10–30	57–170
Steam and		
Water (liquid)	250–400	1420–2270
Fuel oil (light)	60–90	340–510
Fuel oil (heavy)	15–25	85–140
Compressed air	5–50	28–280
Kerosene and		
Water	25–50	140–280
Oil	20–35	110–200

and

Heat exchanger entropy production rate when both fluids are incompressible liquids

$$\dot{S}_P = \dot{m}_H c_H \ln\left(\frac{T_{out}}{T_{in}}\right)_H + \dot{m}_C c_C \ln\left(\frac{T_{out}}{T_{in}}\right)_C \qquad (9.25)$$

And, if both fluids are ideal gases with constant specific heats, then Eqs. (6.22) and (7.37) give

$$\dot{m}_H (c_p)_H (T_{in} - T_{out})_H = \dot{m}_C (c_p)_C (T_{out} - T_{in})_C$$

and

Heat exchanger entropy production rate when both fluids are ideal gases

$$\dot{S}_P = \dot{m}_H\left[(c_p)_H \ln\left(\frac{T_{out}}{T_{in}}\right)_H - R_H \ln\left(\frac{p_{out}}{p_{in}}\right)_H\right] + \dot{m}_C\left[(c_p)_C \ln\left(\frac{T_{out}}{T_{in}}\right)_C - R_C \ln\left(\frac{p_{out}}{p_{in}}\right)_C\right] \qquad (9.26)$$

EXAMPLE 9.5

A single-tube, single-pass heat exchanger is used to cool a compressed air flow of 0.200 kg/s from 90.0 to 75.0°C. The cooling fluid is liquid water that enters the heat exchanger at 20.0°C and leaves at 40.0°C. If the overall heat transfer coefficient is 140. W/(m²·K) and all flow streams have negligible pressure drop, determine the required heat exchanger area and the entropy production rate for (a) parallel flow and (b) counterflow. Assume the compressed air behaves as an ideal gas with constant specific heats.

(Continued)

EXAMPLE 9.5 *(Continued)*

Solution

First, draw a sketch of the system (Figure 9.9).

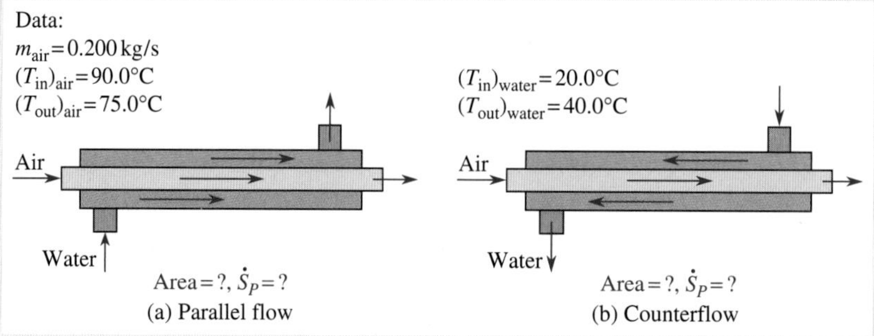

FIGURE 9.9
Example 9.5.

The unknowns are the heat exchanger area and the entropy production rate for parallel flow and counterflow conditions.

The heat exchanger area, from Eq. (9.21), is

$$A = \frac{\dot{Q}}{U(\Delta T)_{\text{LMTD}}}$$

where the log mean temperature difference has different values for the parallel and counterflow arrangements. From Eq. (9.22), we have

a. Parallel flow:

$$(\Delta T)_{\text{LMTD}} = \frac{(75.0-40.0)-(90.0-20.0)}{\ln\dfrac{75.0-40.0}{90.0-20.0}} = 50.5\text{ K}$$

b. Counterflow:

$$(\Delta T)_{\text{LMTD}} = \frac{(75.0-20.0)-(90.0-40.0)}{\ln\dfrac{75.0-20.0}{90.0-40.0}} = 52.5\text{ K}$$

For each case, the heat transfer rate $\dot{Q}$ is obtained by applying the modified energy rate balance equation to only the air flow stream. Then, using the ideal gas assumption and the fact that the value of $\dot{Q}$ must be positive for use in Eq. (9.21),

$$\dot{Q} = |\dot{m}_{\text{air}}(h_{\text{out}}-h_{\text{in}})_{\text{air}}| = |\dot{m}_{\text{air}}(c_p)_{\text{air}}(T_{\text{out}}-T_{\text{in}})_{\text{air}}|$$
$$= |(0.200\text{ kg/s})[1004\text{ J/(kg·K)}](75.0-90.0\text{ K})| = |-3010\text{ J/s}| = 3010\text{ J/s}$$

then, the corresponding heat exchanger areas are

$$A_{\substack{\text{parallel}\\\text{flow}}} = \frac{3010\text{ J/s}}{[140.\text{W/(m}^2\text{·K)}](50.5\text{ K})} = 0.426\text{ m}^2$$

and

$$A_{\text{counterflow}} = \frac{3010\text{ J/s}}{[140.\text{W/(m}^2\text{·K)}](52.5\text{ K})} = 0.410\text{ m}^2$$

Also, in this case, one of the heat transfer fluids is an ideal gas and one is an incompressible liquid. Combining Eqs. (7.33) and (7.37) into Eq. (9.24) with the condition $(p_{\text{in}})_{\text{air}} = (p_{\text{out}})_{\text{air}}$ (i.e., a negligible pressure drop) gives

$$\dot{S}_P = \dot{m}_{\text{air}}(c_p)_{\text{air}}\ln\left(\frac{T_{\text{out}}}{T_{\text{in}}}\right)_{\text{air}} + \dot{m}_{\text{water}}c_{\text{water}}\ln\left(\frac{T_{\text{out}}}{T_{\text{in}}}\right)_{\text{water}} \quad \text{(a)}$$

Now, $\dot{m}_{air}$ is given and $\dot{m}_{water}$ can be found from the modified energy rate balance equation by combining Eq. (6.19) with $(p_{in})_{water} = (p_{out})_{water}$ and Eq. (6.22) with Eq. (9.23) as

$$\dot{m}_{water} = \dot{m}_{air}\left\{\left[(c_p)_{air}/c_{water}\right](T_{in} - T_{out})_{air}/(T_{out} - T_{in})_{water}\right\}$$

Our calculations begin with this last equation:

$$\dot{m}_{water} = (0.2\text{ kg/s})\left[\frac{1.004\text{ kJ/(kg·K)}}{4.186\text{ kJ/(kg·K)}}\right]\left[\frac{(90.0 - 75.0\text{ K})}{(40.0 - 20.0\text{ K})}\right] = 0.036\text{ kg/s}$$

Then, the entropy production rate equation (a) gives

$$\dot{S}_P = (0.200\text{ kg/s})[1004\text{ J/(kg·K)}]\ln\left(\frac{75.0 + 273.15}{90.0 + 273.15}\right)$$

$$+ (0.036\text{ kg/s})[4186\text{ J/(kg·K)}]\ln\left(\frac{40.0 + 273.15}{20.0 + 273.15}\right) = 1.48\text{ W/K}$$

Notice that the entropy production rate in the previous example is independent of whether the heat exchanger is parallel or counterflow, since the same amount of heat transfer occurs in each case. We would see a difference if we had included the effect of the viscous pressure drop in the entropy production rate equation. The counterflow arrangement requires less heat transfer area and therefore produces a smaller pressure drop than the parallel flow arrangement. Then, the counterflow heat exchanger has a smaller entropy production rate than a parallel flow heat exchanger with the same $\dot{Q}$, $U(\Delta T)_H$, and $(\Delta T)_C$ values.

Exercises

13. Determine the entropy production rate in Example 9.5 if the air mass flow rate is increased from 0.200 kg/s to 0.500 kg/s. Keep the values of all the other variables except $\dot{m}_{water}$ the same as they are in Example 9.5. **Answer**: $\dot{S}_P = 3.67\text{ W/K}$.
14. If the cooling water mass flow rate in Example 9.5 is decreased so that it exits the heat exchanger at 60.0°C rather than 40.0°C, then determine the new entropy production rate for the heat exchanger. Keep the values of all the variables except $\dot{m}_{water}$ the same as they are in Example 9.5. **Answer**: $\dot{S}_P = 1.16\text{ W/K}$.
15. Suppose the air in Example 9.5 is to be cooled to 35.0°C rather than 75.0°C. Determine the new entropy production rate for this system. Keep the values of all the variables except $\dot{m}_{water}$ the same as they are in Example 9.5. **Answer**: $\dot{S}_P = 3.47\text{ W/K}$.

9.7 MIXING

A *mixer* normally has two or more inlet flow streams but only one outlet flow stream. Often mixers are used simply to mix different chemical species to produce a final product. When all the entering fluids have the same composition but are at different temperatures, the mixer becomes a type of simple heat exchanger.

Consider the dual-inlet, single-exit mixer shown in Figure 9.10. The steady state, steady flow energy rate balance equation (neglecting any change in kinetic and potential energy) is

$$\dot{Q} - \dot{W} + \dot{m}_1 h_1 + \dot{m}_2 h_2 - \dot{m}_3 h_3 = 0$$

and the similar mass rate balance equation is

$$\dot{m}_1 + \dot{m}_2 - \dot{m}_3 = 0$$

Combining these two equations and introducing the mass fraction y as

$$\dot{m}_1/\dot{m}_3 = y \quad (9.27)$$

or

$$\dot{m}_2/\dot{m}_3 = 1 - y$$

where y is always bound by $0 \le y \le 1$, which gives

$$\dot{Q} - \dot{W} + \dot{m}_3[y(h_1 - h_2) + (h_2 - h_3)] = 0$$

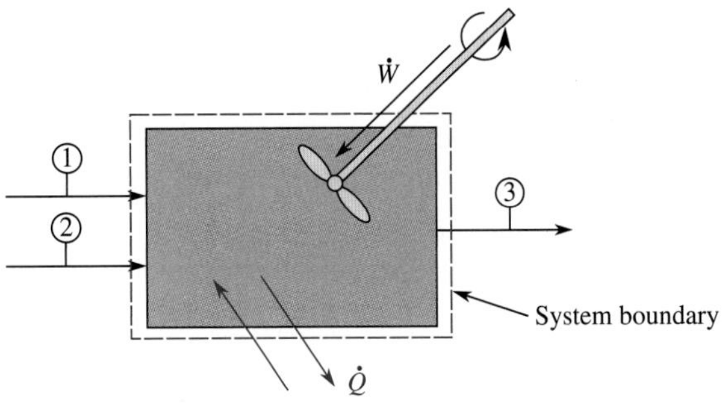

FIGURE 9.10
A simple mixing system.

Similarly, the appropriate entropy rate balance equation for an isothermal system boundary is

$$\begin{aligned}(\dot{S}_P)_{\text{mixing}} &= \dot{m}_3 s_3 - \dot{m}_1 s_1 - \dot{m}_2 s_2 - \frac{\dot{Q}}{T_b}\\ &= \dot{m}_3[y(s_2 - s_1) + (s_3 - s_2)] - \frac{\dot{Q}}{T_b}\end{aligned}$$

Many mixing systems use the inlet flow streams to induce mixing and thus tend to be isobaric (i.e., $p_1 = p_2 = p_3$).[4] To simplify the preceding results somewhat, consider the adiabatic, aergonic, isobaric mixing of two flow streams of the same material but at different temperatures. Then, these formulae reduce to

$$y(h_1 - h_2) + (h_2 - h_3) = 0 \tag{9.28}$$

and

Entropy production rate in mixing two nonreacting flow streams

$$(\dot{S}_P)_{\text{mixing}} = \dot{m}_3[y(s_2 - s_1) + (s_3 - s_2)] > 0 \tag{9.29}$$

If these materials are identical incompressible liquids with negligible mixer pressure loss or identical ideal gases with constant specific heats, then Eqs. (6.19) and (7.33) or Eqs. (6.22) and (7.37) can be used to give

$$y(T_1 - T_2) + T_2 - T_3 = 0 \tag{9.30}$$

and

$$(\dot{S}_P)_{\text{mixing}} = \dot{m}_3 c\left(y \ln\frac{T_2}{T_1} + \ln\frac{T_3}{T_2}\right) \tag{9.31}$$

where $c = c_p$ in the case of ideal gases. Combining these two equations by eliminating T_3 gives

Entropy production rate in mixing two flow streams of the same incompressible liquid or the same ideal gas (in this case, $c = c_p$)

$$(\dot{S}_P)_{\text{mixing}} = \dot{m}_3 c \ln\left\{\left[1 + y\left(\frac{T_1}{T_2} - 1\right)\right]\left(\frac{T_1}{T_2}\right)^{-y}\right\} \tag{9.32}$$

For a given T_1/T_2 ratio, there is a critical mass fraction (y_c) that produces a *maximum* rate of entropy production.[5] The critical mass fraction can be found from Eq. (9.32) by setting $d\dot{S}_P/dy = 0$ and solving for $y = y_c$. The result is

$$y_c = \frac{(1 - T_1/T_2) + \ln(T_1/T_2)}{(1 - T_1/T_2)\ln(T_1/T_2)} \tag{9.33}$$

which gives

$$(\dot{S}_P)_{\substack{\text{mixing}\\ \text{(max)}}} = \dot{m}_3 c \ln\left\{\left[1 + y_c\left(\frac{T_1}{T_2} - 1\right)\right]\left(\frac{T_1}{T_2}\right)^{-y_c}\right\} \tag{9.34}$$

Figure 9.11 shows the variation in y_c with the absolute temperature ratio T_1/T_2, and Figure 9.12 shows the resulting $(\dot{S}_P)_{\text{mixing (max)}}$ vs. T_1/T_2 relation.

Note that y_c is limited by its definition to be less than or equal to 1 in Figure 9.11 and that, in Figure 9.12, it is impossible to have a mixer whose entropy production rate falls in the region *above* the curve shown.

Finally, the analysis of the adiabatic, aergonic, isobaric (i.e., viscousless) mixing of identical ideal gases with constant specific heats produces formulae identical to Eqs. (9.32), (9.33), and (9.34) except c is replaced with the constant pressure specific heat c_p.

---

[4] Isobaric mixing also requires negligible viscous friction losses.

[5] It is a maximum because $d^2\dot{S}_P/dy^2 < 0$ when evaluated at $y = y_c$. Also, the minimum value of $\dot{S}_P$ is always zero, which occurs here when $T_1 = T_2$.

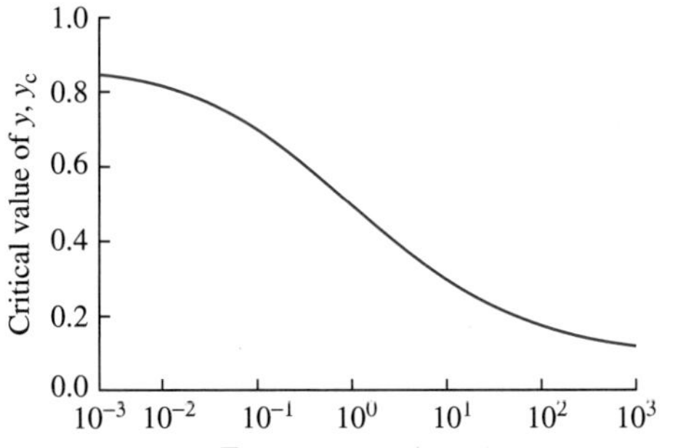

FIGURE 9.11
The critical mass fraction required to produce a maximum entropy production rate in adiabatic, aergonic, isobaric (i.e., viscousless) mixing of identical incompressible liquids or ideal gases with constant specific heats (note that $y_c = 1.0$ at $T_1/T_2 = 0$, and $y_c \to 0$ as $T_1/T_2 \to \infty$).

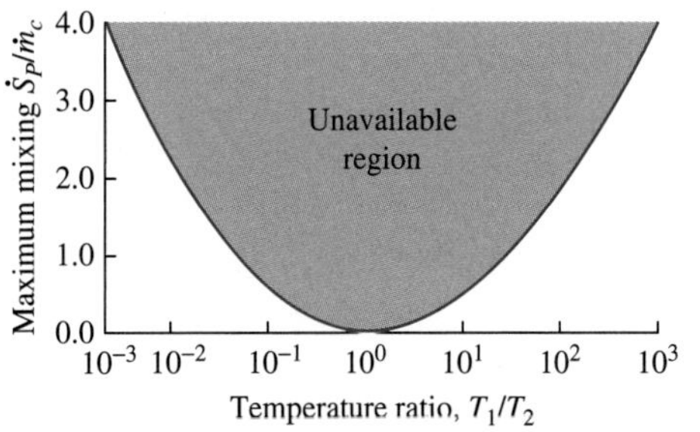

FIGURE 9.12
The maximum entropy production rate vs. temperature ratio for a simple mixing process.

EXAMPLE 9.6

A bathroom shower is set with equal hot and cold mass flow rates of 0.300 lbm/s. The hot water is at 140.°F and the cold water is 50.0°F. Determine

a. The shower mixture temperature and its entropy production rate.
b. The critical mass fraction y_c and the value of the maximum entropy production rate.

Solution

First, draw a sketch of the system (Figure 9.13).

The unknowns are the shower mixture temperature and its entropy production rate, the critical mass fraction y_c, and the value of the maximum entropy production rate. The material is liquid water.

Assume the water is incompressible with a constant specific heat and the mixing takes place adiabatically, aergonically, and isobarically.

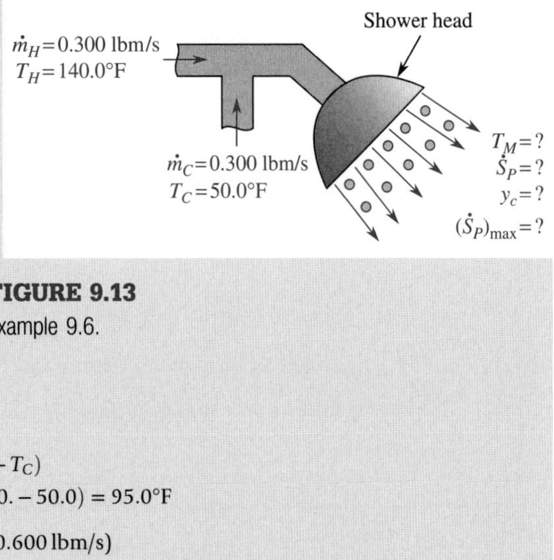

FIGURE 9.13
Example 9.6.

a. From Eq. (9.27), since $\dot{m}_H = \dot{m}_C$ and $\dot{m}_M = \dot{m}_H + \dot{m}_C$, $y = \dot{m}_H/\dot{m}_M = 1/2 = 0.500$ and, with $T_1 = T_H$, $T_2 = T_C$, and $T_3 = T_M$,

$$\frac{T_1}{T_2} = \frac{T_H}{T_C} = \frac{140.+459.67}{50.0+459.67} = 1.18$$

Then, from Eq. (9.30), we have

$$T_3 = T_M = T_C + y(T_H - T_C)$$
$$= 50.0 + 0.500(140. - 50.0) = 95.0°\text{F}$$

and, from Eq. (9.32), we have (using $\dot{m}_3 = \dot{m}_M = \dot{m}_H + \dot{m}_C = 0.600$ lbm/s)

$$(\dot{S}_P)_{\text{mixing}} = (0.600\,\text{lbm/s})[1.0\,\text{Btu/(lbm}\cdot\text{R)}]\ln\left\{[1+0.5(0.18)](1.18)^{-0.5}\right\}$$
$$= [2.051\times 10^{-3}\,\text{Btu/(s}\cdot\text{R)}](778.17\,\text{ft}\cdot\text{lbf/Btu})$$
$$= 1.60\,\text{ft}\cdot\text{lbf/(s}\cdot\text{R)}$$

b. From Eq. (9.33), we have

$$y_c = \frac{(1-T_H/T_C)+\ln(T_H/T_C)}{(1-T_H/T_C)\ln(T_H/T_C)} = \frac{(1-1.18)+\ln(1.18)}{(1-1.18)\ln(1.18)} = 0.486$$

(*Continued*)

EXAMPLE 9.6 *(Continued)*

Then, from Eq. (9.34),

$$(\dot{S}_P)_{\substack{\text{mixing}\\ (\text{max})}} = (0.600\,\text{lbm/s})[1.0\,\text{Btu/(lbm·R)}]\ln\{[1+0.486(0.18)](1.18)^{-0.486}\}$$

$$= [2.056\times 10^{-3}\,\text{Btu/(s·R)}](778.17\,\text{ft·lbf/Btu}) = 1.60\,\text{ft·lbf/(s·R)}$$

This example illustrates that mixing identical materials at almost the same absolute temperatures with equal mass flow rates produces nearly the maximum possible entropy production rate. Less entropy is produced if the mixing fraction is either $y < 0.5$ or $y > 0.5$ when $T_1 \approx T_2$.

Exercises

16. If the hot water mass flow rate is 0.500 lbm/s and the cold water mass flow rate is 0.300 lbm/s in Example 9.6, determine the entropy production rate of the mixing process. Keep the values of all the variables except T_M and y the same as they are in Example 9.6. **Answer:** $(\dot{S}_P)_{\text{mixing}} = 1.90\,\text{ft·lbf/(s·R)}$.

17. Recalculate parts a and b in Example 9.6 when the hot water temperature is reduced from 140.°F to 120.°F. Keep the values of all the variables except T_M and y_c the same as they are in Example 9.6. **Answer:** (a) $(\dot{S}_P)_{\text{mixing}} = 0.966\,\text{ft·lbf/(s·R)}$, (b) $y_c = 0.489$ and $(\dot{S}_P)_{\text{mixing max}} = 0.966\,\text{ft·lbf/(s·R)}$.

18. If the cold water temperature is increased from 50.0°F to 60.0°F, recalculate parts a and b in Example 9.6. Keep the values of all the variables except T_M and y_c the same as they are in Example 9.6. **Answer:** (a) $(\dot{S}_P)_{\text{mixing}} = 1.196\,\text{f·lbf/(s·R)}$, (b) $y_c = 0.488$, and $(\dot{S}_P)_{\text{mixing max}} = 1.196\,\text{ft·lbf/(s·R)}$.

9.8 SHAFT WORK MACHINES

A shaft work machine is defined in Chapter 6 as any device that has work (or power) input or output through a rotating or reciprocating shaft. These devices are normally steady state, steady flow, single-inlet, single-outlet systems, and their resulting modified energy rate balance equation is

$$\dot{Q}_{\text{actual}} - \dot{W}_{\text{actual}} = \dot{m}\left[h_2 - h_1 + (V_2^2 - V_1^2)/2g_c + (g/g_c)(Z_2 - Z_1)\right] \tag{6.30}$$

Assuming an isothermal system boundary, the modified entropy rate balance equation is

$$(\dot{Q}/T_b)_{\text{actual}} + \dot{m}(s_1 - s_2) + \dot{S}_P = 0$$

or

$$\dot{Q}_{\text{actual}} = \dot{m}\,T_b(s_2 - s_1) - T_b\dot{S}_P$$

Using this expression for $\dot{Q}$ in the preceding equation and solving for $\dot{W}_{\text{actual}}$ produces

$$\dot{W}_{\text{actual}} = \dot{m}\left[(h_1 - T_b s_1) - (h_2 - T_b s_2) + (V_1^2 - V_2^2)/2g_c + (g/g_c)(Z_1 - Z_2)\right] - T_b\dot{S}_P \tag{9.35}$$

and this can be rearranged to provide an expression for the entropy production rate of a shaft work machine as

Entropy production in a shaft work machine

$$\dot{S}_P\Big|_{\substack{\text{shaft work}\\ \text{machine}}} = \left(\frac{\dot{m}}{T_b}\right)\left[(h_1 - T_b s_1) - (h_2 - T_b s_2) + \frac{V_1^2 - V_2^2}{2g_c} + \frac{g(Z_1 - Z_2)}{g_c}\right] - \frac{\dot{W}_{\text{actual}}}{T_b} \tag{9.36}$$

For a reversible process, $\dot{S}_P = 0$, and Eq. (9.35) reduces to

$$\dot{W}_{rev} = \dot{m}\left[(h_2 - T_b s_2) - (h_1 - T_b s_1) + (V_2^2 - V_1^2)/2g_c + (g/g_c)(Z_2 - Z_1)\right] \tag{9.37}$$

and since, from Chapter 7,

$$\dot{W}_{\text{actual}} = \dot{W}_{\text{rev}} + \dot{W}_{\text{irr}}$$

by comparing Eqs. (9.35) and (9.37), we see that

$$\dot{W}_{\text{irr}} = T_b\dot{S}_P$$

The work transport energy efficiency η_W is defined in Chapter 4 for a work producing system (e.g., an engine or motor) by Eq. (4.72) as

$$(\eta_W)_{\substack{\text{work}\\\text{producing}}} = \frac{\dot{W}_{\text{act}}}{\dot{W}_{\text{rev}}} \tag{4.72}$$

Inserting Eqs. (9.35) and (9.37) gives

$$\begin{aligned}(\eta_W)\Big|_{\substack{\text{work}\\\text{producing}}} &= \frac{\dot{m}[(h_1 - T_b s_1) - (h_2 - T_b s_2) + (V_1^2 - V_2^2)/(2g_c) + (g/g_c)(Z_1 - Z_2)] - T_b\dot{S}_P}{\dot{m}[(h_1 - T_b s_1) - (h_2 - T_b s_2) + (V_1^2 - V_2^2)/(2g_c) + (g/g_c)(Z_1 - Z_2)]}\\ &= 1 - \frac{T_b\dot{S}_P}{\dot{m}[(h_1 - T_b s_1) - (h_2 - T_b s_2) + (V_1^2 - V_2^2)/(2g_c) + (g/g_c)(Z_1 - Z_2)]}\end{aligned} \tag{9.38}$$

and, for a work absorbing system (pump, compressor, etc.), Eq. (4.71) gives

$$(\eta_W)\Big|_{\substack{\text{work}\\\text{absorbing}}} = \frac{\dot{m}[(h_1 - T_b s_1) - (h_2 - T_b s_2) + (V_1^2 - V_2^2)/(2g_c) + (g/g_c)(Z_1 - Z_2)]}{\dot{m}[(h_1 - T_b s_1) - (h_2 - T_b s_2) + (V_1^2 - V_2^2)/(2g_c) + (g/g_c)(Z_1 - Z_2)] - T_b\dot{S}_P} \tag{9.39}$$

EXAMPLE 9.7

Determine the maximum (reversible) power that could be produced by expanding 0.500 kg/s of steam in a shaft work machine from 8.00 MPa, 300.°C to a saturated vapor at 100.°C. Neglect any kinetic and potential energy effects. The temperature of the system boundary is T_b = 20.0°C

Solution

First, draw a sketch of the system (Figure 9.14).

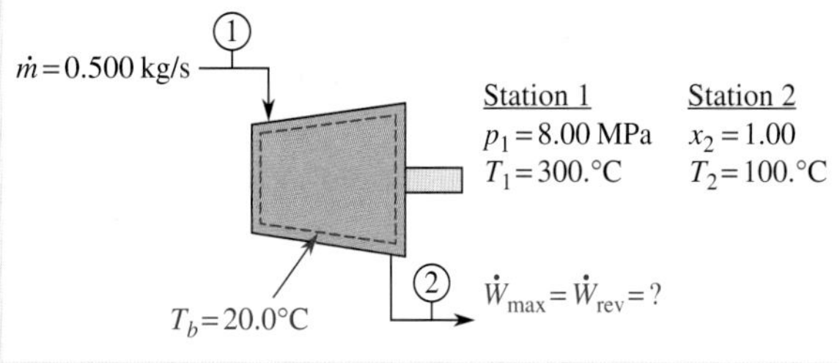

FIGURE 9.14
Example 9.7.

The unknown is the maximum (reversible) power for this system. For this open system, we have the following thermodynamic station data:

Station 1	Expansion →	Station 2
p_1 = 8.00 MPa		x_2 = 1.00
T_1 = 300.°F		T_2 = 100.°C

Neglecting all kinetic and potential energy effects, Eq. (9.37) reduces to

$$\dot{W}_{\text{max}} = \dot{W}_{\text{rev}} = \dot{m}\,[(h_1 - T_b s_1) - (h_2 - T_b s_2)]$$

and, using the values just given, we get

$$\begin{aligned}\dot{W}_{\text{max}} &= (0.50\ \text{kg/s})\{[2785.0 - (20.0 + 273.15)(5.7914)] - [2676.0 - (20 + 273.15)(7.3557)]\}\\ &= 284\ \text{kW}\end{aligned}$$

Exercises

19. If the mass flow rate of steam in Example 9.7 is increased from 0.500 kg/s to 0.850 kg/s with all the inlet and exit steam properties remaining constant, what is the maximum (reversible) power output? **Answer**: $\dot{W}_{\text{max}} = \dot{W}_{\text{rev}}$ = 482 kW.
20. Determine the maximum (reversible) power output in Example 9.7 when the steam exits with a quality of 90.0% rather than 100.%. **Answer**: $\dot{W}_{\text{max}} = \dot{W}_{\text{rev}}$ = 308 kW.
21. Suppose the inlet steam pressure and temperature in Example 9.7 is increased to 10.0 MPa and 800.°C. Determine the maximum (reversible) power output. **Answer**: $\dot{W}_{\text{max}} = \dot{W}_{\text{rev}}$ = 712 kW.

9.9 UNSTEADY STATE PROCESSES IN OPEN SYSTEMS

In Chapter 6, we analyze the energy transport requirements in the emptying and filling of a rigid container. In this section, we carry out an entropy analysis of the filling of a rigid container and determine whether it is more efficient to fill it adiabatically or isothermally.

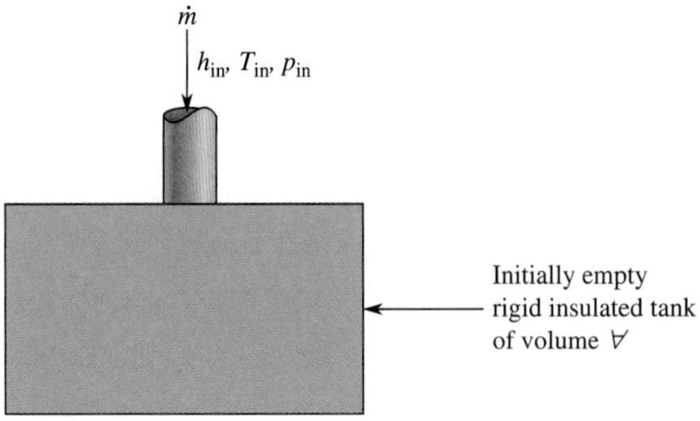

FIGURE 9.15
The filling of an insulated, rigid container.

The energy rate balance analysis of the adiabatic, aergonic system shown in Figure 9.15 is done in Chapter 6. The result is

$$u_2 = h_{in} \tag{6.34}$$

and, in the case of an incompressible, constant specific heat liquid, this means that the final temperature T_2 is

$$T_2 = T_{in} + (v/c)p_{in} \tag{6.35}$$

For an ideal gas with constant specific heats, it means that

$$T_2 = (c_p/c_v)T_{in} = kT_{in} \tag{6.36}$$

The entropy rate balance for this unsteady state system is

$$\dot{S} = \frac{dS}{dt} = \underbrace{\frac{\dot{Q}}{T_b}}_{0\,\text{(insulated)}} + \dot{m}s_{in} + \dot{S}_P$$

Integrating this equation from its initial empty state 1 to its final filled state 2 with S_{in} equal to a constant and solving for the entropy production gives

$${}_1(S_P)_2 = S_2 - S_1 - (m_2 - m_1)s_{in} = m_2s_2 - m_1s_1 - (m_2 - m_1)s_{in}$$

Now, for simplicity, assume that the container is initially evacuated, $m_1 = 0$, and

$${}_1(S_P)_2 = m_2(s_2 - s_{in}) \tag{9.40}$$

then, Eqs. (7.33), (7.37), and (9.40) give

Entropy production in adiabatically filling a rigid tank with an incompressible liquid

$${}_1(S_P)_2\Big|_{\substack{\text{incomp.}\\ \text{liquid}\\ \text{(adiabatic)}}} = m_2c\ln\frac{T_2}{T_{in}} = m_2c\ln\left[1 + \frac{(pv)_{in}}{cT_{in}}\right] \tag{9.41}$$

and

Entropy production in adiabatically filling a rigid tank with an ideal gas

$${}_1(S_P)_2\Big|_{\substack{\text{ideal}\\ \text{gas}\\ \text{(adiabatic)}}} = m_2c_p\ln\frac{T_2}{T_{in}} - m_2R\ln\frac{p_2}{p_{in}} = m_2c_p\ln k \tag{9.42}$$

where, for the ideal gas case, $m_2 = p_2\forall_2/(RT_2) = p_2\forall/(kRT_{in})$, where $\forall$ is the volume of the container and where we use $p_2 = p_{in}$.

A similar energy rate balance on the same system except now uninsulated and kept isothermal at $T_1 = T_{in} = T_2$ gives

$$\dot{Q} + \dot{m}_{in}h_{in} = \frac{d}{dt}(mu)$$

and multiplying this equation by dt and integrating gives

$${}_1Q_2 + (m_2 - m_1)h_{in} = m_2u_2 - m_1u_1$$

again setting $m_1 = 0$ for an initially evacuated container produces

$$u_2 = h_{in} + \frac{{}_1Q_2}{m_2} = u_{in} + (pv)_{in} + \frac{{}_1Q_2}{m_2}$$

For an incompressible isothermal liquid, this reduces to

$$u_2 - u_{\text{in}} = c(T_2 - T_{\text{in}}) = (pv)_{\text{in}} + \frac{{}_1Q_2}{m_2} = 0$$

or

$$ {}_1Q_2 = -m_2(pv)_{\text{in}} \tag{9.43}$$

and, for an isothermal ideal gas, it becomes

$$ {}_1Q_2 = -m_2RT_{\text{in}} = -m_2c_vT_{\text{in}}(k-1)$$

The integrated entropy rate balance for this system with both S_{in} and T_b constant is

$$ {}_1(S_P)_2 = m_2s_2 - m_1s_1 - (m_2 - m_1)s_{\text{in}} - \frac{{}_1Q_2}{T_b}$$

and, when $m_1 = 0$, this becomes

$$ {}_1(S_P)_2 = m_2(s_2 - s_{\text{in}}) - \frac{{}_1Q_2}{T_b} \tag{9.44}$$

For an isothermal incompressible liquid with $T_2 = T_{\text{in}} = T_b = T$, Eqs. (7.33), (9.43), and (9.44) give

Entropy production in isothermally filling a rigid tank with an incompressible liquid

$$ {}_1(S_P)_2\Big|_{\substack{\text{incomp.}\\ \text{liquid}\\ \text{(isothermal)}}} = m_2(pv)_{\text{in}}/T \tag{9.45}$$

and, for an isothermal ideal gas with $p_2 = p_{\text{in}}$ and $T_2 = T_{\text{in}} = T_b = T$,

Entropy production in isothermally filling a rigid tank with an ideal gas

$$ {}_1(S_P)_2\Big|_{\substack{\text{ideal}\\ \text{gas}\\ \text{(isothermal)}}} = m_2c_v(k-1) = m_2R = p_2\forall/T \tag{9.46}$$

where $R = c_p - c_v = c_v(c_p/c_v - 1) = c_v(k - 1)$ and $m_2 = p_2\forall/(RT)$.

Now, comparing Eqs. (9.41) and (9.45) for filling the container with the same amount of an incompressible liquid (i.e., m_2 is the same in each case), we see that, since $\ln(1 + x) < x$ for all $x > 0$,

$$ {}_1(S_P)_2\Big|_{\substack{\text{incomp.}\\ \text{liquid}\\ \text{(adiabatic)}}} < {}_1(S_P)_2\Big|_{\substack{\text{incomp.}\\ \text{liquid}\\ \text{(isothermal)}}} \quad \text{(for adding the same amount of mass in each case)} \tag{9.47}$$

but, on comparing Eqs. (9.42) and (9.46) for ideal gases, we find that we cannot add the same amount of mass in each case because

$$(m_2)\Big|_{\substack{\text{adiabatic}\\ \text{filling}}} = p_2\forall/(RT_2) = p_2\forall/(kRT_{\text{in}}) = (m_2)/k\Big|_{\substack{\text{isothermal}\\ \text{filling}}} \tag{9.48}$$

Since

$$k\left(\frac{R}{c_p}\right) = \frac{c_p - c_v}{c_v} = k - 1$$

and the series expansion for the logarithm of k for $0 < k < 2$ is

$$\ln k = (k-1) - (k-1)^2/2 + (k-1)^3/3 - \cdots$$

clearly $kR/c_p > \ln k$, which produces the following result when the tanks are filled to the same pressure (but not with the same amount of mass):

$${}_1(S_P)_2\Big|_{\substack{\text{ideal}\\\text{gas}\\\text{(adiabatic)}}} < {}_1(S_P)_2\Big|_{\substack{\text{ideal}\\\text{gas}\\\text{(isothermal)}}}$$

In the dissipative hydraulic flows studied earlier in this chapter, we find that less entropy is produced in both incompressible and ideal gas systems if the flows are carried out adiabatically as opposed to isothermally. The preceding analysis shows that this is also true in the filling of a rigid vessel. Note, however, that these two filling processes do not normally produce the same final system state. For example, adiabatic filling clearly produces a higher final temperature than isothermal filling when starting from the same initial temperature. A complete entropy production analysis has to include any additional processes required to reduce both systems to the same final state.

EXAMPLE 9.8

A 3.00 ft^3 rigid container is filled with oxygen entering at 70.0°F to a final pressure of 2000. psia. Assuming the container is initially evacuated and that the oxygen behaves as an ideal gas with constant specific heats, determine the amount of entropy produced when the container is filled

a. Adiabatically by insulating it.
b. Isothermally by submerging it in a water bath at 70.0°F while filling.

Solution

First, draw a sketch of the system (Figure 9.16).

The unknowns are the amount of entropy produced when the container is filled adiabatically by insulating it and isothermally by submerging it in a water bath at 70.0°F while filling. The material is the oxygen gas in the tank.

From Table C.13a of *Thermodynamic Tables to accompany Modern Engineering Thermodynamics*, we find for oxygen $c_p = 0.219$ Btu/(lbm·R), $R = 48.29$ ft·lbf/(lbm·R), $k = 1.39$. The final temperature after filling adiabatically is given by Eq. (6.36) as

$$(T_2)_{\substack{\text{adiabatic}\\\text{filling}}} = kT_{\text{in}} = 1.39(70.0 + 459.67\text{ R}) = 736\text{ R} = 277°\text{F}$$

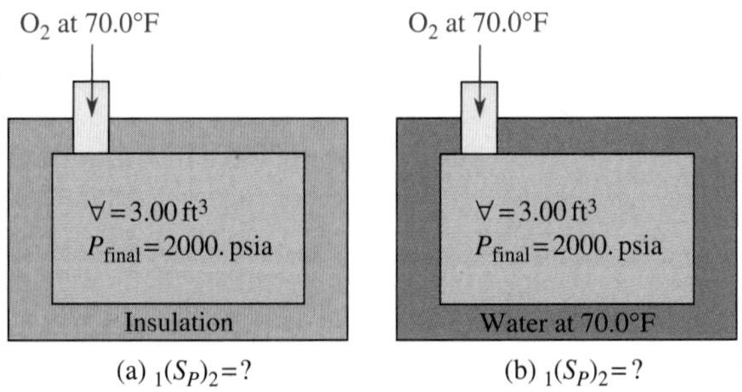

FIGURE 9.16
Example 9.8.

and, for isothermal filling,

$$(T_2)_{\substack{\text{isothermal}\\\text{filling}}} = T_{\text{in}} = 70.0 + 459.67 = 530.\text{ R}$$

The final mass of oxygen in the container can be found from the ideal gas equation of state as

$$m_2 = \frac{p_2\forall_2}{RT_2}$$

so

$$(m_2)\Big|_{\substack{\text{adiabatic}\\\text{filling}}} = \frac{p_2\forall_2}{R(T_2)_{\substack{\text{adiabatic}\\\text{filling}}}} = \frac{(2000.\text{ lbf/in.}^2)(144\text{ in.}^2/\text{ft}^2)(3.00\text{ ft}^3)}{[48.29\text{ ft·lbf/(lbm·R)}](736\text{ R})} = 24.3\text{ lbm}$$

and

$$(m_2)\Big|_{\substack{\text{isothermal}\\\text{filling}}} = \frac{p_2\forall_2}{R(T_2)_{\substack{\text{isothermal}\\\text{filling}}}} = \frac{(2000.\text{ lbf/in.}^2)(144\text{ in.}^2/\text{ft}^2)(3.00\text{ ft}^3)}{[48.29\text{ ft·lbf/(lbm·R)}](530.\text{ R})} = 33.8\text{ lbm}$$

a. From Eq. (9.42) for adiabatic filling, we have

$$[{}_1(S_P)_2]_{\substack{\text{adiabatic}\\\text{filling}}} = (m_2)_{\substack{\text{adiabatic}\\\text{filling}}} c_p \ln(k) = (24.3\,\text{lbm})[0.219\,\text{Btu/(lbm}\cdot\text{R)}]\ln(1.39) = 1.75\,\text{Btu/R}$$

b. From Eq. (9.46) for isothermal filling, we have

$$[{}_1(S_P)_2]_{\substack{\text{isothermal}\\\text{filling}}} = (m_2)_{\substack{\text{isothermal}\\\text{filling}}} R = (33.8\,\text{lbm})\left(\frac{48.29\,\text{ft}\cdot\text{lbf/(lbm}\cdot\text{R)}}{778.16\,\text{ft}\cdot\text{lbf/Btu}}\right) = 2.10\,\text{Btu/R}$$

which is about 20% greater than the entropy produced in part a.

Exercises

22. If the volume of the rigid container in Example 9.8 is reduced from 3.00 ft^3 to 2.00 ft^3, what is the new entropy production for the filling processes described in parts a and b? Keep the values of all the other variables the same as they are in Example 9.8. **Answer**: (a) $[{}_1(S_P)_2]_{adiabatic}$ = 1.17 Btu/R, (b) $[{}_1(S_P)_2]_{isothermal}$ = 1.40 Btu/R.
23. The filling pressure in Example 9.8 is to be reduced from 2000. psia to 1500. psia. Determine the new entropy production for the filling processes described in parts a and b. Keep the values of all the other variables the same as they are in Example 9.8. **Answer**: (a) $[{}_1(S_P)_2]_{adiabatic}$ = 1.31 Btu/R, (b) $[{}_1(S_P)_2]_{isothermal}$ = 1.57 Btu/R.
24. If the rigid container discussed in Example 9.8 is filled with air instead of oxygen, determine the new entropy production for the filling described in parts a and b. Keep the values of all the other variables the same as they are in Example 9.8. **Answer**: (a) $[{}_1(S_P)_2]_{adiabatic}$ = 1.76 Btu/R, (b) $[{}_1(S_P)_2]_{isothermal}$ = 2.10 Btu/R.
25. Determine the amount of entropy produced as 15.7 kg of argon gas is isothermally compressed into a 1.75 m^3 rigid container at 25.0°C. **Answer**: $[{}_1(S_P)_2]_{isothermal}$ = 3.27 kJ/K.

Equations (9.42), (9.46), and (9.48) can be combined to give

$$\left[\frac{[{}_1(S_P)_2]_{\text{isothermal}}}{[{}_1(S_P)_2]_{\text{adiabatic}}}\right]_{\substack{\text{ideal}\\\text{gas}}} = \frac{k-1}{\ln k}$$

which is greater than 1.0 for $k > 1.0$. This result is shown in Figure 9.17.

However, the specific entropy production ratio on a per unit mass of charge basis, that is, as ${}_1(S_P)_2/m_2 = {}_1(s_P)_2$, is

$$\left[\frac{[{}_1(s_P)_2]_{\text{isothermal}}}{[{}_1(s_P)_2]_{\text{adiabatic}}}\right]_{\substack{\text{ideal}\\\text{gas}}} = \frac{k-1}{k\ln k}$$

which is less than 1.0 for $k > 1.0$, because the two processes do not take on the same total charge of gas (see Eq. (9.48)).

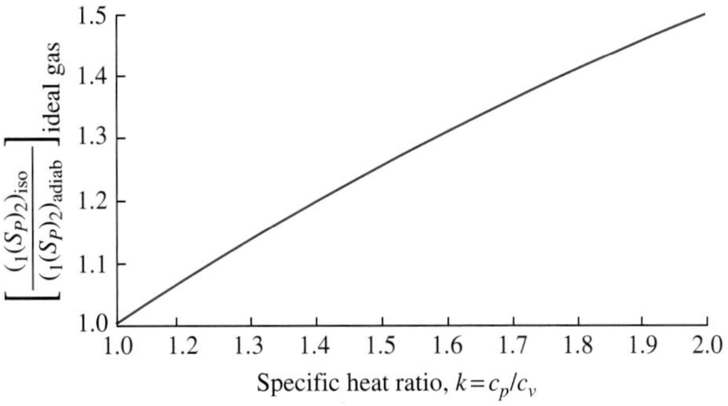

FIGURE 9.17
Entropy production ratio for an ideal gas.

CASE STUDIES OF ENTROPY PRODUCTION IN OPEN SYSTEMS

The following examples are typical case studies in open system applied engineering thermodynamics. They are included here to provide the student with an exposure to a second law analysis of complex systems typical of 21st century engineering technology. Energy and entropy rate balance equations are used as tools to understand these technologies.

Case study 9.1. Temperature magic

A *vortex tube* is a seemingly magical device that separates the temperature of its inlet flow stream into hot and cold outlet flow streams. Remarkably, it contains no moving parts other than the flowing gas itself.

The inlet flow enters the vortex tube chamber tangentially, and the resulting swirling motion causes the gaseous core along the chamber centerline to become extremely cold while the gas near the chamber wall becomes very hot. Internal baffles allow the core gas to exit through one tube (cold outlet) while the wall gas exits through the other (hot outlet), thus producing flow stream temperature separation. The illustration in Figure 9.18 shows the operation of this simple device.

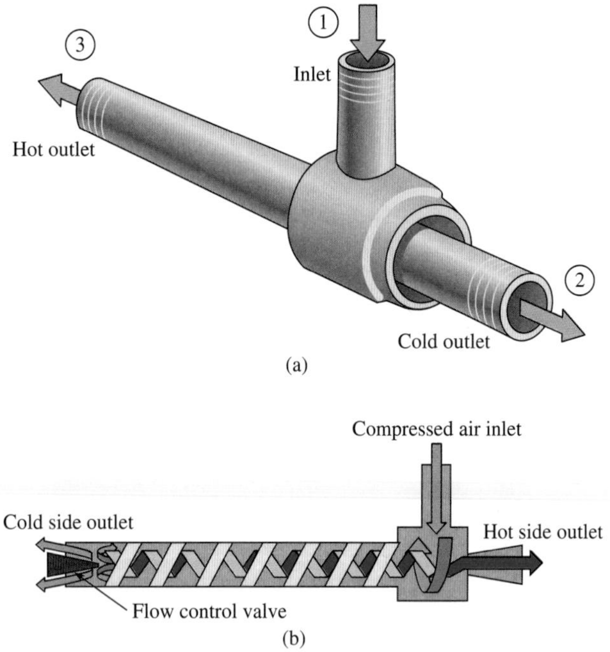

FIGURE 9.18
(a) A typical vortex tube. (b) A schematic showing how the off-center inlet causes the flow to form a swirl (vortex) inside the tube. The center of the vortex becomes cold and exits one end of the tube, and the outer part of the vortex becomes hot and exits the other end on the tube.

This remarkable device was discovered by Georges Joseph Ranque and was first described in a French patent in 1931.[6] In 1933, Ranque presented a paper to the Societe Francaise de Physique on this device, and nothing more was heard about it until 1945, when a vortex tube was found by an American and British investigation team at the end of World War II in the laboratory of Rudolph Hilsch at the University of Erlangen, Germany. Hilsch had begun research on the vortex tube in 1944, after reading Ranque's paper, and he published his results in Germany in 1946 and in the United States in 1947. Since then, interest in the vortex tube has remained high, and it is now frequently used in industry for inexpensive localized cooling applications.

Applying the energy rate balance to the adiabatic and aergonic vortex tube shown in Figure 9.18 and assuming ideal gas behavior with constant specific heats yields

$$y(T_1 - T_2) + T_2 - T_3 = 0$$

where $y = \dot{m}_3/\dot{m}_1 = \dot{m}_{\text{hot}}/\dot{m}_{\text{cold}}$ is the hot-side mass fraction defined by Eq. (9.27). Note that this result is exactly the same as Eq. (9.30), which was developed for the mixing operation. Thus, the first law of thermodynamics is insensitive to whether the fluids are being mixed or separated. It yields the same result in either case.

However, application of the entropy rate balance to the same system produces

$$(\dot{S}_P)_{\text{vortex tube}} = \dot{m}_3[y(s_1 - s_2) + s_2 - s_3] > 0 \qquad (9.49)$$

and comparing this with Eq. (9.29) yields the remarkable result that

$$(\dot{S}_P)_{\text{vortex tube}} = -(\dot{S}_P)_{\text{mixing}} \qquad (9.50)$$

yet both entropy production rates must be positive. This means that these two processes cannot simply be the reverse of each other. They cannot both follow the same thermodynamic path. The vortex tube separation phenomena must occur by a process that is unavailable to the simple mixing process and vice versa; otherwise, one of these devices violates the second law of thermodynamics.

To produce temperature separation, the vortex tube must have a significant pressure drop between the inlet and the outlet flow streams. It does not work isobarically. This pressure drop is not necessary in the mixing operation. Mixing is usually nearly isobaric, and emulation of the vortex tube separation operation requires a higher mixer outlet pressure than inlet pressure. This cannot be done without introducing heat or work energy transport into the system, which would alter the basic nature of the simple mixing device. Therefore, it is clear that the vortex tube inlet pressure is the source of the energy needed to produce the observed temperature separation. It is also the source of the entropy generation needed to allow Eq. (9.50) to be valid, as shown in Example 9.9, which follows.

Though the vortex tube is not an isobaric device, its two exit pressures are essentially equal (i.e., $p_1 \approx p_2$). Combining this condition with Eq. (7.37) for ideal gases and substituting the result into Eq. (9.49) gives

$$(\dot{S}_P)_{\text{vortex tube}} = \dot{m}_3\left[c_p \ln \frac{(T_1/T_2)^y}{1 + y(T_1/T_2 - 1)} + R \ln \frac{p_3}{p_2}\right] \qquad (9.51)$$

[6] *In 1932, he applied for a U.S. patent, which was awarded March 27, 1934 (U.S. patent number 1,952,281).*

EXAMPLE 9.9

The Vortec Corporation manufactures a vortex tube to provide hot and cold air from a standard compressed air system. For an equally split mass flow rate ($y = 0.500$), Table 9.2 lists the hot and cold outlet temperatures for various inlet pressures when the inlet temperature is 70.0°F. Assuming the exit pressure is atmospheric, determine the entropy production rate per unit mass flow rate for each pressure shown and plot the results.

Table 9.2 Hot and Cold Outlet Temperatures

Inlet pressure		Outlet temperatures		$\frac{T_H(°F)+460}{T_C(°F)+460}$
(psig)	(psia)	T_{hot}(°F)	T_{coldt}(°F)	
0.000	14.7	70.0	70.0	1.000
20.00	34.7	119.0	19.5	1.209
40.00	54.7	141.0	–3.00	1.315
60.00	74.7	150.0	–14.0	1.368
80.00	94.7	156.0	–22.0	1.406
100.0	114.7	161.0	–29.0	1.441
120.0	134.7	164.0	–34.0	1.465
140.0	154.7	166.0	–39.0	1.487

Solution

First, draw a sketch of the system (Figure 9.19).

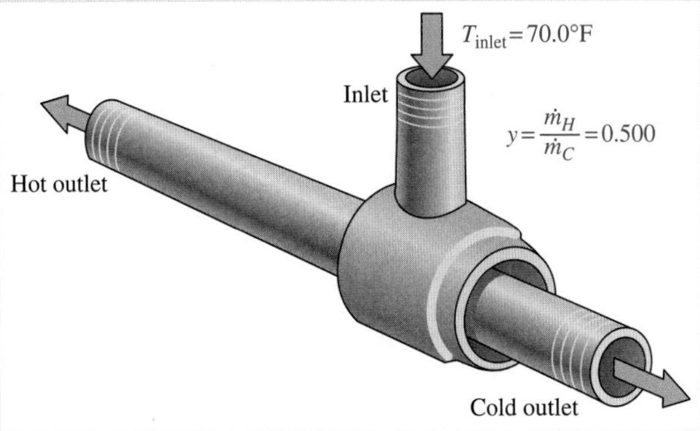

FIGURE 9.19
Example 9.9, system.

The unknowns are the entropy production rate per unit mass flow rate for each pressure shown; plot the results. The material is air.

Table 9.3 Remaining Results for Example 9.9

Inlet pressure psig	T_1/T_2	$\dot{S}_P/\dot{m}_3$ Btu/(lbm · R)
0.000	1.000	0.0000
20.00	1.209	0.0577
40.00	1.315	0.0878
60.00	1.368	0.1084
80.00	1.406	0.1241
100.0	1.441	0.1367
120.0	1.465	0.1474
140.0	1.487	0.1565

(*Continued*)

EXAMPLE 9.9 (*Continued*)

Using Eq. (9.51) for air, with R = 0.0685 Btu/(lbm·R), c_p = 0.240 Btu/(lbm·R), and y = 0.500, to calculate $\dot{S}_P/\dot{m}_3$ gives the following results. At 20 psig = 34.7 psia, $p_2 = p_3$ = 14.7 psia, and from Table 9.2, we find $T_H/T_C = T_3/T_2 = 1.209$. Then,

$$(\dot{S}_P/\dot{m}_3)_{\substack{\text{vortex}\\\text{tube}}} = [0.240\,\text{Btu/(lbm·R)}]\ln\frac{(1.209)^{0.5}}{1+0.500(0.209)} + [0.0685\,\text{Btu/(lbm·R)}]\ln\frac{34.7}{14.7}$$

$$= -0.0011 + 0.0588 = 0.0577\,\text{Btu/(lbm·R)}$$

Notice that the first term (which corresponds to isobaric separation) in this calculation is negative while the second term (resulting from the pressure loss) is positive and dominant. Therefore, if the vortex tube is required to be isobaric, it could not work, because to do so would violate the second law of thermodynamics. However, isobaric mixing is possible because then the lead term in this equation is positive and the second term is zero. The remaining results are in Table 9.3. These values are plotted Figure 9.20.

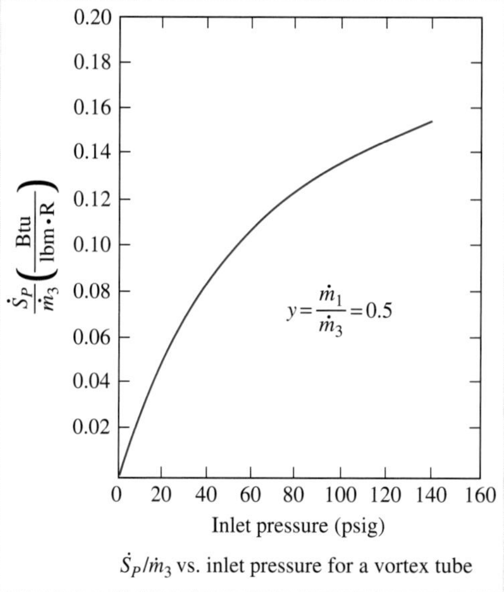

FIGURE 9.20
Example 9.9, results.

CAN VORTEX TUBES BE USED FOR AIR CONDITIONING?

In factories where compressed air is readily available, vortex tubes are often used to cool parts during machining or to provide cool air to workers in enclosed environments. But, when you take into account the energy needed to compress the air, vortex tubes are not very efficient cooling devises.

Vortex tubes also work with liquids as well as gases, but only at very high inlet pressures. Using the incompressible liquid equations of state in the entropy balance equation, you can show that indeed a temperature separation occurs in a vortex tube, providing the inlet pressure is very high.[7]

[7] *See: R.T. Balmer, "Pressure-Driven Ranque-Hilsch Temperature Separation in Liquids," ASME* Journal of Fluids Engineering *(1988), pp. 161–164.*

CASE STUDY 9.2. HYDRODYNAMIC FLOW SYSTEMS

A variety of hydrodynamic flow situations can be effectively analyzed with the entropy rate balance. Consider a steady state, steady flow, single-inlet, single-outlet system. The MERB for this system using the definition of enthalpy, $h = u + pv$, and the definition of density, $\rho = 1/v$, can be written as

$$\dot{Q} - \dot{W} = \dot{m}\left[u_2 - u_1 + (p/\rho)_2 - (p/\rho)_1 + (V_2^2 - V_1^2)/2g_c + (g/g_c)(Z_2 - Z_1)\right]$$
$$= \dot{m}\{u_2 - u_1 - (g/g_C)\left[{}_1(h_L)_2\right]\}$$

where, using the specific weight $\gamma = \rho g$,

$$_1(h_L)_2 = (pg_c/\gamma)_1 - (pg_c/\gamma)_2 + (V_1^2 - V_2^2)/2g + Z_1 - Z_2 \qquad (9.52)$$

is the *head loss* between the inlet station 1 and the exit station 2. In fluid mechanics texts, Eq. (9.52) is known as the *Bernoulli equation*, named after the Swiss mathematician and hydrodynamist Daniel Bernoulli (1700–1782). The MSRB for this system is

$$\dot{S}_P = \dot{m}(s_2 - s_1) - \dot{Q}/T_b$$

If the flowing fluid is a constant specific heat incompressible liquid,

$$\rho_1 = \rho_2 = \text{constant}$$

$$u_2 - u_1 = c(T_2 - T_1)$$

$$s_2 - s_1 = c\ \ln\frac{T_2}{T_1}$$

and, if it is a constant specific heat ideal gas,

$$\rho = p/(RT) = 1/v$$

$$u_2 - u_1 = c_v(T_2 - T_1)$$

$$s_2 - s_1 = c_p \ln\frac{T_2}{T_1} - R\ln\frac{p_2}{p_1}$$

Now, let us compare the rate of entropy production for two cases (i.e., two thermodynamic paths), adiabatic flow and isothermal flow.

Case A. SS, SF, SI, SO, aergonic, adiabatic flow[8]

Incompressible, constant specific heat liquids. In this case, the MERB gives

$$T_2 = T_1 + g\left[{}_1(h_L)_2\right]/(cg_c)$$

and the MSRB gives

$$\dot{S}_P\Big|_{\substack{\text{incomp.}\\\text{liquid.}\\\text{(adiabatic)}}} = \dot{m}c\ln\left\{1 + g\left[{}_1(h_L)_2\right]/(cg_cT_1)\right\} \qquad (9.53)$$

Ideal gases with constant specific heats. In this case, the MERB gives

$$T_2 = T_1 + g\left[{}_1(h_L)_2\right]/(c_vg_c)$$

and the MSRB gives

$$\left(\dot{S}_P\right)\Big|_{\substack{\text{ideal}\\\text{gas}\\\text{(adiabatic)}}} = \dot{m}c_p\ln\left[1 + g\left[{}_1(h_L)_2\right]/(c_vg_cT_1)\right] - \dot{m}R\ln\frac{p_2}{p_1} \qquad (9.54)$$

Case B. SS, SF, (SI, SO) aergonic, isothermal flow

Incompressible, constant specific heat liquids. In this case, the MERB gives

$$\dot{Q} = -\dot{m}g\left[{}_1(h_L)_2\right]/g_c$$

and the MSRB gives

$$\left(\dot{S}_P\right)\Big|_{\substack{\text{incomp.}\\\text{liquid}\\\text{(isothermal)}}} = \dot{m}g\left[{}_1(h_L)_2\right]/(T_bg_c) \qquad (9.55)$$

Ideal gases with constant specific heats. In this case, the MERB again gives

$$\dot{Q} = -\dot{m}g\left[{}_1(h_L)_2\right]/gc$$

and the MSRB gives

$$\left(\dot{S}_P\right)\Big|_{\substack{\text{ideal}\\\text{gas}\\\text{(isothermal)}}} = \dot{m}g\left[{}_1(h_L)_2\right]/(T_bg_c) - \dot{m}R\ln\frac{p_2}{p_1} \qquad (9.56)$$

The question is this: Which process, adiabatic or isothermal, is the more efficient by producing less entropy? For the incompressible liquids, we must compare Eqs. (9.53) and (9.55). Using a series expansion for the logarithm, we find that, for all $x > 0$, $\ln(1 + x) < x$, so that for $T_b = T_1$, we have

$$c\ln\left\{1 + g\left[{}_1(h_L)_2\right]/(cg_cT_1)\right\} < g\left[{}_1(h_L)_2\right]/(T_1g_c)$$

and, consequently,

$$\left(\dot{S}_P\right)\Big|_{\substack{\text{incomp.}\\\text{liquid}\\\text{(adiabatic)}}} < \left(\dot{S}_P\right)\Big|_{\substack{\text{incomp.}\\\text{liquid}\\\text{(isothermal)}}}$$

Similarly, for the ideal gases, we compare Eqs. (9.54) and (9.56) and again find that.

$$\left(\dot{S}_P\right)\Big|_{\substack{\text{ideal}\\\text{gas}\\\text{(adiabatic)}}} < \left(\dot{S}_P\right)\Big|_{\substack{\text{ideal}\\\text{gas}\\\text{(isothermal)}}}$$

In both of these cases, the adiabatic process produces less entropy. This is always true in this type of comparison, because an adiabatic system eliminates the entropy production due to heat transfer across the system boundary.

(*Continued*)

HYDRODYNAMIC FLOW SYSTEMS *Continued*

Table 9.4 Hydraulic Flow Systems

System	Heat Loss Formula
Flow in a straight pipe	${}_1(h_L)_2 = f(L/D)(V^2/2g)$, where f is the Darcy-Weisbach friction factor
Flow through valves, fittings, etc. ("minor losses")	${}_1(h_L)_2 = K_M(V^2/2g)$, where K_M is the minor loss coefficient
Flow through sudden contractions or expansions	$[{}_1(h_L)_2]_{\text{contraction}} = K_C(V^2/2g)$, where K_C is the contraction coefficient and V is the contraction outlet velocity or expansion inlet velocity. Also $[{}_1(h_L)_2]_{\text{expansion}} = (1 - D_1^2/D_2^2)^2(V^2/2g)$
Flow through a hydraulic jump	${}_1(h_L)_2 = (y_2 - y_1)^3/(4y_1y_2)$, where $y_2 - y_1$ is the jump height

Note: Values for the coefficients f, K_M, and K_C can be found in standard fluid mechanics textbooks.

Combining the MERB and the MSRB equations for any type of material produces a general formula for the entropy production rate inside a steady state, steady flow, single-inlet, single-outlet system with an isothermal boundary as

$$\dot{S}_P = \dot{m}\{s_2 - s_1 + [(g/g_c)[{}_1(h_L)_2] - u_2 + u_1]/T_b\} \quad (9.57)$$

Table 9.4 shows typical head loss formulations for a few common hydrodynamic flow situations. With these formulae, the entropy production rates can be calculated for many different hydraulic or pneumatic flow systems.

The hydraulic jump is a very effective phenomenon for dissipating energy. It commonly appears at the end of chutes or spillways to dissipate the kinetic energy of the flow. It is also an effective mixing process due to the violent dissipative agitation that takes place. The following example illustrates the entropy production rate in a hydraulic jump.

[8] *Recall that these terms mean we assume that the flow is steady state (SS) and steady flow (SF), that the flow has a single inlet and a single outlet (SI, SO), and that there is no work (aergonic) or heat transfer (adiabatic).*

EXAMPLE 9.10

At a rate of 500. lbm/s, water at 50.0°F flows down a spillway onto a horizontal floor. A hydraulic jump 1.80 ft high appears at the bottom of the spillway. The jump has an inlet velocity of 8.00 ft/s and an inlet height of 1.00 ft. Determine the energy dissipation rate and entropy production rate.

Solution

First, draw a sketch of the system (Figure 9.21).

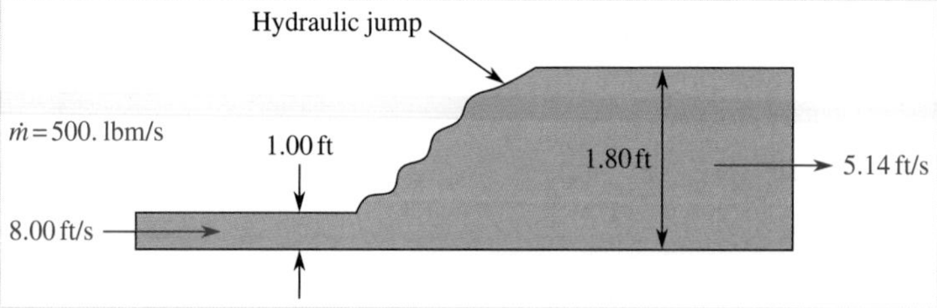

FIGURE 9.21
Example 9.10.

The unknowns are the energy dissipation rate and entropy production rate. The material is liquid water.

From the formula in Table 9.4, we get

$${}_1(h_L)_2 = (1.80 - 1.00)^3/[4(1.0)0(1.80)] = 0.0711\text{ ft}$$

and

$$\dot{m}(g/g_c)[{}_1(h_L)_2] = (500.\text{ lbm/s})\{(32.174\text{ ft/s}^2)/[32.174\text{ lbm}\cdot\text{ft}/(\text{lbf}\cdot\text{s}^2)]\}(0.0711\text{ ft})$$
$$= (35.55\text{ ft}\cdot\text{lbf/s})/(778.17\text{ ft}\cdot\text{lbf/Btu}) = 0.0457\text{ Btu/s}$$

Assuming the flow to be incompressible and adiabatic, the entropy production rate is given by Eq. (9.53) as

$$\dot{S}_P = (500.\,\text{lbm/s})[1.00\,\text{Btu/(lbm}\cdot\text{R)}]$$

$$\times \ln\left[1 + \frac{(32.174\,\text{ft/s}^2)(0.0711\,\text{ft})}{\left(1.00\dfrac{\text{Btu}}{\text{lbm}\cdot\text{R}}\right)\left(32.174\dfrac{\text{lbm}\cdot\text{ft}}{\text{lbf}\cdot\text{s}^2}\right)(50.0 + 459.67\,\text{R})}\right]$$

$$= 0.0697\,\text{Btu/(s}\cdot\text{R)}$$

Exercises

26. Determine the entropy production rate in the hydraulic jump discussed in Example 9.10 if the mass flow rate of the water is reduced from 500. lbm/s to 50.0 lbm/s. Keep the values of all the other variables the same as they are in Example 9.10. **Answer**: $\dot{S}_P = 0.00697$ Btu/(s·R).
27. If the exit water height (y_2) in Example 9.10 is 1.50 ft rather than 1.80 ft, what is the entropy production rate of the hydraulic jump. Keep the values of all the other variables the same as they are in Example 9.10. **Answer**: $\dot{S}_P = 0.0204$ Btu/(s·R).
28. If the inlet water temperature in Example 9.10 is increased from 50.0°F to 80.0°F, what is the new entropy production rate of the hydraulic jump? Keep the values of all the other variables the same as they are in Example 9.10. **Answer**: $\dot{S}_P = 0.0659$ Btu/(s·R).

The next example illustrates the use of the more complex *direct method* of determining the entropy production rate in a simple laminar hydrodynamic flow situation. In the design of hydraulic systems, the viscous losses in the flow stream oil can be very large. For this reason, it is usually desirable to keep the flow of hydraulic oil laminar rather than turbulent.

EXAMPLE 9.11

The velocity profile in the steady isothermal laminar flow of an incompressible Newtonian fluid in a horizontal circular tube of radius R is given by

$$V = V_m\left[1 - (x/R)^2\right]$$

where V_m is the maximum (i.e., centerline) velocity of the fluid, and x is the radial coordinate measured from the centerline of the tube.

Determine the entropy production rate due to laminar viscous losses in water at 20.0°C flowing in a 2.50 cm diameter pipe with a centerline velocity of 0.500 m/s. The viscosity of the water is 10.1 × 10^{-3} kg/(m·s), and the length of the pipe is 10.0 m.

Solution

First, draw a sketch of the system (Figure 9.22).

The unknown is the entropy production rate due to laminar viscous losses. The material is liquid water.

Here we use Eq. (7.72), which is the *direct method*, to determine the entropy production rate due to fluid viscosity. By differentiating the velocity formula given in the problem statement, we get

$$\frac{dV}{dx} = -2V_m(x/R^2)$$

then

$$(\sigma_W)_{\text{vis}} = \frac{\mu}{T}\left(\frac{dV}{dx}\right)^2 = \frac{4\mu V_m^2 x^2}{R^4 T}$$

and Eq. (7.72) gives

$$(\dot{S}_P)_{W_{\text{vis}}} = \int_{\mathcal{V}} (\sigma_W)_{\text{vis}} d\mathcal{V}$$

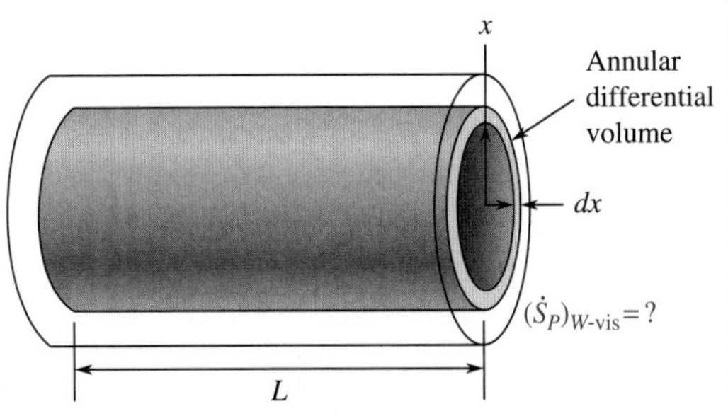

FIGURE 9.22
Example 9.11.

(*Continued*)

EXAMPLE 9.11 *(Continued)*

For the differential volume element $d\forall$, we use the volume of an annulus of thickness dx, or $d\forall = 2\pi Lx\,dx$. Then, we can evaluate Eq. (7.72) as

$$(\dot{S}_P)_{\substack{W\\ \text{vis}}} = \frac{8\pi\mu L V_m^2}{R^4 T}\int_0^R x^3\,dx = \frac{2\pi\mu L V_m^2}{T}$$

For this problem,

$$\mu = 10.1\times 10^{-3}\ \text{kg/(m}\cdot\text{s)}$$
$$L = 10.0\ \text{m}$$
$$V_m = 0.500\ \text{m/s}$$
$$T = 20.0^\circ\text{C} = 293\ \text{K}$$

then

$$(\dot{S}_P)_{\substack{W\\ \text{vis}}} = \frac{2\pi[10.1\times 10^{-3}\ \text{kg/(m}\cdot\text{s)}](10.0\ \text{m})(0.500\ \text{m/s})^2}{20.0+273.15\ \text{K}} = 5.41\times 10^{-4}\ \text{W/K}$$

In this example, we again have a very low entropy production rate. This is due, in this case, to the fact that laminar flow is a very energy efficient type of flow. Turbulent flow (which occurs spontaneously here at higher flow velocities) is much more dissipative and consequently is a much less energy efficient flow.

Exercises

29. If the fluid being pumped through the pipe in Example 9.11 is SAE-30 motor oil with a viscosity of 0.400 kg/(m·s) instead of water, determine the entropy production rate due to viscosity. Keep the values of all the other variables the same as they are in Example 9.11. **Answer:** $(\dot{S}_P)_{W\text{-vis}} = 0.0214\ \text{W/K}$.
30. If we increase the maximum (centerline) velocity of the water in Example 9.11 from 0.500 m/s to 3.00 m/s, determine the new entropy production rate due to viscosity. Keep the values of all the other variables the same as they are in Example 9.11. **Answer:** $(\dot{S}_P)_{W\text{-vis}} = 1.95\times 10^{-3}\ \text{W/K}$.
31. If we increase the length of the pipe in Example 9.11 from 10.0 m to 1000. m, determine the new entropy production rate due to viscosity. Keep the values of all the other variables the same as they are in Example 9.11. **Answer:** $(\dot{S}_P)_{W\text{-vis}} = 5.41\times 10^{-3}\ \text{W/K}$.
32. The diameter of the pipe in Example 9.11 is increased from 2.50 to 3.75 cm, but the mass flow rate is maintained constant. Determine the new entropy production rate due to the fluid's viscosity. Keep the values of all the variables except V_m the same as they are in Example 9.11. **Answer:** $(\dot{S}_P)_{W\text{-vis}} = 1.07\times 10^{-5}\ \text{W/K}$.

SUMMARY

In this chapter, we investigate a series of open systems and carry out a second law analysis of them using the entropy balance or the entropy rate balance. The primary purpose of this material is to stimulate your thinking by addressing modern technologies from a second law point of view. Several new equations are introduced in this chapter that deal with these technologies. Some of the more important equations follow.

1. The general open system entropy rate balance (SRB) equation is

$$\int_\Sigma \left(\frac{\dot{q}}{T_b}\right)_{\text{act}} dA + \sum_{\text{inlet}} \dot{m}s - \sum_{\text{outlet}} \dot{m}s + \dot{S}_P = \dot{S}_{\text{system}} \tag{9.4}$$

and, when the system boundaries are isothermal, this equation reduces to

$$\left(\frac{\dot{Q}}{T_b}\right)_{\text{act}} + \sum_{\text{in}} \dot{m}s - \sum_{\text{out}} \dot{m}s + \dot{S}_p = \dot{S}_{\text{system}} \tag{9.5}$$

This equation is the *actual* heat transfer rate, and T_b is the temperature where the system boundary occurs. When you have a steady state (SS), steady flow (SF), single inlet, single outlet (SI, SO), isothermal boundary (IB) system, Eqn. (9.5) reduces to the *modified* entropy rate balance:

$$\frac{\dot{Q}}{T_b} + \dot{m}(s_{\text{in}} - s_{\text{out}}) + \dot{S}_P = 0 \tag{9.10}$$

Multiplying this equation through by dt and integrating over time from system state 1 to state 2 gives the open system *modified entropy balance* (MSB) equation as

$$\frac{{}_1Q_2}{T_b} + \int_1^2 \dot{m}(s_{\text{in}} - s_{\text{out}})dt + {}_1(S_P)_2 = 0 \tag{9.11}$$

2. For flow in nozzles, diffusers, or throttles, we have the entropy production rate of an incompressible fluid in an adiabatic nozzle, diffuser, or throttle:

$$(\dot{S}_P)\Big|_{\substack{\text{adiabatic}\\\text{incompressible}\\\text{fluid}}} = \dot{m}c\ln\frac{T_{\text{out}}}{T_{\text{in}}} > 0 \tag{9.14}$$

and the entropy production rate of an incompressible fluid in a nozzle, diffuser, or throttle with heat transfer

$$\dot{S}_P\Big|_{\substack{\text{incompressible}\\\text{fluid}}} = \dot{m}c\ln\frac{T_{\text{out}}}{T_{\text{in}}} - \frac{\dot{Q}}{T_b} > 0 \tag{9.15}$$

3. For flow in heat exchangers, we have the heat exchanger entropy production rate:

$$\dot{S}_P = \dot{m}_H(s_{\text{out}} - s_{\text{in}})_H + \dot{m}_C(s_{\text{out}} - s_{\text{in}})_C \tag{9.24}$$

See Eqs. (9.25) and (9.26) when the hot and cold fluids are known to be incompressible liquids or ideal gases.

4. When two flow streams, 1 and 2, are combined to form a mixed outlet flow stream, 3, the entropy production rate is

$$(\dot{S}_P)_{\text{mixing}} = \dot{m}_3[y(s_2 - s_1) + (s_3 - s_2)] > 0 \tag{9.29}$$

See Eq. (9.32) for the entropy production rate in mixing identical incompressible liquids or identical ideal gases.

5. The entropy production rate for shaft work machines is

$$\dot{S}_P\Big|_{\substack{\text{shaft work}\\\text{machine}}} = \left(\frac{\dot{m}}{T_b}\right)\left[(h_1 - T_b s_1) - (h_2 - T_b s_2) + \frac{V_1^2 - V_2^2}{2g_c} + \frac{g(Z_1 - Z_2)}{g_c}\right] - \frac{\dot{W}_{\text{actual}}}{T_b} \tag{9.36}$$

6. The entropy production in adiabatically filling a rigid tank with an incompressible liquid is

$${}_1(S_P)_2\Big|_{\substack{\text{incomp.}\\\text{liquid}\\\text{(adiabatic)}}} = m_2 c\ln\frac{T_2}{T_{\text{in}}} = m_2 c\ln\left[1 + \frac{(pv)_{\text{in}}}{cT_{\text{in}}}\right] \tag{9.41}$$

7. The entropy production in adiabatically filling a rigid tank with an ideal gas is

$${}_1(S_P)_2\Big|_{\substack{\text{ideal}\\\text{gas}\\\text{(adiabatic)}}} = m_2 c_p \ln\frac{T_2}{T_{\text{in}}} - m_2 R\ln\frac{p_2}{p_{\text{in}}} = m_2 c_p \ln k \tag{9.42}$$

8. The entropy production in isothermally filling a rigid tank with an incompressible liquid is

$${}_1(S_P)_2\Big|_{\substack{\text{incomp.}\\\text{liquid}\\\text{(isothermal)}}} = m_2(pv)_{\text{in}}/T \tag{9.45}$$

9. The entropy production in isothermally filling a rigid tank with an ideal gas is

$${}_1(S_P)_2\Big|_{\substack{\text{ideal}\\\text{gas}\\\text{(isothermal)}}} = m_2 c_v(k-1) = m_2 R = p_2\forall/T \tag{9.46}$$

FINAL COMMENTS ON THE SECOND LAW

In Chapter's 7, 8, and 9, we deal with the fundamentals of the second law of thermodynamics for nonequilibrium systems. This law introduces the new thermodynamic property *entropy*, which is *not* conserved in any real engineering process. The second law of thermodynamics states that entropy must always be *produced* in any irreversible process; however, a positive entropy production does not mean that the net system entropy must necessarily increase, because entropy may be transported out of a system faster than it is produced within the system; therefore, *the entropy level of the entire system can either increase or decrease in any real process.*

In Chapter 8 we investigate closed system applications of the second law of thermodynamics for both reversible and irreversible processes. In so doing, we expand the first law examples given in Chapter 5 to include an entropy balance analysis. Since entropy production is a direct consequence of system losses that lead to diminished operating efficiency, many of the examples in this chapter focus on determining the entropy production or its rate to gain further insight into the causes of system inefficiency. This is done by using either the auxiliary entropy production equations (the *direct method*) or an appropriate entropy balance equation (the *indirect method*).

In this chapter, we use the second law of thermodynamics in the analysis of open systems. We use the entropy balance to determine the entropy production rates for a variety of common engineering devices (nozzles, diffusers, throttles, heat exchangers, etc.). We also look at how different processes that achieve the same end states affect the amount of entropy produced by the system during those processes. This allows us to choose the process or method of changing the state that is the least dissipative and consequently the most efficient. Determining processes that minimize the entropy production produces economic and productivity rewards to the user.

Problems (* indicates problems in SI units)

1. An inventor claims to have a steady state, steady flow system with an isentropic flow stream (i.e., $s_{in} = s_{out}$) that requires heat addition. Show whether or not this system violates the second law of thermodynamics.
2. The inventor in Example 9.1 now claims to have a system that generates heat by pumping water through a pipe isothermally (i.e., $T_{in} = T_{out}$). Show whether or not this system violates the second law of thermodynamics.
3.* Suppose the inventor in Example 9.1 filed a new patent claim that said the heater was 20.0 kW and the heat transfer boundary temperature was 100°C. What are the water flow rate and entropy production rate under these conditions?
4.* The inventor in Example 9.1 needs to know the heat transfer rate and entropy production rate required to heat 5.00 kg/s from 15°C to 50°C. You are the engineer, so what are the answers?
5.* The inventor in Example 9.1 now wants to patent a water cooling system that cools 0.500 kg/s of water from 50°C to 15°C by removing heat with a heat transfer boundary temperature of 10.0°C. What is (a) the heat transfer rate and (b) the entropy production rate for this process.
6. An inventor reports that she has a refrigeration compressor that receives saturated Refrigerant-134a vapor at 0.00°F and delivers it at 150. psia and 120.°F. The compression process is adiabatic. Determine whether or not this process violates the second law of thermodynamics.
7.* A 1.00 MW steam power plant operates on the simple reversible thermodynamic cycle shown Figure 9.23.
   a. What is its thermal efficiency?
   b. What is the steam mass flow rate in this system?
   c. What is its thermal efficiency if it is operated on a Carnot cycle?

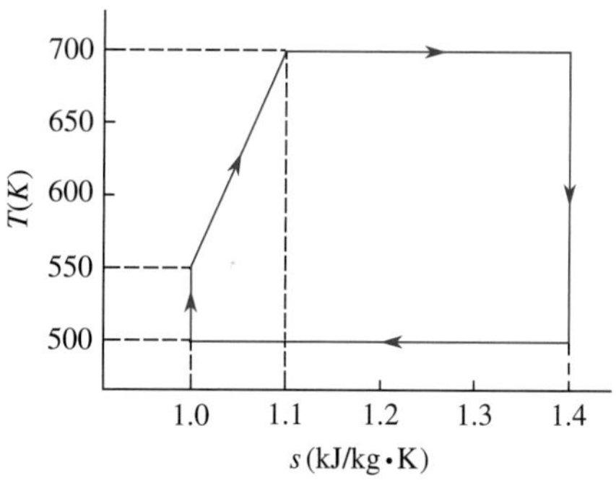

FIGURE 9.23
Problem 7.

8.* A steam turbine is limited to a maximum inlet temperature of 800.°C. The exhaust pressure is 0.0100 MPa, and the moisture in the turbine exhaust is not to exceed 9.00%.
   a. What is the maximum allowable turbine inlet pressure if the flow is adiabatic and reversible?
   b. What is the maximum power output per unit mass flow rate?
9.* A steam turbine receives steam at 1.00 MPa and 700.°C and exhausts at 0.100 MPa. If the turbine can be considered to operate as a steady flow, reversible, adiabatic machine, what is the work done per pound of steam flowing? Neglect any changes in kinetic or potential energy.
10. Saturated mercury vapor enters a steady flow turbine of a high-pressure auxiliary power system at 600. psia and emerges as a mixture of liquid and vapor at a pressure of 1.00 psia. What must be the flow rate if the power output is to be 10.0 kW? Assume the turbine is reversible and adiabatic, and neglect any changes in kinetic or potential energy.
11.* In the year 2138, a law requires certain limits on the production of entropy of any marketable piece of technology. This law is similar in nature to the old air pollution laws of the 20th century. It sets an upper limit of 1.00×10^{-3} kJ/(kg · K · s) on the

entropy production rate density of any new technology. This becomes known simply as the EPRD number, determined by dividing the entire mass of the system generating the entropy into its total entropy production rate. Determine the steady state EPRD of an insulated steam turbine that has a total mass of 2000. kg, takes in steam at 3.50 MPa, 400.°C at a rate of 2.00 kg/s and exhausts it at 5.00 kPa, 90.0% quality.

12. Determine the entropy production rate as 5.00 lbm/s of saturated water vapor at 14.696 psia is condensed isothermally and aergonically in a steady flow, steady state process to a saturated liquid. Ignore all kinetic and potential energy changes. Explain the significance of your answer.

13.* Air is throttled from 1.00 MPa and 30.0°C to 0.100 MPa in a steady flow, adiabatic process. Assuming constant specific heat ideal gas behavior and ignoring any changes in kinetic and potential energy, determine
  a. The change in flow stream entropy.
  b. The entropy produced per kg of air flowing.

14. Determine the nozzle outlet diameter in Example 9.2 required to increase the nozzle efficiency by decreasing the entropy production rate by 25.0%.

15.* Determine the final temperature and the entropy production per unit mass of air at 1.00 MPa, 25.0°C, and 2.00 m/s that expands adiabatically through a horizontal nozzle to 0.100 MPa and 100. m/s. Assume constant specific heat ideal gas behavior, and ignore any changes in kinetic and potential energy.

16. Refrigerant-134a enters an insulated nozzle at 25.0 psia, 80.0°F, and 10.0 ft/s. The flow accelerates and reaches 15.0 psia and 60.0°F just before it exits the nozzle. The process is adiabatic and steady flow.
  a. What is the exit velocity?
  b. Is the flow reversible or irreversible?

17. Air is expanded in an insulated horizontal nozzle from 100. psia, 100.°F to 26.0 psia, 70.0°F. Neglecting the inlet velocity and any change in potential energy, determine (a) the outlet velocity and (b) the entropy production rate per unit mass flowing. Assume the air behaves as an ideal gas with constant specific heats.

18.* Steam at 40.0 MPa, 800.°C expands through a heated nozzle to 0.100 MPa and 90.0% quality at a rate of 100. kg/h. Neglect the inlet velocity and any change in potential energy, and take the entropy production rate to be 10.0% of the magnitude of the entropy transport rate due to heat transfer. Determine
  a. The entropy production rate if the surface temperature of the nozzle is 450.°C.
  b. The exit velocity.
  c. The exit area of the nozzle.

19. Refrigerant-134a flows steadily through an adiabatic throttling valve. At the inlet to the valve, the fluid is a saturated liquid at 110.°F. At the valve outlet, the pressure is 20.0 psia. Neglecting any changes in kinetic and potential energy, determine
  a. The quality of the fluid at the valve outlet.
  b. The entropy production per pound of R-134a flowing through the valve.

20.* Carbon dioxide (CO_2) at 50 MPa and 207°C is expanded isothermally through an uninsulated nozzle to 1.50 MPa in a steady state, steady flow process. There is no change in potential energy across the nozzle, and the surface temperature of the nozzle is 307°C. The entropy production rate magnitude in this problem can be taken to be 10% of the absolute value of the heat transfer rate. Assuming the CO_2 to be a constant specific heat ideal gas, determine
  a. The heat transfer rate of the nozzle per kg of CO_2 flowing.
  b. The change in kinetic energy of the CO_2 across the nozzle per kg of CO_2 flowing.

21. Refrigerant-134a is throttled irreversibly through an insulated, horizontal, constant diameter tube. Saturated liquid R-134a enters the tube at 80.0°F and exits the tube at 10.0°F.
  a. What is the increase in entropy per lbm of R-134a flowing through the tube?
  b. What is the entropy production rate per unit mass flow rate of R-134a?
  c. Show the initial and final states on a *T-s* diagram.
  d. Determine the average Joule-Thomson coefficient for this process.

22. Saturated liquid Refrigerant-134a is expanded irreversibly in a refrigerator expansion valve from 100.°F to 0.00°F. Determine (a) the entropy of the R-134a after the expansion and (b) the entropy production rate of the expansion process per unit mass flow rate. Assume the process is adiabatic and aergonic, and neglect any changes in kinetic and potential energy.

23.* A steady state desuperheater (a type of mixing heat exchanger) adiabatically mixes superheated vapor and liquid with the properties shown in Figure 9.24. Complete vaporization of the liquid reduces the enthalpy of the vapor to h = 1390 kJ/kg at the exit of the desuperheater. Compute the rate of entropy production in the desuperheater.

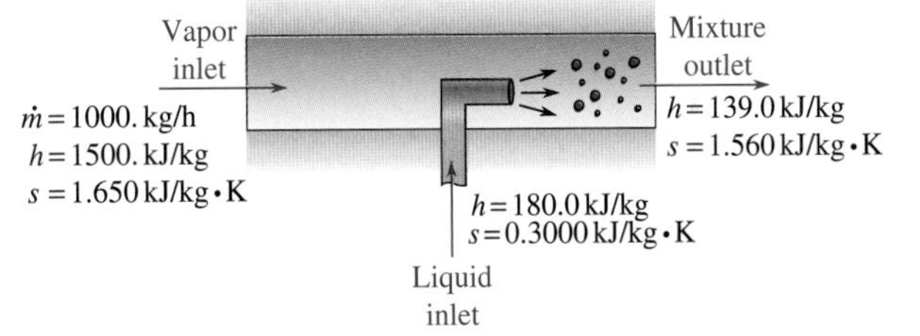

FIGURE 9.24
Problem 23.

24. A solar concentrating heat exchanger system directs sunlight onto a long, straight pipe. The pipe receives 153.616 Btu/h per foot of length. If water enters the pipe at 50.0 lbm/h as saturated liquid at 300.°F and is heated isothermally so that it leaves as vapor at 20.0 psia, then
  a. How long is the pipe for steady state, steady flow conditions.
  b. What is the rate of entropy production.
  c. Show whether this system violates the second law of thermodynamics.

25.* Consider a simple constant pressure boiler that converts 3.00 kg/min of saturated liquid water at 1.00 atm pressure into saturated vapor at 1.00 atm in a steady state, steady flow, single-inlet, single-outlet process.
  a. What is the heat transfer rate into the boiler?
  b. What is the entropy production rate inside the boiler?

26.* A brilliant young engineering student just invented a new chrome-plated digital heat exchanger that has water flowing through it at 14.41 kg/s. At the inlet, the water is a saturated vapor at 200.°C. The water passes isothermally through the heat exchanger while it absorbs heat from the environment. At the outlet, the pressure is 1.00 MPa. Determine (a) the heat transfer rate, and (b) the entropy production rate. (c) Show whether this device violates the second law of thermodynamics. Assume that the system boundary temperature is isothermal at 200.°C.

27. A contact feedwater heat exchanger for heating the water going into a boiler operates on the principle of mixing steam with liquid water. For the steady flow adiabatic process shown in Figure 9.25, calculate

a. The rate of change of entropy of the entire heater.
b. The rate of entropy production inside the heater.

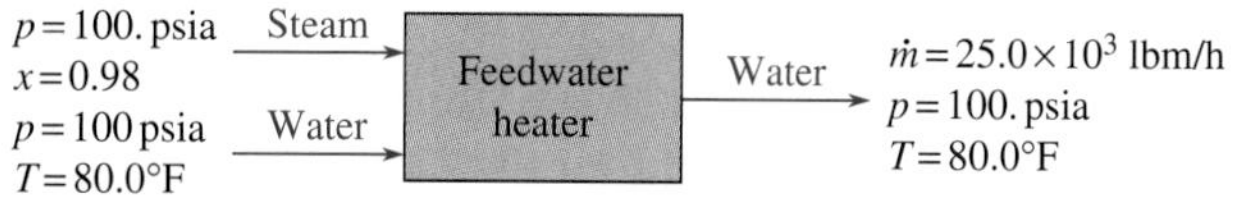

FIGURE 9.25
Problem 27.

28. As an engineering consultant, you are asked to review a design proposal in which an electric resistance heater is to be used in conjunction with a precision air bearing (Figure 9.26). The heater uses 100 W of electrical power and the air bearing has a constant surface temperature of 160.°F. The heater is well insulated on the outside and air enters the bearing at 40.0°F, 35.0 psia and exits at 80.0°F' 40.0 psia. The bearing is a steady flow, steady state device with an air flow rate of 35.55 lbm/h. Determine

a. Whether this device violates the first law of thermodynamics.
b. Whether this device violates the second law of thermodynamics.

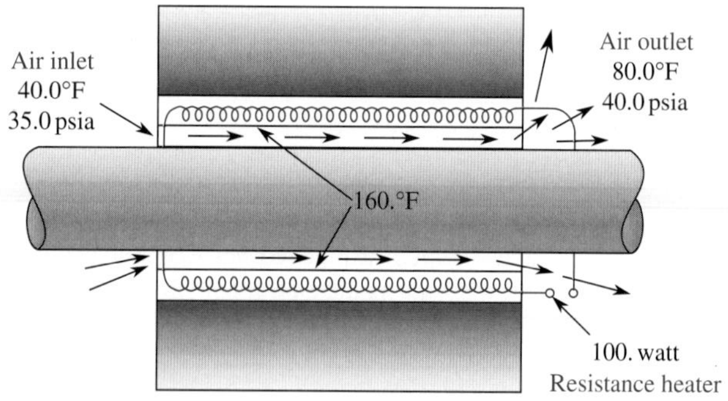

FIGURE 9.26
Problem 28.

29.* A slide projector contains a 500. W lightbulb cooled by an internal fan that blows room air across the bulb at a rate of 1.00 kg/min. If the equilibrium surface temperature of the bulb is 350.°C and the inlet temperature (the room air) is at 20.0°C, then determine (a) the outlet temperature of the cooling air and (b) the rate of entropy production in the air passing through the projector. Assume the air is an ideal gas with constant specific heats and that it undergoes an aergonic process.

30. Determine the total entropy production rate for the heat exchanger shown in Figure 9.27. In addition to the air-water heat transfer within the heat exchanger, the air also loses an unknown amount of heat to the surroundings while the water receives an additional 10.0 Btu/s from the surroundings. Assume the internal air-water interface is isothermal at 100.°F and the outer surface of the heat exchanger is isothermal at 70.0°F. Neglect all flow stream pressure losses.

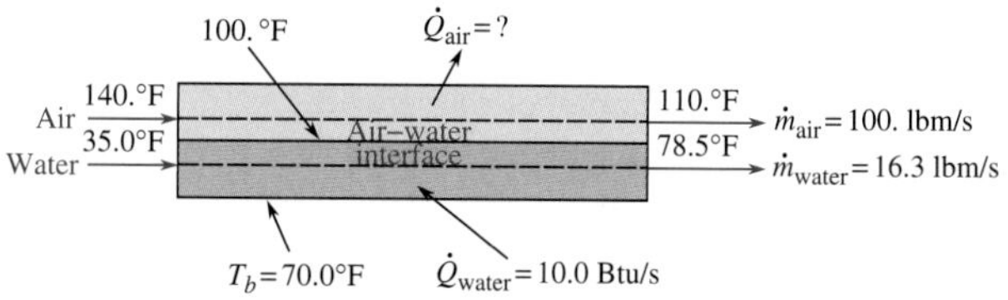

FIGURE 9.27
Problem 30.

31.* A new Yo Yo Dyne propulsion system has three flow streams, as shown in Figure 9.28. It mixes 0.500 kg/s of saturated water vapor at 100.°F with 0.200 kg/s of saturated liquid water at 100.°C in a steady flow, steady state, isobaric process. This system is cheaply made and uninsulated; consequently, it loses heat at the rate of 75.0 kJ/s to the surroundings. Assuming the system boundary temperature is isothermal at 100.°C, determine

a. The quality of the outlet mixture.
b. The entropy production rate of the system.

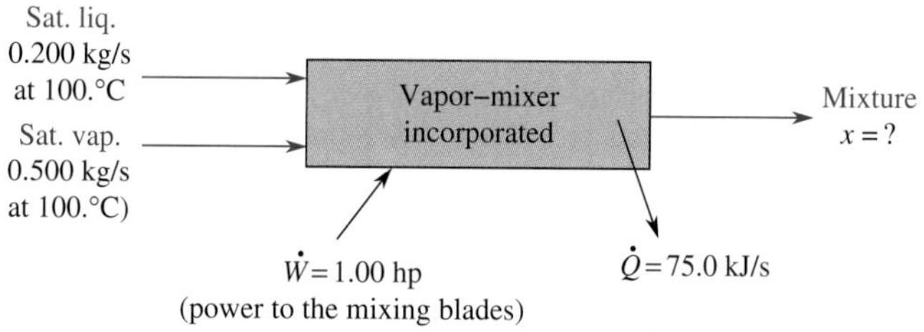

FIGURE 9.28
Problem 31.

32.* In a steady flow, adiabatic, aergonic desuperheater (a kind of mixing heat exchanger), water is sprayed into superheated steam in the proper amount to cause the superheated steam to become saturated.

a. Calculate the mass flow rate of water necessary for desuperheating.
b. What is the entropy production rate of this system?
c. Show whether this process violates the second law of thermodynamics.

Given that

- Steam mass flow rate = 200. kg/h.
- Steam entering state = 10.0 MPa, 600.°C.
- Water entering state = 10.0 MPa, 100.°C.
- Steam outlet state = 10.0 MPa, saturated vapor.

33. A steady state, steady flow steam mixer consists of a box with two inlet pipes and one outlet pipe. One inlet pipe carries saturated water vapor at 50.0 lbm/s and 20.0 psia. The other inlet pipe carries saturated liquid water at 10.0 lbm/s and 20.0 psia. The mixing process is isobaric. In addition, 9602 Btu/s of heat is added to the box by heat transfer from an external source. The surface temperature of the box is isothermal at 300.°F, and no work is done on or by the mixing box. Determine

a. The quality (or temperature if superheated) of the exit flow.
b. The rate of entropy production inside the mixing box.

34. You are now a world famous energy researcher commissioned by the National Entropy Foundation to determine the effect of vapor generation on the entropy production rate in a two-phase mixture. Your experiment consists of a steady state, closed loop flow system in which a liquid-vapor mixture of Refrigerant-134a flows through a test section consisting of a stainless steel tube 0.319 in. in diameter and 4.00 ft long. A constant wall heat flux is imposed along the length of the tube of 450. Btu/(h · ft²) The inlet quality is 0.00% and the outlet quality is 72.6%. The wall surface temperature along the length of the tube is given in °F by

$$T_w = 199.28 - 0.007217 \times T + 0.39449 \times T^2$$

for $0 \le x \le 4.00$ ft. The mass flow rate of the R-134a is 1000. lbm/h, and the inlet and outlet temperatures are 10.0°F and 0.00°F, respectively. Determine the total entropy production rate in the test section of Figure 9.29.

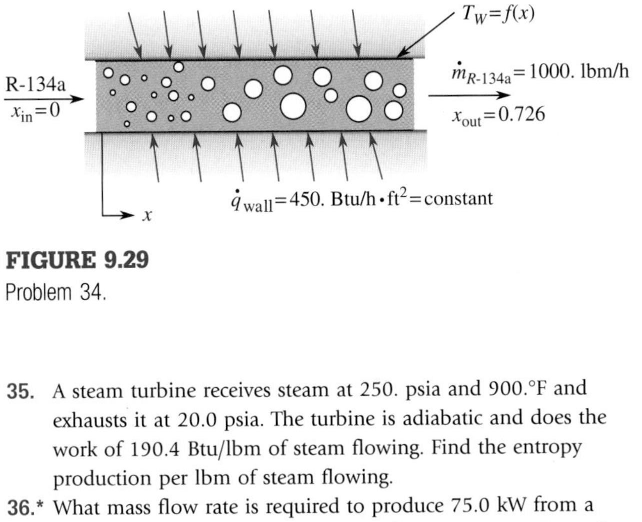

FIGURE 9.29
Problem 34.

35. A steam turbine receives steam at 250. psia and 900.°F and exhausts it at 20.0 psia. The turbine is adiabatic and does the work of 190.4 Btu/lbm of steam flowing. Find the entropy production per lbm of steam flowing.

36.* What mass flow rate is required to produce 75.0 kW from a steam turbine with inlet conditions of 2.00 MPa, 900.°C and exit conditions of 0.100 MPa, 200.°C if the turbine is reversible but not adiabatic? The heat transfer from the turbine occurs at a surface temperature of 50.0°C. Neglect any changes in kinetic and potential energy.

37. Calculate the isentropic efficiency of a continuous flow adiabatic compressor that compresses 20.0 lbm/min of a constant specific heat ideal gas. The test data for this compressor are inlet state = 1.00 atm and 25.0°C; outlet state = 1.00 MPa and 350.°C; c_p = 1.00 KJ/(kg · K), R = 0.250 kJ/(kg · K). The isentropic efficiency of a compressor is defined as

$$\eta_s = \frac{\dot{W}_{\text{isentropic compression}}}{\dot{W}_{\text{actual compression}}}$$

38. A steady flow, steady state air compressor with a surface temperature of 80.0°F handles 4000. ft³/min measured at the intake state of 14.1 psia, 30.0°F and a velocity of 70.0 ft/s. The discharge is at 45.0 psia and has a velocity of 280. ft/s. Both the inlet and exit stations are located 4.00 ft above the floor. Determine the discharge temperature and the power required to drive the compressor for

a. A reversible adiabatic process.
b. An irreversible adiabatic process with a compressor work transport efficiency of 80.0%.

39.* Determine the power required to compress 15.0 kg/min of superheated steam in an uninsulated, reversible compressor from 0.150 MPa, 600.°C to 1.50 MPa, 500.°C in a steady state, steady flow process. Neglect any changes in kinetic and potential energy. The boundary temperature is 20.0°C.

40. An adiabatic, steady flow compressor is designed to compress superheated steam at a rate of 50.0 lbm/min. At the inlet to the compressor, the state is 100. psia and 400.°F; and at the compressor exit, the state is 200. psia and 600.°F. Neglecting any kinetic or potential energy effects, calculate

a. The power required to drive the compressor.
b. The rate of entropy production of the compressor.

41.* A steady flow air compressor takes in 5.00 kg/min of atmospheric air at 101.3 kPa and 20.0°C and delivers it at an exit pressure of 1.00 MPa. The air can be considered an ideal gas with constant specific heats. Potential and kinetic energy effects are negligible. If the process is not reversible but is adiabatic and polytropic with a polytropic exponent of n = 147, calculate

a. The power required to drive the compressor.
b. The entropy production rate of the compressor.

42. An uninsulated, irreversible steam engine whose surface temperature is 200.°F produces 50.0 hp with a steam mass flow rate of 15.0 lbm/min. The inlet steam is at 400.°F, 100. psia; and it exits at 14.7 psia, 90.0% quality. Determine (a) the rate of heat loss from the engine and (b) its entropy production rate.

43.* A design for a turbine has been proposed involving the adiabatic steady flow of steam through the turbine. Saturated vapor at 300.°C enters the turbine and the steam leaves at 0.200 MPa with a quality of 95.0%. (a) Draw a *T-s* diagram for the turbine, and (b) determine the work and entropy production per kilogram of steam flowing through the turbine. The turbine's boundary temperature is 25.0°C.

44. An uninsulated, warp drive steam turbine on a Romulan battle cruiser has a surface temperature of 200.°F. It produces 50.0 hp with a steam mass flow rate of 150. lbm/min. The inlet steam is at 400.°F, 100. psia, and it exits at 16.0 psia, 90.0% quality. Determine

a. The heat transfer rate from the engine.
b. The entropy production rate of the engine.
c. Show whether the Romulans have discovered how to build steam engines that violate the second law of thermodynamics.

45.* Steam enters a turbine at 1.50 MPa and 700.°C and exits the turbine at 0.200 MPa and 400.°C. The process is steady flow, steady state, and adiabatic. The system boundary temperature is 35.0°C. Determine the following on the basis of a steam flow rate of 6.30 kg/s:

a. The entropy production rate of the turbine.
b. The work transport energy efficiency of the turbine.
c. The turbine's actual output power.

46. Steam at 400. psia and 50.0% quality is heated in a steady flow, isobaric heat exchanger until it becomes a saturated vapor. It is then expanded adiabatically through a turbine to 1.00 psia and 98.0% quality. This is followed by isobaric cooling to a saturated liquid in a second heat exchanger then compression and heating to the initial state. For the turbine alone, determine

a. The *net* power output per unit mass flow rate of steam.

b. The entropy production rate per unit mass flow rate of steam flowing in the system.

47.* Through a clerical error, our purchasing department ordered a finely crafted but mysterious device from a foreign manufacturer (Figure 9.30). The manuals are all in a foreign language, and the only intelligible information is in the form of some numbers printed next to the entrance and exit ports and the rotating shaft. The 10.0 kW rating on the shaft may mean either a work input or a work output. Determine

a. The flow direction, *a* to *b* or *b* to *a*.

b. The mass flow rate.

c. The entropy production rate of the device.

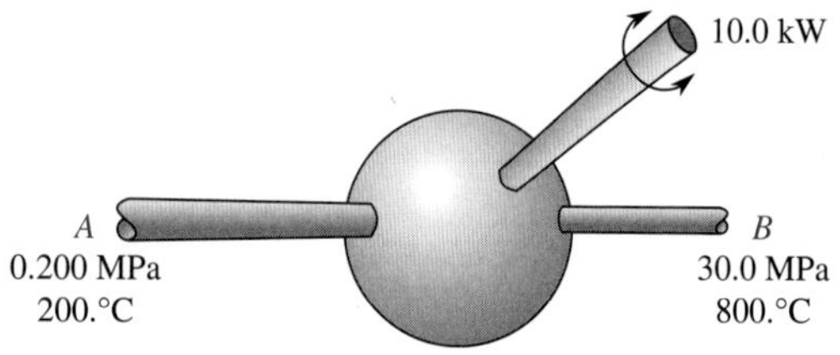

FIGURE 9.30
Problem 47.

Assume a steady state, adiabatic device and that the working substance is H_2O. Assume also that the kinetic and potential energies of outlet and inlet flow streams are negligible.

48. An irreversible, steady state, steady flow steam turbine that has no thermal insulation has an isothermal surface temperature of 100.°F. It operates with a steam mass flow rate of 15.0 lbm/s. The turbine inlet is at 300. psia, 1200.°F, and the exit is at 14.7 psia, 300.°F. If the turbine's entropy production rate is 0.500 Btu/(s · R), then determine

a. The turbine's heat transfer rate.

b. The turbine's work rate (power).

49.* A dog food manufacturer wishes to carry out the canning and sterilization process of a new pet food product at 100.°C, 0.0100 MPa. After crawling around in the rafters of the plant, an engineer finds a 1.60 MPa wet steam line. A pipe is run from this line to the canning area, where it produces 1.40 MPa steam with 2.00% moisture. Now, instead of just throttling down to the 0.0100 MPa state needed and wasting all that energy, the engineer decides to drop the pressure through a small adiabatic steam turbine.

a. What steam mass flow rate must be used to obtain 1.00 hp from the turbine?

b. What is the turbine's entropy production rate under these conditions?

50. A nuclear reactor heats a fluid for the steady flow power plant shown in Figure 9.31. The mass flow rate is 10.0 lbm/s. Determine

a. The horsepower input to the adiabatic pump.

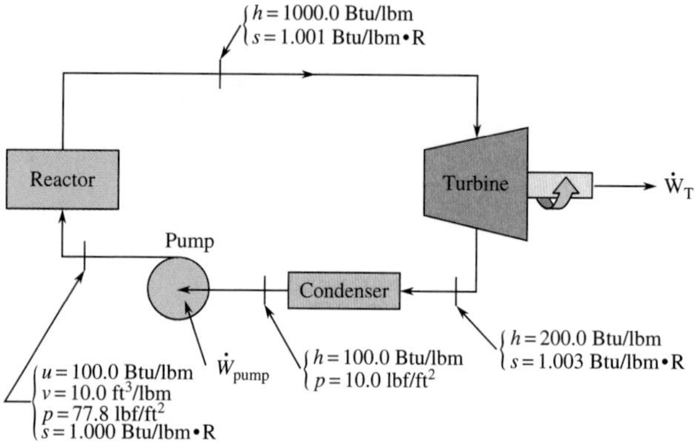

FIGURE 9.31
Problem 50.

b. The entropy production rate in the insulated reactor.

c. The entropy production rate in the insulated turbine.

51. The sales literature for the device shown in Figure 9.32 claims that the outlet temperature is slightly higher than the inlet temperature due to the presence of the vortex tube (see Case Study 9.1).

a. Assuming the vortex tube and the rest of the system are isentropic, determine the outlet temperature (T_4) from the data given in Figure 9.32.

b. Explain how the temperature rise claimed by the manufacturer could in fact exist.

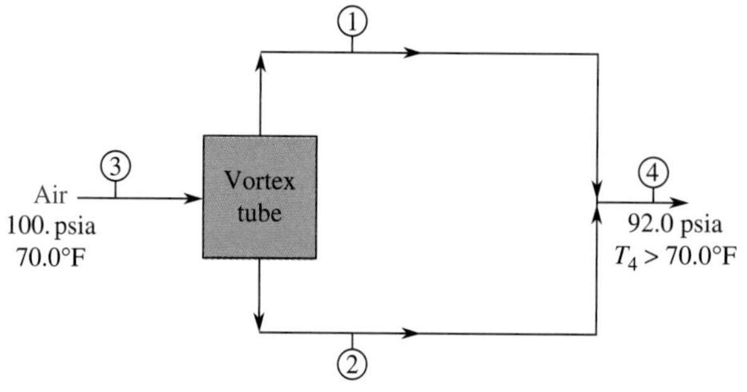

FIGURE 9.32
Problem 51.

52.* Determine the entropy production rate as 0.0500 kg/s of air flows through a vortex tube from an inlet pressure of 1.00 MPa. Both hot and cold side exit pressures are 101.3 kPa, and the hot side temperature is 50.0°C while the cold side temperature is –40.0°C. Two thirds of the inlet mass flow rate passes through the hot side exit. Assume constant, specific heat, ideal gas behavior and neglect any changes in kinetic and potential energy (see Case Study 9.1).

53. Using Eqs. (9.41), (9.29), and (7.33), discuss the possibility of having a temperature separation occur in a constant, specific heat, incompressible liquid flowing through a vortex tube (see Case Study 9.1).

54. A company claims to be able to manufacture a vortex tube using air that reaches –250.°F cold side and +250.°F hot side (both at atmospheric pressure) with an inlet pressure of 20.0 psig and a hot side mass flow fraction of 50.0%. Does their vortex tube violate the second law of thermodynamics? Assume constant specific heat ideal gas behavior and neglect any changes in kinetic and potential energy.

55.* Determine the maximum possible hot side exit temperature in a vortex tube using air when the cold side temperature is 0.00°C, the hot side mass flow fraction is 50.0%, the inlet pressure is 0.800 MPa, and both exits are at atmospheric pressure. Assume constant, specific heat, ideal gas behavior and neglect any changes in kinetic and potential energy.

56. Determine the heat transfer rate and the entropy production rate for the steady state, steady flow of water flowing through a straight horizontal 1.0 in. inside diameter pipe 10.0 ft long at a rate of 5.00 ft^3/min. The Darcy-Weisbach friction factor for this flow is 0.0320. The flow is isothermal at 75.0°F, and the mass density of the water is 62.26 lbm/ft^3.

57.* Determine the entropy production rate and heat transfer in a newly designed valve with 1.10 kg/s of an incompressible hydraulic oil flowing through it. The minor loss coefficient of the valve is 26.3, and the inlet and outlet oil temperatures are 53.0 and 49.0°C, respectively. The surface temperature of the valve is constant at 46.0°C. The flow velocity through the valve is constant at 3.00 m/s, and the specific heat of the oil is 1.13 kJ/(kg · K).

58.* A valve in an air handling system has a mass flow rate of 2.90 kg/min and a minor loss coefficient of 11.56. The valve inlet temperature is 20.0°C and the valve is insulated. The inlet and exit pressures of the valve are 1.66 and 0.300 MPa, respectively. The air velocity through the valve is 2.70 m/s. Assuming air to be an ideal gas with constant specific heats, determine the exit temperature and entropy production rate of the valve.

59. Air enters a sudden contraction in a pipe at a rate of 0.300 lbm/s. The contraction coefficient is 0.470, the exit temperature is 156°F and the exit diameter is 2.00 in. The pipe is insulated. The pressure across the contraction drops from 185 to 50.0 psia. Determine the entrance temperature and the entropy production rate of the contraction. Assume the air to be an ideal gas with constant specific heats.

60.* Liquid mercury enters an insulated sudden expansion with a velocity of 1.15 m/s at 20.0°C. The inlet and exit areas are 1.00×10^{-3} and 0.00×10^{-2} m^2, respectively. Assuming the flow is incompressible and adiabatic, determine the exit temperature and the entropy production rate of the expansion. The specific heat and density of the mercury are 0.1394 kJ/(kg · K) and 13,579 kg/m^3.

61.* It is required to dissipate 3.30 kJ/s of water flow energy in a spillway with a hydraulic jump. An amount of 1000. kg/s of water enters at 15.0°C and a depth of 0.500 m. The water passes through the jump fast enough to be considered adiabatic. Determine

a. The required hydraulic jump depth, $y_2 - y_1$ (see Table 9.4).
b. The exit water temperature.
c. The entropy production rate of the hydraulic jump.

62.* Suppose 0.730 kg/s of oil at 1.20 MPa and 20.0°C enters a system at 8.00 m/s and exits the system at 0.800 MPa, 40.0°C, and 4.00 m/s. The exit is 4.00 m below the inlet. The specific weight of the oil is 6000. N/m^3, which is a constant throughout the system. The specific heat of the oil is constant at 1.21 kJ/(kg · K). Determine (a) the heat transfer rate and (b) the entropy production rate of this system if its boundary temperature is maintained constant at 40.0°C.

63. Determine the entropy production rate for the isothermal steady laminar flow of a constant specific heat, incompressible power law non-Newtonian fluid in a horizontal circular tube of radius R whose velocity profile is given by

$$V = V_m\left[1 - (x/R)^{(n+1)/n}\right]$$

where V_m is the maximum (i.e., centerline) velocity, x is the radial coordinate measured from the tube's centerline, and n is the power law exponent (a positive constant). Sketch a plot of $\dot{S}_P$ vs. n and determine the value of n that minimizes $\dot{S}_P$.

64. The steady laminar flow of a constant specific heat, incompressible Newtonian fluid through a horizontal circular tube of radius R has a velocity profile given by

$$V = V_m\left[1 - (x/R)^2\right]$$

where V_m is the maximum (i.e., centerline) velocity and x is the radial coordinate measured from the tube's centerline. If there is a uniform heat flux $\dot{q}_s$ at the tube wall, then the temperature profile within the fluid (neglecting axial conduction) is given by

$$T = T_o + (\dot{q}_s R/k_t)\left[\alpha - (x/R)^2 + 0.25(x/R)^4\right] \quad \text{(a)}$$

where T_o and α are constants, and k_t is the thermal conductivity of the fluid.

a. Combine the heat transfer and viscous work entropy production rates per unit volume to show that the total entropy production rate per unit volume for this system is given by

$$\sigma = \frac{4\mu V_m^2 x^2}{R^4 T} + \frac{1}{k_t}\left\{\frac{\dot{q}_s}{T}\left[2x/R - (x/R)^3\right]\right\}^2 \quad \text{(b)}$$

b. Comment on the integration of σ over the system volume. Remember that the T in Eq. (b) is given by Eq. (a), and in a polar cylindrical coordinate system, $d\mathcal{V} = 2\pi Lx\,dx$. Carry out the integration analytically, if you can.
c. What factors in Eq. (b) can be manipulated to minimize $\dot{S}_P$?

65. The velocity and temperature profiles established in the free convection of a fluid contained between two flat parallel vertical walls maintained at different isothermal temperatures T_1 and T_2 are

$$V = \rho\beta g b^2 (T_2 - T_1)(x/b)[(x/b)^2 - 1]/12\mu$$

and

$$T = (T_1 + T_2)/2 - (T_2 - T_1)(x/2b)$$

where x is measured from the centerline between the plates, β is the coefficient of volume expansion, ρ is the density, and $2b$ is the distance between the plates (Figure 9.33). Determine a formula for the entropy production rate per unit depth and length of this flow due to viscous effects.

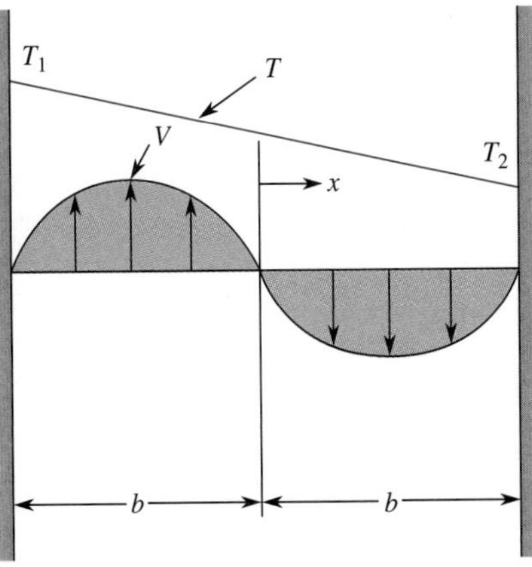

FIGURE 9.33
Problem 65.

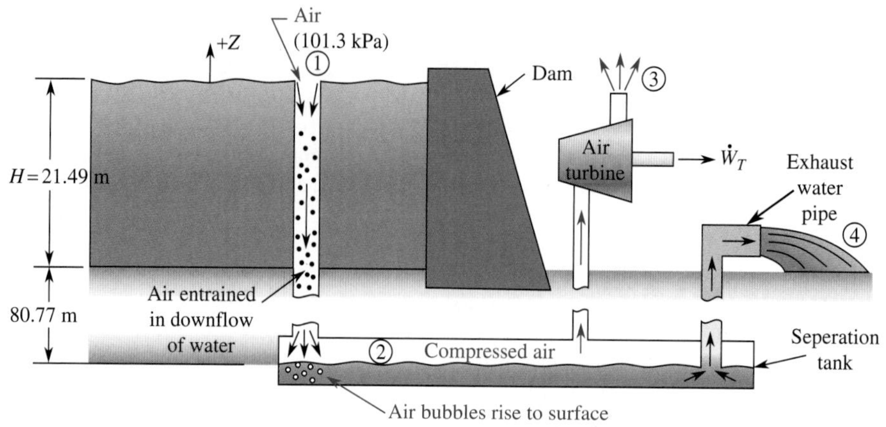

FIGURE 9.34
Problem 66.

66.* Figure 9.34 is a schematic of a novel hydraulic air compressor built in Michigan in 1906 at the Victoria Copper Mining Company and operated until 1921. This plant used the power of falling water from the Ontonagon River to produce high-pressure air with no moving mechanical parts. The falling water compressed entrained air bubbles, which were then separated from the water in a large underground separation tank. The efficiency of this system is defined to be rate of energy removed

$$\eta = \frac{\text{Rate of energy removed from the compressed air}}{\text{Net hydraulic power of the water}} = \frac{\dot{m}_a(h_2 - h_3)_a}{\dot{m}_w(gH/g_c)}$$

where the a subscript refers to the air and the w subscript refers to the water. Using the data given in Table 9.5, determine

a. The separation tank temperature, assuming the air and water are in constant thermal equilibrium in an adiabatic downflow.
b. The air turbine output power.
c. The entropy production rate for the compressor, assuming it is completely adiabatic.

In your analysis, assume air to be a constant specific heat ideal gas and water to be an incompressible liquid. Neglect any changes in kinetic energy of the air and the water. Note that the air turbine discharges to the atmosphere, so $p_3 = p_{atm} = 101.3$ kPa.

67.* A bottle of beer is emptied adiabatically at a rate of 0.100 kg/s. What is the rate of change of entropy of the contents of the beer bottle if the properties of the beer are

$$h = 53.0\text{ kJ/kg} \quad T = 10.0°\text{C}$$
$$p = 0.1013\text{ MPa} \quad s = 1.000\text{ kJ/(kg}\cdot\text{K)}$$

and the entropy production rate is 0.0100 kJ/(s · K).

Table 9.5 Data for the Victoria Hydraulic Air Compressor

$\eta = 57.4\%$	$\dot{m}_w = 6119.1$ kg/s
$p_{2a} = 882.6$ kPa	$T_{1a} = T_{1w} = 20.0°$C
$\dot{m}_a = 5.430$ kg/s	$H = 21.49$ m

Source: Data from W. Rice, "Performance of Hydraulic Gas Compressors." ASME Journal of Fluids Engineering (December 1976), pp. 645–653.

68. Helium at 70.0°F enters a 3.00 ft^3 rigid tank that is filled to a final pressure of 2200. psia. Assuming the tank is initially evacuated and that the helium behaves as an ideal gas with constant specific heats, determine the amount of entropy produced when the tank is filled (a) adiabatically and (b) isothermally at 70.0°F.

69. Plot the isothermal to adiabatic entropy production ratio for filling an initially evacuated container with equal amounts of an incompressible liquid vs. the dimensionless ratio $pv/(cT)$ as this ratio ranges from 0.00 to 10.0.

70.* A 0.0370 m^3 hydraulic cylinder is filled with 30.0 kg of oil (an incompressible liquid) from a supply at 20.0°C and 50.0 MPa. The specific heat of the oil is 1.83 kJ/(kg · K). Determine

a. The entropy produced if the cylinder is filled isothermally at 20.0°C.
b. The final temperature and the entropy produced if the cylinder is filled adiabatically.

71. Equation (9.47) shows that filling a rigid container adiabatically with an incompressible liquid produces less entropy than filling it isothermally. However, an adiabatic filling process produces a temperature rise in the vessel (see Eq. (6.35)) such that $T_{final} = T_2 > T_1$.

a. Develop a formula for the additional entropy produced when the rigid container adiabatically filled with an incompressible liquid is cooled back to the initial temperature by submerging it into a large isothermal bath at temperature T1.
b. Compare the total entropy produced by the adiabatic filling plus the cooling process described in part a with that produced during an isothermal filling process at temperature T_1.

Design problems

The following are elementary open-ended design problems. The objective is to carry out a preliminary thermal design as indicated. A detailed design with working drawings is not expected unless otherwise specified. These problems do not have specific answers, so each student's design is unique.

72. A liquid *cavitates* when the local pressure drops below the saturation pressure and the liquid begins to vaporize, or boil.

The cavitation process is the formation of these vapor bubbles. Carry out the preliminary thermodynamic design of a closed loop apparatus that illustrates this phenomenon by having a liquid pumped through a transparent nozzle wherein the pressure drops below the local saturation pressure and the vapor bubbles become visible. Choose a suitable liquid and provide an engineering sketch containing all the major dimensions and materials of your system. Estimate the pump size and power required, and the entropy production rate of your nozzle.

73. The Engine Test Facility at the Arnold Air Force Base in Tennessee requires a test cell 48.0 ft in diameter and 85.0 ft long with an air flow rate of 4300. ft^3/s at atmospheric pressure and temperature. Carry out the preliminary thermodynamic design of a compressor-nozzle system that meets these requirements. The compressor inlet is at atmospheric conditions, and the inlet temperature of the nozzle must be 70.0°F (this means an intercooling system must be used). Determine the horsepower required to drive the compressor, the nozzle outlet temperature, and the entropy production rate of your nozzle. Assume the compression process is polytropic with $n = 1.30$ and the pressure loss through the nozzle is 15.0% of the inlet pressure.

74. The Aeropropulsion System Test Facility at the Arnold Air Force Base in Tennessee requires a heat exchanger capable of cooling 2750. lbm/s of gas turbine exhaust from 3500. to 80.0°F before discharging it to the atmosphere. Carry out a preliminary thermodynamic design of a suitable heat exchanger using water as the second fluid. Determine the amount of cooling water needed and recommend an appropriate source. Also determine the entropy production rate of your heat exchanger.

75.* Carry out the preliminary thermodynamic design of a fuel mixing valve for a furnace that efficiently mixes inlet flow streams of air at 2.50 kg/s and methane at 0.500 kg/s, both at atmospheric pressure and temperature, and produces one outlet flow stream. Assume these gases behave as constant specific heat ideal gases. Provide a dimensioned engineering sketch and estimate the entropy production rate of your valve.

76.* Carry out the preliminary thermodynamic design of a system that uses a vortex tube to cool a full body suit for a firefighter. The suit must be able to reject up to 1300. kJ/h, and this cooling rate must be easily adjustable by the wearer. Use the data given in Case Study 9.1 or from appropriate industrial literature.

77. Carry out the preliminary thermodynamic design of a system that uses a vortex tube to heat a skintight suit for an underwater diver by bleeding off part of the air supply to the diver. The suit must be able to supply 800. Btu/h, and this heating rate must be easily adjustable by the diver. Use the data given in Case Study 9.1 or from appropriate industrial literature.

78.* Carry out the preliminary thermodynamic design of a system that isothermally fills an initially evacuated, rigid, cylindrical tank with air at 20.0°C to 20.0 MPa with a minimum amount of entropy production. The tank is 0.250 m in diameter and 1.50 m high. Assume the air behaves as a constant specific heat ideal gas. Discuss the technology and economics of how the tank is maintained isothermal during the filling process. Note that there will beentropy production in the pipes and valves used to connect the tank to the air supply.

79. The Von Karman Gas Dynamics Facility at the Arnold Air Force Base in Tennessee has need of a blowdown wind tunnel, to be supplied from a tank containing compressed air initially at 1000. psia and 70.0°F. The wind tunnel is to be at the end of a nozzle attached to the tank and must be 8.00 ft in diameter and have a velocity of 10.0×10^3 ft/s at 2000. R. The tank must be of sufficient size to sustain these wind tunnel conditions for a minimum of 30.0 min of testing. Carry out the preliminary thermodynamic design of such a facility and estimate the size of the storage tank required and the power required to compress the air in filling the tank if it must be done overnight (i.e., in 8.00 h). Determine the entropy production associated with both the filling and emptying of the tank if both are done isothermally.

Computer problems

The following computer assignments are designed to be carried out on a personal computer using a spreadsheet or equation solver. They are meant to be exercises using some of the basic formulae of this chapter. They may be used as part of a weekly homework assignment.

80.* Develop a program that allows you to plot the entropy production rate discussed in Example 9.1 vs. the heat transfer rate. Let the heat transfer rate range from 100. to 1000. watts. Assume all the remaining variables are as given in Example 9.1.

81. Develop a program that determines the entropy production rate of an incompressible fluid or a constant specific heat ideal gas of your choice flowing through an adiabatic nozzle. Input all the variables with proper units. Allow the choice of working in either Engineering English or SI units.

82. Develop a program that determines the entropy production rate of an incompressible fluid or constant specific heat ideal gas of your choice from an adiabatic diffuser. Input all the variables with proper units. Allow the choice of working in either Engineering English or SI units.

83. Develop a program that determines the entropy production inside a heat exchanger having two inlets and two outlets, when the mass flow rates, temperatures, and fluid properties of both of the flow stream fluids are known. Assume the fluids do not mix inside the heat exchanger, and allow either flow stream to be an incompressible liquid or a constant specific heat ideal gas at the user's discretion. Input all the variables with proper units. Allow the choice of working in either Engineering English or SI units.

84. Develop a program that performs an energy rate balance on a gas turbine engine. Input the appropriate gas properties (in the proper units), the turbine's heat loss or gain rate, and the input mass flow rate, and the inlet and exit temperatures. Output to the screen the turbine's output power. Assume the gas behaves as a constant specific heat ideal gas, and neglect all kinetic and potential energy terms. Allow the choice of working in either Engineering English or SI units.

85. Curve fit the hot and cold vortex tube outlet temperature data given in Case Study 9.1 vs. the inlet absolute pressure. Then develop a program that returns these outlet temperatures plus the COP of this device when it is used as a Carnot heat pump and when it is used as a Carnot refrigerator or air conditioner when the user inputs the inlet pressure. Use this program to generate enough data to plot the values of these two COPs vs. the inlet absolute pressure.

86. Develop a program that outputs the entropy production rate for the filling of an initially evacuated rigid vessel with an incompressible liquid or an ideal gas of your choice when the vessel is filled either adiabatically or isothermally (again, your choice). Input all necessary information in proper units. Allow the user to work in either Engineering English or SI units.

Create and solve problems

Engineering education tends to focus on the process of solving problems. It ignores teaching the process of formulating solvable problems. However, working engineers are never given a well-phrased problem statement to solve. Instead, they need to react to situational information and organize it into a structure that can then be solved using the methods learned in college.

These "Create and Solve" problems are designed to help you learn how to formulate solvable thermodynamics problems from engineering data. Since you provide the numerical values for some of the variables, these problems do not have unique solutions. Their solutions depend on the assumptions you need to make and how you set them up to create a solvable problem.

87. You are a new engineer at a small paper making company. The company's chief engineer wants to install a new drum dryer that produces heat at a rate of 500,000 Btu/h with a drum surface temperature of 175°F. The factory has a steam boiler and can produce saturated vapor at any desired pressure and flow rate. Write and solve a thermodynamics problem that satisfies the chief engineer's needs. Choose a steam mass flow rate and inlet conditions, then determine the exit quality and the entropy production rate of the dryer.

88.* Your boss wants to know how much power is required to compress 15.0 kg/min of air (an ideal gas) in an insulated compressor from 0.100 MPa, 20.0°C to 1.50 MPa, in a steady state, steady flow process. Write and solve a thermodynamics problem that provides her with the answer plus the final temperature of the air.

89. Your new job at a domestic refrigerator manufacturing company involves the design of expansion valves. The liquid refrigerant passes through the expansion valve (also called a *throttle valve*), where its pressure abruptly decreases, causing flash evaporation of 35% of the liquid. The resulting mixture of liquid and vapor, at a lower temperature and pressure, then travels through the refrigerator's evaporator coil and is completely vaporized by cooling the warm air of the space being refrigerated. The resulting refrigerant vapor returns to the refrigerator's compressor inlet to complete the thermodynamic cycle. For a new refrigerator design, saturated liquid Refrigerant 134a enters the expansion valve at 80.0°F and exits at 10.0°F. Your supervisor needs you to determine the following items: (a) The increase in entropy per lbm of R-134a flowing through the expansion valve. (b) The entropy production rate per unit mass flow rate of R-134a flowing through the expansion valve. (c) The average Joule-Thomson coefficient for this process. Write and solve a thermodynamics problem to provide the answers.

90.* The corporate vice president of the company that employs you needs to know what the entropy production and the heat transfer rates are in a newly designed valve that has been designed to control the flow of saturated liquid ammonia. At the normal flow rate 0.55 kg/s, the minor loss coefficient of the valve is 18.3, and the inlet and outlet oil temperatures are 3.00 and 19.0°C, respectively. The surface temperature of the valve is constant at 20.0°C. Under normal conditions, the flow velocity through the valve is constant at 7.50 m/s. Write and solve a thermodynamics problem that gives your corporate vice president the answers to his questions.

CHAPTER 10

Availability Analysis

CONTENTS

10.1 WHAT IS AVAILABILITY?

Availability[1] is a measure of how useful the energy within a system can be. It is the name given to the amount of energy within a system that is "available" to do useful work, and consequently, it is a measure of the "quality" of the energy present within a system. Some forms of energy within a system are more available to do useful work than others. Consequently, these energy forms have a higher value, or availability, than the others. In addition, a system may contain forms of energy that cannot do any useful work at all because they do not have a high enough potential, so they are called *unavailable* energy. Figure 10.1 illustrates how a system can have available and unavailable energy.

For example, one gallon of gasoline contains about 158 MJ (150,000 Btu) of energy and currently sells for about \$2–\$4 dollars. Yet, if you purchase 158 MJ of electrical energy from your local electrical power company, it costs you about \$5. Why does electrical energy cost more than an equivalent amount of chemical energy? The answer is simply that electrical energy is more available to do useful work than chemical fuel. An electric motor can convert about 90% of the electrical energy supplied to it into useful output work, whereas an internal

[1] Availability is called *exergy* and *essergy* (essence of energy) in European textbooks. The term *exergy* was coined to be similar in form to the words energy and entropy, and the term *essergy* is a contraction of "essence of energy." Both terms lack any obvious connotation and consequently are not used here.

CRITICAL THINKING

Is some of the energy with a system "unavailable"? Does the magnitude of the available energy within a system depend on the accessible technology? What about nuclear energy, is it "available" to do useful work? If not, could future technologies make it available?

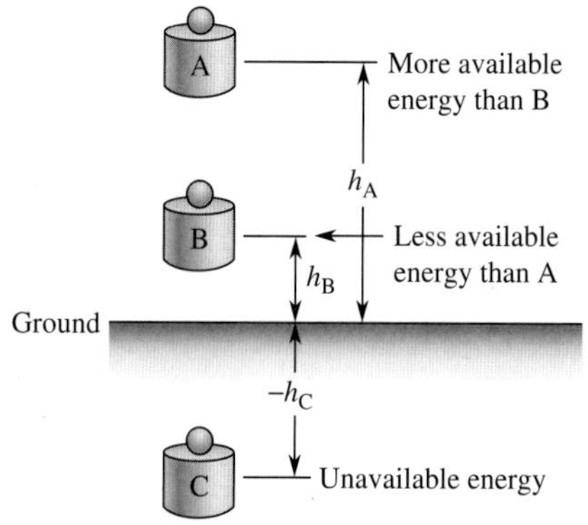

FIGURE 10.1
A system's available and unavailable energy. The potential energy of weight *A* has more "available" potential energy to do useful work relative to the ground than does weight *B*. However, the potential energy of weight *C* is negative relative to the ground and therefore is "unavailable" to do useful work relative to the ground.

combustion engine can convert only about 20–30% of the chemical energy in its fuel into useful output work. Therefore, electrical energy is more valuable to our society than chemical fuel (notice that we have not taken into account how the electrical energy is generated). Being more valuable, it costs more because it has a higher "quality" of work "availability."

Much of the world around us today is driven by various energy conversion technologies. Automobiles, power plants, manufacturing facilities, and computers are all important elements in our society today. Since we know that energy is conserved in all processes, energy would seem to be an inexhaustible resource to be used over and over again, powering our technological needs. However, though energy is conserved, it is *degraded* through use to a form that is not useful in technological systems. For example, the kinetic energy of a bouncing rubber ball is continuously degraded by internal friction produced by the deformation of the ball during impact. This causes the height to which the ball rebounds after each impact to progressively decrease. Using the second law of thermodynamics, we can pinpoint where energy degradation occurs within an engineering system; and by redesigning the system to minimize the energy degradation, we can improve the overall energy conversion performance of the system.

With increasing material costs and decreasing resources, it is apparent that the efficient use of energy resources will be of primary importance to engineers in the future. In engineering design, the concept of pure energy in somewhat misleading. *What a designer needs to know is how much of the energy present in a given system can be used for a particular process.* That is, of the total energy contained within a system, how much is available to do useful work? Therefore, it is important for engineers to understand how to determine the consequence (i.e., the amount of *useful* energy available) of the energy within a given system. The manipulation of the amount of energy present is the subject of the first law of thermodynamics. The second law of thermodynamics tells us where the irreversibilities within the system are located (through the entropy production term). If we combine these two laws and define a new thermodynamic property (called *availability*) that is a measure of the useful energy, we can then identify where energy is being degraded or lost within a system and decide how to modify the system to reduce these losses.

This chapter provides a basic introduction to the availability property and the availability balance. Open and closed system availability balances are developed, and a new system efficiency based on the second law of thermodynamics is developed. Examples are presented illustrating the use of this material for power plants, refrigeration systems, heat pumps, internal combustion engines, and heat exchangers. A summary at the end of the chapter reiterates the main concepts and equations developed in the chapter.

Useful work is associated with a quantity called the *potential of a conservative force*. The following section on fields and forces provides a background on the mathematical nature of fields, potentials, and conservative forces.

10.2 FUN WITH SCALAR, VECTOR, AND CONSERVATIVE FIELDS

10.2.1 Scalar and Vector Fields

A scalar (or vector) function is a mathematical expression defined at each point in a region of space whose *resultant* value is a scalar (or vector) quantity that depends only on the point where the expression was evaluated and not on the coordinate system used. The region of space over which a scalar function is defined is called the *scalar field* of the scalar function, and the region of space over which a vector function is defined is called the *vector field* of the vector function. For example, the scalar pressure, temperature, and density functions at each point within an object define the pressure, temperature, and density scalar fields of the object. Similarly, the velocity vector at each point within a flowing fluid defines the velocity vector field of the fluid.

10.2.2 Conservative Fields

A vector field is said to be *conservative* if it has a vanishing line integral around every closed path c in its region of definition, or

$$\oint_c \vec{A} \cdot d\vec{x} = 0$$

Since the line integral of a conservative vector field $\vec{A}$ around any closed path is always zero, the value of its integral between any two arbitrary points $\vec{x}_1$ and $\vec{x}_2$ depends only on the end points themselves and is independent of the path taken between these points (i.e., the integral is a *point function*). Further, any vector field $\vec{A}$ that obeys the preceding equation must also obey the equation $\vec{A} = -\vec{\nabla}P$, where $\vec{\nabla}$ is the differential operator[2] (called the *del* operator) and $P(\vec{x})$ is called the *potential*[3] of $\vec{A}$ at $\vec{x}$. The term $\vec{\nabla}P$ is called the *gradient* of P, and for a point $\vec{x}$ and an arbitrarily chosen reference point $\vec{x}_0$, we can write

$$P(\vec{x}) = P(\vec{x}_0) - \int_{\vec{x}_0}^{\vec{x}} \vec{A} \cdot d\vec{x}$$

The *gradient* of a scalar function P is usually written as $\vec{\nabla}P = \vec{A}$, where $\vec{A}$ is called the *gradient vector.*

10.3 WHAT ARE CONSERVATIVE FORCES?

Any force produced by a reversible process is a point function (i.e., it depends only on the end points of the process) and is therefore called a *conservative force*. The term *conserve* means to preserve from loss or decay. The "conservation" laws of physics state that certain measurable quantities (e.g., mass, energy, momentum) do not change with time in an isolated system. Similarly, a conservative force is one whose magnitude is not diminished in time by its own action (i.e., it is nondissipative). Thus, the total work done by a conservative force is independent of the path producing the displacement and is equal to zero when its path is a closed loop, or

$$W_{\substack{\text{conserative}\\\text{force}}} = \oint_C \vec{F} \cdot d\vec{r} = 0.$$

Since conservative forces are reversible (i.e., nondissipative), they form a conservative vector field and have an energy *potential* (or stored energy), Φ, defined as $\vec{F} = -\vec{\nabla}\Phi$. Nonconservative (irreversible) forces (such as friction) that depend on other forces (such as sliding velocity) are dissipative, and no energy potential can be defined for them.

OK, BUT WHAT DOES ALL THIS REALLY MEAN?

In simpler terms, this means that we can divide forces into two categories: conservative forces and nonconservative forces. If the "net" work done by a force acting on a system is always zero, the force is said to be conservative. In other words, if the work done by a force depends only on the initial and final states of a system and not on the path taken by the force, then it is a conservative force. Otherwise, it is non-conservative.

Examples of conservative forces:

- The force of gravity.
- Coulomb's force in electrostatics.
- A completely elastic deformation.

Examples of nonconservative forces:

- Friction (both sliding and viscous).
- Inelastic (or plastic) deformation.
- Electrical resistance.

---

[2] This operator has the following form in Cartesian coordinates

$$\vec{\nabla}(\,) = \frac{\partial(\,)}{\partial x}\vec{i} + \frac{\partial(\,)}{\partial y}\vec{j} + \frac{\partial(\,)}{\partial z}\vec{k}.$$

[3] Note that the potential P is not uniquely determined by $\vec{A}$, since any other potential of the form $P' = P +$ a constant also satisfies this equation. Consequently, a conservative force field $\vec{A}$ can always be written as the negative gradient of its potential P.

10.4 MAXIMUM REVERSIBLE WORK

Any mathematical point function can be interpreted as the potential of a conservative vector. Since all thermodynamic properties (both extensive and intensive) are point functions, they must also represent such potentials. Some of these vector potential relations are well known. For example, Fourier's law relates the *conduction heat transfer rate per unit area* vector to its potential, the local temperature T, as $\vec{q} = -k_t \vec{\nabla}(T)$, where k_t is the thermal conductivity.

Since the potential of a conservative force is equal to the reversible work done on or by a system, perhaps it would be useful to know the potential for the *maximum possible* reversible work that a system could produce from any given state. Using a combination of the energy and entropy balance, we can compute the reversible work produced or absorbed by a closed system with isothermal boundaries at temperature T_b, as the system changes from state 1 to state 2, as

$$({}_1W_2)_{\text{rev}} = E_1 - E_2 + ({}_1Q_2)_{\text{rev}} = E_1 - E_2 - T_b(S_1 - S_2)$$

where $E = U + mV^2/2g_c + mgZ/g_c$. To determine the maximum possible reversible work we must define a *minimal energy reference state*, which we call the *ground state* of the system.[4] We denote quantities in the ground state with a zero subscript (e.g., p_0 and T_0 denote the pressure and temperature of the ground state). Then, for an arbitrary starting state, the maximum reversible work that a closed system can perform is

$$(W)_{\substack{\text{maximum}\\\text{reversible}}} = E - E_0 - T_0(S - S_0) \tag{10.1}$$

where the system boundary is now assumed to be at the ground state temperature T_0 (see Figure 10.2). Note that the change in entropy in Eq. (10.1) is solely due to a heat transfer by the system and not as a result of irreversibilities (either internal or external) that may occur during the process of bringing the system to the ground state.

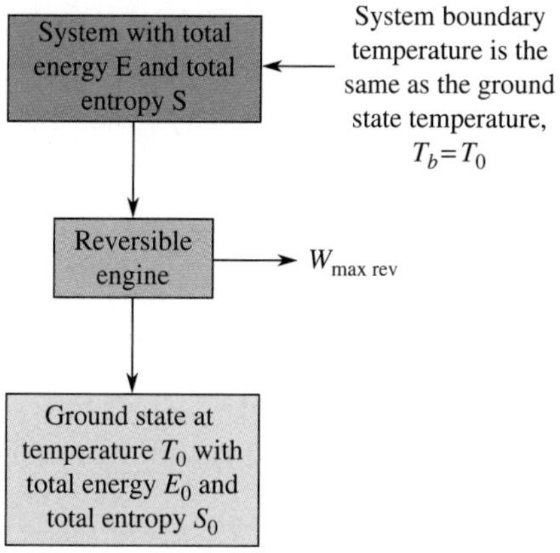

FIGURE 10.2
The maximum reversible work a system can produce. Note that the system boundary must be at the same temperature as the ground state to produce the maximum reversible work.

Now, what constitutes a suitable *ground* state? If we are to have a minimum system energy in the ground state, then clearly the kinetic and potential energies of this state should be zero, so we set $V_0 = Z_0 = 0$. Beyond this, the remaining properties of the ground state can be arbitrarily chosen.

10.5 LOCAL ENVIRONMENT

If the maximum reversible work is to be a useful concept in engineering analysis, a practical ground state must be chosen. It is easy to see that the most convenient ground state for any given system is its *local environment*. Recall that the term *surroundings* refers to everything outside the system boundaries, consequently the surroundings must include the local environment. We therefore define the *local environment* and the *ground state* as in the display boxes. Since the ground state and the local environment are essentially identical, we denote all the properties of the ground state and local environment with a zero subscript (see Figure 10.3).

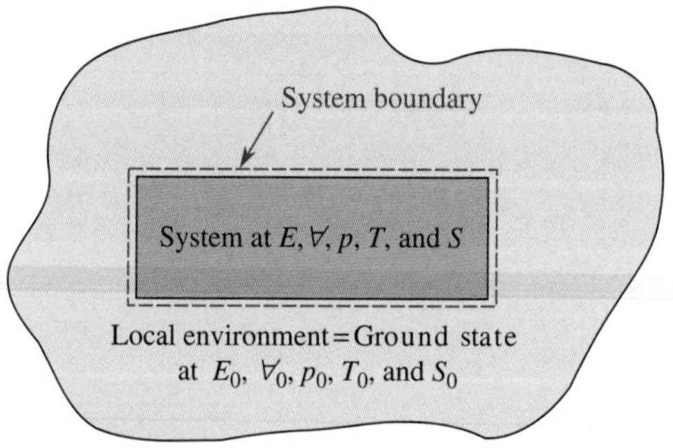

FIGURE 10.3
The surroundings, local environment, and the ground state.

WHAT IS THE *LOCAL ENVIRONMENT*?

The local environment is a portion of the total surroundings in contact with the system boundaries. It must be large enough for all of its intensive properties to be constant, and it must be insensitive to state changes of the system.

---

[4] Some authors call this the *dead state*. However, the term *ground state* is more easily understood, since it is conceptually similar to the zero energy *ground* level reference state commonly used in gravitational potential energy analysis.

WHAT IS A SYSTEM'S *GROUND STATE*?

The "ground state" of a system is the state of a system that has zero kinetic and potential energies and that is in thermodynamic equilibrium with its local environment. Consequently, the local environment of a system is at the system's ground state.

WHAT IS THE GROUND STATE NOTATION?

Zero subscripts are used to denote the ground state and the local environment thermodynamic properties as T_0, p_0, $\mathcal{V}_0$, v_0, U_0, u_0, H_0, h_0, S_0, and s_0.

10.6 AVAILABILITY

Equation (10.1) represents the maximum possible reversible work that a system could produce as it changed states to the ground state, but it is not the maximum possible *useful* reversible work that could be produced. To obtain an expression for the *useful* work, we must subtract the work associated with moving the local environment as the system volume changes from its initial volume $\mathcal{V}$ to its final volume $\mathcal{V}_0$ in the ground state. Since the local environment is at a constant pressure p_0, Eq. (4.26) gives the reversible moving boundary work as

$$(W)_{\substack{\text{reversible}\\\text{moving}\\\text{boundary}}} = p_0(\mathcal{V}_0 - \mathcal{V}) = -p_0(\mathcal{V} - \mathcal{V}_0)$$

Then, we can compute:

$$(W)_{\substack{\text{maximum}\\\text{reversible}\\\text{useful}}} = (W)_{\substack{\text{maximum}\\\text{reversible}}} - (W)_{\substack{\text{reversible}\\\text{moving}\\\text{boundary}}} = E - E_0 + p_0(\mathcal{V} - \mathcal{V}_0) - T_0(S - S_0) \tag{10.2}$$

Since we are dealing with "reversible" work modes here, the maximum useful reversible work for a closed system given by Eq. (10.2) must be the *potential* for some conservative force vector, and we could call this work the *energy potential* (or *work potential*) of that force. However, the term *energy potential* is easily confused with the term *potential energy* (the potential of the gravitational force vector) and consequently is not a good choice for the name of this new potential. Therefore, we choose the term *availability* for the name of this new potential, since this term is synonymous with the word *useful* and it represents the energy available within the system to do useful work. Figure 10.4 illustrates this new concept.

Therefore, the *total availability*, A, of a closed system is defined as

$$\begin{aligned} A \equiv (W)_{\substack{\text{maximum}\\\text{reversible}\\\text{useful}}} &= E - E_0 + p_0(\mathcal{V} - \mathcal{V}_0) - T_0(S - S_0) \\ &= m[u - u_0 + p_0(v - v_0) - T_0(s - s_0) + V^2/2g_c + gZ/g_c] \end{aligned} \tag{10.3}$$

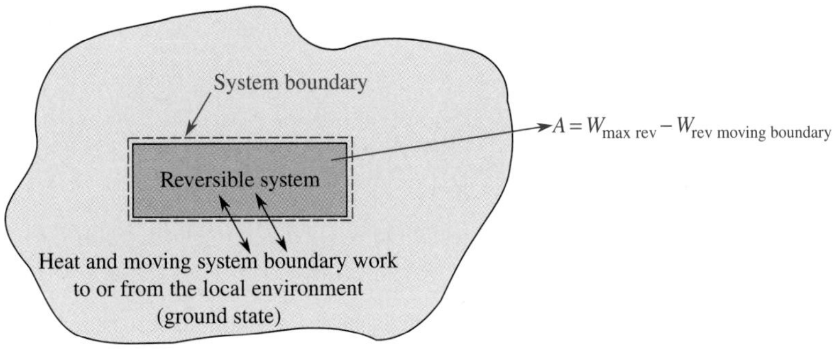

FIGURE 10.4
Availability as the maximum, reversible, useful work that can be produced.

WHERE DID THE NAME "AVAILABILITY" COME FROM?

The potential we call *availability* today has been known by various names over the years. In 1873, Gibbs called it the *available energy of a body and a medium*; in 1889, Gouy called it the *utilizable energy*; and in 1953, Rant named it *exergy*. Joseph Keenan introduced the term *availability* in 1941, and it is still used.

where the kinetic and potential energy terms have been moved to the end of the last expression for convenience. The *specific availability*, *a*, is now defined as

Specific availability, *a*

$$a = A/m = u - u_0 + p_0(v - v_0) - T_0(s - s_0) + V^2/2g_c + gZ/g_c \tag{10.4}$$

Most processes do not take the system to its ground state. In that case, as a closed system undergoes a process that carries it from state 1 to state 2, the change in total and specific availabilities are

$$\begin{aligned} A_2 - A_1 &= E_2 - E_1 + p_0(\mathcal{V}_2 - \mathcal{V}_1) - T_0(S_2 - S_1) \\ &= m\left[u_2 - u_1 + p_0(v_2 - v_1) - T_0(s_2 - s_1) + (V_2^2 - V_1^2)/2g_c + g(Z_2 - Z_1)/g_c\right] \end{aligned} \tag{10.5}$$

$$a_2 - a_1 = u_2 - u_1 + p_0(v_2 - v_1) - T_0(s_2 - s_1) + (V_2^2 - V_1^2)/2g_c + g(Z_2 - Z_1)/g_c \tag{10.6}$$

The following examples illustrate the calculation of these availability functions for various closed systems.

EXAMPLE 10.1

A cylindrical drinking glass 0.0700 m in diameter and 0.150 m high is three-fourths full of cold liquid water at 10.0°C and is placed at 0.762 m above the floor on a table in a room. Determine the total availability of the water in the glass relative to the floor. Take the local environment (ground state) to be at $p_0 = 0.101$ MPa and $T_0 = 20.0°C = 293$ K.

Solution

First, draw a sketch of the system (Figure 10.5).

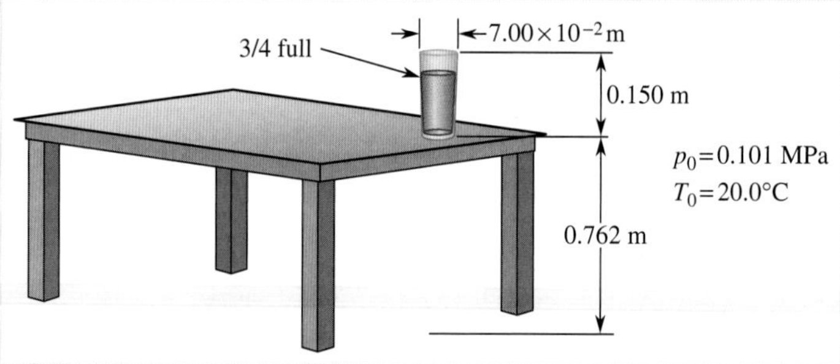

FIGURE 10.5
Example 10.1.

The unknown is the total availability of the water in the glass relative to the floor. The material is water, and the system is closed.

The total availability of a system is given by Eq. (10.3) as

$$A = m\left[u - u_0 + p_0(v - v_0) - T_0(s - s_0) + \frac{V^2}{2g_c} + \frac{gZ}{g_c}\right]$$

Water at 10.0°C and atmospheric pressure is a slightly compressed liquid. However, the amount of compression is very small, and we can use the *u*, *v*, *s*, and u_0, v_0, s_0 values of saturated liquid water at 10.0°C and 20.0°C, respectively from Table C.1b of *Thermodynamic Tables to accompany Modern Engineering Thermodynamics* in the calculation. The glass is not moving, so $V = 0$, but it does have a potential energy defined by its height at $Z = 0.762$ m. The mass of water in the glass is

$$m = \pi R^2 L\rho = \pi(0.0350\,\text{m})^2\,(3/4 \times 0.150\,\text{m})\,(1000.\,\text{kg/m}^3) = 0.433\,\text{kg}$$

and the total available energy is then

$$A = (0.433\,\text{kg})[(42.0 - 83.9\,\text{kJ/kg}) + (101\,\text{kN}/m^2)(0.001000 - 0.001002\,\text{m}^3/\text{kg})$$
$$- (293\,\text{K})(0.1510 - 0.2965\,\text{kJ/kg}\cdot\text{K}) + 0 + (9.81\,\text{m/s}^2)(0.762\,\text{m})/1] = 3.56\,\text{kJ}$$

Exercises

1. In Example 10.1, the total available energy was positive even though the water is colder than the local environment and the contribution from the potential energy term is quite small. Determine the contribution of each term in the total availability equation in this example and comment on its significance. **Answer**: $m(u - u_0) = -18.1$ kJ, $mp_0(v - v_0) \approx 0$ kJ, $-mT_0(s - s_0) = 18.5$ kJ, and $mgZ/g_c = 3.23$ kJ. The entropy change dominates.
2. Recalculate the total availability in Example 10.1 when the water is at 20.0°C instead of 10.0°C. **Answer**: 3.23 kJ. (The only nonzero term here is potential energy.)
3. If the mass of water in the glass in Example 10.1 is doubled, but all the other variables remain unchanged, determine the new availability of the water. **Answer**: A = 7.12 kJ.

CRITICAL THINKING

Why is some of the energy within a system unavailable for use? Does its availability or unavailability depend on the technology being used? Could energy that is unavailable in one system be considered as available in another? Give some examples.

EXAMPLE 10.2

Determine the specific available energy in a stationary, rigid tank containing air (an ideal gas) at 20.0°C and 1.500 MPa. Take the local environment (ground state) to be at $p_0 = 0.101$ MPa and $T_0 = 20.0°\text{C} = 293$ K.

Solution

First, draw a sketch of the system (Figure 10.6).

The unknown is the specific available energy in a stationary, rigid tank, The material is air and the system is closed.

The specific availability is given by Eq. (10.4) as

$$a = u - u_0 + p_0(v - v_0) - T_0(s - s_0) + V^2/2g_c + gZ/g_c$$

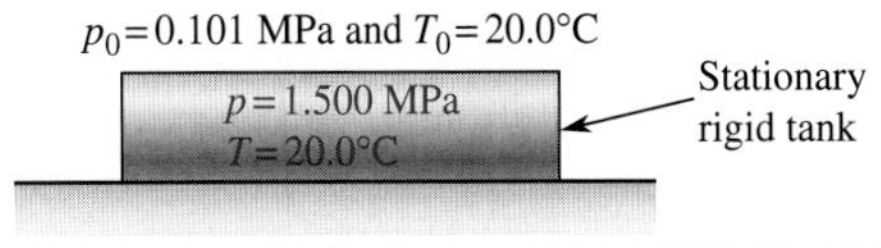

FIGURE 10.6
Example 10.2.

For an ideal gas, $u - u_0 = c_v(T - T_0)$, $v - v_0 = R(T/p - T_0/p_0)$, and $s - s_0 = c_p\ln(T/T_0) - R\ln(p/p_0)$. Then,

$$a = c_v(T - T_0) + p_0R(T/p - T_0/p_0) - T_0c_p\ln(T/T_0) + V^2/2g_c + gZ/g_c$$

or

$$a = (0.781\,\text{kJ/kg}\cdot\text{K})(20.0 - 20.0°\text{C}) + (101\,\text{kN/m}^2)(0.286\,\text{kJ/kg}\cdot\text{K})\left(\frac{293\,\text{K}}{1500.\,\text{kN/m}^2} - \frac{293\,\text{K}}{101\,\text{kN/m}^2}\right)$$
$$- (293\,\text{K})\left[1.004\,\text{kJ/kg}\cdot\text{K})\ln\left(\frac{293\,\text{K}}{293\,\text{K}}\right) - (0.286\,\text{kJ/kg}\cdot\text{K})\ln\left(\frac{1500.\,\text{kN/m}^2}{101\,\text{kN/m}^2}\right)\right] + 0 + 0 = 148\,\text{kJ/kg}$$

Exercises

4. If the tank temperature in Example 10.2 is increased from 20.0°C to 200.°C and all the other variables remain unchanged, determine the new specific availability of the gas in the tank. **Answer**: a = 140. kJ/kg.
5. If the pressure in the tank in Exercise 10.2 is reduced from 1.500 MPa to 0.500 MPa and all the other variables remain unchanged, determine the new specific availability of the gas in the tank. **Answer**: a = 67.0 kJ/kg.
6. Recalculate the specific availability in Example 10.2 when the tank contains carbon dioxide at 50.0°C and 2.250 MPa. Assume all the other variables remain unchanged. **Answer**: a = 115 kJ/kg.

EXAMPLE 10.3

The instrument cooling system on an aircraft flying at 30.0 × 10^3 ft at 500. mph contains 5.00 lbm of saturated liquid Refrigerant-22 at 50°F. As the aircraft lands, a malfunction occurs that causes the refrigerant to be heated to 400.°F and 100. psia. Determine

a. The total availability of the refrigerant before and after the aircraft lands.
b. The change in total availability during the landing.

Take the local environment (ground state) to be saturated liquid Refrigerant-22 at T_0 = 70.0°F (so that $p_0 = p_{sat}(70.0°F)$ = 136.12 psia).

Solution

First, draw a sketch of the system (Figure 10.7).

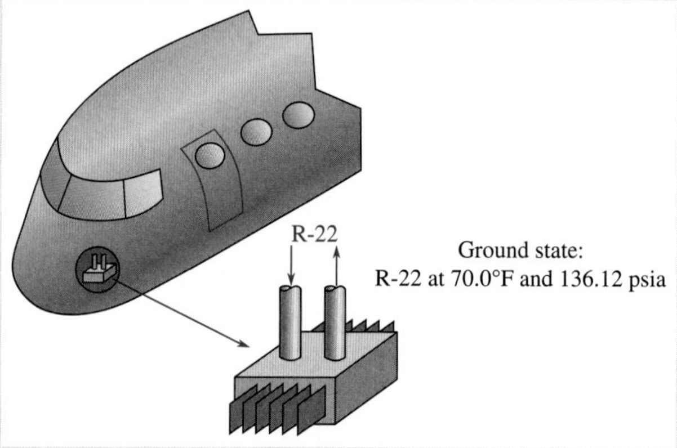

FIGURE 10.7
Example 10.3.

The unknowns are the total availability of the refrigerant before and after the aircraft lands and the change in total availability during the landing. The material is R-22, and the system is closed.

a. The total availability of a system is given by Eq. (10.3) as

$$A = m\left[u - u_0 + p_0(v - v_0) - T_0(s - s_0) + \frac{V^2}{2g_c} + \frac{gZ}{g_c}\right]$$

Tables C.9a and C.10a give the thermodynamic properties of the refrigerant as

State 1 (flying)	State 2 (landed)	Ground state
$x_1 = 0.00$	$p_2 = 100.\text{psia}$	$x_0 = 0.00$
$T_1 = 50.0°F$	$T_2 = 400.°F$	$T_0 = 70.0°F$
$v_1 = 0.01281\ \text{ft}^3/\text{lbm}$	$v_2 = 1.046\ \text{ft}^3/\text{lbm}$	$v_0 = 0.01325\ \text{ft}^3/\text{lbm}$
$u_1 = 24.04$ Btu/lbm	$u_2 = 154.77$ Btu/lbm	$u_0 = 29.78$ Btu/lbm
$s_1 = 0.0519$ Btu/lbm·R	$s_2 = 0.31464$ Btu/lbm·R	$s_0 = 0.06296$ Btu/lbm·R
$V_1 = 500$ mph	$V_2 = 0$	$p_0 = 136.12$ psia
$Z_1 = 30{,}000$ ft	$Z_2 = 0$	

The total availability is given by Eq. (10.5) as

$$A = m\left[u - u_0 + p_0(v - v_0) - T_0(s - s_0) + V^2/2g_c + gZ/g_c\right]$$

so

$$A_1 = (5.00\,\text{lbm})\left[\left(24.04 - 29.78\,\frac{\text{Btu}}{\text{bm}}\right) + \left(136.12\,\frac{\text{lbf}}{\text{in}^2}\right)\left(\frac{144\,\text{in}^2/\text{ft}^2}{778.16\,\text{ft}\cdot\text{lbf/Btu}}\right)\left(0.01218 - 0.01325\,\frac{\text{ft}^3}{\text{lbm}}\right)\right]$$

$$-(5.00\,\text{lbm})(530\,\text{R})\left(0.0519 - 0.06296\,\frac{\text{Btu}}{\text{lbm}\cdot\text{R}}\right)$$

$$+(5.00\,\text{lbm})\left[\frac{[(500.\,\text{mph})(5280\,\text{ft/mi})(1\,\text{h}/3600\,\text{s})]^2}{2(32.174\,\text{lbm}\cdot\text{ft/lbf}\cdot\text{s}^2)(778.16\,\text{ft}\cdot\text{lbf/Btu})} + \frac{(32.174\,\text{ft/s}^2)(30.0\times10^3\,\text{ft})}{(32.174\,\text{lbm}\cdot\text{ft/lbf}\cdot\text{s}^2)(778.16\,\text{ft}\cdot\text{lbf/Btu})}\right]$$

$$= 247\,\text{Btu}$$

and

$$A_2 = (5.00\,\text{lbm})\left[\left(154.77 - 29.78\frac{\text{Btu}}{\text{bm}}\right) + \left(136.12\frac{\text{lbf}}{\text{in}^2}\right)\left(\frac{144\,\text{in}^2/\text{ft}^2}{778.16\,\text{ft}\cdot\text{lbf/Btu}}\right)\left(1.046 - 0.01325\frac{\text{ft}^3}{\text{lbm}}\right)\right.$$

$$\left. -(530\,\text{R})\left(0.31464 - 0.06296\frac{\text{Btu}}{\text{lbm}\cdot\text{R}}\right) + 0 + 0\right] = 88.1\,\text{Btu}$$

b. Now, from part a, we have $A_2 - A_1 = 88.1 - 247 = -159$ Btu. Note that the availability is higher in the first state.

Exercises

7. Determine the change in total availability in Example 10.3 if the initial (flying) state of the R-22 is a saturated vapor rather than a saturated liquid. Assume all the other variables remain unchanged. **Answer:** $A_2 - A_1 = -161$ Btu.
8. If the state of the R-22 in the final (landed) state of Example 10.3 is a saturated liquid at 100.°F rather than a superheated vapor, recompute the change in total availability for this system. **Answer:** $A_2 - A_1 = -246$ Btu.
9. To illustrate the impact of choosing the local environment (ground state), compute the total availabilities and the change during landing in Example 10.3 if the ground state is changed from a saturated liquid at 70.0°F, where $p_0 = 136.122$ psia, to a saturated liquid at −40.0°F, where $p_0 = p_{sat}(-40.0°\text{F}) = 15.222$ psia, which is close to atmospheric pressure. **Answer:** $A_1 = 258$ Btu, $A_2 = 129$ Btu, and $A_2 - A_1 = -129$ Btu.

10.7 CLOSED SYSTEM AVAILABILITY BALANCE

Since availability is a function of the system's thermodynamic properties, it is therefore a thermodynamic property itself. Perhaps, we can gain more insight into its engineering use if we carry out an availability balance for a closed system. Using the general balance equation, Eq. (2.11), we can write

$$A_{\text{transport}} + A_{\text{production}} = A_{\text{gain}} \tag{10.7}$$

If availability were conserved like energy, we would be able to set $A_{\text{production}} = 0$, but since availability is defined for only *reversible* processes and most systems undergo irreversible processes, we can expect that $A_{\text{production}} \neq 0$ and availability is not conserved.

Since it is difficult to decide heuristically how total availability is transported across a system boundary, it is easier to develop Eq. (10.7) from the definition of the gain in total availability given by Eq. (10.5) as

$$A_{\text{gain}} = A_2 - A_1 = E_2 - E_1 + p_0(\forall_2 - \forall_1) - T_0(S_2 - S_1)$$

From an energy balance on a system undergoing an actual irreversible process from state 1 to state 2, we have (recall that we can choose to follow either an actual irreversible path or a hypothetical reversible path to evaluate the change in the total system energy E, because energy is a point function whose integral is independent of the integration path)

$$E_2 - E_1 = ({}_1Q_2)_{\text{act}} - ({}_1W_2)_{\text{act}} = {}_1Q_2 - {}_1W_2$$

where we drop the subscripts on heat and work transfer and allow them to represent the actual (irreversible) process values from here on. The entropy balance for this situation is

$$S_2 - S_1 = \int_1^2 \frac{đQ}{T_b} + {}_1(S_P)_2$$

Combining the last three equations, the gain in availability for a closed system undergoing an irreversible process from state 1 to state 2 becomes

$$_1(A_{gain})_2 = \int_1^2 \left(1 - \frac{T_0}{T_b}\right) \bar{d}Q - {}_1W_2 + p_0(\forall_2 - \forall_1) - T_0\, {}_1(S_p)_2 \tag{10.8}$$

Comparing Eqs. (10.7) and (10.5), we see that the heat transport of total availability can be identified as

$$A_{\substack{\text{heat}\\\text{transport}}} = \int_1^2 \left(1 - \frac{T_0}{T_b}\right) \bar{d}Q \tag{10.9}$$

and, if the boundary temperature is constant over all the heat transfer surfaces, then Eq. (10.9) reduces to

$$A_{\substack{\text{heat}\\\text{transport}}} = \int_1^2 \left(1 - \frac{T_0}{T_b}\right) \bar{d}Q = \sum \left(1 - \frac{T_0}{T_{bi}}\right) ({}_1Q_2)_i \tag{10.10}$$

where the summation is over all the heat transfer surfaces at temperature T_{bi}, where heat transfer $({}_1Q_2)_i$ occurs. Again, comparing Eqs. (10.7), (10.8), and (10.9), we see that the work transport of total availability can be identified as

$$A_{\substack{\text{work}\\\text{transport}}} = -{}_1W_2 + p_0(\forall_2 - \forall_1) \tag{10.11}$$

and, finally, that leaves the net production of total availability as the remaining term:

$$A_{\text{production}} = {}_1(A_P)_2 \equiv -T_0\, {}_1(S_P)_2 \tag{10.12}$$

Note that, since the second law of thermodynamics requires that ${}_1(S_P)_2 \geq 0$, Eq. (10.12) dictates that

$$_1(A_P)_2 \leq 0 \tag{10.13}$$

for all real irreversible processes. Since the negative of production is destruction, Eq. (10.13) tells us that *availability is always destroyed in any irreversible process*. Since negative availability production is a somewhat contradictory and perhaps confusing phrase, we give this term a name that accurately represents its function; we call it the *irreversibility* of the process and give it the symbol I:

$$\text{Irreversibility} = {}_1I_2 = {}_1(A_{\text{production}})_2 = {}_1(A_{\text{destruction}})_2 = T_0\, {}_1(S_P)_2 \geq 0 \tag{10.14}$$

and the irreversibility rate $\dot{I}$ is

$$\text{Irreversibility rate} = \dot{I} = T_0 \dot{S}_P \geq 0 \tag{10.15}$$

Finally, substituting Eqs. (10.9), (10.11), (10.12), and (10.14) into Eq. (10.7) gives the closed system total availability balance (AB) as

$$\int_1^2 \left[1 - \frac{T_0}{T_b}\right] \bar{d}Q - {}_1W_2 + p_0(\forall_2 - \forall_1) - {}_1I_2 = (A_2 - A_1)_{\text{system}} = [m(a_2 - a_1)]_{\text{system}} \tag{10.16}$$

and the corresponding total availability *rate* balance (ARB) is obtained by differentiating Eq. (10.16) with respect to time to yield

$$\int_\Sigma \left[1 - \frac{T_0}{T_b}\right] \dot{q}\, d\Sigma - \dot{W} + p_0 \frac{d\forall}{dt} - \dot{I} = \left(\frac{dA}{dt}\right)_{\text{system}}$$

where Σ is the surface area of the system and T_b is the boundary temperature at the differential surface $d\Sigma$. In this equation, $\dot{q}$ is the heat flux, defined in Chapter 7 as $\dot{q} = \bar{d}^2 Q/d\Sigma dt$. For most systems, the surface integral can be reduced to a summation over a finite number of isothermal heat transfer surface areas at temperature T_{bi}, with each subjected to a heat transfer rate $\dot{Q}_i$ as

$$\int_\Sigma \left[1 - \frac{T_0}{T_b}\right] \dot{q}\, d\Sigma = \sum_i \left(1 - \frac{T_0}{T_{bi}}\right) \dot{Q}_i$$

Then the availability rate balance for a closed system becomes

$$\sum_i \left(1 - \frac{T_0}{T_{bi}}\right) \dot{Q}_i - \dot{W} + p_0 \frac{d\forall}{dt} - \dot{I} = \left(\frac{dA}{dt}\right)_{\text{system}} \tag{10.17}$$

CRITICAL THINKING

How do Eqs. (10.17) and (10.19) change for steady state processes? Does a system have to be steady state for $d\forall/dt$ to be zero? If a system is steady state, does that necessarily mean that $d\forall/dt$ and dA/dt are both zero?

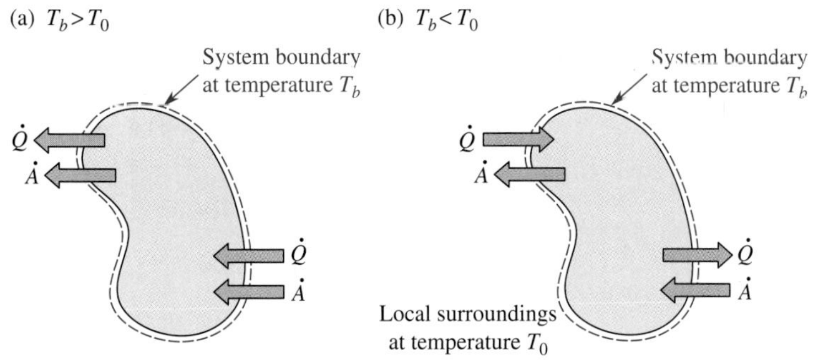

FIGURE 10.8
(a) When $T_b > T_0$, the heat transfer and the associated availability transfer are both in the same direction (either both into or both out of the system). (b) When $T_b < T_0$, the heat transfer and the associated availability transfer are in opposite directions.

If the system has a single heat transfer mode occurring at a constant system boundary temperature T_b, then Eqs. (10.16) and (10.17) reduce to

$$\left[1 - \frac{T_0}{T_{bi}}\right](_1Q_2) - {_1W_2} + p_0(\forall_2 - \forall_1) - {_1I_2} = (A_2 - A_1)_{\text{system}} = [m(a_2 - a_1)]_{\text{system}} \tag{10.18}$$

and

$$\left(1 - \frac{T_0}{T_b}\right)\dot{Q} - \dot{W} + p_0\frac{d\forall}{dt} - \dot{I} = \left(\frac{dA}{dt}\right)_{\text{system}} \tag{10.19}$$

Since availability represents the maximum useful reversible work that a system produces, each term in the availability balance must have this same meaning. Note that the heat transport of availability (Eqs. (10.9) and (10.10)) represents the maximum reversible work that a Carnot engine produces while operating between isothermal reservoirs at temperatures T and T_0 (see Eqs. (7.9) and (7.16)). The work transport of availability (Eq. (10.11)) corresponds to the difference between the actual work and the reversible work necessary to move the local environment if the system changes volume during a process. The irreversible loss of available energy associated with the *actual* heat and work transports of availability are contained in the irreversibility term I. This term accounts for the difference between an actual irreversible process and a hypothetical reversible process. Note that the availability gain can be either positive or negative (a system can gain or lose available energy during a process), and the heat and work availability transports can also be either positive or negative. However, Eq. (10.14) requires that the irreversibility must *always be positive* ($I \geq 0$) because of its relation to the entropy production and the second law of thermodynamics (i.e., $S_P \geq 0$).

When T_b is greater than T_0, the heat transfer and the associated availability transfer are in the same direction (either both into or both out of the system). However, when T_b is less than T_0, they are in opposite directions. Figure 10.8 illustrates this point and the following examples illustrate the use of this material.

EXAMPLE 10.4

A constant pressure piston-cylinder apparatus contains 1.00 kg of saturated liquid water at 120.°C. Heat is added to this system until the contents reach a quality of 50.0%. The surface temperature of the cylinder is constant at 130.°C. Determine the irreversibility of this process. The local environment (ground state) is at $p_0 = 0.101$ MPa and $T_0 = 20.0°\text{C} = 293$ K.

(*Continued*)

EXAMPLE 10.4 (*Continued*)

Solution

First, draw a sketch of the system (Figure 10.9).

The unknown is the irreversibility of this process. The material is water, and the system is closed.

The irreversibility for this system can be computed from Eq. (10.14) using an entropy balance to calculate the entropy production or directly from Eq. (10.18). We use Eq. (10.18) for this example. Solving this equation for the irreversibility gives

$$ {}_1I_2 = \left[1 - \frac{T_0}{T_{bi}}\right]({}_1Q_2) - {}_1W_2 + p_0 m(v_2 - v_1) - [m(a_2 - a_1)]_{\text{system}} $$

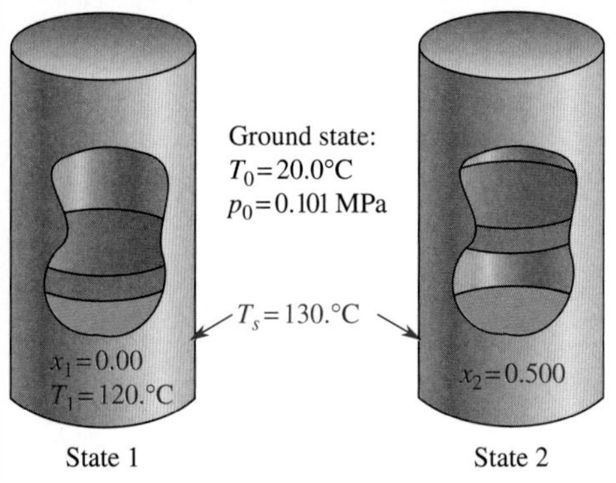

FIGURE 10.9
Example 10.4.

and, since the entire process takes place at constant pressure, we can write

$$ {}_1W_2 = m\int p dv = mp(v_2 - v_1) $$

Applying the closed system energy balance equation to this system gives

$$ {}_1Q_2 = m(u_2 - u_1) + {}_1W_2 $$

The state properties are

State 1	State 2	Ground State
$x_1 = 0$	$x_2 = 0.500$	$T_0 = 20.0°\text{C} = 293\text{ K}$
$T_1 = 120.°\text{C}$	$p_2 = p_1 = p_{\text{sat}}(120.°\text{C}) = 198.5\text{ kN/m}^2$	$p_0 = 0.101\text{ MPa}$
$v_1 = v_f(120.°\text{C})$	$v_2 = v_f(120.°\text{C}) + x_2 v_{fg}(120.°\text{C})$	
$= 0.001060\text{ m}^3/\text{kg}$	$= 0.44648\text{ m}^3/\text{kg}$	
$u_1 = u_f(120.°\text{C}) = 503.5\text{ kJ/kg}$	$u_2 = u_f + x_2 u_{fg} = 1516.4\text{ kJ/kg}$	
$s_1 = s_f(120.°\text{C}) = 1.5280\text{ kJ/kg}\cdot\text{K}$	$s_2 = s_f + x_2 s_{fg} = 4.3292\text{ kJ/kg}\cdot\text{K}$	

The change in specific availability of the system is calculated from Eq. (10.4) with $V_1 = V_2 = 0$ and $Z_1 = Z_2$ as

$$ \begin{aligned} a_2 - a_1 &= u_2 - u_1 + p_0(v_2 - v_1) - T_0(s_2 - s_1) \\ &= (1516.4 - 503.5\text{ kJ/kg}) + (101\text{ N/m}^2)(0.44648 - 0.001060\text{ m}^3/\text{kg}) \\ &\quad - (293\text{ K})(4.3292 - 1.5280\text{ kJ/kg}\cdot\text{K}) = 237\text{ kJ/kg} \end{aligned} $$

Then, we can compute the work using $p_2 = p_1 = p$ as

$$ \begin{aligned} {}_1W_2 &= mp(v_2 - v_1) \\ &= (1.00\text{ kg})(198.5\text{ kN/m}^2)(0.44648 - 0.001060\text{ m}^3/\text{kg}) = 88.4\text{ kJ} \end{aligned} $$

and the heat transfer becomes

$$ \begin{aligned} {}_1Q_2 &= m(u_2 - u_1) + {}_1W_2 \\ &= (1.00\text{ kg})(1516.4 - 503.5\text{ kJ/kg}) + 88.4\text{ kJ} = 1100\text{ kJ} \end{aligned} $$

Then, Eq. (10.15) gives the irreversibility of this process as

$$ \begin{aligned} {}_1I_2 &= \left(1 - \frac{293\text{ K}}{403\text{ K}}\right)(110\text{ kJ}) - 88.4\text{ kJ} + (101\text{ kN/m}^2)(1.00\text{ kg})(0.44648 - 0.001060\text{ m}^3/\text{kg}) \\ &\quad - (1.00\text{ kg})(236.9\text{ kJ/kg}) = 20.3\text{ kJ} \end{aligned} $$

Exercises

10. Suppose the initial state in Example 10.4 is a saturated vapor at 120.°C instead of a saturated liquid at 120.°C and the system boundary temperature is lowered from 130.°C to 100.°C. Determine the irreversiblilty in the process under these conditions, assuming all the other variables remain unchanged. **Answer**: ${}_1I_2 = 44.0$ kJ.
11. Determine the irreversibility of the process in Example 10.4 if the system boundary temperature is (a) 150.°C, and (b) 110.°C instead of 130.°C. Assume all the other variables remain unchanged. **Answer**: (a) 58.3 kJ, (b) −21.4 kJ. (The fact that ${}_1I_2 < 0$ in part b means that it would be impossible to have a surface temperature of only 110.°C for this process.)
12. What is the minimum possible system boundary temperature for the process described in Example 10.4? (Hint: Find T_b that produces ${}_1I_2 = 0$.) **Answer**: $T_b = 120.°\text{C}$.

EXAMPLE 10.5

A small room has a single 100. W lightbulb hanging from the ceiling. The walls are not insulated, so a steady state condition is reached where the room walls are at 24.0°C. Determine the irreversibility rate within the room. The local environment (ground state) outside the room is at $p_0 = 0.101$ MPa and $T_0 = 15.0$°C.

Solution

First, draw a sketch of the system (Figure 10.10).

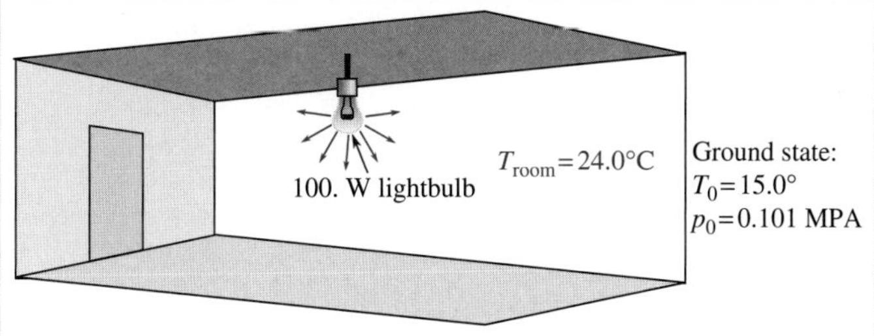

FIGURE 10.10
Example 10.5.

The unknown is the irreversibility rate within the room. The irreversibility rate for this closed system can be determined from Eq. (10.19) as

$$\dot{I} = \left(1 - \frac{T_0}{T_b}\right)\dot{Q} - \dot{W} + p_0\frac{d\forall}{dt} - \left(\frac{dA}{dt}\right)_{\text{system}}$$

In this problem, the system has a fixed volume so $d\forall/dt = 0$, and since the system is steady state, we also have $(dA/dt)_{\text{system}} = 0$. The energy rate balance applied to this closed system tells us that $\dot{Q} = \dot{W} = 100\ W$. Then, Eq. (10.19) gives

$$\dot{I} = \left(1 - \frac{15.0 + 273\text{ K}}{24.0 + 273\text{ K}}\right)(-100\text{ W}) - (-100\text{ W}) + p_0(0) - 0 = 97.0\text{ W}$$

Note that, of the 100. W that enters the room as electrical power, only 3% remains as available energy.

Exercises

13. Suppose the local environment (ground state) temperature in Example 10.5 is lowered from 15.0°C to 0.00°C. Determine the irreversibility rate within the room assuming all the other variables remain unchanged. **Answer**: $\dot{I} = 91.9$ W.
14. If we change the system in Example 10.5 from the room to the lightbulb itself, then the boundary temperature increases from 24.0°C to 40.0°C. Assuming all other variables remain unchanged, determine the irreversibility rate of the lightbulb. **Answer**: $\dot{I} = 92.0$ W
15. The basal metabolic rate of the human body is about 400. Btu/h. This means that the body gives off this much heat, or $\dot{Q}_{\text{body}} = -400.$ Btu/h, when it is resting (i.e., when $\dot{W}_{\text{body}} = 0$). Suppose the human body is a reversible ($\dot{I} = 0$) closed system with a surface boundary temperature of 98.6°F in an environment with a temperature of 70.0°F. What is the rate of change of total availability of the human body? **Answer**: $(dA/dt)_{\text{body}} = -20.5$ Btu/h.

10.8 FLOW AVAILABILITY

Before we extend the availability balance to open systems, we must evaluate the availability transport associated with mass flowing across the system boundary. Consider an incremental amount of mass dm moving with velocity V crossing the system boundary at height Z in a time interval dt. This mass carries with it an incremental amount of available energy, dA, defined as

$$(dA)_{\substack{\text{mass}\\\text{flow}}} = a(dm) = [u - u_0 + p_0(v - v_0) - T_0(s - s_0) + V^2/2g_c + gZ/g_c](dm)$$

and the associated availability transport rate for the incremental mass dm is

$$(dA/dt)_{\substack{\text{mass}\\\text{flow}}} = \dot{A}_{\substack{\text{mass}\\\text{flow}}} = a(dm/dt) = a\dot{m} = [u - u_0 + p_0(v - v_0) - T_0(s - s_0) + V^2/2g_c + gZ/g_c](\dot{m})$$

Figure 6.1 in Chapter 6 indicates that the amount of incremental "flow work" associated with moving the mass dm across the system boundary is

$$(\bar{d}W)_{\substack{\text{mass}\\\text{flow}}} = pv(dm)$$

and the corresponding flow work *rate* is

$$\left(\frac{\bar{d}W}{dt}\right)_{\substack{\text{mass}\\\text{flow}}} = pv\left(\frac{dm}{dt}\right) = \dot{m}pv$$

If we assume that the incremental mass dm does not undergo any heat transfer or change in kinetic or potential energy as it moves through the incremental distance dP required to cross the system boundary, then Eq. (10.11) gives the incremental work mode availability transport as

$$(\bar{d}W)_{\substack{\text{mass}\\\text{flow}}} - p_0(d\forall) = (\bar{d}W)_{\substack{\text{mass}\\\text{flow}}} - p_0(vdm)$$

where $d\forall$ is the volume of the incremental mass dm. Then,

$$(dA)_{\substack{\text{mass}\\\text{flow}}} = a_W dm = pvdm - p_0 vdm = v(p - p_0)dm$$

and the work mode net availability transport *rate* is

$$(\dot{A})_{\substack{\text{mass}\\\text{flow}}} = a_W \dot{m} = v(p - p_0)\dot{m}$$

Combining these equations with the nonwork mass flow availability transport gives the total flow availability transport, $a_f dm$, as

$$adm + a_W dm = (a + a_W)dm = a_f dm$$

or

$$a_f = a + a_w = u - u_0 + p_0(v - v_0) - T_0(s - s_0) + pv - pv_0 + \frac{V^2}{2g_c} + \frac{gZ}{g_c}$$
$$= (u + pv) - (u_0 + p_0 v_0) - T_0(s - s_0) + \frac{V^2}{2g_c} + \frac{gZ}{g_c}$$

or

The specific flow availability of a flow stream

$$a_f = h - h_0 - T_0(s - s_0) + \frac{V^2}{2g_c} + \frac{gZ}{g_c} \tag{10.20}$$

where a_f is the *specific flow availability* of the mass crossing the system boundary, and the *total flow availability* of the mass crossing the system boundary is

$$\dot{A}_f = \dot{m}(a_f) = \dot{m}\left(h - h_0 - T_0(s - s_0) + \frac{V^2}{2g_c} + \frac{gZ}{g_c}\right)$$

The concept of flow availability is illustrated in the following example.

EXAMPLE 10.6

Compute the specific flow availability at the exit of a garden hose used to fill a child's wading pool. The hose is held horizontally 4.00 ft above the ground and the water exits the hose at 3.00 ft/s at 50.0°F. Take the local environment (ground state) to be atmospheric temperature and pressure of 70.0°F and 14.7 psia.

Solution

First, draw a sketch of the system (Figure 10.11).

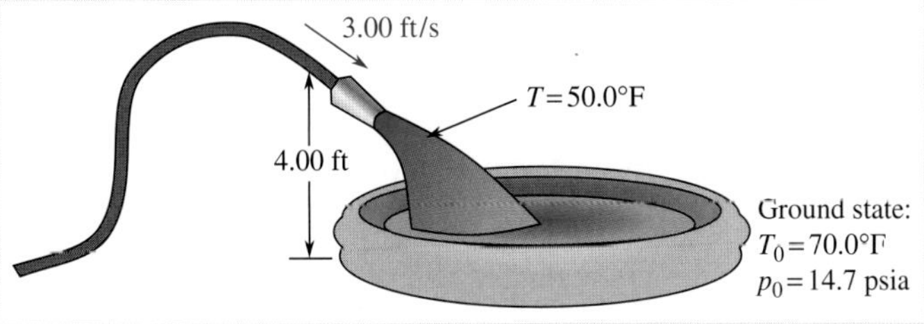

FIGURE 10.11
Example 10.6.

The unknown is the specific flow availability at the exit of a garden hose. The material is water, and the system is open.

The specific flow availability is defined by Eq. (10.16) as

$$a_f = h - h_0 - T_0(s - s_0) + \frac{V^2}{2g_c} + \frac{gZ}{g_c}$$

The liquid water exiting the garden hose fits our definition of an incompressible fluid, so Eqs. (3.62) and (7.33) can be used to compute

$$h - h_0 = c(T - T_0) + v(p - p_0)$$

and

$$s - s_0 = c\ln\left(\frac{T}{T_0}\right)$$

Using the subscript w for the water, we can write the specific flow availability of the stream of water exiting the garden hose as

$$(a_f)_w = c_w(T_w - T_0) + v_w(p_w - p_0) - c_w T_0 \ln\left(\frac{T_w}{T_0}\right) + \frac{V_w^2}{2g_c} + \frac{gZ_w}{g_c}$$

where $T_0 = T_{air} = 70.0°F$ and $p_0 = p_{air} = 14.7$ psia. The liquid water is a slightly compressed liquid; however, the amount of compression is only a few psia ($p_{actual} - p_{sat}(50.0°F) = 14.7 - 0.1780 = 14.52$ psia), and the effect of this small amount of compression is negligible. Thus, for the specific volume v we use the value of saturated liquid at the actual temperature of the water, or $v = v_{sat}(50.0°F) = 0.01602$ ft³/lbm. Substituting in all the appropriate numerical values into the specific flow availability equation, we get

$$a_f = (1.00\,\text{Btu/lbm})(50.0 - 70.0\,\text{R}) + 0.01602\,\text{ft}^3/\text{lbm})(14.7 - 14.7\,\text{lbf/in}^2)\left(\frac{144\,\text{lbf/in}^2}{778.16\,\text{ft}\cdot\text{lbf/Btu}}\right)$$
$$- (70.0 + 459.67\,\text{R})\ln\left(\frac{50.0 + 459.67}{70.0 + 459.67}\right) + \frac{(3.00\,\text{ft/s}^2)^2}{2(32.174\,\text{lbm}\cdot\text{ft/lbf}\cdot\text{s}^2)(778.16\,\text{ft}\cdot\text{lbf/Btu})}$$
$$+ \frac{(32.174\,\text{ft/s}^2)(4.00\,\text{ft})}{(32.174\,\text{lbm}\cdot\text{ft/lbfgs}^2)(778.16\,\text{ft}\cdot\text{lbf/Btu})} = 0.393\text{Btu/lbm}$$

Exercises

16. Determine the value of the flow availability in Example 10.6 when the garden hose is lying on the ground. Assume that the other variables remain unchanged. **Answer**: $(a_f)_w = 0.388$ Btu/lbm.
17. Suppose you put a nozzle on the end of the garden hose in Example 10.6 and increase the velocity of the water leaving the hose from 3.00 ft/s to 15.0 ft/s. Determine the new flow availability, assuming all the other variables remain unchanged. **Answer**: $(a_f)_w = 0.397$ Btu/lbm.
18. While you are spraying the garden hose in Example 10.6, a weather front comes through and lowers the atmospheric temperature from 70.0°F to 50.0°F. Determine the new flow availability assuming all the other variables remain unchanged. **Answer**: $(a_f)_w = 0.00500$ Btu/lbm.

10.9 OPEN SYSTEM AVAILABILITY RATE BALANCE

The open system availability rate balance is obtained from the closed system availability rate balance simply by adding the flow availability resulting from all the inlet and outlet flow streams to Eq. (10.17) to yield the open system availability rate balance (ARB):

Open system availability rate balance

$$\sum_{i=1}^{n}\left(1-\frac{T_0}{T_{bi}}\right)\dot{Q}_i - \dot{W} + \sum_{\text{inlet}} \dot{m}a_f - \sum_{\text{outlet}} \dot{m}a_f + p_0\dot{V} - \dot{I} = \left(\frac{dA}{dt}\right)_{\text{system}} \tag{10.21}$$

As in the case of the open system energy rate balance, several specific cases merit special consideration. These cases are described in detail next.

■ Case 1 Steady State

A steady state is reached when all system properties are independent of time, then $d(\textit{any system property})/dt = 0$. Thus, at a steady state, we have

$$\frac{dA}{dt} = \frac{d}{dt}\left[E - E_0 + p_0(V - V_0) - T_0(S - S_0)\right]_{\text{system}} = 0$$

and, since E_0, p_0, V_0, T_0, and S_0 are all constants, this reduces to

$$\frac{dA}{dt} = \left(\frac{dE}{dt} + p_0\frac{dV}{dt} - T_0\frac{dS}{dt}\right)_{\text{system}} = 0$$

But since E, V, and S are also system properties, their individual time derivatives must also vanish, so that, in a steady state, we must have the following system conditions:

$$\textbf{Steady state means:}\ \frac{dA}{dt} = \frac{dE}{dt} = \frac{dV}{dt} = \frac{dS}{dt} = 0 \tag{10.22}$$

■

■ Case 2 Steady Flow

Equation (6.22) defines steady flow as

$$\sum_{\text{inlet}} \dot{m} = \sum_{\text{outlet}} \dot{m}$$

so that there is no accumulation or depletion of mass within the system during steady flow. ■

■ Case 3 Single Inlet, Single Outlet

In an open system with a single inlet and a single outlet, the summation signs on the flow availability terms can be dropped. Then,

$$\sum_{\text{inlet}} \dot{m}a_f - \sum_{\text{outlet}} \dot{m}a_f = (\dot{m}a_f)_{\text{inlet}} - (\dot{m}a_f)_{\text{outlet}} \tag{10.23}$$

If the system is also at a *steady flow*, then Eq. (10.22) tells us that $\dot{m}_{\text{inlet}} = \dot{m}_{\text{outlet}}$, and the flow availability terms for a steady flow system with a single inlet and single outlet can then be written as

$$\sum_{\text{inlet}} \dot{m}a_f - \sum_{\text{outlet}} \dot{m}a_f = \dot{m}[(a_f)_{\text{inlet}} - (a_f)_{\text{outlet}}] \tag{10.24}$$

■

■ Case 4 Isothermal Boundaries

If the system has a constant (isothermal) system boundary temperature, T_b, then all the heat transports can be combined into a single net heat transport term, $\sum_i \dot{Q}_i = \dot{Q}_{net} = \dot{Q}$, where we drop the subscript *net* for convenience. Then the (net) heat transport of availability for isothermal boundaries can be written as

$$\sum_{i=1}^{n}\left[1-\frac{T_0}{T_{bi}}\right]\dot{Q}_i = \left[1-\frac{T_0}{T_b}\right]\dot{Q}_{net} = \left[1-\frac{T_0}{T_b}\right]\dot{Q} \tag{10.25}$$

■

■ Case 5 Constant Volume

If the volume of space enclosed by the system boundary is constant in time, then $d\forall/dt = 0$, and the loss or gain of available energy associated with the moving environment vanishes. Then, $p_0(\forall_2 - \forall_1) = 0$ in Eq. (10.18), and $p_0(d\forall/dt) = 0$ in Eq. (10.19). Note that Eq. (10.22) tells us that, if the system is at a steady state, then it also must have a constant volume. ■

10.10 MODIFIED AVAILABILITY RATE BALANCE (MARB) EQUATION

Now, we can refer to the open system availability rate balance for a *steady state, steady flow, single-inlet, single-outlet, constant volume, isothermal boundary* system (or SS, SF, SI, SO, CV, IB system) as:

The modified availability rate balance for a SS, SF, SI, SO, CV, IB open system

$$\left[1-\frac{T_0}{T_b}\right]\dot{Q} - \dot{W} + \dot{m}[(a_f)_{in} - (a_f)_{out}] - \dot{I} = 0 \tag{10.26}$$

where Eq. (10.15) requires that $\dot{I} \geq 0$.

EXAMPLE 10.7

A horizontal pipe carrying superheated steam at 5000. psia and 1000.°F suddenly develops a small crack. Steam enters the crack at 50.0 ft/s and passes through it in a steady state, adiabatic ($\dot{Q} = 0$), aergonic ($\dot{W} = 0$) process to exit at 14.696 psia with a velocity of 300. ft/s. Determine the specific flow availabilities at the inlet and outlet of the crack, and calculate the irreversibility per unit mass of steam exiting the crack. Take the local environment (ground state) to be saturated liquid water at 70.0°F.

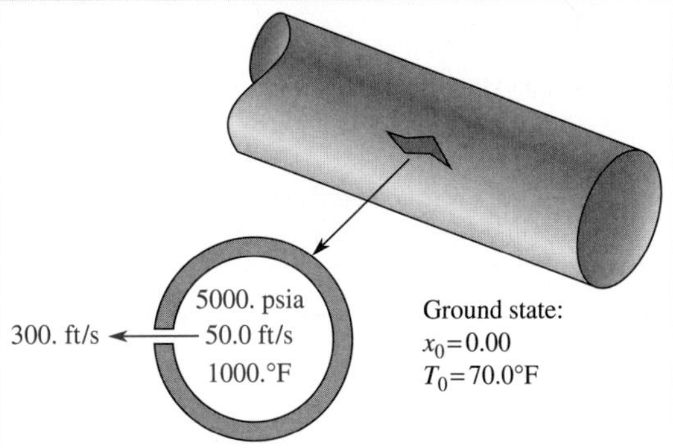

FIGURE 10.12
Example 10.7.

Solution

First, draw a sketch of the system (Figure 10.12).

The unknowns are the specific flow availabilities at the inlet and outlet of the crack and the irreversibility per unit mass of steam exiting the crack. The material is steam, and this is an open system. The data at the inlet and outlet flow stations are

Station 1	Station 2
$p_1 = 5000.$ psia	$p_2 = 14.696$ psia
$T_1 = 1000.$°F	?
$h_1 = 1363.4$ Btu/lbm	
$s_1 = 1.3990$ Btu/lbm·R	

(*Continued*)

EXAMPLE 10.7 *(Continued)*

For a steady state adiabatic, aergonic process $\dot{Q} = \dot{W} = 0$, and the resulting energy rate balance gives (this is a type of throttling process, see Chapter 4)

$$h_2 = h_1 + \frac{V_1^2 - V_2^2}{2g_c} = 1363.4\frac{\text{Btu}}{\text{lbm}} + \frac{(50.0\,\text{ft/s})^2 - (300.\,\text{ft/s})^2}{2\left(32.174\frac{\text{lbm}\cdot\text{ft}}{\text{lbf}\cdot\text{s}^2}\right)\left(778.16\frac{\text{ft}\cdot\text{lbf}}{\text{Btu}}\right)} = 1360\frac{\text{Btu}}{\text{lbm}}$$

The values $p_2 = 14.696$ psia and $h_2 = 1361.7$ Btu/lbm fix the exit state, and the remaining properties at the exit station can now be determined from the steam tables (by interpolation) as $T_2 = 655°\text{F}$ and $s_2 = 1.9981$ Btu/lbm·R. Then, Eq. (10.20) gives the specific flow availability as

$$a_f = h - h_0 - T_0(s - s_0) + \frac{V^2}{2g_c} + \frac{gZ}{g_c}$$

The local environment (ground state) values are $h_0 = h_f(70.0°\text{F}) = 38.1$ Btu/lbm, and $s_0 = s_f(70.0°\text{F}) = 0.0746$ Btu/lbm·R, and inserting the numerical values with $Z_1 = Z_2 = 0$, we get

$$a_{f1} = (1363.4 - 38.1\,\text{Btu/lbm}) - (70.0 + 459.67\,\text{R})(1.3990 - 0.0746\,\text{Btu/lbm}\cdot\text{R}) + \frac{(50.0\,\text{ft/s})^2}{2\left(32.174\frac{\text{lbm}\cdot\text{ft}}{\text{lbf}\cdot\text{s}^2}\right)\left(778.16\frac{\text{ft}\cdot\text{lbf}}{\text{Btu}}\right)} + 0 = 624\frac{\text{Btu}}{\text{lbm}}$$

and

$$a_{f2} = (1361.7 - 38.1\,\text{Btu/lbm}) - (70.0 + 459.67\,\text{R})(1.9970 - 0.0746\,\text{Btu/lbm}\cdot\text{R}) + \frac{(300.\,\text{ft/s})^2}{2\left(32.174\frac{\text{lbm}\cdot\text{ft}}{\text{lbf}\cdot\text{s}^2}\right)\left(778.16\frac{\text{ft}\cdot\text{lbf}}{\text{Btu}}\right)} + 0 = 307\frac{\text{Btu}}{\text{lbm}}$$

The irreversibility rate is found from Eq. (10.26), and if we divide through by $\dot{m}$, we have the irreversibility per unit mass of steam flowing, or

$$\dot{I}/\dot{m} = I/m = \left[1 - \frac{T_0}{T_b}\right]\dot{Q}/\dot{m} - \dot{W}/\dot{m} + a_{f1} - a_{f2}$$

and, since $\dot{Q} = \dot{W} = 0$ here, this equation reduces to

$$\frac{I}{m} = a_{f1} - a_{f2} = 624 - 307 = 317\frac{\text{Btu}}{\text{lbm}}$$

Exercises

19. Determine the irreversibility rate in Example 10.7 if the mass flow rate out of the pipe crack is 0.100 lbm/s. **Answer**: $\dot{I} = 31.7$ Btu/s.
20. If the crack in the pipe in Example 10.7 is horizontal at a height of 15.0 ft from the floor, determine the inlet and exit flow availabilities of the steam relative to the floor. **Answer**: Only $gZ/g_c = 0.0193$ Btu/lbm is added to a_{f1} and a_{f2} in Example 10.7.
21. If the leak in Example 10.7 is not adiabatic but has a heat transfer rate per unit flow rate of $\dot{Q}/\dot{m} = -316$ Btu/lbm and a boundary temperature of 500.°F, determine the new irreversibility per unit mass of steam exiting the crack. Assume all other variables remain unchanged (i.e., assume this small value of $\dot{Q}/\dot{m}$ does not significantly change the values of h_2 and s_2). **Answer**: $I/m = 316$ Btu/lbm.

EXAMPLE 10.8

Superheated steam at 2.80 lbm/s, 100. psia, and 500.°F enters a horizontal, stationary, insulated nozzle with a negligible inlet velocity and expands to 10.0 psia. The friction and other irreversibilities within the nozzle cause the exit velocity to be only 95.0% of that produced by an isentropic expansion. Taking the local environment (ground state) to be saturated liquid water at 70.0°F, determine

a. The inlet specific flow availability.
b. The exit specific flow availability.
c. The irreversibility rate inside the nozzle.

Solution

First, draw a sketch of the system (Figure 10.13).

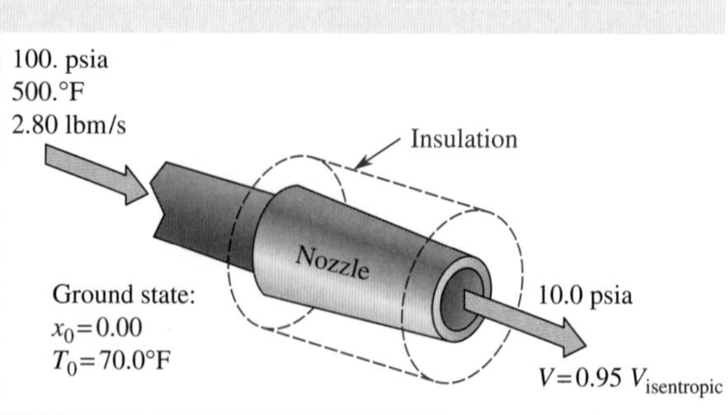

FIGURE 10.13
Example 10.8.

The unknowns are the inlet specific flow availability, the exit specific flow availability, and the irreversibility rate inside the nozzle. The system is open, and material is steam.

a. The station data at the inlet and outlet of the nozzle are

Station 1	Station 2s	Ground State
$p_1 = 100.\text{ psia}$	$p_{2s} = p_2 = 10.0\text{ psia}$	$x_0 = 0$
$T_1 = 500.°\text{F}$	$s_{2s} = s_1 = 1.7087\text{ Btu/lbm}\cdot\text{R}$	$T_0 = 70.0°\text{F}$
$h_1 = 1279.1\text{ Btu/lbm}$	$x_{2s} = 0.94735$	$s_0 = 0.0746\text{ Btu/lbm}\cdot\text{R}$
$s_1 = 1.7087\text{ Btu/lbm}\cdot\text{R}$	$h_{2s} = 161.2 + 0.94735(982.1)$ $= 1091.6\text{ Btu/lbm}$	

Since this an aergonic, adiabatic, steady state, steady flow, single-inlet, single-outlet system, we can use the energy rate balance with $V_1 = 0$ and $Z_1 = Z_2 = 0$ to determine the isentropic exit velocity as

$$\begin{aligned} V_{2s} &= [2g_c(h_1 - h_{2s})]^{1/2} \\ &= [2(32.174\text{ lbm}\cdot\text{ft/lbf}\cdot\text{s}^2)(1279.1 - 1091.6\text{ Btu/lbm})(778.16\text{ ft}\cdot\text{lbf/Btu})]^{1/2} \\ &= 3064\text{ ft/s} \end{aligned}$$

and the actual exit velocity is $V_2 = 0.95 \times V_{2s} = 0.95(3064) = 2911$ ft/s. We can use the energy rate balance to calculate the actual exit specific enthalpy as

$$h_2 = h_1 - \frac{V_2^2}{2g_c} = 1279.1\text{ Btu/lbm} - \frac{(2911\text{ ft/s})^2}{2\left(32.174\frac{\text{lbm}\cdot\text{ft}}{\text{lbf}\cdot\text{s}^2}\right)\left(778.16\frac{\text{ft}\cdot\text{lbf}}{\text{Btu}}\right)} = 1110\text{ Btu/lbm}$$

Then, we can determine the actual quality of the exit steam as

$$x_2 = (h_2 - h_{2f})/h_{2fg} = (1110 - 161.2)/982.1 = 0.9660$$

from which we can calculate the actual entropy at the exit as

$$s_2 = s_{2f} + x_2 s_{2fg} = 0.2836 + (0.9660)(1.5043) = 1.7368\text{ Btu/lbm}\cdot\text{R}$$

The specific flow availability is defined by Eq. (10.20) as

$$a_f = h - h_0 - T_0(s - s_0) + \frac{V^2}{2g_c} + \frac{gZ}{g_c}$$

from which we can now calculate the entrance specific flow availability as

$$\begin{aligned} a_{f1} &= (1279.1 - 38.1\text{ Btu/lbm}) - (70.0 + 459.67\text{ R})(1.7087 - 0.0746\text{ Btu/lbm}\cdot\text{R}) \\ &\quad + 0 + 0 = 375\text{ Btu/lbm} \end{aligned}$$

b. The specific flow availability at the exit can now be determined from Eq. (10.20) as

$$\begin{aligned} a_{f2} &= (1109.9 - 38.1\text{ Btu/lbm}) - (70.0 + 450.67\text{ R})(1.7368 - 0.0746\text{ Btu/lbm}\cdot\text{R}) \\ &\quad + \frac{(2910.9\text{ ft}^a\text{s})^2}{2\left(32.174\frac{\text{lbm}\cdot\text{ft}}{\text{lbf}\cdot\text{s}^2}\right)\left(778.16\frac{\text{ft}\cdot\text{lbf}}{\text{Btu}}\right)} + 0 = 361\text{Btu/lbm} \end{aligned}$$

c. The irreversibility rate for this system is given by Eq. (10.26) as

$$\dot{I} = \left(1 - \frac{T_0}{T_b}\right)\dot{Q} - \dot{W} + \dot{m}(a_{f1} - a_{f2})$$

and, since $\dot{Q} = \dot{W} = 0$ here, this equation reduces to

$$\dot{I} = \dot{m}\,(a_{f1} - a_{f2})$$

or

$$\dot{I} = (2.80\text{ lbm/s})(375 - 361\text{ Btu/lbm}) = 39.2\text{ Btu/s}$$

(Continued)

EXAMPLE 10.8 (*Continued*)

Exercises

22. Determine the actual exit enthalpy in Example 10.8 if the nozzle efficiency is 85.0% instead of 95.0%. Assume all the other variables remain unchanged. **Answer:** $h_2 = 1144$ Btu/lbm.

23. In other thermodynamics problems, we often neglect the effect of flow stream velocity in our analysis. Recompute the irreversibility rate in Example 10.8 neglecting the exit velocity and observe the effect flow stream velocity has on the answer. Note that both h_2 and s_2 change, and that station 2 is now superheated. **Answer:** = 373 Btu/s (a very significant error is produced by neglecting the flow stream velocity).

24. If the inlet pressure is increased from 100. psia to 400. psia, determine the new actual exit velocity. Assume all the other variables remain unchanged. **Answer:** $V_2 = 2640$ ft/s (to three significant figures).

One further case needs to be introduced. To be able to deal with the inequality that appears in the entropy production term of the entropy balance, we introduce the simplifying assumption of a *reversible* process. Even though reversible processes do not occur in practical engineering problems, it is often a useful simplifying assumption to invoke. However, when it is used, one should not expect calculated values to closely match the results of experimental measurements.

■ Case 6 Reversible Processes

In any system undergoing reversible processes, $\dot{S}_P = \dot{I} = 0$. Then, the resulting SS, SF, SI, SO, CV, IB availability rate balance for a system undergoing an internally *reversible* processes is

The modified availability rate balance for a SS, SF, SI, SO, CV, IB, *reversible* system

$$\left[1 - \frac{T_0}{T_b}\right]\dot{Q} - \dot{W} + \dot{m}\left[(a_f)_{\text{in}} - (a_f)_{\text{out}}\right] = 0 \qquad (10.27)$$

Note that the first term on the left-hand side of this equation represents external heat transfer irreversibilities, even though we made the system internally reversible. ■

EXAMPLE 10.9

A large uninsulated steam turbine receives superheated steam at 18.0 kg/s, 500.°C, and 3.00 MPa and exhausts it to 0.0100 MPa with a quality of 96.0%. If the turbine is assumed to be internally reversible, determine the rate of heat loss from the surface of the turbine if the power output is 20.0×10^3 kW. The surface temperature of the turbine is uniform at 350.°C, and the local environment (ground state) is taken to be saturated liquid water at $T_0 = 20.0$°C. Neglect all flow stream kinetic and potential energies in this problem.

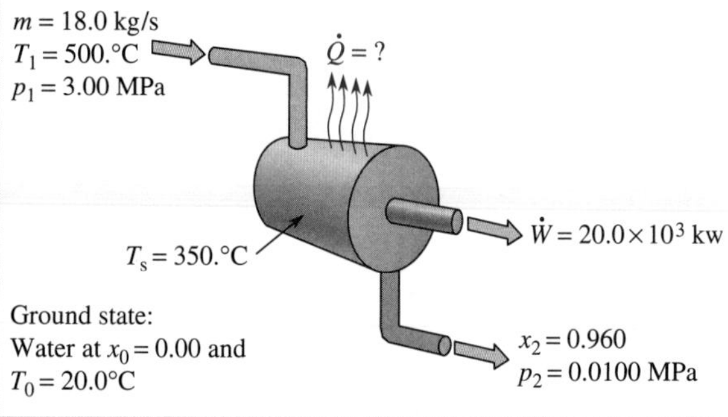

FIGURE 10.14
Example 10.9.

Solution

First, draw a sketch of the system (Figure 10.14).

The unknown is the rate of heat loss from the surface of the turbine. This is an open system, and the material is steam. The station data at the inlet and outlet of the turbine are

Station 1	Station 2	Ground State
$p_1 = 3.00$ MPa	$p_2 = 0.0100$ MPa	$x_0 = 0.00$
$T_1 = 500.$°C	$x_2 = 0.960$	$T_0 = 20.0$°C
$h_1 = 3456.5$ kJ/kg	$h_2 = 191.8 + 0.96(2392.8)$	$h_0 = 83.9$ kJ/kg
$s_1 = 7.2346$ kJ/kg·K	$= 2488.9$ kJ/kg	$s_0 = 0.2965$ kJ/kg·K
	$s_2 = 0.6491 + 0.96(7.5019)$	
	$= 7.8509$ kJ/kg·K	

The entrance and exit specific flow availabilities are given by Eq. (10.20) (neglecting all flow stream kinetic and potential energy terms as per the problem statement) as

$$a_f = h - h_0 - T_0(s - s_0)$$

so

$$a_{f1} = h_1 - h_0 - (T_0)(s_1 - s_0) = (3456.5 - 83.9\,\text{kJ/kg}) - (20.0 + 273.15\,\text{K})(7.2346 - 0.2965\,\text{kJ/kg}\cdot\text{K}) = 1339\,\text{kJ/kg}$$

and

$$a_{f2} = h_2 - h_0 - (T_0)(s_2 - s_0) = (2488.9 - 83.9\,\text{kJ/kg}) - (20.0 + 273.15\,\text{K})(7.8509 - 0.2965\,\text{kJ/kg}\cdot\text{K}) = 190.4\,\text{kJ/kg}$$

Then, Eq. (10.27), for a reversible process, gives the required heat transfer as

$$\dot{Q} = \frac{\dot{W} + \dot{m}(a_{f2} - a_{f1})}{1 - \frac{T_0}{T_b}}$$

or

$$\dot{Q} = \frac{20.0 \times 10^3\,\text{kJ/s} + (18.0\,\text{kg/s})(190.4 - 1339\,\text{kJ/kg})}{1 - \frac{20.0 + 273.15\,\text{K}}{350. + 273.15\,\text{K}}} = -1260\,\text{kJ/s} = -1260\,\text{kW}$$

Exercises

25. Determine the heat loss from the turbine in Example 10.9 if it is not reversible but instead has a power output of 15.0×10^3 kW with an isentropic efficiency of 88.0%. Assume all the other variables remain unchanged. Answer: $\dot{Q} = -6840\,\text{kJ/s}$.

26. If we lower the surface temperature on the turbine in Example 10.9 from 350.°C to 100.°C, determine the new heat loss rate from the turbine, assuming all the other variables remain unchanged. **Answer:** $\dot{Q} = -3120\,\text{kJ/s}$.

27. An error was made in reporting the steam mass flow rate in Example 10.9. The actual mass flow rate is 20.0 kg/s not 18.0 kg/s. Determine the heat loss rate now, assuming all the other variables remain unchanged. Answer: $\dot{Q} = -5600\,\text{kJ/s}$.

10.11 ENERGY EFFICIENCY BASED ON THE SECOND LAW

The common work and thermal energy conversion efficiencies defined in Chapters 4 and 7 for fluid pumps and compressors, heat engines, heat pumps, air conditioners, refrigerators, and so forth are based on energy transport ratios for these technologies. Such efficiencies do not reveal the source of the losses within these devices, because they have no term containing the irreversibilities within the system. Consequently, these efficiencies are often called *first law energy conversion efficiencies* and are described by Eq. (4.70).

10.11.1 First Law (Energy) Efficiency

$$\eta_E \equiv \frac{\text{Desired energy result}}{\text{Required energy input}} \tag{4.70}$$

For example, in Chapter 7, we introduce the first law (thermal) energy efficiency of a heat engine as

$$\eta_T \equiv \frac{\text{Net work output}}{\text{Total heat input}} = \frac{(W_{\text{out}})_{\text{net}}}{(Q_{\text{in}})_{\text{total}}} = \frac{(\dot{W}_{\text{out}})_{\text{net}}}{(\dot{Q}_{\text{in}})_{\text{total}}} \tag{7.5}$$

Equation (4.70) tells us that the energy conversion efficiency is the ratio of the magnitude of the energy that has been converted divided by the magnitude of energy initially present that could be converted. Since energy is conserved, if the energy conversion process is not 100%, then some of the energy initially present must have been converted into a form different from that desired. This energy is said to be "lost" to the system, since it did not end up in the proper energy form.

All technology can be categorized into four broad genres:

1. Those that output some form of energy (such as an engine) as their primary function.
2. Those that absorb energy (such as a pump) as their primary function.

3. Those whose primary function involves energy transfers internal to the system (such as heat exchangers).
4. All other technologies.

For example, given a technology whose primary function is energy output, we can write

$$\text{Actual useful energy output} = \text{Maximum reversible useful energy output} - \text{Lost energy}$$

and, given a technology with a primary function of absorbing energy, we have

$$\text{Actual useful energy input} = \text{Maximum reversible useful energy input} + \text{Lost energy}$$

where we recognize that, in each case, the lost energy is not destroyed but only in a form not used by the technology.

A different and more revealing type of energy conversion efficiency can now be defined, based on ratios of *available* energy rather than total energy. This new efficiency contains terms representing the irreversibilities within the system, so that we can see exactly where the loss of efficiency occurs within a system. To distinguish it from the ordinary first law efficiencies already discussed, this new type of energy conversion efficiency is given the symbol ε and is called *second law* efficiency.

10.11.2 Second Law (Availability) Efficiency

Of the available energy initially resident in (or the net available energy put into) a system, some of it leaves the system with the "desired result" of the operation of the system, some of it is lost to the local environment (usually through unwanted heat transfer), and some of it is destroyed by the irreversibilities operating inside the system. This can be written in equation form as

$$A_{\substack{\text{initial or}\\ \text{net input}}} = A_{\substack{\text{desired}\\ \text{result}}} + A_{\text{loss}} + A_{\text{destruction}} \tag{10.28}$$

We can now define a meaningful second law efficiency (or effectiveness) ε based on availability as

Second law efficiency

$$\varepsilon = \frac{A_{\substack{\text{desired}\\ \text{result}}}}{A_{\substack{\text{initial or}\\ \text{net input}}}} = \frac{\dot{A}_{\substack{\text{desired}\\ \text{result}}}}{\dot{A}_{\substack{\text{initial or}\\ \text{net input}}}} \tag{10.29}$$

EXAMPLE 10.10

Determine the first and second law efficiencies for heating a liquid from temperature T to temperature $T + \Delta T$ in a *closed*, uninsulated tank. The temperature change ΔT is due to an external heat transfer Q_{in}, and since the tank is not insulated, there is also a heat loss to the local environment, Q_{loss}. The ground state temperature is T_0.

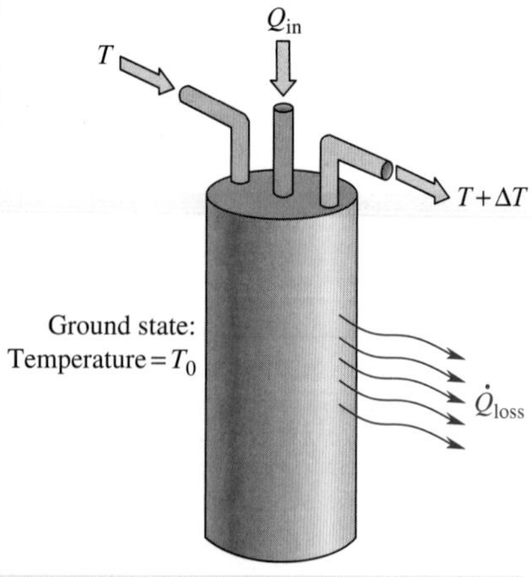

FIGURE 10.15
Example 10.10.

Solution

First, draw a sketch of the system (Figure 10.15).

The unknowns are the first and second law efficiencies for heating a liquid from temperature T to temperature $T + \Delta T$ in a *closed*, uninsulated tank.

An energy balance for this system with $W = 0$ and $\Delta pe = \Delta ke = 0$ is

$$Q_{in} - |Q_{loss}| - 0 = m(u_2 - u_1 + 0 + 0)$$

or

$$Q_{in} = m(u_2 - u_1) + |Q_{loss}|$$

For an incompressible liquid, we can write

$$u_2 - u_1 = c(T_2 - T_1) = c[(T + \Delta T) - T] = c(\Delta T)$$

The "desirable result" here is the increase in internal energy of the liquid in the tank, $m(u_2 - u_1)$, so the first law (thermal) efficiency is

$$\eta_T = \frac{\text{Desirable result}}{\text{Cost}} = \frac{m(u_2 - u_1)}{Q_{in}} = \frac{mc(\Delta T)}{Q_{in}}$$

The second law efficiency is given by Eq. (10.29) as

$$\varepsilon = \frac{A_{\text{desirable result}}}{A_{\text{initial or net input}}}$$

where

$$A_{\text{desirable result}} = A_2 - A_1 = m\left[u_2 - u_1 + p_0(v_2 - v_1) - T_0(s_2 - s_1) + (V_2^2 - V_1^2)/2g_c + g(Z_2 - Z_1)/g_c\right]$$

since $V_1 = V_2$, $Z_1 = Z_2$, and $v_1 = v_2$ here, the this equation reduces to

$$A_{\text{desirable result}} = A_2 - A_1 = m[u_2 - u_1 - T_0(s_2 - s_1)]$$

For an incompressible liquid, $u_2 - u_1 = c(T_2 - T_1) = c(\Delta T)$, and $s_2 - s_1 = c\ \ln(T_2/T_1)$, then

$$A_{\text{desirable result}} = mc\left[T_2 - T_1 - T_0 \ln\left(\frac{T_2}{T_1}\right)\right] = mc\left[\Delta T - T_0 \ln\left(\frac{T + \Delta T}{T}\right)\right]$$

so

$$\varepsilon = mc\left[\frac{\Delta T - T_0 \ln\left(\frac{T + \Delta T}{T}\right)}{Q_{\text{in}}}\right] = mc\left(\frac{\Delta T - T_0 \ln\left(1 + \frac{\Delta T}{T}\right)}{Q_{\text{in}}}\right)$$

But, from the first law efficiency defined earlier, $Q_{\text{in}} = mc\,(\Delta T)/\eta_T$, so

$$\varepsilon = \eta_T\left[1 - \frac{T_0}{\Delta T}\ln\left(1 + \frac{\Delta T}{T}\right)\right]$$

For example, if $T_0 = 70.0°\text{F} = 530.$ R and $T = 120.°\text{F} = 580.$ R, then

$$\varepsilon = \eta_T\left[1 - \frac{530.}{50.0} \times \ln\left(1 + \frac{50.0}{580.}\right)\right] = 0.124 \times \eta_T$$

so that, even if the first law thermal efficiency of the heating process were 100%, the corresponding second law efficiency would be only 12.4%. This is due to the fact that ΔT is very small in comparison with T and T_0.

Exercises

28. Determine the first and second law efficiencies for the closed liquid heater in Example 10.10 when 300. lbm of liquid water at 50.0°F is heated to 100.°F using 16.5×10^3 Btu. The temperature of the local environment is 50.0°F. **Answer**: $\eta_T = 90.9\%$, and $\varepsilon = 4.20\%$.

29. If the closed liquid heater in Example 10.10 contains 150. kg of liquid water at 15.0°C and is heated to 40.0°C, determine the first and second law efficiencies of this process. The amount of heat added to the heater is 23.0×10^3 kJ, and the local environmental temperature is 10.0°C. **Answer**: $\eta_T = 68.3\%$, and $\varepsilon = 6.20\%$.

30. Show that, when $\Delta T \ll T$, the second law efficiency defined in Example 10.10 for a closed liquid heating system can be written as $\varepsilon = \eta_T(1 - T_0/T)$. (Hint: If ΔT is small, i.e., $\Delta T \ll T$, then the logarithm term can be expanded as $\ln(1 + \Delta T/T) \approx \Delta T/T$.) Note that as $T \to T_0$ (or as $\Delta T \to 0$) in this case, then $\varepsilon \to 0$. This is due to the fact that there is no available energy in the desired result (i.e., $A_2 \to A_1$).

Example 10.10 shows that heating a liquid in a closed container is not a very efficient use of energy. What would happen if the heating were done in an open system, such as a domestic hot water heater? Would that be any better? Example 10.11 answers this question and shows that the process of heating a liquid has a poor second law efficiency no matter how it is done.

EXAMPLE 10.11

In this example, we determine the first and second law efficiencies for heating a liquid from temperature T to temperature $T + \Delta T$ in an *open*, uninsulated heating tank with a negligible pressure drop and compare the results to those obtained in Example 10.10 for heating in a *closed* tank. As in Example 10.10, the temperature change ΔT is due to the external heat transfer rate $\dot{Q}_{\text{in}}$, and since the tank is not insulated there is also a heat loss rate to the local environment, $\dot{Q}_{\text{loss}}$. Again, take the ground state temperature to be T_0.

(*Continued*)

EXAMPLE 10.11 (*Continued*)

Solution

First, draw a sketch of the system (Figure 10.16).

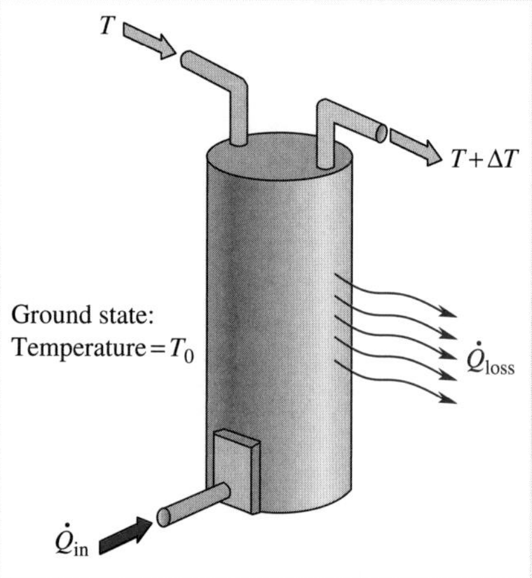

FIGURE 10.16
Example 10.11.

The unknowns are the first and second law efficiencies for heating a liquid from temperature T to temperature $T + \Delta T$ in an *open* uninsulated heating tank.

An energy balance on this system with $\dot{W} = 0$ and $\Delta \text{pe} = \Delta \text{ke} = 0$ is

$$\dot{Q}_{\text{in}} - \dot{Q}_{\text{loss}} - 0 = \dot{m}(h_2 - h_1 + 0 + 0)$$

or

$$\dot{Q}_{\text{in}} = \dot{m}(h_2 - h_1) + \dot{Q}_{\text{loss}}$$

We can treat the liquid as incompressible with a negligible pressure drop and write

$$h_2 - h_1 = u_2 - u_1 + v(p_2 - p_1) = c(T_2 - T_1) + 0 = c[(T + \Delta T) - T] = c(\Delta T)$$

The "desirable result" in this problem is the increase in enthalpy of the liquid flowing through the system, $\dot{m}(h_2 - h_1)$, so the first law (thermal) efficiency becomes

$$\eta_T = \frac{\text{Desirable result}}{\text{Cost}} = \frac{\dot{m}(h_2 - h_1)}{\dot{Q}_{\text{in}}} = \frac{\dot{m}c(\Delta T)}{\dot{Q}_{\text{in}}}$$

The second law efficiency is given by Eq. (10.29) as

$$\varepsilon = \frac{\dot{A}_{\text{desired result}}}{\dot{A}_{\text{initial or net input}}}$$

where

$$\dot{A}_{\text{desirable result}} = \dot{m}(a_{f2} - a_{f1}) = \dot{m}\left[h_2 - h_1 - T_0(s_2 - s_1) + (V_2^2 - V_1^2)/2g_c + g(Z_2 - Z_1)/g_c\right]$$

and, since $V_1 = V_2$ and $Z_1 = Z_2$ here, this equation reduces to

$$\dot{A}_{\text{desirable result}} = \dot{m}[h_2 - h_1 - T_0(s_2 - s_1)]$$

For a constant pressure incompressible liquid, $h_2 - h_1 = u_2 - u_1 + v(p_2 - p_1) = c(T_2 - T_1) + 0 = c(\Delta T)$ and $s_2 - s_1 = c\ \ln(T_2/T_1)$, then

$$\dot{A}_{\text{desirable result}} = \dot{m}c\left[T_2 - T_1 - T_0 \ln\left(\frac{T_2}{T_1}\right)\right] = \dot{m}c\left[\Delta T - T_0 \ln\left(\frac{T + \Delta T}{T}\right)\right]$$

so

$$\varepsilon = \dot{m}c\left[\frac{\Delta T - T_0 \ln\left(\frac{T + \Delta T}{T}\right)}{\dot{Q}_{\text{in}}}\right] = \dot{m}c\left(\frac{\Delta T - T_0 \ln\left(1 + \frac{\Delta T}{T}\right)}{\dot{Q}_{\text{in}}}\right)$$

But, from the first law efficiency defined previously, $\dot{Q}_{\text{in}} = \dot{m}c(\Delta T)/\eta_T$, so we have

$$\varepsilon = \eta_T\left[1 - \frac{T_0}{\Delta T}\ln\left(1 + \frac{\Delta T}{T}\right)\right]$$

as was found in Example 10.10. If we again set $T_0 = 70.0°\text{F} = 530.\ \text{R}$ and $T = 120.°\text{F} = 580.\ \text{R}$, then this equation gives

$$\varepsilon = \eta_T\left[1 - \frac{530.}{50.0}\ln\left(1 + \frac{50.0}{580.}\right)\right] = 0.124 \times \eta_T$$

so that, even if the first law thermal efficiency of a domestic hot water heater were 100%, the corresponding second law efficiency would be only 12.4%. This is again due to the fact that the ΔT here is very small in comparison with the values of T and T_0.

Exercises

31. Determine the first and second law efficiencies for the open liquid heater in Example 10.11 when 3.00 lbm/s of liquid water at 50.0°F is heater to 100.°F with a heat input of 174 Btu/s. The temperature of the local environment is 50.0°F. **Answer:** $\eta_T = 86.2\%$, and $\varepsilon = 4.0\%$.
32. If the open liquid heater in Example 10.11 has a flow rate of 1.50 kg/s of liquid water at 15.0°C that is heated to 40.0°C, determine the first and second law efficiencies of this process. The rate of heat added to the heater is 180. kJ/s, and the local environmental temperature is 10.0°C. **Answer:** $\eta_T = 87.2\%$, and $\varepsilon = 7.9\%$.
33. Show that, when ΔT is small (i.e., $\Delta T \ll T$), the second law efficiency derived for the open liquid heating system in Example 10.11 can be written as $\varepsilon \approx \eta_T(1 - T_0/T)$. (Hint: When $\Delta T \ll T$, the logarithm can be expanded as ln $(1 + \Delta T/T) \approx \Delta T/T$.)

Consider the operation of a heat engine. The "desired result" of the operation of a heat engine is the net power output of the engine $\dot{W}$, and the availability of the desired result is again just the net power output $\dot{W}$. The availability input rate to the engine is that due to the net heat transfer rate, which is the difference between the heat transfer availability rate input from the high temperature source at T_H and the heat transfer availability rate loss to the sink at T_L, or

$$\dot{A}_{\substack{\text{desired}\\\text{result}}} = \dot{W}$$

and

$$\dot{A}_{\substack{\text{net}\\\text{input}}} = \left(1 - \frac{T_0}{T_H}\right)\dot{Q}_H - \left(1 - \frac{T_0}{T_L}\right)|\dot{Q}_L|$$

and the resulting second law availability efficiency of a heat engine is

$$\varepsilon_{HE} = \frac{\dot{W}}{\left(1 - \frac{T_0}{T_H}\right)\dot{Q}_H - \left(1 - \frac{T_0}{T_L}\right)|\dot{Q}_L|} \quad (10.30)$$

If $T_0 = T_L$, the second law efficiency of a heat engine that rejects heat to the local environment reduces to

Second law efficiency of a heat engine that rejects heat to the local environment

$$\text{When } T_0 = T_L,\ \varepsilon_{HE} = \frac{\dot{W}/\dot{Q}_H}{1 - \frac{T_L}{T_H}} = \frac{\eta_T}{\eta_{\text{Carnot}}} \quad (10.31)$$

where, from Eq. (7.16), we use $\eta_{\text{Carnot}} = 1 - T_L/T_H$.

A power plant is a type of modern heat engine. The following example illustrates the application of this material to a power plant.

EXAMPLE 10.12

The electric power plant at Mount Etna has a heat input to the boiler of 1.00×10^6 kJ/s at 700.°C and rejects 7.00×10^5 kJ/s of heat to the condenser at 40.0°C while producing 3.00×10^5 kJ/s of electrical power. The local environment (ground state) is at 0.101 MPa and 5.00°C. Determine

a. The first law thermal efficiency of the power plant.
b. The rate at which available energy enters the boiler.
c. The rate at which available energy enters the condenser.
d. The second law efficiency of the power plant.

Solution

First, draw a sketch of the system (Figure 10.17).

(*Continued*)

EXAMPLE 10.12 *(Continued)*

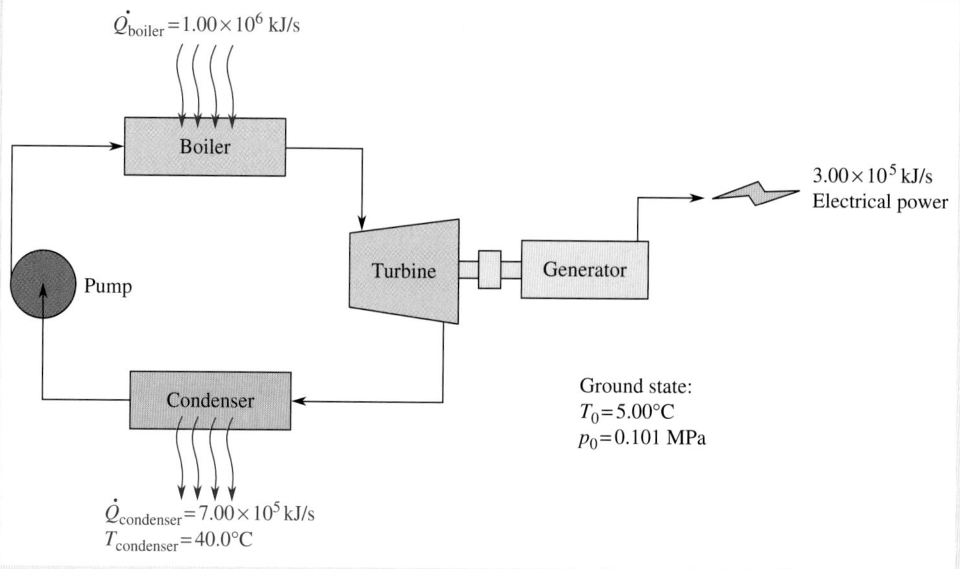

FIGURE 10.17
Example 10.12.

The unknowns are the first law thermal efficiency of the power plant, the rate at which available energy enters the boiler, the rate at which available energy enters the condenser, and the second law efficiency of the power plant.

a. The first law thermal efficiency of a heat engine is given by Eq. (7.5) as

$$\eta_T = \frac{\text{Net work output}}{\text{Total heat input}} = \frac{|W_{\text{out}}|_{\text{net}}}{|Q_{\text{input}}|_{\text{total}}} = \frac{|\dot{W}_{\text{out}}|_{\text{net}}}{|\dot{Q}_{\text{input}}|_{\text{total}}}$$

so, the overall thermal efficiency of the power plant is

$$\eta_T = \frac{3.00 \times 10^5 \text{ kJ/s}}{1.00 \times 10^6 \text{ kJ/s}} = 0.300 = 30.0\%$$

b. The rate at which available energy enters the boiler is given by

$$\dot{A}_{\substack{\text{boiler}\\\text{input}}} = \left(1 - \frac{T_0}{T_H}\right)\dot{Q}_H = \left(1 - \frac{5.00 + 273.15 \text{ K}}{700. + 273.15 \text{ K}}\right)(1.00 \times 10^6 \text{ kJ/s}) = 7.14 \times 10^5 \text{ kJ/s}$$

c. The rate at which available energy enters the condenser is given by

$$\dot{A}_{\substack{\text{condenser}\\\text{input}}} = \left(1 - \frac{T_0}{T_H}\right)|\dot{Q}_L| = \left(1 - \frac{5.00 + 273.15 \text{ K}}{40.0 + 273.15 \text{ K}}\right)(7 \times 10^6 \text{ kJ/s}) = 0.78 \times 10^5 \text{ kJ/s}$$

d. The second law availability efficiency of the power plant is given by Eq. (10.30) as

$$\varepsilon_{HE} = \frac{\dot{W}}{\left(1 - \frac{T_0}{T_H}\right)\dot{Q}_H - \left(1 - \frac{T_0}{T_L}\right)|\dot{Q}_L|} = \frac{\dot{W}}{\dot{A}_{\substack{\text{turbine}\\\text{input}}} - \dot{A}_{\substack{\text{condenser}\\\text{input}}}}$$

$$= \frac{3.00 \times 10^5 \text{ kJ/s}}{\left(1 - \frac{5.00 + 273.15 \text{ K}}{700. + 273.15 \text{ K}}\right)(1.00 \times 10^6 \text{ kJ/s}) - \left(1 - \frac{5.00 + 273.15 \text{ K}}{40.0 + 273.15 \text{ K}}\right)(7.00 \times 10^5 \text{ kJ/s})}$$

$$= 0.472 = 47.2\%$$

Exercises

34. Calculate the second law thermal efficiency of the power plant in Example 10.12 if the boiler outlet temperature is increased from 700.°C to 1000.°C. Assume all the other variables remain unchanged. **Answer:** $\varepsilon_{HE} = 42.7\%$.
35. Suppose we lower the condenser temperature in Example 10.12 from 40.0°C to 15.0°C. Assuming all the other variables remain unchanged, determine the new second law thermal efficiency of the power plant. **Answer:** $\varepsilon_{HE} = 43.5\%$.
36. Determine the second law thermal efficiency of the power plant in Example 10.12 when we set $T_L = T_0 = 5.00°C$. **Answer:** $\varepsilon_{HE} = 42.0\%$.
37. Why is the value of ε_{HE} calculated in Exercise 36 less than the value calculated in the Example 10.12? (Hint: Does it include the heat transfer irreversibilities and the lost available energy in the heat transfer between the condenser and the environment?)

CRITICAL THINKING

The combustion temperature of the burning fuel in the boiler of Example 10.12 is 1800.°C. Suppose this temperature is taken as the heat source temperature T_H. Would the value of the second law availability efficiency ε_{HE} calculated in Example 10.12 increase or decrease? Explain what additional irreversibilities are introduced by this change in system boundary that would produce the change you observe in ε_{HE}.

In the case of a heat pump, the net available energy input rate is again just the pump work rate (power). The available energy rate of the "desired result" (heating) is $(1 - T_0/T_H)_H$. Then, the second law availability efficiency of a heat pump is

$$\varepsilon_{HP} = \frac{\left(1 - \frac{T_0}{T_H}\right)\dot{Q}_H}{|\dot{W}_{in}|} = \left(1 - \frac{T_0}{T_H}\right)\left(\frac{\dot{Q}_H}{|\dot{W}_{in}|}\right) = \left(1 - \frac{T_0}{T_H}\right)\mathrm{COP}_{\substack{\text{actual}\\ \text{heat pump}}} \tag{10.32}$$

and, if $T_0 = T_L$, the second law efficiency of a heat pump that absorbs heat from the local environment reduces to

Second law efficiency of a heat pump that absorbs heat from the local environment

$$\text{When } T_0 = T_L,\ \varepsilon_{HP} = \frac{\mathrm{COP}_{\substack{\text{actual}\\ \text{heat pump}}}}{\mathrm{COP}_{\substack{\text{Carnot}\\ \text{heat pump}}}} \tag{10.33}$$

where the coefficient of performance (COP) of a Carnot heat pump given in Eq. (7.18) is used. The following example illustrates the use of this material.

EXAMPLE 10.13

A heat pump is designed to provide 30.0×10^3 Btu/h of heat to a small house at 70.0°F when the outside temperature is 30.0°F. The electric motor driving the heat pump draws 1.50 hp. Determine

a. The first law thermal efficiency (i.e., the COP) of the heat pump.
b. The second law availability efficiency of the heat pump.

Solution

First, draw a sketch of the system (Figure 10.18).

The unknowns are the first law thermal efficiency (i.e., the COP) of the heat pump and the second law availability efficiency of the heat pump.

a. The first law thermal efficiency of the heat pump is given by Eq. (4.70) as

$$\eta_T = \mathrm{COP}_{\substack{\text{actual}\\ \text{heat pump}}} = \frac{\text{Desired energy result}}{\text{Required energy input}} = \frac{\dot{Q}_H}{|\dot{W}_{in}|} = \frac{30.0 \times 10^3\ \text{Btu/h}}{(1.50\ \text{hp})(2545\ \text{Btu/hp}\cdot\text{h})} = 7.86$$

(Continued)

EXAMPLE 10.13 (*Continued*)

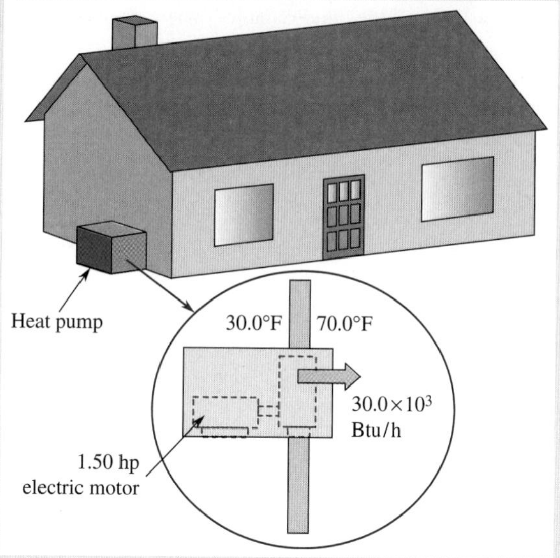

FIGURE 10.18
Example 10.13.

b. The second law availability efficiency is given by Eq. (10.32) with $T_0 = T_L = 30.0°\text{F}$ as

$$\varepsilon_{HP} = \left(1 - \frac{T_0}{T_H}\right)\left(\text{COP}_{\substack{\text{actual}\\ \text{heat pump}}}\right) = \left(1 - \frac{30.0 + 459.67\text{ R}}{70.0 + 459.67\text{ R}}\right)(7.86) = 0.594 = 59.4\%$$

Note that, since $T_0 = T_L$ here, Eq. (10.33) could have been used with

$$\text{COP}_{\substack{\text{Carnot}\\ \text{heat pump}}} = \frac{T_H}{T_H - T_L} = \frac{70.0 + 459.67}{70.0 - 30.0} = 13.241$$

to find

$$\varepsilon_{HP} = \frac{\text{COP}_{\substack{\text{actual}\\ \text{heat pump}}}}{\text{COP}_{\substack{\text{Carnot}\\ \text{heat pump}}}} = \frac{7.86}{13.24} = 0.594 = 59.4\%$$

Exercises

38. If the house temperature in Example 10.13 increases from 70.0°F to 85.0°F and all the other variables remain unchanged, determine the new second law efficiency of the heat pump. **Answer**: $\varepsilon_{HP} = 79.4\%$.

39. When the outside temperature in Example 10.13 increases from 30.0°F to 40.0°F, determine the new second law thermal efficiency of the heat pump. Assume all the other variables remain unchanged. **Answer**: $\varepsilon_{HP} = 44.5\%$.

40. The heat provided to the house in Example 10.13 is suddenly reduced from 30.0×10^3 Btu/h to 25.0×10^3 Btu/h. What is the new second law thermal efficiency of the heat pump? Assume all the other variables remain unchanged. **Answer**: $\varepsilon_{HP} = 49.5\%$.

In the case of a refrigeration or air conditioning system, the net available energy input rate is also $|\dot{W}_{\text{in}}|$, but now the available energy rate of the "desired result" (cooling) is $|(1 - T_0/T_L)\dot{Q}_L|$, where the absolute value has been used to keep the efficiency a positive number. Then, the second law availability efficiency of a refrigeration or air conditioning system is

$$\varepsilon_{R/AC} = \frac{\left(1 - \frac{T_0}{T_L}\right)\dot{Q}_L}{|\dot{W}_{\text{in}}|} = \left(1 - \frac{T_0}{T_L}\right)\left(\frac{\dot{Q}_L}{|\dot{W}_{\text{in}}|}\right) = \left(1 - \frac{T_0}{T_L}\right)\text{COP}_{\substack{\text{actual}\\ \text{ref or air cond}}} \qquad (10.34)$$

and, if $T_0 = T_H$, the second law efficiency of a refrigerator or air conditioner that rejects heat to the local environment reduces to

Second law efficiency of a refrigerator or air conditioner that rejects heat to the local environment

$$\text{When } T_0 = T_H,\ \varepsilon_{R/AC} = \frac{\text{COP}^{\text{actual}}_{\text{ref or air cond}}}{\text{COP}^{\text{Carnot}}_{\text{ref or air cond}}} \tag{10.35}$$

where the coefficient of performance of a Carnot refrigerator or air conditioner given in Eq. (7.20) is used. The following example illustrates the use of this material.

EXAMPLE 10.14

A common window air conditioner has an actual COP of 8.92. If the inside temperature is $T_0 = T_L = 20.0°\text{C}$ and the outside temperature is $T_H = 35.0°\text{C}$, determine the second law availability efficiency of this air conditioner.

Solution

First draw a sketch of the system (Figure 10.19).

The unknown is the second law availability efficiency of this air conditioner.

Equation (7.20) gives the coefficient of performance of a Carnot refrigerator or air conditioner as

$$\text{COP}^{\text{Carnot}}_{\text{ref. or air cond.}} = \frac{T_L}{T_H - T_L} = \frac{20.0 + 273.15\ \text{K}}{(35.0 + 273.15\ \text{K}) - (20.0 + 273.15\ \text{K})} = 19.6$$

then, Eq. (10.35) can be used to determine the second law efficiency as

$$\varepsilon_{R/AC} = \frac{\text{COP}^{\text{actual}}_{\text{ref. or air cond.}}}{\text{COP}^{\text{Carnot}}_{\text{ref. or air cond.}}} = \frac{8.92}{19.6} = 0.455 = 45.5\%$$

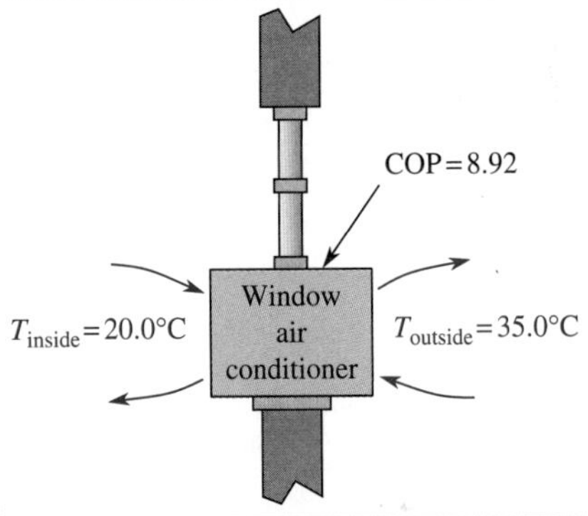

FIGURE 10.19
Example 10.14.

Exercises

41. Determine the second law thermal efficiency of the air conditioner in Example 10.14 if the outside air temperature is increased from 35.0°C to 40.0°C. Assume all the other variables remain unchanged.
Answer: $\varepsilon_{R/AC} = 60.9\%$.

42. If the inside air temperature in Example 10.14 increases from 20.0°C to 22.0°C, determine the new second law thermal efficiency of the window air conditioner. Assume all the other variables remain unchanged.
Answer: $\varepsilon_{R/AC} = 39.3\%$.

43. When the outside temperature in Example 10.14 increases from 35.0°C to 40.0°C, the actual COP of the air conditioner decreases from 8.92 to 7.5. Determine the new second law thermal efficiency of the air conditioner.
Answer: $\varepsilon_{R/AC} = 51.2\%$.

When we deal with heat exchangers in which the fluids do not mix, the desirable result is the increase in temperature of the cold stream (we could choose the desired result to be the decrease in temperature of the hot stream if we wished). Thus, the corresponding $\dot{A}_{\text{desired result}}$ is the increase in the available energy rate of the cold stream, or from Figure 10.20a,

$$\dot{A}_{\text{desirable result}} = \dot{m}_C(a_{f4} - a_{f3})$$

and the source availability is the decrease in the available energy rate of the hot stream:

$$\dot{A}_{\substack{\text{initial or}\\ \text{net input}}} = \dot{m}_H(a_{f1} - a_{f2})$$

Then, the second law availability efficiency for a nonmixing heat exchanger is the ratio of these two terms:

Second law efficiency of a nonmixing heat exchanger

$$\varepsilon_{\text{nonmixing HX}} = \frac{\dot{m}_C(a_{f4} - a_{f3})}{\dot{m}_H(a_{f1} - a_{f2})} \tag{10.36}$$

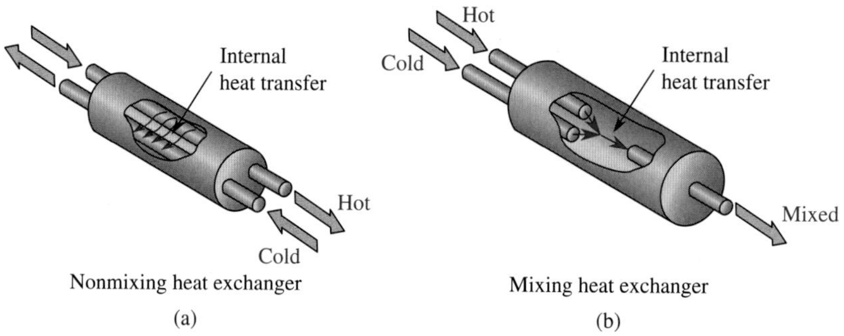

FIGURE 10.20
Schematic examples of nonmixing and mixing heat exchangers.

EXAMPLE 10.15

The inlet air to a gas turbine engine is preheated with the engine's exhaust gases. The preheater is insulated so that all heat transfer is internal. The engine's exhaust gas enters the preheater at 500.°C and 1.10 atmospheres pressure and exits at 400.°C and 1.00 atm. The inlet air enters the preheater at 20.0°C and 1.50 atm and exits at 1.40 atm. The mass flow rates of the inlet air and the exiting exhaust are approximately the same at 0.800 kg/s. Both gases can be treated as constant specific heat ideal gases. The specific heat and gas constant of the exhaust gas are $(c_p)_{exh.} = 0.990$ kJ/(kg·K) and $R_{exh.} = 0.272$ kJ/(kg·K). Neglecting all kinetic and potential energies and taking local environment (ground state) as $p_0 = 1.00$ atm and $T_0 = 20.0$°C, determine

a. The exit temperature of the inlet air.
b. The second law availability efficiency of the preheater.

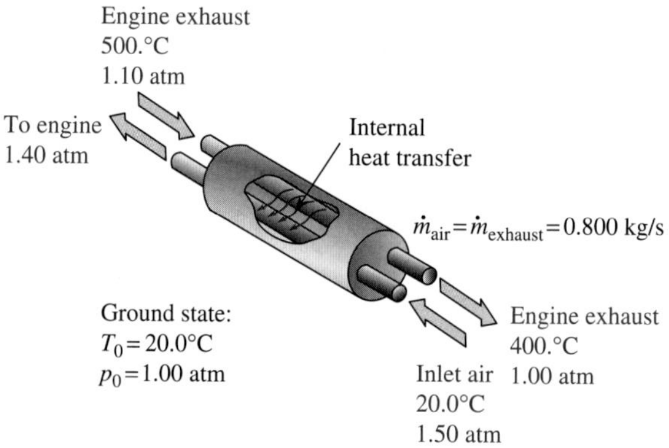

FIGURE 10.21
Example 10.15.

Solution

First, draw a sketch of the system (Figure 10.21).

The unknowns are the exit temperature of the inlet air and the second law availability efficiency of the preheater.

a. The air exit temperature can be obtained from an energy rate balance on the steady state preheater with $\dot{Q} = \dot{W} = 0$ and, with the kinetic and potential energy terms neglected, as

$$\dot{m}_{exh}(h_{in} - h_{out})_{exh} = \dot{m}_{air}(h_{out} - h_{in})_{air}$$

Since both sides of the heat exchanger are ideal gases, this can be written as

$$\dot{m}_{exh}(c_p)_{exh}(T_{in} - T_{out})_{exh} = \dot{m}_{air}(c_p)_{air}(T_{out} - T_{in})_{air}$$

from which we can solve for the exit air temperature as

$$(T_{out})_{air} = (T_{in})_{air} + \left(\frac{\dot{m}_{exh}(c_p)_{exh}}{\dot{m}_{cold}(c_p)_{air}}\right)(T_{in} - T_{out})_{exh}$$

or

$$(T_{out})_{air} = 20.0°\text{C} + \left(\frac{(0.800\text{ kg/s})(0.990\text{ kJ/kg}\cdot\text{K})}{(0.800\text{ kg/s})(1.004\text{ kJ/kg}\cdot\text{K})}\right)(500. - 400.°\text{C}) = 119°\text{C}$$

b. The second law availability efficiency of the preheater is given by Eq. (10.36), where the flow availability is given by Eq. (10.20):

$$a_f = h - h_0 - T_0(s - s_0) + \frac{V^2}{2g_c} + \frac{gZ}{g_c}$$

and since both flow streams are constant specific heat ideal gases, this equation reduces to

$$a_f = c_p(T - T_0) - T_0\left[c_p \ln\left(\frac{T}{T_0}\right) - R\ln\left(\frac{p}{p_0}\right)\right] + \frac{V^2}{2g_c} + \frac{gZ}{g_c}$$

We neglect kinetic and potential energy terms here, so we can evaluate this equation for each of the four flow streams as

$$\begin{aligned}(a_f)_{\text{in-exh}} &= \left(0.990\frac{\text{kJ}}{\text{kg}\cdot\text{K}}\right)(500. - 20.0\ \text{K})\\ &- (20.0 + 273.15\ \text{K})\left[\left(0.990\frac{\text{kJ}}{\text{kg}\cdot\text{K}}\right)\ln\left(\frac{500. + 273.15}{20.0 + 273.15}\right) - \left(0.272\frac{\text{kJ}}{\text{kg}\cdot\text{K}}\right)\ln\left(\frac{1.10}{1.00}\right)\right]\\ &+ 0 + 0 = 201\ \text{kJ/kg}\end{aligned}$$

$$\begin{aligned}(a_f)_{\text{out-exh}} &= \left(0.990\frac{\text{kJ}}{\text{kg}\cdot\text{K}}\right)(400. - 20.0\ \text{K})\\ &- (20.0 + 273.15\ \text{K})\left[\left(0.990\frac{\text{kJ}}{\text{kg}\cdot\text{K}}\right)\ln\left(\frac{400. + 273.15}{20.0 + 273.15}\right) - \left(0.272\frac{\text{kJ}}{\text{kg}\cdot\text{K}}\right)\ln\left(\frac{1.00}{1.00}\right)\right]\\ &+ 0 + 0 = 135\ \text{kJ/kg}\end{aligned}$$

$$\begin{aligned}(a_f)_{\text{in-air}} &= \left(1.004\frac{\text{kJ}}{\text{kg}\cdot\text{K}}\right)(20.0 - 20.0\ \text{K})\\ &- (20.0 + 273.15\ \text{K})\left[\left(1.004\frac{\text{kJ}}{\text{kg}\cdot\text{K}}\right)\ln\left(\frac{20.0 + 273.15}{20.0 + 273.15}\right) - \left(0.286\frac{\text{kJ}}{\text{kg}\cdot\text{K}}\right)\ln\left(\frac{1.50}{1.00}\right)\right]\\ &+ 0 + 0 = 34.0\ \text{kJ/kg}\end{aligned}$$

$$\begin{aligned}(a_f)_{\text{out-air}} &= \left(1.004\frac{\text{kJ}}{\text{kg}\cdot\text{K}}\right)(118.6 - 20.0\ \text{K})\\ &- (20.0 + 273.15\ \text{K})\left[\left(1.004\frac{\text{kJ}}{\text{kg}\cdot\text{K}}\right)\ln\left(\frac{118.6 + 273.15}{20.0 + 273.15}\right) - \left(0.286\frac{\text{kJ}}{\text{kg}\cdot\text{K}}\right)\ln\left(\frac{1.40}{1.00}\right)\right]\\ &+ 0 + 0 = 41.9\ \text{kJ/kg}\end{aligned}$$

Then Eq. (10.36) gives the second law availability efficiency as

$$\varepsilon_{\text{nonmixing HX}} = \frac{\dot{m}_{\text{air}}[(a_f)_{\text{out air}} - (a_f)_{\text{in air}}]}{\dot{m}_{\text{exh}}[(a_f)_{\text{in exh}} - (a_f)_{\text{out exh}}]} = \frac{(0.800\ \text{kg/s})(41.9 - 34.0\ \text{kJ/kg})}{(0.800\ \text{kg/s})(201 - 135\ \text{kJ/kg})} = 0.119 = 11.9\%$$

Exercises

44. Determine the second law efficiency of the air preheater in Example 10.15 when the exhaust inlet temperature is increased from 500.°C to 750.°C. Assume all the other variables (except T_4) remain unchanged. **Answer:** $\varepsilon_{HX} = 48.0\%$.
45. Calculate the second law efficiency of the air preheater in Example 10.15 when the exhaust exit temperature is decreased from 400.°C to 350.°C. Assume all the other variables (except T_4) remain unchanged. **Answer:** $\varepsilon_{HX} = 24.0\%$.
46. The mass flow rates in Example 10.15 are changed from 0.800 kg/s to $\dot{m}_{\text{exhaust}} = 0.900$ kg/s and $\dot{m}_{\text{air}} = 0.700$ kg/s. Assuming all the other variables remain unchanged, determine the new second law thermal efficiency of the air preheater. **Answer:** $\varepsilon_{HX} = 18.4\%$.

If the hot and cold fluids mix inside the heat exchanger (see Figure 10.20b), then the second law availability efficiency becomes

Second law efficiency of a heat exchanger in which the hot and cold fluids mix

$$\varepsilon_{\text{mixing HX}} = \frac{(1 - y)(a_{f3} - a_{f2})}{y(a_{f1} - a_{f3})} \tag{10.37}$$

where y is the hot mass fraction defined as $y = \dot{m}_H/\dot{m}_m = \dot{m}_1/\dot{m}_3$.

EXAMPLE 10.16

A student is using a bathroom sink with separate hot and cold water faucets. The student turns on both faucets and adjusts them to create a pool of warm water in the sink. The sink's drain is open, so the faucets must be kept running to maintain the pool of water. The hot water faucet provides 130.°F water at 0.180 lbm/s, and the cold water faucet provides 60.0°F water at 0.270 lbm/s. The local environment (ground state) is at $p_0 = 14.7$ psia at $T_0 = 55.0$°F. Neglecting all flow stream kinetic and potential energy and assuming the sink itself is insulated, determine

a. The temperature of the mixed water in the sink.
b. The second law availability efficiency of the sink as a mixing-type heat exchanger.

Solution

First, draw a sketch of the system (Figure 10.22).

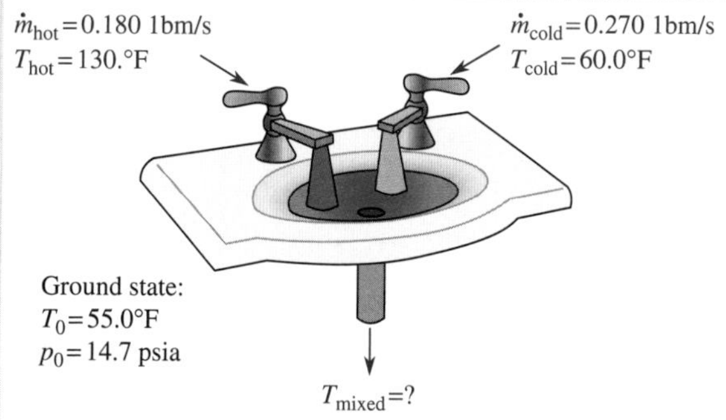

FIGURE 10.22
Example 10.16.

The unknowns are the temperature of the mixed water in the sink and the second law availability efficiency of the sink as a mixing-type heat exchanger.

a. Since the sink is insulated, the mixing process can be taken as adiabatic, with all the heat transfer occurring inside the system. The energy rate balance for an adiabatic, aergonic, steady state, steady flow, double-inlet, single-outlet system with negligible kinetic and potential energies is

$$\dot{m}_H h_H + \dot{m}_C h_C = \dot{m}_M h_M = (\dot{m}_H + \dot{m}_C) h_M$$

or

$$\dot{m}_H (h_H - h_M) = \dot{m}_C (h_M - h_C)$$

where we use the conservation of mass, $\dot{m}_M = \dot{m}_H + \dot{m}_C$. The thermodynamic state of the water here is a slightly compressed liquid, but since the compressibility of liquid water is so small, it can be treated as an incompressible liquid with a constant specific heat c. Assuming $p_H = p_C = p_m$, then Eq. (3.34) can be used in the previous equation to produce

$$\dot{m}_H c (T_H - T_M) = \dot{m}_C c (T_M - T_C)$$

or

$$T_M = \frac{\dot{m}_H T_H + \dot{m}_C T_C}{\dot{m}_H + \dot{m}_C} = \frac{(0.180\,\text{lbm/s})(130. + 459.67\,\text{R}) + (0.270\,\text{lbm/s})(60.0 + 459.67\,\text{R})}{0.180\,\text{lbm/s} + 0.270\,\text{lbm/s}}$$

$$= 548\,\text{R} = 88.0°\text{F}$$

b. The specific flow availability is defined by Eq. (10.20) as

$$a_f = h - h_0 - T_0(s - s_0) + \frac{V^2}{2g_c} + \frac{gZ}{g_c}$$

The liquid water exiting the faucet fits our definition of an incompressible fluid so that we can use Eq. (3.34)

$$h - h_0 = c(T - T_0) + v(p - p_0)$$

and Eq. (7.33)

$$s - s_0 = c \ln\left(\frac{T}{T_0}\right)$$

Combining these equations for the flow of an incompressible fluid gives

$$a_f = c(T - T_0) + v(p - p_0) - cT_0 \ln\left(\frac{T}{T_0}\right) + \frac{V^2}{2g_c} + \frac{gZ}{g_c}$$

In this problem, we have $p_H = p_C = p_m = p_0$ and we neglect all kinetic and potential energy terms. Then, the flow availabilities become

$$a_{fH} = a_{f1} = c_w(T_1 - T_0) - c_w T_0 \ln\left(\frac{T_1}{T_0}\right) = \left(1.00 \frac{\text{Btu}}{\text{lbm}\cdot\text{R}}\right)(130. - 55.0\,\text{R})$$
$$- \left(1.00 \frac{\text{Btu}}{\text{lbm}\cdot\text{R}}\right)(55.0 + 459.67\,\text{R}) \ln\left(\frac{130. + 459.67\,\text{R}}{55.0 + 459.67\,\text{R}}\right)$$
$$= 4.99\,\text{Btu/lbm}$$

$$a_{fC} = a_{f2} = c_w(T_2 - T_0) - c_w T_0 \ln\left(\frac{T_2}{T_0}\right) = \left(1.00 \frac{\text{Btu}}{\text{lbm}\cdot\text{R}}\right)(60.0 - 55.0\,\text{R})$$
$$- \left(1.00 \frac{\text{Btu}}{\text{lbm}\cdot\text{R}}\right)(55.0 + 459.67\,\text{R}) \ln\left(\frac{60.0 + 459.67\,\text{R}}{55.0 + 459.67\,\text{R}}\right)$$
$$= 0.0240\,\text{Btu/lbm}$$

$$a_{fM} = a_{f3} = c_w(T_3 - T_0) - c_w T_0 \ln\left(\frac{T_3}{T_0}\right) = \left(1.00 \frac{\text{Btu}}{\text{lbm}\cdot\text{R}}\right)(88.0 - 55.0\,\text{R})$$
$$- \left(1.00 \frac{\text{Btu}}{\text{lbm}\cdot\text{R}}\right)(55.0 + 459.67\,\text{R}) \ln\left(\frac{88.0 + 459.67\,\text{R}}{55.0 + 459.67\,\text{R}}\right)$$
$$= 1.02\,\text{Btu/lbm}$$

The second law availability efficiency is now given by Eq. (10.37) with $y = \dot{m}_H/\dot{m}_m = \dot{m}_1/\dot{m}_3 = 0.180/(0.180 + 0.270) = 0.400$. Then,

$$\varepsilon_{\text{mixing HX}} = \frac{(1-y)(a_{f3} - a_{f2})}{y(a_{f1} - a_{f3})} = \frac{(1 - 0.400)\left(1.02 - 0.0240 \frac{\text{Btu}}{\text{lbm}\cdot\text{R}}\right)}{0.400\left(4.99 - 1.02 \frac{\text{Btu}}{\text{lbm}\cdot\text{R}}\right)} = 0.374 = 37.4\%$$

Exercises

47. The hot water temperature in Example 10.16 is reduced from 130.°F to 120.°F. Determine the new mixture temperature in the sink and the second law efficiency of the mixing in the sink. Assume all the other variables (except T_m) remain unchanged. **Answer**: T_m = 84°F, and ε_{HX} = 38.2%.

48. If the hot and cold mass flow rates in Example 10.16 are equalized at 0.225 lbm/s each, determine the mixture temperature in the sink and new second law efficiency of this mixing process. Assume all the other variables (except T_m) remain unchanged. **Answer**: T_m = 95.0°F, and ε_{HX} = 41.5%.

49. Explain how the second law efficiency in Example 10.16 might be increased. (Hint: Describe how the mass flow rates and the temperatures could be altered to increase the value of ε.)

50. According to Figure 10.8, when $T_b > T_0$, the heat and availability transports are in the same direction, but when $T_b < T_0$, they are in opposite directions. In a heat exchanger that has $T_C < T_H < T_0$, show that the hot flow stream gains availability during the internal heat transfer process and the cold flow stream loses it. (Hint: Look at the difference between the inlet and exit flow stream specific flow availabilities.)

SUMMARY

In this chapter, we study a new concept in applied thermodynamics called *available energy*. The importance of this material is discussed in the Introduction, and necessary background material is presented in the sections on scalar and vector fields, conservative fields, and conservative forces. The concept of availability is based on the *maximum reversible work possible in a system*, limited by the conditions present in the *local environment*. Once the necessary background material is presented, we are able to define *availability* as the maximum possible *useful* reversible work that a system could supply relative to its local environment as a ground state. At this point, we could develop a *closed system availability balance* and carry out the solution of several example problems. Before we could extend this to open systems, we had to define the concept of *flow availability*. With this as the concept of how available energy crosses the system boundary, we are able to develop an *open system availability balance* and modify the general form of this equation for several typical conditions, such as steady state and steady flow. The chapter concludes with a discussion of a new type of efficiency, the *second law efficiency*. This efficiency tells us how efficiently the available energy within the system is used. This is a very important concept in the design of energy conversion systems.

Some of the more important equations introduced in this chapter are (recall that the thermodynamic properties of the local environment or ground state are written as T_0, p_0, $\forall_0$, v_0, U_0, u_0, H_0, h_0, S_0 and s_0) follow.

1. The maximum reversible work that a closed system can perform is given by Eq. (10.1) as

$$(W)_{\substack{\text{maximum}\\\text{reversible}}} = E - E_0 - T_0(S - S_0)$$

and the maximum reversible *useful* work that a closed system can perform is given by Eq. (10.2) as

$$(W)_{\substack{\text{maximum}\\\text{reversible}\\\text{useful}}} = (W)_{\substack{\text{maximum}\\\text{reversible}}} - (W)_{\substack{\text{reversible}\\\text{moving}\\\text{boundary}}} = E - E_0 + p_0(\forall - \forall_0) - T_0(S - S_0)$$

2. The total availability A of a closed system is given by Eq. (10.3) as

$$A = (W)_{\substack{\text{maximum}\\\text{reversible}\\\text{useful}}} = E - E_0 + p_0(\forall - \forall_0) - T_0(S - S_0)$$

$$= m\left[u - u_0 + p_0(v - v_0) - T_0(s - s_0) + V^2/2g_c + gZ/g_c\right]$$

and the specific availability, a, of a closed system is given by Eq. (10.4) as

$$a = A/m = u - u_0 + p_0(v - v_0) - T_0(s - s_0) + V^2/2g_c + gZ/g_c$$

3. The change in total and specific availability of a closed system as it passes from state 1 to state 2 is given by Eqs. (10.5) and (10.6) as

$$A_2 - A_1 = E_2 - E_1 + p_0(\forall_2 - \forall_1) - T_0(S_2 - S_1)$$

$$= m[u_2 - u_1 + p_0(v_2 - v_1) - T_0(s_2 - s_1) + (V_2^2 - V_1^2)/2g_c + g(Z_2 - Z_1)/g_c]$$

and

$$a_2 - a_1 = u_2 - u_1 + p_0(v_2 - v_1) - T_0(s_2 - s_1) + (V_2^2 - V_1^2)/2g_c + g(Z_2 - Z_1)/g_c$$

4. The irreversibility I and the irreversibility rate $\dot{I}$ that occur inside a system are given by Eqs. (10.14) and (10.15) as

$$_1I_2 = T_0\, {_1(S_P)_2} \geq 0$$

and

$$\dot{I} = T_0 \dot{S}_P \geq 0$$

5. The availability balance (AB) and availability rate balance (ARB) for a closed system with a single heat transfer mode occurring at a constant system boundary temperature T_b is given by Eqs. (10.18) and (10.19) as

$$\left[1 - \frac{T_0}{T_{bi}}\right](_1Q_2) - {_1W_2} + p_0(\forall_2 - \forall_1) - {_1I_2} = (A_2 - A_1)_{\text{system}} = [m(a_2 - a_1)]_{\text{system}}$$

and

$$\left(1 - \frac{T_0}{T_b}\right)\dot{Q} - \dot{W} + p_0\frac{d\forall}{dt} - \dot{I} = \left(\frac{dA}{dt}\right)_{\text{system}}$$

6. The specific flow availability of mass crossing a system boundary is defined by Eq. (10.20) as

$$a_f = h - h_0 - T_0(s - s_0) + \frac{V^2}{2g_c} + \frac{gZ}{g_c}$$

7. The general open system availability rate balance is given by Eq. (10.21) as

$$\sum_{i=1}^{n}\left(1-\frac{T_0}{T_{bi}}\right)\dot{Q}_i-\dot{W}+\sum_{\text{inlet}}\dot{m}a_f-\sum_{\text{outlet}}\dot{m}a_f+p_0\dot{V}-\dot{I}=\left(\frac{dA}{dt}\right)_{\text{system}}$$

and the modified availability rate balance (MARB) for steady state, steady flow, single-inlet, single-outlet, constant volume, isothermal boundary (SS, SF, SI, SO, CV, IB) open systems is given by Eq. (10.26) as

$$\left[1-\frac{T_0}{T_b}\right]\dot{Q}-\dot{W}+\dot{m}[(a_f)_{\text{in}}-(a_f)_{\text{out}}]-\dot{I}=0$$

8. The second law efficiency (or effectiveness) is defined in Eq. (10.29) as

$$\varepsilon=\frac{A_{\substack{\text{desired}\\\text{result}}}}{A_{\substack{\text{initial or}\\\text{net input}}}}=\frac{\dot{A}_{\substack{\text{desired}\\\text{result}}}}{\dot{A}_{\substack{\text{initial or}\\\text{net input}}}}$$

9. The second law availability efficiency of a *heat engine* is given by Eq. (10.30) as

$$\varepsilon_{HE}=\frac{\dot{W}}{\left(1-\frac{T_0}{T_H}\right)\dot{Q}_H-\left(1-\frac{T_0}{T_L}\right)|\dot{Q}_L|}$$

which becomes $\varepsilon_{HE}=\eta_T/\eta_{\text{Carnot}}$ when $T_0=T_L$.

10. The second law availability efficiency of a *heat pump* is given by Eq. (10.32) as

$$\varepsilon_{HP}=\frac{\left(1-\frac{T_0}{T_H}\right)\dot{Q}_H}{|\dot{W}_{\text{in}}|}=\left(1-\frac{T_0}{T_H}\right)\left(\frac{\dot{Q}_H}{|\dot{W}_{\text{in}}|}\right)=\left(1-\frac{T_0}{T_H}\right)\text{COP}_{\substack{\text{actual}\\\text{heat pump}}}$$

which becomes $\varepsilon_{HP}=(\text{COP})_{\text{actual HP}}/(\text{COP})_{\text{Carnot HP}}$ when $T_0=T_L$.

11. The second law availability efficiency of a *refrigerator or air conditioner* is given by Eq. (10.34) as

$$\varepsilon_{R/AC}=\frac{\left(1-\frac{T_0}{T_L}\right)\dot{Q}_L}{|\dot{W}_{\text{in}}|}=\left(1-\frac{T_0}{T_L}\right)\left(\frac{\dot{Q}_L}{|\dot{W}_{\text{in}}|}\right)=\left(1-\frac{T_0}{T_L}\right)\text{COP}_{\substack{\text{actual}\\\text{ref or air cond}}}$$

which reduces to $\varepsilon_{R/AC}=(\text{COP})_{\text{actual R/AC}}/(\text{COP})_{\text{Carnot R/AC}}$ when $T_0=T_H$.

12. The second law availability efficiency of a four flow stream nonmixing heat exchanger is given by Eq. (10.36) as

$$\varepsilon_{\text{nonmixing HX}}=\frac{\dot{m}_C(a_{f4}-a_{f3})}{\dot{m}_H(a_{f1}-a_{f2})}$$

and, for a three flow stream heat exchanger in which the fluids mix inside its boundaries, it is given by Eq. (10.37) as

$$\varepsilon_{\text{mixing HX}}=\frac{(1-y)(a_{f3}-a_{f2})}{y(a_{f1}-a_{f3})}$$

where y is the hot mass fraction, $y=\dot{m}_{\text{hot}}/\dot{m}_{\text{mixed}}$.

Problems (*indicates problems in SI units)

1. Determine the potential for the following vector field: $\vec{A} = y\vec{i} + x\vec{j} + 0\vec{k}$.
2. Compute the vector field for the following potential: $P = ax + by + cz$.
3. Compute the vector field for the following potential: $P = 3x^3 + 2y^2 - z + 7$.
4. Compute the vector field for the following potential: $P = 5x^2y/z + \sin(x)$.
5. Coulomb's law of the electrostatic force of attraction between two isolated charges, Q_1 and Q_2, separated by a distance r is $\vec{F} = (Q_1Q_2/4\pi\varepsilon r^2)\,\vec{i}_r$, where $\vec{i}_r$ is the unit vector pointing along the line of centers of the charges and ε is the dielectric constant of the medium containing the charges. Show that this force has a potential given by $P = Q_1Q_2/4\pi\varepsilon r$.
6. Newton's law for the gravitational force between to bodies of mass m_1 and m_2 separated by a distance r is $\vec{F} = (Gm_1m_2/r^2)\,\vec{i}_r$, where G is the gravitational constant and $\vec{i}_r$ is a unit vector pointing along the line of centers of the masses. Determine the gravitational potential for this force and show that it satisfies the Laplace equation in cylindrical coordinates,

$$\nabla^2\vec{F} = \frac{\partial^2\vec{F}}{\partial r^2} + \frac{1}{r}\frac{\partial\vec{F}}{\partial r} + \frac{1}{r^2}\frac{\partial^2\vec{F}}{\partial\theta^2} + \frac{\partial^2\vec{F}}{\partial z^2}$$

7.* The total energy contained in a closed rigid system is 550. kJ and its total entropy is 2.7521 kJ/K. The ground state total energy, total entropy, and absolute temperature of the system are 50.0 kJ, 1.0000 kJ/K, and 273 K, respectively. Determine the maximum reversible work this system can produce.

8.* The total energy contained in a closed, sealed, rigid can of tuna is 3000. J with a total entropy of 2.8330 J/K. The ground state total energy, total entropy, and temperature of the system are 100.0 J, 0.0275 J/K, and 20.0°C, respectively. Determine the maximum reversible work available from the can of tuna.

9. The total energy contained in a closed, sealed, rigid can of carbonated soda is 10.0 Btu with a total entropy of 0.0739 Btu/R. The ground state total energy, total entropy, and temperature of the system are 1.00 Btu, 0.0619 Btu/K, and 70.0°F, respectively. Determine the maximum reversible work available from the can of soda.

10.* An inventor claims to have developed a new closed, sealed battery that can produce a maximum reversible work of 3.80 kJ. The total energy contained in the battery is 2.00 kJ with a total entropy of 7.5150 J/K. If the ground state total energy and total entropy are 100. J and 0.5660 J/K, respectively, what ground state temperature is required to meet the inventor's maximum reversible work claim?

11. As a designer, you are required to develop a new closed, sealed thermal energy storage cell that has a maximum reversible work output of 500. Btu. The ground state total energy, total entropy, and temperature are 5.00 Btu, 0.11690 Btu/R, and 50.0°F, respectively. If the total entropy of the cell must be ten times its ground state value, what should the total energy of the cell be?

12. An open bucket containing 30.0 lbm of liquid water at 70.0°F is sitting 6 ft above the floor on a ladder. Determine the total availability of the water in the bucket relative to the floor. The local environment is at 14.7 psia and 70.0°F.

13.* An open bucket containing 14.0 kg of liquid water at 20.0°C is spun on a rope in a horizontal plane 2.00 m above the floor with a tangential velocity of 5.00 m/s. Determine the total availability of the water in the bucket relative to the floor. The local environment is at 0.101 MPa and 20.0°C.

14.* Determine the total availability of a 7.00×10^{-3} kg incompressible lead bullet traveling vertically at 1000. m/s at a height of 50 m above the ground. The temperature of the bullet is 150.°C and its specific heat is 0.167 kJ/kg·K. The local environment is at 0.101 MPa and 20.0°C.

15.* A stationary tethered balloon contains helium gas (an ideal gas here) at 0.00°C and 0.0700 MPa at a height of 1000. m. Determine the specific availability of the helium in the balloon relative to the ground, where the local environment is at $p_0 = 0.101$ MPa and $T_0 = 20.0$°C.

16.* The air (an ideal gas here) in the ballast tanks of a submarine is at 10.0°C and 1.50 MPa when the submarine is cruising at 3.00 m/s, 100. m below sea level. Determine the specific availability of the air in the ballast tanks relative to sea level where the local environment is at 0.101 MPa and 20.0°C.

17. Integrate Eq. (7.52) and use Eqs. (10.13a) and (10.14a) to determine the irreversibility and total availability change for an aergonic closed system in which the temperature increases from $T_1 = 70.0$°F to $T_2 = 200.$°F for the cases where the heat transfer varies with the system absolute temperature according to the relationships
   a. $Q = K_1T$ (convection).
   b. $Q = K_2T^4$ (radiation)

   where $K_1 = 3.70$ Btu/R and $K_2 = 5.40 \times 10^{-4}$ Btu/R^4. The system boundary is maintained isothermal at 350.°F and the local environment is at 14.7 psia and 70.0°F.

18. The temperature distribution due to conduction heat transfer inside a flat plate with an internal heat generation is given by $T = T_0 + (T_s - T_0)(x/L)^2$, where T_s is the surface temperature at $x = L$, and T_0 is the centerline temperature at $x = 0$ (Figure 10.23). Use Eqs. (7.66) and (10.13b) to determine a formula for the steady state irreversibility rate for this system.

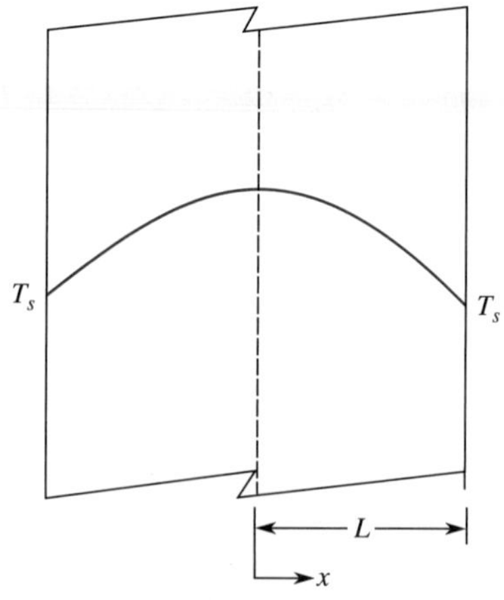

FIGURE 10.23
Problem 18.

19. A current of 100. A is passed through a 6.00 ft long stainless steel wire 0.100 inch in diameter. The electrical resistivity of the wire is 1.97×10^{-5} Ω·in, and its thermal conductivity is

12.5 Btu/hr·ft·R. If the outer surface temperature of the wire is maintained constant at 300.°F and the temperature profile inside the wire is given by $T = T_w + \rho_e J_e^2(R^2 - r^2)/(4k_t)$, where T_w is the wall temperature of the wire, R is its outside radius, and r is measured from the center of the wire. Use Eqs. (7.75) and (10.13b) to determine the steady state irreversibility production rate within the wire due to the flow of electricity through it. Assume all the physical properties are independent of temperature.

20. A constant pressure piston-cylinder apparatus contains 1.30 lbm of saturated liquid water at 212°F. Heat is added to this system until the contents reach a quality of 85.0%. The surface temperature of the cylinder is constant at 250.°F. Determine the irreversibility of this process if the local environment is at 14.7 psia and 70.0°F.

21. 1.00 lbm of saturated water vapor at 212°F is condensed in a closed, nonrigid system to saturated liquid at 212°F in a constant pressure process by a heat transfer across a system boundary with a constant temperature of 80.0°F. What is (a) the irreversibility and (b) the change in total availability for this process if the local environment is at 14.7 psia and 70.0°F?

22.* A closed, rigid container encloses 1.50 kg of air at 0.100 MPa and 20.0°C. We wish to increase the temperature to 40.0°C by heat transfer. Assuming constant specific heat ideal gas behavior, determine (a) the irreversibility and (b) the change in total availability of the air when this change of state is accomplished using a constant system boundary temperature of 100.°C and the local environment is at 0.101 MPa and 20.0°C.

23. A sealed, rigid kitchen pressure cooker with a volume of 1.00 ft^3 contains 2.20 lbm of a mixture of liquid plus vapor water at 14.7 psia. The pressure cooker is then heated until its internal pressure reaches 20.0 psia. Determine (a) the heat transfer during the process, (b) the irreversibility of the process if the inner surface of the pressure cooker is constant at 250.°F, and (c) the change in total availability of the water if the local environment is at 14.7 psia and 70.0°F.

24. A closed, sealed, rigid container is filled with 0.05833 ft^3 of liquid water and 0.94167 ft^3 of water vapor in equilibrium at 1.00 psia. The vessel is then heated until its contents become a saturated vapor. If this heating process is done irreversibly, determine (a) the total irreversibility for this process if the surface temperature of the vessel is maintained constant at 300.°F and (b) the change in total availability of the water if the local environment is at 14.7 psia and 70.0°F.

25.* Determine the irreversibility of a 4.00 × 10^{-3} kg, 80.0°C lead bullet traveling at 900. m/s that impacts a perfectly rigid surface aergonically and adiabatically. The specific heat of lead at the mean temperature of the bullet is 167 J/kg·K, and the local environment is at 0.101 MPa and 20.0°C.

26. A small room has a single 60.0 W lightbulb hanging from the ceiling. The walls are not insulated, so a steady state condition is reached where the room walls are at 55.0°F. Determine the irreversibility rate within the room if the local environment is at 14.7 psia and 40.0°F.

27.* The surface temperature of a 100. W incandescent lightbulb is 60.0°C. The surface temperature of a 20.0 W fluorescent tube producing the same amount of light as the 100. W incandescent lightbulb is 30.0°C. Determine the steady state irreversibility rate of each light source when the local environmental temperature is 20.0°C, and comment on which is the more efficient.

28.* An automobile engine heater is an electrical resistance heater that is plugged into a 110. V ac outlet and inserted into the oil dipstick tube of the engine. Its purpose is to keep the engine oil warm during the winter when the car is not in use, thus allowing the engine to start easier. Determine the steady state irreversibility produced during an 8.00 h period by a 100. W steady state engine heater whose surface is isothermal at 90.0°C.

29. Determine the irreversibility produced when 3.00 lbm of carbon dioxide at 70.0°F and 30.0 psia are adiabatically mixed with 7.00 lbm of carbon dioxide at 100.°F and 15.0 psia. The final mixture pressure is 17.0 psia. Assume the carbon dioxide behaves as a constant specific heat ideal gas and that the local environment is at 14.7 psia and 70.0°F.

30.* Determine the irreversibility produced as 10.0 kg of liquid water at 10.0°C is adiabatically mixed with 20.0 kg of liquid water at 80.0°C. The specific heat of the water is 4.20 kJ/kg·K, and the local environment is at 0.101 MPa and 20.0°C.

31. Here is the classical coffee and cream problem. Which of the following processes produces less irreversibility?

a. Mixing cream with hot coffee and then letting the mixture cool to the drinking temperature.

b. Letting the coffee cool to a temperature such that, when the cream is added, the mixture will be at the drinking temperature.

Do not ignore the cooling heat transfer irreversibility.

32.* An engine operating on a Carnot cycle extracts 10.0 kJ of heat per cycle from a thermal reservoir at 1000.°C and rejects a smaller amount of heat to a low-temperature thermal reservoir at 10.0°C. Determine the net change in availability of the engine per cycle of operation when the local temperature is 0.00°C.

33. An engine operating on a Carnot cycle extracts heat at a rate of 500. Btu/s from a thermal reservoir at 1000.°F and rejects heat to a low-temperature thermal reservoir at 100.°F. Determine the rate of change in availability of the engine when the local environmental temperature is 70.0°F.

34. A fire hose is used to extinguish a building fire. The end of the hose is held 50.0 ft from the ground on a ladder and sprays 50.0°F water at a velocity of 12.0 ft/s. Assuming the water is an incompressible liquid, determine the specific flow availability at the exit of the hose when the local environment (ground state) is at 14.7 psia and 70.0°F.

35. Recompute the specific flow availability in Problem 34 using Table C.1a of *Thermodynamic Tables to accompany Modern Engineering Thermodynamics* when the water exits the fire hose as a *saturated liquid* at 50.0°F and the local environment (ground state) is *saturated liquid* water at 70.0°F.

36.* A garden hose is used to fill a swimming pool. The hose is laid on the ground and the water exits at 1.00 m/s at 15.0°C. Assuming the water is an incompressible liquid, determine the specific flow availability at the exit of the hose when the local environment (ground state) is at 0.101 MPa and 20.0°C.

37.* Recompute the specific flow availability in Problem 36 using Table C.1b when the water exits the garden hose as a *saturated liquid* at 15.0°C and the local environment (ground state) is *saturated liquid* water at 20.0°C.

38.* Superheated steam at 1000.°C and 1.00 MPa flows through a pipe located 20.0 m above the floor in a power plant with a velocity of 50.0 m/s. Determine the specific flow availability of the steam in the pipe when the local environment (ground state) is saturated liquid water at 20.0°C.

39.* Saturated liquid ammonia at 0.00°C flows through a pipe located 10.0 m below the floor in a refrigeration system with a velocity of 3.30 m/s. Determine the specific flow availability of the ammonia in the pipe when the local environment (ground state) is saturated liquid ammonia at 0.00°C.

40. Superheated Refrigerant-22 at 100. psia and 100.°F flows through a tube in a large industrial air conditioner 50.0 ft above the ground at a velocity of 20.0 ft/s. Determine the specific flow availability of the refrigerant in the tube when the local environment (ground state) is saturated liquid Refrigerant-22 at 0.00°F.

41.* Saturated liquid Refrigerant-134a at 12.0°C flows through a tube in an automobile air conditioning system at a velocity of 0.300 m/s. The tube is 1.00 m above the ground. Determine the specific flow availability of the refrigerant in the tube when the local environment (ground state) is saturated liquid Refrigerant-134a at 0.00°C.

42.* Superheated mercury vapor at 2.00 MPa flows at a velocity of 7.50 m/s through a pipe in a portable nuclear power plant. The pipe is 0.500 m above the ground. Determine the specific flow availability of the mercury in the pipe. Use saturated liquid mercury at 0.100 MPa as the local environment (ground state).

43. Air (an ideal gas here) at 100. psia and 150.°F flows through a pipe in a factory with a velocity of 10.0 ft/s. The pipe is located 75.0 ft above the floor. Determine the specific flow availability of the air in the pipe when the local environment (ground state) is at 14.7 psia and 70.0°F.

44.* Saturated water vapor enters an isentropic turbine of a power plant at 4.00 MPa and exits at 1.00×10^{-3} MPa. Neglecting kinetic and potential energy effects, determine the difference in specific flow availability between the entrance and exit of the turbine. Use saturated liquid water at 20.0°C as the local environment (ground state).

45. Steam enters the isentropic turbine of power plant at 200. psia and 500.°F. How much does the change in specific flow availability between the inlet and exit of the turbine increase if the exit pressure is lowered from 14.7 psia to 1.00 psia? Neglect all kinetic and potential energy effects. Use saturated liquid water at 80.0°F as the local environment (ground state).

46. A small portable nuclear-powered steam turbine has an inlet state of 200. psia, 600.°F and an outlet temperature of 95.0°F. Assuming that the exit state is also a saturated vapor, determine the change in specific flow availability between the inlet and exit of the turbine. Neglect all kinetic and potential energy effects. Use saturated liquid water at 70.0°F as the local environment (ground state).

47.* A horizontal pipe carrying superheated steam at 30.0 MPa and 1000.°C suddenly develops a small crack. Steam enters the crack at 30.0 m/s and passes through it in a steady state, adiabatic, aergonic process to exit at 0.101 MPa with a velocity of 250. m/s. Determine the specific flow availabilities at the inlet and outlet of the crack, and calculate the irreversibility per unit mass of steam exiting the crack. Use saturated liquid water at 20.0°C as the local environment (ground state).

48.* Superheated steam at 8.30 kg/s, 1.00 MPa, and 400.°C enters a horizontal, stationary, insulated nozzle with a negligible velocity and expands to 10.0×10^{-3} MPa. The friction and other irreversibilities within the nozzle cause the exit velocity to be only 85.0% of that produced by an isentropic expansion. Taking the local environment (ground state) to be that of saturated liquid water at 20.0°C, determine

a. The inlet specific flow availability.
b. The exit specific flow availability.
c. The irreversibility rate inside the nozzle.

49. A large, uninsulated steam turbine receives superheated steam at 40.0 lbm/s, 1000.°F, and 800. psia and exhausts it to 1.00 psia with a quality of 92.0%. If the turbine is assumed to be internally reversible, determine the heat loss from the surface of the turbine if the power output is 30.0×10^3 kW. The surface temperature of the turbine is uniform at 225°F and the local environment (ground state) is saturated liquid water at 20.0°C. Neglect all flow stream kinetic and potential energies in this problem.

50. A steady flow, steady state air compressor handles 4000. ft^3/min measured at the intake state of 14.1 psia, 30.0°F and a velocity of 70.0 ft/s. The discharge is at 45.0 psia and has a velocity of 280. ft/s. Both the inlet and exit stations are located 4.00 ft above the floor. Using the specific flow availability relative to the local environmental (ground state) temperature of 80.0°F and a pressure of 14.7 psia, determine

a. The discharge temperature and the power required to drive the compressor if the process is reversible and adiabatic.
b. The discharge temperature and the power required to drive the compressor if the process is irreversible and adiabatic with a compressor work transport efficiency of 80.0%.

51.* Determine the work required to compress 15.0 kg/min of superheated steam in an uninsulated, reversible compressor from 0.150 MPa, 600.°C to 1.50 MPa, 500.°C in a steady state, steady flow process. Neglect any changes in kinetic and potential energy. Use the flow availability approach to calculate the specific flow availabilities at the inlet and exit if the environmental temperature is 20.0°C. Choose the local environment (ground state) to be saturated liquid water at the environmental temperature.

52. An adiabatic, steady flow compressor is designed to compress superheated steam at a rate of 50.0 lbm/min. At the inlet to the compressor, the state is 100. psia and 400.°F; and at the compressor exit, the state is 200. psia and 600.°F. Neglecting any kinetic or potential energy effects, calculate (a) the power required to drive the compressor and (b) the rate of availability destruction by the compressor. Use saturated liquid water at 80.0°F as the local environment (ground state).

53.* A steady flow air compressor takes in 5.00 kg/min of atmospheric air at 101.3 kPa and 20.0°C and delivers it at an exit pressure of 1.00 MPa. The air can be considered an ideal gas with constant specific heats. Potential and kinetic energy effects are negligible. If the process is not reversible but is adiabatic and polytropic with a polytropic exponent of $n =$ 1.47, calculate

a. The power required to drive the compressor.
b. The entropy production rate of the compressor.
c. The entrance and exit specific flow availabilities if the ground state local environmental temperature and pressure are 20.0°C and 101.3 kPa.

54. An uninsulated, irreversible steam engine whose surface temperature is 200.°F produces 50.0 hp with a steam mass

flow rate of 15.0 lbm/min. The inlet steam is at 400.0°F, 100. psia and it exits at 14.7 psia, 90.0% quality. Determine
a. The rate of heat loss from the engine.
b. Its entropy production rate.
c. Its entrance and exit specific flow availabilities. Use saturated liquid water at 80.0°F as the local environment (ground state).

55.* A design for a turbine has been proposed involving the adiabatic, steady flow of steam through the turbine. Saturated vapor at 300.°C enters the turbine and the steam leaves at 0.200 MPa with a quality of 95.0%.
a. Draw a *T-s* diagram for the turbine.
b. Determine the work and entropy production per kilogram of steam flowing through the turbine.
c. If the atmospheric temperature is 25.0°C, determine the entrance and exit specific flow availabilities.
Neglect any kinetic and potential energy effects and take the local environment (ground state) to be saturated liquid water at 80.0°F.

56.* Steam enters a turbine at 2.00 MPa and 700.°C and exits the turbine at 0.200 MPa and 400.°C. The process is steady flow, steady state, and adiabatic. Using saturated liquid water at 20.0°C as the local environment (ground state), determine the following items on the basis of a steam flow rate of 6.30 kg/s:
a. The irreversibility rate of the turbine.
b. The turbine's entrance and exit specific flow availabilities.
c. The turbine's actual output power.

57. Saturated liquid water enters a badly worn boiler feed pump at 1.00 psia and exits the pump as a saturated liquid at 600 psia in a steady flow, steady state, adiabatic process. Using saturated liquid water at 70.0°F as the local environment (ground state), determine the following items on the basis of a steam flow rate of 75.0 lbm/s:
a. The irreversibility rate of the pump.
b. The pump's entrance and exit specific flow availabilities.
c. The pump's actual input power.

58.* Saturated liquid ammonia enters a boiler in a refrigeration system at −20.0°C and exits the boiler as a superheated vapor at 100. kPa and 10.0°C in a steady flow, steady state, process. Using saturated liquid ammonia at 0.00°C as the local environment (ground state), determine the following items on the basis of an ammonia flow rate of 12.0 kg/s:
a. The irreversibility rate of the boiler.
b. The boiler's entrance and exit specific flow availabilities.
c. The boiler's heat input rate.

59. Saturated Refrigerant-22 vapor enters the condenser of a large refrigeration system at 80.0°F and exits as a saturated liquid at 80.0°F in a steady flow, steady state process. Using saturated liquid Refrigerant-22 at 0.00°F as the local environment (ground state), determine the following items on the basis of an ammonia flow rate of 80.0 lbm/s:
a. The irreversibility rate of the condenser.
b. The condenser's entrance and exit specific flow availabilities.
c. The condenser's heat rejection rate.

60. Air (an ideal gas) enters a throttle at 70.0°F and 150. psia and exits at 14.7 psia in a steady flow, steady state, adiabatic, aergonic process. Determine the irreversibility rate per unit mass of air flowing through the throttle. The local environment (ground state) is 14.7 psia and 70.0°F.

61.* Superheated steam at 1.30 kg/s, 0.0100 MPa, and 400.°C enters a horizontal, stationary, insulated diffuser with a velocity of 100. m/s. The friction and other irreversibilities within the nozzle cause the exit pressure to be 115% of that produced by an isentropic expansion. Taking the local environment (ground state) to be that of saturated liquid water at 20.0°C, determine
a. The inlet specific flow availability.
b. The exit specific flow availability.
c. The irreversibility rate inside the diffuser.

62.* A new solar collection system has a net input availability rate of 600. kJ/s and an availability destruction rate of 80.0 kJ/s. Determine the second law availability efficiency of this system if it loses availability at a rate of 345 kJ/s to the surroundings.

63. A closed domestic gas hot water heater contains 415 lbm of water at 45.0°F. Determine the first and second law efficiencies as this water is heated to 130.°F using 39,589 Btu from a gas burner. The temperature of the local environment (ground state) is 55.0°F.

64.* A large, closed, industrial electrical hot water heater contains 2000. kg of water at 15.0°C. Determine the first and second law efficiencies as this water is heated to 75.0°C using 0.550 MJ of electrical energy. The temperature of the local environment (ground state) is 20.0°C.

65. In Example 10.10, the following equation was developed for the second law availability efficiency ε of heating a liquid in a *closed* uninsulated tank:

$$\varepsilon = mc\left[\frac{\Delta T - T_0 \ln\left(1 + \frac{\Delta T}{T}\right)}{Q_{in}}\right]$$

Is there a value of ΔT that maximizes ε? Hint: Set $d\varepsilon/d(\Delta T) = 0$ and solve for ΔT.

66.* An electric heater is used to increase the temperature of liquid water in an open tank from 10.0 to 80.0°C in a steady state, steady flow process. The mass flow rate of the water through the tank is 3.70 kg/s, and the electric heater adds 1500. kW to the water. Determine the first and second law efficiencies of this system. The temperature of the local environment (ground state) is 10.0°C.

67.* A solar water heater is used to increase the temperature of liquid water in an open cattle watering tank from 3.00°C to 18.0°C in a steady state, steady flow process. The mass flow rate of the water through the tank is 1.50 kg/s, and the solar heater adds 100. kW to the water. Determine the first and second law efficiencies of this system. The temperature of the local environment (ground state) is 0.00°C.

68. In Example 10.11, the following equation was developed for the second law availability efficiency ε of heating a liquid in an *open*, uninsulated tank:

$$\varepsilon = \dot{m}c\left[\frac{\Delta T - T_0 \ln\left(1 + \frac{\Delta T}{T}\right)}{\dot{Q}_{in}}\right]$$

Is there a value of ΔT that maximizes ε? Hint: Set $d\varepsilon/d(\Delta T) =$ 0 and solve for ΔT.

69. An automobile engine has a Carnot thermal efficiency of 56.0% and an actual thermal efficiency of 21.0%. Determine the second law availability efficiency of this engine.

70. The turbine of a large power plant receives 1.00×10^8 Btu/h of heat from the boiler at 900.°F and rejects 5.50×10^6 Btu/h of heat to the condenser at 60.0°F while producing 22.0×10^3 kW of electrical power. The local environment (ground state) is at 14.7 psia and 50.0°F. Determine
   a. The first law thermal efficiency.
   b. The rate at which available energy enters the boiler.
   c. The rate at which available energy enters the condenser.
   d. The second law availability efficiency of the power plant.

71. The Heat-Master is a new heat pump design that has a Carnot COP (coefficient of performance) of 15.5. However, its actual coefficient of performance is only 6.90. Determine the second law availability efficiency of this heat pump.

72.* A heat pump is designed to provide 9.00 kW of heat to a small house at 20.0°C when the outside temperature is 5.00°C. The electric motor driving the heat pump draws 1.20 kW. Determine:
   a. The first law thermal efficiency (i.e., COP) of the heat pump.
   b. The second law availability efficiency of the heat pump.

73. The Cool-Master is a new window air conditioner design that has a Carnot coefficient of performance (COP) of 18.9. However, its actual coefficient of performance is only 3.30. Determine the second law availability efficiency of this air conditioner.

74. A common window air conditioner has an actual COP of 7.88. If the inside temperature is $T_0 = T_L = 75.0$°F and the outside temperature is 95.0°F, then determine the second law availability efficiency of this air conditioner.

75. Liquid water is to be heated in a nonmixing heat exchanger with waste steam. The liquid water (an incompressible liquid) flows through the heat exchanger at 15.0 lbm/s with an inlet temperature of 50.0°F and an exit temperature of 75.0°F. The steam flows through the heat exchanger at 3.00 lbm/s with an inlet state of 600.°F and 60.0 psia and an exit state of 400.°F and 40.0 psia. Determine the second law availability efficiency of this heat exchanger. Neglect all kinetic and potential energy effects, and use saturated liquid water at 50.0°F as the local environment (ground state).

76. The inlet air to a gas turbine engine is preheated with the engine's exhaust gases. The preheater is insulated so that all heat transfer is internal. The engine's exhaust gas enters the preheater at 800.°F and 1.30 atm pressure and exits at 500.°F and 1.00 atm. The inlet air enters the preheater at 70.0°F and 1.40 atm and exits at 1.30 atm. The mass flow rates of the inlet air and the exiting exhaust are approximately the same at 2.10 lbm/s. Both gases can be treated as constant specific heat ideal gases. The specific heat and gas constant of the exhaust gas are $(c_p)_{\text{exh.}} = 0.238$ Btu/(lbm·R) and $R_{\text{exh.}} = 0.0640$ Btu/(lbm·R). Neglecting all kinetic and potential energies, and taking the local environment (ground state) as $p_0 = 1.00$ atm, and $T_0 = 70.0$°F, determine:
   a. The exit temperature of the inlet air.
   b. The second law availability efficiency of the preheater.

77.* Can you believe that 0.800 kg/s of air at 60.0°C and 1.50 MPa is mixed in an open heat exchanger with 1.50 kg/s of air at 20.0°C and 1.50 MPa to produce an outlet mixture at 30.0°C at 1.50 MPa? Determine the second law availability efficiency of this heat exchanger. Neglect all kinetic and potential energy effects, and take the local environment (ground state) to be 0.101 MPa and 20.0°C.

78.* A student is using a bathtub with separate hot and cold water faucets. The student turns on both faucets and adjusts them to create a pool of warm water in the tub. The tub's drain is open, so the faucets must be kept running to maintain the pool of water. The hot water faucet provides 40.0°C water at 0.100 kg/s, and the cold water faucet provides 15.0°C water at 0.250 kg/s. The local environment (ground state) is at 0.101 MPa and 18.0°C. Neglecting all flow stream kinetic and potential energy and assuming the tub itself is insulated, determine
   a. The temperature of the mixed water in the tub.
   b. The second law availability efficiency of the tub as a mixing type heat exchanger.

Design Problems

The following are open-ended design problems. The objective is to carry out a preliminary thermal design as indicated. A detailed design with working drawings is not required unless otherwise specified by your instructor. These problems do not have specific answers, so each student's design is unique.

79.* Carry out the preliminary design of a closed domestic hot water heater that has a first law efficiency of at least 95.0% and a second law efficiency of at least 10.0%. The inlet water temperature is fixed at 10.0°C. The remaining variables (including the ground state) are unrestrained and can be chosen to fit the needs of the designer.

80. You are to prepare the preliminary design of a commercial, open, liquid water heater that has a first law efficiency of at least 88.0% and a second law efficiency of at least 15.0%. The inlet water temperature is fixed at 50.0°F. The remaining variables (including the ground state) are unrestrained and can be chosen to fit the needs of the designer.

81.* As chief engineer of a large heat exchanger company, you are to prepare the preliminary design of a nonmixing heat exchanger that has a second law availability efficiency of at least 15.0%. The two flow streams are to be liquid water, with one flow stream having a mass flow rate of 10 kg/s, entering at 10.0°C and leaving at 30.0°C. The remaining variables (including the ground state) are not specified and are left to the discretion of the designer.

82. Design a benchtop apparatus that illustrates the basic principles of a mixing heat exchanger. You may use either liquids or gases or a combination of liquids and gases as the flow streams. Determine the measurements that must be made to compute the second law availability of the heat exchanger.

83. Carry out the preliminary design of an instrument that provides a readout of the specific availability of any fluid (liquid or gas) in which it is immersed. Determine the necessary sensors and any calibration procedure required.

Computer Problems

The following open-ended computer problems are designed to be done on a personal computer using a spreadsheet or equation solver.

84. Plot the specific availability of the air in the tank in Example 10.2 as a function of air pressure. Assume all the remaining variables are constant.

85. Create a specific availability vs. temperature curve for saturated liquid and saturated vapor water. Neglect any kinetic or potential energy effects.

86. Using the data and situation described in Example 10.6, plot the specific flow availability vs. (a) water inlet temperature, (b) water velocity, and (c) height above the ground. Use this information to create a three-dimensional surface with availability on the vertical axis and water velocity and height as the other two coordinates. For each part of this problem, assume all the variables except those under consideration are constant at their values given in Example 10.6.

87. Example 10.9 contains a situation involving a large steam turbine. Using the data provided there and assuming the values of all the remaining variables are constant, plot the turbine's heat loss as a function of
   a. The steam flow rate.
   b. The surface temperature of the turbine.
   c. The ground state temperature (note that changing T_0 changes the a_f values).

88. The second law efficiency of heating a liquid in a closed container is evaluated in Example 10.10. Using the results obtained there, plot the ratio of second law availability efficiency to the first law energy efficiency (ε/η_T) vs. ΔT for a variety of T and T_0 values. Comment on the general trends of these curves.

Writing to Learn Problems

Provide a coherent 500-word written response to the following questions on 8½ by 11 in. paper, double spaced, with 1-inch margins on all sides. Unless your instructor indicates otherwise, your response should include the following items:

   a. An opening thesis statement containing the argument you wish to support.
   b. A body of supporting material.
   c. A conclusion section in which you use the supporting material to substantiate your thesis statement.

89. *Availability* is the name given to the amount of energy within a system that can produce useful work. Describe in your own words what constitutes "useful work" and provide at least three representative examples.

90. Is the definition of what constitutes *useful work* a cultural variable (e.g., could work considered useful in one culture not be considered useful in another)? Provide specific arguments and at least three examples to support your contention.

91. If the numerical value of availability represents the amount of energy within a system that can be converted into useful work, does this value depend on existing energy conversion technology accessible for use with the system? Provide specific arguments and at least three examples to support your contention.

92. Useful work is associated with the *potential* of a *conservative force*. Provide a definition of *potential* using only words and arguments that your (hypothetical) nine-year-old sister would understand. Provide at least three examples she could relate to.

93. The concept of a *conservative force* is central to defining useful work. Describe the differences between conservative and nonconservative forces using only words and arguments a second-year college music major would understand. Provide examples of at least three forces of each type.

94. In the text, it is stated that electrical energy is more available to do useful work through a rotating shaft of an 90.0% efficient electrical motor than is an equivalent amount of fuel chemical energy used to power the rotating shaft of an internal combustion engine with an energy conversion efficiency of 20.0%. However, if the chemical fuel is supplied to a fuel cell that converts it directly into electrical energy with a 90.0% efficiency, would the chemical energy of the fuel now have a higher availability than an equivalent amount of electrical energy?

95. Write a letter to your (hypothetical) ten-year-old brother in which you describe in words he would understand the concept of a *local environment*. Be sure to distinguish it from the complete surroundings, and provide at least three physical examples he would be able to understand.

Create and Solve Problems

Engineering education tends to focus on the process of solving problems. It ignores teaching the process of formulating solvable problems. However, working engineers are never given a well-phrased problem statement to solve. Instead, they need to react to situational information and organize it into a structure that can then be solved using the methods learned in college.

These "Create and Solve" problems are designed to help you learn how to formulate solvable thermodynamics problems from engineering data. Since you provide the numerical values for some of the variables, these problems do not have unique solutions. Their solutions depend on the assumptions you need to make and how you set them up to create a solvable problem.

96. You have been hired as a thermal engineer at a company that manufactures domestic cookware appliances. Your first job is to analyze a new design for a pressure cooker. It has a volume of 0.15 ft^3 and initially contains 2.50 lbm of a mixture of liquid plus vapor water at 14.7 psia. When the pressure cooker is heated electrically, its internal pressure reaches 35.0 psia. To understand the design, you need to know the heat transfer during the process, the irreversibility of the process if the inner surface of the pressure cooker is constant at 250.°F, and the change in total availability of the water if the local environment is at 14.7 psia and 70.0°F. Write and solve a thermodynamics problem that answers these questions.

97. As the resident thermal engineer at a new company developing inventive new ideas, you are faced with a scenario where 1.00 lbm of saturated water vapor at 212°F is condensed inside a flexible balloon to saturated liquid at 212°F. This occurs in a constant pressure process by a heat transfer from the balloon to the environment. The balloon has an average surface temperature during this process of 125°F. Your job is to determine the irreversibility and the change in total availability for this process if the local environment is at 14.7 psia and 70.0°F. Write and solve a thermodynamics problem that provides the answers to these questions.

98.* It is now 1923, and you are working for Thomas Edison in his New Jersey research laboratory. The surface temperature of his 50.0 W incandescent lightbulb is 60.0°C. But the surface temperature of a Westinghouse 50.0 W incandescent lightbulb is 50.0°C. Tom asks you to determine the steady state irreversibility rate of each lightbulb when the local environmental temperature is 20.0°C and comment on why the Westinghouse bulb is more efficient. Write and solve a thermodynamics problem that answers these questions.

99.* As a chief engineer at the local fire station, you are asked to use one of your fire hoses to fill a large swimming pool. The hose will be laid on the ground and the water will exit at 3.50 m/s at 18.0°C. Before you can respond to this request, you need to know the specific flow availability at the exit of the hose when the local environment is at 0.101 MPa and 20.0°C. Write and solve a thermodynamics problem that answers this question.

100.* You received a promotion to production engineer at a food-processing factory. A particularly critical process involves the mixing of ingredients with warm water in a large insulated vat. The vat's drain is kept open, so water must be continuously added to maintain the proper mix. The water comes from a hot water source at 38.0°C at 3.10 kg/s and a cold water source at 15.0°C at 2.50 kg/s. The local environment is at 0.101 MPa and 20.0°C. To keep control of this process, you need to know the temperature of the mixed water in the tub and the second law availability efficiency of the tub as a mixing-type heat exchanger. Write and solve a thermodynamics problem that provides the answers to these questions.

CHAPTER 11

More Thermodynamic Relations

CONTENTS

11.1 KYNNING (INTRODUCTION)

We have a problem. It turns out that no meters, gauges, or instruments of any kind can be used to directly measure the internal energy or the enthalpy or the entropy of a system. How, then, do you get numerical values for thermodynamic properties that are not directly measurable? In Chapter 3, we discuss this subject briefly and find that numerical values of properties that are not directly measurable (e.g., u, h, and s) can sometimes be calculated from the numerical values of properties that are measurable (e.g., p, v, and T).

For elementary materials, such as incompressible solids (or liquids) and ideal gases, we have relatively simple equations of state that provide the necessary relations. For example, the specific internal energy, specific enthalpy, and specific entropy of an incompressible material are related to its temperature, pressure, and specific volume by

$$(u_2 - u_1)_{\text{incomp}} = c(T_2 - T_1),\ (h_2 - h_1)_{\text{incomp}} = c(T_2 - T_1) + v(p_2 - p_1),\ \text{and}\ (s_2 - s_1)_{\text{incomp}} = c\ln(T_2/T_1)$$

where c is the specific heat of the material. And, in the case of an ideal gas, these properties are related by

$$(u_2 - u_1)_{\text{ideal gas}} = c_v(T_2 - T_1),\ (h_2 - h_1)_{\text{ideal gas}} = c_p(T_2 - T_1),\ \text{and}\ (s_2 - s_1)_{\text{ideal gas}} = c_p\ln(T_2/T_1) - R\ln(p_2/p_1)$$

Complex materials require more sophisticated equations of state plus a knowledge of various mathematical property interrelationships to be able to evaluate their unmeasurable thermodynamic properties. In this chapter, we build on the equations introduced in Chapter 3 to formulate new property relations that can be used to compute numerical values for u, h, and s for complex real materials.

We begin this chapter by rounding out our list of useful thermodynamic properties by defining two new properties, the Helmholtz and Gibbs functions. We then move on to develop a series of general mathematical results, called the *Maxwell equations*, that relate a number of thermodynamic properties. We end this chapter by using the principle of corresponding states to develop a set of generalized thermodynamic property charts that are valid for many real gases.

11.2 TWO NEW PROPERTIES: HELMHOLTZ AND GIBBS FUNCTIONS

If we consider a stationary closed system containing a pure substance subjected only to a moving boundary mechanical work mode, then the combined energy and entropy balance is given by Eq. (7.30) as

$$du = T\,ds - p\,dv \tag{11.1}$$

Since any two independent properties fix the thermodynamic state of a pure substance subjected to only one work mode (see Chapter 4), we can take the two independent properties here to be s and v, or

$$u = u(s, v)$$

The total differential of this composite function, then, has the form

$$du = \left(\frac{\partial u}{\partial s}\right)_v ds + \left(\frac{\partial u}{\partial v}\right)_s dv \tag{11.2}$$

Comparing Eqs. (11.1) and (11.2) we see that

$$T = \left(\frac{\partial u}{\partial s}\right)_v$$

and

$$p = -\left(\frac{\partial u}{\partial v}\right)_s$$

For this system, we can also write, from Eq. (7.31),

$$dh = T\,ds + v\,dp \tag{11.3}$$

and, in this case, we take the two independent properties to be s and p, so that

$$h = h(s, p)$$

whose total differential is

$$dh = \left(\frac{\partial h}{\partial s}\right)_p ds + \left(\frac{\partial h}{\partial p}\right)_s dp \tag{11.4}$$

On comparing Eqs. (11.3) and (11.4), we see that

$$T = \left(\frac{\partial h}{\partial s}\right)_p$$

and

$$v = \left(\frac{\partial h}{\partial p}\right)_s$$

EXAMPLE 11.1

To illustrate the relation between the constant volume and constant pressure specific heats and entropy, begin with Eqs. (11.1) and (11.3) and show that the constant volume and constant pressure specific heats are related to specific entropy by:

$$c_v = T\left(\frac{\partial s}{\partial T}\right)_v$$

and

$$c_p = T\left(\frac{\partial s}{\partial T}\right)_p$$

Solution

The constant volume specific heat is defined by Eq. (3.15) as

$$c_v = \left(\frac{\partial u}{\partial T}\right)_v$$

and the constant pressure specific heat is defined by Eq. (3.19) as

$$c_p = \left(\frac{\partial h}{\partial T}\right)_p$$

Equation (11.1) is $du = Tds - pdv$, and if we it divide through by dT, we get

$$\frac{du}{dT} = T\frac{ds}{dT} - p\frac{dv}{dT}$$

If we now require the specific volume v to be constant during this operation, then this equation becomes

$$\left.\frac{du}{dT}\right|_v = T\left.\frac{ds}{dT}\right|_v - p\left.\frac{dv}{dT}\right|_v \tag{a}$$

Now, a total derivative restrained with a constant parameter is just a partial derivative, or

$$\left.\frac{du}{dT}\right|_v = \left(\frac{\partial u}{\partial T}\right)_v = c_v$$

$$\left.\frac{ds}{dT}\right|_v = \left(\frac{\partial s}{\partial T}\right)_v$$

$$\left.\frac{dv}{dT}\right|_v = \left(\frac{\partial v}{\partial T}\right)_v = 0 \text{ (since } v \text{ is to be held constant here)}$$

Then, substituting these results into Eq. (a) gives one of the desired relations:

$$\left(\frac{\partial u}{\partial T}\right)_v = c_v = T\left(\frac{\partial s}{\partial T}\right)_v$$

Similarly, beginning with Eq. (11.3), we have $dh = T\,ds + v\,dp$, and dividing this through by dT gives

$$\frac{dh}{dT} = T\frac{ds}{dT} + v\frac{dp}{dT} \tag{b}$$

Again, imposing the condition that p must be constant during this operation, we get

$$\left.\frac{dh}{dT}\right|_p = \left(\frac{\partial h}{\partial T}\right)_p = c_p$$

$$\left.\frac{ds}{dT}\right|_p = \left(\frac{\partial s}{\partial T}\right)_p$$

$$\left.\frac{dp}{dT}\right|_p = \left(\frac{\partial p}{\partial T}\right)_p = 0 \text{ (since } p \text{ is to be held constant here)}$$

Then, substituting these results into Eq. (b) gives the other desired relation:

$$\left(\frac{\partial h}{\partial T}\right)_p = c_p = T\left(\frac{\partial s}{\partial T}\right)_p$$

The following exercises are designed to strengthen your understanding of the thermodynamics and the mathematics of the material presented in this part of the chapter.

Exercises

1. Use the results of Example 11.1 to show that a material that has an entropy function of the form $s(T, v) = A + B(\ln T) + C(\ln v)$, where A, B, and C are constants, has a constant volume specific heat given by $c_v = B$. Hint: Substitute the given function into the relation $c_v = T(\partial s/\partial T)_v$.
2. If the specific internal energy of a material is found to depend on its specific entropy and specific volume according to the relation $u(s, v) = A + Bs + Cv^2 + Ds/v$, where A, B, C, and D are all constants, then determine an expression for $p(s, v)$ for this material. **Answer**: $p(s, v) = -(2Cv - Ds/v^2)$.
3. If the specific enthalpy of a material depends on specific entropy and pressure according to $h(s, p) = A + Bs + Cp^2 + Ds^3p$, where A, B, C, and D are all constants, then determine an expression for $T(s, p)$ for this material. **Answer**: $T(s, p) = B + 3Ds^2p$.

We now introduce two new thermodynamic properties. The first is the total Helmholtz function F, named after the German physicist and physiologist Hermann Ludwig Ferdinand von Helmholtz (1821–1894), defined as

$$F = U - TS$$

Dividing by the system mass gives the specific Helmholtz function f as

$$f = u - Ts$$

Differentiating this equation gives

$$df = du - T\,ds - s\,dT$$

but from Eq. (11.1) we have

$$du - T\,ds = -p\,dv$$

so that

$$df = -p\,dv - s\,dT \tag{11.5}$$

If we presume the existence of a functional relation of the form

$$f = f(v, T)$$

then its total differential is

$$df = \left(\frac{\partial f}{\partial v}\right)_T dv + \left(\frac{\partial f}{\partial T}\right)_v dT \tag{11.6}$$

and, on comparing Eqs. (11.5) and (11.6), we see that

$$p = -\left(\frac{\partial f}{\partial v}\right)_T$$

and

$$s = -\left(\frac{\partial f}{\partial T}\right)_v$$

The second new thermodynamic function is the total Gibbs function G, named after the American physicist Josiah Willard Gibbs (1839–1903), defined as

$$G = H - TS$$

Dividing by the system mass gives the specific Gibbs function g as

$$g = h - Ts \tag{11.7}$$

Differentiating Eq. (11.7) gives

$$dg = dh - T\,ds - s\,dT$$

but from Eq. (11.3), we have

$$dh = T\,ds + v\,dp$$

so that

$$dg = v\,dp - s\,dT \tag{11.8}$$

If we presume a functional relation of the form

$$g = g(p, T)$$

then its total differential is

$$dg = \left(\frac{\partial g}{\partial p}\right)_T dp + \left(\frac{\partial g}{\partial T}\right)_p dT \tag{11.9}$$

and comparing Eqs. (11.8) and (11.9) gives

$$v = \left(\frac{\partial g}{\partial p}\right)_T$$

and

$$s = -\left(\frac{\partial g}{\partial T}\right)_p$$

Table 11.1 Summary of Thermodynamic Property Relations

New Property Relations
$T = \left(\frac{\partial u}{\partial s}\right)_v = \left(\frac{\partial h}{\partial s}\right)_p$
$v = \left(\frac{\partial h}{\partial p}\right)_s = \left(\frac{\partial g}{\partial p}\right)_T$
$p = -\left(\frac{\partial u}{\partial v}\right)_s = -\left(\frac{\partial f}{\partial v}\right)_T$
$s = -\left(\frac{\partial f}{\partial T}\right)_v = -\left(\frac{\partial g}{\partial T}\right)_p$

Table 11.1 summarizes these results. The importance of this set of partial differential equations lies in the fact that they relate easily measurable properties (p, v, T) to nonmeasurable properties (u, h, s, f, and g). Therefore, accurate p, v, T data on any pure substance can be used to generate information about u, h, s, f, and g for that substance. However, they do not provide a direct method for calculating u, h, or s from p, v, and T information. We must look for additional information to complete this task. But, first, we take a short diversion into phase change processes for which we can determine important results based on what we already know.

EXAMPLE 11.2

The design of a new Happy Food fast-food processing system requires the values of the specific Helmholtz and Gibbs functions for superheated water vapor at 200. psia and 400.°F. Since most thermodynamic tables do not list these properties directly, you are asked to calculate them for these conditions.

Solution

First, draw a sketch of the system (Figure 11.1).

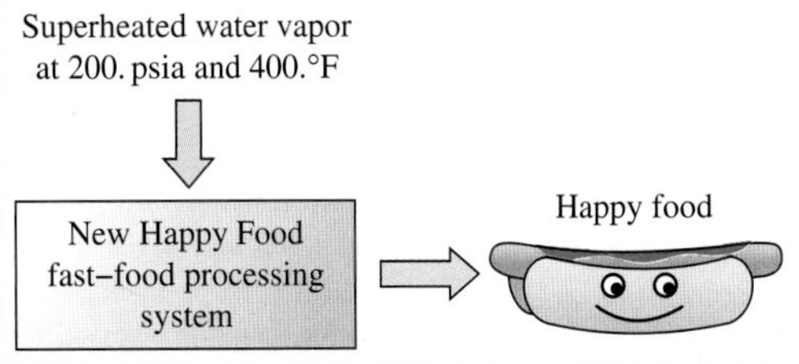

FIGURE 11.1
Example 11.1.

The unknowns are the values of the specific Helmholtz and Gibbs functions for superheated water vapor at 200. psia and 400.°F. The specific Helmholtz and Gibbs functions are defined in the text as

$$\text{Specific Helmholtz function: } f = u - Ts$$

and

$$\text{Specific Gibbs function: } g = h - Ts$$

Therefore, we need u, h, and s information at the state defined by $p = 200.$ psia and $T = 400.°\text{F} = 860.\text{R}$. From Table C.3a in *Thermodynamic Tables to accompany Modern Engineering Thermodynamics*, we find that, at this state,

$$u(p = 200.\,\text{psia}, T = 400.°\text{F}) = 1123.5\,\text{Btu/lbm}$$
$$h(p = 200.\,\text{psia}, T = 400.°\text{F}) = 1210.8\,\text{Btu/lbm},$$

and

$$s(p = 200.\,\text{psia}, T = 400.°\text{F}) = 1.5602\,\text{Btu/lbm}\cdot\text{R}$$

Then,

$$f = 1123.5\,\text{Btu/lbm} - (400. + 459.67\,\text{R})(1.5602\,\text{Btu/lbm}\cdot\text{R}) = -218\,\text{Btu/lbm}$$

and

$$g = 1210.8\,\text{Btu/lbm} - (400. + 459.67\,\text{R})(1.5602\,\text{Btu/lbm}\cdot\text{R}) = -131\,\text{Btu/lbm}$$

The following exercises consider different pressures and temperatures and explore why the specific Helmholtz and Gibbs functions were negative in Example 11.2.

Exercises

4. Determine the value of the specific Helmholtz function of the superheated water vapor in Example 11.2 if the pressure is maintained at 200. psia but the temperature is increased to 1000.°F. **Answer:** $f = -1320$ Btu/lbm.
5. Determine the specific Helmholtz function of saturated liquid water at 200. psia. **Answer:** $f = -103$ Btu/lbm.
6. The Helmholtz and Gibbs functions calculated in Example 11.2 are both negative. Though this has no particular significance at this point, use Table C.3a to determine the temperature and pressure at which the Helmholtz function is zero for water. **Answer:** $f = 0$ for saturated liquid water at 0.0887 psia and 32.018°F (the triple point of water).

11.3 GIBBS PHASE EQUILIBRIUM CONDITION

During a phase change process, the system pressure and temperature are *not* independent properties. This means that, if we hold one of them constant during a phase change, the other must also remain constant. Under the condition of constant pressure and temperature, $dp = dT = 0$, and Eq. (11.9) gives $dg = 0$. Since $g = g_f + xg_{fg}$, we can then write

$$dg = dg_f + x\,dg_{fg} + g_{fg}\,dx = 0$$

Again, Eq. (11.8) can be used to evaluate $dg_f = dg_{fg} = 0$. Since x can vary during the phase equilibrium, dx cannot be zero. Therefore, we are forced to conclude from the preceding equation that $g_{fg} = 0$ at phase equilibrium, or $g_f = g_g$. Using the definition of the Gibbs function, Eq. (11.7), we see that, at phase equilibrium,

$$g_f = g_g = h_f - (T_{sat})s_f = h_g - (T_{sat})s_g$$

or

$$h_g - h_f = h_{fg} = (T_{sat})(s_g - s_f) = (T_{sat})s_{fg}$$

or

$$s_{fg} = h_{fg}/T_{sat} \tag{11.10}$$

This gives us an important relation between the entropy and the enthalpy of a phase change, but we need much more information to complete the process of determining the nonmeasurable properties u, h, and s from the measurable properties p, v, and T. The following example illustrates the use of Eq. (11.10).

EXAMPLE 11.3

Use Eq. (11.10) to calculate the phase change entropy s_{fg} for water at exactly 1.00 MPa and compare the result with the value for s_{fg} at exactly 1.00 MPa listed in Table C.2b in *Thermodynamic Tables to accompany Modern Engineering Thermodynamics*.

Solution[1]

The unknown is the phase change entropy of water. From (Eq. 11.10), we have

$$s_{fg} = \frac{h_{fg}}{T_{sat}}$$

and from Table C.2b at $p = 1.00$ MPa, we find that

$$h_{fg} = 2015.3\ \text{kJ/kg}$$

and

$$T_{sat} = 179.90°\text{C}$$

then, Eq. (11.10) gives

$$s_{fg} = \frac{h_{fg}}{T_{sat}} = \frac{2015.3\ \text{kJ/kg}}{179.90 + 273.15\ \text{K}} = 4.4482\ \text{kJ/kg}\cdot\text{K}$$

Comparing this with the value for s_{fg} listed in Table C.2b at $p = 1.00$ MPa, we find that it is exactly the same.

The following exercises illustrate some of the many uses of Eq. (11.10).

Exercises

7. Use Eq. (11.10) to compute the values of s_{fg} for water at 0.0100 MPa and compare the result with the values listed in Table C.2b. Find h_{fg} and T_{sat} at 0.01 MPa from Table C.2b. **Answer**: $(s_{fg})_{calc} = 7.5021$ kJ/kg·K.
8. Use Eq. (11.10) to compute the value of h_{fg} for water at 100.°F and compare the result with the value listed in Table C.1a. Find values for s_{fg} and T_{sat} at 100.°F from Table C.1a. **Answer**: $(h_{fg})_{calc} = 1036.96$ Btu/lbm.
9. Use the values for h_{fg} and s_{fg} found in Table C.2b for water at 10.0 MPa and Eq. (11.10) to calculate the value of T_{sat} at this state. **Answer**: $(T_{sat})_{calc} = 584.2$ K = 311.06°C.

[1] *To achieve the desired result, we need to carry a lot more significant figures than usual.*

11.4 MAXWELL EQUATIONS

Two sets of equations are named after the Scottish physicist James Clerk Maxwell (1831–1879): the electromagnetic field equations and the thermodynamic property equations. The thermodynamic Maxwell equations allow additional numerical information to be obtained about the nonmeasurable properties u, h, and s from accurately measured p, v, and T data.

Consider an arbitrarily continuous function of the form

$$z = z(x, y)$$

Then, we can write its total differential as

$$dz = \left(\frac{\partial z}{\partial x}\right)_y dx + \left(\frac{\partial z}{\partial y}\right)_x dy = Mdx + Ndy \tag{11.11}$$

where we set

$$M = \left(\frac{\partial z}{\partial x}\right)_y$$

and

$$N = \left(\frac{\partial z}{\partial y}\right)_x$$

If we now differentiate M with respect to y while holding x constant and differentiate N with respect to x while holding y constant, we get

$$\left(\frac{\partial M}{\partial y}\right)_x = \frac{\partial^2 z}{\partial y\, \partial x}$$

and

$$\left(\frac{\partial N}{\partial x}\right)_y = \frac{\partial^2 z}{\partial x\, \partial y}$$

Since we require $z(x, y)$ to be a continuous function, it follows that

$$\frac{\partial^2 z}{\partial y\, \partial x} = \frac{\partial^2 z}{\partial x\, \partial y}$$

or that

$$\left(\frac{\partial M}{\partial y}\right)_x \equiv \left(\frac{\partial N}{\partial x}\right)_y \tag{11.12}$$

Recall that the thermodynamic state of any pure substance is fixed by any pair of independent intensive thermodynamic properties of that substance. That is, any property of a pure substance can be written as a function of any other two independent properties of that substance. Consequently, if x and y are such independent properties, then z also is a property, provided that Eq. (11.12) is satisfied.

EXAMPLE 11.4

Suppose we make a series of measurements in the laboratory and think we discovered a new thermodynamic property, call it z. Our experimental data provide an empirical equation of the form: $dz = p\,dv + v^2 dp$. Is z a new property?

Solution

The unknown is whether or not z is a new thermodynamic property. Equation (11.11) here has the form

$$dz = Mdx + Ndy = p\,dv + v^2 dp$$

so $M = p$, $N = v^2$, $x = v$, and $y = p$. The cross differentials in Eq. (11.12) are

$$\left(\frac{\partial M}{\partial y}\right)_x = \left(\frac{\partial p}{\partial p}\right)_v = 1$$

(Continued)

EXAMPLE 11.4 (*Continued*)

and

$$\left(\frac{\partial N}{\partial x}\right)_y = \left(\frac{\partial (v^2)}{\partial v}\right)_p = 2v \neq \left(\frac{\partial M}{\partial y}\right)_x$$

Since Eq. (11.12) is not satisfied here, then z cannot be a thermodynamic property.

Exercises

10. Suppose the expression experimentally discovered in Example 11.4 is $dz = p^2 dv + v^2 dp$, would z be the thermodynamic property? **Answer**: No, because $2p \neq 2v$.
11. If the expression experimentally discovered in Example 11.4 is $dz = p\,dv - v\,dp$, would z be a thermodynamic property? **Answer**: No, $1 \neq -1$.
12. If the expression reported in Example 11.4 is $dz = p\,dv + v\,dp$, would z be a thermodynamic property? **Answer**: Yes, $dz = d(pv)$.

If we now look at our four basic property relationships as differential equations of the form of Eq. (11.11),

$$du = T\,ds - p\,dv \quad (11.1)$$

$$dh = T\,ds + v\,dp \quad (11.3)$$

$$df = -p\,dv - s\,dT \quad (11.5)$$

$$dg = v\,dp - s\,dT \quad (11.8)$$

then, Eq. (11.12) must be valid for these equations, since we already know that all of these functions are thermodynamic properties. Applying Eq. (11.12) to each of these equations yields a new set of equations, known as the *Maxwell thermodynamic equations*:

Maxwell thermodynamic equations

$$\left(\frac{\partial T}{\partial v}\right)_s = -\left(\frac{\partial p}{\partial s}\right)_v \quad (11.13)$$

$$\left(\frac{\partial T}{\partial p}\right)_s = \left(\frac{\partial v}{\partial s}\right)_p \quad (11.14)$$

$$\left(\frac{\partial p}{\partial T}\right)_v = \left(\frac{\partial s}{\partial v}\right)_T \quad (11.15)$$

$$\left(\frac{\partial v}{\partial T}\right)_p = -\left(\frac{\partial s}{\partial p}\right)_T \quad (11.16)$$

While the Maxwell thermodynamic equations provide additional information about u, h, and s in terms of p, v, and T, they cannot be solved to produce the direct functional relations between these properties that we seek. However, these relations are used a little later in this chapter in conjunction with other material to provide the desired u, h, and s relations from experimental p, v, T data.

EXAMPLE 11.5

Suppose we have the ideal gas equation of state, $pv = RT$, but know nothing about the entropy of this type of gas. Use the appropriate Maxwell equations to determine a mathematical relation for the entropy of an ideal gas during an isothermal process.

Solution

The ideal gas equation of state is $pv = RT$, and so we know a p, v, T relation for our material. Perusing the Maxwell equations, we see two, Eqs. (11.15) and (11.16), that involve only p, v, T variables on one side of the equation. We can choose either of these equations to satisfy the problem statement, so we select Eq. (11.16):

$$\left(\frac{\partial v}{\partial T}\right)_p = -\left(\frac{\partial s}{\partial p}\right)_T$$

Solving for v from the ideal gas equation of state gives $v = RT/p$, so the partial derivative we need is

$$\left(\frac{\partial v}{\partial T}\right)_p = R/p$$

then,

$$\left(\frac{\partial s}{\partial p}\right)_T = -R/p$$

so that $ds_T = -R(dp/p)_T$, where the subscript T is used to indicate that the temperature is to be held constant. This can be integrated for the constant temperature condition to give

$$(s_2 - s_1)_T = -R\int_1^2 (dp/p)_T + \text{func}(T) = -R\ln(p_2/p_1) + \text{func}(T)$$

where func(T) is an arbitrary *function of integration*. The function of integration here depends on the temperature T, and for an isothermal process, it is treated as a constant. Since we happen to know that the entropy relation for an ideal gas is in fact $s_2 - s_1 = c_p \ln(T_2/T_1) - R\ln(p_2/p_1)$, it is easy to see that the function of integration here is simply $c_p \ln(T_2/T_1)$.

The following exercises reinforce the concepts presented in Example 11.5.

Exercises

13. Use the other Maxwell equation (Eq. 11.15) available for the solution of Example 11.5 to find a different ideal gas entropy relation. **Answer**: $(s_2 - s_1)_T = R\ln(v_2/v_1) + \text{func}(T)$.
14. Show that $(\partial s/\partial)_p = (\partial p/\partial T)_s$. **Answer**: Invert Eq. (11.14).
15. Show that $(s_2 - s_1)_v = -\int_1^2 (\partial T/\partial v)_s\, dp + \text{func}(v)$. **Answer**: Use Eq. (11.13).

Before we continue with our search for the illusive u, h, s equations in terms of p, v, T variables, the following example shows that the form taken by the Maxwell equations depends on the type of reversible work mode present in the system.

EXAMPLE 11.6

The equation of state for a nonlinear rubber band is given by

$$F = KT(L/L_o - 1)^2$$

where F is the stretching force, L is the stretched length, L_o is the initial length, K is the elastic constant, and T is the absolute temperature of the material. Then,

a. Determine the Maxwell equations for this material.
b. Show that the internal energy of this material is a function of temperature only.
c. Determine the heat transfer required when the rubber band is stretched isothermally and reversibly from $L_o = 0.0700$ m to $L = 0.200$ m at $T = 20.0°\text{C}$ when $K = 0.150$ N/K.

Solution

The unknowns here are the Maxwell equations for this material, showing that the internal energy of this material is a function of temperature only and the heat transfer required when the rubber band is stretched isothermally and reversibly between two states.

a. Since the reversible work mode involved in the stretching process is

$$(\bar{d}W)_{\text{rev}} = -F\,dL$$

the Maxwell equations for this material can be easily obtained from those derived in the text by replacing p with $-F$ and v with $L/m = \ell$, the *specific length* of the material. Then Eqs. (11.13) to (11.16) become

$$\left(\frac{\partial T}{\partial \ell}\right)_s = \left(\frac{\partial F}{\partial s}\right)_\ell$$

$$\left(\frac{\partial T}{\partial F}\right)_s = -\left(\frac{\partial \ell}{\partial s}\right)_F$$

$$\left(\frac{\partial F}{\partial T}\right)_\ell = -\left(\frac{\partial s}{\partial \ell}\right)_T$$

and

$$\left(\frac{\partial \ell}{\partial T}\right)_F = \left(\frac{\partial s}{\partial F}\right)_T$$

(Continued)

EXAMPLE 11.6 (*Continued*)

b. The combined energy and entropy balance for this material is

$$du = T\,ds + F\,d\ell$$

so that

$$\left(\frac{\partial u}{\partial \ell}\right)_T = T\left(\frac{\partial s}{\partial \ell}\right)_T + F$$

From the third Maxwell equation for this substance listed in part a and the given equation of state, we have

$$\left(\frac{\partial s}{\partial \ell}\right)_T = -\left(\frac{\partial F}{\partial T}\right)_\ell$$
$$= -K(L/L_o - 1)^2$$

and

$$T\left(\frac{\partial s}{\partial \ell}\right)_T = -KT(L/L_o - 1)^2 = -F$$

Therefore,

$$\left(\frac{\partial u}{\partial \ell}\right)_T = -F + F = 0$$

If we now set $u = u(T, \ell)$ and differentiate it, we get

$$du = \left(\frac{\partial u}{\partial T}\right)_\ell dT + \left(\frac{\partial u}{\partial \ell}\right)_T d\ell$$
$$= c_\ell dT + \left(\frac{\partial u}{\partial \ell}\right)_T d\ell$$

where c_ℓ is the constant length specific heat. Now since $(\partial u/\partial \ell)_T = 0$ here, then this equation reduces to $du = c_\ell dT$, so u is only a function of T.

c. A closed system energy balance applied to this material for an isothermal process with $(u_2 - u_1)_T = 0$ gives

$${}_1Q_2 = {}_1W_2 = -\int_{L_o}^{L} F\,dL$$
$$= -KT\int_{L_o}^{L} (L/L_o - 1)^2\,dL$$
$$= -KTL_o(L/L_o - 1)^3/3$$
$$= -(0.150\,\text{N/K})(293\,\text{K})(0.0700\,\text{m})\left(\frac{0.200}{0.0700} - 1\right)^3\Big/3$$
$${}_1Q_2 = -6.57\,\text{N}\cdot\text{m}$$

Consequently, there is a heat transfer out of the system equal in magnitude to the work input.

Exercises

16. Determine the heat transfer and work required to stretch the rubber band in Example 11.6 if the elastic constant of the rubber is increased from 0.150 N/K to 10.0 N/K. **Answer**: ${}_1Q_2 = {}_1W_2 = -438$ N·m.
17. If the temperature of the rubber band in Example 11.6 is increased from 20.0°C to 60.0°C, determine the heat transfer and work required to stretch the rubber band assuming all the other variables remain unchanged. **Answer**: ${}_1Q_2 = {}_1W_2 = -7.47$ N·m.
18. How much heat transfer and work is required to stretch the rubber band in Example 11.6 twice as far, to $L = 0.400$ m instead of 0.200 m, if everything else remains constant? **Answer**: ${}_1Q_2 = {}_1W_2 = -107$ N·m.

11.5 THE CLAPEYRON EQUATION

Benoit Pierre Emile Clapeyron (1799–1864) was a French mining engineer and a contemporary of Carnot who, in the, 1830s, took an interest in studying the physical behavior of gases and vapors. He was able to derive a relation for the enthalpy change of the liquid to vapor phase transition (h_{fg}) in terms of pressure, temperature, and specific volume, thus providing one of the first equations for calculating a property that is not directly measurable in terms of properties that are directly measurable. Today, this relation is most easily derived from one of the Maxwell equations, Eq. (11.15). For an isothermal phase change from a saturated liquid to a saturated vapor, the pressure and temperature are independent of volume. Then, Eq. (11.15) becomes

$$\left(\frac{\partial p}{\partial T}\right)_v = \left(\frac{dp}{dT}\right)_{\text{sat}} = \frac{s_g - s_f}{v_g - v_f} = s_{fg}/v_{fg}$$

and, using Eq. (11.10), we obtain the Clapeyron equation as[2]

$$\left(\frac{dp}{dT}\right)_{\text{sat}} = h_{fg}/(T_{\text{sat}} v_{fg}) \tag{11.17}$$

For most substances, $v_g \gg v_f$, so we can approximate $v_{fg} \approx v_g$. Also, for vapors at very low pressures, the saturated vapor curve can be accurately approximated by the ideal gas equation of state, so we can write $v_g = RT_{\text{sat}}/p_{\text{sat}}$. Then Eq. (11.17) becomes

$$\left(\frac{dp}{dT}\right)_{\text{sat}} = p_{\text{sat}} h_{fg}/(RT_{\text{sat}}^2)$$

or

$$\left(\frac{dp}{p}\right)_{\text{sat}} = \left[h_{fg}/(RT_{\text{sat}}^2)\right] dT_{\text{sat}} \tag{11.18}$$

This equation is often called the *Clapeyron-Clausius equation*. For small pressure and temperature changes, h_{fg} can be assumed to be constant and Eq. (11.18) can be integrated from a reference state to any other state to give

$$\ln(p/p_0)_{\text{sat}} = (h_{fg}/R)\frac{T_{\text{sat}} - T_0}{T_{\text{sat}} T_0}$$

or

$$p_{\text{sat}} = p_0 \exp\left[(h_{fg}/R)\frac{T_{\text{sat}} - T_0}{T_{\text{sat}} T_0}\right]$$

where p_0 and T_0 are reference state values. An exponential relation between p_{sat} and T_{sat} fits experimental data quite well for most substances at low pressure.

EXAMPLE 11.7

In 1849, William Rankine proposed the following relation between the saturation pressure and saturation temperature of water:

$$\ln p_{\text{sat}} = 14.05 - \frac{6289.78}{T_{\text{sat}}} - \frac{913{,}998.92}{T_{\text{sat}}^2}$$

where p_{sat} is in psia, and T_{sat} is the temperature in °F + 461.2 (at that time –461.2°F was Rankine's best estimate of absolute zero temperature). Determine h_{fg} at 212.0°F from the Rankine equation and compare the result with that listed in the steam tables in *Thermodynamic Tables to accompany Modern Engineering Thermodynamics*.

Solution

Differentiating Rankine's equation, we obtain

$$\left(\frac{1}{p}\frac{dp}{dT}\right)_{\text{sat}} = \frac{6289.78}{T_{\text{sat}}^2} + \frac{1{,}827{,}997.8}{T_{\text{sat}}^3}$$

then, using Eq. (11.18), we get

$$h_{fg} = \left[\frac{RT^2}{p}\left(\frac{dp}{dT}\right)\right]_{\text{sat}} = R(6289.78 + 1{,}827{,}997.8/T_{\text{sat}})$$

(Continued)

[2] The Clapeyron equation is valid for any type of phase change in a simple substance. For example, if we let the i subscript denote the solid phase, then for melting we can write

$$\left(\frac{dp}{dT}\right)_{\substack{\text{solid–}\\\text{liquid}\\\text{saturation}}} = h_{if}/(T_{\text{sat}} v_{if})$$

and, for sublimation,

$$\left(\frac{dp}{dT}\right)_{\substack{\text{solid–}\\\text{vapor}\\\text{saturation}}} = h_{ig}/(T_{\text{sat}} v_{ig})$$

EXAMPLE 11.7 (*Continued*)

From Table C.13a in *Thermodynamic Tables to accompany Modern Engineering Thermodynamics*, we find $R = 85.78$ ft·lbf/(lbm·R) = 0.1102 Btu/(lbm·R). Then, at 212.0°F,

$$h_{fg}(212.0°\text{F}) = [6289.78\,\text{R} + (1{,}827{,}997.8\,\text{R}^2)/(461.2 + 212.0\,\text{R})] \times [0.1102\,\text{Btu/(lbm}\cdot\text{R)}] = 992.37\,\text{Btu/lbm}$$

Table C.1a gives $h_{fg}(212.0°\text{F}) = 970.4$ Btu/lbm. Thus, the value obtained from Rankine's equation is in error by only +2.26%.

Exercises

19. Determine p_{sat} from the Rankine equation given in Example 11.7 when $T_{sat} = 212.0°\text{F}$ and compare it with the value of T_{sat} given in Table C.1a at 212.0°F. **Answer:** $(p_{sat})_{calc} = 14.73$ psia, and from Table C.1a, $p_{sat}(212.0°\text{F}) = 14.696$ psia.
20. Using the relations given in Example 11.7, find the value of T_{sat} for water when $h_{fg} = 1037$ Btu/lbm and compare your result with the value given in Table C.1a. **Answer:** $(T_{sat})_{calc} = 124.6°\text{F}$, and from Table C.1a, $T_{sat} = 100.0°\text{F}$.
21. Determine h_{fg} in Example 11.7 if the temperature is increased from 212.0°F to 500.0°F and compare your result with the value given in Table C.1a. **Answer:** $(h_{fg})_{calc} = 902.7$ Btu/lbm, and from Table C.1a, $h_{fg}(500.0°\text{F}) = 714.8$ Btu/lbm.

11.6 DETERMINING *u*, *h*, AND *s* FROM *p*, *v*, AND *T*

We are now ready to combine the previous results to produce u, h, and s relations from p, v, and T data. For a simple substance, any two independent intensive properties fix its thermodynamic state. Consider the specific internal energy described by a function of temperature and specific volume. We can write this as $u = u(T, v)$. Differentiating this function, we get

$$du = \left(\frac{\partial u}{\partial T}\right)_v dT + \left(\frac{\partial u}{\partial v}\right)_T dv$$

From Eq. (11.1), we can write

$$\left(\frac{\partial u}{\partial v}\right)_T = T\left(\frac{\partial s}{\partial v}\right)_T - p$$

and using the Maxwell Eq. (11.15), this becomes

$$\left(\frac{\partial u}{\partial v}\right)_T = T\left(\frac{\partial p}{\partial T}\right)_v - p$$

In Chapter 3, we introduce the constant volume specific heat c_v as

$$c_v = \left(\frac{\partial u}{\partial T}\right)_v \tag{3.15}$$

and our equation for the total differential du then becomes

$$du = c_v\,dT + \left[T\left(\frac{\partial p}{\partial T}\right)_v - p\right] dv \tag{11.19}$$

Therefore, the change in specific internal energy for any simple substance can be determined by integrating Eq. (11.19):

$$u_2 - u_1 = \int_{T_1}^{T_2} c_v\,dT + \int_{v_1}^{v_2} \left[T\left(\frac{\partial p}{\partial T}\right)_v - p\right] dv \tag{11.20}$$

Here, we achieved what we set out to do. Equation (11.20) has u cast completely in terms of the measurable quantities p, v, T, and c_v.

Similarly, we can consider the specific enthalpy to be given by a continuous function of temperature and pressure, $h = h(T, p)$. Then, its total differential is

$$dh = \left(\frac{\partial h}{\partial T}\right)_p dT + \left(\frac{\partial h}{\partial p}\right)_T dp$$

In Chapter 3, we introduce the constant pressure specific heat c_p as

$$c_p = \left(\frac{\partial h}{\partial T}\right)_p \tag{3.19}$$

Introducing the definition of specific enthalpy into Eq. (11.1) gives

$$du = dh - p\,dv - v\,dp = T\,ds - p\,dv$$

or

$$dh = T\,ds + v\,dp \tag{11.21}$$

and, from this equation, we can deduce that

$$\left(\frac{\partial h}{\partial p}\right)_T = T\left(\frac{\partial s}{\partial p}\right)_T + v$$

Using the Maxwell Eq. (11.16), we get

$$\left(\frac{\partial h}{\partial p}\right)_T = -T\left(\frac{\partial v}{\partial T}\right)_p + v$$

and our total differential dh becomes

$$dh = c_p\,dT + \left[v - T\left(\frac{\partial v}{\partial T}\right)_p\right]dp \tag{11.22}$$

The change in specific enthalpy for any simple substance is then given by

$$h_2 - h_1 = \int_{T_1}^{T_2} c_p\,dT + \int_{p_1}^{p_2}\left[v - T\left(\frac{\partial v}{\partial T}\right)_p\right]dp \tag{11.23}$$

Again, we are successful. Equation (11.23) has h cast completely in terms of the measurable quantities, p, v, T, and c_p. Also note that Eqs. (11.20) and (11.23) are related by the fact that

$$h_2 - h_1 = u_2 - u_1 + p_2v_2 - p_1v_1$$

Finally, we can carry out the same type of analysis for the specific entropy of a simple substance. If we let $s = s\,(T, v)$, then

$$ds = \left(\frac{\partial s}{\partial T}\right)_v dT + \left(\frac{\partial s}{\partial v}\right)_T dv$$

From Eqs. (11.1) and (3.15), we can deduce that

$$\left(\frac{\partial s}{\partial T}\right)_v = \left(\frac{1}{T}\right)\left(\frac{\partial u}{\partial T}\right)_v = \frac{c_v}{T}$$

and, using the Maxwell Eq. (11.15), we can write the total differential ds as

$$ds = \left(\frac{c_v}{T}\right)dT + \left(\frac{\partial p}{\partial T}\right)_v dv \tag{11.24}$$

Integrating this gives a relation for the change in specific entropy of a pure substance based completely on measurable quantities:

$$s_2 - s_1 = \int_{T_1}^{T_2}\frac{c_v}{T}dT + \int_{v_1}^{v_2}\left(\frac{\partial p}{\partial T}\right)_v dv \tag{11.25}$$

By assuming $s = s(T, p)$, we can also show that (see Problem 27 at the end of this chapter)

$$ds = \frac{c_p}{T}dT - \left(\frac{\partial v}{\partial T}\right)_p dp \tag{11.26}$$

and

$$s_2 - s_1 = \int_{T_1}^{T_2}\frac{c_p}{T}dT - \int_{p_1}^{p_2}\left(\frac{\partial v}{\partial T}\right)_p dp \tag{11.27}$$

This completes the process of discovering relations for the unmeasurable u, h, and s properties in terms of the measurable properties p, v, and T. The following example illustrates the use of these results.

EXAMPLE 11.8

In Chapter 3, an equation of state developed in 1903 by Pierre Berthelot (1827–1907) was briefly discussed. Using this equation of state, develop equations based on measurable properties for the changes in (a) specific internal energy, (b) specific enthalpy, and (c) specific entropy for an isothermal process.

Solution

The Berthelot equation is given in Eq. (3.46) as

$$p(v-b) = RT - a(v-b)/(Tv^2)$$

where a and b are constants. Solving this equation for p gives

$$p = RT/(v-b) - a/(Tv^2)$$

a. The change in specific internal energy is given by Eq. (11.20), for which we need

$$\left(\frac{\partial p}{\partial T}\right)_v = R/(v-b) + a/(T^2v^2)$$

Then, for an isothermal process ($T_1 = T_2$), Eq. (11.20) gives

$$(u_2-u_1)_T = \int_{v_1}^{v_2}\left[RT/(v-b) + a/(Tv^2) - RT/(v-b) + a/(Tv^2)\right]dv$$

$$= -(2a/T)(1/v_2 - 1/v_1) = 2a(v_2-v_1)/(Tv_1v_2)$$

b. To find the change in specific enthalpy, we could use Eq. (11.23). However, to evaluate this equation, we need to be able to determine the relation $(\partial v/\partial T)_p$. Since the Berthelot equation is not readily solvable for $v = v\,(T, p)$, we choose instead to use the simpler approach, utilizing only the definition of specific enthalpy, $h = u + pv$. Then,

$$(h_2-h_1)_T = (u_2-u_1)_T + p_2v_2 - p_1v_1 = \frac{2a(v_2-v_1)}{Tv_1v_2} + p_2v_2 - p_1v_1$$

$$= \frac{3a(v_2-v_1)}{Tv_1v_2} + RT\left(\frac{v_2}{v_2-b} - \frac{v_1}{v_1-b}\right)$$

c. Finally, since we already evaluated the relation $(\partial p/\partial T)_v$, we choose to use Eq. (11.25) for the isothermal specific entropy relation:

$$(s_2-s_1)_T = \int_{v_1}^{v_2}\left(\frac{\partial p}{\partial T}\right)_v dv$$

$$= \int_{v_1}^{v_2}\left[R/(v-b) + a/(T^2v^2)\right]dv$$

$$= R\ln\left[(v_2-b)/(v_1-b)\right] + a(v_2-v_1)/(T^2v_1v_2)$$

Exercises

22. Setting $a = 0$ in the Berthelot equation of state used in Example 11.8 produces the Clausius equation of state, $p(v-b) = RT$ (see Eq. (3.43)). Determine equations for the change in specific internal energy, specific enthalpy, and specific entropy for a Clausius gas undergoing an isothermal process. **Answer**: $(u_2-u_1)_T = 0$, $(h_2-h_1)_T = RT\left[(v_2/(v_2-b) - v_1/(v_1-b)\right]$, and $(s_2-s_1)_T = R\ln[(v_2-b)/(v_1-b)]$.
23. Evaluate the change in specific internal energy of water vapor when it is modeled as a Berthelot gas, with $a = 4.30\ \text{MN}\cdot\text{m}^4\cdot\text{K/kg}^2$ and $b = 4.50 \times 10^{-3}\ \text{m}^3/\text{kg}$, and undergoes an isothermal compression from a specific volume 40.0 m^3/kg to a specific volume of 5.00 m^3/kg at a constant temperature of 100.°C. **Answer**: $(u_2-u_1)_T = -4.03$ kN·m/kg = −4.03 kJ/kg.
24. Evaluate the change in specific enthalpy of water vapor when it is modeled as a Berthelot gas, with $a = 4.30\ \text{MN}\cdot\text{m}^4\cdot\text{K/kg}^2$ and $b = 4.50 \times 10^{-3}\ \text{m}^3/\text{kg}$, and undergoes an isothermal expansion from a specific volume of 10.0×10^{-3} m^3/kg to a specific volume of 1.00 m^3/kg at a constant temperature of 500.°C. Use $R_{water} = 461$ N·m/(kg·K). **Answer**: $(h_2-h_1)_T = 261$ kN·m/kg = 261 kJ/kg.

Note that, for an ideal gas undergoing an isothermal process,

$$(u_2-u_1)_T = (h_2-h_1)_T = 0 \quad \text{and} \quad (s_2-s_1)_T = R\ln(v_2/v_1)$$

Therefore, the equations developed in Example 11.8 can be considered to be Berthelot corrections to ideal gas behavior.

Equation (11.24) has the same $Mdx + Ndy$ form as Eq. (11.11), so that we can utilize Eq. (11.12) to produce the property relation

$$\left[\frac{\partial(c_v/T)}{\partial v}\right]_T = \left[\frac{\partial}{\partial T}\left(\frac{\partial p}{\partial T}\right)_v\right]_v$$

or

$$\left(\frac{\partial c_v}{\partial v}\right)_T = T\left(\frac{\partial^2 p}{\partial T^2}\right)_v \tag{11.28}$$

Similarly, Eq. (11.26) has the same $Mdx + Ndy$ form and application of Eq. (11.12), so it gives

$$\left[\frac{\partial (c_p/T)}{\partial p}\right]_T = -\left[\frac{\partial}{\partial T}\left(\frac{\partial v}{\partial T}\right)_p\right]_p$$

or

$$\left(\frac{\partial c_p}{\partial p}\right)_T = -T\left(\frac{\partial^2 v}{\partial T^2}\right)_p \tag{11.29}$$

Both Eqs. (11.28) and (11.29) give specific heat information from measurable p, v, and T properties.

EXAMPLE 11.9

Using the Berthelot equation of state given in Example 11.8, determine an equation for the isothermal variation in the constant volume specific heat with volume change.

Solution

From Eq. (11.28) we have what we seek,

$$\left(\frac{\partial c_v}{\partial v}\right)_T = T\left(\frac{\partial^2 p}{\partial T^2}\right)_v$$

and from Example 11.8, the Berthelot equation of state can be written as

$$p = \frac{RT}{v-b} - \frac{a}{Tv^2}$$

so that

$$\left(\frac{\partial p}{\partial T}\right)_v = \frac{R}{v-b} + \frac{a}{T^2v^2}$$

and

$$\left(\frac{\partial^2 p}{\partial T^2}\right)_v = -\frac{2a}{T^3v^2}$$

then,

$$\left(\frac{\partial c_v}{\partial v}\right)_T = -\frac{2a}{T^2v^2}$$

and, to find an explicit $c_v = c_v(T, v)$ equation, the preceding equation can be integrated from a reference state specific volume v_o to give

$$c_v = -\frac{2a}{T^2}\int_{v_0}^{v}\frac{dv}{v^2} = 2a(v_0 - v)/(T^2 v_0 v) + f(T)$$

where $f(T)$ is a function of integration. Note that c_v is independent of v only in the case where $a = 0$ in the Berthelot equation of state.

Exercises

25. The Clausius equation of state, $p(v - b) = RT$ (see Eq. (3.43)), can be obtained by setting $a = 0$ in the Berthelot equation of state. Rework Example 11.9 to determine how the constant volume specific heat of a Clausius gas undergoing an isothermal process depends on the specific volume of the gas. **Answer**: For a Clausius gas undergoing an isothermal process, c_v does not depend on the specific volume of the gas.
26. For the Berthelot equation of state used in Example 11.9, determine an expression for the mixed partial derivative

$$\left[\frac{\partial}{\partial T}\left(\frac{\partial c_v}{\partial v}\right)_T\right]_v = ?$$

Answer: $4a/(T^3v^2)$.

27. Evaluate the constant volume specific heat relation developed in Example 11.9 for a material in which $a = 2.30\ \text{MN}\cdot\text{m}^4\cdot\text{K/kg}^2$. Use $v_0 = 0.100\ \text{m}^3/\text{kg}$ and $v = 0.0200\ \text{m}^3/\text{kg}$. **Answer**: $c_v = 1.321\ \text{kN}\cdot\text{m/kg} = 1.321\ \text{kJ/kg}$.

Finally, if Eqs. (11.24) and (11.26) are set equal to each other,

$$\left(\frac{c_v}{T}\right)dT + \left(\frac{\partial p}{\partial T}\right)_v dv = \left(\frac{c_p}{T}\right)dT - \left(\frac{\partial v}{\partial T}\right)_p dp$$

and, if we solve for dT,

$$dT = \left(\frac{T}{c_p - c_v}\right)\left(\frac{\partial v}{\partial T}\right)_p dp + \left(\frac{T}{c_p - c_v}\right)\left(\frac{\partial p}{\partial T}\right)_v dv$$

Then, writing the general relation $T = T(p, v)$ and differentiating it, we get

$$dT = \left(\frac{\partial T}{\partial p}\right)_v dp + \left(\frac{\partial T}{\partial v}\right)_p dv$$

and, by comparing coefficients of dp and dv in these two equations, it is clear that

$$\left(\frac{\partial T}{\partial p}\right)_v = \left(\frac{T}{c_p - c_v}\right)\left(\frac{\partial v}{\partial T}\right)_p$$

and

$$\left(\frac{\partial T}{\partial v}\right)_p = \left(\frac{T}{c_p - c_v}\right)\left(\frac{\partial p}{\partial T}\right)_v$$

or

$$c_p - c_v = T\left(\frac{\partial v}{\partial T}\right)_p\left(\frac{\partial p}{\partial T}\right)_v = T\left(\frac{\partial p}{\partial T}\right)_v\left(\frac{\partial v}{\partial T}\right)_p$$

Using Eq. (3.3), we can write

$$\left(\frac{\partial p}{\partial T}\right)_v = -\left(\frac{\partial v}{\partial T}\right)_p\left(\frac{\partial p}{\partial v}\right)_T$$

which, when substituted into the previous equation, yields

$$c_p - c_v = -T\left(\frac{\partial v}{\partial T}\right)_p^2\left(\frac{\partial p}{\partial v}\right)_T$$

In Chapter 3, we define the isobaric coefficient of volume expansion β as

$$\beta = \frac{1}{v}\left(\frac{\partial v}{\partial T}\right)_p \tag{3.5}$$

and the isothermal coefficient of compressibility κ as

$$\kappa = -\frac{1}{v}\left(\frac{\partial v}{\partial p}\right)_T \tag{3.6}$$

Substituting these two relations into the previous equation gives the final result:

$$c_p - c_v = T\beta^2 v/k \tag{11.30}$$

This equation reveals several important results. First of all, $c_p = c_v$ for all simple substances at absolute zero temperature. Second, since $\beta^2/\kappa = 0$ for incompressible materials, then $c_p = c_v$ for all incompressible materials. In this case, the p and v subscripts are normally dropped and we write $c_p = c_v = c$ for all incompressible materials. Finally, since T, β, v, and κ are always positive, then $c_p \geq c_v$ for all simple substances.

EXAMPLE 11.10

Using the data in Table 3.2, determine the difference between c_p and c_v for saturated liquid water at 20.0°C.

Solution

From Table 3.2, for water, we find that

$$\beta = 0.207 \times 10^{-6}\ \text{K}^{-1}$$
$$\kappa = 45.9 \times 10^{-11}\ \text{m}^2/\text{N}$$

and, from Table C.1b in *Thermodynamic Tables to accompany Modern Engineering Thermodynamics*, we find that

$$v = v_f(20.0°\text{C}) = 0.001002\ \text{m}^3/\text{kg}$$

Then, from Eq. (11.30), we have

$$c_p - c_v = \frac{T\beta^2 v}{\kappa} = \frac{(293\ \text{K})(0.207\times10^{-6}\ \text{K}^{-1})^2(0.001002\ \text{m}^3/\text{kg})}{45.9\times10^{-11}\ \text{m}\cdot\text{s}^2/\text{kg}}$$

$$= 2.74\times10^{-5}\ \text{J}/(\text{kg}\cdot\text{K}) = 2.74\times10^{-8}\ \text{kJ}/(\text{kg}\cdot\text{K})$$

In most applications, this difference is clearly negligible, since the value of c_p for liquid water at standard temperature and pressure is 4.18 kJ/(kg·K).

Exercises

28. Show that, for an ideal gas, defined by $pv = RT$, the isobaric coefficient of volume expansion is $\beta = 1/T$ and the isothermal coefficient of compressibility is $\kappa = 1/p$. Then, show that, for an ideal gas, Eq. (11.30) gives $c_p - c_v = R$.
29. Using the methods of Example 11.10, determine the difference between the constant pressure specific heat c_p and the constant volume specific heat c_v for liquid mercury at 20.0°C. For liquid mercury, $v = 7.4 \times 10^{-5}$ m^3/kg. **Answer:** $(c_p - c_v)_{\text{mercury}} = 1.79 \times 10^{-5}$ J/(kg·K).
30. Rework Example 11.10 for liquid benzene at 20°C. For liquid benzene, $v = 1.15 \times 10^{-3}$ m^3/kg. **Answer:** $(c_p - c_v)_{\text{benzene}} = 5.46 \times 10^{-4}$ J/(kg·K).

Now we need to develop a strategy for the integration of Eqs. (11.19), (11.22), (11.24), and (11.26) for arbitrary states. Since u, h, and s are *point* functions, the integration results are independent of the actual integration path, so we should pick a path that is easy to evaluate. In addition, since u and h lack well-defined absolute zero values, the integration path must begin at an arbitrary reference state.

Since all equations of state should reduce to the ideal gas equation of state at low pressures, we can postulate that they must all be reducible to the following form:

$$pv = RT + f(v, T)$$

where the function $f(v, T)$ must be on the order of $1/v$ so that it vanishes as $p \to 0$ and $v \to \infty$. Consequently, the reference state is usually taken to be at some arbitrary reference temperature T_0 and at essentially zero pressure $p_0 = 0$, and zero density or infinite specific volume, $v = \infty$.

To generate a numerical value from Eqs. (11.19), (11.22), (11.24), and (11.26) for u, h, and s, we start the integration process at this reference state. Now, the choice of the easiest integration path from the reference state to the actual state depends on the form of the arguments in the integrals. For example, in evaluating Eq. (11.19) for the specific internal energy, the easiest path to follow is a constant specific volume line for the first integral (since c_v is defined for a constant volume process) and to follow a constant temperature line for the second integral. Since the first integration line is at zero pressure, the constant volume specific heat along this path is c_v^0 (the superscript zero indicates a *zero pressure* constant volume specific heat). Note that the integration path must be only *piecewise* continuous; therefore, it can have "kinks."

Since all of our equations of state must have the form $pv = RT + f(v, T)$,

$$p = \frac{RT}{v} + \frac{f(v,T)}{v}$$

$$\left(\frac{\partial p}{\partial T}\right)_v = \frac{R}{v} + \frac{1}{v}\left(\frac{\partial f}{\partial T}\right)_v$$

and

$$T\left(\frac{\partial p}{\partial T}\right)_v = R\frac{T}{v} + \frac{T}{v}\left(\frac{\partial f}{\partial T}\right)_v$$

then,

$$T\left(\frac{\partial p}{\partial T}\right)_v - p = \frac{RT}{v} + \frac{T}{v}\left(\frac{\partial f}{\partial T}\right)_v - \left(\frac{RT}{v} + \frac{f}{v}\right) = \frac{T}{v}\left(\frac{\partial f}{\partial T}\right)_v - \frac{f}{v}$$

so that Eq. (11.19) now gives

$$u - u_0 = \int_{T_0}^{T} c_v^0 dT + \int_{v_0=\infty}^{v}\left[T\left(\frac{\partial p}{\partial T}\right)_v - p\right]dv = \int_{T_0}^{T} c_v^0 dT + \int_{v_0=\infty}^{v}\left[\frac{T}{v}\left(\frac{\partial f}{\partial T}\right)_v - \frac{f}{v}\right]dv$$

Note that following a path of constant T for the second integral in this equation is very logical since f depends on only v and T and we are integrating over v.

For example, if we had an equation of state of the form $pv = RT + \alpha T^2/v$, then

$$p = \frac{RT}{v} + \frac{\alpha T^2}{v^2}$$

$$\left(\frac{\partial p}{\partial T}\right)_v = \frac{R}{v} + \frac{2\alpha T}{v^2}$$

and

$$T\left(\frac{\partial p}{\partial T}\right)_v = \frac{RT}{v} + \frac{2\alpha T^2}{v^2}$$

then,

$$T\left(\frac{\partial p}{\partial T}\right)_v - p = \frac{RT}{v} + \frac{2\alpha T^2}{v^2} - \left(\frac{RT}{v} + \frac{\alpha T^2}{v^2}\right) = \frac{\alpha T^2}{v^2}$$

Equation (11.19) now gives

$$\begin{aligned} u_1 - u_0 &= \int_{T_0}^{T_1} c_v^0 dT + \int_{v_0=\infty}^{v_1} \left(\frac{\alpha T^2}{v^2}\right) dv \\ &= \int_{T_0}^{T_1} c_v^0 dT + \alpha T^2 \left(-\frac{1}{v}\right)\Big|_{\infty}^{v_1} \\ &= \int_{T_0}^{T_1} c_v^0 dT + \frac{\alpha T_1{}^2}{v_1} \end{aligned}$$

and if the zero pressure specific heat c_v^0 is constant over the temperature range T_0 to T_1, this equation reduces to

$$u_1 - u_0 = c_v^0(T_1 - T_0) - \frac{\alpha T_1^2}{v_1}$$

A similar equation can be easily developed for a second state, and we can then combine them to produce an equation for the change in specific internal energy between these two states for this material as

$$u_2 - u_1 = c_v^0(T_2 - T_1) - \alpha\left(\frac{T_2^2}{v_2} - \frac{T_1^2}{v_1}\right)$$

and the reference state values have completely cancelled out.

11.7 CONSTRUCTING TABLES AND CHARTS

We are now able to use Eqs. (11.19), (11.22), (11.24), and (11.26) to construct thermodynamic tables and charts. The construction of thermodynamic tables and charts like the ones in *Thermodynamic Tables to accompany Modern Engineering Thermodynamics* require, first of all, that a great deal of accurate experimental p, v, T, and c_v (or c_p) data be obtained. These data are reduced to mathematical equations through curve-fitting techniques. The resultant mathematical equations are used to derive equations for u, h, and s using the thermodynamic property relations discussed previously. One of the simplest methods for generating saturation and superheat tables is carried out as follows.

A. The following four data sets must be developed from appropriate experiments:
   Data set 1. Saturation temperature and saturation pressure (T_{sat}, p_{sat}).
   Data set 2. Pressure, specific volume, and temperature in the superheated vapor region and along the saturated vapor curve (p, v, T).
   Data set 3. Saturated liquid specific volume (or density) and saturation temperature (v_f, T_{sat}).
   Data set 4. Low- (or zero) pressure constant volume specific heat, c_v^0 and temperature T in the superheated vapor region and along the saturated vapor curve. The superscript 0 is used to denote the fact that the c_v^0 values are measured at essentially zero pressure.
B. Once these four data sets have been obtained, a mathematical equation is curve fit to each of them to obtain four mathematical equations of the form

$$\text{Curve fit 1: } p_{sat} = p_{sat}(T_{sat}) \tag{11.31a}$$

$$\text{Curve fit 2: } p = p(v, T) \text{ (for superheated and saturated vapor)} \tag{11.31b}$$

$$\text{Curve fit 3: } v_f = v_f(T_{\text{sat}}) \tag{11.31c}$$

$$\text{Curve fit 4: } c_v^0 = c_v^0(T) \text{ (for superheated and saturated vapor)} \tag{11.31d}$$

C. If a very low-pressure reference state (p_0, v_0, T_0) is chosen such that $(c_v)_0 = c_v^0$, then Eqs. (11.19), (3.17), and (11.24) are used to calculate values for u, h, and s relative to this reference state as

$$u = u_0 + \int_{T_0}^{T} c_v^0 dT + \int_{v_0}^{v} \left[T\left(\frac{\partial p}{\partial T}\right)_v - p \right] dv \tag{11.32}$$

$$h = u + pv \tag{3.17}$$

and

$$s = s_0 + \int_{T_0}^{T} \left(\frac{c_v^0}{T}\right) dT + \int_{v_0}^{v} \left(\frac{\partial p}{\partial T}\right)_v dv \tag{11.33}$$

where u_0 and s_0 are the internal energy and entropy values of the reference state. Note that these reference state properties always cancel out in a typical internal energy change ($u_2 - u_1$) or entropy change ($s_2 - s_1$) calculation, so their values can be arbitrarily chosen and need not be made known to the user of the table or chart. Typically u_0 and s_0 are chosen so as to make h_f and s_f zero at the reference temperature T_0, and T_0 is often taken to be the triple point temperature (see, for example, the first row of values for water in Table C.1a), because the triple point is a well-defined and easily reproducible reference state. Therefore, u_0 and s_0 are seldom chosen to be zero themselves. The generation of the tables can now be carried out as follows.

11.7.1 Saturation tables

A temperature entry saturation table can be constructed as follows:

1. A temperature $T = T_{\text{sat}}$ is chosen at which the properties are to be determined.
2. Next, p_{sat} is calculated from Eq. (11.31a).
3. Equation (11.31b), which must be valid for saturated vapor as well as superheated vapor, is used to calculate v_g at these p_{sat} and T_{sat} values.
4. Chose a reference temperature T_o and assign arbitrary values to u_o and s_o.
5. The expression for $(dp/dT)_{\text{sat}}$ is determined by differentiating Eq. (11.31a).[3] The values of u_g, h_g, and s_g are then calculated from Eqs. (11.32), (3.17), and (11.33) by setting $(\partial p/\partial T)_v = (dp/dT)_{\text{sat}}$.
6. Equation (11.31c) is used to calculate $v_{fg} = v_g - v_f$ at the T_{sat} value.
7. The remaining saturated liquid properties are determined from the Clapeyron and Gibbs Eqs. (11.17) and (11.10) as follows:

$$h_f = h_g - h_{fg} = h_g - T_{\text{sat}}(v_{fg})\left(\frac{dp}{dT}\right)_{\text{sat}}$$

$$u_f = u_g - u_{fg} = u_g - (h_{fg} - p_{\text{sat}} v_{fg})$$

and

$$s_f = s_g - s_{fg} = s_g - h_{fg}/T_{\text{sat}}$$

This sequence of operations is repeated for a variety of T_{sat} values, and the compilation of all these results gives a temperature entry saturation table like Table C.1a or C.1b in *Thermodynamic Tables to accompany Modern Engineering Thermodynamics*.

Beginning the calculation sequence with a $p = p_{\text{sat}}$ value and calculating the corresponding T_{sat} value from Eq. (11.31a) and continuing as just described produces a pressure entry saturation table like Table C.2a or C.2b.

---

[3] Note that Eq. (11.31b) should yield the same values of p_{sat} and T_{sat} as Eq (11.31a). However, they both are both empirical equations and may not yield the same values of $(dp/dT)_{\text{sat}}$. In this case, Eq. (11.31a) is preferable.

11.7.2 Superheated vapor tables

Superheated vapor tables are somewhat easier to construct.

1. Begin by choosing a pair of pressure-temperature (p, T) values and calculate the corresponding specific volume v from Eq. (11.31b).
2. Values are then calculated for u, h, and s from Eqs. (11.32), (3.17), and (11.33) utilizing $(\partial p/\partial T)_v$ determined from Eq. (11.31b).
   The compilation of v, u, h, and s for each set of p and T values chosen forms a superheated vapor table like Table C.3a or C.3b.

Many tables do not list both u and h values, since these properties are simply related to each other through $h = u + pv$. Therefore, if only one is listed in a table (it is usually h), the other can be easily calculated.

11.8 THERMODYNAMIC CHARTS

When accurate values for p, T, v, u, h, and s have been determined for the construction of saturated and superheated property tables, it is a relatively simple task to plot these values to form thermodynamic charts. Two-dimensional plots allow only two independent properties to be plotted, and the remaining properties have to be added to the plot as parametric families of lines representing constant property values (isotherms, isobars, etc.). For example, the Mollier diagram (see Figure 7.15a) has h and s as coordinate axes. This means that all the remaining property information must be displayed as families of lines of constant p, constant T, constant v, and so forth.

Because of the large number of variables to choose from and the lack of any standard thermodynamic chart format, the charts found in *Thermodynamic Tables to accompany Modern Engineering Thermodynamics* have many coordinate axes (h-s, T-s, p-h, p-v, v-u, etc.).

Although all thermodynamic tables and charts up to about 1950 were generated from manual calculations, the use of a modern digital computer can substantially reduce the amount of human labor involved. Most of the required software programming is straightforward, simply by following the steps outlined previously. However, one aspect of this process that is not so obvious involves solving Eq. (11.31b) for v when p and T are known. These equations are often so algebraically complex that v cannot be determined explicitly in terms of p and T.

EXAMPLE 11.11

A new substance has the following equations of state corresponding to Eqs. (11.31a) through (11.31d). Here, we just letter the equations (a) through (d) to avoid any confusion. That is, Eq. (11.31a) is just called (a) here.

$$p_{\text{sat}} = \exp\left[A_1 - \frac{A_2}{T_{\text{sat}}}\right] \tag{a}$$

$$p = \frac{RT}{v} - \left(\frac{T}{v^2}\right)\exp\left[B_1 - \frac{B_2}{T}\right] \tag{b}$$

$$v_f = \frac{1}{C_1 + C_2 T_{\text{sat}}} \tag{c}$$

$$c_v^0 = D_1 = \text{constant} \tag{d}$$

where A_1, A_2, B_1, B_2, C_1, C_2, and D_1 are all empirical constants. Determine the equations for u_g, u_f, h_g, h_f, s_g, and s_f for this material in the saturated region.

Solution

For the saturation tables, let $A = A_1 - A_2/T_{\text{sat}}$ and $B = B_1 - B_2/T$, then $p_{\text{sat}} = \exp[A]$ and

$$p = \frac{RT}{v} - \left(\frac{T}{v^2}\right)\exp[B]$$

then,

$$\left(\frac{dp}{dT}\right)_{\text{sat}} = \left(\frac{A_2}{T_{\text{sat}}^2}\right)\exp[A]$$

so that

$$T_{sat}\left(\frac{dp}{dT}\right)_{sat} - p_{sat} = T_{sat}\left(\frac{A_2}{T_{sat}^2}\right)\exp[A] - \exp[A] = \left(\frac{A_2}{T_{sat}} - 1\right)\exp[A]$$

and

$$\int_{v_0}^{v_g}\left[T_{sat}\left(\frac{dp}{dt}\right)_{sat} - p_{sat}\right]dv = \left\{\left(\frac{A_2}{T_{sat}} - 1\right)\exp[A]\right\}(v_g - v_0)$$

Since Eq. (b) must be valid to the saturated vapor line, it can be solved to find v_g. For this problem, Eq. (b) is a quadratic equation in v_g with the following solution:

$$v_g = \frac{RT_{sat}}{2p_{sat}} + \sqrt{\left(\frac{RT_{sat}}{2p_{sat}}\right)^2 - \frac{T_{sat}}{p_{sat}}\exp[B]}$$

From Eq. (c) and the preceding result for v_g, we can now calculate $v_{fg} = v_g - v_f$.

Then, with c_v^0 equal to a constant (D_1), we get

$$u_g - u_0 + \int_{T_0}^{T_{sat}}\left[T\left(\frac{dp}{dT}\right)_{sat} - p\right]dv = u_0 + c_v^0(T_{sat} - T_0) + \left\{\left(\frac{A_2}{T_{sat}} - 1\right)\exp[A]\right\}(v_g - v_0)$$

then, with v_g from the preceding equation, we can easily find $h_g = u_g + p_{sat}v_g$.

To find the saturated vapor entropy, we need to determine

$$\int_{v_0}^{v_g}\left(\frac{dp}{dT}\right)_{sat} dv = \int_{v_0}^{v_g}\left(\frac{A_2}{T_{sat}^2}\right)\exp[A]dv = \left\{\left(\frac{A_2}{T_{sat}^2}\right)\exp[A]\right\}(v_g - v_0)$$

Then, with c_v^0 equal to a constant, we get

$$s_g = s_0 + \int_{T_0}^{T_{sat}}\frac{c_v^0}{T}dT + \int_{v_0}^{v}\left(\frac{dp}{dT}\right)_{sat} dv$$

$$= s_0 + c_v^0\ln\left(\frac{T_{sat}}{T_0}\right) + \left\{\left(\frac{A_2}{T_{sat}^2}\right)\exp[A]\right\}(v_g - v_0)$$

Now, we can determine the remaining saturated liquid properties as

$$h_f = h_g - h_{fg} = h_g - T_{sat}\,v_{fg}\left(\frac{dp}{dT}\right)_{sat} = h_g - v_{fg}\left(\frac{A_2}{T_{sat}}\right)\exp[A]$$

$$u_f = u_g - u_{fg} = u_g - (h_{fg} - p_{sat}\,v_{fg}) = u_g - \left\{v_{fg}\left(\frac{A_2}{T_{sat}}\right)\exp[A] - p_{sat}v_{fg}\right\}$$

and

$$s_f = s_g - s_{fg} = s_g - \frac{h_{fg}}{T_{sat}} = s_g - v_{fg}\exp[A]\left(\frac{A_2}{T_{sat}^2}\right)$$

Inserting the empirical values for the constants A_1 through D_1 into these equations provides the desired set of property relations for this material in the saturated region.

Exercises

31. Determine an expression for $T(\partial p/\partial T)_v - p$ for the material described in Example 11.11 in the superheated region. **Answer:**

$$\left(\frac{\partial p}{\partial T}\right)_v = \frac{R}{v} - \left(\frac{1}{v^2}\right)\exp[B] - \frac{T}{v^2}\left(-\frac{B_2}{T^2}\right)\exp[B] = \frac{R}{v} - \frac{\exp[B]}{v^2}\left(1 - \frac{B_2}{T}\right)$$

so that

$$T\left(\frac{\partial p}{\partial T}\right)_v - p = T\left\{\frac{R}{v} - \frac{\exp[B]}{v^2}\left(1 - \frac{B_2}{T}\right)\right\} - \left(\frac{RT}{v} - \frac{T}{v^2}\exp[B]\right) = \left(\frac{B_2}{v^2}\right)\exp[B]$$

where $B = B_1 - B_2/T$.

32. Determine an expression for the specific internal energy of the material in Example 11.11 in the superheated region. **Answer:**

$$u = u_0 + \int_{T_0}^{T} c_v^0 dT + \int_{v_0}^{v}\left[T\left(\frac{\partial p}{\partial T}\right)_v - p\right]dv = u_0 + c_v^0(T - T_0) - B_2\exp[B]\left(\frac{1}{v} - \frac{1}{v_0}\right)$$

where $B = B_1 - B_2/T$.

(*Continued*)

EXAMPLE 11.11 (*Continued*)

33. Determine an expression for the specific entropy of the material in Example 11.11 in the superheated region. **Answer:**

$$s = s_0 + \int_{T_0}^{T} \frac{c_v^0}{T} dT + \int_{v_0}^{v} \left(\frac{\partial p}{\partial T}\right)_v dv = s_0 + c_v^0 \ln\left(\frac{T}{T_0}\right) + R \ln\left(\frac{v}{v_0}\right) + \exp[B]\left(1 - \frac{B_2}{T}\right)\left(\frac{1}{v} - \frac{1}{v_0}\right)$$

where $B = B_1 - B_2/T$.

11.9 GAS TABLES

We now move from the region of the vapor dome and its immediate surroundings out into the region where the material behaves as an "ideal gas." However, we need to incorporate the temperature dependence of the ideal gas's specific heats.

When the concept of an ideal gas was introduced in Chapter 3, it was noted that ideal gases generally do not have constant specific heats. Therefore, one of the simplest steps we can take to make the ideal gas equations more accurate is to take into account their temperature-dependent specific heats. This is what is done for you in Tables C.16 and C.17 in *Thermodynamic Tables to accompany Modern Engineering Thermodynamics*.

The specific internal energy and enthalpy values listed in the gas tables are determined from an integration of Eqs. (3.37) and (3.40) by incorporating an accurate specific heat vs. temperature data curve fit as

$$u = u_o + \int_{T_o}^{T} c_v \, dT$$

and

$$h = h_o + \int_{T_o}^{T} c_p \, dT$$

where u_o, h_o, and T_o are the arbitrarily chosen reference state values, all of which are set equal to zero in these tables.

Since the ideal gas specific entropy depends on more than just temperature, it is not listed in these tables. However, the temperature-dependent part of the specific entropy is listed as the ϕ function, where

$$\phi = \int_{T_o}^{T} (c_p/T) dT \tag{11.34}$$

and, from Eq. (7.35), we have

$$\begin{aligned} s &= s_o + \int_{T_o}^{T} (c_p/T) dT - R \ln (p/p_o) \\ &= s_o + \phi - R \ln (p/p_o) \end{aligned} \tag{7.35}$$

In the gas tables, s_o and T_o are both arbitrarily set equal to zero, and p_o is set equal to 1 atm. As always, the arbitrarily chosen reference states for u, h, and s cancel out when we calculate the changes $u_2 - u_1$, $h_2 - h_1$, and $s_2 - s_1$. Such changes are calculated directly from values taken from the gas tables for u and h, but the change in entropy is given by combining the previous equations as

$$s_2 - s_1 = \phi_2 - \phi_1 - R \ln (p_2/p_1) \tag{11.35}$$

or an isentropic process, $s_2 = s_1$; Eq. (11.35) then gives

$$(\phi_2 - \phi_1)/R = \ln (p_2/p_1) = \ln \frac{p_2 p_o}{p_1 p_o} = \ln (p_{r2}/p_{r1})$$

where p_r is the *relative pressure,* defined as

$$p_r = p/p_o = \exp(\phi/R)$$

consequently,

$$p_2/p_1 = p_{r2}/p_{r1} \tag{11.36}$$

Equation (7.34) expresses the specific entropy in terms of temperature and specific volume as

$$s_2 - s_1 = \int_{T_1}^{T_2} (c_v/T)dT + R\ln(v_2/v_1) \tag{7.34}$$

Again, for an isentropic process, $s_2 = s_1$ and

$$-\frac{1}{R}\int_{T_1}^{T_2} (c_v/T)dT = \ln(v_2/v_1) = \ln\frac{v_2 v_o}{v_1 v_o} = \ln(v_{r2}/v_{r1})$$

where v_r is the *relative volume,* defined as

$$v_r = v/v_0 = \exp\left[-\frac{1}{R}\int_{T_0}^{T} (c_v/T)dT\right]$$

consequently,

$$v_2/v_1 = v_{r2}/v_{r1} \tag{11.37}$$

The p_r and v_r columns of the gas tables are to be used *only for isentropic processes,* and their values are to be used *only in Eqs. (11.36) and (11.37).*

EXAMPLE 11.12

A diesel engine has a compression ratio of 19.2 to 1. Air at 60.0°F and 14.7 psia is drawn into the engine during the intake stroke and compressed isentropically during the compression stroke. Using the gas tables, determine the final temperature and pressure of the air at the end of the compression stroke and the work required per lbm of air present.

Solution

First, draw a sketch of the system (Figure 11.2).

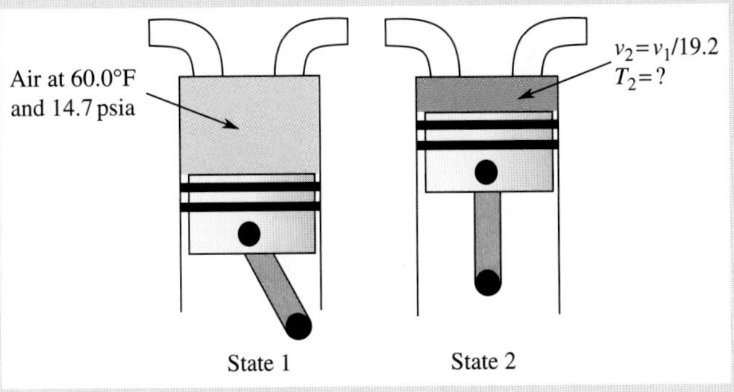

FIGURE 11.2
Example 11.12.

The unknowns here are the final temperature and pressure of the air at the end of the compression stroke and the work required per lbm of air present.

The piston-cylinder arrangement of a diesel engine forms a closed system for the air being compressed. The unknowns are T_2, p_2, and ${}_1W_2/m$. The energy balance for this system (neglecting any changes in the potential and kinetic energies of the air) is

$${}_1Q_2 - {}_1W_2 = m(u_2 - u_1)$$

(Continued)

EXAMPLE 11.12 (*Continued*)

Now, ${}_1Q_2 = 0$ (isentropic processes are also adiabatic), so

$${}_1W_2/m = u_1 - u_2$$

The gas tables are to be used for the thermodynamic properties of air here, because they are more accurate than the standard constant specific heat ideal gas equations. From Table C.16a in *Thermodynamic Tables to accompany Modern Engineering Thermodynamics*, we find that, at 60.0°F = 520. R,

$$u_1 = 88.62\ \text{Btu/lbm}$$

$$p_{r1} = 1.2147$$

and

$$v_{r1} = 158.58$$

For a compression ratio of 19.2 to 1, $v_2/v_1 = 1/19.2$. Then, from Eq. (11.37),

$$v_{r2} = v_{r1}(v_2/v_1) = 158.58/19.2 = 8.26$$

scanning down the v_r column in Table C.16a, we find that $v_r = 8.26$ at about

$$T_2 = 1600\ \text{R} = 1140°\text{F}$$

$$u_2 = 286.06\ \text{Btu/lbm}$$

and

$$p_{r2} = 71.73$$

Then, from Eq. (11.36),

$$p_2 = p_1(p_{r2}/p_{r1}) = (14.7\ \text{psia})(71.73/1.2147)$$
$$= 868.1\ \text{psia}$$

Finally, from the preceding energy balance,

$${}_1W_2/m = u_1 - u_2 = 88.62 - 286.06 = -197.44\ \text{Btu/lbm}$$

Exercises

34. Determine the final temperature and pressure in Example 11.12 if the compression ratio is 10.7 to 1 instead of 19.2 to 1. **Answer:** T_2 = 1300. R and p_2 = 392. psia.

35. Rework Example 11.12 for a compression ratio of 19.3 to 1 when air at 300. K and 0.100 MPa is drawn into the engine during the intake stroke. **Answer:** T_2 = 920. K, p_2 = 5.92 MPa, and ${}_1W_2/m = -477$ kJ/kg.

36. Use the relations for an isentropic process for an ideal gas,

$$\frac{T_2}{T_1} = \left(\frac{v_2}{v_1}\right)^{1-k} = \left(\frac{p_2}{p_1}\right)^{\frac{k-1}{k}}$$

to compute the final temperature T_2 and pressure p_2 of the compression process described in Example 11.12. Then, assuming constant specific heats, compute the work required per unit mass for the compression ${}_1W_2/m = u_1 - u_2 = c_v(T_1 - T_2)$. **Answer:** T_2 = 1236°F, p_2 = 920 psia, and ${}_1W_2/m = -200.$ Btu/lbm.

11.10 COMPRESSIBILITY FACTOR AND GENERALIZED CHARTS

In Chapter 3, we discussed the van der Waals, Dieterici, Berthelot, Beattie-Bridgeman, and Redlich-Kwong equations of state as possible models of nonideal, or real, gas behavior. Several of these equations are just variations on the basic ideal gas equation, $pv = RT$. We now introduce one of the most powerful engineering real gas equations of state: the *compressibility factor*.

In 1880, the Dutch physicist Johannes Diderik van der Waals (1837–1923) reasoned that, if the p-v-T equation of state could be nondimensionalized, then all gases might be found to fit the same dimensionless p-v-T equation. Further, he noted that, since every substance has a vapor dome and every vapor dome has a unique critical point (its peak), perhaps the critical point properties (p_c, T_c, and v_c) could be used to create a dimensionless equation of state. He defined the dimensionless variables *reduced pressure* p_R, *reduced temperature* T_R, and *reduced specific volume* v_r as[4]

$$p_R = p/p_c \qquad (11.38a)$$

$$T_R = T/T_c \qquad (11.38b)$$

$$v_R = v/v_c \qquad (11.38c)$$

where p and T are the *actual* pressure and temperature of the gas, and p_c and T_c are the *critical state* pressure and temperature of the gas (see Table C.12 in *Thermodynamic Tables to accompany Modern Engineering Thermodynamics*). Then, he hypothesized that $p_R = p_R(v_R, T_R)$ would define a generalized dimensionless equation of state that would be valid for all substances. Today, this hypothesis is called van der Waals's *law of corresponding states*, but unfortunately, it has been found to be valid only for materials with similar molecular structures.

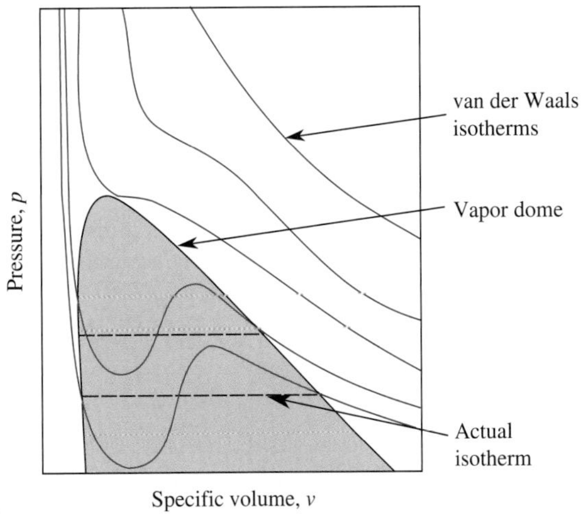

FIGURE 11.3
Schematic of van der Waals isotherms near the vapor dome.

In 1883, van der Waals introduced his now classical equation of state:

$$p = RT/(v - b) - a/v^2 \qquad (3.44)$$

in which the constants a and b are corrections for intermolecular forces and molecular volume, respectively. This is a cubic equation in v (see Example 11.11) and has isotherms shaped as shown in Figure 11.3. For given p and $T < T_c$ values, there are three real roots of this equation. One root corresponds to v_f, another corresponds to v_g, and the third is a meaningless root between v_f and v_g.

Van der Waals noted that the critical temperature isotherm seemed to have an inflection point at the critical point. If this were generally true, then all equations of state would have to obey the mathematical constraints of an inflection point, or

$$\left(\frac{\partial p}{\partial v}\right)\bigg|_{T_c, v_c} = \left(\frac{\partial^2 p}{\partial v^2}\right)\bigg|_{T_c, v_c} = 0$$

Applying these conditions to the van der Waals equation yields

$$\left(\frac{\partial p}{\partial v}\right)\bigg|_{T_c, v_c} = RT_c/(v_c - b)^2 + 2a/v_c^3 = 0$$

and

$$\left(\frac{\partial^2 p}{\partial v^2}\right)\bigg|_{T_c, v_c} = 2RT_c/(v_c - b)^3 - 6a/v_c^4 = 0$$

and solving these two equations simultaneously for a and b while using Eq. (3.44) evaluated at the critical point gives

$$a = 9RT_c v_c/8 = \frac{27}{64}\left(\frac{R^2 T_c^2}{p_c}\right)$$

and

$$b = v_c/3 = \frac{RT_c}{8p_c}$$

Thus, if the van der Waals equation accurately represented universal material behavior, the constants a and b could be determined from a single experimental measurement at the critical point.

---

[4] Note that the *reduced* properties p_R and v_R are not the same as the *relative* properties p_r and v_r introduced in the previous section.

Substituting this formula for a and b back into the van der Waals equation and dividing both sides by p_c gives

$$p/p_c = p_R = 8T_R/(3v_R - 1) - 3/v_R^2$$

which is the type of dimensionless equation of state that he was seeking. Unfortunately, the molecular interactions in real substances are more complex than the elementary corrections used in the van der Waals equation, so it does not work well over large pressure and temperature ranges.

A serious flaw in van der Waals's law of corresponding states is that it breaks down at low pressures, where ideal gas behavior is expected. Further, his equation of state (Eq. (3.44)) has the curious property that (from the preceding equation for b) the ratio $p_c v_c/(RT_c) = 3/8 = 0.375$, whereas for an ideal gas, $pv/(RT)$ is always equal to unity. This similarity led many researchers to investigate $pv/(RT)$ data, and this grouping is now called the *compressibility factor Z*, as

$$Z = pv/(RT) \tag{11.39}$$

where $Z = 1$ for an ideal gas. Figure 11.4 shows that the experimental data for the compressibility factor for many different gases fall together when Z is plotted against *reduced pressure* p_R while holding the *reduced temperature* T_R constant. Figures 11.4, 11.5, 11.6, and 11.7 constitute a set of compressibility charts that can be used to solve real gas compressibility factor problems.

The van der Waals equation predicts that $Z_c = p_c v_c/(RT_c) = 0.375$, but many experiments on a large number of substances have shown that $0.23 \leq Z_c \leq 0.375$. So Z_c is not the same for all substances. Though many people tried to correlate Z with p or T for various substances, it was not until 1939 that H. C. Weber correlated Z with p_R and T_R and thus produced the first generalized compressibility chart of the form $Z = Z(p_R, T_R)$. There was, however, a problem with this chart, in that lines of constant reduced specific volume could not be added because the v_R data were inconsistent. In 1946, Gouq-Jen Su solved this problem, as shown in Figure 11.5, by choosing the product $v_R Z_C$ as a "pseudo" reduced specific volume v'_R, defined as

$$v'_R = v_R Z_c = (v/v_c)Z_c = v/v'_c$$

where a new critical state specific volume v'_c has been defined as

$$v'_c = v_c/Z_c = RT_c/p_c$$

This change produced a much better correlation of the experimental data and lines of constant v'_R could be accurately added to the chart. The resulting $Z = Z(p_R, T_R, v'_R)$ plot is now called the *generalized compressibility chart* and is shown in Figures 11.5, 11.6, and 11.7. Values for p_c and T_c for various substances can be found in Table C.12.

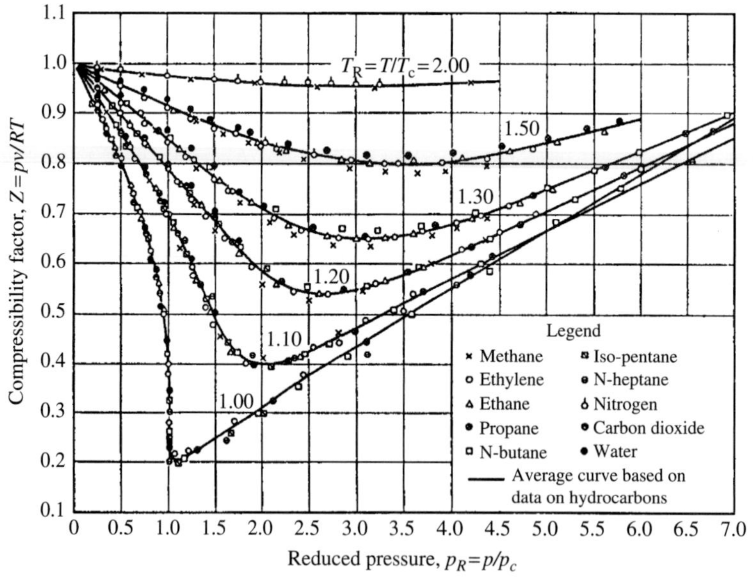

FIGURE 11.4

A generalized compressibility chart for various gases. (*Sources: Reprinted with permission from Su, G.-J. "Modified law of corresponding states for real gases." Ind. Eng. Chem. 38 (8), 1948, p. 804. Also reprinted with permission of the publisher from Reynolds, W. C., Perkins, H. C., 1977. Engineering Thermodynamics. McGraw-Hill, New York.*)

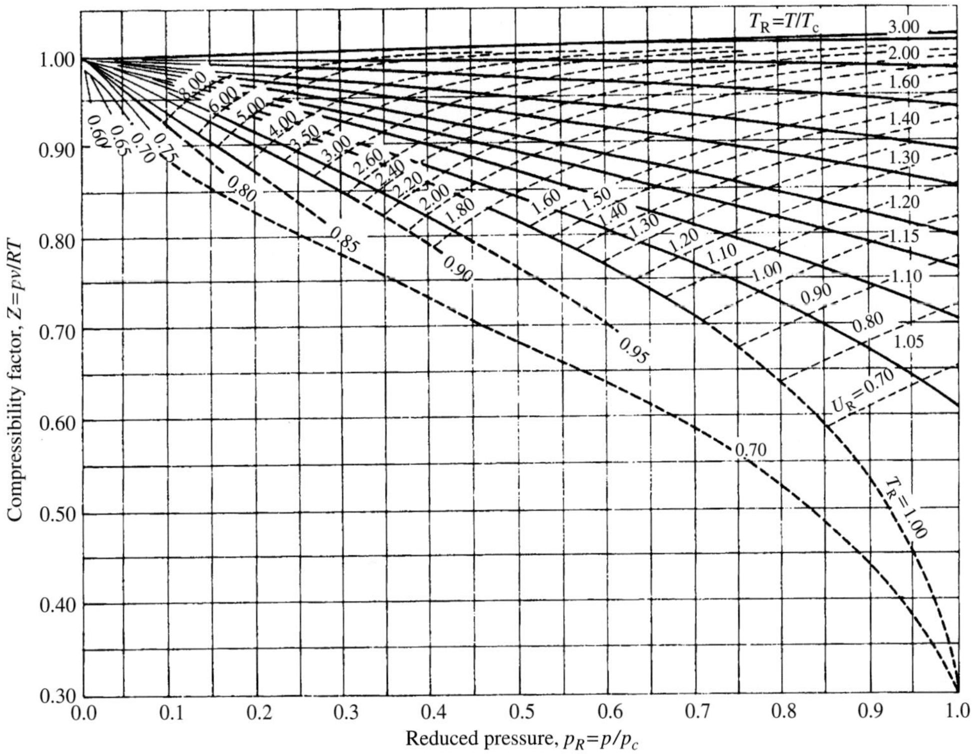

FIGURE 11.5

The generalized (Nelson-Obert) compressibility chart—low-pressure range, $0 \leq p_R \leq 1.0$. Note that $v'_R = v/v'_c = vp_c/RT_c$. (*Source: "Nelson-Obert Compressibility Charts," adapted courtesy of Professor E. F. Obert, University of Wisconsin, Madison, from Obert, E. F., 1960. Concepts of Thermodynamics. McGraw-Hill, New York.*)

Though the van der Waals equation is not very accurate as a universal equation of state, Su's modified compressibility factor formulation as an approximate approach to the law of corresponding states has found universal acceptance within the engineering community.

EXAMPLE 11.13

Using the compressibility charts, find the pressure exerted by 8.20 lbm of carbon monoxide in a 1.00 ft³ rigid tank at –78.0°F.

Solution

From Table C.12a, we find that

$$T_c = 240.\ \text{R}$$

$$p_c = 507\ \text{psia}$$

and

$$v_c = \frac{1.49\ \text{ft}^3/\text{lbmole}}{28.011\ \text{lbm/lbmole}} = 0.053\ \text{ft}^3/\text{lbm}$$

Also, from Table C.13a, we find that $R = 0.0709$ Btu/lbm·R. Then, we have

$$T_R = \frac{T}{T_c} = \frac{-78.0 + 460.}{240.} = 1.60$$

and

$$v = \frac{1.00\ \text{ft}^3}{8.20\ \text{lbm}} = 0.122\ \text{ft}^3/\text{lbm}$$

(Continued)

EXAMPLE 11.13 *(Continued)*

with

$$v_c' = \frac{RT_c}{p_c} = \frac{[0.0709\ \text{Btu/(lbm}\cdot\text{R)}](240.\ \text{R})(778.16\ \text{ft}\cdot\text{lbf/Btu})}{(507\ \text{lbf/in}^2)(144\ \text{in}^2/\text{ft}^2)} = 0.181\ \text{ft}^3/\text{lbm}$$

so that

$$v_R' = \frac{v}{v_c'} = \frac{0.122}{0.181} = 0.67\ \text{(notice that we do } \textit{not} \text{ use the actual critical specific volume } v_c \text{ here)}$$

Using $T_R = T/T_c = 1.60$ and $v_R' = v/v_c' = 0.67$, we find from Figure 11.6 that

$$p_R = \frac{p}{p_c} = 2.10 \quad \text{and} \quad Z = 0.850$$

Then, we can calculate

$$p = p_c p_R = 507(2.10) = 1070\ \text{psia}$$

Exercises

37. Determine the pressure exerted by the carbon monoxide in Example 11.13 if it is at a temperature of 100.°F and all the other variables remain unchanged. **Answer**: p = 1670 psia.

38. Suppose the tank containing the carbon monoxide in Example 11.13 was isothermally crushed by a giant winged wombat, causing the pressure in the tank to increase to 2000. psia. Determine the final volume of the tank. **Answer**: $\forall_{\text{final}} = 0.516\ \text{ft}^3$.

39. Management has just informed us that the tank in Example 11.13 really contains carbon dioxide rather than carbon monoxide and that the tank was really at 178°F, not −78.0°F. Assuming all the remaining variables are unchanged, determine the pressure of the CO_2 in the tank. **Answer**: p = 964 psia.

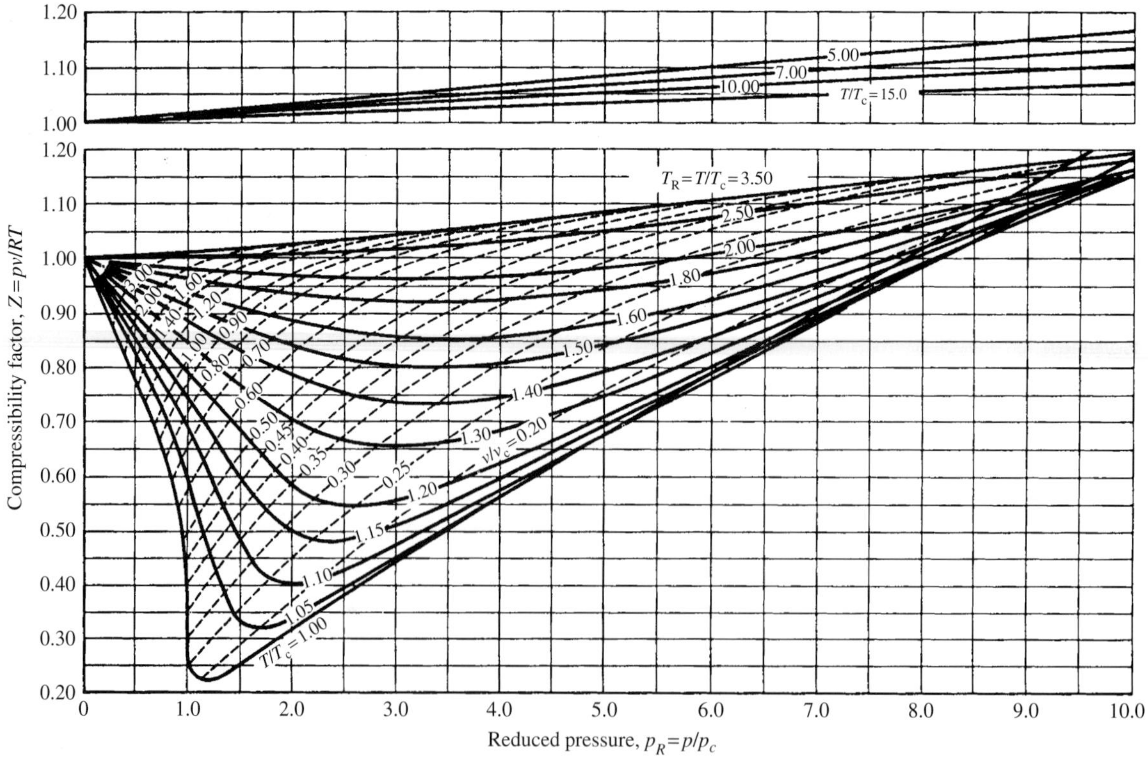

FIGURE 11.6

The generalized (Nelson-Obert) compressibility chart—low-pressure range, $0 \le p_R \le 10.0$. Note that $v_R' = v/v_c' = vp_c/RT_c$. *(Source: "Nelson-Obert Compressibility Charts," reprinted courtesy of Professor E. F. Obert, University of Wisconsin, Madison, from Obert, E. F., 1948. Thermodynamics. McGraw-Hill, New York.)*

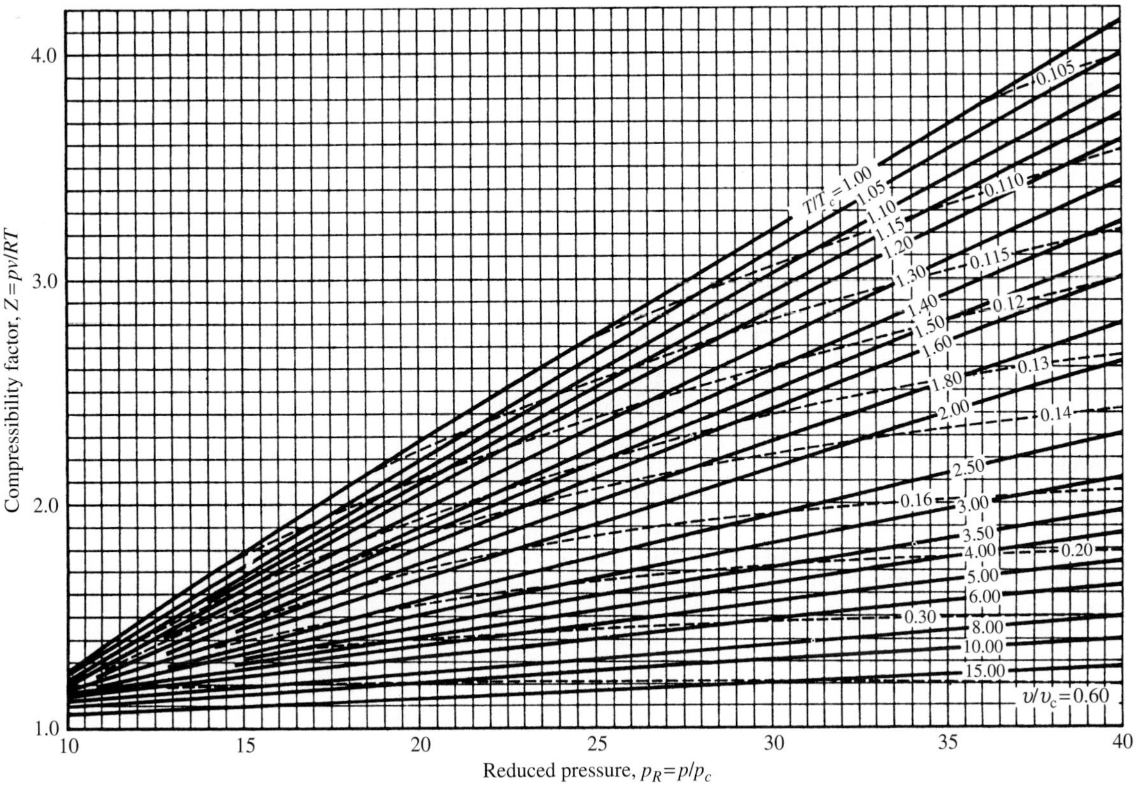

FIGURE 11.7
The generalized (Nelson-Obert) compressibility chart—high-pressure range. Note that $v'_R = v/v'_c = vp_c/RT_c$. (*Source: "Nelson-Obert Compressibility Charts," adapted courtesy of Professor E. F. Obert, University of Wisconsin, Madison, from Obert, E. F., 1960. Concepts of Thermodynamics. McGraw-Hill, New York.*)

EXAMPLE 11.14

Compressed natural gas (CNG) is essentially methane (CH_4). CNG is currently being used as a replacement fuel for gasoline in some automobiles. This requires replacing the automobile's gasoline tank with a high-pressure 0.100 m^3 cylinder filled with CNG. Under normal conditions, the tank pressure is no more than 20.0 MPa when the tank is filled with a maximum of 15.6 kg of CNG. However, the worst case condition would be if the automobile were consumed by fire and the tank temperature reached 1000.°C. Using the compressibility charts, determine the maximum pressure in the CNG tank at this worst case temperature.

Solution

First, draw a sketch of the system (Figure 11.8).

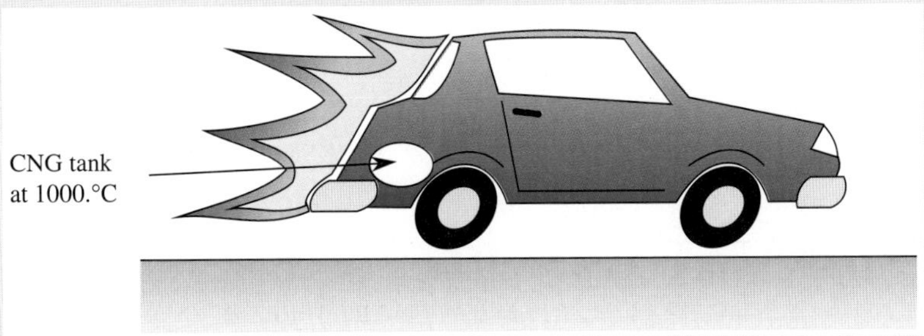

FIGURE 11.8
Example 11.14

The unknown is the maximum pressure in the CNG tank. From Table C.12b, we find the critical state properties of methane to be

$$T_c = 191.1\text{ K} \quad \text{and} \quad p_c = 4.64\text{ MPa}$$

(Continued)

EXAMPLE 11.14 (*Continued*)

Since this is a closed fixed volume system,

$$v_1 = v_2 = v = \mathcal{V}/m = 0.100\,\text{m}^3/15.6\,\text{kg} = 6.40 \times 10^{-3}\text{m}^3/\text{kg}$$

Table C.13b, gives the gas constant for methane as $R = 0.518$ kJ/kg·K, then

$$v'_R = v/v'_c = vp_c/RT_c = (6.40 \times 10^{-3}\text{m}^3/\text{kg})(4640\ \text{kPa})/[(0.518\,\text{kJ/kg}\cdot\text{K})(191.1\,\text{K})] = 0.300$$

and

$$T_R = T/T_c = (1000. + 273.15)/191.1 = 6.66$$

Using these values of v'_R and T_R in Figure 11.7, we find that $p_R = p_2/p_c \approx 32.0$, and the worst case pressure is $(p_2)_{\text{worst case}} = p_R p_c = 32.0(4.64) = 148$ MPa.

Exercises

40. Determine the final (worst case) pressure in the CNG tank in Example 11.14 if the tank temperature reached only 500.°C. **Answer:** $(p_2)_{\text{worst case}} = 90.$ MPa.
41. If we decrease the size of the CNG tank in Example 11.14 from 0.100 m^3 to 0.0500 m^3 and decrease its temperature in the fire from 1000.°C to 200.°C, determine the final (worst case) pressure in the tank. **Answer:** $(p_2)_{\text{worst case}} = 8.00$ MPa.

Compressibility factor data can be used to estimate the specific enthalpy and entropy pressure dependence of substances as follows. Integrating the specific enthalpy total differential given in Eq. (11.22) from a reference state at p_o, v_o, T_o, and h_o to any other state at p, v, T, and h gives

$$h = h_o + \int_{T_o}^{T} c_p dT + \int_{p_o}^{p}\left[v - T\left(\frac{\partial v}{\partial T}\right)_p\right]dp$$

Then, arbitrarily choosing h_o, T_o, and p_o to be zero gives

$$h = \int_0^T c_p dT + \int_0^p\left[v - T\left(\frac{\partial v}{\partial T}\right)_p\right]dp$$

$$= h^* + \int_0^p\left[v - T\left(\frac{\partial v}{\partial T}\right)_p\right]dp$$

where h^* is the ideal gas specific enthalpy defined earlier in the discussion of the gas tables. From Eq. (11.39), we can write

$$v = ZRT/p$$

so that

$$\left(\frac{\partial v}{\partial T}\right)_p = ZR/p + (RT/p)(\partial Z/\partial T)_p \tag{11.40}$$

Then,

$$v - T\left(\frac{\partial v}{\partial T}\right)_p = ZRT/p - ZRT/p - (RT^2/p)(\partial Z/\partial T)_p$$

$$= -(RT^2/p)(\partial Z/\partial T)_p$$

and

$$h = h^* - R\int_0^p (T^2/p)(\partial Z/\partial T)_p\,dp$$

Nondimensionalizing this equation with $T = T_c T_R$ and $p = p_c p_R$ and rearranging it gives

$$(h^* - h)/T_c = R\int_0^{p_R} (T_R^2/p_R)(\partial Z/\partial T_R)_{p_R} dp_R$$

This equation still depends on the substance under consideration, because the value of R is substance dependent. This dependence can be removed by multiplying both sides of this equation by the molecular mass M (in lbm/lbmoles or kg/kgmole) and thus converting it into molar units:

$$(\bar{h}^* - \bar{h})/T_c = \Re\int_0^{p_R} (T_R^2/p_R)(\partial Z/\partial T_R)_{p_R} dp_R$$

where $\bar{h}^* = h^* M$ is the ideal gas molar enthalpy, $\bar{h} = hM$ is the real substance molar enthalpy, and $\Re = RM$ is the universal gas constant. Using compressibility factor data, this equation has been integrated and the results are shown in Figure 11.9.

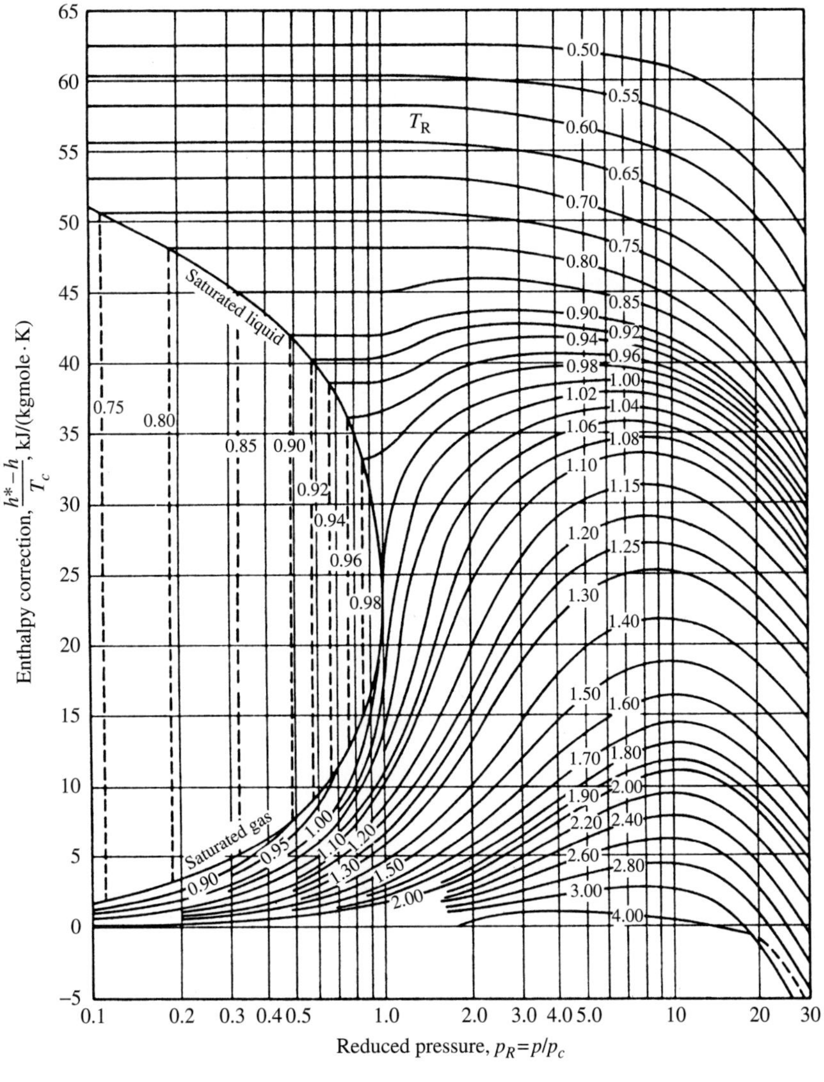

FIGURE 11.9

Generalized chart for enthalpy correction. Note: To convert this figure to English units, use the factor 1 Btu/(lbmole · R) = 4.1865 kJ/(kgmole · K). (*Source: Van Wylen, G. J., Sonntag, R. E., 1973. Fundamentals of Classical Thermodynamics. Wiley, New York. Copyright © John Wiley & Sons. Reprinted by permission of John Wiley & Sons.*)

To find the change in specific enthalpy between states 1 and 2 using this figure, values of p_R and T_R are calculated and values of $(\overline{h}^* - \overline{h})/T_c$ are then read from the figure. The change in specific enthalpy between states 1 and 2 is then determined from

$$h_2 - h_1 = (h_2^* - h_1^*) - \left[\left(\frac{\overline{h}^* - \overline{h}}{T_c}\right)_2 - \left(\frac{\overline{h}^* - \overline{h}}{T_c}\right)_1\right]\left(\frac{T_c}{M}\right) \tag{11.41}$$

where $h_2^* - h_1^*$ is determined from the gas tables (Table C.16 in *Thermodynamic Tables to accompany Modern Engineering Thermodynamics*) or by assuming constant specific heats over the temperature range from T_1 to T_2 and using $h_2^* - h_1^* = c_p(T_2 - T_1)$.

EXAMPLE 11.15

As chief engineer at Precision Throttles Inc., you are responsible for designing an adiabatic, aergonic, steady state, steady flow device to throttle carbon dioxide (CO_2) from 20.0 MPa, 150.°C to 0.101 MPa. As part of this design, you must accurately predict the exit temperature of the throttle. Neglecting all kinetic and potential energies, use the generalized charts under the assumption of constant specific heats to determine the exit temperature of the throttle.

Solution

First, draw a sketch of the system (Figure 11.10).

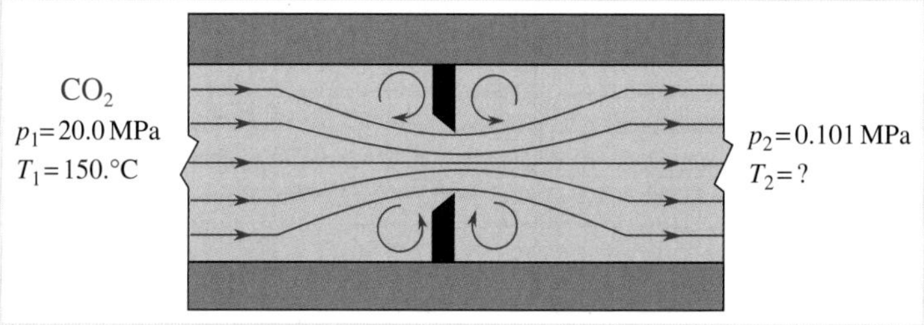

FIGURE 11.10
Example 11.15.

The unknown is the exit temperature of the throttle. The energy rate balance for an adiabatic, aergonic, steady state, steady flow, single-inlet, single-outlet throttle is

$$\dot{Q} - \dot{W} + \dot{m}(h_1 - h_2) = 0$$

and since $\dot{Q} = \dot{W} = 0$ here, we get $h_2 - h_1 = 0$. From Table C.12b, we find the critical temperature and pressure for CO_2 are

$$T_c = 304.2\text{ K} \quad \text{and} \quad p_c = 7.39\text{ MPa}$$

and that the molecular mass of CO_2 is 44.01 kg/kgmole. From Table C.13b, we now find the value of the constant pressure specific heat of CO_2 as

$$c_p = 0.845\text{ kJ/kg}\cdot\text{K}$$

Then,

$$p_{R1} = 20/7.39 = 2.71$$

and

$$T_{R1} = (150. + 273.15\text{ K})/(304.2\text{ K}) = (423.15\text{ K})/(304.2\text{ K}) = 1.39.$$

Then, Figure 11.9 gives $[(h^* - h)/T_c]_1 \approx 14.0$ kJ/kgmole·K. At the exit of the throttle, we have

$$p_{R2} = (0.101\text{ MPa})/(7.39\text{ MPa}) = 0.0135$$

and Figure 11.9 gives $[(h^* - h)/T_c]_2 \approx 0$. Then, using Eq. (11.41), we have

$$h_2 - h_1 = 0$$

$$= \left(0.845\frac{\text{kJ}}{\text{kg}\cdot\text{K}}\right)(T_2 - 423.15\text{ K}) - \left(0 - 14.0\frac{\text{kJ}}{\text{kgmole}\cdot\text{K}}\right)\left(\frac{304.2\text{ K}}{44.01\text{ kg/kgmole}}\right)$$

Solving for T_2 gives

$$T_2 = 423.15\,\text{K} - \frac{14.0\,\text{kJ/kgmole}\cdot\text{K}}{0.845\,\text{kJ/kg}\cdot\text{K}}\left(\frac{304.2\,\text{K}}{44.01\,\text{kg/kgmole}}\right) = 308.6\,\text{K} = 35.5°\text{C}$$

Exercises

42. Determine the exit temperature in the throttle described in Example 11.15 if the inlet pressure is reduced from 20.0 MPa to 10.0 MPa. **Answer**: $T_2 = 117°C$.

43. If the temperature of the CO_2 entering the throttle in Example 11.15 is reduced from 150.°C to 100.°C, determine the exit temperature of the CO_2. **Answer**: $T_2 = -90.0°C$.

44. Suppose the CO_2 in Example 11.15 is replaced with air using the same inlet and exit conditions. Determine the exit temperature of the air from the throttle. **Answer**: $T_{2\text{-air}} = 141°C$.

Similarly, integration of Eq. (11.26) between a zero value reference state ($p_o = T_o = s_o = 0$) and a final state at p, v, T, and s gives

$$s = \int_0^T (c_p/T)dT - \int_0^p (\partial v/\partial T)_p dp$$

and, for an ideal gas, we have, from Eq. (7.35),

$$s^* = \int_0^T (c_p/T)dT - R\int_0^p dp/p$$

Combining these results, we get

$$s = s^* + \int_0^p \left[R/p - (\partial v/\partial T)_p\right]dp$$

Now, using Eq. (11.40),

$$\begin{aligned} R/p - (\partial v/\partial T)_p &= R/p - ZR/p - (RT/p)(\partial Z/\partial T)_p \\ &= (R/p)\left[1 - Z - T(\partial Z/\partial T)_p\right] \end{aligned}$$

Then,

$$s^* - s = R\int_0^p \left[T(\partial Z/\partial T)_p + Z - 1\right](dp/p)$$

Again, nondimensionalizing with $T = T_c T_R$ and $p = p_c p_R$ and multiplying both sides by the molecular mass M to remove substance dependence from the equation, we get the molar results:

$$\bar{s}^* - \bar{s} = \mathfrak{R}\int_0^{p_R} \left[T_R(\partial Z/\partial T_R)_{p_R} + Z - 1\right](dp_R/p_R)$$

Compressibility data have been used to integrate this equation, and the results are shown in Figure 11.11. To find the change in specific entropy between states 1 and 2 using this figure, calculate the values of p_R and T_R and obtain values for $\bar{s}^* - \bar{s}$ from the figure. Then, we can compute

$$s_2 - s_1 = (s_2^* - s_1^*) - \left[(\bar{s}^* - \bar{s})_2 - (\bar{s}^* - \bar{s})_1\right](1/M) \tag{11.42}$$

where $s_2^* - s_1^* = \phi_2 - \phi_1 - R\ln(p_2/p_1)$ from the gas tables, or by assuming constant specific heats over the temperature range from T_1 to T_2 and using $s_2^* - s_1^* = c_p \ln(T_2/T_1) - R\ln(p_2/p_1)$.

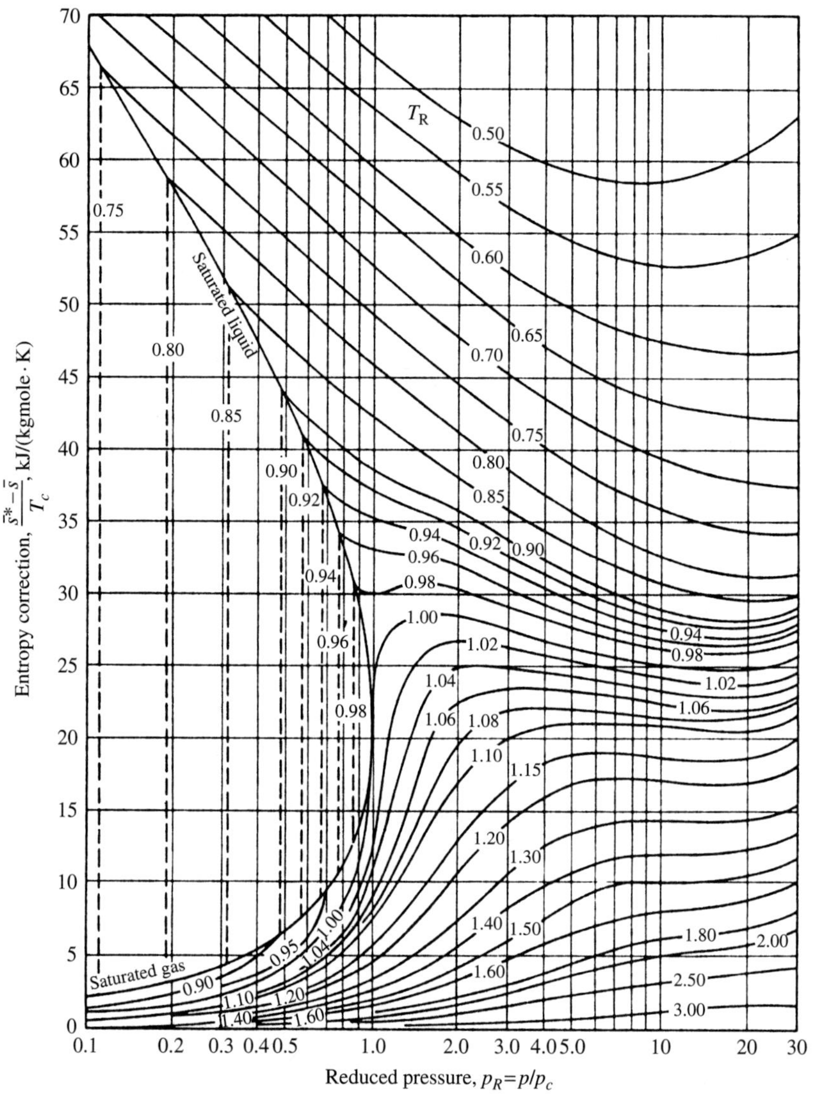

FIGURE 11.11

Generalized chart for the entropy correction. Note: To convert this figure to English units use the factor 1 Btu/(lbmole·R = 4.1865 kJ/(kgmole·K). *(Source: Van Wylen, G. J., Sonntag, R. E., 1973. Fundamentals of Classical Thermodynamic. Wiley, New York. Copyright © John Wiley & Sons. Reprinted by permission of John Wiley & Sons.)*

EXAMPLE 11.16

Ethylene (C_2H_4) gas is to be isothermally compressed from 150. psia to 15.0 × 10^3 psia at 80.0°F. Using the compressibility charts, determine

a. The change in specific enthalpy.
b. The change in specific internal energy.
c. The change in specific entropy of the ethylene.

Solution

a. The change in specific enthalpy of the ethylene is given by Eq. (11.41) as

$$h_2 - h_1 = (h_2^* - h_1^*) - \left[\left(\frac{\bar{h}^* - \bar{h}}{T_c}\right)_2 - \left(\frac{\bar{h}^* - \bar{h}}{T_c}\right)_1\right]\left(\frac{T_c}{M}\right)$$

and since this is an isothermal process, $T_2 = T_1$ and the ideal gas portion of this equation for a constant specific heat gas is $h_2^* - h_1^* = c_p(T_2 - T_1) = 0$. The properties of ethylene at its critical state and its molecular mass are found in Table C.12a as $T_c = 508.3$ R, $p_c = 742$ psia, and $M = 28.05$ lbm/lbmoles. Then,

$$p_{R1} = \frac{150.}{742} = 0.202 \quad \text{and} \quad T_{R1} = \frac{80.0 + 459.67}{508.3} = 1.06$$

$$p_{R2} = \frac{15.0 \times 10^3}{742} = 20.2 \quad \text{and} \quad T_{R2} = T_{R1} = 1.06$$

Using $p_{R1} = 0.202$ and $T_{R1} = 1.06$, Figure 11.9 gives the enthalpy correction for state 1 as

$$\left(\frac{\bar{h}^* - \bar{h}}{T_c}\right)_1 = 1.50 \frac{\text{kJ}}{\text{kgmole} \cdot \text{K}} = \left(1.50 \frac{\text{kJ}}{\text{kgmole} \cdot \text{K}}\right)\left(\frac{1\ \text{Btu}/(\text{lbmole} \cdot \text{R})}{4.1865\ \text{kJ}/(\text{kgmole} \cdot \text{K})}\right) = 0.360 \frac{\text{Btu}}{\text{lbmole} \cdot \text{R}}$$

and using $p_{R2} = 20.2$ and $T_{R2} = 1.06$, Figure 11.9 gives the enthalpy correction for state 2 as

$$\left(\frac{\bar{h}^* - \bar{h}}{T_c}\right)_2 = 31.5 \frac{\text{kJ}}{\text{kgmole} \cdot \text{K}} = \left(31.5 \frac{\text{kJ}}{\text{kgmole} \cdot \text{K}}\right)\left(\frac{1\ \text{Btu}/(\text{lbmole} \cdot \text{R})}{4.1865\ \text{kJ}/(\text{kgmole} \cdot \text{K})}\right) = 7.52 \frac{\text{Btu}}{\text{lbmole} \cdot \text{R}}$$

Then, Eq. (11.41) gives

$$\begin{aligned} h_2 - h_1 &= (h_2^* - h_1^*) - \left[\left(\frac{\bar{h}^* - \bar{h}}{T_c}\right)_2 - \left(\frac{\bar{h}^* - \bar{h}}{T_c}\right)_1\right]\left(\frac{T_c}{M}\right) \\ &= 0 - [7.52 - 0.360\ \text{Btu}/(\text{lbm} \cdot \text{R})]\left(\frac{508.3\ \text{R}}{28.05\ \text{lbm/lbmole}}\right) \\ &= -130. \frac{\text{Btu}}{\text{lbm}} \end{aligned}$$

Note that, since the compressibility charts cannot be read to more than two or three significant figures, our final calculation results are limited to this accuracy as well.

b. The compressibility charts do not give values for the specific internal energy, so it must be calculated from the definition of enthalpy as $u = h - pv$, or $u_2 - u_1 = h_2 - h_1 - (p_2v_2 - p_1v_1)$, where $v_1 = Z_1RT_1/p_1$ and $v_2 = Z_2RT_2/p_2$. For $p_{R1} = 0.202$ and $T_{R1} = 1.06$, Figure 11.5 gives $Z_1 = 0.940$, and for $p_{R2} = 20.2$ and $T_{R2} = T_{R1} = 1.06$, Figure 11.7 gives $Z_2 = 2.15$. The gas constant R for ethylene can be found in Table C.13a as $R = 55.1$ ft·lbf/(lbm·R). Then,

$$v_1 = \frac{Z_1RT_1}{p_1} = \frac{0.940[55.1\ \text{ft} \cdot \text{lbf}/(\text{lbm} \cdot \text{R})](80.0 + 459.67\ \text{R})}{(150.\ \text{lbf/in}^2)(144\ \text{in}^2/\text{ft}^2)} = 1.29 \frac{\text{ft}^3}{\text{lbm}}$$

and

$$v_2 = \frac{Z_2RT_2}{p_2} = \frac{2.15[55.1\ \text{ft} \cdot \text{lbf}/(\text{lbm} \cdot \text{R})](80 + 459.67\ \text{R})}{(15.0 \times 10^3\ \text{lbf/in}^2)(144\ \text{in}^2/\text{ft}^2)} = 0.030 \frac{\text{ft}^3}{\text{lbm}}$$

Then,

$$\begin{aligned} u_2 - u_1 &= h_2 - h_1 - (p_2v_2 - p_1v_1) \\ &= 130. \frac{\text{Btu}}{\text{lbm}} - \left(15.0 \times 10^3 \frac{\text{lbf}}{\text{in}^2}\right)\left(144 \frac{\text{in}^2}{\text{ft}^2}\right)\left(0.0300 \frac{\text{ft}^3}{\text{lbm}}\right)\left(\frac{1\ \text{Btu}}{778.16\ \text{ft} \cdot \text{lbf}}\right) \\ &\quad - \left(150. \frac{\text{lbf}}{\text{in}^2}\right)\left(144 \frac{\text{in}^2}{\text{ft}^2}\right)\left(1.29 \frac{\text{ft}^3}{\text{lbm}}\right)\left(\frac{1\ \text{Btu}}{778.16\ \text{ft} \cdot \text{lbf}}\right) \\ &= -180. \frac{\text{Btu}}{\text{lbm}} \end{aligned}$$

c. Finally, from Eq. (11.42), we have the change in specific entropy as

$$s_2 - s_1 = (s_2^* - s_1^*) - [(\bar{s}^* - \bar{s})_2 - (\bar{s}^* - \bar{s})_1]\left(\frac{1}{M}\right)$$

(Continued)

EXAMPLE 11.16 *(Continued)*

where, for a constant specific heat ideal gas,

$$s_2^* - s_1^* = c_p \ln(T_2/T_1) - R\ln(p_2/p_1) - s_1^*$$

and, since $T_2 = T_1$ here, this becomes

$$s_2^* - s_1^* = 0 - \left[\frac{55.1\ \text{ft}\cdot\text{lbf}/(\text{lbm}\cdot\text{R})}{778.16\ \text{ft}\cdot\text{lbf}/\text{Btu}}\right]\ln\left(\frac{15.0\times10^3}{150.}\right) = -0.326\ \frac{\text{Btu}}{\text{lbm}\cdot\text{R}}$$

Using $p_{R1} = 0.202$ and $T_{R1} = 1.06$, Figure 11.11 gives the entropy correction for state 1 as

$$(\bar{s}^* - \bar{s})_1 = 1.50\frac{\text{kJ}}{\text{kgmole}\cdot\text{K}} = \left(1.50\frac{\text{kJ}}{\text{kgmole}\cdot\text{K}}\right)\left(\frac{1\ \text{Btu}/(\text{lbmole}\cdot\text{R})}{4.1865\ \text{kJ}/(\text{kgmole}\cdot\text{K})}\right) = 0.360\frac{\text{Btu}}{\text{lbmole}\cdot\text{R}}$$

and using $p_{R2} = 20.2$ and $T_{R2} = 1.06$, Figure 11.11 gives the entropy correction for state 2 as

$$(\bar{s}^* - \bar{s})_2 = 22.2\frac{\text{kJ}}{\text{kgmole}\cdot\text{K}} = \left(22.2\frac{\text{kJ}}{\text{kgmole}\cdot\text{K}}\right)\left(\frac{1\ \text{Btu}/(\text{lbmole}\cdot\text{R})}{4.1865\ \text{kJ}/(\text{kgmole}\cdot\text{K})}\right) = 5.30\frac{\text{Btu}}{\text{lbmole}\cdot\text{R}}$$

Then, Eq. (11.42) gives

$$\begin{aligned} s_2 - s_1 &= (s_2^* - s_1^*) - \left[(\bar{s}^* - \bar{s})_2 - (\bar{s}^* - \bar{s})_1\right]\left(\frac{1}{M}\right) \\ &= -0.326\frac{\text{Btu}}{\text{lbm}\cdot\text{R}} - \left[5.30 - 0.360\frac{\text{Btu}}{\text{lbm}\cdot\text{R}}\right]\left(\frac{1}{28.05\ \text{lbm/lbmole}}\right) \\ &= -0.500\frac{\text{Btu}}{\text{lbm}\cdot\text{R}} \end{aligned}$$

The following exercises illustrate some of the elements of Example 11.16.

Exercises

45. Determine the values of Z_1 and Z_2 in Example 11.16 if the isothermal compression occurs at 1060°F instead of 80.0°F and all the remaining variables are unchanged. **Answer**: $Z_1 = 1.0$ and $Z_2 = 2.0$.

46. Determine the changes in specific enthalpy and specific entropy in Example 11.16 if the final pressure is 7500. psia instead of 15.0×10^3 psia and all the remaining variables are unchanged. **Answer**: $h_2 - h_1 = -145$ Btu/lbm and $s_2 - s_1 = -0.509$ Btu/(lbm·R).

47. Rework Example 11.16 when the ethylene is replaced by air but all the other variables are unchanged. **Answer**: $h_2 - h_1 = -4.85$ Btu/lbm, $u_2 - u_1 = -51.1$ Btu/lbm, and $s_2 - s_1 = -0.357$ Btu/(lbm·R).

11.11 IS STEAM EVER AN IDEAL GAS?

One of the most unforgivable mistakes that a thermodynamics student can make is to use the ideal gas equations to calculate the values of the properties *u*, *h*, and *s* of superheated steam. Yet we do just that in the next chapter when we discuss the thermodynamics of water vapor and air mixtures (i.e., humidity). Where did this great academic fear of steam as an ideal gas come from, and is it really justified?

When the term *steam vapor* was introduced into engineering jargon in the 19th century, it originally meant only visible, or "wet" steam. That is, steam whose state was far enough under the vapor dome to contain tiny visible foglike liquid water droplets (sometimes called *water dust*). As soon as the steam became a saturated or superheated vapor, it became invisible to the naked eye and was called *steam gas*, and for a long time, its properties were actually calculated from the ideal gas equations. By the end of the 19th century, it had become clear to thermodynamicists that the ideal gas equations did not accurately describe the behavior of high-pressure steam, and during the first half of the 20th century, considerable effort was devoted to developing new empirical equations for steam properties over the full range of pressures and temperatures of industrial interest.

However, these empirical equations were generally too complex and time consuming for ordinary engineering work, so they were used instead to generate elaborate saturation and superheated steam tables that were accurate to within 1% or less over their full range. Those tables could be used easily and quickly by working engineers with at most a simple linear interpolation required between table entries. These tables were widely distributed and continuously improved through a series of annual International Conferences on the Properties of Steam, which began in 1929. Today, the full steam tables have small pressure and temperature increment listings and fill an entire book.

To provide engineering students with a working knowledge of these new tables, a condensed version was appended to all thermodynamics textbooks. Authors and professors attempted to encourage the use of these tables and discourage the use of ideal gas equations for steam by extending the definition of a vapor to include

WHERE DID THE STEAM TABLES COME FROM?

There is no record of who first took interest in measuring the p-v-T properties of steam, but the development of the steam engine and the associated Industrial Revolution it produced created a strong practical need for such information. By 1683, Samuel Morland (1625–1695) is said to have acquired data on the pressure and temperature of saturated steam near atmospheric pressure. In 1662, Robert Boyle (1627–1691) developed the equation $p\mathcal{V} = \text{constant}$ for isothermal "elastic fluids" (i.e., compressible gases), and this equation was used over 100 years later for steam by James Watt (1736–1819) in his improved steam engine patent of 1782. In about 1787, Jacques Charles (1746–1823) developed the equation for isobaric gas behavior, $\mathcal{V}/T = \text{constant}$, and soon thereafter the laws of Boyle and Charles were combined into the ideal gas equation[5] that we have today, $p\mathcal{V} = mRT$.

The combined Boyle-Charles ideal gas equation continued to be used for steam engine design until the end of the 19th century, when engines were operating at sufficiently high pressures and temperatures to render the equation noticeably inaccurate.

In the 1840s, the French scientist Henri Victor Regnault (1810–1878) was sponsored by his government to carry out a series of precise measurements of the saturation properties of various substances, including water. He found that the Boyle-Charles ideal gas equation was only approximately true for real substances. By 1847, he had correlated his experimental results for the saturation pressure and temperature of steam with the formula given in the problem set at the end of this chapter. Regnault's data was considered to be an accurate authoritative source for over 60 years, and many others made mathematical correlations from it. By the end of the 19th century, many steam tables based on various correlations of increasing complexity of Regnault's data had become available for engineering use.

[5] *Boyle's law was independently discovered in 1676 by Edme Mariotte (1620–1684) and is sometimes known as Mariotte's law. Also, Charles's law was independently discovered in 1802 by Joseph Louis Gay-Lussac (1778–1850) and is sometimes known as Gay-Lussac's law. Carnot, Clapeyron, and many others were using the ideal gas equation in the form $pv = R(T + A)$, where T was in °C and A = 273 °C was then an empirical constant. Later, in 1848, William Thomson (Lord Kelvin) recognized that −273 °C corresponded to absolute zero temperature.*

any state near the vapor dome and below the critical point. Under this definition, superheated steam was now a vapor, not a gas, and it would be unforgivable for a student to apply ideal gas equations to a vapor.

For many years, this subterfuge was successful. But growing student computer literacy and the availability of complex software containing all the equations necessary to generate accurate steam properties will eventually make the use of printed tables obsolete. Yet, there remains the nagging question of whether or not the ideal gas equations can be used to describe the thermodynamic properties of steam with reasonable accuracy in low-pressure or high-temperature situations. If so, then engineering students could write relatively simple computer programs to solve challenging thermodynamic steam (or any other "vapor") problems without using elaborate software for generating property values.

It was the pragmatic Scottish engineer William John Macquorn Rankine (1820–1872), in his *Manual of the Steam Engine and Other Prime Movers,*[6] who noted that

> Steam attains a condition which is sensibly that of a perfect gas, by means of a very moderate extent of superheating; and it may be inferred that the formulae for the relations between heat and work which are accurate for steam-gas are not materially erroneous for actual superheated steam; while they possess the practical advantage of great simplicity.

The concept of a region of steam ideal gas behavior is illustrated in Figure 11.12. This figure is a Mollier diagram that shows the regions in which the equation's $v_g = RT_{sat}/p_{sat}$, $v = RT/p$, $h = h_g + C_p(T - T_{sat})$ and $s = s_g + c_p \ln(T/T_{sat}) - R \ln(p/p_{sat})$ are accurate to within about 1% or less of the actual steam table values (where h_g and s_g are evaluated at T_{sat}). The use of these ideal gas equations for steam in the regions shown produces errors of only a few percent in an analysis, which is often quite acceptable for many engineering thermodynamic applications. The reader can easily define the regions shown in Figure 11.12 by using the preceding equations and a steam table.

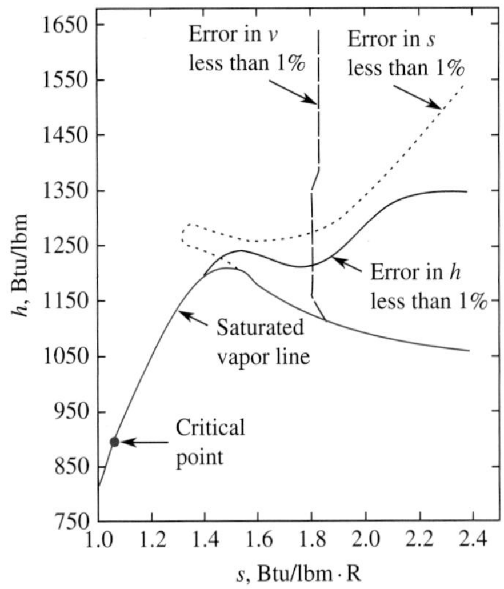

FIGURE 11.12
The ideal gas equations are accurate to 1% or less in the regions to the *right* of the dashed and dotted lines shown on this Mollier diagram.

[6] This book has the honor of being the first comprehensive engineering thermodynamics textbook. It was first printed in 1859 and went through 17 editions.

EXAMPLE 11.17

Suppose 1.00 lbm of superheated steam at 1.00 psia and 200.°F is mixed adiabatically and aergonically with 5.00 lbm of superheated steam at 5.00 psia and 400.°F in a closed, rigid system. Determine the final temperature and pressure.

Solution

The unknowns here are T_2 and p_2, and since this is a closed system, the energy balance (neglecting any changes in system kinetic and potential energy) is

$$\underset{\substack{0\\(\text{adiabatic})}}{{}_1Q_2} - \underset{\substack{0\\(\text{aergonic})}}{{}_1W_2} = m(u_2 - u_1) = U_2 - U_1$$

or $U_2 = U_1$, or $u_2 = u_1 = U_1/m_1 = (m_A u_A + m_B u_B)/(m_A + m_B)$. Also, for a closed, rigid system, the total volume and mass are constant, so

$$v_2 = v_1 = (m_A v_A + m_B v_B)/(m_A + m_B)$$

This problem is very difficult to solve using the steam tables or the Mollier diagram. It requires the construction of lines of constant u and constant v for various combinations of pressure and temperature, then finding the intersection of the $u = u_1 = u_2$ and $v = v_1 = v_2$ lines. However, the steam states given in the problem statement fall within the ideal gas region of Figure 11.12, so we can solve this problem with reasonable accuracy using the ideal gas equations of state. Since the reference state values ultimately cancel out here, we can simplify the algebra by taking $T_o = 0$ R and $u_o = 0$ Btu/lbm, then we can write $u_A = c_v T_A$, $u_B = c_v T_B$, and $u_2 = c_v T_2$. Also, from the Boyle-Charles ideal gas equation, we have that $v_A = RT_A/p_A$, $v_B = RT_B/p_B$, and $v_2 = RT_2/p_2$. Then,

$$u_2 = c_v T_2 = u_1 = c_v(m_A T_A + m_B T_B)/(m_A + m_B)$$

or

$$T_2 = (m_A T_A + m_B T_B)/(m_A + m_B)$$

and

$$v_2 = \frac{RT_2}{p_2} = \frac{R(m_A T_A/p_A + m_B T_B/p_B)}{m_A + m_B}$$

or

$$p_2 = \frac{(m_A + m_B)T_2}{m_A T_A/p_A + m_B T_B/p_B}$$

For the values given in the problem statement, we get

$$T_2 = \frac{(1.00\,\text{lbm})(659.67\,\text{R}) + (5.00\,\text{lbm})(859.67\,\text{R})}{6.00\,\text{lbm}} = 827\,\text{R} = 367°\text{F}$$

and

$$p_2 = \frac{(6.00\,\text{lbm})(827\,\text{R})}{(1.00\,\text{lbm})(659.67\,\text{R})/(1.00\,\text{psia}) + (5.00\,\text{lbm})(859.67\,\text{R})/(5.00\,\text{psia})} = 3.26\,\text{psia}$$

Note that, since the specific heat and gas constant cancel out in the equations for the final temperature and pressure in Example 11.17, they are independent of the material being mixed. The following exercises illustrate the use of this material.

Exercises

48. Determine the final temperature and pressure in Example 11.17 if the 1.00 lbm of superheated steam is at 5.00 psia and 200.°F instead of 1.00 psia and 200.°F. **Answer:** T_2 = 367°F and p_2 = 5.00 psia.
49. Suppose we increase the mass and temperature of one of the components in Example 11.17 so that we are mixing 5.00 lbm of superheated steam at 1.00 psia and 800.°F with 5.00 lbm of superheated steam at 5.00 psia and 400.°F. Determine the final temperature and pressure of this mixture. **Answer:** T_2 = 600.°F and p_2 = 1.48 psia.
50. Suppose the superheated steam components in Example 11.17 are replaced with superheated R-22 at exactly the same temperatures and pressures. Determine the final temperature and pressure of this mixture. **Answer:** T_2 = 367°F and p_2 = 3.26 psia.

SUMMARY

In this chapter, we discuss a series of generalized thermodynamic property relations. We also introduce two thermodynamic properties, the Helmholtz and Gibbs functions, and develop a series of differential property relations, known as the *Maxwell thermodynamic property equations*. The Clapeyron equation, Gibbs phase equilibrium equation, and a series of differential relations for the nonmeasurable u, h, s properties cast in terms of the measurable p, v, T properties allow us to develop a general procedure for constructing thermodynamic tables and charts.

In the concluding section of this chapter, the history and philosophy of modeling steam with the ideal gas equations is discussed. It is seen that this is a reasonably accurate approximation for saturated vapor at low pressure and superheated vapor (steam gas) at low pressure or high temperature.

Some of the more important equations introduced in this chapter follow. Do not attempt to use them without understanding their limitations.

1. Two new thermodynamic properties: the specific Helmholtz function, $f = u - Ts$, and the specific Gibbs function, $g = h - Ts$.
2. The Gibbs phase equilibrium condition:

$$s_{fg} = \frac{h_{fg}}{T_{\text{sat}}}$$

3. The Maxwell equations:

$$\left(\frac{\partial T}{\partial v}\right)_s = -\left(\frac{\partial p}{\partial s}\right)_v \quad \left(\frac{\partial T}{\partial p}\right)_s = \left(\frac{\partial v}{\partial s}\right)_p \quad \left(\frac{\partial p}{\partial T}\right)_v = \left(\frac{\partial s}{\partial v}\right)_T \quad \left(\frac{\partial v}{\partial T}\right)_p = -\left(\frac{\partial s}{\partial p}\right)_T$$

4. The Clapeyron equation:

$$\left(\frac{dp}{dT}\right)_{\text{sat}} = \frac{h_{fg}}{T_{\text{sat}} v_{fg}}$$

5. General u, h, and s property relations:

$$u_2 - u_1 = \int_{T_1}^{T_2} c_v dT + \int_{v_1}^{v_2} \left[T\left(\frac{\partial p}{\partial T}\right)_v - p\right] dv$$

$$h_2 - h_1 = \int_{T_1}^{T_2} c_p dT + \int_{p_1}^{p_2} \left[v - T\left(\frac{\partial v}{\partial T}\right)_p\right] dp$$

$$s_2 - s_1 = \int_{T_1}^{T_2} \frac{c_v}{T} dT + \int_{v1}^{v2} \left(\frac{\partial p}{\partial T}\right)_v dv$$

$$= \int_{T_1}^{T_2} \frac{c_p}{T} dT - \int_{p_1}^{p_2} \left(\frac{\partial v}{\partial T}\right)_p dp$$

6. Specific heat relations:

$$\left(\frac{\partial c_v}{\partial v}\right)_T = T\left(\frac{\partial^2 p}{\partial T^2}\right)_v \left(\frac{\partial c_p}{\partial p}\right)_T = -T\left(\frac{\partial^2 v}{\partial T^2}\right)_p c_p - c_v = T\beta^2 v/k$$

7. Using the gas tables to find changes in entropy: $s_2 - s_1 = \phi_2 - \phi_1 - R \ln(p_2/p_1)$
9. The compressibility factor: $Z = pv/(RT)$.
10. Relations for use with the Generalized charts:

$$h_2 - h_1 = (h_2^* - h_1^*) - \left[\left(\frac{\overline{h}^* - \overline{h}}{T_c}\right)_2 - \left(\frac{\overline{h}^* - \overline{h}}{T_c}\right)_1\right]\left(\frac{T_c}{M}\right)$$

$$s_2 - s_1 = (s_2^* - s_1^*) - \left[(\overline{s}^* - \overline{s})_2 - (\overline{s}^* - \overline{s})_1\right](1/M)$$

where $h_2^* - h_1^*$ and $s_2^* - s_1^*$ are the equivalent ideal gas changes in specific enthalpy and specific entropy of the real gas.

Problems (* indicates problems in SI units)

1.* Calculate the specific Gibbs and Helmholtz functions of saturated water vapor at 100.°C.

2. You are given an unknown material whose boiling point at atmospheric pressure is 50.0°F. You are also given the following property values at this pressure and temperature: $S_f = 0.310$ Btu/(lbm·R), $s_g = 1.76$ Btu/(lbm·R), and $h_g = 940.$ Btu/lbm. Calculate the following quantities for this material at this state (a) h_{fg}, (b) h_f, (c) g_f, and (d) g_g.

3. Using Eq. (11.3), show that

$$c_p = \left(\frac{\partial h}{\partial T}\right)_p = T\left(\frac{\partial s}{\partial T}\right)_p$$

4. Beginning with Eq. (11.1), show that

$$\frac{c_v}{T} = \left(\frac{\partial s}{\partial T}\right)_v = -\left(\frac{\partial^2 f}{\partial T^2}\right)_v$$

5. Beginning with Eq. (11.3), show that

$$\frac{c_p}{T} = \left(\frac{\partial s}{\partial T}\right)_p = -\left(\frac{\partial^2 g}{\partial T^2}\right)_p$$

6. Using Eq. (11.1), show that another definition of thermodynamic pressure is

$$p = T\left(\frac{\partial s}{\partial v}\right)_u$$

7. Using Eqs. (11.8) and (11.10) and the Gibbs phase equilibrium conditions, show that

$$\left(\frac{dp}{dT}\right)_{\text{sat}} = \frac{s_{fg}}{v_{fg}} = \frac{h_{fg}}{T_{\text{sat}} v_{fg}}$$

8.* Calculate s_{fg} for water at 100.°C using the Gibbs phase equilibrium conditions and compare it with the value listed in the steam tables.

9. Using the h_{fg} data from the steam tables at p_{sat} = 14.7, 100., 200., and 300. psia, calculate the value of s_{fg} at these temperatures from the Gibb's phase equilibrium condition and compare your results with the s_{fg} values listed in the steam tables.

10. Determine whether or not any of the following are properties
   a. $dM = \frac{7}{3}u^3 s\,du + \frac{1}{2}u^2 s^3 ds$
   b. $dN = (h/T)dT + \ln(1/T)dh$
   c. $X = \int [(p/2)\,ds - (s/2)\,dp]$
   d. $Y = \int (v\,dp)$

11. Beginning with Eq. (11.1) and the condition that $s = s(u, v)$, show that, for an ideal gas,

$$\left(\frac{\partial T}{\partial v}\right)_u = 0$$

and thus that $T = T(u)$ or $u = u(T)$ only.

12. Beginning with Eq. (11.1) and using the appropriate Maxwell equation, show that

$$\left(\frac{\partial u}{\partial v}\right)_T = T\left(\frac{\partial p}{\partial T}\right)_v - p$$

13. Using the results of Problem 12, show that it can be further reduced to
   a. $\left(\frac{\partial u}{\partial v}\right)_T = T^2\left[\frac{\partial(p/T)}{\partial T}\right]_v$
   b. $\left(\frac{\partial u}{\partial v}\right)_T = -\left[\frac{\partial(p/T)}{\partial(1/T)}\right]_v$

14. Beginning with Eq. (11.3) and using the appropriate Maxwell equation, show that

$$\left(\frac{\partial h}{\partial p}\right)_T = -T\left(\frac{\partial v}{\partial T}\right)_p + v$$

15. Using the results of Problem 14, show that it can be further reduced to
   a. $\left(\frac{\partial h}{\partial p}\right)_T = -T^2\left[\frac{\partial(v/T)}{\partial T}\right]_p$
   b. $\left(\frac{\partial h}{\partial p}\right)_T = \left[\frac{\partial(v/T)}{\partial(1/T)}\right]_p$

16. Let the isentropic exponent k for an arbitrary substance be defined by the process pv^k = constant.
   a. Show that $k = -(v/p)(\partial p/\partial v)_s$.
   b. Using Eqs. (11.24) and (11.26) and the classical definition of an isentropic process (s = constant) along with the appropriate Maxwell equations, show that part a reduces to $k = -(v/p)(\partial p/\partial v)_T(c_p/c_v)$.
   c. Show that, for an ideal gas, part b reduces to $k = c_p/c_v$.

17. An empirical equation of state has been proposed of the form

$$pv = RT + pA(T) + p^2 B(T)$$

where $A(T)$ and $B(T)$ are empirically determined functions of temperature. Beginning with Eq. (11.1) and using the appropriate Maxwell thermodynamic property equation, show that, for this material,

$$\left(\frac{\partial u}{\partial p}\right)_T = -T\frac{dA}{dT} - p\left(B + T\frac{dB}{dT}\right)$$

18. A simple magnetic substance has the following differential equation of state:

$$du = T\,ds + \mu_o v \mathbf{H}\cdot d\mathbf{M}$$

where **H** is the strength of the applied magnetic field, **M** is the magnetization vector, and μ_o is the magnetic permeability of free space (a constant). For this substance, show that the thermodynamic temperature is defined by

$$T = \left(\frac{\partial u}{\partial s}\right)_{\mathbf{M}}$$

and that the Maxwell equation analogous to Eq. (11.13) is

$$\left(\frac{\partial T}{\partial \mathbf{M}}\right)_s = \left(\frac{\partial \mu_o v \mathbf{H}}{\partial s}\right)_{\mathbf{M}}$$

19. A system involves both reversible expansion work ($-\int p\,dv$) and reversible electrochemical work ($\int \phi dq$, where ϕ is the voltage and q is the charge per unit mass). For such a system, its specific enthalpy is now defined as $h = u + pv - \phi q$.
   a. Find an expression for the differential change in specific Gibbs free energy, dg, in terms of p, v, s, T, ϕ, and q.
   b. Find the Maxwell equation $(\partial q/\partial T)_{p,\phi}$ = ?

20.* Estimate h_{fg} for water at 10.0°C using Eq. (11.17) and compare your answer with the steam table value.

21.* Estimate v_{fg} for water at 10.0°C using Eq. (11.17) and compare your answer with the steam table value.

22.* Using the triple point of water (0.0100°C, 611.3 Pa) as a reference state, estimate the saturation pressure of ice in equilibrium with water vapor at −20.0°C if h_{ig} = 2834.8 kJ/kg is a constant over this range.

23.* If saturated solid ice at −10.0°C is subjected to an isothermal compression process to 200. MPa, will it melt? Use the triple point (0.0100°C, 611.3 Pa) as the reference state, where h_{if} = −333.41 kJ/kg and $v_{if} = 9.08 \times 10^{-5}$ m^3/kg. Sketch this process on a p-T diagram.

24. At very low pressures, a substance has a saturation curve given by $p_{sat} = \exp(A_1 - A_2/T_{sat})$, where A_1 and A_2 are constants. Show that h_{fg} is constant for this substance.

25. At very low pressures, the saturation curve for a particular substance is given by $p_{sat} = \exp[A_1 + A_2/T_{sat} + A_3(\ln T_{sat})]$, where A_1, A_2, and A_3 are constants. Show that h_{fg} varies linearly with T_{sat} for this substance.

26. Using Eq. (11.22),
   a. Show that the Joule-Thomson coefficient defined by Eq. (6.25) can be written as

$$\mu_J = (\partial T/\partial p)_h = \left[T(\partial v/\partial T)_p - v\right]/c_p$$

   b. Use this result to evaluate the Joule-Thomson coefficient for an ideal gas.

27. Assuming $s = s(T, p)$ for a simple substance, derive Eqs. (11.26) and (11.27) of this chapter.

28. The following equation of state has been proposed for a gas

$$pv = RT + A/T - B/T^2$$

where A and B are constants. Beginning with this equation, develop equations based on the measurable properties p, v, and T for the property changes $u - u_o$, $h - h_o$, and $s - s_o$, where u_o, h_o, and s_o are reference state properties at p_o, v_o, and T_o.

29. Develop an equation based on measurable properties p, v, and T for the property changes $u_2 - u_1$, $h_2 - h_1$, and $s_2 - s_1$ of a van der Waals gas with constant specific heats. The van der Waals equation of state, Eq. (3.44), can be written as $p = RT/(v - b) - a/v^2$, where a and b are constants.

30. Develop equations based on measurable properties p, v, and T for the isothermal property changes $(u_2 - u_1)_T$, $(h_2 - h_1)_T$, and $(s_2 - s_1)_T$ of a Dieterici gas (see Eq. (3.45)).

31. Develop equations based on measurable properties p, v, and T for the isothermal property changes $(u_2 - u_1)_T$, $(h_2 - h_1)_T$, and $(s_2 - s_1)_T$ of a Beattie-Bridgeman gas (see Eq. (3.47)).

32. Develop equations based on measurable properties p, v, and T for the isothermal property changes $(u_2 - u_1)_T$, $(h_2 - h_1)_T$, and $(s_2 - s_1)_T$ of a Redlich-Kwong gas (see Eq. (3.48)).

33. Determine $(\partial c_v/\partial v)_T$ for a Redlich-Kwong gas (see Eq. (3.48)) and integrate this to find the function $c_v = c_v(T, v, v_o)$, where v_0 is a reference state specific volume.

34.* Using Eq. (11.30), calculate the difference between c_p and c_v for (a) copper at 300°C, (b) mercury at 20°C, (c) glycerin at 20°C. Use Tables 3.1 and 3.2 for compressibility values. The densities at 20°C are ρ_{Cu} = 8954 kg/m^3, ρ_{Hg} = 13,579 kg/m^3, and ρ_{glyc} = 1264 kg/m^3.

35. Determine $(\partial c_v/\partial v)_T$ for a van der Waals gas (see Eq. (3.44)).

36. Determine $(\partial c_v/\partial v)_T$ for a Beattie-Bridgeman gas (see Eq. (3.47)) and integrate this to find the function $c_v = c_v(T, v, v_o)$, where v_o is a reference state specific volume.

37. Saturated mercury vapor has an equation of state of the form

$$p_{sat} = \left(\frac{RT}{v}\right)_{sat} - \left(\frac{T}{v^2}\right)_{sat} \exp(A_1 + A_2/T_{sat} + A_3 \ln T_{sat})$$

where A_1, A_2, and A_3 are constants. The constant volume specific heat c_v is also constant for this material. Determine equations that allow u_g, h_g, and s_g to be calculated relative to a reference state at p_o, v_o, u_o, and s_o in terms of the measurable quantities p, v, and T.

38.* A standard spark ignition piston-cylinder automobile engine has a compression ratio of 8.60 to 1, and the intake air is at 0.100 MPa, 17.0°C. For an isentropic compression process, use the gas tables (Table C.16b) to determine
   a. The work required per unit mass of air compressed.
   b. The temperature at the end of the compression stroke.
   c. The pressure at the end of the compression stroke.

39. Air enters an isentropic, steady flow, axial compressor at 14.7 psia and 60.0°F and exits at 197 psia. Determine the exhaust temperature and the input power per unit mass flow rate. Use Table C.16a in *Thermodynamic Tables to accompany Modern Engineering Thermodynamics* in your solution.

40.* An engineer claims to have designed an uninsulated diffuser that expands 3.00 kg/s of air from 1.00 MPa, 37.0°C to 0.100 MPa, 17.0°C. The inlet and exit air velocities are 80.0 and 5.00 m/s, respectively. Use the gas tables (Table C.16b) to determine the heat transfer rate and entropy production rate for the diffuser, if the average wall temperature is 27.0°C. Will the diffuser work as designed?

41.* Determine the final pressure, temperature, and required work per unit mass when 1.00 m^3 of air is isentropically compressed from 0.150 MPa, 300. K to 0.100 m^3 using
   a. Constant specific heat ideal gas equations.
   b. The gas tables for air (Table C.16b).

42.* An insulated axial flow air compressor for a gas turbine engine is being tested in a laboratory. The inlet conditions are 0.090 MPa and −3.00°C, and the outlet is at 0.286 MPa and 217°C. Use the gas tables to determine the ratio of the power input for an isentropic process to the actual adiabatic power input. This ratio is defined to be the compressor's *isentropic efficiency*.

43.* An insulated air compressor with an isentropic efficiency of 78.0% compresses air from 0.100 MPa, 290. K to 10.0 MPa. Use the gas tables to determine the power required per unit mass flow rate and the exit air temperature.

44. Use the gas tables (Table C.16a) to determine the final temperature and the minimum possible power required to compress 3.00 lbm/s of air from 14.7 psia, 40.0°F to 10.0 atm. in a steady flow, adiabatic process.

45.* Air is compressed in an adiabatic, steady flow process from 0.081 MPa, 400. K, to 2.50 MPa with an isentropic efficiency of 85.0%. Use the gas tables to determine
   a. The power required per unit mass flow rate.
   b. The actual inlet temperature.
   c. The entropy production rate per unit mass flow rate.

46.* An uninsulated piston-type air compressor operates in a steady flow process from 0.100 MPa, 300. K to 2.00 MPa, 540. K. Use the gas tables to determine per unit mass flow rate,
   a. The power required.
   b. The heat transfer rate.

when the entropy production rate per unit mass flow rate is 0.538 kJ/(kg·K) and the mean cylinder external wall temperature is 432 K.

47. An inventor claims that 700. ft·lbf was used to compress 0.450 lbm of air isothermally in a closed piston-cylinder apparatus from 14.7 psia, 70.0°F to 2000. psia. Assuming ideal gas behavior,

a. Is this process possible?

b. If not, what is the maximum possible compression pressure that could be reached with this process?

48.* Determine the compressibility factor for methane at 20.0 MPa and 0.00C.

49.* In 1879, the French physicist Emile Amagat generated experimental data in a mine shaft at Verpilleux, France, for his research on the compressibility of gases. There he used a vertical column of mercury 327 m high to measure the compressibility of nitrogen at a pressure of 430. atm. Assuming the temperature at the bottom of the mine shaft was 30.0°C, use the compressibility charts to determine the value of the compressibility factor for nitrogen under these conditions.

50.* For air at 20.0°C, there is a unique pressure above p_R = 1.00 at which the compressibility factor is the same as that of an ideal gas. Use the compressibility charts to determine this pressure.

51.* For air at 20.0°C, use the compressibility charts to determine the low pressure range in which the compressibility factor of air differs from that of an ideal gas by no more than 2.00% (i.e., $1.0 \leq Z \leq 0.980$). Is it reasonable to assume ideal gas behavior for air at pressures up to 3.45 MPa (500. psia)?

52. 200. lbm of carbon dioxide is to be put into a rigid 3.00 ft^3 tank at 87.5°F. Use the compressibility factor to determine the final pressure.

53. Determine the ratio of v_c'/v_c for the following substances: (a) water vapor, (b) nitrogen, (c) propane, and (d) methane.

54.* Using the generalized charts, determine the sum of the heat transfer rate and power required to isothermally compress 0.300 kg/s of hydrogen in a steady flow process from 2.00 to 20.0 atm at 50.0 K. Is it possible to carry out this process adiabatically?

55.* Using the generalized charts, determine the entropy change as 0.730 kg of carbon monoxide is expanded from 35.0 MPa to 0.100 MPa in an isothermal process at 100. K.

56. Compare the specific volumes of water vapor obtained from the steam tables to those obtained from the compressibility factor charts (Figures 11.5–11.7) at the following states

a. 14.7 psia, 300.°F.

b. 6000. psia, 1000 °F.

c. 8000. psia, 2000.°F.

57. Compare the values of $h_2 - h_1$ and $s_2 - s_1$ for water vapor obtained from (a) the gas tables (Table C.16c) and (b) the generalized charts (Figures 11.9 and 11.11) with those obtained from the steam tables for the following conditions: 14.7 psia, 300.°F (state 1) and 6000. psia, 1000.°F (state 2).

58. Use the generalized charts to calculate the heat transfer rate required to cool 7.00 lbm/s of argon gas from 500.°F, 2000. psia to 300.°F in a steady flow, constant pressure heat exchanger. Assume the specific heats of argon are constant over this temperature range.

59. Methane is throttled adiabatically with negligible velocity change from 1500. psia, 70.0°F to atmospheric pressure. Assuming constant specific heats and using the generalized charts, determine the exit temperature.

60. Carbon dioxide is throttled adiabatically with negligible velocity change from 2500. psia, 800. R to atmospheric pressure. Use the generalized charts to determine the exit temperature by

a. Assuming constant specific heats.

b. Using the gas tables (Table C.16c).

61.* Helium in an external storage tank on a spacecraft is expanded through an isentropic attitude control nozzle with a negligible inlet velocity from 2.00 MPa, 10.0 K to 0.0100 MPa. Assuming constant specific heats and using the generalized charts, determine the exit temperature and velocity.

62.* Sulfur dioxide with a negligible inlet velocity is expanded through an isentropic nozzle from 20.0 MPa, 500. K to 0.200 MPa in a chemical processing unit. Assuming constant specific heats and using the generalized charts, determine the exit temperature and velocity.

63. Hydrogen is cooled in an isobaric heat exchanger from 5000. to 527 R at 20.0 psia. The heat transfer occurs across an isothermal wall at 500. R inside the heat exchanger. Use the generalized charts to determine the hydrogen's heat transfer and entropy production rates per unit mass flow rate.

64. Use the generalized charts to determine the changes in specific enthalpy and specific entropy of nitrogen as it undergoes an isobaric cooling process from 2000. to 1000. R at 14.7 psia assuming

a. Constant specific heats.

b. Temperature dependent specific heats.

c. Compute the percentage difference between the results of parts a and b.

65.* According to Dalton's law of partial pressures, the partial pressure exerted by the water vapor in a mixture of air and water vapor is equal to the pressure the water vapor would exert if it alone occupied the total volume of the mixture. If 1.00 m^3 of humid air at 20.0°C contains 10.3 g of water vapor, determine

a. The partial pressure of the water vapor.

b. The maximum partial pressure of water vapor at this temperature.

c. The ratio of the answer in part a to that of part b (this ratio is called the *relative humidity* of the air).

66. Steam is throttled from 100. psia, 500.°F to 14.7 psia in an isenthalpic process. Determine the change in specific entropy and the exit temperature of the steam using

a. Ideal gas equations.

b. The steam tables.

c. Compute the percent error in assuming ideal gas behavior.

67.* In the warp drive system of an intergalactic spacecraft, 13.0 kg/s of water vapor is reversibly and isothermally expanded from 500. to 125 Pa at 100.°C. Determine the heat transfer rate and the power produced. Assume ideal gas behavior.

68. Water vapor is heated from 300. to 400.°F in a steady flow isobaric process. Determine the percent error in calculating the heat transfer rate per unit mass flow rate by using the ideal gas equations for system pressures of (a) 20.0 psia, (b) 2.00 psia, and (c) 0.200 psia.

69.* Saturated water vapor at 10.0 kPa is expanded reversibly and isentropically in a steady flow process in a doorknob heat treating plant to 5.00 kPa. Determine the final temperature, the heat transfer rate, and the power produced per unit mass flow rate. Assume the steam is an ideal gas.

Design Problems

The following are open-ended design problems. The objective is to carry out a preliminary thermal design as indicated. A detailed design with working drawings is not expected unless otherwise specified. These problems do not have specific answers, so each student's design is unique.

70. Design a system to liquefy nitrogen by repeatedly expanding it until it reaches the saturation temperature. (Hint: Consult your library about the Linde gas liquefaction process.)
71. Design a system to cut ice into various two-dimensional shapes using localized pressure to produce a phase change and thus locally "melting" out the desired shape. (Suggestion: Try a high-pressure "cookie-cutter" technique.)
72.* Design a fire extinguisher system that expands liquid carbon dioxide under a suitable pressure at ambient temperature (which can vary by ±50.0°C) and produce a fine spray of solid carbon dioxide particles at high velocity and low temperature.
73. Design a 3.00 ft^3 cylindrical tank that safely contains oxygen gas at 2500. psia under ambient temperature conditions (which can vary by ±100.°F). (Suggestion: Consult the ASME Pressure Vessel Design Codes in your library.)
74.* Design an experiment that illustrates and accurately measures the difference in p-v-T behavior between water vapor and a suitable ideal gas (such as air) over the temperature range from 0.00 to 150.°C.

Computer Problems

The following computer assignments are designed to be carried out on a personal computer using a spreadsheet or equation solver. They may be used as part of a weekly homework assignment.

75.* Develop a computer program, spreadsheet, equation solver, or the like to determine the percent error in the pressure predicted by the van der Waals equation of state for water along the T = 500°C isotherm, $p = RT/(v - b) - a/v^2$. Use Table C.3b or other appropriate tables for the actual values of p for various values of v at T = 500.°C. The van der Waals coefficients for water vapor can be found in Table C.15b.
76.* Develop a computer program, spreadsheet, equation solver, or the like to determine the percent error in the pressure predicted by the Beattie-Bridgeman equation of state for water along the T = 500.°C isotherm, $p = [(1 - \varepsilon)(v + B)RT - A]/v^2$, where $A = A_0(1 - a/v)$, $B = B_0(1 - b/v)$, and $\varepsilon = c/(vT^3)$. Use Table C.3b or other appropriate tables for the actual values of p for various values of v at T = 500.°C. The Beattie-Bridgeman coefficients for water vapor can be found in Table C.15b.
77.* Develop a computer program, spreadsheet, equation solver, or the like to determine the percent error in the pressure predicted by the Redlich-Kwong equation of state for water along the T = 500.°C isotherm, $p = RT/(v - b) - a/[v\,(v + b)T^{1/2}]$. Use Table C.3b or other appropriate tables for the actual values of p for various values of v at T = 500.°C. The Redlich-Kwong coefficients for water vapor can be found in Table C.15b.
78. Plot a minimum of 100 points along the T = 500. R isotherm on p-v coordinates for air using
    a. The ideal gas equation of state.
    b. The Clausius equation of state (use the van der Waals value for b).
79. Plot a minimum of 100 points along the T = 500. R isotherm on p-v coordinates for hydrogen using
    a. The ideal gas equation of state.
    b. The van der Waals equation of state.
80. Plot a minimum of 100 points along the T = 500. R isotherm on p-v coordinates for methane using
    a. The ideal gas equation of state.
    b. The Beattie-Bridgeman equation of state.
81. Using the steam tables as a guide, find the regions on a Mollier diagram or a p-v diagram where the ideal gas equations with constant specific heats are accurate to within ±1.00% for
    a. Specific volume v.
    b. Enthalpy h.
    c. Entropy s.
82. Expand Problem 81 by adding temperature-dependent ideal gas specific heats.
83. Expand Problem 81 by using the van der Waals equation of state in place of the ideal gas equation of state.
84. Develop an interactive computer program for ammonia using the Beattie-Bridgeman equation of state to produce the following results from responses to appropriate screen prompts:
    a. Output p when v and T are input.
    b. Output T when p and v are input.
    c. Output v when p and T are input.
85. Develop an interactive computer program that replaces the gas tables Tables C.16a and C.16b. Do this in two steps:
    a. First have the program return u, h, ϕ, p_r, and v_r when T is input by assuming constant specific heats.
    b. Modify the program developed in step a to include the temperature-dependent specific heats given in Table C.14 in *Thermodynamic Tables to accompany Modern Engineering Thermodynamics*.
86.* The purpose of this assignment is to investigate the accuracy of several historically important p-T relations for saturated water vapor. Using the tables in the tables book or some other source for accurate saturation p-T data, calculate, tabulate, and plot the percent error in saturation pressure for each of the following cases using % error = (CP – TP)/TP, where CP is the calculated saturation pressure and TP is the saturation pressure found in the steam tables.
    a. By 1847, Henri Regnault had developed an equation from his experimental $p_{sat} - T_{sat}$ results for saturated steam. It was valid in the range of –33.0 to 232°C and had the form

$$\log_{10} p_{sat} = A - BD^n - CE^n$$

where p_{sat} is in mm Hg and

$$A = 6.2640348 \quad \log_{10} D = 9.994049292 - 10$$
$$\log_{10} B = 0.1397743 \quad \log_{10} E = 9.998343862 - 10$$
$$\log_{10} C = 0.6924351 \quad n = T_{sat} + 20.0°C$$

Make your % error calculations every 20.0°C between 20.0 and 220.°C.

    b. In 1849 Williams Rankine fit his own equation to Regnault's data and came up with the following relation:

$$\log_{10} p_{sat} = A - B/T_{sat} - C/T_{sat}^2$$

where p_{sat} is in psia and T_{sat} = T_{sat}(in °F) + 461.2 (–461.2°F was Rankine's best estimate of absolute zero), and

$$A = 6.1007 \quad \log_{10} B = 3.43642 \quad \log_{10} C = 5.59873$$

Make your percent error calculations here at 20.0°F intervals between 40.0 and 700.°F.

c. By 1899, very careful $p_{sat} - T_{sat}$ measurements had been made, and K. Thiesen curve fit these data to the following equation:

$$(T_{sat} + 459.6)\log_{10} p_{sat} = 5.409(T_{sat} - 212) - (8.71\times 10^{-10})\left[(689 - T_{sat})^4 - (477)^4\right]$$

where p_{sat} is in atmospheres and T_{sat} is in °F. Make your percent error calculations here at intervals of 20.0°F between 20.0 and 700.°F.

d. By 1915, G. A. Goodenough had developed the following more complex equation, which he claimed fit steam quite well over the entire range from 32.0°F to the critical point (705°F).

$$\log_{10} p_{sat} = A - B/T_{sat} - C\log_{10} T_{sat} - DT_{sat} + ET_{sat}^2 - F$$

where $F = 0.0002\{10 - 10[(T_{sat} - 829.6)/100]^2 + [(T_{sat} - 829.6)/100]^4\}$. Here, p_{sat} is in psia, T_{sat} is in R, and

$$A = 10.5688080 \quad \log_{10} D = 7.6088020 - 10$$
$$\log_{10} B = 3.6881209 \quad \log_{10} E = 4.1463000 - 10$$
$$C = 0.0155$$

Make your percent error calculations at any convenient temperature interval between 32.0 and 705°F. Note: The use of base 10 logarithms in these equations is the way these equations were originally written, and it has been continued here for historical accuracy.

87.* Equations (11.31) for mercury are given by

1. $\ln p_{sat} = 23.6321 - 7042.6208/T_{sat} - 0.1207\,(\ln T_{sat}) - 58{,}060.290/T^2_{sat}$.
2. $p = RT/v - (T/v^2)\exp[10.3338 - 312.0954/T - 2.0795\,(\ln T)]$.
3. $v_f = [12{,}813.6070 - 2.4531(T_{sat} - 600) - 0.000267(T_{sat} - 600)^2]^{-1}$.
4. $c_v^o = 62.168\ \text{J/(kg}\cdot\text{K)} = \text{constant}$.

where p and p_{sat} are in Pa, T and T_{sat} are in K, and v and v_f are in m^3/kg, and $R = 41.4453$ J/(kg·K). Develop a computer spreadsheet that returns v, u, h, and s values when p and T are input. Make sure your program checks to see what region (saturated or superheated) your input data are in. For $T_{sat} = 750.$ K, you should get $h_{fg} = 291$ kJ/kg and $s_{fg} = 0.388$ kJ/kg. These values are independent of your reference state values. Note: More ambitious programs can now be produced by adding subroutines that return the remaining properties when any pair of independent properties (T-s, p-h, etc.) are input.

CHAPTER 12

Mixtures of Gases and Vapors

CONTENTS

12.1 WPROWADZENIE (INTRODUCTION)

In this chapter, we deal with the problem of generating thermodynamic properties for homogenous mixtures of gases and vapors that are not involved in chemical reactions. Properties of chemically reacting mixtures are discussed in detail in Chapter 15.

We can define the composition of any mixture based on how we physically create the mixture. For example, we can create a mixture by combining measured *masses* (or *weights*) of things or by combining measured *volumes* of things. The items or things that make up the mixture are called the *components* of the mixture, and knowing how much of each one is present defines the *composition* of the mixture.

This all seems very simple, so why do we need to dwell on it here? First of all, there are two ways to measure mass, the regular mass (lbm or kg) and the molar mass (lbmole or kgmole), and a composition based on the regular mass is not the same as a composition based on the molar mass. Second, since the conservation of mass law tells us that the total mass of the mixture is simply the sum of the masses of all the components in the mixture, is the total volume of the mixture the same as the sum of the volumes of all the components in the mixture? Well, that depends. If the components are immiscible, then the total volume is the sum of the individual component volumes. But gases and vapors are not immiscible, so how do the component volumes affect the total volume?

So, whereas the basic definitions of mixture composition for insoluble solids and liquids seems very easy, the practicality of implementing these definitions for mixtures of soluble gases and vapors is not so easy. However, the mixture composition is not really our primary goal. *Our primary goal is to be able to determine the thermodynamic properties of a mixture so that we can apply the first and second laws to an engineering system containing a mixture.* Are mixture thermodynamic properties just the sum of the thermodynamic properties of their components? No! With the exception of mass, the extensive thermodynamic properties (e.g., mixture total volume $\forall_m$, mixture total

CRITICAL THINKING

Whereas the volumes of insoluble solids and liquids seem easily defined, because their molecules do not intermingle, the same is not true for gases and vapors, where their molecules do intermingle. But, what happens when you mix two *soluble* liquids? Is the mixture volume in this case just the sum of the volumes of the mixture components before they were mixed? Can you find an example of soluble liquids where the final mixture volume is not equal to the sum of their premixed component volumes?

internal energy U_m, mixture total enthalpy H_m, and mixture total entropy S_m) of a mixture are *not* generally equal to the sum of the extensive thermodynamic properties of the components. It turns out that the value of any thermodynamic property of a mixture is just the mass weighted sum of the *partial specific properties* of the mixture's components. Therefore, to determine the numeric value of a thermodynamic property of a mixture we need to know (a) the exact composition of the mixture and (b) the values of the partial specific properties of all the components in the mixture. This is what the first half of this chapter is all about.

The second half of this chapter deals with the application of this material to a very special mixture of gases and vapors, air and water vapor (atmospheric air). This is normally the domain of heating, ventilating, and air conditioning (HVAC) engineers, but since the atmosphere affects all of us in our daily life, it provides a good textbook application of gas and vapor mixture theory. The basic elements of HVAC involve applying the first and second laws to systems designed to cool and dehumidify or heat and humidify atmospheric and building air. To carry out this analysis we need the numerical values for specific internal energy, specific enthalpy, and specific entropy for various mixtures of air and water vapor. Since atmospheric air is a fairly complex mixture, it is more convenient to refer to industry prepared tables and charts for accurate thermodynamic property values for this mixture.

12.2 THERMODYNAMIC PROPERTIES OF GAS MIXTURES

Unfortunately, there is no single measure of mixture composition. A mixture composition often is given simply in percent, but the percent is calculated on a mass (or weight) basis,[1] a molar basis, a volume basis, or a pressure basis; and the numerical values depend on which basis is used in the calculation. This ambiguity leads us to define four composition percentages or fraction measures for mixtures of gases.

Consider a homogeneous mixture made up of N distinct gases, each of which has a unique molecular mass, M_i. Let the mass of each gas present in the mixture be m_i. Then the mass balance gives the total mass of the mixture m_m as

$$m_m = m_1 + m_2 + \cdots + m_N = \sum_{i=1}^{N} m_i \tag{12.1}$$

The corresponding number of moles n_i of gas i with molecular mass M_i can be determined from Eq. (1.9) as

$$n_i = m_i/M_i \tag{1.12}$$

and because the mole unit is just another measure of mass, the total number of moles of mixture n_m is simply

$$n_m = n_1 + n_2 + \cdots + n_N = \sum_{i=1}^{N} n_i \tag{12.2}$$

With these two mass measures, we can define two different mass-based composition measures or fractions as

The *mass fraction* w_i of chemical species i in the mixture is

$$w_i = \frac{m_i}{m_m} \tag{12.3}$$

and

The *mole fraction* χ_i of chemical species i in the mixture is

$$\chi_i = \frac{n_i}{n_m} \tag{12.4}$$

[1] This is also called a *gravimetric* basis.

With the exception of system mass, the extensive properties of a system are not generally conserved in any thermodynamic process, so that their mixture values are not normally equal to the sum of their constituent values. The changes in extensive or intensive properties of a mixture with composition must always be determined experimentally. However, the extensive properties of gases are mathematically homogeneous functions of the first degree in mass.[2] For example, the total volume $\forall_m$ of a homogeneous mixture of gases can be written as a function of the mixture total pressure p_m, mixture temperature T_m, and the mass composition of the mixture $m_1, m_2, \ldots, m_N$ as

$$\forall_m = \forall_m(p_m, T_m, m_1, \ldots, m_N)$$

and, when the mixture pressure and temperature are held constant, this is a homogeneous function of the first degree in the masses m_i. This means that, if all the remaining variables (the m_i) are multiplied by an arbitrary constant λ, then the mixture volume also is multiplied by λ, or

$$\lambda\forall_m = \forall_m(p_m, T_m, \lambda m_1, \ldots, \lambda m_N)$$

where λ is an arbitrary variable. Differentiating this equation with respect to λ while holding the pressure and temperature constant and setting $\lambda = 1$ gives

$$\forall_m\big|_{p_m,T_m} = \left(\frac{\partial\forall_m}{\partial m_1}\right)m_1 + \cdots + \left(\frac{\partial\forall_m}{\partial m_N}\right)m_N = \sum_{i=1}^{N} m_i\hat{v}_i \tag{12.5}$$

where

$$\hat{v}_i = \left(\frac{\partial\forall_m}{\partial m_i}\right)_{p_m,T_m,m_j} \tag{12.6}$$

here $\hat{v}_i$ is defined to be the *partial specific volume* of gas i in the mixture and $\forall_i = m_i\hat{v}_i = n_i\hat{\bar{v}}_i$ is the *partial volume* of gas i in the mixture.[3] Equation (12.5) leads us to a third common composition measure, the volume fraction:

The *volume fraction* ψ_i of gas i in a mixture of gases is

$$\psi_i = \frac{\forall_i}{\forall_m} \tag{12.7}$$

Even though pressure is not an extensive property, the implicit function theorem from calculus tells us that, if $\partial\forall_m/\partial p_m \neq 0$ in Eq. (12.5), we can write the total pressure p_m of a homogenous mixture of N gases as a function of the mixture volume $\forall_m$, mixture temperature T_m, and the mass composition of the mixture $m_1, m_2, \ldots, m_N$ as

$$p_m = p_m(\forall_m, T_m, m_1, m_2, \ldots, m_N)$$

and, when the total volume and temperature of the mixture are constant, this too is a homogenous function of the first degree in the masses m_i. Following the development of Eq. (12.5), we can write

$$p_m = \sum m_i\left(\frac{\partial p_m}{\partial m_i}\right)_{\forall_m,T_m,m_j} = \sum_{i=1}^{N} m_i\hat{p}_i = \sum p_i = p_1 + p_2 + \cdots + p_N \tag{12.8}$$

where $\hat{p}_i = (\partial p_m/\partial p_i)_{\forall_m,T_m,m_j}$ is the *partial specific pressure* of gas i in the mixture and $p_i = m_i(\partial p_m/\partial p_i)_{\forall_m,T_m,m_j}$ is the *partial pressure* of gas i in the mixture. Equation (12.8) provides our fourth common composition measure, the pressure fraction:

The *pressure fraction* π_i of gas i in a mixture of gases is

$$\pi_i = \frac{p_i}{p_m} \tag{12.9}$$

---

[2] See, for example, Kestin, J., 1979. *A Course in Thermodynamics*. Hemisphere Publishing Corporation, McGraw-Hill Book Company, New York, vol. 1, pp. 326–327.

[3] The concept of "partial properties" was introduced by Lewis, G. N., 1907. A new system of thermodynamic chemistry, in Proceedings of the American Academy, 43, p. 273.

Table 12.1 summarizes the four composition measures thus far defined.

Dividing both sides of Eq. (12.5) by the mixture mass m_m produces the specific volume of the mixture as

$$v_m = \frac{\forall_m}{m_m} = \sum_{i=1}^{N} \frac{m_i}{m_m} \hat{v}_i = \sum w_i \hat{v}_i$$

where w_i is the mass fraction m_i/m_m. The concepts of total, specific, and partial specific properties can be extended to other extensive properties, as shown in Table 12.2. On a molar basis, total, molar specific, and partial molar specific properties are defined in Table 12.3.

Finally, the constant volume and constant pressure specific heats for the mixture are defined as

$$c_{v_m} = \left(\frac{\partial u_m}{\partial T_m}\right)_{v_m} \quad \text{and} \quad c_{p_m} = \left(\frac{\partial h_m}{\partial T_m}\right)_{v_m}$$

These specific heats can also be written on a molar basis and in terms of the partial specific heats of the component gases, as shown in Table 12.4.

The mass, mole, volume, and pressure composition fractions have the characteristic that they always sum to unity; that is,

$$\sum_{i=1}^{N} w_i = \sum_{i=1}^{N} \chi_i = \sum_{i=1}^{N} \Psi_i = \sum_{i=1}^{N} \pi_i = 1.0 \tag{12.10}$$

Table 12.1 Four Composition Measures

Name of Composition Fraction	Defining Equation
Mass fraction w_i	$w_i = m_i/m_m$
Mole fraction χ_i	$\chi_i = n_i/n_m$
Volume fraction ψ_i	$\psi_i = \forall_i/\forall_m$
Partial pressure fraction π_i	$\pi_i = p_i/p_m$

Table 12.2 Total, Specific, and Partial Specific Properties of a Mixture of Gases

Total Property of the Mixture	Specific Property of the Mixture	Partial Specific Property of Gas *i* in the Mixture
$\forall_m = \sum_{i=1}^{N} m_i \hat{v}_i$	$v_m = \forall_m/m_m = \sum_{i=1}^{N} \frac{m_i}{m_m} \hat{v}_i = \sum_{i=1}^{N} w_i \hat{v}_i$	$\hat{v}_i = \left(\frac{\partial \forall_m}{\partial m_i}\right)_{p_m, T_m, m_j}$
$U_m = \sum_{i=1}^{N} m_i \hat{u}_i$	$u_m = U_m/m_m = \sum_{i=1}^{N} \frac{m_i}{m_m} \hat{u}_i = \sum_{i=1}^{N} w_i \hat{u}_i$	$\hat{u}_i = \left(\frac{\partial U_m}{\partial m_i}\right)_{p_m, T_m, m_j}$
$H_m = \sum_{i=1}^{N} m_i \hat{h}_i$	$h_m = H_m/m_m = \sum_{i=1}^{N} \frac{m_i}{m_m} \hat{h}_i = \sum_{i=1}^{N} w_i \hat{h}_i$	$\hat{h}_i = \left(\frac{\partial H_m}{\partial m_i}\right)_{p_m, T_m, m_j}$
$S_m = \sum_{i=1}^{N} m_i \hat{s}_i$	$s_m = S_m/m_m = \sum_{i=1}^{N} \frac{m_i}{m_m} \hat{s}_i = \sum_{i=1}^{N} w_i \hat{s}_i$	$\hat{s}_i = \left(\frac{\partial S_m}{\partial m_i}\right)_{p_m, T_m, m_j}$

CRITICAL THINKING

Is it possible to define a *partial temperature* of gas i in a mixture of N gases? Since the mixture temperature varies *inversely* with the system mass in most equations of state for gases, is temperature a homogeneous function of the first degree in the masses m_i? Could the partial temperature of gas i in the mixture be the temperature exhibited by gas i when it alone occupies the volume of the mixture $\forall_m$ at the pressure of the mixture p_m? (See Problem 67 at the end of this chapter.)

Table 12.3 Total, Molar Specific, and Partial Molar Specific Properties of a Mixture of Gases

Total Property of the Mixture	Molar Specific Property of the Mixture	Partial Molar Specific Property of Gas *i* in the Mixture
$\forall_m = \sum_{i=1}^{N} n_i \hat{\bar{v}}_i$	$\bar{v}_m = \forall_m/n_m = \sum_{i=1}^{N} \frac{n_i}{n_m}\hat{\bar{v}}_i = \sum_{i=1}^{N} \chi_i \hat{\bar{v}}_i$	$\hat{\bar{v}}_i = \left(\frac{\partial \forall_m}{\partial n_i}\right)_{p_m, T_m, n_j}$
$U_m = \sum_{i=1}^{N} n_i \hat{\bar{u}}_i$	$\bar{u}_m = U_m/n_m = \sum_{i=1}^{N} \frac{n_i}{n_m}\hat{\bar{u}}_i = \sum_{i=1}^{N} \chi_i \hat{\bar{u}}_i$	$\hat{\bar{u}}_i = \left(\frac{\partial U_m}{\partial n_i}\right)_{p_m, T_m, n_j}$
$H_m = \sum_{i=1}^{N} n_i \hat{\bar{h}}_i$	$\bar{h}_m = H_m/n_m = \sum_{i=1}^{N} \frac{n_i}{n_m}\hat{\bar{h}}_i = \sum_{i=1}^{N} \chi_i \hat{\bar{h}}_i$	$\hat{\bar{h}}_i = \left(\frac{\partial H_m}{\partial n_i}\right)_{p_m, T_m, n_j}$
$S_m = \sum_{i=1}^{N} n_i \hat{\bar{s}}_i$	$\bar{s}_m = S_m/n_m = \sum_{i=1}^{N} \frac{n_i}{n_m}\hat{\bar{s}}_i = \sum_{i=1}^{N} \chi_i \hat{\bar{s}}_i$	$\hat{\bar{s}}_i = \left(\frac{\partial S_m}{\partial n_i}\right)_{p_m, T_m, n_j}$

Table 12.4 Specific Heats of a Mixture and the Partial Specific Heats of the Gases in the Mixture

Specific Heat of the Mixture	Partial Specific Heat of Gas *i* in the Mixture
$c_{v_m} = \left(\frac{\partial u_m}{\partial T_m}\right)_{v_m} = \sum_{i=1}^{N} w_i \hat{c}_{vi}$	$\hat{c}_{vi} = \left(\frac{\partial \hat{u}_i}{\partial T_m}\right)_{v_m}$
$c_{p_m} = \left(\frac{\partial h_m}{\partial T_m}\right)_{v_m} = \sum_{i=1}^{N} w_i \hat{c}_{pi}$	$\hat{c}_{pi} = \left(\frac{\partial \hat{h}_i}{\partial T_m}\right)_{p_m}$
$\bar{c}_{v_m} = \left(\frac{\partial \bar{u}_m}{\partial T_m}\right)_{v_m} = \sum_{i=1}^{N} \chi_i \hat{\bar{c}}_{vi}$	$\hat{\bar{c}}_{vi} = \left(\frac{\partial \hat{\bar{u}}_i}{\partial T_m}\right)_{v_m}$
$\bar{c}_{p_m} = \left(\frac{\partial \bar{h}_m}{\partial T_m}\right)_{v_m} = \sum_{i=1}^{N} \chi_i \hat{\bar{c}}_{pi}$	$\hat{\bar{c}}_{pi} = \left(\frac{\partial \hat{\bar{h}}_i}{\partial T_m}\right)_{v_m}$

therefore, when either w_i, χ_i, ψ_i, or π_i is multiplied by 100, it represents the composition *percentage* of gas i present on a mass, molar, volume, or pressure basis, respectively. Note, however, that w_i and χ_i do not have the same numerical values; therefore, a *mass based percentage analysis of the composition depends on which fractional base is used in its determination.*

If we consider the mixture to be a unique substance, then we can compute its *equivalent molecular mass* M_m from Eqs. (12.1), (1.9), and (12.4) as

$$M_m = \frac{m_m}{n_m} = \frac{1}{n_m}\left(\sum_{i=1}^{N} m_i\right) = \sum_{i=1}^{N} \frac{n_i M_i}{n_m} = \sum_{i=1}^{N} \chi_i M_i \tag{12.11}$$

and using Eqs. (12.2), (1.9), and (12.3) as

$$M_m = \frac{m_m}{n_m} = \frac{m_m}{\left(\sum_{i=1}^{N} \frac{m_i}{M_i}\right)} = \frac{1}{\left(\sum_{i=1}^{N} \frac{(m_i/m_m)}{M_i}\right)} = \frac{1}{\left(\sum_{i=1}^{N} \frac{w_i}{M_i}\right)} \tag{12.12}$$

Using Eqs. (12.11) and (12.12), we can now easily convert back and forth between mass and mole fractions with

$$w_i = \frac{m_i}{m_m} = \frac{n_i M_i}{n_m M_m} = \chi_i\left(\frac{M_i}{M_m}\right) \tag{12.13}$$

and

$$\chi_i = \frac{n_i}{n_m} = w_i\left(\frac{M_m}{M_i}\right) \tag{12.14}$$

Last, if the mixture is a gas or a vapor, we can determine its equivalent gas constant R_m from the universal gas constant $\Re = 8.313\ \text{kJ/(kgmole}\cdot\text{K)} = 1545.35\ \text{ft}\cdot\text{lbf/(lbm}\cdot\text{R)}$ and Eq. (12.11) or (12.12) as

$$R_m = \frac{\Re}{M_m} \tag{12.15}$$

At this point, we have developed general formulae for determining the values of thermodynamic properties and other important characteristics of mixtures of substances in their solid, liquid, vapor, or gaseous states. Before we can continue further, we need to know how these mixture thermodynamic properties are related to each other. Since the number of possible mixture compositions is infinite, the construction of thermodynamic tables or charts for all possible mixtures is impractical except for very common mixtures, such as pure air and air–water vapor mixtures.

EXAMPLE 12.1

A new gas furnace requires a mixture of 50.0% propane and 50.0% air on a mass basis. Determine (a) the equivalent molecular mass of the mixture, (b) the mixture composition on a molar basis, and (c) the equivalent gas constant of the mixture.

Solution

a. The mixture composition on a mass basis is $w_{\text{propane}} = 0.500$, and $w_{\text{air}} = 0.500$. The molecular masses of the components are found in Table C.13 in *Thermodynamic Tables to accompany Modern Engineering Thermodynamics* as

$$M_{\text{propane}} = 44.09\ \text{kg/kgmole} \quad \text{and} \quad M_{\text{air}} = 28.97\ \text{kg/kgmole}$$

Then, the equivalent molecular mass of the mixture is given by Eq. (12.12) as

$$M_m = \frac{1}{\sum_{i=1}^{N} \frac{w_i}{M_i}} = \frac{1}{\frac{0.500}{44.09} + \frac{0.500}{28.97}} = 35.0\frac{\text{kg}}{\text{kgmole}}$$

b. The equivalent molar values are given by Eq. (12.14) as

$$\chi_i = \frac{n_i}{n_m} = w_i\left(\frac{M_m}{M_i}\right)$$

and this is used as follows:

$$\chi_{\text{propane}} = w_{\text{propane}}\left(\frac{M_m}{M_{\text{propane}}}\right) = 0.500\left(\frac{35.0\ \text{kg/kgmole}}{44.1\ \text{kg/kgmole}}\right) = 0.397$$

and the remaining component is

$$\chi_{\text{air}} = w_{\text{air}}\left(\frac{M_m}{M_{\text{air}}}\right) = 0.500\left(\frac{35.0\ \text{kg/kgmole}}{28.97\ \text{kg/kgmole}}\right) = 0.603$$

so that the composition is 50.0% propane and 50.0% air on a mass basis, but it is 39.7% propane and 60.3% air on a molar basis. (Note that, once we knew the propane molar concentration, we could have determined the air molar concentration through Eq. (12.10), since $w_{\text{propane}} + w_{\text{air}} = 1.0$ and $\chi_{\text{propane}} + \chi_{\text{air}} = 1.0$, so $\chi_{\text{air}} = 1.0 - \chi_{\text{propane}} = 1.0 - 0.397 = 0.603$.)

c. The equivalent gas constant for this mixture is given by Eq. (12.15) as

$$R_m = \frac{\Re}{M_m} = \frac{8.3143\ \text{kJ/kgmole}\cdot\text{K}}{34.97\ \text{kg/kgmole}} = 0.238\frac{\text{kJ}}{\text{kg}\cdot\text{K}}$$

The following example is slightly more complicated than Example 12.1, because it deals with a mixture containing four components. The solution method, however, is the same as in Example 12.1.

EXAMPLE 12.2

The molar composition of air that is normally used to determine the thermodynamic properties of air at standard temperature and pressure is

Component	Molar %
Nitrogen	78.09
Oxygen	20.95
Argon	0.930
CO_2 and trace gases	0.0300
Total	100.00%

Determine (a) the equivalent molecular mass, (b) the gas constant for this mixture, and (c) the composition of air on a mass (or weight) basis.

Solution

a. Since we are given the molar composition for air, we can find its equivalent molecular weight from Eq. (12.11). Assuming that argon and carbon dioxide are the only minor components present, Table C.13 provides the necessary molecular masses as

$$M_{\text{nitrogen}} = 28.02\ \text{kg/kgmole}$$
$$M_{\text{oxygen}} = 32.00\ \text{kg/kgmole}$$
$$M_{\text{argon}} = 39.94\ \text{kg/kgmole}$$
$$M_{\text{carbon dioxide}} = 44.01\ \text{kg/kgmole}$$

Then, Eq. (12.11) gives

$$\begin{aligned} M_{\text{air}} &= \sum_{i=1}^{4} \chi_i M_i = \chi_{N_2} M_{N_2} + \chi_{O_2} M_{O_2} + \chi_{Ar} M_{Ar} + \chi_{CO_2} M_{CO_2} \\ &= 0.7809(28.02\ \text{kg/kgmole}) + 0.2095(32.00\ \text{kg/kgmole}) \\ &\quad + 0.00930(39.94\ \text{kg/kgmole}) + 0.000300(44.10\ \text{kg/kgmole}) \\ &= 28.97\ \text{kg/kgmole} \end{aligned}$$

b. Equation (12.15) gives the gas constant of this mixture as

$$R_{\text{air}} = \frac{\Re}{M_{\text{air}}} = \frac{8.3143\ \text{kJ/(kgmole}\cdot\text{K)}}{28.97\ \text{kg/kgmole}} = 0.287\frac{\text{kJ}}{\text{kg}\cdot\text{K}}$$

which agrees with the values given in Table C.13b in *Thermodynamic Tables to accompany Modern Engineering Thermodynamics*.

c. Equation (12.13) can be used to determine the corresponding mass or weight fraction composition as

$$w_{N_2} = \chi_{N_2}\left(\frac{M_{N_2}}{M_{\text{air}}}\right) = 0.7809\left(\frac{28.02\ \text{kg/kgmole}}{28.97\ \text{kg/kgmole}}\right) = 0.7553 = 75.53\%\ \text{by mass}$$

$$w_{O_2} = \chi_{O_2}\left(\frac{M_{O_2}}{M_{\text{air}}}\right) = 0.2095\left(\frac{32.00\ \text{kg/kgmole}}{28.97\ \text{kg/kgmole}}\right) = 0.2314 = 23.14\%\ \text{by mass}$$

$$w_{Ar} = \chi_{Ar}\left(\frac{M_{Ar}}{M_{\text{air}}}\right) = 0.00930\left(\frac{39.94\ \text{kg/kgmole}}{28.97\ \text{kg/kgmole}}\right) = 0.0128 = 1.28\%\ \text{by mass}$$

$$w_{CO_2} = \chi_{CO_2}\left(\frac{M_{CO_2}}{M_{\text{air}}}\right) = 0.000300\left(\frac{44.01\ \text{kg/kgmole}}{28.97\ \text{kg/kgmole}}\right) = 0.00046 = 0.0456\%\ \text{by mass}$$

Note the difference between the mass and mole fraction composition values.

Exercises

1. Research has suddenly discovered that the composition given in Example 12.1 was wrong. It should have been 30.0% propane and 70.0% air on a mass basis. Determine the proper molar basis composition for this new mixture. **Answer**: $\chi_{\text{propane}} = 0.220$ and $\chi_{\text{air}} = 0.780$.

(*Continued*)

EXAMPLE 12.2 *(Continued)*

2. Oops, Research discovered still another error; it turns out that the composition given in Example 12.1 was correct except that it is 50.0% propane and 50.0% air on a molar basis. Now , Research wants you to determine the corresponding mass based composition. **Answer**: $w_{\text{propane}} = 0.603$ and $w_{\text{air}} = 0.397$.
3. A more detailed composition for air than that given in Example 12.2 is

Component	Molar %	(Answer: Mass %)
Nitrogen (N_2)	78.084	75.519
Oxygen (O_2)	20.948	23.143
Argon (Ar)	0.934	1.288
Carbon dioxide (CO_2)	0.0314	0.0477
Neon (Ne)	0.00182	0.00127
Helium (He)	0.000520	0.0000720
Methane (CH_4)	0.000200	0.000110
Krypton (Kr)	0.000110	0.000320
Hydrogen (H_2)	0.0000500	0.00000350
Dinitrogen monoxide (N_2O)	0.0000500	0.0000620
Xenon (Xe)	0.00000800	0.0000360
	Total = 100.000%	100.000%

Determine the equivalent molecular mass and the corresponding composition on a mass (or weight) basis. **Answer:** $M_m = 28.97$ kg/kgmole = 28.97 lbm/lbmole, and see the table.

12.3 MIXTURES OF IDEAL GASES

A mixture of ideal gases behaves as a unique ideal gas with an equivalent molecular mass M_m and gas constant R_m given by Eqs. (12.11) or (12.12) and (12.15). Ideal gas mixtures obey all of the ideal gas equations of state:

$$p_m \forall_m = m_m R_m T_m$$

$$u_{m2} - u_{m1} = \int_{T_{m1}}^{T_{m2}} c_{vm}\, dT_m$$

$$h_{m2} - h_{m1} = \int_{T_{m1}}^{T_{m2}} c_{pm}\, dT_m$$

and

$$s_{m2} - s_{m1} = \int_{T_{m1}}^{T_{m2}} (c_{vm}/T_m)\, dT_m + R_m \ln(v_{m2}/v_{m1})$$

$$= \int_{T_{m1}}^{T_{m2}} (c_{pm}/T_m)\, dT_m - R_m \ln(p_{m2}/p_{m1})$$

where p_m and T_m are the mixture pressure and temperature, respectively. If the mixture can be considered to have constant specific heats, then these equations reduce to

$$u_{m2} - u_{m1} = c_{vm}(T_{m2} - T_{m1})$$

$$h_{m2} - h_{m1} = c_{pm}(T_{m2} - T_{m1})$$

$$s_{m2} - s_{m1} = c_{vm} \ln(T_{m2}/T_{m1}) + R_m \ln(v_{m2}/v_{m1})$$

$$= c_{pm} \ln(T_{m2}/T_{m1}) - R_m \ln(p_{m2}/p_{m1})$$

From $p_m \forall_m = m_m R_m T_m$ and Eqs. (12.3), (12.12), and (12.15), we find that, for a mixture of ideal gases,

$$\forall_m = \frac{m_m R_m T_m}{p_m} = m_m \frac{\Re}{M_m}\left(\frac{T_m}{p_m}\right)$$

$$= m_m \Re \left(\sum_{i=1}^{N} \frac{w_i}{M_i}\right)\left(\frac{T_m}{p_m}\right) = \left(\frac{\Re T_m}{p_m}\right)\sum_{i=1}^{N} \frac{m_i}{M_i}$$

Then, Eq. (12.6) can be used to find

$$\hat{v}_i = \left(\frac{\partial \forall_m}{\partial m_i}\right)_{p_m, T_m, m_j} = \left(\frac{\Re}{M_i}\right)\left(\frac{T_m}{p_m}\right) = \frac{R_i T_m}{p_m} = v_i \tag{12.16}$$

where v_i is the specific volume of gas i at the pressure and temperature of the mixture. Similarly, using the appropriate equations for the partial specific internal energy $\hat{u}_i$, enthalpy $\hat{h}_i$, and entropy $\hat{s}_i$, we can show that, for a mixture of ideal gases with constant specific heats, the partial specific properties of gas i in the mixture are the same as the corresponding specific properties; that is,

$$\hat{u}_i = c_{vi}(T_m - T_0)$$
$$\hat{h}_i = c_{pi}(T_m - T_0)$$

and

$$\hat{s}_i = s_i = c_{pi} \ln(T_m/T_0) - R_i \ln(p_m/p_0)$$

where T_0 and p_0 are arbitrary reference state values.

These relations were also discovered experimentally in the 19th century and are now known as the Gibbs-Dalton and Amagat laws. In 1801, John Dalton (1766–1844) carried out a series of experiments that led him to conclude that the total pressure p_m of a mixture of ideal gases was equal to the sum of the *partial pressures* of the individual component gases in the mixture, where the partial pressure p_i of gas i in a mixture of ideal gases is the pressure gas i would exert if it alone occupied the volume of the mixture at the temperature of the mixture. This is known as *Dalton's law of partial pressures,* and it can be written as

$$p_m = \sum_{i=1}^{N} p_i = p_1 + p_2 + p_3 + \cdots + p_N \tag{12.17}$$

where

$$p_i = \frac{m_i R_i T_m}{\forall_m} \tag{12.18}$$

Later, Emile Amagat (1841–1915) discovered experimentally that the total volume $\forall_m$ of a mixture of ideal gases was equal to the sum of the *partial volumes* $\forall_i$ of the individual component gases in the mixture, where the partial volume $\forall_i$ of gas i in a mixture of ideal gases is the volume gas i would occupy if it alone was at the pressure and temperature of the mixture. This is known as *Amagat's law of partial volumes,* and it can be written as

$$\forall_m = \sum_{i=1}^{N} \forall_i = \forall_1 + \forall_2 + \forall_3 + \cdots + \forall_N \tag{12.19}$$

where

$$\forall_i = \frac{m_i R_i T_m}{p_m} \quad \left(\text{or } v_i = \frac{R_i T_m}{p_m}\right) \tag{12.20}$$

WHAT IS *DALTON'S LAW OF PARTIAL PRESSURES*?

Dalton's law of partial pressures states that the *partial pressure* p_i of ideal gas i in a mixture of ideal gases is equal to the pressure gas i would exert if it alone occupied the volume of the mixture at the temperature of the mixture.

WHAT IS *AMAGAT'S LAW OF PARTIAL VOLUMES*?

Amagat's law of partial volumes states that the *partial volume* $\forall_i$ of ideal gas i in a mixture of ideal gases is equal to the volume gas i would occupy if it alone was at pressure and temperature of the mixture.

Finally, the thermodynamic description of a mixture of ideal gases was completed through the work of Josiah Willard Gibbs (1838–1903), who generalized Dalton's law to define all the partial properties (except volume) of the components in the mixture to be equal to the values that those properties would have if each component gas alone occupied the volume of the mixture at the temperature of the mixture and at the partial pressure of that component. The Gibbs-Dalton ideal gas mixture law presumes that no molecular interactions take place between the components of the mixture, because each component is presumed to behave as though the other components were not present.

Under these conditions, we conclude that all the extensive properties of a mixture of ideal gases are conserved, and the mixture value of any extensive property can be determined by summing the contributions made by each gas present in the mixture.

Therefore, for ideal gases only,

$$\forall_m = \sum_{i=1}^{N} \forall_i = \sum_{i=1}^{N} m_i v_i \tag{12.21}$$

and the specific volume of this mixture can be calculated from

$$v_m = \frac{\forall_m}{m_m} = \sum_{i=1}^{N} \frac{m_i}{m_m} v_i = \sum_{i=1}^{N} w_i v_i \tag{12.22}$$

where w_i is the mass fraction of ideal gas i in the mixture and v_i is the specific volume of that gas determined at the pressure and temperature of the mixture (i.e., $v_i = R_i T_m/p_m$). Table 12.5 lists the mass and molar equations for all the total and specific properties of a mixture of ideal gases.

The mass fraction w_i, the mole fraction χ_i, the volume fraction ψ_i, and the pressure fraction π_i make four composition measures that can be used to describe a mixture. However, for ideal gases, there is a simple relation between these four quantities. From Eqs. (12.16) and (12.18), we can write the pressure and volume fractions as

$$\pi_i = \frac{p_i}{p_m} = \frac{m_i R_i T_m/\forall_m}{m_m R_m T_m/\forall_m} = \frac{w_i R_i}{R_m} = \frac{w_i M_m}{M_i} = \frac{n_i}{n_m} = \chi_i$$

and

$$\psi_i = \frac{\forall_i}{\forall_m} = \frac{m_i R_i T_m/p_m}{m_m R_m T_m/p_m} = \frac{w_i R_i}{R_m} = \frac{w_i M_m}{M_i} = \frac{n_i}{n_m} = \chi_i$$

Table 12.5 Mass and Molar Total and Specific Thermodynamic Properties of Ideal Gas Mixtures

Total Property	Mass Specific Property	Molar Specific Property
$\forall_m = \sum_{i=1}^{N} \forall_i = \sum_{i=1}^{N} m_i v_i$	$v_m = \sum_{i=1}^{N} w_i v_i$	$\bar{v}_m = \sum_{i=1}^{N} \chi_i \bar{v}_i$
$U_m = \sum_{i=1}^{N} U_i = \sum_{i=1}^{N} m_i u_i$	$u_m = \sum_{i=1}^{N} w_i u_i$	$\bar{u}_m = \sum_{i=1}^{N} \chi_i \bar{u}_i$
$H_m = \sum_{i=1}^{N} H_i = \sum_{i=1}^{N} m_i h_i$	$h_m = \sum_{i=1}^{N} w_i h_i$	$\bar{h}_m = \sum_{i=1}^{N} \chi_i \bar{h}_i$
$S_m = \sum_{i=1}^{N} S_i = \sum_{i=1}^{N} m_i s_i$	$s_m = \sum_{i=1}^{N} w_i s_i$	$\bar{s}_m = \sum_{i=1}^{N} \chi_i \bar{s}_i$
	$c_{vm} = \sum_{i=1}^{N} w_i c_{vi}$	$\bar{c}_{vm} = \sum_{i=1}^{N} \chi_i \bar{c}_{vi}$
	$c_{pm} = \sum_{i=1}^{N} w_i c_{pi}$	$\bar{c}_{pm} = \sum_{i=1}^{N} \chi_i \bar{c}_{pi}$

Consequently, for a mixture of ideal gases, we have

$$\frac{p_i}{p_m} = \pi_i = \frac{V_i}{V_m} = \psi_i = \frac{n_i}{n_m} = \chi_i = w_i\left(\frac{M_m}{M_i}\right)$$

or

$$\pi_i = \psi_i = \chi_i = w_i\left(\frac{M_m}{M_i}\right) \tag{12.23}$$

which relates all four composition measures. Thus, the pressure fraction, volume fraction, and mole fraction are all equal and differ from the mass fraction by only a molecular mass ratio.

EXAMPLE 12.3

An analysis of the exhaust gas from an experimental engine produces the following results on a molar basis:

Carbon dioxide = 9.51%
Water = 19.01%
Nitrogen = 71.48%

Assuming ideal gas behavior, (a) determine the volume fraction, pressure fraction, and mass fraction composition of the mixture, and (b) if the total pressure of the mixture is 14.7 psia, determine the partial pressure of the water vapor in the exhaust gas mixture.

Solution

a. For ideal gas behavior, Eq. (12.23) tells us that the mole fractions, volume fractions, and the pressure fractions are all the same, or

$$\chi_{CO_2} = \psi_{CO_2} = \pi_{CO_2} = 9.51\%$$
$$\chi_{H_2O} = \psi_{H_2O} = \pi_{H_2O} = 19.01\%$$
$$\chi_{N_2} = \psi_{N_2} = \pi_{N_2} = 71.48\%$$

The equivalent molecular mass of this ideal gas mixture is given by Eq. (12.11) as

$$\begin{aligned} M_m &= \sum_{i=1}^{N} \chi_i M_i = \chi_{CO_2} M_{CO_2} + \chi_{H_2O} M_{H_2O} + \chi_{N_2} M_{N_2} \\ &= 0.0951(44.01\text{ kg/kgmole}) + 0.1901(18.02\text{ kg/kgmole}) + 0.7148(28.02\text{ kg/kgmole}) \\ &= 27.64\text{ kg/kgmole} \end{aligned}$$

and the corresponding mass fractions are given by Eq. (12.23) as

$$w_{CO_2} = \psi_{CO_2}\left(\frac{M_{CO_2}}{M_m}\right) = (9.51\%)\left(\frac{44.01\text{ kg/kgmole}}{27.64\text{ kg/kgmole}}\right) = 15.14\%$$

$$w_{H_2O} = \psi_{H_2O}\left(\frac{M_{H_2O}}{M_m}\right) = (19.01\%)\left(\frac{18.02\text{ kg/kgmole}}{27.64\text{ kg/kgmole}}\right) = 12.39\%$$

$$w_{N_2} = \psi_{N_2}\left(\frac{M_{N_2}}{M_m}\right) = (71.48\%)\left(\frac{28.02\text{ kg/kgmole}}{27.64\text{ kg/kgmole}}\right) = 72.46\%$$

b. If the total pressure is 14.7 psia, then the partial pressure of the water vapor in the exhaust gas mixture is given by Eq. (12.23) as

$$p_{H_2O} = p_m \chi_{H_2O} = (14.7\text{ psia})(0.1901) = 2.79\text{ psia}$$

Exercises

4. Find the partial volume of the nitrogen in the exhaust gas in Example 12.3 if the total volume of the gas mixture is 8.00 ft^3. **Answer:** $V_{N_2} = 2.14\text{ ft}^3$.
5. If there are 20.0 moles of the exhaust gas mixture in Example 12.3, how many moles of carbon dioxide would be in the mixture? **Answer:** $n_{CO_2} = 14.3$ moles.
6. If there is 0.650 lbm of the exhaust gas mixture in Example 12.3, how many lbm of water vapor would be in the mixture? **Answer:** $m_{\text{water vapor}} = 0.081$ lbm.

EXAMPLE 12.4

Though oxygen is necessary to sustain life, breathing oxygen at elevated pressure has toxic effects. It causes changes in lung tissue and affects the liver and brain. Acute oxygen poisoning at high pressures can cause convulsions that can lead to death (even at atmospheric pressure, pure oxygen can be breathed safely for only two hours). Oxygen poisoning at elevated environmental pressure can be avoided by maintaining the oxygen partial pressure equal to that of atmospheric air at standard temperature and pressure (STP). Also, since atmospheric nitrogen is very soluble in blood and body tissue, rapid depressurization causes nitrogen bubbles to form in the blood and tissue (nitrogen embolisms) producing a condition commonly called the *bends*.

Divers going to great depths in the sea are able to circumvent this problem somewhat by breathing a compressed helium-oxygen mixture in which the oxygen partial pressure is adjusted so that it is always equal to its value in atmospheric air at STP. Helium, being much less soluble in body tissue than nitrogen, decreases the time required for depressurization when the diver returns to the surface.

The engineering problem that we must solve is stated as follows:

a. For a deep water diver, determine the proper helium-oxygen breathing mixture composition for a dive to 100. m below the surface of the water, where the pressure is 1.08 MN/m^2. Give your answer in mole, volume, and mass fractions.
b. Determine the effective gas constant, the specific heats, and the specific heat ratio for this mixture.

Assume helium and oxygen behave as ideal gases with constant specific heats.

Solution

a. From Example 12.3 and Eq. (12.23), we find that the partial pressure of oxygen in air at STP is

$$p_{O_2} = \chi_{O_2} p_m = 0.2095(0.1013\,\text{MN/m}^2) = 0.0212\,\text{MN/m}^2$$

Therefore, at a total pressure in 100. m of water of 1.08 MN/m^2, this same partial pressure requires a mole and volume fraction of oxygen of only

$$\chi_{O_2} = \psi_{O_2} = \pi_{O_2} = p_{O_2}/p_m = 0.0212/1.08 = 0.0196$$

and, from Eq. (12.10), the helium mole and volume fractions are

$$\chi_{He} = \psi_{He} = 1 - \chi_{O_2} = 1 - 0.0196 = 0.980$$

The equivalent mass fractions are given by Eq. (12.13), where the mixture equivalent molecular mass can be computed from Eq. (12.11), as

$$M_m = \chi_{O_2} M_{O_2} + \chi_{He} M_{He} = 0.0196(32.00) + 0.980(4.003) = 4.55\,\text{kg/kgmole}$$

then,

$$w_{O_2} = \chi_{O_2}(M_{O_2}/M_m) = 0.0196(32.00/4.55) = 0.138$$

and

$$w_{He} = 1 - w_{O_2} = 0.862$$

Therefore, the required oxygen concentration is only 1.96% on a volume or molar basis, but it is 13.8% on a mass or weight basis.

b. The mixture equivalent gas constant can be computed from Eq. (12.15) as

$$R_m = \frac{\Re}{M_m} = \frac{8.3143\,\text{kJ/(kgmole}\cdot\text{K)}}{4.55\,\text{kg/kgmole}} = 1.86\,\text{kJ/(kg}\cdot\text{K)}$$

and the mixture specific heats can be determined from the equations in Table 12.5 and Table C.13b as

$$c_{vm} = w_{O_2} c_{vO_2} + w_{He} c_{vHe} = 0.138(0.657) + 0.862(3.123) = 2.78\,\text{kJ/(kg}\cdot\text{K)}$$

and

$$c_{pm} = w_{O_2} c_{pO_2} + w_{He} c_{pHe} = 0.138(0.917) + 0.862(5.200) = 4.61\,\text{kJ/(kg}\cdot\text{K)}$$

Finally, the specific heat ratio of the mixture is

$$k_m = \frac{c_{pm}}{c_{vm}} = \frac{4.61}{2.78} = 1.66$$

(Note that $k_m \neq \sum w_i k_i$ because of its definition as a ratio.)

Exercises

7. Rework Example 12.4 for a dive to 200. m below the surface of the water, where the pressure is 2.063 MN/m^2. **Answer:** $\chi_{O_2} = 0.0103$, $\chi_{He} = 0.990$, $R_m = 1.94$ kJ/kg·K, $c_{vm} = 2.93$ kJ/kg·K, $c_{pm} = 4.87$ kJ/kg·K, and $k = 1.66$.
8. Rework Example 12.4 for an argon-oxygen mixture subject to the same condition of $p_{O_2} = 0.0212$ MN/m^2. **Answer:** $\chi_{O_2} = 0.0196$, $\chi_{He} = 0.980$, $R_m = 0.209$ kJ/kg·K, $c_{vm} = 3.08$ kJ/kg·K, $c_{pm} = 5.13$ kJ/kg·K, and $k = 1.66$.

Ideal gas mixture equations are used to produce property values in thermodynamic problems just as though the mixture is a single unique gas. This is illustrated in the next example.

EXAMPLE 12.5

Determine the power per unit mass flow rate required to isentropically compress the helium-oxygen mixture described in Example 12.2 from atmospheric pressure (0.101 MN/m^2) and 20.0°C to 1.08 MN/m^2 in a steady flow, steady state process. Assume the mixture has constant specific heats.

Solution

The unknown here is the power per unit mass flow rate $\dot{W}/\dot{m}_m$. Since this is an open system, the energy rate balance for a steady state, steady flow, single-inlet, single-outlet system is (neglecting kinetic and potential energy effects)

ERB (SS, SF, SI, SO)

$$\dot{Q} - \dot{W} + \dot{m}_m(h_{m1} - h_{m2}) = 0$$

and, since an isentropic process is normally also adiabatic,

$$\dot{W}/\dot{m}_m = h_{m1} - h_{m2} = c_{pm}(T_{m1} - T_{m2})$$

For an ideal gas in an isentropic process, Eq. (7.38) gives

$$T_{m2} = T_{m1}(p_{m2}/p_{m1})^{(k_m-1)/k_m}$$

and, using the results of Example 12.4, this gives

$$T_{m2} = (20.0 + 273.15\,\text{K})(1.08/0.101)^{0.66/1.66} = 803\,\text{K} = 530°\text{C}$$

Then,

$$\dot{W}/\dot{m}_m = c_{pm}(T_{m1} - T_{m2}) = [4.61\,\text{kJ/(kg·K)}](293 - 803\,\text{K}) = -2350\,\text{kJ/kg}$$

Exercises

9. Determine the isentropic power per unit mass flow rate required in Example 12.5 if the inlet temperature is 10.0°C rather than 20.0°C. **Answer:** $\dot{W}/\dot{m}_m = -2030$ kJ/kg.
10. If the exit pressure in Example 12.5 is increased to 2.06 MN/m^2, determine the required isentropic power input per unit mass flow rate. **Answer:** $\dot{W}/\dot{m}_m = -3110$ kJ/kg.
11. Determine the aergonic (zero work) heat transfer required to cool the compressed mixture in Example 12.5 from 530°C back to 20.0°C again. **Answer:** $\dot{Q}/\dot{m}_m = -2350$ kJ/kg.

12.4 PSYCHROMETRICS

Psychrometrics is the study of *atmospheric air*, which is a mixture of pure air and water vapor at atmospheric pressure.[4] The pure air portion of an air–water vapor mixture is commonly called *dry air*; consequently, atmospheric air is said to consist of a mixture of dry air and water vapor. Both the air and the water vapor in

[4] The term *psychrometer* is from the Greek *psychros* (cold) and *meter* (measure).

CRITICAL THINKING

On page 77 of the July/August 1993 issue of *Family Handyman* magazine, a helpful hint is given on how to determine the amount of propane remaining a cook stove tank. According to this magazine you just "Pour a cup of hot water over the outside of the tank. A condensation line will appear on the tank surface at the level of the remaining propane."

Since the formation of a "condensation line" requires reducing the liquid-vapor interface inside the tank to a temperature below the dew point temperature, can you explain how this test works? Are there any conditions under which this test would not work? (Hint: Look at the thermodynamics of the propane's evaporation and condensation processes that result from the heating by the hot water and the subsequent cooling by the local atmosphere.)

this mixture are treated as ideal gases (even though we say *water vapor* and not *water gas*). This particular mixture of ideal gases is important because of its meteorological and environmental comfort implications.

To begin this discussion, we define two new composition measures for the amount of water vapor present in the mixture. Both measures are a type of *humidity*, as is shown.[5]

1. The *relative humidity* ϕ is the ratio of the actual partial pressure of the water vapor present in the mixture to the saturation pressure of the water vapor at the temperature of the mixture, or

$$\text{Relative humidity} = \phi = \frac{p_w}{p_{\text{sat}}} \tag{12.24}$$

The value of p_{sat} can be found in Table C.1 in *Thermodynamic Tables to accompany Modern Engineering Thermodynamics* at the temperature of the mixture. Since $0 \leq \phi \leq 1$, the relative humidity is normally reported as a percentage. This is the common meteorological humidity measure.

2. The *humidity ratio* ω is the ratio of the mass of water vapor present in the mixture divided by the mass of dry air present in the mixture, or

$$\text{Humidity ratio } \omega = \frac{m_w}{m_a} \tag{12.25}$$

where $m_m = m_a + m_w$, and $p_m = p_a + p_w$ = atmospheric pressure. Assuming ideal gas behavior for both the air and water vapor, we can write $m_w = p_w\forall_m/(R_wT_m)$ and $m_a = p_a\forall_m/(R_aT_m)$, then

$$\omega = \frac{p_wR_a}{p_aR_w} = \frac{p_wM_w}{p_aM_a} = \frac{18.016p_w}{28.97p_a} = 0.622\frac{p_w}{p_a} = 0.622\left[\frac{p_w}{p_m - p_w}\right] \tag{12.26a}$$

From Eq. (12.24), we find that $p_w = \phi p_{\text{sat}}$, and substituting this into Eq. (12.26a) provides a formula that relates the two humidity measures:

$$\omega = 0.622\phi\frac{p_{\text{sat}}}{p_a} = 0.622\left[\frac{\phi p_{\text{sat}}}{p_m - p_w}\right] \tag{12.26b}$$

A colorful term from the meteorological profession is the *dew point* temperature T_{DP}, which is the temperature at which liquid water (dew) condenses out of the atmosphere at constant atmospheric pressure (and consequently at constant water vapor partial pressure):

$$T_{DP} = T_{\text{sat}}\ (\text{evaluated at } p_w) \tag{12.27}$$

If the partial pressure of the water vapor (p_w) is known, then the dew point temperature can be found in Table C.2. Figure 12.1 illustrates these concepts on a pressure-specific volume schematic.

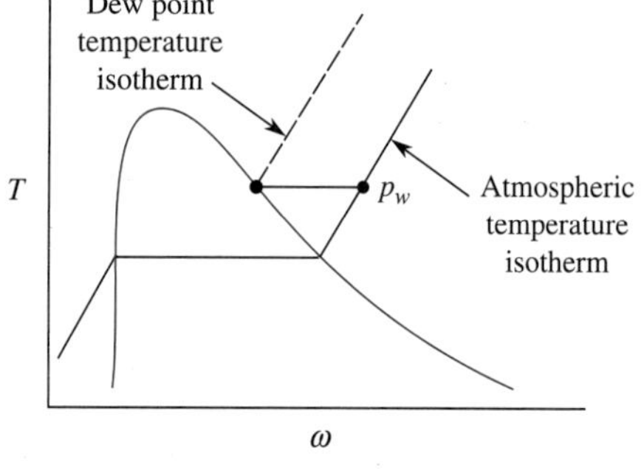

FIGURE 12.1
The partial pressure and dew point temperature of a mixture of water vapor and dry air.

[5] Since neither of these two humidity measures corresponds to any of the four composition measures previously discussed, this brings the number of composition measures used in this chapter to six.

EXAMPLE 12.6

On a particular day, the weather forecast states that the relative humidity is 56.8% when the atmospheric temperature and pressure are 25.0°C and 0.101 MPa, respectively. Determine:

a. The partial pressure of the water vapor in the atmosphere.
b. The humidity ratio of the atmosphere.
c. The dew point temperature of the atmosphere.

Solution

a. From Table C.1b, we find that

$$p_{\text{sat}}(25.0°\text{C}) = 0.003169\,\text{MPa}$$

and, from Eq. (12.24), we can calculate the partial pressure of the water vapor present in the mixture as

$$p_w = \phi p_{\text{sat}} = 0.568(0.003169\,\text{MPa}) = 0.00180\,\text{MPa} = 1.80\,\text{kPa}$$

b. From Dalton's law for partial pressure, we can find the partial pressure of the dry air in the mixture as

$$p_a = p_m - p_w = 101 - 1.8 = 99.2\,\text{kPa}$$

then, Eq. (12.26a) gives the humidity ratio ω as

$$\omega = 0.622(p_w/p_a) = 0.622(1.80/99.2) = 0.0113\,\text{kg H}_2\text{O per kg of dry air}$$

Note that, since the value of ω is not constrained to lie between 0 and 1, it is not reported as a percentage.

c. Using Eq. (12.27) and Table C.2b, we find the dew point temperature to be

$$T_{DP} = T_{\text{sat}}(0.00180\,\text{MPa}) = 15.8°\text{C}$$

Exercises

12. If the relative humidity in Example 12.6 is 45.0% rather than 56.8% and all the remaining variables are the same, determine the new dew point temperature. **Answer**: T_{DP} = 12.1°C.
13. Suppose the atmospheric temperature in Example 12.6 is 20.0°C rather than 25.0°C and all other variables remain the same. Determine the humidity ratio of this mixture. **Answer**: ω = 0.00830 kg H_2O per kg of dry air.
14. Rework Example 12.6 for a relative humidity of 35.0%, an atmospheric temperature of 20.0°C, and an atmospheric pressure of 0.101 MPa. **Answer**: (a) p_w = 0.820 kPa, (b) $\omega = 5.00 \times 10^{-3}$ kg H_2O per kg of dry air, (c) T_{DP} = 4.00°C.

The steady state, steady flow, isothermal boundary energy and entropy rate balances for a mixture of dry air and water vapor with negligible flow stream kinetic and potential energies can be written either on an unmixed *component* basis as

$$\dot{Q} - \dot{W} + \dot{m}_a(h_1 - h_2)_a + \dot{m}_w(h_1 - h_2)_w = 0 \tag{12.28}$$

and

$$\dot{Q}/T_b + \dot{m}_a(s_1 - s_2)_a + \dot{m}_w(s_1 - s_2)_w + \dot{S}_p = 0 \tag{12.29}$$

or on a premixed *mixture* basis as

$$\dot{Q} - \dot{W} + \dot{m}_m(h_1 - h_2)_m = 0$$

and

$$\dot{Q}/T_b + \dot{m}_m(s_1 - s_2)_m + \dot{S}_p = 0$$

where the mixture enthalpy and entropy changes are given by

$$(h_1 - h_2)_m = w_a(h_1 - h_2)_a + w_w(h_1 - h_2)_w$$

and

$$(s_1 - s_2)_m = w_a(s_1 - s_2)_a + w_w(s_1 - s_2)_w$$

In these formulae, h_a is found in the gas tables (Table C.16), h_w is found in the superheated steam tables, and w_a and w_w are the mass fractions of the dry air and water vapor. However, since psychrometrics involves only a two-component mixture, there is no particular advantage to using the complicated premixed *mixture* formula. Therefore, we confine our attention to the simpler unmixed *component* form illustrated in Eqs. (12.28) and (12.29).

12.5 THE ADIABATIC SATURATOR

Evaporative humidification processes normally occur without external heat transfer and are therefore adiabatic. If the outlet of an evaporative humidifier is saturated with water vapor ($\phi = 100\%$), then the device is known as an *adiabatic saturator*. A simple adiabatic saturator is shown in Figure 12.2. It consists of an inlet air–water vapor flow stream at temperature T_1, a liquid makeup water flow stream at temperature T_2, and an outlet air–water vapor flow stream. If the unit is insulated and made long enough, the outlet flow stream will be saturated with water vapor, and the temperature T_3 of the outlet flow stream is then called the *adiabatic saturation* temperature.

Since this device is adiabatic and aergonic, its steady state, steady flow energy rate balance (ERB) reduces to (neglecting changes in flow stream kinetic and potential energy)

$$\dot{m}_{a1}h_{a1} + \dot{m}_{w1}h_{w1} + \dot{m}_{w2}h_{w2} - \dot{m}_{a3}h_{a3} - \dot{m}_{w3}h_{w3} = 0$$

where we have chosen to separate the contributions from the air and water components according to Eq. (12.28). From the conservation of mass, $\dot{m}_{a1} = \dot{m}_{a3} = \dot{m}_a$ and $\dot{m}_{w2} = \dot{m}_{w3} - \dot{m}_{w1}$.

Then, the ERB becomes

$$\dot{m}_a(h_{a1} - h_{a3}) + (\dot{m}_{w3} - \dot{m}_{w1})h_{w2} + \dot{m}_{w1}h_{w1} - \dot{m}_{w3}h_{w3} = 0$$

or

$$\dot{m}_a(h_{a1} - h_{a3}) + \dot{m}_{w1}(h_{w1} - h_{w2}) + \dot{m}_{w3}(h_{w2} - h_{w3}) = 0$$

Dividing by $\dot{m}_a$ and introducing the humidity ratios $\omega_1 = \dot{m}_{w1}/\dot{m}_a$ and $\omega_3 = \dot{m}_{w3}/\dot{m}_a$, and solving for ω_1 gives

$$\omega_1 = \frac{(h_{a3} - h_{a1}) + \omega_3(h_{w3} - h_{w2})}{h_{w1} - h_{w2}} \tag{12.30}$$

Since we can treat the air here as an ideal gas and assuming $T_3 = T_2$,

$$h_{a3} - h_{a1} = c_{pa}(T_3 - T_1) = c_{pa}(T_2 - T_1)$$

and since the liquid makeup water is only a slightly compressed liquid, we can write

$$h_{w2} \approx h_f(T_2) = h_{f2}$$

Finally, since the outlet state contains saturated water vapor at the adiabatic saturation temperature, $T_3 = T_2$,

$$h_{w3} = h_g(T_3) = h_g(T_2) = h_{g2}$$

The water vapor in the inlet region is superheated. A quick check of the Mollier diagram (Figure 7.15) reveals that the isotherms in the low-pressure superheated region are very nearly horizontal. Therefore, the enthalpy of water vapor in this region depends only on temperature, so we can take

$$h_{w1} = h_g(T_1) = h_{g1}$$

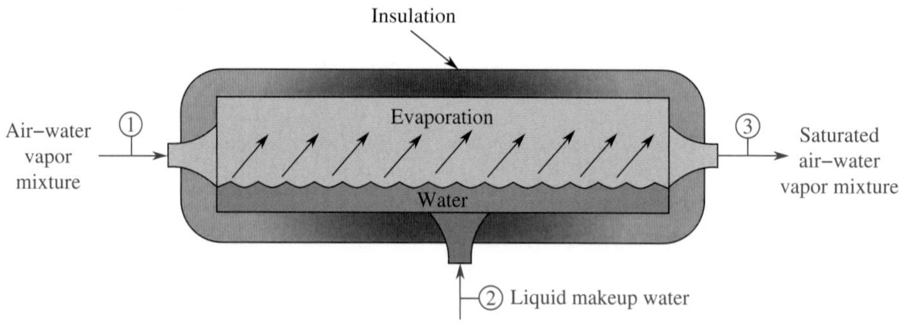

FIGURE 12.2
An adiabatic saturator.

Then, Eq. (12.30) becomes

Adiabatic saturator inlet humidity ratio

$$\omega_1 = \frac{c_{pa}(T_2 - T_1) + \omega_3(h_{fg2})}{h_{g1} - h_{f2}} \tag{12.31}$$

Thus, by simply measuring the inlet temperature T_1 and the outlet adiabatic saturation temperature $T_2 = T_3$, we can calculate the inlet humidity ratio of the air–water vapor mixture, ω_1, directly from Eq. (12.31). However, an adiabatic saturator must be extremely long to obtain 100% relative humidity at the outlet. This difficulty is overcome by the sling psychrometer discussed next.

12.6 THE SLING PSYCHROMETER

Figure 12.3 illustrates a simple device for determining air humidity, called a *sling psychrometer*. It contains two thermometers, one of which is covered with a wick saturated with ambient temperature liquid water. These two thermometers are called *dry bulb* and *wet bulb*. When the sling psychrometer is spun rapidly in the air, the evaporation of the water from the wick causes the wet bulb thermometer to read lower than the dry bulb thermometer. After the psychrometer has been spun long enough for the thermometers to reach equilibrium temperatures, the unit is stopped and the two thermometers are quickly read. A psychrometric chart (or table) is then used to convert the dry bulb temperature T_{DB} and the wet bulb temperature T_{WB} into humidity information. The wet bulb temperature is approximately equal to the adiabatic saturation temperature, so $T_{WB} \approx T_2 = T_3$ in Eq. (12.31).

Figure 12.4 illustrates the major characteristics of a psychrometric chart. Larger charts of professional engineering quality can be found in Charts D.5 and D.6 of *Thermodynamic Tables to accompany Modern Engineering*

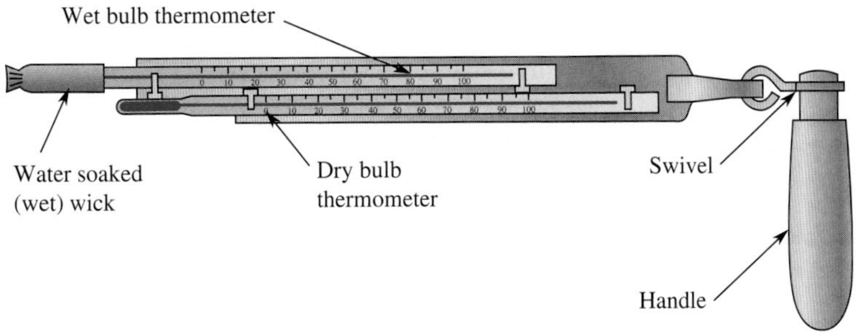

FIGURE 12.3
A sling psychrometer.

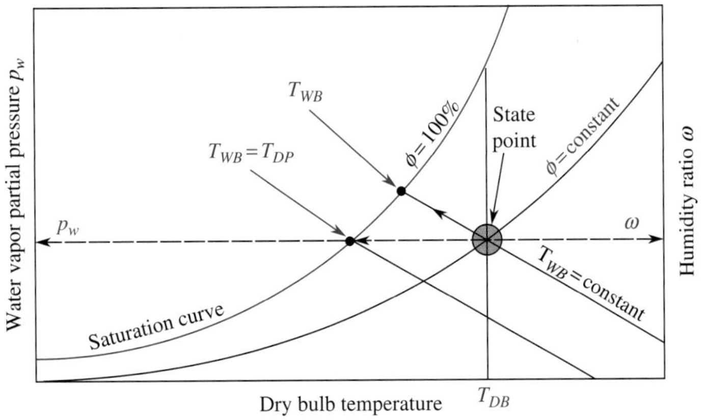

FIGURE 12.4
The elements of a psychrometric chart. The intersection of the dry bulb and wet bulb constant temperature lines determine the state of the water vapor in the system, from which T_{DP}, p_w, ϕ, and ω can then be found.

CRITICAL THINKING

A sling psychrometer can be used to determine the humidity of the surrounding air because its wet bulb temperature is nearly the same as the adiabatic saturation temperature, and Eq. (12.31) can then be used to find $\omega_1 = \omega_{air}$ in Figure 12.2. The combined heat and mass transfer rate analysis of a wet bulb thermometer yields the following result:

$$\frac{\omega_{adiabatic} - \omega_{air}}{T_{air} - T_{adiabatic}} = \left(\frac{\omega_{WB} - \omega_{air}}{T_{air} - T_{WB}}\right)\left(\frac{\text{Pr}}{\text{Sc}}\right)^{2/3}$$

where $\omega_{adiabatic} = \omega_3$ in Figure 12.2, ω_{WB} is the humidity ratio in the vicinity of the wet bulb, $T_{air} = T_{DB} = T_1$ in Figure 12.2, $\text{Pr} = c_p\mu/k$ is the Prandtl number, and $\text{Sc} = \mu/(\rho D)$ is the Schmidt number. Pr and Sc are traditional dimensionless numbers composed of viscosity μ, constant pressure specific heat c_p, specific heat ratio k, density ρ, and mass diffusivity D. From the equation, we see that $T_{WB} = T_{adiabatic}$ only if $(\text{Pr/Sc})^{2/3} = 1$. It turns out that, for water vapor in air, the Prandtl to Schmidt number ratio is about 1.0, so the wet bulb and adiabatic saturation temperatures are about the same. However, for any other chemical vapor–air mixture, $(\text{Pr/Sc})^{2/3}$ generally is not equal to 1, and the wet bulb and adiabatic saturation temperatures are not equal.

Suppose you had a serious leak or spill of a dangerous liquid chemical, like a refrigerant that subsequently evaporated into the air of a closed room, and you wanted to use a sling psychrometer to estimate the resulting concentration of the chemical in the air. Is it possible to use this equation to correct the wet bulb temperature (with the bulb wetted with the spilled liquid) so that it could be used with a psychrometric chart (or Eq. (12.31)) to give an estimate of the concentration of the spilled chemical in the air?

Thermodynamics. Note that the dry bulb temperature is just the temperature registered on any ordinary thermometer and the psychrometric chart is just part of the p-T diagram for saturated and superheated water vapor in the low-pressure region. When the mixture is saturated with water vapor (ϕ = 100%), no water can evaporate from the wet bulb wick and $T_{WB} = T_{DB} = T_{DP}$. Note also that a psychrometric chart is drawn for a fixed total pressure, thus Charts D.5 and D.6 are valid only for mixtures at 1 atm total pressure.

EXAMPLE 12.7

Wet and dry bulb temperature measurements made outside on a cold day reveal that T_{DB} = 5.0°C and T_{WB} = 4.0°C. Using the psychrometric chart, determine

a. ϕ, ω, T_{DP}, and p_w for the outside air
b. The values of ϕ, ω, T_{WB}, and p_w if this mixture is heated at constant pressure to 25.0°C.

Solution

a. From Chart D.6 at T_{DB1} = 5.0°C and T_{WB1} = 4.0°C, we read ϕ_1 = 80.%, ω_1 = 0.004 kg of water vapor per kg of dry air, T_{DP1} = 20°C, and p_{w1} = 700. N/m^2.
b. Now the mixture is heated at constant pressure until its dry bulb temperature increases to 25.0°C. Note that, when the temperature is stated without a modifier (i.e., "wet" or "dry"), we presume it is the ordinary, or dry bulb, temperature. Then, Chart D.6 gives $\phi_2 \approx$ 20.%, $\omega_2 = \omega_1$, T_{WB2} = 13°C, $T_{DP2} = T_{DP1}$, and $p_{w2} = p_{w1}$.

This is shown in Figure 12.5.

Notice that, under these conditions, the relative humidity and the wet and dry bulb temperatures change, but none of the other characteristics change. This is because the amount of water vapor and the amount of air present do not change.

Exercises

15. Determine the values of ϕ, ω, T_{WB}, and p_w when the mixture in Example 12.7 is reheated to 20.0°C rather than 25.0°C and all the other variables remain the same. **Answer:** ϕ = 28%, ω = 0.004 kg H_2O per kg of dry air, T_{WB} = 11°C, and p_w = 700. N/m^2.
16. If the dry bulb temperature in Example 12.7 is 8.0°C rather than 5.0°C and all the other variables remain the same, determine ϕ, ω, and p_w for the outside air. **Answer:** ϕ = 50%, ω = 0.0035 kg H_2O per kg of dry air, and p_w = 550 N/m^2.
17. Rework Example 12.7 for a dry bulb temperature of 10.0°C and a wet bulb temperature of 8.0°C. **Answer:** (a) ϕ = 75%, ω = 0.006 kg H_2O per kg of dry air, T_{DP} = 6.0°C, and p_w = 900 N/m^2; and (b) ϕ = 30%, ω = 0.006 kg H_2O per kg of dry air, T_{WB} = 14°C, and p_w = 900 N/m^2.

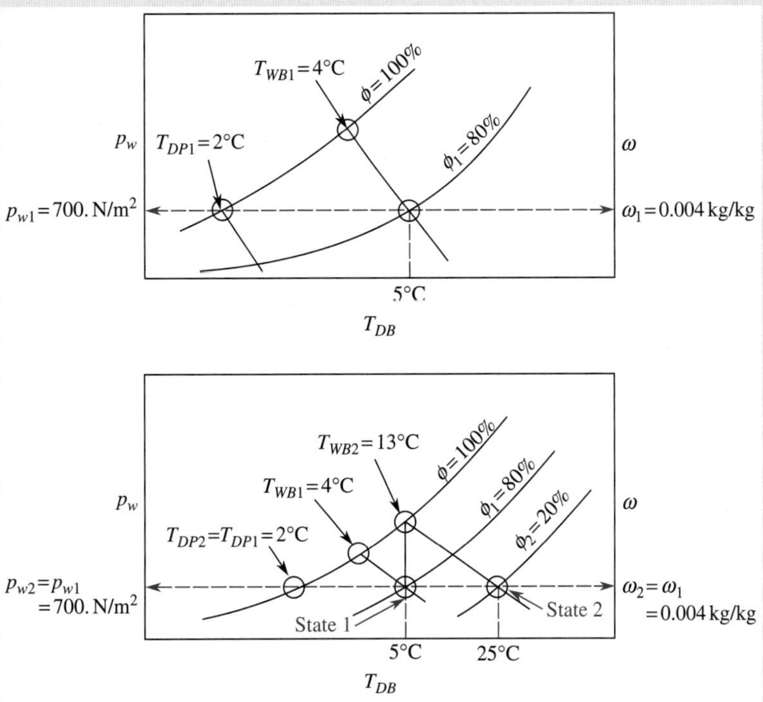

FIGURE 12.5
Example 12.7.

If a sling psychrometer is spun such that the air velocity over the wick is greater than 3.0 m/s, then the wet bulb temperature is essentially equal to the adiabatic saturation temperature T_2 in Eq. (12.31). The following example illustrates this point.

EXAMPLE 12.8

The wet and dry bulb temperatures measured in a dormitory room are 60.0°F and 70.0°F, respectively, when the barometric pressure is 14.7 psia. Assuming that the wet bulb temperature is equal to the adiabatic saturation temperature, use Eq. (12.31) to find the humidity ratio (ω) in the room and compare your answer with that obtained from the psychrometric chart, Chart D.5.

Solution

Here, we have T_{WB} = 60.0°F and T_{DB} = 70.0°F. Then, from Table C.1a in *Thermodynamic Tables to accompany Modern Engineering Thermodynamics*, we find

$$h_{g1} = h_g(70.0°F) = 1092.0 \text{ Btu/lbm}$$
$$h_{fg2} = h_{fg}(60.0°F) = 1059.6 \text{ Btu/lbm}$$
$$h_{f2} = h_f(60.0°F) = 28.1 \text{ Btu/lbm}$$

and

$$p_{w3} = p_{sat}(60.0°F) = 0.2563 \text{ psia}$$

Then, Eq. (12.26a) gives

$$\omega_3 = 0.622(0.2563)/(14.7 - 0.2563) = 0.0110 \text{ lbm water per lbm of dry air}$$

and, from Eq. (12.31), we get[6]

$$\omega_1 = \frac{0.240(60-70) + 0.0110(1059.6)}{1092.0 - 28.1} = 0.00874 \text{ lbm water per lbm of dry air}$$
$$= 0.00874(7000) = 61.2 \text{ grains of water per lbm of dry air}$$

(Continued)

EXAMPLE 12.8 (*Continued*)

which, with $T_{WB} = 60.0°F$ and $T_{DB} = 70.0°F$, the psychrometric chart, Chart D.5, gives approximately

$$\omega_1 = 61 \text{ grains of water per lbm of dry air}$$

which is essentially the same as that calculated from Eq. (12.31).

Exercises

18. If the dry bulb temperature in the room discussed in Example 12.8 is 80.0°F rather than 70.0°F, calculate the humidity ratio in the room and compare your answer with that obtained from the psychrometric chart. **Answer:** $\omega_1 = 45.2$ grains of water per lbm of dry air.

19. Rework Example 12.8 for wet and dry bulb temperatures of 65.0°F and 85.0°F, respectively. **Answer:** $\omega_1 = 60.2$ grains of water per lbm of dry air.

20. Using the method of your choice, determine the relative humidity ratio in the room discussed in Example 12.8 when the wet and dry bulb temperatures are 18.0°C and 22.0°C, respectively. **Answer:** $\omega = 0.0115$ kg of water per kg of dry air.

[6] *The grain is the smallest of the ancient Egyptian measures of weight (see Chapter 1) and originally represented the average weight of a grain of barley corn. Today, it is still used in some engineering fields (e.g., in the heating, ventilating, and air conditioning field) as a mass unit, with 7000 grains = 1 lbm.*

Equation (12.31) gives essentially the same values as obtained from the psychrometric chart, but the chart is much easier and quicker to use.

12.7 AIR CONDITIONING

Complete air conditioning involves producing an environment with desired pressure, temperature, humidity, purity, and circulation characteristics. In this section, we are concerned only with altering the temperature and the humidity in typical air conditioning applications.

In Example 12.8, we see how winter air is severely dehumidified if it is simply heated up to room temperature. Water vapor must be added to bring its humidity up into the 40 to 50% relative humidity range. This can easily be done by blowing the heated air across a moist surface, as shown in Figure 12.6. This is the technique used in a common room humidifier.

The humidification process 2–3 shown in Figure 12.6 is also an example of evaporative cooling. When unsaturated air is brought into contact with liquid water at the same (dry bulb) temperature, some of the water evaporates (thus cooling the mixture) and the resulting air–water vapor mixture moves upward along the T_{WB} = constant line, as shown in Figure 12.6b. The minimum dry bulb temperature that can be produced by evaporative cooling occurs when the outlet air becomes saturated with water vapor ($\phi = 100\%$), then $T_{DB} = T_{WB}$. This concept is illustrated in the following example.

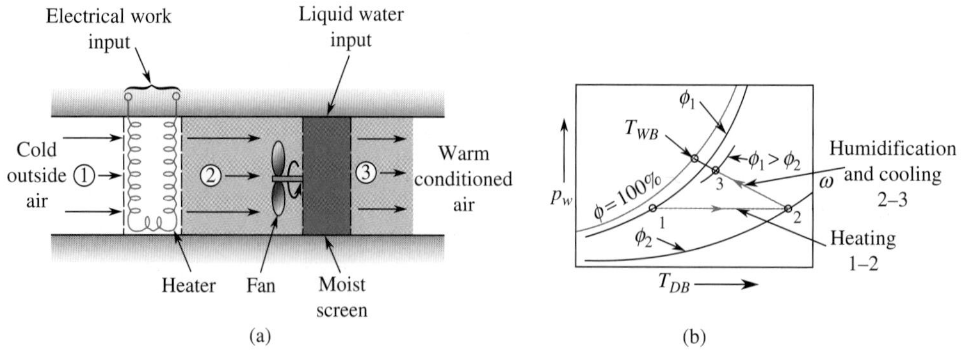

FIGURE 12.6
Temperature and humidity conditioning of cold winter air.

WHAT ENVIRONMENTAL CONDITIONS MAKE PEOPLE COMFORTABLE?

Humans are essentially isothermal open systems with complex temperature-regulating mechanisms. Body temperature (98.6°F, 37.0°C) is normally above the surrounding environmental temperature so that the excess heat generated by the irreversibilities inside the body can be removed by normal convection, conduction, and radiation heat transfer mechanisms. During periods of physical stress or high environmental temperature, the body produces a surface layer of water, called *perspiration*, whose evaporation into the atmosphere helps cool the body. This is one of the body's primary temperature-regulating mechanisms. When the relative humidity of the surrounding atmosphere is high, the evaporation of body perspiration is low and the body automatically tries to minimize its internal heat generation, resulting in the person's feeling lethargic and becoming inactive. Because the sensation of human comfort is so subjective, attempts to define a "comfortable" atmosphere have met with only limited success. Tests have shown that a relative humidity below 15% produces a dried (or *parched*) condition of the membranes in the mouth, nose, and lungs and an increased susceptibility to disease germs. However, a relative humidity above 70% causes an accumulation of moisture in the clothing and a general "sticky" or "muggy" feeling. For best health and comfort conditions, it has been found that the relative humidity should range from 40 to 50% during cold winter weather and from 50 to 60% during warm summer weather.

EXAMPLE 12.9

Desert air at 110.°F and 10.0% relative humidity is to be cooled and humidified by using evaporative cooling only. Determine the minimum outlet mixture temperature and its relative humidity.

Solution

The minimum outlet temperature associated with the evaporation process is the wet bulb temperature corresponding to a dry bulb temperature of 110.°F and 10.0% relative humidity. From Chart D.5, we find that this is approximately

$$(T_{DB})_{\min} = T_{WB} = 69°\text{F}$$

and, of course, the relative humidity at this new dry bulb temperature is 100%.

Exercises

21. Determine the minimum outlet dry bulb temperature that could be realized through evaporation of the liquid water in part b of Example 12.7. **Answer**: $(T_{DB})_{\min} = T_{WB}(25.0°\text{C}, 20.0\%$ relative humidity$) = 13°\text{C}$.
22. Determine the minimum outlet dry bulb temperature that could be realized using the evaporation of liquid water into air with an inlet dry bulb temperature of 20.0°C and an inlet relative humidity of 50.0%. **Answer**: $(T_{DB})_{\min} = T_{WB}(20.0°\text{C}$ and 50.0% relative humidity$) = 14°\text{C}$.
23. Suppose we wanted to produce air with a dry bulb temperature of 60.0°F and a relative humidity of 100.% simply by allowing liquid water to evaporate into the air. What is the maximum dry bulb temperature of the inlet air? Note: In this case, $(T_{DB})_{\max}$ occurs when $\phi = 0\%$ (i.e., when $\omega = 0$). **Answer**: $(T_{DB})_{\max} = 107°\text{F}$.

Hot, humid air can be easily cooled and dehumidified by cooling it to below its dew point (saturation) temperature, condensing out some of the water, then reheating the remaining air–water vapor mixture to the desired temperature. This is illustrated in Figure 12.7. The water in the cooling section condenses at various temperatures, but it is assumed to exit the system at temperature T_2 in Figure 12.7.

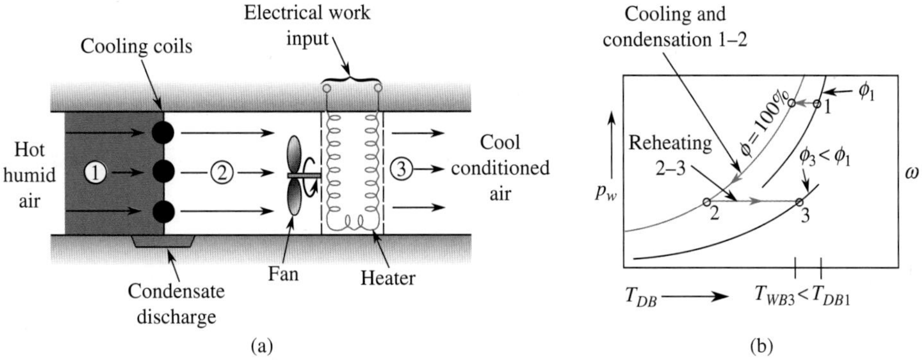

FIGURE 12.7
Dehumidification by cooling, condensing, and reheating again.

12.8 PSYCHROMETRIC ENTHALPIES

The psychrometric chart also contains enthalpy information that is useful in energy balance calculations. Though water vapor may be added or removed from the mixture by an air conditioning system, the mass flow rate of the dry air component is usually constant throughout the system. This makes it convenient to define the mixture's specific enthalpy on a *per unit mass of dry air* basis, rather than on a per unit mass of mixture basis. Specific enthalpies so constructed are referred to as *psychrometric enthalpies*, denoted by $h^{\#}$ to distinguish them from the ordinary form of the specific enthalpy. Thus, the specific psychrometric enthalpy is defined as

$$h^{\#} = H_m/m_a$$

whereas the ordinary mixture specific enthalpy is defined as

$$h_m = H_m/m_m$$

Note that, since $H_m = m_a h^{\#} = m_m h_m = (m_a + m_w)h_m$, we have

$$h^{\#} = (1+\omega)h_m$$

Further, since

$$H_m = H_a + H_w = m_a h_a + m_w h_w$$

then,

$$H_m/m_a = h^{\#} = h_a + (m_w/m_a)h_w = h_a + \omega h_w \tag{12.32}$$

Values of $h^{\#}$ are given on the psychrometric charts in Charts D.5 and D.6.[7]

EXAMPLE 12.10

Air has a dry bulb temperature of 50.0°C and a relative humidity of 40.0% at a total pressure of 0.101 MPa. Calculate the value of $h^{\#}$ from its definition and compare this value with that found in Chart D.6 for these conditions.

Solution

The basic definition of $h^{\#}$ is given by Eq. (12.32) as $h^{\#} = h_a + \omega h_w$, where $h_a = c_p(T_{DB} - T_{ref})$. When the psychrometric Charts D.5 and D.6 were developed, the reference temperature used in the h_a ideal gas equation was chosen to be 0°C. From Table C.13b, we find for air that $c_p = 1.004$ kJ/(kg·K). Then,

$$h_a = [1.004\ \text{kJ/(kg}\cdot\text{K)}][(50.0+273.15\ \text{K}) - (0+273.15\ \text{K})] = 50.2\ \text{kJ/(kg dry air)}$$

Equation (12.26b) gives the humidity ratio ω for this mixture as

$$\omega = 0.622\left(\frac{\phi p_{sat}}{p_m - \phi p_{sat}}\right)$$

where, from Table C.1b, $p_{sat} = p_{sat}(50.0°\text{C}) = 0.01235$ MPa. Then,

$$\omega = 0.622\left(\frac{0.400 \times 0.01235}{0.101 - 0.400 \times 0.01235}\right) = 0.0319\ \frac{\text{kg water vapor}}{\text{kg dry air}}$$

Since the water vapor in the mixture is superheated, its specific enthalpy h_w is determined from Table C.3b at the temperature of the mixture and the partial pressure of the water vapor. From Eq. (12.24), we have $p_w = \phi p_{sat} = 0.400 \times 0.01235 = 5.00 \times 10^{-3}$ MPa. Then, from Table C.3b at 50.0°C and 5.00×10^{-3} MPa, we find that $h_w = 2593.6$ kJ/kg water vapor. Using the definition of $h^{\#}$ given in Eq. (12.32), we now have

$$h^{\#} = h_a + \omega h_w = 50.2\ \frac{\text{kJ}}{\text{kg dry air}} + \left(0.0319\ \frac{\text{kg water vapor}}{\text{kg dry air}}\right) \times \left(2593.6\ \frac{\text{kJ}}{\text{kg water vapor}}\right)$$
$$= 133\ \frac{\text{kJ}}{\text{kg dry air}}$$

Chart D.6 also gives this value for ω at $T_{DB} = 50.0°\text{C}$ and $\phi = 40.0\%$.

---

[7] Note that h_a has a zero reference state at 0°F (not 0 R) in Chart D.6a and 0°C in Chart D.6b with the h_w value coming from the appropriate steam table in each case. Recall that the choice of a reference state is arbitrary, so long as property differences are used in the calculations.

Exercises

24. Use the defining equation for $h^{\#}$ to calculate the psychrometric enthalpy of air with a dry bulb temperature of 100.°C and a relative humidity of 81.0% at a total pressure of 0.101 MPa. **Answer:** $h^{\#}$ = 227 kJ/(kg dry air).

25. Use the defining equation for $h^{\#}$ to calculate its value for air at 100.°C and a relative humidity of 69.1% at a total pressure of 0.200 MPa. **Answer:** $h^{\#}$ = 998 kJ/(kg dry air).

26. Use the defining equation for $h^{\#}$ to calculate its value for air at 100.°F and a relative humidity of 52.6% at atmospheric pressure. Compare your value with that found in Chart D.5. **Answer:** $h^{\#}$ = 48.2 Btu/(lbm dry air).

Using the psychrometric enthalpy values in the energy rate balance on the adiabatic saturator (see Figure 12.2) gives

$$\dot{m}_a h_1^{\#} + \dot{m}_{w2} h_{w2} - \dot{m}_a h_3^{\#} = 0$$

or

$$h_1^{\#} + \omega_2 h_{w2} = h_3^{\#}$$

Typically, $\omega_2 h_{w2} \ll h_1$, so that we obtain $h_1^{\#} \approx h_3^{\#}$, and since both h_a and h_w depend only on temperature at low pressures, lines of constant $h^{\#}$ are parallel to lines of constant T_{WB}, as shown in Figure 12.8. The following example illustrates the processes of dehumidification in an air conditioning application.

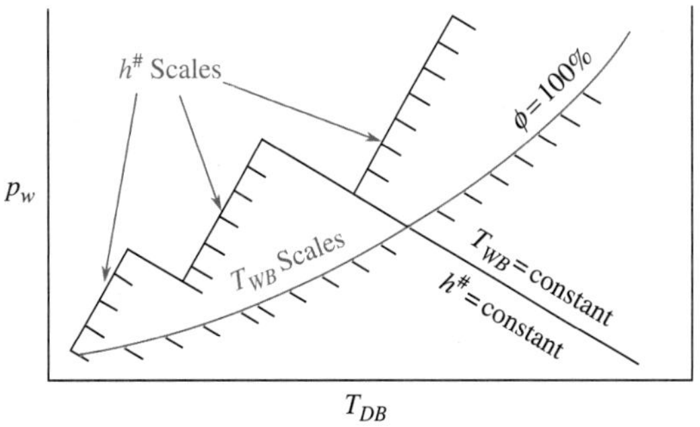

FIGURE 12.8
Reading the psychrometric enthalpy from the psychrometric chart.

EXAMPLE 12.11

A new paint-drying system requires air with a dry bulb temperature of 35.0°C and a relative humidity of 80.0% to be cooled and dehumidified to a dry bulb temperature of 20.0°C and a relative humidity of 40.0%. Determine the heat transfer rate per unit mass flow rate of dry air required to carry out this process.

Solution

The cooling and dehumidification process is illustrated in Figure 12.9.

For an aergonic (zero work) system, an energy rate balance on the air in the paint booth gives

$$\dot{Q} = \dot{m}_a h_3^{\#} - \dot{m}_a h_1^{\#} - \dot{m}_{w2} h_{w2}$$

Neglecting the magnitude of the thermal energy associated with the condensate water (i.e., setting $\dot{m}_{w2} h_{w2} = 0$), we get

$$\dot{Q}/\dot{m}_a = h_3^{\#} - h_1^{\#}$$

Values for psychrometric enthalpy are easily found on the psychrometric chart, Chart D.6. Looking at this chart, we find the intersection of the lines for T_{DB1} = 35.0°C and for ϕ_1 = 80.0%. Then, we follow the diagonal line upward and to the left until we intersect the psychrometric enthalpy axis and read $h_1^{\#} \approx 110$ kJ/kg dry air. Using a similar technique at T_{DB3} = 20.0°C and ϕ_3 = 40.0%, we find that $h_3^{\#} \approx 35$ kJ/kg dry air. The heat transfer rate per unit mass flow rate of dry air is

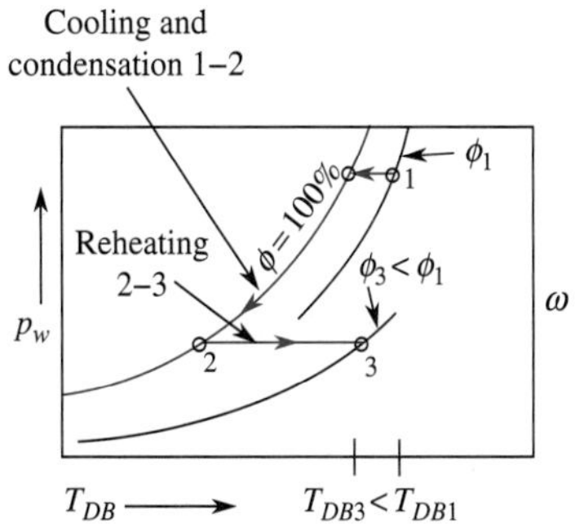

FIGURE 12.9
Example 12.11.

$$\dot{Q}/\dot{m}_a = h_3^{\#} - h_1^{\#} - 35 - 110 = -75 \text{ kJ/kg dry air}$$

This means that 75 kJ of thermal energy must be removed from every kilogram of (dry) air that passes through the system.

Exercises

27. Use Chart D.6 to determine the initial state psychrometric enthalpy $h_1^{\#}$ of the air in Example 12.11 if it is at 40.0°C instead of 35.0°C at the same relative humidity. **Answer:** $h_1^{\#} \approx 140$ kJ/kg dry air.

28. If the final state in Example 12.11 is at a dry bulb temperature of 22.0°C and a relative humidity of 20.0%, use Chart D.6 to find the psychrometric enthalpy of this state. **Answer:** $h_2^{\#} \approx 30$ kJ/kg dry air.

29. Rework Example 12.11 for an inlet dry bulb temperature of 100.°F at a relative humidity of 70.0% and an exit dry bulb temperature of 70.0°F at a relative humidity of 10.0%. **Answer:** $\dot{Q}/\dot{m}_a \approx -37.5$ Btu/lbm dry air.

EXAMPLE 12.12

Moist air at atmospheric pressure and 25.0°C with a relative humidity of 80.0% is to be cooled and dehumidified to 20.0°C with a relative humidity of 40.0%. On a per unit mass of dry air basis, determine (a) the amount of water to be removed, (b) the cooling heat transfer rate, and (c) the reheating heat transfer rate.

Solution

The schematic for this process is the same as Figure 12.7. From the psychrometric chart (Chart D.6), we can find the following information:[8]

State 1	State 2	State 3
$T_{DB1} = 25.0°C$	$T_{WB2} = 6.0°C$	$T_{DB3} = 20.0°C$
$\phi_1 = 80.0\%$	$\phi_2 = 100.\%$	$\phi_3 = 40.0\%$
$h_1^\# = 67$ kg/(kg da)	$h_1^\# = 21$ kg/(kg da)	$h_3^\# = 35$ kg/(kg da)
$\omega_1 = 0.016$ kg H_2O/(kg da)	$\omega_2 = 0.0056$ kg H_2O/(kg da)	$\omega_3 = \omega_2$

a. The amount of water removed per unit mass of dry air is

$$\omega_1 - \omega_2 = 0.016 - 0.0056 = 0.010 \text{ kg } H_2O/(\text{kg dry air})$$

b. The amount of cooling required per unit mass of dry air is given by an energy rate balance on the cooling section as

$$\dot{Q}_{\text{cooling}}/\dot{m}_{\text{dry air}} = h_2^\# - h_1^\# + (\omega_1 - \omega_2)h_f(T_2)$$
$$= 21 - 67 + 0.010(25.2) = -45.7 \text{ kJ/(kg dry air)}$$

c. The reheating heat transfer rate is given by an energy rate balance on the reheating section:

$$\dot{Q}_{\text{reheating}}/\dot{m}_{\text{dry air}} = h_3^\# - h_2^\# = 35 - 21 = 14 \text{ kJ/(kg dry air)}$$

Exercises

30. Suppose the inlet air in Example 12.12 has a relative humidity of 100.% instead of 80.0% and all the other variables remain the same. How much water then has to be removed per unit mass of dry air to achieve the outlet state of T_{DB2} = 20.0°C and ϕ_2 = 40.0%? **Answer**: $\omega_1 - \omega_2 = 0.0144$ kg H_2O per kg of dry air
31. If the final dry bulb temperature in Example 12.12 is 25.0°C (and $\phi_3 \approx 30.0\%$) rather than 20.0°C and all the other parameters remain the same (i.e., the mixture is dehumidified but has no net cooling), determine the cooling and reheating heat transfer rates per unit mass flow rate of dry air. **Answer**: $\dot{Q}_{\text{cooling}}/\dot{m}_{\text{dry air}} = -45.7$ kJ/(kg dry air) and $\dot{Q}_{\text{reheating}}/\dot{m}_{\text{dry air}} = 24.0$ kJ/(kg dry air)
32. Rework Example 12.12 for an inlet condition of T_{DB1} = 100.°F and ϕ_1 = 80.0% and an outlet condition of T_{DB2} = 60.0°F and ϕ_2 = 40.0%. **Answer**: (a) $\omega_1 - \omega_2 = 0.0247$ lbm H_2O/(lbm dry air), (b) $\dot{Q}_{\text{cooling}}/\dot{m}_{\text{dry air}} = -41.9$ Btu/(lbm dry air), and (c) $\dot{Q}_{\text{reheating}}/\dot{m}_{\text{dry air}} =$ 5.5 Btu/(lbm dry air)

[8] *Values taken from reading a chart are typically accurate to only two significant figures.*

Another common air conditioning design problem where the psychrometric chart is put to good use is in the mixing of two or more wet airstreams. This normally involves determining how to mix the inlet airstreams to produce a desired output conditional airstream or predicting the outlet airstream properties when all the inlet airstream properties are known.

The conservation of mass equation for water when wet airstreams 1 and 2 are adiabatically and aergonically mixed to form wet airstream 3 is

$$\dot{m}_{w3} = \dot{m}_{w1} + \dot{m}_{w2}$$

or

$$\dot{m}_{w3} = \dot{m}_{a3}\omega_3 = \dot{m}_{a1}\omega_1 + \dot{m}_{a2}\omega_2$$

or

$$\omega_3 = (\dot{m}_{a1}/\dot{m}_{a3})\omega_1 + (\dot{m}_{a2}/\dot{m}_{a3})\omega_2 \quad (12.33)$$

From the energy rate balance applied to this process, we get

$$\underbrace{\dot{Q} - \dot{W}}_{0} + \dot{m}_{a1}h_1^\# + \dot{m}_{a2}h_2^\# - \dot{m}_{a3}h_3^\# = 0$$

or

$$h_3^\# = (\dot{m}_{a1}/\dot{m}_{a3})h_1^\# + (\dot{m}_{a2}/\dot{m}_{a3})h_2^\# \qquad (12.34)$$

If the states of the inlet flow streams are known, then Eqs. (12.33) and (12.34) allow the calculation of two independent thermodynamic properties (ω_3 and $h_3^\#$) that fix the state of the outlet flow stream. The following example illustrates this type of problem.

EXAMPLE 12.13

Suppose 2000. ft^3/min of air at 14.7 psia, 50.0°F, ϕ = 80.0% is adiabatically and aergonically mixed with 1000. ft^3/min of air at 14.7 psia, 100.°F, and ϕ = 40.0%. Determine the dry bulb temperature and the relative humidity of the outlet mixture.

Solution

First, draw a sketch of the system (Figure 12.10).

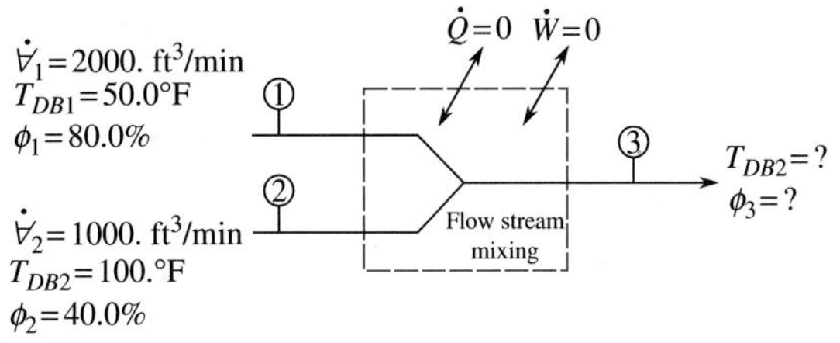

FIGURE 12.10
Example 12.13.

The unknowns are the dry bulb temperature and the relative humidity of the outlet mixture.

At 50.0°F, $p_{w1} = \phi_1 p_{\text{sat}}(50.0°\text{F}) = 0.800(0.178) = 0.142$ psia. Then,

$$\begin{aligned} v_{a1} &= R_a T_m / p_{a1} \\ &= [53.34\ \text{ft}\cdot\text{lbf}/(\text{lbm}\cdot\text{R})](50.0 + 459.67\ \text{R})/[(14.7 - 0.142\ \text{lbf/in}^2)(144\ \text{in}^2/\text{ft}^2)] \\ &= 13.0\ \text{ft}^3/(\text{lbm dry air}) \end{aligned}$$

Similarly,

$$p_{w2} = \phi_2 p_{\text{sat}}(100.°\text{F}) = 0.400(0.9503) = 0.380\ \text{psia}$$

and

$$\begin{aligned} v_{a2} &= 53.34(100. + 459.67)/[(14.7 - 0.380)(144)] \\ &= 14.5\ \text{ft}^3/(\text{lbm dry air}) \end{aligned}$$

Also, since $\dot{m}_a = \dot{V}_a / v_a$,

$$\begin{aligned} \dot{m}_{a1} &= (2000.\ \text{ft}^3/\text{min})/[13.0\ \text{ft}^3/(\text{lbm dry air})] = 154\ \text{lbm dry air/min} \\ \dot{m}_{a2} &= (1000.\ \text{ft}^3/\text{min})/[14.5\ \text{ft}^3/(\text{lbm dry air})] = 69.0\ \text{lbm dry air/min} \end{aligned}$$

and, using the conservation of mass applied to the air,

$$\dot{m}_{a3} = \dot{m}_{a1} + \dot{m}_{a2} = 154 + 69.0 = 223\ \text{lbm dry air/min}$$

Then, from the psychrometric chart (Chart D.5), we find[9]

$$\begin{aligned} \omega_1 &= [44\ \text{grains of water vapor/(lbm dry air)}]/(7000.\ \text{grains/lbm}) \\ &= 0.0063\ \text{lbm water vapor/(lbm dry air)} \\ \omega_2 &= [115\ \text{grains of water vapor/(lbm dry air)}]/(7000.\ \text{grains/lbm}) \\ &= 0.0164\ \text{lbm water vapor/(lbm dry air)} \\ h_1^\# &= 19\ \text{Btu/(lbm dry air)} \\ h_2^\# &= 42\ \text{Btu/(lbm dry air)} \end{aligned}$$

From the water conservation of mass equation, Eq. (12.33), we can now calculate

$$\begin{aligned} \omega_3 &= (154/223)(0.0063) + (69.0/223)(0.0164) \\ &= 0.0094\ \text{lbm water vapor/(lbm dry air)} \\ &= 0.0094\,(7000.) = 66\ \text{grains of water vapor/(lbm dry air)} \end{aligned}$$

and the resulting energy rate balance equation, Eq. (12.34), gives

$$\begin{aligned} h_3^\# &= (154/223)(19) + (69.0/223)(42) \\ &= 26\ \text{Btu/(lbm dry air)} \end{aligned}$$

(Continued)

EXAMPLE 12.13 *(Continued)*

From the point where the lines ω = 65.8 grains/(lbm dry air) = constant and $h^{\#}$ = 26 Btu/(lbm dry air) = constant intersect on the psychrometric chart, we can read from this chart that

$$T_{DB} = 63°\text{F},\ T_{WB} = 59°\text{F},\ \phi = 75\%,\ \text{and}\ T_{DP} = 56°\text{F}$$

Exercises

33. If the 2000. ft^3/min of air entering station 1 in Example 12.13 is at a dry bulb temperature of 70.0°F instead of 50.0°F with all the remaining parameters unchanged, determine its new mass flow rate. **Answer**: $\dot{m}_{a1}$ = 147 lbm dry air/min.
34. If the 1000. ft^3/min of air entering station 2 in Example 12.13 is at a relative humidity of 10.0% rather than 40.0% with all the remaining parameters unchanged, determine its new mass flow rate. **Answer**: $\dot{m}_{a2}$ = 70.5 lbm dry air/min.
35. Determine the dry bulb temperature and relative the humidity of the outlet mixture in Example 12.13 if the volume flow rate at station 2 is increased from 1000. ft^3/min to 2000. ft^3/min and all the remaining variables remain unchanged. **Answer**: T_{DB} = 74°F and ϕ = 60.%.

[9] *The examples and problems throughout this text have been written such that at least three significant figures appear in the variables. Unfortunately, small textbook charts can often be read to only two significant figures; however, psychrometric computer programs can be found on the Internet that provide much more accuracy and are recommended for student use.*

12.9 MIXTURES OF REAL GASES

If the components of an ideal gas mixture interact in any way or one or more of the gases is not ideal, the resulting mixture is not ideal and does not obey the Gibbs-Dalton and Amagat laws. Then, its partial properties must be determined from accurate pressure, volume, temperature, and specific heat data by the techniques discussed in the previous chapter.

Though Amagat's law, Eqs. (12.16) and (12.19), may not hold for a mixture of real gases, the definition of partial specific volumes, Eqs. (12.5) and (12.6), is always valid. The difference is that, for real gases,

$$\hat{v}_i = (\partial \forall_m/\partial m_i)_{p_m,\,T_m,m_j} \neq v_i = R_i T_m/p_m$$

For a binary mixture of gases A and B, $\hat{v}_A$ and $\hat{v}_B$ can be determined at any composition from experimental data of v_m vs. w_A, as shown in Figure 12.11.

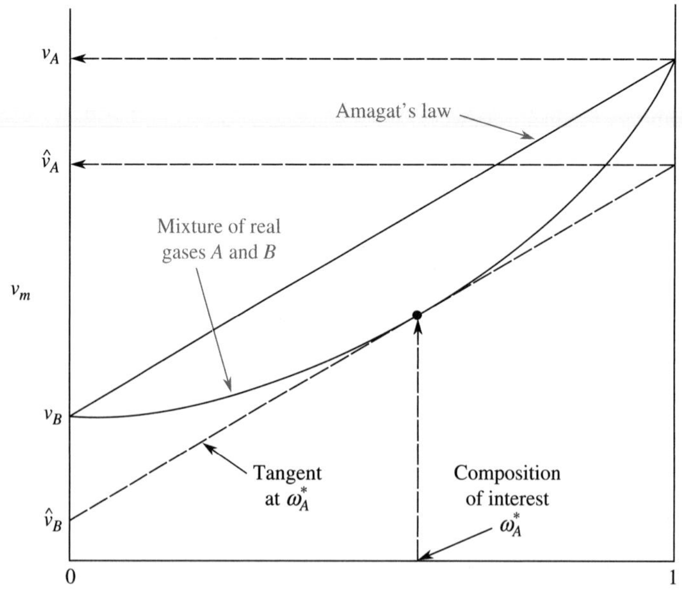

FIGURE 12.11
Determining $\hat{v}_A$ and $\hat{v}_B$ at $w_A = w_A^*$ from real gas data by the method of tangents.

When p, v, T, and c_v data on the gas mixture of interest are not available, engineering approximations can be obtained by combining either Dalton's or Amagat's law with the simplified compressibility factor equation of state $p\forall = ZmRT$. For example, using Dalton's law,

$$p_m = \sum_{i=1}^{N} p_{Di} = Z_{Dm} m_m R_m T_m / \forall_m \tag{12.35}$$

where p_{Di} is the Dalton compressibility factor partial pressure, defined by

$$p_{Di} - Z_{Di} m_i R_i T_m / \forall_m$$

and Z_{Di} and Z_{Dm} are the Dalton species i and mixture compressibility factors, respectively. Substituting the latter equation into the former and solving for Z_{Dm} gives

Dalton compressibility factor

$$Z_{Dm} = \sum_{i=1}^{N} \left(\frac{w_i M_m}{M_i} \right) Z_{di} = \sum_{i=1}^{N} \chi_i Z_{Di} \tag{12.36}$$

For each gas i, the Dalton Z_{Di} compressibility factor is determined from the compressibility charts (Figures 7.5, 7.6, and 7.9 in Chapter 7) by using the reduced temperature T_{Ri} and reduced pseudospecific volume v'_{Ri} for gas i at the temperature and volume of the mixture, or

$$\begin{aligned} T_{Ri} &= T_m / T_{ci} \\ v'_{Ri} &= v_{Di} p_{ci} / (R_i T_{ci}) \\ &= (\forall_m / m_i)(m_m / m_m)(p_{ci}) / (R_i T_{ci}) = (v_m / w_i)[p_{ci} / (R_i T_{ci})] \\ &= (\forall_m / n_i)(n_m / n_m)(p_{ci}) / (\Re T_{ci}) = (\bar{v}_m / x_i)[p_{ci} / (\Re T_{ci})] \end{aligned}$$

where $v_{Di} = \forall_m / m_i = v_m / w_i$ is the Dalton specific volume of gas i, and $v_m = \forall_m / m_m$ and $\bar{v}_m = \forall_m / n_m$ are the mixture mass and molar specific volumes, respectively. Note that we cannot use the reduced pressure $p_{Ri} = p_{Di}/p_{ci}$ in this case because $p_{Di} = Z_{Di} m_i R_i T_m / \forall_m$, and Z_{Di} is not usually known in advance.

EXAMPLE 12.14

A new type of fuel for the camping stove shown in Figure 12.12 is made by mixing 3.00 lbm of methane (CH_4) with 4.00 lbm of propane (C_3H_8) and compressing the mixture into a 1.00 ft^3 portable storage tank. Use the Dalton compressibility factor to determine the total pressure in the tank when it is exposed to the hot summer sun and its internal temperature reaches 240.°F.

FIGURE 12.12
Example 12.14.

Solution

The unknown is the total pressure in the tank. Here, we use Eq. (12.35) to determine the mixture pressure as

$$p_m = \frac{Z_{Dm} m_m R_m T_m}{\forall_m}$$

where Z_{Dm} is the Dalton compressibility factor, determined from Eq. (12.36) as

$$Z_{Dm} = \sum_{i=1}^{N} \left(\frac{w_i M_m}{M_i} \right) Z_{Di}$$

First, we find the mixture composition and molecular mass. The mass of the entire mixture is $m_m = m_{methane} + m_{propane} = 3.00 + 4.00 = 7.00$ lbm. The mass fractions are

$$\begin{aligned} w_{methane} &= 3.00/7.00 = 0.429 \\ w_{propane} &= 4.00/7.00 = 0.571 \end{aligned}$$

(*Continued*)

EXAMPLE 12.14 *(Continued)*

The molecular masses of the components are found in Table C.12a in *Thermodynamic Tables to accompany Modern Engineering Thermodynamics* as

$$M_{\text{methane}} = 16.043\ \text{lbm/lbmole}$$
$$M_{\text{propane}} = 44.097\ \text{lbm/lbmole}$$

then, Eq (12.12) gives the molecular mass of the mixture as

$$M_m = \frac{1}{\sum_{i=1}^{2} \frac{w_i}{M_i}} = \frac{1}{\frac{w_{\text{methane}}}{M_{\text{methane}}} + \frac{w_{\text{propane}}}{M_{\text{propane}}}} = \frac{1}{\frac{0.429}{16.043} + \frac{0.571}{44.097}} = 25.2\ \frac{\text{lbm}}{\text{lbmole}}$$

The Dalton compressibility factors for the components, Z_{Di}, are found from the reduced temperature T_{Ri} and the reduced pseudospecific volume v'_{Ri} for each component. From Tables C.12a and C.13a, we find that

$$(p_c)_{\text{methane}} = 673\ \text{psia},\ (T_c)_{\text{methane}} = 343.9\ \text{R, and } R_{\text{methane}} = 96.3\ \text{ft}\cdot\text{lbf}/(\text{lbm}\cdot\text{R})$$
$$(p_c)_{\text{propane}} = 617\ \text{psia},\ (T_c)_{\text{propane}} = 665.9\ \text{R, and } R_{\text{propane}} = 35.0\ \text{ft}\cdot\text{lbf}/(\text{lbm}\cdot\text{R})$$

The specific volume of the mixture is

$$v_m = \frac{\forall_m}{m_m} = \frac{1.00\ \text{ft}^3}{7.00\ \text{lbm}} = 0.143\ \text{ft}^3/\text{lbm}$$

then, the reduced temperature and reduced pseudospecific volume for methane are

$$(T_R)_{\text{methane}} = \frac{T_m}{(T_c)_{\text{methane}}} = \frac{240. + 459.67\ \text{R}}{343.9\ \text{R}} = 2.03$$

and

$$(v'_R)\text{methane} = \left(\frac{v_m}{w_{\text{methane}}}\right)\left(\frac{(p_c)_{\text{methane}}}{R_{\text{methane}}(T_c)_{\text{methane}}}\right)$$
$$= \left(\frac{0.143\ \text{ft}^3/\text{lbm mixture}}{0.429\ \text{lbm methane/lbm mixture}}\right) \times \left(\frac{(673\ \text{lbf/in}^2)(144\ \text{in}^2/\text{ft}^2)}{(96.3\ \text{ft}\cdot\text{lbf/lbm}\cdot\text{R})(343.9\ \text{R})}\right)$$
$$= 0.975$$

For propane, they are

$$(T_R)_{\text{propane}} = \frac{T_m}{(T_c)_{\text{propane}}} = \frac{240. + 459.67\ \text{R}}{665.9\ \text{R}} = 1.05$$

and

$$(v'_R)\text{propane} = \left(\frac{v_m}{w_{\text{propane}}}\right)\left(\frac{(p_c)_{\text{propane}}}{R_{\text{propane}}(T_c)_{\text{propane}}}\right)$$
$$= \left(\frac{0.143\ \text{ft}^3/\text{lbm mixture}}{0.571\ \text{lbm methane/lbm mixture}}\right) \times \left(\frac{(617\ \text{lbf/in}^2)(144\ \text{in}^2/\text{ft}^2)}{(35.0\ \text{ft}\cdot\text{lbf/lbm}\cdot\text{R})(665.9\ \text{R})}\right)$$
$$= 0.955$$

Using Figure 7.6 with $T_R = 2.03$ and $v'_R = 0.975$, we find that the Dalton compressibility factor for methane is $(Z_D)_{\text{methane}} = 0.975$; and using $T_R = 1.05$ and $v'_R = 0.95$, we find that the Dalton compressibility factor for propane is $(Z_D)_{\text{propane}} = 0.720$. Then, Eq. (12.36) gives the mixture Dalton compressibility factor as

$$Z_{Dm} = \sum_{i=1}^{2}\left(\frac{w_i M_m}{M_i}\right) Z_{Di} = \left(\frac{w_{\text{methane}} M_m}{M_{\text{methane}}}\right)(Z_D)_{\text{methane}} + \left(\frac{w_{\text{propane}} M_m}{M_{\text{propane}}}\right)(Z_D)_{\text{propane}}$$
$$= \left(\frac{(0.429)(25.2)}{16.043}\right)(0.975) + \left(\frac{(0.571)(25.2)}{44.097}\right)(0.720) = 0.892$$

The mixture gas constant can be easily calculated from

$$R_m = \frac{\Re}{M_m} = \frac{1545.35\ \text{ft}\cdot\text{lbf}/(\text{lbmole}\cdot\text{R})}{25.2\ \text{lbm/lbmole}} = 61.3\ \frac{\text{ft}\cdot\text{lbf}}{\text{lbm}\cdot\text{R}}$$

and Eq. (12.35) gives the mixture pressure as

$$p_m = \frac{Z_{Dm}\, m_m\, R_m\, T_m}{V_m}$$
$$= \frac{(0.892)(7.00\,\text{lbm})(61.3\,\text{ft}\cdot\text{lbf/lbm}\cdot\text{R})(240.+459.67\,\text{R})}{1.00\,\text{ft}^3}$$
$$= (268\times 10^3\,\text{lbf/ft}^2)\div(144\,\text{in.}^2/\text{ft}^2) = 1860\,\text{psia}$$

Exercises

36. Determine the total pressure in the tank in Example 12.14 when the mixture is changed to 2.00 lbm of methane and 5.00 lbm of propane with all of the other variables remaining unchanged. **Answer:** $p_{total} = 1470$ psia.

37. Determine the total pressure in the tank in Example 12.14 if the tank volume is reduced from 1.00 ft^3 to 0.500 ft^3 with all the other variables remaining unchanged. **Answer:** $p_{total} = 3400$ psia.

38. Determine the total pressure in the tank in Example 12.14 when the propane is replaced by the same mass of ethane (C_2H_6). **Answer:** $p_{total} = 2200$ psia.

Alternatively, we could use Amagat's law and incorporate a real gas compressibility factor as

$$V_m = \sum_{i=1}^{N} V_{Ai} = \frac{Z_{Am} m_m R_m T_m}{p_m} \tag{12.37}$$

where V_{Ai} is the Amagat compressibility factor partial volume defined by

$$V_{Ai} = Z_{Ai} m_i R_i T_m / p_m$$

and Z_{Ai} and Z_{Am} are Amagat species i and mixture compressibility factors, respectively. Substituting the latter equation into the former gives

Amagat compressibility factor

$$Z_{Am} = \sum_{i=1}^{N} (w_i M_m / M_i) Z_{Ai} = \sum_{i=1}^{N} x_i Z_{Ai} \tag{12.38}$$

In this case, for each gas i, the Amagat Z_{Ai} compressibility factor is determined from the compressibility charts using the reduced temperature T_{Ri} and reduced pressure p_{Ri} for gas i at the temperature and pressure of the mixture, or

$$p_{Ri} = p_m / p_{ci}$$

and

$$T_{Ri} = T_m / T_{ci}$$

Note that we cannot use the reduced pseudospecific volume $v'_{Ri} = v_{Ai} p_m / (R_i T_m)$ in this case because the Amagat specific volume is given by $v_{Ai} = V_{Ai}/m_i = Z_{Ai} R_i T_m / p_m$, and Z_{Ai} is not usually known in advance.

EXAMPLE 12.15

The test chamber atmosphere for a new electrical device requires a mixture of 1.00 kg of each of the following gases: ammonia (NH_3), chlorine (Cl_2), and nitrous oxide (N_2O). Use the Amagat compressibility factor to determine the volume occupied by this mixture at a total pressure of 20.0 MPa and a mixture temperature of 500. K.

Solution

Here, we use Eq. (12.37) to determine the mixture total volume as

$$V_m = \frac{Z_{Am}\, m_m\, R_m\, T_m}{p_m}$$

where Z_{Am} is the Amagat compressibility factor, determined from Eq. (12.38) as

$$Z_{Am} = \sum_{i=1}^{N} \left(\frac{w_i M_m}{M_i}\right) Z_{Ai}$$

(Continued)

EXAMPLE 12.15 *(Continued)*

First, we find the mixture composition and molecular mass. The mass of the entire mixture is $m_m = m_{\text{ammonia}} + m_{\text{chlorine}} + m_{\text{nitrous oxide}} = 1.00 + 1.00 + 1.00 = 3.00$ kg. The mass fractions are

$$w_{\text{ammonia}} = w_{\text{chlorine}} = w_{\text{nitrous oxide}} = 1.00/3.00 = 0.333$$

The molecular masses of the components are found in Table C.12b as

$$M_{\text{ammonia}} = 17.030 \text{ kg/kgmole}$$

$$M_{\text{chlorine}} = 70.906 \text{ kg/kgmole}$$

$$M_{\text{nitrous oxide}} = 44.013 \text{ kg/kgmole}$$

Equation (12.12) gives the molecular mass of the mixture as

$$m_m = \frac{1}{\sum_{i=1}^{3} \frac{w_i}{M_i}} = \frac{1}{\frac{w_{\text{ammonia}}}{M_{\text{ammonia}}} + \frac{w_{\text{chlorine}}}{M_{\text{chlorine}}} + \frac{w_{\text{nitrous oxide}}}{M_{\text{nitrous oxide}}}} = \frac{1}{\frac{0.333}{17.030} + \frac{0.333}{70.906} + \frac{0.333}{44.013}} = 31.4 \frac{\text{kg}}{\text{kgmole}}$$

The Amagat compressibility factors for the components, Z_{Ai}, are found from the reduced pressure and reduced temperature for each component. From Table C.12b, we find that

$$(p_c)_{\text{ammonia}} = 11.280 \text{ MPa} \quad \text{and} \quad (T_c)_{\text{ammonia}} = 405.5 \text{ K}$$

$$(p_c)_{\text{chlorine}} = 7.710 \text{ MPa} \quad \text{and} \quad (T_c)_{\text{chlorine}} = 417.0 \text{ K}$$

$$(p_c)_{\text{nitrous oxide}} = 7.270 \text{ MPa} \quad \text{and} \quad (T_c)_{\text{nitrous oxide}} = 309.7 \text{ K}$$

and the component gas constants can be computed as

$$R_{\text{ammonia}} = \Re/M_{\text{ammonia}} = 8.3143 \text{ kJ/(kgmole}\cdot\text{K)}/17.030 \text{ kg/kgmole} = 0.488 \text{ kJ/kg}\cdot\text{K}$$

$$R_{\text{chlorine}} = \Re/M_{\text{chlorine}} = 8.3143 \text{ kJ/(kgmole}\cdot\text{K)}/70.906 \text{ kg/kgmole} = 0.117 \text{ kJ/kg}\cdot\text{K}$$

$$R_{\text{nitrous oxide}} = \Re/M_{\text{nitrous oxide}} = 8.3143 \text{ kJ/(kgmole}\cdot\text{K)}/44.013 \text{ kg/kgmole} = 0.189 \text{ kJ/kg}\cdot\text{K}$$

Then, the reduced temperatures and pressures are

$$(T_R)_{\text{ammonia}} = \frac{T_m}{(T_c)_{\text{ammonia}}} = \frac{500.\text{ K}}{405.5 \text{ K}} = 1.23$$

$$(p_R)_{\text{ammonia}} = \frac{p_m}{(p_c)_{\text{ammonia}}} = \frac{20.0 \text{ MPa}}{11.280 \text{ MPa}} = 1.77$$

$$(T_R)_{\text{chlorine}} = \frac{T_m}{(T_c)_{\text{chlorine}}} = \frac{500.\text{ K}}{417.0 \text{ K}} = 1.20$$

$$(p_R)_{\text{chlorine}} = \frac{p_m}{(p_c)_{\text{chlorine}}} = \frac{20.0 \text{ MPa}}{7.710 \text{ MPa}} = 2.59$$

$$(T_R)_{\text{nitrous oxide}} = \frac{T_m}{(T_c)_{\text{nitrous oxide}}} = \frac{500.\text{ K}}{309.7 \text{ K}} = 1.61$$

$$(p_R)_{\text{nitrous oxide}} = \frac{p_m}{(p_c)_{\text{nitrous oxide}}} = \frac{20.0 \text{ MPa}}{7.270 \text{ MPa}} = 2.75$$

and using these values on Figure 7.6 gives the following Amagat compressibility factors:

$$(Z_A)_{\text{ammonia}} = 0.64$$

$$(Z_A)_{\text{chlorine}} = 0.55$$

$$(Z_A)_{\text{nitrous oxide}} = 0.86$$

Then, Eq. (12.38) gives the Amagat compressibility factor for the mixture as

$$Z_{Am} = \sum_{i=1}^{3}\left(\frac{w_i M_m}{M_i}\right) Z_{Ai}$$

$$= \left(\frac{w_{\text{ammonia}} M_m}{M_{\text{ammonia}}}\right)(Z_A)_{\text{ammonia}} + \left(\frac{w_{\text{chlorine}} M_m}{M_{\text{chlorine}}}\right)(Z_A)_{\text{chlorine}} + \left(\frac{w_{\text{nitrous oxide}} M_m}{M_{\text{nitrous oxide}}}\right)(Z_A)_{\text{nitrous oxide}}$$

$$= \left(\frac{(0.333)(31.4\text{ kg/kgmole mixture})}{17.030\text{ kg/kgmole ammonia}}\right)(0.64)$$

$$+ \left(\frac{(0.333)(31.4\text{kg/kgmole mixture})}{70.906\text{ kg/kgmole chlorinee}}\right)(0.55)$$

$$+ \left(\frac{(0.333)(31.4\text{ kg/kgmole mixture})}{44.013\text{ kg/kgmole nitrous oxide}}\right)(0.86) = 0.68$$

The mixture gas constant can now be calculated from

$$R_m = \frac{\Re}{M_m} = \frac{8.3143\text{ kJ/(kgmole}\cdot\text{K)}}{31.4\text{ kg/kgmole}} = 0.265\frac{\text{kJ}}{\text{kg}\cdot\text{K}}$$

Then, Eq. (12.37) gives the total volume of the mixture as

$$\forall_m = \frac{Z_{Am}\, m_m\, R_m\, T_m}{p_m} = \frac{(0.68)(3.00\text{ kg})(0.265\text{ kJ/(kg}\cdot\text{K))}(500.\text{ K})}{(20.0\text{ MPa})(1000.\text{ kPa/MPa})} = 0.0135\text{ m}^3$$

Exercises

39. Determine the volume in Example 12.15 if the composition is changed to 1.00 kg of ammonia and chlorine but 5.00 kg of nitrous oxide. **Answer:** $\forall = 0.0297\text{ m}^3$.
40. If the mixture temperature in Example 12.15 is increased from 500. K to 1000. K, what volume would then be required to hold the mixture if all the other variables remain unchanged? **Answer:** $\forall = 0.0401\text{ m}^3$.
41. If the nitrous oxide gas is eliminated from the mixture in Example 12.15 and the masses of ammonia and chlorine are each increased to 1.50 kg, what volume would then be required to hold this new mixture if all the other variables remain unchanged? **Answer:** $\forall = 0.0141\text{ m}^3$.

It should be clear from the preceding formulae that $Z_{Di} \neq Z_{Ai}$; therefore, the resultant Dalton and Amagat mixture compressibility factors also in general are not equal, or $Z_{Dm} \neq Z_{Am}$.

Dalton's law for mixtures of real gases (Eq. (12.35)) is based on the premise that each gas in the mixture acts as though it alone occupies the entire volume of the mixture at the temperature of the mixture. Therefore, the gases are assumed not to interact in any manner. We do not find this to be true experimentally except at very low pressures or very high temperatures.

Amagat's law for mixtures of real gases (Eqs. (12.37)), on the other hand, incorporates the resultant mixture pressure and, therefore, automatically takes gas molecular interactions into account. Consequently, it tends to be more accurate than Dalton's law at high pressures and low temperatures.

A third method of incorporating the compressibility factor charts into predicting the behavior of real gas mixtures involves defining a *pseudocritical pressure* and a *pseudocritical temperature* for the mixture as

Kay's law

$$P_{cm} = \sum_{i=1}^{N} x_i P_{ci} \tag{12.39}$$

and

$$T_{cm} = \sum_{i=1}^{N} x_i T_{ci} \tag{12.40}$$

This was introduced by W. B. Kay in 1936 and is now known as *Kay's law*. The reduced pressure and temperature of the mixture can then be computed from

$$p_{Rm} = p_m/p_{cm}$$

and

$$T_{Rm} = T_m/T_{cm}$$

and these values are used to find the mixture's compressibility factor, called the *Kay's compressibility factor*, Z_{Km}, directly from the compressibility charts. This compressibility factor is then used in the normal way, as, for example, in the equation $p_m \forall_m = Z_{Km} m_m R_m T_m$, where $R_m = \Re/M_m$.

EXAMPLE 12.16

Determine the critical pressure and temperature for air using Kay's law. Use the composition information for air given in Example 12.2.

Solution

Using Eqs. (12.39) and (12.40), the composition data given in Example 12.2 and the critical point data given in Table C.12b in *Thermodynamic Tables to accompany Modern Engineering Thermodynamics* give

$$(p_c)_{air} = \chi_{N_2}(p_c)_{N_2} + \chi_{O_2}(p_c)_{O_2} + \chi_{Ar}(p_c)_{Ar} + \chi_{CO_2}(p_c)_{CO_2}$$
$$= 0.7809(3.39) + 0.2095(5.08) + 0.0093(4.86) + 0.0003(7.39)$$
$$= 3.76 \text{ MPa}$$

and

$$(T_c)_{air} = \chi_{N_2}(T_c)_{N_2} + \chi_{O_2}(T_c)_{O_2} + \chi_{Ar}(T_c)_{Ar} + \chi_{CO_2}(T_c)_{CO_2}$$
$$= 0.7809(126.2) + 0.2095(154.8) + 0.00930(151) + 0.0003(304.2)$$
$$= 133 \text{ K}$$

These values agree quite well with the values of 3.774 MPa and 132.4 K for air given in Table C.12b.

Exercises

42. If the composition of air is simplified to 79.0% nitrogen and 21.0% oxygen on a molar basis, use Kay's law to determine the critical pressure and temperature of this mixture and compare these results with those given in Example 12.16 for the more accurate composition of air. **Answer:** p_c = 3.74 MPa, T_c = 133 K, both less than 1% from the values determined in Example 12.14
43. Determine the critical pressure, temperature, and compressibility factor for the mixture given in Example 12.14 using Kay's law. **Answer:** $(p_c)_{mix}$ = 655 psia, $(T_c)_{mix}$ = 450. R, and Z_{Km} = 0.84
44. Determine the critical pressure, temperature, and compressibility factor for the mixture given in Example 12.15 using Kay's law. **Answer:** $(p_c)_{mix}$ = 9.80 MPa, $(T_c)_{mix}$ = 384 K, and Z_{Km} = 0.70

Note that, in general, $Z_{Dm} \neq Z_{Am} \neq Z_{Km}$. Which one of these three is the most accurate in a specific instance depends on the molecular characteristics and thermodynamic state of the gas under consideration. A demonstration of the accuracy of these three methods of modeling real gas behavior is provided in the following example.

EXAMPLE 12.17

The molar specific volume of a mixture of 30.0% nitrogen and 70.0% methane (on a molar basis) at 1500. psia and −100.°F is measured and found to be 1.315 ft^3/lbmole. Calculate the molar specific volume of this mixture under these conditions, using

a. Ideal gas mixture behavior.
b. The Dalton compressibility factor.
c. The Amagat compressibility factor.
d. Kay's law.

Compute the percent error in each case.

Solution

a. For ideal gas mixture behavior,

$$v_m = \frac{\Re T_m}{p_m} = \frac{[1545.35\,\text{ft}\cdot\text{lbf}/(\text{lbmole}\cdot\text{R})](-100.+459.67\,\text{R})}{(1500.\,\text{lbf/in}^2)(144\,\text{in}^2/\text{ft}^2)} = 2.57\,\text{ft}^3/\text{lbmole}$$

and

$$\%\,\text{error} = \left(\frac{2.57-1.315}{1.315}\right)(100) = 95.4\%\,\text{high}$$

b. From Table C.12a, we find

$$(p_c)_{N_2} = 492\,\text{psia}$$

$$(T_c)_{N_2} = 227.1\,\text{R}$$

and

$$(P_c)_{CH_4} = 673\,\text{psia}$$

$$(T_c)_{CH_4} = 343.9\,\text{R}$$

Since the mixture specific volume is the unknown here, it must be determined by a trial and error method, using the reduced pseudospecific volume v'_{Ri}, which, as was shown earlier, can be written in a variety of forms, for example, as

$$v'_{Ri} = \bar{v}_m p_{ci}/(x_i \Re T_{ci})$$

We assume values for $\bar{v}_m$, find $(Z_{Dm})_N$, $(Z_D)_{CH}$, and Z_{Dm}, then check the assumption with $\bar{v}_m = Z_{Dm}/\Re T_m/p_m$. Assume $\bar{v}_m = 1.51\text{ft}^3/\text{lbmole}$, then

$$(v'_R)_{N_2} = \frac{(1.51)(492)(144)}{(0.300)(1545.35)(227.1)} = 1.02$$

and

$$(v'_R)_{CH_4} = \frac{(1.51)(673)(144)}{(0.700)(1545.35)(343.9)} = 0.393$$

Then,

$$(T_R)_{N_2} = (-100.+459.67)/227.1 = 1.58$$

and

$$(T_R)_{CH_4} = (-100.+459.67)/343.9 = 1.05$$

From Figure 7.6 in Chapter 7, we find that, for these values,

$$(Z_D)_{N_2} = 0.91$$

and

$$(Z_D)_{CH_4} = 0.39$$

Then, from Eq. (12.36), we have

$$Z_{Dm} = 0.300(0.91) + 0.700(0.39) = 0.59$$

Now, checking the $\bar{v}_m$ assumption,

$$\bar{v}_m = Z_{Dm}/\Re T_m/p_m = \frac{0.59(1545.35)(-100.+459.67)}{(1500.)(144)}$$

$$= 1.52\,\text{ft}^3/\text{lbmole}$$

which is close enough to our original assumption. Then,

$$\%\,\text{error} = \left(\frac{1.52-1.315}{1.315}\right)(100) = 15.6\,\%\,\text{high}$$

(Continued)

EXAMPLE 12.17 *(Continued)*

c. Using the Amagat compressibility factor method, we have

$$(p_R)_{N_2} = p_m/(p_c)_{N_2} = \frac{1500.}{492} = 3.05$$

$$(T_R)_{N_2} = T_m/(T_c)_{N_2} = (-100. + 459.67)/227.1 = 1.58$$

$$(p_R)_{CH_4} = \frac{1500.}{673} = 2.23$$

$$(T_R)_{CH_4} = (-100. + 459.67)/343.9 = 1.05$$

Using these values in Figure 7.6 of Chapter 7, we find that

$$(Z_A)_{N_2} = 0.84$$

and

$$(Z_A)_{CH_4} = 0.35$$

Then, Eq. (12.38) gives

$$Z_{Am} = 0.300(0.84) + 0.700(0.35) = 0.50$$

and

$$\bar{v}_m = Z_{Am}\Re T_m/p_m = \frac{0.50(1545.35)(-100. + 459.67)}{1500.\,(144)} = 1.29\ \text{ft}^3/\text{lbmole}$$

or an error of

$$\left(\frac{1.29 - 1.315}{1.315}\right)(100) = -1.90\%\ \text{low}$$

d. Using Kay's law, Eqs. (12.39) and (12.40), we get

$$p_{cm} = 0.300(492) + 0.700(673) = 619\ \text{psia}$$

and

$$T_{cm} = 0.300(227.1) + 0.700(343.9) = 309\ \text{R}$$

Then,

$$p_{Rm} = 1500./619 = 2.42$$

and

$$T_{Rm} = (-100. + 459.67)/309 = 1.17$$

For these reduced values, Figure 7.6 of Chapter 7 gives $Z_{Km} = 0.51$. Then, the mixture molar specific volume is

$$\bar{v}_m = Z_{Km}\Re T_m/p_m = \frac{0.51(1545.35)(-100. + 459.67)}{1500.\,(144)} = 1.31\ \text{ft}^3/\text{lbmole}$$

which has a negligible error from the measured value of 1.315 ft^3/lbmole.

SUMMARY

In this chapter, we deal with the problem of generating thermodynamic properties for homogeneous, nonreacting mixtures. Because of their engineering value, we focus our analysis on gases and vapors, but the theory extending beyond Eq. (12.15) can be easily modified to cover mixtures of liquids and solids.

We find that, if the mixture components and ultimately the mixture itself behave as an ideal gas, then all extensive properties are additive and the partial specific properties reduce to the component specific properties. This produces simple working equations for all the intensive properties (v, u, h, and s) of the mixture.

Last, we combine Dalton's and Amagat's laws with the compressibility factor technique to produce methods for dealing with the *p-v-T* mixture properties of real gases and vapors.

Some of the more important equations introduced in this chapter follow. Do not attempt to use them blindly without understanding their limitations. Please refer to the text material where they are introduced to gain an understanding of their use and limitations.

1. Mass, mole, volume, and pressure fractions of gas mixtures are defined as follows:
   a. w_i is the *mass fraction* of gas i in a mixture of N gases whose total mass is $m_m = \sum m_i$

$$w_i = \frac{m_i}{m_m}$$

   b. χ_i is the *mole fraction* of gas i in a mixture of N gases whose total molar mass is $n_m = \sum n_i$

$$\chi_i = \frac{n_i}{n_m}$$

   c. ψ_i is the *volume fraction* of gas i in a mixture of N ideal gases that occupy a total volume of $\forall_m = \sum_{i=1}^{N} \forall_i$ (where $\forall_i$ is the *partial volume* of ideal gas i in the mixture)

$$\psi_i = \frac{\forall_i}{\forall_m}$$

   d. π_i is the *pressure fraction* of gas i in a mixture of N ideal gases that are subjected to a total pressure of $p_m = p_i$ (where p_i is the *partial pressure* of ideal gas i in the mixture)

$$\pi_i = \frac{p_i}{p_m}$$

   and the mass, mole, volume, and pressure fractions always sum to unity, or

$$\sum_{i=1}^{N} w_i = \sum_{i=1}^{N} \chi_i = \sum_{i=1}^{N} \psi_i = \sum_{i=1}^{N} \pi_i = 1.0$$

2. The effective molecular mass of a mixture of N gases is given by

$$M_m = \sum_{i=1}^{N} M_i \quad \text{and} \quad M_m = \frac{1}{\sum_{i=1}^{N} \frac{w_i}{M_i}}$$

3. The equivalent gas constant of a mixture of N gases is then given by

$$R_m = \frac{\Re}{M_m}$$

   where $\Re = 8.3143\,\text{kJ/(kgmole}\cdot\text{K)} = 1545.35\,\text{ft}\cdot\text{lbf/(lbm}\cdot\text{R)}$ is the universal gas constant.
4. If all the gases in the mixture are ideal gases, then the composition measures are related as follows:

$$\chi_i = \frac{n_i}{n_m} = \psi_i = \frac{\forall_i}{\forall_m} = \pi_i = \frac{p_i}{p_m} = w_i\left(\frac{M_m}{M_i}\right)$$

5. The changes in specific internal energy, specific enthalpy, and specific entropy for mixtures of ideal gases are given by

$$u_{m2} - u_{m1} = c_{vm}(T_{m2} - T_{m1})$$
$$h_{m2} - h_{m1} = c_{pm}(T_{m2} - T_{m1})$$
$$s_{m2} - s_{m1} = c_{vm}\ln(T_{m2}/T_{m1}) + R_m\ln(v_{m2}/v_{m1})$$
$$= c_{pm}\ln(T_{m2}/T_{m1}) - R_m\ln(p_{m2}/p_{m1})$$

where the mixture specific heats are given by

$$c_{v_m} = \sum_{i=1}^{N} w_i \hat{c}_{vi} \quad \text{and} \quad c_{p_m} = \sum_{i=1}^{N} w_i \hat{c}_{pi}$$

6. The study of the special mixture of ideal gases of dry air and water vapor is called *psychrometrics*, and the two important composition measures used in the field of meteorology are
   a. The *relative humidity* ϕ of the mixture, defined as

   $$\phi = \frac{p_w}{p_{\text{sat}}}$$

   where p_w is the actual partial pressure of the water vapor in the mixture and p_{sat} is the saturation pressure of the water vapor at the temperature of the mixture.
   b. The *humidity ratio* ω of the mixture, defined as

   $$\omega = \frac{m_w}{m_a}$$

   where m_w is the mass of water vapor present in the mixture and m_a is the mass of dry air present in the mixture. These two composition measures are related by

   $$\omega = 0.622\left(\frac{\phi p_{\text{sat}}}{p_m - \phi p_{\text{sat}}}\right)$$

7. The *dew point temperature* T_{DP} is defined as

   $$T_{DP} = T_{\text{sat}}\ (\text{evaluated at } p_w)$$

   where p_w is the partial pressure of the water vapor present, $p_w = \phi p_{\text{sat}}$.
8. The humidity ratio at the inlet to an *adiabatic saturator* is given by

   $$\omega_1 = \frac{c_{pa}(T_2 - T_1) + \omega_3(h_{fg2})}{h_{g1} - h_{f2}}$$

9. The *psychrometric enthalpy* $h^{\#}$ of the mixture is defined as

   $$h^{\#} = \frac{H_m}{m_a} = h_a + \omega h_w$$

10. For mixtures of *real gases,* we can use
   a. The *Dalton* compressibility factor, Z_{Dm},

   $$p_m \forall_m = Z_{Dm} m_m R_m T_m$$

   where

   $$Z_{Dm} = \sum_{i=1}^{N}\left(\frac{w_i M_m}{M_i}\right) Z_{Di} = \sum_{i=1}^{N} \chi_i Z_{Di}$$

   b. the *Amagat* compressibility factor, Z_{Am},

   $$p_m \forall_{Am} = Z_{Am} m_m R_m T_m$$

   where

   $$Z_{Am} = \sum_{i=1}^{N} (w_i M_m / M_i) Z_{Ai} = \sum_{i=1}^{N} x_i Z_{Ai}$$

   c. the *Kay* compressibility factor, Z_{Km},

   $$p_m \forall_m = Z_{Km} m_m R_m T_m$$

   where Z_{Km} is found from the pseudocritical pressure and temperature of the mixture

   $$p_{cm} = \sum_{i=1}^{N} x_i p_{ci} \quad \text{and} \quad T_{cm} = \sum_{i=1}^{N} x_i T_{ci}$$

Problems (* indicates problems in SI units)

1.* 1.00 kg of CH_4, 6.30 kg of O_2, and 13.2 kg of N_2 are combined to form a gas mixture. Find the molecular mass of the mixture.

2. Ammonia gas is flowing in a tube whose cross-sectional area is 1.00 ft^2. The density of the ammonia is 0.100 lbm/ft^3. At some point in the system, CO_2 is added at a rate of 1.00×10^{-3} lbm/h in a steady flow process. At a point further downstream, a detector indicates that the concentration of CO_2 on a mass basis is 1.00×10^{-3}%. Determine the inlet velocity of the ammonia.

3. The following is the gravimetric analysis of a gaseous mixture:

Constituent	% by mass
N_2	60.0
CO_2	22.0
CO	11.0
O_2	7.00

Find the mole fractions of the components in the mixture.

4. If the amount of helium produced by the large-scale liquefaction of air is 1.30 lbm per 100. tons of air, determine the mass, mole, and volume fractions of helium in air. Assume air is a mixture of ideal gases.

5.* A mixture of air and water vapor at 20.0°C and 0.101 MPa has a relative humidity of 100.%. Using the information given in Example 12.2, determine
 a. The molar composition.
 b. The effective molecular mass.
 c. The mass concentration.
 d. The effective gas constant of the mixture, whose components are nitrogen, oxygen, water vapor, argon, and carbon dioxide.

6. A gas bulb of volume 0.100 ft^3 contains hydrogen at a pressure of 10.0 psia and a temperature of 50.0°F. Nitrogen is introduced into the bulb such that the final pressure is 20.0 psia and the final temperature is 80.0°F. Find the mole fraction and the mass fraction of the hydrogen in the final state.

7. Using Eqs. (12.11), (12.12), and (12.15), show that the equivalent gas constant of a mixture R_m can be determined directly from the mass and mole fraction compositions (w_i and x_i) and the species gas constants ($R_i = \Re/M_i$) as
 a. $R_m = \sum w_i R_i$
 b. $R_m = \dfrac{1}{\sum (x_i/R_i)}$

8. A mixture of ideal gases at a total pressure of 40.0 psia and 70.0°F contains 0.600 lbm of hydrogen (H_2) and 4.80 lbm of oxygen (O_2). Determine
 a. The mole fraction of the hydrogen in the mixture.
 b. The equivalent molecular mass of the mixture.
 c. The total volume occupied by the mixture under the conditions stated.

9.* (a) How many kg of nitrogen must be mixed with 5.00 kg of carbon dioxide to produce an ideal gas mixture that is 50.0% by volume of each component? (b) For the resulting mixture of part a, determine the mixture molecular mass (M_m), gas constant (R_m), and the partial pressure of the nitrogen if that of the carbon dioxide is 0.0700 MPa.

10. A furnace exhaust stack is instrumented so that a sample of the stack gas can be analyzed for composition. The analysis gives the following volume fractions: 70.0% N_2, 20.0% CO_2, 8.00% CO, 1.00% Ar, and 1.00% H_2O.
 a. What is the molecular mass of the stack gas?
 b. Calculate the mass fraction of each of the gases.
 c. If the measured pressure in the stack is 15.0 psia, what is the partial pressure of the carbon monoxide?

11.* The Department of Homeland Security discovered that 10.0 kg of nitrogen (N_2) was mixed with 3.82 kg of a possibly toxic unknown gas. The resulting mixture occupies a volume of 2.00 m^3 at 0.800 MPa and 65.0°C. Both gases and the mixture are ideal gases. Determine
 a. The molecular mass of the gas mixture.
 b. The volume fraction of each gas present in the mixture.

12. Onboard a starship, a demented alien creature releases a mixture of ideal gases made up of 4.00 lbm of molecular oxygen (O_2) and 6 lbm of an unknown and possibly toxic gas. Before the gas mixture was released, you noticed that its original container had a volume of 10.0 ft^3, and at 150.°F, it had a pressure of 114.3 psia. Because you are a line officer engineer, the captain asks you for your assessment of the situation. Is the gas lethal? To draw a proper conclusion, you must determine
 a. The molecular mass of the unknown gas.
 b. The probable name of this gas.
 c. The volume fraction of each gas present in the mixture.

13. The measured molecular masses of many naturally occurring chemical elements differ appreciably from integer values due to the presence of isotopes of the element in the test sample. Commercially available neon has a measured molecular mass of 20.183. The gas is known to be a mixture of two isotopes whose molecular masses are 20.0 and 22.0. Determine the mole and mass fractions of each of the neon isotopes present in commercial neon.

14. 1.00 ft^3 of steam at 300.°F and 14.7 psia is mixed with 3.00 ft^3 of methane at 80.0°F and 14.7 psia. Assuming ideal gas behavior for both of these substances, determine
 a) The mass fractions.
 b) The mole fractions.
 c) The pressure fractions.
 d) Both the constant pressure and constant volume specific heats of the mixture.

15. A rigid insulated tank is divided into two compartments by a partition. Initially, 5.60 lbm of nitrogen is introduced into one compartment at a pressure of 30.0 psia and a temperature of 140.°F. At the same time, 13.2 lbm of carbon dioxide is introduced into the other compartment at a pressure of 15.0 psia and a temperature of 60.0°F. The partition is then removed and the gases are allowed to mix. Assuming ideal gas behavior, find the pressure in the tank after the mixing.

16.* A perfect gas mixture consists of 3.00 kg of nitrogen (N_2) and 5.00 kg of carbon dioxide (CO_2) at a pressure of 1.00 MPa and a temperature of 30.0°C. If the mixture is heated at constant pressure to 40.0°C, find the work and the heat transfer required for this process.

17.* We invented a new process in which 2.00 kg/min of hydrogen at 5.00°C and 1.00 atm is continuously aergonically mixed with

1.00 kg/min of nitrogen at 30.0°C and 1.00 atm. The mixture leaves the mixing chamber at 60.0°C and 0.800 atm. Find the heat transfer rate and indicate whether it is into or out of the mixture.

18. On planet 3M4G6 in the subsystem Zeta-12, the atmosphere consists of a binary mixture of sewer gas (methane) and an unknown gas, called *Esh-nugim Marookee Moo* by the local mushfoot natives. An extremely accurate spectral scan reveals that the mass fraction to mole fraction ratio for methane is 0.9341875 and the mole fraction of methane is 0.4281378. Determine the name of the unknown gas (in English).
19. Methane, ethane, and propane, all ideal gases, are mixed together in equal parts by mass to create a new "super" fuel gas. Then, it is adiabatically compressed from 40.0°F to 1.75 ft^3 at 300. psia, 80.0°F. Determine the work required.
20.* A chemical processing facility produces exhaust gases (46.0% N_2, 43.0% CO_2, and 11.0% SO_2 by volume) at a total pressure of 0.320 MPa at 1000.°C. It is proposed that energy be recovered from this gas by expanding it through a turbine to atmospheric pressure. Assuming ideal gas behavior, determine the maximum possible power output per unit mass flow rate for this system.
21.* Two parts of molecular hydrogen gas are mixed with one part of molecular oxygen gas (on a molar basis) at 2.00 MPa, 0.00°C and expanded through a reversible nozzle from a negligible inlet velocity to 292 m/s at the entrance to the combustion zone of a rocket engine. Through a preheating process in the nozzle, the gas mixture receives 1325.5 kJ per kg of mixture of heat at 500.°C. Determine the exit pressure of the gas mixture. Assume ideal gas behavior.
22. A diving experiment is to be performed with a mixture of helium and air at a total pressure of 50.0 psia. The composition must be such that the partial pressure of oxygen in the compressed mixture is the same as that in air at standard pressure and temperature (14.7 psia and 70.0°F). Assuming a closed system and ideal gas behavior, determine
    a. The work required to isentropically compress 2.70 lbm of the mixture from 14.7 psia, 70.0°F to 50.0 psia.
    b. The heat transfer required to aergonically cool the compressed mixture back to 70.0°F again.
23.* Acetylene and oxygen are drawn from pressurized storage tanks and mixed together in an oxyacetylene torch in a ratio of 5.00 parts of oxygen per 1 part acetylene on a volume basis. This mixture flows reversibly through the torch from 0.140 MPa, 20.0°C, to atmospheric pressure at 173°C, at which point it is ignited by the flame. The mean surface temperature of the torch is 30.0°C, and it uses 0.100 m^3/s of oxygen at STP (0.101325 MPa, 20.0°C). Assuming ideal gas behavior, determine the work and heat transfer rates.
24.* 18.0 m^3/s of methane are mixed with 10.0 m^3/s of isobutane in a test of a new furnace gas. The mixture is preheated before being ignited by passing it through an adiabatic, isobaric heat exchanger. The second fluid in the heat exchanger is condensing steam at 200.°C flowing at 8.30 kg/s. The steam enters as a saturated vapor and exits with a quality of 21.0%. The gas mixture enters at 20.0°C and exits at 150.°C (Figure 12.13). Determine the entropy production rate of the heat exchanger.

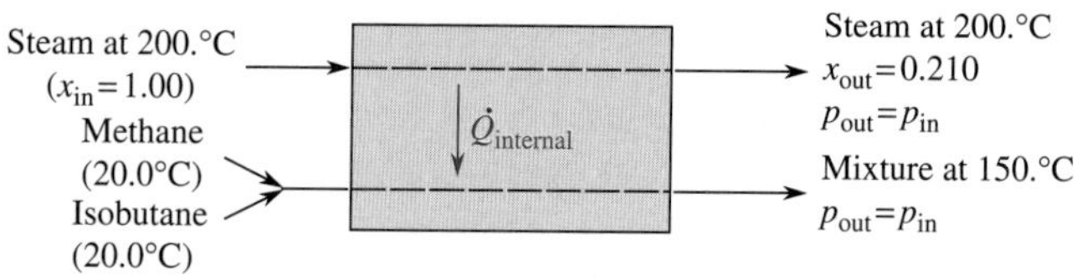

FIGURE 12.13
Problem 24.

25.* Air at 0.101 MPa and 50.0°C is saturated with water vapor. It is to be aergonically heated to 80.0°C in a steady flow, isobaric process by putting it in contact with an isothermal reservoir at 100.°C. Determine the heat transfer and entropy production rates per unit mass flow rate if
    a. The presence of the water vapor is ignored.
    b. The presence of the water vapor is considered (as an ideal gas).
    c. Determine the percent error in the answer a due to the effect of the water vapor.
26.* The combustion of 1.00 mole of octane, C_8H_{18}, yields 8 moles of CO_2, 9.00 moles of H_2O, and 47.0 moles of N_2. The exhaust gases at 811 K, 0.172 MPa from a spark ignition engine are to be expanded through a turbocharger used to compress the incoming air charge to the engine. The incoming air is at 20.0°C and atmospheric pressure, and the turbocharger turbine exhausts to the atmosphere (Figure 12.14). Find the compressor's isentropic outlet temperature T_2 and its isentropic power input per unit mass flow rate $(\dot{W}_C/\dot{m})_s$.

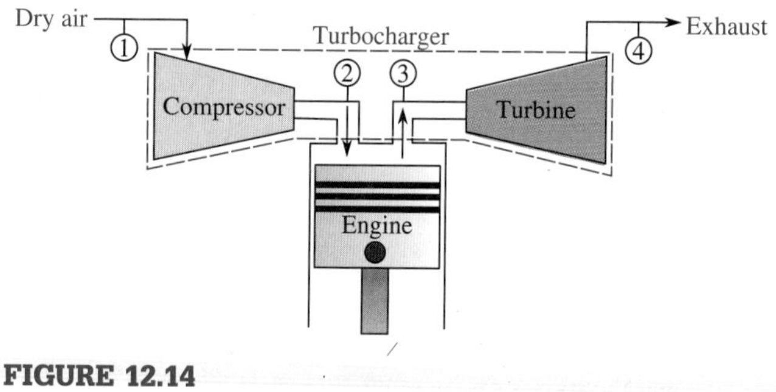

FIGURE 12.14
Problem 26.

27. A pneumatic motor in a highly explosive environment uses a mixture of the ideal gases argon and helium in equal parts on a mass basis. The motor has inlet conditions of 150. psia at 500. R and an exit pressure of 14.7 psia. If the motor must produce 3.00 hp of output power while operating in a steady state, steady flow, reversible, and adiabatic manner, then
    a. What is the exit temperature of the gas mixture?
    b. What mass flow rate of the mixture is required?
28.* An insulated gas turbine is attached to the exhaust stack of an oil-fired boiler in a power plant. The pressure and temperature at the inlet to the turbine are 0.500 MPa absolute, 1000.°C and the exit pressure is atmospheric. The exhaust gas analysis by volume is 12.0% CO_2, 2.00% CO, 4.00% O_2, and 82.0% N_2. What is the maximum possible power output from this turbine per kg of exhaust gas flowing through it? Assume that the

exhaust gas is an ideal gas and ignore all kinetic and potential energy effects.

29. An air–water vapor mixture at 14.7 psia, 100.°F, and 40.4% relative humidity is contained in a 10.0 ft^3 closed tank. The tank is cooled until the water just begins to condense. Determine the temperature at which condensation begins.

30. When the dew point of atmospheric air is between 60.0°F and 70.0°F, the weather is said to be *humid*, and when it is above 70.0°F it is said to be *tropical*. If the partial pressure of water vapor in the air is 0.400 psia when the dry bulb temperature is 80.0°F, determine the relative humidity and whether the weather is humid or tropical.

31. The volume fractions of the gases in the atmosphere of Mars measured at the surface of the planet, where the total atmospheric pressure is 0.112 psia, were found by the early Viking I mission to be 95.0% CO_2, 2.50% N_2, 2.00% Ar, 0.400% O_2, and 0.100% H_2O. Determine
   a. The partial pressure of the water vapor in the Martian atmosphere.
   b. The mass fractions of all the gases in the Martian atmosphere.
   c. The humidity ratio of the Martian atmosphere (consider the Martian dry "air" to be everything in its atmosphere except the water vapor).

32. An engineer at a party is handed a cold glass of beverage with ice in it. The engineer estimates the outside temperature of the glass to be 35.0°F and the room temperature to be 70.0°F. What is the relative humidity in the room when moisture just begins to condense on the outside of the glass?

33.* The list of cities in Table 12.6 includes representative wet and dry bulb temperatures. Use the psychrometric chart to determine their corresponding relative humidity, humidity ratio, dew point temperature, and water vapor partial pressure.

34. People often clean eyeglasses by holding them near their mouths and exhaling heavily on them. This usually causes the lenses to "fog"; the moisture is then wiped off, and this process cleans the lenses. Assuming the air in the lungs has a relative humidity of 75.0% at a dry bulb temperature of 100.°F, determine the maximum temperature of the glasses that just causes moisture droplets (i.e., "fog") to form when cleaned in this manner.

35.* 7000 m^3/min of air at 28.0°C and 0.101 MPa with a relative humidity of 60.0% is to be cooled at constant total pressure to its dew point temperature. Determine the required heat transfer rate and indicate its direction.

36. 30 ft^3/min of air with a dry bulb temperature of 90.0°F and a relative humidity of 80.0% is to be cooled and dehumidified to a dry bulb temperature of 65.0°F and a relative humidity of 50.0%. Determine (a) the wet bulb temperature of the air before dehumidification, (b) the dew point temperature of the air after dehumidification, (c) the amount of moisture removed during the dehumidification process, and (d) the amount of heat removed during the cooling part of the dehumidification process.

37.* A room containing 275 m^3 of air at 1.00 atm pressure is to be humidified. The initial conditions are T_{DB} = 24.0°C and ϕ = 20.0%, and the final conditions are T_{DB} = 20.0°C and ϕ = 60.0%. Determine the mass of water that must be added to the room air.

38. 1000. ft^3/h of moist air at atmospheric pressure, 80.0°F and 70.0% relative humidity is to be cooled to 50.0°F at constant total pressure. Find whether or not this can be done without the removal of water from the air. If it cannot, determine the minimum amount of water that must be removed in lbm/h.

39. A classroom contains 6000. ft^3 of air–water vapor mixture at 1.00 atm total pressure. The dry bulb temperature is 70.0°F and the wet bulb temperature is 65.0°F. Assuming a closed constant total pressure system, determine the following:
   a. The relative humidity.
   b. The partial pressure of the water vapor.
   c. The dew point.
   d. The amount of water that must be added to or removed from the air in the room to achieve 40.0% relative humidity at the same dry bulb temperature.

40. 10,000. ft^3/h of moist air at 14.7 psia and 75.0°F is to be cooled to 45.0°F at constant total pressure. Find the amount (lbm/h) of water condensed if the mole fraction of water in the inlet mixture is 0.0260.

41.* Atmospheric air can be dehumidified by cooling the air at constant total pressure until the moisture condenses out. Suppose that air with a humidity ratio of 5.00×10^{-3} kg water per kg of dry air must be achieved by cooling incoming atmospheric air with a dry bulb temperature of 25.0°C and a wet bulb temperature of 20.0°C.
   a. To what temperature must the incoming air–water vapor mixture be cooled to achieve a humidity ratio of 5.00×10^{-3} kg water per kg dry air?
   b. How much water must be removed per kg of dry air to achieve this state?

42. Outside atmospheric air with a dry bulb temperature of 90.0°F and a wet bulb temperature of 85.0°F is to be passed through an air conditioning device so that it enters a house at 71.0°F and 40.0% relative humidity. The process consists of two steps.

Table 12.6 Problem 33

City	T_{WB}	T_{DB}	ϕ	ω	T_{DP}	P_w
Berlin	21.0°C	32.0°C				
Chicago	75.0°F	97.0°F				
Hong Kong	28.0°C	33.0°C				
New York City	75.0°F	93.0°F				
Paris	20.0°C	31.0°C				
Rome	23.0°C	36.0°C				
Tokyo	79.0°F	92.0°F				

First, the air passes over a cooling coil, where it is cooled below its dew point temperature and the water condenses out until the desired humidity ratio is reached. Then, the air is passed over a reheating coil until its temperature reaches 71.0°F. Determine

a. The amount of water removed per pound of dry air passing through the device.
b. The heat removed by the cooling coil in Btu/(lbm dry air).
c. The heat added by the reheating coil in Btu/(lbm dry air).

43.* An airstream with a mass flow rate of 2.00 kg/s, a dry bulb temperature of 10.0°C, and a wet bulb temperature of 8.00°C is mixed with an airstream having a mass flow rate of 1.00 kg/s, a dry bulb temperature of 40.0°C, and a wet bulb temperature of 35.0°C. For the resulting mixture, determine the (a) humidity ratio, (b) psychrometric enthalpy, (c) relative humidity, (d) dry bulb temperature, (e) wet bulb temperature, and (f) dew point temperature.

44.* Exactly 200 m^3/min of air with a dry bulb temperature of 7.00°C and a wet bulb temperature of 5.00°C is continuously mixed with 500. m^3/min of air with a dry bulb temperature of 32.0°C and a relative humidity of 60.0%. The mixing chamber is at atmospheric pressure and is electrically heated with a power consumption of 3.00 kW. For the resulting mixture, determine (a) the dry bulb temperature, (b) the wet bulb temperature, (c) the dew point temperature, and (d) the relative humidity.

45.* 100. kg of atmospheric air (whose composition is given in Example 12.2) is cooled in a 0.500 m^3 constant volume container to 200. K. Determine the mixture pressure using Dalton's compressibility factor. Compare this result with that obtained by assuming ideal gas (i.e., $Z_m = 1.00$) behavior.

46. Show that the ratio of the Amagat specific volume, $v_{Ai} = \forall_{Ai}/m_i$, to the Dalton specific volume, $v_{Di} = \forall_m/m_i$, of gas i can be written as $v_{Ai}/v_{Di} = Z_{Ai}p_{Di}/Z_{Di}p_m$.

47.* Atmospheric air, whose composition is given in Example 12.2, is compressed to 1000. atm at 0.00°C. Determine the density of the compressed air using Amagat's compressibility factor and compare this result with that obtained by assuming ideal gas (i.e., $Z_m = 1.00$) behavior.

48.* In normal psychrometric analysis at or below atmospheric pressure, both the air and the water vapor are treated as ideal gases. However, at high pressures, this assumption is no longer valid. Determine the relative humidity (ϕ) that results when 10.0 g of water are added to 1.00 kg of dry air at a total pressure of 10.0 MPa at 350.°C. Assume Amagat's compressibility factor is valid here and compare your result with that obtained by assuming ideal gas behavior.

49.* 0.500 m^3 of a mixture of 35.0% acetylene (C_2H_2), 25.0% oxygen (O_2), 20.0% hydrogen (H_2), and 20.0% sulfur dioxide (SO_2) on a mass basis is to be adiabatically compressed in a piston-cylinder arrangement from 1.00 atm, 20.0°C to 100. atm and 300.°C.

a. Using Kay's law, find the final volume of the mixture.
b. Assuming constant specific heats, determine the work required per unit mass of mixture.

50.* A mixture of 80.0% methane and 20.0% ethane on a molar basis is contained in an insulated 1.00 m^3 tank at 3.00 MPa and 50.0°C. An automatic flow control valve opens, causing the tank pressure to drop quickly to 2.00 MPa before it closes again.

a. Calculate the mass that escaped from the tank using Kay's law.
b. Determine the equilibrium tank temperature when the control valve closes, assuming that the gas remaining in the tank underwent a reversible process.

51.* 6.00 kg of hydrogen gas (H_2) is mixed with 28.0 kg of nitrogen gas (N_2) and compressed to 40.5 MN/m^2 at 300.°C. At this state, the specific volume of the mixture is measured and found to be 0.0160 m^3/kg. Determine the specific volume of this state as predicted by each of the following models and calculate the percent deviation from the measured value.

a. Ideal gas model.
b. Dalton's law compressibility factor.
c. Amagat's law compressibility factor.
d. Kay's law.

52.* Determine the total mixture volume when 1 kgmole each of hydrogen (H_2) and helium (He) gases are mixed at 15.0 atm and 40.0 K, using

a. Dalton's law compressibility factor.
b. Amagat's law compressibility factor.
c. Kay's law.
d. Which of the three results do you believe is the most accurate, and why?

Design Problems

The following are open-ended design problems. The objective is to carry out a preliminary thermal design as indicated. A detailed design with working drawings is not expected unless otherwise specified. These problems do not have specific answers, so each student's design is unique.

53. Design an electrically driven sling psychrometer to produce the wet and dry bulb temperatures on digital readouts. The finished product must cost less than 30.0 h of minimum wage pay and be battery powered. If possible, fabricate and test your design. (Suggestion: Try designing around inexpensive, "off the shelf" components.)

54. Design an apparatus to measure the dew point of an air sample based on the cooling of a mirrored surface until it fogs. If possible, build and test this apparatus. (Suggestion: Consider thermoelectric cooling of a polished metal plate.)

55.* Design a system to remove the respiration carbon dioxide from inside a spacecraft and replace it with oxygen. Use a living quarters volume of 10.0 m^3 with the crew generating a maximum of 2.00×10^{-5} m^3/s of CO_2. Assume the mixture enters your system at 30.0°C and exits it at 20.0°C. Maintain the same oxygen partial pressure in your mixture as that in atmospheric air at 0.1013 MPa and 20.0°C.

56.* Design a system to remove the respiration carbon dioxide from inside a submarine and replace it with oxygen. The air volume of the submarine is 1000. m^3, and the crew can generate a maximum of 1.30×10^{-3} m^3/s of CO_2. The submarine must be able to achieve a depth of 300. m. Assume the air mixture enters your system at 30.0°C and exits at 20.0°C. Maintain the oxygen partial pressure at all times in your mixture equal to that in atmospheric air at 0.1013 MPa and 20.0°C.

57.* Cooling towers are large evaporative cooling systems that can be used to transfer heat from warm water to the atmosphere by evaporation of the water to be cooled. Prepare a preliminary design for a cooling tower that will cool 30,000 kg/s of water from 40.0°C to 30.0°C. Atmospheric air enters at 20 ± 10°C with a relative humidity of 45 ± 15%. Establish the overall physical dimensions of the cooling tower, air flow rate, water pumping power, fan power (if forced convection is used), makeup water requirements, air exit conditions, and so forth (Figure 12.15).

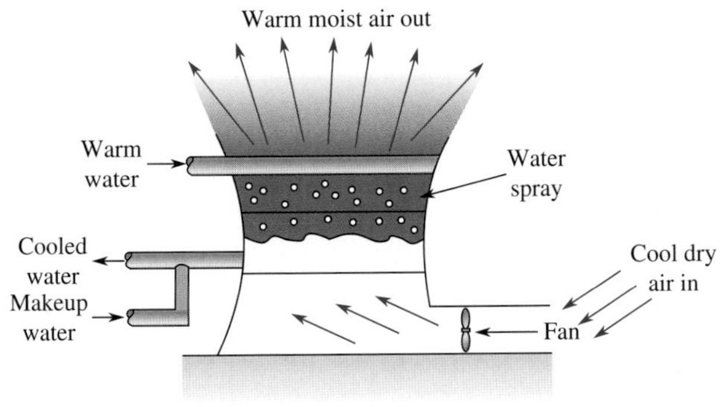

FIGURE 12.15
Problem 57.

Computer Problems

The following open-ended computer problems are designed to be done on a personal computer using a spreadsheet or equation solver.

58. Develop a computer program that returns all four ideal gas mixture composition fractions when any one composition fraction is input for an arbitrary mixture of ideal gases. Have the user input the values of the gas constant, molecular masses, number of gases in the mixture, and anything else you need to make the appropriate calculations. Make sure the user enters the input variables in the proper units, and make sure that correct units appear with the output values.

59.* Develop a simple computer version of the gas tables (Table C.16 in *Thermodynamic Tables to accompany Modern Engineering Thermodynamics*) for an arbitrary mixture of ideal gases with constant specific heats. Input the composition (on a mass, molar, or volume basis), specific heats (in proper units), and temperature from the keyboard. Output, in a properly formatted manner, the values of u, h, ϕ, p_r, and v_r with correct units. Use reference levels of $h_0 = \phi_0 = 0$ at $T_0 = 300.$ K and $p_0 = 1.00$ atm.

60.* Modify Problem 59 by adding the mixture pressure to the list of keyboard input variables and output the specific entropy s instead of ϕ. Use $s_0 = \phi_0 = 0$ at $T_0 = 300.$ K and $p_0 = 1.00$ atm.

61.* Create an accurate expanded version of the gas tables (Table C.16) for an arbitrary mixture of ideal gases with temperature-dependent specific heats. Allow the user to choose the gases in the mixture from those you list in a screen menu. Have the user input the composition (on a mass, molar, or volume basis), the pressure, and the temperature from the keyboard in response to screen prompts. Output the values of u, h, s, p_r, and v_r with correct units. Use reference levels of $h_0 = s_0 = 0$ at $T_0 = 300.$ K and $p_0 = 1.00$ atm.

62. The saturation pressure curve for ammonia (NH_3) can be approximated with $p_{sat} = \exp[A - B/T_{DB} - C(\ln T_{DB}) - D(T_{DB}) + E(T_{DB}^2)]$, where p_{sat} is in psia, T_{DB} is in R, and $A = 58.88706$, $B = 8.58730 \times 10^3$, $C = 6.40125$, $D = 9.55176 \times 10^{-4}$, and $E = 3.39860 \times 10^{-6}$. Equation (12.28) can be modified to give the humidity ratio ω of an air-ammonia mixture as follows:

$$\omega = (M_{NH3}/M_a)[p_{NH3}/(p_m - p_{NH3})] = 0.588[p_{NH3}/(p_m - p_{NH3})]$$

Using these equations and Eq. (12.24), develop an interactive computer program in English units that returns properly formatted values for p_m, T_{DB}, ϕ, and ω for an air-ammonia mixture, when

a. p_m, T_{DB}, and ϕ are input from the keyboard.
b. p_m, T_{DB}, and ω are input from the keyboard.

Make sure the user is prompted for the input variables in the proper units, and that correct units appear with the output values.

63. Modify the program of Problem 62 to allow the user to separately choose either English or Metric (SI) units for the input and the output values.

64.* Using Eqs. (12.24) and (12.26a) and the four equations that follow,[10] develop an interactive computer program in metric (SI) units that replaces the psychrometric chart of Chart D.6. Prompt the user for keyboard input of atmospheric pressure p_m, atmospheric temperature T_{DB}, and either the relative humidity ϕ or the humidity ratio ω. Return to the screen properly formatted values (with units) for p_m, T_{DB}, T_{DP}, ω, ϕ, $h^{\#}$, and v_a.

a. $p_{sat} = 0.1 \exp[14.4351 - 5333.3/(T_{DB} + 273.15)]$, for $0 \leq T_{DB} \leq 38°C$
b. $T_{DP} = 5333.3/[14.4351 - \ln(p_w/0.1)] - 273.15$, for $0 \leq T_{DP} \leq 38°C$
c. $h^{\#} = 1.005(T_{DB}) + \omega[2501.7 + 1.82(T_{DB})]$
d. $v_a = (0.286{\times}10^{-3})(T_{DB} + 273.15)/(p_m - p_w)$

where p_{sat}, p_m, and p_w are in MPa; T_{DB} and T_{DP} are in °C; $h^{\#}$ is in kJ/(kg dry air); and v_a is in m^3/(kg dry air).

65.* Modify the program of Problem 64 to allow the user to separately choose either metric (SI) or English units for the input and the output values.

66.* Expand Problem 65 by adding Eq. (12.31) to your program. In Eq. (12.31), let $\omega_1 = \omega$, $T_1\omega = T_{DB}$, $T_2 = T_{WB}$, and use $c_{pa} = 1.005$ kJ/(kg · K). Also, use[11]

e. $h_{g1} = 2501.7 + 1.82(T_{DB})$
f. $h_{f2} = 4.194(T_{WB})$
g. $h_{fg2} = 2501.7 - 2.374(T_{WB})$
h. $\omega_3 = 0.622 p_{sat}/(p_m - p_{sat})$

where h_{g1}, h_{f2}, and h_{fg2} are in kJ/kg, T is in °C, and p_{sat} is evaluated at T_{WB} and obtained from Eq. (a) in Problem 64 by

[10] These equations are from Liley, P. E., 1980. Approximations for the thermodynamic properties of air and steam useful in psychrometric calculations. Mech. Eng. News 17 (4), 19–20.
[11] These equations also are from Liley, P. E. Approximations for the thermodynamic properties of air and steam useful in psychrometric calculations.

replacing T_{DB} with T_{WB}. The wet bulb temperature can now be used as an input parameter. Prompt for inputs of p_m, T_{DB}, and either T_{WB}, ϕ, or ω. Return to the screen properly formatted values (with appropriate units) of p_m, T_{DB}, T_{WB}, T_{DP}, ω, ϕ, $h^{\#}$, and v_a.

Special Problems

67. The implicit function theorem from calculus tells us that if $M\forall_m/MT_m \neq 0$ in Eq. (12.5), then we can write the temperature T_m of a mixture of N gases as a function of the mixture total volume $\forall_m$, mixture total pressure p_m, and the mass composition of the mixture $m_1, m_2, \ldots, m_N$ as

$$T_m = T_m(\forall_m, p_m, m_1, m_2, \ldots, m_N)$$

However, when the total volume and pressure of the mixture are constant, this generally is *not* a homogenous function of the first degree in the masses m_i. The temperature usually varies inversely with the system mass in most equations of state for gases and vapors. Consequently, when the mixture masses are multiplied by an arbitrary constant λ, the mixture temperature is multiplied by $1/\lambda$, or

$$(1/\lambda)T_m = T_m(p_m, \forall_m, \lambda m_1, \lambda m_2, \ldots, \lambda m_N)$$

Show that differentiating this equation with respect to λ while holding the mixture pressure and volume constant gives

$$\left(\frac{\partial(T_m/\lambda)}{\lambda}\right)_{p_m,\forall_m} = -\frac{T_m}{\lambda^2}\Big|_{p_m,\forall_m}$$

$$= \left(\frac{\partial T_m}{\partial \lambda m_1}\right)\left(\frac{\partial \lambda m_1}{\partial \lambda}\right) + \cdots + \left(\frac{\partial T_m}{\partial \lambda m_N}\right)\left(\frac{\partial \lambda m_N}{\partial \lambda}\right)$$

$$= \left(\frac{\partial T_m}{\partial \lambda m_1}\right) m_1 + \cdots + \left(\frac{\partial T_m}{\partial \lambda m_N}\right) m_N$$

and setting $\lambda = 1$ gives

$$T_m\big|_{p_m,\forall_m} = -\left(\frac{\partial T_m}{\partial m_1}\right) m_1 - \cdots - \left(\frac{\partial T_m}{\partial m_N}\right) m_N = \sum_{i=1}^{N} m_1 \hat{T}_i = \sum_{i=1}^{N} T_i$$

where

$$\hat{T}_i = -\left(\frac{\partial T_m}{\partial m_i}\right)_{p_m,\forall_m,m_j}$$

Can we then define Ti to be the *partial specific temperature* of gas i in the mixture and define $T_i = m_i\hat{T}_i$ to be the *partial temperature* of gas i in the mixture (Figure 12.16)? Does this lead us to a fourth composition measure, the temperature fraction?

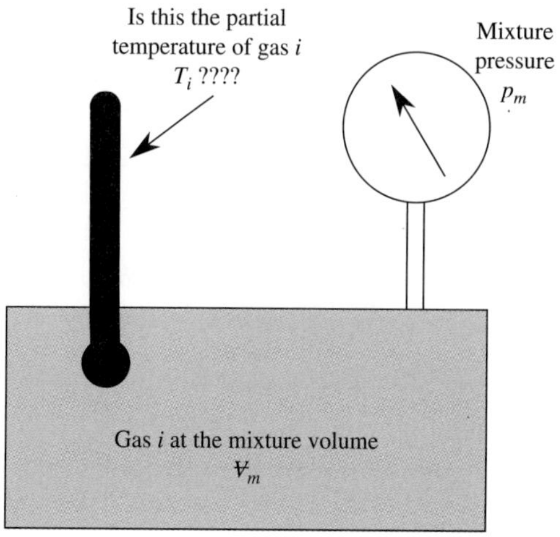

FIGURE 12.16
Problem 67.

Is this the *temperature fraction* of gas i in a mixture of gases?

$$\tau_i = T_i/T_m$$

Also, in the case of a mixture of ideal gases, can the following interpretation be developed similar to the Dalton and Amagat law ideal gas partial pressure and partial volume? Can the *partial temperature of ideal gas i* (T_i) be defined as the temperature exhibited by ideal gas i when it alone occupies the volume of the mixture $\forall_m$ at the pressure of the mixture p_m? Or,

$$T_i = \frac{p_m \forall_m}{m_i R_i}$$

CHAPTER 13

Vapor and Gas Power Cycles

CONTENTS

13.1 BEVEZETÉSÉNEK (INTRODUCTION)

The material in this chapter embodies the heart and soul of applied thermodynamics. Because of the human and cultural impact this technology has had on society, the material is presented chronologically to provide a historical framework for its study. It is important that engineers today develop an understanding of the relative effect the technology they create can have on society. Therefore, the normally dry technical aspects of this enormously powerful technology have been augmented with a small amount of relevant humanistic information in the hope of giving it an interesting perspective.

The first part of this chapter deals with vapor power cycles and their associated technology. Were it not for this technology, the Industrial Revolution that began in the mid 18th century would not have taken place, and the world would be a much different place today. The section begins with basic definitions of engines, machines, and heat, then moves on to a detailed discussion of the Rankine cycle and its various attributes. The section ends with a discussion of modern power plant thermodynamic cycles and technology.

The second part of this chapter focuses on gas power cycles. These power cycles were originally developed as an alternative to dangerous vapor cycle engines. The Stirling, Ericsson, and Brayton cycles are *external combustion* cycles, meaning that the combustion process takes place outside the prime mover. The Lenoir, Otto, and Diesel cycles, on the other hand, are *internal combustion* cycles, with the combustion process taking place inside the engine itself. Gas power cycle technology has also had a profound impact on our culture. Gas power cycles have come to dominate portable power systems, such as automobiles, ships, trains, and airplanes. A great deal of the petroleum imported into the United States goes to fuel the machines powered by these cycles. Part II of this chapter covers the thermodynamics of these cycles in detail. As engineers, we are particularly concerned with understanding how the cycle thermal efficiency depends on design variables, such as compression ratio and combustion temperature.

13.2 PART I. ENGINES AND VAPOR POWER CYCLES

We need to start out with a clear understanding of some of the terminology we use in this chapter. By looking at the historical roots of words like *machine* and *engine*, we discover the true meaning of our profession—engineering.

13.2.1 What Is a Machine?

The English word *machine* is from the Greek word μηχανη, meaning a device consisting of interrelated parts with separate functions. Until the mid 19th century, the terms *engine* and *machine* were used interchangeably (though an "engine" is more than just a machine—it is an *ingenious* machine). These machines were normally driven by either animal, wind, or water power and would be referred to as animal (or horse) engines, wind (or air) engines, or water (or hydraulic) engines, respectively.

13.2.2 What Is an Engine?

The English words *engine* and *ingenious* are derived from the same Latin root word *ingignere*, meaning "to create." About AD 200 Tertullian[1] referred to a military battering ram in Latin as an *ingenium* or product of genius, and soon thereafter the word *ingen* was used in Latin to describe all military machines (catapults, assault towers, etc.). The Latin word *ingen* then became assimilated into English but its spelling was changed to *engine*.

---

[1] Quintus Septimius Florens Tertullianus (ca. AD 150–230) was an early Christian author who helped establish Latin (rather than Greek) as the language of Christianity. He coined many new Latin words and phrases as he wrote about the moral and practical problems facing the early Christians of his time (proper dress, military service, marriage and divorce, arts, theater, etc.).

13.2.3 What Is an Engineer?

In English, the *–er* ending to a word often means someone who does something. For example, someone who sings is a sing*er*, someone who builds is a build*er*, and someone who writes is a writ*er*. So, naturally, someone who creates ingenious things is called an *engineer*. That is where the name of our profession comes from. In French and German, the word engineer is translated as *ingenieur*, and in Italian, it is *ingegnere*. So, engineers are the people who create ingenious solutions to society's problems. We are the engineers.

13.2.4 What Is a Heat Engine?

The vast number of different engines (or machines) developed over time are usually classified either *generally* according to their source of power (animal, wind, water, etc.) or *specifically* according to their function. For example, beginning in the medieval period, an ingenious machine (an *ingen* or engine) was simply called a *gin* (a contraction applicable to either *ingen* or *engine*).[2] So when, in 1793, Eli Whitney (1765–1825) built an ingenious engine (or machine) for removing the seeds from raw cotton, his "cotton engine" became known as a *cotton gin*. A calculating engine is a machine whose function is to make calculations (e.g., an adding machine), whereas a pneumatic engine is an engine whose source of power is air pressure.

A *heat engine* is any machine whose source of power is heat (fire, steam, solar, etc.), and whose specific function is undefined (i.e., it simply produces work, which can be mechanical, electrical, chemical, etc., in nature).

Initially, heat engines simply produced mechanical work that was used directly (e.g., in manufacturing) or else they were connected to other engines. For example, they were often connected to pumping engines (pumps) to move water and later to electrical engines (generators) to generate electricity. Most heat engines use either a vapor or a gas as the internal energy transfer medium (even a thermoelectric device can be thought of as transporting energy via an internal electron gas). Typical heat engine characteristics are shown in Table 13.1.

We call the device that actually produces the heat engine's output work the *prime mover*. A prime mover can be a reciprocating piston-cylinder steam engine, a steam turbine, an internal combustion engine, and so forth. Figure 13.1 illustrates these terms.

It is common to use the words *engine* and *prime mover* interchangeably when referring to reciprocating piston-cylinder devices. We follow this custom in this textbook when no confusion is likely to occur.

The reason heat engines are so important to the study of thermodynamics is that the history of heat engine technology is essentially the history of the Industrial Revolution. The heat engine whose prime mover was the reciprocating piston steam engine was the first large-scale source of portable mechanical power. It was a source of power that did not depend on wind or river and could therefore be located anywhere. The heat engine is still the primary source of power for travel and electricity today, and it is likely to remain so for the foreseeable future.

Table 13.1 Some Typical Heat Engine Characteristics

Heat Sources	Heat Sinks	Working Fluid	Work Output Prime Movers	Cycle Types	Uses
Combustion	Atmosphere	Gas	Engine	Power	Transportation
Nuclear	Oceans, lakes	Vapor	Turbine	Refrigeration	Generate electricity
Atmosphere	Rivers	Vapor	Reversed engine	Heat pump	Heating and cooling
Ocean	Groundwater	Vapor	Solid state (e.g., thermoelectric)	Power	Generate electricity

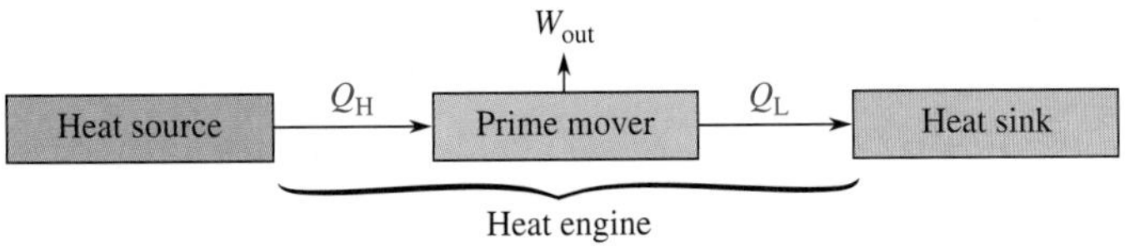

FIGURE 13.1
Heat engine terminology.

[2] The alcoholic beverage gin (a spirit distilled from grain and originally flavored with the juice of juniper berries) is a contraction of the word *geneva*, which comes from the Latin *juniperus*.

THE EVOLUTION OF STEAM POWER

Heat engine steam power technology began in 1698, when Thomas Savery (1650–1715) was granted a patent by King William III of England for "an engine for raising water by the impellant force of fire." Savery's machine was designed to pump water from flooded English mines, and he called it a *fire engine* because it drew its power from the fire under the boiler rather than from horses or wind. Its only moving mechanical parts were hand operated valves and automatic check valves. It drew water by suction when a partial vacuum was created as steam condensed inside a pumping chamber. The operation of this engine is shown in Figure 13.2.

Savery's engine could not draw water from a depth of more than about 20 ft (6.1 m), and it was somewhat dangerous because of the relatively high boiler pressure required to push the water out through the discharge pipe. Boiler technology had not advanced beyond that of the brewing industry then in existence, and many boiler explosions are known to have occurred. Nonetheless, Savery's engine was an enormous economic success. It used fire to pump water, which was something that no other engine could do, and it was the first technologically successful heat engine. In spite of its large size and rather primitive use of steam, its simplicity made it a popular means of pumping water through short distances. Engines of this design were in continuous use in England until 1830.

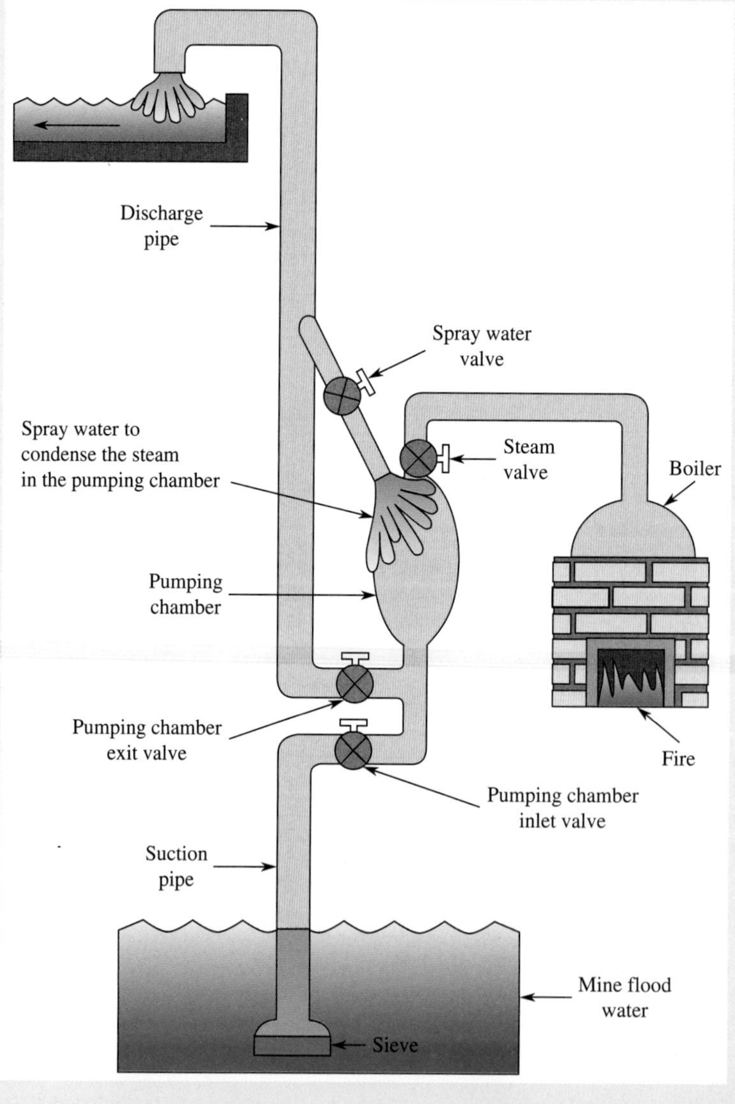

FIGURE 13.2

A schematic of the operation of Savery's fire engine. Steam was generated in the boiler at 100–150 psig. To operate the engine, the manual control valve was opened to fill the pumping chamber with steam. This valve was then closed and cold water was allowed to flow over the outside of the pumping chamber, causing the steam inside to condense and form a partial vacuum. Atmospheric pressure (acting on the surface of the mine water) then forced the water from the flooded mine up through the suction pipe and into the pumping chamber.

By 1712, Thomas Newcomen (1663–1729), an English blacksmith, had devised a better steam engine, which could pump water from very great depths while simultaneously eliminating the need for a high boiler pressure. A schematic of the operation of his engine is shown in Figure 13.3. He introduced a piston-cylinder arrangement in place of Savery's pumping chamber, with the piston attached to one end of a walking beam and a positive displacement piston-cylinder pump, located deep in the mine, attached to the other end of the walking beam. The pump end was counterweighted so as to hold the driving piston at the top of its stroke. When low-pressure (3 to 5 psig) steam was introduced under the driving piston and condensed using a cold water spray, the resulting partial vacuum allowed atmospheric pressure to push the driving piston down (thus charging the pump with water). Venting the driving piston to the atmosphere allowed the counterweight to drop, thus pumping the water. These engines typically ran at a speed of about ten cycles per minute.

Opening the manual valve again allowed steam from the boiler to force this water out through the vertical discharge pipe. A skilled operator could run this engine at about five cycles per minute.

Since both the Savery and Newcomen engines depended on creating a partial vacuum by condensing steam and using atmospheric pressure to cause the necessary motion, they are called *atmospheric* engines. Also, since both engines alternately heated and cooled large metal chambers (Savery's pumping chamber and Newcomen's piston-cylinder) during each cycle of operation, both had very low thermal efficiencies (a fraction of 1%). But, since the concept of thermal efficiency had not yet been developed, their poor thermal performance went undetected. However, as the technology advanced and more engines came into use, it became clear to their owners that they required an enormous amount of fuel (wood or coal) to keep them operating.

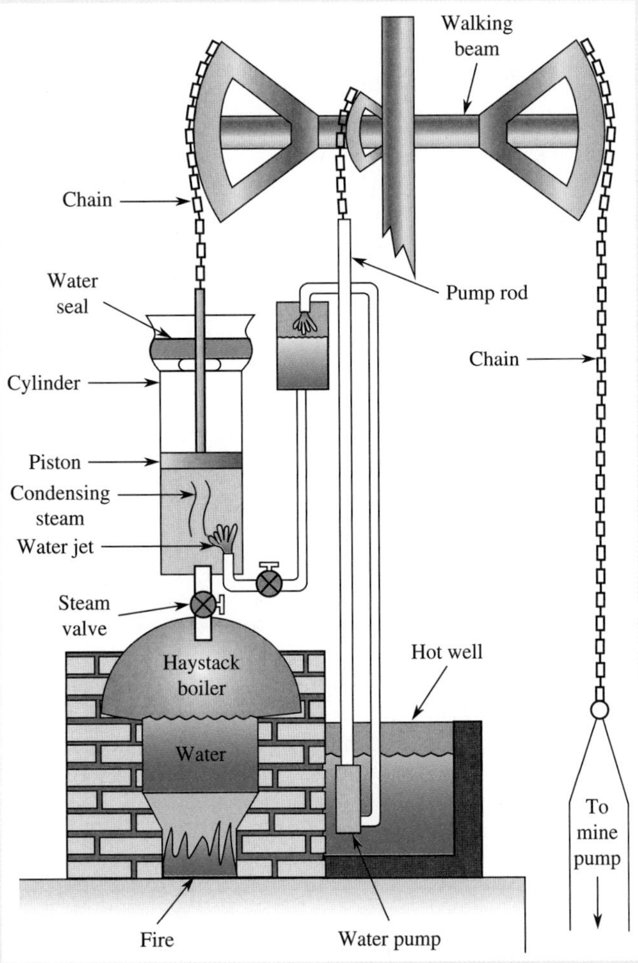

FIGURE 13.3

A schematic of the operation of Newcomen's steam engine. When the cylinder has been filled with steam, the driving piston rises and the weight of the pump rod forces a stroke on the pump (not shown) at the bottom of the mine shaft. The steam valve is then closed and the injection water valve opened. The cold water jet condenses the steam in the cylinder, producing a partial vacuum. Atmospheric pressure (acting on the top of the driving piston) forces the driving piston down, and the walking beam lifts the pump rod and makes the engine ready for another pumping cycle.

By the late 1760s, John Smeaton (1724–1792)[3] had undertaken a study of the fuel efficiencies of a number of England's Newcomen steam engines. He called his measure of fuel efficiency the *duty* of the engine, and he defined it as follows:

> The *duty* of a pumping engine is equal to the number of pounds of water that are raised one foot in height by the engine's pump when one 84.0 lbm bushel of coal is burned in the boiler.

Because Smeaton's efficiency (duty) was applied to the total energy conversion process from chemical input (coal) to work output (water pumped), it constituted a measure of the overall system efficiency that included the boiler and pump efficiencies and the efficiency of the piston-cylinder operation. By assuming the average energy content of coal to be 1.30×10^3 Btu/lbm, we can easily convert Smeaton's duty measurements into overall thermal efficiencies. One 84.0 lbm bushel (bu) of coal = $(1.30 \times 10^3$ Btu/lbm)(84.0 lbm/bu)(778.16 ft·lbf/Btu) = 8.50×10^8 ft·lbf of energy. Defining thermal efficiency η_T as the ratio of net output (pounds of water raised 1 ft) to net input (ft·lbf of energy of coal consumed), we obtain (in percent)

$$\eta_T(\text{in}\%) = \frac{\text{Duty}}{8.50 \times 10^8} \times 100 \tag{13.1}$$

EXAMPLE 13.1

In 1765, John Smeaton built a small steam engine with a 10.0 in (0.254 m) diameter piston having a 38.0 in (0.965 m) stroke and found that, when it was used to drive a water pump, it could pump 291,900 lbf of water 10.0 ft high when one 84.0 lbm bushel of coal was burned in the boiler. Determine the duty and thermal efficiency of this engine.

Solution

The work required by Smeaton's engine to raise 291,900 lbf of water 10.0 ft in height was $mg(\Delta h)$ = 2,919,000 ft·lbf. Therefore, it could have raised 2,919,000 lbf of water 1 ft in height using the same amount of work. Since the *duty* of any engine is the amount of water it can raise 1 ft when one 84.0 lbm bushel of coal is burned in the boiler, the duty of Smeaton's engine is 2,919,000. Note that, because of the way *duty* is defined, it is expressed without units or dimensions. Then, Eq. (13.1) gives the thermal efficiency of Smeaton's engine as

$$\eta_T = \frac{\text{Duty}}{8.5 \times 10^8} \times 100 = \frac{2{,}919{,}000}{8.5 \times 10^8} \times 100 = 0.344\%$$

His engine was not very efficient.

Exercises

1. Suppose Smeaton's engine in Example 13.1 had pumped 291,900 lbf of water 15.0 ft high. What would its duty and thermal efficiency be in this case? **Answer**: Duty = 4,378,500 and $\eta_T = 0.515\%$.
2. How high would Smeaton's engine in Example 13.1 have to pump 1.00 lbf of water to have a duty of 5,000,000? **Answer**: Δh = 5,000,000 ft.
3. What would be the duty of Smeaton's engine in Example 13.1 if it had a thermal efficiency of 10%? **Answer**: Duty = 85.0×10^6.

For five years, Smeaton collected duty measurements for over 30 Newcomen pumping engines and found they had an average duty of 5,590,000, which corresponds to an average overall thermal efficiency of 0.65%. With these results he was able to conclude that large diameter pistons with short strokes made the most efficient engines. Using this result and improved cylinder boring techniques (necessary to reduce piston leakage), he was able to build an engine with a 52.0 in (1.32 m) diameter piston and a 7.00 ft (2.13 m) stroke in 1772 that had a duty of 9,450,000 and an overall thermal efficiency of 1.11%. Thus, he was able to produce an engine with *double* the efficiency of the average Newcomen engine simply by using his experimental observations to optimize its design.

---

[3] John Smeaton was a successful English engineer. In about 1750, he introduced the name *civil engineer* for any nonmilitary engineer (*civil* being simply a contraction of the word *civilian*). In 1771, he started the British Institution of Civil Engineers (the world's first professional engineering society). This was followed by the founding of the British Institution of Mechanical Engineers in 1847 by George Stephenson (1781–1848). In America, the American Society of Civil Engineers (ASCE) was founded in 1852. This was followed by the American Institute of Mining Engineers (AIME) in 1871, the American Society of Mechanical Engineers (ASME) in 1880, the American Institute of Electrical Engineers (AIEE) in 1884, the Society of Automotive Engineers (SAE) in 1904, American Institute of Chemical Engineers (AIChE) in 1908, and many others.

WHO WAS JAMES WATT?

James Watt (1736–1819) was a young Scottish machinist at the College of Glasgow and, in 1764, was given the job of repairing a classroom teaching scale model of a Newcomen engine. The engine was not actually broken. Its problem was that it consumed so much steam that its boiler was empty after only a few cycles of operation and consequently it soon stopped running. Watt discovered that this was due to the alternate heating and cooling of the piston-cylinder unit during the condensation and reheat portions of each cycle. Part of the thermal energy contained in the steam had to be used to reheat the piston and cylinder after the condensation process was brought about by the cold water jet. This energy therefore became unavailable for doing mechanical work. The small size of the scale model engine had magnified this effect to the point where the engine was so inefficient that it would run for only a few cycles before using up all the steam available in the boiler.[4] Several months later, he realized that, if the steam was condensed not *inside* the piston-cylinder unit, but *outside* of it in a separate condenser chamber, then the piston and cylinder could be continuously kept at the high temperature of the steam. This way, the piston-cylinder unit would not have to be reheated during each cycle of the engine and the steam consumption would be greatly reduced. From this simple observation was born one of the most significant technological innovations of the 18th century, the separate steam condenser.

With the addition of a condenser unit, the efficiency of a full-size Newcomen engine was increased several-fold. This meant that these engines could be reduced in size somewhat (most Newcomen engines at that time were as big as a two-story house) and adapted to other uses than pumping water out of flooded mines. The smaller Watt engines provided the medium-scale power sources necessary to bring about the onset of centralized manufacturing, which was the beginning of the Industrial Revolution. How Watt added a separate condenser to a standard Newcomen pumping engine is shown in Figure 13.4.

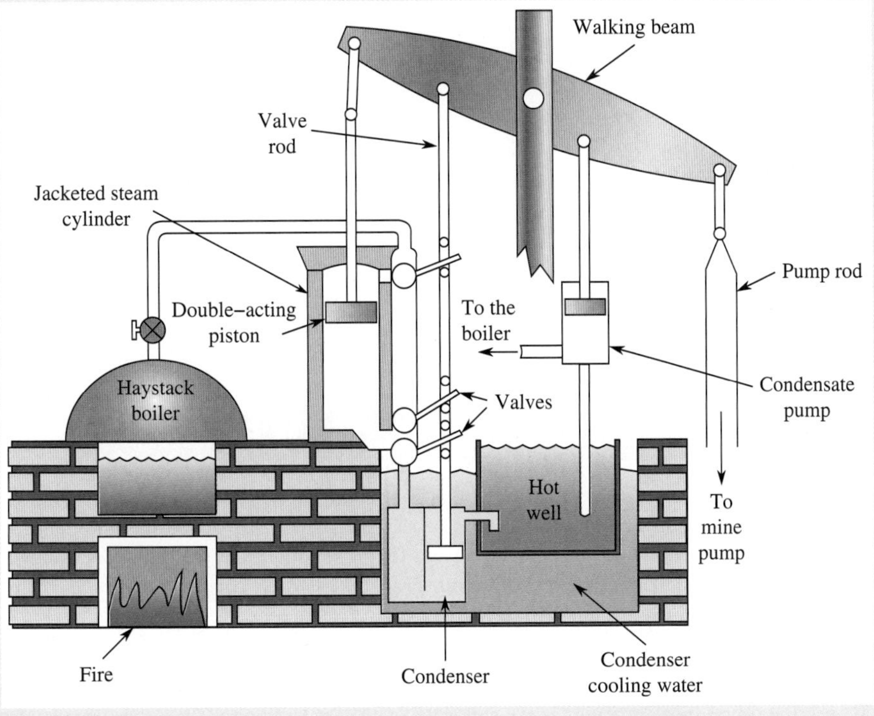

FIGURE 13.4
Watt's condenser added to a Newcomen pumping engine. Automatic valves alternately admit steam from the boiler to the proper side of a double-acting piston inside a steam jacketed cylinder, then into a separate condenser unit.

In 1775, Watt entered into a business partnership with the British industrialist Matthew Boulton (1728–1809) for the purpose of manufacturing his version of the external condenser Newcomen engine. Boulton provided the financial support (and consequently took two thirds of the patent rights) and Watt was responsible for the engineering and manufacturing activities.

(*Continued*)

WHO WAS JAMES WATT? *Continued*

The Boulton and Watt company was very successful and manufactured steam engines for many years, including the engine used in 1807 by Robert Fulton (1765–1815) on the first American steamboat, the *Clermont* (Figure 13.5).

FIGURE 13.5
An illustration of Robert Fulton's steamboat *Clermont.*

Over the years Watt made many important technological advances including automatic controls (the centrifugal governor), safety devices (the pressure gauge), and efficiency improvements (the double-acting cylinder).

The term *horsepower* was introduced by Savery as a measure of how many horses driving a mechanical pump were replaced by his fire engine. However, as a unit of power measurement, it lacked a precise definition until Watt carried out experiments in about 1780 to determine how much power an average horse could deliver on a continuous basis. He then multiplied this value by a factor of 2 to ensure that, when he sold engines rated at a given horsepower, the purchasers would have no complaints as to their performance. Using this technique, he finally arrived at the figure of 33,000 ft·lbf/min as his horsepower definition. In addition to his conservative two-horse horsepower definition, some other equivalences that Watt felt were valid are 1 horsepower = 2 average horses = 3 powerful oxen = 12 men working cranks = 396 gallons of water falling 10 ft in 1 min. Much later, the electrical unit of power was named after him, 1 watt = 1 volt·ampere (and 1 hp = 746 W).

By 1800, Watt's improvements had increased the thermal efficiency of a full-size Newcomen steam engine by about a factor of 4. But, even then, the overall thermal efficiency was only around 4 or 5%. This was to be the upper limit of atmospheric engine thermal efficiency, because they were soon to be replaced by a new technological breakthrough: the high-pressure, expansion steam engine.[5]

[4] *This occurred because small-scale models magnify the inefficiencies that arise from heat loss from the engine's surface to the atmosphere. Since the surface area to volume ratio of a given geometric shape always increases as the physical dimensions of the shape decrease, small heat engine models are inherently much less efficient than their full-scale counterparts.*

[5] *It has been estimated that the upper limit of the thermal efficiency of the reciprocating atmospheric steam engine using the technology available in 1915 was less than 15%.*

BOILER EXPLOSIONS AND EXPANSION STEAM ENGINES

Before the early 19th century, high-pressure steam (anything over about 10 psig) was considered extremely dangerous. Explosions of the primitive boilers of Savery's time caused much damage and loss of life and traumatized steam engine manufacturers. But new boiler materials and manufacturing and testing techniques in the early 1800s allowed operational steam pressures many times the single atmosphere of pressure available to Watt's engines. Engines of 100 psig were common by 1840, and 200 psig was in use by 1880. High-pressure engine technology brought another quantum leap in thermal efficiency, another factor of 3, from Watt's 4 to 5% in 1800 to 12 to 15% by 1850.

High-pressure engines used the pressure of the steam to *push* the piston by expanding against it, rather than having atmospheric pressure push the piston into a vacuum as the atmospheric engines had done. Thus, they were known as *expansion* engines. By the 1820s, high-pressure engines were sufficiently efficient that Watt's condenser unit was no longer considered to be essential, and consequently the steam was often exhausted from the cylinder directly into the atmosphere. The condenser always increased the engine's thermal efficiency somewhat, but if the piston-cylinder was not cooled during each cycle (as it was in the Newcomen atmospheric engines but not in the newer expansion engines), this effect was minimal. The elimination of the condenser and its attendant pump simplified the engine's construction and further reduced its cost and size, with only a small loss in thermal efficiency. At this point, the steam engine was finally small enough to become truly portable and its application to locomotion on land (railroads) and water (steamboats) produced new transportation technologies that changed the face of the world.

EXAMPLE 13.2

In 1807, Robert Fulton (1765-1815) successfully piloted his walking beam paddlewheel steamboat *Clermont* from New York City to Albany, New York (see Figure 13.5). His single-cylinder engine was made by the Boulton and Watt steam engine manufacturing company in England and produced 20.0 hp with a piston diameter of 2.00 ft and a piston stroke of 4.00 ft. The two side paddlewheels were 15.0 ft in diameter. The boiler was made of copper and it weighed 4,000 lbf dry. If the engine had a duty of 35.0 million and ran at 18.0 strokes per minute, then determine:

a. The average pressure of the cycle.
b. The actual thermal efficiency of the engine.
c. The heat rate produced by the boiler.

Solution

a. The average cylinder pressure of the engine can be determined by setting the calculated power produced by the piston equal to the actual power produced by the engine, or

$$\begin{aligned}\dot{W}_{out} &= (p_{avg}) \times (\text{Piston displacement}) \times (\text{Piston strokes/min}) \\ &= (p_{avg}) \times \left[\left(\frac{\pi (D_{piston})^2}{4}\right)(\text{Piston stroke})\right] \times (\text{Strokes/min}) \\ &= (p_{avg}) \times \left[\left(\frac{\pi (2.00)^2}{4}\text{ft}^2\right)(4.00\text{ ft/Stroke})\right] \times (18.0\text{ Strokes/min}) \\ &= (p_{avg}) \times (226\text{ ft}^3/\text{min})\end{aligned}$$

Now, $\dot{W}_{out} = 20$ hp = (20 hp)(33,000 ft·lbf/hp·min) = 660,000 ft·lbf/min, so

$$p_{avg} = \frac{660{,}000\text{ ft}\cdot\text{lbf/min}}{226\text{ ft}^3/\text{min}} = 2918\text{ lbf/ft}^2 = \frac{2918\text{ lbf/ft}^2}{144\text{ in}^2/\text{ft}^2} = 20.3\text{ lbf/in}^2$$

b. From Eq. (13.1), we have

$$\eta_T(\%) = \frac{\text{Duty}}{8.5 \times 10^8} \times 100 = \frac{35.0 \times 10^6}{8.50 \times 10^6} = 4.12\%$$

c. The heat rate produced by the boiler can be found from the thermodynamic definition of the thermal efficiency as $\eta_T = \dot{W}_{out}/\dot{Q}_{boiler}$. Then,

$$\dot{Q}_{boiler} = \frac{\dot{W}_{out}}{\eta_T} = \frac{(20.0\text{ hp})(2545\text{ Btu/hp}\cdot\text{h})}{0.0412} = 1.24 \times 10^6\text{Btu/h}$$

Exercises

4. Determine the average cycle pressure in Example 13.2 if the engine produces 30.0 instead of 20.0 hp and all the other variables remain unchanged. **Answer:** $p_{avg} = 27.3$ psi.
5. Determine the actual thermal efficiency of the engine in Example 13.2 if it has a duty of 26.0 × 10^6 instead of 35.0 × 10^6 and all the other variables remain unchanged. **Answer:** $\eta_T = 3.06\%$.
6. Determine the heat rate produced by the boiler in Example 13.2 if the engine produces 25.0 instead of 20.0 hp, and all the other variables remain unchanged. **Answer:** $\dot{Q}_{boiler} = 1.55 \times 10^6$ Btu/h.

WHAT IS A STEAM ENGINE *INDICATOR*?

In an effort to continue to improve the performance of his steam engines, Watt wanted to know how the pressure varied with piston position inside the cylinder as the engine was running. About 1790, he developed an ingenious device for this purpose (Figure 13.6), which he called a *steam engine indicator*. This device drew the actual pressure–volume diagram of the steam inside the cylinder as the engine was running. Such p–$\forall$ diagrams soon became known as *indicator diagrams,* and the area enclosed by these diagrams represented the reversible work produced inside the engine.

(*Continued*)

WHAT IS A STEAM ENGINE *INDICATOR*? *Continued*

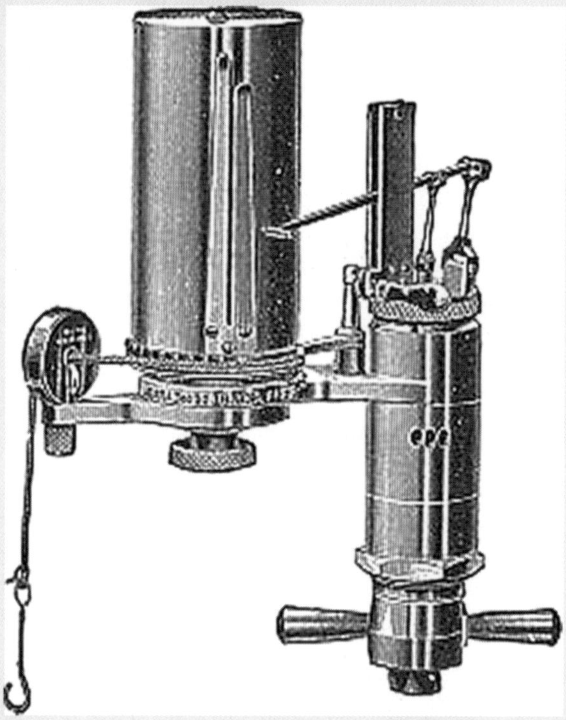

FIGURE 13.6
Steam engine indicator.

13.3 CARNOT POWER CYCLE

In Chapter 7, we discuss how the young French military engineer Sadi Carnot (1796–1832) came to understand the rudiments of heat engine theory in the 1820s using a water wheel analogy. His theory was based on the caloric (or fluid) theory of heat, and he believed that heat passed through a heat engine undiminished (like water passes through a water wheel undiminished), and in so doing, the heat engine could perform work. Today, we know that this is incorrect. The heat flow through an engine is diminished (i.e., reduced) by its conversion into work. Carnot's ideas were so revolutionary that they were largely ignored. Soon after Carnot's death from scarlet fever, Emile Clapeyron (1799–1864), in 1834, strengthened Carnot's ideas by using more precise mathematical derivations. From Carnot's description of a reversible heat engine, Clapeyron constructed its thermodynamic cycle. He deduced that it must be composed of two isothermal processes and two reversible adiabatic processes. Using the pressure-volume steam engine indicator diagram format common at that time, he deduced the cycle shape shown in Figure 13.7a. This cycle is still known as *Carnot's cycle*, but because it is defined to be a reversible cycle, no heat engine can ever be made to operate using it.

The Carnot cycle is important because it was the first heat engine cycle ever to be properly conceptualized and because no other heat engine, reversible or irreversible, can ever be more efficient than a Carnot cycle heat engine (though even a reversible heat engine *may* be less efficient), thus it can be used as a benchmark or standard for comparison, for gauging both real and reversible (ideal) heat engine performance.

When Rudolph Clausius (1822–1888) formalized the second law of thermodynamics and defined entropy in 1860, Carnot's reversible adiabatic processes became an isentropic process, and the Carnot cycle was defined by two T = constant processes and two s = constant processes. The Carnot cycle then took on its characteristic rectangular shape on a T–s diagram, as shown in Figure 13.7b.

In Chapter 7, we also discuss how Carnot's ideas led to the development of the Kelvin absolute temperature scale and, finally, to an expression for the thermal efficiency of a reversible heat engine (see Eq. (7.16)), which we call in this chapter the *Carnot thermal efficiency*:

$$(\eta_T)_{\text{Carnot}} = 1 - T_L/T_H \tag{13.2}$$

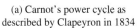

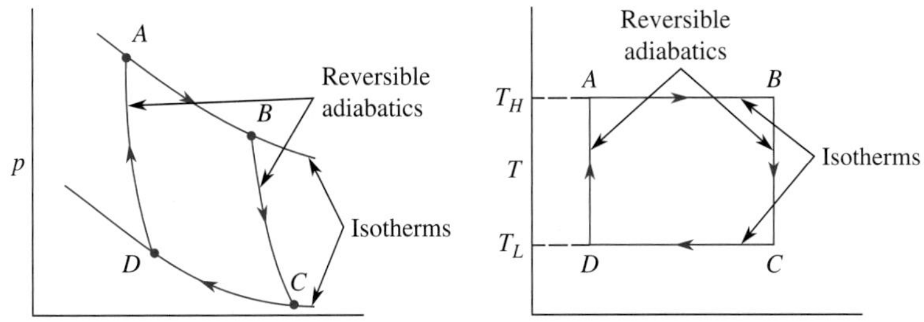

(a) Carnot's power cycle as described by Clapeyron in 1834

(b) Carnot's power cycle using Clausius' entropy concept of 1860

FIGURE 13.7
(a) Carnot's power cycle as described by Clapeyron in 1834. (b) Carnot's power cycle using Clausius' entropy concept of 1860. The reversible Carnot p–$\forall$ cycle diagram was chosen as the logo for the Mechanical Engineering Honor Society, ΠΤΣ (Pi Tau Sigma).

where T_H and T_L are the absolute isothermal temperatures of the high-temperature heat addition and low-temperature heat rejection reservoirs, respectively.

13.4 RANKINE CYCLE

In the period from 1850 to 1880, the subject of thermodynamics was formally developed. One of its early practical goals was to provide a scientific foundation for the empirical steam technology that had by then grown to dominate the economy of the Western World. By 1850, it had been determined that heat was a form of energy, and by 1860, the first and second laws of thermodynamics had been accurately formulated by Clausius, Kelvin, Joule, and others. But, it was the Scottish engineer William John Macquorn Rankine (1820–1872) who first worked out the thermodynamic cycle for a steam engine with an external condenser.[6] Because Rankine was the first person to understand how this type of steam engine worked thermodynamically, the thermodynamic cycle for adiabatic cylinder engines is called the *Rankine cycle* today.

Though the Newcomen cycle is obsolete, the Rankine cycle is still in common use. Therefore, we carry out an analysis of the thermal efficiency of the Rankine cycle and focus on its further development throughout the remainder of this chapter. The difference between the Newcomen and Rankine cycles can be seen by comparing parts a and b of Figure 13.8.

The Rankine cycle is a thermodynamic representation of a high-pressure or expansion type of steam engine cycle, and because of the shape of the *T-s* saturation curve for water, the ideal or reversible Rankine cycle without superheat is very close to the (reversible) Carnot cycle. The difference between the Rankine and Carnot cycles is shown in Figure 13.8c. Because it is very difficult to efficiently pump wet vapor back into the boiler in the Rankine cycle, the vapor is completely condensed into a liquid in the condenser and, using a common liquid pump, pumped into the boiler as a compressed liquid. The Rankine cycle boiler feed pump raises the pressure of the liquid condensate to only that of the boiler (state 4), leaving its temperature nearly equal to that of the condenser (state 3).

The thermal efficiency η_T of a heat engine is defined in Eq. (7.5) as

$$\eta_T = \frac{\text{Net work output}}{\text{Total heat input}} = \frac{\text{Engine work output} - \text{Pump work input}}{\text{Boiler heat input}} \tag{7.5}$$

[6] An adiabatic cylinder steam engine is any expansion engine or any atmospheric engine with an external condenser. Rankine's work on this subject was published in his classic *Manual of the Steam Engine and Other Prime Movers*, first published in 1859. This text went through 17 editions and was in print for over 50 years. It is considered to be the first comprehensive engineering thermodynamics textbook.

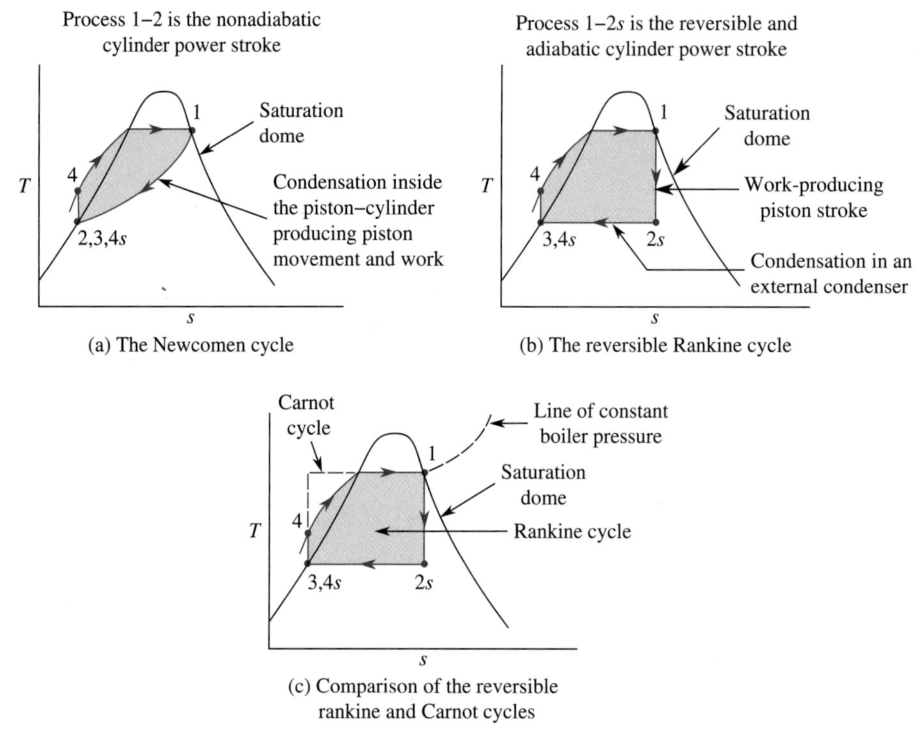

FIGURE 13.8
The Newcomen, Rankine, and Carnot cycles.

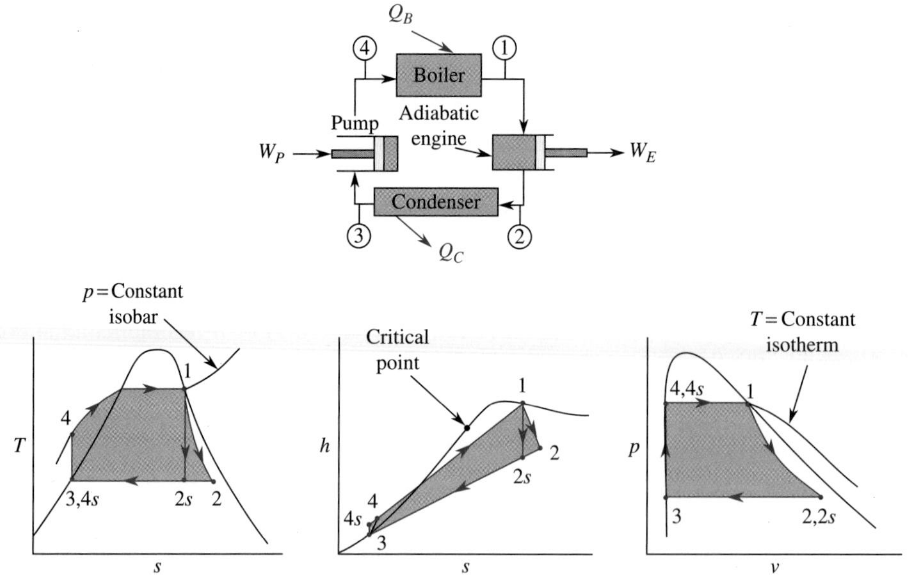

FIGURE 13.9
Reciprocating heat engines operating on Newcomen and Rankine cycles.

Consider the Rankine cycle shown in Figure 13.9. An energy rate balance on the steady state boiler, the condensate pump, and the piston-cylinder prime mover (neglecting any changes in flow stream kinetic or potential energy) gives

$$\text{Heat transport rate into the boiler} = \dot{Q}_B = \dot{m}(h_1 - h_4)$$

$$\text{The magnitude of the power } \textit{into} \text{ the condensate pump} = |\dot{W}_{\text{pump}}| = \dot{m}(h_4 - h_3)$$

$$\text{Power } \textit{out of} \text{ the piston-cylinder prime mover} = \dot{W}_{pm} = \dot{m}(h_1 - h_2) + \dot{Q}_{pm}$$

For a Rankine cycle, the prime mover can be assumed to be insulated ($\dot{Q}_{pm} = 0$), then the prime mover power output becomes

$$(\dot{W}_{pm})_{\text{Rankine}} = \dot{m}(h_1 - h_2)$$

and Eq. (7.5) gives the thermal efficiency of a Rankine cycle heat engine as

$$(\eta_T)_{\text{Rankine}} = \frac{\dot{W}_{pm} - |\dot{W}_p|}{\dot{Q}_{\text{boiler}}} = \frac{h_1 - h_2 - (h_4 - h_3)}{h_1 - h_4} \tag{13.3}$$

13.5 OPERATING EFFICIENCIES

Since no machine is really reversible, we need to develop a method of making accurate power consumption or production calculations for real irreversible machines. This is usually done through the introduction of an empirically determined performance measure, called an operating *efficiency*. Because of the manner in which this type of technology evolved, several different types of efficiency measures are in common use today.

The physical meaning of an operating efficiency depends on where the system boundaries are drawn. If the system under consideration consists of only the working fluid, then this efficiency represents the effect of the irreversibilities that occur only within the working fluid. However, if the system under consideration consists of the entire work-producing or work-absorbing machine (including the working fluid), then this efficiency represents the effect of the irreversibilities within the working fluid as well as those within the machine itself (such as bearing friction). To be effective, an operating efficiency should apply to a system consisting of the device plus the working fluid it contains.

13.5.1 Mechanical Efficiency

The first important measure of the performance of any device is the work transport energy efficiency η_W, which is defined in Chapter 4 as the ratio of the actual work to the reversible work for a work-producing device:

$$(\eta_m)_{\substack{\text{work-}\\\text{producing}\\\text{device}}} = \frac{W_{\text{actual}}}{W_{\text{reversible}}} = \frac{\dot{W}_{\text{actual}}}{\dot{W}_{\text{reversible}}}$$

or the ratio of the reversible work to the actual work for a work-absorbing device:

$$(\eta_m)_{\substack{\text{work-}\\\text{absorbing}\\\text{device}}} = \frac{W_{\text{reversible}}}{W_{\text{actual}}} = \frac{\dot{W}_{\text{reversible}}}{\dot{W}_{\text{actual}}}$$

This efficiency compares the actual performance of a device with what would occur if the device were reversible (but not adiabatic), and it is commonly known as the *reversible efficiency* of the device. However, in mechanical devices, the source of the internal irreversibilities is primarily mechanical friction. Consequently, it is customary to refer to the reversible efficiency of a mechanical device as simply the *mechanical efficiency* of the device and to use the notation η_m instead of η_W. The mechanical efficiency of work-producing and work-absorbing devices is defined mathematically in Table 13.2.

13.5.2 Isentropic Efficiency

Next, we define the *isentropic efficiency* η_s as the ratio of the actual work to the isentropic work for a work-producing device:

$$(\eta_s)_{\substack{\text{work-}\\\text{producing}\\\text{device}}} = \frac{W_{\text{actual}}}{W_{\text{isentropic}}} = \frac{\dot{W}_{\text{actual}}}{\dot{W}_{\text{isentropic}}}$$

or the ratio of the isentropic work to the actual work for a work-absorbing device:

$$(\eta_s)_{\substack{\text{work-}\\\text{absorbing}\\\text{device}}} = \frac{W_{\text{isentropic}}}{W_{\text{actual}}} = \frac{\dot{W}_{\text{isentropic}}}{\dot{W}_{\text{actual}}}$$

It is similar to the work transport energy efficiency η_W, defined in Chapter 4, but whereas η_W was based on comparing the actual performance of the device with what would occur if the device were reversible, the

Table 13.2 Definitions of Some Common Efficiencies

Efficiency	Work-producing device	Work-absorbing device
Mechanical efficiency (η_m)	$(\eta_m)_{\text{work-producing device}} = \dfrac{W_{\text{actual}}}{W_{\text{reversible}}} = \dfrac{\dot{W}_{\text{actual}}}{\dot{W}_{\text{reversible}}}$	$(\eta_m)_{\text{work-absorbing device}} = \dfrac{W_{\text{reversible}}}{W_{\text{actual}}} = \dfrac{\dot{W}_{\text{reversible}}}{\dot{W}_{\text{actual}}}$
Isentropic efficiency (η_s)	$(\eta_s)_{\text{work-producing device}} = \dfrac{W_{\text{actual}}}{W_{\text{isentropic}}} = \dfrac{\dot{W}_{\text{actual}}}{\dot{W}_{\text{isentropic}}}$	$(\eta_s)_{\text{work-absorbing device}} = \dfrac{W_{\text{isentropic}}}{W_{\text{actual}}} = \dfrac{\dot{W}_{\text{isentropic}}}{\dot{W}_{\text{actual}}}$
Relative efficiency (η_r)	$(\eta_r)_{\text{work-producing device}} = \dfrac{W_{\text{reversible}}}{W_{\text{isentropic}}} = \dfrac{\dot{W}_{\text{reversible}}}{\dot{W}_{\text{isentropic}}}$	$(\eta_r)_{\text{work-absorbing device}} = \dfrac{W_{\text{isentropic}}}{W_{\text{reversible}}} = \dfrac{\dot{W}_{\text{isentropic}}}{\dot{W}_{\text{reversib1e}}}$
Thermal efficiency (η_T)	$(\eta_T)_{\text{isentropic}} = \dfrac{(\dot{W}_{\text{out}})_{\text{isentropic}}}{\dot{Q}_{\text{in}}}$	$(\eta_T)_{\text{reversible}} = (\eta_T)_{\text{indicated}} = \dfrac{(\dot{W}_{\text{out}})_{\text{reversible}}}{\dot{Q}_{\text{in}}}$
	$(\eta_T)_{\text{actual}} = (\eta_T)_{\text{brake}} = \dfrac{(\dot{W}_{\text{out}})_{\text{actual}}}{\dot{Q}_{\text{in}}} = \dfrac{(\dot{W}_{\text{out}})_{\text{brake}}}{\dot{Q}_{\text{in}}}$	

where $\dot{W}_{\text{out}}$ represents the magnitude of the *net* power output in each case.

isentropic efficiency η_s is based on comparing the actual performance of the device with that which would occur if the device were adiabatic as well as reversible (i.e., isentropic). Since most prime movers and pumps are thermally insulated, *we always assume that they are adiabatic when their heat loss is not given*. The isentropic efficiency of work-producing and work-absorbing devices is defined mathematically in Table 13.2.

13.5.3 Relative Efficiency

It is also possible to define a *relative efficiency* (or *efficiency ratio*) η_r for these devices that relates the mechanical and isentropic efficiencies. For a work-producing device, the relative efficiency is defined as the ratio of the reversible work to the isentropic work:

$$(\eta_r)_{\substack{\text{work-}\\\text{producing}\\\text{device}}} = \frac{W_{\text{reversible}}}{W_{\text{isentropic}}} = \frac{\dot{W}_{\text{reversible}}}{\dot{W}_{\text{isentropic}}}$$

and for a work-absorbing device, it is defined as

$$(\eta_r)_{\substack{\text{work-}\\\text{absorbing}\\\text{device}}} = \frac{W_{\text{isentropic}}}{W_{\text{reversible}}} = \frac{\dot{W}_{\text{isentropic}}}{\dot{W}_{\text{reversib1e}}}$$

Then, with a little algebra, we can write $\eta_s = \eta_m \eta_r$ for either a work-producing or work-absorbing device. Notice that, if a device is insulated and is therefore adiabatic, then $\eta_s = \eta_m$ and $\eta_r = 1.0$.

The terms *shaft* and *brake* are also commonly used to describe the *actual* work or power produced or absorbed by a device.[7] These terms are interchangeable, but for clarity, the term *actual* is used most often throughout this chapter. Further, the terms *reversible* and *indicated* are synonymous because the reversible work or power produced inside a device can be determined from p–$\forall$ data provided by an indicator diagram; therefore, we can write $\dot{W}_{\text{reversible}} = \dot{W}_{\text{indicated}}$. Consequently, for a work-producing device, we can always write

$$\dot{W}_{\text{actual}} = \dot{W}_{\text{shaft}} = \dot{W}_{\text{brake}} = (\eta_s)(\dot{W}_{\text{isentropic}}) = (\eta_m)(\dot{W}_{\text{reversible}})$$

and for a work-absorbing device, we can always write

$$\dot{W}_{\text{actual}} = \dot{W}_{\text{shaft}} = \dot{W}_{\text{brake}} = (\dot{W}_{\text{isentropic}})/(\eta_s) = (\dot{W}_{\text{reversible}})/(\eta_m)$$

[7] The term *brake* is a descriptive term that comes from an early method of measuring the power output of machines using a friction band brake dynamometer called a *Prony brake*. It was named after Baron Gaspard Clair Francois Marie Riche de Prony (1755–1839) and was developed in the 1830s to measure the power output of water wheels and steam engines.

13.5.4 Thermal Efficiency

We also have to deal with three types of thermal efficiencies for work-producing or work-absorbing systems: *isentropic*, *reversible* (or indicated), and *actual* (or brake). These thermal efficiencies are defined mathematically in Table 13.2.

Using these definitions, the isentropic efficiency of the adiabatic piston-cylinder work-producing prime mover of the Rankine cycle heat engine shown in Figure 13.9 is

$$(\eta_s)_{\substack{\text{prime}\\ \text{mover}}} = (\eta_s)_{pm} = \frac{(h_1 - h_2)_{\text{actual}}}{(h_1 - h_2)_{\text{isentropic}}} = \frac{h_1 - h_2}{h_1 - h_{2s}} \tag{13.4a}$$

or

$$h_1 - h_2 = (h_1 - h_2)_{\text{actual}} = (h_1 - h_{2s})(\eta_s)_{pm} \tag{13.4b}$$

where h_{2s} is determined from p_2 (but not T_2) and the condition $s_{2s} = s_1$, as shown in Figure 13.9. Similarly, the isentropic efficiency of the adiabatic work-absorbing condensate pump in the heat engine shown in Figure 13.9 is

$$(\eta_s)_{\text{pump}} = (\eta_s)_p = \frac{(h_4 - h_3)_{\text{isentropic}}}{(h_4 - h_3)_{\text{actual}}} = \frac{h_{4s} - h_3}{h_4 - h_3} \tag{13.5}$$

where h_{4s} is determined from p_4 (but *not* T_4) and the condition $s_{4s} = s_3$, as shown in Figure 13.9. If the fluid being pumped is an incompressible liquid (v = constant) with a constant specific heat c, then Eq. (7.33) of Chapter 7 clearly shows that any isentropic process that it undergoes must also be isothermal. That is, $T_{4s} = T_3$ and consequently $u_{4s} = u_3$. Then, for $v_{4s} = v_4 = v_3$ and (note that, for an isentropic pump, points 3 and 4*s* coincide on a *T*–*s* diagram but not on a *p*–*v* diagram, see Figure 13.9) $p_{4s} = p_4$ and

$$\begin{aligned} h_{4s} - h_3 &= u_{4s} - u_3 + p_{4s}v_{4s} - p_3 v_3 \\ &= c(T_{4s} - T_3) + v_3(p_{4s} - p_3) \\ &= v_3(p_4 - p_3) \end{aligned} \tag{13.6}$$

Equation (13.5) can now be written as

$$(\eta_s)_{\substack{\text{incompressible}\\ \text{liquid}\\ \text{pump}}} = \frac{v_3(p_4 - p_3)}{h_4 - h_3} = \frac{v_3(p_4 - p_3)}{c(T_4 - T_3) + v_3(p_4 - p_3)} \tag{13.7}$$

or

$$h_4 - h_3 = (h_4 - h_3)_{\text{actual}} = v_3(p_4 - p_3)/(\eta_s)_p \tag{13.8}$$

Substituting Eqs. (13.4b) and (13.8) into the Rankine cycle thermal efficiency, Eq. (13.3) gives

$$(\eta_T)_{\text{Rankine}} = \frac{(h_1 - h_{2s})(\eta_s)_{pm} - v_3(p_4 - p_3)/(\eta_s)_p}{h_1 - h_3 - v_3(p_4 - p_3)/(\eta_s)_p} \tag{13.9a}$$

where $(\eta_s)_{pm}$ is the isentropic efficiency of the prime mover. The maximum possible Rankine cycle thermal efficiency occurs when both the prime mover and the condensate pump are isentropic: $(\eta_s)_{pm} = (\eta_s)_p = 1.0$, or

$$(\eta_T)_{\substack{\text{maximum}\\ \text{Rankine}}} = (\eta_T)_{\substack{\text{isentropic}\\ \text{Rankine}}} = \frac{h_1 - h_{2s} - v_3(p_4 - p_3)}{h_1 - h_3 - v_3(p_4 - p_3)} \tag{13.9b}$$

The phrase *isentropic Rankine cycle thermal efficiency* is used here to denote that the prime mover and the condensate pump are both isentropic (i.e., reversible *and* adiabatic). Clearly, the entire cycle is not isentropic, since thermal irreversibilities are associated with reheating the cold condensate returned to the hot boiler. This notation is necessary to distinguish between *reversible* Rankine cycles, in which the prime movers and pumps are modeled as reversible but are not adiabatic, and those Rankine cycles in which these items are modeled as both reversible and adiabatic (i.e., isentropic). This same notation is used in referring to other power and refrigeration cycles that contain isentropic components.

EXAMPLE 13.3

The boiler on the toy steam engine shown in Figure 13.10 is heated by a 300. watt electrical heater and produces saturated vapor at 20.0 psia. If the exhaust steam from the engine is condensed at 14.7 psia, determine the maximum possible thermal efficiency and net power output of the engine, assuming it operates on (a) a Carnot cycle and (b) a Rankine cycle.

FIGURE 13.10
Example 13.3, toy steam engine.

Solution

a. A Carnot cycle is defined to be a reversible cycle, so it automatically represents the maximum possible performance. Equation (13.2) gives the Carnot cycle thermal efficiency as

$$(\eta_T)_{\text{Carnot}} = 1 - T_L/T_H$$

where

$$T_L = T_{\text{sat}}(14.7\text{ psia}) = 212°\text{F} = 671.67\text{ R}$$

and

$$T_H = T_{\text{sat}}(20.0\text{ psia}) = 228.0°\text{F} = 687.67\text{ R}$$

Then,

$$(\eta_T)_{\text{Carnot}} = 1 - 671.67/687.67 = 0.0233 = 2.33\%$$

and since

$$(\eta_T)_{\text{Carnot}} = (\dot{W}_{\text{net}})_{\text{Carnot}}/\dot{Q}_{\text{boiler}}$$

then

$$(\dot{W}_{\text{net}})_{\text{Carnot}} = (\eta_T)_{\text{Carnot}} \times \dot{Q}_{\text{boiler}} = 0.0233(300.) = 6.98\text{ watts}$$

b. The maximum possible Rankine cycle thermal efficiency occurs when the piston-cylinder unit is reversible and the boiler feed pump is isentropic (Figure 13.11):

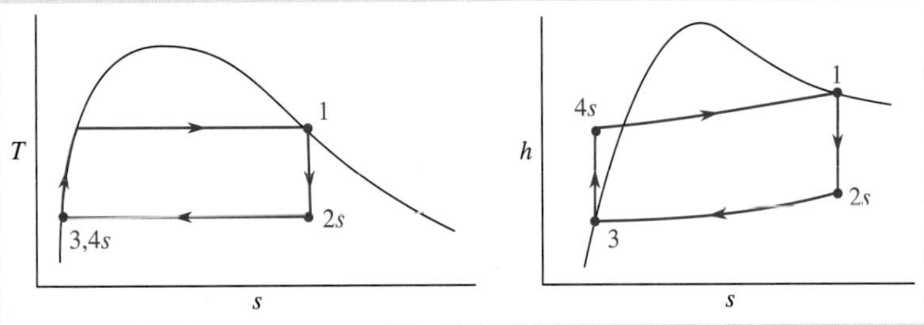

FIGURE 13.11
Example 13.3, T–s diagram.

Station 1—Engine inlet

$p_1 = 20.0$ psia

$x_1 = 1.00$

$h_1 = h_g(20.0 \text{ psia}) = 1156.4$ Btu/lbm

$s_1 = s_g(20.0 \text{ psia}) = 1.7322$ Btu/lbm·R

Station 2s—Engine exit

$p_{2s} = 14.7$ psia

$s_{2s} = s_1 = 1.7322$ Btu/lbm·R

$$x_{2s} = [s_{2s} - s_f(14.7 \text{ psia})]/s_{fg}(14.7 \text{ psia}) = [1.7322 - 0.3122]/1.4447 = 0.9829$$

$$h_{2s} = h_f(14.7 \text{ psia}) + x_{2s}h_{fg}(14.7 \text{ psia}) = 180.1 + 0.9829(970.4) = 1133.9 \text{ Btu/lbm}$$

Station 3—Condenser exit

$p_3 = p_{2s} = 14.7$ psia

$x_3 = 0$

$h_3 = h_f(14.7 \text{ psia}) = 180.1$ Btu/lbm

$v_3 = v_f(14.7 \text{ psia}) = 0.01672 \text{ ft}^3/\text{lbm}$

Station 4s—Boiler inlet

$p_{4s} = p_1 = 20.0$ psia

$s_{4s} = s_3$

Then, the isentropic efficiency of this system is given by Eq. (13.9b) as

$$(\eta_T)_{\substack{\text{maximum}\\ \text{Rankine}}} = \frac{h_1 - h_{2s} - v_3(p_4 - p_3)}{h_1 - h_3 - v_3(p_4 - p_3)}$$

$$= \frac{(1156.4 - 1133.9 \text{ Btu/lbm}) - (0.01672 \text{ ft}^3/\text{lbm})(20.0 - 14.7 \text{ lbf/in}^2)\left(\dfrac{144 \text{ in}^2/\text{ft}^2}{118.16 \text{ ft}\cdot\text{lbf/Btu}}\right)}{(1156.4 - 180.1 \text{ Btu/lbm}) - (0.01672 \text{ ft}^3/\text{lbm})(20.0 - 14.7 \text{ lbf/in}^2)\left(\dfrac{144 \text{ in}^2/\text{ft}^2}{118.16 \text{ ft}\cdot\text{lbf/Btu}}\right)}$$

$$= 0.0230 = 2.30\%$$

Exercises

7. Determine the maximum Carnot cycle net power output in Example 13.3 if the boiler pressure is increased to 30.0 psia from 20.0 psia. Assume all the other variables remain unchanged. **Answer**: $(\dot{W}_{\text{net}})_{\text{Carnot}} = 16.2$ W.
8. Determine the maximum Rankine cycle net power output in Example 13.3 if the boiler pressure is increased to 40.0 psia from 20.0 psia. Assume all the other variables remain unchanged. **Answer**: $(\dot{W}_{\text{net}})_{\text{max. Rankine}} = 22.2$ W.
9. Determine the minimum power required to operate the boiler feed pump per unit mass flow of steam in Example 13.3 if the boiler pressure is 35.0 psia instead of 20.0 psia. Assume all the other variables remain unchanged. **Answer**: $|\dot{W}/\dot{m}|_{\text{pump}} = 0.0184$ W·h/(lbm).

The thermal efficiency calculated in part b of the previous example was for a reversible Rankine cycle engine and, thus, is impossible to achieve with a real Rankine cycle engine. Considering the mechanical and thermal inefficiencies of these relatively crude toy engines, it is easy to see why their actual operating thermal efficiencies are in the range of 1%. Also note that the ratio of pump power to engine power in this example is only

$$|\dot{W}_p/\dot{W}_{pm}| = \frac{(0.01672)(20.0 - 14.7)(144/778.16)\text{ Btu/lbm}}{(1156.4 - 1133.9) - (0.01672)(20.0 - 14.7)(144/778.16)\text{ Btu/lbm}}$$

$$= 0.00073 = 0.073\%$$

and consequently the pump input power could be safely neglected in comparison to the engine output power in the thermal efficiency calculation. Even so, for clarity and completeness, the pump power is included in all subsequent thermal efficiency calculations carried out in the examples given in this chapter.

EXAMPLE 13.4

In 1876, George H. Corliss (1817–1888) built what was then the largest steam engine ever made (Figure 13.12), for the United States Centennial Exposition in Philadelphia, Pennsylvania. It had two cylinders, each with a 40.0-inch bore and a 10.0 ft stroke. Steam entered the engine as a saturated vapor at 100. psia, producing 1400. hp at only 36.0 rpm. The engine's flywheel was 30.0 ft in diameter and weighed 56.0 tons. If the steam was condensed at 14.7 psia, determine the Rankine cycle thermal efficiency of this engine assuming (a) isentropic prime mover and pump, (b) an engine isentropic efficiency of 55.0% and a pump isentropic efficiency of 65.0%, and (c) the steam mass flow rate required to produce 1400. hp.

FIGURE 13.12
Example 13.4, The Corliss engine.

Solution

a. The thermodynamic states at the four main points around the Rankine cycle for this system are (Figure 13.13)

Station 1—Engine inlet	Station 2s—Engine exit
$p_1 = 100.$ psia	$p_{2s} = 14.7$ psia
$x_1 = 1.00$	$s_{2s} = s_1$
$h_1 = h_g(100.\text{ psia}) = 1187.8$ Btu/lbm	$x_{2s} = [s_{2s} - s_f(14.7\text{ psia})]/s_{fg}(14.7\text{ psia})$
$s_1 = s_g(100.\text{ psia}) = 1.6036$ Btu/lbm·R	$= [1.6036 - 0.3122]/1.4447 = 0.8939$
	$h_{2s} = h_f(14.7\text{ psia}) + x_{2s}h_{fg}(14.7\text{ psia})$
	$= 180.1 + 0.8939\ (970.4)$
	$= 1047.5$ Btu/lbm

Station 3—Condenser exit	Station 4s—Boiler inlet
$p_3 = p_{2s} = 14.7$ psia	$p_{4s} = p_1 = 100$ psia
$x_3 = 0$	$s_{4s} = s_3$
$h_3 = h_f(14.7 \text{ psia}) = 180.1$ Btu/lbm	
$v_3 = v_f(14.7 \text{ psia}) = 0.01672$ ft^3/lbm	

Then, the isentropic efficiency of this system is given by Eq. (13.9b) as

$$(\eta_T)_{\substack{\text{maximum}\\\text{Rankine}}} = \frac{h_1 - h_{2s} - v_3(p_4 - p_3)}{h_1 - h_3 - v_3(p_4 - p_3)}$$

$$= \frac{(1187.8 - 1047.5 \text{ Btu/lbm}) - (0.01672 \text{ ft}^3\text{/lbm})(100. - 14.7 \text{ lbf/in}^2)\left(\frac{144 \text{ in}^2\text{/ft}^2}{118.16 \text{ ft}\cdot\text{lbf/Btu}}\right)}{(1187.8 - 180.1 \text{ Btu/lbm}) - (0.01672 \text{ ft}^3\text{/lbm})(100. - 14.7 \text{ lbf/in}^2)\left(\frac{144 \text{ in}^2\text{/ft}^2}{118.16 \text{ ft}\cdot\text{lbf/Btu}}\right)}$$

$$= 0.139 = 13.9\%$$

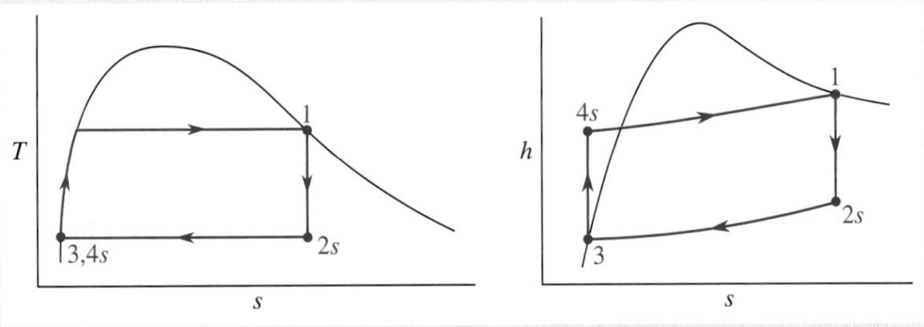

FIGURE 13.13
Example 13.4, *T–s* diagram.

b. Here, we use Eq. (13.9a) with $(\eta_s)_{pm} = 0.550.$ and $(\eta_s)_p = 0.650.$

$$(\eta_T)_{\text{Rankine}} = \frac{(h_1 - h_{2s})(\eta_s)_{pm} - v_3(p_4 - p_3)/(\eta_s)_p}{h_1 - h_3 - v_3(p_4 - p_3)/(\eta_s)_p}$$

$$= \frac{(1187.8 - 1047.5 \text{ Btu/lbm})(0.550) - (0.01672 \text{ ft}^3\text{/lbm})(100. - 14.7 \text{ lbf/in}^2)\left(\frac{144 \text{ in}^2\text{/ft}^2}{118.16 \text{ ft}\cdot\text{lbf/Btu}}\right)\left(\frac{1}{0.650}\right)}{(1187.8 - 180.1 \text{ Btu/lbm}) - (0.01672 \text{ ft}^3\text{/lbm})(100. - 14.7 \text{ lbf/in}^2)\left(\frac{144 \text{ in}^2\text{/ft}^2}{118.16 \text{ ft}\cdot\text{lbf/Btu}}\right)\left(\frac{1}{0.650}\right)}$$

$$= 0.0762 = 7.62\%$$

c. If the actual power output from the engine is 1400 hp and the isentropic efficiency of the engine is 55%, then the mass flow rate of steam required is

$$\dot{m} = \frac{\dot{W}_{\text{actual}}}{(h_1 - h_{2s})(\eta_s)_{pm}} = \frac{(1400. \text{ hp})(2545 \text{ Btu/hp}\cdot\text{h})}{(1187.8 - 1047.5 \text{ Btu/lbm})(0.550)} = 46{,}200 \text{ lbm/h}$$

Exercises

10. Determine the mass flow rate of steam required for the Corliss engine in Example 13.4 if it produced only 1000. hp instead of 1400. hp. Assume all the other variables remain unchanged. **Answer**: $\dot{m} = 33{,}000$ lbm/h.

11. If the isentropic efficiency of the Corliss engine in Example 13.4 were increased from 55.0% to 75.0%, determine the mass flow rate then required to produce 1400. hp. Assume all the other variables remain unchanged. **Answer**: $\dot{m} = 33{,}860$ lbm/h.

12. If the condenser pressure in Example 13.4 were reduced from 14.7 psia to 1.00 psia, recalculate items a, b, and c in Example 13.4. Assume all the other variables remain unchanged. **Answer**: a. $(\eta_T)_{\text{max Rankine}} = 26.1\%$, b. $(\eta_T)_{\text{Rankine}} = 14.3\%$, and c. $\dot{m} = 22{,}3000$ lbm/h.

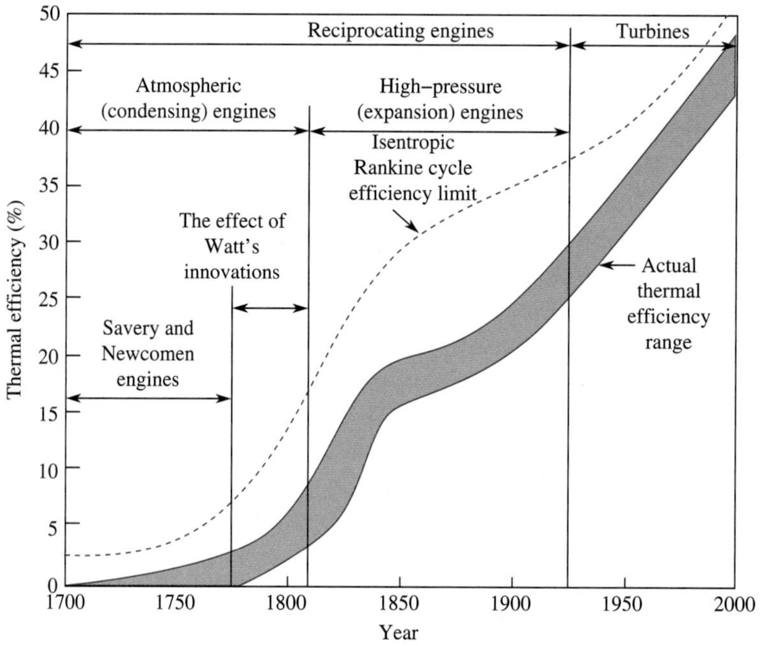

FIGURE 13.14
A chronology of steam engine thermal efficiency, 1700–2000.

The actual thermal efficiencies achieved by these early engines were naturally considerably less than that predicted through an isentropic analysis. Figure 13.14 illustrates the growth of both the actual and the isentropic efficiencies for the past three centuries. Since the early steam engines were very large and expensive, the firm of Boulton and Watt devised a creative marketing scheme based on the superior thermal efficiency of their engine. They let the purchaser pay for his engine by giving the company one third of the value of the fuel saved with the new engine as compared with the fuel consumption of a standard Newcomen engine of the same size.

13.6 RANKINE CYCLE WITH SUPERHEAT

Between 1850 and 1890, a variety of mechanical complexities were added to the reciprocating steam engine to improve its thermal efficiency. For example, the cylinders were often staged in series, so that the steam was first expanded in a high-pressure cylinder then exhausted to lower-pressure cylinder stages. Series staging of an engine's cylinders with the steam expanding only partially in each stage was then called *compounding*. Two-stage (duplex) expansion was introduced in 1811, three-stage (triplex) in 1871, and four-stage (quadruplex) in 1875.

By 1880, it was recognized that initially dry saturated steam became wet when condensation occurred during the expansion stroke of the piston (process 1 to 2 in Figure 13.8b). The water droplets thus formed inside the cylinder tended to cool it slightly, and they promoted corrosion. This meant that the cylinder walls were being alternately cooled and heated (as in the original Newcomen engine) slightly with each cycle of the engine, and this reduced the engine's thermal efficiency. However, if the steam entered the cylinder in a superheated state, then the amount of moisture produced during the expansion stroke was greatly reduced or eliminated altogether. Originally, the term *surcharging* was used to denote the use of superheated steam. This term has since been replaced by the more direct term *superheating*.

Superheating the steam at the entrance to the cylinder alters the equivalent Carnot cycle by raising T_H to the superheating temperature and consequently increases the equivalent Carnot efficiency considerably. This, in turn, makes the Rankine cycle appear less desirable by comparison, as shown in Figure 13.15. But all vapor cycle heat engines operate on the Rankine cycle, and using the Carnot cycle for engineering comparison purposes is purely academic. A more realistic comparison would be between the *isentropic* Rankine cycle and the *actual* Rankine cycle.

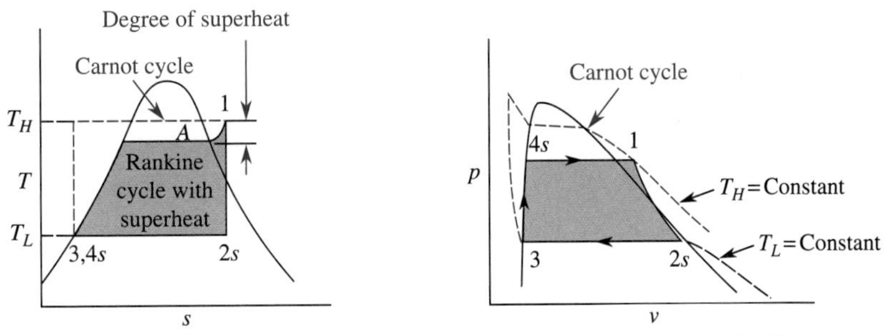

(a) Temperature–entropy diagram for Carnot and Rankine cycles with superheat

(b) Pressure–volume diagram for Carnot and Rankine cycles with superheat

FIGURE 13.15
A comparison between the Carnot cycle and the isentropic Rankine cycle with superheat.

The *degree of superheat* is defined to be the difference between the actual superheated vapor temperature and the saturation temperature at the pressure of the superheated vapor; that is,

$$\left\{\begin{array}{l}\text{Degree of superheat of superheated} \\ \text{vapor at temperature } T \text{ and pressure } p\end{array}\right\} = T - T_{sat}(p) \tag{13.10}$$

For example, in Figure 13.15a, the degree of superheat is $T_1 - T_A$. The degree of superheat that can be used in any particular heat engine design is limited only by the engine's ability to resist high temperatures. This has led to the industrial development and use of high-temperature alloys and ceramics for critical heat engine components.

EXAMPLE 13.5

In the 1890s, the Lancashire Steam Motor Company (now called British Leyland) manufactured the lawn mower shown in Figure 13.16, which was powered by a Rankine cycle steam engine. If it had a superheated boiler outlet state of 100. psia and 500.°F, and a condenser pressure of 1.00 psia, then find

a. The degree of superheat at the boiler outlet.
b. The equivalent Carnot cycle thermal efficiency of the lawn mower.
c. The isentropic Rankine cycle thermal efficiency of the lawn mower.

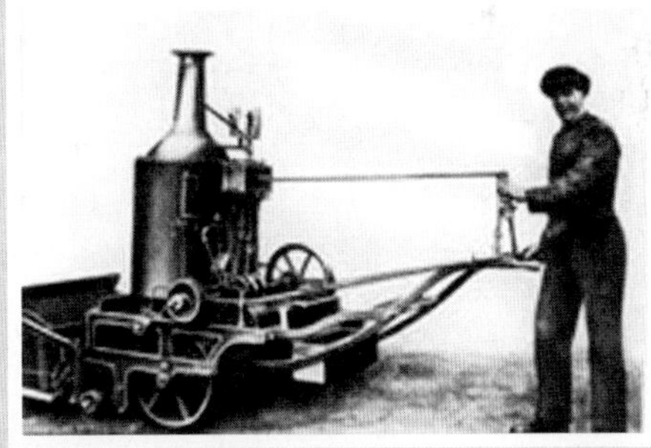

FIGURE 13.16
Example 13.5, steam-powered lawn mower.

Solution

A system sketch is shown in Figure 13.17.

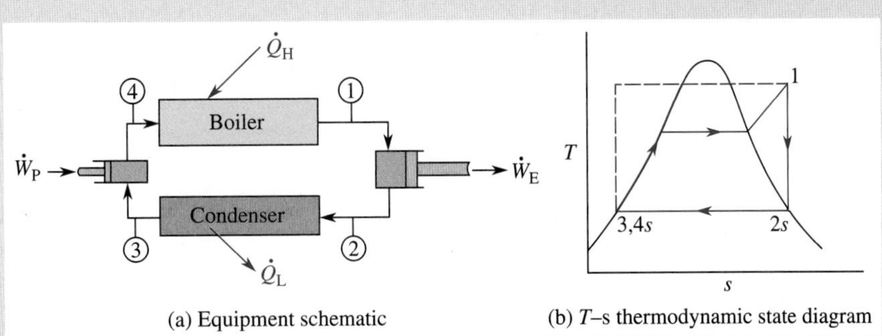

(a) Equipment schematic

(b) T–s thermodynamic state diagram

FIGURE 13.17
Example 13.5, system sketch.

(Continued)

EXAMPLE 13.5 *(Continued)*

The thermodynamic states of the steam at the four monitoring stations shown in the equipment schematic are

Station 1	Station 2s
$p_1 = 100.$ psia	
$T_1 = 500.°F$	
$h_1 = 1279.1$ Btu/lbm	
$s_1 = 1.7087$ Btu/(lbm·R)	

Station 3	Station 4s
$p_3 = 1.00$ psia	$p_{4s} = p_4 = 100.$ psia
$x_3 = 0.00$	$s_{4s} = s_3 = 0.1326$ Btu/(lbm·R)
$h_3 = h_f = 69.7$ Btu/lbm	$h_{4s} = h_3 + v_3(p_4 - p_3)$
$s_3 = s_f = 0.1326$ Btu/(lbm·R)	$= 69.7 + (0.01614)(100. - 1.00)(144/778.16)$
$v_3 = v_f(1.00 \text{ psia}) = 0.01614 \text{ ft}^3/\text{lbm}$	$= 70.0$ Btu/lbm

where we have calculated the following items:

$$x_{2s} = (s_2 - s_{f2})/s_{fg2} = (s_1 - s_{f2})/s_{fg2}$$
$$= (1.7087 - 0.1326)/1.8455 = 0.8540$$
$$h_{2s} = h_{f2} + x_{2s}h_{fg2} = 69.7 + (0.8540)(1036.0)$$
$$= 954.4 \text{Btu/lbm}$$

a. The degree of superheat at the outlet of the boiler is determined from Table C.2a in *Thermodynamic Tables to accompany Modern Engineering Thermodynamics* and Eq. (13.10) as

$$\text{Degree of superheat} = 500. - T_{sat}(100. \text{ psia})$$
$$= 500. - 327.8 = 172°\text{F}$$

b. Here, the highest temperature in the cycle is $T_H = 500. + 459.67. = 960.$ R, and the lowest temperature in the cycle is $T_L = T_{sat}(1.00 \text{ psia}) = 101.7 + 459.67 = 561.4$ R. Then, Eq. (7.16) gives the Carnot cycle thermal efficiency as

$$(\eta_T)_{\text{Carnot}} = 1 - \frac{T_L}{T_H} = 1 - \frac{561.4}{960.} = 0.415 = 41.5\%$$

c. Equation (13.9b) is used to determine the isentropic Rankine cycle thermal efficiency as

$$(\eta_T)_{\substack{\text{isentropic}\\ \text{Rankine}}} = \frac{h_1 - h_{2s} - v_3(p_4 - p_3)}{h_1 - h_3 - v_3(p_4 - p_3)} = \frac{1279.1 - 954.4 - (0.01614)(100. - 1)(144/778.16)}{1279.1 - 69.7 - (0.01614)(100. - 1)(144/778.16)}$$
$$= 0.268 = 26.8\%$$

Exercises

13. Determine the degree of superheat in Example 13.5 if the boiler outlet pressure were 120. psia instead of 100. psia. Assume all the other variables remain unchanged. **Answer**: Degree of superheat = 159°F.
14. Find the equivalent Carnot cycle thermal efficiency of the engine in Example 13.5 if the condenser pressure were 2.00 psia instead of 1.00 psia. Assume all the other variables remain unchanged. **Answer**: $(\eta_T)_{\text{Carnot}} = 39.0\%$.
15. Determine the actual Rankine cycle thermal efficiency of the system described in Example 13.5 if the prime mover and boiler feed pump isentropic efficiencies were 0.600 and 0.700, respectively. Assume all the other variables remain unchanged. **Answer**: $(\eta_T)_{\text{Rankine}} = 16.1\%$.

Initially, the purpose of superheating the vapor was simply to eliminate or at least reduce the amount of moisture in the engine's low-pressure stages. However, at the higher boiler pressures and temperatures of the 20th century, the effect of superheating the steam can add as many as 5 percentage points to the isentropic Rankine cycle thermal efficiency.

WHAT IS WRONG WITH USING WET STEAM?

High-efficiency modern steam turbines are designed to expand the steam into the *wet* vapor region in the low-pressure stage of the turbine. When the steam quality is less than 1.0, some of the vapor $(1-x)$ condenses into tiny liquid water droplets. Some of these droplets are deposited on the turbine walls, promoting corrosion, and some are entrained in the steam flow, eventually coalescing into large droplets. When large water droplets strike the front of high-speed moving blades, they can cause impact erosion, as shown by Figure 13.18.

FIGURE 13.18
Front edge of turbine blades. *(Source: Photo courtesy of Sanders, W. P. Turbo-Technic Services, Inc., Aurora, Canada.)*

FIGURE 13.19
Trailing edge of turbine blades. *(Source: Photo courtesy of Sanders, W. P. Turbo-Technic Services, Inc., Aurora, Canada.)*

This type of impact damage occurs where blades have a tip velocity around 800 ft/s and a steam quality of 97% (3.0% moisture) or less. Water droplets can also damage blade trailing edges, as shown in Figure 13.19. This is more of a gouging operation and produces blade thinning and stress concentration at positions where cracks and blade failure can occur. Superheating the turbine inlet steam minimizes the exhaust steam moisture that produces these detrimental effects.

13.7 RANKINE CYCLE WITH REGENERATION

A *regeneration* process in a system is a feedback process whereby energy is transferred internally from one part of a system to a different part of the system to improve the system's overall energy conversion efficiency. This internal transport of energy within a system is called *regeneration*, and the associated equipment is called a *regenerator*. In the case of heat engines, regeneration usually involves utilizing otherwise waste exhaust thermal energy to preheat fluid in another part of the same system. More specifically, in the case of the Rankine cycle heat engine, a certain percentage of the vapor passing through the prime mover is removed and used to preheat the boiler feedwater to a temperature between the condenser outlet temperature and the boiler outlet temperature. This significantly reduces the thermal irreversibility that occurs when relatively cold condenser outlet water is pumped back into a much hotter boiler. By reducing a major irreversibility of the cycle, the overall thermal efficiency of the cycle is increased.

Regenerative feedwater heating is shown schematically in Figure 13.20. These regenerators are simply heat exchangers. There are two common types:

a. *Open loop* (or direct contact) heat exchangers, in which the regeneration vapor mixes directly with the boiler feedwater.

b. *Closed loop* heat exchangers, in which the regeneration vapor and the boiler feedwater do not mix until after the regeneration vapor has been condensed into a liquid.

As an example of regeneration thermodynamics, consider the open loop regenerative feedwater heater shown in Figure 13.20a. These units were normally adiabatic (all the heat transfer occurs internally) and aergonic. A steady state, steady flow mass and energy rate balance (neglecting any changes in potential or kinetic energy) on the regenerator yields

$$\dot{m}_4 + \dot{m}_5 - \dot{m}_6 = 0$$

and

$$\dot{m}_4 h_4 + \dot{m}_5 h_5 - \dot{m}_6 h_6 = 0$$

Combining these two equations to eliminate $\dot{m}_4$ gives

$$h_6 = (1 - y)h_4 + yh_5$$

where we use $y = \dot{m}_5/\dot{m}_6 = \dot{m}_5/\dot{m}_1$ as the mass fraction of regeneration vapor extracted from the prime mover. Solving for y produces

$$y = \frac{h_6 - h_4}{h_5 - h_4} \tag{13.11}$$

where

$$h_4 = h_3 + v_3(p_4 - p_3)/(\eta_s)_{\text{pump1}}$$

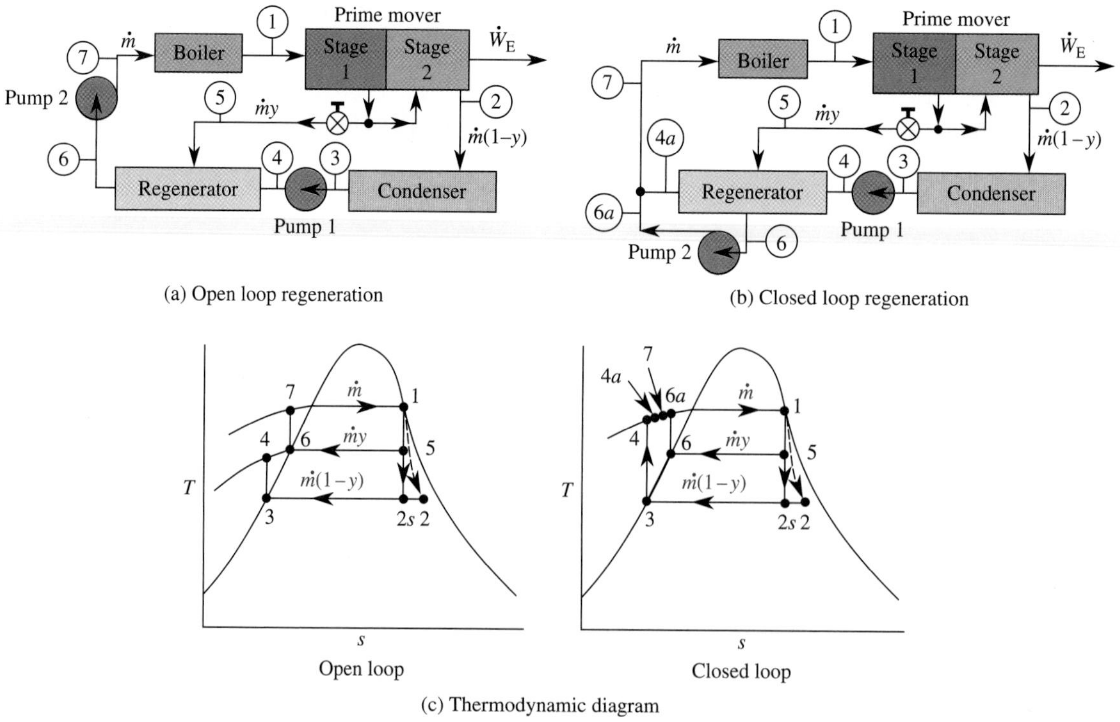

FIGURE 13.20
The Rankine cycle with regeneration.

and

$$h_5 = h_1 - (h_1 - h_{5s})(\eta_s)_{\text{stage1}}$$

This equation allows us to calculate the mass fraction of vapor that must be removed from the prime mover and added to the open loop regenerator to achieve a desired saturated liquid state at station 6.

Suppose we have a total of N regenerators (either open, closed, or any combination) in a system. Let the fraction of vapor removed for the first regenerator be $\dot{m}_{\text{regen 1}}/\dot{m}_{\text{total}} = y_1$, the amount removed for the second be y_2, and so forth, with y_N being fed to the Nth regenerator. Now, the general equation for thermal efficiency is

$$\eta_T = \frac{\dot{W}_{\text{net}}}{\dot{Q}_{\text{boiler}}} = \frac{\dot{Q}_{\text{boiler}} - |\dot{Q}_{\text{condenser}}|}{\dot{Q}_{\text{boiler}}} = 1 - \frac{|\dot{Q}_{\text{condenser}}|}{\dot{Q}_{\text{boiler}}}$$

so that the thermal efficiency of a Rankine cycle with N regenerators can be written as

$$\begin{aligned}(\eta_T)_{\substack{\text{Rankine cycle}\\ \text{with } N \text{ regenerators}}} &= 1 - \frac{|\dot{Q}_{\text{condenser}}|}{\dot{Q}_{\text{boiler}}} = 1 - \frac{\dot{m}_{\text{condenser}}(h_{\text{in}} - h_{\text{out}})_{\text{condenser}}}{\dot{m}_{\text{boiler}}(h_{\text{out}} - h_{\text{in}})_{\text{boiler}}} \\ &= 1 - \frac{\dot{m}_{\text{total}}(1 - y_1 - y_2 - \ldots - y_N)(h_{\text{in}} - h_{\text{out}})_{\text{condenser}}}{\dot{m}_{\text{total}}(h_{\text{out}} - h_{\text{in}})_{\text{boiler}}} \\ &= 1 - \left(\frac{(h_{\text{in}} - h_{\text{out}})_{\text{condenser}}}{(h_{\text{out}} - h_{\text{in}})_{\text{boiler}}}\right)(1 - y_1 - y_2 - \ldots - y_N)\end{aligned}$$

Therefore, the thermal efficiency of a Rankine cycle with N open or closed regenerators can always be determined from

$$(\eta_T)_{\substack{\text{Rankine cycle}\\ \text{with } N \text{ regenerators}\\ \text{(either open or closed)}}} = 1 - \left[\frac{(h_{\text{in}} - h_{\text{out}})_{\text{condenser}}}{(h_{\text{out}} - h_{\text{in}})_{\text{boiler}}}\right](1 - y_1 - y_2 - \ldots - y_N) \tag{13.12a}$$

This equation provides a simple solution to a complex problem when multiple open or closed regenerators are used. For the open and closed regenerators shown in Figure 13.20, Eq. (13.12a) reduces to

$$(\eta_T)_{\substack{\text{Rankine cycle}\\ \text{with 1 regenerator}\\ \text{(either open or closed)}}} = 1 - \left(\frac{h_2 - h_3}{h_7 - h_6}\right)(1 - y) \tag{13.12b}$$

where

$$\begin{aligned}h_2 &= h_5 - (h_5 - h_{2s})(\eta_s)_{\text{stage2}} \\ h_5 &= h_1 - (h_1 - h_{5s})(\eta_s)_{\text{stage1}} \\ h_7 &= h_6 + v_6(p_7 - p_6)/(\eta_s)_{\text{pump2}}\end{aligned}$$

The only property in Eq. (13.12a) that is directly affected by the presence of the regenerators is the boiler inlet enthalpy, $(h_{\text{in}})_{\text{boiler}}$. Notice that maximizing this value also maximizes the cycle's thermal efficiency. Therefore, an optimum set of regenerator flows exists that maximizes the cycle's thermal efficiency. This is illustrated in the next example problem.

EXAMPLE 13.6

A two-stage steam turbine receives dry saturated steam at 200. psia. It has an interstage pressure of 80.0 psia and a condenser pressure of 1.00 psia. Determine

a. The isentropic Rankine cycle thermal efficiency of the system without regeneration present.
b. The isentropic Rankine cycle thermal efficiency of the system and the mass fraction of regeneration steam required with an open loop boiler feedwater regenerator at a pressure of 80.0 psia.

(Continued)

EXAMPLE 13.6 *(Continued)*

Solution

a. First, draw a sketch of the configuration of the isentropic Rankine cycle system before boiler feedwater regeneration is added (Figure 13.21).

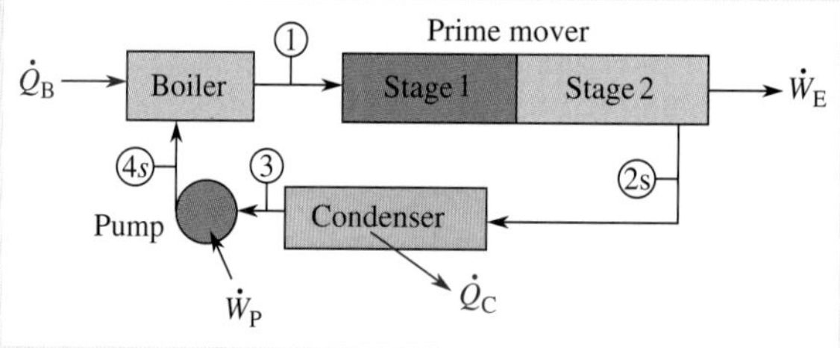

FIGURE 13.21
Example 13.6, part a.

Its thermal efficiency is

$$(\eta_T)_{\substack{\text{isentropic}\\ \text{Rankine}}} = \frac{(h_1 - h_{2s}) - (h_{4s} - h_3)}{(h_1 - h_{4s})}$$

where, assuming an incompressible liquid condensate, Eq. (13.6) gives $h_{4s} = h_3 + v_3(p_4 - p_3)$ and $v_3 = v_4$ (1.00 psia) = 0.01614 ft^3/lbm. The monitoring station data are as follows:

Station 1	Station 2s
$p_1 = 200.$ psia	
$x_1 = 1.00$	$p_{2s} = p_2 = 1.00$ psia
$h_1 = 1199.3$ Btu/lbm	$s_{2s} = s_1 = 1.5466$ Btu/(lbm · R)
$s_1 = 1.5466$ Btu/(lbm · R)	$h_{2s} = 863.5$ Btu/lbm
Station 3	**Station 4s**
$p_3 = 1.00$ psia	$p_{4s} = p_4 = 200.$ psia
$x_3 = 0.00$	$s_{4s} = s_3 = 0.1326$ Btu/(lbm · R)
$h_3 = 69.7$ Btu/lbm	$h_{4s} = h_3 + v_3(p_4 - p_3)$
$s_3 = 0.1326$ Btu/(lbm · R)	$= 69.70 + (0.01614)(200. - 1.00)(144/778.16)$
	$= 69.70 + 0.594 = 70.3$ Btu/lbm

where, at station 2s, we use

$$x_{2s} = \frac{s_{2s} - s_{f2}}{s_{fg2}} = \frac{s_1 - s_{f2}}{s_{fg2}} = \frac{1.5466 - 0.1326}{18455} = 0.7662$$

then,

$$h_{2s} = 69.70 + (0.7662)(1036.0) = 863.5 \text{ Btu/lbm}$$

The thermal efficiency is now

$$(\eta_T)_{\substack{\text{isentropic}\\ \text{Rankine}}} = \frac{1199.3 - 863.5 - 0.594}{1199.3 - 70.3} = 0.297 = 29.7\%$$

b. Now, draw a sketch of the configuration of the isentropic Rankine cycle system with one open loop boiler feedwater regenerator (Figure 13.22).

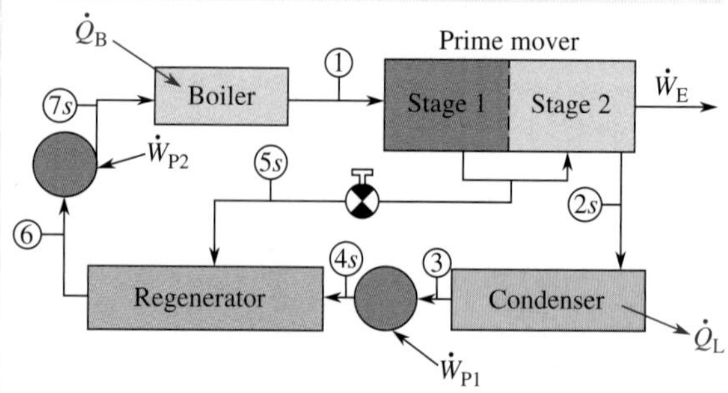

FIGURE 13.22
Example 13.6, part b.

The properties at monitoring stations 1, 2, and 3 are the same as they were in part a, but pump 1 brings the condensate pressure up to only 80.0 psia (to match the vapor inlet pressure), and pump 2 brings the pressure the rest of the way up from 80.0 to 200. psia. Then, $v_3 = v_f(1.00 \text{ psia}) = 0.01614 \text{ ft}^3/\text{lbm}$ and $v_6 = v_f(80.0 \text{ psia}) = 0.01757 \text{ ft}^3/\text{lbm}$. The additional monitoring station data needed are

Station 4s	Station 5s
$p_{4s} = p_4 = 80.0$ psia	$p_{5s} = p_4 = 80.0$ psia
$s_{4s} = s_3 = 0.1326$ Btu/(lbm·R)	$s_{5s} = s_1 = 1.5466$ Btu/(lbm·R)
$h_{4s} = h_3 + v_3(p_4 - p_3)$	$x_{5s} = 0.9358$
$= 69.7 + (0.01614)(80.0 - 1.00)(144/778.16)$	$h_{5s} = 1125.7$ Btu/lbm
$= 69.7 + 0.236 = 69.9$ Btu/lbm	

Station 6	Station 7s
$p_6 = 80.0$ psia	$p_{7s} = p_7 = 200.$ psia
$x_6 = 0.00$	$s_{7s} = s_6 = 0.4535$ Btu/(lbm·R)
$h_6 = 282.2$ Btu/lbm	$h_{7s} = h_6 + v_6(p_7 - p_6)$
$s_6 = 0.4535$ Btu/(lbm·R)	$= 282.2 + (0.01757)(200. - 80.0)(144/778)$
	$= 282.2 + 0.390 = 282.6$ Btu/lbm

where, at station 5s, we determine that $x_{5s} = (1.5466 - 0.4535)/1.1681 = 0.9358$ and $h_{5s} = 282.2 + (0.9358)(901.4) = 1125.7$ Btu/lbm. Equation (13.11) now gives the value of y as

$$y = \frac{h_6 - h_{4s}}{h_{5s} - h_{4s}} = \frac{282.6 - 69.9}{1125.7 - 69.9} = 0.201$$

then, the isentropic thermal efficiency of the cycle is given by Eq. (13.12b) with all the $\eta_s = 1.0$ as

$$(\eta_T)_{\substack{\text{Rankine cycle}\\ \text{with 1 regenerator}}} = 1 - \left(\frac{h_{2s} - h_3}{h_1 - h_{7s}}\right)(1 - y) = 1 - \left(\frac{863.5 - 69.7}{1199.3 - 282.6}\right)(1 - 0.201) = 0.308 = 30.8\%$$

By altering the value of y (the amount of steam bled from the turbine) and recomputing the value of h_{7s}, it becomes clear that the cycle thermal efficiency is maximized at 31.2% when $y = 0.138$. This is shown in Figure 13.23.

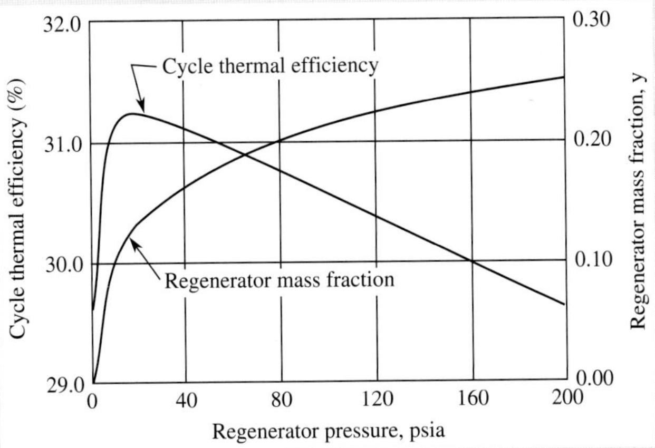

FIGURE 13.23
Example 13.6, plot of solution.

Exercises

16. The operator of the plant described in Example 13.6 changes the boiler outlet state from dry saturated steam at 200. psia to dry saturated steam at 400. psia. Determine the mass fraction of regeneration steam that is now required to produce saturated liquid water at 80.0 psia at the end of the open loop regenerator. Assume all the other variables remain unchanged. **Answer:** $y = 0.210$

17. If the interstage steam pressure in Example 13.6 is increased from 80.0 to 150. psia, determine the new isentropic Rankine cycle thermal efficiency. Assume all the other variables remain unchanged. **Answer:** $(\eta_T)_{\text{isentropic Rankine}} = 31.0\%$.

18. The power plant engineer mentioned in Exercise 16 just informed us that the first and second stage prime mover isentropic efficiencies are 75.0% and 68.0%, respectively, and that the boiler feed pump isentropic efficiency is 83.0%. Now determine the actual Rankine cycle thermal efficiency of this system with the 80.0 psia open loop regenerator. **Answer:** $(\eta_T)_{\text{Rankine}} = 25.8\%$.

Note that the effect of a single regeneration unit in the previous example was to increase the thermal efficiency by only 1.1%. The extra expense of the regenerators and the additional pumps cannot be economically justified unless the system is large enough to make such a small thermal efficiency increment produces a significant savings in fuel costs.

This type of Rankine cycle regeneration was first seriously proposed in 1890 and first implemented in 1898 with a four-stage, quadruple-expansion (quadruplex), reciprocating piston-cylinder steam engine. This system achieved an actual thermal efficiency of 22.8%, which was remarkably high for its day. After about 1910, the production of large reciprocating piston-cylinder steam engines decreased rapidly. They were being replaced by a new prime mover technology, the steam turbine. Consequently, regeneration was temporarily discontinued until about 1920, at which point the steam turbine had been established as the preferred prime mover for large stationary power plants. After 1920, regeneration became standard practice in the design of large, vapor cycle, turbine, prime mover power plants.

WHAT IS THE WORLD'S SMALLEST STEAM ENGINE?

The world's smallest steam engine was built using nanotechnology (the ability to create objects at the atomic scale). It is about 5 microns wide (approximately the size of a red blood cell), and was developed by Dr. Jeff Sniegowski at Sandia National Laboratories in Albuquerque, New Mexico. Steam is produced by a small electrical current that boils the water in a tiny boiler. The engine was built using computer chip technology by photographing and reducing an image to a very small size, then etching it on a silicon wafer. The etchings are done in layers to build up the three-dimensional working engine. However, you need an electron microscope to see it.

13.8 THE DEVELOPMENT OF THE STEAM TURBINE

By the end of the 19th century, the large, slow-speed reciprocating piston-cylinder steam engine had reached its upper limit in size and complexity. When the maximum practical piston speed had been reached in reciprocating steam engine design, the only way to increase the work output further was to increase the physical size of the engine. The largest reciprocating steam engine ever built in the United States was constructed in 1891 by the E. P. Allis Company (renamed the Allis Chalmers Company in 1901) of Milwaukee, Wisconsin. It was installed as a pumping engine at the Chapin mine in Iron Mountain, Michigan, in 1892. It was a duplex steeple compound condensing engine with high-pressure cylinders 50. inches in diameter and low-pressure cylinders 100. inches in diameter, both with strokes of 10 ft. It was 54 ft high, 75 ft long, weighed 725 tons, and had a flywheel 40. ft in diameter. At its maximum speed of 10. rpm it could produce over 1200 hp.

Meanwhile, a new heat engine prime mover technology was quickly being developed: the steam turbine. The word *turbine* was coined in 1822 from the Latin root word *turbo*, for "that which spins." When it was coined it was applied only to water wheels (as in hydraulic or water turbines).

A turbine is a prime mover in which mechanical rotating shaft work is produced by a steady flow of fluid through the system. The output work is produced by changing the *momentum* of the working fluid as it passes through the system (the turbine). Reciprocating prime mover output work, on the other hand, is produced by changing the *pressure* of a fixed mass of working fluid within the system (the piston-cylinder apparatus).

There are two basic types of turbine designs: *impulse* and *reaction*. Both the impulse and reaction turbine concepts date from antiquity. The paddle-type water wheels developed in Italy in about 70 BC (and used throughout the world, well into the 20th century AD) were of the impulse type (Figure 13.24, left). Also, in the first century AD, a Greek known today only as Heron (or, in Latin, Hero) of Alexandria (Egypt) devised a simple reaction steam turbine (called an *aeolipile*) in which a hollow copper sphere was made to rotate by steam jetting out of four nozzles mounted perpendicular to the axis of rotation (see Figure 13.24, right). No practical use was then made of this device. However, it was known to James Watt and his contemporaries because they experimented with steam-driven reaction turbines of the Heron type and still found them impractical due to the extremely high rotational speeds required to make them efficient enough to be competitive with existing reciprocating steam engine technology.

In the impulse turbine, high-velocity fluid jets from stationary nozzles impinge on a set of blades on a rotor. The impulse force generated by the momentum change of the fluid passing through these blades causes the

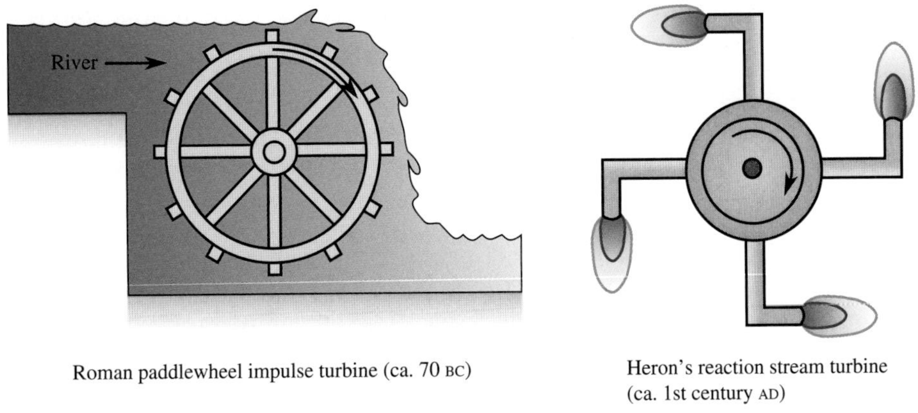

FIGURE 13.24
Roman paddlewheel impulse turbine (ca. 70 BC) and Heron's reaction steam turbine (ca. 1st century AD).

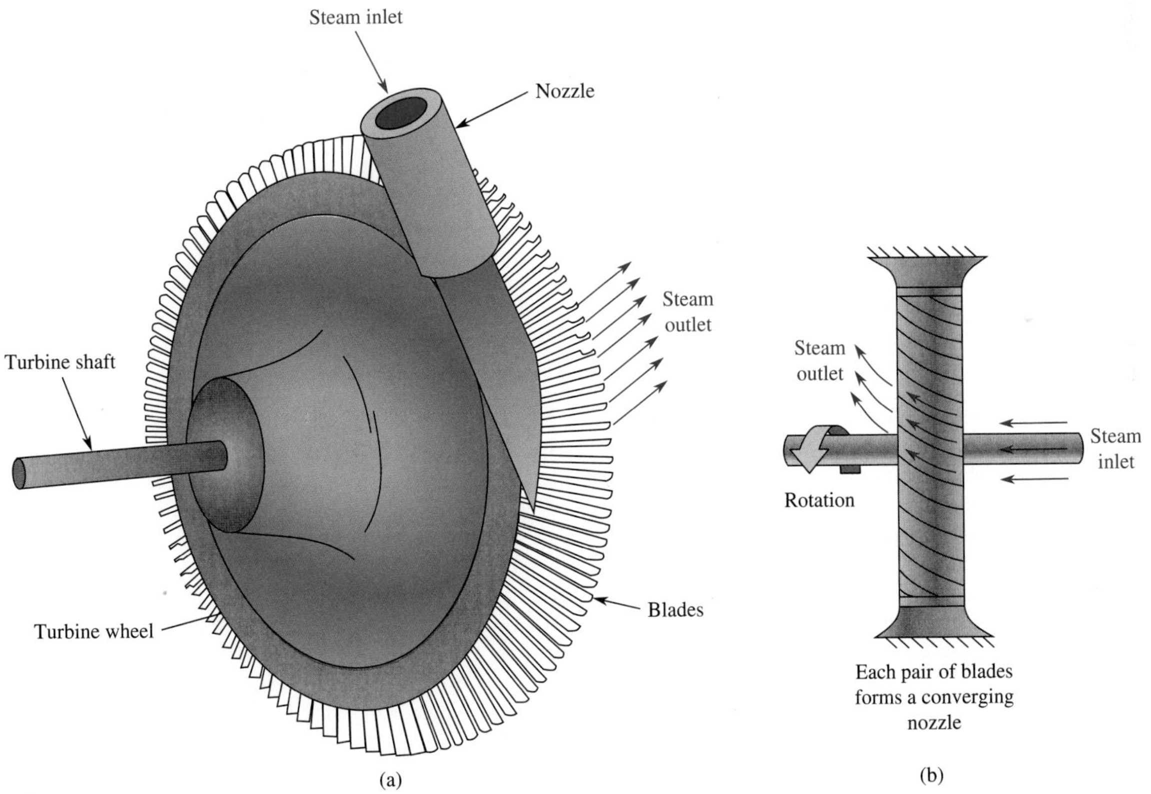

FIGURE 13.25
Characteristics of (a) impulse and (b) reaction turbines.

rotor to spin rapidly, like blowing on a pinwheel. In a reaction turbine, the rotation is caused by a reaction force generated by the momentum change of the fluid accelerating through nozzles attached to the rotor itself (like the nozzles on a lawn sprinkler). The nozzles in a rotor-type reaction turbine are not the same as the simple axi-symmetric cylindrical jet producing nozzles of the impulse turbine. Instead, they are two-dimensional nozzlelike channels formed in the passage between the blades of each row. The characteristics of impulse and reaction turbines are shown in Figure 13.25.

It can be easily shown from the momentum balance equations of fluid mechanics that the maximum energy conversion efficiency of an impulse turbine occurs when the fluid enters the rotor blades parallel to the direction of motion of the blades and with a velocity equal to exactly twice the blade average velocity and that this efficiency drops off quickly when these conditions are deviated from. Similarly, it can be shown that a reaction turbine has a maximum energy conversion efficiency when the fluid enters the rotor blades parallel to the direction of motion of the blades and with a velocity exactly equal to the blade average velocity. The effect of the turbine's nozzles is to convert static pressure energy into dynamic kinetic energy, whose momentum can then be manipulated by the turbine's geometry to drive the rotor. Nozzles are analyzed in Chapter 6, and the outlet velocity of an adiabatic nozzle with a negligible inlet velocity is given by Eq. (6.16) as

$$V_{\text{out}} = \sqrt{2g_c(h_{\text{in}} - h_{\text{out}})} \tag{6.16}$$

A nozzle receiving steam at 200. psia, 700.°F and exhausting to 1.00 psia has an enthalpy drop of, say, 400. Btu/lbm. The resulting nozzle outlet velocity is supersonic and can be calculated from Eq. (6.16) as

$$\begin{aligned} V_{\text{out}} &= \{2[32.174\,\text{lbm}\cdot\text{ft}/(\text{lbf}\cdot\text{ft})](400.\,\text{Btu/lbm})(778.16\,\text{ft}\cdot\text{lbf/Btu})\}^{1/2} \\ &\approx 4500\,\text{ft/s} \end{aligned}$$

Then, a reaction turbine operating at its most efficient speed requires a blade average velocity of about 4500 ft/s. If we assume a mean rotor radius of 1.0 ft, then the angular velocity of the rotor at its most efficient operating speed is 4500 radians per second, or about 43,000 rpm. This is an extremely high rotational speed and is very dangerous, due to the high centrifugal stresses and the high bearing loads produced by unbalanced forces. Also, few auxiliary turbine driven devices (e.g., an electrical generator) could be made to operate at these speeds. Therein was the major problem with early turbine development. How could they be slowed down while still maintaining their good energy conversion and thermal efficiencies?

This problem remained unsolved until the end of the 19th century, when it was discovered that certain types of turbine *staging* significantly decrease the turbine's operating speed while maintaining its energy conversion efficiency. The region between stationary nozzles in an impulse turbine is referred to as an *impulse stage*. It was discovered that the effect of adding two extra rows of blades, one stationary and one moving, to each stage of an impulse turbine reduced its most efficient operating speed by about a factor of 2. In a reaction turbine, every other row of blades is stationary, and the combination of a stationary row and a moving row forms a *reaction stage* of the turbine. Large reaction turbines typically have 30 to 100 or more stages, whereas impulse turbines, normally, have fewer than 10 stages.

In 1883, the Swedish engineer Carl Gustaf Patrik DeLaval (1845–1913) built and ran the first practical single-stage impulse steam turbine. It had a mean rotor diameter of 3 inches (0.076 m) and produced about 1.5 hp with a shaft speed of 40,000 rpm. In 1889, he discovered that, if the pressure ratio across his stationary nozzles were less than about 0.55, he could increase the nozzle exit velocity to supersonic speeds by making the nozzles with a converging-diverging internal profile. The high shaft velocity of the DeLaval impulse turbine required a gearbox to reduce the output rotational speed to a usable value. Since high-speed gear reduction is very inefficient, it became necessary to find other ways of reducing the turbine's speed effectively. The addition of multiple stages of fixed nozzles and moving blades to the shaft of a DeLaval impulse turbine was first carried out by the Frenchman Auguste Camille Edmond Rateau (1863–1930) in 1899. By having a sufficient number of these stages in series, he was able to reduce the efficient rotating speed enough to allow electrical generators to be driven directly from the output shaft. Also, in 1898, the American engineer Charles Gordon Curtis (1860–1953) introduced multiple sets of stationary and moving blade rows downstream from a single stationary nozzle. This technique exposed each set of moving vanes to a different mean velocity and also had the effect of slowing down an impulse turbine while maintaining its energy conversion efficiency. Rateau staging exposes each row of moving blades to nearly constant pressure, usually called *pressure compounding* or *pressure staging*. Curtis staging exposes each row of stationary blades to nearly constant velocity, usually called *velocity compounding* or *velocity staging*. Figure 13.26 illustrates these impulse turbine staging concepts.

The practice of putting several different constant pressure stages in series in a reaction turbine was introduced in 1884 by the Englishman Charles Algernon Parsons (1854–1931). His first turbine had 14 stages (14 pairs of stationary and moving blade rows) and produced about 10 hp at 18,000 rpm. This was still too fast for direct coupling to the existing electrical generators, so Parsons designed a new high-speed generator that could be driven directly by his turbine.

By the early 20th century, the pressure and velocity staged DeLaval impulse turbine and the multiple pressure staged Parsons reaction turbine became nearly equal rivals in terms of cost and efficiency. A lot was at stake at this point in time, because electrical power generation was just coming into existence, and it was clear that it had the potential for creating a second Industrial Revolution, based on electricity rather than steam.

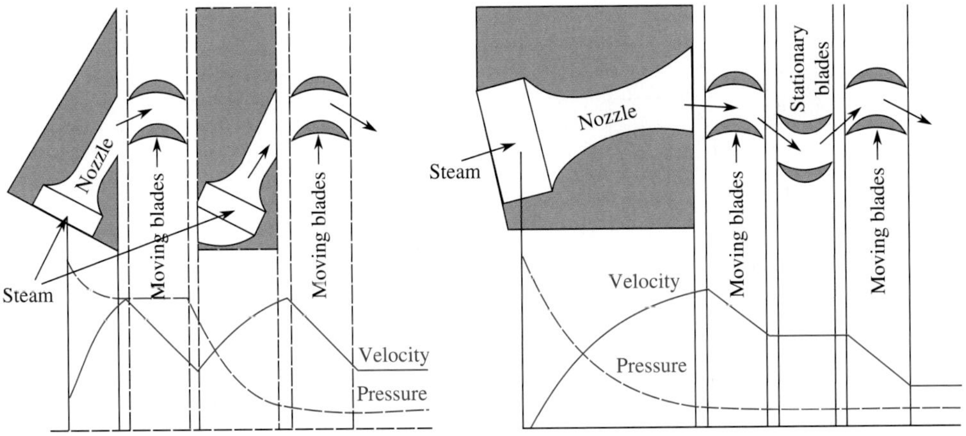

FIGURE 13.26
Pressure (Rateau) and velocity (Curtis) staging in the DeLaval impulse turbine.

Thomas Alva Edison's (1847–1931) development of a practical electric lightbulb in 1879 opened the doors of a remarkable new technology, electricity. To make his lightbulb marketable, he had to develop and produce a means of putting electricity directly into the home. He had to conceive and build an entire electrical power plant and electrical distribution network. This he did, and in 1882, he opened the Pearl Street power station in New York City, the first such station in the world. By 1890, several electrical power stations were in place in major cities across the United States, and they were rapidly growing in size and complexity. Initially, reciprocating steam engines drove the electrical generators, but it became clear rather quickly that this type of prime mover was not going to be able to meet the needs of this growing industry for very long. Reciprocating steam engines were too slow, too large, and too unreliable to carry the burden. The steam turbine was cultivated as a viable replacement prime mover.

By 1900, the Westinghouse Electric Company was manufacturing multistage reaction steam turbines of the Parson's type for the electrical power generation industry, and the General Electric Company was developing an impulse turbine of the DeLaval type with Curtis velocity staging for the same market.

Thus, the search for a suitable prime mover for large-scale electrical generators was the motivation that led to the successful commercial development of the steam turbine. Though early steam turbines were actually less energy efficient than their reciprocating counterparts, their potential for improvement was enormous. In addition, they were about *ten times smaller* than a reciprocating engine with the same power output. Also, even very large steam turbines could be made to run efficiently at generator speeds (1800 or 3600 rpm), they were quiet, and they required little maintenance. It was for these reasons, not the reasons of improved thermal or mechanical efficiency, that by 1920, the steam turbine had replaced virtually all large-scale reciprocating steam engines. By 1960, virtually all small- and medium-scale reciprocating steam engines had been replaced by electric motors or internal combustion engines.

13.9 RANKINE CYCLE WITH REHEAT

By 1920, boiler technology had advanced to the point where steam at 650°F, 250 psia was generally available. In the early 1920s, the regenerative process, initially developed in the late 1890s to improve the thermal efficiency of reciprocating steam engines, was reintroduced as a means of improving steam turbine power plant thermal efficiency. Regeneration using steam turbine prime movers required that steam be extracted from between one or more of the turbine stages and used to preheat the boiler feedwater. During the 1920s, boiler technology continued to increase rapidly, and by 1930, steam was commonly supplied at 725°F and 550 psia. This led to the commercial use of steam *reheat,* in which steam is extracted from the outlet of a turbine stage, returned to the boiler to be reheated, then brought back to the inlet of the next turbine stage for further expansion. After its introduction in the mid 1920s, reheat technology became unpopular during the Depression due to technical and economic difficulties. Single reheat cycles were again introduced in the 1940s, and double reheat cycles were introduced in the 1950s. This prevents excessive moisture levels from occurring in the low-pressure stages and has the effect of slightly increasing the thermal efficiency of the cycle. A simple power plant utilizing reheat (but no regeneration) is shown schematically in Figure 13.27.

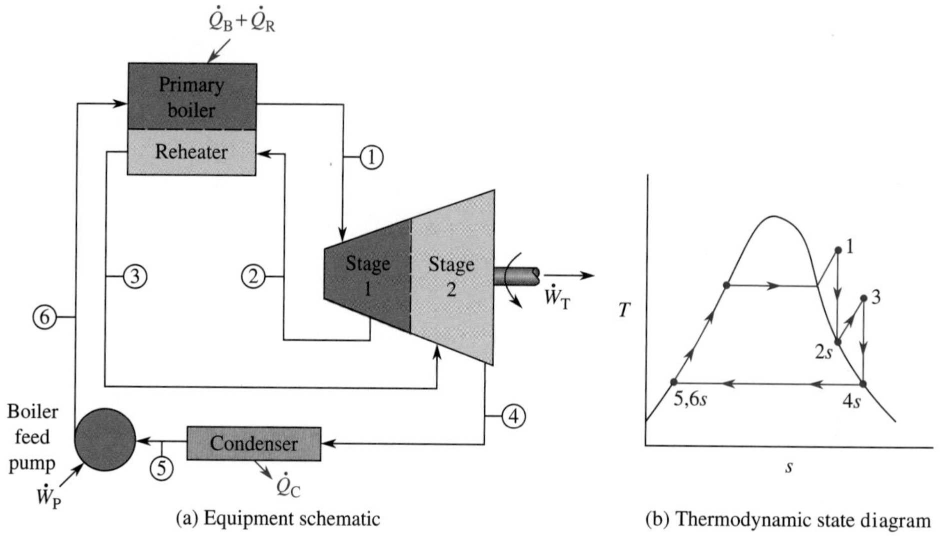

FIGURE 13.27
A Rankine cycle power plant with reheat.

The thermal efficiency of the Rankine cycle power plant with reheat shown in Figure 13.27 can be computed from the general thermal efficiency definition as

$$(\eta_T)_{\substack{\text{Rankine}\\\text{cycle with}\\\text{one reheat unit}}} = \frac{\dot{W}_{pm} - |\dot{W}_p|}{\dot{Q}_B + \dot{Q}_R} = \frac{(h_1 - h_2) + (h_3 - h_4) - (h_6 - h_5)}{(h_1 - h_6) + (h_3 - h_2)}$$

and, if the liquid condensate is considered to be incompressible, then Eq. (13.7b) can be used to give

$$(\eta_T)_{\substack{\text{Rankine}\\\text{cycle with}\\\text{one reheat unit}}} = \frac{(h_1 - h_2) + (h_3 - h_4) - v_5(p_6 - p_5)/(\eta_s)_p}{(h_1 - h_6) + (h_3 - h_2)}$$

where $(\eta_s)_p$ is the isentropic efficiency of the boiler feed pump. Using Eq. (13.4a), we can introduce the isentropic efficiencies of the two turbine stages $(\eta_s)_{pm1}$ and $(\eta_s)_{pm2}$ as

$$(\eta_s)_{pm1} = \frac{h_1 - h_2}{h_1 - h_{2s}} \tag{13.13}$$

and

$$(\eta_s)_{pm2} = \frac{h_3 - h_4}{h_3 - h_{4s}} \tag{13.14}$$

Finally, the thermal efficiency of the cycle can be written as

$$(\eta_T)_{\substack{\text{Rankine}\\\text{cycle with}\\\text{one reheat unit}}} = \frac{(h_1 - h_{2s})(\eta_s)_{pm1} + (h_3 - h_{4s})(\eta_s)_{pm2} - v_5(p_6 - p_5)/(\eta_s)_p}{(h_1 - h_6) + (h_3 - h_2)} \tag{13.15}$$

where the values of h_2 and h_6 in the denominator of this equation are calculated from Eqs. (13.13) and (13.7b) as

$$h_2 = h_1 - (h_1 - h_{2s})(\eta_s)_{pm1} \tag{13.16}$$

and

$$h_6 = h_5 + v_5(p_6 - p_5)/(\eta_s)_p \tag{13.17}$$

EXAMPLE 13.7

The first Rankine cycle steam turbine prime mover with reheat used in the United States was at the Crawford Avenue power station of the Commonwealth Edison Company of Chicago, Illinois, which went into operation in September 1924. The primary steam was at 700.°F, 600. psia, with reheat to 700.°F at 100. psia. The isentropic efficiencies of the first and second turbine stages and the boiler feed pump were 84.0, 80.0, and 61.0%, respectively. The condenser pressure was 1.00 psia with saturated liquid being produced at its outlet.

Determine

a. The Rankine cycle thermal efficiency of the plant with reheat.
b. The Rankine cycle thermal efficiency of the plant without reheat (assume a turbine isentropic efficiency of 82.0% for this calculation).

Solution

a. Using the notation of Figure 13.27, the monitoring station data for this problem are

Station 1	Station 2s
$p_1 = 600.$ psia	$p_{2s} = p_2 = 100.$ psia
$T_1 = 700.°F$	$s_{2s} = s_1 = 1.5874$ Btu/(lbm · R)
$h_1 = 135.6$ Btu/lbm	$x_{2s} = 0.9856$
$s_1 = 1.5874$ Btu/(lbm · R)	$h_{2s} = 1175.0$ Btu/lbm
Station 3	**Station 4s**
$p_3 = p_{2s} = 100.$ pisa	$p_{4s} = p_4 = 1.00$ psia
$T_3 = 700.°F$	$s_{4s} = s_3 = 1.8035$ Btu/(lbm · R)
$h_3 = 1379.2$ Btu/lbm	$x_{4s} = 0.9054$
$s_3 = 1.8035$ Btu/(lbm · R)	$h_{4s} = 1007.7$ Btu/lbm
Station 5	**Station 6s**
$p_5 = 1.00$ psia	$p_{6s} = p_6 = 600.$ psia
$x_5 = 0.00$	$s_{6s} = s_5 = 0.1326$ Btu/(lbm · R)
$h_5 = 69.7$ Btu/lbm	$h_{6s} = 72.5$ Btu/lbm
$s_5 = 0.1326$ Btu/(lbm · R)	

where the following calculations have been used:

$$x_{2s} = \frac{s_{2s} - s_{f2}}{s_{fg2}} = \frac{s_1 - s_{f2}}{s_{fg2}} = \frac{1.5874 - 0.4745}{1.1291} = 0.9856$$

Then,

$$h_{2s} = h_{f2} + x_{2s}(h_{fg2}) = 298.6 + (0.9856)(889.2) = 1175.0 \text{ Btu/lbm}$$

$$x_{4s} = \frac{s_3 - s_{f4}}{s_{fg4}} = \frac{1.8035 - 0.1326}{1.8455} = 0.9054$$

and

$$h_{4s} = h_{f4} + x_{4s}(h_{fg4}) = 69.7 + (0.9054)(1036.0) = 1007.7 \text{ Btu/lbm.}$$

Since $v_5 = v_f$ (1.0 psia) = 0.01614 ft^3/lbm, Eqs. (13.16) and (13.17) can now be used to give

$$h_2 = 1350.6 - (0.840)(1350.6 - 1175.0) = 1203.1 \text{ Btu/lbm}$$

and

$$h_6 = 69.7 + (0.01614)(600. - 1.00)(144/778.16)/(0.610)$$
$$= 69.7 + 2.93 = 72.6 \text{ Btu/lbm}$$

Then, finally, Eq. (13.15) yields

$$\eta_T = \frac{(1350.6 - 1175.0)(0.840) + (1379.2 - 1007.7)(0.800) - 2.93}{(1350.6 - 72.6) + (1379.2 - 1203.1)}$$
$$= 0.304 = 30.4\%$$

(*Continued*)

EXAMPLE 13.7 *(Continued)*

b. Here, we remove the reheat loop by simply eliminating the pipes with monitoring stations 2 and 3 in Figure 13.27. The power plant's thermal efficiency becomes

$$\eta_T = \frac{(h_1 - h_{4s})(\eta_s)_{pm} - (h_{6s} - h_5)/(\eta_s)_{pm}}{h_1 - h_6}$$

where all the enthalpy values are the same as they were in part a except for h_{4s}. Here, $s_{4s} = s_1$, and

$$x_{4s} = \frac{s_1 - s_{f4}}{s_{fg4}} = \frac{1.5874 - 0.1326}{1.8455} = 0.7883$$

so

$$h_{4s} = 69.7 + (0.7883)(1036.0) = 886.4 \text{ Btu/lbm}$$

Then,

$$\eta_T = \frac{(1350.6 - 886.4)(0.82) - 2.93}{1350.6 - 72.6} = 0.296 = 29.6\%$$

Exercises

19. Determine the isentropic Rankine cycle thermal efficiency of the Crawford Avenue power plant without reheat discussed in Example 13.7b. Assume all the variables except the turbine and boiler feed pump isentropic efficiencies remain unchanged. **Answer**: $(\eta_T)_{\text{isentropic Rankine}} = 36.2\%$.
20. If the isentropic efficiency of the boiler feed pump in the Crawford Avenue power plant with reheat in Example 13.7(a) is increased from 61.0% to 85.0%, determine the power plant's new Rankine cycle thermal efficiency. Assume all the other variables remain unchanged. **Answer**: $(\eta_T)_{\text{Rankine}} = 30.4\%$ (there is no change in $(\eta_T)_{\text{Rankine}}$ to three significant figures).
21. If the reheat pressure in the Crawford Avenue power plant with reheat in Example 13.7a is decreased from 100. psia to 80.0 psia, determine the power plant's new Rankine cycle thermal efficiency. Assume all the other variables remain unchanged. **Answer**: $(\eta_T)_{\text{Rankine}} = 30.4\%$ (there is no change in $(\eta_T)_{\text{Rankine}}$ to three significant figures).

Note that the interstage reheating used in Example 13.7 increases the Rankine cycle thermal efficiency by only 0.8%. However, it has the much more important effect of reducing the moisture content at the turbine exit. Wet steam with a moisture content of more than 8 to 10% can produce serious blade erosion problems in the low-pressure region of a turbine. The effect of reheating in this example keeps the exit moisture level within this range, whereas without reheating, part b of the example shows that the exit moisture level would be (1 – 0.7883)(100) = 21.2%, which is much too high.

13.10 MODERN STEAM POWER PLANTS

In the years since the 1930s, the advancements in boiler technology have been as dramatic as those in turbine technology. Turbine inlet pressures and temperatures continued to increase over the years, mainly due to significant improvements in high-temperature strength properties of various metal alloys. The simultaneous use of superheat, reheat, and regeneration, along with improved turbine isentropic and mechanical efficiencies at higher turbine inlet temperatures and pressures, allowed actual operating power station thermal efficiencies to reach percentages in the low 40s by the 1980s (see Figure 13.9). In the 1930s, the turbine-generator unit output reached 200 MW, and by the 2000s, it had surpassed 2000 MW. Figure 13.28 shows the 2800 MW combined cycle gas turbine (CCGT) power plant built in Chiba, Japan.

FIGURE 13.28
Chiba Japan 2800 MW CCGT power plant with turbines supplied by GE and Mitsubishi. The facility cost $1.5 billion and was completed in 2000.

HOW DANGEROUS ARE BOILER EXPLOSIONS?

Between 1898 and 1902, there were 1600 boiler explosions in the United States, in which 1184 people were killed. On March 10, 1905, a boiler explosion in a shoe factory in Brockton, Massachusetts, killed 58 people and injured an additional 117 people (Figure 13.29); and on December 6, 1906, a similar explosion occurred in a shoe factory in Lynn, Massachusetts. As a result of the ensuing public outcry, the state of Massachusetts enacted the first legal code of rules for the construction of steam boilers, in 1907. From 1908 to 1910, Ohio and various other states enacted similar legislation, but because no two states had exactly the same code, boiler manufacturers had great difficulty satisfying all the varying and occasionally conflicting rules. In 1911, the American Society of Mechanical Engineers (ASME) joined with the boiler manufacturers to formulate a set of uniform standards for the design and construction of safe boilers. The first edition of the resulting ASME Boiler and Pressure Vessel Code was produced in 1914, and by the 1930s, many advancements in boiler technology had been made. An updated and modernized version of this code is used throughout industry today.

FIGURE 13.29
Brockton shoe factory explosion.

Modern power plant performance can be expressed in four ways:

1. The actual thermal efficiency $(\eta_T)_{\text{actual}}$, where

$$(\eta_T)_{\text{actual}}\left((\dot{W}_{\text{out}})_{\text{net}}/\dot{Q}_{\text{in}}\right)_{\text{actual}}$$

Typical ranges of this efficiency are shown in Figure 13.30 for various power-producing technologies.

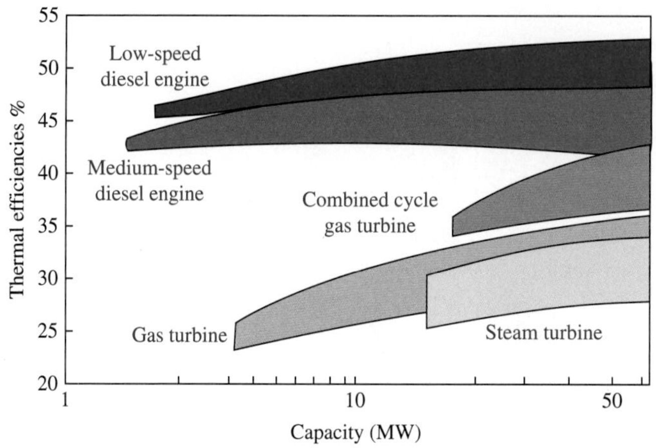

FIGURE 13.30

Thermal efficiency ranges of various power-producing technologies.

2. A similar performance measure, the *heat rate* of the power plant, which is defined to be the inverse of the thermal efficiency but in mixed units (e.g., Btu/(kW·h)). These two measures are related by

$$\text{Heat rate in Btu/(kW}\cdot\text{h)} = \frac{3412\ \text{Btu/(kW}\cdot\text{h)}}{(\eta_T)_{\text{actual}}}$$

where the decimal form (*not* percent) of $(\eta_T)_{\text{actual}}$ is used.
3. The actual steam flow rate divided by the actual plant electrical output power, $\dot{m}_{\text{steam}}/\dot{W}_{\text{elect}}$, in mixed units of (lbm steam)/(kW·h).
4. The ratio of the actual thermal efficiency of the power plant to the isentropic Rankine cycle thermal efficiency, $(\eta_T)_{\text{actual}}/(\eta_T)_{\text{isentropic}}$. This ratio is often expressed as a percentage and is commonly called by the misleading term *engine efficiency*. More accurately, it is an overall heat engine *thermal efficiency ratio.*

By the 1930s, it had been realized that water was not necessarily the best working fluid for a vapor cycle heat engine. Since the deviations between the Carnot cycle and the isentropic Rankine cycle are due to the characteristics of the working fluid, clearly, the *ideal* working fluid for a heat engine should make the Rankine cycle as close to the Carnot cycle as possible. More specifically, the ideal working fluid should have the following characteristics (see Figure 13.31):

1. It should have a critical temperature well above the metallurgical limit of the boiler and turbine, so that efficient isothermal high-temperature heat transfer can occur in the boiler.
2. It should have a relatively low saturation vapor pressure at high temperatures, so that high mechanical stresses are not produced in the boiler or turbine.
3. It should have an ambient temperature saturation pressure slightly above atmospheric pressure, so that the condenser does not have to be operated at a vacuum.
4. It should have large phase change enthalpies (h_{fg}) and low liquid specific heats, so that the heat required to bring the liquid condensate up to the vaporization temperature is a small percentage of the vaporization heat (this reduces boiler heat transfer irreversibilities and ensures that regeneration devices are effective).
5. It should have the slope of its saturated vapor and liquid lines as nearly vertical as possible on a *T–s* diagram.
6. It should have a triple-point temperature well below the ambient temperature to prevent the formation of solids (i.e., freezing) within the system.
7. It should be chemically stable (i.e., not dissociate at high temperatures), nontoxic, noncorrosive, and inexpensive.

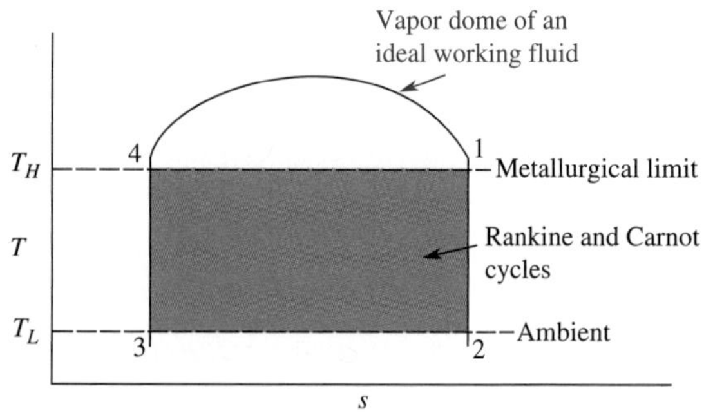

FIGURE 13.31

The Rankine cycle with the ideal working fluid becomes a Carnot cycle.

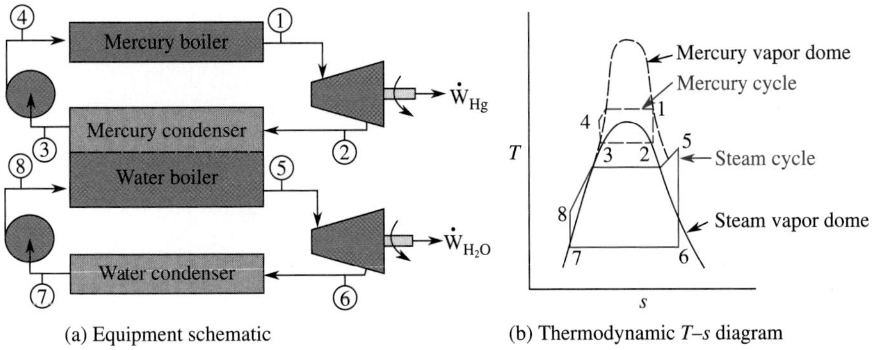

FIGURE 13.32
A mercury-water binary power plant.

No known fluid meets all seven of these conditions quite as well as water. But other fluids meet some of these conditions significantly better than water. For example, the critical state of mercury is at 1649°F and 2646 psia, which meets item 1 much better than water, whose critical state is 705°F and 3204 psia. However, the saturation pressure of mercury at 100°F is a very high vacuum (thus violating item 3). At 1000°F, the saturation pressure of mercury is about 180 psia, which makes it attractive for use in a dual working fluid or *binary cycle* system, as shown in Figure 13.32. Here, the mercury condenser also serves as the steam generator, and the combined binary cycle thermal efficiency is much higher (in the range of 50–60% for isentropic systems) than either one operating alone (in the range of 30–40% for isentropic systems).

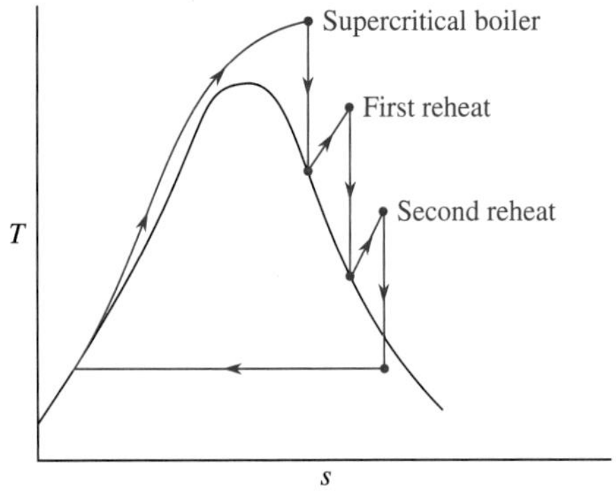

FIGURE 13.33
A supercritical Rankine cycle with two stages of reheat.

Between the 1930s and 1960s, several mercury-water binary cycle power plants were put into commercial operation. But, despite their superior thermal efficiencies, the problems of high initial cost, mercury toxicity, and numerous operating and maintenance problems prevented such plants from being commercially successful. However, the use of two or more working fluids within the same power plant still holds promise for significantly improving overall thermal performance in the future.

Supercritical power plants were developed after 1950 with boiler pressures as high as 5000 psia at 1200°F. Figure 13.33 illustrates a supercritical Rankine cycle with two stages of reheat. However, the high operating and maintenance costs of supercritical plants often offset the cost benefits due to their increased thermal efficiencies.

EXAMPLE 13.8

The Philadelphia Electric Power Company Eddystone Power Plant has the highest operating conditions of any electrical generating facility in the world. The boiler has a supercritical outlet state of 5000. psia at 1200.°F. After expansion in the first and second stages of the turbine, the steam is reheated to 1000.°F at 1000. psia and to 1000.°F at 300. psia, respectively. The condenser pressure is 0.400 psia, and the power plant has eight regenerators. The steam mass flow rate is 1.50×10^6 lbm/h and the power plant produces 325 MW of electrical power. Neglecting the eight regenerators in this plant, determine

a. The isentropic thermal efficiency of this power plant.
b. The isentropic efficiency of the turbine-generator power unit.

Solution
First, draw a sketch of the system (Figure 13.34).

(*Continued*)

EXAMPLE 13.8 *(Continued)*

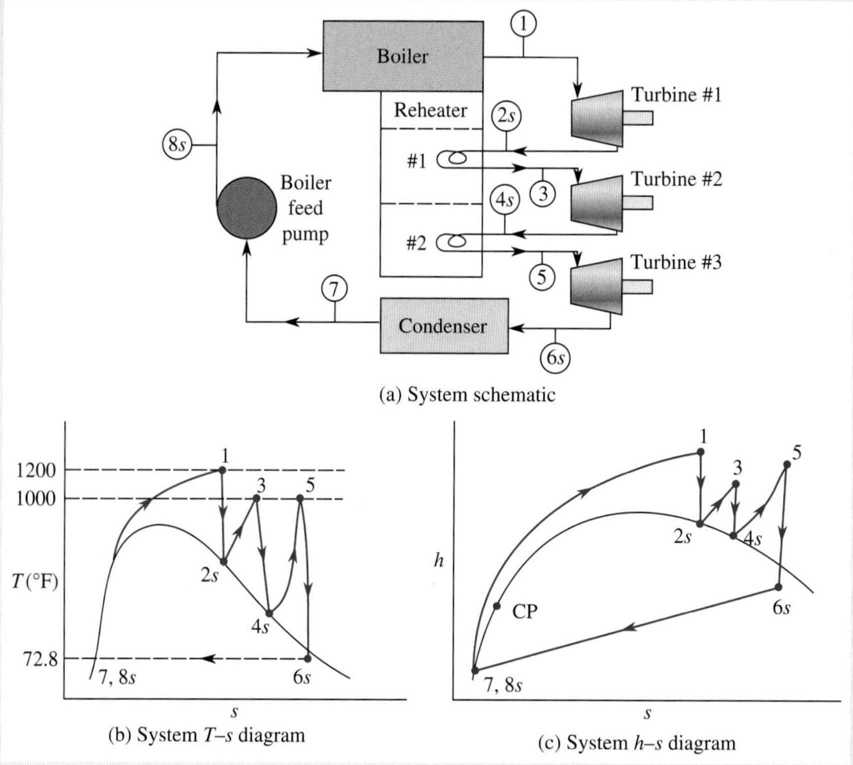

FIGURE 13.34
Example 13.8.

The thermodynamic data at the various points around this cycle are as follows:

Station 1—Turbine 1 inlet	Station 2s—Turbine 1 exit
$p_1 = 5000.$ psia	$p_{2s} = 1000.$ psia
$T_1 = 1200.$°F	$s_{2s} = s_1 = 1.5068$ Btu/lbm·R
$h_1 = 1530.8$ Btu/lbm	$h_{2s} = 1316.9$ Btu/lbm
$s_1 = 1.5068$ Btu/lbm·R	(by interpolation in Table C.3a)
Station 3—Turbine 2 inlet	**Station 4s—Turbine 2 exit**
$p_3 = 1000.$ psia	$p_{4s} = 300.$ psia
$T_3 = 1000.$°F	$s_{4s} = s_3 = 1.6532$ Btu/lbm·R
$h_3 = 1505.9$ Btu/lbm	$h_{4s} = 1343.8$ Btu/lbm
$s_3 = 1.6532$ Btu/lbm·R	(by interpolation in Table C.3a)
Station 5—Turbine 3 inlet	**Station 6s—Turbine 3 exit**
$p_5 = 300.$ psia	$p_{6s} = 0.400$ psia
$T_5 = 1000.$°F	$s_{6s} = s_5 = 1.7966$ Btu/lbm·R
$h_5 = 1526.4$ Btu/lbm	$x_{6s} = (1.7966 - 0.0799)/1.9762 = 0.8687$
$s_5 = 1.7966$ Btu/lbm·R	$h_{6s} = 40.9 + 0.8687(1052.4)$
	$= 955.1$ Btu/lbm
Station 7—Condenser exit	**Station 8s—Boiler inlet**
$p_7 = 0.400$ psia	$p_{8s} = p_1 = 5000.$ psia
$x_7 = 0.00$	$s_{8s} = s_7$
$h_7 = h_f(0.4 \text{ psia}) = 40.9$ Btu/lbm	$h_{8s} = h_7 + v_7(p_{8s} - p_7) = 40.9 + 0.01606(5000. - 0.400)(144/778.16)$
$v_7 = v_f(0.4 \text{ psia}) = 0.01606$ ft^3/lbm	$= 55.76$ Btu/lbm

a. The isentropic thermal efficiency of this Rankine cycle power plant is given by Eq. (13.15) with $(\eta_s)_{pm1} = (\eta_s)_{pm2} = (\eta_s)_p = 1.0$ as

$$(\eta_T)_s = \frac{(h_1 - h_{2s}) + (h_3 - h_{4s}) + (h_5 - h_{6s}) - v_7(p_{8s} - p_7)}{(h_1 - h_{8s}) + (h_3 - h_{2s}) + (h_5 - h_{4s})}$$

where the numerator in this equation is

$$\text{Numerator} = (1530.8 - 1316.9) + (1505.9 - 1343.8) + (1526.4 - 955.1) - 0.01606(5000. - 0.400)\left(\frac{144}{778.16}\right) = 932.4 \text{ Btu/lbm}$$

and the denominator is

$$\text{Denominator} = (1530.8 - 55.76) + (1505.9 - 1316.9) + (1526.4 - 1343.8) = 1847 \text{ Btu/lbm}$$

then, the isentropic thermal efficiency is

$$(\eta_T)_s = \frac{932.4 \text{ Btu/lbm}}{1847 \text{ Btu/lbm}} = 0.505 = 50.5\%$$

b. The isentropic efficiency of the complete power generating unit is

$$(\eta_s)_{\text{turbine-generator}} = \frac{(\dot{W}_{\text{out}})_{\text{net actual}}}{(\dot{W}_{\text{out}})_{\text{net isentropic}}}$$

where the actual net power output is given as

$$(\dot{W}_{\text{out}})_{\text{net actual}} = 325 \text{ MW} = (325{,}000 \text{ kW})(3412 \text{ Btu/kW}\cdot\text{h}) = 1.11 \times 10^9 \text{ Btu/h}$$

and the isentropic net power output can be calculated from

$$\begin{aligned}(\dot{W}_{\text{out}})_{\text{net isentropic}} &= \dot{W}_{\text{turbine 1}} + \dot{W}_{\text{turbine 2}} + \dot{W}_{\text{turbine 3}} - |\dot{W}|_{\text{pump}} \\ &= \dot{m}\{(h_1 - h_{2s}) + (h_3 - h_{4s}) + (h_5 - h_{6s}) - v_7(p_7 - p_{8s})\} \\ &= (1.50 \times 10^6 \text{ lbm/h})\{(1530.8 - 1316.9) + (1505.9 - 1343.8) \\ &\quad + (1526.4 - 955.1) - 0.01606(5000. - 0.400)(144/778.16)\} \text{ Btu/lbm} \\ &= 1.40 \times 10^9 \text{ Btu/h}\end{aligned}$$

Then,

$$(\eta_s)_{\text{turbine-generator}} = \frac{1.11 \times 10^9 \text{ Btu/h}}{1.40 \times 10^9 \text{ Btu/h}} = 0.793 = 79.3\%$$

The isentropic efficiencies calculated in parts a and b are somewhat low because we choose to omit the eight regenerator units in this analysis.

Exercises

22. Determine the isentropic thermal efficiency of the Eddystone Power Plant discussed in Example 13.8 if the boiler outlet state temperature is increased from 1200.°F to 1500.°F. Assume all the other variables remain unchanged. **Answer:** $(\eta_T)_s = 52.1\%$.

23. If the second reheat pressure in Example 13.8 is increased from 300. psia to 600. psia, what would be the new isentropic thermal efficiency of the power plant? Assume all the other variables remain unchanged. **Answer:** $(\eta_T)_s = 50.2\%$.

24. If the operating conditions of the power plant in Example 13.8 remain unchanged, but the generator power output drops from 325 MW to 315 MW, determine the new isentropic efficiency of the power generating unit. **Answer:** $(\eta_s)_{\text{turbine-generator}} = 76.8\%$.

The introduction of nuclear power in the 1960s added a new facet to heat source technology. Nuclear safety restrictions require the use of a double-loop heat transfer system to keep the radioactive reactor cooling fluid from entering the turbine, and this effectively limits the maximum nuclear power plant secondary loop temperature and pressure to around 1000°F and 800 psia. This has the effect of limiting nuclear plant thermal efficiencies to the low 30%s, while the thermal efficiencies of fossil fueled plants reached the low 50%s.

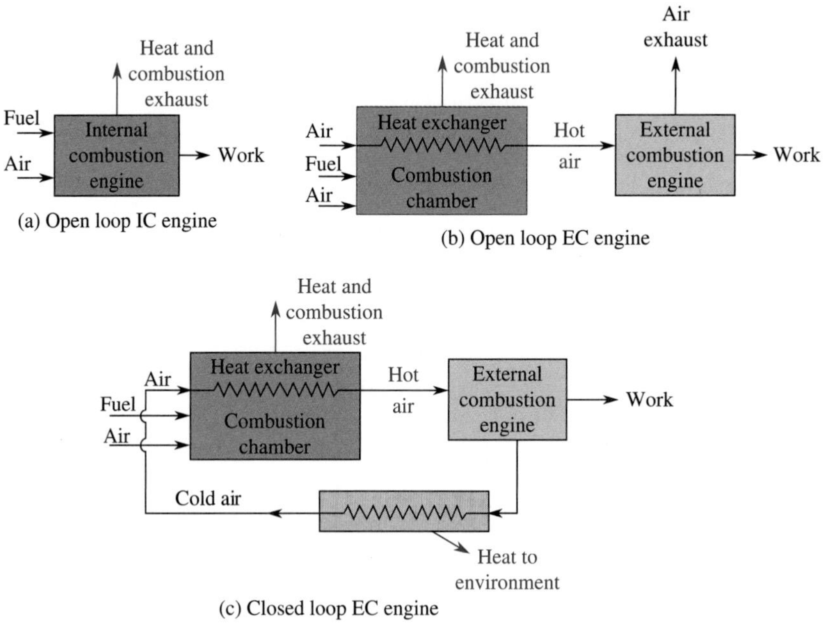

FIGURE 13.35
Internal and external combustion engine thermodynamic loop classifications.

13.11 PART II. GAS POWER CYCLES

Heat engines whose working fluid is a gas rather than a vapor undergo gas power thermodynamic cycles. Like steam power, gas prime movers fall into two broad mechanical design categories: reciprocating and turbine. In addition, unlike steam power prime movers, they fall into two heat source categories: external combustion (EC) and internal combustion (IC). In external combustion engines, the working fluid does not enter into the combustion process. Combustion, if it occurs at all, occurs outside the engine, with the resulting heat being transferred into the working fluid at some point. All steam engines, therefore, are external combustion prime movers. Since the working fluid of an internal combustion engine always enters into the heat-generating combustion process, thus depleting the fuel and oxygen supply of the working fluid, the combustion products must be removed and fresh fuel and oxygen added during each thermodynamic cycle. Consequently, all internal combustion engines operate on an open loop process, whereas external combustion engines can operate on either an open loop or closed loop process (see Figure 13.35).

Gas power cycle prime movers (engines) developed slightly later than their steam engine counterparts. They evolved largely as an alternative to steam power technology, and by 1900, they were already very competitive with small to medium power steam engines. By the mid 20th century, they had replaced all steam power within the transportation industry and the small to medium electrical generating industry, leaving only large electrical power plants as the major commercial users of steam.

13.12 AIR STANDARD POWER CYCLES

Most modern gas power cycles involve the use of open loop internal or external combustion engines. The working fluid has highly variable physical and chemical properties throughout these engines, and this makes their thermodynamic cycle very difficult to analyze. Since the most abundant chemical constituent of the working fluid of air-breathing engines is nitrogen, which is largely chemically inert within the engine, it is possible to devise an effective *closed loop* engine model in which air alone is considered to be the working fluid. Such an approximation to real engine thermodynamics is called the *air standard cycle*, ASC for short. The ASC allows a simple but highly idealized closed loop thermodynamic analysis to be carried out on an otherwise very complex open loop system. The assumptions embodied in an ASC analysis of an IC or EC engine are as follows:

1. The engine operates on a closed loop thermodynamic cycle and the working fluid is a fixed mass of atmospheric air.
2. This air behaves as an ideal gas throughout the cycle.

3. The combustion process within the engine is replaced by a simple heat addition process from an external heat source.
4. The intake and exhaust processes of the engine are replaced by an external heat rejection process to the environment.
5. All processes within the thermodynamic cycle are assumed to be reversible.

Note that item 5 implies that all processes within a cycle that have no associated heat transfer are also isentropic processes (i.e., they are reversible *and* adiabatic). Since the numerical results of an analysis using the ASC depends on how the ideal gas specific heat issue is handled, an ASC analysis is further characterized as either a *cold air standard cycle*, if the specific heats of air are assumed to be constant and evaluated at room temperature, or a *hot air standard cycle*, if the specific heats of air are assumed to be temperature dependent. When a more complex analysis is done, in which the actual fuel-air mixture and exhaust gases are used, it is usually called a *real mixture standard cycle*.

An IC or EC engine operating on an ASC can be represented schematically as shown in Figure 13.36. Since the actual operating thermal efficiency of an IC engine is often compared to its ideal ASC thermal efficiency to evaluate the impact of real world irreversibilities on the engine's performance, the ASC analysis serves the same type of idealized benchmark comparison function as the isentropic Rankine cycle analysis of vapor power cycles.

If the working fluid in the Carnot cycle shown in Figure 13.37 is air (functioning as an ideal gas), then we would have a Carnot ASC. Since (by definition) a Carnot cycle is thermodynamically reversible, using the notation of Figure 13.37, $s_1 = s_2$ and $s_3 = s_4$.

Isentropic ideal gas compression and expansion processes with constant specific heats are discussed in Chapter 7, and the *p-v-T* relation for these processes is given by Eq. (7.38), which shows that it can be written directly in terms of either the temperature ratio, pressure ratio, or compression (volume) ratio.

For an isentropic expansion process, $s_1 = s_2 = s_{2s}$,

$$\frac{T_{2s}}{T_1} = \left(\frac{p_{2s}}{p_1}\right)^{(k-1)/k} = \left(\frac{v_{2s}}{v_1}\right)^{1-k}$$

For an isentropic compression process, $s_3 = s_4 = s_{4s}$,

$$\frac{T_3}{T_{4s}} = \left(\frac{p_3}{p_{4s}}\right)^{(k-1)/k} = \left(\frac{v_3}{v_{4s}}\right)^{1-k}$$

where

$$T_1 = T_{4s} = T_H$$

and

$$T_{2s} = T_3 = T_L$$

We further define

$$v_{2s}/v_1 = v_3/v_{4s} = \text{isentropic compression ratio, CR}$$

and

$$p_1/p_{2s} = p_{4s}/p_3 = \text{isentropic pressure ratio, PR}$$

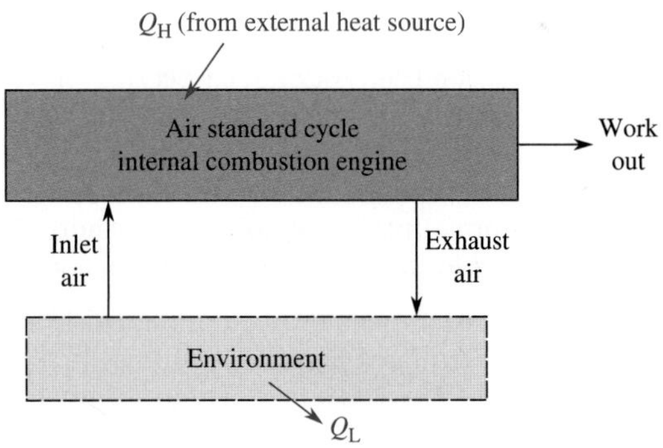

FIGURE 13.36
IC or EC engine operating on a closed loop air standard cycle.

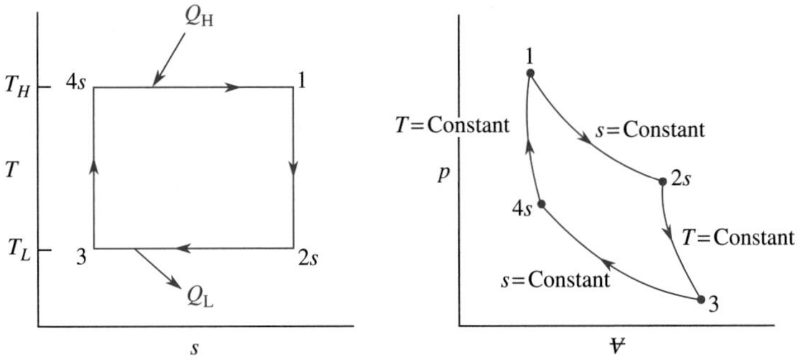

FIGURE 13.37
The Carnot air standard cycle.

CRITICAL THINKING

When we say that an engine or compressor is *isentropic*, we mean that the entropy at the exit is the same as the entropy at the inlet. But this condition alone is not enough to fix the exit state, since we must specify two independent thermodynamic properties to fix a state. We could choose the actual exit pressure or the actual exit temperature as the second independent property at the exit. However, in practice, we *always* choose the actual exit pressure as the second independent property. Why do we do this?

The answer is simply that the exit pressure is more often known or easily specified than the exit temperature. For example, when an engine exhausts into the atmosphere, the exhaust pressure is always atmospheric pressure, independent of the inlet pressure and the processes that occur inside the engine. But the exhaust temperature always depends on the inlet temperature and the complex processes that occur between the inlet and the exit inside the engine.

Therefore, *the isentropic pressure is always taken to be equal to the actual pressure at the exit of an isentropic process*, and the *isentropic pressure ratio* is always the same as the *actual pressure ratio* for these systems.

The equation for the thermal efficiency of the Carnot cold ASC can now be written as

$$(\eta_T)_{\substack{\text{Carnot}\\ \text{cold ASC}}} = 1 - T_L/T_H = 1 - \text{PR}^{(1-k)/k} = 1 - \text{CR}^{1-k} \tag{13.18}$$

Most vapor power cycles fall at least partially under the vapor dome and can therefore be modeled with a single practical thermodynamic cycle, the Rankine cycle. Unfortunately, outside the vapor dome, no one thermodynamic cycle models all possible practical gas power cycles. In the next sections, we discuss a few commercially valuable gas power cycles and evaluate their ASC thermal efficiencies. While these cycles do not cover all possible cycles, they are the ones that have had significant economic success over the years. We discuss them in the chronological order in which they were developed.

13.13 STIRLING CYCLE

Many early steam boilers exploded because of weak materials, faulty design, and poor construction. The resulting loss in human life and property inspired many people to attempt to develop engines that did not need a high-pressure boiler. In 1816, the Scottish clergyman Robert Stirling (1790–1878) patented a remarkable closed loop external combustion engine in which a fixed mass of air passed through a thermodynamic cycle composed of two isothermal processes and two isochoric (constant volume) processes. Figure 13.38 shows the *T–s* and $p-\mathcal{V}$ diagrams for this cycle, along with an equipment schematic.

Stirling's engine was remarkable, not only in its mechanical and thermodynamic complexity, but also because he originated a thermal regeneration process in which the heat released during the isochoric expansion process from state 1 to state 2 is stored within the system (in the regenerator) and reintroduced into the working fluid (air) during the isochoric compression process from state 3 to state 4. This was the first use of thermal regeneration in a power cycle and predates its use in steam engines by many years.[8] The complexity of construction and high cost limited production of Stirling's engine to small units (0.5 to 10 hp). Generally known as *hot air engines*, they found extensive use on small farms between 1820 and 1920 for pumping water and other light duties.

The thermal efficiency of the Stirling cycle is given by

$$(\eta_T)_{\text{Stirling}} = \frac{(\dot{W}_{\text{out}})_{\text{net}}}{\dot{Q}_{\text{H}}} = \frac{\dot{Q}_{\text{H}} - |\dot{Q}_{\text{L}}|}{\dot{Q}_{\text{H}}} = 1 - \frac{|\dot{Q}_{\text{L}}|}{\dot{Q}_{\text{H}}}$$

where (see Figure 13.38a) for a reversible ASC engine, we can write

$$\dot{Q}_{\text{H}} = \dot{m}\, T_H(s_1 - s_4)$$

and

$$|\dot{Q}_{\text{L}}| = \dot{m}\, T_L(s_2 - s_3)$$

[8] Note that this is a completely different type of "regeneration" than that used with the Rankine vapor cycle.

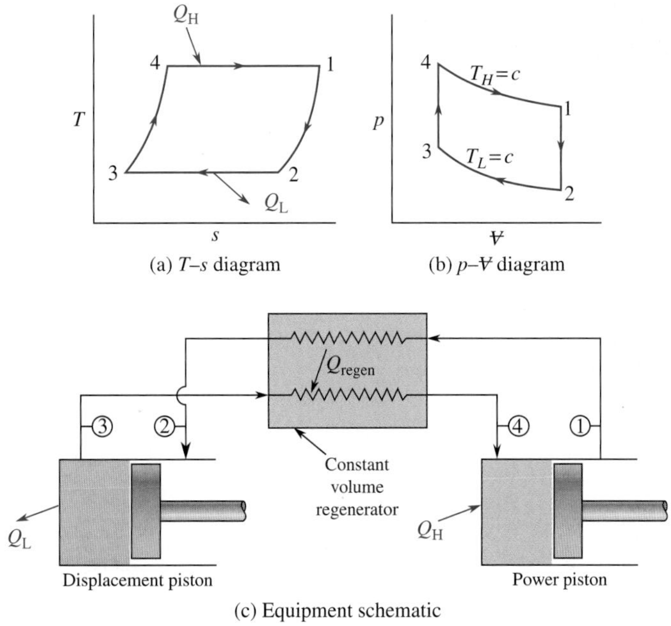

FIGURE 13.38
The Stirling cycle.

These entropy changes can be evaluated for the cold ASC using Eq. (7.36) of Chapter 7 as

$$s_1 - s_4 = c_v \ln(T_1/T_4) + R\ln(v_1/v_4) = R\ln(v_1/v_4)$$

and

$$s_2 - s_3 = c_v \ln(T_2/T_3) + R\ln(v_2/v_3) = R\ln(v_2/v_3)$$

Figures 13.38a and 13.38b show that $T_2 = T_3 = T_L$, $T_1 = T_4 = T_H$, and $v_2/v_3 = v_1/v_4$, so that we have $s_1 - s_4 = s_2 - s_3$ and the preceding equations give

$$(\eta_T)_{\substack{\text{Stirling}\\ \text{cold ASC}}} = 1 - T_L/T_H = 1 - T_2/T_1 = 1 - T_3/T_4 \tag{13.19}$$

which is the same as the thermal efficiency of the Carnot cold ASC operating between the same two temperature limits.

EXAMPLE 13.9

A new Stirling cycle engine using air as the working fluid is being designed to power the small portable electric generator shown Figure 13.39. The piston displacement is to be 0.0110 m^3 and the minimum volume in the cylinder is to be $\forall_3 = \forall_4 = 1.00 \times 10^{-3}$ m^3. The design calls for a maximum power piston pressure of $p_1 = 0.300$ MPa, a minimum displaced piston pressure of $p_2 = 0.100$ MPa, and a displacer inlet temperature of $T_2 = 30.0$°C. Use the Stirling cold ASC analysis shown in Figure 13.38 to complete the design, determining

a. The displacer piston maximum pressure (p_3).
b. The power piston maximum pressure (p_4).
c. The mass of air in the engine (m).
d. The heat addition temperature ($T_1 = T_4$).
e. The Stirling cold ASC thermal efficiency of the engine.

FIGURE 13.39
Example 13.9.

(Continued)

EXAMPLE 13.9 *(Continued)*

Solution

Using the Stirling cycle diagram shown in Figure 13.38, we can carry out the following analysis.

a. Since $T_2 = T_3$, the ideal gas equation of state gives $p_3 = p_2(\forall_2/\forall_3)$, where $\forall_2 = \forall_1$ = Piston displacement $- \forall_3 = 0.011 - 0.001 = 0.010\ \text{m}^3$. Then, $p_3 = (0.100\ \text{MPa})(0.0100/0.00100) = 1.00\ \text{MPa}$.

b. Since $T_1 = T_4$, the ideal gas equation of state gives $p_4 = p_1(\forall_1/\forall_4)$, where $\forall_4 = \forall_3 = 0.001\ \text{m}^3$. Then, $p_4 = (0.300\ \text{MPa}) \times (0.0100/0.00100) = 3.00\ \text{MPa}$.

c. Again using the ideal gas equation of state, we find that $m = p_2\forall_2/RT_2$ and

$$m = \frac{(0.100\,\text{MPa})(1000\,\text{kPa/MPa})(0.0100\,\text{m}^3)}{(0.286\,\text{kJ/kgK})(30.0 + 273.15\,\text{K})} = 0.0115\,\text{kg}$$

d. The ideal gas equation of state gives

$$T_1 = \frac{p_1\forall_1}{mR} = \frac{(0.300\,\text{MPa})(1000\,\text{kPa/MPa})(0.0100\,\text{m}^3)}{(0.0115\,\text{kg})(0.286\,\text{kJ/kg}\cdot\text{K})} = 912\,\text{K}$$

e. Equation (13.19) gives the Stirling cold ASC thermal efficiency of this engine as

$$(\eta_T)_{\substack{\text{Stirling}\\ \text{cold ASC}}} = 1 - \frac{T_L}{T_H} = 1 - \frac{T_2}{T_1} = 1 - \frac{30.0 + 273.15\,\text{K}}{912\,\text{K}} = 0.668 = 66.8\%$$

Exercises

25. Determine what the maximum displaced piston pressure would be in Example 13.9 if the piston displacement is increased from $0.0110\ \text{m}^3$ to $0.0200\ \text{m}^3$. Assume all the other variables remain unchanged. **Answer:** $p_3 = 1.90$ MPa.

26. If the maximum power piston design pressure is increased from 0.300 MPa to 0.800 MPa in Example 13.9, determine the corresponding design heat addition temperature. Assume all the other variables remain unchanged. **Answer:** $T_1 = T_4 = 2430$ K.

27. Determine the Stirling cold ASC thermal efficiency in Example 13.9 if the piston displacement is reduced from $0.0110\ \text{m}^3$ to $8.00 \times 10^{-3}\ \text{m}^3$. Assume all the other variables remain unchanged. **Answer:** $(\eta_T)_{\text{Stirling cold ASC}} = 66.7\%$ (no change).

Though the Stirling cycle engine did not compete well with alternate gas power cycle engines after about 1880, its potential for high thermal efficiency (and consequently low fuel consumption) plus the low noise and low air pollution traits of an external combustion process caused renewed interest in the late 20th century in its applicability for automotive use.

13.14 ERICSSON CYCLE

In 1833, the Swedish-born engineer John Ericsson (1803–1889) developed a different type of hot air, reciprocating, external combustion engine, which could operate on either an open or closed loop cycle. Ericsson's engine also used a thermal regenerator, but it differed from Stirling's in that the constant volume regeneration process was replaced by a constant pressure regeneration process. Therefore, the Ericsson cycle consists of two isothermal processes and two isobaric (constant pressure) processes, as shown in Figure 13.40.

In 1839, Ericsson moved to America and continued to develop his engine. His large engines (up to 300 hp, with pistons 14 ft in diameter) were very inefficient and could not compete economically with existing steam engine technology. However, his small engines were reasonably successful and several thousand were sold by 1860. By 1880, the popularity of his engine had dropped off, and it was considered to be obsolete technology until modern gas turbine power plants came into being in the mid 20th century. The Ericsson cycle is approximated by an open loop gas turbine that has multistage compressor intercooling (to approximate T_L = constant) and multistage turbine reheating (to approximate T_H = constant) along with thermal regeneration, as shown in Figure 13.41.

The Ericsson cycle thermal efficiency is given by

$$(\eta_T)_{\text{Ericsson}} = \frac{(\dot{W}_{\text{out}})_{\text{net}}}{\dot{Q}_{\text{H}}} = \frac{\dot{Q}_{\text{H}} - |\dot{Q}_{\text{L}}|}{\dot{Q}_{\text{H}}} = 1 - \frac{|\dot{Q}_{\text{L}}|}{\dot{Q}_{\text{H}}}$$

where (see Figure 13.40a), for a reversible ASC engine, we can write

$$\dot{Q}_{\text{H}} = \dot{m}\,T_H(s_1 - s_4)$$

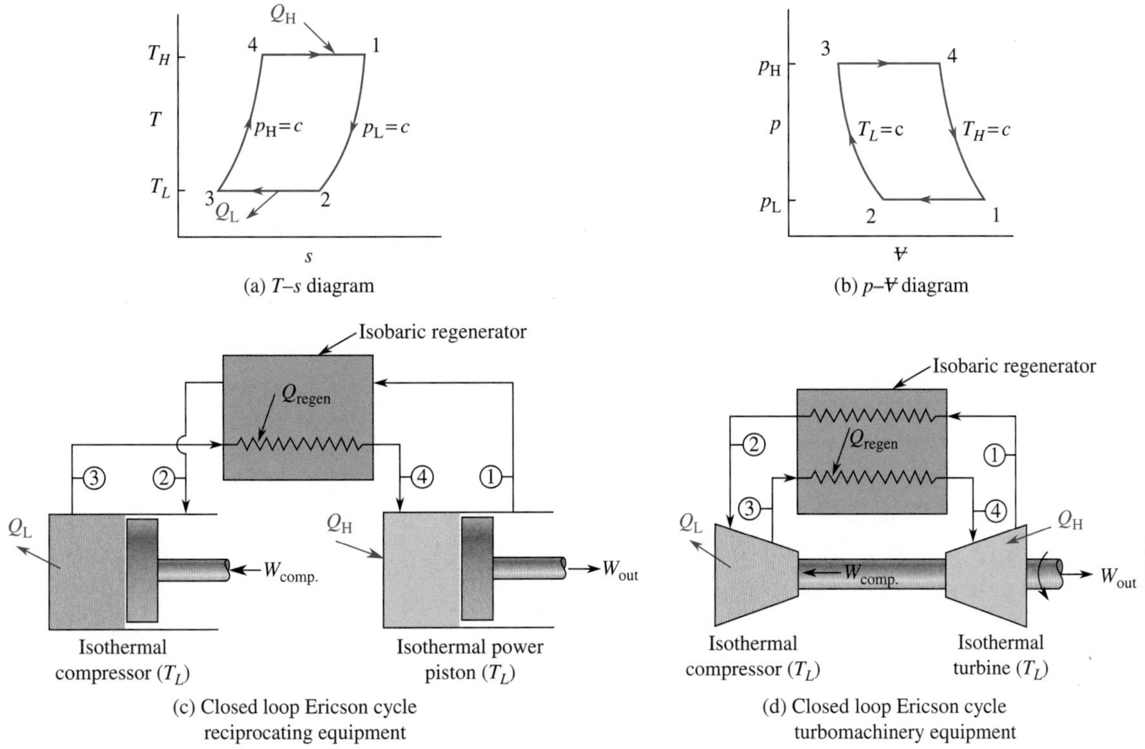

FIGURE 13.40
The Ericsson cycle.

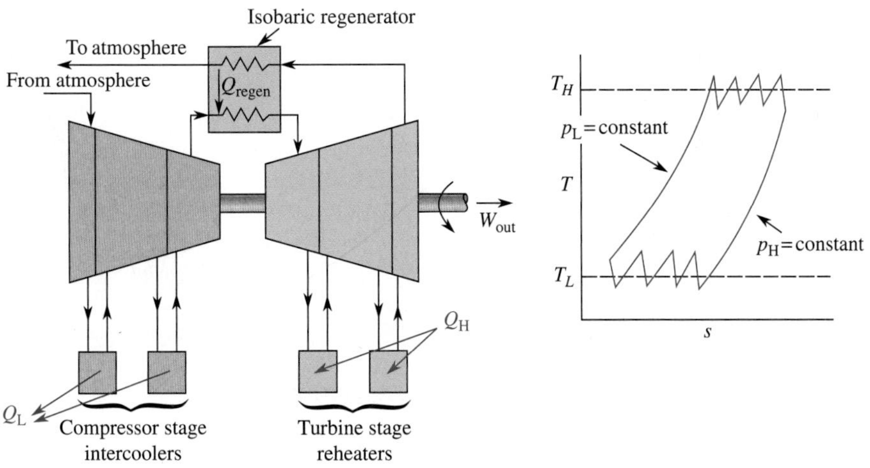

FIGURE 13.41
An open loop gas turbine power plant as an approximation to the Ericsson cycle.

and

$$|\dot{Q}_L| = \dot{m}\, T_L(s_2 - s_3)$$

Equation (7.37) of Chapter 7 gives these entropy changes for the cold ASC as

$$s_1 - s_4 = c_p \ln(T_1/T_4) + R\ln(p_1/p_4) = R\ln(p_1/p_4)$$

and

$$s_2 - s_3 = c_p \ln(T_2/T_3) + R\ln(p_2/p_3) = R\ln(p_2/p_3)$$

Figure 13.40b shows that $T_2 = T_3 = T_L$, $T_1 = T_4 = T_H$, and $p_1/p_4 = p_2/p_3$, so that the previous equations yields $s_1 - s_4 = s_2 - s_3$ and the thermal efficiency becomes

$$(\eta_T)_{\substack{\text{Ericsson} \\ \text{cold ASC}}} = 1 - T_L/T_H = 1 - T_2/T_1 = 1 - T_3/T_4 \tag{13.20}$$

Thus, like the Stirling cold ASC, the Ericsson cold ASC has the same thermal efficiency as the Carnot cold ASC operating between the same two temperature limits. The Stirling and Ericsson cycle engines of the 19th century were large and costly, and their actual operating thermal efficiencies were quite poor. They were ultimately replaced by a newer, more efficient engine technology introduced in the second half of the 19th century, the internal combustion Otto and Diesel cycles, discussed later in this chapter. However, the high thermal efficiency potential of the Stirling and Ericsson cycles produces periodically renewed interest in utilizing these cycles within the framework of modern technology.

EXAMPLE 13.10

A new Ericsson cycle engine is being designed with a pressure ratio of PR = $p_4/p_1 = 2.85$, a power piston outlet pressure of $p_1 = 0.500$ MPa, a maximum volume of $\forall_1 = 0.0110$ m^3, and a minimum volume of $\forall_3 = 3.00 \times 10^{-3}$ m^3. The engine will contain 0.0500 kg of air. Use the Ericsson cold ASC analysis shown in Figure 13.40 to complete this design, determining

a. The compressor inlet pressure and volume (p_2 and $\forall_2$).
b. The power piston outlet pressure and inlet volume (p_4 and $\forall_4$).
c. The compressor outlet pressure (p_3).
d. The temperatures at the inlet and outlet of the power and displacer pistons (T_1, T_2, T_3, and T_4).
e. The Ericsson cold ASC thermal efficiency of this engine.

Solution

Using the Ericsson cycle diagram shown in Figure 13.40 we can carry out the following analysis.

a. For this cycle, the compressor inlet pressure is the same as the power piston outlet pressure (see Figure 13.40), or $p_2 = p_1 = 0.500$ MPa. The compressor inlet volume is $\forall_2 = \forall_3 \times (\text{CR})$, where the isentropic compression ratio CR = $\forall_1/\forall_4$ and $\forall_4 = mRT_4/p_4$. Now, from Figure 13.40,

$$T_4 = T_1 = \frac{p_1\forall_1}{mR} = \frac{(0.500\,\text{MPa})(1000\,\text{kPa/MPa})(0.0110\,\text{m}^3)}{(0.0500\,\text{kg})(0.286\,\text{kJ/kg}\cdot\text{K})} = 385\,\text{K}$$

and

$$p_4 = p_3 = p_2(\text{PR}) = p_1(\text{PR}) = (0.500\,\text{MPa})(2.85) = 1.43\,\text{MPa}$$

so

$$\forall_4 = \frac{mRT_4}{p_4} = \frac{(0.0500\,\text{kg})(0.286\,\text{kJ/kg}\cdot\text{K})(385\,\text{K})}{(1.43\,\text{MPa})(1000\,\text{kPa/MPa})} = 0.00385\,\text{m}^3$$

and the isentropic compression ratio is

$$\text{CR} = \frac{\forall_1}{\forall_4} = \frac{0.0110\,\text{m}^3}{0.00385\,\text{m}^3} = 2.86$$

Then, the compressor inlet volume is $\forall_2 = \forall_3 \times (\text{CR}) = (3.00 \times 10^{-3})(2.86) = 0.00858$ m^3.

b. For this cycle, the power piston outlet pressure is the same as the compressor inlet pressure (see Figure 13.40): $p_4 = p_3 = 1.43$ MPa, and from part a, $\forall_4 = 0.00385$ m^3.

c. Since $p_4/p_1 = p_3/p_2$ = PR, the compressor outlet pressure is $p_3 = p_2(\text{PR})$, where the isentropic pressure ratio is given as PR = 2.86. Then, $p_3 = 0.500(2.86) = 1.43$ MPa.

d. Using the ideal gas equation of state, we have

$$T_3 = T_2 = \frac{p_2\forall_2}{mR} = \frac{(0.500\,\text{MPa})(1000\,\text{kPa/MPa})(0.00858\,\text{m}^3)}{(0.0500\,\text{kg})(0.286\,\text{kJ/kg}\cdot\text{K})} = 300.\,\text{K}$$

e. The Ericsson cold ASC thermal efficiency for this engine is given by Eq. (13.20) as

$$(\eta_T)_{\substack{\text{Ericsson} \\ \text{cold ASC}}} = 1 - \frac{T_L}{T_H} = 1 - \frac{T_2}{T_1} = 1 - \frac{300.\,\text{K}}{385\,\text{K}} = 0.221 = 22.1\%$$

Exercises

28. If the isentropic pressure ratio in Example 13.10 is increased from 2.85 to 3.10, determine the resulting compressor inlet and exit pressures. Assume all other variables remain unchanged. **Answer:** $p_2 = 0.500$ MPa and $p_3 = 1.55$ MPa.

29. If the minimum volume $\forall_3$ in the Ericsson cycle engine in Example 13.10 is decreased from 3.0×10^{-3} m^3 to 1.00×10^{-3} m^3, determine the new compressor inlet pressure and volume (p_2 and $\forall_2$). Assume all other variables remain unchanged. **Answer:** $p_2 = 0.500$ MPa and $\forall_2 = 0.00285$ m^3.

30. A modification of the Ericsson cycle engine proposed in Example 13.10 calls for increasing the power piston outlet pressure and the compressor piston inlet pressure from $p_1 = p_2 = 0.500$ MPa to $p_1 = p_2 = 1.00$ MPa. Determine the new Ericsson cold ASC thermal efficiency resulting from this design change. Assume all other variables remain unchanged. **Answer:** $(\eta_T)_{\text{Ericsson cold ASC}} = 22.3\%$ (no change).

13.15 LENOIR CYCLE

Both the Stirling and the Ericsson cycles are for external combustion with thermal regeneration. Initially, an appropriately sized furnace was used as their heat source. This made these engines rather large and awkward, and even though they could theoretically achieve high (reversible) thermal efficiencies, the mechanical and thermal irreversibilities of the early engines were very large; consequently, they had rather low actual operating thermal efficiencies. In 1860, the French engineer Jean Joseph Etienne Lenoir (1822–1900) made the first commercially successful internal combustion engine. He converted a reciprocating steam engine to admit a mixture of air and methane during the first half of the piston's outward (suction) stroke, at which point it was ignited with an electric spark and the resulting combustion pressure acted on the piston for the remainder of the outward (expansion) stroke. The following inward stroke of the piston was used to expel the exhaust gases, then the cycle began over again. This cycle is (ideally) composed of only three effective processes: constant volume (combustion), constant entropy (power), and constant pressure (exhaust), as shown in Figure 13.42.

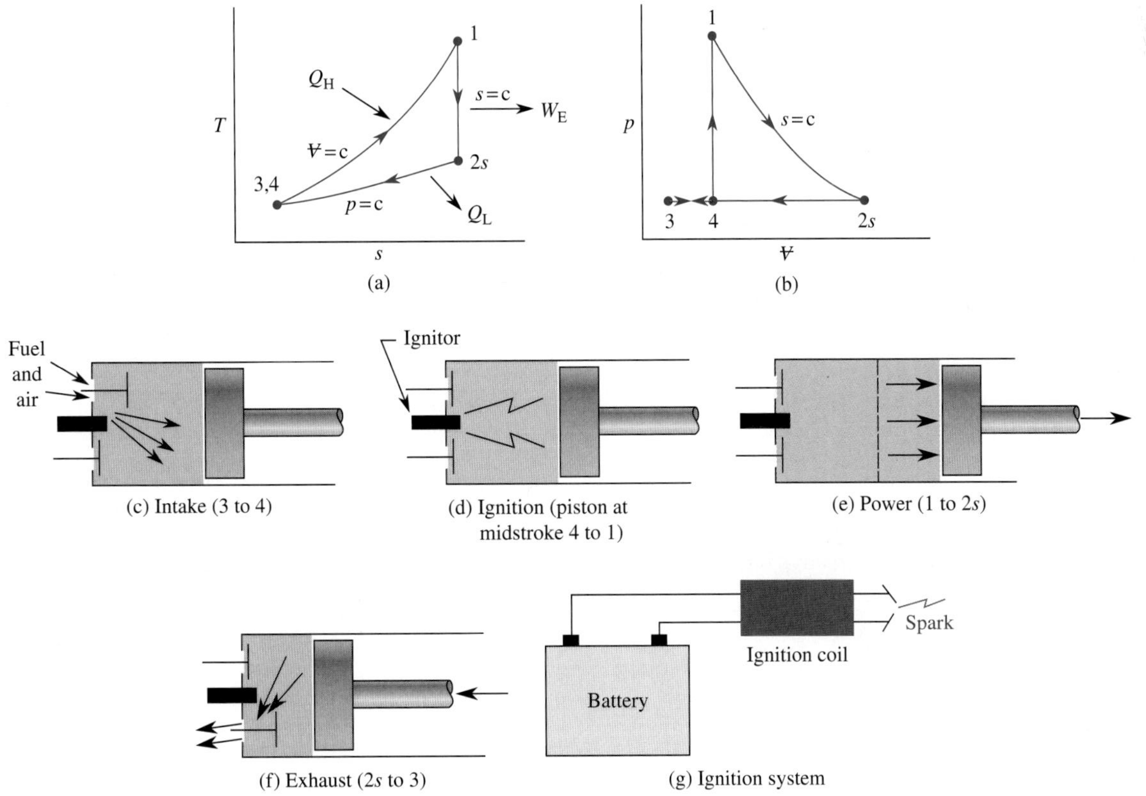

FIGURE 13.42
The Lenoir cycle.

The fuel-air mixture was ignited by an electric spark inside the cylinder. The spark was generated by a battery and an induction coil, which was basically the same technique used with all spark ignition engines through the middle of the 20th century.

The Lenoir engine ran smoothly, but because the air-fuel mixture was not compressed before ignition, the engine had a very low actual thermal efficiency (less than 4%).[9] Consequently, Lenoir engines became popular only in small sizes (0.5 to 3 hp) because their fuel consumption was very high.

The thermal efficiency of the Lenoir cycle is given by

$$(\eta_T)_{\text{Lenoir}} = \frac{(\dot{W}_{\text{out}})_{\text{net}}}{\dot{Q}_{\text{H}}} = \frac{\dot{Q}_{\text{H}} - |\dot{Q}_{\text{L}}|}{\dot{Q}_{\text{H}}} = 1 - \frac{|\dot{Q}_{\text{L}}|}{\dot{Q}_{\text{H}}}$$

where, from Figure 13.42a, for a cold ASC with an isentropic expansion from 1 to 2*s*, we have

$$|\dot{Q}_{\text{L}}| = |{}_{2s}\dot{Q}_3| = \dot{m}\,(u_{2s} - u_3) + \dot{m}\,p(v_{2s} - v_3)$$
$$= \dot{m}(h_{2s} - h_3) = \dot{m}\,c_p(T_{2s} - T_3)$$

and

$$\dot{Q}_{\text{H}} = {}_4\dot{Q}_1 = \dot{m}\,c_v(T_1 - T_4)$$

Because the intake air comes from an isothermal source (the atmosphere), $T_3 = T_4$. However, the exhaust gas is confined to a fixed mass, so the condition $p_{2s} = p_3$ requires that $T_{2s}/T_3 = v_{2s}/v_3$, where v_{2s}/v_3 is the isentropic compression ratio, CR. Then, the thermal efficiency becomes

$$\begin{aligned}(\eta_T)_{\substack{\text{Lenoir}\\ \text{cold ASC}}} &= 1 - \frac{c_p(T_{2s} - T_3)}{c_v(T_1 - T_4)} = 1 - kT_3\left(\frac{T_{2s}/T_3 - 1}{T_1 - T_4}\right)\\ &= 1 - kT_3\left(\frac{v_{2s}/v_3 - 1}{T_1 - T_4}\right) = 1 - kT_3\left(\frac{\text{CR} - 1}{T_1 - T_4}\right)\end{aligned} \quad (13.21)$$

EXAMPLE 13.11

The small model airplane jet engine shown in Figure 13.43 operates on the Lenoir cycle. It has a maximum temperature of T_1 = 800. R and an intake temperature of $T_3 = T_4$ = 530. R. The expansion, exhaust, and intake pressures are all $p_{2s} = p_3 = p_4$ = 14.7 psia, and the engine contains 1.00 × 10^{-3} lbm of air. For this engine, determine

a. The combustion pressure p_1.
b. The isentropic compression ratio CR = v_{2s}/v_3.
c. The Lenoir cold ASC thermal efficiency.

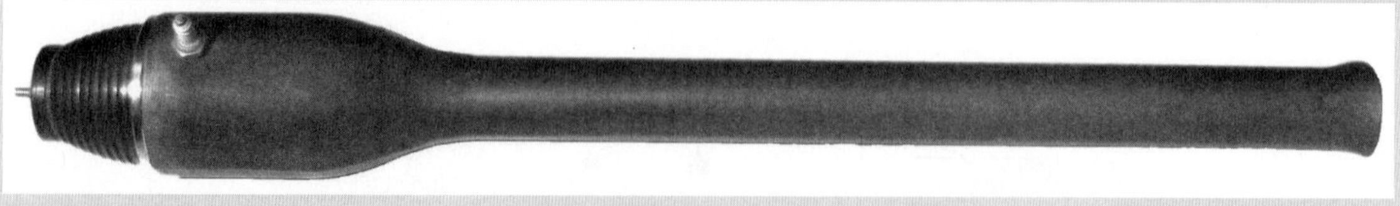

FIGURE 13.43
Example 13.11.

[9] These engines were often called *atmospheric* gas engines for the same reason.

Solution

Using the Lenoir cycle diagram shown in Figure 13.42, we can carry out the following analysis.

a. From the ideal gas equation of state, we can calculate $p_1 = mRT_1/\forall_1$. For this cycle, we have

$$\forall_1 = \forall_4 = \frac{mRT_4}{p_4} = \frac{(1.00 \times 10^{-3}\ \text{lbm})\left(53.34\,\frac{\text{ft}\cdot\text{lbf}}{\text{lbm}\cdot\text{R}}\right)(530\ \text{R})}{\left(14.7\,\frac{\text{lbf}}{\text{in}^2}\right)\left(144\,\frac{\text{in}^2}{\text{ft}^2}\right)} = 0.0134\ \text{ft}^3$$

Then,

$$p_1 = \frac{mRT_1}{\forall_1} = \frac{(1.00 \times 10^{-3}\ \text{lbm})\left(53.34\,\frac{\text{ft}\cdot\text{lbf}}{\text{lbm}\cdot\text{R}}\right)(800.\ \text{R})}{(0.0134\ \text{ft}^3)\left(144\,\frac{\text{in}^2}{\text{ft}^2}\right)} = 22.2\ \text{psia}$$

b. For the Lenoir cycle, the isentropic compression ratio is $\text{CR} = v_{2s}/v_3 = T_{2s}/T_3$. From Eq. (7.38), we have

$$T_{2s} = T_1\left(\frac{p_{2s}}{p_1}\right)^{\frac{k-1}{k}} = (800.\ \text{R})\left(\frac{14.7\ \text{psia}}{22.2\ \text{psia}}\right)^{\frac{0.4}{1.4}} = 711\ \text{R}$$

Then, the isentropic compression ratio for this engine is

$$\text{CR} = \frac{v_{2s}}{v_3} = \frac{T_{2s}}{T_3} = \frac{711\ \text{R}}{530.\ \text{R}} = 1.34$$

c. Equation (13.21) gives the Lenoir cold ASC thermal efficiency as

$$(\eta_T)_{\substack{\text{Lenoir}\\ \text{cold ASC}}} = 1 - \frac{kT_3(\text{CR}-1)}{T_1 - T_4} = 1 - \frac{1.40(530.\ \text{R})(1.34-1)}{800. - 530.\ \text{R}} = 0.0656 = 6.56\%$$

Exercises

31. If the maximum combustion temperature of the Lenoir model airplane jet engine in Example 13.11 is increased from 800. R to 1000. R, determine the corresponding combustion pressure, p_1. Assume all the other variables remain unchanged. **Answer:** $p_1 = 27.7$ psia

32. If the air intake temperature T_3 in Example 13.11 is lowered from 530. R to 500. R, determine the new isentropic compression ratio for the engine. Assume all the other variables remain unchanged. **Answer:** CR = 1.40.

33. If the combustion temperature in Example 13.11 is increased from 800. R to 1500. R, determine the corresponding Lenoir cold ASC thermal efficiency of the engine. **Answer:** $(\eta_T)_{\text{Lenoir cold ASC}} = 15.7\%$.

Its relatively simple construction and good reliability, and the fact that methane was readily available and inexpensive in many urban areas (where it was already being used extensively for illumination), made the Lenoir cycle engine quite successful from about 1860 to 1890. After the turn of the century, the Lenoir engine lost popularity and became obsolete. However, the Lenoir cycle appeared briefly again in the German V-l rocket engines ("buzz bombs"), used during World War II.

13.16 BRAYTON CYCLE

The main reason why the Lenoir cycle had such a poor thermal efficiency was that the fuel-air mixture was not compressed before it was ignited. Many people recognized this fact, but it was not until 1873 that a more efficient internal combustion engine, utilizing preignition compression, was developed. George B. Brayton (1830–1892), an American engineer, adopted the dual reciprocating piston technique of Stirling and Ericsson but used one piston only as a compressor and the second piston only to deliver power. A combustion chamber was inserted between the two pistons to provide a constant pressure heat addition process. Thus, the Brayton ASC consists (ideally) of two isentropic processes (compression and power) and two isobaric processes (combustion and exhaust), as shown in Figure 13.44.

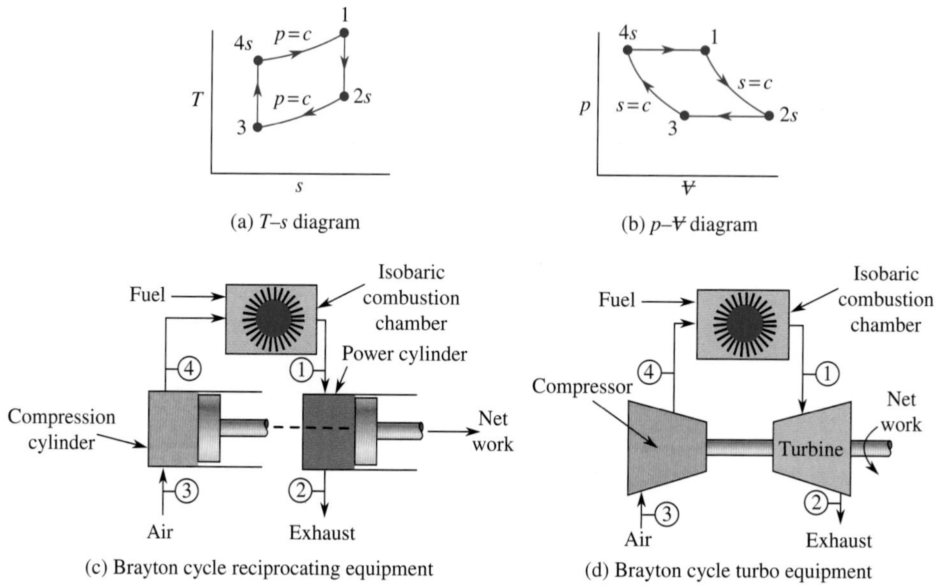

FIGURE 13.44
The open loop Brayton cycle.

BRAYTON CYCLE GAS TURBINE ENGINE

The original Brayton cycle was conceived as a closed loop external combustion hot air engine like those of Stirling and Ericsson. However, it was found to run more reliably when it was converted into an open loop internal combustion engine, as shown in Figure 13.44. Much later, it was discovered to be an adequate model for gas turbine engines.

The thermal efficiency of the Brayton cycle is given by

$$(\eta_T)_{\text{Brayton}} = \frac{(\dot{W}_{\text{out}})_{\text{net}}}{\dot{Q}_{\text{H}}} = \frac{\dot{W}_{pm} - |\dot{W}_c|}{\dot{Q}_{\text{H}}}$$

For the cold ASC, both the compressor and the prime mover (either reciprocating piston-cylinder or turbine) are considered to be isentropic, and we can write

$$\dot{W}_{pm} = \dot{m}(h_1 - h_{2s}) = \dot{m}c_p(T_1 - T_{2s})$$

and

$$|\dot{W}_c| = \dot{m}(h_{4s} - h_3) = \dot{m}c_p(T_{4s} - T_3)$$

Since the combustion chamber is isobaric,

$$\dot{Q}_{\text{H}} = \dot{m}(h_1 - h_{4s}) = \dot{m}c_p(T_1 - T_{4s})$$

Then,

$$(\eta_T)_{\substack{\text{Brayton}\\\text{cold ASC}}} = \frac{(T_1 - T_{2s}) - (T_{4s} - T_3)}{T_1 - T_{4s}} = \frac{(T_1 - T_{4s}) - (T_{2s} - T_3)}{T_1 - T_{4s}} = 1 - \frac{T_{2s} - T_3}{T_1 - T_{4s}} \tag{13.22}$$

Now, from the ideal gas isentropic formula used earlier, Eq. (7.38), we see from Figure 13.44b that

$$T_1/T_{2s} = (p_1/p_{2s})^{(k-1)/k} = (p_{4s}/p_3)^{(k-1)/k} = T_{4s}/T_3$$

so that $T_{2s}/T_3 = T_1/T_{4s}$, and Eq. (13.22) becomes

$$(\eta_T)_{\substack{\text{Brayton}\\\text{cold ASC}}} = 1 - \frac{(T_{2s}/T_3 - 1)T_3}{(T_1/T_{4s} - 1)T_{4s}} = 1 - T_3/T_{4s}$$

Equation (7.38) also allows us to write

$$T_3/T_{4s} = (p_{4s}/p_3)^{(1-k)/k} = (v_3/v_{4s})^{1-k}$$

so that

$$(\eta_T)_{\substack{\text{Brayton}\\ \text{cold ASC}}} = 1 - T_3/T_{4s} = 1 - \text{PR}^{(1-k)/k} = 1 - \text{CR}^{1-k} \quad (13.23)$$

where PR is the isentropic pressure ratio p_{4s}/p_3, and CR is the isentropic compression ratio v_3/v_{4s}. Thus, the thermal efficiency of the Brayton cold ASC can be written as a function of the isentropic pressure or compression ratio and the specific heat ratio of the working fluid.

EXAMPLE 13.12

The average power cylinder and compression cylinder pressures for the early 1878 Brayton cycle engine shown in Figure 13.45 were $p_1 = p_{4s} = 0.210$ MPa and $p_{2s} = p_3 = 0.190$ MPa, respectively. For this engine, determine

a. The isentropic pressure ratio PR.
b. The isentropic compression ratio CR.
c. The Brayton cold ASC thermal efficiency.

Solution

Using the Brayton cycle diagram shown in Figure 13.44, we can carry out the following analysis.

a. The isentropic pressure ratio of a Brayton cycle engine is given by

$$\text{PR} = \frac{p_{4s}}{p_3} = \frac{0.210\ \text{MPa}}{0.190\ \text{MPa}} = 1.11$$

b. The isentropic compression ratio of a Brayton cycle engine is given by

$$\text{CR} = (\text{PR})^{1/k} = (1.11)^{1/1.40} = 1.07$$

c. Equation (13.23) gives the Brayton cold ASC thermal efficiency as

$$(\eta_T)_{\substack{\text{Brayton}\\ \text{cold ASC}}} = 1 - \frac{T_3}{T_{4s}} = 1 - \text{PR}^{(1-k)/k} = 1 - (1.11)^{(1-1.40)/1.40} = 0.0294 = 2.94\%$$

FIGURE 13.45
Example 13.12.

Exercises

34. Determine the isentropic pressure ratio for the Brayton cycle engine in Example 13.12 if the power piston pressure is increased from 0.210 MPa to 0.300 MPa. Assume all the other variables remain unchanged. **Answer**: PR = 1.58.
35. If the compression pressure of the Brayton cycle engine in Example 13.12 is lowered from 0.190 MPa to 0.100 MPa, determine the corresponding isentropic compression ratio of the engine. Assume all the other variables remain unchanged. **Answer**: CR = 1.70.
36. If operating conditions of the Brayton cycle engine in Example 13.12 are altered such that the isentropic temperature at the exit of the compressor is $T_{4s} = 35.0$°C when the inlet air temperature is 22.0°C, determine the Brayton cycle cold ASC thermal efficiency of this engine under these operating conditions. **Answer**: $(\eta_T)_{\text{Brayton cold ASC}} = 4.22\%$.

Since $T_3 = T_L$ but $T_{4s} < T_1 = T_H$, the Brayton cold ASC thermal efficiency is *less* than that of the Carnot cold ASC working between the same temperature limits (T_1 and T_3). However, for fixed values of the temperature limits T_1 and T_3, there is an optimum value for the compressor outlet temperature T_{4s} that maximizes the net output work. This can be determined as follows for an isentropic turbine and compressor:

$$(\dot{W}_{\text{out}})_{\substack{\text{net}\\ \text{isentropic}}} = \dot{W}_{pm} - |\dot{W}_c| = \dot{m}c_p(T_1 - T_{2s} - T_{4s} + T_3)$$

Now, holding T_1 and T_3 fixed and replacing T_2 with its isentropic equivalent, $T_{2s} = T_1T_3/T_{4s}$, we get

$$(\dot{W}_{\text{out}})_{\substack{\text{net}\\ \text{isentropic}}} = \dot{m}c_p(T_1 - T_1T_3/T_{4s} - T_{4s} + T_3)$$

The optimum value of T_{4s} that causes the net isentropic output work to be a maximum can be found by differentiating $(\dot{W}_{out})_{net\ isentropic}$ with respect to T_{4s} and setting the result equal to zero, or

$$\frac{d(\dot{W}_{out})_{net\ isentropic}}{dT_{4s}} = \dot{m}c_p\left(0 + T_1T_3/T_{4s}^2 - 1 + 0\right) = 0$$

Then, solving for $T_{4s} = (T_{4s})_{opt}$ gives

$$(T_{4s})_{opt} = \sqrt{T_1\,T_3} \tag{13.24}$$

The corresponding optimum pressure and compression ratios are

$$PR_{opt} = \left[(T_{4s})_{opt}/T_3\right]^{k/(k-1)} = (T_1/T_3)^{k/[2(k-1)]} \tag{13.25}$$

and

$$CR_{opt} = \left[(T_{4s})_{opt}/T_3\right]^{1/(k-1)} = (T_1/T_3)^{1/[2(k-1)]} \tag{13.26}$$

while the thermal efficiency at the maximum net isentropic work output is

$$(\eta_T)_{\substack{\text{max work}\\ \text{Brayton}\\ \text{cold ASC}}} = 1 - T_3/(T_{4s})_{opt} = 1 - (T_3/T_1)^{1/2} = 1 - (T_L/T_H)^{1/2} \tag{13.27}$$

HOW DID THE BRAYTON CYCLE BECOME A GAS TURBINE ENGINE?

The reciprocating piston-cylinder Brayton cycle engine, while more efficient than the Lenoir cycle engine, was at the same time mechanically more complex and costly. Its relatively low compressor pressure ratio limited its efficiency and its ability to compete effectively with existing reciprocating steam engine economics. These factors stifled the development of the reciprocating Brayton cycle engine, and the cycle might have quickly become obsolete if it had not been for a new prime mover technology being developed for steam, the turbine. By replacing steam with gas, a new type of gas-powered prime mover, the gas turbine, was produced.

Because gas and steam turbines have many characteristics in common, several gas turbine engines were under development at the same time steam turbines were being developed. One characteristic that they do not have in common, however, is that gas turbine power plants require gas compressors, while vapor turbine power plants condense the working fluid to the liquid phase before compressing it with pumps. Early liquid pumps were fairly efficient, but early gas compressors were very inefficient due to a lack of understanding of the dynamics of high-speed compressible flow. This single fact proved to be a major stumbling block in the development of gas turbine engine technology.

This problem is illustrated by Eq. (13.29). For a gas turbine to have a net work output, its thermal efficiency obviously has to be a positive number. This means that both the turbine (the prime mover) and the compressor need high enough isentropic efficiencies for Eq. (13.28) to be obeyed. One of the early major problems in compressible fluid mechanics was to understand how to compress a gas efficiently in a rotary compressor. Turbine prime movers, on the other hand, had already undergone considerable development within the steam power industry and were already 70 to 90% isentropically efficient.

A gas compressor is not simply a turbine running backward, and since compressible flow theory had not yet been completely developed, gas compressor development was carried out largely by trial and error. By 1900, most compressors had isentropic efficiencies of less than 50%, so that the product $(\eta_s)_{pm}(\eta_s)_c$ in Eq. (13.28) was on the order of 0.4. Since typical early gas turbines operated with very small compressor pressure ratios, say, PR = 1.5, and relatively small combustion chamber temperatures, say, 700°F = 1150 R, for an ambient inlet temperature of 70°F = 530 R, Eq. (13.28) requires that $(\eta_s)_{pm}(\eta_s)_c \geq (530/1160)(1.5)^{0.286} = 0.513$. But this was impossible for early units, because while they may have had good turbine isentropic efficiencies of around 90%, they also had very poor compressor isentropic efficiencies of around 50% or less. Thus, many early prototype gas turbine test engines failed to operate under their own power.

The first Brayton cycle gas turbine unit to produce a net power output (11 hp) was built in 1903. It had a very low actual thermal efficiency (about 3%) and could not compete economically with the other prime movers of its time. Compressor efficiency problems continued to plague gas turbine technology, and many new prototype engines were designed and built as late as the 1930s that still could not produce a net power output. Since the thrust produced by an aircraft engine is not considered to be part of the engine's work output (thrust is force, not work), aircraft engines do not necessarily need high thermal efficiencies to be effective. It was in this industry that the gas turbine engine first became successful.

When the isentropic efficiencies of the compressor and prime mover are taken into account, the Brayton cycle thermal efficiency becomes

$$(\eta_T)_{\text{Brayton}} = \frac{(\dot{W}_s)_{pm}(\eta_s)_{pm} - |(\dot{W}_s)_c/(\eta_s)_c|}{\dot{Q}_H} = \frac{(T_1 - T_{2s})(\eta_s)_{pm} - (T_{4s} - T_3)/(\eta_s)_c}{T_1 - T_4}$$

where $T_4 = T_3 + (T_{4s} - T_3)/(\eta_s)_c$. It is clear that this efficiency has a positive value only if

$$(\eta_s)_{pm}(\eta_s)_c \geq (T_{4s} - T_3)/(T_1 - T_{2s}) = T_{4s}/T_1$$

or

$$(\eta_s)_{pm}(\eta_s)_c \geq (T_3/T_1)\text{PR}^{(k-1)/k} = (T_L/T_H)\text{PR}^{(k-1)/k} = (T_3/T_1)\text{CR}^{k-1} = (T_L/T_H)\text{CR}^{k-1} \tag{13.28}$$

13.17 AIRCRAFT GAS TURBINE ENGINES

The major function of an aircraft jet engine is to produce a high-velocity exhaust jet whose thrust is large enough to propel the aircraft. The engine's thrust T is given by

$$T = \dot{m}(V_{\text{exhaust}} - V_{\text{inlet}})/g_c \tag{13.29}$$

where both the inlet and exhaust velocities are measured in a coordinate system fixed to the engine, and the mass flow rate is $\dot{m} = \dot{m}_{\text{fuel}} + \dot{m}_{\text{air}} = \dot{m}_{\text{exhaust}}$. A jet engine needs to produce only enough net output power to drive the aircraft's accessories (fuel pump, hydraulics, generator, etc.), and consequently, it need not have a very high thermal efficiency (the exhaust kinetic energy is considered to be lost energy in a thermal efficiency analysis). This was an ideal application for the inherently inefficient gas turbine engine of the 1930s. The pressures of World War II caused intense research and development in aircraft gas turbine *turbojet* engine development. The first successful turbojet aircraft was the German Heinkel-178, which flew for the first time on August 27, 1939. The engine weighed 800 lbf (364 kg) and produced a thrust of 1100 lbf (4890 N) at 13,000 rpm. As a result of intense wartime technological development, axial flow compressors with pressure ratios of 3.0 and isentropic efficiencies of 75–80% were available by the end of World War II.

Modern aircraft gas turbine engines have compressor pressure ratios as high as 25, and ceramic-coated super alloys have allowed turbine inlet temperatures to approach 3000 R. Their turbine isentropic efficiencies are typically in the range of 85 to 95%, and their compressor isentropic efficiencies usually fall in the range of 80 to 90%. Figure 13.46 illustrates the construction of a modern gas turbine engine.

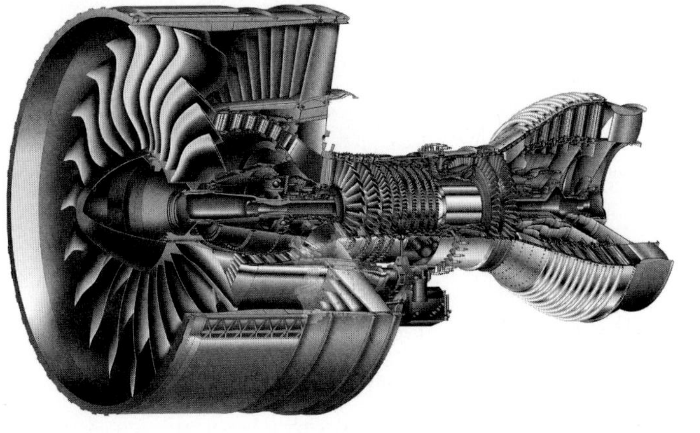

FIGURE 13.46
World class design. The GP7000 was designed by a 50/50 joint venture between GE and Pratt & Whitney for the Airbus A380.
(Source: United Technologies, Pratt & Whitney Aircraft.)

When regeneration, interstage compressor cooling, and interstage turbine reheat are added to the Brayton cycle, it approximates the more efficient Ericsson cycle, as shown in Figure 13.39. The major focus of modern gas turbine development centers on increasing the turbine inlet temperature through the development of new high-strength, high-temperature materials. Modern gas turbine heat engines are used mainly in small- to medium-size stationary power-generating stations and as prime movers throughout the transportation industry.

EXAMPLE 13.13

Under static ground testing at sea-level conditions, the Pratt & Whitney JT3D-3B Turbofan engine has the actual internal temperatures and pressures as shown in Figure 13.47.

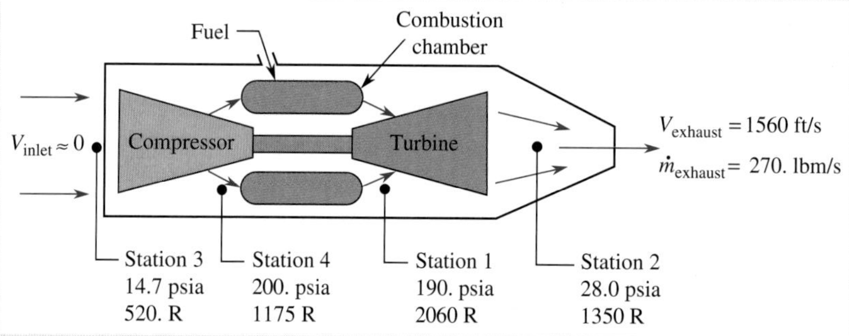

FIGURE 13.47
Example 13.13.

Determine

1. The engine's static thrust.
2. The compressor and turbine isentropic efficiencies for
   a. The Brayton cold air standard cycle.
   b. The Brayton hot air standard cycle using the gas tables for air, Table C.16a.
3. The ASC and actual thermal efficiencies for
   a. The Brayton cold air standard cycle.
   b. The Brayton hot air standard cycle using the gas tables for air, Table C.16a.
   c. The maximum work Brayton cold ASC thermal efficiency.

Solution

1. The engine's static thrust is given directly by Eq. (13.29) as

$$T = \dot{m}(V_{\text{exhaust}} - V_{\text{inlet}})/g_c$$
$$= (370.\,\text{lbm/s})(1560.\,\text{ft/s} - 0)[32.174\,\text{lbm}\cdot\text{ft/(lbf}\cdot\text{s}^2)]$$
$$= 17{,}900\,\text{lbf}$$

2a. The compressor's isentropic efficiency is given by

$$(\eta_s)_c = \frac{(\dot{W}_c)_{\text{isentropic}}}{(\dot{W}_c)_{\text{actual}}} = \frac{T_{4s} - T_3}{T_4 - T_3}$$

and, using $k = 1.40 =$ constant for the cold ASC, we have

$$T_{4s} = T_3(p_{4s}/p_3)^{(k-1)/k} = (520.)(200./14.7)^{(1.4-1)/1.4} = 1100\,\text{R} = 640.°\text{F}$$

so that the compressor's isentropic efficiency using constant specific heats is

$$(\eta_s)_{\substack{\text{compressor}\\ \text{(constant}\\ \text{specific heats)}}} = \frac{1100 - 520.}{1175 - 520.} = 0.886 = 88.6\%$$

Similarly, the turbine's (prime mover) isentropic efficiency is given by

$$(\eta_s)_{pm} = \frac{(\dot{W}_{pm})_{\text{actual}}}{(\dot{W}_{pm})_{\text{isentropic}}} = \frac{T_1 - T_2}{T_1 - T_{2s}}$$

where, using the constant specific heats, we obtain

$$T_{2s} = T_1(p_{2s}/p_1)^{(k-1)/k} = (2060)(28.0/190.)^{(1.40-1)/1.40} = 1190\,\text{R} = 730.°\text{F}$$

Then,

$$\underset{\substack{\text{(constant}\\\text{specific heats)}}}{(\eta_s)_{pm}} = \frac{2060 - 1350}{2060 - 1190} = 0.816 = 81.6\%$$

3a. The Brayton cold ASC thermal efficiency is given by

$$\underset{\substack{\text{Brayton}\\\text{cold ASC}}}{(\eta_T)} = \frac{T_1 - T_{2s} - (T_{4s} - T_3)}{T_1 - T_{4s}}$$

$$= \frac{2060 - 1190 - (1100 - 520.)}{2060 - 1100} = 0.302 = 30.2\%$$

but the actual thermal efficiency of the engine, based on constant specific heats and the data provided in the schematic, is

$$\underset{\substack{\text{(actual, constant}\\\text{specific heats)}}}{(\eta_T)_{\text{Brayton}}} = \frac{T_1 - T_2 - (T_4 - T_3)}{T_1 - T_4}$$

$$= \frac{2060 - 1350 - (1175 - 520.)}{2060 - 1175} = 0.062 = 6.2\%$$

2b. We can easily take into account the temperature-dependent specific heats by using Table C.16a in *Thermodynamic Tables to accompany Modern Engineering Thermodynamics*. For the compressor, $p_{r4} = (p_{4s}/p_3) = (1.2147)(200./14.7) = 16.5$ and, by interpolation in Table C.16a, we find that $T_{4s} = 1084$ R = 624°F. Then,

$$\underset{\substack{\text{(variable}\\\text{specific heats)}}}{(\eta_s)_c} = \frac{T_{4s} - T_3}{T_4 - T_3} = \frac{1084 - 520.}{1175 - 520.} = 0.861 = 86.1\%$$

Similarly, for the turbine,

$$p_{r2} = p_{r1}(p_{2s}/p_1) = (196.16)(28.0/190.) = 28.9$$

and, by interpolation in Table C.16a, we find that $T_{2s} = 1261$ R = 801°F. Then,

$$\underset{\substack{\text{(variable}\\\text{specific heats)}}}{(\eta_s)_{pm}} = \frac{T_1 - T_2}{T_1 - T_{2s}} = \frac{2060 - 1350}{2060 - 1261} = 0.889 = 88.9\%$$

3b. Finally, the Brayton hot ASC can be easily determined from

$$\underset{\text{hot ASC}}{(\eta_T)_{\text{Brayton}}} = \frac{h_1 - h_{2s} - (h_{4s} - h_3)}{h_1 - h_{4s}}$$

where, from Table C.16a,

$$h_3 = 124\,\text{Btu/lbm (at 520.R)}$$

$$h_{4s} = 262\,\text{Btu/lbm (by interpolation at 1084 R)}$$

$$h_1 = 521\,\text{Btu/lbm (at 2060 R)}$$

$$h_{2s} = 307\,\text{Btu/lbm (by interpolation at 1261 R)}$$

Then,

$$\underset{\text{hot ASC}}{(\eta_T)_{\text{Brayton}}} = \frac{521 - 307 - (262 - 124)}{521 - 262} = 0.293 = 29.3\%$$

(*Continued*)

EXAMPLE 13.13 (*Continued*)

and the engine's actual thermal efficiency, based on temperature-dependent specific heats, is

$$(\eta_T)_{\substack{\text{Brayton}\\ \text{(actual, variable}\\ \text{specific heats)}}} = \frac{h_1 - h_2 - (h_4 - h_3)}{h_1 - h_4}$$

where $h_4 = 284.9$ Btu/lbm(at 1175 R) and $h_2 = 329.9$ Btu/lbm(at 1350 R). Then,

$$(\eta_T)_{\substack{\text{Brayton}\\ \text{(actual, variable}\\ \text{specific heats)}}} = \frac{521 - 329.9 - (284.9 - 124)}{521 - 284.9} = 0.128 = 12.8\%$$

3c. The maximum work Brayton cold ASC thermal efficiency is given by Eq. (13.27) as

$$(\eta_T)_{\substack{\text{max work}\\ \text{Brayton}\\ \text{cold ASC}}} = 1 - \sqrt{\frac{T_3}{T_1}} = 1 - \sqrt{\frac{520.\text{ R}}{2060\text{ R}}} = 0.502 = 50.2\%$$

This is much greater than the actual thermal efficiency for this engine, because an aircraft engine need produce only enough work output to drive the engine's auxiliary equipment (generator, fuel pump, etc.), and most of the engine's energy output is in the kinetic energy of its exhaust (which produces thrust).

Exercises

37. Determine the optimum compressor outlet temperature of the Pratt & Whitney jet engine analyzed in Example 13.13 that maximizes the net output work of the engine. **Answer**: $(T_{4s})_{\text{opt}} = 1035$ R.

38. Determine the isentropic efficiency of the Brayton cycle compressor in Example 13.13 if the outlet pressure is increased from 200. psia to 210. psia. Assume all the other variables remain unchanged. **Answer**: $(\eta_s)_{\text{compressor}} = 90.3\%$.

39. If the temperature at the entrance to the turbine in Example 13.13 is increased from 2060 R to 2460 R, determine the new Brayton cycle cold ASC thermal efficiency of the engine. Assume all the other variables remain unchanged. **Answer**: $(\eta_T)_{\text{Brayton cold ASC}} = 33.8\%$.

Note that, whereas the Brayton hot ASC cycle thermal efficiency of Example 13.13 is relatively high (about 30%), the actual thermal efficiency of an aircraft turbojet engine is normally quite low. This is not because of poor engine design, but because most of the combustion energy is put into the kinetic energy of the exhaust gas rather than into the mechanical shaft work output. In aircraft engine design, the thrust to weight ratio of the engine is a key parameter, and the engine's thermal efficiency is secondary.

13.18 OTTO CYCLE

The Stirling and Ericsson external combustion gas power cycles were originally developed to combat the dangerous high-pressure boilers of the early steam engines. The Lenoir internal combustion engine was simpler, smaller, and used a more convenient fuel than either of these engines, but it had a very poor thermal efficiency. Brayton managed to increase the thermal efficiency of the internal combustion engine by providing a compression process before combustion using the two-piston Stirling and Ericsson technique with a separate combustion chamber. But the ultimate goal of commercial internal combustion engine development was to combine all the basic processes of intake, compression, combustion, expansion (power), and exhaust within a single piston-cylinder apparatus. This was finally achieved in 1876 by the German engineer Nikolaus August Otto (1832–1891). The basic elements of the ASC model of the Otto cycle are shown in Figure 13.48. It is composed of two isochoric processes and two isentropic processes.

After several years of experimentation, Otto finally built a successful internal combustion engine that allowed all the basic processes to occur within a single piston-cylinder arrangement. The thermodynamic cycle of Otto's engine required four piston strokes and two crankshaft revolutions to complete, but it ran smoothly, was relatively quiet, and was very reliable and efficient. Otto's engine was an immediate success, and by 1886, more than 30,000 had been sold. They became the first serious competitor to the steam engine in the small- and medium-size engine market.

Initially, Otto's engine used illuminating gas (methane) as its fuel, but by 1885, many Otto cycle engines were already being converted into liquid hydrocarbon (gasoline) burning engines. The development of the ingenious float-feed carburetor for vaporizing liquid fuel in 1892 by the German Wilhelm Maybach (1847–1929) heralded

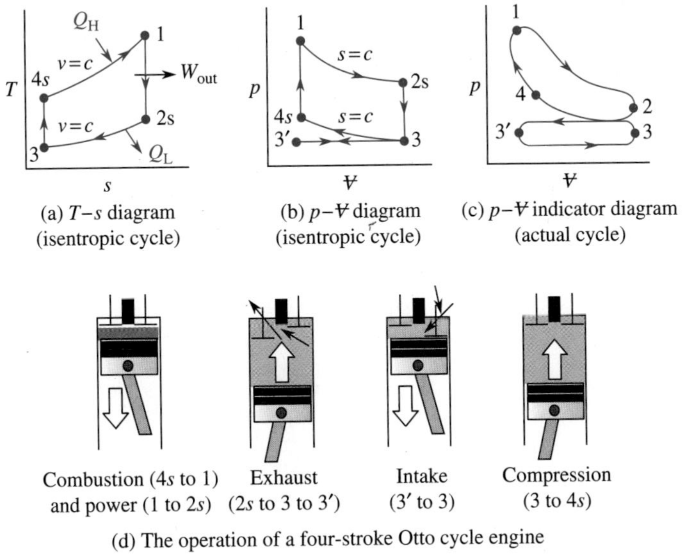

FIGURE 13.48
The Otto air standard cycle.

the dawn of the automobile era. The German engineer Karl Friedrich Benz (1844–1929) is generally credited with building the first practical automobile, using a low-speed Otto cycle engine running on liquid hydrocarbon fuel, in 1885. He used engine exhaust heat to vaporize the fuel before it was fed into the engine.

WHO INVENTED THE "OTTO" CYCLE?

Unknown to Nikolaus Otto, the four-stroke cycle IC engine had already been patented in the 1860s by the French engineer Alphonse Eugene Beau de Rochas (1815–1893). However, Rochas did not actually build and test the engine he patented. Since Otto was the first to actually construct and operate the engine, the cycle is named after him rather than Rochas.

In 1878, the Scottish engineer, Dugald Clerk (1854–1932) developed a two-stroke version of the Otto cycle, producing one crankshaft revolution per thermodynamic cycle (it was like the Lenoir engine but with preignition compression). In 1891, Clerk went on to develop the concept of IC engine supercharging. This increased the thermal efficiency of the engine by further compressing the induction charge before ignition.

Although Clerk's two-stroke engine was inherently less fuel efficient than Otto's four-stroke cycle engine, it gave a more uniform power output (which is important only for single- or dual-cylinder engines) and had almost double the power to weight ratio of the Otto engine. The two-stroke Otto cycle (it never became known as the Clerk cycle) engine became successful as a small, lightweight engine for boats, lawn mowers, saws, and so forth.

The thermal efficiency of the Otto cycle is given by

$$(\eta_T)_{\text{Otto}} = \frac{(\dot{W}_{\text{out}})_{\text{net}}}{\dot{Q}_H} = \frac{\dot{Q}_H - |\dot{Q}_L|}{\dot{Q}_H} = 1 - \frac{|\dot{Q}_L|}{\dot{Q}_H}$$

where, from Figure 13.48, $|\dot{Q}_L| = \dot{m}(u_{2s} - u_3)$ and $\dot{Q}_H = \dot{m}(u_1 - u_{4s})$.

Then, the thermal efficiency of the Otto hot ASC is

$$(\eta_T)_{\substack{\text{Otto}\\ \text{hot ASC}}} = 1 - \frac{u_{2s} - u_3}{u_1 - u_{4s}}$$

For the Otto *hot ASC*, Table C.16a or C.16b in *Thermodynamic Tables to accompany Modern Engineering Thermodynamics* are used to find values for the specific internal energies. Since the processes from 1 to 2*s* and from 3 to 4*s* are isentropic, we use the v_r columns in these tables to find

$$\frac{v_3}{v_{4s}} = \frac{v_{r3}}{v_{r4}} = \frac{v_{2s}}{v_1} = \frac{v_{r2}}{v_{r1}} = \text{CR}$$

where $CR = v_3/v_{4s}$ is the isentropic compression ratio. If the intake temperature and pressure (T_3 and p_3) are known, we can find u_3 and v_{r3} from the table. Then, if we know the compression ratio (CR), we can find

$$v_{r4} = \frac{v_{r3}}{CR} \text{ and } v_{r2} = v_{r1} \times CR$$

We can now find u_{4s} and T_{4s} from the tables. However, to find u_1, T_1, u_{2s}, and T_{2s}, we need to know more information about the system. Consequently, the heat of combustion ($Q_H/m = \dot{Q}_H/\dot{m}$), maximum pressure ($p_1$), or maximum temperature (T_1) in the cycle is usually given to complete the analysis.

For the Otto *cold ASC*,

$$|\dot{Q}_L| = \dot{m}(u_{2s} - u_3) = \dot{m}\, c_v(T_{2s} - T_3) \text{ and } \dot{Q}_H = \dot{m}(u_1 - u_{4s}) = \dot{m}\, c_v(T_1 - T_{4s}).$$

Then,

$$(\eta_T)_{\substack{\text{Otto}\\ \text{cold ASC}}} = 1 - \frac{T_{2s} - T_3}{T_1 - T_{4s}} = 1 - \left(\frac{T_3}{T_{4s}}\right)\left(\frac{T_{2s}/T_3 - 1}{T_1/T_{4s} - 1}\right)$$

The process 1 to 2*s* and process 3 to 4*s* are isentropic, so

$$T_1/T_{2s} = T_{4s}/T_3 = (v_1/v_{2s})^{1-k} = (v_{4s}/v_3)^{1-k}$$
$$= (p_1/p_{2s})^{(k-1)/k} = (p_{4s}/p_3)^{(k-1)/k}$$

Since $T_1/T_{4s} = T_{2s}/T_3$,

$$(\eta_T)_{\substack{\text{Otto}\\ \text{cold ASC}}} = 1 - T_3/T_{4s} = 1 - PR^{(1-k)/k} = 1 - CR^{1-k} \tag{13.30}$$

where $CR = v_3/v_{4s}$ is the isentropic compression ratio and $PR = p_{4s}/p_3$ is the isentropic pressure ratio.

Since $T_3 = T_L$ but $T_{4s} < T_1 = T_H$, the Otto cold ASC thermal efficiency is less than that of a Carnot cold ASC operating between the same temperature limits (T_1 and T_3). Because the Otto cycle requires a constant volume combustion process, it can be carried out effectively only within the confines of a piston-cylinder or other fixed volume apparatus by a nearly instantaneous rapid combustion process.

EXAMPLE 13.14

The isentropic compression ratio of a new lawn mower Otto cycle gasoline engine is 8.00 to 1, and the inlet air temperature is $T_3 = 70.0°F$ at a pressure of $p_3 = 14.7$ psia. Determine

a. The air temperature at the end of the isentropic compression stroke T_{4s}.
b. The pressure at the end of the isentropic compression stroke before ignition occurs p_{4s}.
c. The Otto cold ASC thermal efficiency of this engine.

Solution

a. The isentropic compression ratio for an Otto cycle engine is defined as

$$CR = \frac{v_3}{v_{4s}} = \left(\frac{T_3}{T_{4s}}\right)^{\frac{1}{1-k}}$$

from which we have

$$T_{4s} = \frac{T_3}{CR^{1-k}} = T_3 \times CR^{k-1} = (70.0 + 459.67\,R)(8.00)^{0.40} = 1220\,R$$

b. For the Otto cycle, the isentropic pressure and compression ratios are related by $PR = CR^k$, where $PR = p_{4s}/p_3$ and $CR = v_3/v_{4s}$. Then,

$$p_{4s} = p_3 CR^k = (14.7\,\text{psia})(8.00)^{1.40} = 270.\,\text{psia}$$

c. Equation (13.30) gives the Otto cold ASC thermal efficiency as

$$(\eta_T)_{\substack{\text{Otto}\\ \text{cold ASC}}} = 1 - \frac{T_3}{T_{4s}} = 1 - PR^{\frac{1-k}{k}} = 1 - CR^{1-k} = 1 - (8.00)^{1-1.40} = 0.565 = 56.5\%$$

Exercises

40. If the lawn mower in Example 13.14 is left outside on a cold day when T_3 is reduced from 70.0°F to 30.0°F, determine the new temperature at the end of the isentropic compression stroke. Assume all the other variables remain unchanged. **Answer:** $T_{4s} = 1130$ R.

41. If the clearance volume on the lawn mower in Example 13.14 is decreased such that the compression ratio is increased from 8.00 to 8.50 to 1, determine the new pressure at the end of the isentropic compression stroke. Assume all the other variables remain unchanged. **Answer:** $p_{4s} = 294.1$ psia.

42. If the maximum temperature in the cycle (T_{4s}) is 2400 R, determine the Otto cycle *hot ASC* thermal efficiency of this engine. Assume all the other variables remain unchanged. **Answer:** $(\eta_T)_{\text{Otto hot ASC}} = 52.8\%$.

The actual pressure–volume diagram from an engine operating on a gas or vapor power cycle is called an *indicator diagram*,[10] and the enclosed area is equal to the net reversible work produced inside the engine. The *mean effective pressure* (mep) of a reciprocating engine is the *average net pressure* acting on the piston during its displacement. The *indicated* (or reversible) work output $(W_I)_{\text{out}}$ of the piston is the net positive area enclosed by the indicator diagram, as shown in Figure 13.49, and is equal to the product of the mep and the piston displacement, $\mathcal{V}_2 - \mathcal{V}_1 = \pi/4\,(\text{Bore})^2(\text{Stroke})$, or

$$(W_I)_{\text{out}} = \text{mep}\,(\mathcal{V}_2 - \mathcal{V}_1) \tag{13.31}$$

The *indicated* power output $(\dot{W}_I)_{\text{out}}$ is the net (reversible) power developed *inside* all the combustion chambers of an engine containing n cylinders and is

$$(\dot{W}_I)_{\text{out}} = \text{mep}(n)(\mathcal{V}_2 - \mathcal{V}_1)(N/C) \tag{13.32}$$

where N is the rotational speed of the engine and C is the number of crankshaft revolutions per power stroke ($C = 1$ for a two-stroke cycle and $C = 2$ for a four-stroke cycle). The *actual* power output of the engine as measured by a dynamometer is called the *brake* power $(\dot{W}_B)_{\text{out}}$, and the difference between the indicated and brake power is known as the *friction* power (i.e., the power dissipated in the internal friction of the engine) $\dot{W}_F$, or

$$(\dot{W}_I)_{\text{out}} = (\dot{W}_B)_{\text{out}} + \dot{W}_F$$

therefore, the engine's mechanical efficiency η_m is simply (see Table 13.2)

$$\eta_m = \frac{\dot{W}_{\text{actual}}}{\dot{W}_{\text{reversible}}} = \frac{(\dot{W}_B)_{\text{out}}}{(\dot{W}_I)_{\text{out}}} = 1 - \frac{\dot{W}_F}{(\dot{W}_I)_{\text{out}}} \tag{13.33}$$

From Eq. (13.31), we can write

$$\begin{aligned} \text{mep} &= (W_I)_{\text{out}}/(\mathcal{V}_2 - \mathcal{V}_1) = ((W_I)_{\text{out}}/m_a)/v_2 - v_1 \\ &= \left[(\dot{W}_I)_{\text{out}}/\dot{m}_a\right]/(v_2 - v_1) \end{aligned}$$

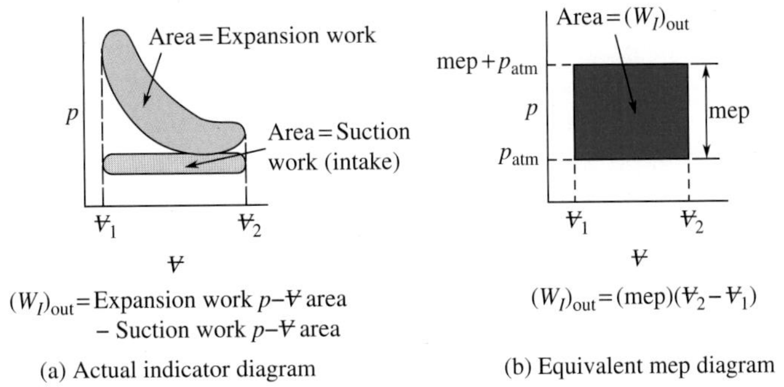

FIGURE 13.49
Mean effective pressure (mep) and indicator diagram relation.

[10] The term *indicator diagram* dates from about 1790, when James Watt developed an apparatus to continuously record (i.e., *indicate*) the variations in pressure within a steam engine cylinder. It is used today to denote any p–$\mathcal{V}$ diagram that is constructed from actual pressure–volume data.

where m_a and $\dot{m}_a$ are the mass of air in the cylinder and the cylinder's air mass flow rate, respectively. The ASC (i.e., reversible or indicated, see Table 13.2) thermal efficiency of *any* internal or external combustion engine can now be written as

$$(\eta_T)_{\text{ASC}} = \frac{(\dot{W}_{\text{out}})_{\text{reversible}}}{\dot{Q}_{\text{in}}} = \frac{(\dot{W}_1)_{\text{out}}}{\dot{Q}_{\text{fuel}}} = \frac{(\dot{W}_1)_{\text{out}}/\dot{m}_a}{\dot{Q}_{\text{fuel}}/\dot{m}_a}$$

where $\dot{Q}_{\text{in}} = \dot{Q}_{\text{fuel}}$ is the heating value of the fuel. Combining these equations gives

$$\text{mep} = \frac{(\eta_T)_{\text{ASC}}(\dot{Q}_{\text{fuel}}/\dot{m}_a)}{v_2 - v_1} = \frac{(\eta_T)_{\text{ASC}}(\dot{Q}_{\text{fuel}}/\dot{m}_{\text{fuel}})}{(A/F)(v_2 - v_1)}$$

where $A/F = \dot{m}_a/\dot{m}_{\text{fuel}}$ is the air–fuel ratio of the engine. Now,

$$v_2 - v_1 = v_1(v_2/v_1 - 1) = RT_1(\text{CR} - 1)/p_1$$

so Eq. (13.32) becomes

$$(\dot{W}_1)_{\text{out}} = \frac{(\eta_T)_{\text{ASC}}(\dot{Q}/\dot{m})_{\text{fuel}}(DNp_1/C)}{(A/F)(RT_1)(\text{CR} - 1)} \tag{13.34}$$

where $D = n(\forall_2 - \forall_1) = \pi/4(\text{Bore})^2 \times (\text{Stroke}) \times (\text{Number of cylinders})$ is the total piston displacement of the engine. Equation (13.34) allows us to determine the horsepower output of an ideal frictionless internal combustion engine, and when actual dynamometer test data are available, Eq. (13.33) allows us to determine the engine's mechanical efficiency.

EXAMPLE 13.15

A six-cylinder, four-stroke Otto cycle internal combustion engine has a total displacement of 260. in^3 and a compression ratio of 9.00 to 1. It is fueled with gasoline having a specific heating value of 20.0×10^3 Btu/lbm and is fuel injected with a mass-based air-fuel ratio of 16.0 to 1. During a dynamometer test, the intake pressure and temperature were found to be 8.00 psia and 60.0°F while the engine was producing 85.0 brake hp at 4000. rpm. For the Otto cold ASC with $k = 1.40$, determine the

a. Cold ASC thermal efficiency of the engine.
b. Maximum pressure and temperature of the cycle.
c. Indicated power output of the engine.
d. Mechanical efficiency of the engine.
e. Actual thermal efficiency of the engine.

Solution

a. From Eq. (13.30), using $k = 1.40$ for the cold ASC,

$$(\eta_T)_{\substack{\text{Otto}\\\text{cold ASC}}} = 1 - \text{CR}^{1-k} = 1 - 9.00^{-0.40} = 0.585 = 58.5\%$$

b. From Figure 13.48*a*,

$$\dot{Q}_{\text{H}} = \dot{Q}_{\text{fuel}} = (\dot{m}c_v)_a(T_1 - T_{4s}) = \dot{m}_{\text{fuel}}(A/F)(c_v)_a(T_1 - T_{4s})$$

and

$$T_1 = T_{\text{max}} = T_{4s} + \frac{(\dot{Q}/\dot{m})_{\text{fuel}}}{(A/F)_{\text{mass}}(c_v)_a}$$

Since process 3 to 4*s* is isentropic, Eq. (7.38) gives

$$T_{4s} = T_3\text{CR}^{k-1} = (60.0 + 459.67)(9.00)^{0.40} = 1250\ \text{R}$$

Then,

$$T_{\text{max}} = \frac{20.0 \times 10^3\ \text{Btu/lbm fuel}}{(16.0\ \text{lbm air/lbm fuel})[0.172\ \text{Btu/(lbm air} \cdot \text{R)}]} + 1250\ \text{R} = 8520\ \text{R}$$

Since process 4*s* to 1 is isochoric, the ideal gas equation of state gives

$$p_{\text{max}} = p_1 = p_{4s}(T_1/T_{4s})$$

and, since the process 3 to 4s is isentropic,

$$T_{4s}/T_3(p_{4s}/p_3)^{(k-1)/k}$$

or

$$p_{4s} = p_3(T_{4s}/T_3)^{k/(k-1)} = (8.00\,\text{psia})\left(\frac{1250\,\text{R}}{520\,\text{R}}\right)^{1.40/0.40} = 172\,\text{psia}$$

then,

$$p_{max} = (172\,\text{psia})[(8520\,\text{R})/1250\,\text{R}] = 1170\,\text{psia}$$

c. Equation (13.34) gives the indicated power as

$$|\dot{W}_I|_{out} = \frac{(0.585)(20.0\times10^3\,\text{Btu/lbm})(260.\,\text{in}^3/\text{rev})(4000.\,\text{rev/min})(1170\,\text{lbf/in}^2)/2}{(16.0)[0.0685\,\text{Btu/(lbm}\cdot\text{R)}](8520\,\text{R})(9.00-1)(12\,\text{in/ft})(60\,\text{s/min})}$$

$$= (132{,}00\,\text{ft}\cdot\text{lbf/s})\left(\frac{1\,\text{hp}}{550\,\text{ft}\cdot\text{lbf/s}}\right) = 241\,\text{hp}$$

d. Equation (13.33) gives the mechanical efficiency of the engine as

$$\eta_m = \frac{(\dot{W}_B)_{out}}{(\dot{W}_I)_{out}} = \frac{85.0\,\text{hp}}{241\,\text{hp}} = 0.353 = 35.3\%$$

e. Finally, the actual thermal efficiency of the engine can be determined from Eqs. (7.5) and (13.33) as

$$(\eta_T)_{\substack{\text{Otto}\\ \text{actual}}} = \frac{(\dot{W}_B)_{out}}{\dot{Q}_{fuel}} = \frac{(\eta_m)(\dot{W}_I)_{out}}{\dot{Q}_{fuel}} = (\eta_m)(\eta_T)_{\substack{\text{Otto}\\ \text{cold ASC}}}$$

$$= (0.353)(0.585) = 0.207 = 20.7\%$$

Exercises

43. If the Otto cycle engine discussed in Example 13.15 has its compression ratio increased to 10.0 to 1, what would be its new Otto cold ASC thermal efficiency? Assume all other variables remain unchanged. **Answer:** $(\eta_T)_{\text{Otto cold ASC}} = 60.2\%$.

44. Find p_{max} and T_{max} for the Otto cycle engine discussed in Example 13.15 when the compression ratio is decreased from 9.00 to 8.00 to 1. Assume all other variables remain unchanged. **Answer:** p_{max} = 1040 psia and T_{max} = 8460 R.

45. Determine the indicated horsepower in Example 13.15 if the engine's displacement is increased from 260. in^3 to 300. in^3. Assume all other variables remain unchanged. **Answer:** $(\dot{W}_I)_{out}$ = 280. hp.

46. Determine the mechanical efficiency of the Otto cycle engine in Example 13.15 if the actual brake horsepower is 88.0 hp instead of 85.0 hp. Assume all other variables remain unchanged. **Answer:** η_m = 36.3%.

The previous example illustrates that the Otto cold ASC analysis generally predicts thermal efficiencies that are far in excess of the actual thermal efficiencies. Typical Otto cycle IC engines have actual operating thermal efficiencies in the range of 15–25%. The large difference between the cold ASC (which contains at least one isentropic process) thermal efficiency and the actual thermal efficiency is due to the influence of the second law of thermodynamics through the large number of thermal and mechanical irreversibilities inherent in this type of reciprocating piston-cylinder engine. To improve its actual thermal efficiency, the combustion heat losses and the number of moving parts in the engine must be reduced.

WHAT IS THE SMALLEST INTERNAL COMBUSTION ENGINE?

The Cox Tee Dee .010 model airplane engine (Figure 13.50) has the smallest internal combustion engine ever put into production. This amazing little engine weighs just under an ounce and runs at 30,000 rpm. The fuel is 10–20% castor oil plus 20–30% nitromethane mixed with methanol. With a bore of 0.237 in (6.02 mm) and a stroke of 0.226 in (5.74 mm), it has a power output of about 5 W.

(Continued)

WHAT IS THE SMALLEST INTERNAL COMBUSTION ENGINE? *Continued*

FIGURE 13.50
Cox Tee engine.

13.19 ATKINSON CYCLE

In Otto cycle engines, the pressure in the cylinder at the end of the expansion (power) stroke is still 3 to 5 atm when the exhaust valve opens. The British engineer James Atkinson (1846–1914) realized that the efficiency of the Otto cycle could be improved if the combustion gases could be expanded to near atmospheric pressure before being exhausted from the engine. In 1882, he invented a piston engine that allowed the intake, compression, power, and exhaust strokes of the four-stroke cycle to occur in a single crankshaft revolution. The crankshaft was mounted on a separate axis from the piston rods and was connected by a series of levers that allowed all four strokes of the cycle to occur in a single crankshaft revolution (Figure 13.51). The complex crankshaft also produced a power stroke that was longer than the compression stroke, which allowed the engine to achieve a greater efficiency than an equivalent Otto cycle engine.

The *Atkinson cycle* engines were slightly more efficient than comparable Otto cycle engines of the day, but they were also larger and more expensive. Consequently, Atkinson's engine did not achieve market success and soon disappeared.

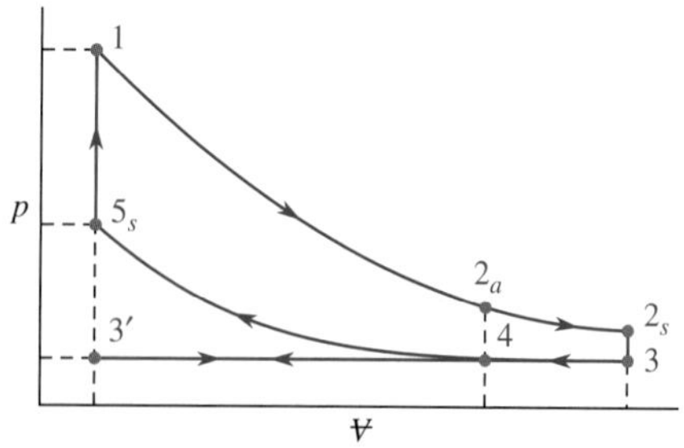

FIGURE 13.51
The ideal Atkinson cycle consists of the following operations: $1–2_s$, isentropic (reversible and adiabatic) expansion; $2_s–3$: isochoric (constant volume) cooling; 3–4–3′, isobaric (constant pressure) exhaust and intake; 3–4, isobaric (constant pressure) cooling; $4–5_s$, isentropic compression; and $5_s–1$, isochoric heating (combustion).

The additional work produced by the Atkinson cycle over the equivalent Otto cycle is the area enclosed by the area $2a \rightarrow 2s \rightarrow 3 \rightarrow 4 \rightarrow 2a$. The thermal efficiency of the Atkinson cold ASC is given by

$$(\eta_T)_{\substack{\text{Atkinson}\\ \text{cold ASC}}} = 1 - \frac{k(\text{ER} - \text{CR})}{\text{ER}^k - \text{CR}^k}$$

where $\text{ER} = v_{2s}/v_1$ is the isentropic expansion ratio and $\text{CR} = v_4/v_{5s}$ is the isentropic compression ratio. Note that, as the expansion ratio ER approaches the compression ratio CR, the Atkinson cold ASC thermal efficiency should approach the Otto cycle cold ASC thermal efficiency. To be effective, the expansion ratio should be greater than the compression ratio. Typical values are compression ratio = 8:1 and expansion ratio = 10:1.

Ilmor Engineering, a firm that is co-owned by Roger Penske and that supplies Honda engines to the Indy Racing League, is developing a new three-cylinder engine that simulates the Atkinson cycle. Two cylinders operate on the conventional four-stroke cycle and empty their exhaust into a third, low-pressure expansion cylinder, which allows the expansion and compression processes to operate independently. A prototype engine was first displayed at the 2009 Stuttgart Engine Exposition.

13.19.1 Modern Atkinson Cycle

In 1947, an American engineer named Ralph Miller patented an ingenious variation of the original Atkinson cycle. Rather than varying the actual length of the intake stroke, he realized that you could simply delay closing the intake valve past the end of the intake stroke. Then, as the piston traveled back up the cylinder, it simply pushed air back out into the intake manifold. The compression began only when the intake valve was finally closed, and by altering when the intake valve closed, you could effectively change the compression ratio of the engine.

Today, the term *Atkinson cycle* is used to describe Miller's modified four-stroke Otto cycle, in which the intake valve is held open longer than normal to allow the piston to push some of the intake air back out of the cylinder. This reduces the compression ratio, but the subsequent expansion ratio is unchanged. This means that the compression ratio is smaller than the expansion ratio, meeting one of the essential features of the Atkinson cycle.

The goal of the modern Atkinson cycle is to allow the pressure in the combustion chamber at the end of the power stroke to be as close to atmospheric pressure as possible. This maximizes the energy obtained from the combustion process.

Because an Atkinson cycle engine does not compress as much air as a similar size Otto cycle engine, it has a lower power density (power output per unit of engine mass). Four-stroke Atkinson cycle engines with the addition of a turbocharger or supercharger to make up for the loss of power density are known as *Miller cycle* engines.

While an Otto cycle engine modified to run on the Atkinson cycle provides good fuel economy, it has a lower power output than a traditional Otto cycle engine. However, the power of the engine can be supplemented by an electric motor during times when more power is needed. This forms the basis of Atkinson cycle hybrid electric automobiles. Their electric motors can be used independently of, or in combination with, the Atkinson-cycle engine, to provide the most efficient means of producing the desired power. Toyota, Ford, Chevrolet, Lexus, and Mercedes have all produced hybrid electric automobiles using Atkinson cycle engines in recent years. For more information, see the animated Atkinson cycle engine at http://www.animatedengines.com/atkinson.shtml.

13.20 MILLER CYCLE

By closing the intake valve when the piston is at the bottom of its stroke, the Otto cycle engine begins compressing the air when the crankshaft has no leverage to push it up. A flywheel is often necessary to keep the engine running. In the 1940s, the American engineer R. H. Miller (1890–1967) realized that the crankshaft would have a much easier time pushing the piston up if it did not start the compression stroke until it had rotated part way up, so he used a longer lever arm (see Figure 13.52).

When a Miller cycle engine has delayed (late) intake valve closure, it reduces the load on the piston as it rises to begin the compression stroke. This can produce significant horsepower with the addition of a turbocharger or supercharger, which compresses the air for the engine. A modern-day Atkinson cycle engine, on the other hand, delays its intake valve closure simply to cause the compression stroke to be shorter than the power stroke, thus realizing some of the same efficiency benefits of the original Atkinson cycle engine.

A Miller cycle engine is very similar to a modern Atkinson cycle engine. However, there are two big differences:

1. A Miller cycle engine depends on a turbocharger or supercharger.
2. A Miller cycle engine has either an early or late intake valve closing during the compression stroke. When the intake valve closes late, the piston travels 20 to 30% of the way back up to the top of the cylinder before the intake valve finally closes, so that the engine compression starts at the pressure of the turbocharger or supercharger.

The effect is increased efficiency, up to about 15%. This type of engine was first used in ships and stationary power-generating plants, but in the 1990s, it was adapted by Mazda for use in the Mazda Millennia.

To be effective, the Miller cycle turbocharger or supercharger must be able to compress the air with less energy than with the engine's pistons. This occurs only at low pressures, so the Miller cycle uses the turbocharger or supercharger for the first part of the compression process and uses the piston for the remainder. Successful production versions of this cycle use variable valve timing to control the Miller cycle to maximize the engine's efficiency.

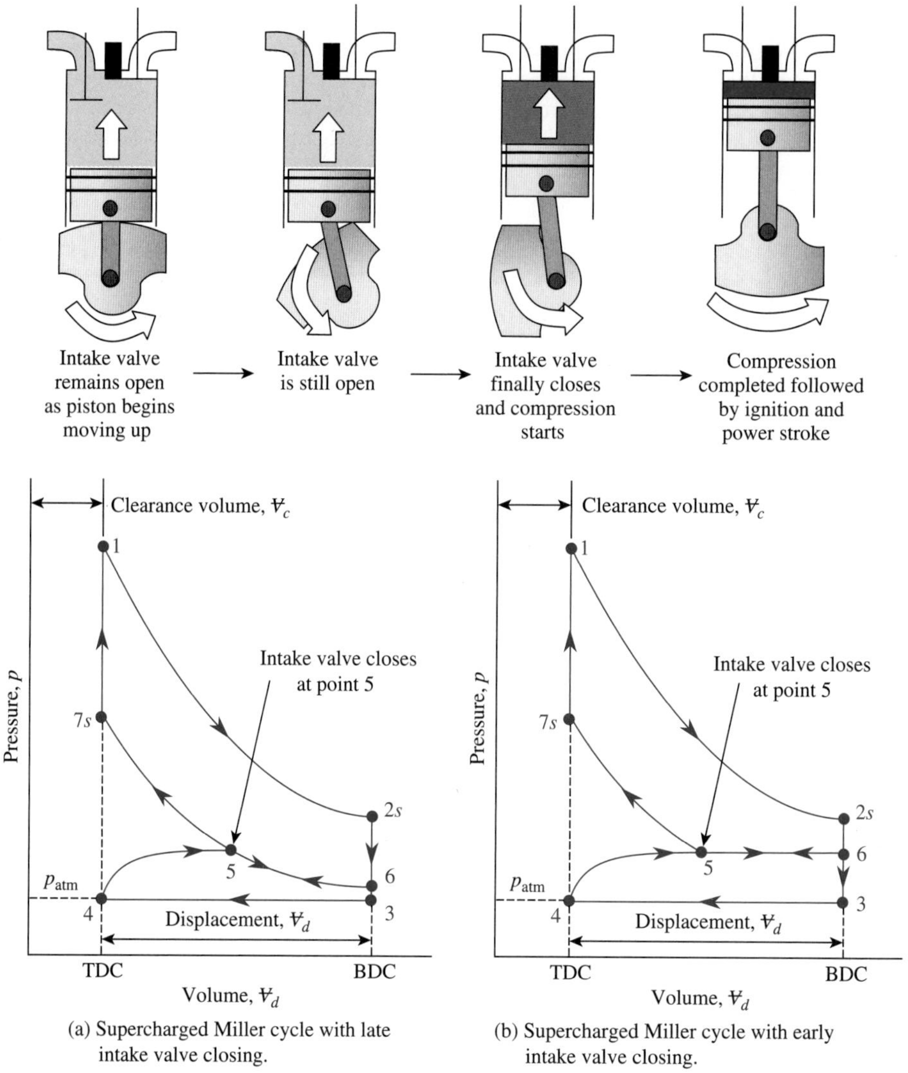

FIGURE 13.52
The ideal supercharged Miller cycle with early and late intake valve closing.

The Miller cycle has one additional benefit: When the intake air is compressed by a supercharger then cooled by an intercooler, it has a higher density and a lower temperature than that obtained by a piston compression alone, further increasing the engine's efficiency.

EXAMPLE 13.16

A four-cylinder, 3.50 liter automobile engine operates on an ideal Miller cycle with early closing intake valves shown in Figure 13.52b. It has a compression ratio of 8.00 to 1 and an expansion ratio of 10.0 to 1. The turbocharger provides air at 200. kPa and 40.0°C when the intake valve closes. The air–fuel ratio is 15.0 to 1 and the fuel has a heating value of 43,300 kJ/kg. Using a cold ASC with $k = 1.35$, determine the temperature and pressure at all points of the cycle.

Solution

For each cylinder, the displacement volume is $V_d = 3.50/4 = 0.875 \text{ L} = 8.75 \times 10^{-4} \text{ m}^3$. The clearance volume, V_c, is calculated from the expansion ratio as $\text{ER} = (V_c + V_d)/V_c$, or

$$V_c = V_d/(\text{ER} - 1) = (8.75 \times 10^{-4})/(10.0 - 1) = 9.72 \times 10^{-5} \text{ m}^3$$

Then, from Figure 13.52b,

$$\forall_c = \forall_1 = \forall_{7s} = \forall_4 = 9.72 \times 10^{-5}\ \text{m}^3$$

Also,

$$\forall_{6s} = \forall_{2s} = \forall_3 = \forall_d + \forall_c = 8.75 \times 10^{-4}\ \text{m}^3 + 9.72 \times 10^{-5}\ \text{m}^3 = 9.72 \times 10^{-4}\ \text{m}^3$$

and

$$\forall_5 = \forall_{7s} \times \text{CR} = (9.72 \times 10^{-5}\ \text{m}^3) \times 8.00 = 7.78 \times 10^{-4}\ \text{m}^3$$

Now, we can compute the temperature and pressure at each point in the cycle by starting at state 5 in Figure 13.52b, the closing of the intake valve, where we know the pressure and temperature.

State 5 in Figure 13.52b

This is where the intake valve closes and the following information is given in the problem statement: p_5 = 200 kPa and T_5 = 40.0°C = 313 K.

State 6 in Figure 13.52b

The process from states 5 to 6 is isentropic, so

$$p_{6s} = p_5\left(\frac{\forall_5}{\forall_{6s}}\right)^k = (200.\ \text{kPa})\left(\frac{7.78 \times 10^{-4}\ \text{m}^3}{9.72 \times 10^{4}\ \text{m}^3}\right) = 148\ \text{kPa}$$

and

$$T_{6s} = T_5\left(\frac{\forall_5}{\forall_{6s}}\right)^{k-1} = (313\ \text{K})\left(\frac{7.78 \times 10^{-4}\ \text{m}^3}{9.72 \times 10^{-4}\ \text{m}^3}\right) = 648\ \text{K}$$

State 7s in Figure 13.52b

The process from states 5 to 7*s* is also isentropic, so

$$p_{7s} = p_5(\forall_5/\forall_{7s})^k = p_5(\text{CR})^k = (200.\ \text{kPa})(8.00)^{1.35} = 3310\ \text{kPa}$$

and

$$T_{7s} = T_5(\forall_5/\forall_{7s})^{k-1} = T_5(\text{CR})^{k-1} = (313\ \text{K})(8.00)^{0.35} = 648\ \text{K}$$

State 1 in Figure 13.52b

The mass of air in the cylinder is

$$m_{\text{air}} = \frac{p_{6s}\forall_{6s}}{RT_{6s}} = \frac{(148\ \text{kPa})(9.72 \times 10^{4}\ \text{m}^3)}{(0.287\ \text{kJ/kg} \cdot \text{K})(648\ \text{K})} = 1.73 \times 10^{-3}\ \text{kg}$$

Since we have an air–fuel ratio of 15.0, the mass of fuel in the cylinder is $m_{\text{fuel}} = \dfrac{m_{\text{air}}}{AF+1} = \dfrac{1.73 \times 10^{-3}\ \text{kg}}{15.0+1} = 1.08 \times 10^{-4}\ \text{kg}$

The heat produced by the combustion process is then

$$Q_{\text{comb}} = m_{\text{fuel}} \times \text{Fuel heating value} = (1.08 \times 10^{-4}\ \text{kg})(43300\ \text{kJ/kg-fuel}) = 4.69\ \text{kJ}.$$

Also, $Q_{\text{comb}} = m_{\text{air}}c_{\text{v-air}}(T_1 - T_{7s}) = (1.73 \times 10^{-3}\ \text{kg})(T_1 - 648\ \text{K}) = 4.69$ kJ. Solving for T_1 gives T_1 = 3940 K. Then, since the process from state 7*s* to 1 is a constant volume process, $p_1 = p_{7s}(T_1/T_{7s})$ = (3310 kPa)(3940 K/624 K) = 20.2 MPa.

State 2s in Figure 13.52b

The process from 1 to 2*s* is isentropic, so

$$p_{2s} = p_1\left(\frac{\forall_1}{\forall_{2s}}\right)^k = (20.2 \times 10^3\ \text{kPa})\left(\frac{9.72 \times 10^{-5}\ \text{m}^3}{9.72 \times 10^{-4}\ \text{m}^3}\right)^{1.35} = 901\ \text{kPa}$$

and

$$T_{2s} = T_1\left(\frac{\forall_1}{\forall_{2s}}\right)^{k-1} = (3920\ \text{K})\left(\frac{9.72 \times 10^{-5}\ \text{m}^3}{9.72 \times 10^{-4}\ \text{m}^3}\right) = 1760\ \text{K}$$

(Continued)

EXAMPLE 13.16 *(Continued)*

State 3 in Figure 13.52b

$p_3 = p_{exhaust} = 101$ kPa, and since the process from 2*s* to 3 is a constant volume process, we have $T_3 = T_{2s}(p_3/p_{2s}) = (1760 \text{ K})(101 \text{ kPa}/901 \text{ kPa}) = 196$ K.

State 4 in Figure 13.52b

$p_4 = p_3 = 101$ kPa and T_4 = atmospheric temperature.

Exercises

47. Determine the clearance volume in Example 13.16 if the compression ratio is 9.00 to 1 and the expansion ratio is 12.0 to 1. **Answer:** Clearance volume = 7.96×10^{-5} m^3.

48. Determine the pressure and temperature at the end of the compression stroke in Example 13.16 if the supercharger boost pressure is only 100. kPa instead of 200. kPa. **Answer:** p_{7s} =1660 kPa and T_{7s} = 648 K.

49. If the fuel used in Example 13.16 is changed to a hotter burning fuel with a heating value of 51,700 kJ/kg-fuel, determine the temperature at the end of the combustion process. **Answer:** T_1 = 4580 K.

13.21 DIESEL CYCLE

Rudolf Christian Karl Diesel (1858–1913) was a well-educated linguist and social theorist, but most of all, he was a remarkable engineer. He was born in Paris, but he received his technical education in Munich under Karl von Linde (1842–1934), a renowned pioneer in mechanical refrigeration.

Though the actual thermal efficiency of Otto's engine was many times better than that of Lenoir's, it was still barely competitive with the ever improving Rankine cycle steam engine. Diesel felt that he could eliminate the electrical ignition system of the Otto cycle engine if he could compress the air to the point where its temperature would be high enough to cause the fuel to ignite spontaneously. This would raise the maximum temperature of the cycle and consequently improve its thermal efficiency. He also felt that a higher combustion temperature would allow cheaper, heavier hydrocarbon fuels (such as kerosene, a common lamp oil in the late 19th century) to be used. On August 10, 1893, Diesel's first compression ignition engine ran under its own power for the first time, and by 1898, Diesel had become a millionaire simply by selling franchises for the industrial use of his engine.[11]

Diesel had originally intended to create an isothermal combustion process in the cylinder, so as to eliminate the heat transfer irreversibilities and thus approach the Carnot cycle thermal efficiency. He was not able to do this; instead, the ASC model of his cycle consists of two isentropic processes (compression and power), one isobaric process (combustion), and one isochoric process (exhaust), as shown in Figure 13.53.

DR. DIESEL VANISHES FROM A STEAMSHIP

Inventor of Oil Engine Missing after a Journey from Antwerp to Harwich

By Marconi Transatlantic Wireless Telegraph to *The New York Times*

LONDON, Sept. 30.—Dr. Rudolf Diesel, the famous inventor of the Diesel oil engine, has disappeared in most mysterious circumstances. He left Antwerp yesterday to attend in London the annual meeting of the Consolidated Diesel Engine Manufacturers. He embarked on the steamer *Dresden*, accompanied by a fellow Director, George Carels, and Herr Luckmann, Chief Engineer of the company.

Dr. Diesel had a cabin to himself. On the arrival of the vessel at Harwich at 6 o'clock this morning he was missing. His bed had not been slept in, though his night attire was laid out on it.

It is conjectured by his friends that Dr. Diesel fell overboard. He complained to a friend some time ago that he was occasionally troubled with insomnia, and it is possible that when his friends retired to their cabins he decided to continue

[11] Diesel's 1893 test engine compressed air to 80 atm, a pressure never before achieved by a machine. He was nearly killed when the engine subsequently exploded.

to stroll on deck. He was in the best of health, in very cheerful spirits, and had expressed most sanguine expectations as to the future of his engine and the development of the company.[12]

---

[12] *Ten days later, the crew of the Dutch boat Coertsen came upon the corpse of a man floating in the sea. The body was in such an advanced state of decomposition that they did not bring it aboard. Instead, the crew retrieved personal items (pill case, wallet, pocket knife, eyeglass case) from the clothing of the dead man and returned the body to the sea. On October 13, these items were identified by Rudolf's son as his father's. No one knows for sure how or why Diesel went overboard. Some believe he committed suicide because of his numerous "breakdowns," and some believe he was murdered by either fellow Germans (who resented his lack of nationalism) or by coal industrialists (who resented his engine).*

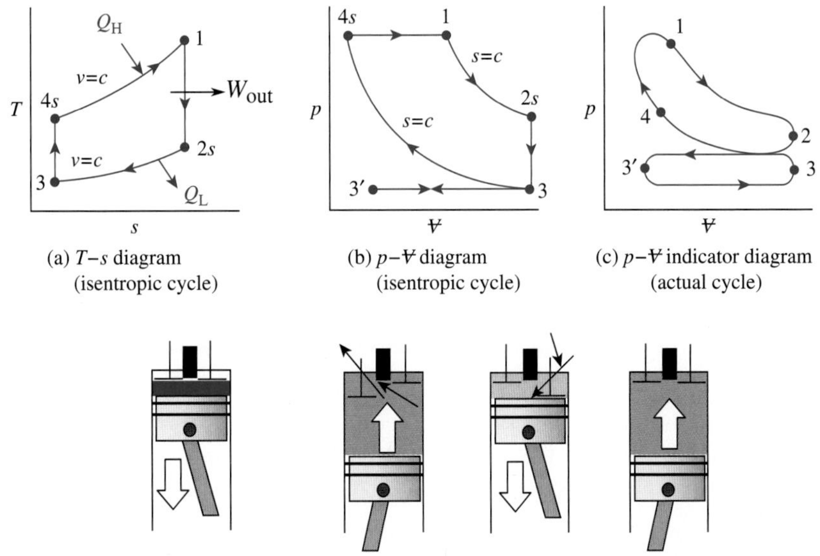

(a) T–s diagram (isentropic cycle)

(b) p–$\forall$ diagram (isentropic cycle)

(c) p–$\forall$ indicator diagram (actual cycle)

Combustion (4s to 1) and power (1 to 2s)

Exhaust (2s to 3 to 3′)

Intake (3′ to 3)

Compression (3 to 4s)

(d) The operation of a four–stroke Diesel cycle engine

FIGURE 13.53
The Diesel air standard cycle.

The thermal efficiency of the Diesel cycle is

$$(\eta_T)_{\text{Diesel}} = \frac{(\dot{W}_{\text{out}})_{\text{net}}}{\dot{Q}_{\text{H}}} = \frac{\dot{Q}_{\text{H}} - |\dot{Q}_{\text{L}}|}{\dot{Q}_{\text{H}}} = 1 - \frac{|\dot{Q}_{\text{L}}|}{\dot{Q}_{\text{H}}}$$

where (see Figure 13.53a), for the hot ASC,

$$|\dot{Q}_{\text{L}}| = \dot{m}(u_{2s} - u_3) \quad \text{and} \quad \dot{Q}_{\text{H}} = \dot{m}[(u_1 - u_{4s}) + p(v_1 - v_{4s})] = \dot{m}(h_1 - h_{4s})$$

Then, the thermal efficiency of the Diesel hot ASC is

$$(\eta_T)_{\substack{\text{Diesel}\\ \text{hot ASC}}} = 1 - \frac{u_{2s} - u_3}{h_1 - h_{4s}}$$

For the Diesel *cold ASC*, this becomes

$$(\eta_T)_{\substack{\text{Diesel}\\ \text{cold ASC}}} = 1 - \frac{c_v(T_{2s} - T_3)}{c_p(T_1 - T_{4s})} = 1 - \frac{T_3(T_{2s}/T_3 - 1)}{kT_{4s}(T_1/T_{4s} - 1)} \tag{13.35}$$

For the isentropic processes from 1 to 2s and from 3 to 4s, we have

$$T_{2s}/T_1 = (v_{2s}/v_1)^{1-k}$$

and

$$T_3/T_{4s} = (v_3/v_{4s})^{1-k} = \text{CR}^{1-k}$$

and, for the isobaric process from 4*s* to 1, we have

$$T_1/T_{4s} = v_1/v_{4s} = \text{CO}$$

where CO is called the *cutoff* ratio of the engine. For the isochoric process from 2*s* to 3, it can easily be shown that $T_{2s}/T_3 = \text{CO}^k$, then

$$(\eta_T)_{\substack{\text{Diesel}\\ \text{cold ASC}}} = 1 - \frac{\text{CR}^{1-k}(\text{CO}^k - 1)}{k(\text{CO} - 1)} \tag{13.36}$$

WHAT DOES *CUTOFF* MEAN?

The term *cutoff* is another archaic steam engine jargon term that has been absorbed into modern IC engine terminology. It was introduced in the 1780s by James Watt, when he realized that, if the steam entering the cylinder was "cut off" (i.e., shut off) when the piston had completed only a portion of its stroke and the natural expansion of the steam was allowed to complete the stroke, then the engine's thermal efficiency increased significantly. Today, this term is used to indicate where the combustion process "cuts off" (i.e., stops) in a compression ignition internal combustion engine. It is determined by the geometry of the combustion chamber and the fuel charge.

EXAMPLE 13.17

The Wärtsilä RT-flex96C, the two-stroke, turbocharged, low-speed Diesel engine shown in Figure 13.54, was manufactured by the Finnish manufacturer Wärtsilä. With 14 cylinders, it is one of the largest reciprocating engines in the world. Determine the following items for the Diesel cold ASC with $k = 1.4$. Assume that, when the engine is producing 80,080 kW of output power, it has a compression ratio of 18.0 to 1 with a cutoff ratio of 2.32, and it uses fuel with a heating value of 45.5×10^3 kJ/kg with a fuel flow rate of rate 3.35 kg/s.

a. The Diesel cold ASC thermal efficiency of the engine.
b. The actual thermal efficiency of the engine.
c. The mechanical efficiency of the engine.

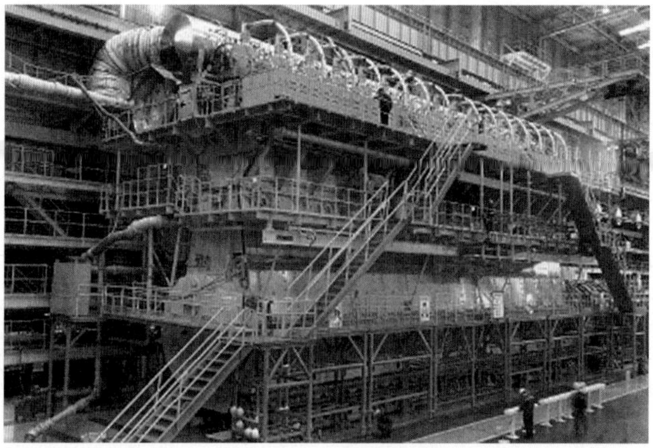

FIGURE 13.54
Example 13.17.

The 14-cylinder Wärtsilä RT-flex96C marine engine was put into service in September 2006 aboard the *Emma Mærsk*. Its maximum continuous power output was 80,080 kW (108,920 bhp) at 102 rpm. Measuring 27.3 m long and 13.5 m high, its overall weight is 2300 tons.

Solution

a. The compression ratio CR = 18.0, and the cutoff ratio CO = 2.32, so the Diesel cold ASC thermal efficiency is given by Eq. (13.36) as

$$(\eta_T)_{\substack{\text{Diesel}\\ \text{cold ASC}}} = 1 - \frac{(18.0)^{-0.40}[(2.32)^{1.40} - 1]}{1.40(2.32 - 1)} = 0.617 = 61.7\%$$

b. The rate of energy provided by burning the fuel is

$$\dot{Q}_{\text{fuel}} = (\text{Fuel heating value in kJ/kg-fuel}) \times (\text{Fuel flow rate in kg-fuel/s})$$

$$= (45.5 \times 10^3 \text{ kJ/kg-fuel}) \times (3.35 \text{ kg-fuel/s}) = 152 \times 10^3 \text{ kJ/s} = 152{,}000 \text{ kW}$$

Then the actual thermal efficiency of the engine is given by Eqs. (7.5) and (13.33) as

$$(\eta_T)_{\substack{\text{Diesel}\\ \text{actual}}} = \frac{(\dot{W}_B)_{\text{out}}}{\dot{Q}_{\text{fuel}}} = \frac{80{,}080}{152{,}000} = 0.527 = 52.7\%$$

c. Since $(\eta_T)_{\text{actual}} = (\eta_m)(\eta_T)_{\text{cold ASC}}$,

$$\eta_m = \frac{(\eta_T)_{\text{actual}}}{(\eta_T)_{\text{cold ASC}}} = \frac{0.525}{0.617} = 0.851 = 85.1\%$$

Exercises

50. Determine the Diesel cold ASC efficiency discussed in Example 13.17 if the compression ratio is decreased to 14.0 to 1. Assume all other variables remain unchanged. **Answer:** $(\eta_T)_{\text{Diesel cold ASC}} = 61.6\%$.

51. Find the Diesel cold ASC thermal efficiency of the engine in Example 13.17 if the cutoff ratio is increased to 4.65. **Answer:** $(\eta_T)_{\text{Diesel cold ASC}} = 61.1\%$.

52. Calculate the indicated horsepower output for the Diesel engine in Example 13.17. **Answer:** $(\dot{W}_I)_{\text{out}} = (\eta_T)_{\substack{\text{Diesel}\\ \text{cold ASC}}} \times \dot{Q}_{\text{fuel}} = 0.661 \times 152{,}000 \text{ kW} = 100{,}500 \text{ kW}$.

53. Determine the actual thermal efficiency of the diesel engine in Example 13.16 if it produces 85,500 kW rather than 80,080 kW. Assume all other variables remain unchanged. **Answer:** $(\eta_T)_{\text{Diesel actual}} = 56.1\%$.

NO RAY OF LIGHT ON DIESEL MYSTERY

German Inventor Was a Millionaire and His Home Was Happy

Special Cable to *The New York Times*

LONDON, Oct. 1—Today brought no fresh tidings in regard to Dr. Rudolf Diesel, the oil engine inventor, and that he met his death by drowning on his way from Antwerp to Harwich on Monday night appears to be certain.

Dr. Diesel's friends are greatly mystified. While on the one hand the probabilities of an accidental fall overboard seem remote, on the other hand they cannot conceive any motive which might prompt a suggestion of suicide.

Dr. Diesel was a wealthy man. His patent rights in the Diesel engine were sold for huge sums in various countries, and, having amassed a fortune, he had to all intents and purposes retired from active business. It is believed that he amassed a fortune of $2,500,000 in a few years.

Sidney Whitman, the oldest of Dr. Diesel's English friends, says Dr. Diesel lived in Munich in one of the most palatial modern houses in Bavaria, which cost $250,000 to build. Mr. Whitman says that when he was in America last year Dr. Diesel struck up a great friendship with Thomas A. Edison. The latter told him not to eat too much and he would live to be a hundred.

NO LIGHT ON DIESEL'S FATE

Special Cable to *The New York Times*

LONDON, Oct. 2.—The mystery of the disappearance on Monday night of Dr. Rudolf Diesel on the cross-channel steamer *Dresden* is still unexplained. A report that Dr. Diesel did not sail on the *Dresden* receives no confirmation. The inventor was seen on deck after the vessel left Flushing.

The British Board of Trade inquiry has elicited nothing, and an examination of Dr. Diesel's private papers at Munich has been equally without result.

Baron Schmidt, Dr. Diesel's son-in-law, declares that the theory of suicide in a sudden fit of aberration is entirely unsupported.

When the inventor went to see his daughter at Frankfurt after several days' shooting in the Bavarian highlands he complained that he had overstrained himself and that this had accentuated the weakness of the heart from which he had suffered in recent years. Nevertheless, he was in the best possible spirits.

DIESEL WAS BANKRUPT

He Owed $375,000—Tangible Assets Only about $10,000
By Marconi Transatlantic Wireless Telegraph to *The New York Times*

MUNICH, Oct. 14.—The deplorable state of the late Dr. Rudolf Diesel's finances was revealed at today's meeting of his creditors here.

The meeting found itself unable to take definite action regarding the administration of Dr. Diesel's wrecked fortune, as the exact state of affairs remains to be cleared up. It is declared that the inventor's liabilities are approximately $375,000, against which the tangible assets are only $10,000.

Figures laid before the meeting showed a state of confusion in the tabulation of Dr. Diesel's supposed assets. In the case of some houses at Hamburg and Munich it was found that there was an overvaluation of $125,000.

Diesel cycle engines operate with a much higher compression ratio than Otto cycle engines (12 to 24 vs. 8 to 11) and therefore are more efficient than Otto cycle engines. Typical Otto cycle engine actual thermal efficiencies are in the range from 15 to 25%, whereas for Diesel cycle engines, they normally fall in the range from 30 to 45%. As the previous example illustrates, very large Diesel engines can have efficiencies greater than 50%.

Because combustion takes place intermittently in internal combustion engines and therefore the cylinder is alternately heated by combustion and cooled by the intake stroke (plus the fact that most of these engines have water jacket cooling), they have the same heat transfer irreversibilities as the early 18th century Newcomen steam engines. A similar cyclic cylinder heating and cooling process in the Newcomen steam engine led James Watt to develop the external steam condenser that improved the engine's thermal efficiency fourfold.

13.22 MODERN PRIME MOVER DEVELOPMENTS

Solving for the power from an energy rate balance (ERB) on a steady state, steady flow, single-inlet, single-outlet prime mover (neglecting any changes in kinetic or potential energy) yields

$$\dot{W}_{\text{out}} = \dot{m}(h_{\text{in}} - h_{\text{out}}) - |\dot{Q}|_{\text{loss}}$$

Since the primary objective of any prime mover is to produce power ($\dot{W}_{\text{out}} > 0$), any heat loss ($|\dot{Q}|_{\text{loss}} > 0$) clearly reduces both the power output and the thermal efficiency of the prime mover. Consequently, most external combustion prime movers are heavily insulated to minimize their heat loss and maximize their thermal efficiency. When there is no heat loss or gain by a prime mover, it can properly be called an *adiabatic prime mover*.

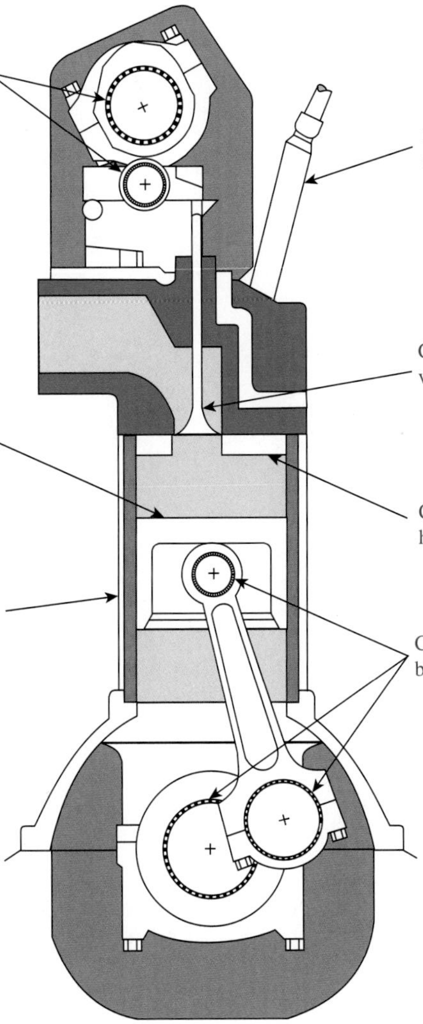

FIGURE 13.55
Typical ceramic components of a modern IC engine.

Current internal combustion engines are not only uninsulated, they are intentionally cooled. This is done to prevent the buildup of excessively high internal temperatures that in turn would cause material failure due to loss of strength. Consequently, about 80% of the chemical energy originally contained in the fuel leaves an IC engine as thermal energy in the coolant and exhaust gases.

The development of high-temperature, high-strength ceramic or *superalloy* engine components is a step in the direction of creating a truly adiabatic IC engine. In addition to reducing engine heat loss, these new components allow higher internal operating temperatures to be achieved, and this in turn increases the maximum theoretical thermal efficiency of the engine. For example, doubling the operating temperature from 2000°F (1093°C) to 4000°F (2204°C) increases the Carnot isentropic efficiency by about 10%. Figure 13.55 illustrates some of the current uses of ceramics in reciprocating IC engines. Similar advances are being made in gas turbine engine technology.

Other thermal efficiency–increasing technology, such as supercharging, turbocharging, and turbocompounding, can be used to extract some of the thermal energy from the engine's exhaust gases (see Figure 13.56). However, it must be remembered that all cyclic heat engines (including all IC engines when their cycle is closed by the environment) must have a heat loss rate to the environment, dictated by the second law of thermodynamics, of

$$|\dot{Q}|_{\text{loss}} \geq (T_{\text{environment}}/T_{\text{engine max}})\,|\dot{Q}|_{\text{fuel}}$$

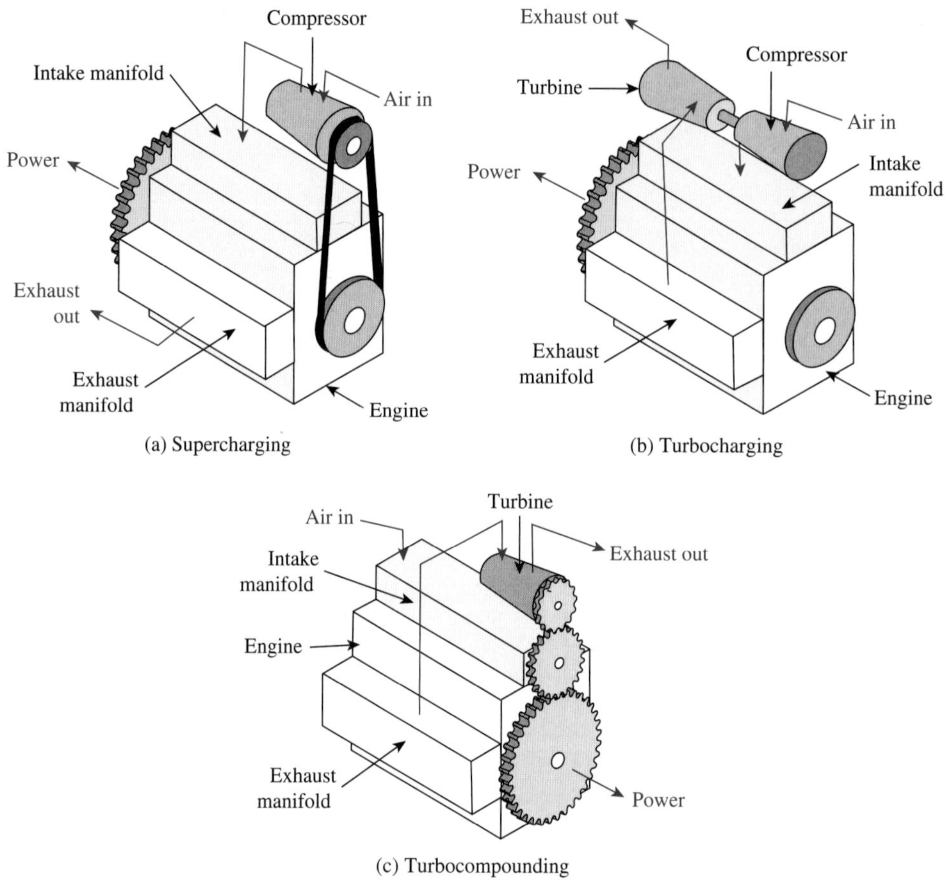

FIGURE 13.56
Air and exhaust flow arrangements for supercharging, turbocharging, and turbocompounding an internal combustion engine.

13.23 SECOND LAW ANALYSIS OF VAPOR AND GAS POWER CYCLES

The difference between the isentropic thermal efficiency and the actual thermal efficiency of a system is due to the effects of the second law of thermodynamics. The second law can be used to determine viscous irreversibilities leading to pressure losses in pipes, valves, and fittings; to determine heat transfer irreversibilities due to incomplete insulation and large temperature gradients; and to determine mechanical and chemical irreversibilities in pumps, compressors, and prime movers due to friction and chemical reactions. However, piping and ancillary viscous losses (irreversibilities) are normally determined through the empirical *friction factor* material introduced in Chapter 9 (and found in most fluid mechanics textbooks), and mechanical and thermal losses in machinery are globally lumped into the empirically determined isentropic efficiency, η_s. The application of the second law to complex engines and turbines is so difficult today that it is not normally used in the engineering design stage of product development. This will, no doubt, change as technology and engineering analysis advance in the future.

The thermal efficiency η_T is essentially a *first law* energy conversion efficiency, in that it is concerned with the effectiveness of an energy conversion process as the ratio of a desired output to a required input. The isentropic efficiency η_s, on the other hand, can be viewed as a *second law* energy conversion efficiency in that it compares the actual energy conversion performance of a real (irreversible) device with its idealized reversible counterpart. Consequently, the primary role of the second law of thermodynamics in the analysis of vapor and gas power cycles today is through the (largely empirical) determination and use of the isentropic efficiency. Many of the practical aspects of engineering courses in heat transfer, fluid mechanics, electrical circuit theory, and machine design today are the result of the consequences of the second law of thermodynamics.

CASE STUDIES IN APPLIED THERMODYNAMICS

The following are examples of typical case studies in the field of applied thermodynamics. They are meant to demonstrate the practical use of the material presented in this chapter. The examples are chosen from a wide variety of well-known technologies, and the thermodynamic analysis has been presented essentially as a diagnostic tool. In this way, we can develop a quick understanding of how some simple and some complex items behave from a thermodynamic point of view.

Case study 13.1. The Stanley Steamer automobile

In 1897, a Rankine cycle steam-powered automobile was introduced by the twin brothers Francis Edgar Stanley (1849–1918) and Freelan Oscar Stanley (1849–1940). Their automobiles were affectionately known as *Stanley Steamers* (Figure 13.57), and most had a two-cylinder, 30.0 hp reciprocating steam engine with a 4.0 in bore and a 5.0 in stroke. The boiler operated at 600. psia and 600.°F and was fueled with gasoline or kerosene. The engine was mounted in the rear of the automobile and connected directly to the drive wheels. Therefore, Stanley Steamers did not require a drive shaft, transmission, or differential. The engine (and the automobile's) speed was controlled simply by altering the amount of steam reaching the engine with a hand-operated *throttle* valve. The Stanley Steamers did not have a condenser until 1917. Before then, they exhausted their spent steam directly into the atmosphere. When a condenser was finally added to the engine, its main function was to conserve and recycle water and not to improve the efficiency of the power plant. Consequently, the condenser looked and operated very much like a standard automobile radiator, condensing at atmospheric pressure instead of a vacuum. Under these conditions, the engine produced 30.0 hp with an isentropic efficiency of about 80.0%. Using basic engineering thermodynamics, we can estimate the steam flow rate required for the engine and the amount of water consumed in traveling 1 mile at 55.0 mph.

Station 1—Engine inlet	Station 2s—Condenser inlet
$p_1 = 600.$ psia	$p_{2s} = p_2 = 14.7$ psia
$T_1 = 600°\text{F}$	$s_{2s} = s_1 = 1.5322\ \text{Btu/lbm}\cdot\text{R}$
$h_1 = 1289.5\ \text{Btu/lbm}$	$x_{2s} = (1.5322 - 0.3122)/1.4447 = 0.8445$
$s_1 = 1.5322\ \text{Btu/lbm}\cdot\text{R}$	$h_{2s} = 180.1 + 0.8445(970.4) = 999.6\ \text{Btu/lbm}$

Station 3—Pump inlet	Station 4s—Boiler inlet
$p_3 = p_{2s} = 14.7$ psia	$p_{4s} = p_1 = 600.$ psia
$x_3 = 0.00$	$s_{4s} = s_3 = 0.3122\ \text{Btu/lbm}\cdot\text{R}$
$h_3 = 180.1\ \text{Btu/lbm}$	$h_{4s} = h_3 + v_3(p_{4s} - p_3)$
$s_3 = 0.3122\ \text{Btu/lbm}\cdot\text{R}$	$= 180.1 + 0.01672(600. - 14.7)(144/778.16)$
$v_3 = 0.01672\ \text{ft}^3/\text{lbm}$	$= 181.9\ \text{Btu/lbm}$

Then, since $\dot{W} = (30.0\ \text{hp})(2545\ \text{Btu/hp}\cdot\text{h}) = 76{,}400\ \text{Btu/h}$, and $(\eta_s)_{\text{engine}} = 0.800$, the required steam mass flow rate is

$$\dot{m} = \frac{\dot{W}}{(h_1 - h_{2s})(\eta_s)_{\text{engine}}} = \frac{76{,}400\ \text{Btu/h}}{(1289.5 - 999.6\ \text{Btu/lbm})(0.800)} = 329.\ \text{lbm/h}$$

and, if the vehicle is traveling at 55.0 mph, it uses

$$\frac{329\ \text{lbm/h}}{55.0\ \text{mi/h}} = 5.98\ \text{lbm/mi}$$

or about $\frac{3}{4}$ of a gallon of water per mile.

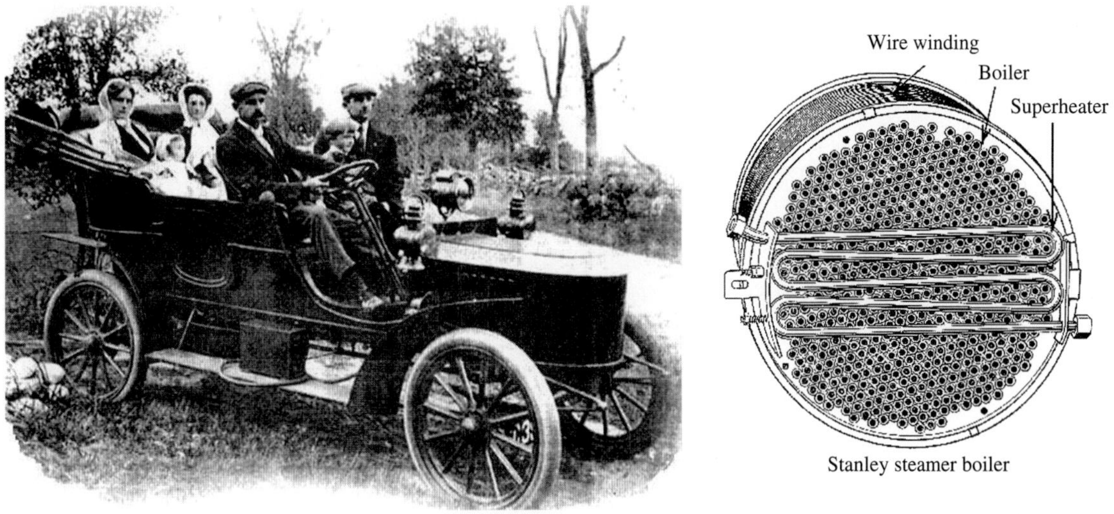

FIGURE 13.57
1909 Stanley Steamer car and boiler.

(Continued)

CASE STUDIES IN APPLIED THERMODYNAMICS *Continued*

Note that, if we take the isentropic efficiency of the boiler feed pump to be 70.0%, then the thermal efficiency of the entire Stanley power plant is quite respectable at

$$(\eta_T)_{\substack{\text{Stanley}\\ \text{Steamer}}} = \frac{(h_1 - h_{2s})(\eta_s)_{\text{engine}} - v_3(p_{4s} - p_3)/(\eta_s)_{\text{pump}}}{h_1 - h_4}$$

$$= \frac{(1289.5 - 999.6)(0.80) - 0.01672(600. - 14.7)\left(\frac{144}{778.16}\right)\left(\frac{1}{0.700}\right)}{1289.5 - \left(180.1 + 0.01672(600. - 14.7)\left(\frac{144}{778.16}\right)\left(\frac{1}{0.700}\right)\right)}$$

$$= 0.207 = 20.7\%$$

In Chapter 7, we learned that the amount of entropy production inside a system is a key factor in limiting its energy conversion efficiency. For example, the more moving parts and friction there are inside an engine, the lower is its isentropic efficiency. Since Stanley engines used double-acting cylinders with two power strokes occurring with every revolution of the engine's crankshaft, they produced the same power as an equivalent eight-cylinder Otto cycle internal combustion engine that has only one power stroke occurring in every two revolutions of the engine's crankshaft.

A typical eight-cylinder Otto cycle engine has 50–100 moving parts with very significant mechanical friction, whereas the two-cylinder, double-acting Stanley engine had only 15–25 moving parts. In fact, an entire Stanley automobile had only 37 moving parts. Consequently, the Stanley automobiles were very effective energy conversion devices.

In 1906, the Stanley brothers set a new world land speed record of 127.66 mph at the Dewar Cup Race in Omond Beach, Florida, with a steam-powered race car called the Rocket (Figure 13.58). The car's body was made by a canoe factory and looked like an upside-down boat. The engine was the same two-cylinder Stanley steam engine used in their production vehicles, but it had been enhanced with a 4.50 in bore and a 6.50-inch stroke. The standard Stanley boiler had been enlarged to a 30.0 inches in diameter and was 18.0 inches high, and operated at 1000 psia and about 700°F. Under these conditions the little two-cylinder Stanley steam engine produced a whopping 250 hp.

FIGURE 13.58
The Rocket race car.

In 1907, the Stanleys again tried to set a new world land speed record. This attempt ended in disaster, when the car became airborne at about 190 mph and crashed. The driver survived the crash, but a spectator in the crowd had to reinsert the driver's right eye, which had been dislodged from its socket by the force of the impact. At the time of the crash, the boiler pressure had been increased to an incredible 1300 psia.

Case study 13.2. The drinking bird as a heat engine

Many novelty stores carry a toy called the *drinking bird*, which bobs up and down, apparently drinking from a glass of water (Figure 13.59). This toy is really a small heat engine that uses the evaporation of water from its head as the power source for its operation (Figure 13.60).

The head must be lightly covered with something that will hold a small amount of water (e.g., a light fuzz) and the beak is simply a wick that keeps the head wet. As the water evaporates from the head, it cools the vapor inside the head causing a slight vacuum to form. The liquid in the bottom of the bird is then drawn up into the bird's neck, shifting the center of gravity of the bird forward, and ultimately causing it to tip. When the bird tips, the beak is rewetted and a vapor bubble passes through the neck equalizing the pressures between the ends of the bird and restoring the bird's center of gravity. The bird then returns to its original upright position. This cycle is continuously repeated until the water source is exhausted.

FIGURE 13.59
Drinking bird.

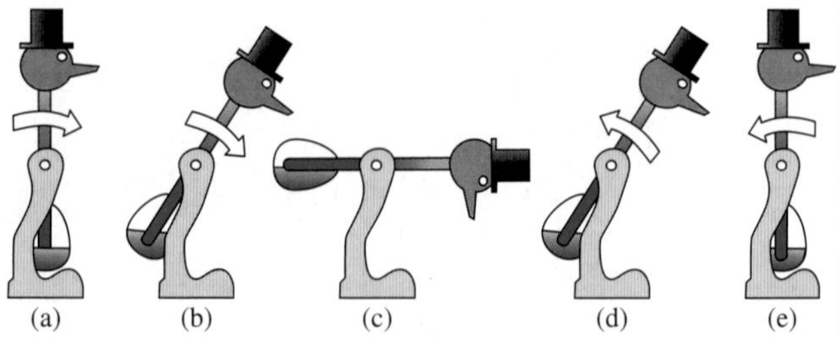

FIGURE 13.60
The operation of the drinking bird.

Since the motion of the bird could easily be connected to a shaft to produce work, the drinking bird is clearly some form of an engine. So how does this engine work?

By simple observation, we conclude that the evaporation of liquid water from the head of the bird causes the physical changes in the liquid inside the bird. Also, our observations tell us that the relative humidity in the room must be below 100% for the bird to operate. Therefore, it would seem that the bird is some form of steam engine, where the *steam* is just the water vapor that evaporates from the bird's head, and the partial pressure of this water vapor must be lower than the saturation pressure of water vapor in the room air for the bird to operate. Since the bird's head is constantly moving, it is not unreasonable to assume that its temperature is approximately equal to the wet bulb temperature of the room. If we then complete the steam cycle by assuming the evaporated water is condensed in the atmosphere at the dew point temperature, we can construct the *T–s* vapor power cycle shown Figure 13.61.

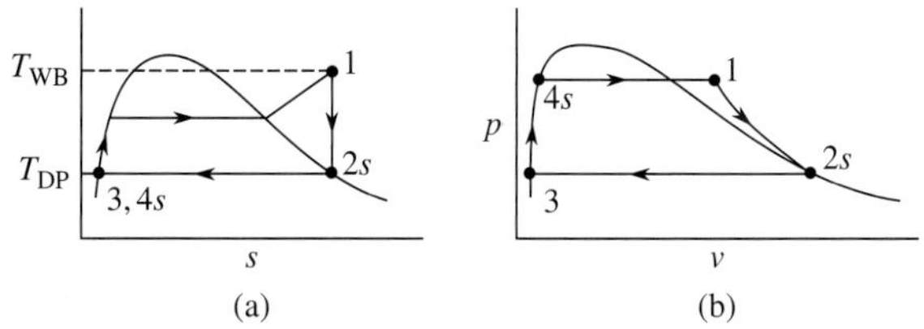

FIGURE 13.61
T–s and *p–v* diagrams for the drinking bird.

We can estimate its power-producing potential and operational thermal efficiency as follows. The equivalent Carnot cycle thermal efficiency is given by

$$(\eta_T)_{\substack{\text{Carnot}\\ \text{drinking bird}}} = 1 - \frac{T_L}{T_H} = 1 - \frac{T_{DP}}{T_{WB}}$$

and the isentropic Rankine cycle thermal efficiency is

$$(\eta_T)_{\substack{\text{Isentropic}\\ \text{Rankine}\\ \text{drinking bird}}} = \frac{(h_1 - h_{2s}) - (h_{4s} - h_3)}{h_1 - h_{4s}}$$

where the states are identified using the relative humidity ϕ, the dry bulb temperature T_{DB}, the wet bulb temperature T_{WB}, and the dew point temperature T_{DP} as follows:

Station 1—Bird's head

$p_1 = \phi p_{sat}(T_{DB})$
$T_1 = T_{WB}$
$h_1 = h(p_1, T_1)$
$s_1 = s(p_1, T_1)$

Station 3—End of condensation

$p_3 = p_{2s}$
$x_3 = 0.0$
$h_3 = h(p_3, x_3)$
$s_3 = s(p_3, x_3)$

Station 2s—Beginning of condensation

$T_{2s} = T_{DP}$
$s_{2s} = s_1$
$h_{2s} = h(T_{2s}, s_{2s})$
$p_{2s} = p(T_{2s}, s_{2s})$

Station 4s—Liquid brought back to pressure p_1

$p_{4s} = p_1$
$s_{4s} = s_3$
$h_{4s} = h(p_{4s}, s_{4s})$

Figure 13.62 shows how these thermal efficiencies vary with room relative humidity. Clearly, the drier the air in the room, the more efficient the drinking bird becomes.

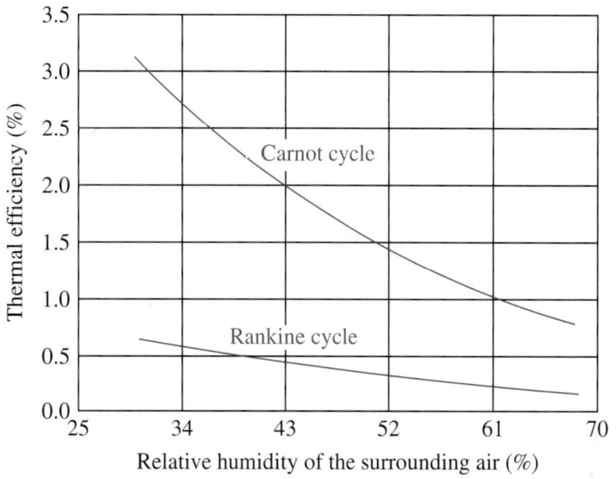

FIGURE 13.62
Thermal efficiency of a drinking bird.

The drinking bird thermal efficiency is roughly equivalent to the low thermal efficiencies of the original Newcomen and Watt steam engines. However, a scaled-up version could be effective as an engine in dry climates.

Case study 13.3. Aircraft engine development

The world's largest aircraft engine manufacturers developed a new generation of gas turbine engines that provide 60,000 to 100,000 lbf of thrust. Pratt & Whitney has its PW4000, Rolls Royce has its Trent, and General Electric has its GE90. The PW4000 has an isentropic pressure ratio of 31.5 and a bypass ratio of 5:1, producing 60,000 lbf of thrust. The GE90 has an isentropic pressure ratio of 50.0, and a bypass ratio of 10.0 to 1, producing a thrust in the 80,000 lbf range. This case study focuses on the GE90 turbofan engine development.

Before we can discuss aircraft engine design, we need to do some background work. The thrust produced by an aircraft engine is primarily the difference between the rate of momentum imparted to the gases exiting the engine and the momentum rate of the inlet air. Neglecting the small momentum produced by the fuel flow rate, the thrust can be written as

$$\text{Thrust} = T = \dot{m}_a(V_{\text{exhaust}} - V_{\text{aircraft}})/g_c$$

where, in this case, $V_{\text{aircraft}} = V_{\text{inlet air}}$ and $\dot{m}_a = \dot{m}_{\text{air}} = \dot{m}_{\text{exhaust}}$. Since the exhaust velocity is always greater than the aircraft velocity, we see from this equation that there are two ways to increase the thrust of an aircraft engine:

1. Increase the engine's exhaust gas velocity.
2. Increase the engine's air mass flow rate.

A *turbojet* engine is a Brayton cycle gas turbine engine that produces all of its thrust by generating a very high exhaust gas velocity. This is done by forcing all the turbine's hot exhaust gases through a flow nozzle to maximize the exhaust velocity (see Figure 13.63a). A *turbofan* engine is also a Brayton cycle gas turbine engine. But it

(*Continued*)

CASE STUDIES IN APPLIED THERMODYNAMICS *Continued*

produces thrust by a combination of high-speed exhaust gas plus low-speed "bypass" air produced by an external "fan" (or a low-pressure ratio compressor stage) directly connected to the engine's turbine to increase the value of the air mass flow rate $\dot{m}_{air}$ (see Figure 13.63b). Finally, a *turboprop* engine is a Brayton cycle gas turbine engine that produces some thrust with a high-speed exhaust gas, like the turbojet, with the remaining thrust coming from a standard low-speed propeller driven by the engine's turbine through a speed-reducing gearbox (see Figure 13.63c).

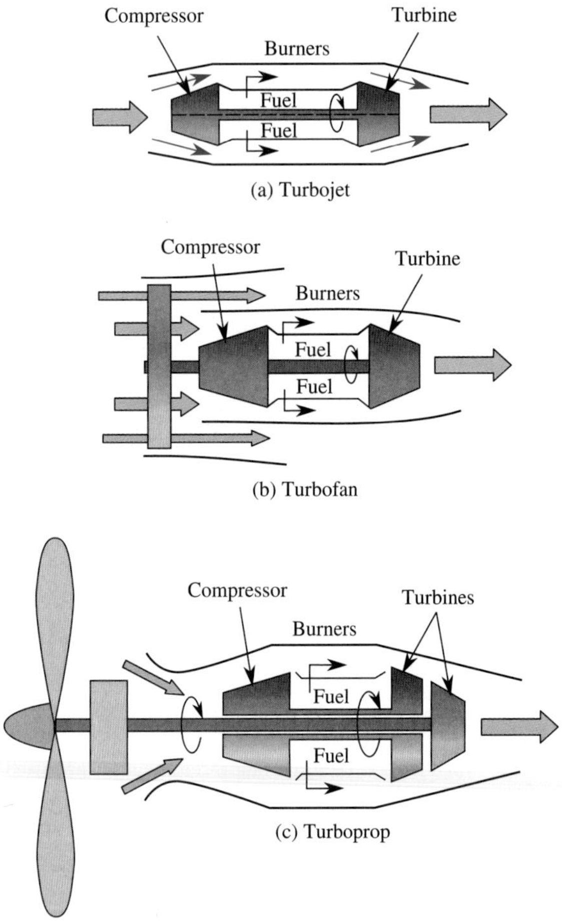

FIGURE 13.63
Brayton cycle gas turbine engines.

Each of these engine designs has its advantages and disadvantages. A turbojet engine is ideal for very high-speed flight, but it does not perform well at low speeds or low altitudes. A turbofan engine performs well at moderate to high speeds, and a turboprop engine performs well at low aircraft speeds but not at high speeds.

The propulsion efficiency of these three engine designs is given by

$$\eta_{propulsion} = \frac{\text{Thrust power output}}{\text{Thrust power output} + \text{Lost kinetic energy rate}}$$

Neglecting the momentum associated with the mass flow rate of the fuel, the thrust of a turbojet engine is given by

$$T_{turbojet} = \dot{m}_a(V_{exhaust} - V_{aircraft})/g_c$$

and the thrust of a turbofan or turboprop engine is given by

$$T_{\substack{turbofan \\ or\ turboprop}} = \dot{m}_a(\overline{V}_{exhaust} - V_{aircraft})/g_c$$

where
$\dot{m}_a$ = air mass flow rate,

$V_{aircraft} = V_{inlet\ air}$ (the air inlet velocity is the same as the aircraft velocity),

$V_{exhaust}$ = turbojet exhaust gas velocity,

$\overline{V}_{exhaust}$ = turbofan or turboprop average exhaust gas velocity, defined as

$$\overline{V}_{exhaust} = (\dot{m}_{aH}V_{eH} + \dot{m}_{aC}V_{eC})/(\dot{m}_{aH} + \dot{m}_{aC}) = (\dot{m}_{aH}V_{eH} + \dot{m}_{aC}V_{eC})/\dot{m}_a,$$

where $\dot{m}_{aH}$ and V_{eH} are the mass flow rate and exhaust gas velocity of that portion of the exhaust passing through the hot combustion chamber, and $\dot{m}_{aC}$ and V_{eC} are the mass flow rate and exhaust velocity of that portion of the exhaust gases that "bypass" the combustion chamber to produce additional thrust by increasing the magnitude of $\dot{m}_a$.

Then, the propulsion efficiencies can be written as

$$(\eta_{propulsion})_{turbojet}$$

$$= \frac{\dot{m}_a(V_{exhaust} - V_{aircraft})(V_{aircraft}/g_c)}{\dot{m}_a(V_{exhaust} - V_{aircraft})(V_{aircraft}/g_c) + (\dot{m}_a/2g_c)(V_{exhaust} - V_{aircraft})^2}$$

$$= \frac{2}{1 + V_{exhaust}/V_{aircraft}}$$

and

$$(\eta_{propulsion})_{\substack{turbofan \\ or\ turboprop}}$$

$$= \frac{\dot{m}_a(\overline{V}_{exhaust} - V_{aircraft})(V_{aircraft}/g_c)}{\dot{m}_a(\overline{V}_{exhaust} - V_{aircraft})(V_{aircraft}/g_c) + (\dot{m}_a/2g_c)(\overline{V}_{exhaust} - V_{aircraft})^2}$$

$$= \frac{2}{1 + \overline{V}_{exhaust}/V_{aircraft}}$$

Note that $\eta_{propulsion}$ increases as aircraft speed approaches the engine's exhaust velocity ($V_{aircraft} \to V_{exhaust}$), but the engine's thrust vanishes as $V_{aircraft} \to V_{exhaust}$. So, to maintain the thrust while we increase $V_{aircraft}$, we must increase $\dot{m}_a$. Therefore, an engine with a large $\dot{m}_a$ and a small $V_{exhaust}$ is more efficient than an engine with the same thrust moving a small $\dot{m}_a$ at a high $V_{exhaust}$.

Another measure of aircraft performance is its *thrust efficiency*, defined as

$$\eta_{thrust} = \frac{\text{Thrust power}}{\text{Fuel power}} = \frac{T \times V_{aircraft}}{\dot{Q}_{fuel}}$$

where $\dot{Q}_{fuel}$ is the rate of heat produced by the burning fuel. For turbojet engines, this is

$$(\eta_{thrust})_{turbojet} = \frac{\dot{m}_a(V_{exhaust} - V_{aircraft})V_{aircraft}}{\dot{Q}_{fuel}}$$

and, for turbofan and turboprop engines, this it

$$(\eta_{thrust})_{\substack{turbofan\\ or\, turboprop}} = \frac{\dot{m}_a(\overline{V}_{exhaust} - V_{aircraft})V_{aircraft}}{\dot{Q}_{fuel}}$$

We can now maximize the thrust efficiency by differentiating it with respect to the aircraft's velocity V while holding all the other variables constant, then setting the result equal to zero to find the optimum aircraft velocity as

$$V_{optimum} = \left(\frac{V_{exhaust}}{2}\right)_{turbojet} = \left(\frac{\overline{V}_{exhaust}}{2}\right)_{\substack{turbofan\\ or\, turboprop}}$$

Inserting this result into the turbojet, turbofan, and turboprop propulsion efficiency equations produces a common optimum propulsion efficiency for all three engines of

$$(\eta_{propulsion})_{optimum} = \frac{2}{3} = 0.667 = 66.7\%$$

The GE90 engine development

In early 1990, the Aircraft Engines Division of General Electric launched the design of the GE90, a high-thrust, turbofan, gas turbine engine (Figure 13.64) designed to meet the needs of the emerging superjumbo wide-body passenger aircraft market. It was clear at that time that there would be increasing air traffic occurring over longer distances as world markets continued to evolve in the Pacific Rim, Eastern Europe, and South America. More passengers traveling longer distances meant more air congestion or larger planes. Also, in the 1990s, the aging fleet of over 3000 Boeing 747s could be replaced with new wide-body, twin-engine passenger aircraft capable of carrying 300 passengers over 6000 miles. Such an aircraft would have a wingspan of about 200 ft and a gross weight of about 500,000 lbf. This meant that each engine required a takeoff thrust of about 100,000 lbf. The resulting GE90 engine specifications were

Bypass ratio $(\dot{m}_{aC}/\dot{m}_{aH}) = 10.0$ to 1
Compressor compression ratio $p_{2s}/p_1 = \text{PR} = 50.0$ to 1
Engine thrust $= T = 100{,}000$ lbf
Fan diameter $= D_{fan} = 123$ in. $= 10.25$ ft
$\dot{m}_a = 2150$ lbm/s

FIGURE 13.64
Cross-section of the GE90 engine.

For a static (takeoff) thrust of 100,000 lbf, the average exit velocity must be

$$\overline{V}_{exhaust} = \frac{Tg_c}{\dot{m}_a} = \frac{(100{,}000\,\text{lbf})(32.174\,\text{lbm}\cdot\text{ft/lbf}\cdot\text{s}^2)}{2150\,\text{lbm/s}} = 1500\,\text{ft/s}$$

and for a cruising speed of 500. mph = 733 ft/s, this exhaust velocity yields a propulsion efficiency of

$$\eta_{propulsion} = \frac{2}{1+\dfrac{1500.}{733}} = 0.657 = 65.7\%$$

Note that, since $V_{aircraft} \approx V_{exhaust}/2$ here, this is very close to the optimum propulsion efficiency (66.7%) for this engine.

The GE-IA: The first U.S. turbojet engine

In January 1941, the U.S. National Academy of Sciences reported that gas turbine engines were impractical for aircraft propulsion because their power to weight ratio was too low. However, on August 24, 1939, Germany flew its first turbojet-powered aircraft (the Heinkel-178), and by June 1944, German combat jet aircraft (Messerschmitt ME-262 *Swallow*, powered by two Junkers Juno turbojet engines with an air speed of 540 mph at 20,000 ft) had entered World War II. Also, on May 15, 1941, the British first flew a Gloster *Meteor* jet aircraft powered by a single Whittle W-1X turbojet engine as part of their war research and development program. These events prompted the U.S. government to issue a contract to General Electric in September 1941 to build and test 15 gas turbojet engines based on the designs of the British aircraft engineer Frank Whittle. On March 18, 1942, the first GE type I-A (pronounced "eye-A") turbojet engine was completed and tested at GE's River Works facility in Lynn, Massachusetts. It produced about 1250 lbf of static thrust at 15,000 rpm. On October 1, 1942, the first U.S. turbojet-powered aircraft (the Bell XP-59A *Aircomet*) was flown at Muroc Dry Lake (now called Edwards Air Force Base) in California. It had an air speed of 400 mph, 140 mph less than the German ME-262.

The tests with the Bell XP-59A did not result in U.S. combat aircraft during World War II, but it did play a key role in the later development of the XP-80 *Shooting Star*, used in the Korean War in the early 1950s.

If the air mass flow rate into the GE-IA engine is 18.0 lbm/s and it produces 1,250 lbf of static thrust, then we can compute the jet exit velocity from the relation given in the GE90 case study as

$$V_{exhaust} = \frac{T \times g_c}{\dot{m}_a} = \frac{(1250\,\text{lbf})[32.174\,\text{lbm}\cdot\text{ft/(lbf}\cdot\text{s}^2)]}{18.0\,\text{lbm/s}} = 2230\,\text{ft/s}$$

Then, at a flight speed of 400. mph = 587 ft/s, the propulsion efficiency is

$$\eta_{propulsion} = \frac{2}{1+\dfrac{2230\,\text{ft/s}}{587\,\text{ft/s}}} = 0.417 = 41.7\%$$

Case study 13.4. Model Stirling engine projects

Though the mechanism and thermodynamic cycle of a Stirling engine are not easy to understand, you can make a working Stirling engine at home. Numerous model Stirling engine designs are available (for example, see *Making Stirling Engines*, by Andy Ross,

(*Continued*)

CASE STUDIES IN APPLIED THERMODYNAMICS *Continued*

published by Ross Experimental, 1660 W. Henderson Rd., Columbus, Ohio); however, most require machinist skills. The Stirling engine described here can be made from commonly available components and requires minimum manual skill to assemble.

Stirling engine 1. A test tube engine

This project has been designed for any engineering student with access to simple hand-tools. It was originally developed in 1992 by Wilfried Schlagenhauf in Germany. If you follow the instructions carefully, you can make a simple Stirling cycle engine that runs under its own power. Figure 13.65 shows the eight basic parts. You need a standard test tube (about ¾ inch in diameter and 6 inches long) (1), and five glass marbles (2) that roll freely inside the test tube and function as a free piston displacer. The power cylinder (5) is a 1-inch diameter plastic snap-cap pill bottle with the bottom cut off. Cut the neck off a small balloon (6) and stretch the opening over the uncut end of the pill bottle and snap the cap in place to hold and seal the balloon. Then, cut a small hole in the pill bottle top to insert the air tube and mount it upside down on the base (8). The power piston assembly (4) consists of another plastic bottle cap, small enough to fit inside the power cylinder without jamming when at an angle. A 3 inch long nail is pushed through the center of the bottle cap to provide the piston connecting rod. The test tube support assembly (3) consists of an angle bracket mounted to the base (8), and an adjustable (erector set) member attached to the test tube pivot joint. The air tube (9) is 2 or 3 mm ID pliable plastic or rubber tubing. Finally, counterweights (10) need to be added to the cold end of the test tube to control its motion. The engine is driven by the heat of a single candle (7). A complete engine kit is available from Ginsberg Scientific (ginsbergscientific.com).

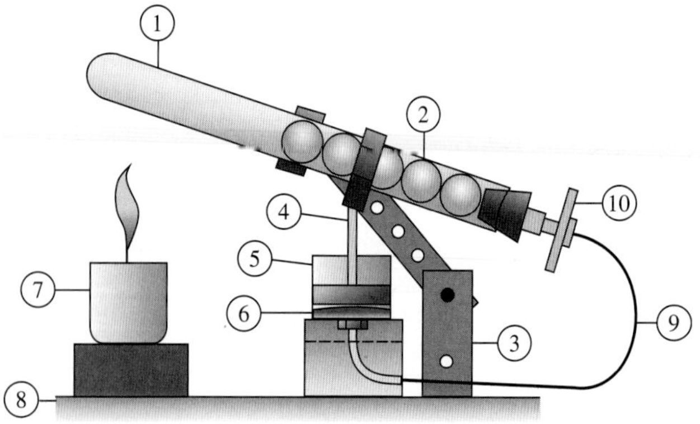

FIGURE 13.65
A test tube Stirling engine.

Here is how the cycle works (Figure 13.66): (a) When the free end of the test tube is pointing up, the air it contains is heated. The heated air expands and inflates the balloon. (b) The expanding balloon pushes the power piston assembly up, which in turn causes the test tube to pivot its free end downward. Then the marbles roll towards the free end of the test tube, displacing the hot air to the higher, cold end, where it cools, causing the balloon to retract, lowering the piston assembly, and the cycle begins again. When properly adjusted, the engine operates at about one to three cycles per second, and the engine generates enough power to drive a small external device.

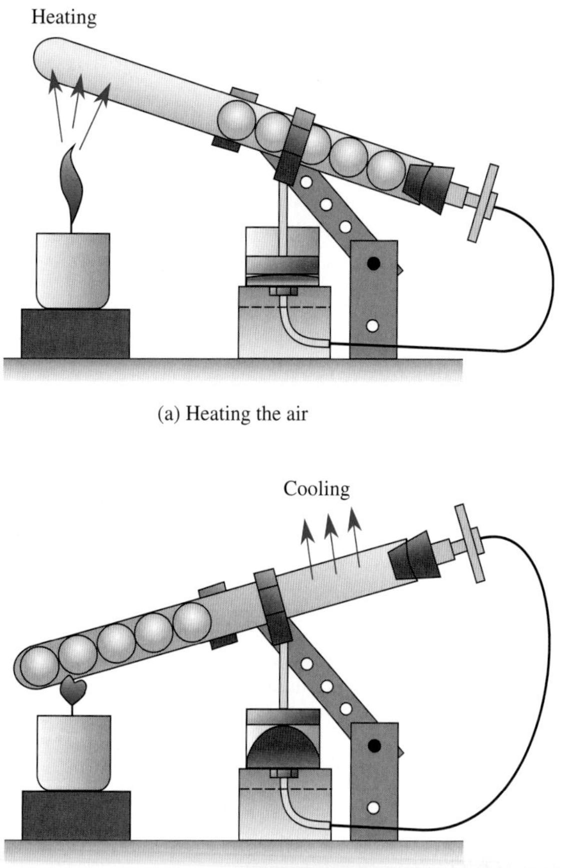

(a) Heating the air

(b) Cooling the air

FIGURE 13.66
How this engine works.

Stirling engine 2. A liquid piston fruit-jar engine

A number of easily made liquid piston Stirling cycle engines can be found in the text *Liquid Piston Stirling Engines*, by C. D. West (New York: Van Nostrand Reinhold, 1983). One of the most intriguing is the fruit-jar engine, shown in Figure 13.67, connected to a water pump. This unit should pump about 5 gallons per hour when properly adjusted. In some cases, the water level in the cylinders and the length of the tuning line inserted into the hot cylinder may require careful adjustment. Some very interesting thermodynamic experiments can be done with this design. For best results, see the details provided in the text by C. D. West.

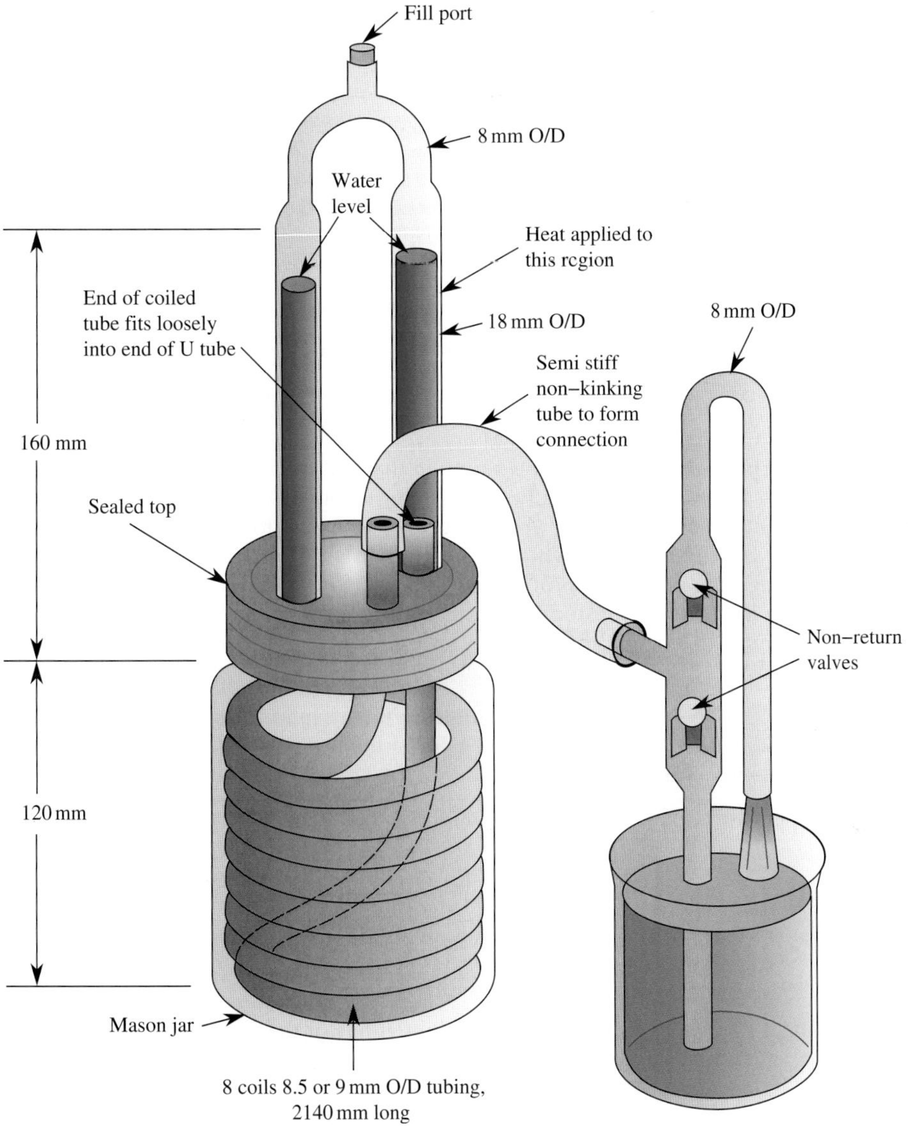

FIGURE 13.67
Fruit-jar Stirling engine.

SUMMARY

Vapor and gas power cycles are the heart and soul of applied thermodynamics. Their commercialization brought humanity from a meager medieval cottage industry to an advanced technological society with a high standard of living. The empirical development of power-producing technology that began in the 17th century provided the basis and motivation for the theoretical understanding of thermodynamics as an intellectual pursuit in the mid 19th century. Most of the technology described in this chapter was developed by inventors who did not understand the theoretical principles involved. Only after they created their ingenious machines did others come along to interpret their work in the light of a complete thermodynamic analysis. Until now, technology was often developed by this simple discovery method. The future holds less of this empirical approach to technological development. Most modern technology is far too complicated to continue to be developed by trial and error methods. We need to know much more about exactly what we are dealing with to discover new technological advances that will benefit humankind. Table 13.3 summarizes the basic elements of the power cycles presented in this chapter.

Table 13.3 Summary of Power Cycle Thermodynamic Processes

Cycle Name	Cycle Process
Carnot (1820)	Two constant entropy and two constant temperature processes (2*s* and 2*T*)
Rankine (1859)	Two constant entropy and two constant pressure processes (2*s* and 2*p*), mostly under the vapor dome
Stirling (1816)	Two constant temperature and two constant volume processes (2*T* and 2$\forall$)
Ericsson (1833)	Two constant temperature and two constant pressure processes (2*T* and 2*p*)
Lenoir (1860)	One constant entropy, one constant volume, and one constant pressure process (1*s*, 1$\forall$, and 1*p*)
Brayton (1873)	Two constant entropy and two constant pressure processes (2*s* and 2*p*)
Otto (1876)	Two constant entropy and two constant volume processes (2*s* and 2$\forall$)
Atkinson (1885) and Miller (1947)	Two constant entropy, one constant volume, and one constant pressure processes (2*s*, 1$\forall$, and 1*p*)
Diesel (1893)	Two constant entropy, one constant volume, and one constant pressure processes (2*s*, 1$\forall$, and 1*p*)

The field of power plant thermodynamics is so broad that it is difficult to present it adequately in just one chapter. In this chapter, we attempt to find a new course for its presentation by carefully charting the chronology of its development. Our goal is to provide you, the reader, with a historical perspective on this important technology and historical benchmarks with which to judge the significance of its impact on society and to broaden your understanding of your chosen profession. Contrary to the opinion held by most historians, the history of the human race is primarily a history of its technological development, with the social *faux pas* of the ruling aristocracy being much less significant than the concurrent advances in mathematics, metallurgy, mechanics, and so forth.

Some of the more important equations introduced in this chapter follow. Do not attempt to use them blindly without understanding their limitations. Please refer to the text material where they were introduced to gain an understanding of their use and limitations.

1. The thermal efficiency of a steam engine whose output is rated in *duty* is

$$\eta_T(\text{in}\%) = \frac{\text{Duty}}{8.50 \times 10^8} \times 100$$

2. The thermal efficiency of an engine operating on a Carnot vapor power cycle is

$$(\eta_T)_{\text{Carnot}} = 1 - T_L/T_H$$

where T_L and T_H are the high and low temperature limits of the cycle.

3. The thermal efficiency of a Rankine cycle power plant without regeneration or reheat is

$$(\eta_T)_{\text{Rankine}} = \frac{(h_1 - h_{2s})(\eta_s)_{pm} - v_3(p_4 - p_3)/(\eta_s)_p}{h_1 - h_3 - v_3(p_4 - p_3)/(\eta_s)_p}$$

Note that the maximum (or isentropic) thermal efficiency of this cycle occurs when $(\eta_s)_{pm} = (\eta_s)_p = 1.0$.

4. The thermal efficiency of a Rankine cycle with one stage of regeneration is

$$(\eta_T)_{\substack{\text{Rankine cycle}\\ \text{with 1 regenerator}\\ \text{(either open or closed)}}} = 1 - \left(\frac{h_2 - h_3}{h_7 - h_6}\right)(1 - y)$$

5. The thermal efficiency of a Rankine cycle with one stage of reheat is

$$(\eta_T)_{\substack{\text{Rankine}\\ \text{cycle with}\\ \text{one reheat unit}}} = \frac{\dot{W}_{pm} - |\dot{W}_p|}{\dot{Q}_B + \dot{Q}_R} = \frac{(h_1 - h_2) + (h_3 - h_4) - (h_6 - h_5)}{(h_1 - h_6) + (h_3 - h_2)}$$

where $h_2 = h_1 - (h_1 - h_{2s})(\eta_s)_{pm1}$ and $h_6 = h_5 + v_5(p_6 - p_5)/(\eta_s)_p$, and of course $p_6 = p_{6s}$.

6. The thermal efficiency of an engine operating on a Carnot gas power cycle is

$$(\eta_T)_{\substack{\text{Carnot}\\ \text{cold ASC}}} = 1 - T_L/T_H = 1 - \text{PR}^{(1-k)/k} = 1 - \text{CR}^{1-k}$$

where PR is the isentropic pressure ratio and CR is the isentropic compression ratio.

7. The thermal efficiency of a Stirling cold ASC external combustion engine is

$$(\eta_T)_{\substack{\text{Stirling}\\ \text{cold ASC}}} = 1 - T_L/T_H = 1 - T_2/T_1 = 1 - T_3/T_4$$

8. The thermal efficiency of an Ericsson cold ASC external combustion engine is

$$(\eta_T)_{\substack{\text{Ericsson}\\ \text{cold ASC}}} = 1 - T_L/T_H = 1 - T_2/T_1 = 1 - T_3/T_4$$

9. The thermal efficiency of a Lenoir cold ASC internal combustion engine is

$$(\eta_T)_{\substack{\text{Lenoir}\\ \text{cold ASC}}} = 1 - kT_3(\text{CR} - 1)/(T_1 - T_4)$$

10. The thermal efficiency of a Brayton cold ASC external combustion engine is

$$(\eta_T)_{\substack{\text{Brayton}\\ \text{cold ASC}}} = 1 - T_3/T_{4s} = 1 - \text{PR}^{(1-k)/k} = 1 - \text{CR}^{1-k}$$

11. The thermal efficiency of an Otto cold ASC internal combustion engine is

$$(\eta_T)_{\substack{\text{Otto}\\ \text{cold ASC}}} = 1 - T_3/T_{4s} = 1 - \text{PR}^{(1-k)/k} = 1 - \text{CR}^{1-k}$$

12. The thermal efficiency of a Diesel cold ASC internal combustion engine is

$$(\eta_T)_{\substack{\text{Diesel}\\ \text{cold ASC}}} = 1 - \frac{\text{CR}^{1-k}(\text{CO}^k - 1)}{k(\text{CO} - 1)}$$

where CO is the *cutoff* ratio of the engine.

13. The indicated cold ASC power output of either an Otto or Diesel cycle engine is

$$(\dot{W}_1)_{\text{out}} = \frac{(\eta_T)_{\text{ASC}}(\dot{Q}/\dot{m})_{\text{fuel}}(DNp_1/C)}{(A/F)(RT_1)(\text{CR} - 1)}$$

where D is the engine's total displacement, N is the engine speed in revolutions per time, C is the number of crankshaft revolutions per power stroke, and A/F is the air/fuel ratio of the engine.

14. The thermal efficiency of an Atkinson cold ASC internal combustion engine is

$$(\eta_T)_{\substack{\text{Atkinson}\\ \text{cold ASC}}} = 1 - \frac{k(\text{ER} - \text{CR})}{\text{ER}^k - \text{CR}^k}$$

where ER is the isentropic expansion ratio v_{2s}/v_1 and CR is the isentropic compression ratio v_3/v_{4s}.

15. The actual thermal efficiency of any of the cycles discussed in this chapter is

$$(\eta_T)_{\text{actual}} = (\eta_m)(\eta_T)_{\text{ASC}}$$

where η_m is the mechanical efficiency of the engine.

Problems (* indicates problems in SI units)

1. The duty of a 1718 Newcomen engine was found to be 4.30 million. Determine its thermal efficiency (%).
2. In 1767, John Smeaton measured the performance of a particularly efficient Newcomen engine that had a 42.0-inch diameter piston and found that it produced 16.7 net horsepower with a duty of 7.44 million. Determine (a) the thermal efficiency of the engine and (b) the boiler heat input rate.
3. In 1767 John Smeaton measured the performance of a Newcomen engine with a 60.0-inch diameter piston and found that it produced 40.8 net horsepower with a duty of 5.88 million. Determine (a) the thermal efficiency of this engine and (b) its boiler heat input rate.
4. In 1767, John Smeaton measured the performance of a Newcomen engine with a 75.0-inch diameter piston and found that it produced 37.6 net horsepower with a duty of 4.59 million. Determine (a) the thermal efficiency of this engine and (b) the boiler heat input rate.
5. In 1772, John Smeaton used the results of his tests on various existing Newcomen cycle engines to design and build his own atmospheric steam engine. It had a 52.0-inch diameter piston with a 7.0 ft stroke and operated at 12.5 strokes per minute with a mean effective pressure of 7.50 lbf/in^2. It produced a remarkably high duty of 9.45 million. Determine the (a) thermal efficiency, (b) the horsepower output, and (c) the boiler

heat input rate of this engine. Ignore the boiler feed pump power requirement.

6. In 1790 John Curr of Sheffield, England, made an atmospheric steam engine with a 61.0-inch diameter piston and an 9.50 ft stroke. The engine operated with a mean effective piston pressure of 7.00 lbf/in^2 and ran at 12.0 strokes per minute. It produced a duty of 9.38 million. Determine (a) the thermal efficiency, (b) the horsepower output, and (c) the boiler heat input rate of this engine. Ignore the boiler feed pump power requirement.
7.* An engine operating on a Carnot cycle extracts 10.0 kJ of heat per cycle from a thermal reservoir at 1000.°C and rejects a smaller amount of heat to a low-temperature thermal reservoir at 10.0°C. Determine the net work produced per cycle of operation.
8.* We need to get 1.00 kW of power from a heat engine operating on a Carnot cycle. 3.00 kW of heat is supplied to the engine from a thermal reservoir at 600. K. What is the required temperature of the low-temperature reservoir, and how much heat must be rejected to it?
9. For the thermal reservoir temperatures shown in Table 13.4, determine (a) the Carnot cycle heat engine thermal efficiency and (b) which is the more effective method of increasing this efficiency, increasing T_H by an amount ΔT or lowering T_L by an amount ΔT, and why.

Table 13.4 Problem 9

No.	T_H (°F)	T_L (°F)
1	4000.	500.
2	4000.	100.
3	2000.	500.
4	2000.	100.

10. Steam enters the turbine of a power plant at 200. psia, 500.°F. How much will the Carnot cycle thermal efficiency increase if the condenser pressure is lowered from 14.7 to 1.00 psia?
11. Steam enters the piston cylinder of a Newcomen cycle steam engine at 14.7 psia, 212°F and condenses at 6.00 psia. Plot the thermal efficiency of the reversible cycle vs. the average cylinder wall temperature over the range 170.°F $\leq (T_b)_{avg} \leq$ 212°F. Neglect the power used by the boiler feed pump.
12. Determine the maximum possible thermal efficiency of an atmospheric steam engine with a boiler temperature of 212°F and a condensation temperature of 70.0°F if it operates on
   a. A Carnot cycle.
   b. A Newcomen cycle assuming an average cylinder wall temperature of 141°F.
   c. A Rankine cycle.
13. Steam with a quality of 1.00 enters the turbine of a Rankine cycle power plant at 400. psia and exits at 1.00 psia. Neglecting pumping power, determine the Rankine cycle isentropic thermal efficiency.
14. Determine the decrease in Carnot and isentropic Rankine cycle thermal efficiencies if the condenser is removed from the engine discussed in text Example 13.5 and the steam is allowed to exhaust directly into the atmosphere at 14.7 psia. Assume the cycles are still closed loop.
15. Show that the Carnot and isentropic Rankine cycle thermal efficiencies for an engine whose boiler produces dry saturated steam at 100. psia and whose condenser operates at 1.00 psia are 29.7% and 26.1%, respectively. Draw the appropriate *T–s* and *p–v* diagrams for these cycles.
16. Rework Problem 15 for an engine without a condenser, with the steam exhausting directly into the atmosphere at 14.7 psia. Determine the Carnot and isentropic Rankine cycle thermal efficiencies and their percent decrease due to the removal of the condenser. Assume the cycles are still closed loop.
17. Steam enters the turbine of a Rankine cycle power plant at 200. psia and 500.°F. How much will the isentropic Rankine cycle thermal efficiency increase if the condenser pressure is lowered from 14.7 psia to 1.00 psia? Neglect the pump work.
18. A small portable nuclear-powered Rankine cycle steam power plant has a turbine inlet state of 200. psia, 600.°F, and a condenser temperature of 80.0°F. The steam mass flow rate is 0.500 lbm/s. Assuming that the condenser exit state is a saturated liquid, that the pump and turbine isentropic efficiencies are 55.0% and 75.0%, respectively, and that there are no pressure losses across the boiler or condenser, determine
   a. The actual power required to drive the pump.
   b. The actual power output of the plant.
   c. The actual thermal efficiency of the plant.
19. A Rankine cycle power plant is to be used as a stationary power source for a polar research station. The working fluid is Refrigerant-22 at a flow rate of 9.00 lbm/s, and the turbine inlet state is saturated vapor at 200.°F. The condenser is air cooled and has an internal temperature of 0.00°F. Assuming a turbine and pump isentropic efficiency of 85.0% and that the refrigerant leaves the condenser as a saturated liquid, determine the overall thermodynamic efficiency and the net power output of the system.
20.* A solar-powered Rankine cycle power plant uses 18.5 × 10^3 m^2 of solar collectors. Refrigerant-22 is used as the working fluid at a flow rate of 1.00 kg/s and is transformed to a saturated vapor in the solar collectors (which function as the boiler) at a temperature of 40.0°C. The condenser for the system operates at 20.0°C and has a quality of 0.00 at the exit. The pump and prime mover isentropic efficiencies are 65.0% and 75.0%, respectively. Determine the prime mover power output and system thermal efficiency when the incident solar flux is 8.00 W/m^2.
21.* Saltwater oceans have subsurface stratification layers called *thermoclines* across which large temperature differences can exist. A 1.00 MW Rankine cycle power plant using ammonia as the working fluid is being designed to operate on a thermocline temperature difference. The ammonia exits the boiler as a saturated vapor at 28.0°C and exits the condenser as a saturated liquid at 10.0°C. For an isentropic system, determine
   a. The thermal efficiency of the power plant.
   b. The pump to turbine power ratio.
   c. The required mass flow rate of ammonia.
22. The condensation that can occur in the low-pressure end of a steam turbine is undesirable because it can cause corrosion and blade erosion, thus reducing the turbine's isentropic efficiency. This can be avoided by superheating the steam before it enters the turbine. What degree of superheat would be required if the steam entered an isentropic turbine at 300. psia and exited as a saturated vapor at 1.00 psia?
23.* Steam enters the turbine of a Rankine cycle power plant at 12.0 MPa and 400.°C. How much does the isentropic thermal

efficiency increase if the condenser pressure is lowered from 0.100 to 1.00 × 10^{-3} MPa?

24. Steam enters the turbine of a Rankine cycle power plant at 200. psia and 500.°F. How much does the isentropic thermal efficiency increase if the condenser pressure is lowered from 14.7 to 1.00 psia?

25. Steam leaves the boiler of a Rankine cycle power plant at 3000. psia and 1000.°F. It is isenthalpically throttled to 600. psia before it enters the turbine. It then exits the turbine at 1.00 psia. Neglecting the pump work, determine
   a. The maximum thermal efficiency of this plant.
   b. Its maximum thermal efficiency if the boiler is operated at 600. psia and 1000.°F and no throttling occurs at the entrance to the turbine.

26. Steam exits a boiler and enters a turbine at 200. psia, 600.°F with a mass flow rate of 30.0 × 10^3 lbm/h. It exits the turbine at 1.00 psia and is condensed. The condensate then reenters the boiler as a saturated liquid at 90.0°F. The turbine drives an electrical generator that delivers a net 3000. kW of power. Determine
   a. The Rankine cycle actual thermal efficiency of the entire system.
   b. The isentropic efficiency of the combined turbine–generator unit.

27. An isentropic Rankine cycle using steam is shown in Figure 13.68. For the given data, determine
   a. The quality of the turbine exhaust steam.
   b. The cycle thermal efficiency.
   c. The mass flow rate required to produce 10,000 Btu/s net output power.

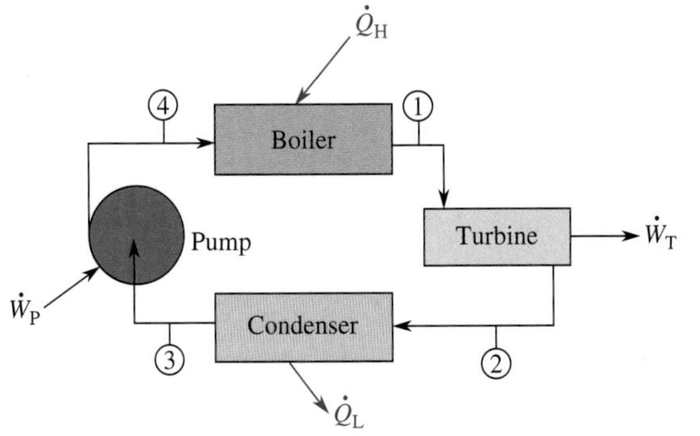

FIGURE 13.68
Problem 27.

Station 1	Station 2	Station 3	Station 4
p_1 = 1000. psia T_1 = 1000.°F	p_2 = 3.00 psia $s_2 = s_1$	p_3 = 3.00 psia x_3 = 0.00	p_4 = 1000. psia $s_4 = s_3$

The process path from station 1 to 2 is an isentropic expansion and that from station 3 to 4 is an isentropic compression.

28.* For a constant steam boiler temperature of 400.°C, a constant turbine exhaust moisture content of 5.00%, and a constant condenser pressure of 3.00 kPa, plot the Rankine cycle thermal efficiency vs. boiler pressure over the boiler pressure range from 1.00 to 20.0 MPa. Ignore the power required by the boiler feed pump.

29.* Steam is supplied to a turbine at a rate of 1.00 × 10^6 kg/h at 500.°C and 10.0 MPa, and it exhausts to a condenser at 2.00 kPa. A single open loop regenerator is used to heat the boiler feedwater with steam extracted from the turbine at 6.00 MPa. The condensate exits the regenerator at 6.00 MPa as a saturated liquid. Neglecting the pump power, determine
   a. The mass flow rate of steam extracted from the turbine.
   b. The system's isentropic Rankine cycle thermal efficiency.
   c. The isentropic Rankine cycle thermal efficiency of the same system without a regenerator.
   d. The percent increase in thermal efficiency due to the regenerator.

30. Steam enters a turbine with an isentropic efficiency of 83.0% at 300. psia, 800.°F and exhausts to a condenser at 0.250 psia. The boiler feedwater is heated in a single, open loop regenerator with steam extracted from the turbine at 100. psia. Saturated liquid leaves the regenerator at the pressure of the extraction steam. Neglecting all pump work, determine
   a. The percent mass flow of steam extracted from the turbine.
   b. The turbine power output per unit mass flow of steam.
   c. The system thermal efficiency.
   d. The turbine power output per unit mass flow of steam when the regenerator is not in use.
   e. The system thermal efficiency when the regenerator is not in use.
   f. The percent increase in system thermal efficiency produced by the regenerator.

31. Repeat items a, b, and c of Problem 30 for steam extraction pressures of 75.0, 50.0, 25.0, and 10.0 psia and plot these results vs. the extraction pressure.

32. A turbine having an isentropic efficiency of 86.0% receives steam at 4.00 × 10^6 lbm/h, 300. psia, 1000.°F and exhausts it to a condenser at 1.00 psia. Steam is extracted from the turbine at 788,000 lbm/h and 200. psia to heat the boiler feedwater in a single, closed loop regenerator. The extract steam then exits the regenerator as a saturated liquid. Neglecting the pump power, determine
   a. The pressure and temperature of the extract steam as it leaves the regenerator.
   b. The Rankine cycle thermal efficiency of this system.
   c. The Rankine cycle thermal efficiency of this system without regeneration.
   d. The percent increase in thermal efficiency due to the regenerator.

33. Determine the blade tip velocity of DeLaval's first steam turbine. It had a rotor diameter of 3.00 inches and ran at 40.0 × 10^3 rpm.

34.* Determine the isentropic exit velocity from a reaction turbine nozzle if steam enters the nozzle at 30.0 MPa, 500.°C and exits at 1.00 MPa.

35. Steam at 600. psia and 800.°F enters the high-pressure stage of an isentropic turbine and is reheated to 60.0 psia and 700.°F before entering the low-pressure stage. The steam then exhausts to a condenser at 1.00 psia. Neglecting pump power, determine
   a. The isentropic Rankine cycle thermal efficiency.
   b. The isentropic Rankine cycle thermal efficiency that would occur if the steam were not reheated.
   c. The percent increase in thermal efficiency due to the reheating operation.

36. Steam enters the high-pressure turbine of a Rankine cycle power plant at 1200. psia and 700.°F and exits as a saturated vapor. It is then reheated to 600.°F before it enters the low-pressure turbine, which exhausts to a condenser at 1.00 psia. The isentropic efficiencies of the high- and low-pressure turbines and the boiler feed pump are 88.0%, 79.0%, and 65.0%, respectively. Determine the thermal efficiency and the net power per unit mass flow rate of steam for this plant.

37.* Consider a steam turbine with a constant inlet temperature of 500.°C connected to a constant pressure condenser at 1.00 kPa. When the steam expands to a saturated vapor in the turbine, it is removed and reheated in the boiler to 500.°C then returned to the turbine to continue to expand until it reaches the condenser pressure. The turbine isentropic efficiency is constant at 80.0%. Ignoring the boiler feed pump power, plot

- **a.** The Rankine cycle thermal efficiency.
- **b.** The percent moisture in the turbine exhaust vs. the boiler pressure over a boiler pressure range of 1.00 to 20.0 MPa. Computerized steam tables are recommended for this problem.

38. The first steam turbine used in an American electrical power plant was a Westinghouse reaction turbine of the Parson's type installed at the Hartford, Connecticut, Electric Light Company in 1902. The turbine inlet state was 200. psig and 400.°F, the generator produced 2.00 MW, and the plant had a heat rate of 35.0×10^3 Btu/(kW · h). Determine its thermal efficiency.

39. In 1903, a General Electric Curtis impulse steam turbine was installed at the Fisk Street Station of the Commonwealth Electric Company in Chicago, Illinois, and was at that time the most powerful steam turbine in the world.[13] The turbine inlet state was 175 psig with 150.°F of superheat, and the condenser pressure was 1.50 in of mercury. When the generator produced a net 5000. kW, the steam flow rate per unit of electrical power produced ($\dot{m}/\dot{W}_{\text{elect.}}$) was 22.5 lbm/(kW · h). For this unit, determine

- **a.** The isentropic power output of the turbine.
- **b.** The isentropic efficiency of the turbine-generator unit.
- **c.** The isentropic Rankine cycle thermal efficiency of the power plant, assuming saturated liquid exits the condenser and neglecting pump work.

40. In 1939, the Port Washington, Wisconsin, power plant of the Milwaukee Electric Railway and Light Company[14] had an unusually high heat rate of 10,800 Btu/(kW · h). Determine its thermal efficiency.

41. Refrigerant-22 is used as the working fluid in a 1.00 MW Rankine bottoming cycle for a steam power plant. The bottoming cycle turbine inlet state is saturated vapor at 210.°F, and the condenser outlet is saturated liquid at 70.0°F. The turbine and pump isentropic efficiencies are 85.0% and 70.0%, respectively. Determine

- **a.** The thermal efficiency of the bottoming cycle.
- **b.** The ratio of the pump to turbine power.
- **c.** The required mass flow rate of refrigerant.

42. It is common to model hot ASC performance with the same formula used in cold ASC analysis except that a specific heat ratio (k) typical of high-temperature gas is used. Determine the Carnot ASC thermal efficiency of an engine with an 8.00 to 1 compression ratio, using

- **a.** A cold ASC analysis with $k = 1.40$.
- **b.** A hot ASC analysis with $k = 1.30$.
- **c.** Determine the percent decrease in the Carnot thermal efficiency between the cold and hot ASC analysis.

43. Air enters an engine at 40.0°F and is compressed isentropically in a 9.00 to 1 compression ratio. Determine the Carnot ASC thermal efficiency of this engine, using

- **a.** A cold ASC analysis with $k = 1.40$.
- **b.** A hot ASC analysis using the gas tables (Table C.16a in *Thermodynamic Tables to accompany Modern Engineering Thermodynamics*).
- **c.** Determine the percent decrease in the Carnot thermal efficiency between the cold and hot ASC analysis.

44.* Air enters an engine at atmospheric pressure and 17.0°C and is isentropically compressed to 871.4 kPa. Determine the Carnot ASC thermal efficiency of the engine, using

- **a.** A cold ASC analysis with $k = 1.40$.
- **b.** A hot ASC analysis using the gas tables (Table C.16b).
- **c.** Determine the percent decrease in the Carnot thermal efficiency between the cold and hot ASC analysis.

45. Determine the mechanical efficiency of a Stirling cycle engine operating with a 1300.°F heater and a 100.°F cooler. The engine produces a net 10.0 hp output with a heat input of 80.0×10^3 Btu/h.

46. In 1964, an experimental Stirling engine was installed in a modified Chevrolet Corvair at the General Motors Research Laboratory. Alumina (aluminum oxide) heated to 1200.°F served as the heat source for the engine, while the atmosphere at 100.°F served as the heat sink (because of the use of alumina in the engine, the car was dubbed the *Calvair* by GM researchers). Assuming a mechanical efficiency of 67.0%, determine the actual thermal efficiency of the engine based on a cold ASC analysis.

47. A Stirling cycle engine uses 0.0800 lbm of air as the working fluid. Heat is added to this air isothermally at 1500.°F and is rejected isothermally at 200.°F. The initial volume of the air before the heat addition ($\mathcal{V}_4$ in Figure 13.38) is 0.750 ft^3 and the final volume after the heat addition ($\mathcal{V}_1$ in Figure 13.38) is 1.00 ft^3. For the cold ASC, determine

- **a.** The air pressure at the beginning and end of the expansion stroke (p_4 and p_1 in Figure 13.38).
- **b.** The air pressure at the beginning and end of the compression stroke (p_2 and p_3 in Figure 13.38).
- **c.** The cold ASC thermal efficiency of the engine.
- **d.** The net reversible work produced inside the engine per cycle of operation.

48. In 1853, John Ericsson constructed a huge 300. hp hot air engine that ran on the Ericsson cycle. It had pistons 14.0 ft in diameter and it consumed 2.00 lbm of coal per indicated

---

[13] On May 28, 1975, this turbine-generator unit was designated as the seventh National Historic Mechanical Engineering Landmark by the American Society of Mechanical engineers.

[14] In 1980, this power plant was designated as the 48th National Historic Mechanical Engineering Landmark by the American Society of Mechanical Engineers.

horsepower hour. Assuming a heating value for coal of 13.0×10^3 Btu/lbm and that the engine was reversible, determine the thermal efficiency of this engine.

49. The air standard Ericsson cycle (see Figure 13.40) is made up of an isothermal compressor ($T_2 = T_3 = T_L$), an isothermal prime mover ($T_1 = T_4 = T_H$), and an isobaric regenerator ($p_1 = p_2$ and $p_3 = p_4$). Show that the compressor and prime mover must have identical pressure ratios, that is, $p_2/p_3 = p_1/p_4$, and show that this also requires that $v_3/v_2 = v_4/v_1$.

50.* An Ericsson cycle operates with a compressor inlet pressure and volume of 1.00 MPa, 0.0200 m^3 and a turbine inlet pressure and volume of 5.00 MPa, 0.0400 m^3. For a reversible cycle, determine
   a. The heat added.
   b. The heat rejected.
   c. The work done.
   d. The thermal efficiency of the cycle.

51.* An inventor claims to have developed a Lenoir engine with an isentropic compression ratio of 8.00 to 1 that produces a combustion temperature of 1500.°C when the intake temperature is 20.0°C. Assuming $k = 1.40$ and that the engine operates on a cold ASC, show whether or not the inventor's claim is possible.

52.* A World War II Lenoir cycle "buzz bomb" has an air intake at 10.0°C, a combustion temperature of 1000.°C, and a compression ratio of 2.30. Determine its cold ASC thermal efficiency.

53.* Plot the cold ASC thermal efficiency of a Lenoir engine having an air intake at 15.0°C and a combustion temperature of 2000.°C vs. the isentropic compression ratio over the range $1.01 \leq \text{CR} \leq 5.50$.

54. A Brayton cold ASC has a turbine isentropic efficiency of 92.0% and a compressor isentropic pressure ratio of 13.37. The compressor and turbine inlet temperatures are 500. and 2200. R, respectively. Determine the value of the compressor isentropic efficiency that causes the overall thermodynamic thermal efficiency of this system to be exactly zero.

55. A Brayton cold ASC has a turbine that is 80.0% isentropically efficient and a compressor with an isentropic pressure ratio of 7.00. The compressor inlet temperature is 530. R and the turbine inlet temperature is 2460 R. Determine the compressor isentropic efficiency that causes the entire cycle thermal efficiency to become exactly zero. Assume $k = 1.40$.

56. Show that the product of the compressor and turbine isentropic efficiencies must be greater than $(T_L/T_H)^{1/2}$ if a Brayton cycle gas turbine unit is to operate at its maximum power output.

57. A test on an open loop Brayton cycle gas turbine produced the following results:

    Net power output = 180.5 hp
    Air mass flow rate = 20.0×10^3 lbm/h
    Inlet air temperature = 80.0°F
    Inlet air pressure = 14.5 psia
    Compressor exit pressure = 195 psia
    Compressor isentropic efficiency = 85.0%
    Combustion chamber heat addition = 4.00×10^6 Btu/h

   Using a cold ASC analysis and assuming $k = 1.40$, determine
   a. The cycle thermal efficiency.
   b. The isentropic efficiency of the turbine.

58. On July 29, 1949, the first gas turbine installed in the United States for generating electric power went into service at the Belle Isle station of the Oklahoma Gas and Electric Company.[15] It had a 15-stage compressor with an isentropic pressure ratio of 6 to 1, a 2-stage turbine with overall entrance and exit temperatures of 1400°F and 780°F, respectively, and the turbine-generator unit was rated at 3500 kW. Assuming a Brayton cold ASC, determine
   a. The isentropic efficiency of the turbine.
   b. The Brayton cold ASC thermal efficiency of the entire turbine compressor unit.

59.* The regenerator in a Brayton cycle is simply a heat exchanger designed to transfer heat from hot exhaust gas to cool inlet gas. In an "ideal" regenerator, the exit temperature of the inlet (heated) gas is equal to the entrance temperature of the exhaust (cooled) gas. Since this is not normally the case in practice, *regenerator* (or *heat exchanger*) *efficiency* can be defined as

$$\eta_{\text{regeneration}} = \frac{(\dot{Q}_{\text{regeneration}})_{\text{actual}}}{(\dot{Q}_{\text{regeneration}})_{\substack{\text{maximum}\\ \text{possible}}}} = \frac{(h_{\text{out}} - h_{\text{in}})_{\text{heated}}}{(h_{\text{in}})_{\text{cooled}} - (h_{\text{in}})_{\text{heated}}}$$

and, for constant specific heats, this reduces to

$$\eta_{\text{regeneration}} = \frac{(T_{\text{out}} - T_{\text{in}})_{\text{heated}}}{(T_{\text{in}})_{\text{cooled}} - (T_{\text{in}})_{\text{heated}}}$$

Note that regeneration is practical only when the engine exhaust temperature is greater than the compressor exhaust temperature. Therefore, as the compression and expansion ratios of the compressor and prime mover increase, the effectiveness of regeneration decreases. Determine an expression for the limiting isentropic pressure ratio (PR) in terms of T_1, T_3, and k for which regeneration is no longer useful in the Brayton ASC with regeneration as shown in Figure 13.69. Evaluate this expression to find the limiting pressure ratio when $T_1 = 1500.$°C, $T_3 = 10.0$°C, and $k = 1.40$.

60.* An aircraft gas turbine engine operating on a Brayton cycle has a cold ASC thermal efficiency of 25.0% when the intake air is at 20.0°C and the combustion chamber outlet temperature is at 1200.°C. Assuming $k = 1.40$, determine
   a. The isentropic pressure ratio of the engine.
   b. The isentropic compression ratio of the engine.
   c. The isentropic outlet temperature of the engine's compressor.
   d. The optimum isentropic pressure ratio for maximum isentropic power output from the engine.
   e. The optimum isentropic compression ratio for maximum isentropic power output from the engine.
   f. The engine's thermal efficiency when operated at the maximum isentropic power output.

61. In Professor John L. Krohn's laboratory at Arkansas Tech University, air enters the compressor of an ideal Brayton cycle at $p_1 = 14.5$ psi, $T_1 = 70.0$°F with a volumetric flow rate of 20.0×10^3 ft^3/min. The compressor pressure ratio is 12.0

[15] On November 8, 1984, the Belle Isle gas turbine was designated as the 73rd National Historic Mechanical Engineering Landmark by the American Society of Mechanical Engineers.

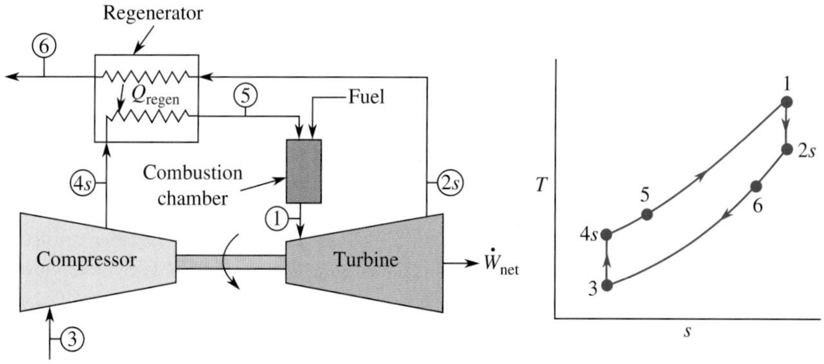

FIGURE 13.69
Problem 59.

and the turbine inlet temperature is 2500.°F. Using the *hot air standard cycle* analysis, determine the:

a. Net work output of the cycle in Btu/hr and MW.
b. Hot ASC thermal efficiency of the cycle.

62. An internal combustion engine operating on the Otto cycle has a pressure and temperature of 13.0 psia and 70.0°F at the beginning of the compression stroke (state 3 in Figure 13.48) and a pressure at the end of the compression stroke of 200. psia. For the cold ASC with k = 1.40, determine

a. The compression ratio, CR.
b. The temperature at the end of the compression stroke.
c. The thermal efficiency of the cycle.

63. An eight-cylinder, four-stroke Otto cycle racing engine has a 4.00-inch bore and a 4.00-inch stroke with a compression ratio of 10.0 to 1. Find the mean effective pressure in the cylinders when the engine is running at 5000. rpm and burning fuel at a rate of $\dot{Q}_{fuel}$ = 1.00 × 10^3 Btu/s. Assume k = 1.4.

64. Determine the actual brake horsepower produced by a four-stroke Otto cycle internal combustion engine operating with a 7.50 to 1 compression ratio and a mechanical efficiency of 30.0% when the combustion of the fuel is producing 225,000 Btu/h inside the engine. Assume k = 1.40.

65.* Air enters an Otto cycle internal combustion engine at 90.0 kPa and 15.0°C. The engine has a compression ratio of 8.00 to 1. During the combustion process, 3000. kJ per kg of air is added to the air. Assuming a reversible engine, determine

a. The pressure and temperature at the end of each process of the cycle.
b. The engine's cold ASC thermal efficiency.
c. The mean effective pressure of the engine.

66. Determine the output brake horsepower of a small two-stroke Otto cycle internal combustion engine that has the following characteristics:

Displacement = 5.00 in^3
Speed = 2000. rpm
Compression ratio = 8.00 to 1
Air–fuel ratio = 15.0 to 1
Mechanical efficiency = 30.0%
Fuel heating value = 18.0 × 10^3 Btu/lbm
Ambient conditions = 14.7 psia and 70.0°F

67. A dynamometer test of a six-cylinder Otto four-stroke cycle engine with a 231 in^3 displacement gave the following results at 4000. rpm:

Indicated power output = 250. hp
Actual (or brake) power output = 75.0 hp
Heating value of the fuel being used = 20.0 × 10^3 Btu/lbm
Fuel consumption rate = 54.0 lbm/h

Determine

a. The mechanical efficiency η_m of the engine.
b. The ASC thermal efficiency of the engine.
c. The isentropic compression ratio (CR) of the engine, assuming an Otto cold ASC.
d. The mean effective pressure (mep) inside the combustion chamber.

68. Professor John L. Krohn at Arkansas Tech University is running an engine test. At the beginning of the compression process in a *hot air standard cycle,* Otto cycle, the conditions are p_1 = 14.7 psi, T_1 = 77.0°F. The compression ratio is 8.50 and the pressure doubles during the constant volume heat addition. For this cycle, use the air tables (Table C.16a) to determine the

a. Heat addition per unit mass.
b. Net work per unit mass.
c. Hot ASC thermal efficiency.
d. The maximum temperature reached in the cycle.

69.* At the beginning of the compression process in a hot air standard Diesel cycle, the conditions are p_1 = 1.00 bar, T_1 = 25.0°C, V_1 = 700. cm^3. The engine has a compression ratio of 18.0 and the heat addition per unit mass is 920. kJ/kg. For this cycle, Professor John L. Krohn at Arkansas Tech University wants you to use the air tables (Table C.16b) to determine the

a. Maximum temperature reached.
b. Cutoff ratio.
c. Net work.
d. The hot ASC thermal efficiency.

70. A Diesel cycle internal combustion engine has a compression ratio of 18.0 to 1 and a cutoff ratio of 2.20. Determine the cold ASC thermal efficiency of this engine. Assume k = 1.40.

71. Show that, as the cutoff ratio of the Diesel cycle approaches 1.00, the Diesel cold ASC thermal efficiency becomes equal to that of the Otto cycle with the same compression ratio.

72. Determine the value of the cutoff ratio that causes the Diesel cold ASC thermal efficiency to become zero for an engine with a 20.0 to 1 compression ratio. Assume $k = 1.40$.

73. A Diesel cycle internal combustion engine has a compression ratio of 15.0 to 1. At the beginning of the compression process, the pressure is 14.7 psia and the temperature is 520. R. The maximum temperature of the cycle is 4868 R. Determine the cold ASC thermal efficiency. Assume $k = 1.40$.

74.* A two-cylinder, two-stroke Diesel cycle internal combustion engine with a 16.2 to 1 compression ratio and a total displacement of 1.50 L burns kerosene having a heating value of 40.0×10^3 kJ/kg when using an air–fuel ratio of 25.0 to 1. The intake temperature and pressure are 100. kPa and 15.0°C. When the engine is running at 1200. rpm and producing 6.00 kW of power, determine (assuming $k = 1.40$) the engine's

a. Cutoff ratio.
b. Cold ASC thermal efficiency.
c. Indicated power output.
d. Mechanical efficiency.
e. Actual thermal efficiency.

75. The Atkinson cycle is similar to the Otto cycle except that the constant volume exhaust-intake stroke at the end of the Otto cycle power stroke has been replaced by a constant pressure process in the Atkinson cycle, as shown in Figure 13.70. Q_H occurs during process 4*s* to 1, and Q_L occurs during process 2*s* to 3 in each case.

a. Sketch the *T–s* diagram for the Atkinson cycle numbering and labeling all the process path lines as in the *p–V* diagram of Figure 13.70.
b. Determine the Atkinson cold ASC thermal efficiency for $k = 1.40$, $T_1 = 8000.$ R, $T_3 = 520.$ R, and $CR = v_3/v_{4s} = 8.0$.

Design Problems

The following are open-ended design problems. The objective is to carry out a preliminary thermal design as indicated. A detailed design with working drawings is not expected unless otherwise specified. These problems have no specific answers, so each student's design is unique.

76. Design a small, single-cylinder, piston-type steam engine and boiler that can be used to power a toy vehicle, such as a train or a tractor. Choose a convenient fuel such as alcohol, and design the boiler so that it can supply enough steam to your engine. Make sure the boiler has a pressure relief valve, and pay close attention to other safety considerations.

77. Develop a preliminary design for a closed loop Rankine cycle steam power plant to be used in a compact automobile. The prime mover may be either a reciprocating piston or turbine and must produce a net output of 40.0 hp with a thermal efficiency in excess of 35.0%.

a. Specify inlet and outlet states for the boiler, prime mover, condenser, and boiler feed pump.
b. Choose typical values for the isentropic efficiencies of the prime mover and boiler feed pump, and calculate the overall thermal efficiency of the power plant.
c. Specify the fuel to be used in the boiler.
d. Estimate the overall power plant weight, including fuel storage.
e. Specify conditions needed to meet part-load operation and prime mover speed control during vehicle acceleration and deceleration.
f. Specify all additional equipment needed to connect the prime mover output shaft to the vehicle drive wheels.

78. Design a small, single-cylinder Stirling or Ericsson cycle external combustion engine that can be used to demonstrate the operation of this type of engine in the classroom. Choose a convenient fuel such as alcohol. Provide detailed working drawings and a thermodynamic analysis.

79. Design a Brayton cycle power system to propel a small drone aircraft that will be used for military target practice. The fueled drone must weigh less than 500. kg, and since these aircraft are not reusable, they must be produced at minimum cost. Determine or specify the air mass flow rate, pressure ratio, compressor and turbine isentropic efficiencies, turbine inlet temperature, thrust, thrust to weight ratio, exhaust temperature, and hot ASC thermal efficiency. A computer program will help you carry out parametric studies of the variables involved.

80. Design a personal exercise machine that uses the otherwise dissipated human exercise energy in some productive way. For example, the exercise energy could be converted into mechanical, chemical, or electrical energy, which could then be used in some domestic device (for example, to power a computer, TV set, or kitchen appliance). Another solution would be to design a system that would feed the exercise energy directly into the local electrical power grid for credit against the

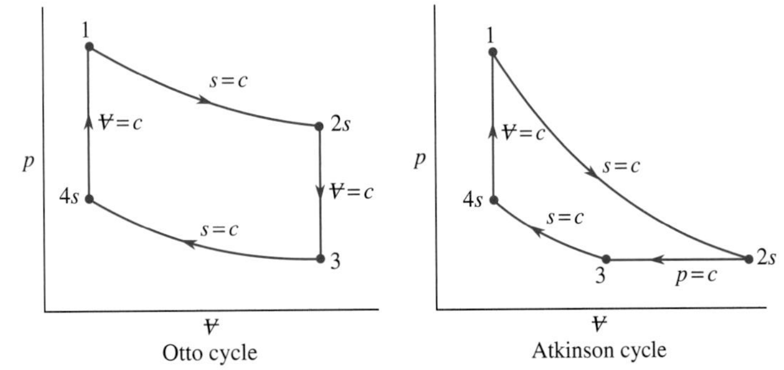

FIGURE 13.70
Problem 75.

user's electrical bill (like wind power devices today). If 100 million people exercised for 15 min every day of the year at a power level of 100. W each, how many barrels of crude oil (at 5.80×10^6 Btu per barrel) and how many tons of coal (at 26.0×10^6 Btu per ton) could be saved annually?

Computer Problems

The following computer programming assignments are designed to be carried out on any personal computer using a spreadsheet or equation solver. They are meant to be exercises in the manipulation of some of the basic formulae of this chapter. They may be used as part of a weekly homework assignment.

81. Write a computer program or use an equation solver or spreadsheet to carry out a parametric study of the isentropic Rankine cycle with regeneration given in Example 13.6, part b. Allow the regenerator pressure p_{5s} to vary from 1.00 to 200. psia (this simulates opening a valve in the regenerator steam line from the turbine). Maintain the regenerator outlet state as a saturated liquid (i.e., set $h_6 = h_f(p_{5s})$), because if $h_6 < h_f(p_{5s})$, the regenerator is not operating as effectively as possible; and if $h_6 > h_f(p_{5s})$, there will be vapor at the entrance of the boiler feed pump and its performance will be seriously degraded.

a. Determine the values of p_{5s} and mass fraction y that maximize the cycle efficiency.

b. Produce a plot of cycle efficiency η_T and mass fraction y on the y-axis vs. regenerator pressure p_{5s} on the x-axis. As part of this solution, note that, when p_{5s} is equal to the turbine inlet pressure (200. psia) or the condenser inlet pressure (1.00 psia), the regenerator has no effect and the cycle efficiency is the same as without regeneration (i.e., 29.7%).

82. Repeat Problem 78 using the cycle data from Example 13.6, except now let the boiler pressure range from 100. to 1000. psia and keep the regenerator pressure at 25.0% of the boiler pressure. Plot the cycle efficiency and the regenerator mass fraction vs. boiler pressure. Does the cycle efficiency have a maximum value under these conditions?

83. Develop a computer program or use a spreadsheet or equation solver to calculate the Otto ASC thermal efficiency as the compression ratio ranges from 1.00 to 12.0 for the following values of k: 1.40, 1.35, 1.30, 1.25, and 1.20 (note that values of k less than 1.40 give an approximate hot ASC result). Plot these results (either manually or with a computer) with the thermal efficiency on the vertical axis and the compression ratio on the horizontal axis, using k as a parameter for the family of curves. Utilize at least 25 points per curve.

84. Develop a computer program or use a spreadsheet or equation solver to calculate the Diesel cold ASC thermal efficiency as the compression ratio (CR) ranges from 12.0 to 30.0 for the following cutoff ratios: 1.50, 2.00, 2.50, 3.00, and 3.50. Plot these results (either manually or with a computer) with the thermal efficiency on the vertical axis and the compression ratio on the horizontal axis, using the cutoff ratio as a parameter for the family of curves. Utilize at least 25 points per curve.

85. Develop a computer program or use a spreadsheet or equation solver that determines the value of the cutoff ratio that causes the Diesel ASC thermal efficiency to become zero as the compression ratio ranges from 12.0 to 30.0 for the following values of k: 1.40, 1.35, 1.30, 1.25, and 1.20. Plot these results (either manually or with a computer) with the cutoff ratio on the vertical axis and the compression ratio on the horizontal axis, using k as a parameter for the family of curves.

CHAPTER 14

Vapor and Gas Refrigeration Cycles

CONTENTS

14.1 INTRODUKSJON (INTRODUCTION)

This chapter is divided into two parts, one dealing with common vapor refrigeration cycles and one dealing with common gas refrigeration cycles. Most of these cycles are reversed power cycles, and their analysis amounts to a reapplication of the power cycle material presented in Chapter 13. Reversed vapor (Rankine) cycles are commonly called *vapor-compression* refrigeration cycles and reversed gas cycles are normally referred to by the cycle name (e.g., a *reversed Brayton* refrigeration cycle).

Like power cycle technology, refrigeration technology has had an enormous impact on our culture and the way we live. It changed our diet, the architecture of our buildings, the agriculture on our farms, and many other items that touch our everyday life. Just as we would find it very difficult to return to a time without the portable power produced by engines, we would also find life much less comfortable in a time without refrigeration and air conditioning.

14.2 PART I. VAPOR REFRIGERATION CYCLES

The basic concepts of refrigeration, air conditioning, and heat pumps were introduced in Chapter 7. This technology is usually modeled as a backward-running heat engine. When a heat engine runs backward (or in reverse), it receives a net input of work W that causes an amount of heat Q_L to be removed from a low-temperature region and an amount of heat Q_H to be added to a high-temperature region. So, it actually cools the low-temperature region and heats the high-temperature region.

A backward-running heat engine is a refrigeration machine, but its exact technical name depends on exactly how it is being used. For example, if food occupies the low-temperature region, then the device is indeed called a *refrigerator*, but if people occupy the low-temperature region, then the device is called an *air conditioner*.[1] On the other hand, if people occupy the high-temperature region and utilize Q_H for space-heating purposes, then the device is called a *heat pump*. Though the details of their design and operation differ slightly, refrigerators, air conditioners, and heat pumps can all be modeled as backward-running heat engines. These distinctions are shown in Figure 14.1.

HOW CAN AN ENGINE RUN BACKWARD?

To get an engine to run backward, you need to put work *into* it where the work normally comes *out*. If the engine has an output shaft, you simply attach a motor or something to the shaft to turn it so that you are putting work into the engine instead of having the engine produce work. A heat engine converts some of the heat from a high-temperature source into work and rejects the remaining heat to a low-temperature sink. A "backward-running" heat engine draws heat from a low-temperature source, adds work energy to it, and rejects everything to a high-temperature sink (see Figure 14.1).

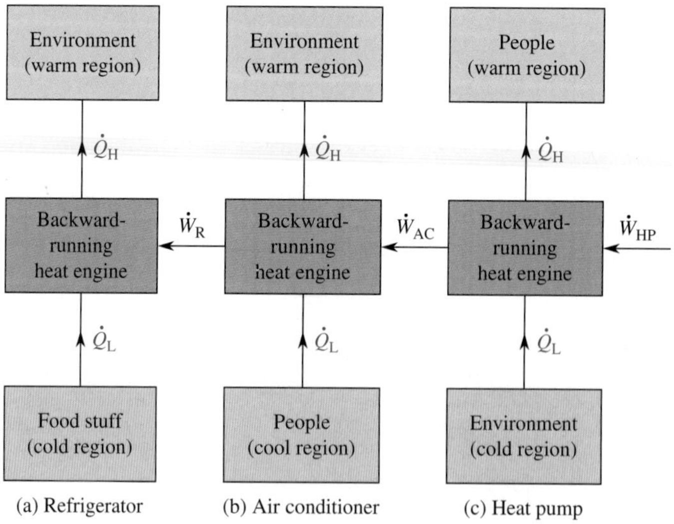

FIGURE 14.1
Characteristics of refrigerators, air conditioners, and heat pumps.

[1] In addition to lowering room or building air temperature, air conditioners also usually filter the air and alter its humidity (see Chapter 8).

14.3 CARNOT REFRIGERATION CYCLE

In Chapter 7, we discovered that refrigerators, air conditioners, and heat pumps usually have actual thermal efficiencies in excess of 100%. This is due simply to the mathematical way in which their thermal efficiencies are defined (the desired energy output divided by the required energy input) and does not imply the violation of any physical law. However, claims of thermal efficiency in excess of 100% cause obvious credibility problems in the public domain, so the term *efficiency* is not often used with this technology. Instead, we simply rename the thermal efficiency the *coefficient of performance* (COP), which is expressed as a pure number, usually between 1 and 10, rather than as a percentage. The COP definitions have been given in Eqs. (7.17) and (7.19) (recall that work *into* and heat *out of* a system are both negative quantities in our sign convention; as in previous chapters, we use their absolute values in these equations and assign algebraic signs to the symbols to avoid confusion):

$$(\eta_T)_{\substack{\text{heat}\\\text{hump}}} = \text{COP}_{\text{HP}} = \frac{|Q_H|}{|W_{\text{HP}}|} = \frac{|\dot{Q}_H|}{|\dot{W}_{\text{HP}}|} = \frac{|Q_H|}{|Q_H| - Q_L} = \frac{|\dot{Q}_H|}{|\dot{Q}_H| - \dot{Q}_L} \tag{7.17}$$

and

$$(\eta_T)_{\substack{\text{refrigerator or}\\\text{air conditioner}}} = \text{COP}_{\text{R/AC}} = \frac{Q_L}{|W_{\text{R/AC}}|} = \frac{\dot{Q}_L}{|\dot{W}_{\text{R/AC}}|} = \frac{Q_L}{|Q_H| - Q_L} = \frac{\dot{Q}_L}{|\dot{Q}_H| - \dot{Q}_L} \tag{7.19}$$

and it is easily shown that

$$\text{COP}_{\text{HP}} = \text{COP}_{\text{R/AC}} + 1 \tag{14.1}$$

From Figure 14.1 and Eq. (14.1), it is evident that Eqs. (7.17) and (7.19) can also be written as

$$\text{COP}_{\text{HP}} = \frac{|\dot{Q}_H|}{|\dot{W}_{\text{in}}|_{\text{net}}} \tag{7.17a}$$

and

$$\text{COP}_{\text{R/AC}} = \frac{\dot{Q}_L}{|\dot{W}_{\text{in}}|_{\text{net}}} = \text{COP}_{\text{HP}} - 1 \tag{7.19a}$$

Note that Eq. (7.17a) is simply the inverse of the general forward-running heat engine thermal efficiency equation (see Eq. (7.5)); that is,

$$\text{COP}_{\text{HP}} \equiv \frac{1}{(\eta_T)_{\substack{\text{forward-running}\\\text{heat engine}}}} \tag{14.2}$$

and Eq. (7.17a) then gives

$$\text{COP}_{\text{R/AC}} \equiv \frac{1}{(\eta_T)_{\substack{\text{forward-running}\\\text{heat engine}}}} - 1 \tag{14.3}$$

Therefore, the COP for any of the heat engines discussed in Chapter 13 operating on a reversed thermodynamic cycle as a heat pump, refrigerator, or air conditioner can be easily obtained through the use of Eqs. (14.2) and (14.3).[2]

For example, Eq. (7.16) gives the Carnot thermal efficiency as

$$(\eta_T)_{\text{Carnot}} = 1 - \frac{T_L}{T_H} = \frac{T_H - T_L}{T_L}$$

Then, Eqs. (14.2) and (14.3) can be used directly to give the COP of a Carnot engine running backward as a heat pump, refrigerator, or air conditioner as

$$\text{COP}_{\text{Carnot HP}} = \frac{T_H}{T_H - T_L} \tag{14.4}$$

$$\text{COP}_{\text{Carnot R/AC}} = \frac{T_L}{T_H - T_L} \tag{14.5}$$

Also, it is easy to show that Eq. (14.1) remains valid for these systems.

---

[2] However, it is difficult to imagine an internal combustion engine (like the Otto and Diesel cycle) running backward because it would require a heat-absorbing (endothermic) combustion reaction.

EXAMPLE 14.1

A new die-casting operation has a large amount of waste heat available at 200.°C in a location where the local environmental temperature is 20.0°C. As chief engineer in charge of thermal energy management, investigate the possibility of recovering some of this waste heat by determining

a. The thermal efficiency of a Carnot engine operating between these temperatures.
b. The coefficient of performance of a Carnot heat pump operating between these temperatures.
c. The coefficient of performance of a Carnot refrigerator or air conditioner operating between these temperatures.

Solution

First, draw a sketch of the system (Figure 14.2).

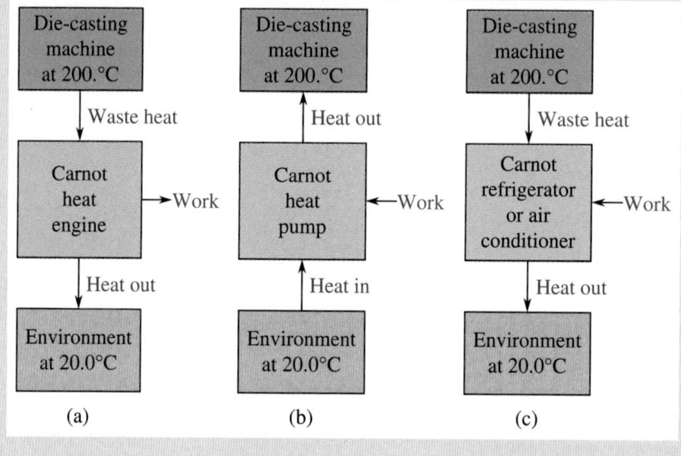

FIGURE 14.2
Example 14.1.

a. Equation (13.2) gives the thermal efficiency of a Carnot engine operating between the temperature limits of 200. + 273.15 = 473.15 K and 20.0 + 273.15 = 293.15 K as

$$(\eta_T)_{\text{Carnot}} = 1 - \frac{T_L}{T_H} = 1 - \frac{293.15\ \text{K}}{473.15\ \text{K}} = 0.380 = 38.0\%$$

b. Equation (14.4) gives the coefficient of performance of the same Carnot engine running backward as a heat pump as

$$\text{COP}_{\text{Carnot HP}} = \frac{T_H}{T_H - T_L} = \frac{473.15\ \text{K}}{473.15 - 293.15\ \text{K}} = 2.63$$

c. Equation (14.5) gives the coefficient of performance of the same Carnot engine running backward as a refrigerator or air conditioner as

$$\text{COP}_{\text{Carnot R/AC}} = \frac{T_L}{T_H - T_L} = \frac{293.15\ \text{K}}{473.15 - 293.15\ \text{K}} = 1.63$$

Note that the $\text{COP}_{\text{Carnot HP}} = \text{COP}_{\text{Carnot R/AC}} + 1$ as Eq. (14.1) requires, and that Eqs. (14.2) and (14.3) are also satisfied here.

Thermal energy management is a serious problem in the industrial environment. Lost thermal energy often reflects poor process design and lost money. It can be remedied by considering the waste heat as an energy source and applying a technology that can utilize it in some fashion.

Exercises

1. Suppose the waste heat in Example 14.1 is available at 35.0°C instead of 200.°C while the environmental temperature remains at 20.0°C. Determine the thermal efficiency of a Carnot engine operating between these temperatures and the coefficient of performance of a Carnot air conditioning unit. **Answer:** $(\eta_T)_{\text{Carnot}} = 4.90\%$, and $\text{COP}_{\text{Carnot AC}} = 20.5$.
2. During winter, the environmental temperature in the die-casting facility in Example 14.1 drops to 0.00°C. Recompute the thermal efficiency and coefficient of performance, assuming the waste heat temperature remains at 200.°C. **Answer:** $(\eta_T)_{\text{Carnot}} = 42.3\%$, $\text{COP}_{\text{Carnot HP}} = 2.37$, and $\text{COP}_{\text{Carnot R/AC}} = 1.37$.

3. A salesperson from a waste heat recovery company visits you and claims to have a new engine that can convert 50.% of the waste heat in Example 14.1 into useful shaft work. How would you evaluate this claim? **Answer**: No engine can be more efficient than a (hypothetical) Carnot engine, and since a Carnot engine is only 38% efficient for converting this waste heat into useful work, the salesperson's claim of a 50.% conversion is impossible to achieve.

Development of refrigeration technology

Natural refrigeration			Artificial refrigeration			
Ice and snow	Surface evaporation	Radiation cooling	Vapor-compression	Expanding gas	Absorption	Miscellaneous
Used from aniquity. Harvested during the winter and stored in pits and ravines covered with straw. Extensive ice harvesting and storage industries developed over the ages to provide ice during the summer months and in southern regions.	Used from aniquity. Evaporation from the surface of porous containers will keep the contents cool.	Used by the Egyptians. Shallow pans filled with water and exposed to the night sky will cause the water to freeze by radiation heat transfer to the sky even when the surrounding air temperature is above freezing.	Originated by Jacob Perkins in 1834. Commercialized by James Harrison in 1856.	Originated by John Gorrie in 1844. Commercialized by Alexander Kirk in 1862.	Originated and commericalized by Ferdinand Carré in 1859.	Refrigerating mixtures – known from antiquity. Reduced pressure – Originated by William Cullen, 1755. Thermoelectric – Originated by Jean-Charles Peltier, 1834. Joule–Thomson – Originated by James Prescott Joule and William Thomson, 1850. Vortex Tube – Originated by Georges Ranque, 1931.

FIGURE 14.3
The development of natural and artificial refrigeration technologies.

Two primary types of refrigeration are available, *natural* (e.g., ice) and artificial. Artificial refrigeration has been subdivided in this chapter into vapor cycles (specifically vapor-compression and absorption cycles) and gas expansion (specifically reversed Stirling and Brayton cycles). These refrigeration methods are illustrated in Figure 14.3 and discussed in detail in this chapter.

14.4 IN THE BEGINNING THERE WAS ICE

The use of natural ice for refrigeration spread throughout the world in prehistoric times. China had ice houses for storing winter ice and snow by 1100 BC. The early Greeks and Romans are known to have used ice and snow for cooling drinks but not for preserving foods. In about 300 BC, the king of Macedon had several trenches dug and filled with snow to cool kegs of wine given to his troops on the eve of a major battle, hoping it would make them more courageous. Ice and snow were harvested by farmers during the winter throughout the United States, Europe, and Asia (see Figure 14.4). Ice was stored in special *icehouses*, underground, or in pits and ravines and covered with straw to insulate it from the daytime sun.

Cutting ice at Moundsville, West Virginia, circa 1900 · Source: Church of God Archives, Anderson University, Anderson, Indiana

FIGURE 14.4
Ice harvesting in the 19th century.

Initially, natural ice refrigeration was merely a convenience, providing a cool drink or preserving food a bit longer. However, the development and extensive use of ice as a refrigeration technology had a very significant social impact, in that it allowed whole populations to change to a healthier diet. In the distant past, people used salting and drying as the main technology for preserving

IS IT CALLED AN *ICEBOX* OR A *REFRIGERATOR*?

An *icebox* was a wooden box that contained both ice and food to be preserved (Figure 14.5). Originally, the ice was put on the bottom and the food on the top. But eventually it was realized that this was inefficient because the cold air is heavier than warm air. Thereafter, the ice was put in the top of the icebox and the food was always placed below it, so that the heavier cold air could circulate around the food. The icebox was invented in 1803 and manufactured in the United States until 1953. Ice needed to be added every day or two in the original iceboxes. But, by 1923, improved thermal insulation design required ice to be added only every five to seven days.

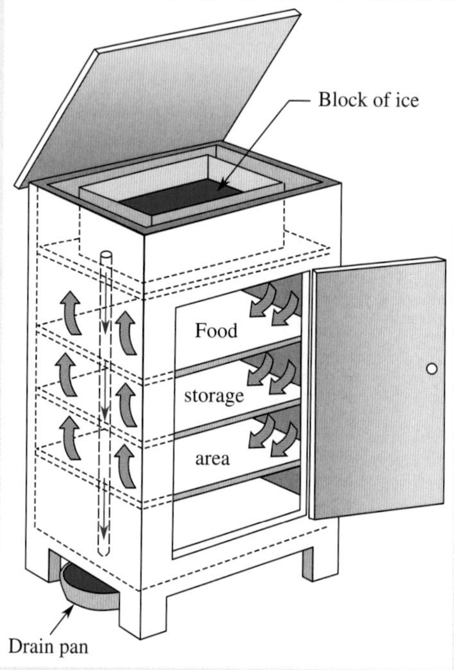

FIGURE 14.5
Domestic icebox.

The term *refrigerator* is reserved for a device that does not use ice to produce cold temperatures, even though the device may be used to preserve food (Figure 14.6). There are vapor-compression refrigerators, gas expansion refrigerators, thermoelectric refrigerators, and so forth.

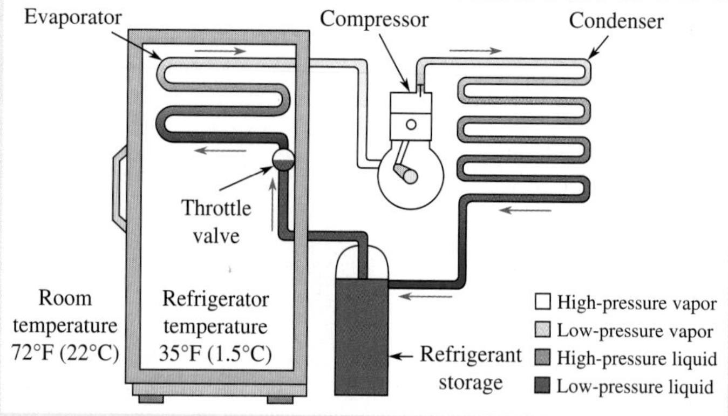

FIGURE 14.6
Domestic refrigerator.

meat and fish. By 1830, the use of ice to preserve food in American iceboxes was quite common. The standard of living of average Americans improved between 1830 and 1860, as their diet changed from one of bread and salted or dried meat or fish to one that regularly included refrigerated fresh meat, fruits, and vegetables.

Preserving food in a sterilized metal can was patented in 1825, but it did not become a commercial success until 1875, about the same time mechanical refrigeration systems were being marketed.

As the food refrigeration industry grew, the demand for natural ice increased dramatically. Natural ice was harvested from ponds, lakes, and rivers in rural communities around the world by farmers during the winter months. New York City used 12,000 tons of natural ice in 1843, 100,000 tons in 1856, and 1 million tons in 1879. Natural ice harvesting in the United States reached its peak in 1886 at 25 million tons. From 1845 to 1860, the mechanical refrigeration systems of Perkins, Gorrie, and Carré were used primarily for making ice to replace natural (winter) ice.

A common unit of commercial and household refrigeration or air conditioning is the *ton*. The following example illustrates the use of this old unit of measurement.

IS IT DANGEROUS TO STUFF A CHICKEN WITH SNOW?

The great British philosopher and statesman Sir Francis Bacon (1561–1626) was keenly interested in the possibility of using snow to preserve meat. In March 1626, he stopped in the country on a trip to London and purchased a chicken. He had the chicken killed and cleaned on the spot, then he packed it with snow and took it with him to London (Figure 14.7). Unfortunately, the experiment only caused his own death a few weeks later. The 65-year-old statesman apparently caught a chill while stuffing the chicken with snow and came down with terminal bronchitis. Refrigeration was clearly not something to be taken lightly.

FIGURE 14.7
The price of experimentation.

WHAT IS A "TON" OF REFRIGERATION?

A ton of refrigeration or air conditioning is the amount of heat that must be removed from 1 ton (2000 lbm) of water in one day (24 hours) to freeze it at 32°F at 1 atmosphere pressure. It is also the amount of heat absorbed by the melting of 1 ton of ice in 24 hours at 32°F at atmospheric pressure. Using more conventional units,

$$1 \text{ ton of refrigeration or air conditioning} = 200.\ \text{Btu/min} = 12.0 \times 10^3\ \text{Btu/h} = 214.\ \text{kJ/min} = 12{,}600\ \text{kJ/h}$$

EXAMPLE 14.2

The earth's polar ice caps contain about 2.50 × 10^{16} m^3 of ice. Determine the tons of refrigeration produced if all this ice were to melt at 0.00°C in a 24.0 h period. The density of ice at 0.00°C is 917 kg/m^3.

Solution

The mass of ice present in the polar ice caps is (2.50 × 10^{16} m^3)(917 kg/m^3)(2.2046 lbm/kg) = 5.05 × 10^{19} lbm = 2.53 × 10^{16} tons. Since a *ton* of refrigeration is equal to the amount of heat absorbed by melting 1 ton of ice at 0.00°C at atmospheric pressure in one 24.0 h day, melting the earth's polar ice caps at 0.00°C in a 24.0 h period would produce 2.53 × 10^{16} tons of refrigeration.

Exercises

4. Suppose the polar ice caps in Example 14.2 melt over a period of 14.0 years. Then, how many tons of refrigeration would be produced? **Answer**: 6.85 × 10^{12} tons.
5. How many Btu per hour would be produced by the melting of the polar ice caps in Example 14.2? **Answer**: 3.04 × 10^{20} Btu/h.
6. How long would it take to melt the polar ice caps in Example 14.2 if the Earth receives an extra 10^{15} kJ per year from the sun? **Answer**: 7.56 million years.

14.5 VAPOR-COMPRESSION REFRIGERATION CYCLE

Like the steam engine, refrigeration technology had a significant impact on society and the way we live. First of all, it changed the way we process food; it created large new agricultural markets and provided a healthier diet for many people. Later, it was applied to making our living environment more comfortable and productive. Initially, it was a spinoff technology from steam engine and gas power cycle prime movers that were simply made to operate thermodynamically backward. Then, it became a powerful force in shaping our culture.

The first vapor-compression refrigeration system using a closed cycle process was patented in 1834 by the American Jacob Perkins (1766–1849). He chose ethyl ether (or, more accurately, diethyl ether, $C_2H_5OC_2H_5$) as the refrigerant, because at low pressures, its temperature was low enough to freeze water on the outside of the evaporator. The ether vapor was compressed in a piston-cylinder apparatus and condensed into a liquid at a higher saturation pressure and temperature. Finally, the liquid ether was throttled through a valve back into the low-pressure evaporator. This system is illustrated in Figure 14.8. Since this process occurs beneath the vapor dome of the working fluid (ether), it is clearly a reversed Rankine cycle device.

All vapor-compression cycle refrigeration systems operate essentially on a reversed Rankine cycle, as shown in Figure 14.8b. In these systems, the boiler is normally called the *evaporator* and the prime mover is replaced by a *compressor*. Also, it would seem reasonable to replace the boiler feed pump of the forward-running Rankine cycle with some form of prime mover in the reversed Rankine or vapor-compression cycle, whose work output could be used to offset the work input to the compressor. Unfortunately, this is not economically feasible in most small- to medium-scale refrigeration systems, as the following example illustrates.

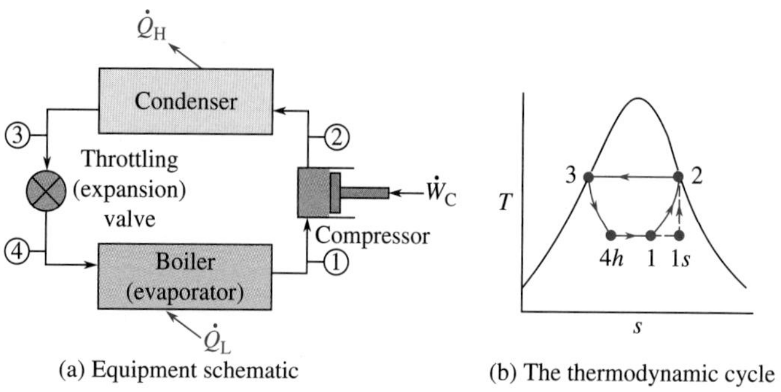

FIGURE 14.8
Jacob Perkins's closed-loop vapor-compression refrigeration cycle.

WHERE DID "MECHANICAL REFRIGERATION" COME FROM?

The first vapor-compression refrigeration system was patented by Jacob Perkins (1766–1849) in 1834 (Figure 14.9). Though Perkins was an American, his refrigerator was made in England and was not an economic success. A similar machine was made in the United States in 1856 by Alexander Catlin Twinning (1801–1884), again with little financial success. In each case, the evaporator was immersed in a salt brine solution and the cold brine was used to make ice, but it attracted little attention for more than 20 years, after which natural refrigeration had begun to cause changes in people's dietary habits. In 1855, James Harrison (1816–1893), a Scotsman who emigrated to Australia, produced a commercially successful refrigerator similar to Perkins's for the manufacture of ice. Since natural ice is difficult to find in Australia, Harrison's *artificial* ice machine was an instant success.

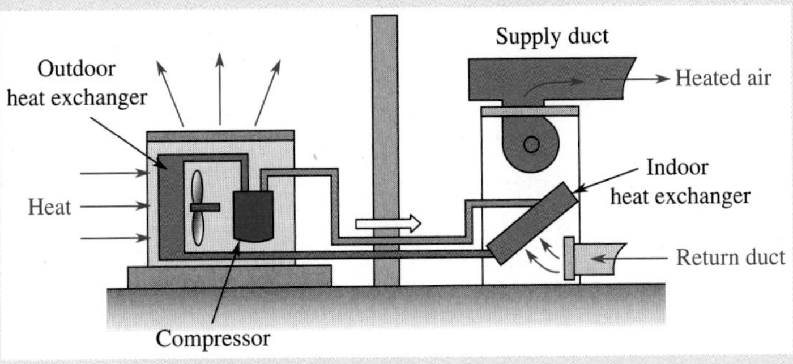

FIGURE 14.9
Jacob Perkins's 1834 refrigeration apparatus.

EXAMPLE 14.3

A refrigeration system for a supermarket is to be designed using R-22 to maintain frozen food at –15.0°C while operating in an environment at 20.0°C. The refrigerant enters the condenser as a saturated vapor and exits as a saturated liquid. Determine the COP for this refrigerator, using

a. A reversed Carnot cycle operating between these temperature limits.
b. An isentropic vapor-compression cycle with an isentropic expansion turbine installed between the high-pressure condenser and the low-pressure evaporator.
c. An isentropic vapor-compression cycle with an aergonic, adiabatic, throttling expansion valve installed between the high-pressure condenser and the low-pressure evaporator.

Solution

First, draw a sketch of the system (Figure 14.10).

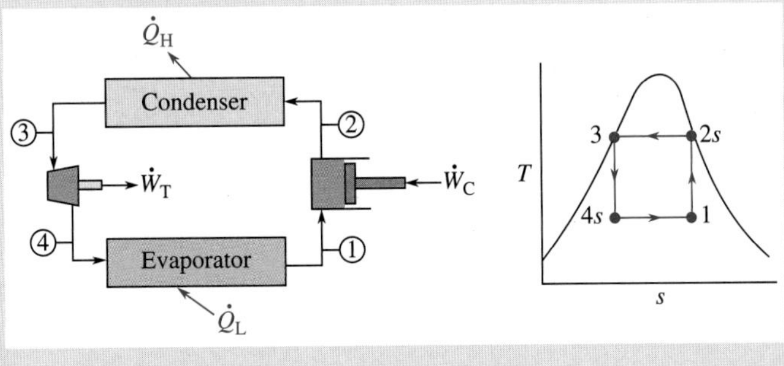

FIGURE 14.10
Example 14.3, system sketch.

a. Here, $T_H = 20.0 + 273.15 = 293.15$ K, and $T_L = -15.0 + 273.15 = 258.15$ K. Then, Eq. (14.5) gives

$$\text{COP}_{\substack{\text{Carnot}\\\text{refrigerator}}} = \frac{T_L}{T_H - T_L} = \frac{258.15}{293.15 - 258.15} = 7.38$$

(Continued)

EXAMPLE 14.3 (*Continued*)

b. From Table C.9b in *Thermodynamic Tables to accompany Modern Engineering Thermodynamics*, the thermodynamic data at the monitoring stations shown in the schematic are

Station 1	Station 2s
$T_1 = -15.0°C$	$T_{2s} = 20.0°C$
$s_1 = s_{2s} = 0.89973$ kJ/(kg·K)	$x_{2s} = 1.00$
$x_1 = 0.9395$	$h_{2s} = 256.5$ kJ/kg
$h_1 = 231.0$ kJ/kg	$s_{2s} = 0.89973$ kJ/(kg·K)
	$p_{2s} = 909.9$ kPa

Station 3	Station 4s
$T_3 = 20.0°C$	$T_{4s} = T_1 = -15.0°C$
$x_3 = 0.00$	$s_{4s} = s_3 = 0.25899$ kJ/(kg·K)
$h_3 = 68.67$ kJ/kg	$x_{4s} = 0.1765$
$s_3 = 0.25899$ kJ/(kg·K)	$h_{4s} = 65.6$ kJ/kg
$p_3 = p_2 = 909.9$ kPa	

where we have calculated

$$x_1 = \frac{s_1 - s_{f1}}{s_{fgx1}} = \frac{s_{2s} - s_{f1}}{s_{fg1}} = \frac{0.89973 - 0.11075}{0.83977} = 0.9395$$

$$h_1 = h_{f1} + x_1(h_{fg1}) = 27.33 + (0.9395)(216.79) = 231.0\ \text{kJ/kg}$$

$$x_{4s} = \frac{s_3 - s_{f4}}{s_{fg4}} = \frac{0.25899 - 0.11075}{0.83977} = 0.1765$$

and

$$h_{4s} = h_{f4} + x_{4s}(h_{fg4}) = 27.33 + (0.1765)(216.79) = 65.59\ \text{kJ/kg}$$

Then,

$$\text{COP}_{\substack{\text{isentropic}\\ \text{vapor-compression}\\ \text{cycle (with expansion}\\ \text{turbine)}}} = \frac{\dot{Q}_L}{\dot{W}_c - \dot{W}_t} = \frac{h_1 - h_{4s}}{(h_{2s} - h_1) - (h_3 - h_{4s})}$$

$$= \frac{231.0 - 65.59}{(256.5 - 231.0) - (68.67 - 65.59)} = 7.38$$

which is identical to the Carnot efficiency of part a, as it should be, because the Rankine and Carnot cycles are identical in this case (see Figure 14.10).

c. When the isentropic turbine is replaced by an adiabatic, aergonic throttling valve, the process from station 3 to station 4 becomes isenthalpic rather than isentropic, as shown in Figure 14.11.

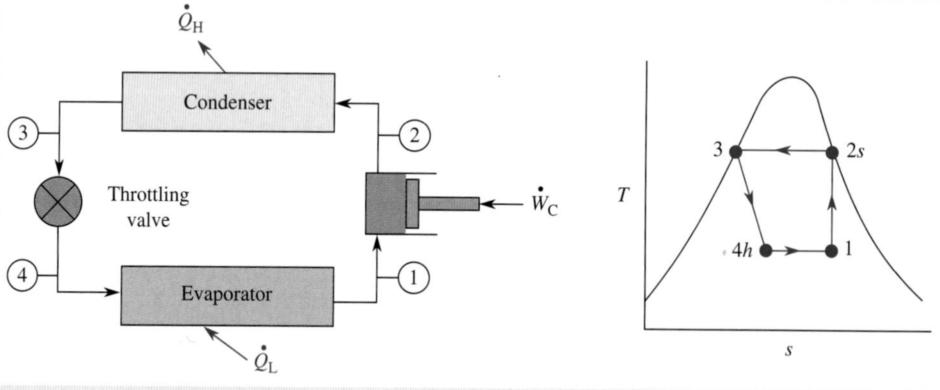

FIGURE 14.11
Example 14.3, Solution, part c.

The thermodynamic data for stations 1, 2*s*, and 3 remain unchanged from part b, but the isenthalpic throttling valve changes the data of station 4*s* to 4*h* as follows.

Station 4*h*

$T_{4h} = T_1 = -15.0°C$

$h_{4h} = h_3 = 68.67$ kJ/kg

$x_{4h} = 0.1910$

$s_{4h} = 0.27081$ kJ/(kg·K)

where we have calculated

$$x_{4h} = \frac{h_{4h} - h_{f4}}{h_{fg4}} = \frac{68.67 - 27.34}{216.79} = 0.1906$$

and

$$s_{4h} = s_{f4} + x_{4h}(s_{fg4}) = 0.11075 + (0.1906)(0.83977) = 0.27081 \text{ kJ/(kg·K)}$$

Finally,

$$\text{COP}_{\substack{\text{isentropie}\\ \text{vapor-compression cycle}\\ \text{(with throttling valve)}}} = \frac{\dot{Q}_L}{\dot{W}_c} = \frac{h_1 - h_{4h}}{h_{2s} - h_1} = \frac{231.0 - 68.67}{256.5 - 231.0} = 6.37$$

Exercises

7. Determine the pressure of the R-22 in the evaporator in Example 14.3. **Answer**: $p_{\text{evaporator}} = p_1 = p_4 = p_{\text{sat}}$(R-22 at −15.0°C) = 295.7 kPa.
8. If ammonia were used in the refrigeration system described in Example 14.3, determine the condenser pressure if all the other variables remain unchanged. **Answer**: $p_{\text{condenser}} = p_2 = p_3 = p_{\text{sat}}$(ammonia at 20.0°C) = 857.12 kPa.
9. The head of your Engineering Department has decided to use R-134a instead of R-22 in the refrigeration system in Example 14.3. Assuming all the other variables remain unchanged, determine the new operating pressure in the evaporator. **Answer**: $p_{\text{evaporator}} = p_1 = p_4 = p_{\text{sat}}$(R-134a at −15.0°C) = 164 kPa.

The decrease in COP from 7.38 to 6.37 (13.7%) in the previous example is not normally sufficient to justify the increased expense of manufacturing, installing, and maintaining a turbine or other prime mover between the condenser and the evaporator in small- and medium-size systems. Also, the working fluid in this part of the cycle contains a mixture of liquid and vapor, and it is difficult to find any prime mover that operates efficiently and reliably with this type of two-phase fluid. Throttling expansion valves, on the other hand, are very inexpensive and reliable under these conditions.

By introducing the isentropic efficiency of the compressor $(\eta_s)_c$, the general formula for the actual thermal efficiency (COP) of a reversed Rankine cycle can be written as

$$\text{COP}_{\substack{\text{vapor-compression cycle}\\ \text{R/AC}}} = \frac{\dot{Q}_L}{\dot{W}_c} = \frac{h_1 - h_{4h}}{(h_{2s} - h_1)/(\eta_s)_c} \tag{14.6}$$

and

$$\text{COP}_{\substack{\text{vapor-compression cycle}\\ \text{HP}}} = \frac{|\dot{Q}_H|}{\dot{W}_c} = \frac{h_2 - h_3}{(h_{2s} - h_1)/(\eta_s)_c} \tag{14.7}$$

where $h_2 = h_1 + (h_{2s} - h_1)/(\eta_s)_c$.

Because throttling processes are ideally isenthalpic, a pressure-enthalpy diagram is often used to describe vapor refrigeration cycles, as shown in Figure 14.12. Process 1 to 2*s* in this figure involves the compression of a liquid-vapor mixture. This is technically more difficult than compressing either a pure vapor or a pure liquid. A method of eliminating this problem is to superheat the vapor, as shown in Figure 14.12b.

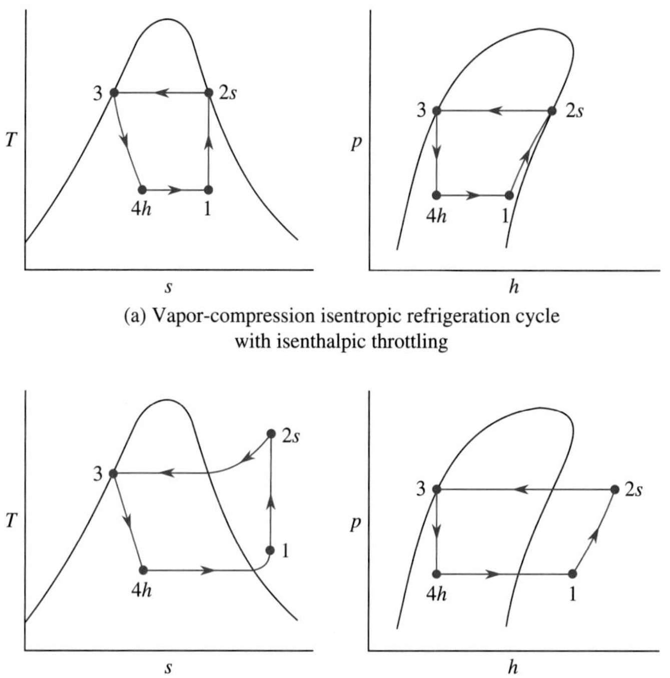

(a) Vapor-compression isentropic refrigeration cycle with isenthalpic throttling

(b) Same as (a) except with superheat

FIGURE 14.12
T–s and *p–h* diagrams for an isentropic vapor-compression cycle.

EXAMPLE 14.4

Repeat part c of Example 14.3 requiring that the evaporator outlet be a saturated vapor at −15.0°C and introduce a compressor isentropic efficiency of 75.0%.

Solution

First, draw a sketch of the system (Figure 14.13). Since the evaporator outlet is a saturated vapor, the compressor outlet is a superheated vapor, as shown in the figure.

The thermodynamic data for the four monitoring stations are (see Example 14.3 for details)

Station 1	Station 2*s*
$T_1 = -15.0°\text{C}$	$p_{2s} = p_2 = 909.9\ \text{kPa}$
$x_1 = 1.00$	$s_{2s} = s_1 = 0.95052\ \text{kJ/(kg·K)}$
$h_1 = 244.13\ \text{kJ/kg}$	$h_{2s} = 271.92\ \text{kJ/kg}$ (from interpolation in Table C.10b)
$s_1 = 0.95052\ \text{kJ/(kg·K)}$	$T_{2s} = 39.3°\text{C}$
Station 3	**Station 4*h***
$T_3 = 20.0°\text{C}$	$T_{4h} = T_1 = -15.0°\text{C}$
$x_3 = 0.00$	$h_{4h} = h_3 = 68.67\ \text{kJ/kg}$
$h_3 = 68.67\ \text{kJ/kg}$	$x_{4h} = 0.1910$
$s_3 = 0.25899\ \text{kJ/(kg·K)}$	$s_{4h} = 0.27088\ \text{kJ/(kg·K)}$

Then, from Eq. (14.7),

$$\text{COP}_{\substack{\text{vapor-compression cycle}\\ \text{R/AC}}} = \frac{\dot{Q}_L}{|\dot{W}|_c} = \frac{h_1 - h_{4h}}{(h_{2s} - h_1)/(\eta_s)_c}$$

$$= \frac{244.13 - 68.67}{(271.92 - 244.13)/0.750} = 4.74$$

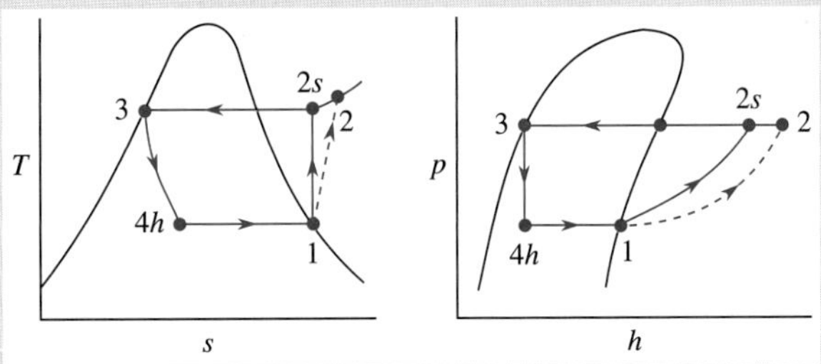

FIGURE 14.13
Example 14.4.

Exercises

10. After several years of use, the isentropic efficiency of the compressor in Example 14.4 decreases from 75.0% to 55.0% due to wear and a lack of maintenance. Determine the new COP for this system. **Answer**: COP = 3.47.
11. Determine the power required to drive the compressor $\dot{W}_c$ in Example 14.4 if the refrigeration system is to produce $\dot{Q}_L = 20$ tons of cooling. Recall that 1 ton of refrigeration is equal to 214. kJ/min. **Answer**: $\dot{W}_c = 19.8$ hp.
12. Determine the mass flow rate of refrigerant required in Example 14.4 if this system produces $\dot{Q}_L = 20$ tons of cooling. Recall that 1 ton of refrigeration is equal to 214. kJ/min. **Answer**: $\dot{m} = 23.9$ kg/min.

Even if the compressor had an isentropic efficiency of 100%, the COP in this example would be only 4.74/0.750 = 6.32, which is still slightly less than the 6.37 of part c in Example 14.3. Thus, adding superheat to the cycle usually does not increase the COP because both $|\dot{W}_c|/\dot{m}$ and $\dot{Q}_L/\dot{m}$ are increased. However, the required mass flow rate $\dot{m}$ is significantly reduced by the addition of superheat. Also, because condensers and evaporators are not 100% effective as heat exchangers, the temperature difference between the working fluid in these devices and their local environment is typically about 15.0°F.

14.6 REFRIGERANTS

Whereas the working fluid of the steam engine (water) was nearly ideal for vapor power cycles, it was totally unsuitable for the refrigeration cycles of commercial interest. The major problem faced by the early developers of refrigeration technology was not the design of the machinery per se but the search for a suitable nontoxic, safe, inexpensive working fluid with satisfactory low-temperature thermodynamic characteristics.

Though water is the cheapest and safest refrigerant available, it is limited to high-temperature applications such as steam-jet refrigeration. Since most refrigeration needs are at temperatures near the freezing point of water, other refrigerants had to be found that boiled at lower temperatures.

Perkins used ethyl ether as his refrigerant. It was a good refrigerant, but it was also toxic and flammable. Also, the entire ethyl ether refrigeration system operated below atmospheric pressure, making it difficult to prevent air from leaking into the system. The danger and complexity of ethyl ether refrigerators caused other inventors to search for alternative refrigeration technologies, which ultimately lead to the rapid development of gas expansion refrigeration cycles between 1860 and 1890.

The French inventor Charles Tellier (1828–1913) introduced methyl ether (CH_3Cl) as a replacement for ethyl ether in 1863. Though methyl ether was also toxic and flammable, it had a higher vapor pressure, and that allowed the entire refrigeration system to operate above atmospheric pressure, thus eliminating the problems caused by air leaking into the system.

IS ETHER A REFRIGERANT OR AN ANESTHETIC?

The *di* in *di*ethyl ether is often dropped, and it is called either *ethyl ether* or simply *ether*. This is the same ether that was first successfully used as an anesthetic in 1846 by the Massachusetts dentist William T. G. Morton. Since the boiling point of ether at atmospheric pressure is 35°C (95°F), slightly below the temperature of the human body, it was common practice in the late 19th and early 20th centuries for physicians to use liquid ether as a local anesthetic by spraying it onto parts of the body where it would then freeze the tissue as it boiled away and consequently numb the local sensations. This is the source of the term *freezing* as a synonym for a local anesthetic (especially in dentistry) today.

HOW DID COMPRESSOR TECHNOLOGY DEVELOP?

By the end of the World War I (1914–1918), reciprocating piston compressors still dominated refrigerant technology, and the primary refrigerants still in use in the Unoted States at that time were ammonia, carbon dioxide, and sulfur dioxide. In 1919, the French engineer Henri Corblin (1867–1947) patented a diaphragm refrigerant compressor in which the oscillating motion of the center of a fixed diaphragm replaced the reciprocating motion of a piston in a cylinder. In 1918, the first hermetically sealed refrigeration compressor was developed by the Australian Douglas Henry Stokes, in which the motor and compressor were sealed together inside a container with the refrigerant. In 1933, Willis Carrier (1876–1950) developed his first centrifugal refrigerant compressor for use with R-11.

During the last half of the 19th century, the development of refrigeration technology flourished in America, especially in the South. In 1866, Thaddeus S. C. Lowe (1832–1913) developed a high-pressure (80 atm) carbon dioxide compressor for manufacturing ice in Dallas, Texas, and Jackson, Mississippi; and in 1872, David Boyle (1837–1891) developed an ammonia compressor (10 atm) for manufacturing ice in Jefferson, Texas. This allowed CO_2 and NH_3 to enter the list of useful refrigerants.

The Swiss physicist Raoul Pierre Pictet (1846–1929) studied the various refrigerants then available and found that sulfur dioxide had suitable thermodynamic properties. In 1874, he developed an SO_2 compressor and refrigerating system that was quite successful. Sulfur dioxide has the advantages of being a natural lubricant for the compressor and it does not burn. Its chief disadvantage is that, on contact with moisture, it forms corrosive sulfuric acid.

In the late 1920s, the American chemist and engineer, Thomas Midgley, Jr. (1889–1944), discovered that certain fluorine compounds were remarkably nontoxic and odorless while simultaneously having the proper thermodynamic properties of a good refrigerant. In the 1930s, the E. I. duPont de Nemours Company became commercially involved in the refrigeration industry by manufacturing and selling Midgley's discovery as a refrigerant. DuPont marketed the product under the commercial trade name *Freon*.

Midgley's refrigerants were halogenated hydrocarbons in which halogen atoms (mainly chlorine and fluorine) were substituted for hydrogen atoms in simple hydrocarbon molecules. Midgley replaced the four hydrogen atoms in methane, CH_4, with two chlorine and two fluorine atoms to produce dichloro-difluoro-methane (or dichlorodifluoromethane, CCl_2F_2). Other common methane based refrigerants are monochlorodifluoromethane $CHClF_2$ and trichloromonofluoromethane CCl_3F. The complex chemical names of these compounds are logical and technically correct, but they are difficult for the nonchemist to pronounce and remember. Consequently, a confusing variety of commercial trade names, such as Freon, Genetron, Isotron, and Frigen, came into popular use during the 1940s. Shortly thereafter, the American Society of Refrigerating Engineers (ASRE)[3] decided to adopt a standard method of refrigerant designation that was based only on the use of numbers.

THE TEFLON CONNECTION!

A young DuPont chemist named Roy J. Plunkett discovered Teflon on April 6, 1938, while experimenting with a halogenated ethylene gas for use as a refrigerant. On this day, Plunkett received a pressurized tank of tetrafluoroethylene (C_2F_4) to study its properties as a nontoxic refrigerant. When he opened the tank nothing came out. After the valve was checked, the tank was weighted and found to be the same weight as when it was full. Something made no sense, so Plunkett had the tank cut open and found a waxy white powder. Being a chemist, Plunkett realized that the gas had somehow spontaneously "polymerized" to form a new material, *polytetrafluoroethylene*. The waxy white powder had some remarkable physical properties: it was not affected by strong acids or bases, was resistant to heat from −450°F to 725°F (−270°C to 385°C), and was very slippery. While these properties were interesting, it was decided that this new material had no particular commercial value. Then came World War II and the top-secret atomic bomb project (the Manhattan Project). A material was needed for gaskets that would resist the terribly corrosive properties of uranium hexafluoride gas. By a chance communication, the director of the Manhattan Project became aware of the new polymeric material that Plunkett had discovered. It was then made into a test gasket and found to be very successful at containing the corrosive gas. After World War II, the new polymer material was not put to any practical use until it began to be used on nonstick cookware in France in 1954. Nonstick cooking utensils were first sold in the United States on December 15, 1960, at Macy's Department Store in New York City. Taking letters from the complicated chemical name *polytetrafluoroethylene*, the new polymer was named *Teflon*.

---

[3] The ASRE merged with the American Society of Heating and Air-Conditioning Engineers (ASHAE) to form the American Society of Heating, Refrigerating and Air-Conditioning Engineers (ASHRAE) in 1959.

14.7 REFRIGERANT NUMBERS

Most halogenated hydrocarbons used in refrigeration have a molecular structure of the form $C_aH_bCl_cF_d$ and the atomic valences require that $c = 2(a + 1) - b - d$. These compounds are given refrigerant R numbers defined by

$$C_aH_bCl_cF_d \text{ is refrigerant number: R-}(a-1)(b+1)d, \text{ where } c = 2(a+1) - b - d \tag{14.8}$$

When $a = 1$, then $a - 1 = 0$ (the methane series of halogenated hydrocarbons) and the zero is omitted in the R number. For example, carbon tetrachloride, CCl_4, has $a = 1$, $b = 0$, $c = 4$, and $d = 0$. Consequently, its R number is R-(0)(0 + 1)0 = R-010 = R-10.

Bromate compounds are indicated with a B after the R number followed by the number of bromine atoms. For example, $CBrF_3$ = R-13B1. Also, ethane and higher hydrocarbon bases can have numerous *isomers* (compounds containing the same number of atoms, but assembled in different ways). In these cases, the most symmetrical atomic arrangement is given the base R number R-$(a - 1)(b + 1)d$, and the remaining arrangements are given the suffixes *a*, *b*, *c*, and so forth as the refrigerant molecule become less and less symmetrical. For example the differences between R-134 (CHF_2-CHF_2) and R-134a (CH_2FCF_3) are illustrated next:

$$\begin{array}{ccccccc} & F & F & & & H & F & \\ H- & C- & C- & H & F- & C- & C- & F \\ & F & F & & & H & F & \end{array}$$

R – 134 R – 134a

Therefore, Midgley's CCl_2F_2 with $a = 1$, $b = 0$, $c = 2$, and $d = 2$ became *Refrigerant-12* (abbreviated R-12), or *Freon-12* if manufactured by DuPont. Similarly, $CHClF_2$ ($a = b = c = 1$, $d = 2$) became *Refrigerant-22* or R-22, CCl_3F ($a = 1$, $b = 0$, $c = 3$, $d = 1$) became *Refrigerant-11* or R-11, and so forth. Ethane-based refrigerants are the 100 number series, and the ethane-based hexachloroethane C_2Cl_6 ($a = 2$, $b = 0$, $c = 6$, $d = 0$) became *Refrigerant-110* or R-110, and so forth. Propane-based refrigerants are the 200 number series, and butane-based refrigerants are assigned the 600 number series. Inorganic (i.e., nonhydrocarbon based) refrigerants are assigned the 700 number series with the last two digits being the molecular mass of the refrigerant. For example, ammonia, NH_3, is *Refrigerant-717* and water, H_2O, is *Refrigerant-718*. Table 14.1 lists the ASHRAE number, chemical formula, and boiling point of some common refrigerants. Figure 14.14 presents typical saturation temperature-pressure curves for some common refrigerants plus a graphical representation of the refrigerant derivatives of methane, CH_4, and ethane, C_2H_6.

Table 14.1 The American Society of Heating, Refrigerating and Air-Conditioning Engineers Refrigerant Numbering System for Some Common Refrigerants

		Boiling Point at Atmospheric Pressure	
Refrigerant Number	**Chemical Formula**	**°F**	**°C**
R-10	CCl_4	170.2	76.8
R-11	CCl_3F	74.9	23.8
R-12	CCl_2F_2	−21.6	−29.8
R-21	$CHCl_2F$	48.1	8.9
R-22	$CHClF_2$	−41.4	−40.8
R-30	CH_2Cl_2	105.2	40.7
R-40	CH_3Cl	−14.8	−23.8
R-50	CH_4 (methane)	−259.0	−161.7
R-110	C_2Cl_6	365.0	185.0
R-111	C_2Cl_5F	279.0	137.2
R-112	$C_2Cl_4F_2$	199.0	92.8
R-123	$CHCl_2CF_3$	81.7	27.6
R-134a	CH_2FCF_3	−15.7	−26.2
R-170	C_2H_6 (ethane)	−127.8	−88.8
R-290	C_3H_8 (propane)	−43.7	−42.1
R-600	C_4H_{10} (butane)	33.1	0.6
R-717	NH_3 (ammonia)	−28.0	−33.3
R-718	H_2O (water)	212.0	100.0

Source: Reprinted by permission from the *ASHRAE Handbook—1985 Fundamentals*.

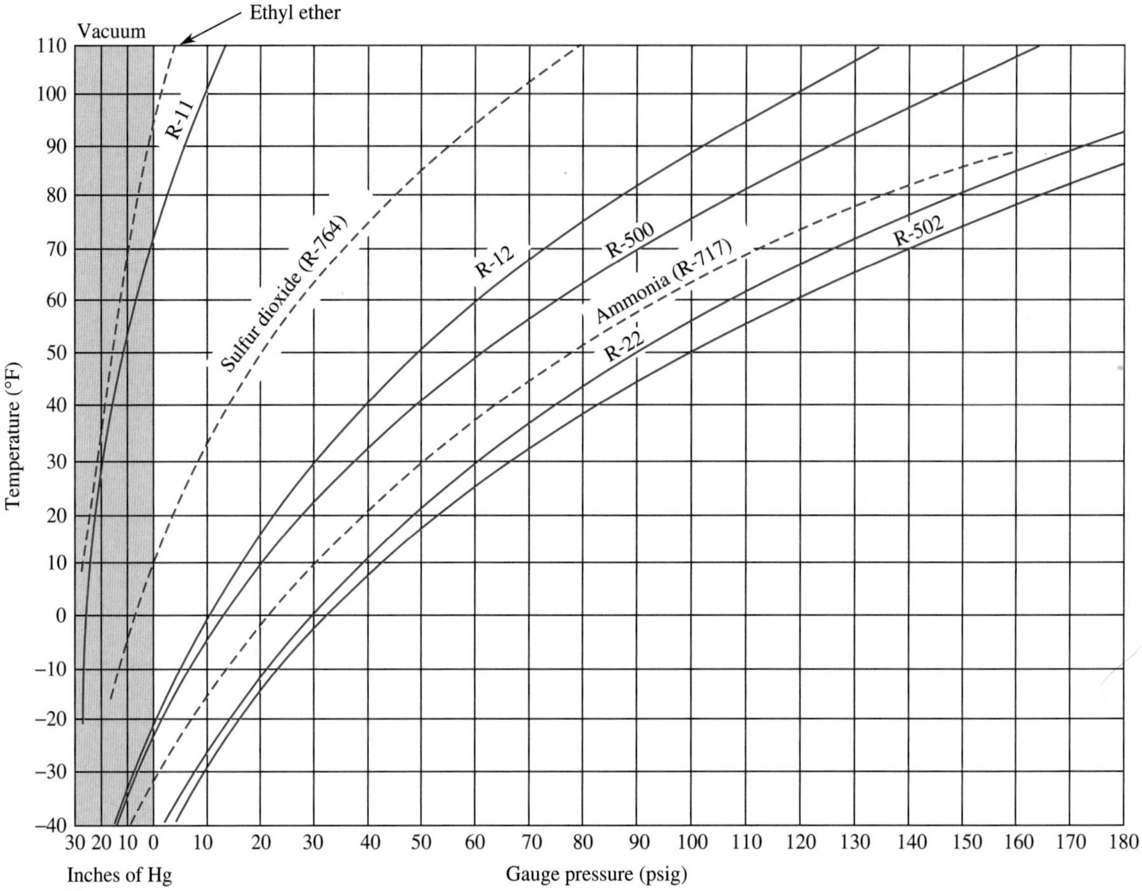

(a)

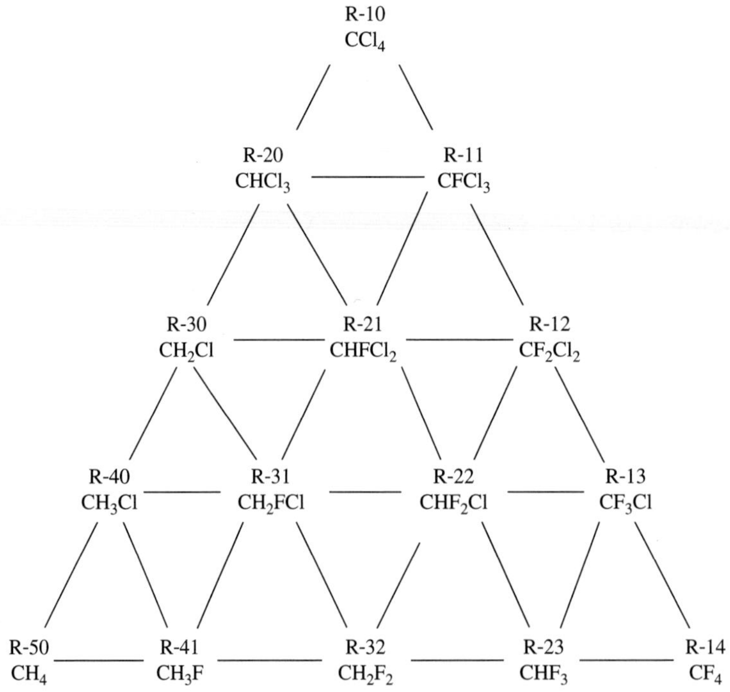

(b) Refrigerant derivatives of methane (CH_4)

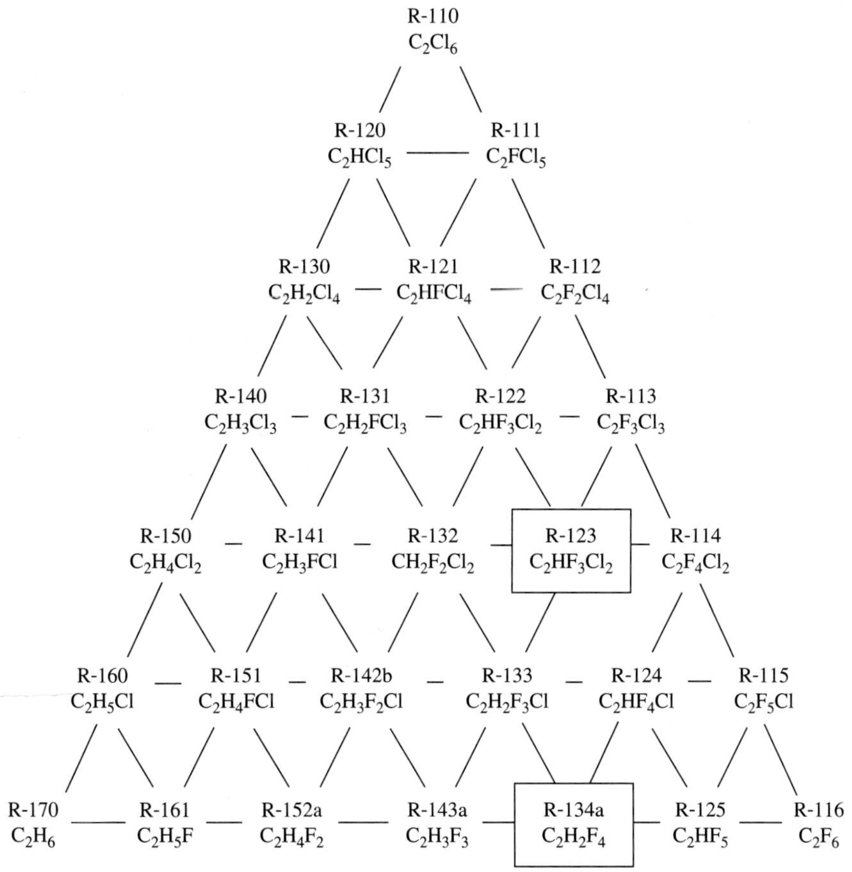

(c) Refrigerant derivatives of ethane (C_2H_6)

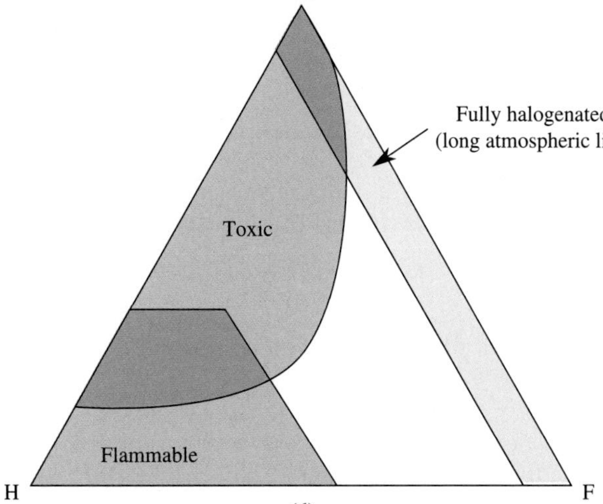

(d)

FIGURE 14.14
(a) Typical saturation temperature–pressure curves for common refrigerants, (b) the refrigerant derivatives of methane, (c) the refrigerant derivatives of ethane, and (d) CFC behavior chart.

EXAMPLE 14.5

As a technical expert in a multibillion-dollar lawsuit, you are asked to determine the refrigerant numbers for the following refrigerants by the prosecuting attorney:

a. Chloroform, $CHCl_3$.
b. Chlorotetrafluoroethane, $CHClFCF_3$.
c. Octafluoropropane, $CF_3CF_2CF_3$.

Solution

Being totally unimpressed by the prosecuting attorney's aggressive questioning, you calmly reply as follows:

a. "Chloroform contains one carbon atom ($a = 1$), one hydrogen atom ($b = 1$), and three chlorine atoms ($c = 3$), and no fluorine atoms ($d = 0$)." Making a quick calculation in your head using Eq. (14.8), you arrive at R-(a – 1)(b + 1)d = R-(1 – 1)(1 + 1)0 = R-020 = R-20 (dropping the leading 0). Then you reply, "So, the refrigerant number for chloroform is R-20."
b. "Chlorotetrafluoroethane, on the other hand, contains two carbon atoms ($a = 2$), one hydrogen atom ($b = 1$), one chlorine atom ($c = 1$), and four fluorine atoms ($d = 4$)." Using Eq. (14.8) again you find R-(2 – 1)(1 + 1)4 = R-124, and you reply, "So, its refrigerant number is: R-124."
c. "Now octafluoropropane is a very interesting compound in that it contains three carbon atoms ($a = 3$), no hydrogen or chlorine ($b = c = 0$), and eight fluorine atoms ($d = 8$)." (Thinking, again using Eq. (14.8), R-(3 – 1)(0 + 1)8 = R-218.) "Consequently its refrigerant number is: R-218."

Exercises

13. Suppose the prosecuting attorney in Example 14.5 asks you for the refrigerant number of carbontetrachloride CCl_4. What would you say then? **Answer**: Your response: "The refrigerant number is: R-14."
14. "Aha!" the prosecutor in Example 14.5 exclaims, "You seem pretty confident of yourself, don't you? Well, then, can you tell me what substance has refrigerant number 720?" (Recall that 700 series refrigerants are inorganic compounds and the last two digits of the R number correspond to the molecular mass of the compound.) **Answer**: Your response: "The compound is neon."
15. The prosecuting attorney in Example 14.5 vociferates, "You don't say, then give me the chemical formula and refrigerant number for trifluoromethane!" **Answer**: Your response: "CHF_3 which is R-23."

14.8 CFCs AND THE OZONE LAYER

Ozone (O_3) in the upper atmosphere absorbs ultraviolet radiation from the sun and prevents much of it from reaching the surface of the Earth. Exposure to ultraviolet radiation is a known source of skin cancer and other biological problems.

All chlorofluorocarbons (CFCs) are combinations of chlorine, fluorine, and carbon atoms. After 1950, the use of chlorofluorocarbons dominated the domestic and automotive refrigeration and air conditioning markets. In the 1950s and 1960s, inexpensive chlorofluorocarbons found use as a propellant in aerosol spray cans for paint, deodorant, hair products, and so forth.

In 1974, Professor Sherry Rowland at the University of California—Irvine and her postdoctorate student Mario Moline postulated that chlorofluorocarbons are so chemically stable that they can exist in the atmosphere for hundreds of years, eventually diffusing into the Earth's stratosphere, where ultraviolet radiation decomposes them to release chlorine atoms. The chlorine atoms then catalyze the conversion of ozone into oxygen as follows:

$$O_3 + Cl \rightarrow O_2 + ClO$$
$$ClO + O \rightarrow O_2 + Cl$$

with the chlorine atom being regenerated. The overall reaction is then

$$O + O_3 \rightarrow 2O_2$$

The CFC production in 1974 was 1 million pounds per year, and the shear volume of CFCs released through spray cans and leaking refrigeration systems could possibly destroy the ozone layer faster than it is created by ultraviolet radiation acting on oxygen molecules. Rowland's hypothesis alluded to a massive global problem, and it had profound impact on CFC use.

But, what will replace the banned CFCs? It was not too difficult to find safe propellants (such as CO_2) for use in aerosol cans, but finding suitable replacements for refrigerants such as R-11 (used in large building air conditioning systems) and R-12 (used in domestic refrigerators and air conditioners and automotive air conditioners) was much less obvious.

HOW WERE CFCs CONTROLLED?

1978: The U.S. Environmental Protection Agency (EPA) banned the use of CFCs in all nonessential aerosol cans. This action alone cut the U.S. consumption of CFCs by 50%.

1980: The European Community limited CFC production and use in aerosols.

1985: The *Ozone Hole* is discovered in the Antarctic.

1987: The Montreal Protocol is signed by 43 nations to decrease overall production of CFCs by 50% by 1999.

1990: Title VI of the Clean Air Act (Stratospheric Ozone Protection) is passed into law in the United States.

1992: The signers of the Montreal Protocol agree to a phase-out schedule for all HCFCs (including R-123) by the year 2030.

The financial investment in existing refrigeration and air conditioning systems is massive, so the replacement refrigerants must have very similar thermodynamic properties to R-11 and R-12, so that they can be used in the same operating equipment with minimal modifications. Today, R-123 is temporarily[4] replacing R-11, and R-134a is replacing R-12. By the end of 1995, EPA banned most production and import of R-12. However, the use of R-12 is still permitted until supplies are depleted. Figures 14.15 and 14.16 show the *p–h* diagrams for these refrigerants and their replacements.

R-123 is $CHCl_2CF_3$ and is called a *hydrochlorofluorocarbon* (HCFC). While it still contains chlorine, it is 50 times less detrimental to the ozone layer than R-11. Consequently, it is viewed as a temporary replacement for R-11, since it too must be phased out by the year 2030. The ultimate replacement for R-11 may be R-245fa ($CF_2HCF_2CFH_2$), a propane-based halocarbon that does not contain chlorine. R-134a is CH_2FCF_3 and is called a *hydrofluorocarbon* (HFC). It contains no chlorine and will not damage the ozone layer. The other common refrigerant in use in large-scale air conditioning and heat pump systems is R-22 ($CHClF_2$). It is also an HCFC, and

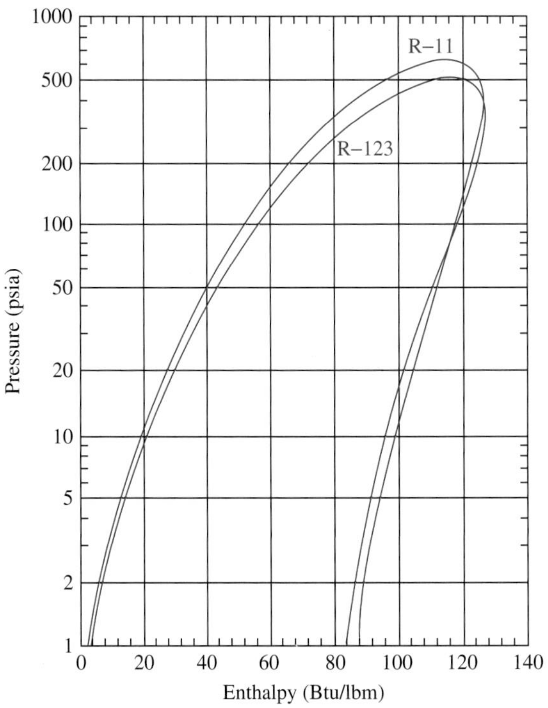

FIGURE 14.15

Superimposed *p–h* diagrams for R-11 and R-123, showing the thermodynamic similarities between these two refrigerants.

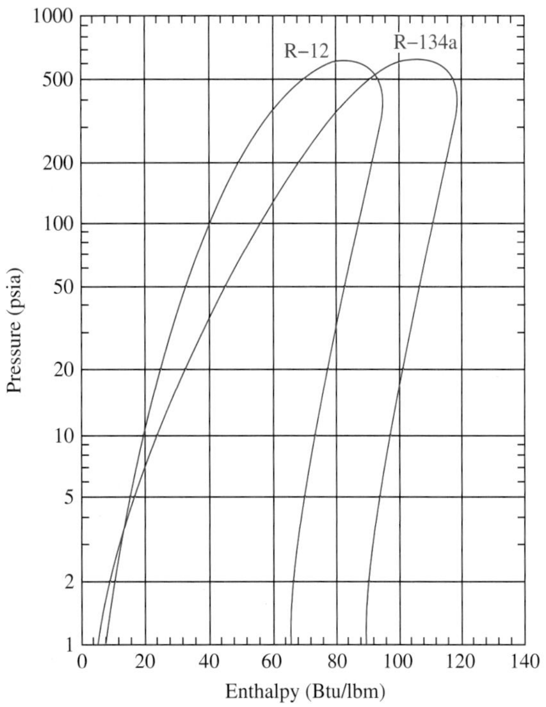

FIGURE 14.16

Superimposed *p–h* diagrams for R-12 and R-134a, showing the thermodynamic similarities between these two refrigerants.

[4] R-123 is scheduled to be phased out in 2020 in new equipment.

CRITICAL THINKING

Replacing refrigerants R-11 and R-12 with R-123 and R-134a in existing equipment is not a matter of simply draining out the old refrigerant and adding the new, because they have different physical and thermodynamic properties. This conversion is very expensive and the owners of the equipment must bear the costs. This is why the final phase-out of all CFCs is not scheduled until the year 2030, when the equipment existing today would be obsolete and need to be replaced anyway. If the CFCs are as dangerous as we think they are, then why are we waiting so long to eliminate them? Who else should share in the conversion costs?

although it still contains chlorine, it is 20 times less detrimental to the ozone layer than R-11 or R-12. However, after January 1, 2010, no virgin R-22 can be used in existing systems, and after January 1, 2015, no recycled refrigerant R-22 can be used in existing systems.

14.9 CASCADE AND MULTISTAGE VAPOR-COMPRESSION SYSTEMS

Refrigeration applications like the quick freezing of processed food or the production of liquefied gases such as liquefied natural gas (LNG, methane) and liquefied petroleum gas (LPG, propane and butane) require moderately cold refrigeration temperatures in the range of −30.°C to −180°C (−22°F to −290°F) with an outside ambient temperature near 20.°C (68°F). This temperature range is too large for a single vapor-compression refrigeration cycle, because it requires a very large pressure ratio across the compressor. To solve this problem, we can connect (or cascade) two or more cycles together to form a *cascade* vapor-compression refrigeration cycle with lower individual compressor pressure ratios, as shown in Figure 14.17. This figure shows a double-cascade

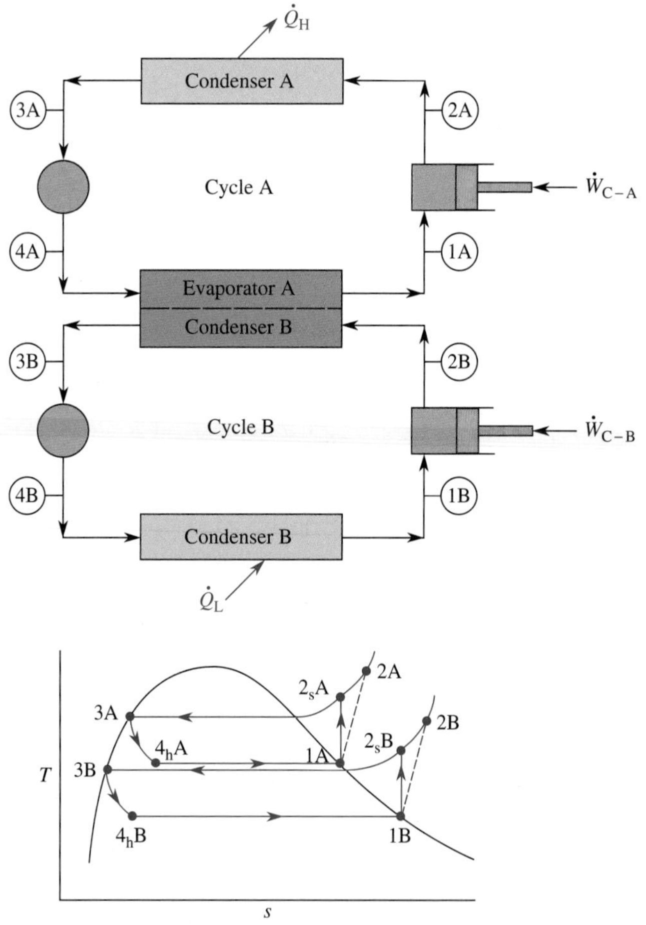

FIGURE 14.17
A dual-cascade, vapor-compression refrigeration system with the same refrigerant used in each cycle.

HOW DO YOU LIQUEFY A GAS LIKE OXYGEN?

Though French engineer Charles Tellier (1828–1913) first suggested the concept of cascade refrigeration in 1867, the Swiss scientist Raoul Pictet (1846–1929) first developed a dual-cascade refrigeration system and used it to produce liquid oxygen in 1877. He used SO_2 in the high-temperature cycle and CO_2 in the low-temperature cycle. He was able to produce only a liquid oxygen mist, but it was the beginning of cryogenic refrigeration. Hydrogen gas was first liquefied in 1898 and helium gas was finally liquefied in 1908.

system using the same refrigerant in each cycle. However, a different refrigerant is often used in each cycle to optimize overall system performance. When different refrigerants are used in the cascaded cycles, separate (or combined) *T–s* diagrams must be used in the analysis. Some industrial systems require three or four cascaded cycles to reach the desired low temperature.

The cascaded cycles are interconnected through insulated closed loop heat exchangers that function as evaporators in the higher temperature cycle (A) and as condensers in the lower temperature cycle (B). An energy balance on an interconnecting heat exchanger provides a relation between the mass flow rates of refrigerant in the two cycles as

$$\frac{\dot{m}_A}{\dot{m}_B} = \frac{h_{2B} - h_{3B}}{h_{1A} - h_{4hA}} \tag{14.9}$$

Where h_{2b} is determined from

$$h_{2B} = (h_{2sB} - h_{1B})/(\eta_s)_{c-B} + h_{1B} \tag{14.10}$$

And the coefficient of performance for the entire cascaded system is

$$\text{COP}_{\text{cascade}} = \frac{\dot{Q}_L}{\dot{W}_{c-A} + \dot{W}_{c-B} + \ldots} = \frac{\dot{Q}_L}{\sum \dot{W}_{\text{compressors}}} \tag{14.11}$$

For a dual-cascade system, Eq. (14.11) becomes

$$\text{COP}_{\substack{\text{dual}\\\text{cascade}}} = \frac{\dot{Q}_L}{\dot{W}_{c-A} + \dot{W}_{c-B}} = \frac{\dot{m}_B(h_{1B} - h_{4hB})}{\dot{m}_A(h_{2sA} - h_{1A})/(\eta_s)_{c-A} + \dot{m}_B(h_{2sB} - h_{1B})/(\eta_s)_{c-B}} \tag{14.12}$$

The following example illustrates that cascading can be used to decrease the individual compressor pressure ratios and increase the coefficient of performance of a system. However, understand that this increase in system COP also requires an increased capital investment and increased maintenance costs.

EXAMPLE 14.6

A food-processing refrigeration unit is required to produce 40.0 tons of refrigeration at an evaporator temperature of −50.0°C and a condenser temperature of 25.0°C. Since this temperature difference is quite large, it was decided to design a dual-cascade unit using. R-22 in both of the cascaded loops. The intermediate heat exchanger connecting the two loops is to operate at −20.0°C, and the isentropic efficiencies of both compressors is 80.0%. The following design specifications were then established for the refrigeration loops shown in Figure 14.17:

Loop A

Station 1A Compressor A inlet	Station 2sA Compressor A outlet	Station 3A Condenser A outlet	Station 4*h*A Expansion valve A outlet
$x_{1A} = 1.00$	$p_{2sA} = 1500.$ kPa	$x_{3A} = 0.00$	$h_{4hA} = h_{3A}$
$T_{1A} = -20.0°C$	$s_{2sA} = s_{1A}$	$T_{3A} = 25.0°C$	

Loop B

Station 1B Compressor B inlet	Station 2sB Compressor B outlet	Station 3B Condenser B outlet	Station 4*h*B Expansion valve B outlet
$x_{1B} = 1.00$	$p_{2sB} = 300.$ kPa	$x_{3B} = 0.00$	$h_{4hB} = h_{3B}$
$T_{1B} = -50.0°C$	$s_{2sA} = s_{1B}$	$T_{3B} = -25.0°C$	

(*Continued*)

EXAMPLE 14.6 (*Continued*)

For this design, determine

a. The mass flow rate of refrigerant in loops A and B.
b. The system's coefficient of performance.
c. The pressure ratios across each of the compressors.

Solution

Use Figure 14.17 as the equipment schematic for this example. From Tables C.9b and C.10b for R-22, we can find the following property values:

Loop A

Station 1A	Station 2sA	Station 3A	Station 4hA
Compressor A in	Compressor A out	Condenser A out	Expansion valve A out
$x_{1A} = 1.00$	$p_{2sA} = 1500.$ kPa	$x_{3A} = 0.00$	$h_{4hA} = h_{3A}$
$T_{1A} = -20.0°C$	$s_{2sA} = s_{1A}$	$T_{3A} = 25.0°C$	
$h_{1A} = 242.05$ kJ/kg	$h_{2sA} = 289.08$ kJ/kg (by interpolation)	$h_{3A} = 74.91$ kJ/kg	$h_{4hA} = h_{3A} = 74.91$ kJ/kg
$s_{1A} = 0.95927$ kJ/kg·K			
$p_{1A} = 244.8$ kPa	$T_{2sA} = 71.07°C$ (by interpolation)		

Loop B

Station 1B	Station 2sB	Station 3B	Station 4hB
Compressor B inlet	Compressor B outlet	Condenser B outlet	Expansion valve B outlet
$x_{1B} = 1.00$	$p_{2sB} = 300.$ kPa	$x_{3B} = 0.00$	$h_{4hB} = h_{3B}$
$T_{1B} = -50.0°C$	$s_{2sB} = s_{1B}$	$T_{3B} = -20.0°C$	
$h_{1B} = 228.51$ kJ/kg	$h_{2sB} = 264.05$ kJ/kg (by interpolation)	$h_{3B} = 21.73$ kJ/kg	$h_{4hB} = h_{3B} = 21.73$ kJ/kg
$s_{1B} = 1.02512$ kJ/kg·K			
$p_{1B} = 63.139$ kPa	$T_{2sB} = 15.0°C$		

a. The mass flow rate in loop B can be found from an energy rate balance on the evaporator as

$$\dot{m}_B = \frac{\dot{Q}_L}{h_{1B} - h_{4hB}} = \frac{(40.0\text{ tons})[210.\text{ kJ/min}/(1\text{ ton})](1\text{ min}/60\text{ s})}{228.51 - 21.73\text{ kJ/kg}} = 0.677\text{ kg/s}$$

Equation (14.10) can now be used to find the actual compressor outlet state in loop B as

$$h_{2B} = (h_{2sB} - h_{1B})/(\eta_s)_{c-B} + h_{1B} = (264.05 - 228.51)/0.80 + 228.51 = 272.9\text{ kJ/kg}$$

Then Eq. (14.9) can be used to find the mass flow rate in loop A as

$$\dot{m}_A = \dot{m}_B\left(\frac{h_{2B} - h_{3B}}{h_{1A} - h_{4hA}}\right) = (0.677\text{ kg/s})\left(\frac{272.94 - 21.73\text{ kJ/kg}}{242.05 - 74.91\text{ kJ/kg}}\right) = 1.02\text{ kg/s}$$

b. Equation (14.12) provides the system COP as

$$\text{COP}_{\substack{\text{dual}\\ \text{cascade}}} = \frac{\dot{m}_B(h_{1B} - h_{4hB})}{\dot{m}_A(h_{2sA} - h_{1A})/(\eta_s)_{c-A} + \dot{m}_B(h_{2sB} - h_{1B})/(\eta_s)_{c-B}}$$

$$= \frac{0.677(228.51 - 21.73)}{1.02(289.08 - 242.05)/0.80 + 0.677(264.05 - 228.51)/0.80}$$

$$= 1.55$$

c. The compressor pressure ratios are obtained form these data as

$$\text{PR}_{\text{compressor A}} = p_{2SA}/p_{1A} = 1500./244.8 = 6.13$$

$$\text{PR}_{\text{compressor B}} = p_{2SB}/p_{1B} = 300./63.139 = 4.75$$

Exercises

16. If the refrigerating capacity of the dual-cascade system described in Example 14.6 is doubled, determine the required mass flow rate of refrigerant in loops A and B. Assume all the other variables remain unchanged. **Answer**: $\dot{m}_A = 2.04\,\text{kg/s}$ and $\dot{m}_B = 1.35\,\text{kg/s}$.

17. We just found another compressor manufacturer that can provide both compressors in Example 14.6 with isentropic efficiencies of 88.0% instead of 80.0%. Determine the new system coefficient of performance, assuming all the other variables remain unchanged. **Answer**: COP = 1.73.

18. Determine the coefficient of performance and the compressor pressure ratio required to produce the refrigeration required in Example 14.6 using a single-stage, vapor-compression R-22 refrigeration system. Use the following states in your calculations: compressor inlet, $x_1 = 1.00$, $T_1 = -50°\text{C}$; compressor outlet, $s_{2s} = s_1$, $p_{2s} = 1500.$ kPa; condenser outlet: $x_3 = 0.00$, $T_3 = 25.0°\text{C}$; expansion valve outlet: $h_{4h} = h_3$. **Answer**: COP = 1.48 and PR = 23.8.

When the same refrigerant is used in two or more cascaded cycles, the interconnecting closed loop heat exchangers can be replaced with more efficient heat exchanger systems consisting of liquid-vapor separators, called *flash chambers,* and open loop direct contact (or mixing) heat exchangers. Such systems are called *multistage* refrigeration systems. A dual-stage refrigeration system is shown in Figure 14.18.

In Figure 14.18, we see that the vapor from the flash chamber mixes directly with vapor from the first-stage compressor. Since the flash vapor is at a lower temperature than the vapor from the first-stage compressor, the flash chamber acts as an intercooler between compressor stages. The liquid from the flash chamber then passes through an expansion valve and flows into the low-temperature evaporator. The entire process is basically a *regeneration* process, similar to the Rankine power cycle with regeneration discussed in Chapter 13, with the

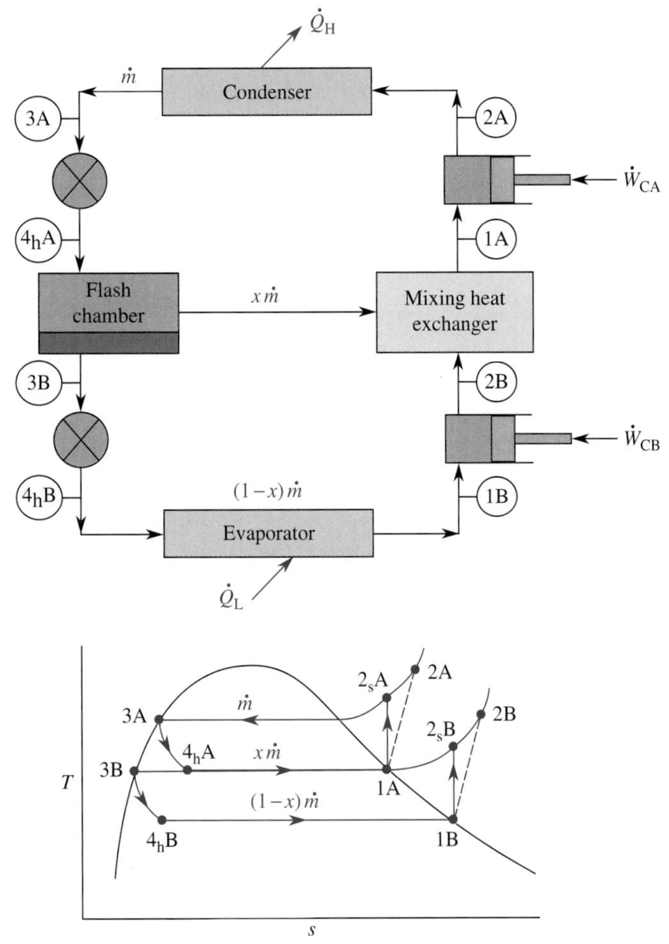

FIGURE 14.18

A dual-stage, vapor-compression refrigeration system. The *flash chamber* functions as a regenerator in this system.

flash chamber functioning as the regenerator. As with the Rankine power cycle, the efficiency of a multistage refrigeration cycle can be optimized through the proper choice of the regenerator (flash chamber) pressure. This pressure then dictates the quality of the vapor entering the direct contact heat exchanger as

$$x_{\text{flash}} = \frac{h_f(\text{at the condenser pressure}) - h_f(\text{at the flash chamber pressure})}{h_{fg}(\text{at the flash chamber pressure})} \tag{14.13}$$

The rate of cooling produced by a dual-stage refrigeration system is given by

$$(\dot{Q}_L)_{\substack{\text{dual}\\\text{stage}}} = \dot{m}_A(1 - x_{\text{flash}})(h_{1B} - h_{4B}) = \dot{m}_B(h_{1B} - h_{4B}) \tag{14.14}$$

and the total power input is

$$\begin{aligned}\sum \dot{W}_{\text{compressors}} &= \dot{W}_A + \dot{W}_B = \dot{m}_A(h_{2A} - h_{1A}) + \dot{m}_B(h_{2B} - h_{1B}) \\ &= \dot{m}_A[(h_{2A} - h_{1A}) + (1 - x_{\text{flash}})(h_{2B} - h_{1B})] \\ &= \dot{m}_A[(h_{2sA} - h_{1A})/(\eta_s)_{c-A} + (1 - x_{\text{flash}})(h_{2B} - h_{1B})/(\eta_s)_{c-B}]\end{aligned} \tag{14.15}$$

The system coefficient of performance can then be computed from Eq. (14.10) as

$$\begin{aligned}\text{COP}_{\substack{\text{dual}\\\text{stage}}} &= \frac{\dot{Q}_L}{\sum \dot{W}_{\text{compressors}}} = \frac{\dot{Q}_L}{\dot{W}_{c-A} + \dot{W}_{c-B}} \\ &= \frac{\dot{m}_{\text{ref}}(1 - x_{\text{flash}})(h_{1B} - h_{4B})}{\dot{m}_{\text{ref}}[(h_{2sA} - h_{1A})/(\eta_s)_{c-A} + (1 - x_{\text{flash}})(h_{2B} - h_{1B})/(\eta_s)_{c-B}]}\end{aligned} \tag{14.16}$$

An energy balance on the mixing heat exchanger in Figure 14.18 gives the value of the specific enthalpy at the inlet of the compression stage in loop A as

$$h_{1A} = x_{\text{flash}} h_5 + (1 - x_{\text{flash}})h_{2B} = x_{\text{flash}} h_g(\text{at } p_{\text{flash}}) + (1 - x_{\text{flash}})h_{2B} \tag{14.17}$$

where we set $h_5 = h_g(\text{at } p_{\text{flash}})$, and we compute h_{2B} from

$$h_{2B} = (h_{2sB} - h_{1B})/(\eta_s)_{c-B} + h_{1B} \tag{14.18}$$

The following example illustrates the effect of flash chamber pressure on the system's overall coefficient of performance.

EXAMPLE 14.7

A large food-processing plant needs a 14.0 ton refrigeration unit with an evaporator pressure of 100. kPa and a condenser pressure of 1600. kPa. We are designing a two-stage, vapor-compression unit using refrigerant R-134a. The flash chamber is to operate at 500. kPa, and the isentropic efficiency of both compressors is 80.0%. The following design specifications have been established for the refrigerant loops shown in Figure 14.18:

Loop A

Station 1A	Station 2sA	Station 3A	Station 4hA
Compressor A inlet	Compressor A outlet	Condenser A outlet	Expansion valve A outlet
$p_{1A} = 500.$ kPa	$p_{2sA} = 1600.$ kPa	$x_{3A} = 0.00$	$h_{4hA} = h_{3A}$
	$s_{2sA} = s_{1A}$	$p_{3A} = 1600.$ kPa	

Loop B

Station 1B	Station 2sB	Station 3B	Station 4hB
Compressor B inlet	Compressor B outlet	Condenser B outlet	Expansion valve B outlet
$x_{1B} = 1.00$	$p_{2sB} = 500.$ kPa	$x_{3B} = 0.00$	$h_{4hB} = h_{3B}$
$p_{1B} = 100.$ kPa	$s_{2sA} = s_{1B}$	$p_{3B} = 500.$ kPa	

We now need to determine

a. The mass flow rate of the two refrigerants.
b. The system's coefficient of performance.
c. The total power required by the compressors.

Solution

Use Figure 14.18 as the equipment schematic for this example. From Tables C.7f and C.8d for R-134a, we can find the following data:

Loop A

Station 1A	Station 2sA	Station 3A	Station 4hA
Compressor A inlet	Compressor A outlet	Condenser A outlet	Expansion valve A outlet
$p_{1A} = 500.\ \text{kPa}$	$p_{2sA} = 1600.\ \text{kPa}$	$x_{3A} = 0.00$	$h_{4hA} = h_{3A}$
$h_{1A} = 265.60\ \text{kJ/kg}$ (see below)	$s_{2sA} = s_{1A} = 0.9486\ \text{kJ/kg·K}$	$p_{3A} = 1600.\ \text{kPa}$	–
$s_{1A} = 0.9486\ \text{kJ/kg·K}$	$h_{2A} = 256.60\ \text{kJ/kg}$ (by interpolation)	$h_{3A} = 134.02\ \text{kJ/kg}$	$h_{4A} = 134.02\ \text{kJ/kg}$

Loop B

Station 1B	Station 2sB	Station 3B	Station 4hB
Compressor B inlet	Compressor B outlet	Condenser B outlet	Expansion valve B outlet
$x_{1B} = 1.00$	$p_{2sB} = 500.\ \text{kPa}$	$x_{3B} = 0.00$	$h_{4hB} = h_{3B}$
$p_{1B} = 100.\ \text{kPa}$	$s_{2sA} = s_{1B} = 0.9395$	$p_{3B} = 500.\ \text{kPa}$	–
$h_{1B} = 231.35\ \text{kJ/kg}$	$h_{2sB} = 264.25\ \text{kJ/kg}$	$h_{3B} = 71.33\ \text{kJ/kg}$	$h_{4hB} = 71.33\ \text{kJ/kg}$
$s_{1B} = 0.9395\ \text{kJ/kg·K}$	(by interpolation)		

The specific enthalpy of station 1A was determined from an energy rate balance on the flash chamber using Eq. (14.17) as

$$h_{1A} = x_{\text{flash}} h_g(\text{at } p_{\text{flash}}) + (1 - x_{\text{flash}}) h_{2B}$$

where h_{2B} was determined from Eq. (14.18) as

$$\begin{aligned} h_{2B} &= (h_{2sB} - h_{1B})/(\eta_s)_{c-B} + h_{1B} \\ &= (264.25 - 231.35)/0.800 + 231.35 = 272.54\ \text{kJ/kg} \end{aligned}$$

The quality of the vapor exiting the flash chamber is given by Eq. (14.13) as

$$x_{\text{flash}} = \frac{h_f(\text{at the condenser pressure}) - h_f(\text{at the flash chamber pressure})}{h_{fg}(\text{at the flash chamber pressure})}$$

$$= \frac{h_f(\text{at } 1600.\ \text{kPa}) - h_f(\text{at } 500.\ \text{kPa})}{h_{fg}(\text{at } 500.\ \text{kPa})} = \frac{134.02 - 71.33}{184.74} = 0.339 = 33.9\%$$

Then

$$h_{1A} = 0.339(252.07) + (1 - 0.339)(272.54) = 265.60\ \text{kJ/kg}$$

a. The mass flow rate in loop B is given by Eq. (14.14) as

$$\dot{m}_B = \frac{(\dot{Q}_L)_{\substack{\text{dual}\\\text{stage}}}}{h_{1B} - h_{4B}} = \frac{(10.0\ \text{tons})[210.\ \text{kJ/min}/(1\ \text{ton})](1\ \text{min}/60\ \text{s})}{231.35 - 71.33\ \text{kJ/kg}} = 218\ \text{kg/s}$$

and, since for the dual-stage system $\dot{m}_B = \dot{m}_A(1 - x_{\text{flash}})$,

$$\dot{m}_A = \dot{m}_B/(1 - x_{\text{flash}}) = (0.218\ \text{kg/s})/(1 - .0339) = 0.330\ \text{kg/s}$$

b. The system COP is given by Eq. (14.16) as

$$\text{COP}_{\substack{\text{dual}\\\text{stage}}} = \frac{\dot{m}_{\text{ref}}(1 - x_{\text{flash}})(h_{1B} - h_{4B})}{\dot{m}_{\text{ref}}[(h_{2sA} - h_{1A})/(\eta_s)_{c-A} + (1 - x_{\text{flash}})(h_{2B} - h_{1B})/(\eta_s)_{c-B}]}$$

$$= \frac{(1 - 0.339)(231.35 - 71.33)}{(292.33 - 265.60)/0.800 + (1 - 0.339)(264.25 - 231.38)/0.800}$$

$$= 1.78$$

(Continued)

EXAMPLE 14.7 *(Continued)*

c. The total compressor power is obtained from Eq. (14.15) as

$$\sum \dot{W}_{\text{compressors}} = \dot{m}_A[(h_{2sA} - h_{1A})/(\eta_s)_{c-A} + (1 - x_{\text{flash}})(h_{2B} - h_{1B})/(\eta_s)_{c-B}]$$
$$= (0.330\text{ kg/s})[(292.33 - 256.60\text{ kJ/kg})/0.800 + (1 - 0.339)(264.25 - 231.35\text{ kJ/kg})/0.800]$$
$$= 23.7\text{ kJ/s} = 23.7\text{ kW}$$

Exercises

19. If the refrigerating capacity of the two-stage system described in Example 14.7 is tripled, determine the required refrigerant mass flow rate. Assume all the other variables remain unchanged. **Answer**: $\dot{m}_{\text{ref}} = 0.654\text{ kg/s}$.

20. We just found another manufacturer that can provide the compressors for the unit described in Example 14.7 with an isentropic efficiency of 90.0% instead of 80.0%. Determine the new system coefficient of performance, assuming all the other variables remain unchanged. **Answer**: COP = 1.88.

21. Using a spreadsheet or equation solver (like EES), develop a plot of system COP vs. flash chamber pressure for the unit discussed in Example 14.7. Note that the maximum COP occurs at a flash chamber pressure of about 500 kPa.

14.10 ABSORPTION REFRIGERATION

Ammonia was discovered in 1774 by the British chemist Joseph Priestley (1733–1804), who noted that his new gas dissolved easily in water (one volume of water dissolves over 1000 volumes of ammonia at STP). The French engineer Ferdinand Carré utilized this property of ammonia's affinity for water to create the first absorption refrigeration system in 1859.

Carré's absorption refrigeration technique is an important vapor refrigeration technology, because it does not require a vapor compressor. It is based on dissolving the refrigerant vapor (ammonia) in a *carrier* liquid (water) and pumping this liquid to a high pressure. A liquid can be pumped more efficiently than a vapor can be compressed, so this technique has a decided advantage over vapor-compression technology. The pressurized liquid is then fed into a *generator*, where the refrigerant vapor is boiled off, now at a much higher pressure, and the carrier liquid returned to the absorber to continue the process. The high-pressure refrigerant vapor then continues through the refrigeration cycle in the normal reversed Rankine manner. Standard and absorption refrigeration systems are shown schematically in Figure 14.19. These are both vapor-compression

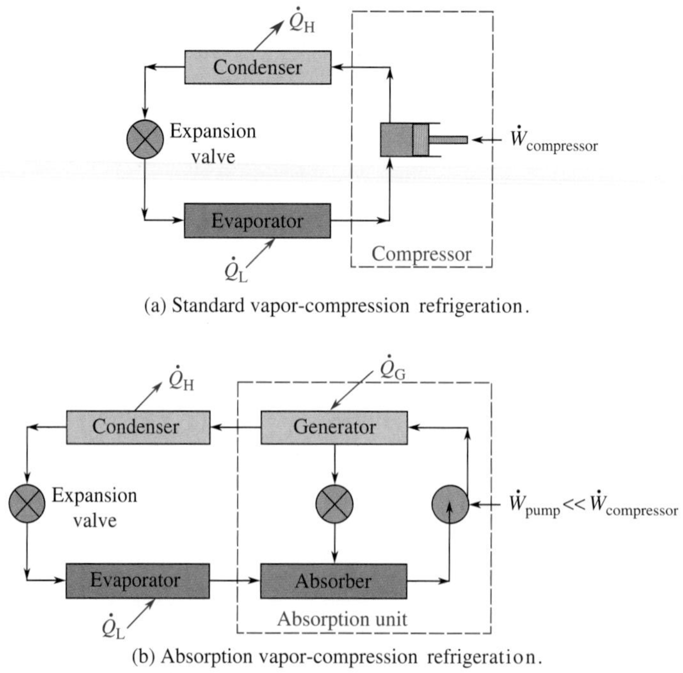

FIGURE 14.19
Basic absorption refrigeration.

cycles, since the compressor unit in the standard system is simply replaced by the absorption unit in the absorption system.

The absorption cycle is a special type of vapor-compression refrigeration cycle, since it is driven by heat as opposed to work. Consequently, the coefficient of performance of an absorption cycle is not computed in the same way as a standard vapor-compression cycle, and the two COPs should not be directly compared. The heat energy necessary to drive an absorption cycle is at a much lower availability than the electrical energy necessary to power a work-driven cycle. There is a need for some work in absorption systems with a generator feed pump, but the amount of pump work required is negligible in comparison to the compressor work needed in a standard vapor-compression cycle. The coefficient of performance of an ideal absorption refrigerator is

$$(\mathrm{COP})_{\substack{\text{absorption}\\ \text{refrigerator}}} = \frac{\text{Refrigeration (evaporator) cooling}}{\text{Generator heat} + \text{Pump work}} = \frac{\dot{Q}_{\text{evaporator}}}{\dot{Q}_{\text{generator}} + \dot{W}_{\text{pump}}} \tag{14.19}$$

A Carnot absorption cycle can be constructed by driving a Carnot refrigerator with a Carnot engine, as shown in Figure 14.20. In this system, both the engine and the refrigerator exhaust heat to the local environment at temperature T_a.

The thermal efficiency of the Carnot engine used here is

$$(\eta_T)_{\substack{\text{Carnot}\\ \text{engine}}} = \frac{\dot{W}_{\text{engine}}}{\dot{Q}_{\text{generator}}}$$

and the thermal efficiency (COP) of the Carnot refrigerator used here is

$$(\eta_T)_{\substack{\text{Carnot}\\ \text{refrigerator}}} = (\mathrm{COP})_{\substack{\text{Carnot}\\ \text{refrigerator}}} = \frac{\dot{Q}_{\text{evaporator}}}{\dot{W}_{\text{refrigerator}}}$$

Since the overall efficiency of a combined system is equal to the product of the efficiencies of its components, the COP of this combined system is equal to the product of the thermal efficiency of the Carnot engine multiplied by the thermal efficiency of the Carnot refrigerator, or

$$(\mathrm{COP})_{\substack{\text{Carnot}\\ \text{absorption}\\ \text{refrigerator}}} = \left(\frac{\dot{W}_{\text{engine}}}{\dot{Q}_{\text{generator}}}\right)\left(\frac{\dot{Q}_{\text{evaporator}}}{\dot{W}_{\text{refrigerator}}}\right) = \frac{\dot{Q}_{\text{evaporator}}}{\dot{Q}_{\text{generator}}} \tag{14.20}$$

as all the work produced by the engine is used to drive the refrigerator, or $\dot{W}_{\text{engine}} = \dot{W}_{\text{refrigerator}} = \dot{W}$. For a Carnot cycle, $\dot{Q}_{\text{generator}} = \dot{W}/(1 - T_a/T_g)$ and $\dot{Q}_{\text{evaporator}} = \dot{W}\,T_e/(T_a - T_e)$. Consequently,

$$(\mathrm{COP})_{\substack{\text{Carnot}\\ \text{absorption}\\ \text{refrigerator}}} = \frac{T_e}{T_g}\left(\frac{T_g - T_a}{T_a - T_e}\right) \tag{14.21}$$

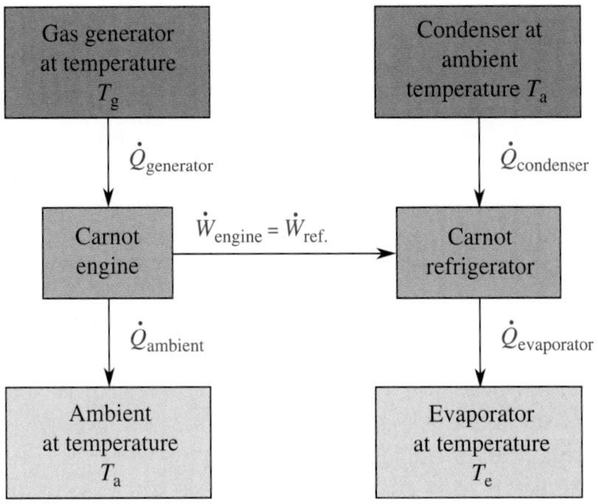

FIGURE 14.20
A Carnot absorption refrigeration cycle.

Since no real system can be more efficient than a Carnot system, Eq. (14.21) represents the maximum possible thermal efficiency of an absorption refrigeration system operating with a generator temperature T_g, an evaporator temperature T_e, and an ambient temperature T_a. Because absorption systems are heat rather than work driven, their practical COP values tend to be around 1.0 or less. Common absorption refrigeration fluid systems are: ammonia (refrigerant)-water (carrier), water (refrigerant)-lithium bromide (carrier), and water (refrigerant)-lithium chloride (carrier). The ammonia-water system was widely used in domestic refrigerators until about 1950. The lithium salt–water systems that use water as the refrigerant cannot go below 32°F (0°C) and consequently are mainly used in air conditioning applications.

Absorption refrigeration dominated the refrigeration market before 1875, which is remarkable considering its inherent complexity and the fact that its design was empirical. A suitable theory for the operation of absorption refrigeration did not appear until 1913, and today multistage regenerative absorption refrigerators can produce temperatures as low as 65 K.

EXAMPLE 14.8

A new absorption refrigeration system with a generator temperature of 100.°C and an evaporator temperature of 5.00°C is being designed to operate in an environment at a temperature of 20.0°C. To provide an upper limit for the operating efficiency, determine the Carnot absorption refrigeration coefficient of performance of this system.

Solution

Use Figure 14.20 as the equipment schematic for this example. Equation (14.21) gives the COP for this system as

$$(\mathrm{COP})_{\substack{\text{Carnot}\\ \text{absorption}\\ \text{refrigerator}}} = \frac{T_e}{T_g}\left(\frac{T_g - T_a}{T_a - T_e}\right) = \left(\frac{5.00 + 273.15}{100. + 273.15}\right)\left(\frac{100. - 20.0}{20.0 - 5.00}\right) = 3.98$$

Since no system can be more efficient than a Carnot system, this represents the maximum COP of any absorption system operating under these conditions.

Exercises

22. The environmental temperature of the absorption refrigerator being designed in Example 14.8 is suddenly increased from 20°C to 30°C. Determine the new Carnot absorption refrigeration COP, assuming all the other variables remain unchanged. **Answer**: $(\mathrm{COP})_{\text{Carnot absorption ref}} = 2.09$.

23. Suppose now we want to convert the absorption refrigeration system design discussed in Example 14.8 into a cryogenic unit with an evaporator temperature of only 65 K. What would be the maximum possible coefficient of performance of this system assuming all the other variables remain unchanged? **Answer**: $(\mathrm{COP})_{\text{Carnot absorption ref}} = 0.06$.

24. Explain why the COP given by Eq. (14.21) goes to zero as the evaporator temperature approaches absolute zero. **Answer**: In this instance, as $T_e \to 0$, the evaporator cooling load $\dot{Q}_e$ also goes to zero. Since the COP is defined as the ratio of the system cooling to the energy input, the COP must vanish as the cooling vanishes.

14.11 COMMERCIAL AND HOUSEHOLD REFRIGERATORS

Commercial and household refrigeration technology essentially developed together, because commercial refrigeration in shops and supermarkets requires the same basic technological advances as household refrigerators. Also, once frozen or chilled food products were purchased by the consumer, similar refrigeration needs were created in the home. Thus. the parallel development of household and commercial refrigeration was advantageous, if the market for chilled and frozen foods was to expand beyond the needs of a single day's food supply.

Throughout the 19th century, mechanical vapor-compression refrigeration systems had been limited to large-scale industrial units powered by steam engines or internal combustion engines. Several major technical bottlenecks prevented small commercial and household vapor-compression refrigerators from being successfully developed. The first problem was the development of a power source suitable for use in a household. The traditional commercial power sources (steam and internal combustion engines) were not suitable for household use. The second problem was the enormous friction in the mechanical seals on the shaft between the power source and the compressor. Without a complex and tight sealing system, refrigerant leaked out, causing serious environmental and maintenance problems. The third problem was the development of an automatic

CRITICAL THINKING

Visit an antique store and find an old icebox. Note the current price. Look it over to understand how it worked. Inspect it for insulation in the walls and measure the size of the food storage compartment. How do you think your diet would change if you had to use an icebox every day? How cold do you think iceboxes were able to keep the food?

WHERE DID BIRDS EYE FROZEN FOOD COME FROM?

Clarence Birdseye (1886–1956) was a very successful businessman and inventor. In 1912, he went to Labrador as a fur trader and discovered that fish caught in weather 50° below zero froze almost instantly and were still fresh months later, when they were thawed out. Slow freezing allows ice crystals to form in the cells of plants and animals, causing them to burst. However, with quick freezing, the cells remain intact, preserving the flavor and nutrition of the food.

After returning from Labrador, Birdseye developed a quick freezing process that preserved the original taste of a variety of foods such as fish, fruits, and vegetables. In 1924, he helped found the General Seafood Company (later to become General Foods Corporation), which successfully marketed his frozen food products, and he became very wealthy. Birds Eye frozen food products are still available in supermarkets.

Birdseye's patented freezing process consisted of placing two flat refrigerated metal plates at −40°F on either side of a food package, causing the food to freeze very quickly. Birdseye was granted nearly 300 patents in his lifetime. In addition to his frozen food patents, he developed infrared heat lamps for home use, a recoilless gun for firing a harpoon, and a method for freeze-drying foods.

refrigerant flow control mechanism that would not let liquid refrigerant enter and subsequently destroy the compressor. Commercial refrigeration systems often required constant refrigerant flow adjustment by a human operator.

The first problem was solved with the development of an effective central electrical power system by Thomas Edison between 1895 and 1920. Inexpensive electric motors then became available as suitable power sources for a household compressor. The second problem was solved with the development of the hermetically sealed motor-compressor unit in 1918. Many of the refrigerants used were dielectrics and did not conduct electricity. Therefore, they could come in direct contact with the motor windings and even act as a coolant for the motor. Numerous float-valve (as found in a toilet tank) refrigerant flow control mechanisms were tried during this period, but the automatic flow control problem was not completely solved until 1926, when Harry Thompson developed the thermostatic expansion valve to automatically control the flow of liquid refrigerant into the evaporator.

The Kelvinator Corporation of Detroit, Michigan, was one of the first companies to build electrical refrigerators. They launched the first household vapor-compression refrigerator in 1918 and sold 67 units that year. By 1921, over 20 manufacturers of home refrigerators sold a total of 5000 units that year. Growth in the domestic refrigeration market was phenomenal, with 75,000 units sold in 1925, 850,000 in 1930, and 1.7 million in 1935. Even during the Depression, sales of home refrigerators remained high. As production increased, the prices fell. In 1920, the average price for a refrigerator was $600; in 1930, it was $275; and in 1935, it was $160. In 1929, as many mechanical refrigerators were produced as old-fashioned iceboxes.

The first commercial refrigeration system appeared in large hotels in the early 1900s for air conditioning and food preservation. By the 1930s, glass-covered self-service display cabinets were available in small groceries and supermarkets for chilled or frozen ice cream, meat, poultry, fish, eggs, and dairy products. Since cold air is heavier than room air, the glass cover disappeared from most horizontal display cases in the 1950s to encourage customer access. Vertical freezer display cases, however, still require glass doors.

At the same time as vapor-compression domestic refrigerators were reaching the market, a major breakthrough was made in absorption refrigeration. In 1925, the Swedish engineers G. Munters and B. von Platen successfully eliminated the mechanical pump between the high- and low-pressure regions by introducing hydrogen gas into the system. Hydrogen lowers the partial pressure of the refrigerant (ammonia) vapor in the evaporator, allowing it to boil at a lower temperature. A percolation or siphon action transports the fluid from the low-pressure absorber into the high-pressure generator (the pressure difference is only a few inches of height in these systems) and gravity forces the fluid to circulate through the remaining system (see Figure 14.22). This refinement opened the way for the development of a household absorption refrigerator that had no moving parts, was completely

WHO INVENTED THE "TV DINNER"?

By the 1930s, General Foods had a few frozen meals on the market (such as Irish stew), but the first individual frozen meals did not appear until World War II. In 1945, Maxson Food Systems Inc. introduced three-part meals called *Strato-Plates* for military airplane passengers. For the next ten years, food engineers worked to make frozen meals more appealing. In 1954, C. A. Swanson & Sons (a Campbell Soup company) introduced the name *TV Dinner* for its new frozen meals that could be eaten while watching television (Figure 14.21). They were an instant success at 98 cents each, with 10 million TV dinners sold in 1955 and 214 million in 1960. In 1990, manufacturers introduced over 650 new frozen dishes, resulting in over 2 billion frozen meals sold per year by the mid 1990s. Today, the TV dinner has morphed into the $4 billion frozen food industry.

The TV dinner allowed families to gather around the television set to share their meals, just as they used to do while gathered around the dinner table. Commercially prepared frozen foods vastly simplified the art of meal preparation and significantly contributed to changing women's role in society by promoting the sharing of meal preparation tasks by all members of the family.

FIGURE 14.21
TV Dinner.

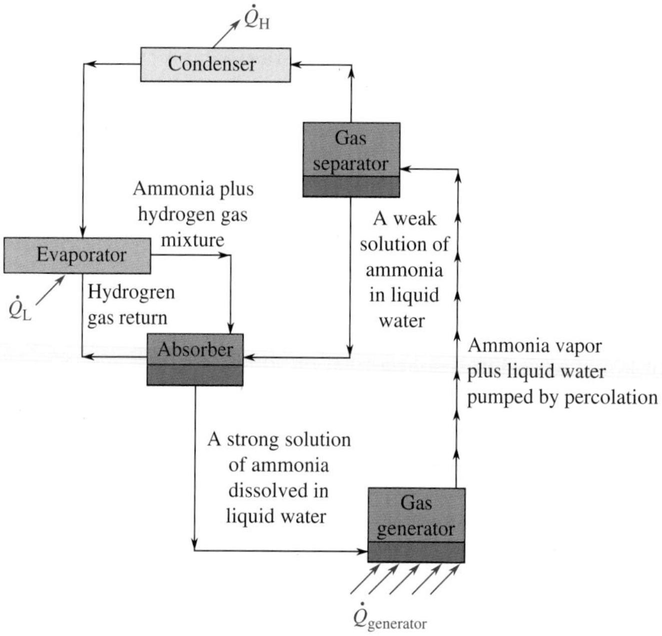

FIGURE 14.22
A gas-powered absorption refrigerator requires no electricity.

silent, and did not require electricity to operate. This technology was first marketed in the United States in 1927 by Electrolux in Evansville, Indiana. Absorption refrigeration was popular in the household until the 1950s, when highly efficient, cascaded, electric-powered, vapor-compression refrigerators dominated the market. Because

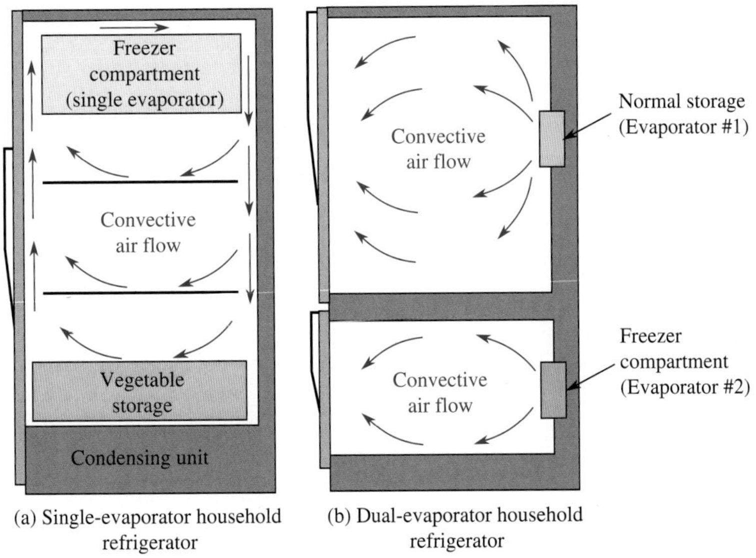

FIGURE 14.23
Illustration of (a) single-evaporator system and (b) dual-evaporator system.

absorption refrigeration has no moving parts and operates from a heat source instead of a work source, it periodically attracts renewed attention. It is particularly attractive from an energy conservation point of view, since it can operate from waste heat or solar energy.

Early vapor-compression and absorption household refrigerators had only one evaporator, located around the freezer compartment. Cooling of the remaining refrigeration space was produced by the natural convection of air passing around the outside of the freezer compartment. Many inexpensive portable refrigerators are still made this way today. The first true "dual-temperature" refrigerator, with separate freezer and refrigeration evaporation coils, appeared in 1939.

In single-evaporator refrigerator-freezers, the evaporator temperature must be colder than the freezer temperature. A dual-evaporator refrigerator allows the pressure and temperature in the evaporators to be controlled separately. The fresh food evaporator is warmer than the freezer evaporator temperature, thus reducing the irreversibilities associated with the heat transfer.

Also, in a single-evaporator system, humid air from the fresh food compartment comes in contact with the cold freezer evaporator, producing frost. When the dehumidified dry air is returned to the fresh food cabinet, it dries the food and reduces food quality. If the fresh food evaporator more closely matches the air temperature in the fresh food compartment, it dehumidifies the air less and builds up less frost. This decreases the need for defrosting with a heater and increases energy efficiency. Figure 14.23 illustrates these differences.

EXAMPLE 14.9

A new household refrigerator-freezer combination unit is being designed with the dual-evaporator system shown in Figures 14.23 and 14.24. The freezer compartment is to be at −18.0°C and the refrigerator compartment is to be at 4.00°C. The outlet of the condenser is at 30.0°C. The cooling capacities of both the refrigeration and the freezer compartments are to be 422 kJ/h each. The system uses refrigerant R-134a with a compressor isentropic efficiency of 80%. Determine

a. The coefficient of performance for this design.
b. The mass flow rate of refrigerant required.
c. The quality at the outlet of the refrigeration evaporator.

(*Continued*)

EXAMPLE 14.9 *(Continued)*

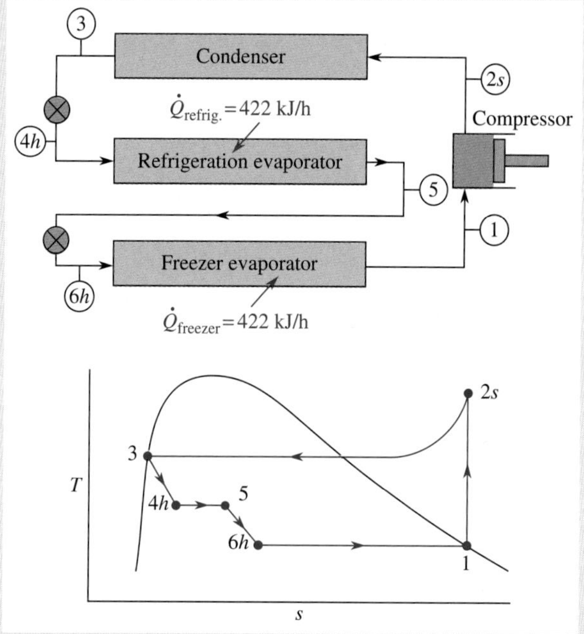

FIGURE 14.24
Example 14.9.

Solution

The properties of the refrigerant at the states shown in Figure 14.24 are as follows (see Tables C.7e, C.7f, and C.8d for the numerical values):

Station 1—Compressor inlet
$x_1 = 1.00$
$T_1 = -18.0°C$
$h_1 = 236.53$ kJ/kg
$s_1 = 0.9315$ kJ/kg·K

Station 2s—Compressor outlet
$s_{2s} = s_1 = 0.9315$ kJ/kg·K
$p_{2s} = p_3 = p_{sat}(30.0°C) = 0.770$ MPa
$h_{2s} = 271.0$ kJ/kg
(interpolation in Table C.7f)

Station 3—Condenser outlet
$x_3 = 0.00$
$T_3 = 30.0°C$
$p_3 = p_{sat}(30.0°C) = 0.770$ MPa
$h_3 = h_f(30.0°C) = 91.49$ kJ/kg

Station 4h—Refrigerator evaporator inlet
$h_{4h} = h_3$
$T_{4h} = 4.00°C$
No further information is needed.

Station 5—Refrigerator evaporator outlet
$x_5 = ?$
$T_5 = T_{4h} = 4.00°C$
$h_5 = ?$

Station 6h—Freezer evaporator inlet
$h_{6h} = h_5 = ?$
$T_{6h} = -18.0°C$
No further information is needed.

The isentropic efficiency of the compressor is 80.0%, or $(\eta_s)_{comp} = 0.800$.

a. The COP for this system is the ratio of the total heat rate removed from the refrigeration plus freezer compartments divided by the power input to the compressor, or

$$\text{COP} = \frac{\dot{Q}_R + \dot{Q}_F}{\dot{W}_C} = \frac{(h_5 - h_{4h}) + (h_1 - h_{6h})}{(h_{2s} - h_1)/(\eta_s)_{comp}}$$

but since $h_5 = h_{6h}$ here, this equation reduces to

$$\text{COP} = \frac{h_1 - h_{4h}}{(h_{2s} - h_1)/(\eta_s)_{comp}} = \frac{236.53 - 91.49}{(271.0 - 236.53)/0.800} = 3.37$$

so that the system COP does not depend on states 5 or 6h.

b. Since $h_5 = h_{6h}$, we can write the refrigeration and freezer compartment cooling rates as $\dot{Q}_R = \dot{m}_{ref}(h_5 - h_{4h})$ and $\dot{Q}_F = \dot{m}_{ref}(h_1 - h_{6h}) = \dot{m}_{ref}(h_1 - h_5)$. Solving these two equations for $\dot{m}_{ref}$ gives

$$\dot{m}_{ref} = \frac{\dot{Q}_R + \dot{Q}_F}{h_1 - h_{4h}} = \frac{(422 + 422\text{ kJ/h})(1\text{ h}/60\text{min})}{236.53 - 91.49\text{ kJ/kg}} = 0.0970\text{ kg/min}$$

c. Now we can find the enthalpy at the exit of the refrigeration evaporator from the energy balance on the refrigeration compartment as

$$h_5 = h_{4h} + \frac{\dot{Q}_R}{\dot{m}_{ref}} = 91.49\ \text{kJ/kg} + \frac{(422\ \text{kJ/h})(1\ \text{h}/60\text{min})}{0.0970\ \text{kg/min}} = 164.0\ \text{kJ/kg}$$

Then, we can find the quality at the exit of the refrigeration evaporator as

$$x_5 = \frac{h_5 - h_f(4.00°\text{C})}{h_{fg}(4.00°\text{C})} = \frac{164.0 - 55.35}{194.19} = 0.559 = 55.9\%$$

Exercises

25. A high-efficiency compressor has just been developed. It has an isentropic efficiency of 95%. Determine the new COP for the refrigerator design presented in Example 14.9. **Answer**: COP = 4.00.
26. Explain why the COP of the dual-evaporator refrigerator in Example 14.9 does not depend on the quality of the refrigerator evaporator's exit. What characteristic of the system depends on this variable? **Answer**: The intermediate state h_5 to h_{6h} cancels out in the COP evaluation. The amounts of cooling in the refrigeration and freezing sections depend on this variable.
27. If the refrigerator in Example 14.9 runs only 50% of the time, determine its annual operating cost if the price of electricity is \$0.15 per kWh. **Answer**: Annual operating cost = \$45.05.

HOW LONG HAVE WE HAD AIR CONDITIONING?

Before 1930, air conditioning for human comfort was available only for large installations, such as movie theaters, department stores, and offices. In 1919, the first air-conditioned movie theater opened in Chicago, Illinois, and the first air-conditioned department store (Abraham and Straus) opened in Brooklyn, New York. The first completely air-conditioned office building was built in San Antonio, Texas, in 1927. By the late 1930s, smaller units became available for restaurants, shops, and hotels.

Air conditioning using cold or chilled water circulating throughout the air handling system of a building was introduced in 1929. The water was chilled by the evaporator in a large central refrigeration unit (the evaporator was usually called the *chiller*).

Small home window air conditioners were not introduced until after 1946, and mass-produced automobiles were not air conditioned until after 1948.

WHERE DID HEAT PUMPS COME FROM?

Lord Kelvin seemed cognizant of the possibility that a "reversed heat engine" could be used for heating and cooling. The first use of the term *heat pump* appears in the British literature in about 1895. The first application of heat pump technology was made by T.G.N. Haldane to heat his London office in 1927. Modern heat pumps (Figure 14.25) use refrigerants R-113 or R-114 to reach hot-side temperatures as high as 150°C.

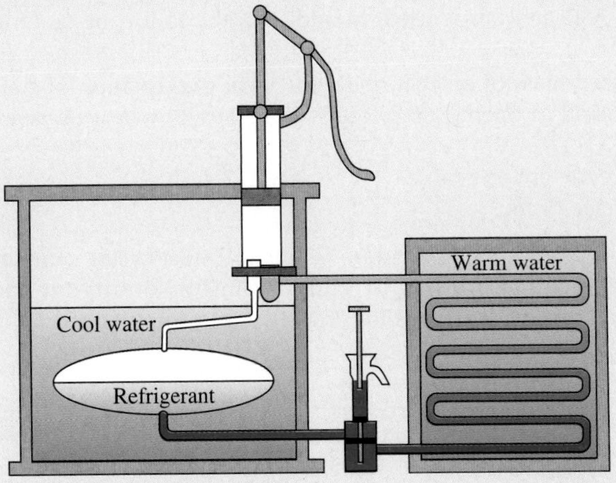

FIGURE 14.25
A heat pump.

WHAT ARE "SOFT" DRINKS AND WHY ARE THEY SERVED COLD?

The *soda fountain* is a unique 1920s American invention at which ice cream and carbonated nonalcoholic (i.e., "soft") drinks were sold. The soft drinks were usually cooled to increase the solubility of the CO_2 used in their carbonation (Figure 14.26). This enhanced their flavor and created a unique tingling feeling in the mouth and throat.

In 1767, Joseph Priestley invented carbonated water, a key component of soft drinks. He added carbon dioxide to water by suspending a container of water above a beer vat at a local brewery.

In 1881, the first cola-flavored beverage was introduced. In 1885, Charles Aderton invented "Dr Pepper"; in 1886, Dr. John S. Pemberton invented "Coca-Cola"; and in 1898, "Pepsi-Cola" was invented by Caleb Bradham. Until 1905, Coca-Cola contained extracts of cocaine (from coca leaves) and caffeine (from the kola nut) and was marketed as a medicine that could cure various ailments.

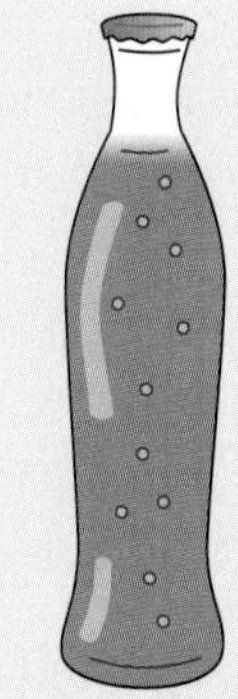

FIGURE 14.26
A soft drink bottle.

14.12 PART II. GAS REFRIGERATION CYCLES

Reversed gas power cycles have the same potential for producing cooling that we see with reversed vapor cycles. But not all gas power cycles have a reversed cycle refrigeration analog. For example, we do not yet know how to reverse internal combustion gas power cycles, because to do so would require the development of rapid endothermic (heat-absorbing) chemical reactions similar to the exothermic combustion reactions used in the power cycles. However, all external combustion gas power cycles have effective reversed cycle refrigeration technologies.

The rapid development of vapor and gas power cycles forever changed the world. They improved productivity and efficiency in agriculture, transportation, textiles, manufacturing, and in many other areas that affect the way we live. The social impact of the reversed power cycles, limited primarily to food preservation and environmental control, was less influential. Though we have known for a long time that many gas power cycles could be reversed to produce cooling, their development into viable technologies had less social impact and consequently grew less rapidly. Part II of this chapter focuses on the technology of gas refrigeration cycles plus a few miscellaneous refrigeration technologies that were not derived from reversed power cycles.

14.13 AIR STANDARD GAS REFRIGERATION CYCLES

The working fluid in gas refrigeration cycles is less complex than in gas power cycles. For example, refrigeration cycles do not involve internal combustion processes that change the working fluid during the cycle, but they can be either open or closed loop cycles. Therefore, there is much less need for a working fluid simplifying model like the air standard cycle (ASC). Nonetheless, we use its simplifying characteristics in analyzing gas refrigeration cycles in which air is the working fluid. In particular, all gas refrigeration cycles are assumed to be closed loop cycles when an ASC analysis is used. In addition, the following assumptions apply to gas refrigeration ASC analysis:

1. The working fluid is a fixed mass of air that obeys the ideal gas equation of state.
2. All inlet or exhaust processes in open loop systems are replaced by heat transfer processes to or from the environment.
3. All processes within the cycle are reversible.
4. The air has constant specific heats.[5]

ASC refrigeration analysis yields reasonably accurate results for most cycles using air as the working fluid. One notable exception is in the area of throttling or Joule-Thomson cooling, in which the amount of cooling depends exclusively on real gas behavior. This is illustrated later in this chapter.

[5] Since the temperature variations within a gas refrigeration cycle are not nearly as large as those within a gas power cycle that contains a combustion process, there is no practical need to distinguish between a hot refrigeration ASC, in which temperature-dependent specific heats are used, and a cold refrigeration ASC, in which constant specific heats are used.

14.14 REVERSED BRAYTON CYCLE REFRIGERATION

In 1844, the American physician John Gorrie (1803–1855) designed and built an air cooling apparatus in Florida to provide air conditioning for his yellow fever patients. His machine had a piston-cylinder apparatus that compressed air that was cooled back to ambient temperature by circulating water. The cooled compressed air was then expanded in a second piston-cylinder apparatus that caused the air to drop to a sufficiently low temperature to produce ice and satisfy other cooling needs. The expanded air was then drawn back into the compressor and the cycle began again. The two piston-cylinder devices were connected together so that the expansion work was used to offset the compression work. This was clearly a reversed, closed loop Brayton cycle, as shown by comparing Figures 13.44 and 14.27.

The COP of an actual reversed Brayton cycle is given by

$$\text{COP}_{\substack{\text{reversed}\\ \text{Brayton cycle}\\ \text{HP}}} = \frac{|\dot{Q}_H|}{|\dot{W}_{in}|_{net}} = \frac{T_2 - T_3}{(T_{2s} - T_1)/(\eta_s)_c - (T_3 - T_{4s})(\eta_s)_c} \tag{14.22}$$

and

$$\text{COP}_{\substack{\text{reversed}\\ \text{Brayton cycle}\\ \text{R/AC}}} = \frac{|\dot{Q}_L|}{|\dot{W}_{in}|_{net}} = \frac{T_1 - T_4}{(T_{2s} - T_1)/(\eta_s)_c - (T_3 - T_{4s})(\eta_s)_c} \tag{14.23}$$

where $T_2 = T_1 + (T_{2s} - T_1)/(\eta_s)_c$ and $T_4 = T_3 - (T_3 - T_4)(\eta_s)_e$. Since the processes 1 to 2*s* and 3 to 4*s* are isentropic and the processes 2*s* to 3 and 4s to 1 are isobaric,

$$T_{2s} = T_1 (p_{2s}/p_1)^{(k-1)/k} = T_1 \text{PR}^{(k-1)/k}$$

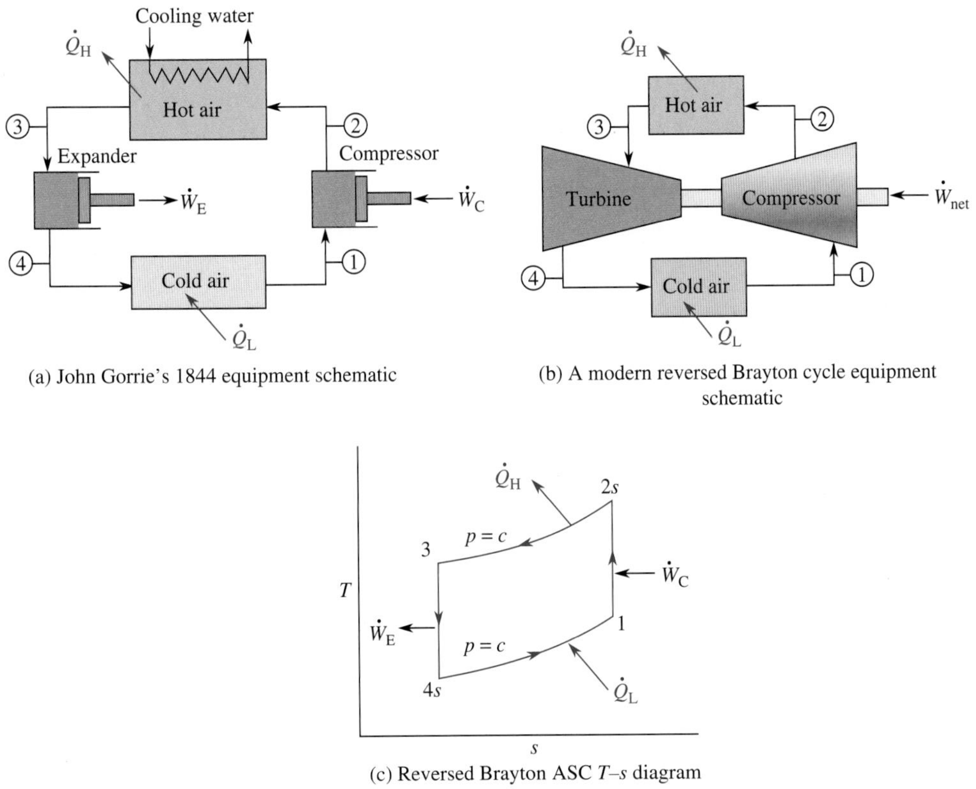

FIGURE 14.27
The reversed Brayton cycle cooling system using air as a working fluid.

WHO INVENTED MECHANICAL REFRIGERATION FIRST, BRAYTON OR GORRIE?

If Gorrie developed his refrigeration cycle in 1844 and Brayton developed his power cycle in 1873, why do we call Gorrie's cycle a *reversed Brayton cycle*? The reversed name order is just tradition. It probably occurred because power cycles have always been more important to the development of societal goals than refrigeration cycles.

The original reversed Brayton cycle using a reciprocating piston compressor and expander (sometimes called *gas expansion with external work*), developed by Dr. Gorrie in 1844, was not commercially successful. He received a patent in 1851 but could not raise capital to produce his refrigerator, and he died a disappointed man.

The reversed Brayton cycle was finally established as a viable refrigeration technology by Sir William Siemens (1823–1883) in 1857. Carl von Linde (1842–1934) unsuccessfully tried to liquefy air using this cycle in 1894. Then, he switched to a Joule-Thompson expansion technique that allowed him to liquefy air at −190°C in 1895. In 1902, the French engineer Georges Claude (1870–1960) finally succeeded in liquefying air using a reversed Brayton cycle. The reciprocating piston technology was finally replaced by turbine technology around 1935 in Germany.

and

$$T_3 = T_{4s}(p_3/p_{4s})^{(k-1)/k} = T_{4s}\text{PR}^{(k-1)/k}$$

where PR is the isentropic pressure ratio. Hence, $T_{2s}/T_3 = T_1/T_{4s}$, and it can be shown that, for an ASC (i.e., $(\eta_s)_c = (\eta_s)_e = 1.0$), Eqs. (14.22) and (14.23) reduce to

$$\text{COP}_{\substack{\text{reversed}\\ \text{Brayton ASC}\\ \text{HP}}} = \frac{T_3}{T_3 - T_{4s}} = \left(1 - \text{PR}^{(1-k)/k}\right)^{-1} \tag{14.24}$$

and

$$\text{COP}_{\substack{\text{reversed}\\ \text{Brayton ASC}\\ \text{R/AC}}} = \frac{T_{4s}}{T_3 - T_{4s}} = \left(\text{PR}^{(k-1)/k} - 1\right)^{-1} \tag{14.25}$$

It is easy to show that these equations can also be obtained directly from Eqs. (13.23), (14.4), and (14.5). These results are illustrated in the following example.

EXAMPLE 14.10

Determine the COP and cycle minimum cooling temperature of Gorrie's 1844 reversed Brayton cycle refrigerator if it has a pressure ratio of 2.00 to 1, a compressor inlet temperature of 70.0°F, and an expander inlet temperature of 80.0°F, using

a. An ASC analysis.
b. An ideal gas analysis that includes typical mid 19th century compressor and expander isentropic efficiencies of 65.0% each.

Solution

Use Figure 14.27 as the system illustration for this example.

a. From the problem statement, we have PR = 2.00, T_1 = 70.0°F = 530. R, and T_3 = 80.0°F = 540. R. Then, Eq. (14.25) gives the COP for a reversed Brayton R/AC ASC as

$$\text{COP}_{\substack{\text{reversed Brayton}\\ \text{ASC R/AC}}} = (2.00^{0.40/1.40} - 1)^{-1} = 4.57$$

and the minimum temperature within the cycle is the cooling temperature T_{4s}, which is given by

$$T_{4s} = T_3/\text{PR}^{(k-1)/k} = (540.\text{ R})/2.00^{0.40/1.40} = 443\text{ R} = -17.0°\text{F}$$

b. In a more realistic analysis, we still assume ideal gas behavior, but now we introduce isentropic compressor and expander efficiencies of $(\eta_s)_c = (\eta_s)_e = 0.650$, and use Eq. (14.23), where the coldest temperature in the cycle is now only

$$T_4 = T_3 - (T_3 - T_{4s})(\eta_s)_e$$
$$= 540. - (540. - 443)(0.650) = 477\text{ R} = 17.0°\text{F}$$

and

$$T_{2s} = T_1 T_3 / T_{4s} = \frac{(530.\text{ R})(540.\text{ R})}{443\text{ R}} = 646\text{ R}$$

Then, Eq. (14.23) gives

$$\text{COP}_{\substack{\text{reversed}\\\text{Brayton cycle}\\\text{R/AC}}} = \frac{T_1 - T_4}{(T_{2s} - T_1)/(\eta_s)_c - (T_3 - T_{4s})(\eta_s)_e}$$
$$= \frac{530. - 477}{[(646 - 530.)/0.650] - (540. - 443)(0.650)}$$
$$= 0.220$$

The results of part b of this example are much more realistic than those of part a due to the large thermodynamic irreversibilities (friction, heat loss, etc.) present in early mechanical equipment.

Exercises

28. If Gorrie had found a way to increase the isentropic efficiency of his compressor in Example 14.10 from 65.0% to 75.0% (but the isentropic efficiency of the expander did not change), determine the new coefficient of performance of the actual (not ASC) system assuming all the other variables remain unchanged. **Answer**: $\text{COP}_{\text{actual}} = 0.244$.

29. A sudden heat wave causes the compressor inlet temperature in Example 14.10 to increase from 70.0°F to 95.0°F. Determine the new actual (not ASC) coefficient of performance of the unit. Assume all the other variables remain unchanged. **Answer**: $\text{COP}_{\text{actual}} = 0.384$.

30. Gorrie found a way to increase the pressure ratio of his compressor in Example 14.10 from 2.00 to 3.00. Determine the new actual (not ASC) coefficient of performance of the unit. Assume all the other variables remain unchanged. **Answer**: $\text{COP}_{\text{actual}} = 0.214$.

Like modern Brayton power cycles, modern reversed Brayton refrigeration cycles can be constructed with regeneration capability (Figure 14.28). Unlike power cycles, however, regeneration in refrigeration cycles does not improve the cycle's thermal efficiency; instead, it *reduces* the COP. However, regeneration does have the advantage of decreasing the minimum cooling temperature T_{4s}. Therefore, the purpose of regeneration in refrigeration cycles is simply to be able to reach lower cooling temperatures.

The use of a modern reversed Brayton refrigeration cycle is illustrated in the following example.

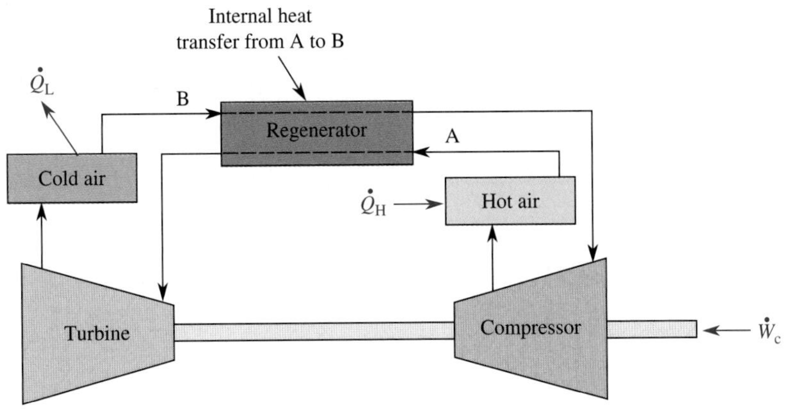

FIGURE 14.28
A modern reversed Brayton cycle with a regenerator.

EXAMPLE 14.11

Suppose 4.00 lbm/s of air at 530. R enters the compressor of a modern reversed Brayton ASC refrigeration unit. The isentropic pressure ratio of the compressor is 3.00 to 1 and the inlet temperature of the expander is 600. R. Determine

a. The expander power.
b. The compressor power.
c. The coefficient of performance of the unit.
d. The refrigeration capacity of the unit in tons.

Solution

Using Figure 14.27 as the illustration for this example, station 1 is the compressor inlet, station 2 is the compressor outlet, station 3 is the expander inlet, and station 4 is the expander outlet. For an ASC, all the processes are reversible, so the η_s of the expander and the compressor are both 1.00.

a. An energy rate balance on the expander gives

$$\dot{W}_{\text{expander}} = \dot{m}\,(h_3 - h_{4s}) = \dot{m}\,c_p(T_3 - T_{4s})$$

where

$$T_{4s} = T_3(p_{4s}/p_3)^{(k-1)/k} = 600.(1/3.00)^{0.40/1.40} = 438\text{ R}$$

Then,

$$\dot{W}_{\text{expander}} = (4.00\text{ lbm/s})[0.240\text{ Btu/(lbm}\cdot\text{R)}](600. - 438.\text{ R}) = 155\text{ Btu/s}$$

b. An energy rate balance on the compressor gives

$$\dot{W}_{\text{compressor}} = \dot{m}\,(h_1 - h_{2s}) = \dot{m}\,c_p(T_1 - T_{2s})$$

where

$$T_{2s} = T_1(p_{2s}/p_1)^{(k-1)/k} = 530.(3.00)^{0.40/1.40} = 725\text{ R.}$$

Then,

$$\dot{W}_{\text{compressor}} = 4.00(0.240)(530. - 725) = -188\text{ Btu/s}$$

c. Equation (14.25) gives the COP of this unit as

$$\text{COP} = \left(\text{PR}^{(k-1)/k} - 1\right)^{-1} = \left(3.00^{0.40/1.40} - 1\right)^{-1} = 2.71$$

d. Finally, the refrigeration capacity of this unit is

$$\begin{aligned}\dot{Q}_L &= \text{refrigeration capacity} = \text{COP}_{\text{R/AC}} \times |\dot{W}_{\text{in}}|_{\text{net}} = 2.71|(188 - 155)| \\ &= (88.0\text{ Btu/s})(60\text{ s/min}) = (5280\text{ Btu/min})(1\text{ ton/200. Btu/min}) \\ &= 26.4\text{ tons of refrigeration}\end{aligned}$$

Exercises

31. If the compressor inlet temperature of the reversed Brayton ASC refrigeration unit in Example 14.11 is reduced from 530. R to 500. R, what is the new refrigeration capacity of this unit? **Answer:** $\dot{Q}_L = 49.8$ tons.
32. The pressure ratio across the compressor of the reversed Brayton ASC refrigeration unit in Example 14.11 is increased from 3.00 to 500. Determine the new ASC coefficient of performance for the new unit. **Answer:** $\text{COP}_{\text{ASC R/AC}} = 1.71$.
33. The reversed Brayton ASC refrigeration unit discussed in Example 14.11 is to be scaled up to handle a refrigeration capacity of 230. tons. Determine the corresponding air mass flow rate required for the new unit. **Answer:** $\dot{m}_{\text{air}} = 34.9$ lbm/s.

14.15 REVERSED STIRLING CYCLE REFRIGERATION

A reversed Stirling cycle refrigerator was first implemented by the Scottish engineer Alexander Carnegie Kirk (1830–1892) in 1862. Kirk was searching for a cooling technology that was safer than the prevailing vapor-compression machines that used explosive ether. He was aware of the engine developed by Robert Stirling in 1816 and felt that, if he put power into the engine instead of letting it produce power, the displacer piston would be cooled. With a compressor pressure ratio of 2.0, he reached an expander temperature of −13°C, and when he increased the pressure ratio to 7.0, he reached −40.°C.

A theoretical analysis of the Stirling cycle was finally carried out by I. A. Wyshnegradski in 1871. Because of its inherent safety, it was extensively used for refrigeration of food products (especially frozen meat) on ships from 1880 to 1900. The safe operation of a reversed Stirling cycle also made it ideal for use in deep mines until about 1930, when synthetic refrigerants (especially R-11 and R-12) made vapor-compression machines safer and more efficient.

The reversed Stirling cycle refrigerator is most effective when hydrogen gas is the working fluid and a very efficient regenerator is employed. Hydrogen has the largest specific heat of all common gases and does not condense at temperatures above 35 K (helium is used for lower temperatures). Under these conditions, Stirling refrigerators are often used to produce temperatures in the −80.0°C to −200.°C (−112°F to −328°F) range.

Recall that the Stirling cycle consists of two constant temperature processes and two constant volume processes (see Figure 14.29). Reciprocating piston technology is still the most effective mechanism used to form this cycle, with the constant volume processes being approximated by the relatively small piston motion near the top and bottom dead center crankshaft positions. Since the Stirling and Carnot cycles have the same thermal efficiency, the reversed Stirling and the reversed Carnot cycles have the same coefficient of performance,

$$\text{COP}_{\substack{\text{reversed Stirling}\\ \text{ASC HP}}} = \frac{T_H}{T_H - T_L} \tag{14.26}$$

and

$$\text{COP}_{\substack{\text{reversed Stirling}\\ \text{ASC R/AC}}} = \frac{T_L}{T_H - T_L} \tag{14.27}$$

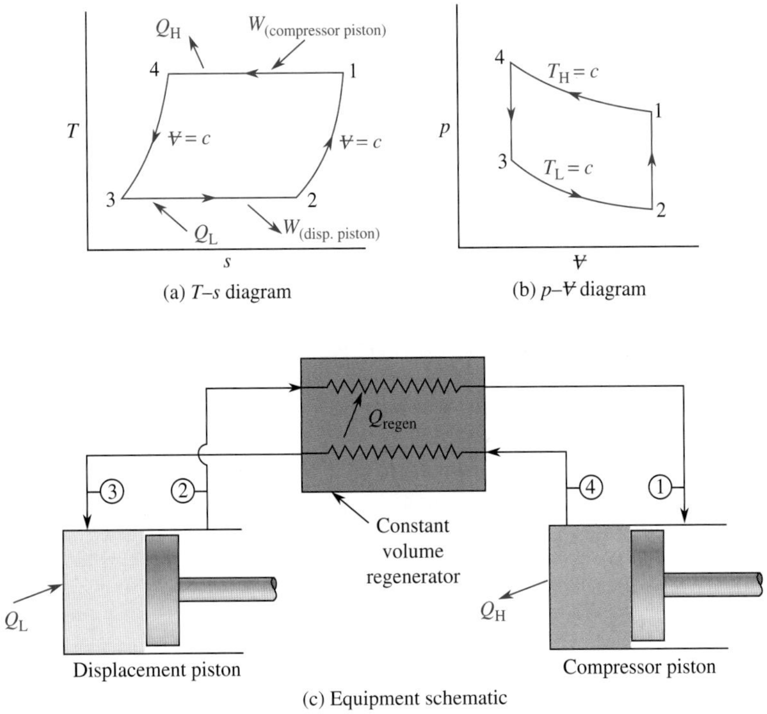

FIGURE 14.29
The reversed Stirling cycle operating as a refrigerator, air conditioner, or heat pump.

CRITICAL THINKING

Stirling engines and refrigerators work best when hydrogen gas is used as the working fluid, because hydrogen has a very large specific heat ($c_{p\text{-hydrogen}} = 14.32$ kJ/kg·K, whereas $c_{p\text{-air}} = 1.004$ kJ/kg·K). Since the hydrogen is not burned, but merely passes through the cycle described in Figure 14.29, then why would having a high specific heat make any difference? Hint: Think about the heat transfer process in the regenerator.

The following example illustrates the use of this material.

EXAMPLE 14.12

A company manufactures the cryogenic Stirling refrigeration cycle microcooler shown in Figure 14.30. The microcooler has a mass of only 0.300 kg and is used to replace liquid nitrogen in infrared thermal imaging cameras. With a power input of only 3.00 W from a 12.0 V battery, the microcooler can reach a temperature of 65.0 K in an environment at 22.0°C. The refrigerating capacity of the microcooler at these conditions is 0.100 J/s. For this design, determine

a. The Stirling ASC coefficient of performance of this refrigeration unit.
b. The actual coefficient of performance of this unit.

FIGURE 14.30
Example 14.12, microcooler.

Solution

a. The Stirling ASC coefficient of performance of a refrigeration unit is given by Eq. (14.27) as

$$\text{COP}_{\substack{\text{reversed Stirling}\\ \text{ASC R/AC}}} = \frac{T_L}{T_H - T_L} = \frac{65.0\,\text{K}}{(273.15 + 22.0\,\text{K}) - 65.0\,\text{K}} = 0.282$$

b. The actual coefficient of performance can be calculated from its definition as

$$\text{COP}_{\substack{\text{reversed Stirling}\\ \text{actual R/AC}}} = \frac{\dot{Q}_L}{\dot{W}_{\text{compressor}}} = \frac{\dot{Q}_{\text{cooling}}}{\dot{W}_{\text{input}}} = \frac{0.100\,\text{J/s}}{3.00\,\text{J/s}} = 0.0333$$

The actual efficiency of Stirling cycle systems is often much lower than the idealized ASC predictions. In addition to the fact that real systems have losses and entropy production that are not accounted for in an ASC analysis, the large dead space within the reciprocating piston mechanism prevents all the gas in the system from passing through the cycle. Also, a reciprocating piston mechanism cannot provide a truly constant volume process at any point in the cycle. It can provide only approximate constant volume conditions when the piston is near the top or bottom dead center positions while the crankshaft is rotating. Figure 14.31 illustrates this volume change for a compression ratio of 8.0.

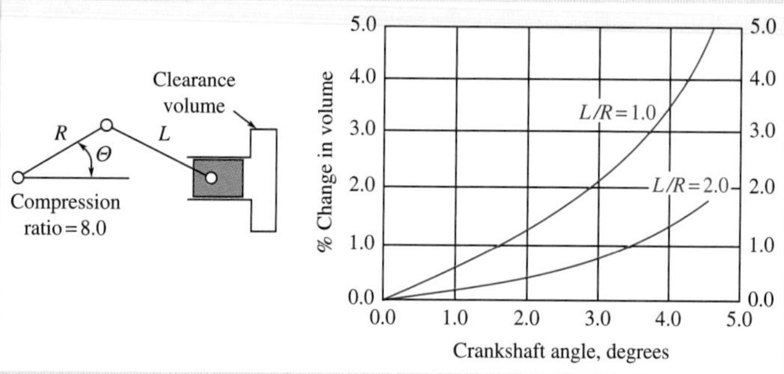

FIGURE 14.31
Example 14.14, volume change.

Exercises

34. If the microcooler described in Example 14.12 is used to produce a temperature of 75.0 K instead of 65.0 K at the same environmental temperature, determine the ASC coefficient of performance of the unit under these conditions. **Answer:** $\text{COP}_{\text{reversed Stirling ASC R/AC}} = 0.341$.

35. The microcooler described in Example 14.12 can also produce 0.250 J/s of cooling at 120. K with a power input of only 1.70 W when the ambient temperature is 22.0°C. Determine the ASC and actual coefficients of performance under these conditions. **Answer:** $COP_{\text{reversed Stirling ASC R/AC}} = 0.685$ and $COP_{\text{actual Stirling R/AC}} = 0.147$.

36. Note that the Stirling ASC coefficient of performance of a refrigerator vanishes as $T_L \to 0$. What happens to the power required to drive these systems as you approach lower and lower temperatures? If a Stirling ASC has the same efficiency as a Carnot ASC and a Carnot ASC is the most efficient cycle possible, then what is the likelihood that we will ever be able to reach absolute zero temperature in the laboratory? **Answer:** As $T_L \to 0$, $\dot{W}_{comp} \to \infty$, consequently we will never be able to reach absolute zero temperature in the laboratory (although temperatures in the range of 0.001 K have been reached).

14.16 MISCELLANEOUS REFRIGERATION TECHNOLOGIES

14.16.1 Joule-Thomson Expansion Cooling

The cooling that results from the expansion or throttling of a gas from a high to a low pressure is called *Joule-Thomson cooling*. A refrigeration ASC analysis predicts that no cooling occurs in this type of throttling, because an ideal gas throttling process is both isenthalpic and isothermal. However, *real* gases and vapors do undergo a temperature change during isenthalpic processes, as dictated by their Joule-Thomson coefficient. This coefficient is discussed in Chapter 6 and defined in Eq. (6.25) as

$$\mu_J = \left(\frac{\partial T}{\partial p}\right)_h \tag{6.25}$$

so that, approximately,

$$(\Delta T)_h = \mu_J(\Delta p)_h \tag{14.28}$$

The value of μ_J can be either positive or negative, and it is usually larger at lower temperatures.[6] For example, the Joule-Thomson coefficient for air at 20°C and several atmospheres is only about 0.3°C/atm (see Figure 6.6). Therefore, throttling air from 100 psig (~6 atm) down to atmospheric pressure produces a temperature drop in the air of only about 2°C. Such a system is shown in Figure 14.32.

The refrigeration or air conditioning COP of a Joule-Thomson expansion throttling device is given by

$$COP_{\substack{J-T\\R/AC}} = \frac{\dot{Q}_L}{|\dot{W}|_c} = \frac{T_1 - T_{4h}}{(|T_1 - T_{2s}|)/(\eta_s)_c}$$

and if $T_1 = T_3$, $p_2 = p_3$, and $p_4 = p_1$, then this reduces to

$$COP_{\substack{J-T\\R/AC}} = \frac{\mu_J(p_2 - p_1)}{T_1\left[(p_2/p_1)^{(k-1)/k} - 1\right]/(\eta_s)_e} \tag{14.29}$$

The vortex tube discussed in Chapter 9 is a variation on this technique. It does not appear to depend on the Joule-Thomson effect, and it can produce cold temperatures in only part of the outlet flow. The remainder of the outlet flow is quite warm (see Figure 9.15).

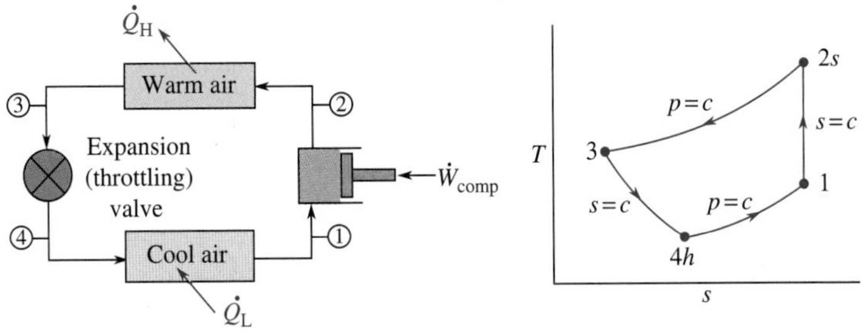

FIGURE 14.32
A Joule-Thomson refrigeration system.

[6] If μ_J is negative, then you have Joule-Thomson heating of the gas on throttling.

EXAMPLE 14.13

Determine the outlet temperature and COP of a Joule-Thomson expansion throttling device using air when the inlet temperature and pressure are 70.0°F and 300. psia and the outlet pressure is 14.7 psia. Assume the Joule-Thomson coefficient for air in this range is 0.0300°F/psi, and that the isentropic efficiency of the air compressor is 90.0%.

Solution

Using Figure 14.32 as the illustration for this example, from Eq. (14.28) we have

$$T_2 - T_1 = \mu_J(p_2 - p_1) = 0.0300(14.7 - 300.) = -8.56°\text{F}$$

Then, $T_2 = 70.0 - 8.56 = 61.4°\text{F}$, and Eq. (14.29) gives the COP as

$$\text{COP}_{\substack{\text{J-T}\\ \text{R/AC}}} = \frac{0.0300(14.7 - 300.)}{(70.0 + 459.67)\left[\left(\frac{14.7}{300.}\right)^{0.40/1.40} - 1\right]/0.90} = 0.0252$$

Therefore, the thermal efficiency of this type of Joule-Thomson expansion throttling refrigeration or air conditioner is only 2.52%.[7]

Exercises

37. The air compressor in Example 14.13 is replaced by one with an isentropic efficiency of 95%. Determine the new Joule-Thomson coefficient of performance of the device. Assume all the other variables remain unchanged. **Answer**: $\text{COP}_{\text{J-T}} = 0.027$.
38. If the inlet air temperature in Example 14.13 is reduced from 70.0°F to 50.0°F, determine the outlet air temperature and the Joule-Thomson coefficient of performance. Assume all the other variables remain unchanged. **Answer**: $T_2 = 41.4°\text{F}$ and $\text{COP}_{\text{J-T}} = 0.0260$.
39. If the initial pressure in Example 14.13 is increased from 300. psia to 3000. psia, determine the new outlet temperature and the Joule-Thomson coefficient of performance of the device. Assume all the other variables remain unchanged. **Answer**: $T_2 = -19.6°\text{F}$ and $\text{COP}_{\text{J-T}} = 0.195$.
40. The Joule-Thomson coefficient for CO_2 at 2.00 MPa is 0.0150°C/kPa. Carbon dioxide initially at 20.0°C is throttled from 2.00 MPa to atmospheric pressure. Determine the outlet temperature and the Joule-Thomson coefficient of performance. **Answer**: $T_2 = -8.50°\text{C}$ and $\text{COP}_{\text{J-T}} = 0.179$.

Though the Joule-Thomson refrigerator is not very effective by itself, it is often used in conjunction with other refrigeration systems. Figure 14.33 shows it being used with a reversed Brayton cycle to liquefy the working fluid. This was the basic technique used by Karl von Linde (1842–1934) to produce liquid air on a large commercial scale in 1895.

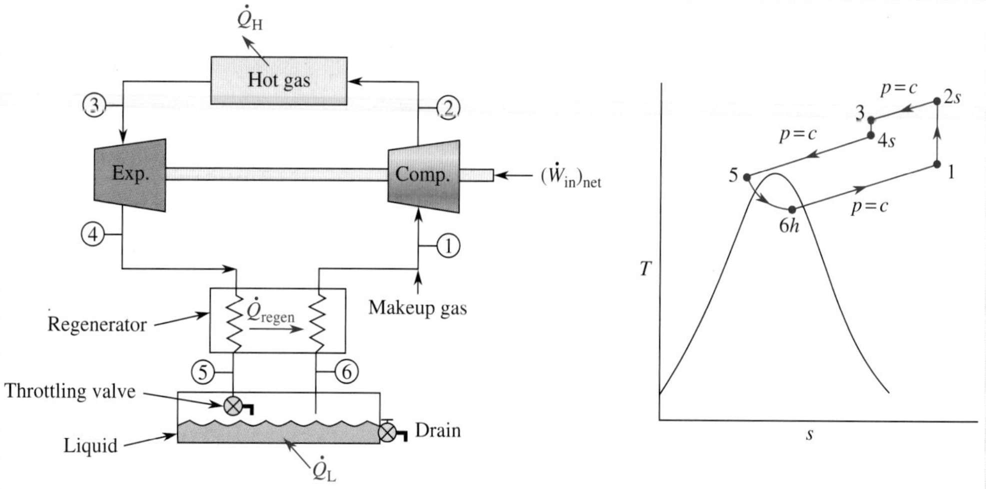

FIGURE 14.33
The basic Linde process for liquefying air.

[7] *Note that the thermal efficiency of a heat pump is always greater than 100% (i.e., its COP > 1.0), but the thermal efficiency of a refrigerator or air conditioner can be less than 100%.*

14.16.2 Refrigerating Mixtures

Any endothermic (heat-absorbing) chemical reaction can be used to produce refrigeration. The simplest refrigerating reactions occur when salts with endothermic heats of solution are dissolved in water. The resulting saltwater (brine) mixtures can become cold enough to freeze ice on the outside of the container and are known as *refrigerating mixtures*. Sodium nitrate ($NaNO_3$), potassium nitrate (KNO_3), ammonium nitrate (NH_4NO_3), and calcium chloride hexahydrate[8] ($CaCl_2 \cdot 6\ H_2O$) were all used as refrigerating mixtures in ancient times. Refrigerating mixtures were used in ancient times to produce artificial ice to cool drinks. More recently, it was discovered that most of these salts also reduce the freezing point of a saltwater mixture, so if one of these salts is added to ice or snow at 0°C (32°F), the salt melts some of the ice or snow, producing a brine solution. Chemical equilibrium requires that thermal energy be added to the endothermic process of dissolving the salt, plus additional thermal energy must be found to melt the ice used to produce the brine. If the mixture container is insulated, then the required thermal energy must come from the mixture itself. This results in lowering the mixture temperature to the freezing point corresponding to the brine concentration. For example, a saturated solution of common hydrated calcium chloride ($CaCl_2 \cdot 6\ H_2O$) can produce an ice-brine mixture temperature of −55°C (−67°F) in an insulated container. This salt is also used to melt ice from sidewalks and roads during winter, so long as the ice temperature is above −55°C (−67°F).

14.16.3 Surface Evaporation

The ancient Egyptians knew that, when water seeped through porous pottery and evaporated from its surface, it would cool the contents. Evaporative cooling is the method warm-blooded animals use to control their body temperature. The evaporation of perspiration from the surface of our skin is one of the main cooling mechanisms of humans. Dogs and cats pant, causing evaporation directly from the surface of their lungs.

HOW DO YOU MAKE ICE CREAM?

Have you ever tried to refreeze melted ice cream? Refreezing changes the texture of the ice cream from smooth and creamy to coarse and unpleasant. This is because refrigerator freezers freeze foods slowly, producing large ice crystals that result in the coarse texture in refrozen ice cream (it also destroys the cellular structure of meats frozen this way).

FIGURE 14.34
A 19th century home ice cream maker.

Cooled or semi-frozen foods containing milk or cream have been known since ancient times. Initially reserved for the wealthy, who could afford such extravagance, they finally became universally available when a Spanish doctor, Blasius Villafranca, discovered that cream could be frozen in a container surrounded by a mixture of saltpeter (potassium nitrate, KNO_3) and snow in 1550. Nancy Johnson, a farmer's wife, developed an ingenious home ice cream freezer using salt and ice in 1846, and in 1851 Jacob Fussel established the first ice cream company in Seven Valleys, Pennsylvania. By the late 1800s, it was commonly known that mixing table salt (NaCl) or hydrated calcium chloride ($CaCl_2 \cdot 6\ H_20$) with solid ice or snow in an insulated container produced subzero temperatures cold enough to make good quality *homemade* ice cream by quick freezing (Figure 14.34). Around 1890, a small ice cream vendor decided to stimulate weekend sales by adding chocolate or fruit syrup to ice cream sold on Sunday. At first this was known as *Sunday ice cream*, but to avoid religious conflict, the name and spelling were subsequently changed to *ice cream sundae*. The first ice cream on a stick was produced commercially in 1904.

[8] Pure anhydrous calcium chloride ($CaCl_2$) has an exothermic heat of solution due to the strong exothermic hydration process that occurs during solution. However, common solid calcium chloride salt is already hydrated as $CaCl_2 \cdot 6\ H_2O$, and has an endothermic heat of solution, since the exothermic hydration process is not required.

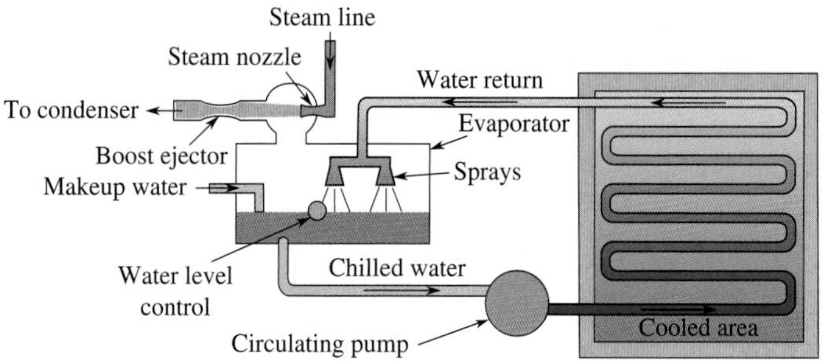

FIGURE 14.35

Steam jet refrigeration. Steam flowing through the steam nozzle and booster ejector produces low pressure in the evaporator. This causes the water to evaporate and cool (or "chill") the water in the evaporator.

14.16.4 Radiation Cooling

The Egyptians were known to use radiation heat loss to a black night sky (which typically has a black-body temperature of about 150 R) from a shallow pan of water, causing it to freeze when the surrounding air temperature is far above freezing.

14.16.5 Reduced Pressure Refrigeration

In 1755, William Cullen discovered that exposing water to a vacuum caused it to boil at a lower temperature. This could then be used to cool a surrounding fluid. Various versions of this technique evolved over the years, including a fascinating steam jet ejector cooling system developed by the French engineer Maurice Leblanc in 1909. Leblanc utilized the Bernoulli effect to evacuate a container of water by attaching it to the throat of a nozzle passing high-velocity steam. The water in the container was thus cooled. This is the technology that was initially used to air condition the Radio City Music Hall in New York City in the 1930s (see Figure 14.35).

14.16.6 Thermoelectric Refrigeration

Thermoelectricity is the direct or spontaneous conversion of heat into electrical energy. In 1834, Jean Charles Peltier (1785–1845) observed that, when an electric current was passed through two different metallic conductors connected in a loop, one of the two junctions between the conductors cooled while the other warmed. When the direction of the current was reversed, the effect was reversed, with the first junction warming and the second cooling. Thermoelectric cooling remained a laboratory curiosity until after 1950, when semiconductors were developed. Semiconductor materials were formulated to produce much more efficient thermoelectric cooling than pure metals. They were used effectively in the U.S. space program in the 1960s and have a growing market today.

14.16.7 Vortex Tube

The technique for separating a flowing fluid into hot and cold outlet flows using a spiral vortex chamber was developed by Georges Ranque (1898–1973) in 1931. Vortex tubes have a limited application in spot cooling in manufacturing operations, electronic cabinet cooling, and body suit cooling. They have no moving parts, and are very effective in manufacturing facilities, since they require only a source of compressed air for their operation (see the Section 9.10 at the end of Chapter 9 for more information).

14.17 FUTURE REFRIGERATION NEEDS

The most pressing need in modern refrigeration, air conditioning, and heat pump technology is improvement in energy efficiency (COP). It is estimated that, in the United States, over 20% of all household electricity used is consumed by refrigerators and about 20% of all the electricity generated in the United States is used for lighting. By the year 2030, it is estimated that the world's energy needs will top 25×10^{12} W (25 TW) per year. Technological developments are needed to make household appliances and industrial operations more efficient, develop heat pumps that operate in colder climates, devise new and more efficient air conditioning cycles and associated technologies, and decrease the overall use of energy in illumination. It was mentioned earlier that refrigeration technology has never had the power and glamour associated with the power-generating technologies, but it will be very important in leading the way to energy conservation in the future.

WHAT IS THE SUPER EFFICIENT REFRIGERATOR PROGRAM?

By the 1990s, refrigeration consumed over 20% of all U.S. household electricity. The Super Efficient Refrigerator Program was a contest sponsored by a consortium of 24 electrical utilities, which offered a prize of $30 million for the manufacture and marketing of a new domestic refrigerator that was at least 50% more efficient than current models and used a chlorine-free refrigerant. Whirlpool Corporation in Evansville, Indiana, easily won the award with a refrigerator that was 80% more efficient than current models, using R-134a as the refrigerant. Its design modifications included

- Foam insulation throughout the cabinet.
- Fuzzy logic controlled defrost cycle, which defrosts only when needed.
- Freezer vacuum insulation panels three times as effective as foam.
- Thicker door with foam insulation.
- Redesigned efficient fan motors.
- Redesigned efficient compressor and drive motor.

Initially, improvement in system efficiency can be had by simply applying what we already know in the way of better thermal insulation, more efficient compressors and pumps, more effective use of proportional control systems, and so forth. The Super Efficient Refrigeration Program (SERP), developed in the 1990s, is a step in this direction (see the box). This program and others like it may produce the energy-efficient products that will be needed in the future.

14.18 SECOND LAW ANALYSIS OF REFRIGERATION CYCLES

A logical method for maximizing (or optimizing) system performance during the design stage of a new product can be based on minimizing the losses (irreversibility rate) within a system. In Chapter 10, we define the *irreversibility rate* of a process to be the product of the local environmental temperature T_0 and the entropy production rate inside the system $\dot{S}_P$ as

$$\text{Irreversibility rate} = \dot{I} = T_0\dot{S}_P \geq 0 \qquad (10.15)$$

This method involves computing the irreversibility rate of each of the components within a system over a range of component parameters (size, efficiency, materials, and so forth) and system operating conditions. Then, you specify the component parameters in the design that minimize the irreversibility and entropy production rates within the system. Equations for the direct calculation of the entropy production rate for various component processes are given in Chapter 7. Some examples follow:

Irreversibility rate due to heat transfer

$$\dot{I}_{\substack{\text{heat}\\\text{transfer}}} = T_0(\dot{S}_P)_{\substack{\text{heat}\\\text{transfer}}} = -T_0\int_{\forall}\frac{\dot{q}}{T^2}\left(\frac{dT}{dx}\right)d\forall$$

Irreversibility rate due to fluid viscosity

$$\dot{I}_{\substack{\text{fluid}\\\text{viscosity}}} = T_0(\dot{S}_P)_{\substack{\text{fluid}\\\text{viscosity}}} = T_0\int_{\forall}\frac{\mu}{T}\left(\frac{dV}{dx}\right)^2 d\forall$$

Irreversibility rate due electrical resistance

$$\dot{I}_{\substack{\text{electrical}\\\text{resistance}}} = T_0(\dot{S}_P)_{\substack{\text{electrical}\\\text{resistance}}} = T_0\int_{\forall}\frac{J_e^2\rho_e}{T}d\forall$$

and so forth. Use of these equations requires a detailed understanding of the operation of the system.

However, if the system contains simply connected flow loops, it is much easier to determine the entropy production and irreversibility rates using a simple entropy rate balance equation (recall that this is called the *indirect method* for determining entropy production). Entropy production rate equations using the indirect method are developed in Chapter 9 for heat exchangers (i.e., boilers, evaporators, and condensers), fluid mixing, and transient operations.

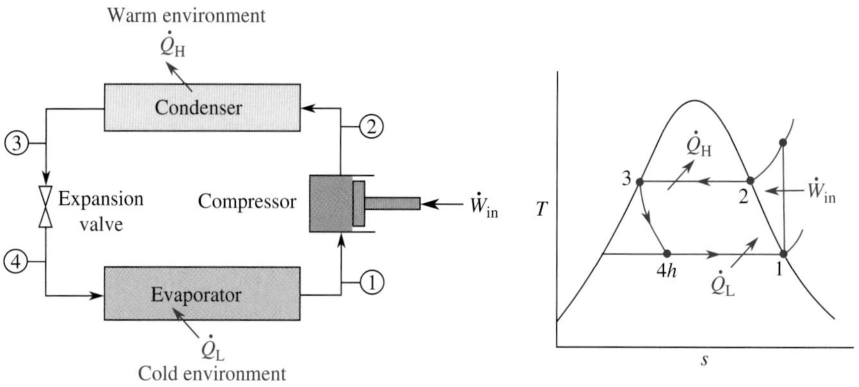

FIGURE 14.36
A simple vapor-compression refrigeration cycle.

Because of the simple flow structure of most refrigeration systems, it is easy to use the indirect method to find the irreversibility rate of various components. For a steady state (SS), steady flow (SF) system component with a single inlet and a single outlet (SI, SO) having isothermal boundaries (IB), the irreversibility rate is

$$\dot{I}_{\substack{\text{SS, SF}\\ \text{SI, SO IB}}} = T_0(\dot{S}_P)_{\substack{\text{SS, SF}\\ \text{SI, SO IB}}} = T_0\left[\dot{m}\left(s_{\text{out}} - s_{\text{in}}\right) - \frac{\dot{Q}}{T_b}\right]$$

For example, for the simple vapor-compression refrigeration system shown in Figure 14.36, the irreversibility rate produced by the adiabatic compressor is

$$\dot{I}_{\substack{\text{adiabatic}\\ \text{compressor}}} = T_0(\dot{S}_P)_{\substack{\text{adiabatic}\\ \text{compressor}}} = \dot{m}_{\text{ref}} T_0(s_2 - s_1) \tag{14.30}$$

and, in the condenser, we have

$$\dot{I}_{\text{condenser}} = T_0(\dot{S}_P)_{\text{condenser}} = T_0\left[\dot{m}_{\text{ref}}(s_3 - s_2) - \frac{\dot{Q}_{\text{condenser}}}{T_{\text{condenser}}}\right] \tag{14.31}$$

where $\dot{Q}_{\text{condenser}} = \dot{Q}_H$, and across the adiabatic expansion valve, we have

$$\dot{I}_{\substack{\text{adiabatic}\\ \text{expansion valve}}} = T_0(\dot{S}_P)_{\substack{\text{adiabatic}\\ \text{expansion valve}}} = \dot{m}_{\text{ref}} T_0(s_{4h} - s_3) \tag{14.32}$$

Finally, the irreversibility rate produced inside the evaporator is

$$\dot{I}_{\text{evaporator}} = T_0(\dot{S}_P)_{\text{evaporator}} = T_0\left[\dot{m}_{\text{ref}}(s_1 - s_{4h}) - \frac{\dot{Q}_{\text{evaporator}}}{T_{\text{evaporator}}}\right] \tag{14.33}$$

where $\dot{Q}_{\text{evaporator}} = \dot{Q}_L$ is the refrigeration or cooling rate of the system. Note that, since a phase change is usually a reversible process (see Chapter 7), the irreversibilities that occur in the condenser and evaporator come from viscous pressure losses between the inlet and outlet and, if the refrigerant exits the evaporator or enters the condenser in a superheated state, then irreversibilities exist for all heat transfer processes that occur outside the vapor dome.

The use of this technique is illustrated in the following example.

EXAMPLE 14.14

The following preliminary design information is available for the vapor-compression refrigeration cycle shown in Figure 14.36 using R-134a:

Station 1 Compressor Inlet	Station 2*s* Compressor Outlet	Station 3 Condenser Outlet	Station 4*h* Expansion Valve Outlet
$x_1 = 1.00$	$p_{2s} = 800.\ \text{kPa}$	$x_3 = 0.00$	$h_{4h} = h_3$
$T_1 = -20.0°\text{C}$	$s_{2s} = s_1$	$p_3 = 725.\ \text{kPa}$	$p_{4h} = 160.\ \text{kPa}$

Note that the designer computes pressure losses of 75.0 kPa in the condenser and 27.0 kPa in the evaporator. The refrigerant mass flow rate is 0.500 kg/s, the local environmental temperature is T_0 = 25.0°C, and the isentropic efficiency of the compressor is 70.0%. Determine

a. The irreversibility rate of each component in the system and the total irreversibility rate of the system.
b. The system COP and ε (the first and second law efficiencies).
c. Select the component with the highest irreversibility rate and suggest ways to improve it.

Solution

Using Figure 14.36 as the illustration for this example, the properties at the four stations can be found in Tables C.7e, C.7f, and C.8d as

Station 1 Compressor inlet	Station 2*s* Compressor outlet	Station 3 Condenser outlet	Station 4*h* Expansion valve outlet
$x_1 = 1.00$	$p_{2s} = 800.$ kPa	$x_3 = 0.0$	$h_{4h} = h_3$
$T_1 = -20.0$°C	$s_{2s} = s_1$	$p_3 = 725$ kPa	$p_{4h} = 160$ kPa
	$= 0.9332$ kJ/kg·K	$h_3 = 87.46$ kJ/kg	$h_{4h} = 87.46$ kJ/kg
$h_1 = 235.31$ kJ/kg	$h_{2s} = 271.10$ kJ/kg	$s_3 = 0.3257$ kJ/kg·K	$x_{4h} = 0.280$
$s_1 = 0.9332$ kJ/kg·K	$T_{2s} = 39.8$°C	$T_3 = 27.9$°C	$s_{4h} = 0.3449$ kJ/kg·K
$p_1 = 132.99$ kPa			$T_{4h} = -15.6$°C

Conditions at stations 2*s*, 3, and 4*h* are determined by interpolation in Table C.7f or with a computer program containing the appropriate properties. Also, note that the condenser outlet temperature T_3 is about 3°C above the environmental temperature of T_0 = 25°C, which is an appropriate temperature drop across the wall of the condenser. A temperature drop of this magnitude also occurs in the evaporator, so the temperature of the refrigerated space is about −12.6°C instead of −15.6°C.

We can now determine the actual conditions at the outlet of the compressor from the two properties $h_2 = (h_{2s} - h_1)/(\eta_s)_{comp} + h_1 = 288.03$ kJ/kg and $p_2 = p_{2s} = 800$ kPa. Interpolation in Table C.7f in *Thermodynamic Tables to accompany Modern Engineering Thermodynamics* (or through the use of an appropriate computer program) gives the following additional properties at this state:

$$s_2 = 0.9814\text{ kJ/kg}\cdot\text{K and } T_2 = 54.97°\text{C}$$

The condenser and evaporator heat transfer rates and the compressor work rate are

$$\dot{Q}_{condenser} = \dot{m}_{ref}(h_3 - h_2) = (0.5\text{ kg/s})(87.46 - 288.03\text{ kJ/kg}) = -100.3\text{ kJ/s}$$

$$\dot{Q}_{evaporator} = \dot{m}_{ref}(h_1 - h_{4h}) = (0.5\text{ kg/s})(235.31 - 87.46\text{ kJ/kg}) = 73.9\text{ kJ/s}$$

$$\dot{Q}_{compressor} = \dot{m}_{ref}(h_2 - h_1) = (0.5\text{ kg/s})(288.03 - 235.31\text{ kJ/kg}) = 26.36\text{ kJ/s}$$

Now we can calculate the desired quantities.

a. The irreversibility rate of the compressor is given by

$$\dot{I}_{\substack{\text{adiabatic}\\ \text{compressor}}} = \dot{m}_{ref}T_0(s_2 - s_1)$$

$$= \left(0.500\frac{\text{kJ}}{\text{s}}\right)(25.0 + 273.15\text{ K})\left(0.9814 - 0.9317\frac{\text{kJ}}{\text{kg}\cdot\text{K}}\right)$$

$$= 7.41\frac{\text{kJ}}{\text{s}} = 7.41\text{ kW}$$

$$\dot{I}_{condenser} = T_0\left(\dot{m}_{ref}(s_3 - s_2) - \frac{\dot{Q}_{condenser}}{T_0}\right)$$

$$= (25.0 + 273.15\text{ K})\left[0.500\frac{\text{kg}}{\text{s}}\left(0.3257 - 0.9814\frac{\text{kJ}}{\text{kg}\cdot\text{K}}\right) - \frac{-100.3\text{ kJ/s}}{25.0 + 273.15\text{ K}}\right]$$

$$= 2.55\frac{\text{kJ}}{\text{s}} = 2.55\text{ kW}$$

$$\dot{I}_{\substack{\text{expansion}\\ \text{valve}}} = \dot{m}_{ref}T_0(s_{4h} - s_3)$$

$$= \left(0.500\frac{\text{kg}}{\text{s}}\right)(25.0 + 273.15\text{ K})\left(0.3449 - 0.3257\frac{\text{kJ}}{\text{kg}\cdot\text{K}}\right)$$

$$= 2.86\frac{\text{kJ}}{\text{s}} = 2.86\text{ kW}$$

(*Continued*)

EXAMPLE 14.14 *(Continued)*

$$\dot{I}_{\text{evaporator}} = T_0\left[\dot{m}_{\text{ref}}(s_1 - s_{4h}) - \frac{\dot{Q}_{\text{evaporator}}}{T_{\text{evaporator}}}\right]$$

$$= (25.0 + 273.15)\left(0.500\frac{\text{kg}}{\text{s}}\right)\left[\left(0.9332 - 0.3449\frac{\text{kJ}}{\text{kg}\cdot\text{K}}\right) - \frac{73.4\ \text{kJ/s}}{-15.6 + 273.15\ \text{K}}\right]$$

$$= 2.15\ \text{kJ/s} = 2.15\ \text{kW}$$

and the total irreversibility rate of the system is

$$\dot{I}_{\text{total}} = \dot{I}_{\text{compressor}} + \dot{I}_{\text{condenser}} + \dot{I}_{\text{expansion valve}} + \dot{I}_{\text{evaporator}}$$

$$= 7.41 + 2.55 + 2.86 + 2.15 = 15.0\ \text{kW}$$

b. The system coefficient of performance is given by

$$\text{COP} = \frac{\dot{Q}_{\text{evaporator}}}{\dot{W}_{\text{compressor}}} = \frac{73.9}{26.4} = 2.80$$

The second law efficiency for a refrigeration system is discussed in Chapter 10. The relevant equation is Eq. (10.34),

$$\varepsilon_{\text{R/AC}} = \frac{\left|\left(1 - \frac{T_0}{T_L}\right)\dot{Q}_L\right|}{\dot{W}} = \left|\left(1 - \frac{T_0}{T_L}\right)\left(\frac{\dot{Q}_L}{\dot{W}}\right)\right| = \left|\left(1 - \frac{T_0}{T_L}\right)\right| \times \text{COP}_{\substack{\text{actual}\\ \text{R/AC}}}$$

$$= \left|\left(1 - \frac{25.0 + 273.15\ \text{K}}{-15.6 + 273.15\ \text{K}}\right)\right| \times 2.85 = 0.494 = 49.4\%$$

c. The component with the highest irreversibility rate in this system is the compressor. It is the weak link in this system. Improving other system components will have only a marginal effect on system performance until a more efficient compressor is designed or found. Note that to reduce the irreversibility rate of the compressor to a value comparable to the irreversibility rates of the other components in the system requires improving the isentropic efficiency of the compressor from 70% to about 88%.

SUMMARY

Refrigeration is a generic term that embodies the topics of refrigeration, air conditioning, and heat pump systems. In this chapter, we divide refrigeration into three broad categories of technology. The first is vapor refrigeration cycles, consisting of vapor-compression cycles and absorption cycles. Vapor-compression cycles are basically reversed Rankine power cycles, whereas absorption refrigeration has no power cycle analog. The second category is gas refrigeration cycles, which consist of reversed versions of external combustion power cycles. The most prominent are the reversed Brayton and reversed Stirling refrigeration cycles. The third category covers all other miscellaneous refrigeration technologies, such as Joule-Thomson cooling; refrigerating mixtures; and evaporation, radiation, reduced pressure, thermoelectric, and vortex tube cooling. The chapter ends with a discussion of future needs in refrigeration technology followed by an example of how the second law of thermodynamics can assist in the design of better refrigeration technologies by minimizing the irreversibility rate within the various system components.

As in the other chapters in this text, a historical timeline is used to develop the material. This is done to provide a perspective on the social and cultural impact produced by the development of refrigeration technology. The reason this approach has been followed throughout this text is to sensitize you, the next generation of engineers, to your responsibility for understanding the enormous potential of your profession to change society.

Some of the more important equations introduced in this chapter follow. Do not attempt to use them blindly without understanding their limitations. Please refer to the text material where they were introduced to gain an understanding of their use.

1. The relation between the coefficient of performance of heat pumps, refrigerators, and air conditioners:

$$\text{COP}_{\text{HP}} = \text{COP}_{\text{R/AC}} + 1$$

2. The coefficient of performance of the reversed Carnot benchmark cycle:

$$COP_{\text{Carnot HP}} = \frac{T_H}{T_H - T_L}$$

and

$$COP_{\text{Carnot R/AC}} = \frac{T_L}{T_H - T_L}$$

3. The equation for assigning refrigerant numbers:

$$\text{R-}(a-1)(b+1)d$$

where a = number of carbon atoms, b = number of hydrogen atoms, c = number of chlorine atoms, and d = number of fluorine atoms in the refrigerant molecule. Note that when $a = 1$, the leading $a - 1 = 0$ number is omitted from the R number designation.

4. The coefficient of performance of a dual-cascade, vapor-compression refrigeration system:

$$COP_{\substack{\text{dual}\\ \text{cascade}}} = \frac{\dot{m}_B(h_{1B} - h_{4hB})}{\dot{m}_A(h_{2sA} - h_{1A})/(\eta_s)_{c-A} + \dot{m}_B(h_{2sB} - h_{1B})/(\eta_s)_{c-B}}$$

5. The coefficient of performance of a dual-stage, vapor-compression refrigeration system:

$$COP_{\text{dual-stage}} = \frac{(1 - x_{\text{flash}})(h_{1B} - h_{4hB})}{(h_{2sA} - h_{1A})/(\eta_s)_{c-A} + (1 - x_{\text{flash}})(h_{2sB} - h_{1B})/(\eta_s)_{c-B}}$$

where x_{flash} is the quality of the vapor leaving the flash chamber.

6. The coefficient of performance of a Carnot absorption refrigeration system:

$$(COP)_{\substack{\text{Carnot}\\ \text{absorption}\\ \text{refrigerator}}} = \frac{T_e}{T_g}\left(\frac{T_g - T_a}{T_a - T_e}\right)$$

where T_e is the evaporator temperature, T_g is the gas generator temperature, and T_a is the ambient temperature.

7. The coefficient of performance of a reversed Brayton cycle heat pump:

$$COP_{\substack{\text{reversed}\\ \text{Brayton cycle}\\ \text{HP}}} = \frac{T_2 - T_3}{(T_{2s} - T_1)/(\eta_s)_c - (T_3 - T_{4s})(\eta_s)_e}$$

and of a reversed Brayton cycle refrigerator or air conditioner:

$$COP_{\substack{\text{reversed}\\ \text{Braytonc ycle}\\ \text{R/AC}}} = \frac{T_1 - T_4}{(T_{2s} - T_1)/(\eta_s)_c - (T_3 - T_{4s})(\eta_s)_e}$$

If the reversed Brayton cycle operates on an air standard cycle (ASC), then these equations become

$$COP_{\substack{\text{reversed}\\ \text{Brayton ASC}\\ \text{HP}}} = \frac{1}{1 - PR^{(1-k)/k}}$$

and

$$COP_{\substack{\text{reversed}\\ \text{Brayton ASC}\\ \text{R/AC}}} = \frac{1}{PR^{(k-1)/k} - 1}$$

where PR is the compressor or expander pressure ratio.

8. The coefficient of performance of a reversed Stirling air standard cycle heat pump and refrigerator or air conditioner is the same as the reversed Carnot cycle:

$$COP_{\substack{\text{reversed Stirling}\\ \text{ASC HP}}} = \frac{T_H}{T_H - T_L}$$

and

$$\text{COP}_{\substack{\text{reversed Stirling}\\ \text{ASC R/AC}}} = \frac{T_L}{T_H - T_L}$$

9. The coefficient of performance of a Joule-Thomson air standard cycle refrigerator or air conditioner is

$$\text{COP}_{\substack{\text{J-T}\\ \text{R/AC}}} = \frac{\mu_J(p_2 - p_1)}{T_1\left[(p_2/p_1)^{(k-1)/k} - 1\right] \Big/ (\eta_s)_c}$$

10. The irreversibility rate produced by an adiabatic compressor is

$$\dot{I}_{\substack{\text{adiabatic}\\ \text{compressor}}} = T_0(\dot{S}_P)_{\substack{\text{adiabatic}\\ \text{compressor}}} = \dot{m}_{\text{ref}} T_0(s_2 - s_1)$$

the irreversibility rate produced by a condenser is

$$\dot{I}_{\text{condenser}} = T_0(\dot{S}_P)_{\text{condenser}} = T_0\left[\dot{m}_{\text{ref}}(s_3 - s_2) - \frac{\dot{Q}_{\text{condenser}}}{T_{\text{condenser}}}\right]$$

the irreversibility rate produced by an adiabatic expansion valve is

$$\dot{I}_{\substack{\text{adiabatic}\\ \text{expansion valve}}} = T_0(\dot{S}_P)_{\substack{\text{adiabatic}\\ \text{expansion valve}}} = \dot{m}_{\text{ref}} T_0(s_{4h} - s_3)$$

and the irreversibility rate produced inside an evaporator is

$$\dot{I}_{\text{evaporator}} = T_0(\dot{S}_P)_{\text{evaporator}} = T_0\left[\dot{m}_{\text{ref}}(s_1 - s_{4h}) - \frac{\dot{Q}_{\text{evaporator}}}{T_{\text{evaporator}}}\right]$$

Problems (* indicates problems in SI units)

1. Using the definition of thermal efficiency, show that the coefficient of performance (COP) of a heat pump is always greater than 1.0.
2. If the coefficient of performance of a window air conditioner in the summer is 5.70, what is the coefficient of performance of this unit if it were used as a window heat pump in the winter?
3. The Rumford Engineering company, a supplier of yours, has just announced that it is offering a new product line for closed electronic cabinet cooling. It is a small refrigeration unit driven by a 0.250 hp electric motor that removes 500. W of heat from the cabinet. What is the coefficient of performance of this unit?
4. A heat pump with a coefficient of performance of 7.30 is used to heat a small house at a rate of 40.0×10^3 Btu/h. What horsepower electric motor is required to drive the heat pump?
5.* It is proposed to operate a reversed Carnot cycle to remove 1500. W of thermal energy from a freezer at −14.0°C and to discharge heat to the environment at 20.0°C. Determine
   a. The coefficient of performance of the system.
   b. The heat transfer rate to the environment.
   c. The required input power.
6.* When the outside environmental temperature is 5.00°C, to what temperature can you heat the interior of a house with a Carnot heat pump that has a coefficient of performance of 8.90?
7.* How much power is required to drive a Carnot refrigeration unit used to remove 15.0 kW of heat from a commercial food storage unit maintained at 3.00°C when the outside temperature is 35.0°C?
8. A new plastic-molding facility has a large amount of waste heat available at 300.°F. The local environmental temperature is 75.0°F. As an entry-level engineer, you are to investigate possible energy savings by determining
   a. The thermal efficiency of a Carnot engine operating between these temperatures.
   b. The coefficient of performance of a Carnot heat pump operating between these temperatures.
   c. The coefficient of performance of a Carnot refrigerator operating between these temperatures.
9. A typical ice cube from your refrigerator contains about 1.50 in^3 of ice. Determine the tons of refrigeration produced if the ice cube melts at 32.0°F in (a) 24 hours and (b) 1 hour. Assume the density of ice at 32.0°F to be 0.0330 lbm/in^3.
10. A large commercial refrigerator has a coefficient of performance of 4.70 and is driven by a 7.00 hp electric motor. Determine the refrigeration capacity of this unit in tons of refrigeration.
11. The air conditioning unit for a building uses a water chiller with a capacity of 12.0 tons of refrigeration. Determine the coefficient of performance of this unit if it is driven by a 14.0 hp electric motor.
12. In 1937, Dr. Willis Haviland Carrier (1876–1950) developed an air conditioning system to cool the air in the Magma Copper Mine in Superior, Arizona.[9] It reduced the air temperature at the 3600 ft level from 133°F to 71.0°F after 4 months of operation. Each air conditioning unit in the system had a capacity of 140. tons of cooling and was driven by a 200. hp electric motor. Determine the coefficient of performance of these units.

[9] In 1976, this air-conditioning system was designated the 13th National Historic Mechanical Engineering Landmark by the American Society of Mechanical Engineers.

13.* A refrigeration unit is to be designed for a meat market using R-22 to maintain meat at 0.00°C while operating in an environment at 30.0°C. The refrigerant enters the condenser as a saturated vapor and exits as a saturated liquid. Determine the coefficient of performance (COP) for this refrigerator, using

a. A reversed Carnot cycle operating between these temperature limits.

b. An isentropic, vapor-compression refrigeration cycle with an aergonic, adiabatic throttling valve installed between the high-pressure condenser and the low-pressure evaporator.

14. A vapor-compression refrigeration unit uses R-22 and has an isentropic efficiency of 88.0% when working between the temperature limits of −20.0°F and 80.0°F. The refrigerant enters the condenser as a saturated vapor and exits as a saturated liquid. Determine the coefficient of performance (COP) of this unit.

15.* A vapor-compression cycle refrigeration system using R-22 has an evaporator temperature of −20.0°C and a condenser temperature of 20.0°C. The refrigerant exits the compressor as a saturated vapor and exits the condenser as a saturated liquid. Determine

a. The isentropic coefficient of performance of this system.

b. The coefficient of performance of a Carnot refrigerator operating between the same temperature limits.

c. The "tons" of isentropic refrigeration per unit mass flow rate of refrigerant.

16.* Repeat Problem 15 using R-134a instead of R-22 as the refrigerant.

17.* A vapor-compression cycle refrigerator uses R-22 as the working fluid. The evaporator temperature is −20.0°C, the condenser temperature is 40.0°C, and the flow rate of R-22 is 0.100 kg/s. If the refrigerant exits the compressor as a saturated vapor and exits the condenser as a saturated liquid, determine

a. The isentropic coefficient of performance.

b. The equivalent reversed Carnot cycle coefficient of performance.

c. The amount of refrigeration (in tons) that this system can provide.

18.* Repeat Problem 17 using R-134a instead of R-22 as the refrigerant.

19.* A vapor-compression cycle refrigerator using R-22 as the working fluid has an evaporator temperature of −30.0°C and a condenser temperature of 50.0°C. The refrigerant enters the compressor at 7.50 kg/min as a saturated vapor and exits the condenser as a saturated liquid. The isentropic efficiency of the compressor is 82.5%. Determine

a. The coefficient of performance for the system.

b. The coefficient of performance for a reversed Carnot cycle with the same temperature limits.

c. The refrigeration (in tons) that this system can provide.

20.* Repeat Problem 19 using R-134a instead of R-22 as the refrigerant.

21. The states in a vapor-compression cycle air conditioner using R-22 as the working fluid are as follows:

Station 1 Compressor inlet	Station 2s Compressor outlet	Station 3 Throttle valve inlet	Station 4h Throttle valve outlet
$T_1 = 20.0°F$	$T_{2s} = 80.0°F$	$T_3 = T_{2s} = 80.0°F$	$T_{4h} = 20.0°F$
$s_1 = s_{2s}$	$x_{2s} = 1.00$	$x_3 = 0.00$	$h_{4h} = h_3$

Determine the actual coefficient of performance of this cycle if the compressor isentropic efficiency is 79.0%.

22. A large vapor-compression cycle refrigeration manufacturer tests its units by directing the condenser heat to the evaporator and adding additional cooling as necessary to fully cool the unit. In the test unit shown Figure 14.37, the refrigerant is R-22 and the refrigerant mass flow rate is 1500. lbm/min.

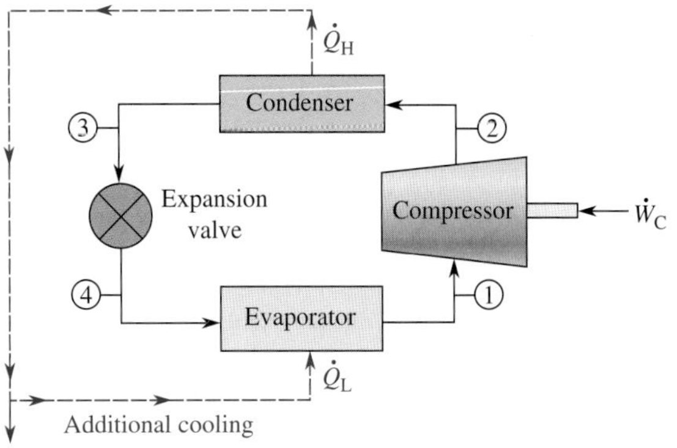

FIGURE 14.37
Problem 22.

Station 1 Compressor inlet	Station 2s Compressor isentropic outlet
$T_1 = 20.0°F$	$p_{2s} = p_3 = 98.7 \approx 100.$ psia
$p_1 = 30.0$ psia	$s_{2s} = s_1$

Station 3 Condenser outlet	Station 4h Expansion valve outlet
$p_3 = 98.87 \approx 100.$ psia	
$x_3 = 0.00$	

Assuming the isentropic efficiency of the compressor is 100.%, determine:

a. The compressor power input.

b. The cooling capacity in tons of refrigeration.

c. The unit's COP.

23. A vapor compression cycle heat pump using R-22 is used to provide 20.0 × 10^3 Btu/h of heat to a house. The evaporator temperature is 14.0°F and the condenser temperature is 70.0°F. The refrigerant exits the compressor as a saturated vapor and exits the condenser as a saturated liquid. The isentropic efficiency of the compressor is 80.0%. Determine

a. The mass flow rate of the refrigerant.

b. The power input to the compressor.

c. The coefficient of performance of this system.

24. Repeat Problem 23 using R-134a instead of R-22 as the refrigerant.

25. A vapor-compression cycle heat pump using low-pressure water as the working fluid is proposed to heat a house. The evaporator is to be buried in the ground below the frost line and consequently will remain at 50.0°F year-round. The condenser is to be inside the house and will operate at a constant 80.0°F. The water enters the condenser as a saturated vapor at 12.6 lbm/min and exits as a saturated liquid. Assume the isentropic efficiency of the compressor is 100.%. Determine

a. The coefficient of performance of this system.

b. The amount of heat (in Btu/h) transferred into the house.

26.* A vapor-compression cycle heat pump using R-134a as the working fluid is to be designed for heating a house. The evaporator is to be buried in the ground below the frost line and consequently will remain at 14.0°C year-round. The condenser is to be inside a house and will operate at a constant 20.0°C. The refrigerant enters the condenser as a saturated vapor at 7.30 kg/min and exits as a saturated liquid. Assume the isentropic efficiency of the compressor is 91.0%. Determine

a. The coefficient of performance of this system.

b. The amount of heat (in kW) transferred into the house.

27.* A vapor-compression cycle heat pump has been developed that uses R-22 as the working fluid. The evaporator is at 0.00°C and the condenser is at 30.0°C. The refrigerant enters the condenser as a saturated vapor at 0.0800 kg/s and exits as a saturated liquid. The isentropic efficiency of the compressor is 83.0%. Determine

a. The coefficient of performance of this system.

b. The power (in kW) required to drive the unit.

28.* A company wants to design a vapor-compression cycle heat pump that uses ammonia as the working fluid. The evaporator is at 0.00°C and the condenser is at 30.0°C. The refrigerant enters the condenser as a saturated vapor at 0.120 kg/s and exits as a saturated liquid. The compressor has an isentropic efficiency of 85.0%. Determine

a. The coefficient of performance of this system.

b. The amount of heat (in kW) transferred from the cold to the warm region.

29.* A special high-temperature vapor-compression cycle heat pump for a spacecraft is to be designed that uses water as the working fluid. The evaporator is at 100.°C and the condenser is at 300.°C. The refrigerant enters the condenser as a saturated vapor at 1.80 kg/min and exits as a saturated liquid. Assume that the isentropic efficiency of the compressor is 91.0%. Determine

a. The coefficient of performance of this system.

b. The amount of heat (in kW) transferred from the cold to the warm region.

30. Use Eq. (14.8) to determine the proper chemical formula and chemical name for the following refrigerants: (a) R-10, (b) R-110, and (c) R-214.

31. Find the chemical formula and chemical name for the following refrigerants using Eq. (14.8): (a) R-30, (b) R-40, and (c) R-50.

32. Use Eq. (14.8) to find the chemical formula and chemical name for the following refrigerants: (a) R-113, (b) R-114, and (c) R-123.

33. Determine the chemical formula and R number of the following refrigerants (see Eq. (14.8)): a) pentachloroethane, b) trichloroethane, and c) octafluoropropane.

34. Use Eq. (14.8) to determine the chemical name and R number of the following refrigerants: (a) $CClF_3$, (b) CHF_3, and (c) CH_2ClF.

35. Provide the chemical name and refrigerant R number of the following materials: (a) NH_3, (b) CO_2, and (c) H_2O.

36.* A dual-cascade system using R-22 in both loops is used to produce 30.0 tons of refrigeration in a large refrigeration unit with an evaporator temperature of −40.0°C and a condenser temperature of 25.0°C. The intermediate heat exchanger between the loops operates at −20.0°C, and the isentropic efficiencies of the compressors in each loop are both 83.0%. The following design specifications have been defined for the loops:

Loop A

Station 1A Compressor A inlet	Station 2sA Compressor A outlet	Station 3A Condenser A outlet	Station 4hA Expansion valve A outlet
$x_{1A} = 1.00$	$p_{2sA} = 1500.\text{ kPa}$	$x_{3A} = 0.00$	$h_{4hA} = h_{3A}$
$T_{1A} = -20.0°C$	$s_{2sA} = s_{1A}$	$T_{3A} = 25.0°C$	

Loop B

Station 1B Compressor B inlet	Station 2sB Compressor B outlet	Station 3B Condenser B outlet	Station 4hB Expansion valve B outlet
$x_{1B} = 1.00$	$p_{2sB} = 300.\text{ kPa}$	$x_{3B} = 0.00$	$h_{4hB} = h_{3B}$
$T_{1B} = -40.0°C$	$s_{2sB} = s_{1B}$	$T_{3B} = -20.0°C$	

For this design, determine

a. The mass flow rate of refrigerant in loops A and B.

b. The system's coefficient of performance.

c. The pressure ratios across both of the compressors.

37.* A new ultralow-temperature, dual-cascade refrigeration system using R-22 in both loops is used to produce 5.00 tons of refrigeration with an evaporator temperature of −60.0°C and a condenser temperature of 25.0°C. The intermediate heat exchanger between the loops operates at −20.0°C and the isentropic efficiencies of the compressors in each loop are both 75.0%. The following operating specifications have been determined for the loops:

Loop A

Station 1A Compressor A inlet	Station 2sA Compressor A outlet	Station 3A Condenser A outlet	Station 4hA Expansion valve A outlet
$x_{1A} = 1.00$	$p_{2sA} = 1500.\text{ kPa}$	$x_{3A} = 0.00$	$h_{4hA} = h_{3A}$
$T_{1A} = -20.0°C$	$s_{2sA} = s_{1A}$	$T_{3A} = 25.0°C$	

Loop B

Station 1B Compressor B inlet	Station 2sB Compressor B outlet	Station 3B Condenser B outlet	Station 4hB Expansion valve B outlet
$x_{1B} = 1.00$	$p_{2sB} = 300.\text{ kPa}$	$x_{3B} = 0.00$	$h_{4hB} = h_{3B}$
$T_{1B} = -60.0°C$	$s_{2sB} = s_{1B}$	$T_{3B} = -20.0°C$	

For this design, determine:

a. The mass flow rate of refrigerant in loops A and B.

b. The system's coefficient of performance.

c. The pressure ratios across both of the compressors.

38.* A dual-cascade system using R-22 in both loops is used to produce 300. tons of refrigeration in a ice-skating rink with an evaporator temperature of −30.0°C and a condenser temperature of 25.0°C. The intermediate heat exchanger between the loops operates at 0.00°C and the isentropic efficiencies of the compressors in each loop are both 85.0%. The following design specifications have been defined for the loops:

Loop A

Station 1A Compressor A inlet	Station 2sA Compressor A outlet	Station 3A Condenser A outlet	Station 4hA Expansion valve A outlet
$x_{1A} = 1.00$	$p_{2sA} = 1500.\text{ kPa}$	$x_{3A} = 0.00$	$h_{4hA} = h_{3A}$
$T_{1A} = 0.00°C$	$s_{2sA} = s_{1A}$	$T_{3A} = 25.0°C$	

Loop B

Station 1B Compressor B inlet	Station 2sB Compressor B outlet	Station 3B Condenser B outlet	Station 4hB Expansion valve B outlet
$x_{1B} = 1.00$	$p_{2sB} = 300.\ \text{kPa}$	$x_{3B} = 0.00$	$h_{4hB} = h_{3B}$
$T_{1B} = -30.0°\text{C}$	$s_{2sB} = s_{1B}$	$T_{3B} = 0.00°\text{C}$	

For this design, determine

a. The mass flow rate of refrigerant in loops A and B.

b. The system's coefficient of performance.

c. The pressure ratios across both of the compressors.

39.* A small, two-stage, vapor-compression refrigeration unit using R-134a produces 15.0 tons of refrigeration with an evaporator pressure of 100. kPa and a condenser pressure of 1200. kPa. The intermediate flash chamber operates at 400. kPa and the isentropic efficiencies of the compressors in each loop are both 89.0%. The following operating specifications have been developed for the stages (loops):

Stage (Loop) A

Station 1A Compressor A inlet	Station 2sA Compressor A outlet	Station 3A Condenser A outlet	Station 4hA Expansion valve A outlet
$p_{1A} = 400.\ \text{kPa}$	$p_{2sA} = 1200.\ \text{kPa}$	$x_{3A} = 0.00$	$h_{4hA} = h_{3A}$
	$s_{2sA} = s_{1A}$	$p_{3A} = 1200.\ \text{kPa}$	

Stage (Loop) B

Station 1B Compressor B inlet	Station 2sB Compressor B outlet	Station 3B Condenser B outlet	Station 4hB Expansion valve B outlet
$x_{1B} = 1.00$	$p_{2sB} = 400.\ \text{kPa}$	$x_{3B} = 0.00$	$h_{4hB} = h_{3B}$
$p_{1B} = 100.\ \text{kPa}$	$s_{2sB} = s_{1B}$	$p_{3B} = 400.\ \text{kPa}$	

For this design, determine

a. The mass flow rate of the refrigerant.

b. The system's coefficient of performance.

c. The total power required by the compressor.

40.* A large, two-stage, vapor-compression packing house refrigeration unit using R-134a produces 55.0 tons of refrigeration with an evaporator pressure of 80.0 kPa and a condenser pressure of 1.00 MPa. The intermediate flash chamber operates at 400. kPa and the isentropic efficiencies of the compressors in each loop are both 92.0%. The following operating specifications have been determined for the stages (loops):

Stage (Loop) A

Station 1A Compressor A inlet	Station 2sA Compressor A outlet	Station 3A Condenser A outlet	Station 4hA Expansion valve A outlet
$p_{1A} = 400.\ \text{kPa}$	$p_{2sA} = 1.00\ \text{MPa}$	$x_{3A} = 0.00$	$h_{4hA} = h_{3A}$
	$s_{2sA} = s_{1A}$	$p_{3A} = 1.00\ \text{MPa}$	

Stage (Loop) B

Station 1B Compressor B inlet	Station 2sB Compressor B outlet	Station 3B Condenser B outlet	Station 4hB Expansion valve B outlet
$x_{1B} = 1.00$	$p_{2sB} = 400.\ \text{kPa}$	$x_{3B} = 0.00$	$h_{4hB} = h_{3B}$
$p_{1B} = 80.0\ \text{kPa}$	$s_{2sB} = s_{1B}$	$p_{3B} = 400.\ \text{kPa}$	

For this design, determine

a. The mass flow rate of the refrigerant.

b. The system's coefficient of performance.

c. The total power required by the compressor.

41.* A very large, two-stage, vapor-compression refrigeration unit using R-134a produces 3000 tons of refrigeration with an evaporator pressure of 100 kPa and a condenser pressure of 2000 kPa. The intermediate flash chamber operates at 800 kPa, and the isentropic efficiencies of the compressors in each loop are both 91%. The following operating specifications have been determined for the stages (loops):

Stage (Loop) A

Station 1A Compressor A inlet	Station 2sA Compressor A outlet	Station 3A Condenser A outlet	Station 4hA Expansion valve A outlet
$p_{1A} = 800.\ \text{kPa}$	$p_{2sA} = 2000.\ \text{kPa}$	$x_{3A} = 0.00$	$h_{4hA} = h_{3A}$
	$s_{2sA} = s_{1A}$	$p_{3A} = 2000.\ \text{kPa}$	

Stage (Loop) B

Station 1B Compressor B inlet	Station 2sB Compressor B outlet	Station 3B Condenser B outlet	Station 4hB Expansion valve B outlet
$x_{1B} = 1.00$	$p_{2sB} = 800.\ \text{kPa}$	$x_{3B} = 0.00$	$h_{4hB} = h_{3B}$
$p_{1B} = 100.\ \text{kPa}$	$s_{2sB} = s_{1B}$	$p_{3B} = 800.\ \text{kPa}$	

For this design, determine

a. The mass flow rate of the refrigerant.

b. The system's coefficient of performance.

c. The total power required by the compressor.

42. Laboratory measurements on an absorption refrigeration system produced the following results: the evaporator absorbs 15.0×10^3 Btu/h while 14.0×10^3 Btu/h of heat is added to the generator. The carrier liquid pump requires 0.500 hp to operate during the test. Determine the coefficient of performance of this unit.

43.* A new absorption refrigeration system is designed to operate in a hazardous environment, where the temperature is 40.0°C. If the generator temperature is 100.0°C and the evaporator temperature is 20.0°C, determine the Carnot absorption refrigeration coefficient of performance of this unit.

44. If the Carnot absorption refrigeration coefficient of performance of a new refrigeration unit is 4.20 and the environmental and generator temperatures are 70.0°F and 300.°F, respectively, determine the cooling (evaporator) temperature.

45.* A new household refrigerator-freezer combination unit is designed with a dual-evaporator system. The freezer compartment is at −20.0°C and the refrigerator compartment is at 4.00°C. The outlet of the condenser is at 28.0°C. The cooling capacities of both the refrigeration and the freezer compartments are to be 400. kJ/h each. The system uses refrigerant R-134a with a compressor isentropic efficiency of 87.0%. Determine (a) the coefficient of performance for this design and (b) the mass flow rate of refrigerant required.

46.* A new commercial refrigerator-freezer combination unit is being designed with a dual-evaporator system. The freezer compartment is to be at −24.0°C and the refrigerator compartment is to be at 8.00°C. The outlet of the condenser is at 26.0°C. The cooling capacities of both the refrigerating and the freezer compartments are to be 800. kJ/h each. The system uses refrigerant R-134a with a compressor isentropic efficiency of 89.0%. Determine

a. The coefficient of performance for this design.

b. The mass flow rate of refrigerant required.

c. The quality at the outlet of the refrigeration evaporator.

47.* A new refrigerator-freezer combination unit is being designed with a dual-evaporator system to be used in a recreational vehicle. The freezer compartment is to be at −14.0°C and the refrigerator compartment is to be at 8.0°C. The outlet of the condenser is at 20.0°C. The cooling capacities of both the refrigerator and the freezer compartments are to be 200. kJ/h each. The system will use refrigerant R-134a with a compressor isentropic efficiency of 70.0%. Determine

a. The coefficient of performance for this design.
b. The mass flow rate of refrigerant required.
c. The quality at the outlet of the refrigeration evaporator.

48.* A reversed Brayton cycle with an isentropic pressure ratio of 1.75 is to be used to refrigerate a food locker. The refrigerant is air. The compressor and expander inlet temperatures are 0.00°C and 14.0°C, respectively. Determine

a. The ASC coefficient of performance of this system.
b. The Carnot coefficient of performance for a refrigerator operating between the same temperature limits.
c. The compressor and expander outlet temperatures.

49. In calculating the reversed Carnot ASC coefficient of performance for comparison with that of a reversed Brayton ASC, the temperature limits are always taken as T_L = "the compressor inlet temperature," and T_H = "the expander inlet temperature." (a) Why is this done? (b) What would happen if the cycle limit temperatures were used instead (i.e., taking T_L = "the expander outlet temperature" and T_H = "the compressor outlet temperature")?

50. A reversed Brayton ASC refrigerator operates with air between 0.00°F (the compressor inlet temperature) and 80.0°F (the expander inlet temperature) with an isentropic pressure ratio of 2.85. Determine the ASC coefficient of performance of this system, assuming (a) constant specific heats and (b) temperature-dependent specific heats (use Table C.16 in *Thermodynamic Tables to accompany Modern Engineering Thermodynamics*).

51. Consider the possibility of converting an automobile turbocharger with radiator intercooling at 100.0°F into a reversed Brayton ASC air conditioning system. For inlet conditions of 14.7 psia at 70.0°F and a compressor pressure ratio of 2.00, determine

a. The ASC coefficient of performance of this system as an air conditioner.
b. The coldest possible air conditioning temperature attainable with this system.

52. The states in a reversed Brayton cycle air conditioner using air (a constant specific heat ideal gas) as the working fluid are as follows:

Station 1 Compressor inlet	Station 2*s* Compressor outlet	Station 3 Expander inlet	Station 4*s* Expander outlet
$T_1 = 80.0°\text{F}$	$T_{2s} = 280.°\text{F}$	$T_3 = 120.°\text{F}$	$T_{4s} = -36.0°\text{F}$
$p_1 = 14.0\,\text{psia}$	$p_{2s} = 42.0\,\text{psia}$	$p_3 = 42.0\,\text{psia}$	$p_{4s} = 14.0\,\text{psia}$

Determine the actual coefficient of performance of this cycle if the compressor and expander isentropic efficiencies are 0.790 and 0.880, respectively. Note: *Not* an ASC analysis.

53. Show that Eqs. (14.22) and (14.23) for the reversed Brayton cycle obey Eq. (14.1).

54.* 1.75 kg/s of air at 300. K enters the compressor of a reversed Brayton cycle heat pump. The isentropic pressure ratio of the compressor is 3.00 to 1, and the inlet temperature of the expander is 335 K. The isentropic efficiencies of the compressor and expander are 91.0% and 85.0%, respectively. Determine

a. The actual power output from the expander.
b. The actual power input to the compressor.
c. The coefficient of performance of the unit.

55. Show that the coefficient of performance of a reversed Brayton ASC heat pump can be written as

$$\text{COP}_{\substack{\text{reversed}\\\text{Brayton ASC}\\\text{HP}}} = \frac{(\text{CR})^{k-1}}{(\text{CR})^{k-1} - 1}$$

where CR is the compression ratio of the system.

56. Using the notation of Figure 14.27 and beginning with the relation

$$\text{COP}_{\substack{\text{reversed}\\\text{Carnot ASC}\\\text{R/AC}}} = \frac{T_L}{T_H - T_L} = \frac{T_1}{T_3 - T_1}$$

show that the minimum possible isentropic pressure ratio of a reversed Brayton ASC refrigerator is

$$\text{PR}_{\text{minimum}} = \left(\frac{T_3}{T_1}\right)^{\frac{k}{k-1}}$$

57.* In 1862, Carnegie Kirk attained a temperature of −40.0°C with a reversed Stirling cycle refrigerator. Assuming the environmental temperature was 20.0°C, determine the reversed Stirling ASC coefficient of performance for his unit.

58.* A reversed Stirling cryogenic refrigerator is used to produce a temperature of −200.°C in an environment at 20.0°C. Determine the reversed Stirling ASC coefficient of performance of this unit.

59. Reversed Stirling cycle air conditioners were used to cool deep mines until about 1930. If the temperature in the mine was 150.°F and the temperature of the air outside the mine was 60.0°F, determine the reversed Stirling ASC coefficient of performance for the air conditioner.

60. A reversed Stirling cycle heat pump is driven by a 14.0 hp electric motor. It produces a heat transfer rate of 75.0×10^3 Btu/h into the high temperature region at 80.0°F from an environment at 50.0°F. Determine

a. The actual coefficient of performance of the heat pump.
b. The reversed Stirling ASC coefficient of performance for this unit.

61.* What power is required to drive a reversed Stirling cycle heat pump with a high-temperature heat transfer rate of 14.0 kW over the temperature difference 30.0°C and 0.00°C, if the actual coefficient of performance is 50.0% of the ASC coefficient of performance for this unit?

62. Determine the cooling temperature and the coefficient of performance of an isenthalpic expansion cooler that expands air from 1500. psia, 70.0°F to 14.7 psia. The isentropic efficiency of the compressor is 85.0%. The mean Joule-Thomson coefficient for this process is 0.0200°F/psi.

63.* When compressed air expands from 20.0°C, 100. atm to 1.00 atm, the mean Joule-Thomson coefficient is 0.150°C/atm. Determine the outlet temperature and the coefficient of performance of an expansion air conditioning system operating under these conditions when the compressor isentropic efficiency is 80.0%.

64* A Joule-Thomson expansion refrigeration system is being considered for use in a meat-storage facility. The working fluid is to be carbon dioxide that is to be expanded from 20.0 atmospheres at 20.0°C to 1.00 atm. The system is to have a compressor with an isentropic efficiency of 75.0%. Determine the cooling temperature and the coefficient of performance of this system.

65. A newly formulated gas is found to have a Joule-Thomson coefficient of 0.500°F/psi. If this gas expands from 70.0°F at 100. psia to atmospheric pressure, determine the exhaust temperature and the coefficient of performance of this system if the compressor has an isentropic efficiency of 88.0%.

66.* A highly toxic gas is discovered to have a negative Joule-Thomson coefficient of −8.00°C/atmosphere. Since this coefficient is negative, the gas heats rather than cools on expansion, and thus it can be made into a heat pump. Determine

- **a.** The exhaust temperature as this gas expands from 20.0°C at 100. atm to 1.00 atm.
- **b.** The coefficient of performance of this unit as a heat pump if it is compressed with an isentropic efficiency of 65.0%.

67* Determine the irreversibility rate produced by an adiabatic compressor that compresses 0.150 kg/s of refrigerant R-134a from 0.00°C at 0.100 MPa to 80.0°C at 0.500 MPa. The local environmental temperature is 20.0°C.

68.* Refrigerant R-22 flows through a condenser at a rate of 2.70 kg/s. It enters the condenser as a superheated vapor at 1.00 MPa and 50.0°C and exits as a saturated liquid at 25.0°C. The average temperature of the condenser is 30°C and the local environmental temperature is 20.0°C. Determine the irreversibility rate for this condenser.

69.* Find the irreversibility rate of an adiabatic expansion valve that reduces 0.0660 kg/s of refrigerant R-22 from a saturated liquid at 25.0°C to a liquid-vapor mixture with a quality of 15.0% at 0.00°C. The local environmental temperature is 20.0°C.

Design Problems

The following are open-ended design problems. The objective is to carry out a preliminary thermal design as indicated. A detailed design with working drawings is not expected unless otherwise specified. These problems have no specific answers, so each student's design is unique.

70. Design a small vapor-compression cycle refrigeration system that can serve as an experimental apparatus for a junior or senior mechanical engineering laboratory course. The system must be instrumented with the proper pressure, temperature, and mass flow transducers, so that its coefficient of performance can be accurately determined. The system should have at least one variable parameter (such as refrigerant mass flow rate) to provide a range of performance to study. You may wish to start by modifying the components of an existing domestic refrigerator. Either construct the apparatus yourself or else provide sufficiently accurate drawings (or sketches) and instructions that it can be made by an engineering technician.

71. Carry out the preliminary thermal design of a vapor-compression cycle heat pump that uses a solar collector as the heat source. Use Refrigerant-134a as the working fluid, and assume an average solar flux of 496 Btu per square foot per day in December (the worst case) with a yearly average solar flux of 1260 Btu per square foot per day. The heat pump must provide $400. \times 10^3$ Btu per day to a house during December and average $200. \times 10^3$ Btu per day during the year. Be sure to determine the following items in your analysis:

- **a.** The required collector surface area for worst case and average conditions. (Is either too large for an average roof?)
- **b.** The resulting system coefficient of performance.
- **c.** The required mass flow rate of R-134a.

Note: Assume a solar flux (sunshine) period of 8 to 10 hours per day.

72. Design a domestic heating system that uses a heat pump to extract heat from the earth to heat a house. The evaporator is to be made of long lengths of plastic pipe buried below the frost line at a constant temperature of 50.0°F. The heat pump system must provide $200. \times 10^3$ Btu/h to the house at 70.0°F. Specify the refrigerant; the length, diameter, and type of plastic to be used for the evaporator piping; the mass flow rate of the refrigerant; and the compressor efficiency. Also determine the pumping losses (i.e., the pressure drop) in the evaporator. Estimate or compute an appropriate temperature differential between the outside and the inside of the condenser and the evaporator.

73. Design a small, laboratory-scale, reversed Brayton cycle air conditioning system that can serve as an experimental apparatus for a junior or senior mechanical engineering laboratory course. The system must be instrumented with the proper pressure, temperature, and mass flow transducers so that its coefficient of performance can be accurately determined. The system should have at least one variable parameter to provide a range of performance to study. You may wish to start by modifying an automotive turbocharger to provide the turbine and compressor stages. Either construct the apparatus yourself or else provide sufficiently accurate and detailed drawings and instructions that it can be made by an engineering technician.

74. Design an inexpensive reversed Brayton cycle air conditioning system for an automobile using air as the working fluid. Convert one of the engine's cylinders into an air compressor or add a separate compressor driven off the fan belt. Determine the amount of cooling required (in tons) when the outside air temperature is 100.°F and the inside temperature is maintained at 70.0°F. Size and locate the components on the automobile. Estimate the unit's coefficient of performance, its input power requirement, and manufacturing cost. The final unit must add no more than $200.00 to the cost of the automobile to the consumer.

Computer Problems

The following computer programming assignments are designed to be carried out on a personal computer using a spreadsheet, equation solver, or programming language. They may be used as part of a weekly homework assignment.

75. Develop a computer program that provides data to plot (either manually or with a computer) the coefficient of performance of a vapor-compression cycle heat pump, refrigerator, or air conditioner vs. the throttling valve outlet quality, x_{4h}. Prompt the user for all the relevant input information, including the compressor's isentropic efficiency.
76. Develop an interactive computer program that determines either the system's coefficient of performance or its cooling capacity in tons (allow the user to choose which) of a vapor-compression cycle air conditioner, refrigerator, or heat pump (again allow the used to choose which). Prompt the user for all necessary information (in proper units) and produce a screen diagram of the system with all the variables and the unknowns shown.
77.* Develop an interactive reversed Brayton ASC computer program that utilizes constant specific heat ideal gas equations of state as the source of the enthalpy values. Allow the user to select either a refrigeration, air conditioning, or heat pump application. Have the user input the appropriate gas constants (or choose the gas from a screen menu), the source and sink temperatures, the mass flow rate of the gas, and the isentropic pressure ratio, PR. Output the coefficient of performance, the source and sink heat transfer rates, and the net power input. Use this program to plot the coefficient of performance vs. the PR for a reversed Brayton cycle refrigerator operating between 20.0°C and −14.0°C. Allow the PR to range from 1.00 to 14.0.
78.* Expand Problem 77 by replacing the ideal gas properties with a gas menu that contains a computerized version of the isentropic gas tables as the source of the enthalpy values.
79. Develop an interactive computer program that determines the coefficient of performance of a Joule-Thomson refrigeration system using air or carbon dioxide (allow the user to choose which). Curve fit the information given in Figure. 6.6 as the source of the proper Joule-Thomson coefficient. Prompt the user for the appropriate temperatures, pressures, and isentropic compressor efficiency.
80. Develop an interactive computer program that determines the coefficient of performance of a reversed Stirling ASC refrigerator or heat pump (allow the user to choose which). Prompt the user for all necessary input information (in the proper units).

Writing to Learn Problems

Provide a coherent 500-word written response to the following questions on $8\frac{1}{2}$ by 11 in paper, double spaced, 12 point font, with 1 in margins on all sides. Unless your instructor indicates otherwise, your response should include the following:

a. An opening thesis statement containing the argument you wish to support.
b. A body of supporting material.
c. A conclusion section in which you use the supporting material to substantiate your thesis statement.

81. Describe how you think the introduction of food refrigeration by artificial ice used in kitchen iceboxes in the mid 19th century could improve the standard of living in a society.
82. While water makes a very good working fluid for power cycles, it does not make a good refrigerant. Describe the reasons for this and discuss the evolution of refrigerant working fluids. What problems are produced by leaking refrigerants?
83. Describe the controversy over CFCs and the ozone layer. Do you believe it is true?
84. Describe the assumptions behind the air standard cycle. Which are the most severe and which are the least severe?
85. Describe what you see as future refrigeration needs in the next 100 years (i.e., what will the refrigeration need be 100 years from now, in your opinion)?

CHAPTER 15

Chemical Thermodynamics

CONTENTS

15.1 EINFÜHRUNG (INTRODUCTION)

This chapter deals with an application of the laws of thermodynamics on which entire textbooks have been written. Chemists call this topic *physical chemistry*, and it forms the basis of much of applied chemistry. It is important to engineers, because it provides a fundamental understanding of the combustion process in engines, power plants, fuel cells, and other chemically based energy conversion processes.

This chapter has three main goals. Our first goal is to be able to calculate the amount of heat produced in the combustion of an organic fuel. The second is to understand the basic elements of chemical equilibrium and dissociation, and the third is to look at the emerging field of fuel cell technology. To be able to discuss these subjects adequately, we need to define what we mean by the term *fuel*, we need to decide how the thermodynamic properties of the products and the reactants are related through a *standard reference state*, and we need to lay the foundations for discussing chemical reaction energy conversion efficiency.

IS IT CHEMISTRY OR ALCHEMY?

A practical chemical technology can be traced to prehistoric times. Primitive metallurgy, medicine, and food preparation are typical examples. They were purely empirical "recipe" driven processes with no form of chemical theory to explain their results. Before the sixth century BC, it was generally believed that all things were composed of a single primitive element. The Greek philosophers Heraclitus (540–480 BC) and Empedocles (490–430 BC) began a new era when they proposed that, instead of a single element, all matter was made up of four elements—air, earth, fire, and water—and that the continual mixing of these elements formed all the objects of the real world.

The Greek philosopher Pythagoras (580–500 BC) is generally credited with recognizing the functional significance of numbers in quantifying the processes of the real world. Pythagoras established an academy of learning in Crotona, Italy, in about 532 BC. The academy prospered long after his death until its destruction in about 390 BC. It is believed that his disciples (known as *Pythagoreans*) working at the academy during this time developed many of the mathematical discoveries now attributed to him (e.g., the Pythagorean theorem for right triangles).

Empedocles adopted Pythagoras' numerology technique in an attempt to quantify the chemistry of his four elements. For example, according to Empedocles, animal bone consisted of two parts water, two parts earth, and four parts fire. Because he believed that all of his four elements were most thoroughly mixed in blood, he concluded that people think mainly with their blood.

From about the second century AD until nearly the 19th century, the world embraced what is considered today to be a scientific and chemical curiosity, *alchemy*. Alchemy was a combination of the occult, astrology, and primitive chemistry. Even its name, derived from Arabic and introduced in the 12th century, is obscure because the root *chem* seems to have no relevant etymological meaning.

The basic function of alchemy was *transmutation*, which was concerned with transmuting age to youth, sickness to health, death to immortality. More notorious was its preoccupation with the physical transmutation of base metals into gold (i.e., transmuting poverty to wealth). Its central elements were mercury (*quicksilver*, the liquid metal), sulfur (the stone that burns), and ammonium chloride (*sal ammoniac*, a source of hydrochloric acid). Successful alchemists tended to be charlatans whose work was shrouded in mystery. From the Medieval period forward, the central focus of alchemy was the making of gold (religion had successfully taken over the immortality issue), and many prominent scientists, including Isaac Newton (1643–1726), experimented with it seriously. (Newton is thought to have contracted mercury poisoning in about 1690 as a result of his alchemy experiments.) The false science of alchemy, which appealed primarily to the human weakness of greed, went without serious intellectual challenge for nearly 2000 years.

In the 17th century, Johann Jochim Becher (1635–1682), an established alchemist at one time engaged in attempting to transmute Danube River sand into gold, proposed that all substances were made up of the classical alchemical elements of mercury, sulfur, and corrosive salts, plus a new fourth weightless element that was *produced* by combustion. In 1697, the German physician Georg Ernst Stahl (1660–1734) named this supposed fourth element *phlogiston* and used it to develop a coherent theory of combustion, respiration, and corrosion. His phlogiston theory quickly won universal scientific approval and was the only scientifically accepted theory of matter for nearly 100 years afterward.

In 1774, the English clergyman and scientist Joseph Priestley (1733–1804) described some of his experimental results in removing phlogiston from air (the "dephlogistication" of air) to the French chemist Antione Laurent Lavoisier (1743–1794), who immediately recognized their importance and subsequently carried out similar experiments himself. Lavoisier soon realized that the dephlogiston that Priestley thought he had been working with was actually a unique chemical and, in 1777, he named it *oxygen* (from the Greek for *acid forming*). By the early 19th century Lavoisier's oxidation theory completely replaced Stahl's phlogiston theory and the era of modem chemistry had begun. Lavoisier had done for chemistry what a century earlier Newton had done for mechanics; he put the subject on a firm analytical foundation. The unfortunate Joseph Priestley was subsequently driven from his home in England because of his public support of the French Revolution, and he settled in America.

WHO KILLED THE GREAT FRENCH CHEMIST LAVOISIER?

Because Lavoisier maintained a career in the French government as well as science (he was on the original 1790 weights and measures committee that led to the development of the metric system we now use), he was caught up in the French Revolution and was accused of political crimes (such as stopping the circulation of air in Paris by a city wall erected at his suggestion in 1787). He was convicted and guillotined on the same day in 1794.

15.2 STOICHIOMETRIC EQUATIONS

In the early 19th century, the English chemist John Dalton (1766–1844) devised a system of chemical symbols and determined the relative masses of some elemental atoms. He also formulated a theory that combinations of different chemical elements occur in simple mass ratios, which led him to the development of a way of writing a chemical formula that mathematically represented chemical reactions. For example, if elements *A* and *B* combine in a two to one ratio by mass to form chemical *C*, Dalton wrote this as

$$2 \text{ atoms of } A + 1 \text{ atom of } B = 1 \text{ atom of } C \tag{15.1}$$

In modern notation, this would simply be

$$\underbrace{2A + B}_{\text{Reactants}} \rightarrow \underbrace{C}_{\text{Product}}$$

where the equality has been replaced by an arrow that indicates the direction of the reaction. The items on the left side of this equation are called the *reactants*, and those on the right side are called the *products* of the reaction.

A description of the net combining properties of atoms and compounds that occur in a chemical reaction is known today as the *stoichiometry* of the reaction. Dalton's chemical equation notation provides a shorthand mathematical version of such a description, and the numerical values that precede the chemical symbols in these equations are called the *stoichiometric coefficients* of the reaction. These coefficients represent the number of atoms or molecules involved in the reaction, and since mass is conserved in ordinary chemical reactions, the number of atoms of each chemical element must be the same in both the reactants and the products. Therefore, a chemical equation can be *balanced* (i.e., mass balanced) by requiring stoichiometric coefficients that produce the same number of atoms of each chemical species on both sides of the reaction equation. For example, the reaction that occurs in burning hydrogen to completion in a pure oxygen atmosphere can be written in modern notation as

$$a(H_2) + b(O_2) \rightarrow c(H_2O)$$

where *a*, *b*, and *c* are the stoichiometric coefficients for the reaction. An individual atomic species balance now gives

$$\text{Atomic hydrogen (H) balance: } 2a = 2c$$

$$\text{Atomic oxygen (O) balance: } 2b = c$$

thus producing two equations in the three unknowns, *a*, *b*, and *c*. Since such reactions are usually of interest per unit mass of fuel supplied, we can arbitrarily set $a = 1$: then, the atomic hydrogen and oxygen balances gives $c = 1$ and $b = \frac{1}{2} = 0.5$. Therefore, our final balanced equation would read

$$H_2 + 0.5(O_2) \rightarrow H_2O \tag{15.2}$$

After an extensive period of experimentation, the Italian chemist Count Amado Avogadro (1776–1856) proposed in 1811 that equal volumes of different gases at the same temperature and pressure contain equal numbers of molecules.

WHAT DOES THE WORD *STOICHIOMETRY* MEAN?

The term *stoichiometry* comes from the Greek words *stoicheion* (component) and *metron* (measure). It was introduced in 1792 by the German chemist Jeremias Benjamin Richter, when he suggested that substances react chemically according to relations that resemble mathematical formulae.

WHAT IS AVOGADRO'S LAW?

Avogadro's law states that equal volumes of different gases at the same temperature and pressure contain equal numbers of molecules. The *Avogadro constant*, N_A (originally called *Avogadro's number*), is the number of atoms in exactly 12 kg of carbon-12. The 2006 value is

$$N_A = 6.022\,1 \times 10^{26} \text{ atoms/kgmole} = 2.731\,6 \times 10^{26} \text{ atoms/lbmole}$$

Although Avogadro introduced this law in 1811, it was not generally accepted by the scientific community until after 1858. Incidentally, André Marie Ampère (1775–1836) popularized the term *molecule* for an assembly of atoms in about 1814.

With this proposition, he was able to show that hydrogen, oxygen, nitrogen, and the like exist as diatomic molecules in nature. Modern experiments have determined that the actual number of molecules in any substance whose mass in kilograms is equal to its (relative) molecular mass M is 6.022×10^{26}. Therefore, the mass of one molecule of this substance is

$$\frac{M}{6.022 \times 10^{26}} \text{ kilograms.}$$

The acceptance of Avogadro's law soon led to the development of the *mole* (sometimes abbreviated *mol,* both being a contraction of the German word *molekul*) as a convenient chemical mass unit. Originally a "mole" was defined as a mass *in grams* that was equal to the molecular mass M of a substance (i.e., originally a "mole" of carbon-12 contained 12 grams of carbon-12). Today, we need to recognize that this "mole" is really a *gram-mole,* or *gmole,* because the "mole" unit is used in so many different units systems today. Now, it must carry a prefix showing the units system being used, such as *gmole, kgmole,* or *lbmole.* But remember that these "moles" are not equal to each other, since 1 kgmole = 1000 gmole = 2.2046 lbmole.

Since equal "moles" of different substances contain the same number of molecules, the stoichiometric coefficients, which initially represented only individual molecules, can also be used to represent the number of moles of each element present. Therefore, Eq. (15.1) can be written in the equivalent form

$$2 \text{ gmole of } A + 1 \text{ gmole of } B = 1 \text{ gmole of } C$$

or

$$2 \text{ kgmole of } A + 1 \text{ kgmole of } B = 1 \text{ kgmole of } C$$

or

$$2 \text{ lbmole of } A + 1 \text{ lbmole of } B = 1 \text{ lbmole of } C$$

and Eq. (15.2) can be interpreted as a reaction between 1 kgmole of H_2 and 0.5 kgmole of O_2 producing 1 kgmole of H_2O or 1 lbmole of H_2 and 0.5 lbmole of O_2 producing 1 lbmole of H_2O and so forth.

Most combustion processes occur in air, not pure oxygen. The composition of air used to determine its thermodynamic properties on a molar or volume basis is 78.09% nitrogen, 20.95% oxygen, 0.93% argon, and 0.03% carbon dioxide and trace elements. For convenience, we round this off to 79.0% N_2 and 21.0% O_2. In doing this, we are essentially dividing air into two components: pure oxygen and a mixture of noncombustibles (N_2, Ar, CO_2). The noncombustible group, called *atmospheric nitrogen,* has a mole fraction composition of

$$\chi_{N_2} = \frac{n_{N_2}}{n_{N_2} + n_{Ar} + n_{CO_2}} = \frac{78.09}{78.09 + 0.93 + 0.03} = \frac{78.09}{79.05} = 0.9879 = 98.79\%$$

$$\chi_{Ar} = \frac{n_{Ar}}{n_{N_2} + n_{Ar} + n_{CO_2}} = \frac{0.93}{79.05} = 0.0118 = 1.18\%$$

and

$$\chi_{CO_2} = \frac{n_{CO_2}}{n_{N_2} + n_{Ar} + n_{CO_2}} = \frac{0.03}{79.05} = 0.00038 = 0.038\%$$

where n_{N_2}, n_{Ar}, and n_{CO_2} are the number of moles of nitrogen, argon, and carbon dioxide present in the mixture. Then, from Eq. (12.11), the equivalent molecular mass of this mixture is

$$(M)_{\substack{\text{atmospheric}\\ \text{nitrogen}}} = \sum \chi_i M_i = (0.9879)(28.016) + (0.0118)(39.944) + (0.00038)(44.01)$$
$$= 28.16 \text{ kg/kgmole} = 28.16 \text{ lbm/lbmole}$$

From this point on, we refer to the atmospheric nitrogen mixture as simply nitrogen, and we assume that air has a molar composition of 21.0% oxygen and 79.0% nitrogen, where the molecular mass of oxygen is still 32.00, but the molecular mass of nitrogen is now 28.16 instead of 28.016. The equivalent molecular mass of air is still 28.97, since the argon and carbon dioxide are now merely grouped with the nitrogen.

For this air composition, each mole of oxygen is accompanied by 79.0/21.0 = 3.76 moles of nitrogen. Thus, if the hydrogen combustion reaction described by Eq. (15.2) were carried out in air instead of pure oxygen, it would be written as

$$H_2 + 0.5[O_2 + 3.76(N_2)] \rightarrow H_2O + 1.88(N_2)$$

In this equation, the nitrogen is assumed to be inert and therefore passes through the reaction unchanged.

CHEMICAL REACTION EQUATIONS AND SIGNIFICANT FIGURES

We have a problem using significant figures in the equations for chemical reactions. For example, in the reaction of hydrogen and oxygen to form water, $H_2 + 0.5\,O_2 \rightarrow H_2O$, we do not want to have to write it as $1.00\,H_2 + 0.500\,O_2 \rightarrow 1.00\,H_2O$. So, in this chapter, we *assume* that the coefficients in an equation for a chemical reaction have at least three significant figures without actually writing them as such in the reaction equations.

The amount of air or oxygen used in a combustion process can be described as

1. The percent of theoretical air or oxygen required to carry out the reaction.
2. The percent of excess or deficit air or oxygen actually used in the reaction.
3. The air/fuel (A/F) or fuel/air (F/A) ratio used in the reaction measured on either a mass or a mole basis.

One hundred percent *theoretical air* is the minimum amount of air that supplies enough oxygen to carry out complete combustion. The percentage of *excess air* is simply the percentage of theoretical air supplied minus 100, and the percentage of *deficit air* is 100 minus the percent of theoretical air supplied. The *air/fuel ratio* (A/F) is the amount of air used per unit of fuel consumed, and it can be expressed either in mass or mole units. The relation between the mass and molar air/fuel ratios is given by

$$(A/F)_{\text{mass}} = [(A/F)_{\text{molar}} \text{ in moles of air/mole of fuel}] \times \left(\frac{\text{Molecular mass of air, } M_{\text{air}}}{\text{Molecular mass of fuel, } M_{\text{fuel}}}\right)$$

The *fuel/air ratio* (F/A) is simply the inverse of the air/fuel ratio.

$$(F/A)_{\text{molar}} = \frac{1}{(A/F)_{\text{molar}}}$$

and

$$(F/A)_{\text{mass}} = \frac{1}{(A/F)_{\text{mass}}}$$

With these definitions, it is easy to see that the hydrogen combustion described in the reaction $H_2 + 0.5[O_2 + 3.76(N_2)] \rightarrow H_2O + 1.88(N_2)$ uses 100% theoretical air and no excess or deficit air is involved. The molar air/fuel ratio for this reaction is

$$(A/F)_{\text{molar}} = \frac{n_{\text{air}}}{n_{\text{fuel}}} = \frac{0.5 \times (1 + 3.76) \text{ moles of air}}{1 \text{ mole of fuel}} = 2.38 \text{ moles of air/mole } H_2$$

and the mass air/fuel ratio is[1]

$$\begin{aligned}(A/F)_{\text{mass}} &= [(A/F)_{\text{molar}} \text{ moles of air/mole of fuel}] \times \left(\frac{\text{Molecular mass of air, } M_{\text{air}}}{\text{Molecular mass of fuel, } M_{\text{fuel}}}\right)\\ &= 2.38 \text{ moles of air/mole of fuel } (H_2) \times \frac{28.97 \text{ Mass of air/mole of air}}{2.016 \text{ Mass of fuel/mole of fuel}}\\ &= 34.2 \text{ grams of air/gram of } H_2 = 34.2 \text{ kg of air/kg of } H_2 = 34.2 \text{ lbm of air/lbm of } H_2\end{aligned}$$

Then, the molar and mass fuel/air ratios are simply

$$(F/A)_{\text{molar}} = (A/F)^{-1}_{\text{molar}} = 0.42 \text{ moles } H_2/\text{mole air}$$

and

$$\begin{aligned}(F/A)_{\text{mass}} = (A/F)^{-1}_{\text{mass}} &= 0.029 \text{ gram } H_2/\text{gram air}\\ &= 0.029 \text{ kg } H_2/\text{kg air}\\ &= 0.029 \text{ lbm } H_2/\text{lbm air}\end{aligned}$$

---

[1] While the mole is in fact a unit of mass, in this section we use the term *mass* to indicate nonmolar units.

If this reaction is carried out with 150% theoretical air (i.e., 50% excess air), it has the form[2]

$$H_2 + 1.5 \times (0.5)[O_2 + 3.76(N_2)] \rightarrow H_2O + 0.25(O_2) + 2.82(N_2)$$

with a molar air/fuel ratio of

$$(A/F)_{molar} = \frac{n_{air}}{n_{fuel}} = \frac{1.5 \times (0.5)(1 + 3.76)}{1} = 3.57 \text{ moles air/mole } H_2$$

Similarly, if it is carried out at 75% theoretical air (i.e., 25% deficit air), it is

$$H_2 + 0.75 \times (0.5)[O_2 + 3.76(N_2)] \rightarrow 0.75(H_2O) + 0.25(H_2) + 1.41(N_2)$$

with a molar air/fuel ratio of

$$(A/F)_{molar} = \frac{n_{air}}{n_{fuel}} = \frac{0.75 \times (0.5)(1 + 3.76)}{1} = 1.785 \text{ moles air/mole } H_2$$

and so forth.

15.3 ORGANIC FUELS

The term *organic* has been used in chemistry since the late 18th century and originally referred only to materials occurring in or derived from living organisms. Today, this term is used to represent *all* compounds of carbon, whether derived from living organisms or not. Other elements frequently found in organic compounds are hydrogen, oxygen, nitrogen, sulfur, and phosphorus. Since the number of organic compounds is very large, they are subdivided into groups having similar properties. Hydrocarbons, alcohols, carbohydrates, proteins, and fats are typical organic compound subdivisions.

Many common organic fuels are made up of only carbon and hydrogen atoms and are consequently called *hydrocarbons*. The hydrocarbon class of organic molecules can be further subdivided into the groups shown in Figure 15.1.

Using the atomic mass balance technique discussed at the beginning of this chapter, it is easily shown that the stoichiometric reaction equation for the combustion of a typical hydrocarbon of the form C_nH_m using 100.% theoretical air is

$$C_nH_m + (n + m/4)[O_2 + 3.76(N_2)] \rightarrow n(CO_2) + (m/2)(H_2O) + 3.76(n + m/4)(N_2) \tag{15.3a}$$

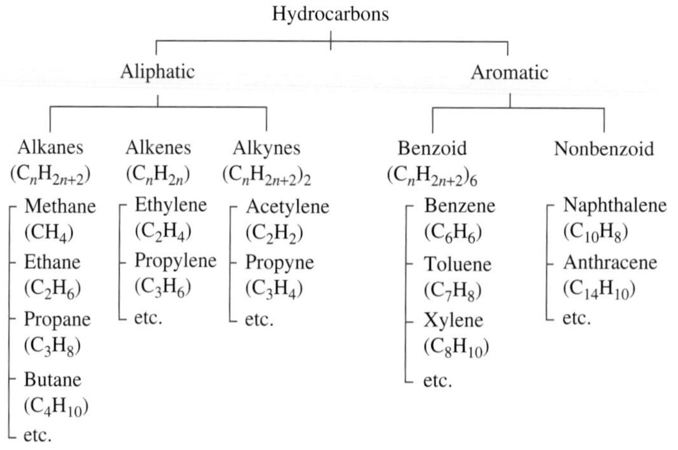

FIGURE 15.1
Classification of hydrocarbons.

[2] Remember, in this chapter we assume each coefficient in an equation for a chemical reaction has at least three significant figures without actually writing them as such in the reaction equation.

ANSWERS SOMETIMES COME IN DREAMS!

In the early part of the 19th century, chemists were puzzled by the fact that it was possible to construct two seemingly different compounds that had dissimilar physical properties yet identical chemical formulae. For example, ethanol and dimethyl ether both have the same chemical formula, C_2H_6O, yet ethanol boils at 79°C while dimethyl ether boils at −24°C. Materials that have the same chemical formula but dissimilar physical properties are called *isomers*. The isomer puzzle was solved by the German chemist Friedrich August Kekulé (1829–1896) in a dream while dozing on the top deck of a horse-drawn bus in London. In his dream, he realized that the atoms of a molecule could be arranged in different geometric structures; and in 1858, he introduced the schematic notation, still used, in which bonds between atoms are represented by lines drawn between their corresponding chemical symbols (e.g., H_2 as H—H). He was then able to show that isomers were simply the result of different bonding patterns. For example, butane (C_4H_{10}) has the isomers *n*-butane, which boils at −0.5°C, and isobutane, which boils at −12°C. (The *n* prefix is always used to denote the *normal*, or chainlike, structure, whereas the *iso* prefix is used to denote the branched structure.) These two isomer bonding patterns are shown in Figures 15.2 and 15.3.

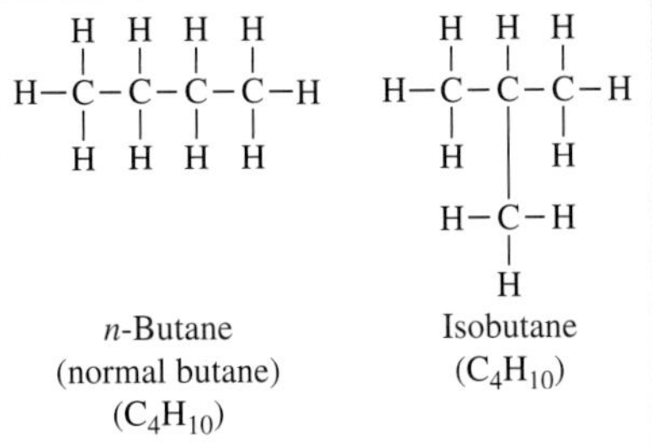

FIGURE 15.2
n-Butane and isobutane.

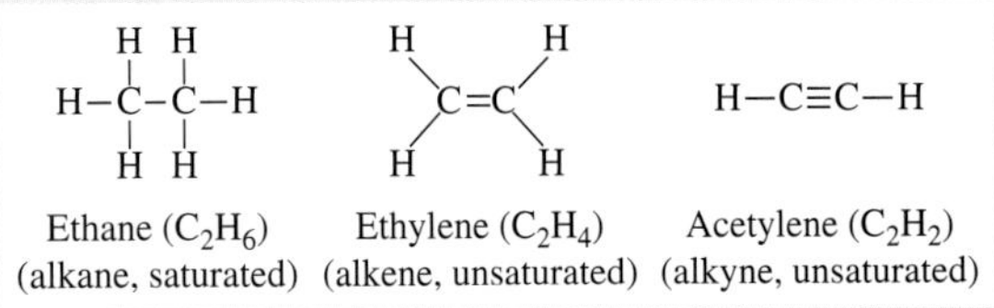

FIGURE 15.3
Ethane (alkane, saturated), ethylene (alkene, unsaturated), and acetylene (alkyne, unsaturated).

Another 19th century hydrocarbon curiosity was the existence of the two classes, *aliphatic* (fatty) and *aromatic* (fragrant) compounds. Aromatic hydrocarbons always had at least six carbon atoms and a smaller proportion of hydrogen atoms than the aliphatic hydrocarbons. In 1865, Kekulé again found the solution in a dream. He envisioned a six-carbon chain closing on itself to form a ring, like a snake biting its own tail, and he concluded that the aromatic compounds contain such rings whereas the aliphatic compounds contain only straight chains.

Within the aliphatic group, the *alkanes* are characterized by having carbon atoms with single bonds between them, while the *alkenes* have carbon atoms with double bonds between them, and *alkynes* have carbon atoms with triple bonds between them. If all the bonds within an organic compound are *single*, then the compound is said to be *saturated*; but if multiple bonds exist between any two carbon atoms in the compound, it is said to be *unsaturated*. Thus, the alkanes are all saturated hydrocarbons, while all the remaining hydrocarbons are unsaturated.

Also, its combustion with excess air using 100.(x) percent theoretical air (i.e., 100.($x-1$) percent excess air), where $x \geq 1.0$, is

$$\begin{aligned} C_nH_m + x(n+m/4)[O_2+3.76(N_2)] \rightarrow\ & n(CO_2)+(m/2)(H_2O) \\ & +(x-1)(n+m/4)(O_2) \\ & +x(3.76)(n+m/4)(N_2) \end{aligned} \tag{15.3b}$$

And its combustion in deficit air using 100.(y) percent theoretical air (i.e., 100.($1-y$) percent deficit air), where $(2n+m)\div(4n+m)\leq y \leq 1.0$, is

$$\begin{aligned} C_nH_m + y(n+m/4)[O_2+3.76(N_2)] \rightarrow\ & [n(2y-1)-m(1-y)/2](CO_2) \\ & +(2n+m/2)(1-y)(CO)+(m/2)(H_2O) \\ & +y(3.76)(n+m/4)(N_2) \end{aligned} \tag{15.3c}$$

In Eq. (15.3c), it has been assumed that the hydrogen is much more reactive than the carbon, so it will take up all the oxygen it needs to be converted into water. This leaves only the carbon subject to incomplete combustion.

EXAMPLE 15.1

Determine the stoichiometric reaction equation for methane (CH_4) burned in

a. 100.% theoretical air.
b. 150.% excess air.
c. 20.0% deficit air.

Solution

We could solve this problem for methane, $CH_4 = C_nH_m$ by setting $n = 1$ and $m = 4$ in Eqs. (15.3a–c). However, since methane is a simple compound, it is more enlightening to carry out the individual atomic balances to obtain the correct reaction equations.

a. The general combustion equation for 1 mole of methane in 100% theoretical air is

$$CH_4 + a[O_2 + 3.76(N_2)] \rightarrow b(CO_2) + c(H_2O) + d(N_2)$$

The element[3] balances are

Carbon (C) balance: $1 = b$
Hydrogen (H_2) balance: $2 = c$
Oxygen (O_2) balance: $a = b + c/2 = 1 + 2/2 = 2$
Nitrogen (N_2) balance: $a(3.76) = d = 2(3.76) = 7.52$

The resulting stoichiometric equation for 100.% theoretical air is

$$CH_4 + 2[O_2 + 3.76(N_2)] \rightarrow CO_2 + 2(H_2O) + 7.52(N_2)$$

b. The 150.% excess air corresponds to 250.% theoretical air. The reaction equation now has O_2 in the products and consequently has the form

$$CH_4 + 2.5(2)[O_2 + 3.76(N_2)] \rightarrow a(CO_2) + b(H_2O) + c(O_2) + d(N_2)$$

and again element balances can be used to find that $a = 1$, $b = 2$, $c = 3$, and $d = 2.5(2)(3.76) = 18.8$; so that, for 150.% excess air, we have

$$CH_4 + 5[O_2 + 3.76(N_2)] \rightarrow CO_2 + 2(H_2O) + 3(O_2) + 18.8(N_2)$$

c. The 20.0% deficit air corresponds to 80.0% theoretical air. Again, assuming all the hydrogen reacts to water, the reaction now has CO in the products and has the form

$$CH_4 + 0.8(2)[O_2 + 3.76(N_2)] \rightarrow a(CO_2) + b(CO) + c(H_2O) + d(N_2)$$

and again the element balances can be used to yield the coefficients $a = 0.2, b = 0.8, c = 2.0$, and $d = 6.016$. Note that these results correspond to the same coefficients one would obtain using Eq. (15.3c). The final reaction equation for 20.0% deficit air is

$$CH_4 + 1.6[O_2 + 3.76(N_2)] \rightarrow 0.2(CO_2) + 0.8(CO) + 2(H_2O) + 6.016(N_2)$$

Exercises

1. Determine the number of kgmoles of water produced in the reaction of Example 15.1 per kgmole of methane burned in 200.% excess air. **Answer**: 2 kgmoles per kgmole of CH_4.
2. Determine the molar and mass A/F ratios for parts a, b, and c in Example 15.1. **Answer:**
   a. $(A/F)_{molar} = 9.52$ lbmole air/lbmole $CH_4 = 9.52$ kgmole air/kgmole CH_4
   $(A/F)_{mass} = 17.24$ lbm air/lbm $CH_4 = 17.24$ kg air/kg CH_4
   b. $(A/F)_{molar} = 23.8$ lbmole air/lbmole $CH_4 = 23.8$ kg air/kg CH_4
   $(A/F)_{mass} = 43.1$ lbm air/lbm $CH_4 = 43.1$ lbm air/lbm CH_4
   c. $(A/F)_{molar} = 7.616$ lbmole air/lbmole $CH_4 = 7.616$ kg air/kg CH_4
   $(A/F)_{mass} = 13.79$ lbm air/lbm $CH_4 = 13.79$ lbm air/lbm CH_4
3. Rework Example 15.1 for combustion in 200% theoretical air. **Answer:** $CH_4 + 4[O_2 + 3.76(N_2)] \rightarrow CO_2 + 2(H_2O) + 2(O_2) + 15.04(N_2)$.

[3] *In modern chemistry, the term* element *refers to the stable form of a substance composed of only one kind of atom. The chemically stable forms of carbon, hydrogen, oxygen, and nitrogen in these reactions are C, H_2, O_2, and N_2, rather than C, H, O, and N. So, even though H_2, O_2, and N_2 are really diatomic molecules, they are considered to be the proper forms for these elements in common chemical reactions.*

Hydrocarbon fuels refined from petroleum normally contain a mixture of many organic components. Gasoline, for example, is a mixture of over 30 compounds. It is, however, convenient to model these fuels as a single *average* hydrocarbon compound of the form C_nH_m, as discussed in the following section.

15.4 FUEL MODELING

It is fairly easy to obtain accurate composition analysis of combustion products with modern gas chromatography or mass spectroscopy techniques. With an accurate combustion analysis of a fuel that is in reality a complex mixture of hydrocarbons, an *equivalent* or *average* hydrocarbon model of the form C_nH_m can be determined from a chemical element balance. For example, if the combustion products contain only CO_2, CO, O_2, H_2O, and N_2, then Eqs. (15.3a–c) could be used to determine the composition parameters n and m when the stoichiometric coefficients of the products are measured. Since the fuel model formula C_nH_m represents an average of all the different hydrocarbon compounds present in the fuel mixture, n and m usually do not turn out to be integers and the resulting model does not represent any real hydrocarbon (except possibly when n and m are rounded to integers).

EXAMPLE 15.2

A new hydrocarbon fuel is being developed that consists of 1.00 kgmole of methane (CH_4) mixed with 3.00 kgmoles of propane (C_3H_8). Determine the hydrocarbon fuel model for this mixture.

Solution

This mixture is assumed to produce 1.00 kgmole of the fuel model C_nH_m, so

$$1\,CH_4 + 1\,C_3H_8 = 1\,C_nH_m$$

An element balance for this equation gives

$$\text{Carbon balance: } 1 + 3(3) = 10 = n$$

$$\text{Hydrogen balance: } 4 + 3(8) = 28 = m$$

Consequently, the hydrocarbon fuel model is $C_{10}H_{28}$.

Exercises

4. Determine the hydrocarbon fuel model in Example 15.2 if only 2.00 kgmoles of propane are used in the mixture. **Answer**: Hydrocarbon fuel model = C_7H_{20}.
5. Determine the hydrocarbon fuel model in Example 15.2 if ethane (C_2H_6) is used in place of the methane. **Answer**: Hydrocarbon fuel model = $C_{11}H_{30}$.
6. 1.00 lbmole each of methane (CH_4), propane (C_3H_8), and butane (C_4H_{10}) are mixed to form 1.00 lbmole of a new super fuel. Determine the hydrocarbon fuel model for this mixture. **Answer**: Hydrocarbon fuel model = C_8H_{22}.

Often, you do not know the exact hydrocarbon composition of a fuel. However, you *can* analyze the products of the fuel's combustion and deduce the fuel model. Several modern instruments produce an accurate exhaust gas analysis. For example, gas chromatography and mass spectrometry are commonly used in exhaust gas analysis today. But perhaps the quickest, simplest, and most inexpensive method of obtaining an approximate combustion analysis is with an Orsat analyzer. This gas analyzer uses a chemical absorption technique to determine the volume fractions (which are equivalent to the mole fractions) of CO_2, CO, and O_2 in the exhaust gas (see Figure 15.4). Since it cannot measure the H_2O content, the exhaust gas sample is always cooled to room temperature, or below the dew point of any water vapor present, so that most of the water in the combustion products condenses out. Therefore, the Orsat technique is said to produce a *dry products* analysis. Also, the Orsat technique cannot detect unburned hydrocarbons (typically CH_4) and free hydrogen (H_2) in the exhaust gas. These are usually small and can normally be neglected. However, studies have shown that the mole fractions of methane and hydrogen in the combustion products of a hydrocarbon can be approximated as $\chi_{CH_4} \approx 0.0022$ and $\chi_{H_2} \approx 0.5(x_{CO})$, and these relations can be used with an Orsat analysis if necessary.

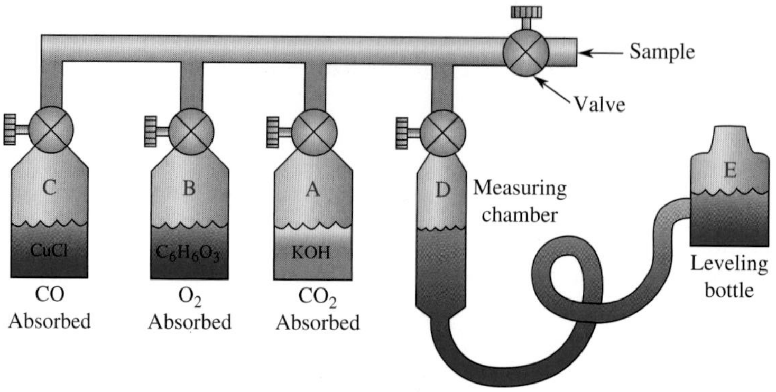

(a)

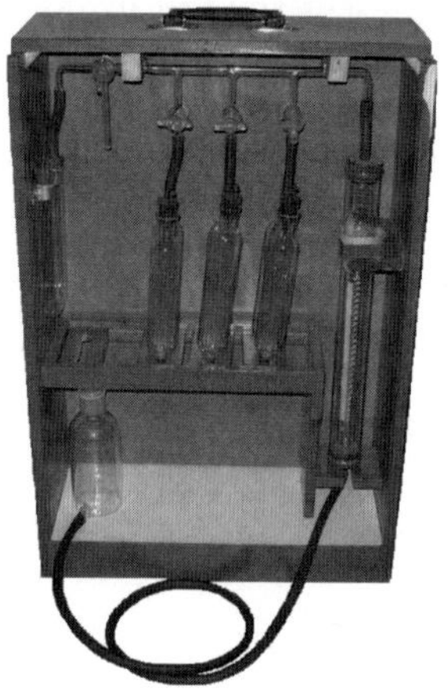

(b)

FIGURE 15.4

A typical Orsat analyzer. (a) Schematic: Vessel *A* contains a potassium hydroxide (KOH) solution that absorbs CO_2. Vessel *B* contains a pyrogallic acid (1,2,3-trihydroxybenzene, $C_6H_6O_3$) solution that absorbs O_2. Vessel *C* contains a cuprous chloride (CuCl) solution that absorbs CO. The remaining gas is assumed to be N_2. Vessel *D* is the measuring chamber, and vessel *E* is the leveling bottle. (b) Photograph of a typical Orsat analyzer.

EXAMPLE 15.3

The exhaust gas of a gasoline-fueled automobile engine is cooled to 20.0°C and subjected to an Orsat analysis. The results (on a volume or mole basis) are

$$\begin{aligned} CO_2 &= 7.10\% \\ CO &= 0.800\% \\ O_2 &= 9.90\% \\ N_2 &= 82.2\% \\ \hline \text{Total} &= 100.\% \end{aligned}$$

Determine

a. The hydrocarbon model (C_nH_m) of the fuel.
b. The composition of the fuel on a molar and a mass basis.
c. The air-fuel ratio on a molar and a mass basis.
d. The % of theoretical air used in the combustion process.

Solution

a. Since the Orsat analysis is carried out at 20.0°C, we assume that virtually all the water of combustion has condensed out and therefore the composition given is on a dry basis. However, the water term must be left in the chemical reaction equation, since it results from the oxidation of the hydrogen in the fuel. For convenience, we write the combustion reaction for 100. moles of dry product formed by burning 1.00 mole of the fuel model C_nH_m. Using the given combustion analysis, we have

$$C_nH_m + a[O_2 + 3.76(N_2)] \rightarrow 7.10(CO_2) + 0.800(CO) + 9.90(O_2) + b(H_2O) + 82.2(N_2)$$

The element balances are

Carbon (C) balance: $n = 7.10 + 0.800 = 7.90$
Hydrogen (H) balance: $m = 2b$
Nitrogen (N_2) balance: $3.76a = 82.2$, or $a = 82.2/3.76 = 21.9$
Oxygen (O_2) balance: $a = 21.9 = 7.10 + 0.800/2 + 9.90 + b/2$, or $b = 9.00$.

Then, from the preceding hydrogen balance, $m = 2b = 18.0$. Consequently, the fuel model is $C_{7.90}H_{18.0}$, which is approximately octane, C_8H_{18}. The final reaction equation is

$$C_{7.90}H_{18.0} + 21.9[O_2 + 3.76(N_2)] \rightarrow 7.10(CO_2) + 0.800(CO) + 9.90(O_2) + 9.00(H_2O) + 82.2(N_2)$$

b. On a molar basis, 1.00 mole of fuel contains 7.90 moles of C and 18.0 moles of H, and on a molar percentage basis, this becomes $[7.90/(7.90 + 18.0)](100.) = 31.0\%$ C and $[18.0/(7.90 + 18.0)](100) = 69.0\%$ H. The molecular mass of the fuel in this model is

$$M_{fuel} = 7.90(12) + 18.0(1) = 113 \text{ kg/kgmole} = 113 \text{ lbm/lbmole}$$

and so the fuel's composition on a mass basis is

$$(7.90 \text{ kgmole C/kgmole fuel})(12.0 \text{ kg C/kgmole C})/(113 \text{ kg fuel/kgmole fuel})$$
$$= 0.840 \text{ kg C/kg fuel} = 0.840 \text{ lbm C/lbm fuel}$$

and

$$(9.00)(2.016)/113 = 0.161 \text{ kg H/kg fuel} = 0.161 \text{ lbm H/lbm fuel}$$

Therefore, the fuel can be said to consist of 31% carbon and 69% hydrogen on a molar basis or 84% carbon and 16% hydrogen on a mass basis.

c. Referring to the final combustion equation determined in part a, the air/fuel ratio on a molar basis is

$$(A/F)_{molar} = \frac{n_{air}}{n_{fuel}} = \frac{21.9 \times (1 + 3.76) \text{ moles of air}}{1 \text{ mole of fuel}} = 104 \text{ moles air/mole fuel}$$

and on a mass basis it is

$$(A/F)_{mass} = (104 \text{ kgmole air/kgmole fuel}) \times \left(\frac{28.97 \text{ kg air/kgmole air}}{113 \text{ kg fuel/kgmole fuel}}\right)$$
$$= 26.7 \text{ kg air/kg fuel} = 26.7 \text{ lbm air/lbm fuel}$$

d. To determine the percent of theoretical air used, we must first determine the minimum air required for complete combustion. The reaction for 100.% theoretical air has the form

$$C_{7.90}H_{18.0} + a[O_2 + 3.76(N_2)] \rightarrow b(CO_2) + c(H_2O) + d(N_2)$$

The element balances are

Carbon (C) balance: $7.90 = b$
Hydrogen (H) balance: $18.0 = 2c$, or $c = 11.0$
Oxygen (O_2) balance: $a = b + c/2 = 7.90 + 9.00/2 = 12.4$
Nitrogen (N_2) balance: $3.76a = d = 3.76(12.4) = 46.6$

Then the theoretical molar air/fuel ratio (for 100% theoretical air) is

$$(A/F)_{\substack{molar\\theoretical}} = \frac{12.4(1 + 3.76)}{1} = 59.0 \text{ mole air/mole fuel}$$

finally, the percent of theoretical air used in the actual combustion process is

$$\% \text{ of theoretical air} = \left[(A/F)_{\substack{molar\\actual}} \Big/ (A/F)_{\substack{molar\\theoretical}}\right] \times 100$$
$$= [104/59.0](100) = 177\%$$

or 77.0% excess air.

(Continued)

EXAMPLE 15.3 (*Continued*)

Exercises

7. Determine the hydrocarbon fuel model in Example 15.3 if the CO and O_2 concentrations are both 0.00%, and the CO_2 and N_2 concentrations are 17.0% and 83.0%, respectively, in the Orsat analysis. **Answer**: Hydrocarbon fuel model = $C_{17}H_{20.3}$.
8. Determine the molar and mass air/fuel ratios in Example 15.3 if the CO and O_2 concentrations are both 0.00%, and the CO_2 and N_2 concentrations are 17.0% and 83.0%, respectively, in the Orsat analysis. **Answer**: $A/F = 10.15$ kgmole air/kgmole fuel = 1.31 kg air/kg fuel.
9. If the CO concentration in Example 15.3 is 0.00% and the N_2 concentration is 83.0%, with all the other concentrations unchanged, determine the hydrocarbon fuel model and % of excess air used in the combustion process. **Answer**: Hydrocarbon fuel model = $C_{7.1}H_{22.28}$ and % of excess air = 74.2%.

In the previous example, we assume that nearly all the water produced by the combustion process condensed out by the time the combustion products cooled to 20.0°C. For this to be a valid assumption, the dew point of the combustion products must be at 20.0°C or higher. The determination of the dew point temperature for this reaction is illustrated in the next example.

EXAMPLE 15.4

Determine the dew point temperature of the combustion products given in Example 15.3 if the total pressure of the mixture is 14.7 psia.

Solution

From Eq. (12.23) of Chapter 12, the volume fractions, mole fractions, and partial pressure ratios are all equal for a mixture of ideal gases. Exhaust products at atmospheric pressure are sufficiently ideal to allow us to determine the water vapor partial pressure at its condensation temperature (i.e., dew point) from this relation. The total number of moles of product, from part a of Example 15.3, is 109 moles. Then, using Eq. (12.23) wherein p_m is the total pressure of the mixture gives

$$\pi_{H_2O} = p_{H_2O}/p_m = \psi_{H_2O} = \chi_{H_2O} = \frac{9.00}{109} = 0.0826$$

where π is the partial pressure ratio, ψ is the volume fraction, and χ is the mole fraction. So,

$$p_{H_2O} = 0.0826(14.7) = 1.21\ \text{psia}$$

The saturation temperature of water vapor at this pressure is defined to be the dew point temperature. By interpolation in Table C.1a in *Thermodynamic Tables to accompany Modern Engineering Thermodynamics*, we find that

$$T_{sat}(1.21\ \text{psia}) = T_{DP} = 108°\text{F} = 42.3°\text{C}$$

Thus, the exhaust products must be cooled to 108°F (42.3°C) or below to condense the water of combustion and have an essentially dry exhaust gas.

Exercises

10. Determine the partial pressure of the water vapor in Example 15.4 if the mixture total pressure is 0.150 MPa. **Answer**: p_{H_2O} = 12.4 kPa.
11. If the dew point temperature in Example 15.4 is 212°F, what is the mixture total pressure? **Answer**: $p_m = 178$ psia.
12. Determine the partial pressure and dew point temperature of the water vapor present in the 100.% theoretical air combustion process given in part d of Example 15.3. Assume the mixture total pressure is 14.7 psia. **Answer**: p_{H_2O} = 2.08 psia, T_{DP} = 128°F.

WHY DO AUTOMOBILE EXHAUST SYSTEMS RUST?

Water condenses in an automobile's exhaust system and drips out the tailpipe until the entire exhaust system has been heated above the dew point temperature by the exhaust gases. This water promotes corrosion and causes the exhaust system to rust out sooner if the vehicle is used for short trips than trips long enough (a half hour or more) to dry out the exhaust system.

If moisture enters the combustion process as humidity in the inlet air, this moisture is carried through the reaction as an inert element and adds to the combustion water in the products. This has the net effect of raising the dew point temperature. This is illustrated in the next example.

EXAMPLE 15.5

During the automobile engine fuel combustion test discussed in Example 15.3, the dry bulb and wet bulb temperatures of the inlet air were measured to be 90.0°F and 75.0°F, respectively. Determine (a) the amount of water carried into the engine in the form of inlet humidity and (b) the new dew point temperature of the exhaust products. Assume the exhaust is at a total pressure of 14.7 psia.

Solution

From the psychrometric chart, Figure D.6a of *Thermodynamic Tables to accompany Modern Engineering Thermodynamics*, we find that, for $T_{DB} = 90°F$ and $T_{WB} = 76°F$, the relative humidity $\phi = 50\%$ and the humidity ratio, $\omega = (105$ grains of H_2O per lbm of dry air) × (1 lbm/7000. grains) = 0.0150 lbm H_2O/lbm dry air. On a molar basis, the humidity ratio is

$$\omega = (0.0150\,\text{lbm}\,H_2O/\text{lbm dry air})\left(\frac{28.97\,\text{lbm dry air/lbmole dry air}}{18.016\,\text{lbm}\,H_2O/\text{lbmole}\,H_2O}\right)$$

$$= 0.0241\,\text{lbmole}\,H_2O/\text{lbmole dry air}$$

From the balanced reaction equation of part a of Example 15.3, we find that the amount of dry air used per mole of fuel is 21.9(1 + 3.76) = 104 moles, and this now carries with it 0.0241(104) = 2.51 moles of water. Assuming this water passes through the reaction unchanged, the total amount of water now in the exhaust is 9.00 + 2.51 = 11.5 moles per mole of fuel. Consequently, the total moles of product are 111.5, the mole fraction of water vapor in the exhaust is now

$$\chi_{H_2O} = \frac{n_{H_2O}}{n_{\text{total}}} = 11.5/111.5 = 0.103$$

and Eq. (12.23) gives the partial pressure of the water vapor in the exhaust as

$$p_{H_2O} = 0.103(14.7) = 1.52\,\text{psia}$$

Again, interpolating in Table C.1a in *Thermodynamic Tables to accompany Modern Engineering Thermodynamics*, we find

$$T_{\text{sat}}(1.52\,\text{psia}) = T_{DP} = 116°\text{F} = 46.5°\text{C}$$

Exercises

13. What happens to the water vapor in the engine's exhaust in Example 15.5 if the surrounding air temperature is 20.0°C? **Answer**: It condenses into liquid water, since the surrounding air temperature is less than the dew point temperature of the water vapor.
14. If the inlet air in Example 15.5 contains 3.214 moles of water vapor per mole of fuel burned (instead of 2.50 moles of H_2O per mole of fuel), determine the new dew point temperature. **Answer**: $T_{DP} = 118°F$.
15. If the inlet air in Example 15.5 has a relative humidity of 100.% and a dry bulb temperature of 90.0°F, what is the new exhaust dew point temperature? **Answer**: $T_{DP} = 123°F$.

By comparing the results of Examples 15.4 and 15.5, we see that combustion air with 50.0% relative humidity has a dew point temperature 7.6°F (4.2°C) higher than that of dry combustion air.

15.5 STANDARD REFERENCE STATE

Because we deal with a variety of elements and compounds in combustion reactions, it is necessary to define a common thermodynamic reference state for all these substances. Recall that, in developing the steam tables, we chose the triple point of water as the reference state and arbitrarily set the specific internal energy of liquid water equal to *zero* at that point. Therefore, the values of u and h in the steam tables are not the *actual* specific internal energies and enthalpies of steam, they are only *relative* values. This is sufficient, since most of our formulae use $u_2 - u_1$ or $h_2 - h_1$ for changes occurring within a system and the effect of the reference state cancels out in the subtraction process. In the case of the gas tables, we take 0 K and 1 atm as the thermodynamic reference state and arbitrarily set the specific internal energy equal to *zero* at this state. However, in the case of combustion processes, a more pragmatic thermodynamic reference state of 25.0°C and 0.100 MPa (approximately 1 atm) is chosen. This is called the *standard reference state* (SRS) for combustion reactions. However, since most of the calorimeters used to study combustion processes are steady state, steady flow, open systems, it is more convenient to set the specific *enthalpy* rather than the specific internal energy of the elements equal to *zero* at this state.

WHAT IS THE STANDARD REFERENCE STATE?

The standard reference state (SRS) is defined by the following temperature and pressure:

$$\text{SRS temperature} = T^\circ = 25.0^\circ\text{C} = 298\,\text{K} = 77.0^\circ\text{F} = 537\,\text{R}$$

$$\text{SRS pressure} = p^\circ = 0.100\,\text{MPa} = 14.5\,\text{psia} \approx 1\,\text{atm}$$

Consequently, the specific internal energies of the elements at the SRS are always negative and computed from $u^\circ = -p^\circ v^\circ$, where $p^\circ = 0.100$ MPa and v° is the corresponding specific volume of the element in question. Thermodynamic properties at the standard reference state are always denoted by a superscript °.

15.6 HEAT OF FORMATION

When a reaction gives off or liberates heat, the reaction is said to be *exothermic*, and when it absorbs heat, it is said to be *endothermic*. Our sign convention for heat transport of energy requires that $Q_{\text{exothermic}} < 0$ whereas $Q_{\text{endothermic}} > 0$. The *heat of formation* of a compound is the heat liberated or absorbed in the reaction when the compound is formed from the stable form of its elements at the standard reference state. For example, if the elements and the resulting compound are both at the standard reference state, then we can write

$$\text{Elements (at the SRS)} \rightarrow \text{Compound (at the SRS)} + \left(\bar{q}_f^\circ\right)_{\text{compound}}$$

where $\left(\bar{q}_f^\circ\right)_{\text{compound}}$ is the molar heat of formation of the compound at the standard reference state.

In 1840, the Swiss chemist Germain Henri Hess (1802–1850) discovered that the total amount of heat liberated or absorbed during a chemical reaction is independent of the thermodynamic path followed by the reaction. This is known as *Hess's law* or the *law of constant heat sums*. It allows us to determine heats of formation for compounds that cannot be synthesized directly from their elements.

For example, the complete combustion of a hydrocarbon compound of the form C_nH_m in pure oxygen, wherein the reactants and the products are both maintained at the standard reference state, can be written as

$$C_nH_m + a(O_2) \rightarrow n(CO_2) + (m/2)(H_2O) + HHV_{C_nH_m}$$

where HHV is the higher heating value of the hydrocarbon (defined later, see Tables 15.2 and 15.3). We also have the following carbon dioxide and water formation reactions:

$$C + O_2 \rightarrow CO_2 - 393.5\,\text{MJ/kgmole}\,CO_2$$

and

$$H_2 + (1/2)(O_2) \rightarrow H_2O - 285.8\,\text{MJ/kgmole}\,H_2O$$

Now Hess's law states that the heats liberated or absorbed in these reactions are independent of the reaction path, so we can rearrange them as

$$CO_2 \rightarrow C + O_2 + 393.5\,\text{MJ/kgmole}\,CO_2$$

WHAT IS HESS'S LAW?

Using caloric theory, Henri Hess tried to extend Dalton's interpretation of chemical reactions by attempting to find examples of the combination of caloric with chemical elements in simple mass ratios. He discovered that, for a given reaction, the total amount of caloric (heat) involved was always the same, independent of the number of intermediate steps contained within the reaction. Today, we know that this is really true only for aergonic, steady state, steady flow, open systems and for isobaric, closed systems where the heat of reaction equals the change in total enthalpy (because enthalpy is a point function and therefore independent of the actual chemical path taken by the reaction).

and

$$H_2O \rightarrow H_2 + (1/2)(O_2) + 285.8\ \text{MJ/kgmole}\ H_2O$$

Then, the combustion equation for the compound C_nH_m can be written as

$$\begin{aligned} C_nH_m + a(O_2) &\rightarrow n(C + O_2 + 393.5) + (m/2)(H_2 + (1/2)O_2 + 285.8) + \text{HHV}_{C_nH_m} \\ &\rightarrow n(C) + (m/2)(H_2) + (n + m/4)(O_2) \\ &\quad + \left[n(393.5) + (m/2)(285.8) + \text{HHV}_{C_nH_m}\right] \end{aligned}$$

Now, an oxygen balance on the original compound combustion equation gives $a = n + m/4$, so the O_2 terms in the previous equation cancel, and again using Hess's law to rearrange this equation, we get

$$\begin{aligned} n(C) + (m/2)(H_2) &\rightarrow C_nH_m - \left[n(393.5) + (m/2)(285.8) + \text{HHV}_{C_nH_m}\right] \\ &\rightarrow C_nH_m + \left(\bar{q}_f^{\circ}\right)_{C_nH_m} \end{aligned} \tag{15.4}$$

where the HHV is in MJ/kgmole of compound. Consequently, the heat of formation in MJ/kgmole of the hydrocarbon fuel C_nH_m at the standard reference state is

$$\left(\bar{q}_f^{\circ}\right)_{C_nH_m} = -\left[n(393.5) + (m/2)(285.5) + \text{HHV}_{C_nH_m}\right] \text{ in MJ/kgmole} \tag{15.5}$$

The use of this relation is illustrated in the following example.

EXAMPLE 15.6

To prevent the Universe from collapsing in a deadly hypergeometric spiral, you must quickly determine the heat of formation of methane gas $CH_4(g)$ at the standard reference state. Normally, you would react carbon and hydrogen gas in your laboratory to form methane and measure its heat of formation during the reaction. Unfortunately, there is no known reaction by which you can form methane by reacting solid carbon with hydrogen gas. How will you save the Universe?

Solution

Even though we do not know how to form $CH_4(g)$ from a direct reaction of solid carbon and hydrogen gas, we can use Eqs. (15.4) and (15.5) with Tables 15.2 and 15.3 to calculate the heat of formation of $CH_4(g)$ at the standard reference state. Tables 15.2 and 15.3 give the higher heating value (HHV) of CH_4 as −890.4 MJ/kgmole. Then Eq. (15.5) gives

$$\begin{aligned} (\bar{q}_f^{\circ})_{CH_4} &= -\left[n(393.5) + (m/2)(285.5) + \text{HHV}_{C_nH_m}\right] \\ &= -\left[1(393.5) + (4/2)(285.5) + \text{HHV}_{CH_4}\right] \\ &= -\left[393.5 + 2(285.8) + (-890.4)\right] = -74.7\ \text{MJ/kgmole of } CH_4 \end{aligned}$$

and then Eq. (15.4) becomes

$$C(s) + 2[H_2(g)] \rightarrow CH_4(g) - 74.7\ \text{MJ/kgmole of } CH_4$$

In this example, we denote the physical state of the substances in parentheses as solid (s), liquid (ℓ), or gas (g). Note that the negative sign on the HHV of methane indicates that the combustion of methane is an exothermic (heat-producing) reaction.

Exercises

16. The methane in Example 15.6 is replaced by acetylene gas. You must now determine the heat of formation of acetylene gas, $C_2H_2(g)$, at the standard reference state to save the Universe. **Answer**: $(\bar{q}_f^{\circ})_{\text{acetylene}} = +227$ MJ/kgmole.
17. Oops, it is not methane or acetylene gas in Example 15.6, it is ethylene gas. So now, to save the Universe, you must determine the heat of formation of ethylene gas, $C_2H_4(g)$, at the standard reference state. **Answer**: $(\bar{q}_f^{\circ})_{\text{ethylene}} = -52.4$ MJ/kgmole.
18. Well, I bet you are surprised to find out that, at the last minute, it was ethane gas that was actually used to generate the deadly hypergeometric collapse of the Universe. To save all life in the Universe, you must now determine the heat of formation of ethane gas, $C_2H_6(g)$, at the standard reference state. **Answer**: $(\bar{q}_f^{\circ})_{\text{ethane}} = -84.5$ MJ/kgmole.

Notice that, in this example, we do not take into account the heat of formation of H_2 from atomic hydrogen H or O_2 from atomic oxygen O. This is because the elements used in the formation of a compound must be in their *stable molecular forms* at the standard reference state. In the case of methane, the elements are solid carbon (graphite), C, and diatomic hydrogen gas, H_2.

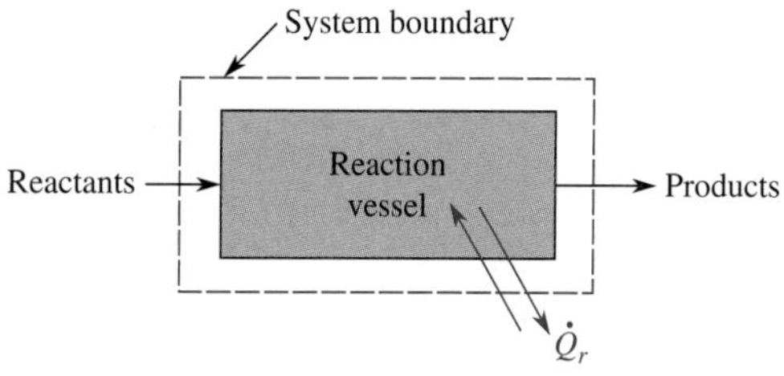

FIGURE 15.5
A steady state, steady flow, aergonic reaction vessel.

Consider a chemical reaction occurring in the steady state, steady flow, aergonic, open system shown in Figure 15.5. The energy rate balance (ERB) applied to this system yields

$$\dot{Q}_r = \sum_P(\dot{m}h) - \sum_R(\dot{m}h) = \sum_P(\dot{n}\bar{h}) - \sum_R(\dot{n}\bar{h}) = \dot{H}_P - \dot{H}_R$$

where $\dot{Q}_r$ is the exothermic or endothermic heat transfer rate of the reaction, and $\dot{H}_P$ and $\dot{H}_R$ are the total enthalpy rates of the products and reactants, respectively.

If this reaction is to be used to determine the standard reference state heat of formation of a compound, then the temperature and pressure of the reactants and the products must be maintained by sufficient heat transfer at 25.0°C and 0.100 MPa. In this case, the specific enthalpies of all the reactant elements are zero (by definition), so $\dot{H}_R = \dot{H}_R^\circ = \dot{H}_{\text{elements}} = \dot{H}_{\text{elements}}^\circ = 0$, and the previous equation reduces to

$$\dot{Q}_r^\circ = \dot{Q}_f^\circ = \dot{H}_P^\circ = \dot{H}_{\text{compound}}^\circ$$

where $\dot{Q}_f^\circ$ is the standard reference state *heat rate of formation* of the compound. In this case, the heat rate of formation, the heat transfer rate of the reaction, and the total enthalpy rate of the products are all equal. We now define the *molar specific enthalpy of formation*, $\bar{h}_f^\circ$, of a compound at the standard reference state as

$$(\bar{h}_f^\circ)_{\text{compound}} = (\dot{H}^\circ/\dot{n})_{\text{compound}} = (\dot{Q}_f^\circ/\dot{n})_{\text{compound}} = \left(\bar{q}_f^\circ\right)_{\text{compound}} \tag{15.6}$$

where $\dot{n}$ and $\bar{q}_f^\circ$ are the molar flow rate and the molar heat of formation of the compound, respectively.

Because of our sign convention that heat energy entering the system is positive while that leaving the system is negative, the heats and enthalpies of formation of *exothermic* reactions are always *negative*, while those of *endothermic* reactions are always *positive*. Table 15.1 gives the specific molar enthalpies (heats) of formation for some common compounds. Heats of formation can also be estimated from the atomic bond energies of the compound.

Table 15.1 Molar Specific Enthalpy of Formation at 25.0°C (77.0°F) and 0.100 MPa

		$\bar{h}_f^\circ$	
Substance	***M*** **kg/kgmole or lbm/lbmole**	**MJ/kgmole**	**Btu/lbmole**
Carbon monoxide, CO(g)	28.011	−110.529	−47,522
Carbon dioxide, CO_2(g)	44.011	−393.522	−169,195
Sulfur dioxide, SO_2(g)	64.07	−296.83	−127,622
Water, H_2O(g)	18.016	−241.827	−103,973
Water, H_2O(ℓ)	18.016	−285.838	−122,896
Methane, CH_4(g)	16.043	−74.873	−32,192
Acetylene, C_2H_2(g)	26.038	+226.731	+97,483
Ethylene, C_2H_4(g)	28.054	+52.283	+22,479
Ethane, C_2H_6(g)	30.070	−84.667	−36,403
Propane, C_3H_8(g)	44.097	−103.847	−44,649
Butane, C_4H_{10}(g)	58.124	−126.148	−54,237
Benzene, C_6H_6(g)	78.114	+82.930	+35,656
Octane, C_8H_{18}(g)	114.23	−208.447	−89,622
Octane, C_8H_{18}(ℓ)	114.23	−2411.952	−107,467
Carbon, C(s)	12.011	0	0
Oxygen, O_2(g)	32.00	0	0
Hydrogen, H_2(g)	2.016	0	0
Nitrogen, N_2(g)	28.013	0	0

Note: Here, (g) indicates gas or vapor state and (ℓ) indicates liquid state.

Source: Van Wylen, G. J., Sonntag, R. E., 1976. Fundamentals of Classical Thermodynamics, SI Version, second ed. Wiley, New York, p. 496 (Table 12.3). Copyright © 1976 John Wiley & Sons. Reprinted by permission of John Wiley & Sons.

EXAMPLE 15.7

The formation of 1 mole of water by the combustion of its elements can be written as

$$H_2 + 0.5 \times O_2 \rightarrow H_2O + \overline{q}_r$$

where $\overline{q}_r$ is the heat transfer that occurs per mole of water formed. Suppose we wanted to make exactly 0.160 kg of liquid water by this combustion reaction. Determine the heat transfer required to keep both the reactants (H_2 and O_2) and the products (H_2O) at the standard reference state (25.0°C and 0.100 MPa) while this reaction takes place.

Solution

For the isothermal (25.0°C) and isobaric (0.100 MPa) combustion of 1.00 kgmole of hydrogen gas with 0.500 kgmole of oxygen gas, Eq. (15.6) gives

$$(\overline{q}_r)_{SRS} = \overline{q}_f^{\circ} = \overline{h}_f^{\circ}$$

and since the reaction is at 25.0°C, the water formed is in the liquid state. Then, from Table 15.1, we find that $(\overline{h}_f^{\circ})_{H_2O(l)} =$ 285.838 MJ/kgmole. So, $(\overline{q}_f^{\circ})_{H_2O(l)SRS} = 285.838$ MJ/kgmole, and the heat transfer required for this reaction on a mass rather than a molar basis is

$$q_r = \frac{\overline{q}_r}{M}$$

where M is the molecular mass. Finally, the total heat transfer required is $Q_r = mq_r$, or

$$Q_r = mq_r = m\frac{\overline{q}_r}{M} = (0.160\,\text{kg})\left(\frac{-285.838\,\text{MJ/kgmole}}{18.016\,\text{kg/kgmole}}\right) = -2.54\,\text{MJ}$$

Exercises

19. Rework Example 15.7 and determine the heat transfer required for the formation of 0.160 kg of water vapor $H_2O(g)$ at the SRS rather than liquid water. **Answer:** $(Q_r)_{H_2O(g)} = -2.15$ MJ.
20. Repeat Example 15.7 and determine the heat transfer required for the formation of 1.00 kg of methane gas $CH_4(g)$ when both the reactants and the products are at the standard reference state. **Answer:** $(Q_r)_{CH_4(g)} = -4.67$ MJ.
21. Use the technique of Example 15.7 to determine the heat transfer required for the formation of 1.00 gallon (6.25 lbm) of liquid octane $C_8H_{18(\ell)}$. **Answer:** $(Q_r)_{C_8H_{18}(\ell)} = -5880$ Btu.

15.7 HEAT OF REACTION

In the previous section, we saw that the heat of formation of a compound was the same as the heat of reaction when that compound was formed from its elements at the standard reference state. An oxidation reaction of a *fuel* is normally called *combustion;* therefore, the heat of reaction of the oxidation of a fuel in air or pure oxygen is also known as the *heat of combustion* or the *heating value* of the fuel.

A *bomb calorimeter* is a closed, rigid (constant volume) vessel that can be used to determine the heat of reaction of a liquid or solid fuel sample (see Figure 15.6). When the final temperature in the bomb has been reduced to its initial standard state temperature of 25.0°C by the water bath, the resulting energy balance on the bomb is

$$Q_r = m(u_P - u_R) = n(\overline{u}_P - \overline{u}_R) = U_P - U_R \tag{15.7}$$

Then, from the definition of enthalpy, we can write

$$H_P - H_R = U_P - U_R + (p\forall)_P - (p\forall)_R$$

and, since the reactants are likely to be solids or liquids with a small volume and a low pressure, we can ignore them and set $(p\forall)_R \approx 0$, then

$$H_P - H_R = U_P - U_R + (p\forall)_P = Q_r + (n\Re T)_P$$

which provides a convenient relation between the constant volume heat of reaction measured by the bomb calorimeter and the total enthalpy change of the reaction occurring inside the bomb calorimeter.

The heat of reaction of gases, liquids, and some solids is more often measured in a steady state, steady flow, aergonic calorimeter, similar to that shown in Figure 15.5. An energy rate balance on this type of calorimeter gives the heat transfer rate of the reaction as

$$\dot{Q}_r = \dot{H}_P - \dot{H}_R$$

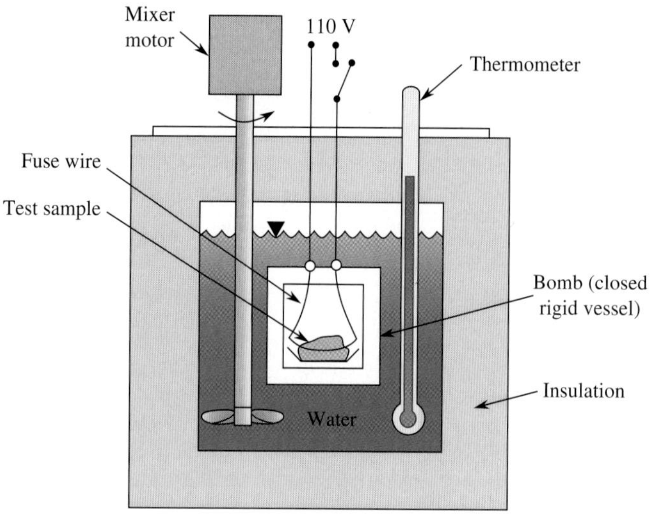

FIGURE 15.6
An adiabatic, constant volume bomb calorimeter.

and dividing through by the fuel molar flow rate $\dot{n}_{\text{fuel}}$ gives the molar heat of reaction $\bar{q}_r$ as

$$\bar{q}_r = \dot{Q}_r/\dot{n}_{\text{fuel}} = \bar{h}_P - \bar{h}_R = \bar{h}_{RP} \tag{15.8}$$

In this equation, we define the quantities

$$\bar{h}_R = \dot{H}_R/\dot{n}_{\text{fuel}} = \sum_R (\dot{n}_i/\dot{n}_{\text{fuel}})\bar{h}_i = \sum_R (n_i/n_{\text{fuel}})\bar{h}_i \tag{15.9}$$

and

$$\bar{h}_P = \dot{H}_P/\dot{n}_{\text{fuel}} = \sum_P (\dot{n}_i/\dot{n}_{\text{fuel}})\bar{h}_i = \sum_P (n_i/n_{\text{fuel}})\bar{h}_i. \tag{15.10}$$

where $i = 1, 2, 3, \ldots, n$, and n is the number of reactants or products. In these equations, $\bar{h}_R$ and $\bar{h}_P$ are the total reactant and product enthalpies per unit mole of fuel consumed. Combining Eqs. (15.9) through (15.11) gives the molar heat of reaction as

$$\bar{q}_r = \sum_P \frac{n_i}{n_{\text{fuel}}}\bar{h}_i - \sum_R \frac{n_i}{n_{\text{fuel}}}\bar{h}_i \tag{15.11}$$

The heating value of a fuel is the heat of reaction produced by the complete combustion of a unit mole (or mass) of the fuel when both the reactants and the products are maintained at the standard reference state (SRS). When the fuel contains hydrogen, the combustion products contain water that can be in either the liquid or vapor phase. The *higher heating value* (HHV) is produced when the water in the combustion products is condensed into the liquid state, and the *lower heating value* (LHV) occurs when this water is in the vapor state:

$$\text{HHV} = \bar{q}_r(\text{at the SRS and liquid } H_2O)$$

$$\text{LHV} = \bar{q}_r(\text{at the SRS and } H_2O \text{ vapor})$$

The relation between these two heating values is simply

$$\text{HHV} = \text{LHV} - \left(\frac{n_{H_2O}}{n_{\text{fuel}}}\right)(\bar{h}^{\circ}_{fg})_{H_2O} = \text{LHV} - \left(\frac{n_{H_2O}}{n_{\text{fuel}}}\right)\left(44.00\frac{\text{MJ}}{\text{kgmole } H_2O}\right) \tag{15.12}$$

where both the HHV and the LHV are in MJ/kgmole of fuel. In this equation, n_{H_2O}/n_{fuel} is the number of moles of water produced per mole of fuel burned, and $(\bar{h}^{\circ}_{fg})_{H_2O} = 44.00$ MJ/kgmole is the phase change molar specific

CRITICAL THINKING

In the calculation of the higher and lower heating values, HHV and LHV, both the reactants and the products are at the SRS. In the HHV calculation, all the water in the combustion products is condensed into a liquid, but in the LHV calculation, all this water is in the vapor phase. Wait a minute, how can you have all the LHV water in the vapor phase at the SRS temperature of 25.0°C, which is clearly below the dew point temperature of water at the SRS pressure of 0.100 MPa ($T_{DP} = T_{sat}(0.100\text{ MPa}) = 100°\text{C}$)? The answer is that you *cannot*, so the LHV state does not really exist. But, if the LHV can never be achieved, then what good is it? The answer is that the LHV is merely a benchmark value used for evaluating combustion processes. For example, the HHV and LHV are most commonly used in the calculation of combustion efficiency $\eta_{combustion}$, defined as

$$\eta_{combustion} = \frac{\text{Actual heat produced by a combustion process}}{\text{HHV or LHV}}$$

In this calculation, the HHV is used if the temperature of the combustion products can be reasonably lowered below 100.°C (as in industrial power plant boilers or domestic home furnaces), and the LHV is used when this is not possible (as in internal combustion engines). When the temperatures of the combustion products are necessarily above 100.°C, the LHV makes a more reasonable benchmark for the calculation of a combustion efficiency (note that, since the LHV is less than the HHV, the combustion efficiency is larger when the LHV is used).

enthalpy of water at the standard reference state temperature (25.0°C or 77.0°F). Note that both the HHV and the LHV are *negative* for heat-producing combustion reactions.

If the reactants *and* the products are both at the standard reference state, then Eq. (15.11) gives the *standard reference state heat of reaction, $\bar{q}_r^\circ$,* as

$$\bar{q}_r^\circ = \sum_P (n_i/n_{fuel})\left(\bar{h}_f^\circ\right)_i - \sum_R (n_i/n_{fuel})\left(\bar{h}_f^\circ\right)_i \tag{15.13}$$

This equation can be used to determine the HHV and LHV heats of combustion tabulated in Tables 15.2 and 15.3, as illustrated in the following example.

Table 15.2 *Molar* Based Higher Heating Values (HHV) for Common Fuels Where Both the Reactants and the Products Are at the SRS and the Water in the Combustion Products Is in the Liquid Phase

	$(\text{HHV})_{molar}$	
Fuel	**MJ/kgmole**	**Btu/lbmole**
Hydrogen, $H_2(g)$	−285.84	−122,970
Carbon, C(s)	−393.52	−169,290
Carbon monoxide, CO(g)	−282.99	−121,750
Methane, $CH_4(g)$	−890.36	−383,040
Acetylene, $C_2H_2(g)$	−1299.60	−559,120
Ethylene, $C_2H_4(g)$	−1410.97	−607,010
Ethane, $C_2H_6(g)$	−1559.90	−671,080
Propylene, $C_3H_6(g)$	−2058.50	−885,580
Propane, $C_3H_8(g)$	−2220.00	−955,070
n-Butane, $C_4H_{10}(g)$	−2877.10	−1,237,800
Benzene, $C_6H_6(g)$	−3270.0	−1,406,000
n-Octane, $C_8H_{18}(g)$	−5454.5	−2,345,000
n-Decane, $C_{10}H_{22}(g)$	−6754.7	−2,904,000
Methyl alcohol, $CH_3OH(g)$	−764.54	−328,700
Ethyl alcohol, $C_2H_5OH(g)$	−1409.30	−606,280

Source: Reprinted by permission of the publisher from Holman, J. P., 1980. Thermodynamics, third ed. McGraw-Hill, New York, p. 466 (Table 11-1).

Table 15.3 *Mass* Higher Heating Values (HHV) for Common Fuels Where Both the Reactants and the Products Are at the SRS and the Water in the Combustion Products Is in the Liquid Phase

	$(HHV)_{mass}$	
Fuel	**MJ/kg**	**Btu/lbm**
Hydrogen, H_2(g)	−142.9	−61,485
Carbon, C(s)	−32,79	−14,108
Carbon monoxide, CO(g)	−10.11	−4,348
Methane, CH_4(g)	−55.65	−23,940
Acetylene, C_2H_2(g)	−49.98	−21,505
Ethylene, C_2H_4(g)	−50.39	−21,679
Ethane, C_2H_6(g)	−52.00	−22,369
Propylene, C_3H_6(g)	−49.01	−21,085
Propane, C_3H_8(g)	−50.45	−21,706
n-Butane, C_4H_{10}(g)	−49.61	−21,341
Benzene, C_6H_6(g)	−41.92	−18,026
n-Octane, C_8H_{18}(g)	−47.85	−20,570
n-Decane, $C_{10}H_{22}$(g)	−47.57	−20,451
Methyl alcohol, CH_3OH(g)	−23.89	−10,272
Ethyl alcohol, C_2H_5OH(g)	−30.64	−13,180

Source: Reprinted by permission of the publisher from Holman, J. P., 1980. Thermodynamics, third ed. McGraw-Hill, New York, p. 466 (Table 11-1).

EXAMPLE 15.8

Determine the higher and lower heating values of methane. Note that, for the determination of the HHV and LHV, the combustion reaction must occur with 100.% theoretical air and both the reactants and the products must be at the standard reference state. For the HHV calculation, the water in the combustion products must be in the liquid phase, and for the LHV calculation, it must be in the vapor phase.

Solution

For 100.% theoretical air, the combustion equation for methane is

$$CH_4 + 2.00[O_2 + 3.76(N_2)] \rightarrow CO_2 + 2.00(H_2O) + 7.52(N_2)$$

Since both the reactants and the products are at the standard reference state, we can use Eq. (15.13) to find the heat of combustion, which is either the HHV or the LHV, depending on how the water term is handled. Then,

$$\bar{h}_R^\circ = \sum_R (n_i/n_{\text{fuel}})(\bar{h}_f^\circ)_i = (\bar{h}_f^\circ)_{CH_4} + 2.00(\bar{h}_f^\circ)_{O_2} + 7.52(\bar{h}_f^\circ)_{N_2}$$

and, from Table 15.1, we find that

$$(\bar{h}_f^\circ)_{CH_4} = -74.873\ \text{MJ/kgmole}\ CH_4$$

Since O_2 and N_2 are the elements of the compound CH_4, in their standard states (see Table 15.1), $(\bar{h}_f^\circ)_{O_2} = (\bar{h}_f^\circ)_{N_2} = 0$. Then, $\bar{h}_R^\circ = -74.873$ MJ/kgmole CH_4. Similarly,

$$\bar{h}_P^\circ = \sum_P (n_i/n_{\text{fuel}})(\bar{h}_f^\circ)_i = (\bar{h}_f^\circ)_{CO_2} + 2.00(\bar{h}_f^\circ)_{H_2O(\ell)} + 7.52(\bar{h}_f^\circ)N_2$$

From Table 15.1, we find that $(\bar{h}_f^\circ)_{N_2} = 0$, and

$$(\bar{h}_f^\circ)_{CO_2} = -393.522\ \text{MJ/kgmole}\ CO_2$$
$$(\bar{h}_f^\circ)_{H_2O(g)} = -241.827\ \text{MJ/kgmole}\ H_2O\ \text{vapor}$$
$$(\bar{h}_f^\circ)_{H_2O(\ell)} = -285.838\ \text{MJ/kgmole}\ H_2O\ \text{liquid}$$

Then,

$$(\overline{h}_P^\circ)_{\text{LHV}} = -393.522 + 2(-241.827) + 0 = -877.176 \text{ MJ/kgmole CH}_4$$

and

$$(\overline{h}_P^\circ)_{\text{HHV}} = -393.522 + 2(-285.838) + 0 = -965.198 \text{ MJ/kgmole CH}_4$$

Finally,

$$\begin{aligned}\text{LHV} &= (\overline{q}_r^\circ)_{\text{LHV}} = (\overline{h}_P^\circ)_{\text{LHV}} - \overline{h}_R^\circ = -877.176 - (-74.873)\\ &= -802.30 \text{ MJ/kgmole CH}_4\end{aligned}$$

and

$$\begin{aligned}\text{HHV} &= (\overline{q}_r^\circ)_{\text{HHV}} = (\overline{h}_P^\circ)_{\text{HHV}} - \overline{h}_R^\circ = -965.198 - (-74.873)\\ &= -890.33 \text{ MJ/kgmole CH}_4\end{aligned}$$

Note that this HHV is essentially the same as that listed in Table 15.2 for methane and that the HHV can be calculated from the LHV using Eq. (15.12) as

$$\begin{aligned}\text{HHV} &= \text{LHV} - \left(\frac{n_{H_2O}}{n_{\text{fuel}}}\right)(\overline{h}_{fg}^\circ)_{H_2O} = \text{LHV} - \left(\frac{n_{H_2O}}{n_{\text{fuel}}}\right)(44.00 \text{ MJ/kgmole H}_2\text{O})\\ &= -802.3 \text{ MJ/kgmole CH}_4 - \left(\frac{2.00 \text{ kgmole H}_2\text{O}}{1.00 \text{ kgmole CH}_4}\right)(44.00 \text{ MJ/kgmole H}_2\text{O})\\ &= -890.3 \text{ MJ/kgmole CH}_4\end{aligned}$$

Exercises

22. Use the methods of Example 15.8 to determine the higher heating value (HHV) of acetylene gas, $C_2H_2(g)$. **Answer**: $\text{HHV}_{C_2H_2(g)} = -1299.6$ MJ/kgmole of $C_2H_2(g)$.

23. Use the technique presented in Example 15.8 to determine the lower heating value (LHV) of propane gas, $C_3H_8(g)$. **Answer**: $\text{LHV}_{C_3H_8(g)} = -2044.0$ MJ/kgmole of $C_3H_8(g)$.

24. Repeat Example 15.8 for *n*-butane gas, $C_4H_{10}(g)$. **Answer**: $\text{HHV}_{C_4H_{10}(g)} = -2877.1$ MJ/kgmole of $C_4H_{10}(g)$, $\text{LHV}_{C_4H_{10}(g)} = -2657.1$ MJ/kgmole of $C_4H_{10}(g)$.

When the reactants or products are *not* at the standard reference state, then their enthalpies are determined from Hess's law by adding to the standard reference state enthalpies the change in enthalpy between the actual temperature and pressure and the standard reference state temperature and pressure. Normally, we can ignore the effect of pressure on enthalpy, so that the molar enthalpy of any compound at temperature T and pressure p is $\overline{h}(T,p)$, given by

$$\overline{h}(T,p) = \overline{h}_f^\circ + [\overline{h}(T) - \overline{h}(T^\circ)] \tag{15.14}$$

If the material can be considered to be an ideal gas with constant specific heats over the temperature range from the standard reference state temperature T° to the actual temperature T, then we can write

$$\overline{h}(T,p) = \overline{h}_f^\circ + \overline{c}_p(T - T^\circ) \tag{15.15}$$

Otherwise, $\overline{h}(T) - \overline{h}(T^\circ)$ must be determined from a more accurate source, such as the gas tables in *Thermodynamic Tables to accompany Modern Engineering Thermodynamics* (Table C.16c). Combining Eqs. (15.8), (15.9), (15.10), and (15.14) gives the general formula for the heat of reaction (or combustion) of a substance *not* at the standard reference state as

$$\overline{q}_r = \sum_P (n_i/n_{\text{fuel}})\left[\overline{h}_f^\circ + \overline{h}(T) - \overline{h}(T^\circ)\right]_i - \sum_R (n_i/n_{\text{fuel}})\left[\overline{h}_f^\circ + \overline{h}(T) - \overline{h}(T^\circ)\right]_i \tag{15.16}$$

The use of Eq. (15.16) is illustrated in the following example.

EXAMPLE 15.9

Compute the heat of reaction of methane when the reactants are at the standard reference state but the products are at 500.°C. Assume that the product gases can be treated as ideal gases with constant specific heats and the combustion water is in its vapor state.

Solution

Here, $\overline{q}_r = \overline{h}_P - \overline{h}_R^\circ$, where, from Example 15.8, $\overline{h}_R^\circ = -74.873$ MJ/kgmole CH_4. Assuming all the components of the products behave as ideal gases with constant specific heats, we can use Eq. (15.15) in Eq. (15.10) to find

$$\begin{aligned}\overline{h}_P &= \sum_P (n_i/n_{\text{fuel}})\left[\overline{h}_f^\circ + \overline{c}_p(T - T^\circ)\right]_i \\ &= \left(\overline{h}_f^\circ\right)_{CO_2} + 2.00\left(\overline{h}_f^\circ\right)_{H_2O(g)} + 7.52\left(\overline{h}_f^\circ\right)_{N_2} \\ &\quad + \left[\left(\overline{c}_p\right)_{CO_2} + 2.00\left(\overline{c}_p\right)_{H_2O(g)} + 7.52\left(\overline{c}_p\right)_{N_2}\right](T - T^\circ) \\ &= -393.522 + 2.00(-241.827) + 0 + [0.03719 + 2(0.03364) \\ &\quad + 7.52(0.02908)](500 - 25) = -723.680 \text{ MJ/kgmole } CH_4\end{aligned}$$

where the molar specific heats are obtained from Table C.13b (note that they are converted from kJ/kgmole to MJ/kgmole for use here). Then,

$$\overline{q}_r = -723.680 - (-74.870) = -648.810 \text{ MJ/kgmole } CH_4$$

Exercises

25. Repeat Example 15.9 for a products temperature of 800.°C rather than 500.°C. Assume all other variables are unchanged. **Answer**: $\overline{q}_r = -552$ MJ/kgmole CH_4.

26. Rework Example 15.9 for the case where the combustion products have been cooled to 30.0°C and the water in the products is in the liquid state. **Answer**: $\overline{q}_r = -889$ MJ/kgmole CH_4.

27. Use the method of Example 15.9 to determine the heat of reaction of acetylene when the reactants are at the standard reference state and the products are at 1000.°C. **Answer**: $\overline{q}_r = -990$ MJ/kgmole C_2H_2.

In Example 15.9, note that, even though the nitrogen does not enter into the chemistry of the reaction, it does enter into the thermodynamics of the reaction, because a significant portion of the heat of combustion went into heating the nitrogen from 25.0°C to 500.°C. Thus, it is easy to see why the use of too much excess air can reduce the net heat production of a combustion reaction and cause significant energy losses to the environment via hot exhaust gases.

EXPLOSION LIMITS AND IGNITION TEMPERATURES OF COMMON FUELS

Fuel burns continuously only when the amount of fuel and air present are within the *explosion* (or *flammability*) *limits* of the reaction. A fuel–air mixture does not ignite when the mixture is below the lower explosion limit (LEL) or when it is above the upper explosion limit (UEL). Both temperature and pressure affect these limits. For example, as the pressure of the mixture decreases below atmospheric pressure, the UEL decreases and the LEL increases. However, as the pressure increases above atmospheric pressure, the UEL increases but the LEL stays nearly constant. Also, as the mixture temperature increases, the UEL increases and the LEL decreases.

The *ignition temperature* is the lowest combustion temperature at which more heat is generated by the combustion process than is lost to the surroundings. A fuel–air mixture does not burn continuously if the combustion temperature is below the ignition temperature, unless heat is supplied to the reaction from the surroundings.

Table 15.4 lists explosion limits and ignition temperatures for some common fuel–air mixtures. These LEL and UEL values are lower and upper fuel to air ratio explosion limit percentages on a molar (or volume) basis.

Note that the LEL for hydrogen is higher than almost all the hydrocarbon fuels shown (including gasoline), making it less dangerous from a LEL explosion point of view.

Table 15.4 Explosion Limits and Ignition Temperatures for Common Fuel–Air Mixtures

Substance	LEL (%)	UEL (%)	Ignition Temp. (°F)
Carbon monoxide (CO)	12.5	74.0	1128
Hydrogen (H_2)	4.00	75.0	968
Methane (CH_4)	5.00	15.0	1301
Acetylene (C_2H_2)	2.50	81.0	763–824
Ethylene (C_2H_4)	2.75	28.6	914
Ethane (C_2H_6)	3.00	12.5	968–1166
Propylene (C_3H_6)	2.00	11.1	856
Propane (C_3H_8)	2.10	10.1	871
n-Butane (C_4H_{10})	1.86	8.4	761
Gasoline	1.12	6.75	495

15.8 ADIABATIC FLAME TEMPERATURE

The maximum possible combustion temperature occurs when combustion takes place inside an adiabatic (i.e., insulated) system. This temperature is called the *adiabatic combustion temperature* or the *adiabatic flame temperature*. In practice, though, the combustion temperature can never reach this temperature, because

1. No system can be made truly adiabatic.
2. The combustion reaction is always somewhat incomplete.
3. The combustion products ionize at high temperatures and thus lower the reaction temperature.

Nonetheless, the adiabatic flame temperature provides a useful upper bound on combustion temperatures and can be used to estimate the thermal effects of combustion on material physical properties and exhaust gas states.

There are actually two types of adiabatic flame temperature, depending on whether the combustion process is carried out under constant volume or constant pressure. The *constant volume adiabatic flame temperature* is the temperature resulting from a complete combustion process that occurs inside of a closed, rigid vessel with no work, heat transfer, or changes in kinetic or potential energy. The *constant pressure adiabatic flame temperature* is the temperature that results from a complete combustion process that occurs at a constant pressure (like an open flame) with no heat transfer or change in kinetic or potential energy. The constant pressure adiabatic flame temperature is lower than the constant volume adiabatic flame temperature, because some of the combustion energy is used to change the volume of the reactants and thus generates work.

For an open, constant pressure, adiabatic system, $\bar{q}_r = 0$ and Eq. (15.9) reduces to $\bar{h}_R = \bar{h}_P$, then,

$$\sum_R (n_i/n_{\text{fuel}})[\bar{h}_f^\circ + \bar{h}(T) - \bar{h}(T^\circ)]_i = \sum_P (n_i/n_{\text{fuel}})[\bar{h}_f^\circ + \bar{h}(T_A) - \bar{h}(T^\circ)]_i$$

where T_A is the adiabatic flame temperature and T is the temperature of the reactants. If the reactants are all at the standard reference state and the products can all be treated as ideal gases with constant specific heats over the temperature range from T° to T_A, then the previous equation reduces to

$$\sum_R (n_i/n_{\text{fuel}})\left(\bar{h}_f^\circ\right)_i = \sum_P (n_i/n_{\text{fuel}})\left[\bar{h}_f^\circ + \bar{c}_p(T_A - T^\circ)\right]_i$$

Now, let us suppose that all of the reactants except the fuel are elements; then, their $\bar{h}_f^\circ$ values are all zero. This equation can now be solved for T_A as

Open system, constant pressure, adiabatic, flame temperature when the reactants are at the SRS:

$$T_A\Big|_{\substack{\text{open}\\ \text{system}}} = T^\circ + \frac{\left(\bar{h}_f^\circ\right)_{\text{fuel}} - \sum_P (n_i/n_{\text{fuel}})\left(\bar{h}_f^\circ\right)_i}{\sum_P (n_i/n_{\text{fuel}})\left(\bar{c}_{pi}\right)_{\text{avg}}} \tag{15.17}$$

Equation (15.17) represents the only method for calculating the adiabatic flame temperature directly. It requires ideal gas behavior, which is usually reasonable, and it requires constant specific heats over the range $T^\circ = 25.0°C$

Table 15.5 Molar Specific Heats Averaged over the Temperature Range from 25 to 3000°C (77 to 5400°F)

	$(\bar{c}_p)_{avg}$		$(\bar{c}_v)_{avg}$	
Substance	**kJ/(kgmole·K)**	**Btu/(lbmole·R)**	**kJ/(kgmole·K)**	**Btu/(lbmole·R)**
CO_2 (g)	58.18	13.90	49.87	11.91
H_2O (g)	42.50	10.15	34.19	8.17
O_2 (g)	32.99	7.88	24.68	5.89
N_2 (g)	31.18	7.45	22.87	5.46

to T_A, which is not so reasonable unless average values are used (as noted in the equation). Average molar specific heats for typical combustion products in the range 25 to 3000°C are given in Table 15.5. This range covers most adiabatic flame temperatures.

In the case of a constant volume adiabatic closed system (Table 15.6), Eq. (15.7) tells us that $\bar{u}_P = \bar{u}_R$, and if the reactants are at the SRS and the products can again be treated as ideal gases with constant (or average) specific heats, it is easy to show that the adiabatic flame temperature in this system is given by

Closed system constant volume adiabatic flame temperature when the reactants at the SRS:

$$T_A\Big|_{\substack{\text{closed}\\\text{system}}} = T^\circ + \frac{(\bar{u}_f^\circ)_{\text{fuel}} - \sum_R (n_i/n_{\text{fuel}})\bar{h}_f^\circ - \Re T^\circ\left[\sum_R (n_i/n_{\text{fuel}}) - \sum_P (n_i/n_{\text{fuel}})\right]}{\sum_P (n_i/n_{\text{fuel}})(\bar{c}_{vi})_{\text{avg}}} \tag{15.18}$$

where we again assume that the reactants contain only the fuel and its combustion elements. Also, we use the definition of enthalpy to find

$$\bar{u}_f^\circ = \bar{h}_f^\circ - (p\bar{v})^\circ = \bar{h}_f^\circ - \Re T^\circ$$

for the ideal gas products and the nonfuel reactants, where T° is the standard reference state absolute temperature (298 K or 537 R). Further, for most liquids and solids at the standard reference state, we can use the approximation $(\bar{u}_f^\circ)_{\text{fuel}} \approx (\bar{h}_f^\circ)_{\text{fuel}}$.

Table 15.6 Constant Volume Adiabatic Flame Temperatures of Common Hydrocarbon Fuels when the Reactants Enter the Combustion Process at 25°C (77°F) and 1 Atm Pressure and the Products Leave the Process at 1 Atm Pressure. The Combustion Is Stoichiometric with No Excess Air

Fuel	Oxidizer	T_{ad} (°C)	T_{ad} (°F)
Acetylene (C_2H_2)	Air	2500	4532
	O_2	3480	6296
Butane (C_4H_{10})	Air	1970	3578
	O_2	3100	5612
Hydrogen (H_2)	Air	2210	4010
	O_2	3200	5792
Methane (CH_4)	Air	1950	3542
	O_2	2810	5090
Propane (C_3H_8)	Air	1980	3596
	O_2	2526	4579
MAPP gas (C_3H_4)	Air	2010	3650
	O_2	2927	5301
Wood	Air	1980	3596

EXAMPLE 15.10

For liquid octane, $C_8H_{18}(\ell)$, determine the following adiabatic flame temperatures when the reactants are in the standard reference state (25°C and 0.100 MPa) and the combustion products are assumed to be ideal gases:

a. The open system (constant pressure) adiabatic flame temperature burning with 100.% theoretical air.
b. The open system (constant pressure) adiabatic flame temperature burning with 200.% theoretical air.
c. The closed system (constant volume) adiabatic flame temperature burning with 100.% theoretical air.

Solution

a. The combustion equation for octane burning with 100.% theoretical air is

$$C_8H_{18} + 12.5[O_2 + 3.76(N_2)] \rightarrow 8(CO_2) + 9(H_2O) + 47(N_2)$$

Since the products can be considered to be ideal gases, we can use Eq. (15.17) and the average specific heat values given in Table 15.5. From Table 15.1, we find,

$$\left(\bar{h}_f^\circ\right)_{\text{fuel}} = \left(\bar{h}_f^\circ\right)_{C_8H_{18}(\ell)} = -249.952\ \text{MJ/kgmole}$$
$$\left(\bar{h}_f^\circ\right)_{CO_2} = -393.522\ \text{MJ/kgmole}$$
$$\left(\bar{h}_f^\circ\right)_{H_2O(g)} = -241.827\ \text{MJ/kgmole}$$

and

$$\left(\bar{h}_f^\circ\right)N_2 = 0 \text{ because it is an element}$$

The open system constant pressure adiabatic flame temperature is given by Eq. (15.17), where

$$\sum_P (n_i/n_{\text{fuel}})\left(\bar{h}_f^\circ\right)_i = 8\left(\bar{h}_f^\circ\right)_{CO_2} + 9\left(\bar{h}_f^\circ\right)_{H_2O} + 47\left(\bar{h}_f^\circ\right)_{N_2}$$
$$= 8(-393.522) + 9(-241.827) + 47(0)$$
$$= -5325\ \text{MJ/kgmole of } C_8H_{18}$$

and

$$\sum_P (n_i/n_{\text{fuel}})(\bar{c}_{pi})_{\text{avg}} = 8\left[(\bar{c}_p)_{CO_2}\right]_{\text{avg}} + 9\left[(\bar{c}_p)_{H_2O}\right]_{\text{avg}} + 47\left[(\bar{c}_p)_{N_2}\right]_{\text{avg}}$$
$$= 8(0.05818) + 9(0.04250) + 47(0.03118)$$
$$= 2.313\ \text{MJ/}\left[(\text{kgmole of } C_8H_{18})\cdot \text{K}\right]$$

Then, Eq. (15.17) gives

$$T_A\Big|_{\substack{\text{open}\\\text{system}}} = 25.0°\text{C} + \frac{-249.952\ \text{MJ/kgmole fuel} - (-5325\ \text{MJ/kgmole fuel})}{2.313\ \text{MJ/(kgmole fuel}\cdot\text{K)}} = 2170°\text{C} = 3940°\text{F}$$

b. The reaction equation when 200.% theoretical air is used is

$$C_8H_{18} + 2(12.5)[O_2 + 3.76(N_2)] \rightarrow 8(CO_2) + 9(H_2O) + 12.5(O_2) + 94(N_2)$$

The numerator in Eq. (15.17) is the same here as it was in part a, since we only added more elements to the reaction side of the equation. The denominator represents the energy required to raise the temperature of all the product gases and is consequently different from part a. In this case,

$$\sum_P (n_i/n_{\text{fuel}})(\bar{c}_{pi})_{\text{avg}} = 8\left[(\bar{c}_p)_{CO_2}\right]_{\text{avg}} + 9\left[(\bar{c}_p)_{H_2O}\right]_{\text{avg}} + 12.5\left[(\bar{c}_p)_{O_2}\right]_{\text{avg}} + 94\left[(\bar{c}_p)_{N_2}\right]_{\text{avg}}$$
$$= 8(0.05818) + 9(0.04250) + 12.5(0.03299) + 94(0.03118)$$
$$= 4.19\ \text{MJ/}\left[(\text{kgmole of } C_8H_{18})\cdot\text{K}\right]$$

Then, Eq. (15.17) gives

$$T_A\Big|_{\substack{\text{open}\\\text{system}\\200\%\ \text{TA}}} = \frac{-249.952 - (-5325)}{4.19} + 25.0 = 1240°\text{C} = 2260°\text{F}$$

(*Continued*)

EXAMPLE 15.10 (Continued)

Thus, the addition of 100.% excess air, a practice sometimes necessary to get complete combustion in high-velocity combustion processes, has the effect of reducing the adiabatic combustion temperature by nearly a factor of 2.

c. For a closed, constant volume system, the adiabatic flame temperature is given by Eq. (15.18). Since the fuel in this example is a liquid, we can assume $(\bar{u}_f^\circ)_{\text{fuel}} \approx (\bar{h}_f^\circ)_{\text{fuel}}$, and Eq. (15.18) becomes

$$T_A\Big|_{\substack{\text{closed}\\\text{system}}} \approx T^\circ + \frac{(\bar{h}_f^\circ)_{\text{fuel}} - \sum_R (n_i/n_{\text{fuel}})\bar{h}_f^\circ - \Re T^\circ\left[\sum_R (n_i/n_{\text{fuel}}) - \sum_P (n_i/n_{\text{fuel}})\right]}{\sum_P (n_i/n_{\text{fuel}})(\bar{c}_{vi})_{\text{avg}}}$$

The numerator is

$$\begin{aligned}(\bar{h}_f^\circ)_{\text{fuel}} - \sum_R (n_i/n_{\text{fuel}})\bar{h}_f^\circ - \Re T^\circ\left[\sum_R (n_i/n_{\text{fuel}}) - \sum_P (n_i/n_{\text{fuel}})\right]\\ = -249.953 - (-5324.62) - 0.0083143(25.0+273)[1+12.4\times 4.76 - (8+9+47)]\\ = 5083.34\ \text{MJ/(kgmole of } C_8H_{18})\end{aligned}$$

and the denominator is

$$\begin{aligned}\sum_P (n_i/n_{\text{fuel}})(\bar{c}_{vi})_{\text{avg}} &= 8\left[(\bar{c}_v)_{CO_2}\right]_{\text{avg}} + 9\left[(\bar{c}_v)_{H_2O}\right]_{\text{avg}} + 47\left[(\bar{c}_v)_{N_2}\right]_{\text{avg}}\\ &= 8(0.04987) + 9(0.03419) + 47(0.02287)\\ &= 1.782\ \text{MJ/[(kgmole of } C_8H_{18})\cdot \text{K]}\end{aligned}$$

Then, the constant volume adiabatic flame temperature is approximately

$$T_A\Big|_{\substack{\text{closed}\\\text{system}}} \approx 25.0 + \frac{5083.34\ \text{MJ/(kgmole of } C_8H_{18})}{1.782\ \text{MJ/[(kgmole of } C_8H_{18})\cdot \text{K]}} = 2880^\circ\text{C} = 5220^\circ\text{F}$$

Note that the constant volume adiabatic flame temperature in Example 15.10 is higher than the constant pressure adiabatic flame temperature due to the energy used in the work performed in a constant pressure process, that is, $p(V_2 - V_1)$.

Exercises

28. Determine the open system constant pressure adiabatic flame temperature for the liquid octane in Example 15.10 when the combustion occurs with 400.% theoretical air. **Answer:** $T_A = 664°\text{C}$.
29. Determine the open system constant pressure adiabatic flame temperature for the liquid octane in Example 15.10 when the combustion occurs with 800.% theoretical air. **Answer:** $T_A = 353°\text{C}$.
30. Determine the closed system constant volume adiabatic flame temperature for the liquid octane in Example 15.10 when the combustion occurs at 200% theoretical air. **Answer:** $T_A = 1610°\text{C}$.

An alternate and somewhat more accurate approach to finding the adiabatic flame temperature is to use the gas tables in *Thermodynamic Tables to accompany Modern Engineering Thermodynamics* (Table C.16c) to determine the thermodynamic properties of CO_2, H_2O, O_2, N_2, and so forth. However, since T_A and the other thermodynamic properties at this state are unknown, T_A must be determined by trial and error as follows:

1. $\bar{h}_R$ is calculated from Eq. (15.9) utilizing Eq. (15.14) or (15.15) if necessary.
2. A trial value for T_A is then assumed.
3. $\bar{h}_P$ is the calculated from the $(\bar{h}_f^\circ)_P$ values and the values of $\bar{h}(T) - \bar{h}(T^\circ)$ in Table C.16c.
4. If the value of $\bar{h}_P$ calculated in step 3 equals that of $\bar{h}_R$ calculated in step 1, then the correct value of T_A is assumed in step 2. Otherwise, a new T_A value is chosen and the process is repeated until $\bar{h}_P \approx \bar{h}_R$.

This manual iteration scheme is rather tedious, and inaccuracies are introduced by the linear interpolations in Table C.16c required to obtain a solution. These inaccuracies can be eliminated by programming accurate molar enthalpy formulae for the products into a microcomputer. The computer can then be programmed to calculate the heat of combustion and iterate to find the adiabatic flame temperature in a small fraction of the time required to carry out these calculations manually. Tables C.14 give accurate correlations for the variation in $\bar{c}_p$ with temperature for various substances. Using this information, we can determine accurate values for

$$\bar{h}(T) - \bar{h}(T^\circ) = \int_{T^\circ}^{T} \bar{c}_p\, dT$$

For example, the reaction for the combustion of liquid octane with Y % theoretical air is

$$C_8H_{18} + (Y/100)12.5\left[O_2 + 3.76(N_2)\right] \rightarrow 8(CO_2) + 9(H_2O) + (Y/100 - 1)O_2 + 47(Y/100)(N_2)$$

To simplify matters, assume that, before the combustion, the reactants are at the SRS. Then, $\bar{h}(T_R) = \bar{h}(T^\circ)$ and $\bar{h}(T_R) - \bar{h}(T^\circ) = 0$ for all the reactants. The heat produced by this reaction when the combustion products are at temperature T_P is

$$\begin{aligned}\bar{q}_r &= \sum_R (n_i/n_{\text{fuel}})(\bar{h}_f^\circ)_i - \sum_P (n_i/n_{\text{fuel}})[\bar{h}_f^\circ + \bar{h}(T_P) - \bar{h}(T^\circ)]_i \\ &= (\bar{h}_f^\circ)_{C_8H_{18}} - 8[\bar{h}_f^\circ + \bar{h}(T_P) - \bar{h}(T^\circ)]_{CO_2} - 9[\bar{h}_f^\circ + \bar{h}(T_P) - \bar{h}(T^\circ)]_{H_2O} \\ &\quad - (Y/100 - 1)[\bar{h}_f^\circ + \bar{h}(T_P) - \bar{h}(T^\circ)]_{O_2} - 47(Y/100 - 1)[\bar{h}_f^\circ + \bar{h}(T_P) - \bar{h}(T^\circ)]_{N_2}\end{aligned}$$

The molar specific heat equations in kJ/(kgmole·K), accurate to at least 0.43% over the range of 300 to 3500 K, can be found in Table C.14b in *Thermodynamic Tables to accompany Modern Engineering Thermodynamics* as

$$\text{Carbon dioxide: } (\bar{c}_p)_{CO_2} = -3.7357 + 30.529\theta^{0.5} - 4.1034\theta + 0.024198\theta^2$$

$$\text{Water: } (\bar{c}_p)_{H_2O} = 143.05 - 183.54\theta^{0.25} + 82.751\theta^{0.5} - 3.6989\theta$$

$$\text{Oxygen: } (\bar{c}_p)_{O_2} = 37.432 + 0.020102\theta^{1.5} - 178.57\theta^{-1.5} + 236.88\theta^{-2}$$

$$\text{Nitrogen: } (\bar{c}_p)_{N_2} = 39.060 - 512.79\theta^{-1.5} + 1072.7\theta^{-2} - 820.40\theta^{-3}$$

where $\theta^\circ = T^\circ/100 = 298/100 = 2.98$, and $\theta_P = T_p/100$. Integrating these equations from the SRS (θ°) to the temperature of the combustion products (θ_P) gives

$$\begin{aligned}[\bar{h}(T_P) - \bar{h}(T^\circ)]_{CO_2} &= 100 \times \int_{\theta^\circ}^{\theta_P} (\bar{c}_p)_{CO_2} d\theta \\ &= -373.57(\theta_P - \theta^\circ) + 2035.3\left[(\theta_P)^{1.5} - (\theta^\circ)^{1.5}\right] \\ &\quad - 205.17\left[(\theta_P)^2 - (\theta^\circ)^2\right] + 0.8066\left[(\theta_P)^3 - (\theta^\circ)^3\right]\end{aligned}$$

$$\begin{aligned}[\bar{h}(T_P) - \bar{h}(T^\circ)]_{H_2O} &= 100 \times \int_{\theta^\circ}^{\theta_P} (\bar{c}_p)_{H_2O} d\theta \\ &= 14{,}305.(\theta_P - \theta^\circ) - 14{,}683.2\left[(\theta_P)^{1.25} - (\theta^\circ)^{1.25}\right] \\ &\quad + 5516.7\left[(\theta_P)^{1.5} - (\theta^\circ)^{1.5}\right] - 184.95\left[(\theta_P)^2 - (\theta^\circ)^2\right]\end{aligned}$$

$$\begin{aligned}[\bar{h}(T_P) - \bar{h}(T^\circ)]_{O_2} &= 100 \times \int_{\theta^\circ}^{\theta_P} (\bar{c}_p)_{O_2} d\theta \\ &= 3743.2(\theta_P - \theta^\circ) + 0.80408\left[(\theta_P)^{2.5} - (\theta^\circ)^{2.5}\right] \\ &\quad + 35.714.\left[(\theta_P)^{-0.5} - (\theta^\circ)^{-0.5}\right] - 23{,}688\left[(\theta_P)^{-1} - (\theta^\circ)^{-1}\right]\end{aligned}$$

$$\begin{aligned}[\bar{h}(T_P) - \bar{h}(T^\circ)]_{N_2} &= 100 \times \int_{\theta^\circ}^{\theta_P} (\bar{c}_p)_{N_2} d\theta \\ &= 3906.0(\theta_P - \theta^\circ) + 102{,}558\left[(\theta_P)^{-1/2} - (\theta^\circ)^{-1/2}\right] \\ &\quad - 107{,}270.\left[(\theta_P)^{-1} - (\theta^\circ)^{-1}\right] + 41{,}020\left[(\theta_P)^{-2} - (\theta^\circ)^{-2}\right]\end{aligned}$$

To simplify the algebra, we define the following terms:

$$\begin{array}{lll} A = \theta_P - \theta^\circ & B = (\theta_P)^{1.25} - (\theta^\circ)^{1.25} & C = (\theta_P)^{1.5} - (\theta^\circ)^{1.5} \\ D = (\theta_P)^2 - (\theta^\circ)^2 & E = (\theta_P)^{2.5} - (\theta^\circ)^{2.5} & F = (\theta_P)^3 - (\theta^\circ)^3 \\ G = (\theta_P)^{-1/2} - (\theta^\circ)^{-1/2} & H = (\theta_P)^{-1} - (\theta^\circ)^{-1} & I = (\theta_P)^{-2} - (\theta^\circ)^{-2} \end{array}$$

Then, the heat of the combustion reaction becomes

$$\begin{aligned}\bar{q}_r &= \sum_R (n_i/n_{\text{fuel}})(\bar{h}_f^\circ)_i - \sum_P (n_i/n_{\text{fuel}})[\bar{h}_f^\circ + \bar{h}(T_P) - \bar{h}(T^\circ)]_i \\ &= (\bar{h}_f^\circ)_{C_8H_{18}} - 8[\bar{h}_f^\circ - 373.57A + 2035.3C - 205.17D + 0.8066F]_{CO_2} \\ &\quad - 9[\bar{h}_f^\circ + 14{,}305.A - 14{,}683.2B + 5516.7C - 184.95D]_{H_2O} \\ &\quad - (Y/100 - 1)[3743.2A + 0.80408E + 35{,}714.G - 23{,}688H]_{O_2} \\ &\quad - 47(Y/100 - 1)[3906.0A + 102{,}558G - 107270.H + 41{,}020I]_{N_2}\end{aligned}$$

From Table 15.1, we find

$$(\bar{h}_f^\circ)_{C_8H_{18}(\ell)} = -249.952 \text{ MJ/kgmole}, \; (\bar{h}_f^\circ)_{CO_2} = -393.522 \text{ MJ/kgmole},$$

$$(\bar{h}_f^\circ)_{H_2O(g)} = -241.827 \text{ MJ/kgmole, and } (\bar{h}_f^\circ)_{N_2} = (\bar{h}_f^\circ)_{O_2} = 0 \text{ because they are elements.}$$

While these equations are difficult to handle using a hand calculator, they are easily solved using an equation solver or a spreadsheet. The spreadsheet in Figure 15.7 illustrates the process for the combustion of liquid octane with 200% theoretical air. Typing control + tilde (Ctrl and Shift + ~) reveals the details, shown in Figure 15.8.

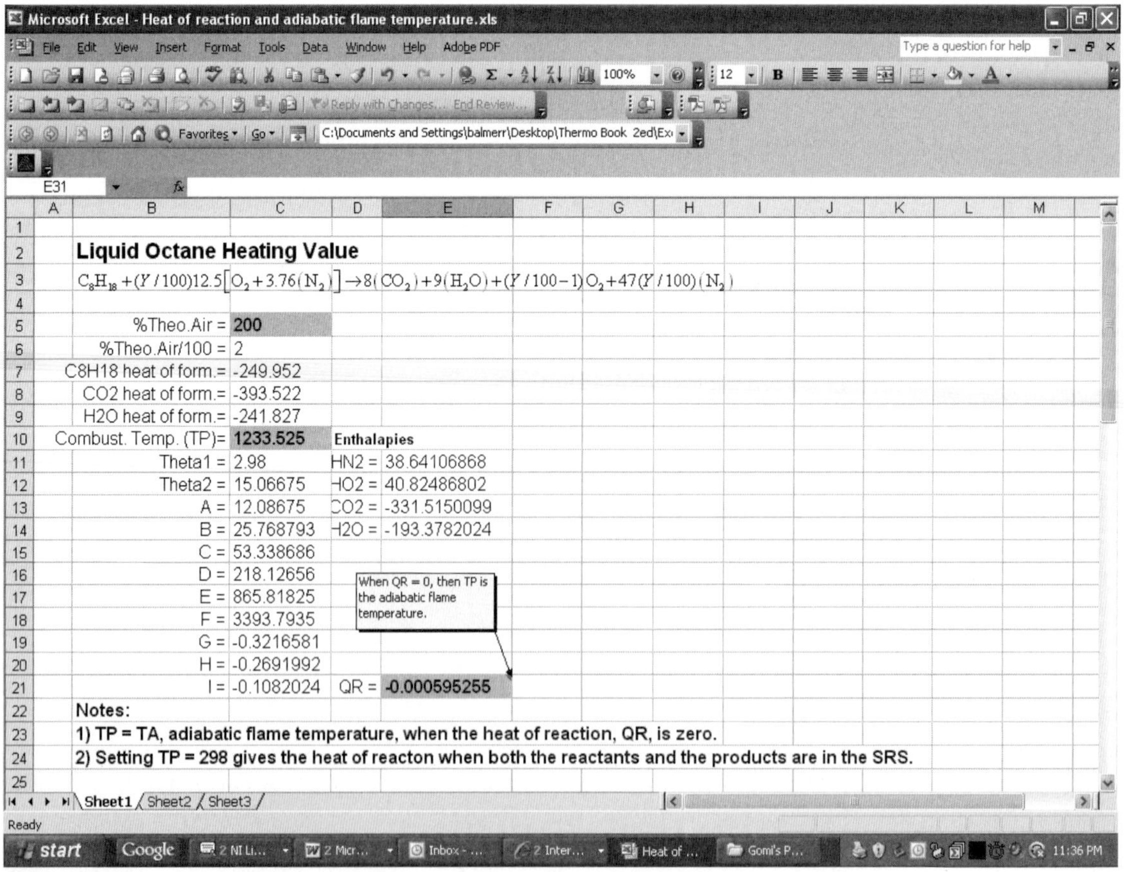

FIGURE 15.7
Solving the equations with a spreadsheet.

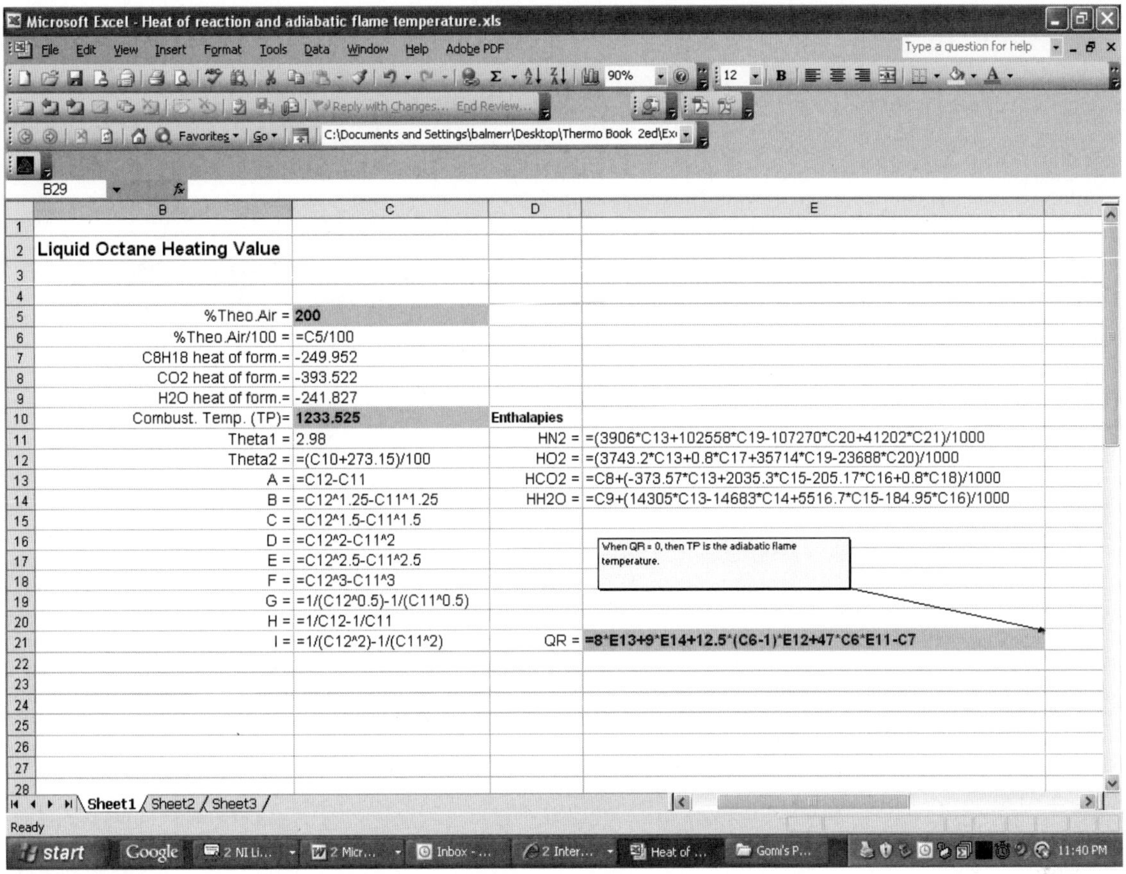

FIGURE 15.8
Accessing the details.

15.9 MAXIMUM EXPLOSION PRESSURE

The maximum possible internal pressure produced by combustion in a closed, rigid system is the pressure that occurs at the adiabatic flame temperature. It is the pressure that occurs when the system is insulated or when the combustion reaction occurs too fast for significant heat transfer to occur (as in explosions). This maximum pressure can be estimated from the ideal gas equation of state as

Maximum explosion pressure:

$$p_{\substack{\max \\ \text{explosion}}} = \frac{n_p \Re T_A}{\forall_p} \tag{15.19}$$

where n_p is the total number of moles of product present in the volume, $\forall_p$, and T_A is the adiabatic flame temperature of the reaction. Even though the adiabatic flame temperature is never reached in practice, this value of $p_{\max}$ is useful as an upper bound in explosion safety calculations.

CRITICAL THINKING

At the SRS, nitroglycerin ($C_3H_5O_9N_3$) is a highly unstable liquid that explodes 25 times faster and with 3 times the energy of gunpowder. In 1867, Alfred Bernhard Nobel (1833–1896) found that clay soaked with nitroglycerin was much more stable and less sensitive to shock than pure nitroglycerin. He called this safer form of nitroglycerin *dynamite* and manufactured it and other explosives for many years. His company was very successful, and with some of the profits, he founded the Nobel Prizes in physics, chemistry, literature, physiology or medicine, and peace. Would the maximum explosion pressure of nitroglycerin be 25 times higher than the explosion pressure of an equal mass of gunpowder?

EXAMPLE 15.11

We want to measure the heat of combustion of liquid octane by burning it in the rigid, sealed, adiabatic bomb calorimeter shown in Figure 15.9. The internal volume is of the combustion bomb is 50.0×10^{-3} ft^3. We insert 10.0 grams of fuel and fill the bomb with enough pure oxygen to have 50.0% excess oxygen for the combustion reaction. Determine the maximum possible explosion pressure inside the bomb when the fuel is ignited.

FIGURE 15.9
Example 15.11.

Solution

The molecular mass of octane (C_8H_{18}) is 114 kg/kgmole, so at 10.0 g (0.0100 kg), it contains

$$(0.0100\,\text{kg})/(114\,\text{kg/kgmole}) = 8.77 \times 10^{-5}\,\text{kgmole}$$

The reaction equation for 50.0% excess pure oxygen is

$$C_8H_{18} + 1.5(12.5)(O_2) \rightarrow 8(CO_2) + 9(H_2O) + 6.25(O_2)$$

and 8.77×10^{-5} kgmole of octane yields

$$(8.77 \times 10^{-5})(8 + 9 + 6.25) = 2.00 \times 10^{-3}\,\text{kgmole of product}$$

or

$$(2.00 \times 10^{-3}\,\text{kgmole})(2.2046\,\text{lbmole/kgmole}) = 4.41 \times 10^{-3}\,\text{lbmole of product}$$

We can estimate the adiabatic flame temperature for this closed system reaction from Eq. (15.18) using the temperature-averaged molar constant volume specific heats found in Table 15.6:

$$T_A\Big|_{\substack{\text{closed}\\\text{system}}} = T^\circ + \frac{(\bar{h}_f^\circ)_{\text{fuel}} - \sum_R (n_i/n_{\text{fuel}})\Re T^\circ - \sum_R (n_i/n_{\text{fuel}})(\bar{h}_f^\circ - \Re T^\circ)_i}{\sum_P (n_i/n_{\text{fuel}})(\bar{c}_{vi})_{\text{avg}}}$$

Since the fuel is in liquid form at a low pressure here, we use the approximation $(\bar{u}_f^\circ)_{\text{fuel}} \approx (\bar{h}_f^\circ)_{\text{fuel}}$. Since $\Re T^\circ = (0.0083143)(298.15) = 2.4789$ MJ/kgmole is a constant, the numerator of Eq. (15.18) becomes

$$\begin{aligned}&-249.952 - 1.5(12.5)(2.4789) - 8(-393.522 - 2.4789)\\&-9(-241.827 - 2.4789) - 6.25(0 - 2.4789) = 5090\,\text{MJ}\end{aligned}$$

and the denominator is

$$8(0.04987) + 9(0.03419) + 6.25(0.02468) = 0.861\,\text{MJ/K}$$

Then, Eq. (15.18) gives

$$(T_A)_{\substack{\text{bomb}\\\text{calorimeter}}} = 25.0 + \frac{5090}{0.861} = 5930^\circ\text{C} = 6210\,\text{K} = 11{,}200\,\text{R}$$

and Eq. (15.19) gives

$$\begin{aligned}p_{\max} &= \frac{n_p \Re T_A}{\forall_p}\\&= \frac{(4.41 \times 10^{-3}\,\text{lbmole})[1545.35\,\text{ft}\cdot\text{lbf/(lbmole}\cdot\text{R)}](11.2 \times 10^3\,\text{R})}{(50.0 \times 10^{-3}\,\text{ft}^3)(144\,\text{in}^2/\text{ft}^2)} = 10{,}600\,\text{psi}\end{aligned}$$

Exercises

31. Determine the maximum explosion pressure in Example 15.11 for a bomb volume of 10.0×10^{-3} ft^3 instead of 50.0×10^{-3} ft^3. Assume all the remaining variables are unchanged. **Answer:** $p_{\text{max explosion}} = 53{,}900$ psi.

32. If the amount of fuel used in the bomb calorimeter in Example 15.11 is 50.0 grams instead of 10.0 grams, determine the maximum explosion pressure in the bomb. Assume all the remaining variables are unchanged.
Answer: $p_{\text{max explosion}} = 53{,}900$ psi.

33. If the bomb calorimeter in Example 15.11 is filled with 100.% excess oxygen instead of 50.0% excess oxygen, determine the maximum explosion pressure in the bomb. Assume all the remaining variables are unchanged.
Answer: $p_{\text{max explosion}} = 11{,}700$ psi.

WHAT IS C-4?

C-4 (or Compound 4) is an off-white plastic explosive that feels like modeling clay. About 91% of C-4 is an explosive called RDX ($C_3H_6N_6O_6$) with the remainder being plasticizer and binder. C-4 is 1.34 times as explosive as trinitrotoluene (TNT) and detonates with a pressure wave of about 8040 m/s (26,400 ft/s or about 18,000 mph).

C-4 is actually very stable. It can be detonated only by a combination of extreme heat and a shockwave brought about by inserting and firing a detonating device. It cannot be detonated by a gunshot, by dropping it onto a hard surface, or even by blowing it up. When ignited with a flame, C-4 burns slowly rather than explodes. Even though soldiers knew that burning C-4 produces poisonous fumes, during the Vietnam War, they used small amounts of it as fuel for cooking.

More accurate calculations that include the energy-absorbing effects of chemical dissociation of the products at high temperatures show that the maximum adiabatic flame temperature of octane with pure oxygen is only about 3100 K (5580 R). Therefore, the maximum explosion temperature calculated in Example 15.11 is high by a factor of about 2, and the actual maximum pressure inside the bomb calorimeter is closer to 5400 psi. Gas pressures at this level can be very dangerous (especially at high temperatures) and the calorimeter must be designed to withstand them.

15.10 ENTROPY PRODUCTION IN CHEMICAL REACTIONS

When the entropy rate balance is applied to a steady state, steady flow, *open system* combustion or reaction chamber with isothermal boundaries at temperature T_b, the total entropy production rate for the reaction is

$$\begin{aligned}(\dot{S}_P)_r &= \sum_{\text{out}} \dot{m}s - \sum_{\text{in}} \dot{m}s - \dot{Q}_r/T_b \\ &= \sum_{\text{out}} \dot{n}\bar{s} - \sum_{\text{in}} \dot{n}\bar{s} - \dot{Q}_r/T_b\end{aligned}$$

where $\dot{Q}_r$ is the heat transport rate of the reaction. In most instances, the products exit the system mixed together in a single flow stream, but the reactants can enter the system either (a) premixed in a single flow stream or (b) in individual flow streams. If the reactants enter through separate flow streams, each carrying a pure substance, then the entropy production rate of the mixing process that must occur inside the system before the reaction can occur is included in the previous equation, which then has the form

$$(\dot{S}_P)_r = \dot{n}_P\bar{s}_P - \sum_R \dot{n}_i\bar{s}_i - \dot{Q}_r/T_b$$

On the other hand, if the reactants enter the system already mixed together in a single flow stream, then this equation becomes

$$(\dot{S}_P)_r = \dot{n}_P\bar{s}_P - \dot{n}_R\bar{s}_R - \dot{Q}_r/T_b$$

where the mixture molar specific entropies are given in Table 12.3 as

$$\bar{s}_P = \sum_P \chi_i\hat{\bar{s}}_i \quad \text{and} \quad \bar{s}_R = \sum_R \chi_i\hat{\bar{s}}_i$$

where χ_i is the mole fraction of substance i, and $\hat{\bar{s}}$ is the partial molar specific entropy, defined in Chapter 12. If both the reactants and the products can be considered mixtures of ideal gases that obey the Gibbs-Dalton ideal gas mixture law discussed in Chapter 12, then $\bar{s}_i = \hat{\bar{s}}_i$, the molar specific entropy of gas i. If one or more of the reactants or products is a liquid or a solid or if the mixture does not obey the Gibbs-Dalton ideal gas mixture law, then a much more complex analysis must be carried out.

Assuming both the reactants and the products to be premixed ideal gases, the total entropy production rate of the reaction is

$$\begin{aligned}(\dot{S}_P)_r &= \dot{n}_P\sum_P \chi_i\bar{s}_i - \dot{n}_R\sum_R \chi_i\bar{s}_i - \dot{Q}_r/T_b \\ &= \sum_P \dot{n}_i\bar{s}_i - \sum_R \dot{n}_i\bar{s}_i - \dot{Q}_r/T_b\end{aligned}$$

and dividing through by the fuel molar flow rate $\dot{n}_{\text{fuel}}$ gives $(\dot{S}_P)_r/\dot{n}_{\text{fuel}} = (S_P)_r/n_{\text{fuel}} = (\bar{s}_P)_r$, where $(\bar{s}_P)_r$ is the specific entropy production per unit mole of fuel consumed:

$$(\bar{s}_P)_r = \sum_P (n_i/n_{\text{fuel}})\bar{s}_i - \sum_R (n_i/n_{\text{fuel}})\bar{s}_i - \bar{q}_r/T_b \tag{15.20}$$

In the case of a *closed system* with an isothermal boundary, the entropy balance equation gives the total entropy production of the reaction as (see Eq. (7.77))

$$\begin{aligned}(S_P)_r &= (S_2 - S_1)_r - {}_1(Q_r)_2/T_b \\ &= n_P\bar{s}_P - n_R\bar{s}_R - (Q_r)/T_b\end{aligned}$$

where the reactants and products are assumed to be mixed at the beginning and end of the reaction, respectively. Again assuming the reactants and the products to be ideal gases and dividing through by the number of moles of fuel present in the reactants gives

$$(S_P)_r/n_{\text{fuel}} = (\bar{s}_P)_r = \sum_P (n_i/n_{\text{fuel}})\bar{s}_i - \sum_R (n_i/n_{\text{fuel}})\bar{s}_i - \bar{q}_r/T_b$$

which is identical to Eq. (15.20). This is as it should be, since the specific entropy production per unit mole of fuel consumed should depend on only the reaction itself and not the analysis frame (i.e., open or closed) used to determine it.

As in the case of enthalpy discussed earlier, we need a common zero point reference state from which to measure the entropies of all of the components in the reaction. Enthalpy and internal energy have no physically well-defined absolute zero values. Even at absolute zero temperature, it can be shown that the enthalpy and internal energy are not generally zero. Entropy, on the other hand, does have an absolute zero point dictated by a state of absolutely perfect molecular order. This is the postulate called the *third law of thermodynamics.*

The primary value of the third law of thermodynamics as far as we are concerned is that it gives us a reference state from which we can construct an absolute entropy scale. This means that we now have three thermodynamic properties with well-defined absolute zero value states: pressure, temperature, and entropy. We assume that all the substances we deal with have ordered crystalline, rather than amorphous solid, phases at absolute zero temperature. Therefore, we can compute the absolute molar specific entropy of an *incompressible* substance at any pressure and temperature from Eq. (7.32) in Chapter 7 as

$$\left[\bar{s}(p,T)_{\text{abs}}\right]_{\substack{\text{incompressible}\\ \text{substance}}} = \int_0^T \bar{c}(dT/T) \tag{15.21}$$

where $\bar{c}$ is the molar specific heat and T is in absolute temperature units (K or R). Similarly, the absolute molar specific entropy in SI units of an *ideal gas* at any pressure and temperature can be determined from Eq. (7.35) in Chapter 7 as

$$\left[\bar{s}(p,T)_{\text{abs}}\right]_{\text{ideal gas}} = \int_0^T \bar{c}_P(dT/T) - \Re \ln(p/0.1\ \text{MPa}) \tag{15.22}$$

where T is in K and p is in MPa. Note that both $\bar{c}$ and $\bar{c}_p \to 0$ as $T \to 0$ therefore, Eqs. (15.21) and (15.22) cannot be integrated by assuming constant specific heats in the temperature range from 0 to T.

WHAT IS THE THIRD LAW OF THERMODYNAMICS?

Unlike the first two laws of thermodynamics, the third law is not a statement about conservation or production. It was developed from quantum statistical mechanics theories in 1906 by Walther Hermann Nernst (1864–1941), for which he won the 1920 Nobel Prize in Chemistry. Basically, it states that the entropy of a perfect crystalline substance vanishes at absolute zero temperature and is independent of the pressure at that point. That is,

$$\lim_{T\to 0}(S)_{\substack{\text{perfect}\\ \text{crystal}}} = \left.\left(\frac{\partial S}{\partial p}\right)\right|_{T=0} = 0$$

Therefore, if we choose absolute zero temperature and any convenient pressure as a reference state for an entropy scale, we have produced an *absolute entropy scale* (i.e., one with an absolute zero point). A pressure of 0.1 MPa (about 1 atm) is usually chosen for the reference state pressure. Consequently, we construct an absolute entropy scale from the point $S = 0$ at 0.1 MPa and 0 K.

Since all of the enthalpy values used thus far in this chapter are based on a standard reference state of $T^\circ = 25.0°C$ and $p^\circ = 0.100$ MPa, it would be convenient to be able to shift our absolute entropy scale to this reference state, T°. To do this, we define $\bar{s}^\circ$ to be the absolute molar specific entropy at the standard reference state and use Eq. (7.35) with $p = p^\circ = constant$ to obtain

$$\bar{s}^\circ = \bar{s}(p^\circ, T^\circ)_{abs} = \int_0^{T^\circ} \bar{c}_P(dT/T)$$

Values for the $\bar{s}^\circ$ of various compounds can be found in Table 15.7. The absolute molar specific entropy at any other state at pressure p and temperature T is given by

$$\bar{s}(p, T)_{abs} = \bar{s}^\circ + \Delta\bar{s}(p^\circ \rightarrow p, T^\circ \rightarrow T)$$

where $\Delta\bar{s}$ represents the change in molar specific entropy between the state at (p°, T°) and that at (p, T). For an *incompressible* substance, this becomes

$$\left[\bar{s}(p, T)_{abs}\right]_{\substack{\text{incompressible}\\ \text{substance}}} = \bar{s}^\circ + \int_{T^\circ}^{T} \bar{c}(dT/T)$$

and for an *ideal gas* it becomes

$$\left[\bar{s}(p, T)_{abs}\right]_{\substack{\text{ideal}\\ \text{gas}}} = \bar{s}^\circ + \int_{T^\circ}^{T} \bar{c}_P(dT/T) - \Re\ln(p/p^\circ)$$

If these substances have constant (or averaged) specific heats in the range of T° to T, then these equations can be integrated to give

$$\left[\bar{s}(p, T)_{abs}\right]_{\substack{\text{incompressible}\\ \text{substance}}} = \bar{s}^\circ + \bar{c}\ln(T/T^\circ) \tag{15.23}$$

and

$$\left[\bar{s}(p, T)_{abs}\right]_{\substack{\text{ideal}\\ \text{gas}}} = \bar{s}^\circ + \bar{c}_p\ln(T/T^\circ) - \Re\ln(p/p^\circ) \tag{15.24}$$

Table 15.7 Molar Specific Absolute Entropy and Molar Specific Gibbs Function of Formation at 25.0°C and 0.100 Mpa

	$\bar{s}^o$		$\bar{g}_f^o$	
Substance	**kJ/(kgmole·K)**	**Btu/(lbmole·R)**	**MJ/kgmole**	**Btu/lbmole**
Carbon monoxide, CO	197.653	47.219	−137.150	−59,003
Carbon dioxide, CO_2	213.795	52.098	−394.374	−169,664
Water, $H_2O(g)$	188.833	45.132	−228.583	−98,333
Water, $H_2O(\ell)$	70.049	16.742	−237.178	−102,036
Methane, CH_4	186.256	44.516	−50.751	−21,834
Acetylene, C_2H_2	200.958	48.030	+2011.234	+90,015
Ethylene, C_2H_4	2111.548	52.473	+68.207	+29,343
Ethane, C_2H_6	2211.602	54.876	−32.777	−14,101
Propane, C_3H_8	270.019	64.361	−23.316	−10,031
Butane, C_4H_{10}	310.227	74.146	−16.914	−7,276
Octane, $C_8H_{18}(g)$	466.835	111.576	+16.859	+7,253
Octane, $C_8H_{18}(\ell)$	360.896	86.256	+6.940	+2,986
Carbon, C(s)	5.740	1.372	0	0
Oxygen, $O_2(g)$	205.138	411.029	0	0
Hydrogen, $H_2(g)$	130.684	31.234	0	0
Nitrogen, $N_2(g)$	191.610	45.796	0	0

Source: Van Wylen, G. J., Sonntag, R. E., 1976. Fundamentals of Classical Thermodynamics, SI Version, second ed. Wiley, New York, p. 496 (Table 12.3). Copyright © 1976 John Wiley & Sons. Reprinted by permission of John Wiley & Sons. Data on C, O_2, H_2, and N_2 are from the Journal of Physical and Chemical Reference Data, 11, Suppl. 2 (1982). Used with permission.

In the case of mixtures of ideal gases that obey the Gibbs-Dalton ideal gas mixture law, the absolute molar specific entropy of chemical species $i, \bar{s}_i$, needed for the entropy balance of Eq. (15.20) is given by

$$\bar{s}_i = \bar{s}_i^\circ + \int_{T^\circ}^{T} \bar{c}_{pi}(dT/T) - \Re \ln(p_i/p^\circ) \tag{15.25}$$

and, when $\bar{c}_{pi}$ is constant (or averaged) over T° to T,

$$\bar{s}_i = \bar{s}_i^\circ + \bar{c}_{pi} \ln(T/T^\circ) - \Re \ln(p_i/p^\circ) \tag{15.26}$$

where p_i is the partial pressure of chemical species i in the mixture.

Accurate integrated values can be found for $\bar{s}_i$ through the use of Table C.16c in *Thermodynamic Tables to accompany Modern Engineering Thermodynamics*. Note that, in this table, we use the condensed notation

$$\bar{\phi}_i = \bar{s}_i^\circ + \int \bar{c}_{pi}(dT/T)$$

so that Eq. (15.25) reduces for use with Table C.16c to

$$\bar{s}_i = \bar{\phi}_i - \Re \ln(p_i/p^\circ) \tag{15.27}$$

Alternatively, a computer program could be easily written to calculate accurate integrated values for $\bar{s}_i$.

The partial pressure p_i of component i in the mixture is determined from the mixture composition via Eq. (12.23) as

$$p_i = \chi_i p_m = w_i(M_m/M_i)p_m$$

where χ_i is the mole fraction, w_i is the mass (or weight) fraction, M_i is the molecular mass of chemical species i in the mixture, M_m is the equivalent molecular mass of the mixture, and p_m is the total pressure of the mixture.

EXAMPLE 15.12

Calculate the entropy produced per mole of fuel when methane is burned with 100.% theoretical air. The reactants are premixed at 25.0°C at a mixture total pressure of 0.100 MPa, and the products are at 200.°C at a total pressure of 0.100 MPa. The molar heating value of methane under these conditions with the water of combustion in the vapor phase is −134.158 MJ per kgmole of methane. Assume constant specific heat ideal gas behavior for all the combustion components.

Solution

Equation (15.20) gives the required entropy production as

$$(\bar{s}_P)_r = \sum_P (n_i/n_{\text{fuel}})\bar{s}_i - \sum_R (n_i/n_{\text{fuel}})\bar{s}_i - \bar{q}_r/T_b$$

where we are given $\bar{q}_r$ = −134.158 MJ/kgmole and we assume that T_b = 200. + 273.15 = 473 K. The reaction equation for 100.% theoretical air is

$$CH_4 + 2[O_2 + 3.76(N_2)] \rightarrow CO_2 + 2(H_2O) + 7.52(N_2)$$

The partial pressures of the reactants can then be found from Eq. (12.23) as

$$\begin{aligned} p_{CH_4} &= (n_{CH_4}/n_R)p_m = [1/(1+2+7.52)](0.100) = (1/10.52)(0.100) \\ &= 9.51\ \text{kPa} \\ p_{O_2} &= (2/10.52)(0.100) = 19.0\ \text{kPa} \\ p_{N_2} &= (7.52/10.52)(0.100) = 71.5\ \text{kPa} \end{aligned}$$

and the partial pressures of the products are

$$\begin{aligned} p_{CO_2} &= (n_{CO_2}/n_P)P_m = (1/10.52)(0.100) = 9.51\ \text{kPa} \\ p_{H_2O} &= (2/10.52)(0.100) = 19.0\ \text{kPa} \\ p_{N_2} &= (7.52/10.52)(0.100) = 71.5\ \text{kPa} \end{aligned}$$

Now,

$$\sum_R (n_i/n_{\text{fuel}})\bar{s}_i = \bar{s}_{CH_4} + 2.00(\bar{s}_{O_2}) + 7.52(\bar{s}_{N_2})$$

where, from Eq. (15.26) with T = 298 K, and using Table 15.7,

$$\begin{aligned}
\bar{s}_{CH_4} &= \bar{s}^o_{CH_4} - \Re \ln[(p_{CH_4} \text{ in MPa})/(0.100 \text{ MPa})] \\
&= 186.256 - 8.3143(\ln 0.0951) \\
&= 205.8 \text{ KJ/(kgmole} \cdot \text{K)} \\
\bar{s}_{O_2} &= 205.138 - 8.3143(\ln 0.190) \\
&= 218.9 \text{ kJ/(kgmole} \cdot \text{K)} \\
\bar{s}_{N_2} &= 191.610 - 8.3143(\ln 0.715) \\
&= 194.4 \text{ kJ/(kgmole} \cdot \text{K)}
\end{aligned}$$

Then,

$$\begin{aligned}
\sum_R (n_i/n_{\text{fuel}})\bar{s}_i &= 205.8 + 2(218.9) + 7.52(194.4) \\
&= 2100 \text{ kJ/(kgmole} \cdot \text{K)}
\end{aligned}$$

Also, for the products,

$$\sum_P (n_i/n_{\text{fuel}})\bar{s}_i = \bar{s}_{O_2} + 2(\bar{s}_{H_2O}) + 7.52(\bar{s}_{N_2})$$

where, again from Eq. (15.26) with $T = 473$ K and using Table 15.7,

$$\begin{aligned}
\bar{s}_{CO_2} &= \bar{s}^o_{CO_2} + (\bar{c}_p)_{CO_2}\left(\ln \frac{473}{298}\right) - \Re\{\ln[(p_{CO_2} \text{ in MPa})/(0.100 \text{ MPa})]\} \\
&= 213.795 + 37.19\left(\ln \frac{473}{298}\right) - 8.3143(\ln 0.0951) \\
&= 250.5 \text{ kJ/(kgmole} \cdot \text{K)} \\
\bar{s}_{H_2O} &= 188.833 + 33.64\left(\ln \frac{473}{298}\right) - 8.3143(\ln 0.190) \\
&= 218.2 \text{ kJ/(kgmole} \cdot \text{K)}
\end{aligned}$$

and

$$\begin{aligned}
\bar{s}_{N_2} &= 191.610 + 29.08\left(\ln \frac{473}{298}\right) - 8.3143(\ln 0.715) \\
&= 207.8 \text{ kJ/(kgmole} \cdot \text{K)}
\end{aligned}$$

Then,

$$\begin{aligned}
\sum_P (n_i/n_{\text{fuel}})\bar{s}_i &= 250.5 + 2(218.2) + 7.52(207.8) \\
&= 2250 \text{ kJ/(kgmole} \cdot \text{K)}
\end{aligned}$$

Finally, the desired result is

$$\begin{aligned}
(\bar{s}_p)_r &= 2250 - 2100 - (-134{,}158/473) \\
&= 434 \text{ kJ/(kgmole} \cdot \text{K)}
\end{aligned}$$

Only about one third of the value of $(\bar{s}_p)_r$ here is due to the reaction itself; the remaining two thirds come from the associated heat transfer. Note that $(\bar{s}_p)_r > 0$, as required by the second law of thermodynamics.

Exercises

34. Determine the molar specific entropy produced in Example 15.12 if the temperature of the surface of the system where the heat transfer occurs T_b is 20.0°C rather than 200.°C. Assume all the remaining variables are unchanged. **Answer**: $(\bar{s}_p)_r = 602$ kJ/(kgmole·K).

35. If the temperature of the products in Example 15.12 is 500.°C instead of 200.°C, determine the specific molar entropy production for this reaction. Assume all the remaining variables are unchanged (including T_b). **Answer**: $(\bar{s}_p)_r = 586$ kJ/(kgmole·K).

36. Determine the specific molar entropy production in Example 15.12 if the pressure of the products of combustion is 0.200 MPa instead of 0.100 MPa. **Answer**: $(\bar{s}_p)_r = 488$ kJ/(kgmole·K).

15.11 ENTROPY OF FORMATION AND GIBBS FUNCTION OF FORMATION

Since the specific molar Gibbs function, $\bar{g} = \bar{h} - T\bar{s}$, depends upon *both* $\bar{h}$ and $\bar{s}$, it does not have an absolute zero reference state. But, since $\bar{s}$ does have an absolute zero reference state, we cannot arbitrarily set the molar specific entropies of the elements of the reaction equal to zero at the standard reference state, as was done earlier with their $\bar{h}_f^o$ values. Therefore, we define the *molar specific entropy of formation*, $\bar{s}_f^o$, of a compound as

$$(\bar{s}_f^o)_{compound} = \bar{s}^o{}_{compound} - \sum_{elements} (n_i/n_{compound})\bar{s}_i^o \quad (15.28)$$

Then, the *molar specific Gibbs function of formation* of a compound, $\bar{g}_f^o$, is given by

$$(\bar{g}_f^o)_{compound} = (\bar{h}_f^o)_{compound} - T^o(\bar{s}_f^o)_{compound} \quad (15.29)$$

where T^o is either 298 K or 537 R depending on whether SI or English units are being used.

Table 15.7 lists values of $\bar{s}^o$ and $\bar{g}_f^o$ for the same substances found in Table 15.1.

Note that, because the elements of the reaction are not considered to be compounds themselves (even though they may be diatomic molecules), they cannot have an entropy of formation. Therefore, for all these elements, we can set $\bar{h}_f^o = \bar{s}_f^o = \bar{g}_f^o = 0$.

EXAMPLE 15.13

As in Example 15.6, the Universe is about to come to an end, except this time it can be saved if you can determine the molar specific entropy of formation and the molar specific Gibbs function of formation for methane gas $CH_4(g)$. Would you please save the Universe again?

Solution

In Example 15.6, we discovered that there is no known reaction by which we can form methane gas by reacting solid carbon with hydrogen gas. However, in Example 15.6, we determined the specific molar heat of formation of CH_4 using the (hypothetical) reaction

$$C(s) + 2(H_2(g)) \rightarrow CH_4$$

Consequently, we can also determine the specific molar entropy of formation and the specific molar Gibbs function of formation using the same reaction. Equation (15.28) gives

$$(\bar{s}_f^o)_{CH_4} = \bar{s}_{CH_4}^o - \left[\left(\frac{n_C}{n_{CH_4}}\right)\bar{s}_C^o + \left(\frac{n_{H_2}}{n_{CH_4}}\right)\bar{s}_{H_2}^o\right]$$

and values for $\bar{s}_{CH_4}^o$, $\bar{s}_C^o$, and $\bar{s}_{H_2}^o$ are found in Table 15.7. Then, the specific molar entropy of formation of methane is

$$(\bar{s}_f^o)_{CH_4} = 186.256[5.740 + 2(130.684)] = 80.852 \text{ kJ/kgmole} \cdot \text{K}$$

Equation (15.29) and Table 15.1 can be used to find the specific molar Gibbs function of formation of methane as

$$\begin{aligned}(\bar{g}_f^o)_{CH_4} &= (\bar{h}_f^o)_{CH_4} - T^o(\bar{s}_f^o)_{CH_4} \\ &= -74.873 \text{ MJ/kgmole} - (298.15 \text{ K})(-80.852 \text{ kJ/kgmole} \cdot \text{K})(1 \text{ MJ}/1000 \text{ kJ}) \\ &= -50.782 \text{ MJ/kgmole}\end{aligned}$$

And you have saved the Universe again. Thanks.

Exercises

37. Using the technique of Example 15.13, determine the molar specific entropy of formation of carbon dioxide, CO_2. **Answer:** $(\bar{s}_f^o)_{CO_2} = 2.92$ kJ/(kgmole·K).

38. Using the results of Exercise 37 and the technique of Example 15.13, determine the molar specific Gibbs function of formation of carbon dioxide, CO_2. **Answer:** $(\bar{g}_f^o)_{CO_2} = -394.4$ MJ/(kgmole·K).

39. Following the technique of Example 15.13, determine the molar specific entropy of formation and the molar specific Gibbs function of formation of water vapor, $H_2O(g)$. **Answer:** $(\bar{s}_f^o)_{H_2O} = -44.400$ kJ/(kgmole·K) and $(\bar{g}_f^o)_{H_2O} = -228.600$ MJ/(kgmole·K).

15.12 CHEMICAL EQUILIBRIUM AND DISSOCIATION

In 1877, the Dutch chemist Jacobus Hendricus van't Hoff (1852–1911) developed the basic principles of chemical equilibrium from the fundamental laws of thermodynamics. For this, among other things, he won the first Nobel Prize for Chemistry in 1901.

Irreversible reaction equations are written as $n_A A + n_B B \rightarrow n_C C + n_D D$ with the implication that A and B are completely and irreversibly consumed in the reaction as C and D are produced. In an irreversible reaction equation, the stoichiometric coefficients represent the actual number of moles of each element present in the reaction vessel. In an equilibrium reaction equation, on the other hand, while the stoichiometric coefficients still represent the number of moles that enter into the equilibrium reaction, they do not necessarily represent the number of moles *present* in the reaction vessel. Therefore, the mole fraction concentrations present cannot be determined from the equilibrium reaction equation alone. Consequently, we continue to use the symbol n_i to represent the total number of moles of species i present in the reaction vessel, but we must now introduce the symbol $v_i \leq n_i$ to represent the number of moles of species i that actually enter into the equilibrium reaction.

For example, at high temperature, the reaction of A and B to form C and D may partially reverse and reform A and B from C and D, and at equilibrium, both the forward and the reverse reactions take place simultaneously, resulting in a reversible "equilibrium" composition containing all four substances. To denote a reversible chemical equilibrium reaction that implies the coexistence of all four substances A, B, C, and D, we use a *double arrow* between the reactants and the products and we use v_i for the stoichiometric coefficients of the reaction, as follows:

$$v_A A + v_B B \leftrightarrows v_C C + v_D D$$

Van't Hoff argued that chemical equilibrium occurs only when a reversible system is in a state of constant uniform pressure and temperature. For a closed system whose only work mode is $p - \forall$, the combined first and second laws in differential form (neglecting any changes in kinetic or potential energy) are

$$\bar{d}Q - \bar{d}W = dU = T\,dS - Td(S_P) - pd\,\forall$$

or

$$T\,dS = dU + p\,d\forall + Td(S_P) = dH - \forall\,dp + Td(S_P) \tag{15.30}$$

where we use the definition of total enthalpy $H = U + p\forall$, and $d(S_P)$ is the differential entropy production rate, which is required to be greater than or equal to zero by the second law of thermodynamics. In Chapter 11, we introduce the Gibbs function G as (see Eq. (11.7)) $G = H - TS$, and on differentiation, it becomes

$$dG = dH - T\,dS - S\,dT$$

Rearranging gives

$$T\,dS = dH - dG - S\,dT$$

and combining this result with Eq. (15.30) gives

$$dG = \forall\,dp - S\,dT - Td(S_P) \tag{15.31}$$

For chemical equilibrium, we require that, T, p, and S_P all be constants; then $dG = 0$ and consequently $G =$ constant. Otherwise, for nonequilibrium chemical reactions that take place at constant temperature and pressure, the second law of thermodynamics requires that

$$dG = -Td(S_P) < 0$$

or

$$G_2 - G_1 = -\int Td(S_P) < 0$$

Consequently, we have the following three results for the Gibbs function of a chemical reaction that occurs at constant temperature and pressure:

a. $dG < 0$ (or $G_2 - G_1 < 0$) implies that the chemical reaction has the *potential to occur* (but this does not imply that it *will* spontaneously occur).
b. $dG = 0$ (or $G_2 = G_1 =$ constant) implies that chemical equilibrium exists and *no further reactions can occur* beyond the equilibrium reactions.
c. $dG > 0$ (or $G_2 - G_1 > 0$) implies that the reaction *cannot occur* at all because to do so would violate the second law of thermodynamics.

For a system that has undergone a chemical reaction at constant temperature and pressure, item b requires that a final state of chemical equilibrium occurs only when $G_{P'} = G_{R'} =$ constant, where P' and R' denote the products and the reactants in the *equilibrium* reaction. Then,

$$G_{R'} = \sum_{R'} v_i \bar{g}_i = \sum_{P'} v_i \bar{g}_i = G_{P'} \tag{15.32}$$

Equation (15.31) can be written on a per unit mole basis for chemical species i with partial pressure p_i and negligible entropy production as

$$d\overline{g}_i = \overline{v}_i\, dp_i - \overline{s}_i\, dT$$

When the system has a constant temperature T and a constant total pressure p_m and the substances involved can be treated as ideal gases, this equation reduces to

$$d\,\overline{g}_i = \overline{v}_i\, dp_i = \Re T(dp_i/p_i)$$

which can be easily integrated from the standard reference state pressure $p°$ to any other state at partial pressure p_i and temperature T as

$$\overline{g}_i(p, T) = \overline{g}_i^{\bullet}(p°, T) + \Re T\, \ln(p_i/p°) \tag{15.33}$$

where $p°$ is the standard reference state pressure of 0.100 MPa, and $\overline{g}_i^{\bullet}$ is known as the molar specific Gibbs function at (unfortunately) a *new* reference state of 0.100 MPa and temperature T.

The new reference temperature is normally chosen to be the mixture temperature T_m rather than the traditional standard reference state temperature of 25.0°C. Consequently, there are new *fourth* reference states for thermodynamic properties discussed in this chapter. They are[4]

1. The *arbitrarily chosen reference state* (e.g., the triple point, as in the steam tables).
2. The *standard reference state* at 0.1 MPa and 25°C.
3. The *absolute value reference state* at 0.1 MPa and 0 K.
4. The mixture temperature reference state at 0.1 MPa and T_m.

Although the introduction of an additional reference state at this point whose temperature is not given a fixed value (like 25.0°C) tends to complicate the logic somewhat, it does simplify the notation and the resulting calculations. Using the definition of the Gibbs function and some simple algebraic manipulation, we can arrive at a working formula for calculating accurate values of $\overline{g}_i(p, T)$ by using property values listed in Table C.16c in *Thermodynamic Tables to accompany Modern Engineering Thermodynamics* and Table 15.7. The required algebraic manipulations are

$$\begin{aligned}\overline{g}_i^{\bullet}(\text{at } p°, T) &= (\overline{g}_f^{\circ})_i + [\overline{g}_i^{\bullet}(\text{at } p°, T) - (\overline{g}_f^{\circ})_i] \\ &= (\overline{g}_f^{\circ})_i + \{[(\overline{h}(\text{at } T) - T\,\overline{s}(\text{at } p°, T)]_i - [(\overline{h}^{\circ}(\text{at } T°) - T°\overline{s}^{\circ}(\text{at } p°, T°)]_i\}\end{aligned}$$

or

$$\overline{g}_i^{\bullet}(\text{at } p°, T) = (\overline{g}_f^{\circ})_i + [\overline{h}(\text{at } T) - \overline{h}^{\circ}(\text{at } T°)]_i - T[\overline{s}(\text{at } p°, T)]_i + T°[\overline{s}^{\circ}(\text{at } p°, T°)]_i \tag{15.34}$$

where $T°$ is the standard reference state temperature of 298 K or 537 R. The superscript ∘ on a quantity implies that it is at the standard reference state, whereas the superscript • implies that it is at the new reference state of $T = T_m$ and 0.100 Mpa. Values for $(\overline{g}_f^{\circ})_i$ and $[\overline{s}^{\circ}(\text{at } p°, T°)]_i$ can be found in Table 15.7, and values for $[\overline{h}(\text{at } T) - \overline{h}^{\circ}(\text{at } T°)]_i$ and $[\overline{s}(\text{at } p°, T)]_i$ (and also $[\overline{s}^{\circ}(\text{at } p°, T°)]_i$) can be found in Table C.16c for various common substances.

Substituting Eq. (15.33) into Eq. (15.32) gives

$$\begin{aligned}\sum_{R'} v_i\overline{g}_i^{\bullet} - \sum_{P'} v_i\overline{g}_i^{\bullet} &= \Re T\left[\sum_{P'} v_i \ln(p_i/p°) - \sum_{R'} v_i \ln(p_i/p°)\right] \\ &= \Re T\left\{\ln\left[\prod_{P'}(p_i/p°)^{v_i}\right] - \ln\left[\prod_{R'}(p_i/p°)^{v_i}\right]\right\} \\ &= \Re T\, \ln\left[\frac{\prod_{P'}(p_i/p°)^{v_i}}{\prod_{R'}(p_i/p°)^{v_i}}\right] = \Re T \ln\left[K_e\right]\end{aligned}$$

[4] Confusing, isn't it?

WHAT DOES THE SYMBOL $\prod()$ MEAN?

In these equations, we use $\prod$ as the symbol for repeated multiplication, just as we use $\sum$ as the repeated summation symbol; that is,

$$\sum_{i=1}^{N}(\alpha_1) = \alpha_1 + \alpha_2 + \alpha_3 + \cdots + \alpha_N,$$

and

$$\prod_{i=1}^{N} = (\alpha_1)\times(\alpha_2)\times(\alpha_3)\times(\cdots)\times(\alpha_N)$$

K_e is the equilibrium constant for the reaction, defined as

$$K_e = \exp\left[\frac{\sum_{R'} v_i\bar{g}_i^\bullet - \sum_{P'} v_i\bar{g}_i^\bullet}{\Re T}\right] \tag{15.35}$$

and, from the previous equation, we also have

The equilibrium constant:

$$K_e = \frac{\prod_{P'}(p_i/p^\circ)^{v_i}}{\prod_{R'}(p_i/p^\circ)^{v_i}} = \frac{(p_C/p^\circ)^{v_C}(p_D/p^\circ)^{v_D}(\cdots)}{(p_A/p^\circ)^{v_A}(p_B/p^\circ)^{v_B}(\cdots)} \tag{15.36}$$

Equation (15.36) indicates that the equilibrium constant is a measure of how much product has been generated by the reaction. Equation (15.36) (or (15.37) later) is normally used to find the actual concentrations v_i if the equilibrium constant K_e is known, whereas Eq. (15.35) is normally used to find K_e if the v_i are known.

EXAMPLE 15.14

Calculate the equilibrium constant for the reversible equilibrium water vapor dissociation reaction equation $H_2O \leftrightarrows H_2 + \frac{1}{2}O_2$ at 0.1 MPa and at the following temperatures: (a) 298 K and (b) 2000. K.

Solution

Here, we do not have an irreversible reaction equation to contend with, so we can find K_e directly from Eq. (15.35). Then,

$$\Re T \ln(K_e) = \sum_{R'} v_i\bar{g}_i^\bullet - \sum_{P'} v_i\bar{g}_i^\bullet = \bar{g}_{H_2O}^\bullet - \bar{g}_{H_2}^\bullet - \tfrac{1}{2}\bar{g}_{O_2}^\bullet$$

where the $\bar{g}_i^\bullet$ are determined from Eq. (15.34).

a. At $T = T^\circ = 298\,\text{K}$, Eq. (15.34) reduces to $\bar{g}_i^\bullet = (\bar{g}_f^\circ)_i$, then

$$\Re T \ln(K_e) = (\bar{g}_f^\circ)_{H_2O} - (\bar{g}_f^\circ)_{H_2} - \tfrac{1}{2}(\bar{g}_f^\circ)_{O_2}$$

and, since H_2 and O_2 are elements, their molar specific Gibbs function of formation is zero. Then, from Table 15.7, we have for water vapor $H_2O(g)$

$$\ln K_e = \left[(\bar{g}_f^\circ)_{H_2O}\right]/\Re T = \frac{-228.583}{0.0083143(298)} = -92.3$$

and

$$K_e = \exp(-92.3) = 8.22\times10^{-41}$$

(Continued)

EXAMPLE 15.14 *(Continued)*

b. At 2000. K = 3600. R and 0.100 MPa, Eq. (15.34) with Tables 15.7 and C.16c in *Thermodynamic Tables to accompany Modern Engineering Thermodynamics* give

$$\begin{aligned}\bar{g}^{\bullet}_{H_2O} &= (\bar{g}^{\circ}_f)_{H_2O} + [(\bar{h}(2000.\text{ K}) - \bar{h}(298\text{ K})]_{H_2O} - 2000.[(\bar{s}(2000.\text{ K})]_{H_2O} + 298[(\bar{s}^{\circ}(298\text{ K})]_{H_2O}\\ &= -228{,}583 + (35{,}540.1 - 4258.3)(2.3258) - 2000.(63.221)(4.1865) + 298(188.833)\\ &= -628900\text{ kJ/kbmole}\end{aligned}$$

$$\begin{aligned}\bar{g}^{\bullet}_{H_2} &= 0 + (26{,}398.5 - 3640.3)(2.3258) - 2000.(44.978)(4.1865) + 298(130.684)\\ &= -285{,}000\text{ kJ/kgmole}\end{aligned}$$

and

$$\begin{aligned}\bar{g}^{\bullet}_{O_2} &= 0 + (29{,}173.9 - 3725.1)(2.3258)\\ &\quad - 2000.(64.168)(4.1865) + 298(205.138) = -417{,}000\text{ kJ/kgmole}\end{aligned}$$

Note: The multipliers 2.3258 and 4.1865 in these equations are necessary to convert the Btu/lbmole and Btu/(lbmole·R) values in Table C.16c into kJ/kgmole and kJ/(kgmole·K), respectively.

Then,

$$\begin{aligned}\Re T \ln K_e &= -629{,}000 - (-285{,}000) - (1/2)(-417{,}000)\\ &= -136{,}000\text{ kJ/kgmole}\end{aligned}$$

so

$$\ln K_e = \frac{-136{,}000}{8.3143(2000.)} = -8.18$$

and

$$K_e = \exp(-8.18) = 2.81 \times 10^{-4}$$

Exercises

40. Find the equilibrium constant for part b of Example 15.14 from the Table C.17 in *Thermodynamic Tables to accompany Modern Engineering Thermodynamics*. Note that the numbers in this table are $\log_{10}(K_e)$, and $\log_{10}(K_e) = 2.30528 \times \log_{10}(K_e) = 2.30528 \times \ln(K_e)$. **Answer**: From Table C.17, we find that $K_e = 10^{-3.531} = 0.000294$.

41. Using the technique of Example 15.14, determine the equilibrium constant for the dissociation of $N_2 \approx 2\,N$ at 2000. K. Compare your answer with that obtained in Table C.17. **Answer**: $\log_{10}(K_e) = 12.02$.

42. Using the technique of Example 15.14, determine the equilibrium constant for the dissociation reaction of water given in Example 15.14 at a temperature of 5300. R ≈ 3000. K. Compare your result with that given in Table C.17. **Answer**: $K_e = 0.0476$.

The magnitude of K_e is a good indicator of the degree to which a reaction goes to completion. Generally, if K_e is less than about 0.01 (or $\ln K_e < -4.6$), then the reaction does not occur to any significant degree. However, if K_e is greater than about 100 (or $\ln K_e > 4.6$), then the reaction essentially goes to completion.

If the components of the reaction are ideal gases that obey the Gibbs-Dalton ideal gas mixture law, then the partial pressures can be expressed in terms of the mole fraction χ_i and the total mixture pressure p_m as (see Eq. (12.23))

$$p_i = \chi_i p_m$$

Then, Eq. (15.36) reduces to

$$K_e = \frac{\prod_{P'} (\chi_i)^{\nu_i}}{\prod_{R'} (\chi_i)^{\nu_i}} \left(\frac{p_m}{p^{\circ}}\right)^{\left(\sum_{P'}\nu_i - \sum_{R'}\nu_i\right)} \tag{15.37}$$

Note that the ν_i and the repeated multiplication ranges (P' and R') in Eqs. (15.36) and (15.37) come from the *equilibrium* reaction equation, but the p_i and χ_i in these equations come from *only the products* of the *irreversible* reaction equation. For example, consider an equilibrium reaction equation for a simple dissociation of the form

$$\nu_A A \leftrightarrows \nu_B B + \nu_C C$$

which is also subject to an overall irreversible reaction equation in which $y\%$ of the A present dissociates into B and C as

$$A \to (1-y)A + y[A_{\text{dissociated}}]$$

Then, the overall irreversible reaction equation can be written as

$$A \to (1-y)A + y\left[(\nu_B/\nu_A)B + (\nu_C/\nu_A)C\right]$$

and the equilibrium constant for this reaction is given by Eq. (15.37) as

$$K_e = \frac{(\chi_B)^{\nu_B}(\chi_C)^{\nu_C}}{(\chi_A)^{\nu_A}}\left(\frac{p_m}{p^\circ}\right)^{(\nu_B+\nu_C-\nu_A)} \tag{15.38}$$

where χ_A, χ_B, and χ_C are determined from the products of the overall irreversible reaction equation as

$$\chi_A = \frac{1-y}{1-y+y(\nu_B/\nu_A+\nu_C/\nu_A)}$$

$$\chi_B = \frac{y(\nu_B/\nu_A)}{1-y+y(\nu_B/\nu_A+\nu_C/\nu_A)}$$

and

$$\chi_C = \frac{y(\nu_C/\nu_A)}{1-y+y(\nu_B/\nu_A+\nu_C/\nu_A)}$$

Table C.17 lists values of the base-10 logarithm of K_e for a variety of simple dissociation reactions of this form at various temperatures.

EXAMPLE 15.15

The overall irreversible carbon dioxide dissociation reaction equation wherein $y\%$ of the CO_2 dissociates into CO and O_2 is

$$CO_2 \to (1-y)(CO_2) + y(CO_2)_{\text{dissociated}}$$

subject to the reversible equilibrium dissociation reaction

$$CO_2 \leftrightarrows CO + \tfrac{1}{2}O_2$$

Then the overall irreversible dissociation reaction equation is

$$CO_2 \to (1-y)(CO_2) + y(CO) + (y/2)(O_2)$$

For this reaction, determine

a. The variation in the degree of dissociation (y) with temperature at a total pressure of 0.100 MPa.
b. The influence of total pressure on the degree of dissociation (y) at 3000. K.

Solution

The total number of moles of product in the overall irreversible reaction equation is $(1-y)+y+y/2 = (2+y)/2$. Then the mole fractions of the products are

$$\chi_{CO_2} = 2(1-y)/(2+y)$$
$$\chi_{CO} = 2y/(2+y)$$
$$\chi_{O_2} = y/(2+y)$$

The reversible equilibrium dissociation reaction equation gives the stoichiometric coefficients, ν_i, as $\nu_{CO_2} = 1.0$, $\nu_{CO} = 1.0$, and $\nu_{O_2} = \frac{1}{2}$. Then, Eq. (15.38) gives the equilibrium constant as

$$K_e = \left[\frac{(\chi_{CO})(\chi_{O_2})^{1/2}}{(\chi_{CO_2})}\right]\left(\frac{p_m}{p^\circ}\right)^{(1+1/2-1)}$$

$$= \left[\frac{\left(\frac{2y}{2+y}\right)\left(\frac{y}{2+y}\right)^{1/2}}{\frac{2(1-y)}{2+y}}\right]\left(\frac{p_m}{p^\circ}\right)^{1/2}$$

$$= \left[\left(\frac{y}{1-y}\right)\left(\frac{y}{2+y}\right)^{1/2}\right]\left(\frac{p_m}{p^\circ}\right)^{1/2}$$

(Continued)

EXAMPLE 15.15 (*Continued*)

a. To determine the required results for part a of this problem, we set $p_m = p° = 0.1$ MPa, and then K_e can be found in Table C.17 for various temperatures. Our task is now to pick several reaction temperatures, look up their corresponding K_e values, and solve the previous equation for y.

Squaring both sides of the preceding reversible equilibrium dissociation reaction equation and rearranging gives a cubic equation in y:

$$y^3 + \frac{3\alpha}{1-\alpha}y - \frac{2\alpha}{1-\alpha} = 0$$

where

$$\alpha = K_e^2\left(\frac{p°}{p_m}\right)$$

A cubic equation in y has a simple algebraic solution. If we let

$$a = \frac{3\alpha}{1-\alpha} \quad \text{and} \quad b = -\frac{2\alpha}{1-\alpha}$$

then the cubic equation becomes $y^3 + ay + b = 0$, and the three roots of this equation are

$$\begin{aligned} y_1 &= A+B \\ y_2 &= -\frac{A+B}{2} + \frac{A-B}{2}\sqrt{-3} \\ y_3 &= -\frac{A+B}{2} - \frac{A-B}{2}\sqrt{-3} \end{aligned}$$

where

$$A = \left(-\frac{b}{2} + \sqrt{\frac{b^2}{4} + \frac{a^3}{27}}\right)^{1/3} \quad \text{and} \quad B = \left(-\frac{b}{2} - \sqrt{\frac{b^2}{4} + \frac{a^3}{27}}\right)^{1/3}$$

Note that, in evaluating these expressions, the cube root of a negative number is computed as the negative of the cube root of the positive value of the number; that is, $(-8)^{1/3} = -(8^{1/3}) = -2$.

If α is less than 1, then $b^2/4 + a^3/27 > 0$ and there is just one real root, given by y_1. But if α is greater than 1, then $b^2/4 + a^3/27 < 0$ and there are three real roots. The following procedure is a simple way of calculating these roots. Let $J = [(-(a/3)^3]^{1/2}$ and $K = J^{1/3}$. Then calculate $L = \arccos[-b/(2J)]$, $M = \cos(L/3)$, and $N = (3^{1/2})\sin(L/3)$. The three roots are then given by

$$\begin{aligned} y_1 &= 2K\cos(L/3) \\ y_2 &= -K(M+N) \\ y_3 &= -K(M-N) \end{aligned}$$

Note that the only valid root here is the one between 0 and 1. Table 15.8 lists several typical results using these solution equations with $p° = 0.1$ MPa.

These results can then be plotted either by hand or with the use of computer software. The final plot showing the variation in the degree of dissociation (y) versus reaction equilibrium temperature is shown in Figure 15.10.

b. With the solution from part a, it is a simple matter to generate reaction equations for various total pressures at $T = 3000$. K. The curve in Figure 15.11 shows the results of this effort and provides a graphical representation of the effect of total pressure on the degree of dissociation (y) at 3000. K.

Table 15.8 Typical Values for the Degree of Dissociation of Carbon Dioxide at Various Temperatures and Pressures with $p° = 0.100$ MPa

Reaction Temperature (K)	Reaction Pressure p_m (MPa)	$\log_{10}(K_e)$	K_e	y
2000. K	0.100 MPa	2.863	0.0014	0.0154
3000. K	0.100 MPa	0.469	0.3396	0.4436
5000. K	0.100 MPa	+1.387	24.378	0.9770
3000. K	1.00 MPa	0.469	0.3396	0.2453
3000. K	10.0 MPa	0.469	0.3396	0.1235

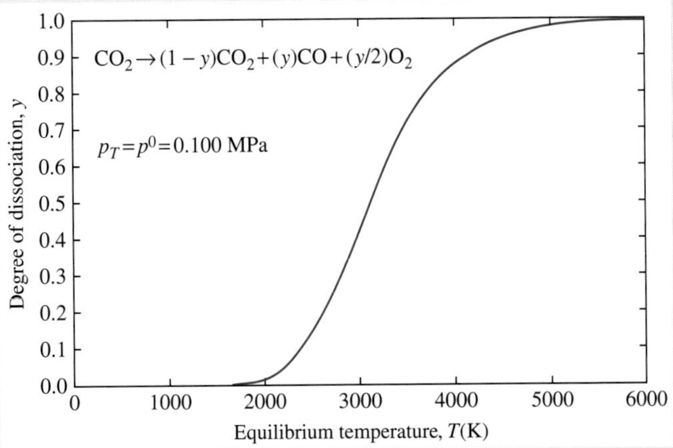

FIGURE 15.10
Example 15.15, Solution, part a.

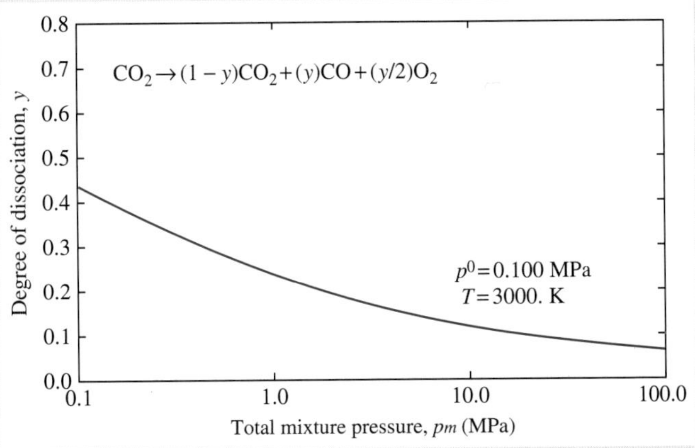

FIGURE 15.11
Example 15.15, Solution, part b.

Exercises

43. Use the solution given in Example 15.15 to determine the dissociation equation for CO_2 at 3000. K and 5.000 MPa. **Answer**: $CO_2 \rightarrow 0.8472(CO_2) + 0.1528(CO) + 0.0764(O_2)$.

44. Use the solution given in Example 15.15 to determine the dissociation equation for CO_2 at 3000. K and 0.01000 MPa. **Answer**: $CO_2 \rightarrow 0.3194(CO_2) + 0.6806(CO) + 0.3403(O_2)$.

45. Use the solution given in Example 15.15 to determine the total mixture pressure that produces the following dissociation reaction for CO_2 at 3000 K: $CO_2 \rightarrow 0.50(CO_2) + 0.50(CO) + 0.25(O_2)$. **Answer**: $p_m = 0.94$ Pa.

An equivalent and perhaps more straightforward way of approaching a dissociation equilibrium reaction is illustrated in the next example.

EXAMPLE 15.16

Determine the amount of H_2 produced as a function of temperature in the thermal dissociation of water vapor at the SRS pressure (i.e., $p_m = p^\circ$).

Solution

The overall irreversible dissociation reaction equation for water vapor is

$$H_2O \rightarrow (1-y)H_2O + y(\nu_{H_2})H_2 + y(\nu_{O_2})O_2$$

(Continued)

EXAMPLE 15.16 (*Continued*)

Table 15.9 Example 15.16, Typical Values

y	$\log K_e$	T(K)
0.001	−4.650	1710
0.010	−3.147	2020
0.100	−1.615	2850
0.500	−0.349	3950

subject to the following reversible equilibrium dissociation reaction equation:

$$H_2O \leftrightarrows H_2 + \frac{1}{2}O_2$$

Thus, $\nu_{H_2} = 1.0$ and $\nu_{O_2} = \frac{1}{2}$, and the overall reaction equation becomes

$$H_2O \rightarrow (1-y)H_2O + yH_2 + (y/2)O_2$$

Thus, $\nu_{H_2O} = 1, \nu_{H_2} = 1$, and $\nu_{O_2} = \frac{1}{2}$, then Eq. (15.38) gives

$$K_e = \frac{\chi_{H_2}\chi_{O_2}^{1/2}}{\chi_{H_2O}}\left(\frac{p_m}{p^\circ}\right)^{(1+1/2-1)}$$

where

$$\chi_{H_2O} = \frac{2(1-y)}{2+y} \quad \chi_{H_2} = \frac{2y}{2+y} \quad \text{and} \quad \chi_{O_2} = \frac{y}{2+y}$$

so that

$$K_e = \frac{y}{1-y}\left(\frac{y}{2+y}\right)^{1/2}(1)^{1/2} = \frac{y}{1-y}\left(\frac{y}{2+y}\right)^{1/2}$$

Arbitrarily choosing values for y and solving for K_e from the previous equation, then looking up the corresponding temperature in Table C.17 in *Thermodynamic Tables to accompany Modern Engineering Thermodynamics* gives the desired relation between the amount of H_2 present (y) and the system temperature (T). Table 15.9 illustrates some typical values.

Exercises

46. Determine the value of the equilibrium constant and the reaction temperature in Example 15.16 when the degree of dissociation is $y = 0.250$. **Answer**: $K_e = 1/9 = 0.1111$, then $T = 3280$ K.
47. If the equilibrium constant in Example 15.16 is 1.0, determine the degree of dissociation y for this reaction. **Answer**: $y = 2/3$.
48. If the degree of dissociation in Example 15.16 is 0.8384, determine the base-10 logarithm of the equilibrium constant and the temperature of the dissociation reaction. **Answer**: $\log_{10}(K_e) = 0.450$ and $T = 5000.$ K.

15.13 RULES FOR CHEMICAL EQUILIBRIUM CONSTANTS

Tables of equilibrium constants are usually limited in size. This limitation can often be overcome by combining reactions available in a table to produce a desired reaction. For example, many tables include equilibrium constants for the water *dissociation* reaction $H_2O \leftrightarrows H_2 + (1/2)O_2$. But what do you do if you need the reverse reaction for the *formation* of water $H_2 + (1/2)O_2 \leftrightarrows H_2O$? Are the equilibrium constants for these two reactions the same? The following three chemical equilibrium constant rules can be used to determine the equilibrium constant for a reaction equation that is not listed in any table but can be constructed from reactions with known equilibrium constants that are listed in a table.

EQUILIBRIUM CONSTANT RULE 1

Let K_{e1} be the equilibrium constant for the reaction $\nu_A A + \nu_B B \leftrightarrows \nu_C C + \nu_D D$ and let K_{e2} be the equilibrium constant for the reverse reaction $\nu_C C + \nu_D D \leftrightarrows \nu_A A + \nu_B B$. Then, these two equilibrium constants are related as follows: $K_{e2} = 1/K_{e1}$.

EQUILIBRIUM CONSTANT RULE 2

Let K_{e1} be the equilibrium constant for the reaction $\nu_A A + \nu_B B \leftrightarrows \nu_C C + \nu_D D$ and let K_{e2} be the equilibrium constant for the reaction $\alpha(\nu_A A + \nu_B B) \leftrightarrows \alpha(\nu_C C + \nu_D D)$, where α is any constant. Then, these two equilibrium constants are related as follows: $K_{e2} = (K_{e1})^{\alpha}$.

EQUILIBRIUM CONSTANT RULE 3

Let K_{e1} be the equilibrium constant for the reaction $\nu_A A + \nu_B B \leftrightarrows \nu_C C + \nu_D D$ and let K_{e2} be the equilibrium constant for a second reaction: $\nu_E E + \nu_F F \leftrightarrows \nu_G G + \nu_H H$. Then, the equilibrium constant for a third reaction, formed by multiplying the first reaction by a constant α and adding it to the second reaction multiplied by a constant β,

$$\alpha(\nu_A A + \nu_B B) + \beta(\nu_E E + \nu_F F) \leftrightarrows \alpha(\nu_C C + \nu_D D) + \beta(\nu_G G + \nu_H H)$$

is

$$K_{e3} = (K_{e1})^{\alpha}(K_{e2})^{\beta}.$$

EXAMPLE 15.17

Determine the equilibrium constants for the following reactions at 5000. K.

a. $H_2 + (1/2)O_2 \leftrightarrows H_2O$
b. $O_2 + N_2 \leftrightarrows 2NO$
c. $O_2 + 3.76\,N_2 \leftrightarrows 2O + 7.52\,N$

Solution

a. Table C.17 gives equilibrium constant values for the spontaneous dissociation of water into hydrogen and water at 5000. K ($H_2O \leftrightarrows H_2 + (1/2)O_2$) as $K_{e1} = 10^{0.450} = 2.82$. The formation of water from hydrogen and oxygen gases is simply the reverse reaction, so rule 1 can be used to determine its equilibrium constant as $K_{e2} = 1/K_{e1} = 1/2.82 = 0.355$.
b. The reaction $O_2 + N_2 \leftrightarrows 2NO$ is the same as the reaction $2[(1/2)O_2 + (1/2)N_2] \leftrightarrows 2\,NO$, which can be obtained from the reaction $(1/2)O_2 + (1/2)N_2 \leftrightarrows NO$ that appears in Table C.17 by multiplying it by $\alpha = 2$. From Table C.17 at 5000. K, we find that the equilibrium constant for the reaction $(1/2)O_2 + (1/2)N_2 \leftrightarrows NO$ is $K_{e1} = 10^{-0.298} = 0.504$. Then, using rule 2, we can calculate the equilibrium constant for the reaction $O_2 + N_2 \leftrightarrows 2NO$ as $K_{e2} = (K_{e1})^{\alpha}$.
c. In Table C.17, we find the reactions $O_2 \leftrightarrows 2O$ and $N_2 \leftrightarrows 2N$. If we multiply the second reaction by 3.76 and add it to the first reaction, we get the given reaction $O_2 + 3.76N_2 \leftrightarrows 2O + 7.52N$ and we can use rule 3 with $\alpha = 1$ and $\beta = 3.76$. At 5000. K, the equilibrium constant for the reaction $O_2 \leftrightarrows 2O$ is $K_{e1} = 10^{-1.719} = 52.4$ and the equilibrium constant for the reaction $N_2 \leftrightarrows 2N$ is $K_{e2} = 10^{-0.570} = 0.269$, then rule 3 gives the equilibrium constant for the combined reaction as $K_{e3} = (K_{e1})^1(K_{e2})^{3.76} = (52.4)^1(0.296)^{3.76} = 0.376$.

Exercises

49. Determine the equilibrium constant for the reaction $H_2 + (1/2)O_2 \leftrightarrows H_2O$ at 2000. K. **Answer:** $K_e = 3396$.
50. Determine the equilibrium constant for the reaction at 5000. K. **Answer:** $K_e = 1/0.2535 = 3.945$.
51. Determine the equilibrium constant for the reaction $O_2 + N_2 \leftrightarrows NO + O + N$ at 5000. K. **Answer:** $K_e = 1.890$.

15.14 THE VAN'T HOFF EQUATION

Both the equilibrium constant K_e and the molar Gibbs function $\bar{g}_i^{\circ}$ depend on the mixture temperature T. To investigate the temperature dependence of K_e, we differentiate Eq. (15.35) with respect to temperature to obtain

$$\frac{dK_e}{dT} = -\frac{1}{\Re T^2}\left[\left(\sum_{R'} \nu_i \bar{g}_i^{\circ} - \sum_{P'} \nu_i \bar{g}_i^{\circ}\right) - T\frac{d}{dT}\left(\sum_{R'} \nu_i \bar{g}_i^{\circ} - \sum_{P'} \nu_i \bar{g}_i^{\circ}\right)\right] K_e$$

From Eq. (11.8), we have the relation $dg = v\,dp - s\,dT$, so that, for a constant pressure process, we can write

$$\frac{d\bar{g}_i^{\bullet}}{dT} = -\bar{s}_i^{\bullet}$$

Introducing this result along with the definition of the molar specific Gibbs function, $\bar{g} = \bar{h} - T\bar{s}$, into the equation for dK_e/dT and rearranging gives

$$\frac{1}{K_e}\left(\frac{dK_e}{dT}\right) = \frac{d(\ln K_e)}{dT} = -\frac{1}{\Re T^2}\left(\sum_R \nu_i \bar{h}_i^{\bullet} - \sum_P \nu_i \bar{h}_i^{\bullet}\right) = \frac{\dot{Q}_r}{\Re T^2}$$

or

The van't Hoff equation:

$$\frac{d(\ln K_e)}{dT} = \frac{\dot{Q}_r}{\Re T^2} \quad (15.39)$$

where $\dot{Q}_r$ is the heat transfer rate of the reaction. This equation is known as the van't Hoff equation. It shows that, for a heat-producing (i.e., exothermic) reaction, the value of K_e decreases when the reaction temperature increases and increases when the reaction temperature decreases. While for a heat-absorbing (i.e., endothermic) reaction, the value of K_e increases when the temperature of the reaction increases and decreases when the reaction temperature decreases. Since the equilibrium constant is a relative measure of the amount of product present, by changing the reaction temperature, we can change the amount of product formed. For example, consider the equilibrium equation for the formation of ammonia from hydrogen and nitrogen gas, $N_2 + 3(H_2) \leftrightarrows 2(NH_3) - 91.25$ kJ. We can increase the amount of ammonia produced by increasing the K_e of the reaction. Since the reaction is exothermic, this can be done by lowering the reaction temperature.

15.15 FUEL CELLS

The highly inefficient heat engine energy conversion technology that provided the mobile power for the technological developments of the 17th through the 20th centuries is coming to an end. The thermal combustion of chemical fuel to propel the heat engines of the past has come to be the source of numerous worldwide social problems today. Chemical and thermal pollution of the Earth's air and water plus a continuously diminishing supply of suitable fuels is clearly signaling the end of the heat engine era. But, what other, less ecologically damaging energy conversion technologies do we have to support the societies of the next few centuries?

What about electrical batteries? Technically, a battery consists of two or more electrochemical *cells* that convert chemical energy directly into electrical energy. The term *battery*, however, is often used to describe a single cell. Figure 15.12 illustrates the basic elements of an electrochemical cell.

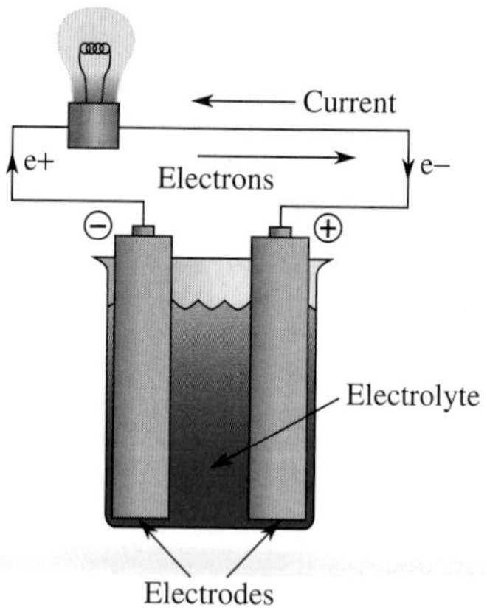

FIGURE 15.12
The operation of a basic electrochemical cell using one electrolyte. Electrical *current* is the time rate of change of electrical charge, originally defined by Benjamin Franklin to flow from the positive to the negative terminal. However, since electrons are the charge carriers for current in conductors and the charge of an electron is negative, the direction of electron flow is actually from the negative to the positive terminal. Nonetheless, we maintain the old concept of current flow from positive to negative and call this the *conventional flow* notation.

Ordinary commercial batteries are *closed systems* because no mass crosses their boundaries; consequently, as they are used, they consume their stored electrical energy and become discharged. If the electrolyte needed to operate the cell were continuously supplied from outside the cell, then it would become an *open system* and would never become discharged. Such an open system electrochemical cell is called a *fuel cell*.

Since a fuel cell (Figure 15.13) does not produce heat as its primary energy conversion mode, it is therefore not a heat engine. Consequently, fuel cells are not subject to the severe limitation of the Carnot (heat engine) efficiency, and fuel cell efficiencies can approach 100% under the proper conditions.

Consider a general steady state, steady flow, open system. Neglecting any changes in kinetic or potential energy, the energy rate balance on this system gives its work transport rate of energy (i.e., power) as

$$\dot{W} = \sum_{\text{in}} \dot{m}_i h_i - \sum_{\text{out}} \dot{m}_i h_i + \dot{Q} = \sum_{\text{in}} \dot{n}_i \bar{h}_i - \sum_{\text{out}} \dot{n}_i \bar{h}_i + \dot{Q}$$

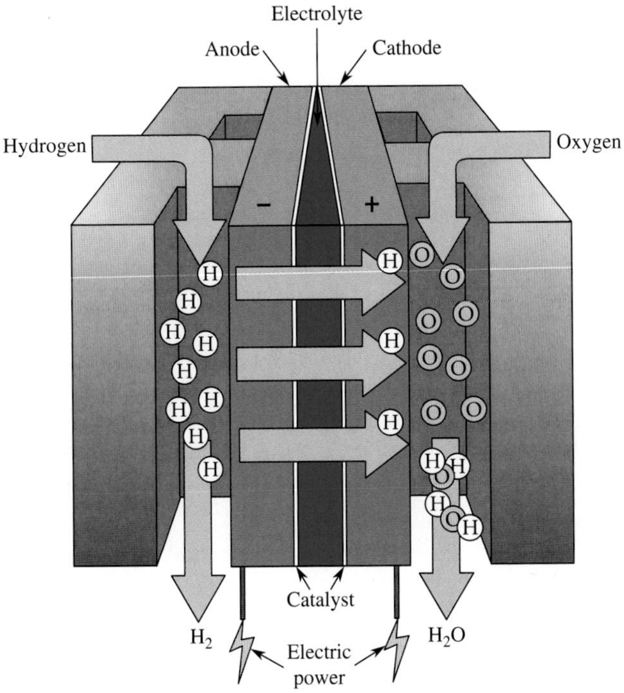

FIGURE 15.13
Fuel cell and fuel battery technology.

WHY IS AN ELECTROCHEMICAL CELL CALLED A *BATTERY*?

The term *battery* was introduced by the American scientist and diplomat Benjamin Franklin (1706–1790) for describing a group of interconnected capacitors. Though it technically means two or more items (like a cannon in an artillery battery), today, single electrochemical cells are often referred to as *batteries* (e.g., the cells in a flashlight should be called *flashlight cells* not *flashlight batteries*).

WHAT IS A FUEL CELL?

A *fuel cell* is an electrochemical open energy conversion system that converts the chemical energy of its fuel *directly* into electrical work energy output by a chemical oxidation reaction.

WHO INVENTED THE FUEL CELL?

Though fuel cells offer an exciting high-efficiency alternative to traditional heat engine energy conversion technology, they are not a new concept. The credit for developing the first fuel cell is given to the British physicist and lawyer Sir William Robert Grove (1811–1896). In 1839, he constructed a fuel battery that consisted of a test tube containing a platinum electrode inverted in a sulfuric acid solution (Figure 15.14). When an electric current was passed through the cell to charge it, the water in the electrolyte decomposed into hydrogen and oxygen inside the inverted test tube. When an electrical load was applied to the electrodes, the hydrogen and oxygen recombined again to form water and produced an electrical current. It is easy to see how this can become a true open system fuel cell if the hydrogen and oxygen are continuously supplied from outside the system.

(*Continued*)

WHO INVENTED THE FUEL CELL? *Continued*

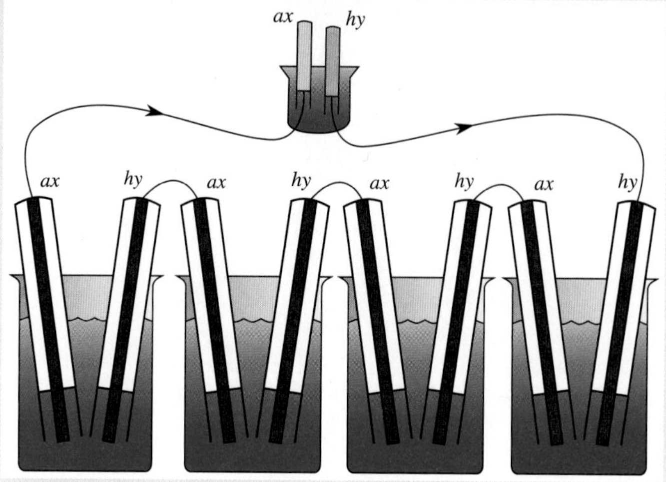

FIGURE 15.14
Grove's fuel cell/battery.

Fuel cell technological development continued sporadically throughout the 19th century until the dramatic technological developments produced by the American inventor Thomas Alva Edison (1847–1931). Edison's development of the light-bulb and associated electrical generating and distribution system technology completely dominated the development of electrical technology for 50 years. Beginning in the 1960s, with the emergence of the United States space program, interest in fuel cell technology was rekindled and its development continues today. The outstanding energy conversion capability of fuel cells plus their low pollution potential and wide variety of operating fuels make fuel cells one of the leading contenders for heat engine replacement in the 21st century.

and if the system has isothermal boundaries at temperature T_b, then the entropy rate balance gives its heat transport rate as

$$\dot{Q} = T_b\left(\sum_{\text{out}} \dot{m}_i s_i - \sum_{\text{in}} \dot{m}_i s_i - \dot{S}_P\right) = T_b\left(\sum_{\text{out}} \dot{n}_i \bar{s}_i - \sum_{\text{in}} \dot{n}_i \bar{s}_i - \dot{S}_P\right)$$

Combining these two equations and using the definition of the molar specific Gibbs function, $\bar{g} = \bar{h} - T\bar{s}$, gives

$$\begin{aligned}\dot{W} &= \sum_{\text{in}} \dot{m}_i (h - T_b s)_i - \sum_{\text{out}} \dot{m}_i (h - T_b s)_i - T_b \dot{S}_P \\ &= \sum_{\text{in}} \dot{n}\left(\bar{h} - T_b \bar{s}\right)_i - \sum_{\text{out}} \dot{n}_i\left(\bar{h} - T_b \bar{s}\right)_i - T_b \dot{S}_P \\ &= \sum_{\text{in}} \dot{n}_i \bar{g}_i - \sum_{\text{out}} \dot{n}_i \bar{g}_i - T_b \dot{S}_P\end{aligned} \tag{15.40}$$

where we assume that the temperature of the fuel cell reactants and products are the same as the system boundary temperature (i.e., $T_R = T_P = T_b$). Note that the second law of thermodynamics requires that $\dot{S}_P \geq 0$.

We can now calculate the "reaction efficiency," η_r, with the following general formula:

$$\eta_r = \frac{\text{The desired result}}{\text{What it costs}} = \frac{\dot{W}}{\sum_{\text{in}} \dot{m}_i h_i - \sum_{\text{out}} \dot{m}_i h_i} = \frac{\dot{W}}{\sum_{\text{in}} \dot{n}_i \bar{h}_i - \sum_{\text{out}} \dot{n}_i \bar{h}_i}$$

or

$$\eta_r = \frac{\sum_{\text{in}} \dot{m}_i g_i - \sum_{\text{out}} \dot{m}_i g_i - T_b \dot{S}_P}{\sum_{\text{in}} \dot{m}_i h_i - \sum_{\text{out}} \dot{m}_i h_i} = \frac{\sum_{\text{in}} \dot{n}_i \bar{g}_i - \sum_{\text{out}} \dot{n}_i \bar{g}_i - T_b \dot{S}_P}{\sum_{\text{in}} \dot{n}_i \bar{h}_i - \sum_{\text{out}} \dot{n}_i \bar{h}_i} \tag{15.41}$$

The maximum power $\dot{W}_{max}$ and maximum reaction efficiency $(\eta_r)_{max}$ occur when the device is completely reversible or $\dot{S}_P = 0$. Then,

$$(\eta_r)_{max} = \frac{\sum_{in} \dot{m}_i g_i - \sum_{out} \dot{m}_i g_i}{\sum_{in} \dot{m}_i h_i - \sum_{out} \dot{m}_i h_i} = \frac{\sum_{in} \dot{n}_i \bar{g}_i - \sum_{out} \dot{n}_i \bar{g}_i}{\sum_{in} \dot{n}_i \bar{h}_i - \sum_{out} \dot{n}_i \bar{h}_i} \tag{15.42}$$

Utilizing Eq. (15.33), the molar forms of Eqs. (15.40), (15.41), and (15.42) are

$$\dot{W} = \sum_R \dot{n}_i \bar{g}_i^\bullet - \sum_P \dot{n}_i \bar{g}_i^\bullet - T_b \dot{S}_P + \Re T \ln\left[\prod_R (p_i/p^\circ)^{\dot{n}_i} / \prod_P (p_i/p^\circ)^{\dot{n}_i}\right] \tag{15.43}$$

$$\eta_r = \frac{\sum_R \dot{n}_i \bar{g}_i^\bullet - \sum_P \dot{n}_i \bar{g}_i^\bullet - T_b \dot{S}_P + \Re T \ln\left[\prod_R (p_i/p^\circ)^{\dot{n}_i} / \prod_P (p_i/p^\circ)^{\dot{n}_i}\right]}{\sum_R \dot{n}_i \bar{h}_i - \sum_P \dot{n}_i \bar{h}_i} \tag{15.44}$$

and

$$(\eta_r)_{\substack{\text{maximum} \\ \text{fuel cell} \\ \text{efficiency}}} = \frac{\sum_R (n_i/n_{\text{fuel}}) \bar{g}_i^\bullet - \sum_P (n_i/n_{\text{fuel}}) \bar{g}_i^\bullet + \Re T \ln\left[\prod_R (p_i/p^\circ)^{(n_i/n_{\text{fuel}})} / \prod_P (p_i/p^\circ)^{(n_i/n_{\text{fuel}})}\right]}{\sum_R (n_i/n_{\text{fuel}}) \bar{h}_i - \sum_P (n_i/n_{\text{fuel}}) \bar{h}_i} \tag{15.45}$$

If the reactants or products are *unmixed* and the pressure of each species (or component) in the reaction is 0.1 MPa, then $p_i = p^\circ$ and $\prod_P (p_i/p^\circ)^{n_i} = \prod_R (p_i/p^\circ)^{n_i} = 1.0$ in these three equations. Otherwise, if the reactants are *premixed* or the product gases are *mixed* (as they normally are) at a total pressure p_m, then the partial pressure of each component gas must be determined. Further, if all the gases present are ideal gases that obey the Gibbs-Dalton ideal gas mixture law, then the partial pressures can be expressed in terms of the mole fractions as $p_i/p^\circ = (\chi_i p_m)/p^\circ$, where x_i is the mole fraction of gas i and $p^\circ = 0.1$ MPa. Then,

$$\ln\left[\prod (p_i/p^\circ)^{n_i}\right] = \ln\left[\prod (\chi_i p_m/p^\circ)^{n_i}\right]$$

in Eqs. (15.43), (15.44), and (15.45).

In the case of a fuel cell, the output power appears as an electrical current I flowing through a potential (voltage) difference ϕ, and using Ohm's law we can calculate the actual power output of the cell as

$$\dot{W} = \phi I = I^2 R_e$$

where ϕ is the cell voltage, I is its current flow, and R_e is the external resistance (recall that work *output* must be positive with our sign convention). Combining this equation with Eq. (15.40) gives the entropy production rate of the fuel cell as

$$(\dot{S}_P)_{\substack{\text{fuel} \\ \text{cell}}} = \left(\sum_P \dot{n}_i \bar{g}_i - \sum_R \dot{n}_i \bar{g} + \phi I\right) / T_b \geq 0 \tag{15.46}$$

Also, it can be shown that the electrical current I produced by a fuel cell is given by

$$I = (\dot{n}_{\text{fuel}}) jF$$

where I is in amperes when $\dot{n}_{\text{fuel}}$ is in kgmole/s. In this equation j is the total valence of the fuel ions in kgmole of electrons per kgmole of fuel, and F is Faraday's constant, defined as

$$\begin{aligned} F &= (6.023 \times 10^{26} \text{electrons/kgmole electrons})(1.602 \times 10^{-19} \text{coulombs/electron}) \\ &= 96.487 \text{ kilocoulombs/kgmole electrons} \end{aligned}$$

but since 1 coulomb = 1 joule/volt = 1 J/V, this can be written as

$$F = 96{,}487 \text{ kJ/(V} \cdot \text{kgmole electrons)}$$

WHAT IS j?

The letter j represents the number of electrons involved in the cell reaction per kgmole of fuel consumed. For example, for any hydrocarbon fuel, C_nH_m, $j = n(4) + m(1) = 4n + m$ kgmole of electrons per kgmole of C_nH_m.

When there is no electron flow in the external circuit, there are no losses within the fuel cell and we can write the maximum (or reversible) power output of a fuel cell as approximately[5]

$$\dot{W}_{\max} \approx \phi_o I = \phi_o(\dot{n}_{\text{fuel}})jF \tag{15.47}$$

where ϕ_o is the open circuit voltage. Then, the maximum reaction efficiency of the fuel cell becomes

$$(\eta_r)_{\substack{\max \\ \text{fuel cell}}} = \frac{(\phi_o jF)}{\sum_R (n_i/n_{\text{fuel}})\bar{h}_i - \sum_P (n_i/n_{\text{fuel}})\bar{h}_i}$$

EXAMPLE 15.18

Determine the maximum theoretical reaction efficiency, open circuit voltage, and maximum theoretical work output per mole of hydrogen consumed for the hydrogen–oxygen fuel cell operating at 25.0°C and 0.100 MPa, shown in Figure 15.15.

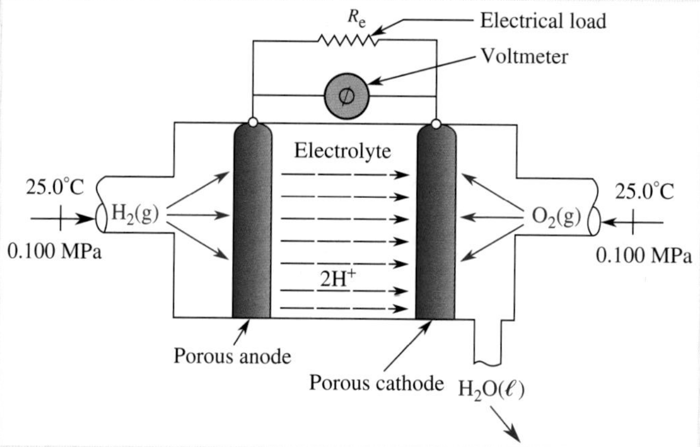

FIGURE 15.15
Example 15.18.

The anode reaction is $H_2(g) \to 2(H^+) + 2(e^-)$.
The cathode reaction is $0.5[O_2(g)] + 2(H^+) + 2(e^-) \to H_2O(\ell)$.
The overall reaction is $H_2(g) + 0.5[O_2(g)] \to H_2O(\ell)$.

Also, $(p_i/p^\circ)_{H_2} = (p_i/p^\circ)_{O_2} = 1.0$.

Solution

Equation (15.45) can be used to calculate the maximum energy conversion reaction efficiency as

$$(\eta_r)_{\max} = \frac{\bar{g}^\bullet_{H_2(g)} + (0.5)(\bar{g}^\bullet_{O_2(g)}) - \bar{g}^\bullet_{H_2O(\ell)} + \Re T \ln 1.0}{\bar{h}_{H_2(g)} + (0.5)(\bar{h}_{O_2(g)}) - \bar{h}_{H_2O(\ell)}}$$

where the $\bar{g}_i^\bullet$ are determined from Eq. (15.34), and the $\bar{h}_i$ are determined from Eq. (15.14). Since the pressure and temperature given in this problem statement are the standard reference state values, Eqs. (15.14) and (15.34) reduce to $\bar{h}_i = (\bar{h}_f^\circ)_i$ and $\bar{g}_i^\bullet = (\bar{g}_f^\circ)_i$. Then, the maximum theoretical reaction efficiency equation becomes

$$(\eta_r)_{\max} = \frac{(\bar{g}_f^\circ)_{H_2(g)} + (0.5)(\bar{g}_f^\circ)_{O_2(g)} - (\bar{g}_f^\circ)_{H_2O(\ell)} + \Re T \ln 1.0}{(\bar{h}_f^\circ)_{H_2(g)} + (0.5)(\bar{h}_f^\circ)_{O_2(g)} - (\bar{h}_f^\circ)_{H_2O(\ell)}}$$

[5] A small amount of internal irreversibility is generated in producing ϕ_o, so ϕ_o is slightly less than $\phi_{rev} = \phi_{max}$.

and, since the enthalpies and Gibbs functions of formation of the elements H_2 and O_2 are always zero, the maximum reaction efficiency becomes (using Tables 15.1 and 15.7)

$$(\eta_r)_{max} = \frac{(\bar{g}_f^\circ)_{H_2O(\ell)}}{(\bar{h}_f^\circ)_{H_2O(\ell)}} = \frac{237.178\ \text{MJ/kgmole}}{285.838\ \text{MJ/kgmole}} = 0.830$$

or 83%. The theoretical open circuit voltage can now be determined from Eq. (15.48) as

$$\begin{aligned}\phi_o &= \left[\sum_R (n_i/n_{fuel})\bar{h}_i - \sum_P (n_i/n_{fuel})\bar{h}_i\right]\left[(\eta_r)_{max}\right]/jF \\ &= \left[-(n_{H_2O}/n_{H_2})(\bar{h}_f^\circ)_{H_2O(\ell)}\right]\left[(\eta_r)_{max}\right]/jF \\ &= \frac{\left[-(1\ \text{kgmole H}_2\text{O/kgmole H}_2)(-285{,}838\ \text{KJ/kgmole H}_2\text{O})\right](0.8298)}{(2.00\ \text{kgmole electrons/kgmole H}_2)\left[96{,}487\ \text{kJ/(V}\cdot\text{kgmole electrons)}\right]} \\ &= 1.23\ \text{V}\end{aligned}$$

where $j = 2.00$ kgmole of electrons per kgmole of H_2 (i.e., the valence of $2H^+$). Finally, Eq. (15.47) can be used to find

$$\begin{aligned}\dot{W}_{max}/\dot{n}_{fuel} &= W_{max}/n_{fuel} = \phi_o jF \\ &= (1.23\ \text{V})(2.00\ \text{kgmole electrons/kgmole H}_2) \\ &\quad\times\left[96{,}487\ \text{KJ/(V}\cdot\text{kgmole electrons)}\right] \\ &= 237{,}000\ \text{kJ/kgmole H}_2 = \left(\bar{g}_f^\circ\right)_{H_2O(\ell)}\ \text{(to significant figures)}\end{aligned}$$

Exercises

52. Determine the power produced by the fuel cell discussed in Example 15.18 per kilogram of H_2 rather than per kgmole of H_2. **Answer:** $\dot{W}/\dot{m} = 118{,}600$ kJ/(kg H_2).

53. Determine the maximum thermal efficiency of the fuel cell discussed in Example 15.18 when the product H_2O is at 116°C = 700. R and 0.100 MPa and the reactants are at the standard reference state. **Answer:** $(\eta_T)_{max} = 90.1\%$.

54. What would the theoretical open circuit voltage be for the fuel cell discussed in Example 15.18 if product H_2O was at 116°C = 700. R and 0.100 MPa and the reactants were at the standard reference state. **Answer:** $\phi_o = 1.32$ V.

Table 15.10 lists the maximum theoretical reaction efficiencies and open circuit voltages for a variety of fuel cell materials at the standard reference state. Those reactions showing efficiencies greater than 100% must absorb heat from the surroundings to maintain steady state operation.

Table 15.10 Fuel Cell Maximum Reaction Efficiency and Open Circuit Voltage for Various Fuels at 25.0°C and 0.100 MPa

Fuel Reaction	ϕ_o(V)	$(\eta_r)_{max}$(%)
$H_2 + 0.5(O_2) \rightarrow H_2O(\ell)$	1.23	83.0
$CO + 0.5(O_2) \rightarrow CO_2$	1.33	90.9
$C(s) + O_2 \rightarrow CO_2$	1.02	100.2
$C_3H_8 + 5(O_2) \rightarrow 3(CO_2) + 4(H_2O(g))$	1.08	101.5
$C_8H_{18}(g) + 12.5(O_2) \rightarrow 8(CO_2) + 9(H_2O(\ell))$	1.10	96.3

Note: Unlabeled elements and compounds are in a gaseous (g) physical state.

15.16 CHEMICAL AVAILABILITY

In Chapter 10, we define the thermodynamic property *availability* as the *maximum reversible useful work that can be produced by a system*. Equation (15.43) describes the rate of work produced or absorbed by a steady state, open system with negligible flow stream kinetic and potential energies. This equation becomes the *maximum reversible* work rate when there are no losses within the system, or $\dot{S}_P = 0$. Since this equation focuses on the energy transport rate due to the chemical species crossing the system boundary, it represents the net chemical flow availability of the system, or

$$\left(\dot{A}_{\substack{\text{flow}\\\text{chemical}}}\right)_{\text{net}} = \sum_R \dot{n}_i[(\bar{a}_f)_i]_{\text{chemical}} - \sum_P \dot{n}_i[(\bar{a}_f)_i]_{\text{chemical}}$$
$$= \sum_R \dot{n}_i\bar{g}_i^{\circ} - \sum_P \dot{n}_i\bar{g}_i^{\circ} + \Re T \ln\left[\frac{\prod_R (p_i/p^{\circ})^{\dot{n}_i}}{\prod_P (p_i/p^{\circ})^{\dot{n}_i}}\right] \tag{15.49}$$

and the specific molar flow availability of chemical species i is

$$[(\bar{a}_f)_i]_{\text{chemical}} = \bar{g}_i^{\circ} + \Re T \ln\left[\frac{p_i}{p^{\circ}}\right] \tag{15.50}$$

The following example illustrates the use of this material.

EXAMPLE 15.19

Determine the net molar specific flow availability of the hydrogen–oxygen fuel cell operating at 25°C and 0.1 MPa analyzed in Example 15.18. Assume that the ground state is the standard reference state, SRS (25°C and 0.1 MPa).

Solution

Recall that the reaction is $H_2 + 0.5\ O_2 \rightarrow H_2O$. The fuel cell has three flow streams (H_2, O_2, and H_2O), all at the SRS pressure $p_i = p^{\circ}$. Also, since the Gibbs function at the SRS reduces to the Gibbs function of formation, Eq. (15.50) gives $[(\bar{a}_f)_i]_{\text{chemical}} = \bar{g}_i^{\circ} + \Re T\ \ln[1] = (\bar{g}_f^{\circ})_i$. Consequently, $[(\bar{a}_f)_{H_2}]_{\text{chemical}} = [(\bar{a}_f)_{O_2}]_{\text{chemical}}$ and $[(\bar{a}_f)_{H_2O}]_{\text{chemical}} = (\bar{g}_f^{\circ})_{H_2O(\ell)} = -237.178$ MJ/kgmole.

Then, the net molar specific flow availability is given by

$$\left(\dot{a}_{\substack{\text{flow}\\\text{chemical}}}\right)_{\text{net}} = \frac{\left(\dot{A}_{\substack{\text{flow}\\\text{chemical}}}\right)_{\text{net}}}{n_f} = \sum_R (n_i/n_f)[(\bar{a}_f)_i]_{\text{chemical}} - \sum_P (n_i/n_f)[(\bar{a}_f)_i]_{\text{chemical}}$$
$$= (n_{H_2}/n_{H_2})[(\bar{a}_f)_{H_2}]_{\text{chemical}} + (n_{O_2}/n_{H_2})[(\bar{a}_f)_{O_2}]_{\text{chemical}} - (n_{H_2O}/n_{H_2})[(\bar{a}_f)_{H_2O}]_{\text{chemical}}$$
$$= 0 + 0 - (1)(-237.178) = 237.178\ \text{MJ/kgmole}\ H_2$$

Exercises

55. Determine the net chemical flow availability of the fuel cell described in Example 15.19 when hydrogen is consumed at a rate of 2.00 kgmole/min. **Answer**: $[\dot{A}(_{\text{flow}})_{\text{net}}]_{\text{chemical}} = 474.4$ MJ/min.
56. Determine the molar specific chemical flow availability of oxygen in air at a total pressure of 3.50 MPa and the SRS temperature of 25.0°C. Assume air consists of a mixture of 21.0% oxygen and 79.0% nitrogen on a molar basis. **Answer**: $[\dot{a}(_{\text{flow}})_{O2}]_{\text{chemical}} = 4940$ kJ/kgmole O_2.
57. If all the flow streams entering and exiting the hydrogen–oxygen fuel cell discussed in Example 15.19 are at 1.00 MPa instead of 0.100 MPa, determine the net chemical flow availability of the fuel cell per kgmole of hydrogen consumed. **Answer**: $[\dot{a}(_{\text{flow}})_{\text{net}}]_{\text{chemical}} = 3090$ kJ/kgmole H_2.

SUMMARY

In this chapter, we deal with the fundamental elements of chemical thermodynamics. Chemistry has its roots in thousands of years of alchemy; its accurate mathematical notation is relatively recent. The 19th century stoichiometric mass balance and the basic concepts of stereochemistry provide a framework on which an accurate combustion analysis of organic fuels can be built. Concepts such as percent of theoretical air, fuel modeling, heat of formation, and the standard reference state plus the first law of thermodynamics applied to a chemical reaction lead to a useful understanding of the heat of combustion of a chemical compound. The adiabatic flame temperature and maximum explosion pressure calculations provide conservative upper bounds for real combustion processes. The introduction of the third law of thermodynamics provides the basis on which to build an absolute entropy scale that can be used to determine chemical reaction irreversibilities via the entropy balance. Also, the Gibbs function from the combined first and second laws is found to be a controlling factor in chemical reactions, chemical equilibrium, and dissociation reactions. Finally, fuel cell analysis provides a means of investigating the maximum possible work that can be produced directly from a chemical reaction.

The often complex formulae of chemical thermodynamics provide an excellent topic for personal computer software development. In this chapter, simple Excel spreadsheet programs are developed for calculating heats of combustion and dissociation equilibrium conditions. The reader is encouraged to utilize and modify this material, to take full advantage of the computer's ability to remove the tedium of calculation and expand the scope of the analysis.

Some of the more important equations introduced in this chapter follow. Do not attempt to use them blindly without understanding their limitations. Please refer to the text material where they were introduced to gain an understanding of their use and limitations.

1. Air/fuel ratio. For a chemical reaction combustion equation of the form

$$\alpha(\text{Fuel}) + \beta(O_2 + 3.76\,N_2) \rightarrow \text{Products}$$

   the molar and mass air/fuel ratios are

$$(A/F)_{\text{molar}} = \frac{\alpha(1+3.76)}{\beta} = \frac{\alpha(4.76)}{\beta}$$
$$(A/F)_{\text{mass}} = \frac{\alpha(4.76)(M_{\text{air}})}{\beta(M_{\text{fuel}})} = \frac{\alpha(4.76)(28.97)}{\beta(M_{\text{fuel}})} = \frac{\alpha(137.9)}{\beta(M_{\text{fuel}})}$$

   where α and β are the stoichiometric coefficients in the reaction equation, and M_{fuel} is the molecular mass of the fuel.
2. Stoichiometric reaction equation for the combustion of a hydrocarbon of the form C_nH_m in 100% theoretical air:

$$C_nH_m + \left(n + \frac{m}{4}\right)[O_2 + 3.76(N_2)]6n(CO_2) + \left(\frac{m}{2}\right)(H_2O) + 3.76\left(n + \frac{m}{4}\right)(N_2)$$

3. The temperature at which water condenses out of the products of combustion:

$$T_{\text{condense}} = T_{\text{sat}}(p_{H_2O})$$

   where $p_{H_2O} = (\chi_{H_2O}/\chi_{\text{total}})p_{\text{total}}$ is the partial pressure of the water vapor in the products.
4. The standard reference state (SRS):

$$\text{Temperature: } T_{\text{SRS}} = T^\circ = 25.0°\text{C} = 298\text{ K} = 77.0°\text{F} = 537\text{ R}$$

$$\text{Pressure: } p_{\text{SRS}} = p^\circ = 0.100\text{ MPa} = 14.5\text{ psia} \approx 1\text{ atmosphere}$$

5. Molar specific heat of formation of a hydrocarbon C_nH_m in MJ/kgmole:

$$(\bar{q}_f^\circ)_{C_nH_m} = -[n(393.5) + (m/2)(285.5) + \text{HHV}_{C_nH_m}]$$

6. Molar specific enthalpy of formation of a compound at the standard reference state is the same as its molar specific heat of formation:

$$(\bar{h}_f^\circ)_{\text{compound}} = (\bar{q}_f^\circ)_{\text{compound}}$$

7. Molar heat of reaction:

$$\bar{q}_r = \sum_P (n_i/n_{\text{fuel}})\bar{h}_i - \sum_R (n_i/n_{\text{fuel}})\bar{h}_i$$

8. Higher and lower heating values of a fuel:

$$\text{HHV} = \bar{q}_r(\text{SRS and } H_2O \text{ in the liquid state})$$

$$\text{LHV} = \bar{q}_r(\text{SRS and } H_2O \text{ in the vapor state})$$

9. Open and closed system adiabatic flame temperature:

$$(T_A)_{\substack{\text{open}\\\text{system}}} = T^\circ + \frac{(\bar{h}_f^\circ)_{\text{fuel}} - \sum_P (n_i/n_{\text{fuel}})(\bar{h}_f^\circ)_i}{\sum_P (n_i/n_{\text{fuel}})(\bar{c}_{pi})_E}$$

and

$$(T_A)_{\substack{\text{closed}\\ \text{system}}} = T^\circ + \frac{(\overline{h}_f^\circ)_{\text{fuel}} - \sum_R (n_i/n_{\text{fuel}})\mathfrak{R}T^\circ - \sum_R (n_i/n_{\text{fuel}})(\overline{h}_f^\circ - \mathfrak{R}T^\circ)_i}{\sum_P (n_i/n_{\text{fuel}})(\overline{c}_{vi})_{\text{avg}}}$$

where $(\overline{c}_{pi})_{\text{avg}}$ and $(\overline{c}_{vi})_{\text{avg}}$ are found in Tables 15.5 and 15.6.

10. Molar specific entropy production for a chemical reaction:

$$(\overline{s}_P)_r = \sum_P (n_i/n_{\text{fuel}})\overline{s}_i - \sum_R (n_i/n_{\text{fuel}})\overline{s}_i - \overline{q}_r/T_b$$

11. Absolute molar specific entropy at temperature T and pressure p:

$$\left[\overline{s}(p,T)_{\text{abs}}\right]_{\substack{\text{incompressible}\\ \text{substance}}} = \overline{s}^\circ + \overline{c}\ln(T/T^\circ)$$

and

$$\left[\overline{s}(p,T)_{\text{abs}}\right]_{\substack{\text{ideal}\\ \text{gas}}} = \overline{s}^\circ + \overline{c}_p \ln(T/T^\circ) - \mathfrak{R}\ln(p/p^\circ)$$

12. Molar specific entropy of formation:

$$(\overline{s}_f^\circ)_{\text{compound}} = \overline{s}^\circ_{\text{compound}} - \sum_{\text{elements}} (n_i/n_{\text{compound}})\overline{s}_i^\circ$$

13. Molar specific Gibbs function of formation:

$$(\overline{g}_f^\circ)_{\text{compound}} = (\overline{h}_f^\circ)_{\text{compound}} - T^\circ(\overline{s}_f^\circ)_{\text{compound}}$$

14. Molar specific Gibbs function at the reference state of 0.1 MPa and the products or reactants mixture temperature T:

$$\overline{g}_i^\bullet(\text{at } p^\circ, T) = (\overline{g}_f^\circ)_i + [(\overline{h}(\text{at } T) - (\overline{h}^\circ(\text{at } T^\circ)]_i - T[\overline{s}(\text{at } p^\circ, T)]_i + T^\circ[\overline{s}^\circ(\text{at } p^\circ, T^\circ)]_i$$

15. Chemical equilibrium constant, based on Gibbs functions:

$$K_e = \exp\left[\frac{\sum_{R'} \nu_i \overline{g}_i^\bullet - \sum_{P'} \nu_i \overline{g}_i^\bullet}{\mathfrak{R}T}\right]$$

and, based on molar concentrations:

$$K_e = \frac{\prod_{P'} (\chi_i)^{\nu_i}}{\prod_{R'} (\chi_i)^{\nu_i}} \left(\frac{p_m}{p^\circ}\right)^{\left(\sum_{P'} \nu_i - \sum_{R'} \nu_i\right)}$$

16. The van't Hoff equation:

$$\frac{d(\ln K_e)}{dT} = \frac{\dot{Q}_r}{\mathfrak{R}T^2}$$

17. Fuel cell maximum reaction efficiency:

$$(\eta_r)_{\substack{\text{maximum}\\ \text{fuel cell}\\ \text{efficiency}}} = \frac{\sum_R (n_i/n_{\text{fuel}})\overline{g}_i^\bullet - \sum_P (n_i/n_{\text{fuel}})\overline{g}_i^\bullet + \mathfrak{R}T\ln\left[\prod_R (p_i/p^\circ)^{(n_i/n_{\text{fuel}})} / \prod_P (p_i/p^\circ)^{(n_i/n_{\text{fuel}})}\right]}{\sum_R (n_i/n_{\text{fuel}})\overline{h}_i - \sum_P (n_i/n_{\text{fuel}})\overline{h}_i}$$

18. Fuel cell open circuit voltage:

$$\phi_o = \frac{(\eta_r)_{\substack{\text{max}\\ \text{fuel cell}}}\left[\sum_R (n_i/n_{\text{fuel}})\overline{h}_i - \sum_P (n_i/n_{\text{fuel}})\overline{h}_i\right]}{jF}$$

where $j = 4n + m$ kgmole electrons per kgmole of C_nH_m and $F = 96{,}487$ kJ/(V·kgmole electrons) is Faraday's constant.

19. Fuel cell electrical current:

$$I = (\dot{n}_{\text{fuel}})jF$$

20. The specific molar flow availability of chemical species i:

$$[(\bar{a}_f)_i]_{\text{chemical}} = \bar{g}_i^{\bullet} + \Re T \ln\left[\frac{p_i}{p^\circ}\right]$$

Problems (* indicates problems in SI units)

1.* Determine the mass in kg of one molecule of water.

2. Determine the mass in lbm of one molecule of methane.

3.* The density of benzene (C_6H_6) is 879 kg/m^3. Determine the volume of
   a. 1 kgmole of benzene.
   b. 1 molecule of benzene.

4. Determine the percentage by mass of carbon in C_8H_{18}.

5. Determine the percentage by mass of aluminum in Al_2O_3.

6. Determine the percentage by mass of oxygen atoms in a molecule of casein of milk, $C_{708}H_{1130}O_{224}N_{180}S_4P_4$.

7. Determine the percentage by mass of carbon, hydrogen, and oxygen in methyl alcohol, $CH_3(OH)$.

8. The spectral analysis of a chemical compound gave the following composition on a mass basis: 29.1% Na, 40.5% S, 30.4% O. Determine the chemical formula of this substance (i.e., find x, y, and z in $Na_xS_yO_z$).

9. How many lbm are in 1 lbmole of
   a. $C_{12}H_{22}O_{11}$ (sucrose, a typical carbohydrate).
   b. $C_{57}H_{110}O_6$ (stearin, a typical fat).
   c. $C_{3032}H_{4816}O_{872}N_{780}S_8Fe_4$ (human hemoglobin, a typical protein).

10.* Convert the following to mass units (kg or lbm): (a) 1.00 kgmole of CO_2, (b) 2.00×10^{-3} lbmole of Fe_2O_3, (c) 6.00 gmoles of SO_2, (d) 0.700 kgmole of CH_4.

11.* Convert the following to molar units (kgmole or lbmole): (a) 14.0 lbm of CO, (b) 0.370 kg of H_2O, (c) 123 g of C_2H_2, (d) 5.00 kg of C_8H_{18}.

12.* When 2.00 kgmoles of $KClO_3$ are heated, 2.00 kgmoles of KCl and 3.00 kgmoles of O_2 are produced. If 50.0 kg of $KClO_3$ are heated, how many kg of KCl and O_2 are produced?

13. Determine the reaction equation and molal analysis of the combustion products for the combustion of carbon disulfide, CS_2, with
   a. 100.% theoretical air.
   b. 150.% theoretical air.
   c. An air/fuel ratio of 47.6 moles of air per mole of fuel to produce CO_2, SO_2, and excess air products.

14. Determine the reaction equation and molal analysis of the combustion products of ammonia, NH_3, on the surface of a heated platinum wire in the presence of
   a. 0.00% excess oxygen.
   b. 125% theoretical oxygen.
   c. 25.0% deficit oxygen to form NO, H_2O, and excess air products.

15. Determine the reaction equation and volumetric analysis of the combustion products for the combustion of methyl alcohol, $CH_3(OH)$, with
   a. 100.% theoretical oxygen.
   b. 150.% theoretical oxygen.
   c. 50.0% theoretical oxygen.

16. Determine the reaction equation and volumetric analysis of the combustion products for the combustion of natural rubber, $[C_3H_8]_{2000}$ or $(C_{6000}H_{16,000})$, with
   a. 100.% theoretical oxygen.
   b. 400.% excess oxygen.
   c. 90.0% deficit oxygen.

17. Determine the reaction equation and molal air/fuel ratio for the combustion of ethyl alcohol, $C_2H_5(OH)$, with
   a. 100.% theoretical air.
   b. 100.% excess air.
   c. 50.0% deficit air.

18. Determine the reaction equation and molal air/fuel ratio for the combustion of dimethyl ketone (acetone), $CO(CH_3)_2$, with
   a. 0.00% excess air.
   b. 100.% excess air.
   c. 30.0% theoretical air.

19. Determine the reaction equation and molar fuel/air ratio for the combustion of wood cellulose, $C_6H_{10}O_5$, with
   a. 100.% theoretical air.
   b. 250.% excess air.
   c. 25.0% deficit air.

20. Determine the reaction equation, molal analysis of the combustion products, and mass air/fuel ratio for the combustion of kerosene, $C_{10}H_{22}$, with 187% excess air.

21. Determine the reaction equation, volumetric analysis of the combustion products, and molal fuel/air ratios for the combustion of tetraethyl lead, $Pb(C_2H_5)_4$, the common antiknock gasoline additive, to form PbO, CO_2, H_2O, and possibly excess air with
   a. 100.% theoretical air.
   b. 10.0% excess air.
   c. An air/fuel ratio of 20.0 lbm of air per lbm of fuel.

22. Develop a reaction equation for the combustion of polyethylene, $[CH_2CH_2]_n$, in 100.% theoretical air.

23. A simple alcohol can be obtained from a hydrocarbon by replacing one of the hydrogen atoms by a hydroxyl group (OH). Develop a reaction equation for the combustion of a generalized

alcohol of this type, $C_nH_{2n+1}(OH)$, in 100.% theoretical air and test it out with

a. Methyl alcohol (also known as methanol or wood alcohol), $CH_3(OH)$.
b. Ethyl alcohol (also known as ethanol or grain alcohol), $C_2H_5(OH)$.
c. Isopropyl alcohol, $C_3H_7(OH)$.

24. Propane, C_3H_8, is burning with 130.% theoretical air in a camp stove. Determine
a. The reaction equation.
b. The molar air/fuel ratio of the combustion.
c. The volumetric analysis of the combustion products.
d. The dew point temperature of the combustion products if the total pressure of the combustion products is 14.7 psia.

25. The combustion of an unknown amount of benzene (*x* moles of C_6H_6) in pure oxygen in a chemical reactor produces the following dry exhaust gas analysis: 44.71% CO_2 and 55.29% O_2. Determine
a. The actual molar and mass fuel/oxygen ratios.
b. The percent of theoretical oxygen used.
c. The molar percentage of water vapor in the exhaust gas before it was dried for this analysis.

26. The combustion of an unknown amount of ethylene (*x* moles of C_2H_4) in pure oxygen in a laboratory experiment produces the following dry exhaust gas analysis on a molar basis: 84.75% CO_2 and 15.25% O_2. Determine
a. The actual molar and mass oxygen/fuel ratios.
b. The percent of excess oxygen used.
c. The molar percentage of water vapor in the exhaust gas before it was dried for this analysis.

27. The combustion of an unknown amount of acetylene (*x* moles of C_2H_2) with pure oxygen in an oxyacetylene torch produces the following dry exhaust gas analysis: 39.14% CO_2 and 60.86% O_2. Determine
a. The actual molar and mass fuel/oxygen ratios.
b. The percent of excess oxygen used.
c. The molar percentage of each combustion product before the exhaust gas was dried for this analysis.

28. An unknown amount of butane (*x* moles of C_4H_{10}) is burned with pure oxygen in a bomb calorimeter. The dry gas molar analysis of the products is 48.72% CO_2 and 51.28% O_2. Determine
a. The actual molar and mass oxygen/fuel ratios.
b. The percent excess oxygen used.
c. The molar percentage of each combustion product before the exhaust gas was dried for this analysis.

29. The combustion of an unknown amount of methane (*x* moles of CH_4) in a furnace results in the following dry exhaust gas molar analysis: 9.52% CO_2, 4.00% O_2, and 86.47% N_2. Determine
a. The actual molar and mass air/fuel ratios.
b. The percent of excess air used.
c. The molar percentage of water vapor in the exhaust gas before it was dried for this analysis.

30. The combustion of an unknown amount of propane (*x* moles of C_3H_8) in an industrial oven produces the following dry exhaust gas analysis on a volume basis: 5.52% CO_2, 12.59% O_2, and 81.89% N_2. Determine
a. The actual molar and mass air/fuel ratios.
b. The percent of theoretical air used.
c. The volume percentage of water vapor in the exhaust gas before it was dried for this analysis.

31. An unknown amount of ethane (*x* moles of C_2H_6) is burned in a combustion chamber with air. The dry gas molal analysis of the exhaust products is 6.64% CO_2, 10.46% O_2, and 82.9% N_2. Determine
a. The actual molar and mass fuel/air ratios.
b. The percent of excess air used.
c. The molar percentage of each combustion product before the exhaust gas was dried for this analysis.

32. The combustion of an unknown amount of propylene (*x* moles of C_3H_6) with air in a prototype space heater produced the following dry exhaust gas molar analysis: 4.27% CO_2, 15.05% O_2, and 80.68% N_2. Determine
a. The actual molar and mass air/fuel ratios.
b. The percent of theoretical air used.
c. The molar percentage of each combustion product before the exhaust gas was dried for this analysis.

33. The following dry exhaust volumetric analysis results from the combustion of an unknown amount of octane (*x* moles of C_8H_{18}) in a spark ignition internal combustion engine: 8.80% CO_2, 8.20% CO, 4.1% H_2, 1.00% NO, 0.200% CH_4, and 77.7% N_2. Determine
a. The actual molar and mass air/fuel ratios.
b. The percent of theoretical air used.
c. The molar percentage of water vapor in the exhaust gas before it was dried for this analysis.

34. The following dry exhaust gas molar analysis results from the combustion of an unknown amount of kerosene (*x* moles of $C_{10}H_{22}$) in a compression ignition internal combustion engine: 6.30% CO_2, 9.40% CO, 4.70% C, 4.70% H_2, 1.50% NO, 0.200% CH_4, and 73.2% N_2. Determine
a. The actual molar and mass air/fuel ratios.
b. The percent deficit air used.
c. The volumetric percentage of water vapor in the exhaust gas before it was dried for this analysis.

35. An unknown hydrocarbon material is burned in a calorimeter with air. An Orsat analysis indicates that the (dry) exhaust gas is made up of only 17.5% CO_2 and 82.5% N_2, with no CO or O_2 present. Determine
a. The fuel model (C_nH_m).
b. The composition of the fuel on a mass basis.
c. The percent theoretical air used.
d. The dew point temperature if the combustion products are at 0.101 MPa.

36.* An unknown hydrocarbon material is burned in air for chemical analysis. An Orsat test indicates that the (dry) exhaust gas contains no CO or O_2 but consists of only 14.9% CO_2 and 85.1% N_2. Determine
a. The fuel model (C_nH_m).
b. The composition of the fuel on a percent mass basis.
c. The molar and mass air/fuel ratios used in the combustion process.
d. The dew point temperature if the combustion products are at 0.101 MPa.

37. An Orsat analysis of the (dry) products of combustion of a gas emanating from the bowels of a creature of immense hypocrisy produces the following composition: 1.00% CO_2, 19.2% O_2, and 79.8% N_2. Determine
a. The chemical formula (C_nH_m) and name of the gas.
b. The composition of the fuel on a percent molar basis.

c. The percent excess air used in the combustion process.
d. The dew point temperature of the products at atmospheric pressure.

38. An Orsat analysis of the (dry) products of combustion of an unknown hydrocarbon indicates that it consists of only 26.1% CO and 73.9% N_2, with no CO_2 or O_2 present. Determine
a. The fuel model (C_nH_m).
b. The composition of the fuel on a percent mass basis.
c. The percent theoretical air used.
d. The dew point temperature of the combustion products at atmospheric pressure.

39.* The combustion of a new fuel that is a mixture of a liquid hydrocarbon and hydrogen gas is to be modeled as a single hydrocarbon. An Orsat analysis of the dry exhaust products of this new fuel shows 6.00% CO, 6.00% O_2, and 88.0% N_2 (no CO_2 is present). Determine
a. The fuel model (C_nH_m) of this mixture.
b. The percentages by mass of carbon and hydrogen present in the mixture.
c. The mass air/fuel ratio used in the combustion process.
d. The dew point temperature of the exhaust products at 0.152 MPa.

40.* An unknown hydrocarbon fuel produced an Orsat dry exhaust gas analysis of 8.30% CO_2, 0.00% CO, 9.10% O_2, and 82.6% N_2. Determine
a. The fuel model (C_nH_m).
b. The composition of the fuel on a percent mass basis.
c. The percent of theoretical air used.
d. The dew point temperature of the exhaust gas when it is at 0.1015 MPa.

41. Amounts of 9.70% CO_2, 1.10% CO, 0.00% O_2, and 87.2% N_2 are found in the Orsat analysis of the dry exhaust gas produced by the cataclysmic combustion of an unknown hydrocarbon substance in air. Determine
a. The fuel model (C_nH_m) of the hydrocarbon.
b. The mass percentages of carbon and hydrogen in the fuel.
c. The percent of deficit air used.
d. The molar percentage of water vapor in the exhaust before it was dried.

42.* The Orsat analysis of the (dry) products of combustion of an unknown hydrocarbon is 9.10% CO_2, 8.90% CO, and 82.0% N_2 (no O_2 is present). Determine
a. The fuel model (C_nH_m).
b. The mass percentages of C and H present in the fuel.
c. The molar air/fuel ratio and percent theoretical air used in the combustion.
d. The dew point temperature at 0.106 MPa.

43.* The Orsat analysis of the (dry) exhaust gas from the combustion of an unknown hydrocarbon is 1.10% CO_2, 1.10% CO, 18.8% O_2, and 79.0% N_2. Determine
a) The fuel model (C_nH_m),
b) The percent mass composition of the fuel,
c) The fuel/air ratio used, and
d) The dew point temperature at 0.644 MPa.

44.* The combustion of a mysterious unknown hydrocarbon fuel in the dimensional stabilization module of the temporal drive unit produces the following Orsat (dry) exhaust gas analysis: 4.50% CO_2, 1.90% CO, 14.1% O_2, and 79.5% N_2. Determine
a. The fuel model (C_nH_m) and its probable name.
b. The percent mass composition of the fuel.
c. The percent of theoretical air used in the combustion.
d. The dew point temperature of the exhaust products at 1.00 MPa.

45. A Wingbarton hydrocarbon bomb explodes in the dry air near an Orsat analyzer. A slightly injured but quick-witted technician quickly carries out a dry gas analysis in the shattered remains of the laboratory to produce the following results: 2.80% CO_2, 0.500% CO, 16.2% O_2, and 80.5% N_2. Determine
a. The fuel model (C_nH_m) and probable name of the mysterious hydrocarbon explosive used in the Wingbarton bomb.
b. The mass percentage composition of the carbon and hydrogen in the bomb.
c. The percentage of excess air available in the laboratory when the bomb exploded.
d. The dew point temperature in the lab after the bomb exploded (assume atmospheric pressure).

46. An international espionage agent uses a computerized pocket-sized Orsat apparatus to grab and analyze a dry sample of the exhaust gases from the new air-breathing, hydrocarbon-burning, Blood-Sucker guided missile. The Orsat readout is 6.20% CO_2, 2.10% CO, 9.90% O_2, and 81.8% N_2. Determine
a. The fuel model C_nH_m and probable name of the Blood-Sucker's fuel.
b. The percentages of carbon and hydrogen by mass in the fuel.
c. The mass fuel/air ratio used by the missile.

47. An ancient internal combustion Mugwump bilge-pump engine burns a mixture of obscure hydrocarbon fuels that can be modeled as a single fuel. An Orsat analysis of the dry exhaust gas from this engine gives the following results: 5.90% CO_2, 5.50% CO, 7.50% O_2, and 81.1% N_2. Determine
a. The single fuel hydrocarbon model (C_nH_m).
b. The carbon and hydrogen composition by mass of this fuel.
c. The percent of theoretical air used in the engine.

48. An Orsat analysis of the combustion of a strange unearthly hydrocarbon gas produces a (dry) result of 5.00% each for CO_2, CO, and O_2, with N_2 making up the remainder. Determine
a. The hydrocarbon fuel model C_nH_m for this strange and somewhat putrid gas.
b. The mass composition of this gas.
c. The percent of excess air used in the combustion.

49. The coagulated remains of a fiendish mutant humanoid creature that evolved on an oxygen-free planet are oxidized and dried in air by a laser beam then processed through a nuclear-powered Orsat analyzer that reports the following composition, using a computerized synthetic voice: "Thee ox-see-gin, car-bon de-oxside and car-bon mo-noxside conzentrations are each exactly seven per-cent. Thee remaining gaz is nitrogen. Zis is very unusual." As chief engineer of the starship Entropy, determine
a. The synthetic formula (C_nH_m) of the mutant's body tissue.
b. The carbon and hydrogen percentages by mass in the tissue.
c. The percent of excess air available in the atmosphere where the tissue was oxidized.

50. An icky, unknown hydrocarbon fuel of the form $(CH_2)_n$ is burned with an unknown amount of excess air x in the following gosh awful reaction:

$$(CH_2)_n + 1.5n(1+x)[O_2 + 3.76(N_2)] \\ \rightarrow n(CO_2) + n(H_2O) + 1.5nx(O_2) + 5.64n(1+x)(N_2)$$

Table 15.11 Problem 50

Component	Molecular Mass	%
N_2	28.0	77.4
O_2	32.0	13.6
H_2O	18.0	4.5
CO_2	44.0	4.5
	Total	100.0

An exhaust gas analysis of the products of this combustion yielded the Table 15.11 percentage composition on a volume basis. Using these data, determine

a. The temperature to which the exhaust must be cooled to cause the water vapor to condense, if the exhaust is at a total pressure of 22.223 psia.

b. The amount of excess air x used in the combustion process.

c. The air/fuel ratio on a mass basis.

In Problems 51 through 55, use Eqs. (15.4), (15.6), and the higher heating value data given in Table 15.2 to compute the molar specific enthalpy of formation at the standard reference state, $(\bar{h}_f^\circ)_{\text{compound}}$, of each compound from its elements. Compare your results with the values given in Table 15.1.

51. Acetylene: $2[C(s)] + H_2(g) \rightarrow C_2H_2(g) + (\bar{q}_f^\circ)_{C_2H_2(g)}, (\bar{h}_f^\circ)_{C_2H_2(g)} = ?$
52. Ethylene: $2[C(s)] + 2[H_2(g)] \rightarrow C_2H_4(g) + (\bar{q}_f^\circ)_{C_2H_4(g)}, (\bar{h}_f^\circ)_{C_2H_4(g)} = ?$
53. Propane: $3[C(s)] + 3[H_2(g)] \rightarrow C_3H_6(g) + (\bar{q}_f^\circ)_{C_3H_6(g)}, (\bar{h}_f^\circ)_{C_3H_6(g)} = ?$
54. Ethane: $2[C(s)] + 3[H_2(g)] \rightarrow C_2H_6(g) + (\bar{q}_f^\circ)_{C_2H_6(g)}, (\bar{h}_f^\circ)_{C_2H_6(g)} = ?$
55. Butane: $4[C(s)] + 5[H_2(g)] \rightarrow C_4H_{10}(g) + (\bar{q}_f^\circ)_{C_4H_{10}(g)}, (\bar{h}_f^\circ)_{C_4H_{10}(g)} = ?$
56. Repeat Example 15.8 using 150.% theoretical air.
57. Repeat Example 15.8 using 90.0% theoretical air. Assume the hydrogen is much more reactive than the carbon and is all converted into water.
58. Repeat Example 15.9 using 150.% theoretical air in the combustion process.
59.* The higher heating value of glucose, $C_6H_{12}O_6(s)$, is −2817.5 MJ/kgmole. Determine the standard reference state molar specific enthalpy of formation of glucose using the reaction

$$C_6H_{12}O_6(s) + 6[O_2(g)] \rightarrow 6[CO_2(g)] + 6[H_2O(\ell)]$$

60. Determine the heat of combustion of propane (C_3H_8) gas with 100.% theoretical air when the reactants are at the standard reference state, but the products are at 2000. R. Use the gas tables in *Thermodynamic Tables to accompany Modern Engineering Thermodynamics* (Table C.16c).
61. Using the gas tables, Table C.16c, determine the heat of combustion of liquid octane, $C_8H_{18}(l)$, with 200.% theoretical air when the reactants are at 537 R and the products are at 4000. R. Explain the significance of your answer.
62. Liquid ethyl alcohol at 77.0°F is burned in 100.% theoretical air. Determine the heat produced per kgmole of fuel when the products are at 540.°F. The molar specific enthalpy of formation of this fuel is − 277.69 MJ/kgmole.
63.* Methane gas (CH_4) at −60.0°C is burned during a severe winter with 200.% theoretical air at the same temperature. The products of combustion are at 300.°C. Assuming constant specific heats, find the heat released per kgmole of fuel.
64. Kerosene (decane, $C_{10}H_{22}$) with a density of 49.3 lbm/ft^3 has a HHV of 20,484 Btu/lbm and costs $0.500 per gallon. Calculate the cost of 1.00 therm (10^5 Btu) obtained by burning kerosene.
65.* The *explosive energy* of a high explosive is defined to be the lower heating value (LHV) of the detonation reaction. The heat of formation of nitroglycerin, $C_3H_5(NO_3)_3$, is −354 MJ/kgmole, and its molecular mass is 227 kg/kgmole.

    a. Find the values of a, b, c, and d in the following reaction describing the detonation of nitroglycerin:

$$C_3H_5(NO_3)_3 \rightarrow a(CO_2) + b(H_2O) + c(O_2) + d(N_2)$$

    b. Determine the explosive energy of nitroglycerin in MJ/kg.

66.* In Problem 65, the explosive energy of a high explosive was defined to be the lower heating value (LHV) of the detonation reaction. The heat of formation of trinitrotoluene (TNT), $C_7H_5(NO_2)_3$, is −54.4 MJ/kgmole, and its molecular mass is 227 kg/kgmole.

    a. Find the values of a, b, c, and d in the following simplified reaction describing the detonation of TNT:

$$C_7H_5(NO_2)_3 \rightarrow a(CO) + b(CH_4) + c(H_2O) + d(N_2)$$

    b. Determine the explosive energy of TNT in MJ/kg.

67. Determine the adiabatic flame temperature of methane (CH_4) burned in 400.% theoretical air in a steady flow process.
68. Determine the adiabatic flame temperature of acetylene (C_2H_2) burned in a steady flow process with (a) 100.% theoretical air, (b) 200.% theoretical air, and (c) 400.% theoretical air.
69. Determine the adiabatic flame temperature of propane (C_3H_6) burned in a steady flow process with (a) 0.00% excess air, (b) 100.% excess air, and (c) 300.% excess air.
70. Determine the adiabatic flame temperature of benzene (C_6H_6) burned in a steady flow process with (a) 0.00% excess air, (b) 100.% excess air, and (c) 300.% excess air.
71.* Determine the adiabatic flame temperature and maximum explosion pressure as 0.001 kgmole of butane (C_4H_{10}) is burned in 100.% theoretical air in a 0.100 m^3 adiabatic bomb calorimeter.
72.* Determine the molar specific entropy production $(\bar{s}_P)_r$ for the reaction $C + O_2 \rightarrow CO_2$, when both the products and the reactants are at the standard reference state. Assume an isothermal boundary at 298 K.
73.* Determine the molar specific entropy production $(\bar{s}_P)_r$ for the combustion of methane in pure oxygen, $CH_4 + 2(O_2) \rightarrow CO_2 + 2(H_2O)$, when both the products and the reactants are at the standard reference state. Assume an isothermal boundary at 298 K.
74.* Determine the molar specific entropy production $(\bar{s}_P)_r$ as ethylene is burned in an adiabatic combustion chamber with 100.% theoretical air. The reactants are at the standard reference state, but the products are at 5.00 MPa. The adiabatic flame temperature is 2291.6°C. Assume constant specific heats (Table C.13b).
75. Using the gas tables (Table C.16c), determine the molar specific entropy production $(\bar{s}_P)_r$ for the combustion of propane with 100.% theoretical air. The reactants are at the standard reference state. The products are at 2000. R and 4.00 MPa, and the heat transfer from the combustion chamber is 571,126 Btu/(lbmole·R). Assume the combustion chamber has isothermal walls at 2000. R.
76. Determine the molar specific entropy of formation $\bar{s}_f^\circ$ for (a) CO, (b) CO_2, and (c) H_2O.

77. Determine the molar specific entropy of formation $\bar{s}_f^\circ$ for (a) methane (CH_4), (b) acetylene (C_2H_2), and (c) propane (C_3H_6).

78. Determine the molar specific entropy of formation $\bar{s}_f^\circ$ of liquid octane $C_8H_{18}(\ell)$, and explain why it is negative.

79.* Determine the specific molar Gibbs function of formation $\bar{g}_f^\circ$ at 25.0°C and 0.100 MPa for the reaction $C + O_2 \rightarrow CO_2$, and compare your result with the value given in Table 15.7.

80.* Use Eq. (15.35) and Table C.17 to find the molar specific Gibbs function of formation at 25.0°C and 0.100 MPa of atomic hydrogen gas from the equilibrium reaction $H_2 \rightleftarrows 2H$.

81. The equilibrium constant for the reaction $0.5(N_2) + 0.5(O_2) \rightleftarrows NO$ is 0.0455 at 4500. R and atmospheric pressure. Assume that air at room temperature and atmospheric pressure contains 21.0% oxygen and 79.0% nitrogen on a molar basis.
   a. As the dissociation occurs, does the total number of moles in the reaction (i) increase, (ii) decrease, or (iii) remain constant?
   b. The equilibrium constant formula (K_e) for the dissociation equation is which of the following? (a) $\frac{p_{NO}}{p_{N_2}p_{O_2}}$ (b) $\frac{p_{N_2}p_{O_2}}{p_{NO}}$ (c) $\frac{p_{NO}}{\sqrt{p_{N_2}p_{O_2}}}$ (d) $\frac{(p_{NO})^2}{p_{N_2}p_{O_2}}$
   c. Air at 4500 R and atmospheric pressure contains what percentage of NO on a molar basis? (a) 0.617, (b) 1.803, (c) 4.55, (d) more than 4.55, (e) less than 0.617.

82.* Determine the equilibrium constant (K_e) for the reaction $CH_4 + H_2O \rightleftarrows CO + 3(H_2)$ at 25.0°C and 0.100 MPa.

83.* Algebraically solve the cubic equation given in Example 15.15 for the degree of dissociation (y) and use Table C.17 to verify the three example computer results given there for
   a. $p_m = 0.100$ MPa, $T = 2000.$ K.
   b. $p_m = 0.100$ MPa, $T = 3000.$ K.
   c. $p_m = 1.00$ MPa, $T = 3000.$ K.

84.* Carbon is burned with 100.% excess oxygen to form an equilibrium mixture of CO_2, CO, and O_2 at 3000. K and 1.00 MPa pressure. Determine the equilibrium composition when only the CO_2 dissociates as $CO_2 \rightleftarrows CO + 0.5(O_2)$. Assume ideal gas behavior.

85.* The equilibrium constant for the water–carbon monoxide reaction $CO + H_2O \rightleftarrows CO_2 + H_2$ at 0.100 MPa and 1000. K is $K_e = 1.442$. Determine the equilibrium mole fraction of each gas present under these conditions. Assume ideal gas behavior.

86. Determine the maximum reversible electrical work output of the fuel cell shown in Figure 15.16, where g is the specific Gibbs free energy.

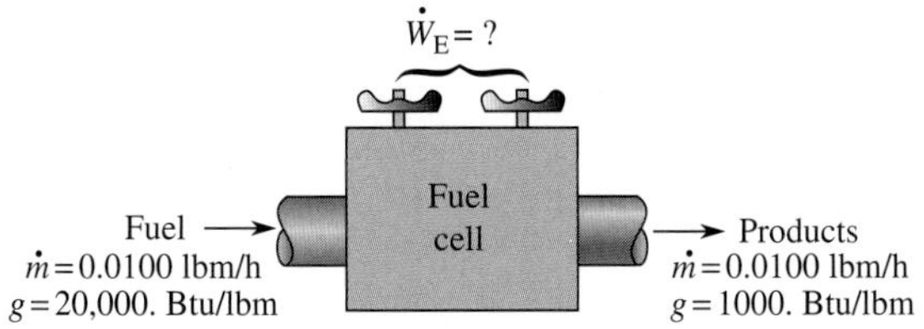

FIGURE 15.16
Problem 86.

87.* Determine the maximum theoretical efficiency, open circuit voltage, and maximum theoretical work output per mole of carbon consumed in a carbon–oxygen fuel cell operating with the reaction $C(s) + O_2 \rightarrow CO_2$, when each component in the reaction is at 25.0°C and 0.100 MPa.

88.* Repeat Example 15.18 for a fuel cell that operates on hydrogen and 100.% theoretical air (instead of pure oxygen), each at 25.0°C and 0.100 MPa.

89.* Determine the maximum efficiency and open circuit voltage for the methane–oxygen fuel cell, $CH_4 + 2(O_2) \rightarrow CO_2 + 2[H_2O(\ell)]$, when each component in the reaction is at 25.0°C and 0.100 MPa.

90.* Repeat Problem 89 for the case where the reactants are premixed to a total pressure of 0.100 MPa at 25.0°C.

91.* Determine the maximum efficiency and open circuit voltage for the propane–oxygen fuel cell, $C_3H_8 + 5(O_2) \rightarrow 3(CO_2) + 4[H_2O(\ell)]$, when each component in the reaction is at 25.0°C and 0.100 MPa.

92.* Repeat Problem 91 for a propane–air fuel cell in which the propane is premixed with 200.% theoretical air at a total pressure of 0.100 MPa and 25.0°C. Assume the combustion products are also mixed and are at a total pressure of 0.100 MPa at 25.0°C.

93.* An inventor claims to have perfected a hydrogen–oxygen fuel cell, $H_2 + 0.5(O_2) \rightarrow H_2O(\ell)$, that produces 300 MJ per kgmole of hydrogen consumed at 25.0°C and 0.100 MPa. Is this possible? If not, what is the maximum possible power output? Assume each component in the reaction is at 25.0°C and 0.100 MPa.

94.* Determine the open circuit, internal entropy production rate per unit molar flow rate of CO in the carbon monoxide–oxygen fuel cell, $CO + 0.5(O_2) \rightarrow CO_2$, when each component in the reaction is at 25.0°C and 0.100 MPa.

95. An inventor claims to have invented a fuel cell that contains a catalyst for the ammonia reaction $N_2 + 3H_2 \rightarrow 2NH_3$ at the standard reference state. The molar enthalpy of formation of ammonia = −19,750 Btu/lbmole and the Gibbs function of formation of ammonia = −7140 Btu/lbmole. Determine
   a. The maximum theoretical reaction efficiency.
   b. The maximum theoretical electrical work output of this fuel cell per lbmole of N_2 consumed.

Design Problems

The following are open-ended design problems. The objective is to carry out a preliminary thermal design as indicated. A detailed design with working drawings is not expected unless otherwise specified. These problems have no specific answers, so each student's design is unique.

96. Design a burner for a furnace that will produce 2.00×10^6 Btu/h at 1500. °F. Choose the fuel, flow rates, air/fuel ratio, burner material, burner geometry, flow controls, and the like.

97. Design a bomb calorimeter to be used to measure the heat of combustion of municipal solid waste. Because of the heterogeneous nature of the waste, the test sample size must be at least 1 lbm. Make the calorimeter either adiabatic or isothermal. Assume all noncombustibles (e.g., metal, glass) have been removed from the waste before it is tested.

98.* Design a system to produce 0.1 kg/h of hydrogen gas from the catalytic reaction of methane and steam, $CH_4 + H_2O(g) \rightarrow 3(H_2) + CO$, at 1500. R and 14.7 psia. Assume the system has an equilibrium composition of $CH_4 + H_2O(g) \rightarrow a(CH_4) + b[H_2O(g)] + c(H_2) + d(CO)$

99.* Design a combustion chamber for a rocket that will oxidize liquid hydrazine, $N_2H_4(\ell)$, with liquid hydrogen peroxide, $H_2O_2(\ell)$, at 1.00 MPa and 2000. K as follows: $N_2H_4(\ell) + 2(H_2O_2(\ell)) \rightarrow 4[H_2O(g)] + N_2$, and that will supply the rocket nozzle with 1000. kg/s of exhaust gas. Data: $(\bar{h}_f^\circ)_{N_2H_4(\ell)} = 50.417$ MJ/kgmole, and $(\bar{h}_f)_{H_2O_2(\ell)} = -187.583$ MJ/kgmole.

100. Design a benchtop laboratory-scale facility to produce methanol from the reaction $CO + 2(H_2) \rightarrow CH_3(OH)$. Determine all flow rates, heat transfers, reaction vessel optimum temperature and pressure, reaction vessel material, and geometry.

Computer Problems

The following open-ended computer problems are designed to be done on a personal computer using a spreadsheet or equation solver.

101. Develop a spreadsheet to compute the LHV of *gaseous* octane and use it to find the adiabatic flame temperature for (a) 100.% and (b) 200.% theoretical air.

102. Develop a spreadsheet to balance the combustion reaction of any simple alcohol of the type $C_nH_{2n+1}(OH)$ with excess or deficit air or oxygen.

103. Develop a spreadsheet with temperature-dependent specific heats for a hydrocarbon fuel of your choice listed in Table 15.1 to do one or more of the following:
    a. Compute the heat of reaction of the fuel when the reaction temperature and percent of excess air or oxygen are input by the user.
    b. Compute the adiabatic flame temperature of the fuel for any excess air or oxygen value.

104. Develop an interactive computer program that outputs the entropy production rate for any (or a series of) reaction(s) of your choice. Input from the keyboard is in response to properly formatted screen prompts for the stoichiometric coefficients, the heat of reaction, the reaction temperature and pressure, and the isothermal heat transfer boundary temperature. Use the program to determine
    a. The maximum pressure for a given isothermal boundary temperature at which the reaction can occur.
    b. The maximum isothermal boundary temperature possible for any given reaction pressure and temperature.

105. Develop an interactive computer program that outputs the equilibrium constant K_e when the stoichiometric coefficients are input from the keyboard in response to properly formatted screen prompts. Assume all the components obey the Gibbs-Dalton ideal gas mixture law.

106. Develop an interactive computer program to replace Table C.17 for the dissociation of H_2O, CO_2, and NO. You have to find the relevant information on H, O, and N in other texts if you wish to also include the dissociation of H_2, O_2, and N_2.

107. Develop an interactive computer program that outputs the molar specific enthalpy $(\bar{h}_i)$, entropy $(\bar{s}_i)$, and Gibbs function $(\bar{g}_i)$ for any substance of your choice at any user input pressure and temperature.

108. Develop an interactive computer program that outputs the power and reaction efficiency of any reaction you desire. Apply this program to a fuel cell analysis and plot the reaction efficiency η_r vs.
    a. The percent of excess oxidizer present.
    b. The reaction temperature.
    c. The reaction pressure.
    d. The entropy production rate.
    e. The fuel cell heat transfer rate $\dot{Q}$.
    Assume isothermal boundaries and input T_b from the keyboard along with all other necessary information.

CHAPTER 16

Compressible Fluid Flow

CONTENTS

16.1 INTRODUCEREA (INTRODUCTION)

This chapter focuses the application of the laws of thermodynamics on the behavior of a very specific type of fluid, a *compressible* fluid. An incompressible fluid has a constant density independent of the magnitude of the applied pressure, but the density of a compressible fluid varies with the applied pressure.

However, a compressible fluid does not exhibit *compressibility effects* whenever it is used in a system. Suppose you have a compressible fluid in a system in which pressure does not change significantly. Then, the effects of compressibility (density changes) do not appear and the compressible fluid behaves like an incompressible fluid. For example, in the heating and ventilating system of a building, the air flow rates are very low and the pressure and density changes are also very small. Consequently, air (by definition, a compressible fluid) behaves

WHAT IS A *COMPRESSIBLE* FLUID?

A compressible fluid is any fluid whose density varies significantly with pressure.

WHEN CAN THE COMPRESSIBILITY OF A FLUID BE IGNORED?

As a general rule of thumb, when $V^2 \ll \Delta p/\Delta \rho$, the compressibility of a fluid is negligible and it behaves like an incompressible fluid.

like an incompressible fluid in this system. Conversely, air flowing at high speed through a gas turbine engine is exposed to large pressure and density changes inside the engine; therefore, it does not behave like an incompressible fluid. In general, if the kinetic energy of a compressible fluid is much less than the ratio of the pressure change to its density change in the flow, then the fluid's compressibility is negligible and it may be treated as an incompressible fluid for engineering analysis purposes.

Consequently, a *compressible flow* is any flow in which the fluid density is not constant in time and space. Though all real substances are compressible to some extent, in normal engineering practice, only gases and vapors are significantly compressible. Liquids and solids are normally considered to be incompressible, except at extremely high pressures, on the order of 10^5 psia (0.7 GPa) or more. Because many modern engineering systems deal with thermodynamic processes involving gases or vapors, it is important to understand the unique flow characteristics of these substances.

In the previous chapters, we discuss the conservation of mass and energy and make extensive use of the one-dimensional mass, energy, and entropy balance equations for open and closed systems. In this chapter, we introduce the conservation of linear momentum law and the corresponding closed and open system one-dimensional momentum balance equations and we apply these equations to systems containing compressible substances. The conservation of linear momentum law, along with the conservation of mass and the two laws of thermodynamics, complete the set of fundamental physical laws and corresponding balance equations necessary for proper engineering design and analysis of all substances.

In this chapter, we characterize the basic properties of a compressible flow and apply them to subsonic and supersonic flows. One of the main areas of application of compressible flows is in converging-diverging nozzles and diffusers. Fluid compressibility produces the additional phenomena of *choked flow* and *shock waves* in these flow systems. In Chapters 6 and 9 we studied the energy and entropy characteristics of nozzles and diffusers. In this chapter, we discover that the conservation of linear momentum adds new facets to nozzle and diffuser efficiency analysis for compressible fluids.

16.2 STAGNATION PROPERTIES

The *stagnation state* of a moving fluid is the state it would achieve if it underwent an adiabatic, aergonic deceleration to zero velocity. The energy rate balance (ERB) for an adiabatic, aergonic, steady state, steady flow, single-inlet, single-outlet open system with negligible change in flow stream potential energy reduces to

$$h_{in} + V_{in}^2/(2g_c) = h_{out} + V_{out}^2/(2g_c)$$

If we let the subscript o refer to the stagnation (or zero velocity) state, then $V_o = 0$, and the preceding equation can be used to define the stagnation specific enthalpy h_o as

$$h_o = h + V^2/(2g_c) \tag{16.1}$$

WHY IS COMPRESSIBLE FLUID FLOW PART OF THERMODYNAMICS?

At the beginning of the 19th century, when the Industrial Revolution was in full swing and the technology of the high-pressure steam engine was in the process of being developed, it became clear that, under certain circumstances, some very peculiar things were happening inside the engine. At that time, engineers were trying to determine how to increase the power output of a given engine while at the same time improving its operating efficiency. The relation between power output and operating conditions of an adiabatic engine can be easily understood by applying the energy rate balance (neglecting any changes in flow stream kinetic and potential energies):

$$\dot{W} = \dot{m}(h_{in} - h_{out})$$

This equation clearly indicates that an effective way to increase the power output $\dot{W}$ is simply to increase the mass flow rate $\dot{m}$ through the engine. This can be done by either increasing the inlet pressure or decreasing the exhaust pressure. As the inlet pressure was increased, engineers found that the power output did in fact increase. But, when the exhaust pressure was decreased, the power also increased, but only up to a certain point. Beyond a certain operating point, something "mysterious" occurred. No matter how much the exhaust pressure was decreased below a certain level, the mass flow rate and consequently the engine's power did not increase further. They called this phenomenon *choked* flow, and it was not fully understood until the study of *compressible fluid flow* was completely developed in the early 20th century. Therefore, compressible fluid flow is of vital importance in the study of applied thermodynamics, because it helps engineers understand the effect that fluid compressibility has on the thermodynamic performance of systems containing high-speed compressible working fluids.

For an ideal gas or a low-pressure vapor with constant specific heats, this equation can be written as

$$V^2/(2g_c) = h_o - h = c_p(T_o - T)$$

or

$$\frac{T_o}{T} = 1 + \frac{V^2}{2g_c c_p T} \tag{16.2}$$

where T_o is the *stagnation temperature* (the temperature at zero velocity).

EXAMPLE 16.1

While driving in your new sports car at 90.0 km/h in still air at 20.0°C, you put your hand out the window with your palm toward the front of the car. What is the air temperature on the center of your palm?

Solution

First, draw a sketch of the system (Figure 16.1).

When your hand is placed perpendicular to the air flow, you should feel the stagnation pressure and temperature of the air flow. The stagnation temperature is given by Eq. (16.2) as

$$T_o = T\left(1 + \frac{V^2}{2g_c c_p T}\right)$$

$$= (20.0 + 273.15)\left(1 + \frac{(90.0\,\text{km/h})(1000\,\text{m/km})(1\,\text{h}/3600\,\text{s})(1\,\text{kJ/kg}/1000\,\text{m}^2/\text{s}^2)}{2(1)(1.004\,\text{kJ/kg}\cdot\text{K})(20.0 + 273.15\,\text{K})}\right)$$

$$= 294\,\text{K} = 20.3°\text{C}$$

So the stagnation temperature rise is not very much at this speed.

Exercises

1. How fast would the sports car in Example 16.1 have to travel to produce a 1.00°C stagnation temperature rise at the center of your hand? **Answer**: V = 161 km/h.
2. Suppose it is winter and the temperature of the air in Example 16.1 is 0.00°C. What would be the stagnation temperature rise at the center of your hand if all the other variables remain the same? **Answer**: $T_o - T$ = 0.310°C (independent of the value of T).
3. Now, you are in an aircraft traveling at 800. km/h in air at 20.0°C. If you put your hand out the window now, what is the air temperature at the center of your hand? **Answer**: T_o = 318 K = 44.6°C.

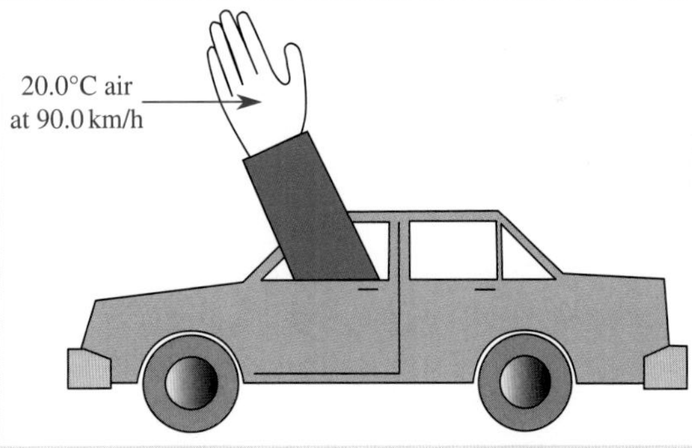

FIGURE 16.1
Example 16.1.

16.3 ISENTROPIC STAGNATION PROPERTIES

If, in addition, we decelerate the flow reversibly (i.e., without friction or other losses) and aergonically, then the entire process becomes isentropic and Eq. (7.38) of Chapter 7 can be combined with Eq. (16.2) to provide equations for the *isentropic stagnation pressure* p_{os} and the *isentropic stagnation density* ρ_{os} based on the *isentropic stagnation temperature* $T_o = T_{os}$ as

$$\frac{T_{os}}{T} = \left(\frac{p_{os}}{p}\right)^{(k-1)/k} = \left(\frac{v_{os}}{v}\right)^{1-k} = \left(\frac{\rho_{os}}{\rho}\right)^{k-1} \tag{7.38}$$

Then, Eq. (16.2) becomes

$$\frac{p_{os}}{p} = \left(1 + \frac{V^2}{2g_c c_p T}\right)^{\frac{k}{k-1}} \tag{16.3}$$

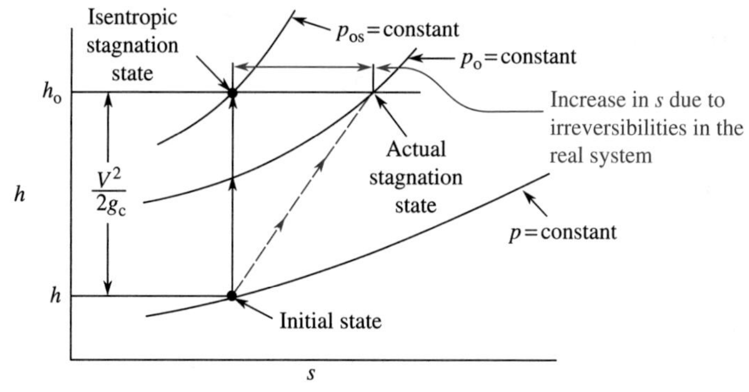

FIGURE 16.2
An isentropic and a real deceleration to the stagnation state on a Mollier diagram.

and

$$\frac{\rho_{os}}{\rho} = \left(1 + \frac{V^2}{2g_c c_p T}\right)^{\frac{1}{k-1}} \tag{16.4}$$

where the os subscript has been added to indicate the *isentropic stagnation state* condition. These states are shown schematically in Figure 16.2.

EXAMPLE 16.2

In Example 16.1, it was determined that the stagnation air temperature for air at 20.0°C traveling at 90.0 km/h = 25.0 m/s was 20.3°C. Now determine the isentropic stagnation pressure and isentropic stagnation density of this air when the atmospheric pressure is 0.101 MPa.

Solution

First, draw a sketch of the system (Figure 16.3).

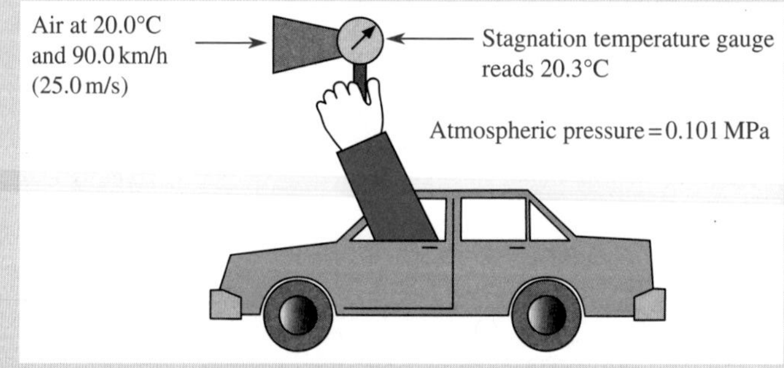

FIGURE 16.3
Example 16.2.

For air, the specific heat ratio is 1.40, and when p = 0.101 MPa, Eq. (16.3) gives the isentropic stagnation pressure as

$$p_{os} = p\left(1 + \frac{V^2}{2g_c c_p T}\right)^{\frac{k}{k-1}}$$

$$= (0.101\text{ MPa})\left(1 + \frac{(25.0\text{ m/s})^2\left(\frac{1\text{ kJ/kg}}{1000\text{ m}^2/\text{s}^2}\right)}{2(1)(1.004\text{ kJ/kg}\cdot\text{K})(20.0 + 273.15\text{ K})}\right)^{\frac{1.40}{1.40-1}}$$

$$= 0.1014\text{ MPa}$$

Note that p_{os} is the same as p to three significant figures. At 20.0°C and 0.101 MPa flow stream conditions, the flow stream density is

$$\rho = \frac{p}{RT} = \frac{101\,\text{kPa}}{(0.286\,\text{kJ/kg}\cdot\text{K})(20.0 + 273.15\,\text{K})} = 1.21\,\text{kg/m}^3$$

Then, Eq. (16.4) gives the isentropic stagnation density as

$$\rho_{os} = \rho\left(1 + \frac{V^2}{2g_c c_p T}\right)^{\frac{1}{k-1}}$$

$$= (1.21\,\text{kg/m}^3)\left(1 + \frac{(25\,\text{m/s})^2\left(\frac{1\,\text{kJ/kg}}{1000\,\text{m}^2/\text{s}^2}\right)}{2(1)(1.004\,\text{kJ/kg}\cdot\text{K})(20.0 + 273.15\,\text{K})}\right)^{\frac{1}{1.40-1}}$$

$$= 1.1213\,\text{kg/m}^3$$

Note that ρ_{os} is the same as ρ to three significant figures. The following example shows that the use of an isentropic stagnation state is not limited to ideal gases.

EXAMPLE 16.3

A new steam turbine design contains a flow nozzle that produces a flow of steam at 14.7 psia and 1000.°F at a velocity of 1612 ft/s. Determine the isentropic stagnation temperature, pressure, and density of this flow.

Solution

Equations (16.2, 16.3), and (16.4) are only for ideal gases. However, we can use the steam tables to solve this problem as follows. Let *station 1* be the exit flow stream from the nozzle and let *station os* be the same flow stream isentropically decelerated to a zero velocity (stagnation) state. From Eq. (16.1), we have $h_{os} = h_1 + V_1^2/2g_c = h_1 + (1612\text{ ft/s})^2/[2(32.174\text{ lbm}\cdot\text{ft/lbf}\cdot\text{s}^2)(778.16\text{ ft}\cdot\text{lbf/Btu})] = h_1 + 51.90$ Btu/lbm. Then, the station data become

Station 1	Station os
$p_1 = 14.7$ psia	$s_{os} = s_1 = 2.1332$ Btu/lbm·R
$T_1 = 1000°\text{F}$	$h_{os} = h_1 + V_1^2/2g_c = 1534.4 + 51.90$ Btu/lbm $= 1586$ Btu/lbm
$h_1 = 1534.4$ Btu/lbm	$p_{os} = ?$ psia
$s_1 = 2.1332$ Btu/lbm·R	$T_{os} = ?$ °F
	$\rho_{os} = ?$ lbm/ft^3

and using these values of s_{os} and h_{os} and the steam table, Table C.3a, in *Thermodynamic Tables to accompany Modern Engineering Thermodynamics* a Mollier diagram for steam, or computerized steam tables, we find that $p_{os} \approx 20.$ psia, $T_{os} \approx 1100°\text{F}$, and $\rho_{os} = 1/v_{os} \approx 1/(46.4\text{ ft}^3\text{/lbm}) = 0.022\text{ lbm/ft}^3$.

Exercises

4. Determine the isentropic stagnation pressure in Example 16.2 if the air stream is moving at 500. km/h instead of 90.0 km/h. Assume all the other variables remain unchanged. **Answer**: $p_{os} = 0.112$ MPa.
5. Use the steam tables or a Mollier diagram for steam to estimate the isentropic stagnation pressure of steam at 1100°F and 600. psia traveling at 1600. ft/s. **Answer**: $p_{os} \approx 800$ psia.
6. If the isentropic stagnation temperature and pressure of the steam in Example 16.3 are 60.0 psia and 1500.°F, determine the corresponding velocity of the steam at station 1, assuming all the other variables remain unchanged. **Answer**: $V \approx 3670$ ft/s.

16.4 THE MACH NUMBER

The Mach number M was introduced in 1929. It was named in honor of Ernst Mach (1838–1916), an Austrian physicist and philosopher who studied high-speed compressible flow in the1870s. It is the dimensionless ratio of the local fluid velocity V to the velocity of sound, c:

$$\text{M} = V/c \tag{16.5}$$

Sound is the propagation of pressure waves in a medium such as air. Sound is created by the motion or vibration of an object, which causes pressure waves in the surrounding medium. A sound wave has the same characteristics as any other type of wave. It has a wavelength, frequency, velocity, and amplitude. The velocity of sound in air at 20°C (70°F) is approximately 344 m/s (1130 ft/s), or 770 miles per hour. The speed of sound is often called the *sonic velocity*.

WHO WAS ERNST MACH?

Ernst Mach (1838–1916) was a physicist and philosopher who established the basic principles of modern scientific thought. In 1864, Mach became professor of mathematics at the University of Graz in Austria. He spent his most productive years as a professor of physics at Charles University in Prague, from 1867 to 1895. In 1881, Mach began a study of the flight of artillery shells using the new technology of photography. In this research, he discovered that the angle θ of the shock cone radiating from the leading edge of a supersonic object was related to the speed of sound c and the velocity of the object V by $\sin\theta = c/V$ (see Figure 16.4), and θ was later called the *Mach angle*. The ratio of the local fluid velocity V to the speed of sound in the fluid c came to be of fundamental value in the study of high-speed aerodynamics, and after 1930, it was called the *Mach number* ($M = V/c$).

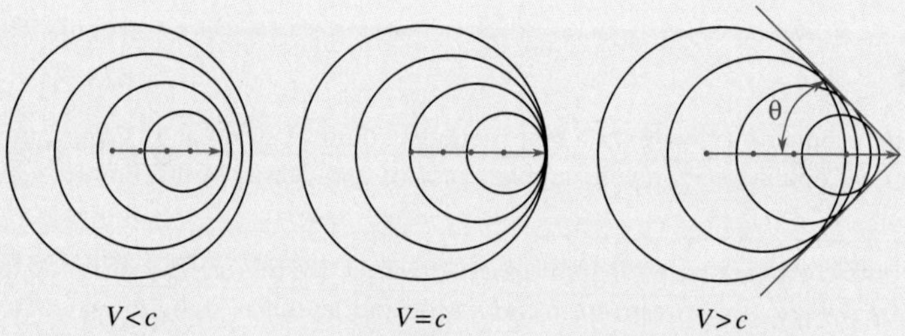

(a)

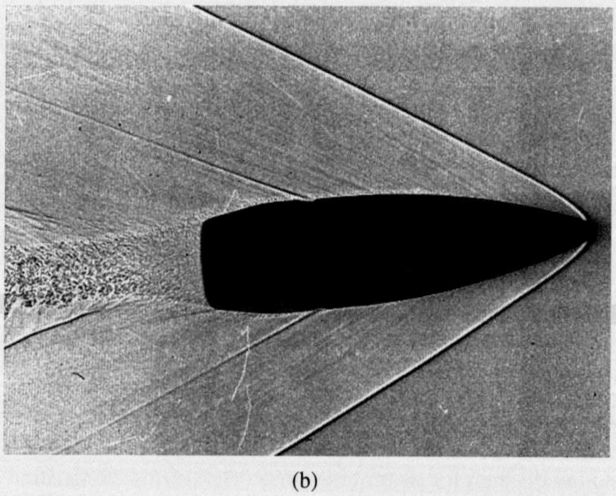

(b)

FIGURE 16.4

(a) The representation of a sound waves produced when an object moves from subsonic ($V < c$) to sonic ($V = c$) to supersonic ($V > c$) velocity. Generally, a conical shock wave sweeps back from the leading edge of the object with the cone angle proportional to the ratio of the sonic velocity of the fluid to the supersonic velocity of the object, c/V. (b) The shadowgraph of a supersonic bullet showing shock waves.

However, Mach would have probably considered his work in aerodynamics insignificant in comparison to his primary contribution in establishing a connection between physics and psychology. He adopted David Hume's position that a phenomenon can be understood from a scientific point of view only in terms of the human observations produced by the phenomenon. This view leads to the very powerful conclusion that no law of physics is acceptable unless it is experimentally verifiable. His strict definition of verifiability allowed the rejection of such long-held concepts as absolute time and space and ultimately provided the framework for the acceptance of Einstein's relativity theory.

Table 16.1 lists the jargon terms developed to describe the different flow regimes. Transonic flow is a flow in which the Mach number fluctuates around 1.0 by a small amount ($\pm\varepsilon$).

We can deduce a relation for the dependence of the isentropic speed of sound (or sonic velocity) c on local thermodynamic properties from a mass and energy balance analysis of a moving acoustical wave. Figure 16.5 shows an open system attached to an isentropic sound wave moving at velocity c through a stationary fluid in a duct that has a constant cross-sectional area A. Our coordinate system is attached to the moving wave, so the fluid appears to be approaching the wave with velocity c. For an adiabatic, aergonic, reversible, steady state, steady flow, single-inlet, single-outlet open system, the mass rate balance (MRB) reduces to

$$\dot{m}_{\text{in}} = \dot{m}_{\text{out}}$$

or

$$\begin{aligned}\rho A c &= (\rho + \partial\rho_s)(A)(c - \partial V_s)\\ &= A(\rho c - \rho\,\partial V_s + c\,\partial\rho_s - \partial\rho_s\,\partial V_s)\end{aligned}$$

Where the subscript s indicates that entropy is held constant during the differentiation process. Neglecting second-order differential terms (i.e., setting $\partial\rho_s\,\partial V_s = 0$), this equation can be rearranged to give

$$\left(\frac{\partial V}{\partial \rho}\right)_s = c/\rho \tag{16.6}$$

Similarly, the energy rate balance for this system becomes

$$h + c^2/2g_c = (h + \partial h_s) + (c - \partial V_s)^2/(2g_c)$$

Again neglecting second-order differential terms, this can be expanded and rearranged to give

$$\left(\frac{\partial h}{\partial V}\right)_s = c/g_c \tag{16.7}$$

Finally, the Gibbs Eq. (7.21) for an isentropic process yields

$$T\partial s_s = \partial h_s - v\partial p_s = 0$$

Table 16.1 Compressible Flow Mach Number Regimes (ε Is a Small Fluctuation)

Flow Regime	Name
$M < 1$	Subsonic
$M = 1$	Sonic
$1-\varepsilon \le M \le 1+\varepsilon$	Transonic
$M > 1$	Supersonic
$M \ge 5$	Hypersonic

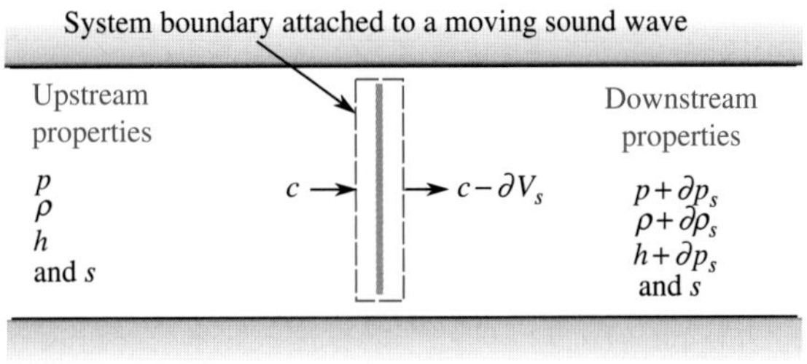

FIGURE 16.5
An isentropic sound wave moving through a stationary fluid in a horizontal duct of constant cross-sectional area A. The coordinate system is fixed to the moving wave so that the surrounding fluid appears to be moving. The s subscript is used to indicate isentropic changes in the properties.

or

$$\left(\frac{\partial h}{\partial p}\right)_s = v = \frac{1}{\rho} \tag{16.8}$$

Multiplying Eq. (16.6) by Eq. (16.7) and dividing by Eq. (16.8) gives a relation for the sonic velocity c in terms of the measurable properties p and ρ:

$$\left(\frac{\partial V}{\partial \rho}\right)_s \left(\frac{\partial h}{\partial V}\right)_s \left(\frac{\partial p}{\partial h}\right)_s = \left(\frac{\partial p}{\partial \rho}\right)_s = \rho(c/g_c)(c/\rho) = c^2/g_c$$

or

$$c = \sqrt{g_c\left(\frac{\partial p}{\partial \rho}\right)_s} \tag{16.9}$$

Equation (16.9) is a valid equation for the isentropic sonic velocity in any compressible substance. In particular, in the case of an ideal gas, Eq. (7.39) relates pressure and density for an isentropic process by

$$pv^k = p\rho^{-k} = \text{constant} \tag{7.39}$$

Solving for $p = \text{constant} \times \rho^k$ and taking the partial differential of p with respect to ρ gives

$$\left(\frac{\partial p}{\partial \rho}\right)_s = kp/\rho = kRT$$

where we use the ideal gas law, $p = \rho RT$. Then, from Eq. (16.9), the speed of sound (sonic velocity) in an ideal gas is

$$c_{\substack{\text{ideal}\\ \text{gas}}} = \sqrt{kg_cRT} \tag{16.10}$$

and from Eq. (16.5),

$$\mathrm{M}_{\substack{\text{ideal}\\ \text{gas}}} = \frac{V}{\sqrt{kg_cRT}} \tag{16.11}$$

EXAMPLE 16.4

Methane (CH_4) gas at 35.0°C flows through a pipe with a velocity of 300. m/s. Determine the Mach number of the methane.

Solution

The velocity of sound in methane at 35.0°C is given by Eq. (16.10). Using Table C.13b in *Thermodynamic Tables to accompany Modern Engineering Thermodynamics* for the values of the specific heat ratio and the gas constant for methane, we get

$$\begin{aligned} c_{\text{methane}} &= \sqrt{(k_{\text{methane}})(g_c)(R_{\text{methane}})(T)} \\ &= \sqrt{(1.30)(1)(518\,\text{J/kgCK})(35.0+273.15\,\text{K})} \\ &= 456\,\text{m/s} \end{aligned}$$

then, Eq. (16.5) (or (16.11)) gives the Mach number as

$$\mathrm{M}_{\text{methane}} = \frac{300.\ \text{m/s}}{456\,\text{m/s}} = 0.658$$

Exercises

7. Determine the Mach number of the methane in Example 16.4 if its velocity is reduced from 300. m/s to 3.00 m/s and all the other variables remained unchanged. **Answer**: $\mathrm{M}_{\text{methane}} = 6.58 \times 10^{-3}$.
8. If the methane in the pipe in Example 16.4 is cooled to 0.00°C and all the other variables remain unchanged, determine the new Mach number of the methane. **Answer**: $\mathrm{M}_{\text{methane}} = 0.700$.
9. If the gas in Example 16.4 is changed from methane to argon, determine the Mach number of the argon assuming all the other variables remain unchanged. **Answer**: $\mathrm{M}_{\text{argon}} = 0.917$.

Since for ideal gases $c_p - c_v = R$ and $k = c_p/c_v$, we can write $c_p = R + c_v = kR/(k-1)$. Also, from Eq. (16.10), we find that $R = c^2/(kg_c T)$. Then, the equations for the isentropic stagnation temperature, pressure, and density (Eqs. (16.2), (16.3), and (16.4)) for an ideal gas become

$$\frac{T_{os}}{T} = 1 + V^2/(2g_c c_p T) = 1 + \frac{k-1}{2}\mathrm{M}^2 \tag{16.12}$$

$$\frac{p_{os}}{p} = \left(1 + \frac{k-1}{2}\mathrm{M}^2\right)^{k/(k-1)} \tag{16.13}$$

and

$$\frac{\rho_{os}}{\rho} = \left(1 + \frac{k-1}{2}\mathrm{M}^2\right)^{1/(k-1)} \tag{16.14}$$

Table C.18 in *Thermodynamic Tables to accompany Modern Engineering Thermodynamics* contains tabulated values of T/T_{os}, p/p_{os}, and ρ/ρ_{os} for air ($k = 1.40$) for various values of M. These tabulations were made using Eqs. (16.12), (16.13), and (16.14) and may be used in place of these equations when convenient. Table 16.2 lists various values of $k = c_p/c_v$ and Figure 16.6 shows k at various pressures.

Table 16.2 Typical Values of the Specific Heat Ratio k

Gas	$k = c_p/c_v$
Monatomic	
Argon, helium, neon, xenon, etc.	1.67
Diatomic	
Air	1.40
Nitrogen	1.40
Oxygen	1.39
Carbon monoxide	1.40
Hydrogen	1.40
Triatomic	
Carbon dioxide	1.29
Sulfur dioxide	1.25

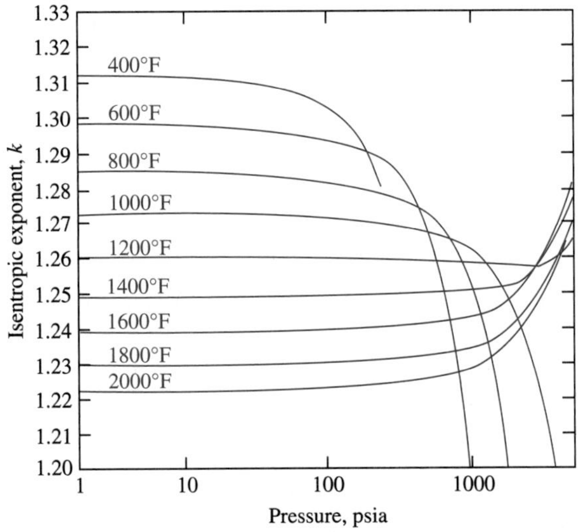

FIGURE 16.6
The variation in the isentropic exponent $k = c_p/c_v$ of steam with pressure and temperature.

EXAMPLE 16. 5

Find the velocity, isentropic stagnation temperature, and isentropic stagnation pressure on an aircraft flying at Mach 0.850 at an altitude where the temperature is –20.0°C and pressure of 0.500 atm. Assume air is an ideal gas with a constant specific heat ratio of $k = 1.40$ and a gas constant of $R = 286\,\text{J/(kg·K)}$.

Solution

From Eq. (16.11), we can calculate the aircraft's velocity as

$$V = \text{M}\sqrt{kg_cRT}$$
$$= (0.850)\{(1.40)(1)[286\,\text{m}^2/(\text{s}^2\cdot\text{K})](-20.0+273.15\,\text{K})\}^{1/2} = 271\,\text{m/s}$$

The isentropic stagnation pressure and temperature can be determined from Eqs. (16.12) and (16.13) as

$$T_{os} = T\left(1+\frac{k-1}{2}\text{M}^2\right)$$
$$= (-20.0+273.15\,\text{K})\left[1+\frac{1.40-1}{2}(0.850)^2\right] = 290.\,\text{K} = 16.6°\text{C}$$

and

$$p_{os} = p\left[1+\frac{k-1}{2}\text{M}^2\right]^{k/(k-1)}$$
$$= (0.500\,\text{atm})\left[1+\frac{1.40-1}{2}(0.850)^2\right]^{1.40/(1.40-1)} = 0.802\,\text{atm} = 81.3\,\text{kPa}$$

Exercises

10. If the aircraft in Example 16.5 has a supersonic Mach number of 2.25 and all the other variables remain unchanged, what would be its velocity? **Answer:** $V = 717$ m/s.
11. Determine the isentropic stagnation temperature of the aircraft in Example 16.5 if it is flying at a supersonic Mach number of 1.30 and all the other variables remain unchanged. **Answer:** $T_{os} = 339$ K = 65.6°C.
12. If the aircraft in Example 16.5 is flying at a hypersonic Mach number of 6.50, determine the isentropic stagnation pressure on the aircraft, assuming all the other variables remain unchanged. **Answer:** $p_{os} = 1300.$ atm.

CRITICAL THINKING

Equations (16.12)–(16.14) are for the isentropic deceleration of an ideal gas to zero velocity (stagnation), and the resulting isentropic stagnation pressure and temperature (p_{os} and T_{os}) are valid only under these conditions. Would you expect the actual values of the stagnation pressure and temperature (p_o and T_o) for a real (nonideal) gas undergoing a real (nonisentropic) deceleration to zero velocity to be larger or smaller than their isentropic counterparts?

Example 16.5 shows that, even at moderate subsonic velocities, there can be a considerable temperature and pressure rise at the stagnation points of moving objects.

16.5 CONVERGING-DIVERGING FLOWS

We now investigate the effect of variations in the flow cross-sectional area on the Mach number of the flow. If we differentiate the mass rate balance equation, $\dot{m} = \rho A v = \text{constant}$, for an isentropic flow, we obtain

$$\rho A\,\partial V_s + \rho V\,\partial A_s + AV\,\partial\rho_s = 0$$
$$\partial V_s/V + \partial A_s/A + \partial\rho_s/\rho = 0$$

or

$$\partial A_s/A = -\partial V_s/V - \partial\rho_s/\rho \tag{16.15}$$

Next, we differentiate Eq. (16.1) for a constant isentropic stagnation enthalpy to get

$$\partial h_{os} = \partial h_s + V\,\partial V_s/g_c = 0$$

or

$$\partial h_s = -V\,\partial V_s/g_c$$

Now, we combine this with Gibbs Eq. (7.21) for an isentropic process to get

$$T\,\partial s_s = \partial h_s - v\,\partial p_s = 0$$

or

$$\partial h_s = v\,\partial p_s = \partial p_s/\rho = -V\,\partial V_s/g_c$$

then,

$$\partial V_s/V = -g_c\,\partial p_s/\rho V^2 \tag{16.16}$$

Substituting Eq. (16.16) into (16.15) and using Eqs. (16.5) and (16.9) gives the desired result

$$\begin{aligned}\partial A_s/A &= -\partial V_s/V - \partial\rho_s/\rho = \left[g_c/V^2 - \left(\frac{\partial\rho}{\partial p}\right)_s\right](\partial p_s/\rho)\\ &= [g_c/V^2 - g_c/c^2](\partial p_s/\rho) = (1-\mathrm{M}^2)(g_c/V^2)(\partial p_s/\rho)\end{aligned}$$

or

$$\left(\frac{\partial A}{\partial p}\right)_s = (1-\mathrm{M}^2)\left(\frac{Ag_c}{\rho V^2}\right) \tag{16.17a}$$

Using Eq. (16.16), we can rewrite Eq. (16.17a) as

$$\frac{\partial A_s}{A} = (1-\mathrm{M}^2)\frac{g_c\partial p_s}{\rho V^2} = (\mathrm{M}^2-1)\frac{\partial V_s}{V} \tag{16.17b}$$

Equations (16.17a) and (16.17b) lead to slightly different but related conclusions about the nature of converging-diverging flows.

Converging subsonic flow: $\partial A_s < 0$ and $\mathrm{M} < 1$

- **Equation (16.17a)**. Converging flows are characterized by the fact that the area becomes smaller in the direction of flow, or $\partial A_s < 0$. Subsonic flows are characterized by the fact that $\mathrm{M} < 1$. Since A, g_c, ρ, and V are all > 0, Eq. (16.17a) shows that $(\partial A/\partial p)_s$ must be > 0. Then, ∂p_s must be < 0 since $\partial A_s < 0$ for converging flows. Consequently, the pressure must decrease in the direction of a subsonic converging flow.
- **Equation (16.17b).** Since $\mathrm{M} < 1$ and $\partial A_s < 0$ for subsonic converging flows (Figure 16.7), then Eq. (16.17b) shows that ∂V_s must be > 0 for these flows. That is, the flow velocity increases in subsonic converging flow. In Chapter 6, we define a nozzle as a flow geometry that converts pressure into kinetic energy, or $\partial p_s < 0$ and $\partial V_s > 0$ in the direction of flow. Consequently, converging subsonic flow corresponds to what we traditionally call *nozzle flow*. Therefore, a converging passage carrying subsonic flow is called a *subsonic nozzle*.

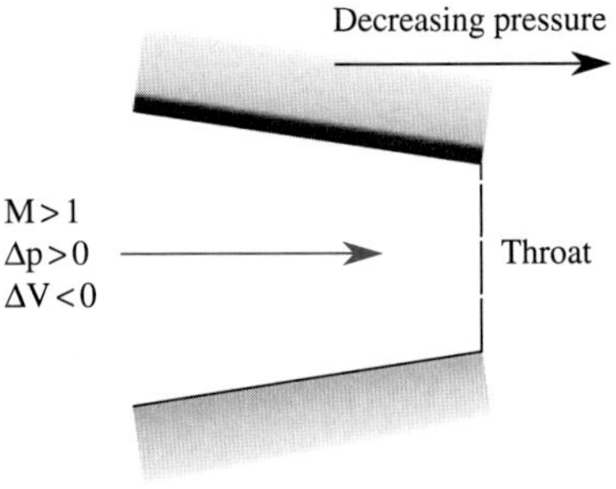

FIGURE 16.7
Converging subsonic nozzle.

Converging supersonic flow: $\partial A_s < 0$ and $\mathrm{M} > 1$

- **Equation (16.17a)**. Here, ∂A_s is still < 0, but now $\mathrm{M} > 1$. Then, Eq (16.17a) tells us that ∂p_s must be > 0 and the pressure increases in the direction of flow.
- **Equation (16.17b)**. If $\partial A_s < 0$ and $\mathrm{M} > 1$, then Eq. (16.17b) tells us that ∂V_s must be < 0, or the flow velocity must decrease in the direction of flow. In Chapter 6, we define a diffuser as a flow geometry that converts kinetic energy into pressure, or $\partial V_s < 0$ and $\partial p_s > 0$ in the direction of flow. Consequently, a converging passage carrying a supersonic flow is called a *supersonic diffuser* (Figure 16.8).

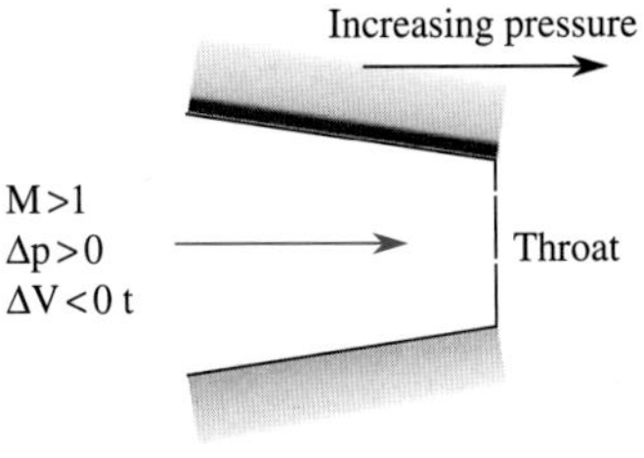

FIGURE 16.8
Converging supersonic diffuser.

Diverging subsonic flow: $\partial A_s > 0$ and $\mathrm{M} < 1$

- **Equation (16.17a)**. Diverging flows are characterized by the fact that the area becomes larger in the direction of flow, or $\partial A_s > 0$. If the flow is

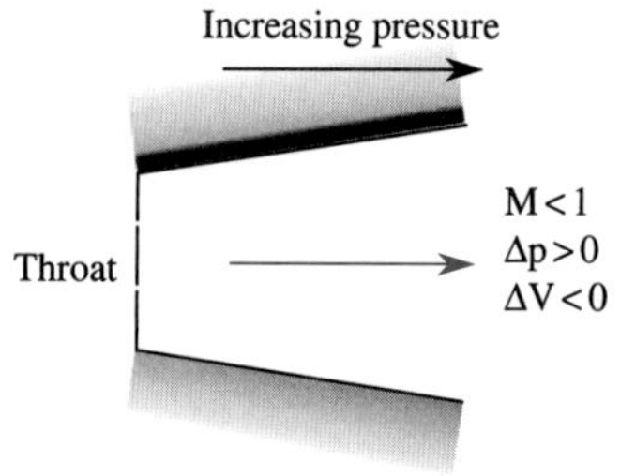

FIGURE 16.9
Diverging subsonic diffuser.

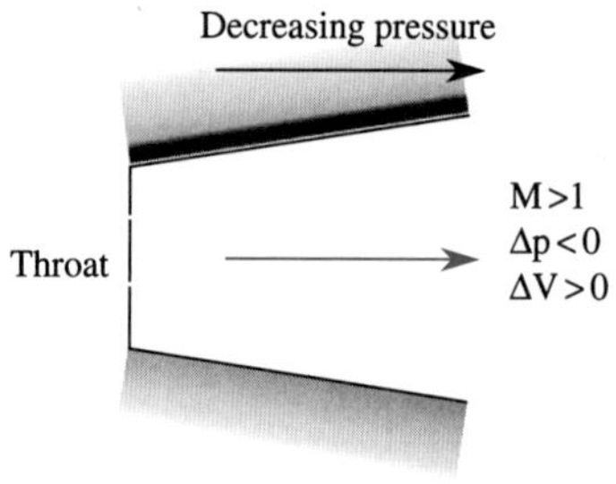

FIGURE 16.10
Diverging supersonic nozzle.

subsonic, $M < 1$, and Eq. (16.17a) tells us that ∂p_s must be > 0. That is, the pressure must increase in the direction of a subsonic diverging flow.

- **Equation (16.17b)**. In the case of diverging subsonic flow, Eq. (16.17b) tells us that, since $\partial A_s > 0$ and $M < 1$, ∂V_s must be < 0, so the flow velocity must decrease in the direction of flow. In Chapter 6, we learned that a device that converts kinetic energy ($\partial V_s < 0$) into pressure ($\partial p_s > 0$) is called a *diffuser*. Consequently, a diverging passage carrying subsonic flow is called a *subsonic diffuser* (Figure 16.9).

Diverging supersonic flow: $\partial A_s > 0$ and $M > 1$

- **Equation (16.17a)**. Since $\partial A_s > 0$ and $M > 1$ here, Eq. (16.17a) tells us that ∂p_s must be < 0 and the pressure must decrease in the direction of flow.
- **Equation (16.17b)**. For $\partial A_s > 0$ and $M > 1$, Eq. (16.17b) tells us that ∂V_s must be > 0 and the flow velocity must increase in the direction of flow. This again corresponds to the definition of a nozzle as a device that converts pressure into kinetic energy, so a diverging passage carrying supersonic flow is called a *supersonic nozzle* (Figure 16.10). This is the type of nozzle used on the space shuttle rocket engine, as shown in the schematic of Figure 16.11.

When $M = 1.0$ in Eq. (16.17b), $(\partial A_s/A) = 0$. This corresponds to a point of minimum cross-sectional area. This point is called the *throat* of the device, characterized by the fact that it can never have a Mach number greater than 1; that is $M_{\text{throat}} \leq 1.0$.

When the Mach number at the throat is equal to 1, we say that the throat is at its *critical* condition and denote its properties in this state with a superscript asterisk. Then, M^*_{throat} is *always* equal to 1.0, and from Eqs. (16.12), (16.13), and (16.14), the critical condition properties at the throat are[1]

$$T^* = T_{os}\left(\frac{2}{k+1}\right) \tag{16.18}$$

$$p^* = p_{os}\left(\frac{2}{k+1}\right)^{k/(k-1)} \tag{16.19}$$

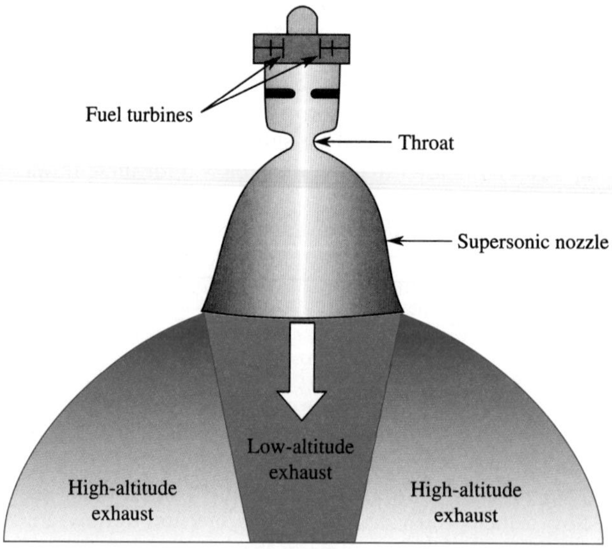

FIGURE 16.11
Space shuttle rocket engine.

[1] These are not the same as the thermodynamic *critical state* properties. The use of the word *critical* here refers to a different type of phenomenon.

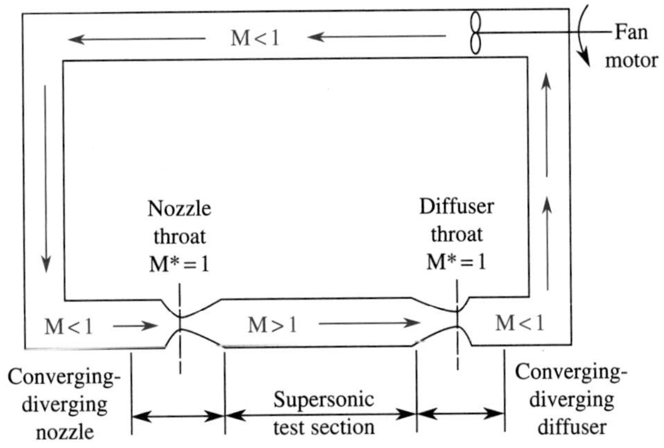

FIGURE 16.12
A supersonic wind tunnel design.

and

$$\rho^* = \rho_{os}\left(\frac{2}{k+1}\right)^{1/(k-1)} \tag{16.20}$$

When a subsonic nozzle and a supersonic nozzle are joined at their throats, they form a converging-diverging nozzle that can be used to generate supersonic velocities. Also, connecting a supersonic diffuser to a subsonic diffuser forms a converging-diverging diffuser that can be used to decelerate a supersonic flow and recover its kinetic energy by converting it into pressure. These two converging-diverging geometries are often combined in the design of a supersonic wind tunnel, as shown in Figure 16.12. Note that the diverging part of the nozzle cannot become supersonic until the throat becomes critical at a Mach number of 1.0. This is called *choked flow* and is discussed in the next section.

EXAMPLE 16.6

A converging-diverging nozzle is attached via a valve to a pipe holding compressed air at 1.00 MPa and 20.0°C. The valve is opened and the air passes through the nozzle and into the atmosphere. Assuming isentropic flow throughout, determine

a. The exit Mach number.
b. The exit temperature.
c. The exit velocity.
d. The pressure at the throat of the nozzle.
e. The temperature at the throat of the nozzle.

Solution
First, draw a sketch of the system (Figure 16.13).

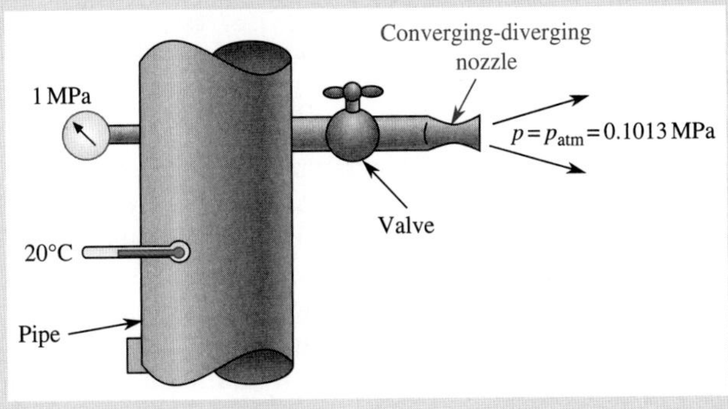

FIGURE 16.13
Example 16.6.

(Continued)

EXAMPLE 16.6 (*Continued*)

a. Assume air is a constant specific heat ideal gas with $k = 1.40$. Since the fluid is air, we can use either Table C.18 in *Thermodynamic Tables to accompany Modern Engineering Thermodynamics* with

$$p/p_{os} = (0.1013\ \text{MPa})/(1.00\ \text{MPa}) = 0.101$$

or Eqs. (16.12) and (16.13). Instead of interpolating in Table C.18, it is easier to solve Eq. (16.13) directly for M:

$$\text{M} = \left\{\frac{2}{k-1}\left[(p_{os}/p)^{(k-1)/k} - 1\right]\right\}^{1/2} = \left\{\frac{2}{1.40-1}\left[\left(\frac{1.00}{0.101}\right)^{(1.40-1/1.40} - 1\right]\right\}^{1/2} = 2.15$$

Note that this agrees with what we would have interpolated from Table C.18.

b. Here, Eq. (16.12) can be used to give

$$T = \frac{T_{os}}{1+\frac{k-1}{2}\text{M}^2} = \frac{20.0+273.15\ \text{K}}{1+\frac{1-1.40}{2}(2.15)^2} = 152\ \text{K} = -121°\text{C}$$

c. Equation (16.11) can now be used to find the exit velocity as

$$V = \text{M}\sqrt{kg_cRT} = (2.15)\{(1.40)(1)[286\ \text{m}^2/(\text{s}^2\cdot\text{K})](152\ \text{K})\}^{1/2} = 530.\ \text{m/s}$$

d. Since the exit velocity is supersonic, the throat must be sonic; therefore, Eq. (16.19) can be used to determine the throat pressure as

$$p_{\text{throat}} = p^* = p_{os}[2/(k+1)]^{k/(k-1)}$$
$$= (1.00\ \text{MPa})\left(\frac{2}{2.40}\right)^{1.4/(1..4-1)} = 0.528\ \text{MPa}$$

e. Similarly, Eq. (16.18) can be used to calculate the throat temperature as

$$T_{\text{throat}} = T^* = T_{os}[2/(k+1)]$$
$$= (20.0+273.15\ \text{K})(2/2.40) = 244\text{K} = -29.2°\text{C}$$

Exercises

13. Determine the exit Mach number in Example 16.6 when the pressure in the pipe is 5.00 MPa and all the other variables remain unchanged. **Answer**: $\text{M}_{\text{exit}} = 3.20$.
14. Determine the exit temperature in Example 16.6 if the temperature in the pipe is 65.0°C. Assume all the other variables remain unchanged. **Answer**: $T_{\text{exit}} = 176$ K = –97.4°C.
15. Determine the temperature and pressure at the throat of the nozzle in Example 16.6 if the temperature and pressure inside the pipe are 65.0°C and 5.00 MPa. Assume all the other variables remain unchanged. **Answer**: $T^* = 8.64$°C and $p^* = 2.64$ MPa.

Example 16.6 demonstrates that a very significant temperature drop can occur inside a supersonic nozzle. If the flowing fluid is a vapor near its saturation state, it is possible that this type of cooling can cause the state to drop through the vapor dome and produce a two-phase mixture inside the nozzle. This is a rather common occurrence for low-temperature steam flow through a nozzle. However, the condensation process that produces this phase change generally requires a longer time to complete than the resident time of the high-speed vapor in the converging part of the nozzle. When this occurs, the vapor exits the nozzle's throat in a nonequilibrium state at a much lower temperature than the proper equilibrium saturation temperature at the exit pressure. This nonequilibrium state is called *supersaturated* and is very unstable. After a sufficient time has passed, the fluid undergoes a rapid condensation downstream from the throat due to a nucleation and growth process of the second phase. This process is irreversible and causes the temperature of the two-phase mixture to rise to the proper equilibrium saturation value (see Figure 16.14). The irreversible condensation process from a' to b shown in Figure 16.14 is sometimes called a *condensation shock*.

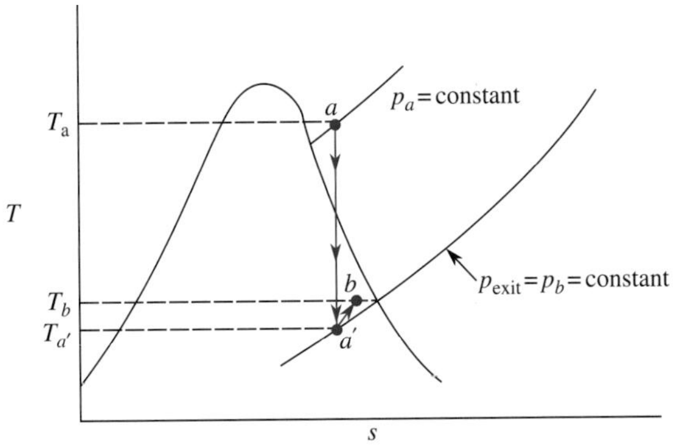

FIGURE 16.14
Isentropic expansion of a vapor near its saturation at state *a* through a nozzle to a nonequilibrium supersaturated state at *a'*. This is followed by an irreversible nucleation and condensation process to equilibrium state *b*.

16.6 CHOKED FLOW

The mass flow rate per unit area in an isentropic nozzle can be determined for an ideal gas from Eqs. (16.11), (16.12), and (16.13) and the ideal gas formula, $\rho = p/RT$, as

$$\begin{aligned}\dot{m}/A &= \rho V = (p/RT)(\mathrm{M}c)\\ &= p\mathrm{M}(\sqrt{kg_cRT})(T_{os}/T_{os})^{1/2}(p_{os}/p_{os})/RT\\ &= p_{os}\mathrm{M}(kg_c/RT_{os})^{1/2}(T_{os}/T)^{1/2}(p/p_{os})\\ &= p_{os}\mathrm{M}\left(\frac{kg_c}{RT_{os}}\right)^{1/2}\left(1+\frac{k-1}{2}\mathrm{M}^2\right)^{(k+1)/2(1-k)}\end{aligned} \qquad (16.21)$$

Figure 16.15 is a schematic of the variation in $\dot{m}/A$ with the ratio of the back pressure p_B to the upstream stagnation pressure p_{os}. When $\mathrm{M}_{throat} = \mathrm{M}^* = 1.0$, the nozzle is passing the maximum possible flow $\dot{m}_{max}$.

Since we always have $\mathrm{M}_{throat} \leq 1.0$, clearly the maximum nozzle mass flow rate occurs when the throat velocity is sonic. Then $\dot{m} = \dot{m}_{max}$, $\mathrm{M}_{throat} = \mathrm{M}^* = 1.0$, $A_{throat} = A^*$, and Eq. (16.21) becomes

$$\dot{m}_{max}/A^* = p_{os}\left(\frac{kg_c}{RT_{os}}\right)^{1/2}\left(\frac{k+1}{2}\right)^{(k+1)/2(1-k)} \qquad (16.22a)$$

Note that the value of $\dot{m}_{max}/A^*$ depends only on the upstream isentropic stagnation properties and is completely independent of the downstream conditions. For air, $k = 1.40$ and, in Engineering English units, this equation reduces to

$$(\dot{m}_{max}/A^*)_{air} = \left[0.532\frac{\mathrm{lbm}\cdot\sqrt{\mathrm{R}}}{\mathrm{lbf}\cdot\mathrm{s}}\right]\left(\frac{p_{os}}{\sqrt{T_{os}}}\right) \qquad (16.22b)$$

where $\dot{m}_{max}$ must be in lbm/s, A^* must be in in^2, p_{os} must be in lbf/in^2, and T_{os} must be in R. This is called Fliegner's (or sometimes Zeuner's) formula and was experimentally discovered in the 1870s. In metric SI units, Eq. (16.22a) reduces to

$$(\dot{m}_{max}/A^*)_{air} = \left[0.0404\frac{\mathrm{kg}\cdot\sqrt{\mathrm{K}}}{\mathrm{N}\cdot\mathrm{s}}\right]\left(\frac{p_{os}}{\sqrt{T_{os}}}\right) \qquad (16.22c)$$

where $\dot{m}_{max}$ must be in kg/s, A^* must be in m^2, p_{os} must be in N/m^2, and T_{os} must be in K.

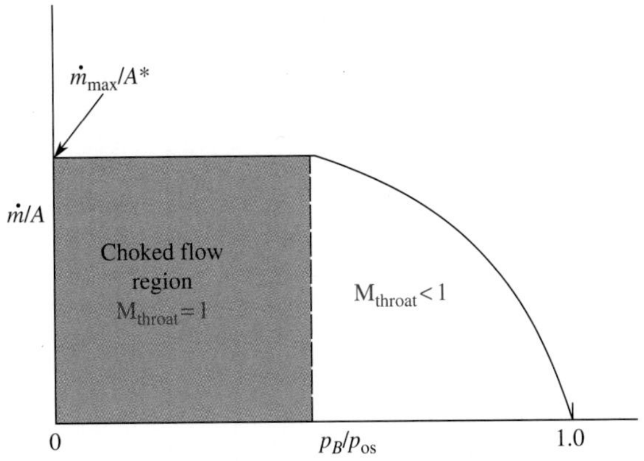

FIGURE 16.15
The relative variation in dimensionless isentropic converging nozzle air mass flow rate with increasing back pressure to upstream stagnation pressure ratio. Since the Mach number in a converging nozzle cannot exceed 1.0, the maximum flow rate through the nozzle occurs when the Mach number at the throat is 1.0.

EXAMPLE 16.7

Automobile safety equipment now includes *air bags*, which are essentially large balloons that inflate very quickly on vehicle impact. To investigate the feasibility of using high-pressure compressed air to inflate bags, we want to determine the minimum tube diameter necessary to completely fill a spherical air bag to a diameter of 3.00 feet in 30.0 milliseconds to a pressure of 15.0 psia using air from a compressed air storage tank maintained at $p_{os} = 1500.$ psia and $T_{os} = 70.0°\text{F}$.

Solution

If the air bags are to be inflated from a high-pressure compressed air storage tank, the maximum air flow rate from the tank corresponds to the minimum required fill-tube diameter operating under choked flow conditions. The average mass flow rate of air into the bag is

$$\dot{m}_{\text{avg}} = \frac{\textit{Mass of air in the bag}}{\textit{Required bag fill time}} = \frac{\rho_{air}\forall_{\text{bag}}}{t_{\text{fill}}}$$

where

$$\forall_{\text{bag}} = \frac{\pi D_{\text{bag}}^3}{6} = \frac{\pi(3.00\text{ ft})^3}{6} = 14.1\text{ ft}^3$$

The temperature of the air entering the bag can be computed from Eq. (16.18) as

$$T_{\text{air}} = T_{os}\left(\frac{2}{k+1}\right) = (70.0 + 459.67\text{ R})\left(\frac{2}{1.40+1}\right) = 441\text{ R}$$

Consequently, the density of the air entering the bag at 15 psia is

$$\rho_{\text{air}} = \frac{p_{\text{air}}}{R_{\text{air}}T_{\text{air}}} = \frac{(15.0\text{ lbf/ft}^2)(144\text{ in}^2/\text{ft}^2)}{(53.34\text{ ft}\cdot\text{lbf/lbm}\cdot\text{R})(441\text{ R})} = 0.0918\frac{\text{lbm}}{\text{ft}^3}$$

The minimum diameter of the bag fill tube may now be determined from Eq. (16.22b) by setting $\dot{m}_{\text{avg}} = \dot{m}_{\text{max}}$:

$$A_{\text{tube}}^* = \frac{\pi D_{\text{tube}}^2}{4} = \frac{\dot{m}_{\text{max}}\sqrt{T_{os}}}{0.532\pi p_{os}}$$

or

$$D_{tube} = \left[\frac{4\dot{m}_{\text{avg}}\sqrt{T_{os}}}{0.532\pi p_{os}}\right]^{1/2} = \left[\frac{4(43.4\text{ lbm/s})\sqrt{70.0+459.67\text{ R}}}{(0.532\text{ lbm}\cdot\sqrt{R}/\text{lbf}\cdot\text{s})(\pi)(1500.\text{ lbf/in}^2)}\right]^{1/2} = 1.26\text{ in}$$

The graph in Figure 16.16 shows the variation in fill-tube diameter with compressed air pressure. Note that, even with a storage tank at 4000. psia, a ¾-inch diameter fill tube is required. Also, since our analysis was for an *isentropic* process, losses in the system would necessitate using a larger diameter to achieve the required inflation.

Considering the size of the fill tube and the financial cost of this type of inflation system, it is not surprising to find that air bags are actually inflated by a rapid chemical reaction (explosion) that releases a large amount of gas in a short period.

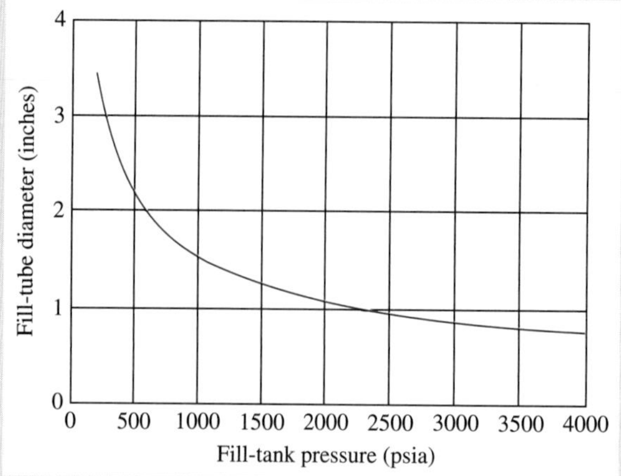

FIGURE 16.16
Example 16.7.

Exercises

16. Determine the fill-tube diameter in Example 16.7 for a compressed air storage (stagnation) pressure of 3000. psia. Assume all the other variables remain unchanged. **Answer:** $D_{tube} = 0.890$ in.
17. Suppose the air bag in Example 16.7 is filled with helium instead of air. Determine the minimum fill-tube diameter required, assuming all the other variables remain unchanged. **Answer:** $D_{tube} = 0.787$ in.
18. Determine the air bag minimum fill-tube diameter in Example 16.7 if the variables are changed to $D_{bag} = 1.0$ m, $T_{os} = 20.0$°C, $p_{os} = 10.0$ MPa, and $p_{bag} = 0.103$ MPa. **Answer:** $D_{tube} = 0.0370$ m.

When $M_{throat} = 1.0$, the nozzle is passing its maximum flow rate and no changes made *downstream* of the throat (such as lowering the exit pressure) will cause the flow to increase. Consequently, the nozzle is said to be *choked* when the velocity of the fluid at its throat reaches sonic velocity.

Dividing Eq. (16.22a) by (16.21) gives the nozzle to throat cross-sectional area ratio for a supersonic nozzle at its maximum flow rate, and for air ($k = 1.4$), this becomes

$$A/A^* = \frac{1}{M}\left[\frac{2}{k+1}\left(1+\frac{k-1}{2}M^2\right)\right]^{(k+1)/2(k-1)} \tag{16.23a}$$

for air ($k = 1.40$), this becomes

The ratio of the nozzle air flow cross-sectional area (A), where the Mach number is M, to the throat cross-sectional area (A^*), where the mach number is $M_{throat} = 1.0$ is:

$$(A/A^*)_{air} = \frac{1}{M}\left(\frac{1+0.2M^2}{1.2}\right)^3 \tag{16.23b}$$

HOW DO AUTOMOBILE AIR BAGS WORK?

Why are air bags inflated with a chemical reaction rather than with compressed air? (See Example 16.7 for the answer.)

Air bags are widely used in automobiles to protect occupants in the event of collision. It is essential that the bag inflate within a few milliseconds. Therefore, the gas must be nontoxic and nonflammable and be produced by a very rapid chemical reaction.

Most air bags contain a mixture of solid sodium azide, NaN_3, and iron oxide, Fe_2O_3, as the source of inflating gas because sodium azide contains 65% nitrogen (by mass) and it decomposes very rapidly at temperatures of 350°C or higher:

$$2NaN_3 \rightarrow 2Na(\ell) + 3N_2(g)$$

and

$$6Na(\ell) + Fe_3O_2 \rightarrow 3Na_2O(s) + 2Fe(s)$$

Because sodium azide is poisonous, all the products and reactants are kept in a sealed container with only the nitrogen gas being allowed to enter the "air" bag.

The decomposition is initiated about 10 milliseconds into a crash by a fuse wire activated by a collision sensor. The bag is completely filled within 30 milliseconds, and porous sections of the bag allow it to deflate in 100 to 200 milliseconds.

CRITICAL THINKING

The term *choked* implies an asphyxiating, suffocating, clogged, or otherwise impeded condition. How would you explain the concept of *choked flow* as used in compressible fluid flow in nontechnical language to a liberal arts friend?

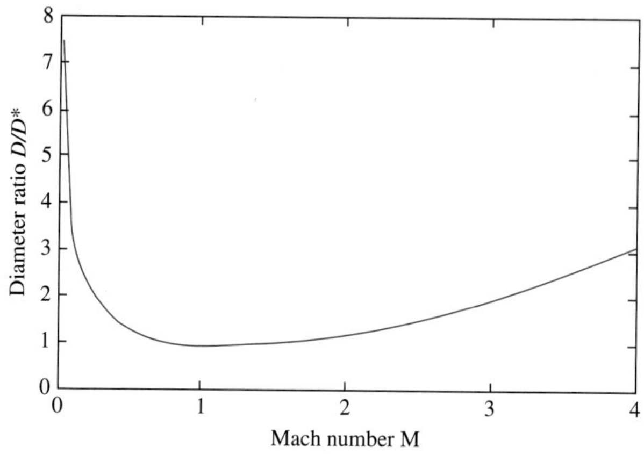

FIGURE 16.17
Diameter ratio variation in a converging-diverging isentropic nozzle for air ($k = 1.4$), from Eq. (16.23b).

For a circular cross-section, $A/A^* = (D/D^*)^2$. Figure 16.17 shows how this diameter ratio varies with Mach number for air according to Eq. (16.23b). If a converging-diverging nozzle does not have the exact shape dictated by Eq. (16.23a), it may still produce supersonic flow; however, the flow will not be isentropic.

EXAMPLE 16.8

To deflate an automobile tire, the valve core must be removed from the tire's valve stem. When the core is removed, the valve stem approximates an isentropic converging nozzle with an internal diameter of 0.0938 in. If the tire is initially at 50.0 psia (35.3 psig) and 70.0°F,

- Is the flow in the open valve stem initially choked?
- If so, at what tire pressure does it unchoke?
- How long does it take to unchoke, if the tire is assumed to have a constant volume of 1.00 ft^3 and a constant internal temperature (T_{os}) of 70°F?

Solution

First, draw a sketch of the system (Figure 16.18).

a. From Eq. (16.19) for air ($k = 1.40$), the flow is choked if

$$p_{exit}/p_{os} < p^*/p_{os} = \left(\frac{2}{k+1}\right)^{k/(k-1)} = \left(\frac{2}{1+1.40}\right)^{1.40/(1.4-1)} = 0.528$$

Here, $p_{exit}/p_{os} = 14.7/50.0 = 0.294$, which is <0.528 therefore, initially, the flow is choked.

b. The flow remains choked until the tire deflates to a pressure of

$$p_{os} = p_{exit}/0.528 = 14.7/0.528 = 27.8 \text{ psia} = 13.1 \text{ psig}$$

c. During the deflation process with a choked flow, $\dot{m} = \dot{m}_{max}$, then Eq. 16.22b gives

$$\dot{m}_{max} = -\left(\frac{dm_T}{dt}\right) = 0.532\left(A^* p_{os}/\sqrt{T_{os}}\right)$$

where

$$A^* = \frac{\pi D^2_{exit}}{4} = \frac{\pi(0.0938 \text{ in})^2}{4(144 \text{ in}^2/\text{ft}^2)} = 4.80 \times 10^{-5} \text{ ft}^2$$

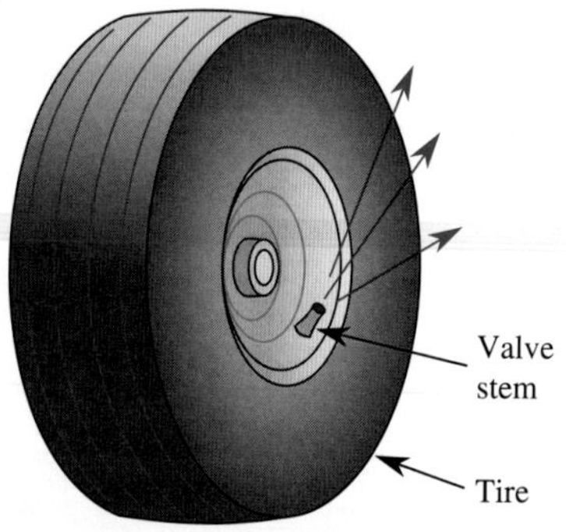

FIGURE 16.18
Example 16.8.

Also, $T_{os} = 70.0°\text{F} = \text{constant}$, and $p_{os} = m_T R T_{os}/V\!\!\!/_T$, where m_T and $V\!\!\!/_T$ are the mass of air in the tire and the volume of the tire, respectively.

Then,

$$\begin{aligned} dm_T/dt &= -0.532\left[A^* m_T R T_{os}/\left(\forall_T \sqrt{T_{os}}\right)\right] \\ &= -0.532\left[A^* R\sqrt{T_{os}}/\forall_T\right] m_T \\ &= -\left[0.532\,\text{lbm}\cdot\sqrt{\text{R}}/(\text{lbf}\cdot\text{s})\right]\left(4.80\times10^{-5}\text{ft}^2\right)[53.34\,\text{ft}\cdot\text{lbf}/(\text{lbm}\cdot\text{R})] \\ &\quad \times \left[(70.0+459.67\,\text{R})^{1/2}/(1.00\,\text{ft}^3)\right] m_T \\ &= -(0.0313\,\text{s}^{-1}) m_T \end{aligned}$$

or

$$\frac{dm_T}{m_T} = -(0.0313)\,dt$$

Integrating this result from the initial mass in the tire m_{Ti} to the mass in the tire $m_{T\tau}$ when the valve stem unchokes at time τ gives

$$\ln(m_{T\tau}/m_{Ti}) = -(0.0313)\tau$$

or

$$\tau = 31.95[\ln(m_{Ti}/m_{T\tau})] \text{ seconds}$$

Now, $m_{Ti}/m_{T\tau} = [p_{os}\forall_T/(RT_{os})]_i/[p_{os}\forall_T/(RT_{os})]_\tau = p_{osi}/p_{os\tau}$, so

$$\tau = 31.95\ln(p_{osi}/p_{os\tau}) = 31.95\ln(50.0/27.8) = 18.7\,\text{s}$$

Exercises

19. Determine the time required for the valve stem in Example 16.8 to unchoke when the tire pressure is increased from 50.0 to 70.0 psia. Assume all the other variables remain unchanged. **Answer:** $\tau = 29.5$ s.
20. If the internal temperature of the tire in Example 16.8 is maintained at 0.00°F rather than 70.0°F during the deflation process, determine the time required for the valve stem to unchoke. Assume all the other variables remain unchanged. **Answer:** $\tau = 20.1$ s.
21. Suppose the valve core is not removed in Example 16.8 but instead the tire has a slow leak through a hole 1.00×10^{-4} inches in diameter. Determine how long it would take for the leak hole to unchoke when all the other variables remain unchanged. **Answer:** $\tau = 1.64 \times 10^7$ s = 190. days.

16.7 REYNOLDS TRANSPORT THEOREM

In Chapter 2, we define a closed system as any system in which mass does not cross the system boundary, but energy (heat and work) may cross the boundary. An open system then is defined to be any system in which both mass and energy may cross the system boundary. In classical mechanics, closed and open systems are called *Lagrangian* and *Eulerian* systems, respectively; and their use in problem solving is referred to as *Lagrangian analysis* and *Eulerian analysis*.

The Lagrangian analysis technique is named after the French mathematician Joseph Louis Lagrange (1736–1813). Basically, it involves solving the equations of energy and motion for a fixed mass (or closed) system. The Eulerian analysis technique is named after the Swiss mathematician Leonhard Euler (1707–1783). It involves solving the equations of energy and motion for a nonconstant spatial volume (or open system). For a given situation, usually one or the other of these techniques is easier to use, but regardless of their ease of application, both must give the same results. Therefore, we must be able to mathematically transform our governing equations back and forth between these two analysis frames. The *Reynolds transport theorem* is a method of carrying out this transformation.

In Chapter 2, we define a simple balance equation for any extensive property X as

$$X_G = X_T + X_P \tag{2.11}$$

and on a rate basis as

$$\dot{X}_G = \dot{X}_T + \dot{X}_P \tag{2.12}$$

where $\dot{X} = dX/dt$ is the rate of change of X within the system (either closed or open) we are analyzing, X_G is the net gain in X by the system, X_T is the net amount of X transported into the system, and X_P is the net amount of X produced inside the system boundaries by some internal process. We note in Chapter 2 that, if X is conserved, then $X_P = \dot{X}_P = 0$.

This balance equation is conceptually accurate. The only problem that arises in its use is that the form of the derivative for the net gain rate of X by the system depends on whether the system is open or closed. We never had to consider this in the past, because these two derivative forms happen to be identical for the simple one-dimensional analysis discussed in this book. However, many times (especially in the study of fluid mechanics), a one-dimensional analysis is not sufficient to solve the problem; and we have to expand to a multidimensional approach, which is often supported by a computer-based numerical technique. Because of the basic power of the Reynolds transport theorem in being able to transform differentiation operations back and forth between multidimensional closed and open systems, because our general rate balance equations have such differentials, and because we deal with both closed and open systems in thermodynamics, the Reynolds transport theorem is being introduced at this point in its full three-dimensional form. We use it *only* to transform the left side of Eq. (2.12), the system rate of gain term. The remaining transport and production rate terms are handled differently.

The Reynolds transport theorem is named after the British engineer and physicist Osborne Reynolds (1842–1912). Its derivation is quite complex and is not presented here, but the interested reader may wish to consult fluid mechanics texts for more information. Because there is a difference between the differentiation operation for a closed system and that for an open system, we must create a new notation that acknowledges this difference.[2]

Therefore, let

$$\dot{X}_G = \frac{DX}{Dt} = \text{the rate of gain of } X \text{ by a closed (Lagrangian) system}$$

where we use the operator symbol D/Dt to denote a fixed mass or closed system time derivative. The same time derivative measured in an open system is given by the Reynolds transport theorem as

$$\underbrace{\frac{DX}{Dt}}_{\substack{\text{Closed system}\\ \text{(Lagrangian) rate}}} = \underbrace{\int_{\forall} \frac{\partial(\rho x)}{\partial t} d\forall + \int_A \rho x(\mathbf{V}\cdot d\mathbf{A})}_{\substack{\text{Open system}\\ \text{(Eulerian) rate}}} \tag{16.24}$$

where $x = X/m$ is the intensive (specific) version of X, and $\mathbf{V}$ is the velocity vector of x as it crosses the surface area element $d\mathbf{A}$, as shown in Figure 16.19. The transport and production rate terms on the right side of Eq. (2.12) can be generalized for a *closed system* to be of the form

$$\dot{X}_T = -\int_A \mathbf{J}_x \cdot dA \tag{16.25}$$

and

$$\dot{X}_P = \int_{\forall} \sigma_x \, d\forall \tag{16.26}$$

where $\mathbf{J}_x$ is the flux (flow per unit area per unit time) of x through the area element dA, and σ_x is the local production rate per unit volume of x inside the volume element $d\forall$. Note that the closed and open systems described by the Reynolds transport theorem do not have the same system boundary and in fact are not the same physical system. A closed system that is to be considered equivalent to a given open system must be much larger than the open system. It must be large enough to include *all* the mass that crosses the boundary of the open system during the analysis period, and it is therefore always much larger than the open system it emulates. Consequently, equivalent closed systems are generally quite awkward and difficult to define, and it was this characteristic that ultimately provided the motivation for developing the open system (or Eulerian) analysis technique.

Combining Eqs. (16.24), (16.25), and (16.26) into the general rate balance Eq. (2.12) and rearranging it slightly gives the generalized *open system* rate balance equation:

$$\int_v \frac{\partial(\rho x)}{\partial t} d\forall = -\int_A \mathbf{J}_x \cdot d\mathbf{A} - \int_A \rho x(\mathbf{V}\cdot d\mathbf{A}) + \int_{\forall} \sigma_x \cdot d\forall \tag{16.27}$$

[2] In fluid mechanics texts, this derivative is often given a special name, such as the *material derivative* or the *substantial derivative*.

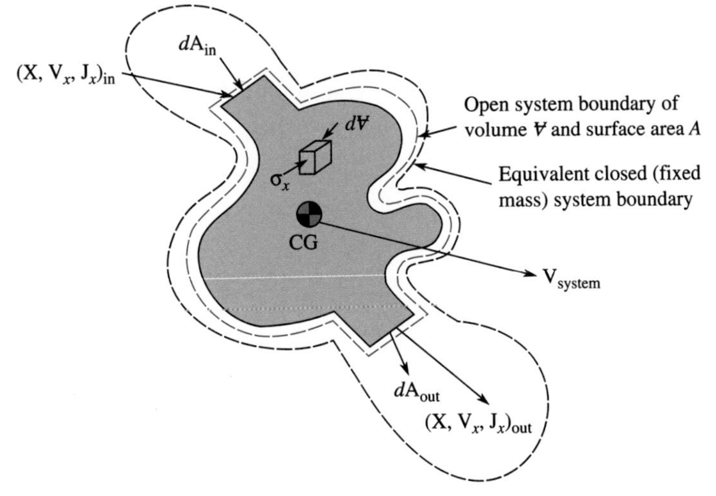

FIGURE 16.19
The vector quantities of Eqs. (16.24), (16.25), and (16.26) and the difference between the open and equivalent closed systems.

In a one-dimensional analysis, ρ, x, $\mathbf{J}_x$, and **V** do not vary across their flow streams, and consequently, they do not depend upon the cross-sectional area **A**. In the following equations, we denote the *magnitude* of a vector quantity by using the same quantity symbol in italic rather than boldface type. For example, the magnitudes of **V**, $\mathbf{J}_x$, and $d\mathbf{A}$ are V, J_x, and dA, respectively. Then, in a one-dimensional flow, the following simplifications occur

$$\begin{aligned}\int_A \mathbf{J}\cdot d\mathbf{A} &= -\Big(J_x\int_A dA\Big)_{\text{in}} + \Big(J_x\int_A dA\Big)_{\text{out}}\\ &= -\sum_{\text{in}} J_xA + \sum_{\text{out}} J_xA\end{aligned} \tag{16.28}$$

and

$$\begin{aligned}\int_A \rho x(\mathbf{V}\cdot d\mathbf{A}) &= -\Big(\rho xV\int_A dA\Big)_{\text{in}} + \Big(\rho xV\int_A dA\Big)_{\text{out}}\\ &= -\sum_{\text{in}} \rho VAx + \sum_{\text{out}} \rho VAx\\ &= -\sum_{\text{in}} \dot{m}x + \sum_{\text{out}} \dot{m}x\end{aligned} \tag{16.29}$$

In Eqs. (16.28) and (16.29), we use the fact that the inflow area vector always points in a direction opposite to the inflow velocity and flux, while the outflow area vector always points in the same direction as the outflow velocity and flux vectors (see Figure 16.19), so that

$$(\mathbf{J}_x\cdot d\mathbf{A})_{\text{in}} = -(J_x\,dA)_{\text{in}}$$

and

$$(\mathbf{V}\cdot d\mathbf{A})_{\text{in}} = -(V\,dA)_{\text{in}}$$

whereas

$$(\mathbf{J}_x\cdot d\mathbf{A})_{\text{out}} = +(J_x\,dA)_{\text{out}}$$

and

$$(\mathbf{V}\cdot d\mathbf{A})_{\text{out}} = +(V\,dA)_{\text{out}}$$

For our one-dimensional analysis, we now require that the system volume $\forall$ not be a function of time; then, we can write

$$\int_v \frac{\partial(\rho x)}{\partial t}\,d\forall = \frac{d}{dt}\int_\forall \rho x\,d\forall$$

CRITICAL THINKING

The Reynolds transport theorem is a very powerful mathematical relation often used in advanced engineering courses. Can you use it to visualize the difference between a Lagrangian (closed system of fixed mass) and a Eulerian (open system of variable mass) fluid momentum rate balance analysis where $\mathbf{X} = m\mathbf{V}$?

and, from the definitions of ρ and x inside the differential volume $d\forall$, we get

$$\rho x\, d\forall = (dm/d\forall)(dX/dm)d\forall = dX$$

so that

$$\frac{d}{dt}\int_{\forall} \rho x\, d\forall = \frac{d}{dt}\int_{\forall} dX = \left(\frac{dX}{dt}\right)_{\text{sys}} \tag{16.30}$$

where X_{sys} is the value of dX integrated over the system volume $\forall$. Substituting Eqs. (16.28), (16.29), and (16.30) into Eq. (16.27) gives the complete one-dimensional open system rate balance equation:

$$\underbrace{(dX/dt)_{\text{sys}}}_{\substack{\text{Net rate}\\ \text{of change}\\ \text{of } X \text{ inside}\\ \text{the open}\\ \text{system}}} = \underbrace{\left(\sum_{\text{in}} J_x A\right) - \sum_{\text{out}} J_x A}_{\substack{\text{Net "conduction"}\\ \text{(i.e., non-mass}\\ \text{flow) transport}\\ \text{rate of } X \text{ into}\\ \text{the open sytem}}} + \underbrace{\left(\sum_{\text{in}} \dot{m}x - \sum_{\text{out}} \dot{m}x\right)}_{\substack{\text{Net mass flow}\\ \text{transport rate}\\ \text{of } X \text{ into the}\\ \text{open system}}} + \underbrace{\int_{\forall} \sigma_x d\forall}_{\substack{\text{Net production}\\ \text{rate of } X \text{ inside}\\ \text{the boundary of}\\ \text{the open system}}} \tag{16.31}$$

If X is a conserved property like mass, energy, or momentum, then $\sigma_x = 0$ and the last term on the right side of Eq. (16.31) vanishes. If X is not conserved, like entropy, then a formula must be found for the variation of σ_x inside $\forall$ so that the integration of σ_x over $\forall$ can be carried out as indicated (this is what was done for the entropy production terms discussed in Chapter 7).

EXAMPLE 16.9

Show that Eq. (16.31) reduces to the standard one-dimensional

a. Mass rate balance, Eq. (6.19).
b. Energy rate balance, Eqs. (6.4) and (6.5).
c. Entropy rate balance, Eq. (9.6).

Solution

a. For mass, $X = m$ and $x = X/m = m/m = 1$. Mass is conserved, so $\sigma_x = \sigma_1 = 0$, and since there are no conduction (i.e., non-mass flow) mechanisms that move mass across a system boundary, $J_x = J_1 = 0$. Then, Eq. (16.31) reduces to

$$(dm/dt)_{\text{sys}} = \sum_{\text{in}} \dot{m} - \sum_{\text{out}} \dot{m}$$

which is identical to the one-dimensional mass rate balance introduced in Eq. (6.19) of Chapter 6.

b. For energy, $X = E$ and $x = E/m = e$. Energy is conserved, so $\sigma_X = \sigma_e = 0$, and the non-mass flow energy fluxes are identified in Chapter 4 as *heat* and *work*, so $J_e = \dot{q} - \dot{w}$. Then, Eq. (16.31) reduces to

$$(dE/dt)_{\text{sys}} = \dot{Q} - \dot{W} + \sum_{\text{in}} \dot{m}e - \sum_{\text{out}} \dot{m}e$$

where

$$\dot{Q} = \sum_{\text{in}} \dot{q}A - \sum_{\text{out}} \dot{q}A \quad \text{and} \quad \dot{W} = \sum_{\text{out}} \dot{W}A - \sum_{\text{in}} \dot{W}A$$

are the *net* heat and work transport rates of energy, and $e = u + V^2/(2g_c) + mgZ/g_c$ is the flow stream specific energy. These results are identical to those for the one-dimensional energy rate balance originally presented in Chapter 6 in Eqs. (6.4) and (6.5).

c. For entropy, $X = S$ and $x = S/m = s$. Entropy is not conserved, so $\sigma_s \neq 0$. The non-mass flow entropy flux is identified in Chapter 7 as $J_s = \dot{q}/T_b$. Then, for an isothermal (i.e., one-dimensional) boundary, Eq. (16.31) reduces to

$$(dS/dt)_{\text{sys}} = \frac{\dot{Q}}{T_b} + \sum_{\text{in}} \dot{m}s - \sum_{\text{out}} \dot{m}s + \dot{S}_P$$

where

$$\dot{Q}/T_b = \sum_{\text{in}} \dot{q}/T_b - \sum_{\text{out}} \dot{q}/T_b$$

is the net non-mass flow entropy transport rate, and

$$\dot{S}_p = \int_{\forall} \sigma_s d\forall$$

is the entropy production rate. These results are identical to those developed in Chapter 9 in for the one-dimensional entropy rate balance originally presented in Eq. (9.6).

Exercises

22. Use Eq. (16.31) to produce a general closed system rate balance equation for an arbitrary property X. **Answer:**

$$\sum_{\text{in}} J_x A - \sum_{\text{out}} J_x A + \int_{\forall} \sigma_x d\forall = \left(\frac{dX}{dt}\right)_{\text{system}}$$

23. Apply the general closed system rate balance equation developed in Exercise 22 to produce the steady state closed system entropy rate balance equation. **Answer:**

$$\sum_{\text{in}} \frac{\dot{q}}{T_b} - \sum_{\text{out}} \frac{\dot{q}}{T_b} + \int_{\forall} \sigma_s d\forall = \left(\frac{dS}{dt}\right)_{\text{system}} = 0$$

or

$$\frac{\dot{Q}}{T_b} + \dot{S}_P = 0$$

where

$$\frac{\dot{Q}}{T_b} = \sum_{\text{in}} \frac{\dot{q}}{T_b} - \sum_{\text{out}} \frac{\dot{q}}{T_b} \quad \text{and} \quad \dot{S}_P = \int_{\forall} \sigma_s d\forall$$

24. Use Eq. (16.31) to produce an availability rate balance for a steady state closed system. **Answer:**

$$\dot{Q} - \dot{W} + \sum_{\text{in}} \dot{m}a_f - \sum_{\text{out}} \dot{m}a_f + \dot{I} = \left(\frac{dA}{dt}\right)_{\text{system}} = 0$$

16.8 LINEAR MOMENTUM RATE BALANCE

The linear momentum rate balance (LMRB) for a one-dimensional open system can be easily developed from Eq. (16.31) by letting $X = m\mathbf{V}$ and $x = m\mathbf{V}/m = \mathbf{V}$. Since momentum is conserved, $\sigma_x = \sigma_V = 0$. External forces are the source of the non-mass flow momentum transport rate across the system boundary, so the one-dimensional momentum flux, $J_x = J_V$, is the external force per unit area, or

$$J_V = (\mathbf{F}_{\text{ext}}/A)g_c$$

Then, Eq. (16.31) gives the LMRB as[3]

$$\frac{d}{dt}(m\mathbf{V})_{\substack{\text{open}\\\text{system}}} = \sum_{\text{net}} (\mathbf{F}_{\text{ext}})g_c + \sum_{\text{in}} \dot{m}\mathbf{V} - \sum_{\text{out}} \dot{m}\mathbf{V} \tag{16.32}$$

[3] The vector nature of momentum necessarily causes Eq. (16.32) to be three dimensional. The one-dimensional restriction on this equation implies only that $\mathbf{V}$ and $\mathbf{F}_{\text{ext}}$ are *area averaged* quantities over the surface of the system. However, components of $\mathbf{V}$ and $\mathbf{F}_{\text{ext}}$ may be in each of the three coordinate directions.

where $\sum_{net}\mathbf{F}_{ext}\, g_c$ is the net sum of all the external forces acting on the system. Note that, for a closed system, $\dot{m} = 0$ and $m_{\text{closed system}} =$ constant. Then, Eq. (16.32) reduces to Newton's second law for a fixed mass closed system:

$$\frac{d}{dt}(m\mathbf{V})_{\substack{\text{closed}\\\text{system}}} = m\left(\frac{d\mathbf{V}}{dt}\right)_{\substack{\text{closed}\\\text{system}}} = m\mathbf{a}_{\substack{\text{closed}\\\text{system}}} = \sum_{\text{net}} \mathbf{F}_{\text{ext}}\, g_c$$

Also, the external forces are normally divided into two categories: *surface* forces, such as pressure and contact forces, and *body* forces, such as gravity and magnetic forces. That is,

$$\sum \mathbf{F}_{\text{ext}} = \sum \mathbf{F}_{\text{surface}} + \sum \mathbf{F}_{\text{body}}$$

For a steady state, steady flow, single-inlet, single-outlet open system, Eq. (16.32) further reduces to

Linear momentum rate balance (SS, SF, SI, SO):

$$\sum \mathbf{F}_{\text{ext}} = \dot{m}(\mathbf{V}_{\text{out}} - \mathbf{V}_{\text{in}})/g_c \tag{16.33}$$

EXAMPLE 16.10

An air jet is used to levitate a 5.00×10^{-3} kg sheet of paper, as shown in Figure 16.20. The diameter of the jet at the nozzle exit is 3.00×10^{-3} m and it is at atmospheric pressure at 20.0°C. Determine the velocity of the jet.

Solution

Assume the flow is steady state and steady flow. Then, the y component of Eq. (16.33) is

$$F_y = -W = \dot{m}(V_{\text{out}} - V_{\text{in}})_y/g_c$$

where W is the weight of the paper, given by

$$W = mg/g_c = (5.00 \times 10^{-3}\,\text{kg})(9.81\,\text{m/s}^2)/(1) = 0.0491\,\text{N}$$

and

$$(V_{\text{out}})_y = 0$$

Also, $\dot{m} = \rho AV = \rho\pi D^2 V/4$, where

$$\rho = \frac{p}{RT} = \frac{101.3 \times 10^3\,\text{kg/(m}\cdot\text{s}^2)}{[286\,\text{m}^2/(\text{s}^2\cdot\text{K})](293.15\,\text{K})} = 1.21\,\text{kg/m}^3$$

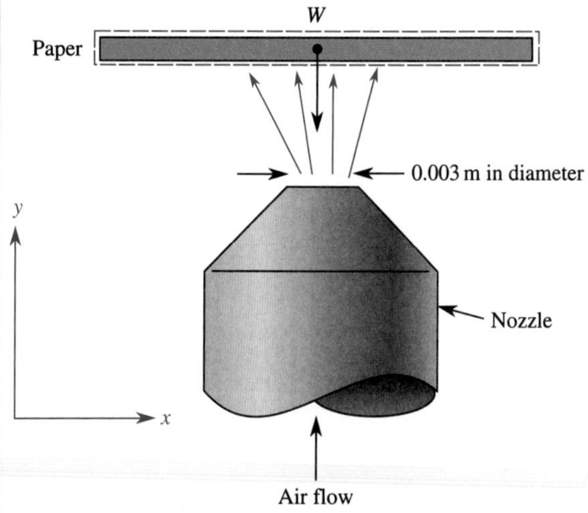

FIGURE 16.20
Example 16.10.

Then,

$$W = \dot{m}(V_{\text{in}})_y/g_c = \rho A V_{\text{in}}^2/g_c = \rho\pi D^2 V_{\text{in}}^2/4g_c$$

So,

$$V_{\text{in}} = \left(\frac{4g_c W}{\rho\pi D^2}\right)^{1/2} = \left[\frac{4(1)(0.0491\,\text{kg}\cdot\text{m/s}^2)}{(1.21\,\text{kg/m}^3)(\pi)(3.00 \times 10^{-3}\,\text{m})^2}\right]^{1/2} = 75.7\,\text{m/s}$$

Exercises

25. The nozzle diameter in Example 16.10 is to be reduced from 3.00×10^{-3} m to 1.50×10^{-3} m. Determine the velocity of the resulting jet required to levitate the sheet of paper, assuming all the other variables remain unchanged. **Answer:** $V_{\text{jet}} = 151$ m/s.
26. Suppose the sheet of paper in Example 16.10 is replaced by a sheet of cardboard with a mass of 5×10^{-2} kg. Determine the velocity of the jet required to levitate the cardboard, assuming all the other variables remain unchanged. **Answer:** $V_{\text{jet}} = 239$ m/s.
27. If the air jet in Example 16.10 is at 0.00°C rather than 20.0°C, determine the velocity of the jet required to levitate the sheet of paper assuming all the other variables remain unchanged. **Answer:** $V_{\text{jet}} = 73.1$ m/s.

16.9 SHOCK WAVES

Shock waves can occur only in compressible substances and are often thought of as strong acoustical (sound) waves. However, they differ from sound waves in two important ways: They travel much faster than normal sound waves, and there is a large and nearly discontinuous change in pressure, temperature, and density across a shock wave. The thickness of a shock wave over which these changes occur is typically on the order of 10^{-7} m (4×10^{-6} in.); consequently, large property gradients occur across the shock wave that make it very dissipative and irreversible. The amplitude of a large shock wave, such as that created by an explosion or a supersonic aircraft, decreases nearly with the inverse square of the distance from the source until it weakens sufficiently to become an ordinary sound wave. The *sonic boom* heard at the surface of the Earth from a high-altitude supersonic aircraft is the weak acoustical remnants of its shock wave.

Since strong shock waves are highly irreversible, they cannot be treated even approximately as isentropic processes. Ordinary sound waves, on the other hand, are very much weaker by comparison and can sometimes be modeled by isentropic processes.

When a shock wave occurs perpendicular (i.e., normal) to the velocity, it is called a *normal* shock, and it can be analyzed with the one-dimensional balance equations. A shock wave that is inclined to the direction of flow is called an *oblique* shock and requires a two- or three-dimensional analysis. We limit our analysis to normal shock waves in this chapter.

The easiest way to generate normal shock waves for laboratory study is to use a supersonic converging-diverging nozzle. Equation (16.13) gives the pressure profile along the nozzle in terms of the isentropic stagnation pressure p_{os} and the local Mach number M as

$$p/p_{os} = \left(1 + \frac{k-1}{2}\mathrm{M}^2\right)^{k/(1-k)}$$

This equation is shown schematically along the nozzle in Figure 16.21. When the back pressure p_B is greater than the throat critical pressure p_* given in Eq. (16.19), the Mach number at the throat is less than 1 and the flow remains subsonic throughout the entire nozzle, with nozzle exit pressure p_{exit} equal to the back pressure p_B. When p_B is less than p_*, the Mach number at the throat is equal to 1 and the flow becomes supersonic in the diverging section of the nozzle. If p_B/p_{os} is between the points *b* and *c* on Figure 16.21, then a normal shock occurs within the diverging section at the point where the flow can isentropically recover to $p_{exit} = p_B$. If we let p_E be the pressure at the exit of the supersonic nozzle when the flow expands isentropically throughout the nozzle (see Figure 16.21), then when p_B/p_{os} is between the points *c* and *d* on Figure 16.21, a normal shock

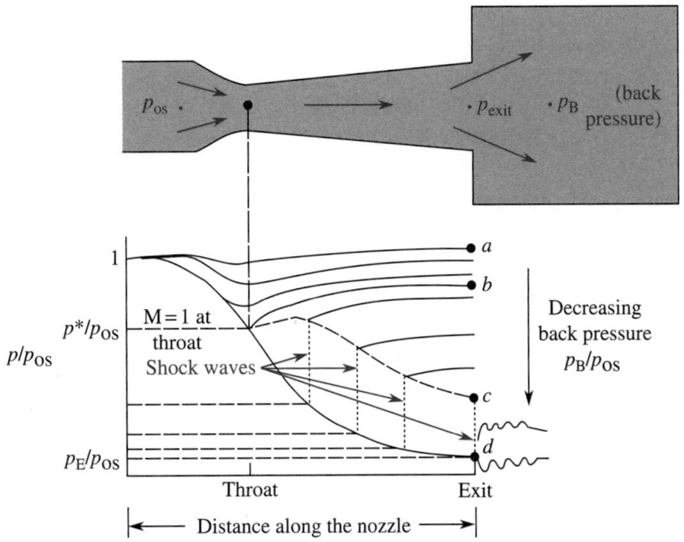

FIGURE 16.21
The pressure distribution in a converging-diverging nozzle when the upstream stagnation pressure is held constant and the downstream back pressure is decreased. Shock waves occur in the diverging section when the flow is supersonic but the back pressure is not low enough to allow complete expansion to the end of the nozzle.

CRITICAL THINKING

Normal shock waves form as a result of a piling up of pressure waves into a strong compression wave front. A similar phenomenon occurs as gravity waves in the ocean approach a beach. The front of each wave steepens as it approaches the beach, but unlike shock waves it eventually topples over forming a *breaker*. Shock waves do not topple over, they continue to grow in strength as their velocity increases. The strength of a normal shock wave is defined as the ratio of the pressure increase across the shock to the original pressure, or $(p_y - p_x)/p_x$. How would you define a similar *strength* for a gravity wave in the ocean?

occurs at the exit plane of the nozzle and $p_{exit} = p_E < p_B$. Finally, when p_B/p_{os} is below point d on Figure 16.21, shock waves occur downstream from the nozzle and $p_{exit} = p_E > p_B$.

If we apply the mass, energy, entropy, and linear momentum balances to the normal shock wave system shown in Figure 16.22, we can relate the upstream (x) and downstream (y) properties across the shock. Assuming a steady state, steady flow, single-inlet, single-outlet, adiabatic, and aergonic system, the mass, energy, and linear momentum rate balances give

Mass rate balance (MRB, SS, SF, SI, SO)

$$\dot{m}_x = \rho_x A_x V_x = \dot{m}_y = \rho_y A_y V_y$$

but, since $A_x = A_y = A$ in Figure 16.22,

$$\rho_x V_x = \rho_y V_y$$

Now, $\rho = p/RT$, so

$$\rho_x V_x = \frac{p_x V_x}{RT_x} = \frac{kg_c p_x V_x}{kg_c RT_x} = p_x \mathrm{M}_x \left(\frac{kg_c}{RT_x}\right)^{1/2}$$

and

$$\rho_y V_y = p_y \mathrm{M}_y \left(\frac{kg_c}{RT_y}\right)^{1/2} = \rho_x V_x$$

or

$$\frac{p_x \mathrm{M}_x}{\sqrt{T_x}} = \frac{p_y \mathrm{M}_y}{\sqrt{T_y}} \tag{16.34}$$

Energy rate balance (ERB, SS, SF, SI, SO, aergonic, adiabatic)

$$h_{ox} = h_x + V_x^2/(2g_c) = h_y + V_y^2/(2g_c) = h_{oy}$$

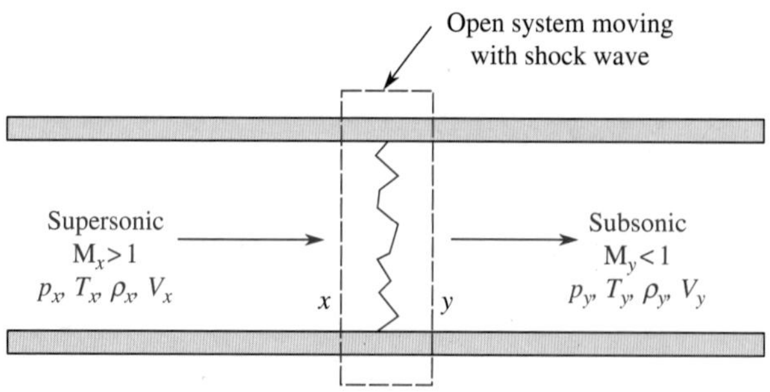

FIGURE 16.22
A normal shock wave moving at supersonic velocity in a constant area adiabatic duct. The coordinate system here is fixed to the shock wave.

and, for an ideal gas with constant specific heats,

$$h_{ox} - h_x = c_p(T_{ox} - T_x) = V_x^2/(2g_c)$$

Since $c_p = kR/(k-1)$ for constant specific heat ideal gases,

$$T_{ox}/T_x = T_{osx}/T_x = 1 + \frac{k-1}{2}\left(\frac{V_x^2}{kg_cRT_x}\right) = 1 + \frac{k-1}{2}M_x^2$$

Similarly, for the downstream region, we can write that

$$\alpha = -\frac{1}{v}\left(\frac{\partial v}{\partial p}\right)_s$$

Since $h_{ox} = h_{oy}$, then $T_{ox} = T_{oy} = T_{osx} = T_{osy}$ and we can divide the preceding two equations to get

$$\frac{T_x}{T_y} = \frac{1 + \frac{k-1}{2}M_y^2}{1 + \frac{k-1}{2}M_x^2} \tag{16.35}$$

Linear momentum rate balance (LMRB, SS, SF, SI, SO)

$$F_x - F_y = (p_x - p_y)A = \dot{m}(V_y - V_x)/g_c$$

or

$$\begin{aligned} p_x - p_y &= (\dot{m}/A)(V_y - V_x)/g_c \\ &= (\rho_y V_y^2 - \rho_x V_x^2)/g_c \end{aligned}$$

Now $\rho = p/RT$, so

$$\begin{aligned} p_x - p_y &= kp_y V_y^2/(kg_cRT_y) - kp_x V_x^2/(kg_cRT_x) \\ &= k(p_y M_y^2 - p_x M_x^2) \end{aligned}$$

or

$$\frac{p_x}{p_y} = \frac{1 + kM_y^2}{1 + kM_x^2} \tag{16.36}$$

Substituting Eqs. (16.35) and (16.36) into Eq. (16.34) yields an equation for M_x, M_y, and k, which can be solved for $M_x \geq 1$ and $M_y \leq 1$ to give

$$M_y^2 = \frac{(k-1)M_x^2 + 2}{2kM_x^2 + 1 - k} \tag{16.37}$$

Because Eq. (16.34) is symmetrical in x and y, the x and y subscripts in Eq. (16.37) can be interchanged to produce an equation for M_x in terms of M_y and k.

EXAMPLE 16.11

A nuclear explosion produces a normal shock wave that travels through still air with a Mach number of 5.50. The pressure and temperature of the air in front of the shock wave are 14.7 psia and 70.0°F. Determine the pressure, temperature, and wind velocity directly behind the shock wave.

Solution

If we attach our reference frame to the moving shock wave, it appears that the air is approaching it with a Mach number of $M_x = 5.50$. The Mach number behind the shock wave can be determined from Eq. (16.37) with $k = 1.40$ as

$$M_y = \left[\frac{(1.40 - 1)(5.50)^2 + 2}{2(1.40)(5.50)^2 + 1 - 1.4}\right]^{1/2} = 0.409$$

(Continued)

EXAMPLE 16.11 *(Continued)*

Then, the temperature directly behind the shock wave is given by Eq. (16.35) as

$$T_y = T_x\left[\frac{1+\frac{k-1}{2}(M_x)^2}{1+\frac{k-1}{2}(M_y)^2}\right] = (70.0+459.67\text{ R})\left[\frac{1+\frac{1.40-1}{2}(5.50)^2}{1+\frac{1.40-1}{2}(0.409)^2}\right] = 3610\text{ R}$$

and the pressure directly behind the shock wave is given by Eq. (16.34) as

$$p_y = p_x\frac{M_x}{M_y}\left(\frac{T_y}{T_x}\right)^{1/2} = (14.7\text{ lbf/in}^2)\left(\frac{5.50}{0.409}\right)\left(\frac{3610\text{ R}}{70.0+459.67\text{ R}}\right)^{1/2} = 516\text{ lbf/in}^2$$

Finally, the wind velocity directly behind the shock wave is just the relative velocity $V_x - V_y$, or

$$V_{\text{wind}} = V_x - V_y = M_xc_x - M_yc_y = M_x\sqrt{kg_cRT_x} - M_y\sqrt{kg_cRT_y}$$

$$= 5.50\sqrt{1.40(32.174\text{ lbm}\cdot\text{ft/lbf}\times\text{s}^2)(53.34\text{ ft}\cdot\text{lbf/lbm}\cdot\text{R})(70.0+459.67\text{ R})}$$

$$-0.409\sqrt{1.40(32.174\text{ lbm}\cdot\text{ft/lbf}\cdot\text{s}^2)(53.34\text{ ft}\cdot\text{lbf/lbm}\cdot\text{R})(3610\text{ R})}$$

$$= 5.00\times10^3\text{ ft/s}$$

These extremely high values for p_y, T_y, and the wind velocity show why large explosions produce so much damage to life and property.

Exercises

28. Determine the Mach number directly behind the shock wave in Example 16.11 if the Mach number of the shock wave itself is 4.0 and all the other variables remain unchanged. **Answer:** $M_y = 0.434$.
29. Use Eq. (16.36) to verify the pressure directly behind the shock wave calculated in Example 16.11. **Answer:** $p_y = 516$ psia (the same as determined in Example 16.11).
30. If the temperature in front of the shock wave in Example 16.11 is 0.00°F rather than 70.0°F and all the other variables remain unchanged, determine the wind velocity directly behind the shock wave. **Answer:** $V_{\text{wind}} = 4570$ ft/s.
31. Determine the constant value assumed by M_y as $M_x \to 4$. Thus, no matter how high M_x is, M_y can never be less than this value. **Answer:** M_y (as $M_x \to 4$) = $[(k-1)/2k]^{1/2} = 0.378$ for air.

Equations (16.35), (16.36), and (16.37) have been tabulated for air (k = 1.4) in Table C.19 in *Thermodynamic Tables to accompany Modern Engineering Thermodynamics*. The reader is encouraged to use this table when its direct entry is convenient. However, rather than interpolating for nondirect entry values, the equations just given can be used to make accurate direct calculations, since they were used to generate the table.

Finally, an entropy rate balance on the shock wave gives

Entropy rate balance (SRB, SS, SF, SI/SO, A)

$$\dot{s}_p = \dot{m}(s_y - s_x) \geq 0$$

So,

$$\begin{aligned}\dot{S}_p/\dot{m} &= s_y - s_x = c_p\ln(T_y/T_x) - R\ln(p_y/p_x)\\ &= R\ln\left[(p_x/p_y)(T_y/T_x)^{k/(k-1)}\right] \geq 0\end{aligned} \tag{16.38}$$

Therefore,

$$p_x/p_y \geq (T_x/T_y)^{k/(k-1)}$$

and by substituting Eqs. (16.35) and (16.36) into this relation, it can be shown that $M_x \geq M_y$. Consequently, the second law of thermodynamics stipulates that shock waves can occur only in supersonic flows, from $M_x \geq 1$ to $M_y \leq 1$ and can never occur in subsonic flows. Equation (16.38) is shown in Figure 16.23 for air (k = 1.40).

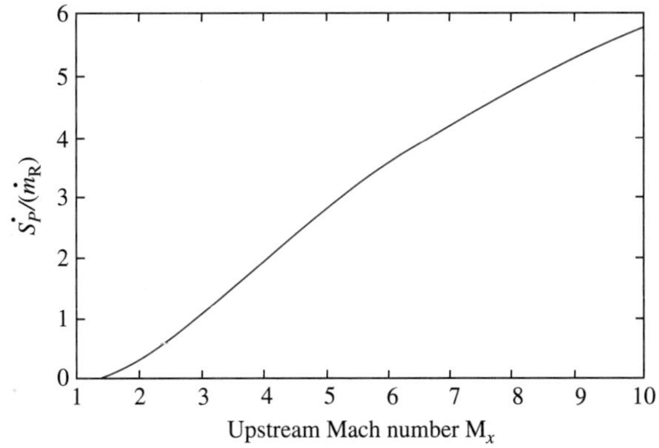

FIGURE 16.23

A plot of Eq. (16.38) for air (k = 1.4) utilizing Eqs. (16.35), (16.36), and (16.37). Note that the second law of thermodynamics requires that $\dot{S}_p/(\dot{m}_R) \geq 0$ for all processes.

EXAMPLE 16.12

A spacecraft directional control thruster is a converging-diverging nozzle that uses high-pressure and high-temperature air. The air enters with isentropic stagnation properties of 7.00 MPa and 2000.°C. The throat diameter is 0.0200 m and the diameter of the exit of the diverging section is 0.100 m. Determine

a. The mass flow rate required for supersonic flow in the diverging section.
b. The Mach number, pressure, and temperature at the exit of the diverging section with this mass flow rate.
c. The outside back pressure required to produce a standing normal shock wave at the exit of the diverging section.

Solution

a. To have supersonic flow in the diverging section of a converging-diverging nozzle, the throat must have a Mach number of unity (i.e., be choked). Therefore, the mass flow rate is the maximum value for air, in SI units, given by Eq. (16.22c), or

$$\dot{m} = \dot{m}_{\text{max}} = 0.0404\left(p_{os}A^*/\sqrt{T_{os}}\right)$$

where $A^* = \pi(D^*)^2/4 = \pi(0.0200\,\text{m})^2/4 = 3.14\times 10^{-4}\,\text{m}^2$. Then,

$$\dot{m} = \frac{[0.0404\,\text{kg}\cdot\sqrt{\text{K}}/(\text{N}\cdot\text{s})](7.00\times 10^6\,\text{N/m}^2)(3.14\times 10^{-4}\,\text{m}^2)}{\sqrt{2000.+273.15\,\text{K}}} = 1.86\,\text{kg/s}$$

b. Here, $A_{\text{exit}}/A^* = (D_{\text{exit}}/D^*)^2 = (0.100/0.0200)^2 = 25.0$. Then, Eq. (16.23) can be inverted to find M_{exit}, and p_{exit} and T_{exit} can be found from Eqs. (16.13) and (16.12), respectively. Unless these equations are programmed into a computer, this can be a tedious set of calculations. Table C.18 in *Thermodynamic Tables to accompany Modern Engineering Thermodynamics* was created to eliminate this tedium by tabulating these equations. The preceding area ratio is a direct entry into this table, so we use it and read

$$M_{\text{exit}} = 5.00$$
$$p_{\text{exit}}/p_{os} = 1.89\times 10^{-3} \text{ and}$$
$$T_{\text{exit}}/T_{os} = 0.16667$$

Then, $p_{\text{exit}} = (1.89\times 10^{-3})(7.00\times 10^6\,\text{N/m}^2) = 13.2\,\text{kN/m}^2$ and

$$T_{\text{exit}} = (0.16667)(2000.+273.15\,\text{K}) = 378.8\,\text{K}$$

The velocity of sound at the exit is

$$c_{\text{exit}} = \sqrt{kg_cRT_{\text{exit}}} = \sqrt{(1.40)(1)[286\,\text{m}^2/(\text{s}^2\cdot\text{K})](378.8\,\text{K})} = 390.\,\text{m/s}$$

then,

$$V_{\text{exit}} = c_{\text{exit}}M_{\text{exit}} = (390.\,\text{m/s})(5.00) = 1950\,\text{m/s}.$$

(Continued)

EXAMPLE 16.12 *(Continued)*

c. An exit plane shock wave is illustrated by region c to d in Figure 16.23. The required back pressure p_B here is equal to the downstream isentropic stagnation pressure p_{osy} necessary to cause a normal shock to occur in the exit plane. Then, $M_x = 5.0$, $p_x = 13.23\ \text{kN/m}^2$, and $T_x = 378.8$ K. The downstream Mach number M_y can be found from Eq. (16.37), and p_y can be found from Eq. (16.36). Finally, Eq. (16.13) can be used to find p_{osy}. However, Table C.19 is a tabular version of these equations, and at $M_x = 5.0$, we again have a direct entry. From this table, we find that

$$M_y = 0.415 \quad p_{osy}/p_{osx} = 0.06172$$

$$p_y/p_x = 29.00 \quad p_{osy}/p_x = 32.654$$

$$T_y/T_x = 5.800$$

Here, $p_{osx} = p_{os} = 7.00$ MPa; therefore, the required back pressure p_B is

$$p_B = p_{osy} = (0.06172)\,p_{osx} = (0.06172)(7.00 \times 10^3\ \text{kN/m}^2) = 432\ \text{kN/m}^2$$

alternatively,

$$p_B = p_{osy} = (32.654)\,p_x = (32.654)(13.23\ \text{kN/m}^2) = 432\ \text{kN/m}^2$$

Exercises

32. Determine the mass flow rate required for supersonic flow in the diverging section of the converging-diverging nozzle in Example 16.12, if the compressed air is at 10.0 MPa and 20.0°C instead of 7.00 MPa and 2000.°C. **Answer**: $\dot{m} = 7.42$ kg/s.

33. Use Table C.18 to find the Mach number at the exit of the diverging section of the converging-diverging nozzle in Example 16.12, if the diameter at the exit of the diverging section is increased to 0.14585 m. Assume all other variables remain unchanged. **Answer**: $M_{exit} = 6.00$.

34. Use Table C.19 to determine the outside back pressure required to produce a standing normal shock wave at the exit of the diverging section of the converging-diverging nozzle in Example 16.12, if the exit Mach number is 6.00 and the upstream stagnation pressure is 7.00 MPa. **Answer**: $p_B = 208$ kPa.

EXAMPLE 16.13

Air enters a converging-diverging nozzle with isentropic stagnation properties of 3.00 atm and 20.0°C and exhausts into the atmosphere (i.e., $p_B = 1.00$ atm). The exit to throat area ratio for the nozzle is 2.00. Determine the pressure, temperature, and velocity at the exit.

Solution

We are given the upstream isentropic stagnation state of 3.0 atm, 20°C, and a back pressure of 1.0 atm. To find the conditions in the exit plane, we must first determine whether or not a shock wave occurs inside the diverging section of the nozzle. This occurs if $p_E < p_B < p^*$, where from Eq.(16.19),

$$p^* = p_{os}\left(\frac{2}{k+1}\right)^{k/(k-1)} = (3.00\ \text{atm})(0.528) = 1.58\ \text{atm}$$

and, from Eq. (16.13),

$$p_E = p_{os}\{1 + [(k-1)/2]M_E^2\}^{k/(1-k)}$$

Since we are given $A_{exit}/A^* = A_E/A^* = 2.00$, we can find M_E by inverting Eq. (16.23b). However, in this case, it is again much easier to use Table C.18 for this area ratio and read (approximately), $M_E = 2.20$ and $p_E/p_{os} = 0.09352$. Then, $p_E = 0.093252 \times (3.0\ \text{atm}) = 0.281$ atm. Thus, $p_E < p_B < p^*$ here and a normal shock must occur somewhere in the diverging section of the nozzle.

Since we now know that a shock wave occurs, we also need to know whether or not it occurs in the exit plane of the nozzle. We could find M_x from the upstream and downstream isentropic stagnation pressures from the relation,

$$p_{osy}/p_{osx} = (p_{osy}/p_y)(p_y/p_x)(p_x/p_{osx})$$

by using Eqs. (16.13), (16.36), and (16.37). This results in the equation

$$\frac{p_{osy}}{p_{osx}} = \frac{\left[\frac{k+1}{2}M_x^2 \Big/ \left(1+\frac{k-1}{2}M_x^2\right)\right]^{k/(k-1)}}{\left(\frac{2k}{k+1}M_x^2 - \frac{k-1}{k+1}\right)^{1/(k-1)}} = \frac{1.00}{3.00} = 0.333$$

However, it is quite tedious to solve this equation for M_x without using a computer, and since we have a direct entry in Table C.19 at this value of p_{osy}/p_{osx}, we use this table in our solution. From Table C.19 at $p_{osy}/p_{osx} = 0.333$, we read $M_x \approx 2.98$ and $M_y \approx 0.476$. From Table C.18 at $M = M_x = 2.98$, we find that $A/A^* \approx 4.16$. But our nozzle only has an $A_e/A^* = 2.00$, so the shock wave must be in the exit plane; therefore, $p_{exit} = p_E = 0.281$ atm, $M_{exit} = M_E = 2.20$, and $T_{exit} = 0.050813(293.15) = 148.96$ K. The pressure readjustment from p_{exit} to p_B occurs outside the exit (see region *c* to *d* in Figure 16.21). Finally, the exit velocity is given by

$$\begin{aligned} V_{exit} &= M_{exit}c_{exit} = M_{exit}\sqrt{kg_cRT_{exit}} \\ &= (2.20)\sqrt{(1.40)(1)[286\,\text{m}^2/(\text{s}^2\cdot\text{K})](148.96\,\text{K})} = 537\,\text{m/s} \end{aligned}$$

Exercises

35. Use Table C.18 to determine the exit Mach number and exit pressure in Example 16.13 for an exit to throat area ratio of the converging-diverging nozzle of 6.78962. Assume all the other variables remain unchanged. **Answer**: $M_{exit} = 3.50$, and $p_{exit} = 39.33$ kPa.

36. Suppose the back pressure in Example 16.13 is increased from 1.00 atm to 2.16261 atm. Use Table C.19 to determine the values of M_x and p_{exit}. **Answer**: $M_x = 2.00$ and $p_{exit} = 1.695$ atm.

37. Determine the exit temperature and velocity in Example 16.13, if the upstream stagnation temperature is reduced from 20.0°C to 0.00°C and all the other variables remain unchanged. **Answer**: $T_{exit} = 139$ K and $V_{exit} = 519$ m/s.

16.10 NOZZLE AND DIFFUSER EFFICIENCIES

Inefficiencies in nozzles and diffusers result from irreversibilities that occur within their boundaries. Shock waves and fluid friction (viscosity) in the wall boundary layer are the most common types of irreversibilities that occur. If a nozzle or diffuser is not designed with exactly the correct wall contour, oblique shocks, boundary layer separation, and turbulence destroy the nozzle's performance.

Because nozzle and diffuser performance depend on their internal irreversibilities, we can base their efficiency equation on the second law of thermodynamics by taking the *isentropic* nozzle and diffuser to be 100% efficient. Then, since the function of a nozzle is to convert pressure (or thermal energy in the case of an ideal gas) into kinetic energy, we can define its efficiency η_N to be (see Figure 16.24)

$$\begin{aligned} \eta_N &= \frac{\text{Actual exit kinetic energy}}{\text{Isentropic exit kinetic energy at the actual exit pressure}} \\ &= \frac{(V_{exit}^2/2g_c)_{actual}}{(V_{exit}^2/2g_c)_{isentropic}} = \frac{(V_{exit}^2/2g_c)_{actual}}{(h_{inlet} - h_{exit})_s} = \frac{(V_{exit}^2/2g_c)_{actual}}{c_p(T_{inlet} - T_{exit})_s} \end{aligned}$$

and using Eqs. (7.38), (16.10), and (7.40), this can be written as

Nozzle efficiency

$$\eta_N = \frac{\frac{(k-1)}{2}(V_{exit}/c_{inlet})^2}{1 - (p_{exit}/p_{inlet})^{(k-1)/k}} \tag{16.39}$$

If the inlet velocity is very slow, then the entrance can be taken to be the isentropic stagnation state, or $p_{inlet} = p_{os}$ and $c_{inlet} = c_{os} = \sqrt{kg_cRT_{os}}$. The p_{exit} and V_{exit} terms in Eq. (16.39) are the *actual* exit pressure and exit velocity values that must be determined from measurements on the actual nozzle. Typical efficiencies for well-designed nozzles vary from 0.90 to 0.99 at high flow rates.

We can also define a *nozzle velocity coefficient* C_v for the nozzle in a similar way:

$$C_v = \frac{\text{Actual exit velocity}}{\text{Isentropic exit velocity at the actual exit pressure}} = \sqrt{\eta_N} \tag{16.40}$$

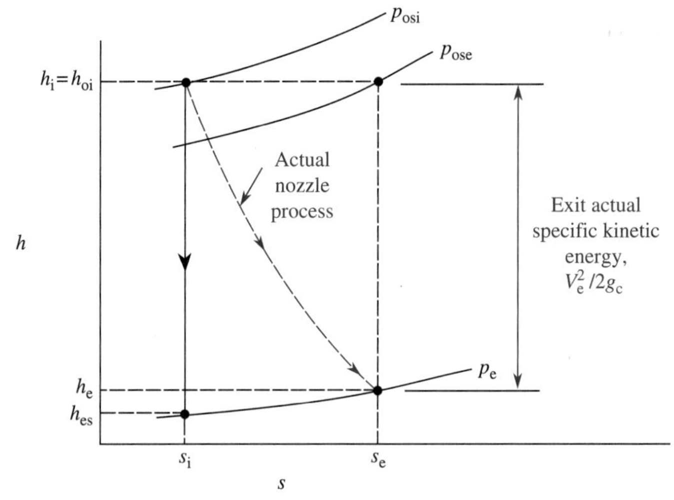

FIGURE 16.24
The thermodynamic process path of a nozzle plotted on h–s coordinates.

Also, it is common to define a *nozzle discharge coefficient* C_d as

$$C_d = \frac{\text{Actual mass flow rate}}{\text{Isentropic mass flow rate}} = \frac{\dot{m}_{\text{actual}}}{\dot{m}_{\text{isentropic}}} = \frac{(\rho AV)_{\text{actual}}}{(\rho AV)_{\text{isentropic}}} \tag{16.41}$$

Typical nozzle discharge coefficients run from 0.60 for sharp-edged nozzles (i.e., orifices) at low flow rates to 0.99 for properly designed nozzles at high flow rates.

EXAMPLE 16.14

Helium enters a newly designed test nozzle at 456.2 kN/m^2 and 283.7 K with a negligible velocity. The exit velocity, temperature, and pressure are measured at the instant when the nozzle first becomes choked and are found to be 474.8 m/s, 370.4 kN/m^2, and 260.1 K, respectively. For these conditions, determine the nozzle's

a. Efficiency.
b. Velocity coefficient.
c. Discharge coefficient.

Solution

- Equation (16.39) gives the nozzle's efficiency η_N as

$$\eta_N = \frac{\frac{k-1}{2}(V_{\text{exit}}/c_{\text{inlet}})^2}{1-(p_{\text{exit}}/p_{\text{inlet}})^{(k-1)/k}}$$

For helium, $k = 1.67$ and $R = 2.007$ kJ/kg·K, and since the flow enters the nozzle with a negligible inlet velocity, we can take $T_{\text{inlet}} \approx T_{osi}$ and $p_{\text{inlet}} \approx p_{osi}$. Then,

$$c_{\text{inlet}} \approx c_{osi} = \sqrt{(1.67)(1)[2077\ \text{m}^2/(\text{s}^2\cdot\text{K})](283.7\ \text{K})} = 992\ \text{m/s}$$

and

$$\eta_N = \frac{\frac{0.67}{2}(474.8/992)^2}{1-(370.4/456.2)^{0.67/1.67}} = 0.957$$

- Equation (16.40) quickly gives the nozzle's velocity coefficient C_v as

$$C_v = \sqrt{\eta_N} = \sqrt{0.957} = 0.978$$

- The nozzle's discharge coefficient C_d is determined from Eq. (16.41) as

$$C_d = \frac{(\rho AV)_{\text{actual}}}{(\rho AV)_{\text{isentropic}}} = \frac{(\rho_e V_e)_{\text{actual}}}{(\rho_e V_e)_{\text{isentropic}}}$$

Now,

$$(\rho_{\text{exit}})_{\text{actual}} = p_{\text{exit}}/RT_{\text{exit}} = \frac{370.4\,\text{kN/m}^2}{[2.077\,\text{kN}\cdot\text{m/(kg}\cdot\text{K)}](260.1\,\text{K})} = 0.686\,\text{kg/m}^3$$

Since the flow is choked, $M_{\text{exit}} = 1.0$, and the isentropic exit temperature and density can be determined from Eqs. (16.18) and (16.20) as

$$(T_{\text{exit}})_s = T^* = T_{os}[2/(k+1)] = (283.7)[2/2.67] = 212.5\,\text{K}$$

and

$$(\rho_{\text{exit}})_s = \rho_{os}[2/(k+1)]^{1/(k-1)} = (p_{os}/RT_{os})[2/(k+1)]^{1/(k-1)}$$

$$= \frac{(456.2\,\text{kN/m}^2)(2/2.67)^{1/0.67}}{[2.077\,\text{kN}\cdot\text{m/(kg}\cdot\text{K)}](283.7\,\text{K})} = 0.503\,\text{kg/m}^3$$

and

$$(V_{\text{exits}})_s = c_{\text{exit}}\,|_s = \sqrt{kg_cR(T_c)}\,|_s$$

$$= \sqrt{(1.67)(1)[2077\,\text{m}^2/(\text{s}^2\cdot\text{K})](212.5\,\text{K})} = 859\,\text{m/s}$$

then,

$$C_d = \frac{(0.686\,\text{kg/m}^3)(474.8\,\text{m/s})}{(0.503\,\text{kg/m}^3)(859\,\text{m/s})} = 0.754$$

Exercises

38. Your technician reports that there was an error in the sensor used to measure the nozzle exit velocity in Example 16.14. The correct exit velocity is 426.3 m/s, not 474.8 m/s. Determine the new values for the nozzle's efficiency, velocity coefficient, and discharge coefficient, assuming all the other variables remain unchanged. **Answer:** $\eta_N = 0.772$, $C_v = 0.878$, and $C_d = 0.677$.
39. Oops, the technician in charge of the nozzle test in Example 16.14 now tells you that the velocity sensor reading is correct, it is the exit temperature sensor reading that was in error. The correct nozzle exit temperature is 271.5 K, not 260.1 K. Determine the new values for the nozzle's efficiency, velocity coefficient, and discharge coefficient, assuming all the other variables remain unchanged. **Answer:** $\eta_N = 0.957$, $C_v = 0.978$, and $C_d = 0.722$.
40. The nozzle in Example 16.14 is retested using air instead of helium. The same inlet conditions are used, but the following new exit conditions are measured: $V_{\text{exit}} = 190.1$ m/s, $p_{\text{exit}} = 430.3$ kPa, and $T_{\text{exit}} = 270.5$ K. Determine the new values for the nozzle's efficiency, velocity coefficient, and discharge coefficient under the air test. **Answer:** $\eta_N = 0.444$, $C_v = 0.666$, and $C_d = 0.964$.

The function of a diffuser, on the other hand, is to convert kinetic energy into pressure. Therefore, we define its efficiency η_D as

$$\eta_D = \frac{\text{Isentropic enthalpy increase at the actual exit stagnation pressure}}{\text{Inlet kinetic energy}}$$

$$= \frac{h_{es} - h_{\text{inlet}}}{V_{\text{inlet}}^2/(2g_c)} = \frac{h_{es} - h_{\text{inlet}}}{h_{oi} - h_{\text{inlet}}}$$

where h_{es} is the enthalpy at the actual exit stagnation pressure but at the same entropy as the inlet state (see Figure 16.25).

For an ideal gas, the diffuser efficiency becomes

$$\eta_D = \frac{c_p(T_{es} - T_{\text{inlet}})}{V_{\text{inlet}}^2/(2g_c)} = \frac{T_{\text{inlet}}(T_{es}/T_{\text{inlet}} - 1)}{V_{\text{inlet}}^2/(2g_c c_p)}$$

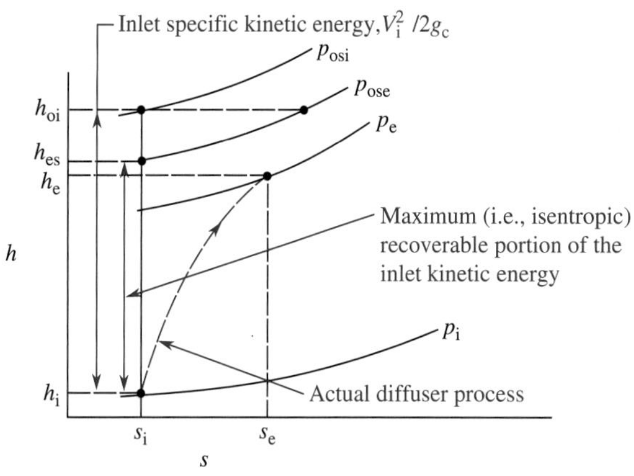

FIGURE 16.25
Thermodynamic process path of a diffuser plotted on *h–s* coordinates.

An ideal diffuser has a negligible exit velocity, so $p_{es} = p_{ose}$ and

$$T_{es}/T_{\text{inlet}} = (p_{es}/p_{\text{inlet}})^{(k-1)/k} = (p_{ose}/p_{\text{inlet}})^{(k-1)/k}$$

Then, the equation for η_D becomes

$$\eta_D = \frac{(p_{ose}/p_{\text{inlet}})^{(k-1)/k} - 1}{(k-1)\text{M}_{\text{inlet}}^2/2}$$

Now, from Eq. (16.13),

$$p_{\text{inlet}} = p_{osi}\left(1 + \frac{k-1}{2}\text{M}_{\text{inlet}}^2\right)^{-k/(k-1)}$$

so that the diffuser efficiency can be written in terms of the inlet Mach number and the isentropic stagnation pressure ratio as

Diffuser efficiency

$$\eta_D = \frac{\left(1 + \frac{k-1}{2}\text{M}_{\text{inlet}}^2\right)(p_{ose}/p_{osi})^{(k-1)/k} - 1}{(k-1)\text{M}_{\text{inlet}}^2/2} \tag{16.42}$$

Therefore, for a constant isentropic stagnation pressure ratio, the diffuser efficiency *decreases* as the inlet Mach number *increases*, asymptotically approaching the value $(p_{ose}/p_{osi})^{(k-1)/k}$ as the inlet Mach number goes to infinity.

A more direct measure of a diffuser's ability to convert kinetic energy into pressure is the *diffuser pressure recovery coefficient* C_p, defined as

$$C_p = \frac{\text{Actual diffuser pressure rise}}{\text{Isentropic diffuser pressure rise}} = \frac{(p_{\text{exit}})_{\text{actual}} - p_{\text{inlet}}}{p_{osi} - p_{\text{inlet}}} \tag{16.43}$$

Because of flow separation from the diffuser wall, C_p values are typically around 0.6.

EXAMPLE 16.15

A subsonic diffuser for a spacecraft attitude control thruster has been tested in the laboratory with an inlet Mach number of 0.890, using pure nitrogen as the working gas. The inlet and exit isentropic stagnation pressures are measured and found to be p_{osi} = 314.7 kPa and p_{ose} = 249.3 kPa. Determine, under these conditions, the diffuser's (a) efficiency and (b) pressure recovery coefficient.

Solution

a. The diffuser's efficiency can be determined directly from the measured conditions using Eq. (16.42) as follows:

$$\eta_D = \frac{\left(1 + \frac{k-1}{2}\text{M}_{\text{inlet}}^2\right)\left(\frac{p_{ose}}{p_{osi}}\right)^{(k-1)/k} - 1}{(k-1)\text{M}_{\text{inlet}}^2/2}$$

$$= \frac{\left(1 + \frac{1.40-1}{2}0.890^2\right)\left(\frac{249.3}{314.7}\right)^{\frac{1.40-1}{1.40}} - 1}{\left(\frac{1.40-1}{2}\right)0.890^2} = 0.529 = 52.9\%$$

b. The diffuser's pressure recovery coefficient can be determined from Eq. (16.43), if we can determine the actual inlet and exit pressures from the given data. Equation (16.13) can be used to relate the actual pressures to their isentropic stagnation values as

$$\frac{p_{ose}}{p_e} = \left(1 + \frac{k-1}{2}\text{M}_{\text{exit}}^2\right)^{\frac{k}{k-1}}$$

and

$$\frac{p_{osi}}{p_i} = \left(1 + \frac{k-1}{2}\text{M}_{\text{inlet}}^2\right)^{\frac{k}{k-1}}$$

Since for a diffuser, $M_{exit} \ll M_{inlet}$, we can assume $(p_{exit})_{actual} = p_e \approx p_{ose} = 309.3$ kPa. The inlet pressure can then be calculated as

$$p_{inlet} = p_i = \frac{p_{osi}}{\left(1+\frac{k-1}{2}M_{inlet}^2\right)^{\frac{k}{k-1}}} = \frac{314.7\,\text{kPa}}{\left(1+\frac{1.40-1}{2}0.890^2\right)^{\frac{1.40}{1.40-1}}} = 188\,\text{kPa}$$

and Eq. (16.43) gives

$$C_p = \frac{(p_{exit})_{actual} - p_{inlet}}{p_{osi} - p_{inlet}} = \frac{249.3\,\text{kPa} - 188\,\text{kPa}}{314.7\,\text{kPa} - 188\,\text{kPa}} = 0.484$$

Exercises

41. If the diffuser shape is modified in Example 16.15 so that the exit stagnation pressure is increased from 249.3 kPa to 261.5 kPa, determine the new values for the diffuser's efficiency and pressure recovery coefficient, assuming all other variables remain unchanged. **Answer:** $\eta_D = 0.623 = 62.3\%$ and $C_p = 0.580$.

42. A supersonic diffuser is also tested with the same inlet and exit conditions as the subsonic diffuser in Example 16.15, except that the inlet Mach number is 2.45 instead of 0.890. Determine the efficiency and pressure recovery coefficient of the supersonic diffuser. **Answer:** $\eta_D = 0.882 = 88.2\%$ and $C_p = 0.778$.

43. What value does the diffuser efficiency η_D approach as the exit isentropic stagnation pressure approaches the inlet isentropic stagnation pressure (i.e., $\eta_D \rightarrow$? as $p_{ose} \rightarrow p_{osi}$)? **Answer:** $\eta_D \rightarrow 1.00 = 100.\%$.

SUMMARY

In this chapter, we investigate the basic phenomena that occur in high-speed compressible flows of gases and vapors. New concepts, such as the stagnation state, Mach number, choked flow, and shock waves, are introduced to fully explain the basic characteristics of these flows. We focus our attention on converging-diverging nozzle and diffuser flow geometries because of their industrial value and their ability to generate supersonic flows and shock waves. Finally, we consider the overall performance of nozzles and diffusers in terms of their actual operating efficiencies.

We also introduce the Reynolds transport theorem in this chapter. This allows us to generalize our open system balance concept and subsequently to easily develop a linear momentum rate balance for open systems.

Some of the more important equations introduced in this chapter follow. Do not attempt to use them blindly without understanding their limitations. Please refer to the text material where they were introduced to gain an understanding of their use and limitations.

1. Stagnation state specific enthalpy h_o of a fluid with specific enthalpy h and velocity V:

$$h_o = h + V^2/(2g_c)$$

2. Stagnation state temperature T_o of an ideal gas with a velocity V at a temperature T and a constant pressure specific heat c_p:

$$T_o = T\left(1 + \frac{V^2}{2g_c c_p T}\right)$$

3. Isentropic stagnation state properties denoted by an os subscript:

$$\frac{T_{os}}{T} = \left(\frac{p_{os}}{p}\right)^{(k-1)/k} = \left(\frac{v_{os}}{v}\right)^{1-k} = \left(\frac{\rho_{os}}{\rho}\right)^{k-1}$$

where

$$\frac{p_{os}}{p} = \left(1 + \frac{V^2}{2g_c c_p T}\right)^{\frac{k}{k-1}} \quad \text{and} \quad \frac{\rho_{os}}{\rho} = \left(1 + \frac{V^2}{2g_c c_p T}\right)^{\frac{1}{k-1}}$$

4. The Mach number M is the ratio of the fluid velocity V to the velocity of sound c in the fluid:

$$M = V/c$$

5. The velocity of sound c in an ideal gas:

$$c_{\substack{\text{ideal}\\\text{gas}}} = \sqrt{kg_cRT}$$

6. The stagnation state properties in terms of the Mach number:

$$\frac{T_{os}}{T} = 1 + \frac{k-1}{2}\text{M}^2, \quad \frac{p_{os}}{p} = \left(1 + \frac{k-1}{2}\text{M}^2\right)^{k/(k-1)}, \quad \frac{\rho_{os}}{\rho} = \left(1 + \frac{k-1}{2}\text{M}^2\right)^{1/(k-1)}$$

7. The properties at the throat of a choked flow nozzle (denoted by an *):

$$T^* = T_{os}\left(\frac{2}{k+1}\right), \quad p^* = p_{os}\left(\frac{2}{k+1}\right)^{k/(k-1)}, \quad \rho^* = \rho_{os}\left(\frac{2}{k+1}\right)^{1/(k-1)}$$

8. The choked flow mass flow rate equations:

$$\dot{m}_{\text{max}}/A^* = p_{os}\left(\frac{kg_c}{RT_{os}}\right)^{1/2}\left(\frac{k+1}{2}\right)^{(k+1)/2(1-k)}$$

and, for air ($k = 1.40$) in the Engineering English units system,

$$(\dot{m}_{\text{max}}/A^*)_{\text{air}} = \left[0.532\frac{\text{lbm}\cdot\sqrt{\text{R}}}{\text{lbf}\cdot\text{s}}\right]\left(\frac{p_{os}}{\sqrt{T_{os}}}\right)$$

where $\dot{m}_{\text{max}}$ is in lbm/s, A^* is in in^2, p_{os} is in psia, T_{os} is in R, and the constant 0.532 has units of $\text{lbm}\cdot\sqrt{\text{R}}/(\text{lbs}\cdot\text{s})$. For air in the SI units system,

$$(\dot{m}_{\text{max}}/A^*)_{\text{air}} = \left[0.0404\frac{\text{kg}\cdot\sqrt{\text{K}}}{\text{N}\cdot\text{s}}\right]\left(\frac{p_{os}}{\sqrt{T_{os}}}\right)$$

where $\dot{m}_{\text{max}}$ is in kg/s, A^* is in m^2, p_{os} is in N/m^2, T_{os} is in K, and the constant 0.0404 has units of $\text{kg}\cdot\sqrt{\text{K}}/(\text{N}\cdot\text{s})$.

9. The general cross-sectional area ratio for a supersonic nozzle at its maximum flow rate:

$$A/A^* = \frac{1}{\text{M}}\left[\frac{2}{k+1}\left(1 + \frac{k-1}{2}\text{M}^2\right)\right]^{(k+1)/2(k-1)}$$

and, for air ($k = 1.40$), this reduces to

$$(A/A^*)_{\text{air}} = \frac{1}{\text{M}}\left(\frac{1+0.2\text{M}^2}{1.2}\right)^3$$

10. Thermodynamic property relations across a shock wave, where the subscript x denotes the upstream ($\text{M}_x > 1$) and the subscript y denotes downstream conditions ($\text{M}_y < 1$):

$$\frac{p_x\text{M}_x}{\sqrt{T_x}} = \frac{p_y\text{M}_y}{\sqrt{T_y}}, \quad \frac{T_x}{T_y} = \frac{1+\frac{k-1}{2}\text{M}_y^2}{1+\frac{k-1}{2}\text{M}_x^2}, \quad \frac{p_x}{p_y} = \frac{1+k\text{M}_y^2}{1+k\text{M}_x^2},$$

$$\text{M}_y^2 = \frac{(k-1)\text{M}_x^2+2}{2k\text{M}_x^2+1-k} \quad (\text{for } \text{M}_x \geq 1)$$

In the last expression M_x and M_y can be interchanged if M_y is known and M_x is the unknown.

11. The nozzle efficiency η_N, velocity coefficient C_v, and discharge coefficient C_d:

$$\eta_N = \frac{\frac{(k-1)}{2}(V_{\text{exit}}/c_{\text{inlet}})^2}{1-(p_{\text{exit}}/p_{\text{inlet}})^{(k-1)/k}}, \quad C_v = \sqrt{\eta_N}, \quad C_d = \frac{\dot{m}_{\text{actual}}}{\dot{m}_{\text{isentropic}}} = \frac{(\rho AV)_{\text{actual}}}{(\rho AV)_{\text{isentropic}}}$$

12. The diffuser efficiency η_D and pressure recovery coefficient C_p:

$$\eta_D = \frac{\left(1+\frac{k-1}{2}\text{M}_{\text{inlet}}^2\right)(p_{ose}/p_{osi})^{(k-1)/k} - 1}{(k-1)\text{M}_{\text{inlet}}^2/2}, \quad C_p = \frac{(p_{\text{exit}})_{\text{actual}} - p_{\text{inlet}}}{p_{osi} - p_{\text{inlet}}}$$

Problems (* indicates problems in SI units)

1. Explain the difference between T_o and T_{os}.
2. Explain why we use both symbols T_o and T_{os} but use only p_{os} and do not refer to p_o at all.
3. The absolute maximum exit velocity from any type of nozzle can be obtained by multiplying Eq. (16.2) by T and setting $T = 0$ R. Then, $(V_{exit})_{max} = \sqrt{2g_c c_p T_o} = \sqrt{2g_c c_p T_{os}}$. Determine the absolute maximum exit velocity for the nozzle in Example 16.6, and determine the percentage of this value achieved by the actual exit velocity.
4.* Saturated water vapor at 150.°C enters an isentropic converging nozzle with a negligible velocity and exits at 0.300 MPa. Determine the exit quality, temperature, and velocity. Do not assume ideal gas behavior.
5. Air flows in a circular tube with a velocity of 275 ft/s at a temperature of 103°F and a pressure of 175 psig. Determine its stagnation pressure and temperature.
6. Calculate the isentropic stagnation temperature and pressure on your hand as you hold it outside the window of an automobile traveling at 55.0 mph on a day when the static temperature and pressure are 70.0°F and 14.7 psia.
7. Steam at 600.°F and 200. psia is traveling at 1500. ft/s. Determine the isentropic stagnation temperature and pressure of the steam by
   a. Assuming steam to be an ideal gas.
   b. Using the steam tables (or Mollier diagram or computer program).
8.* A steam jet with a static pressure and temperature of 10.0 MPa and 400.°C has a velocity of 750. m/s. Determine the isentropic stagnation pressure and temperature of the jet. Do not assume ideal gas behavior.
9. Using Eq. (16.9), show that the speed of sound is infinite in an incompressible substance.
10.* If the speed of sound in saturated liquid water at 90.0°C is 1530 m/s, determine the *isentropic compressibility* α of the water, where

$$\alpha = -\frac{1}{v}\left(\frac{\partial v}{\partial p}\right)_s$$

11.* Determine the Mach number of a meteor traveling at 5000. m/s through still air at 0.00°C.
12. Determine the Mach number of a bullet traveling at 3000. ft/s through still air at 70.0°F.
13. The rotor of an axial flow air compressor has a diameter of 2.30 ft. What is the maximum rpm of the rotor such that its blade tips do not exceed the local sonic velocity when the air in the compressor is at 150.°F?
14.* Determine the stagnation temperature and Mach number of carbon dioxide gas flowing in a 1.00×10^{-2} m diameter circular tube at a rate of 0.100 kg/s. The temperature and pressure are 30.0°C and 0.500 MPa.
15.* The isentropic stagnation–static property formula given in Eqs. (16.12) to (16.14) and (16.18) to (16.20) are valid only for ideal gases. In Chapter 11, the conditions under which steam behaves as an ideal gas are discussed (steam gas). Suppose steam at 4.00 MPa, 400.°C is to be expanded through a converging nozzle under choked flow conditions. Use Eq. (16.18) to calculate T^* and use the steam tables (or Mollier diagram or computer program) with this value of T^* and $s^* = s_{os}$ to find p^* and $\rho^* = 1/v^*$. Then compare these values with the ones calculated from Eqs. (16.19) and (16.20).
16. A nozzle is to be designed to accelerate the flow of air from a Mach number of 0.100 to 1.00. Determine the inlet to exit area ratio of the nozzle assuming isentropic flow.
17. A diffuser is to be designed to reduce the Mach number of air from 0.90 to 0.10. Assuming isentropic flow, determine the exit to inlet area ratio of the diffuser.
18.* Argon escapes into the atmosphere at 0.101325 MPa from a 1.00 m^3 storage tank initially at 5.00 MPa and 25.0°C through a converging-diverging nozzle with a throat area of 1.00×10^{-3} m^2.
   a. Is the flow through the nozzle initially choked?
   b. If so, at what tank pressure does it unchoke?
   c. How long does it take to unchoke if the tank is maintained at 25.0°C?
19. Air at 100. psia and 70.0°F enters a converging nozzle with a negligible velocity and is expanded isentropically until the exit temperature is 32.0°F. Determine the exit Mach number and pressure.
20.* Air at 150. kPa 100.°C enters a converging nozzle with a negligible velocity and is expanded isentropically until the exit pressure reaches 101 kPa. Determine the exit Mach number and temperature.
21.* 1.86 kg/s of air flows through a converging-diverging supersonic wind tunnel whose reservoir isentropic stagnation conditions are 18.0 atm at 300. K and whose exit Mach number is 4.80. If the reservoir isentropic stagnation pressure is raised to 20.0 atm at the same temperature, find the new mass flow rate and exit Mach number if the exit pressure remains constant.
22.* Air enters a converging-diverging isentropic nozzle at 10.0 MPa and 500. K with a negligible velocity and is accelerated to a Mach number of 4.50. Determine the static temperature, pressure, and density at (a) the throat, and (b) the exit. (c) Find the exit to throat area ratio.
23. Low-velocity helium enters a converging-diverging isentropic nozzle at 250. psia and 120.°F. It is accelerated to a Mach number of 2.0. at the exit. Determine the static temperature, pressure, and density at (a) the throat, and (b) the exit. (c) Find the exit to throat area ratio.
24.* A converging-diverging nozzle is attached to a compressed air reservoir at 1.00 MPa and 27.0°C. There are two positions in the nozzle where $A/A^* = 2.00$, one is in the converging section and the other is in the diverging section. Determine the Mach number, pressure, temperature, density, and velocity at each section.
25.* Air at 0.500 MPa and 21.0°C enters an isentropic converging-diverging nozzle with a negligible velocity. The exit to throat area ratio of the nozzle is 1.34. If the throat velocity is sonic, determine the exit static pressure, temperature, and Mach number, if (a) the exit is subsonic and (b) the exit is supersonic.
26. A supersonic converging-diverging nozzle is to be designed to be attached to a standard machine shop air supply having isentropic stagnation conditions of 100. psia and 70.0°F. The throat of the nozzle is to have a diameter of 0.250 in, and the nozzle exhausts into the atmosphere at 14.7 psia. For an isentropic nozzle, determine
   a. The exit Mach number.
   b. The exit temperature.
   c. The mass flow rate of air through the nozzle.
   d. The exit diameter of the diverging section.

27.* Air enters a supersonic isentropic diffuser with a Mach number of 3.00, a temperature of 0.00°C, and a pressure of 1.00×10^{-2} MPa. Assuming the air exits with negligible velocity, determine the exit temperature, pressure, and mass flow rate per unit inlet area.

28.* Propane at 100. kPa 40.0°C is expanded isentropically through a converging-diverging nozzle that has an exit to throat diameter ratio of 2.00. The propane enters with a negligible velocity but reaches sonic velocity at the throat. Determine the exit temperature, pressure, and Mach number, if (a) the exit pressure is high enough so that the exit velocity is subsonic and (b) the exit pressure is low enough that the exit velocity is supersonic.

29. An isentropic converging-diverging nozzle that reaches a Mach number of 4.00 at the exit, when the exit pressure is atmospheric (14.7 psia) and the inlet isentropic stagnation temperature is 70.0°F, is to be built using air as the working fluid. Determine the required inlet isentropic stagnation pressure, the exit static temperature, and the exit to throat area ratio.

30. Acetylene at 50.0 psia, 65.0°F is accelerated through a converging-diverging nozzle isentropically until it reaches an exit pressure of 14.7 psia. Assuming the flow enters the nozzle with a negligibly small velocity, determine the exit Mach number, temperature, and exit to throat area ratio.

31.* Carbon dioxide gas at 13.8 MPa, 20.0°C is expanded isentropically through a converging-diverging nozzle until its exit temperature reaches –100.°C. Assuming the flow enters with a negligible inlet velocity, determine the exit Mach number, pressure, and exit to throat area ratio.

32. Steam at 800. psia, 600.°F expands through an *uninsulated* nozzle to a saturated vapor at 600. psia at a rate of 100. lbm/h. The surface temperature of the nozzle is measured and found to be 450.°F. The entropy production rate of the nozzle is equal to 10.0% of the magnitude of the heat transport of entropy for the nozzle. The potential energies and the inlet velocity can be neglected, but the exit velocity cannot be neglected. Determine
   a. The nozzle's heat transfer rate.
   b. The nozzle's entropy production rate.
   c. The nozzle's exit velocity.
   d. The exit area of the nozzle.

33. Determine the maximum possible mass flow rate of air through a nozzle with a 1.00×10^{-3} m diameter throat and inlet stagnation conditions of 5.00 MPa and 30.0°C.

34.* Determine the maximum flow rate of helium that passes through a nozzle with a 1.00×10^{-2} m diameter throat from an upstream stagnation state of 35.0 MPa at 27.0°C.

35. Determine the minimum throat diameter required for a nozzle to pass 0.250 lbm/s of air from a stagnation state of 100. psia at 70.0°F.

36. Atmospheric air (14.7 psia, 70.0°F) leaks into an initially evacuated 2.00 ft^3 tank through a tiny hole whose area is 1.00×10^{-6} ft^2. Determine the time required for the pressure in the tank to rise to 0.528 times the atmospheric pressure, if the air inside the tank is maintained at 70.0°F.

37.* A tiny leak in a 1.00 m^3 vacuum chamber causes the internal pressure to rise from 1.00 Pa to 10.0 Pa in 3.77 h when the vacuum pump is not operating. Air leaks into the chamber from the atmosphere at 101.3 kPa, 20.0°C, and the air inside the chamber is maintained at 20.0°C by heat transfer with the walls. Determine the diameter of the leak hole.

38. An initially evacuated 1.50 ft^3 tank is to be isothermally filled with air to 50.0 psia and 70.0°F. It is to be filled from a very large constant pressure source at 100. psia and 70.0°F. The tank is connected to the source by a single 0.125-inch inside diameter tube. How long will it take to fill the tank?

39.* An initially evacuated 0.500 m^3 tank is to be isothermally filled with air to 1.00 MPa and 20.0°C. It is to be filled from a very large constant pressure source at 2.50 MPa and 20.0°F. The tank is connected to the source by a single 1.00×10^{-3} m inside diameter tube. How long will it take to fill the tank?

40. Use the Reynolds transport equation (Eq. (16.31)) to develop a formula for a one-dimensional angular momentum rate balance (AMRB), and show that, for steady flow with a single inlet and a single outlet, the AMRB reduces to

$$\sum \mathbf{T}_{ext} = \dot{m}\left[(\mathbf{V}\times\mathbf{r})_{out} - (\mathbf{V}\times\mathbf{r})_{in}\right]/g_c$$

where $\mathbf{T}_{ext} = \mathbf{F}_{ext} \times \mathbf{r}$ is the torque vector due to external forces, $\mathbf{V}$ is the average velocity vector, and $\mathbf{r}$ is the radius vector. Hint: Start with $X = m(\mathbf{V} \times \mathbf{r})$ and utilize the conservation of angular momentum principle.

41. Use the linear momentum rate balance (LMRB) to show that the thrust force F of a rocket engine nozzle (Figure 16.26) is given by

$$F = \dot{m}\,(V_{exit}/g_c) + (p_{exit} - p_a)A_{exit}$$

and show that the absolute maximum thrust produced as $M_{exit} \to \infty$ is given by

$$F_{max} = 2c_p(\rho_{exit}A_{exit}T_{os}) + (p_{exit} - p_a)A_{exit}$$

(Hint: See Problem 3.)

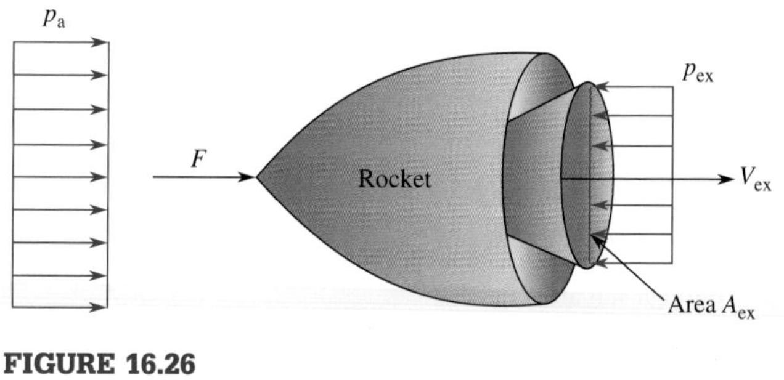

FIGURE 16.26
Problem 41.

42. The thrust F produced by the supersonic flow in the converging-diverging nozzle of a rocket engine is $F = \dot{m}\,(V_{exit}/g_c) + (p_{exit} - p_a)A_{exit}$, where p_{exit} and A_{exit} are the pressure and area at the nozzle exit and p_a is the local atmospheric pressure.
   a. Suppose the stagnation *temperature* is increased by 100% while maintaining the stagnation and exit pressures and nozzle geometry constant. What is the percent increase in thrust?
   b. Suppose the stagnation *pressure* is increased by 100% while maintaining the stagnation and exit temperatures and nozzle geometry constant. What is the percent increase in thrust? Hint: Assume the nozzle is choked in each case, and use air as the exhaust gas.

43. 0.800 lbm/s of air passes through an insulated converging nozzle that has an inlet to exit area ratio of 1.59 to 1. The nozzle is choked, and the stagnation temperature is 80.0°F. The exit pressure is 14.7 psia. Determine the force required to hold the nozzle in place.

44.* Determine the horizontal and vertical forces on the stationary turbine blade shown in Figure 16.27 when it is exposed to a 0.300 kg/s jet of air at 100. m/s.

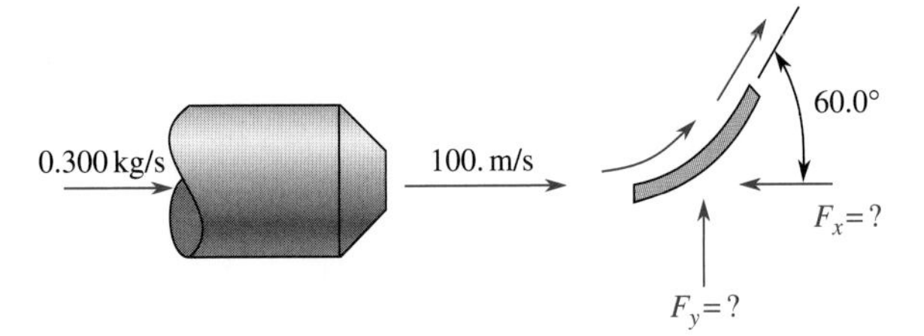

FIGURE 16.27
Problem 44.

45.* Determine the horizontal and vertical restraining forces on the air flow divider shown in Figure 16.28.

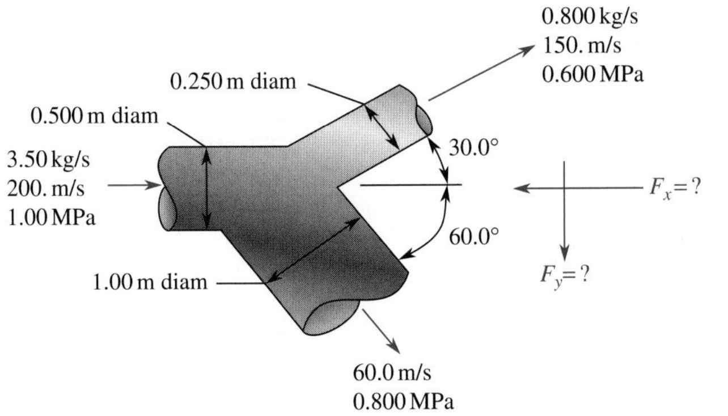

FIGURE 16.28
Problem 45.

46. What two conditions are required of an ideal gas to have $T_{osy} = T_{osx}$ across a normal shock wave?

47.* The flow conditions just downstream of a standing normal shock wave in air in a wind tunnel are $M_y = 0.500$, $p_y = 0.100$ MPa, and $T_y = 450.$ K. Determine the flow conditions just upstream of the shock (i.e., M_x, p_x, and T_x).

48.* A nuclear blast generates a normal shock wave that travels through still air with a Mach number of 5.00. The pressure and temperature in front of the shock (i.e., downstream) are 0.101 MPa and 20.0°C. Determine the air velocity relative to a stationary observer (i.e., the wind velocity), the pressure, and the temperature immediately after the shock wave has passed.

49.* The upstream and downstream temperatures across a normal shock wave in air are measured and found to be 306.3 and 717.6 K, respectively. Determine the upstream and downstream Mach numbers and the pressure ratio across the shock wave.

50.* The upstream and downstream static pressures across a normal shock wave in air are measured and found to be 0.500 and 3.00 MPa, respectively. Determine the upstream and downstream Mach numbers and the temperature ratio across the shock wave.

51.* The upstream and downstream isentropic stagnation pressures across a normal shock wave in air are measured and found to be 124.6801 kPa and 0.101325 MPa, respectively. Determine the upstream and downstream Mach numbers and static pressure and temperature ratios across the shock wave.

52. A converging-diverging nozzle has an exit to throat area ratio of 2.00. The inlet isentropic stagnation air pressure is 2.00 atm and the exit static pressure is 1.00 atm. This flow is supersonic in a portion of the nozzle, terminating in a normal shock inside the nozzle. Determine the local area ratio A/A^* at which the shock occurs.

53.* Air with a velocity of 450. m/s and a static pressure and temperature of 1.00 MPa and 200. K undergoes a normal shock. Determine the velocity and static pressure and temperature after the shock.

54. Use the conservation of mass condition across a shock wave $(\dot{m}_x = \dot{m}_y)$ to show that

$$p_x M_x (T_x)^{-0.5} = p_y M_y (T_y)^{-0.5}$$

55. Using Eqs. (16.34), (16.35), and (16.36), derive Eq. (16.37). (Hint: Use Eqs. (16.35) and (16.36) to eliminate p_x/p_y and $\sqrt{T_y/T_x}$ in Eq. (16.34). Then, square both sides of the resulting equation and solve for M_y^2 in terms of M_x^2.)

56. Using Eqs. (16.13), (16.36), and (16.37), show that

$$\frac{p_{osy}}{p_{osx}} = \left[\frac{(k+1)M_x^2/2}{1+(k-1)M_x^2/2}\right]^{k/(k-1)} \times \left[\frac{2kM_x^2}{k+1} - \frac{k-1}{k+1}\right]^{1/(1-k)}$$

57. It may be shown algebraically that, across a normal shock,

$$\frac{V_x}{c^*}\left(\frac{V_y}{c^*}\right) = 1.0$$

where $c^* = \sqrt{kg_cRT^*}$ is the sonic velocity at the throat. Consequently, it has become customary to use the rather awkward notation $M^* = V/c^*$ so that this relation can be written as $M_x^* M_y^* = 1.0$. Table C.18 in *Thermodynamic Tables to accompany Modern Engineering Thermodynamics* includes an M^* column for this purpose. Verify this relation for the normal shock data given in Example 16.12 by

a. Calculating V_x, V_y, $c^*$and $V_x V_y/(c^*)^2$.

b. Using Table C.18 to calculate $M_x^* M_y^*$.

58. It can be shown, for a normal shock wave, that

$$\frac{\rho_y}{\rho_x} = \frac{V_x}{V_y} = \frac{(k+1)M_x^2}{(k-1)M_x^2+2}$$

Using this relation, determine the maximum density ratio $(\rho_y/\rho_x)_{max}$ that can occur across a normal shock wave in air.

59. Use Eqs. (16.36) and (16.37) to develop the relation

$$\frac{p_y}{p_x} = 1 + \frac{2k}{k+1}(M_x^2 - 1)$$

60. The *strength* of a normal shock wave is defined as $(p_y - p_x)/p_x$. Using the results of Problem 55, show that this can be written as

$$\frac{p_y - p_x}{p_x} = \left(\frac{2k}{k+1}\right)(M_x^2 - 1)$$

61. Use Eqs. (16.35) and (16.37) to develop the relation

$$\frac{T_y}{T_x} = \frac{\left[(k-1)M_x^2+2\right]\left[2kM_x^2-(k-1)\right]}{\left[(k+1)M_x\right]^2}$$

62. In a supersonic wind tunnel utilizing air similar to that shown in Figure 16.12, the converging-diverging nozzle section has an inlet isentropic stagnation pressure of 3.20 atm. The test section has a Mach number of 2.70, and the converging-diverging diffuser section has an exit pressure of 1.00 atm. Determine the efficiency of the converging-diverging diffuser section.

63. Measurements on a prototype nozzle using air produce inlet and exit temperatures of 70.0°F and 60.0°F, respectively, while the exit velocity is 325 ft/s. Determine the nozzle's efficiency and velocity coefficient.

64.* An air diffuser has inlet and exit isentropic stagnation pressures of 3.50 and 3.10 MPa, respectively. The inlet velocity is 300. m/s and the inlet static temperature is 27.0°C. Determine the diffuser efficiency, pressure recovery coefficient, and exit static temperature, if the air leaves the diffuser with a negligible velocity.

65.* 8.00 kg/s of air flows through a diffuser with an inlet diameter of 0.0350 m and a static pressure and temperature of 0.500 MPa and 22.0°C. The air exits through a diameter of 0.900 m at static conditions of 0.540 MPa and 25.0°C. Determine the diffuser's efficiency and pressure recovery coefficient.

66. A diffuser decelerates 15.0 kg/s of carbon dioxide from 200. m/s at 20.0°C and 0.800 MPa to 1.00 m/s at 30.0°C and 1.00 MPa. Determine the diffuser efficiency, pressure recovery coefficient, and the inlet and exit areas.

67.* Experimental measurements on a new methane fuel nozzle for a furnace produce an exit velocity, pressure, and temperature for methane of 335 m/s, 0.100 MPa, and 0.00°C. The upstream stagnation pressure and temperature are 0.150 MPa and 22.0°C. Determine
   a. The nozzle efficiency.
   b. The nozzle's velocity coefficient.
   c. The nozzle's discharge coefficient.

68. A diffuser having an efficiency of 92.0% is to be used to reduce the velocity of an air stream initially at 450. ft/s, 65.0°F, and 50.0 psia down to a Mach number of 0.100. Calculate
   a. The exit to inlet area ratio (A_{exit}/A_{inlet}) required.
   b. The pressure recovery factor for the diffuser.

69. A sonic converging nozzle with a negligible inlet velocity and inlet and throat areas of 2.00 in^2 and 0.500 in^2, respectively, has a velocity coefficient of 0.820 when the upstream stagnation pressure and temperature are 100. psia and 70.0°F. Determine the thrust produced by the nozzle in atmospheric air.

Design Problems

The following are open-ended design problems. The objective is to carry out a preliminary design as indicated. A detailed design with working drawings is not expected unless otherwise specified. These problems do not have specific answers, so each student's design is unique.

70. Design a converging-diverging nozzle system that can be used to demonstrate supersonic flow in the classroom. Choose a convenient gas, inlet conditions, and exit Mach number. Determine the necessary area ratios, pressures, and temperatures throughout the system.

71.* Design an attitude control nozzle for a spacecraft that produces 50.0 N of thrust (see Problem 42) using compressed helium gas stored at 50.0 MPa and 0.00°C. Assume the nozzle discharges into a total vacuum. Specify the nozzle inlet, throat, and exit areas as well as the exit Mach number.

72.* Design a system that has no moving mechanical parts to cool machine shop compressed air at 0.500 MPa, 25.0°C to 0.00°C at 0.101 MPa. The outlet velocity must be at least 10.0 m/s. If possible, fabricate and test your design.

73. Design a small demonstration wind tunnel to be driven from a standard compressed air supply line at 100. psia and 70.0°F. Assume that the maximum volumetric air flow rate available from this supply is 10.0 ft^3/min at 14.7 psia and 70.0°F. The wind tunnel test section must be at least 1.00 inch in diameter and must reach a Mach number of at least 2.25. The air may be exhausted to the atmosphere, but it first must be decelerated to subsonic velocity to minimize noise generation. If possible, fabricate and test your design.

74.* Design a converging-diverging nozzle for a spacecraft thruster that has an exit Mach number of 5.00 when using compressed helium at 50.0 MPa, and 0.00°C. Assume the nozzle exhausts into a total vacuum. Plot the nozzle diameter vs. length along the nozzle, keeping the angle of the diverging wall to less than 10° with the horizontal to prevent flow separation. Show the positions along the nozzle where M = 1, 2, 3, 4, and 5. Determine the mass flow rate through your nozzle.

75. Design a system that produces a constant mass flow rate of 1.00×10^{-2} lbm/s of oxygen from one or more 3.00 ft^3 high-pressure storage bottles initially at 2000. psia and 400. R. The oxygen must be delivered at 50.0 ft/s at 60.0°F and 175 psia. The system must operate continuously for six months and must have a fail-safe backup.

Computer Problems

The following open-ended computer problems are designed to be done on a personal computer using a spreadsheet or problem solver.

76. Develop a computer program that returns values for T^*/T_{os}, p^*/p_{os}, and ρ^*/ρ_{os}, when k and the remaining variables are input in response to a screen prompt.

77. Develop an interactive computer program that returns values for p/p_{os}, T/T_{os}, ρ/ρ_{os}, and A/A^*, when k and M are input from the keyboard in response to a screen prompt.

78. Develop an interactive computer program that returns values for M_y, p_y/p_x, T_y/T_x, ρ_y/ρ_x, $\dot{S}/\dot{m}$, and p_{osy}/p_{osx}, when k and M_x are input from the keyboard in response to a screen prompt.

79. Using Eqs. (16.35), (16.36), (16.37), and (16.38), plot $\dot{S}p/(\dot{m}R)$ vs. M_x for $1 \leq M_x \leq 50$ for (a) air, (b) carbon dioxide, (c) methane, and (d) water vapor (use $k = 1.33$ here).

80.* **Fanno line.** An analysis of the adiabatic aergonic flow of a *viscous* ideal gas with constant specific heats traveling through a constant area duct can be carried out by combining the continuity equation, the energy rate balance (Eq. (16.1)), and the entropy rate balance (using Eq. (7.36)) to obtain the following relation:

$$\frac{\dot{S}_P}{(\dot{m}c_v)} = \frac{s_{out}-s_{in}}{c_v} = \ln\left[\frac{T_{out}}{T_{in}}\left(\frac{T_{os}-T_{out}}{T_{os}-T_{in}}\right)^{(k-1)/2}\right] \geq 0$$

A plot of this function is called the *Fanno line* for the flow. Plot T_{out} vs. $\dot{S}p/(\dot{m}c_v)$ for air using $0 \le T_{out} \le T_{in}$. Take $T_{os} = 300\,\text{K}$ and $T_{in} = 290$ K. Note that $\dot{S}p/(\dot{m}c_v)$ is double valued in T_{out} and its maximum value occurs at M = 1.0. Determine the two values of T_{out} for which $\dot{S}_p = 0$ when $T_{os} = 300\,\text{K}$ and $T_{in} = 290$ K.

81.* **Rayleigh line**. An analysis of the frictionless aergonic flow of an ideal gas with constant specific heats traveling through a constant area duct with heat transfer at the walls can be carried out by combining the continuity equation and the linear momentum rate balance equation to yield the following set of equations:

$$\frac{p_{out}}{p_{in}} = \frac{1 + kM_{in}^2}{1 + kM_{out}^2}$$

$$\frac{T_{out}}{T_{in}} = \left(\frac{M_{out}}{M_{in}}\right)^2 \times \frac{1 + kM_{in}^2}{1 + kM_{out}^2}$$

$$\frac{(T_{os})_{out}}{(T_{os})_{in}} = \left(\frac{M_{out}}{M_{in}}\,\frac{1 + kM_{in}}{1 + kM_{out}}\right)^2 \times \frac{1 + \frac{k-1}{2}M_{out}^2}{1 + \frac{k-1}{2}M_{in}^2}$$

and Eq. (7.37) gives $s_{out} = s_{in} + c_p \ln(T_{out}/T_{in}) - R\ln(p_{out}/p_{in})$. For air, with $(T_{os})_{in} = 100°C$, $p_{in} = 0.5$ MPa, $M_{in} = 0.5$, and $s_{in} = 2.2775$ kJ/(kg·K), generate the following plots for $0 \le M_{out} \le 10$:

a. T_{out} vs. s_{out} (this plot is called the *Rayleigh line*).

b. $\dot{Q}/\dot{m} = c_p[(T_{os})_{out} - (T_{os})_{in}]$ vs. M_{out} (this is the heat transfer per unit mass to or from the air).

c. $\dot{S}_p/(\dot{m}c_p) = (s_{out} - s_{in})/c_p - \frac{\dot{Q}/\dot{m}}{c_p T_w}$ vs. M_{out}, where $T_w = \frac{1}{2}(T_{in} + T_{out})$ is the mean wall temperature. Note that s_{out} is a maximum when $M_{out} = 1.0$.

CHAPTER 17

Thermodynamics of Biological Systems

CONTENTS

17.1 INTRODUÇÃO (INTRODUCTION)

Over the years, thermodynamics has remained essentially an engineering discipline with only infrequent applications elsewhere. In chemistry, courses entitled Physical Chemistry specialize in applying thermodynamics to chemical systems of the type treated in Chapter 15 of this book. Only recently has the developing field of bioengineering begun to apply the macroscopic mass, energy, and entropy balance concepts of classical thermodynamics to living systems. In this chapter, the results of applying these basic laws of thermodynamics to biological systems are reviewed. The conclusions reached will help you understand how your body functions and give you some insight into the operation of the complex molecular phenomena necessary to sustain life on this or any other planet.

In this chapter, the basic thermodynamics of simple living systems (biological cells) is followed by discussions of animal biological energy conversion efficiency, metabolism, nutrition, and exercise. Then, the fascinating subjects of the limits to biological growth and an engineering view of living system mobility are presented. The chapter ends, appropriately, with a thermodynamic discussion of biological aging and death. In this section, an attempt is made to describe how and why living systems age differently from nonliving systems.

17.2 LIVING SYSTEMS

Only in the past few years has science begun to realize how the evolution of life is completely compatible with the laws of physics. A key to this understanding has been the entropic explanation of self-organizing systems and the connection between self-organization and energy flow. Self-organization can exist in both living and nonliving systems, but living systems are self-organizing *and* self-replicating. The origin of life is apparently not an unusual phenomenon. Fairly complex living microscopic creatures existed on Earth within a few hundred million years of its formation.

There is no clear-cut understanding of the scientific concept of *life*. Living systems have six recognized characteristics: (1) molecular organization, (2) metabolism, (3) growth, (4) adaptation, (5) response to stimuli, and (6) reproduction. But perhaps the best definition available today is that a system is said to be "living" if it sustains its low-entropy molecular complexity (i.e., DNA) that contains hereditary information transmitted to offspring on reproduction, by a metabolic energy transport from a high-energy source (food) to a low-energy sink (waste) via catalytic macromolecules, called *enzymes*.

The different living systems on Earth have numerous items in common. For example, they all use the same class of molecules for energy storage, the nucleotide phosphates. Also, of the billions of chemically possible organic compounds, only about 1500 are actually used by living systems. And all these 1500 compounds are made with less than 50 simpler molecular building blocks, utilizing no more than 24 of the available elements. Hydrogen atoms make up 63% of all the atoms in the human body. Oxygen accounts for 25.5%, carbon 9.50%, nitrogen 1.40%, and the 20 remaining elements essential for mammalian life account for only about 0.60%. Only 3 of the 24 elements known to be essential to life on this planet have atomic numbers greater than 34, and these 3 are needed in only trace amounts. Thus, since living systems are made up of the simplest atomic elements, they can be expected to develop early on any planet that has the proper environmental conditions.

Living systems are organized around a *cell* structure of some kind. A cell is like a small factory whose main function is to carry out its metabolic process, and the cell boundaries appear to exist to provide the high enzyme concentration necessary for efficient metabolism. The smallest free-living cell known today is a pleuropneumonialike organism that has a mass of only about 5×10^{-19} kg and has a diameter of only about 10^{-7} m (about one fifth of the wavelength of visible light). This cell contains about 100 enzymes and can be seen only with a high-powered electron microscope. The human body contains about 10^{14} cells with an average diameter of 10^{-5} m. Each cell typically consists of a central nucleus, with the remaining material being the cytoplasm (see Figure 17.1). The chemical activity inside the cell is very high, with each enzyme entering into the synthesis of about 100 molecules per second.

The oldest remnants of life on Earth are cellular microfossils that are over 3.5 billion years old. Since the age of the Earth is only about 4.5 billion years, the thermal and chemical requirements for the evolution of living systems must have developed remarkably fast. All known living systems on Earth are water based and therefore cannot exist far outside the temperature range from 0 to 100°C. It is amazing that the surface of the Earth had regions in this temperature range for at least 80% of its existence.

WHAT IS METABOLISM?

Metabolism is the name given to the processes of breakdown and synthesis of large macromolecules within a cell.

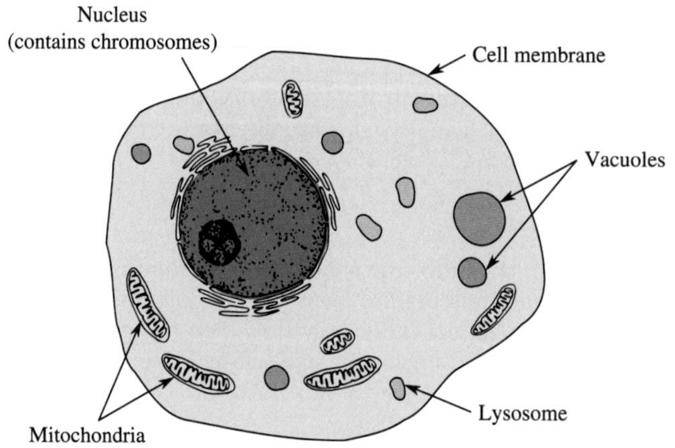

FIGURE 17.1
Schematic of a typical living cell.

17.3 THERMODYNAMICS OF BIOLOGICAL CELLS

It is unlikely that a single energy source was directly responsible for the synthesis of all the organic molecules on the newly formed Earth. In recent decades, laboratory experiments with the elements of carbon, hydrogen, oxygen, and nitrogen have shown that basic organic compounds can be synthesized by a variety of energy sources under early Earth conditions. Table 17.1 lists an estimate of the energy rate per unit area available on the surface of the primitive Earth. Though solar radiation was clearly the largest source of energy, the energy contained in long-wavelength (150 to 200 nm) ultraviolet light is so strong that it decomposes absorbing molecules rather than building them. However, the water of the primitive Earth's oceans protected complex organic molecules from disruptive ultraviolet radiation until the Earth's ozone layer developed. It was not until this protective atmospheric layer had developed that life forms could leave the oceans and populate the dry land.

The most widely used source of energy for the synthesis of primitive organic compounds in the laboratory is an electrical discharge in a mixture of gases. The most common compounds produced by this technique are amino acids, with yields as high as 5%.

As concentrations of organic compounds built up in the primitive oceans, biological life processes began to synthesize and replicate molecules. Enzymes and genetic molecules evolved, but reaction rates were limited by the comparatively low concentrations of these essential building blocks. Specialized molecular barriers then evolved that completely enclosed small volumes of fluid containing complex molecular machinery. These barriers are called *membranes* and the resulting enclosure is called a *cell*. The purpose of biological membranes is to maintain concentration differences that would be advantageous to the molecular operation of the cell. To do this, the membrane must be able to transport certain ions *against* the concentration gradient (this is called *active* transport). This requires that the membrane operate as an energy converter, with some of the internal energy of the cell being used to maintain the various concentration gradients across the membrane. Table 17.2 lists some ion concentrations inside and outside common human cells.

Table 17.1 Estimates of Energy Rates Available for the Formation of Simple Organic Compounds, Averaged over the Surface Area of Primitive Earth

Source	Energy Rates per Unit Area [KJ/($m^2 \cdot a$)]
Electric discharge (lightning, etc.)	170
Solar radiation in the 0–150 nm range	71
Thermal quenching of hot gases from	
Shock waves from meteors and lightning	46
Volcanoes	5.4
Highly ionizing radiation from	
Radioactivity 1.0 km deep in the Earth	33
Solar wind	8.4
Cosmic rays	0.1

Source: Material drawn from Oró, J., Miller, S. L., Urey, H. C. Energy conversion in the context of the origin of life. In: Buvet, R., Allen, M. J., Massué, J.-P. (Eds.), Living Systems as Energy Converters. North-Holland Publishing, 1977, pp. 7–19, New York. Reprinted by permission of Elsevier Science Publishers (Biomedical Division), Amsterdam, and the authors.

Table 17.2 Approximate Ion Concentration Inside and Outside Human Cells

	Concentration in Osmoles per cm^3 of Water	
Ion	Outside the Cell	Inside the Cell
Na^+	144	14.0
K^+	4.1	140
Mg^{2+}	1.5	31
Cl^-	107	4.00
HCO_3^-	27.7	10.0
SO_4^{2-}	0.5	1
HPO_4^{2-}, $H_2PO_4^-$	2.0	11

Note: One osmole is the number of gram moles of the substance that do not diffuse or dissociate in solution. Also, $pH_{outside}$ = 7.4 and pH_{inside} = 7.0, where the concentration of hydrogen ions (H^+) in gmoles/L is 10^{-pH}.

Because cell membranes are molecular machines, their exact structure is not yet completely understood. The most universally accepted model of a membrane is the bimolecular lipid leaflet structure shown schematically in Figure 17.2. In this model, the membrane structure consists of two parallel rows of phospholipid molecules oriented with their hydrophobic chains pointing inward and their hydrophilic (polar) ends pointing outward. The inner and outer surfaces of the membrane are covered with various protein layers, and the membrane thickness is typically 7 to 10 nanometers (10^{-9} m). It is also felt that the membrane must contain a uniform distribution of holes, or *pores*, about 0.8 nm in diameter, through which water and certain hydrated ions can pass. Approximately 0.06% of the membrane area is made up of these pores. The concentration of materials inside the cell is determined exclusively by concentration differences across the membrane.

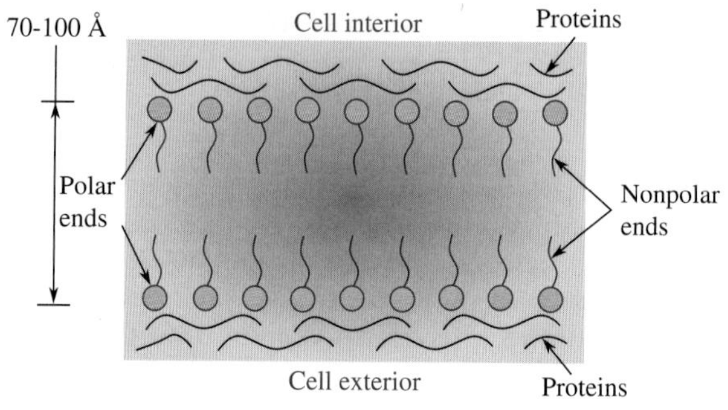

FIGURE 17.2
Schematic of membrane construction.

Membranes of living cells maintain an electrical potential difference between the inside and outside of the cell. With very small electrodes, a reasonably constant current can be continuously drawn from a cell. A cell can produce electricity in this way only if it has a molecular mechanism for maintaining an unequal ion charge difference across its membrane. Such membranes are known to contain a molecular level *ion pump* that transports ion species in only one direction (into or out of the cell).

How much work is required to pump a charged ion from the solution outside the cell into the solution inside the cell? The answer to this question can be developed by considering the transport process to be carried out in two steps. First, consider moving an ion from infinity through a vacuum, through the membrane, and into the cell. Assume in this first step that the cell membrane has no dipole layer (i.e., no net charge on its surface) and the inside of the cell is electrically neutral. Now, as the charged ion moves from infinity to the membrane, it encounters no resistance, so its transport work is zero. As it moves through the cell membrane, it begins to feel electrostatic ion-solvent and ion-ion interactions. We lump all these interactions together and call them *chemical* effects; therefore, the work done against these interactions in moving the charged ion into the cell is called *chemical work*. The chemical work done in moving a mole of ions of chemical species i from infinity into an uncharged cell through a dipole layer–free membrane is equal to the molar *chemical potential* $\bar{\mu}_i$ of ion species i.

The second step in this process is to allow the membrane to have a dipole layer and allow the cell to have a net internal charge. We call the work required to move the ion into this system of net charges the *electrical* work; therefore, the total work required to move the ion from infinity through a dipole-layered membrane into a charged cell is the sum of the chemical work plus the electrical work. This total work is called the *electrochemical* work of the cell.

Define ϕ_k as the electrical potential (in volts) required to transport a unit charge of species i into the cell. Then, the electrical work required to transport one ion of species i with valence z_i kgmole of electrons per kgmole of species i (and thus a charge $z_i e$) is $z_i e\phi_{ic}$, where e is the charge on one electron. The electrical work required to transport 1 mole of species i into the cell is $N_o z_i e\phi_{ic} = z_i F\phi_{ic}$, where N_o is Avogadro's number and $F = N_o e =$ Faraday's constant = 96,487 kilocoulombs/kgmole of electrons. Let $(\overline{w}_{EC})_{ic}$ be the electrochemical work required to transport 1 mole of species i with valence z_i *into* a cell. Then, we can write

$$(\overline{w}_{EC})_{ic} = (W_{EC})_{ic}/n_i = \bar{\mu}_{ic} + z_i F\phi_{ic} \tag{17.1}$$

Unfortunately, neither $\bar{\mu}_{ic}$ or ϕ_{ic} is directly measurable. They were introduced as conceptual quantities for the purpose of separating chemical effects from electrical effects; however, only their combined effect can be observed in the laboratory.

What can we measure? We can measure the electrical potential difference (i.e., voltage) between the inside and outside of the cell. Now, as soon as we introduce an electrode into the cell, we set up a current path, so that the measured potential $\Delta\phi_m$ is not the same as the zero current (no electrode) equilibrium potential $\Delta\phi_e$. Again, $\Delta\phi_e$ cannot be measured, but we can get around that as follows. From Eq. (17.1), we find that the electrochemical work required to move a mole of ions of species i with valence z_i from infinity into the solution *outside* the cell is

$$(\overline{w}_{EC})_{io} = (W_{EC})_{io}/n_i = \bar{\mu}_{io} + z_i F\phi_{io} \tag{17.2}$$

and that the electrochemical work required to move that same mole from infinity *into* the cell is

$$(\overline{w}_{EC})_{ic} = \overline{\mu}_{ic} + z_i F \phi_{ic} \tag{17.3}$$

then, from Eqs. (17.2) and (17.3), we find that the zero current equilibrium electrical potential difference between the inside and outside of the cell due to the presence of species i is

$$(\Delta\phi_e)_i = \phi_{ic} - \phi_{io} = \frac{1}{z_i F}\left[(\overline{w}_{EC})_{ic} - (\overline{w}_{EC})_{io} - (\overline{\mu}_{ic} - \overline{\mu}_{io})\right] \tag{17.4}$$

The molar chemical potential of species i can be written for isothermal, dilute solutions as

$$\overline{\mu}_i = \overline{\mu}_i^o + \Re T \ln c_i \tag{17.5}$$

where $\overline{\mu}_i^o$ is the molar chemical potential when $c_i = 1.0$, $\Re$ is the universal gas constant, T is the absolute temperature, and c_i is the molar concentration of i. Using Eq. (17.5), we can write Eq. (17.4) as

$$(\Delta\phi_e)_i = \frac{1}{z_i F}\left[(\overline{w}_{EC})_{ic} - (\overline{w}_{EC})_{io} - (\overline{\mu}_{ic}^o - \overline{\mu}_{io}^o)\right] + \frac{\Re T}{z_i F} \ln \frac{c_{io}}{c_{ic}} \tag{17.6}$$

To simplify the algebra, we call the first term on the right side of Eq. (17.6) $(\Delta\phi_e^o)_i$, which is the value of $(\Delta\phi_e)_i$ when $c_{ic} = c_{io}$. Then, Eq. (17.6) becomes

$$(\Delta\phi_e)_i = (\Delta\phi_e^o)_i + \frac{\Re T}{z_i F} \ln \frac{c_{io}}{c_{ic}} \tag{17.7}$$

However, we still cannot measure either $(\Delta\phi_e)_i$ or $(\Delta\phi_e^o)_i$. At this point, we arbitrarily assign the electrical potential outside the cell the value zero, and we define the *membrane potential* E_i due to species i as

$$E_i = (\Delta\phi_e)_i - (\Delta\phi_e^o)_i$$

which is given by Eq. (17.7) as

$$E_i = \frac{\Re T}{z_i F} \ln \frac{c_{io}}{c_{ic}} \tag{17.8}$$

At 37°C, Eq. (17.8) becomes (recall that 1 coulomb = 1 joule/volt)

$$\begin{aligned} E_i(\text{at } 37°\text{C}) &= \frac{[8314.3\ \text{J}/(\text{kgmole}\cdot\text{K})](37.0 + 273.15\ \text{K})}{z_i(96{,}487\ \text{kilocoulombs/kgmole})} \ln\left(\frac{c_{io}}{c_{ic}}\right) \\ &= \frac{26.7\ \text{millivolts}\cdot(\text{kgmole electrons/kgmole } i)}{z_i} \ln\left(\frac{c_{io}}{c_{ic}}\right) \end{aligned} \tag{17.9}$$

where z_i is the valence of species i in kgmole of electrons per kgmole of species i. Note that z_i can be either positive or negative in this equation.

WHAT IS SO SPECIAL ABOUT A BODY TEMPERATURE OF 37°C?

As the temperature of an organism increases up to about 40°C, the speed of its enzyme-catalyzed metabolic reactions increases, because the molecules collide more frequently due to thermal agitation. But above 40°C, the weak bonds that control the functional shape of the enzymes begin to break, and they become ineffective at sustaining metabolism. For many years, it was thought that life as we know it could not exist at temperatures above about 40°C.

However, recently hyperthermophilic ("superheat-loving") bacteria have been found in high-temperature environments, such as deep sea volcanic hot vents. They grow at temperatures above 80°C and can survive to temperatures up to 113°C. They are very tough life forms, even surviving temperatures as low as −140°C. It seems possible that they could have been carried through space on meteoroids to populate planets.

EXAMPLE 17.1

Using the concentration data provided in Table 17.2, determine the membrane potential in human cells of sodium, potassium, and chlorine ions at 37.0°C.

Solution

Table 17.2 gives the concentration of sodium ions inside a human cell at 37.0°C as

$$c_{Na_c^+} = 14.0\ \text{osmoles/cm}^3$$

while the concentration of sodium ions outside the cell is

$$c_{Na_o^+} = 144\ \text{osmoles/cm}^3$$

The valence of a sodium ion is 1 kgmole electrons/kgmole Na^+. Then, Eq. (17.9) gives the membrane potential of sodium as

$$E_{Na^+} = \frac{26.7\ \text{mV(kgmole electrons/kgmole Na}^+)}{z_{Na^+}\ \text{kgmole electrons/kgmole Na}^+} \ln\left(\frac{c_{Na_o^+}}{c_{Na_c^+}}\right)$$

$$= \frac{26.7\ \text{mV(kgmole electrons/kgmole Na}^+)}{1\ \text{kgmole electrons/kgmole Na}^+} \ln\left(\frac{144}{14.0}\right) = 62.2\ \text{mV}$$

For potassium, Table 17.2 gives $c_{K_c^+} = 140$ osmoles/cm^3 and $c_{K_o^+} = 4.1$ osmoles/cm^3. The valence of a potassium ion is also 1 kgmole electrons/kgmole K^+, and Eq. (17.9) gives the membrane potential of potassium in a human cell as

$$E_{K^+} = \frac{26.7\ \text{mV(kgmole electrons/kgmole K}^+)}{1\ \text{kgmole electrons/kgmole K}^+} \ln\frac{4.1}{140} = -94.3\ \text{mV}$$

Finally, Table 17.2 gives $c_{Cl_c^-} = 4.00$ osmoles/cm^3 and $c_{Cl_o^-} = 107$ osmoles/cm^3. The valence of a chlorine ion is −1 kgmole electrons/kgmole Cl^-, and Eq. (17.9) gives the membrane potential of chlorine in a human cell as

$$E_{Cl^-} = \frac{26.7\ \text{mV (kgmole electrons/kgmole Cl}^-)}{-1\ \text{kgmole electrons/kgmole Cl}^-} \ln\frac{107}{4.00} = -87.8\ \text{mV}$$

Exercises

1. Determine the membrane potential in human cells of magnesium ions, Mg^{2+}, at 37.0°C. **Answer:** $E_{Mg^{2+}} = -40.4$ mV.
2. Determine the membrane potential in human cells of sulphate ions, SO_4^{2-}, at 37.0°C. **Answer:** $E_{SO_4^{2-}} = 9.3$ mV.
3. Determine the membrane potential in human cells of dihydrogen phosphate ions, $H_2PO_4^-$, at 37.0°C. **Answer:** $E_{H_2PO_4^-} = 45.5$ mV.

Actual measured potentials are generally in the range of −70 to −90 mV and represent the cumulative effect of all the ion species present. However, Na^+, K^+, and Cl^- are the primary high-transport ions in most mammal membranes, and their cell potentials, listed earlier, average out to about the measured value.

Applying the open system energy rate balance equation to a living cell gives

$$\dot{Q} - \dot{W} + \sum_{\text{in}} \dot{m}e - \sum_{\text{out}} \dot{m}e = \left(\frac{dU}{dt}\right)_{\text{cell}}$$

Here, $\dot{Q}$ is the irreversible metabolic heat transfer resulting from the life processes within the cell, $\sum_{\text{in}} \dot{m}e$ is the food energy intake, $\sum_{\text{out}} \dot{m}e$ is the waste product output, and $\dot{W}$ is the total work done on or by the cell. The food taken into the cell can be generalized as glucose and molecular oxygen, and the waste products can be generalized as carbon dioxide and water. The total work done on or by the cell is the electrochemical work done in maintaining the chemical differences across the cell membrane, $(\overline{w}_{EC})_i = (\overline{w}_{EC})_{ic} - (\overline{w}_{EC})_{io}$, and occasional p-$\forall$ work done in enlarging the cell plus γ-A surface tension work done in generating new membrane surface area. Then,

$$\dot{W} = \sum \dot{m}_i \left(\frac{\overline{w}_{EC}}{M}\right)_i + \sum \dot{m}_i \left(\frac{\overline{\mu}}{M}\right)_i + \gamma \dot{A} + p\dot{\forall} \tag{17.10}$$

where we have written all terms on a mass rather than a molar basis (using $\dot{n}_i = \dot{m}_i/M_i$, where M_i is the molecular mass of species i) and the intensive properties have been assumed to be constant in time. Thus, the time rate of change of the cell's total internal energy is

$$\left(\frac{dU}{dt}\right)_{\text{cell}} = \dot{Q} - \sum \dot{m}_i \left(\frac{\overline{w}_{EC}}{M}\right)_i - \sum \dot{m}_i \left(\frac{\overline{\mu}}{M}\right)_i - \gamma \dot{A} - p\dot{\forall} + \sum_{\text{in}} \dot{m}e - \sum_{\text{out}} \dot{m}e$$

Most of the various cellular processes that require energy use adenosine triphosphate (ATP) as the energy source. This compound has about 33 MJ/kgmole of energy stored in each of two phosphate bonds. When these bonds are split by enzyme action to form adenosine diphosphate (ADP), their energy is then made available for other uses. The cell contains many enzymes that can catalyze the splitting of the ATP bonds and utilize the liberated energy.

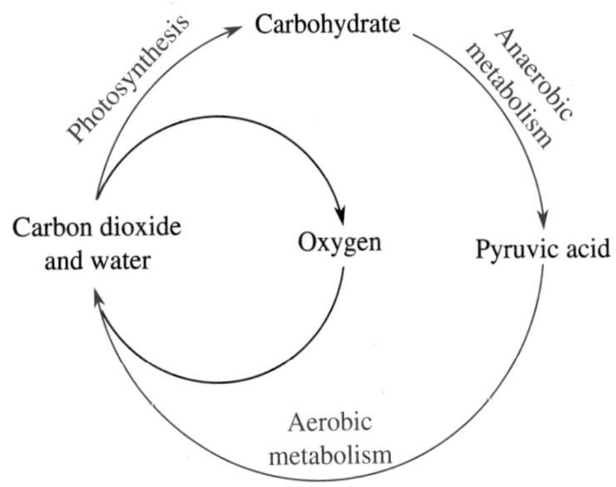

FIGURE 17.3
Energy transport mechanisms in living systems.

Energy storage reactions within the cell, on the other hand, are limited to two basic types: photosynthetic (in plant cells), wherein incoming light is used as the energy source, and metabolism (in animal cells), wherein the food brought into the cell (generally glucose and molecular oxygen) is utilized to reconstitute ATP from ADP, with the production of carbon dioxide and water waste products, which must be expelled from the cell. Figure 17.3 shows how these two energy transport mechanisms are linked together in the life cycle, and Figure 17.4 illustrates the ATP–ADP cycle.

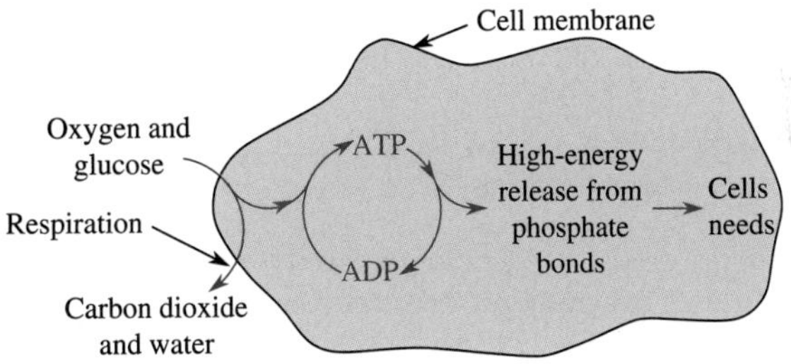

FIGURE 17.4
The ATP–ADP cycle.

An open system entropy rate balance applied to a living cell gives

$$\frac{\dot{Q}}{T_b} + \sum_{\text{in}} \dot{m}s - \sum_{\text{out}} \dot{m}s + \dot{S}_P = \left(\frac{dS}{dt}\right)_{\text{cell}}$$

where T_b is the temperature of the cell boundary (assumed isothermal here). Because the metabolic heat must leave the cell for it to survive, we know that $\dot{Q} < 0$. Also, $\dot{S}_P > 0$ due to the irreversibilities of the life process within the cell. Since food products are brought into the cell and waste products expelled, $\sum_{\text{out}} \dot{m}e > \sum_{\text{in}} \dot{m}e$ (as these two flow streams are at the same temperature and the molecular order of the waste material is less than that of the food). For a cell to grow and continue to maintain its elaborate internal molecular order, we must have $(dS/dt)_{\text{cell}} < 0$, or

$$\left| \sum_{\text{in}} \dot{m}s - \sum_{\text{out}} \dot{m}s + \dot{Q}/T_b \right| \geq \dot{S}_P$$

which is perfectly reasonable so long as the cell remains alive (i.e., $\dot{Q} < 0$ and $\sum_{\text{in}} \dot{m}e < \sum_{\text{out}} \dot{m}e$).

17.4 ENERGY CONVERSION EFFICIENCY OF BIOLOGICAL SYSTEMS

Metabolism is the name given to all anabolic (constructive) and catabolic (destructive) molecular processes within a living system, and it is a direct measure of the energy used by the system. Because a living system is an open system, it is more convenient to speak of its *metabolic rate*, that is, its energy usage per unit time. Part of the metabolic energy can appear as physical work done by the system; part of it can appear as an increase in total system internal energy (as in the case of growth); part of it can appear in the creation of high-energy items, such as eggs, seeds, live offspring, and milk; and virtually all of the irreversibilities associated with these processes appear as heat production within the system.

An open system energy rate balance for the life form shown in Figure 17.5 is

$$\underbrace{\dot{Q}}_{\substack{\text{Metabolic}\\\text{heat transfer}\\(<0)}} - \underbrace{\dot{W}}_{\substack{\text{Work done}\\\text{on or by}\\\text{the system}}} + \underbrace{\sum_{\text{in}} \dot{m}e - \sum_{\text{out}} \dot{m}e}_{\substack{\text{Food, oxygen}\\\text{and}\\\text{waste material}}} = \underbrace{\frac{dU}{dt} + \frac{d}{dt}\left(\frac{mV^2}{2g_c}\right) + \frac{d}{dt}\left(\frac{mgZ}{2g_c}\right)}_{\text{System changes}} \tag{17.11}$$

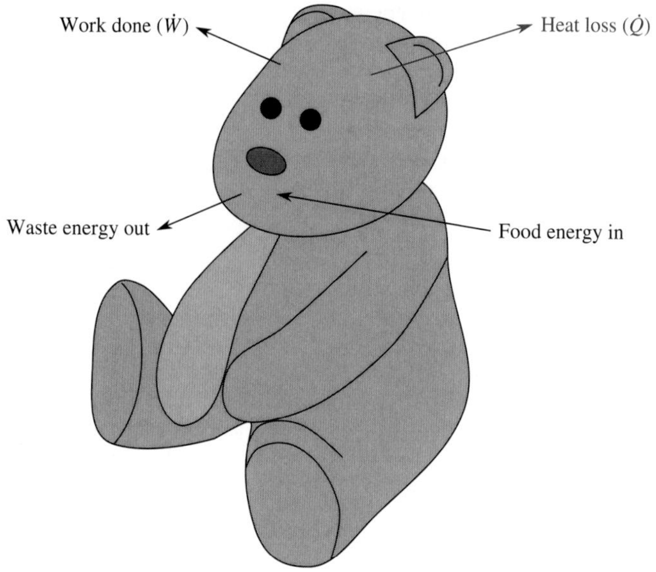

FIGURE 17.5
Energy flows in living systems.

Life processes all have some degree of irreversibility. Therefore, $\dot{Q}$ normally is negative since the internal irreversibilities generally produce internal heat generation, which must be removed from the system if the system is not to overheat.

Classically, the concept of work in thermodynamic analysis has been somewhat ambiguous. As discussed in Chapter 4, during the development of thermodynamics, it was convenient to separate the changes in kinetic and potential energies from the work term. These energy terms are written separately and usually grouped with the system's total internal energy change, as shown in Eq. (17.11). Thus, the work term in the thermodynamic energy balance encompasses all the work transport of energy into or out of a system *except* the work associated with changes in the system's kinetic and potential energy. This can be quite confusing when analyzing biological systems, since one of their major work modes in a social or cultural context is that of mobility, that is, running and climbing, which are the kinetic and potential energy terms we are discussing. Also, whereas a classical thermodynamic system can either do work or have work done on it, in general, a biological system only does work (i.e., the work term is always negative).

As with nonliving work-producing systems, we can define an energy conversion efficiency as

$$\text{Energy conversion efficiency} = \eta_E = \frac{\text{Desired energy result}}{\text{Required energy input}} \tag{17.12}$$

The term *energy* used in this equation must include relevant kinetic and potential energy changes. For example, the energy conversion efficiency of a human climbing a hill could be calculated by choosing the change in potential energy of the person as the desired energy output, while ignoring other types of energy output simultaneously performed (such as aerodynamic drag against the atmosphere). This is acceptable, providing the meaning of the efficiency is clearly defined in each case.

The required energy input part of Eq. (17.12) is more difficult to evaluate. Since, for warm-blooded animals, the net $\dot{Q}$ is always out of the system, it cannot be considered as a source of energy input. Also, one cannot generally input useful energy into a biological system via changes in the system's kinetic or potential energies. Thus, what remains is

$$\text{Required energy input} = -\frac{dU}{dt}$$

Then, we may write Eq. (17.12) as

$$\eta_E = \frac{\dot{W} + \frac{d}{dt}\left(\frac{mV^2}{2g_c}\right) + \frac{d}{dt}\left(\frac{mgZ}{g_c}\right)}{-dU/dt} = 1 + \frac{\dot{Q}}{-dU/dt} \tag{17.13}$$

and since both $\dot{Q}$ and dU/dt are always negative, it is clear that Eq. (17.13) gives an energy conversion efficiency that is always less than 100%.

On the other hand, in the case of most plants and some animals, there is either direct energy conversion of incoming solar radiation or a metabolic reduction resulting from direct body warming from incoming solar radiation. Since radiation is one of the classical heat transfer mechanisms, solar radiation belongs to the $\dot{Q}$ term. In this case, part of the system's $\dot{Q}$ is actually incoming and used within the system and must be considered as part of the total energy input.

The energy conversion efficiency of plant photosynthesis can be defined as

$$(\eta_E)_{\text{photosynthesis}} = \frac{\text{Energy converted to organic molecules by photosynthesis (per unit area)}}{\text{Solar energy input to earth (per unit area)}} \tag{17.14}$$

This is quite low, typically ranging between 0.01 and 1.0%. Part of the reason for the low efficiency is that not all the solar energy incident on a unit area of the Earth is intercepted by a plant. As plants become smaller and more uniformly cover the Earth, this efficiency rises somewhat. For example, in the case of algae (a microscopic one-celled plant), the photosynthetic energy conversion efficiency at a small densely packed test site can be as high as 10%.

The energy conversion efficiency of animals can be defined as

$$(\eta_E)_{\text{animal}} = \frac{\text{Rate of food energy stored in the body as complex organic molecules}}{\text{Rate of energy taken into the body as food}} \tag{17.15}$$

EXAMPLE 17.2

Everyone has heard about the food chain, but few realize how inefficient it is in nature. The energy conversion efficiency from sunlight to plant growth is only about 1.00%, the energy conversion efficiency of the plants eaten by grazing herbivores is about 20.0%, and the energy conversion efficiency of the carnivores who hunt and eat the herbivores is only about 5.00%. So the overall energy conversion efficiency from sunlight to carnivore is about $(0.0100)(0.200)(0.0500) = 1.00 \times 10^{-4} = 0.0100\%$. If the average daily solar energy reaching the surface of the Earth is 15.3 MJ/d · m^2, then how much land is required to grow the plants needed to feed the herbivores eaten by a large carnivore that requires 10.0 MJ/d to stay alive?

Solution

Since our hunting carnivore requires 10.0 MJ of food per day, at a 5.00% food energy conversion rate, it must consume

$$\frac{10.0\ \text{MJ/d}}{0.0500} = 200.\ \text{MJ/d}$$

of herbivore meat. The food energy conversion rate of the grazing herbivores is 20.0%, so they must consume

$$\frac{200.\ \text{MJ/d}}{0.200} = 1000\ \text{MJ/d}$$

in plant food. At a 1.0% energy conversion rate, the plants consumed by the herbivore require

$$\frac{1000\ \text{MJ/d}}{0.0100} = 1.00 \times 10^5\ \text{MJ/d}$$

of solar energy. Since the average solar energy intensity on the surface of the Earth is 15.3 MJ/d · m^2, 100,000 MJ/d of solar energy require an area of

$$\frac{100{,}000\ \text{MJ/d}}{15.3\ \text{MJ/d}\cdot\text{m}^2} = 6540\ \text{m}^2$$

and since 1 acre = 4047 m^2, then

$$6540\ \text{m}^2\left(\frac{1\ \text{acre}}{4047\ \text{m}^2}\right) = 1.62\ \text{acres}$$

of plant food is required to supply the food chain energy required to meet the 10 MJ/d needs on our carnivore.

Exercises

4. Using the results of Example 17.2, determine the number of carnivores that can be supported by herbivores living off 1500 acres of plants. **Answer**: 926 carnivores.
5. If the number of available herbivores in Example 17.2 increases dramatically and the carnivores' hunting energy expenditure is reduced to the point where their food energy conversion efficiency increases from 5.00% to 12.0%, determine the amount of land required to support one carnivore. **Answer**: 0.675 acres/carnivore.
6. If the carnivore in Example 17.2 moves to a tropical climate where the solar intensity and the photosynthetic energy conversion efficiency double, how much land would be required to support its food chain? **Answer**: 0.81 acres.

CRITICAL THINKING

The energy conversion efficiencies for plants and animals defined in Eqs. (17.14) and (17.15) are what we define in Chapter 10 as a *first law* efficiency. Can you use the general definition of a *second law* efficiency given in Chapter 10 to formulate a second law efficiency for plants and animals? What difficulties are encountered in evaluating this new efficiency?

Our conceptual understanding of physiological work is often quite different from our earlier (Chapter 4) definition of thermodynamic work. For example, when an animal walks along a horizontal surface it does *no* net thermodynamic work. There is no net change in kinetic or potential energy, and there is no appreciable sliding friction between the animal's feet and the ground. Only when the animal moves against an external force (such as hydrodynamic drag or inertia forces) is any classical thermodynamic work done. Walking does involve what we culturally call work, but in thermodynamic jargon the energy associated with constant velocity motion along a horizontal plane (in the absence of hydrodynamic drag) merely involves a net conversion of internal energy into heat. Thus, the thermodynamic efficiency of this type of motion in animals (or machines) is zero. If, instead, an animal walks on a horizontal treadmill, then it does do thermodynamic work. This work appears as friction or electricity, depending in the treadmill design. Part of the friction in this case is external to the animal and is measurable as work in the classical thermodynamic sense. The remaining part of the energy expenditure is internal losses within the animal and appears as metabolic heat. Note, however, that the thermodynamic work efficiency of walking on a treadmill returns to zero if the entire treadmill apparatus is included in the system with the animal.

A key element in understanding the thermodynamics of biological systems is comprehending the role of the heat transfer term in the energy rate balance equation of these systems. Since this equation by itself is useful only if you have just one unknown term, since it is not usually satisfactory to simply ignore or set equal to zero those terms for which we do not have values, and since $(dU/dt)_{system}$ is perhaps the most difficult term of all to measure accurately, then it becomes absolutely necessary that a means be found to give accurate measurements of $\dot{Q}$.

17.5 METABOLISM

The metabolic energy in the resting state is called the *basal metabolic rate* (BMR). The BMR is essentially the energy required to keep the molecular machinery of life operating at a zero activity level. Similar measurements at a higher activity level produce *intermediary metabolic rate* results. The basal metabolic rate for humans depends on age, sex, height, general health conditions, and the like. Figure 17.6 shows the variation in the average BMR per unit body surface area for human males and females as a function of age. It is not uncommon to have BMR variations around these normal (or average) values of ±15% for any one individual. Table 17.3 shows the breakdown in energy consumption comprising the BMR in the adult human body. The large energy consumption of the brain is surprising; the brain of a five-year-old child may account for up to 50% of its BMR.

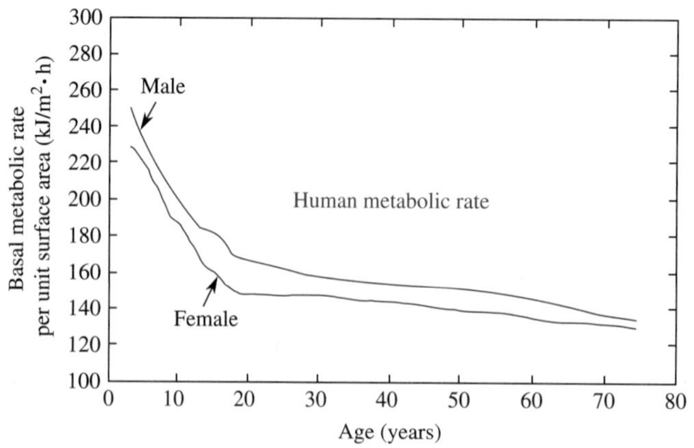

FIGURE 17.6
Average BMR per unit area for humans vs. age.

Table 17.3 Breakdown of the Contributions to the Basal Metabolic Rate of the Various Organs of the Adult Human Body

Organ	Mass (kg)	% of Body Mass	% of BMR
Liver	1.5	2.14	27
Brain	1.4	2.00	20
Heart	0.3	0.43	10
Kidneys	0.3	0.43	8
Muscles	30.0	42.8	26
Remaining body tissue	36.5	52.2	9
Total	70.0	100.0	100.0

Source: Reprinted by permission of the publisher and the author from Margen, S. Energy metabolism. In: McCally, M. (Ed.), Hypodynamics and Hypogravics, 1968 ed. Academic Press, New York.

CRITICAL THINKING

The average basal metabolic rates per unit surface area for male and female humans are shown in Figure 17.6. Why do you think the values for females are less than those for males over their life spans? Also, why do these curves level off at about age 20?

Measuring an animal's metabolic heat transfer directly is called *direct calorimetry*. The technique is very difficult to carry out because the animal's conductive, convective, and radiation heat transport rates must all be measured directly. This is commonly done by putting the animal in a closed box that has water circulating through all six of its sides. If the outside of the box is well insulated, then an energy balance shows that all of the metabolic heat produced by the animal ends up in the circulating water. However, virtually all metabolic measurements done today use a method called *indirect calorimetry*, wherein the CO_2 production and the O_2 consumption are measured instead. Generally, indirect calorimetric techniques are found to be as accurate as the direct techniques and are usually considerably easier and less expensive to use.

The ratio of the number of moles of CO_2 produced to the number of moles of O_2 consumed during an indirect calorimetry test is called the *respiratory quotient* (RQ), and its value depends on the type of food being metabolized. For example, in the metabolism of 1 mole of a typical carbohydrate, glucose,

$$C_6H_{12}O_6 + 6(O_2) \rightarrow 6(CO_2) + 6(H_2O)$$

6 moles of O_2 and 6 moles of CO_2 are involved. Therefore, the RQ of carbohydrate is 1.0. On the other hand, the RQ of protein is 0.8 and that of fat is 0.7. An animal generally consumes a mixture of these substances, so how do we know which value to use as the energy equivalent per liter of O_2 consumed? Tests show that, under basal conditions, the RQ is approximately 0.82 (which is very nearly the average value for the RQs of carbohydrate, protein, and fat), and it can be shown that this gives a mixture composition of these three substances that corresponds to an energy equivalent of 20.2 MJ/m^3 of O_2, or 20.2 kJ/L of O_2. Thus, if we measure the number of liters of O_2 consumed per unit time by an animal in the resting state and multiply this value by 20.2 kJ/L O_2, we obtain the resting energy consumption rate (or basal metabolic rate) of the animal. In the case of a fasting (starving) animal, which is living on the consumption of its own body fat and protein, the energy equivalent per liter of O_2 consumed is 21.3 kJ/L O_2.

WHAT IS KLEIBER'S LAW?

In 1932, Max Kleiber (1893–1976) published a paper, "Body Size and Metabolism," which included a graph (Figure 17.7) that showed that an animal's metabolic rate scales to the three-quarter power of the animal's mass, or $BMR \propto M^{3/4}$. Kleiber's law has been found to hold across 18 orders of size, from microbes to whales.

(*Continued*)

WHAT IS KLEIBER'S LAW? *Continued*

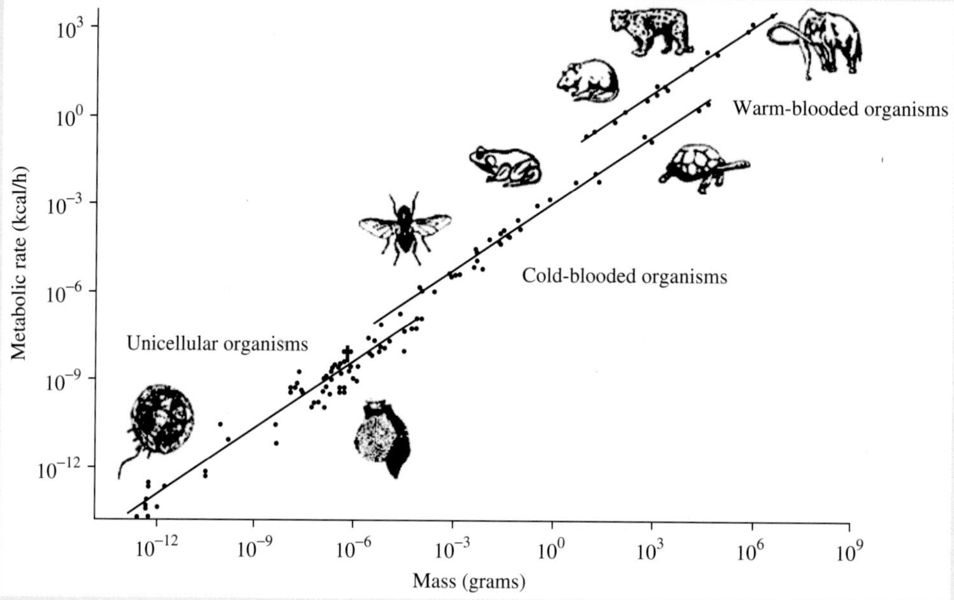

FIGURE 17.7
Kleiber's law graph.

WHAT DO A BANANA, AN ORANGE, AND A PERSON HAVE IN COMMON?

According to a new study, all living organisms share roughly the same resting metabolic rate when body size and temperature are taken into account. The finding suggests that widely diverse species burn energy in predictable patterns. "The [corrected] basal metabolic rate of an apple or tree is remarkably similar to that of bacteria, which is remarkably similar to a fish or person," says James Gillooly, at the University of New Mexico in Albuquerque.

A comparison of the basal metabolic rates for a large number of warm-blooded animals from mice to elephants and birds produced the following empirical correlation, called *Kleiber's law*:

$$\text{BMR} = 293\left(m^{3/4}\right) \tag{17.16}$$

where BMR is the animal's basal metabolic rate in kJ/d, and m is the animal's body mass in kg. Thus, the basal metabolic rate per unit mass of the animal is

$$\text{BMR}/m = 293(m^{-1/4}) = \frac{293}{m^{1/4}} \tag{17.17}$$

and it clearly increases with decreasing body mass.

EXAMPLE 17.3

Determine the basal metabolic rate (BMR) per unit mass of an 80.0 kg adult human and an 8.00 gram mouse. Since both are warm-blooded mammals, explain why there is a difference in these values.

Solution

The basal metabolic rate per unit mass of a warm-blooded animal is given by Eq. (17.17) as

$$\text{BMR}/m = 293(m^{-0.25})$$

where BMR is in kJ/d and m is in kg. Then, the BMR per unit mass of an 80.0 kg human is

$$(\text{BMR}/m)_{\text{human}} = 293(80.0^{-0.25}) = 98.0\ \frac{\text{kJ}}{\text{kg}\cdot\text{d}}$$

and the BMR per unit mass of an 8.00 gram mouse is

$$(\text{BMR}/m)_{\text{mouse}} = 293(0.00800^{-0.25}) = 980.\ \frac{\text{kJ}}{\text{kg}\cdot\text{d}}$$

This calculation shows that the basal metabolic rate per unit mass of a mouse is ten times that of a human. This large difference is primarily due to the difference in surface area to volume ratio of these mammals. Since heat loss from the body is primarily by convection heat transfer, which is proportional to surface area, and internal heat generation inside the body is proportional to its volume, then as the ratio of surface area to volume increases the internal heat generation rate must also increase if a mammal is to maintain its body temperature. To produce higher internal heat generation rates, small animals must feed very often if they are not to starve.

It is a fact that the smaller any object becomes, the larger its surface area to volume ratio becomes. This is easiest to understand with spherical objects. The surface area of a sphere is $4\pi R^2$ whereas its volume is $(4/3)\pi R^3$. Therefore, its surface area to volume ratio is

$$\left(\frac{\text{Surface area}}{\text{Volume}}\right)_{\text{sphere}} = \frac{4\pi R^2}{\frac{4}{3}\pi R^3} = \frac{3}{R}$$

and this ratio decreases inversely with increasing R. Thus, there is a lower limit to the size of warm-blooded animals. The shrew and the hummingbird are the smallest known animals of this kind.

The body temperature of insects and cold-blooded animals is approximately equal to the temperature of their surroundings. Consequently, there is no thermodynamic lower limit to their size.

Exercises

7. Determine the basal metabolic rate (BMR) and the basal metabolic rate per unit mass (BMR/m) of a 1300. kg elephant. **Answers**: $\text{BMR}_{\text{elephant}} = 63{,}400$ kJ/d and $(\text{BMR}/m)_{\text{elephant}} = 48.8$ kJ/kg · d.
8. If a 0.500 gram house fly had to maintain the body temperature of warm-blooded mammal using the same internal heat generation rate mechanisms, what would be its basal metabolic rate (BMR) and its basal metabolic rate per unit mass (BMR/m)? **Answers**: $\text{BMR}_{\text{fly}} = 0.980$ kJ/d and $(\text{BMR}/m)_{\text{fly}} = 1960$ kJ/kg · d.
9. Since whales are aquatic mammals, determine the basal metabolic rate (BMR) and basal metabolic rate per unit mass of a 136,000 kg (150. ton) great blue whale. **Answer**: $\text{BMR}_{\text{whale}} = 2{,}080{,}000$ kJ/d and $(\text{BMR}/m)_{\text{whale}} = 15.3$ kJ/kg · d.

HOW DOES TEMPERATURE AFFECT METABOLISM?

Temperature governs metabolism through its effects on rates of biochemical reactions. Reaction kinetics vary with temperature according to Boltzmann's factor $e^{-E/kT}$, where T is the absolute temperature, E is the activation energy, and k is Boltzmann's constant. The combined effects of body mass (M) and body temperature (T) on the basal metabolic rate can then be written as $\text{BMR} \propto M^{3/4} e^{-E/kT}$, where E is the average activation energy for the enzyme-catalyzed biochemical reactions of metabolism.

17.6 THERMODYNAMICS OF NUTRITION AND EXERCISE

The molecular form of the food we eat can be broken down into the following three categories:

1. **Carbohydrates**. Carbohydrates always contain hydrogen and oxygen atoms in a 2 to 1 ratio, as in water; and they can have very large macromolecules built up from the glucose ($C_6H_{12}O_6$) monomer, with molecular masses as high as 2×10^6 (as in the case of plant starch and glycogen).
2. **Proteins**. Proteins are very large molecules containing carbon, hydrogen, oxygen, and often nitrogen. For example, a single molecule of human hemoglobin ($C_{3032}H_{4816}O_{872}N_{780}S_8Fe_4$) contains a total of 9512 atoms and has a molecular mass of 66,552 kg/kgmole.
3. **Fats (glycerol and fatty acids)**. Fatty acids are much smaller molecules, with typically 16 or 18 carbon atoms per molecule plus attached hydrogen atoms and a carboxyl group (—COOH) at one end. An example of a *saturated* (with hydrogen atoms) fatty acid is shown in Figure 17.8. An example of the same acid *unsaturated* is shown in Figure 17.9.

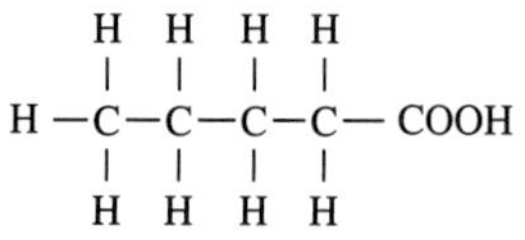

FIGURE 17.8
A saturated fatty acid hydrocarbon chain.

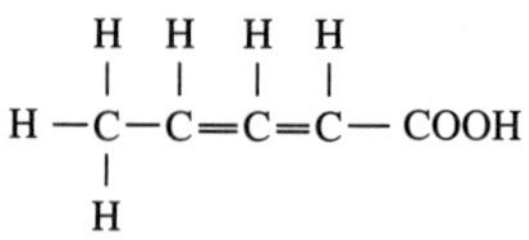

FIGURE 17.9
An unsaturated fatty acid hydrocarbon chain.

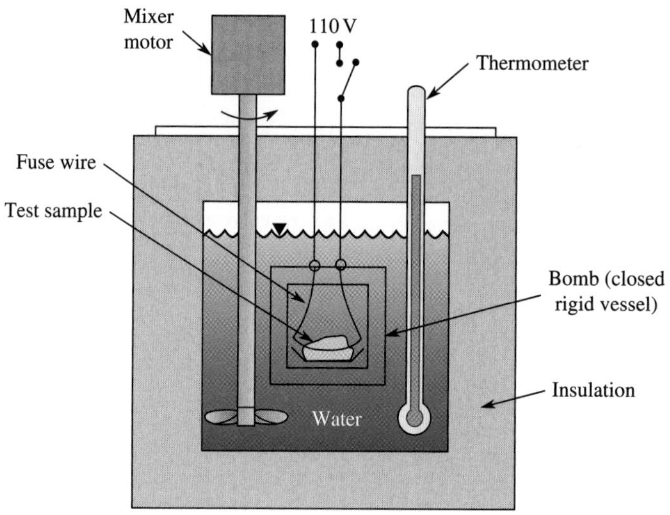

FIGURE 17.10
A schematic of a typical bomb calorimeter.

The energy value of different foods is normally determined by direct calorimetry in a device called a *bomb calorimeter* (see Figure 17.10). In this device, a sample of known mass is ignited in a pressurized atmosphere of excess pure oxygen. The liberated heat of combustion is transferred to water surrounding the combustion chamber, and it can easily be calculated from an energy balance on the calorimeter. The end product of this type of combustion is always CO_2 and H_2O (and nitrogen products when the sample contains bound nitrogen). Since this is exactly the same end state that occurs in the body as a result of enzyme decomposition of food molecules, the same amount of energy must be released in each case. Thus, bomb calorimeter energy measurements represent the total energy available in the sample that can be converted into heat or another form of energy.

Bomb calorimeter studies on dry (water-free) foods give the following averaged results for the specific energies of the basic food components:

$$\left.\begin{array}{ll}\text{Carbohydrate:} & 18.0\ \text{MJ/kg}\\ \text{Protein:} & 22.2\ \text{MJ/kg}\\ \text{Fat:} & 39.8\ \text{MJ/kg}\end{array}\right\}\text{Total energy content (water free)}$$

When these same substances are metabolized in the human body they produce the following specific energy releases:

$$\left.\begin{array}{ll}\text{Carbohydrate:} & 17.2\ \text{MJ/kg}\\ \text{Protein:} & 17.2\ \text{MJ/kg}\\ \text{Fat:} & 38.9\ \text{MJ/kg}\end{array}\right\}\text{Metabolizable energy content (water free)}$$

Using these two sets of values we can compute the food energy conversion efficiency of the human body as

$$\eta_{\text{carbohydrate}} = \frac{17.2}{18.0} \times 100 = 95.5\%$$

$$\eta_{\text{protein}} = \frac{17.2}{22.2} \times 100 = 77.5\%$$

$$\eta_{\text{fat}} = \frac{38.9}{39.8} \times 100 = 97.7\%$$

Thus, 22.5% of the energy in the protein we eat passes through the body unused. The low protein energy conversion efficiency supports the theory that humans were not always meat-eating animals.

These energy content values were for dry or water-free foods. Most foods, especially carbohydrates, contain a large amount of functional water (the mass of the human body, for example, is about 72% water). The energy content of natural or *wet* food is lower that that of dry food due to the dilution effect of the energetically inert water. The average natural state metabolizable specific energy content of the three basic food components is

$$\left.\begin{array}{ll}\text{Carbohydrate:} & 4.20\ \text{MJ/kg}\\ \text{Protein:} & 8.40\ \text{MJ/kg}\\ \text{Fat:} & 33.1\ \text{MJ/kg}\end{array}\right\}\text{Metabolizable energy content natural state foods}$$

Note the extraordinarily large specific energy content of natural state fat.

EXAMPLE 17.4

If the average meal consumed by an adult consists of about 45.0% carbohydrate, 15.0% protein, and 40.0% fat, determine (a) the specific energy content of an average meal with natural state foods and (b) the total mass of an average meal needed to provide 10.5 MJ per day.

Solution

a. The average energy content of natural (or wet) food components is

Carbohydrate:	4.20 MJ/kg
Protein:	8.40 MJ/kg
Fat:	33.1 MJ/kg

Therefore, the specific energy content of the average meal is

$$e_{\text{avg. meal}} = 0.450(4.20) + 0.150(8.40) + 0.400(33.1) = 16.4\text{MJ/kg meal}$$

b. A person who requires a daily food energy intake of 10.5 MJ must then consume

$$\dot{m}_{\text{avg. meal}} = \frac{10.5}{16.4} = 0.640\ \text{kg of average meal/day} = 1.4\ \text{lbm of average meal/day}$$

Exercises

10. Suppose the person in Example 17.4 who needed 10.5 MJ of food energy per day goes on a diet that requires a food energy intake of only 5.0 MJ per day. What total mass of average meal food should this person consume per day? **Answer**: $\dot{m}_{\text{avg. meal}} = 0.300\ \text{kg avg. meal/d} = 0.670.$
11. If the amount of fat described in the average meal in Example 17.4 is reduced from 40.0% to 30.0% and the carbohydrate and protein increase to 50.0% and 20.0%, respectively, determine the new specific energy content of this meal and the total mass of this meal required to produce 10.5 MJ of food energy per day. **Answer**: $\dot{m}_{\text{avg. meal}} = 0.770\ \text{kg avg. meal/d} = 1.69.$
12. People who live in very cold climates usually have diets that have a very high fat content. Suppose their average meal consisted of 20% carbohydrate, 20% protein, and 60% fat. Determine the specific energy content of this meal and the total mass of this meal required to produce 10.5 MJ of food energy per day. **Answer**: $\dot{m}_{\text{avg. meal}} = 0.470\ \text{kg avg. meal/d} = 1.03\ \text{lbm avg. meal/d}.$

Overweight conditions place a greatly increased load on the heart and other organs. For example, each kilogram of body tissue contains 0.885 km of tiny blood vessels. If an individual is 10. kg (22 lbm) overweight, the heart must pump blood through an extra 8.9 km (5.5 miles) of small blood vessels.

EXAMPLE 17.5

People living in affluent societies generally know very little about starvation. Most feel that death is imminent if food is withheld for only a week. However, we know that 1.00 kg of human body fat contains about 33.1 MJ of metabolizable energy, and it would be useful to know:

a. The mass of body fat consumed per day if one uses 10.5 MJ of energy in normal activities.
b. How many days of total fasting are required to lose 10.0 kg of body fat.

Solution

a. A fasting person requiring 10.5 MJ of metabolizable energy per day consumes about

$$\dot{m}_{\text{fat}} = \frac{10.5\ \text{MJ/d}}{33.1\ \text{MJ/kg body fat}} = 0.317\ \text{kg of body fat/d}$$

(Continued)

EXAMPLE 17.5 *(Continued)*

b. The number of fasting days required to lose (consume) 10.0 kg of body fat is

$$t = \frac{m_{fat}}{\dot{m}_{fat}} = \frac{10.0\,\text{kg of body fat}}{0.317\,\text{kg of body fat/d}} = 31.3\,\text{d}$$

Exercises

13. If the person in Example 17.5 requires only 8.40 MJ per day instead of 10.5 MJ per day, how much fat would be lost (consumed) by fasting just one day? **Answer**: $m_{fat} = 0.250$ kg fat.

14. You are stranded in the wilderness without any food (but you have plenty of water to drink). If you have 15.0 kg of excess body fat, how long can you wait to be rescued before your body fat is consumed by your metabolism of 9.00 MJ/d? **Answer**: $t = 55.2$ d.

15. Suppose you have absolutely no excess body fat, and you are stranded without food. Your body then begins to consume your muscle protein, which has a metabolizable value of 8.4 MJ/kg. How much protein would you lose (consume) after being stranded for 30.0 days with a metabolism of 10.5 MJ/da? **Answer**: $m_{protein} = 12.5$ kg protein.

Example 17.5 shows that, if you have 10. kg (220 lbm) of excess body fat, you can theoretically fast for 31.3 days just living on that body fat alone. This also gives you some idea why weight loss by dieting is such a slow process. Fasting for long periods is not a medically sound method of weight loss since the body soon begins to consume its own protein, and this can seriously affect the functioning of the body's organ systems (especially the heart). No one should ever willingly attempt a total fasting diet without consulting a qualified physician.

Whereas most adult humans can survive long periods of fasting, they cannot withstand long periods without water intake. Since the body continually loses water through the skin and lungs, it must be replaced or the body soon becomes dehydrated and death quickly follows. Healthy adults have been known to fast for over 100 days, but no human can survive for more than 10 to 20 days without water.

Tables 17.4 and 17.5 present the metabolizable energy content values for various common foods and the average energy expenditure requirements for various human exercises. Common nutritional tables today have food energy content and exercise energy expenditure levels listed in *Calories*. The capitalization of the word Calorie indicates what nutritionists call a *large calorie*, that is, a *kilocalorie*: 1 Calorie = 1000 calories = 1 kilocalorie. This is confusing notation, since only the capital C tells you that it is not the normal calorie energy unit, a subtle point often overlooked by the publishers of nutrition tables.

Table 17.4 Approximate Energy Content of Some Common Foods

	Metabolizable Energy Content		
Food	**Calories**	**MJ**	**Btu**
Fast foods (average values)			
Hamburger	275	1.15	1090
Cheeseburger	325	1.36	1490
Quarter pound hamburger	450	1.88	1790
With cheese	550	2.30	2180
With cheese and bacon	650	2.72	2580
Fish sandwich	450	1.88	1790
With cheese	500	2.09	1980
Hot dog	300	1.26	1190
With chili or cheese	350	1.47	1390
Regular fries	250	1.05	992
Regular onion rings	350	1.47	1390
Baked potato	250	1.05	992
With sour cream and chives	450	1.88	1790
With chili and cheese	500	2.09	1980
With broccoli and cheese	500	2.09	1980
With bacon and cheese	550	2.30	2180
With cheese	550	2.30	2180

Table 17.4 Approximate Energy Content of Some Common Foods *continued*

Food	Metabolizable Energy Content		
	Calories	MJ	Btu
Pizza (per slice, 8 slices per 13-in. pizza)			
With cheese	350	1.47	1390
With cheese and pepperoni	500	2.09	1980
Salads (1 cup each)			
Lettuce with French dressing	150	0.63	595
Potato with mayonnaise	375	1.57	1490
Chicken and mayonnaise	550	2.30	2180
Egg and mayonnaise	400	1.67	1590
Tuna fish and mayonnaise	500	2.09	1980
Drinks			
Shakes (all flavors, 10 fluid ounces)	350	1.47	1390
Milk (skim, per pint)	180	0.75	714
Cola (all flavors, 10 fluid ounces)	130	0.54	516
Diet cola	0	0.0	0
Beer (per fluid ounce)	8	0.033	30
Whiskey (per fluid ounce)	38	0.16	150
Desserts			
Ice cream (per pint, 10% fat)	600	2.51	2380
Pie (per slice, 8 slices per 9-in. pie)	300	1.26	1190
Chocolate candy (milk, per ounce)	150	0.63	595
Marshmallows (1 large)	25	0.10	99

Source: Jacobsen, M., Fritschner, S. The Fast-Food Guide. Workman Publishing, 1992, New York. Copyright © The Center for Science in the Public Interest. Reprinted by permission of Workman Publishing.

Table 17.5 Approximate Adult Human Energy Expenditure in Exercise

Exercise	Energy Required during Exercise		
	Calories/h	MJ/h	Btu/h
Fast running	910	3.8	3610
Cross-country skiing	910	3.8	3610
Fast swimming	860	3.6	3410
Wrestling	810	3.4	3210
Boxing	690	2.9	2740
Hard cycling	600	2.5	2380
Jogging	600	2.5	2380
Football	600	2.5	2380
Fast dancing	600	2.5	2380
Basketball	550	2.3	2180
Handball	550	2.3	2180
Sawing wood	500	2.1	1980
Shoveling	500	2.1	1980
Tennis	480	2.0	1900
Climbing stairs normally	410	1.7	1630
Baseball	360	1.5	1430
Volleyball	360	1.5	1430
Fast walking	310	1.3	1230
Sexual intercourse	270	1.1	1070
Golf	240	1.0	952
Hoeing	190	0.8	754
Driving a car	140	0.6	556
Card playing	96	0.4	381
Watching TV	72	0.3	286
Basal metabolism	72	0.3	286

IS IT A Calorie OR A calorie?

When the word *calorie* is capitalized, it indicates what nutritionalists call a *large calorie,* or a kilocalorie. That is, in nutrition jargon, 1 Calorie = 1 kilocalorie, yet it takes 1000 calories to equal 1 kilocalorie. Only the capital C distinguishes the two. So, when a nutrition table indicates that your caloric intake should be 2500 Calories per day, it really means 2500 kilocalories per day. Now, since 1 kilocalorie = 4.186 kilojoules, then 2500 Calories/d = 2500 kcal/d = (2500 kcal/d)(4.186 kJ/kcal) = 10,465 kJ/d = 10.456 MJ/d.

From Table 17.5, we find that the energy expenditure required to jog, play football, or fast dance is about 600 Calories per hour, and Table 17.4 tells us that the energy content of milk chocolate is about 150 Calories per ounce. So, if you want to exercise off the energy content of one 1.5-oz milk chocolate candy bar you would have to jog, play football, or fast dance continuously for

$$\frac{(1.5\,\text{oz})(150\,\text{Calories/oz})}{600\,\text{Calories/h}} = 0.375\,\text{hours}$$

This is a lot of hard exercise for one small candy bar.

EXAMPLE 17.6

Suppose you want to exercise off the energy added to your body as a result of eating one pint of ice cream by lifting weights. The external work done by the body equals the change in potential energy of the weights as they are lifted (there is no significant energy recovery within the body, however, when the weights are lowered again). Suppose you are lifting 490. N (110. lbm) a vertical distance of 1.00 m, and you can make one lift in 1.00 second. Approximately how many lifts are required and how long will it take to work off the energy content of the ice cream?

Solution

Each lift requires that an amount of energy be put into the weights of

$$(mgZ/g_c)_{\text{weights}} = (490.\,\text{N})(1.00\,\text{m})/(1) = 490.\,\text{N}\cdot\text{m} = 490.\,\text{J}$$

If we take the human body as the thermodynamic system and apply the energy rate balance and ignore all mass flow energy movements into or out of the system during the exercise period (thus, we are ignoring perspiration energy losses and all O_2 and CO_2 exchanges), then we can write

$$\dot{Q} - \dot{W} = \left[\frac{dU}{dt} + \frac{d}{dt}\left(\frac{mV^2}{2g_c}\right) + \frac{d}{dt}\left(\frac{mgZ}{g_c}\right)\right]_{\text{body}}$$

Since the kinetic and potential energies of the human body do not change significantly during the exercise, we can set

$$\left[\frac{d}{dt}\left(\frac{mV^2}{2g_c}\right) + \frac{d}{dt}\left(\frac{mgZ}{g_c}\right)\right]_{\text{body}} = 0$$

and the energy rate balance becomes

$$\dot{Q} - \dot{W} = \left(\frac{dU}{dt}\right)_{\text{body}} = \dot{U}_{\text{body}}$$

Now, the external work rate that must be done by the system is

$$\dot{W} = -\frac{(mgZ/g_c)_{\text{weights}}}{\Delta t} = \frac{490.\,\text{J}}{1.00s} = 490.\,\text{J/s}$$

It has been shown experimentally that the energy conversion efficiency of animal muscular contraction defined by Eq. (17.12) is about 25%, or

$$(\eta_T)_{\text{muscle}} = \frac{\dot{W}}{\dot{U}_{\text{body}}} = 0.250$$

Then, the rate of total internal energy expenditure within the body is

$$\dot{U}_{\text{body}} = \frac{-\dot{W}}{(\eta_T)_{\text{muscle}}} = \frac{-490\ \text{J/s}}{0.250} = -1960\ \text{J/s}$$

Therefore, $\dot{Q} = \dot{U} + \dot{W} = -1960 + 490 = -1470\ \text{J/s}$. Consequently, the time, τ, required to produce a change in the total internal energy of the system that equals the energy content of one pint of ice cream (see Table 17.4) is

$$\tau = \left(\frac{\Delta U}{\dot{U}}\right)_{\text{body}} = \frac{-(1\ \text{pint})(2.51\ \text{MJ/pint})}{-1.96\times10^{-3}\ \text{MJ/s}} = 1280\ \text{s} = 21.3\ \text{min}$$

Hence, the 490. N weight in this example must be lifted continuously at a rate of one lift per second until a total of 1280 lifts have been made. This is clearly a great deal of physical labor just to overcome the enjoyment of a pint of ice cream. Note that only 25% of the energy in the ice cream gets converted into external work while 75% of its energy is utilized elsewhere within the body to keep the circulatory, respiratory, and other subsystems operating and is ultimately converted into heat inside the body due to the internal irreversibilities of these processes.

Exercises

16. If the muscle efficiency in Example 17.6 is 30.0% instead of 25.0%, how many lifts would be required to work off the energy content of the ice cream? Assume all the other variables remain unchanged. **Answer**: 1540 lifts.
17. If the 490 N weight in Example 17.6 is lifted only 0.50 m instead of 1.00 m, how long would it take to work off the energy content of the ice cream? Assume all the other variables remain unchanged. **Answer**: τ = 2560 s = 42.7 min.
18. Suppose the person in Example 17.6 consumes a cheeseburger instead of a pint of ice cream. Assuming all the other variables remain unchanged, how long would it take to work of the energy content of the cheeseburger? **Answer**: τ = 694 s = 11.6 min.

Physiologically, it is very hard to lose weight by exercising alone. Most of the weight loss that appears after exercising is really water loss due to perspiration. Perspiration is a convection-evaporation heat transfer mechanism that removes the heat generated within the body due to the biological irreversibilities of exercise. Its function is to help maintain a constant body temperature. This type of water loss is quickly replaced in the meals following the exercise and should never be considered as part of a permanent weight loss.

17.7 LIMITS TO BIOLOGICAL GROWTH

For purposes of simplification, consider living systems to have a characteristic length L such that their surface and cross-sectional areas are proportional to L^2 and their volumes are proportional to L^3. The most obvious effect of size on animal evolution is the ability of an animal's skeleton to support its body weight. The ability of a leg bone to withstand direct compression loading is proportional to its yield modulus and to the cross-sectional area of the bone. Hence, the strength of a leg varies with L^2. However, the body weight of the animal is proportional to its volume, which varies with L^3. The ratio of body weight to leg loading then increases with the animal's size, L. Clearly there exists an upper limit (dictated by the elastic properties of bone) to an animal's growth, where its legs can no longer support its weight. The giant dinosaurs of 100 million years ago apparently evolved up to this critical size. Some aquatic dinosaurs were too large to leave the water because without the buoyant supporting force of the water their skeletons could not support their body weight.

Even more crucial to mobile land animals are the bending stresses developed in their bones during walking and running. Small animals can run with very nimble and flexible legs while heavy animals like elephants must walk stiff legged to minimize leg bone bending stresses.

The internal heat generated by biochemical irreversibilities in animals is proportional to the amount of tissue present, and consequently, it varies with L^3. The rate of heat loss by an animal depends on the convective and radiative heat transfer mechanisms, which in turn depend directly on the animal's surface area and, consequently, vary with L^2. Therefore, the ratio of heat generation to heat loss is proportional to $L^3/L^2 = L$, and if an animal's size were to increase indefinitely, a point would eventually be reached where the animal would overheat and die. Thus, at least two mechanisms provide an upper limit to the size of animals: the strength of their supporting tissue and their ability to maintain a moderate body temperature.

The rate at which oxygen and food reach the body's cells depends on the volume of blood in the circulatory system and the pumping capacity of the heart. The volume of blood delivered to the heart is proportional to the cross-sectional area of the aorta and, consequently, varies with L^2, whereas the volume of the heart

WHAT ABOUT BIRDS?

The pulse to breathing rate ratio is about 9.0 for all birds (regardless of their size) because birds have a continuous flow of air in only one direction through their lungs, in contrast to the two-way in-out breathing of mammals. The unidirectional air flow in birds is also countercurrent (in the opposite direction to) the blood flow in the lungs, thus improving the efficiency of gas exchange.

itself is proportional to L^3. Therefore, the ratio of blood flow rate to heart volume varies with $L^2/L^3 = L^{-1}$, and consequently, the heart pulse rate also varies with L^{-1}. For mammals, the heart pulse rate has been correlated with body mass according to

$$\text{Heart pulse rate (in beats per minute)} = 241(m^{-0.25}) = \frac{241}{m^{1/4}} \quad (17.18)$$

where the body mass m is in kilograms.

The same argument can be made for the respiratory system. The ratio of the gas transport rate through the lung wall to the lung volume also varies with L^{-1}, and the breathing rate is also proportional to L^{-1}. Experimentally, we find that the ratio of pulse rate to breathing rate is constant at about 4.5 in all mammals regardless of their size. The breathing rate for mammals has been correlated with body mass as

$$\text{Breathing rate (in breaths per minute)} = 54.0(m^{-0.25}) = \frac{54.0}{m^{1/4}} \quad (17.19)$$

where the body mass m is in kilograms.

EXAMPLE 17.7

As a famous biomedical engineer, you are challenged on your medical exam to determine the heart rate and respiratory rate of a 0.0300 kg mouse, a 70.0 kg human, and a 4000. kg elephant.

Solution

The heartbeat rate for mammals in beats per minute is given by Eq. (17.18) as

$$\text{Heartbeat rate} = 241(m^{-0.25})\ \text{Beats/min}$$

so the heartbeat rates of the mouse, human, and elephant are

$$\begin{aligned}(\text{Heartbeat rate})_{\text{mouse}} &= 241(0.0300^{-0.25}) = 579.\ \text{Beats/min}\\ (\text{Heartbeat rate})_{\text{human}} &= 241(70.0^{-0.25}) = 83.3\ \text{Beats/min}\\ (\text{Heartbeat rate})_{\text{elephant}} &= 241(4000.^{-0.25}) = 30.3\ \text{Beats/min}\end{aligned}$$

The respiratory rate of mammals in breaths per minute is given by Eq. (17.19) as

$$\text{Respiratory (breathing) rate} = 54.0(m^{-0.25})\ \text{Breaths/min}$$

So the breathing rates of the mouse, human, and elephant are

$$\begin{aligned}(\text{Breathing rate})_{\text{mouse}} &= 54.0(0.0300^{-0.25}) = 130.\ \text{Breaths/min}\\ (\text{Breathing rate})_{\text{human}} &= 54.0(70.0^{-0.25}) = 18.7\ \text{Breaths/min}\\ (\text{Breathing rate})_{\text{elephant}} &= 54.0(4000.^{-0.25}) = 6.79\ \text{Breaths/min}\end{aligned}$$

Exercises

19. Determine the heatbeat rate of a 2.80 kg house cat. **Answer**: 186 beats/min.
20. Determine the breathing rate of a 700. kg racehorse. **Answer**: 10.5 breaths/min.
21. Since whales are mammals, determine the heart rate of a 136,000 kg (150. ton) blue whale. **Answer**: 12.6 beats/min.

Because plants lack mobility, their size criteria are generally simpler than those for animals. The main strength concerns in plants center on the buckling of their central trunk and excessive deflections of their cantilevered limbs. Consider a circular cylinder of height h and diameter d. Then, for slender cylinders ($h/d > 25$) the critical height for a cylinder buckling under its own weight can be shown to be

$$h_{\text{critical}} = 0.85\left(\frac{E}{\gamma}\right)^{1/3} d^{2/3} \quad (17.20)$$

where E is the elastic modulus of the trunk and γ is its weight density. It can also be shown that the tallest self-supporting homogeneous tapering conical column with base diameter d is about twice as tall as the critical height given by Eq. (17.20). For live wood, the ratio of $(E/\gamma)^{1/3}$ is approximately 120. $m^{1/3}$ for all trees. Thus, the critical height of trees varies approximately with their base diameter to the 2/3 power according to

$$h_{\text{critical}} = 68.0\left(d^{2/3}\right) \tag{17.21}$$

where h_{critical} and d are in meters and the coefficient 68.0 has units of $m^{1/3}$.

EXAMPLE 17.8

Determine the critical buckling height of a small tree whose base diameter is 5.00×10^{-3} m.

Solution

From Eq. (17.21), we have $h_{\text{critical}} = 68.0(d^{2/3}) = 68.0(5.00 \times 10^{-3})^{2/3} = 1.99$ m.

Exercises

22. Determine the critical buckling height of a tree in your yard that has a base diameter of 8.00 in (0.203 m). **Answer:** $h_{\text{critical}} = 23.5$ m (77.1 ft).
23. A wooden flag pole with a 6.00 in (0.152 m) base diameter is made from fresh wood. Determine its critical buckling height. **Answer:** $h_{\text{critical}} = 19.4$ m (63.6 ft).
24. Determine the critical buckling height of a giant redwood tree with a base diameter of 6.00 m (20.0 ft). **Answer:** $h_{\text{critical}} = 225$ m (738 ft). Note: Such trees seldom exceed a growth height of about 90 m (300 ft).

Tree limbs are sized to withstand the bending forces due to their own weight. If a branch is considered to be a cantilever beam attached at an angle α to the trunk, then there exists a critical length ℓ_{crit} that allows the tip of the branch to extend horizontally. Longer branches droop below the horizontal and shorter branches point upward at an angle approximately the same as their attachment angle α. It can be shown that the equation for ℓ_{crit} is identical in form to Eq. (17.20), except with a different multiplying constant (in this case the diameter d is the limb diameter at the point of attachment). Thus, the shape and size of trees and other plants is proportional to the 2/3 power of the base diameter of the limbs and trunk.

It can be shown that muscular power for animal locomotion is also proportional to the square of the characteristic body dimension. Therefore, the work (i.e., power × time) done by a muscle is proportional to $L^2 \times (L/V)$, or L^3, where V is the locomotion velocity. The kinetic energy of motion at constant velocity is also proportional to L^3 because the animal's mass is proportional to its volume. Since both the work done by the muscle and the system kinetic energy it produces are proportional to L^3 (if we ignore any aerodynamic drag and acceleration effects), we see that there can be no significant size effect in the horizontal locomotion of animals. That is, *all* animals should be able to run at about the same maximum velocity on a horizontal surface.

Consider now an animal running uphill at constant velocity. The rate of energy expenditure in increasing its potential energy (again, ignoring aerodynamic and other effects) is proportional to $L^3 \times (dZ/dt)$. Since its muscular power is always proportional to L^2, an energy rate balance on the animal tells us that its ascent velocity dZ/dt must therefore be proportional to L^{-1}. That is, the speed of an animal running uphill should be inversely proportional to its size. A hill that a rabbit can easily run up may reduce a dog to a trot and a hunter to a walk.

A similar argument can be made for large flying animals. It can be shown with an energy rate balance that the rate of energy expenditure required for hovering or forward flight is proportional to $L^{3.5}$. Since the flight muscles can supply power only proportional to L^2 (again ignoring aerodynamic drag and inertia), the ratio of required power to available power is proportional to $L^{1.5}$. Thus, an upper limit to the size of flying animals is quickly reached. In the case of birds, their aerodynamic design sets this upper limit at about 16 kg (35 lbm).

WHY YOU CANNOT CATCH A RABBIT

Though observations of animals from rabbits to horses show that they can all run about the same maximum speed on a horizontal surface, smaller animals can accelerate and decelerate (i.e., maneuver) much faster than larger animals (people).

17.8 LOCOMOTION TRANSPORT NUMBER

Air, water, and land constitute the three common transport media available on Earth. Accordingly, we assign the following locomotion mechanisms to these media: air, flying; water, swimming; and land, crawling and running. An effective way to study the energy consumption of locomotion is through the dimensionless locomotion transport number *T*, defined as

$$T = \frac{P}{wV} \tag{17.22}$$

where *P* is the animal's total rate of energy (power) expenditure during locomotion (often determined by measuring the rate of O_2 consumption during locomotion), *w* is the animal's weight (not mass), and *V* is its locomotion velocity. At zero velocity, $P = P_o$ (the BMR) and *T* becomes infinite. The most efficient transport velocity is the velocity for which *T* is a minimum. If we ignore aerodynamic drag and assume that *P* is independent of *V*, then $T \to 0$ as $V \to \infty$. The faster the animal moves, the more efficient is its locomotion. This, clearly, is unrealistic, since aerodynamic drag becomes important at even moderate speeds and inertia also becomes important because the animal increases its velocity by flexing its locomotion muscles (legs, wings, etc.) faster. Therefore *P* cannot be independent of *V*.

We can represent total power expenditure during locomotion, *P*, as

$$P = P_o + P_D + P_m$$

HOW EFFECTIVE IS WALKING?

When you walk, some of the kinetic energy of your forward motion is converted into potential energy as you rise on one foot, and a portion of this potential energy is converted back into kinetic energy again as you push forward and drop onto the other foot (Figure 17.11). The efficiency of slow walking is such that only about 65% of your body's kinetic energy is carried forward from one step to another. Internal friction and other irreversibilities in your body consume the remaining 35%. To keep moving, the lost kinetic energy must be replaced by flexing the muscles in your legs, and this energy ultimately comes from the food you eat.

When you get tired walking, you have used up your leg muscles' energy reserve and your body cannot produce it from the stored body fat as fast as it is needed by the muscles. Also, when you carry a load in your arms or on your back when you walk, more energy is lost through increased friction (though the percentage of body plus load kinetic energy lost is about the same). However, African women have somehow learned how to carry loads of up to one fifth of their body weight in baskets on their head without using any additional energy in slow walking.

Since 35% of the kinetic energy is $0.35(mV^2/2g_c) = 0.175(mV^2/g_c)$, and muscle efficiency is only about 25%, then the total energy consumed in slow walking is about $4(0.175)(mV^2/g_c) = 0.7(mV^2/g_c)$. For a 180. lbm person walking at 3.00 ft/s, the energy requirement per step in slow walking is

$$0.7(180.\text{lbm})(3.00\,\text{ft/s})^2/32.2\,\text{lbm.ft/lbf}\cdot\text{s}^2 = 35.2\,\text{ft} \times \text{lbf/step}$$

and, if the walker has a velocity of 2.00 steps/s, then the energy rate is

$$(35.2\,\text{ft}\cdot\text{lbf/step})(2.00\,\text{steps/s}) = 70.4\,\text{ft.lbf/s} \times 1\,\text{Btu}/778\,\text{ft}\cdot\text{lbf} = 0.0905\,\text{Btu/s}$$

Now, 1 Btu = 1.055 kJ, so the energy rate needed is 0.0905 × 1.055 = 0.0955 kJ/s, or 0.0955 kJ/s × (3600 s/h) = 344 kJ/h = 0.344 MJ/h. Table 17.5 lists "fast" walking as requiring 1.3 MJ/h. Since the basal metabolic rate is 0.3 MJ/h, then the fast walking alone is 1.0 MJ/h, which is about three times the slow walking rate just calculated.

FIGURE 17.11
Person walking.

where P_o is the basal metabolic rate (BMR), P_D is the power absorbed by aerodynamic drag, and P_m is the power absorbed by the muscles. From fluid mechanics, we know that the power required to overcome viscous drag is given by

$$P_D = \frac{1}{2}\rho A C_D V^3$$

where ρ is the fluid density, A is the frontal projected area of the animal, and C_D is its drag coefficient. We can therefore determine the most efficient transport velocity by minimizing T as follows:

$$\frac{\partial T}{\partial V} = 0 = -\frac{P_o + P_m}{wV^2} + \frac{\rho A C_D V}{w} + \frac{1}{wV}\left(\frac{\partial P_m}{\partial V}\right) \tag{17.23}$$

where we have assumed the weight to be constant during the locomotion. Equation (17.23) can be rewritten as

$$\rho A C_D V^3 + V\left(\frac{\partial P_m}{\partial V}\right) - (P_o + P_m) = 0 \tag{17.24}$$

which could be solved for the most efficient locomotion velocity if we knew how P_m depends on V. If we assume that P_m increases linearly with V, then we can write

$$P_m = KV$$

where K is a constant. Then, Eq. (17.24) becomes

$$\rho A C_D V^3 - P_o = 0$$

then,

$$V_{\text{most efficient}} = \left(\frac{P_o}{\rho A C_D}\right)^{1/3} \tag{17.25}$$

Figure 17.12 shows T vs. V for a 70 kg human. The minimum value of T occurs at about $V = 1.75$ m/s, which corresponds to a fast walk. Locomotion velocities faster or slower than this value require more energy consumption per distance traveled and are hence less efficient locomotion speeds.

Mechanical locomotion devices have the potential of altering the T vs. V curve by moving its minimum to a higher velocity. Of course, the weight of the locomotion device must be added to the animal's weight such that w in Eq. (17.22) is

$$w = w_{\text{animal}} + w_{\text{device}}$$

therefore, the weight of the locomotion device alone tends to decrease the value of V at minimum T. Bicyclists are willing to carry along the extra weight of their machines, because at their most efficient velocity, their minimum value of T is about 0.064, which is about 25% of their minimum value of T in normal leg locomotion without the bicycle. In fact, the bicyclist has the lowest value of T ever measured for any animal or machine-animal combination.

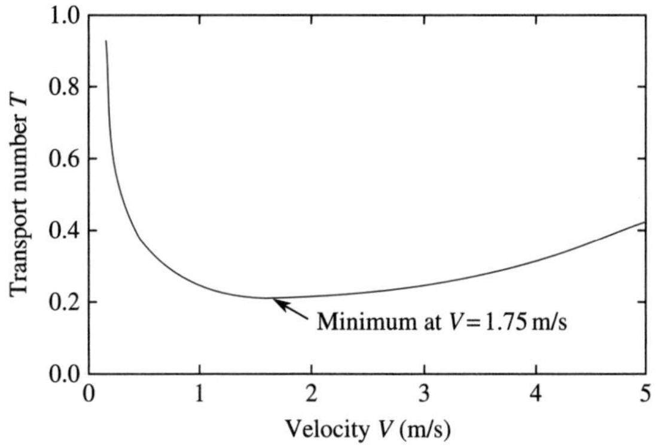

FIGURE 17.12
The locomotion transport number vs. velocity calculated from Eq. 17.22 for a 70-kg human.

EXAMPLE 17.9

Determine the locomotion transport number of a 60.0 kg person traveling at 15.0 miles per hour on a 15.0 kg bicycle while expending 400. W of power pedaling.

Solution

Using Eq. (17.22) with $P = 400.$ W, $w = (60.0 + 15.0)(9.81) = 735$ N, and $V = 15.0$ mph $= (15.0\ \text{miles}/h)(1.609\ \text{km/mile}) = 24.14$ km/h $= 24{,}140$ m/h gives

$$T = \frac{P}{wV} = \frac{(400.\text{N}\cdot\text{m/s})(3600.\text{s/h})}{(735\ \text{N})(24{,}140\ \text{m/h})} = 0.0812$$

Exercises

25. Determine the locomotion transport number of a 0.100 kg fish using 0.150 J/s to swim at a velocity of 0.500 m/s. **Answer:** $T_{fish} = 0.30$.
26. An aircraft weighing 22.0×10^3 lbf uses 3.00×10^3 horsepower to fly at 200. mph. Determine its locomotion transport number. **Answer:** $T_{airplane} = 0.256$.
27. Determine the most efficient velocity for the bicyclist in Example 17.9 if his or her basal metabolic rate is 73.1 J/s, frontal area is 0.750 m^2, aerodynamic drag coefficient is 1.50, and the local density of air is 1.21 kg/m^3. **Answer:** $V_{most\ efficient} = 3.77$ m/s $= 13.6$ km/h.

Figure 17.13 presents data on the dimensionless locomotion transport number T vs. body mass for a large variety of birds, fish, land animals, and machines. The value of the locomotion transport number for an animal of a given mass clearly depends directly on the percentage of its body mass dedicated to locomotion muscles. This percentage is greatest in fish, next largest in birds, and smallest in two- and four-legged runners. Note that fish have the lowest T values and are therefore the most efficient mobile animals. Figure 17.13 also has points for various machines, and the machines that are the most efficient at transport are trains and ships.

The locomotion efficiency for a given animal becomes much lower when it is forced to travel in a different medium. A human consumes 30 times more energy in swimming than does a fish of equivalent mass. Penguins are highly adapted to swimming, but on land they waddle around with a locomotion transport number twice as high as any land animal of equivalent mass.

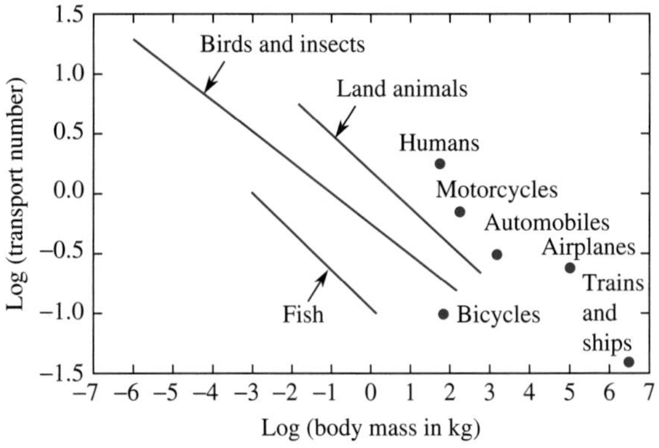

FIGURE 17.13
The average locomotion transport number vs. mass for a variety of animals and machines.

17.9 THERMODYNAMICS OF AGING AND DEATH

There are several important theories of biological aging, but perhaps the most popular is that of molecular error propagation. This theory states that molecular reproduction by enzymes is not perfect. The entropy production of molecular synthesis over a significant period of time cannot be insignificant with regard to the information content (or structure) of the molecule being synthesized. Thus, both evolution *and* aging depend on how the living system responds to error accumulation at the molecular level. Ultimately, the errors build up such that the system can no longer function properly and a catastrophic event leading to death occurs. Equation (17.11) is the energy rate balance applied to the living system of Figure 17.5. It accounts for all the energy flows into and out of the system and the state of the energy within the system at any time. It reveals nothing about the aging process or life span of the system. However, the life span of mammals in captivity has been accurately correlated with body mass as

$$\text{Life span of mamals (in years)} = 11.8\left(m^{0.2}\right) \tag{17.26}$$

where the body mass m is in kilograms. Table 17.6 lists the pulse rates, breathing rates, and life spans of various mammals calculated using Eqs. (17.18), (17.19), and (17.26). Except for human life span, the results of these calculations are reasonably accurate.

Table 17.6 Metabolic Characteristics of Typical Mammals

Mammal	Body Mass (kg)	Pulse Rate (beats/min)	Breathing Rate (breaths/min)	Life Span (years)
Shrew	0.003	1030	230	3.7
Mouse	0.03	580	130	5.9
Rat	0.2	360	80	8.6
Cat	2.8	190	42	14
Dog	15.9	120	27	21
Horse	700	47	10	44
Elephant	4000	30	6.7	62

Even though we know that the metabolic heat generation rate $\dot{Q}$ decreases with increasing age (e.g., see Figure 17.6), this effect must be offset by increasing size (i.e., growth) and eating less or exercising less as age increases. Experiments in which test animals were fed a very low (starvation level) daily diet showed that they generally had a lower metabolic rate and lived longer than did their counterparts who were fed a normal or excessive diet. These particular results, however, occur only when the starvation diet was begun before the animal reached sexual maturity. If it was begun later in life, it had no significant effect on metabolic rate or on life span.

An entropy rate balance on a living system is

$$\frac{\dot{Q}}{T_b} + \sum_{\text{in}} \dot{m}s - \sum_{\text{out}} \dot{m}s + \dot{S}_P = \frac{dS}{dt} \tag{17.27}$$

The first term is the entropy transport due to the metabolic heat transfer, and since $\dot{Q} < 0$, it is negative. The combination of the second and third terms is the net entropy transport into the system via the mass flow of food, respiration, and wastes. Since the entropy of the incoming food is lower than the entropy of the outgoing wastes (both are at the same temperature, but the molecular order of the food is more complex than that of the wastes) and the input and output mass flow rates averaged over a long period of time are essentially the same, these two terms taken together also are negative. The last term on the left side of Eq. (17.27) is the rate of entropy production, which by the second law of thermodynamics must always be positive. The term on the right side is the time rate of change of the entropy of the entire biological system, and it can be either positive or negative depending on the net sign of the left side. Thus, we find that, for any living system,

$$\dot{S}_P > 0$$

$$\frac{\dot{Q}}{T_b} < 0$$

$$\sum_{\text{in}} \dot{m}s - \sum_{\text{out}} \dot{m}s < 0$$

But, these conditions alone are not sufficient to define a living system. The one characteristic that seems to make a living system unique is its peculiar affinity for self-organization, and this characteristic corresponds to a continual decrease in the system's entropy over its life span. As the system "lives," it grows and ages and generally

CRITICAL THINKING

Is life unique in your opinion? That is, are living systems scientifically distinguishable from nonliving systems? Given a sealed box containing an object, how can you tell whether or not the object is living without opening the box? What physical tests could you perform to determine the life state of the object in the box?

A POSSIBLE DEFINITION OF A LIVING SYSTEM (I.E., LIFE)

Living systems are uniquely characterized by a continuously decreasing entropy over their life spans.

$$\left(\frac{dS}{dt}\right)_{\text{living system}} = \left(m\frac{ds}{dt} + s\frac{dm}{dt}\right)_{\text{living system}} < 0$$

becomes more complex at the molecular level. Note that this does not happen with the aging of machines, whose entropy generally increases monotonically with age. Consequently, we can postulate that living systems are defined by the following unique characteristic:

$$\left|\frac{\dot{Q}}{T_b} + \sum_{\text{in}} \dot{m}s - \sum_{\text{out}} \dot{m}s\right| > \dot{S}_P \tag{17.28}$$

and that death occurs when this inequality is violated. What does Eq. (17.28) tell us about the system as it ages? Based on our experiences with the machinery of the Industrial Age, we intuitively feel that old age corresponds to degeneration. In humans, the skin becomes wrinkled, teeth and hair are permanently lost, hearing and sight diminish, joints stiffen—it seems as though people "wear out" as they become older. Actually, what are normally described as degenerative signs of aging are really the result of continued growth, that is, continued systemic molecular organization. Skin becomes wrinkled because the collagen molecules of the skin crosslink to form a more rigid (less elastic) and complex structure. The same thing happens in the lens of the eye, where the macromolecular cross-linking makes the lens so rigid that the eye muscles can no longer change its shape to make it focus properly. Molecular cross-linking also causes loss of hearing sensitivity, and cross-linking within the lubricating fluid of the joints causes this fluid to thicken, which makes the joints arthritic and painful to move. We also see cross-linking and thickening in other biofluids such as blood. It appears that growth in molecular complexity continues long after physical maturity is reached and is the cause of many of the common symptoms of aging. If a biological system were to continue to grow (but not add mass), then its ultimate state would be one of complete rigidity with a very low entropy but with little mobility potential and very low predator survivability. Thus, a living system becomes more delicate as it ages beyond physical maturity and consequently is more prone to death resulting from failure of one of its major subsystems, such as the circulatory or the respiratory system. Cancer is curious in that it represents a reversion to cellular growth and appears to function as a mechanism for preventing the entropy of a living system from becoming too low.

According to the inequality of Eq. (17.28), the life span of a living system could be extended by decreasing the system (or body) temperature. Thus, even though $\dot{Q}$ is decreasing with age, the ratio $\dot{Q}/T_b$ could be made as large as desired by selectively lowering the body temperature. Figure 17.14 presents survival curves for common houseflies raised from birth in environments of different (but constant) temperatures. The longest life spans occur at the lowest environmental temperature (16°C). These insects have also been shown to exhibit increased life spans when they were raised for part of their lives at one temperature and then spent the remainder of their lives at a lower temperature. Similarly, their life spans have been shortened by raising the environmental temperature slightly.

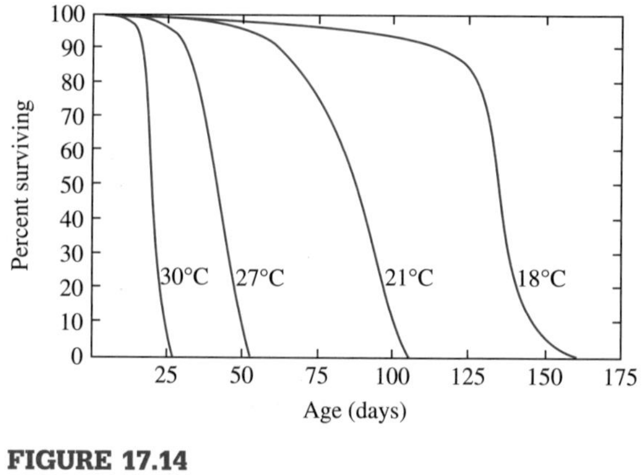

FIGURE 17.14
Survival curves for houseflies raised at different constant environmental temperatures.

Using survival curves such as those of Figure 17.14, researchers developed survival equations similar to those used in describing the kinetics of first-order chemical reactions,

$$\frac{dN}{dt} = -k_d N \tag{17.29}$$

where N is the number of survivors at time t, and k_d is a death rate constant. The constant k_d is often found to be independent of t but dependent on the environmental absolute temperature T. A plot of ln k_d vs. ln T yields a straight line from which the following equation can be obtained:

$$k_d = \alpha T\left[\exp\left(\frac{\bar{s}_d}{\Re} - \frac{\bar{h}_d}{\Re T}\right)\right] \tag{17.30}$$

where α is a constant, T is the absolute temperature, and $\bar{s}_d$ and $\bar{h}_d$ are the specific molar activation entropy and enthalpy of death. It has been shown from data on the death of unicellular organisms and the irreversible thermal denaturization of proteins that $\bar{s}_d$ and $\bar{h}_d$ are related by

$$\bar{s}_d = \frac{\bar{h}_d}{T_c} + \beta \tag{17.31}$$

where $T_c = 330.$ K and is called the *compensation* temperature, and β is a constant equal to –276 kJ/(kgmole · K). Equation (17.31) is often called a *compensation law*, because changes in $\bar{s}_d$ are partially compensated for by changes in $\bar{h}_d$, resulting in a relatively constant value of k_d. The compensation is exact at $T = T_c$ as can be seen by substituting Eq. (17.31) into Eq. (17.30). For the common housefly, the data reduction gives $\bar{h}_d \approx 800$ MJ/kgmole. Since k_d is the death rate constant, the smaller it is, the smaller the death rate becomes and the longer the life span becomes.

Combining Eqs. (17.30) and (17.31) gives

$$k_d = \alpha T\left\{\exp\left[\frac{\bar{h}_d}{\Re}\left(\frac{T - T_c}{T \times T_c}\right) + \frac{\beta}{\Re}\right]\right\} \tag{17.32}$$

Now, because $\bar{h}_d$ is such a large value and $T < T_c$ for most living systems, a small decrease in T can produce a significant decrease in k_d. Using $\Re = 8.3143$ kJ/(kgmole·K), $\bar{h}_d = 800{,}000$ kJ/kgmole, $\beta = -276$kJ/(kgmole·K), and $T_c = 330.$ K, Eq. (17.32) becomes

$$\frac{k_d}{\alpha} = T\left\{\exp\left[9.62 \times 10^4\left(\frac{T - 330.}{330. \times T}\right) - 33.2\right]\right\} \tag{17.33}$$

When $T = 310.$ K (37.0°C), Eq. (17.33) gives $k_d/\alpha = 8.03 \times 10^{-21}$ K. But when T is lowered by just 2 degrees to 308 K (35.0°C), then k_d/α drops by almost a factor of 8 to 1.06×10^{-21} K. If T is dropped all the way down to 293 K (20.0°C), then $k_d/\alpha = 1.15 \times 10^{-28}$ K, a drop of a factor of 10^7. Thus, the death rate constant is *very* sensitive to the body temperature. It has been estimated by some researchers that, if the core temperature of humans were lowered from its present value of 37°C down to 31°C, then the average age at death would increase from about 75 to around 200 years.

EXAMPLE 17.10

The death rate constant for mice at 27.0°C is 0.0350 months^{-1}. Determine the coefficient α in Eq. (17.33) for mice.

Solution

At $T = 27.0°C = 300.$ K, we find, from Eq. (17.33), that

$$k_d/\alpha = 300.\{\exp[9.62 \times 10^4(300. - 330.)/[(330.)(300.)] - 33.2]\} = 2.50 \times 10^{-25}\ \text{K}$$

then,

$$\alpha = 0.035\ \text{months}^{-1}/(2.50 \times 10^{-25}\ \text{K}) = 1.4 \times 10^{23}\text{months}^{-1} \cdot \text{K}^{-1}$$

Exercises

28. Using the value for the coefficient α determined in Example 17.10, determine the death rate constant k_d for mice at 30.0°C (303 K). **Answer**: $k_d = 0.846$ months^{-1}.
29. Integrate Eq. (17.29) to find an expression for the ratio of the number of animals surviving at time t to the initial number of animals present, N/N_0. Then, using the information given in Example 17.10, evaluate this expression to determine the percentage of mice that survive to 5.0 months at 27°C. **Answer**: $N/N_0 \times 100 = 84\%$.
30. Plot log(k_d/α) vs. T for 30.0°C (303 K) $\leq T \leq$ 40.0°C (313 K). **Answer**: See Figure 17.15.

(Continued)

EXAMPLE 17.10 (*Continued*)

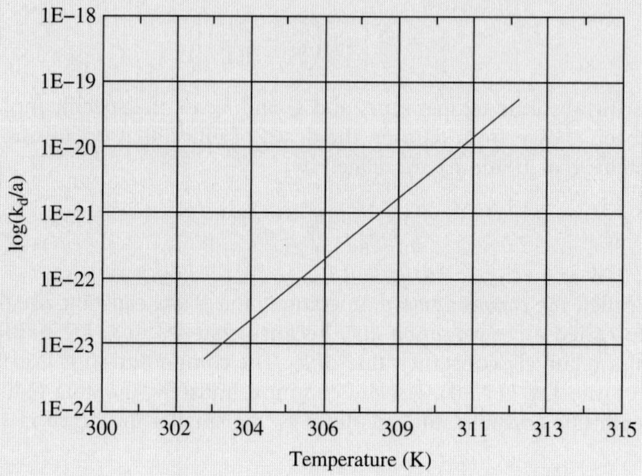

FIGURE 17.15
Example 17.10, Exercise 30.

One of the most interesting unsolved problems in evolutionary biology is that of biological aging and development. What determines the beginning and the end of growth? Why do some cells develop into one kind of organ and other cells into a completely different organ? Also, there is a remarkable similarity between the following biological groups:

- The grouping of elements to form active biochemical entities (such as amino acids).
- The grouping of these entities to form macromolecules.
- The grouping of macromolecules to form cells.
- The grouping of cells to form living creatures (plants and animals).
- The grouping of these living creatures into productive units (families, industries, etc.).
- The grouping of these families into cultures (or societies).
- The grouping of these cultures into nations.

Recently, West, Brown, and Enquist[1] discovered a single universal curve that describes the growth of many diverse species. A plot of a dimensionless mass, $r = (m/M)^{1/4}$, versus a dimensionless time variable, $\tau = (at/4M^{1/4}) - \ln[1 - (m_0/M)^{1/4}]$, for a wide variety of species shows that growth curves for all organisms fall on the same universal curve $r = 1 - e^{-\tau}$ (shown as a solid line in Figure 17.16), where a is a constant, t is time, m_0 is the mass at birth (i.e., at $t = 0$), and M is the maximum body size. Their model identifies r as the proportion of total lifetime metabolic power used for maintenance and other activities and provides the basis for deriving relationships for growth rates and the timing of life history events.

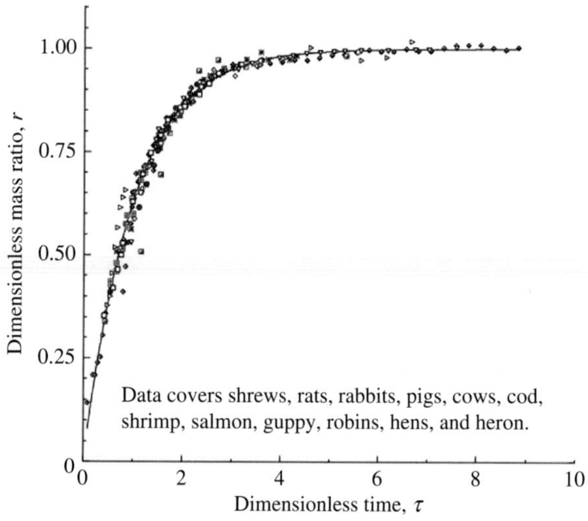

FIGURE 17.16
A plot of the dimensionless mass ratio, $r = (m/M)^{1/4}$, versus the dimensionless time variable, $\tau = (at/4M^{1/4}) - \ln[1 - (m_0/M)^{1/4}]$, for a wide variety of species. (*Source: Geoffrey West, James Brown, and Brian Enquist, "A general model for ontogenetic growth". Adapted by permission from Macmillan Publishers Ltd: Nature 413, no. 201, pp. 628–631, copyright 2001.*)

Thus, there seems to be a common phenomenological driving force that is not only responsible for the organization of molecular structure but is also responsible for the organization of the cultural bonds of nations

[1] *Nature* 413, (October 11, 2001), pp. 628–631.

and beyond. The entropy balance and the second law of thermodynamics may well be the key to understanding the fundamentals of both biomolecular and biosocial phenomena.

SUMMARY

Classical thermodynamics can be used to develop a fundamental understanding of the operation of biological systems. The conservation laws of mass, momentum, and energy are all obeyed by biological systems. The second law of thermodynamics seems to be critical in the understanding of the self-organization, growth, and aging of these systems. It has been argued that evolution via natural selection would be impossible without death; therefore, a death mechanism must be programmed into every living creature. On the other hand, from a thermodynamic point of view, such an argument is not necessary. All one needs to do is to recognize that no real process is completely reversible and that the entropy production for any real process is a positive finite value. Thus, the internal irreversibilities would eventually accumulate to the point of system failure, or death.

The field of biological thermodynamics covers not only individual living plants and animals but also (in ways that we do not yet fully understand) interacting groups of plants and animals, societies, corporations, and nations. Just as a living animal is made up of billions of living cells, each with its own unique function and characteristics, a society is made up of many unique living animals, each having its unique function within the society. Thus, the first and second laws of thermodynamics have the potential to also be the basic laws of social organization, and they may contain the key to the birth, growth, maturity, and decline of social structures.

Some of the more important equations introduced in this chapter follow. Do not attempt to use them blindly without understanding their limitations. Please refer to the text material where they were introduced to gain an understanding of their use and limitations.

1. The membrane potential E_i at 37°C due to the presence of chemical species i is

$$E_i(\text{at } 37°\text{C}) = \frac{26.7\ \text{millivolts}\cdot(\text{kgmole electrons/kgmole } i)}{z_i}\ \ln\left(\frac{c_{io}}{c_{ic}}\right)$$

2. The energy conversion efficiency of biological systems is

$$\eta_E = \frac{\dot{W} + \dfrac{d}{dt}\left(\dfrac{mV^2}{2g_c}\right) + \dfrac{d}{dt}\left(\dfrac{mgZ}{g_c}\right)}{-dU/dt} = 1 + \frac{\dot{Q}}{-dU/dt}$$

3. The basal metabolic rate per unit mass of a mammal of mass m is

$$\text{BMR}/m = 293(m^{-0.25}) = \frac{293}{m^{1/4}}$$

4. The heart rate (pulse) for mammals of mass m is

$$\text{Heart rate (in beats per minute)} = 241(m^{-0.25}) = \frac{241}{m^{1/4}}$$

5. The breathing rate for mammals of mass m is

$$\text{Breathing rate (in breaths per minute)} = 54.0(m^{-0.25}) = \frac{54.0}{m^{1/4}}$$

6. The critical buckling height of a tree with a base diameter d is

$$h_{\text{critical}} = 68.0(d^{2/3})$$

7. The locomotion transport number T is

$$T = \frac{P}{wV}$$

where P is the total power expended in the transportation, w is the weight of the system, and V is its velocity.

8. The velocity that produces the most efficient transport process is given by

$$V_{\text{most efficient}} = \left(\frac{P_o}{\rho A C_D}\right)^{1/3}$$

where P_o is the basal metabolic rate, ρ is the local fluid density, A is the cross-sectional area normal to the direction of motion, and C_D is the drag coefficient of the system.

9. The life span of mammals of mass m is

$$\text{Life span of mamals (in years)} = 11.8\left(m^{0.2}\right)$$

10. The death rate constant k_d for living systems at an environmental temperature T (in K) is

$$\frac{k_d}{\alpha} = T\left\{\exp\left[9.62\times 10^4\left(\frac{T-330.}{330.\times T}\right) - 33.2\right]\right\}$$

where α is a species-specific constant.

Problems (* indicates problems in SI units)

1.* Use Eq. (17.9) to determine the membrane potentials of hydrogen carbonate (HCO_3^-), hydrogen phosphate ($HPO_4^=$), and hydrogen (H^+) ions listed in Table 17.2.

2.* An alien scientist from a galaxy in the star system Luepke has been able to *directly* measure the electrical potential required to transport a unit charge of divalent carbon ions from outside a human cell into the cell and finds it to be 4.10 μV. The alien also measures the value of the chemical potential of the carbon ion and finds it to be 0.520 J/kgmole. Using these measurements, determine the amount of electrochemical work required to transport 1.00 kgmole of carbon ions across the cell membrane.

3.* A muscle contraction is brought about by the release of calcium ions from storage vacuoles, where the calcium concentration is 0.700 mg/mL, into the cell's cytoplasm, where its concentration is 0.100 mg/mL.

a. What is the electrical potential between the vacuole and the cytoplasm?

b. If 1.00 mg of calcium ions is released in a contraction, how much adenosine triphosphate (ATP) must be hydrolyzed from adenosine diphosphate (ADP) according to the reaction $\text{ATP} + \text{water} \rightarrow \text{ADP} + \text{P} + 29.3\ \text{MJ/kgmole}$ to restore the muscle to its initial state?

4.* Assume that an individual adult in our society requires an energy intake of 10.0 MJ/d. Let this energy come exclusively from eating beef that was produced with a 10.0% energy conversion efficiency. Let the beef be fed by corn produced with another 10.0% energy conversion efficiency, and let the corn be produced from sunlight with a 1.00% energy conversion efficiency. Further, let the corn be grown in a region where the solar energy flux is $20.0\ \text{MJ/(m}^2\cdot\text{d)}$.

a. Compute the number of acres (1.000 acre = 4047 m^2) of land necessary to grow the corn required to feed the beef ultimately consumed by one adult person.

b. If some 16.0 × 10^9 acres are available for cultivation on Earth today, estimate the total population that can be supported by this food chain.

c. If the current world population is 6.00 × 10^9 people and the population growth rate is given by $p = p_o\exp(0.0300t)$, where p is the population at time t, p_o is the initial population, and t is time measured in years from the present, determine the number of years into the future when the population calculated in part b will be reached.

5.* Stewart has a BMR of 160. kJ/d and climbs to the 13th floor of his office building in 4.00 min while consuming only 0.0100 kg of carbohydrate having an energy content of 17.2 MJ/kg. The distance between the floors is 7.60 m. What is Stewart's energy conversion efficiency if he does not lose any weight during the climb? Note: Stewart must supply energy to achieve his kinetic energy motion, but this energy is not recovered when he stops.

6. Find the energy conversion efficiency of a 2000. lbf thoroughbred racehorse with a 120. lbf jockey and tack running 1.25 miles on a flat track in 144 s. The horse accelerates to a constant speed at the starting gate and maintains this speed throughout the race. During the race, the horse expends energy at the rate of 33,000. Btu/h. Ignore any aerodynamic effects.

7.* Greg, a professional weightlifter, has the capacity to convert his internal energy into output work at a rate of 2.70 × 10^3 J/s. If the distance from his chest to his extended arms is 0.750 m, how much weight can he bench press in 2.00 s? What is the horsepower output of his arms under these conditions? Assume his arm muscles have an energy conversion efficiency of 25.0%.

8.* During an experiment, it was found that an 80.0 kg man lost 0.260 kg of body fat having an energy content of 33.1 MJ/kg by lifting one 50.0 kg mass from the floor to a 1.50 m high shelf every 5.00 seconds continuously for 4.00 hours. Ignoring respiratory and perspiration losses, determine the energy conversion efficiency of the muscular contractions.

9.* The per capita electrical power consumption in the United States is about 200. kW · h/d. Suppose this power is generated by having mice run in wheels that turn electrical generators. These mice are to be feed Swiss cheese that has a metabolizable energy content of 15.5 MJ/kg and costs $4.00 per kilogram. If the mice have an energy conversion efficiency of 25.0%, how much will it cost to buy the cheese needed to feed the mice who supply the per capita energy needs?

10.* Determine the horsepower corresponding to 1.00 MJ/h. If, in an average 24.0 h day, your energy output is 8.40 MJ, determine your average daily horsepower output.

11.* In 6.00 h, the heat from a guinea pig melts 0.200 kg of ice in an adiabatic calorimeter. Assuming that the heat of fusion of ice is 335 kJ/kg, determine the average metabolic heat production rate of the animal while it is in the calorimeter.

12.* If a person's body has a specific heat equal to that of water and produces 6.28 kJ per min per kilogram of body mass, what is the rate of increase in body temperature in °C per min if the person is suddenly made adiabatic?

13.* Rumor has it that Frankenstein's monster was brought to life by charging it with 1.00 kW of power for 2.00 h, after which it operated with an efficiency of only 25.0%.

a. If the monster consumed its charged energy at a rate of 1.30 MJ/h, how long before it needed to be recharged again?
b. If the monster had a mass of 100. kg, what would be its mean metabolic rate in MJ/d if it were a normal mammal?

14.* The heat of formation of glucose ($C_6H_{12}O_6$) is –996.4 MJ/kgmole. Using the material presented in Chapter 15, determine
a. The amount of heat liberated (i.e., the heat of reaction) at the standard reference state as 1 mole of glucose is metabolized according to the reaction

$$C_6H_{12}O_6 + 6(O_2) \rightarrow 6(H_2O(\ell)) + 6(CO_2) + \text{heat}$$

b. How long does it take to completely metabolize 1.00 kg of pure glucose at a basal metabolic rate of 0.300 MJ/h?

15.* Human blood contains 1.00 gram of glucose ($C_6H_{12}O_6$) per liter, and the average person contains 5.20 L of blood. In metabolizing glucose, 17.1 MJ/kg of energy is released, of which 50.0% is lost as heat and 50.0% is used to make adenosine triphosphate (ATP).
a. How much energy could be stored in the ATP of the blood of an average person?
b. If the metabolic heat of the glucose is removed from the body by the evaporation of perspiration, how much perspiration is evaporated per gram of glucose metabolized? Assume the heat of vaporization of perspiration (H_2O) is 40.6 MJ/kgmole.

16.* During a weekend of fun and frolic, Homer, a humanities student, consumes 8.50 kg of beer with a metabolizable energy content of 1.10 MJ/kg. Homer then went to bed with the intent of sleeping off his entire caloric intake of beer. Assuming he falls asleep at 2:00 AM Wednesday morning, when should he wake up?

17.* In an experiment, the contribution of Joe's brain to his total BMR was found to be 5.00×10^{-2} MJ/h.
a. What is Joe's total mass?
b. How long would it take Joe to metabolize the energy content of one candy bar containing 0.750 MJ while resting?

18.* A serious problem that arises in performing surgery on cats and small dogs is the additional heat loss produced from an open body cavity. For these small animals, this type of surgery effectively doubles the normal heat loss rate. If the anesthetic used depresses the BMR of a 5.00 kg dog by 50.0%, estimate the resulting reduction in the animal's body temperature from its normal body temperature of 102°F (39.2°C) during 2.00 h of
a. Minor surgery *not* requiring the body cavity to be opened.
b. Major surgery requiring an open body cavity procedure. Assume the body of the dog has a specific heat of 4.17 kJ/(kg · K).

19.* What is Superman's top flying speed, if he gets all his energy by eating as many 1.00 MJ chocolate candy bars as he wishes? Neglect his potential energy, aerodynamic drag, and all other losses.

20.* If you consume 25.0 Calories per day more food energy than you use, how many years will it take you to gain 10.0 kg of fat if there are 33.1 MJ per kg of body fat?

21.* During a basal metabolic test, Steve consumes 460. L of O_2.
a. What is Steve's body mass?
b. What is the change in his BMR oxygen intake if he loses 15.0 kg of body mass?

22.* Broiled lobster contains 3.60 MJ/kg of metabolizable food energy. If Sharla has 45.0 min of hard cycling planned later in the day, how much lobster can she eat so that she will be sure not to gain weight?

23.* Suzanne Malaxos, 27, from Perth Australia, won the 12th annual Empire State Building Run-Up in New York City by climbing the 102 story building in 12 min and 24 s. If her body mass is 50.0 kg, and each story is 4.00 m high, determine the amount of body fat she consumed in the race. Assume a muscle energy conversion efficiency of 25.0% and a body fat energy content of 33.1 MJ/kg.

24.* You are at a tailgate party before a baseball game and have just eaten three hot dogs with chili. The mass of the container and the remaining contents of your quarter barrel of beer is 30.3 kg. How many times would you have to lift this barrel 0.500 m in 2.00 s to work off the energy content of the hot dogs? How many hours would it take to do this?

25. Gasoline has a heating value of 20.0×10^3 Btu/lbm. How many 10.0 oz colas must Ted drink to produce enough power by turning a crank to light a 100. W lightbulb for the same number of hours that an internal combustion engine running on 1.00 lbm of gasoline with an overall efficiency of 25.0% could light the same bulb?

26.* Mark decides to build a cabin cruiser. After many hours of sawing wood, it is determined that the sawing required a total of 63.0 MJ of energy from Mark and 15.0% of this energy came from protein (at 17.2 MJ/kg), 60.0% came from fat (at 33.1 MJ/kg), and 25.0% came from carbohydrates (at 17.2 MJ/kg). Determine the mass of protein, fat, and carbohydrates consumed in the process.

27.* A 60.0 kg mountain climber makes a vertical climb of 2000. m in 5.90 h. From a chemical analysis of the urine samples collected during the climb, it is found that 2.00×10^{-2} kg of water-free protein is catabolized. Assuming a 25.0% muscle energy conversion efficiency, find
a. The percentage of the total energy need for the climb that came from protein and the percentage that came from fat.
b. The mass of natural (wet) fat catabolized (i.e., consumed) during the climb.

28.* If you consume two hamburgers, one regular fries, and one 10 oz cola, how many hours on this meal alone can you (a) cross country ski, (b) play tennis, or (c) watch television?

29.* Brian is somewhat overweight and calculates that his excess body fat contains 18.0 miles of extra small blood vessels. Brian's weight is stable, but he consumes 10.5 MJ/d of metabolizable food energy. To reduce his weight, he decides to eat one fewer cheeseburger per day plus jog for one hour per day.
a. How much extra fat does Brian have at the beginning of his diet?
b. How many days does it take for Brian to eliminate this fat?

30.* Steve is jogging at a constant speed of 5.00 mph and encounters a hill that requires an average energy expenditure rate of 3.10 MJ/h. If the hill is 0.400 mi long, determine the mass of natural state foods that Steve must consume to replenish the energy spent climbing the hill, if he consumes
a. Only carbohydrate with an energy content of 4.20 MJ/kg.
b. Only protein with an energy content of 8.40 MJ/kg.
c. Only fat with an energy content of 33.1 MJ/kg.

31.* In a laboratory experiment, two engineering students are asked to determine the caloric value of a commercial brand of diet cocoa. They are provided with a bomb calorimeter, which had to be calibrated by measuring the heat liberated by a substance with a known heat of combustion. They elect to use benzoic acid, which has a known heat of combustion of 6.318 kcal/g. The test consists of igniting a tablet of the test material inside the bomb and measuring the temperature rise of the surrounding water. The energy equivalent (EQ) of a bomb calorimeter is defined to be the product of the mass of the system multiplied by its specific heat. Then, the relation between the heat liberated by a test sample and the measured temperature change of the water is $Q = EQ(T_{final} - T_{initial})$.

a. If the change in temperature of the water was 2.905°C when a 1.1523 g tablet of benzoic acid is tested, determine the energy equivalent of the calorimeter.
b. Then, a 1.0825-g tablet of diet cocoa is tested and produces a temperature change of 1.699°C. Using the results of part a, determine the caloric energy content of the cocoa in kilocalories per gram.

32.* Rob, a young engineer, notices that, over a long period of time, he has added 20.0 lbm (9.07 kg) of excess body fat and decides to lose this extra weight by dieting alone. Rob's activities are such that his caloric intake and energy output are identical. He normally eats *two* of the following meals per day, seven days per week: one cheeseburger, one regular fries, and one 10.0 oz cola. Also, he consumes 2.00 MJ per day of snack food while working. How long will it take him to lose the extra 20.0 lbm of body fat by

a. Eliminating the daily snack food only?
b. Eliminating the daily snack food plus eating only one of these meals per day?
c. Going on a total starvation diet with no food whatsoever being consumed?

33.* Christine, an aspiring lawyer, notices that, over the past year, she has added 10.0 lbm (4.54 kg) of excess body fat, and she decides to work off this extra weight by jogging each evening after work without changing her eating habits. She works 8.00 hours per day (including weekends) with an energy expenditure rate of 0.600 MJ/h. The time spent not working or jogging is spent sleeping or watching television at 0.300 MJ/h. Christine eats *two* of the following meals per day, seven days per week: one baked potato with cheese, one lettuce salad with French dressing, and one pint of skim milk, plus she consumes 1.00 MJ per day of munchies while working. How many hours must she jog *each night* for three weeks to loose the extra 10.0 pounds?

34.* Do you eat like a bird? How much birdseed would a 70.0 kg person have to eat in a day to consume proportionally as much birdseed as does a 0.0120 kg sedentary canary per day? Birdseed contains 60.0% carbohydrate, 12.0% protein, 6.0% fat, and 22.0% water.

35. Jim is a college wrestler weighing 145 lbf and he decides to wrestle in the 132 lbf weight class for the upcoming season. He plans to start his weight-loss program early so that he can be down to the desired weight by the first practice of the season. To accomplish this, he restricts his food intake to 1000. Calories per day and begins an exercise program consisting of jogging for 20.0 min a day plus 10.0 min of other daily exercises that are equivalent to climbing 20 flights of stairs at 12.0 ft per flight. When he is not exercising, his average energy expenditure rate is 500. Btu/h for the remainder of the 24 h day.

a. What is Jim's total energy expenditure during his 30.0 min workout?
b. Assuming his excess weight is all body fat, how many days before his first wrestling practice must he start the program?

36.* The amount of body fat on an average man is 19.0% of his total mass. Tim has a body mass of 80.0 kg and it is determined that 24.0% of his total mass is body fat. He decides to swim 0.500 h each day until his body fat has been reduced to the average. If his daily caloric intake and energy output are equal before he begins swimming and he does not change his caloric input, how many days must he swim to reach his goal?

37. Mary Anne Sorensen's airplane crashed on a mountain in the Yukon wilderness. Sorensen weighed 150. pounds at the time of the crash and 110. pounds when she was rescued 50.0 days later. Assuming that death occurs when 50.0% of her body weight is lost, estimate Sorensen's survival time, assuming

a. A constant weight-loss rate.
b. An exponential weight loss–time relation of the form $w = w_o \exp(-\alpha t)$, where α is a constant and w_o is her weight at the time of the crash.

38.* Tamara Arendt was in the same plane crash as Mary Anne Sorensen (see the previous problem). Tamara was trained in mountaineering and wants to hike down the mountain to safety. However, it will take her 27.0 d of climbing at 15.0 h per day with 9.0 h of rest per day to reach her destination. Her food supply consists of 11.0 MJ of candy bars, 16.2 MJ of peanuts, and 8.30 MJ of soda. The only body tissue she is able to consume during the climb is body fat, which is 20.0% of her initial body weight. When her body fat has been consumed she will die of exhaustion. If she weighs 59.0 kg at the time of the crash, is she better off waiting to be rescued with Mary Anne or climbing down the mountain?

39. In 1638, Galileo estimated that a tree over 300. ft tall would collapse under its own weight. Using the modern theory, determine the diameter of the base of such a tree.

40. What is the maximum height of a California redwood tree whose base diameter is 10.0 ft if its weight density is 40.0 lbf/ft^3 and its elastic modulus is 1.30×10^6 psi?

41. Since the uncertainty in the exponent in Eqs. (17.18) and (17.19) is ±0.0800, show that the total number of heart beats and breaths that occur over a life span is approximately the same for all the mammals listed in Table 17.6.

42.* In a laboratory test, a student's resting pulse rate and lung volume are measured and found to be 60.0 beats per minute and 8.35×10^{-5} m^3, respectively. What is the student's body mass?

43. Compute the locomotion transport number of a 4000. lbf automobile using 60.0 hp to move at a speed of 55.0 mph.

44.* Compute the locomotion transport number of a pedal-powered aircraft whose total mass (including the operator) is 126 kg. The aircraft flies at 15.0 mph when the operator is supplying 1.50 hp to the pedals.

45.* It has been proposed to design a human powered vehicle (HPV) whose total mass (including the 70.0 kg operator) is only 95.0 kg. The vehicle would be capable of traveling at 64.0 km/h while the operator supplies power equal to that of a person running at 5.00 m/s. Determine the locomotion transport number of this vehicle.

46. While Paul is driving his classic 220 hp, 3000. lbf Mustang convertible to his thermodynamics final exam, he is stopped by a state patrolman for traveling 95 mph in a 55 mph speed zone. Paul's excuse to the police officer is that he is performing a locomotion transport number homework experiment for his thermo class. Having heard this excuse countless times before,

the officer asks Paul for the value of his experimental LTN, promising to release him if his answer is correct. Paul replies, "0.29, sir." The officer then consults the state patrolman's guide to locomotion transport numbers for the correct value. Does Paul get arrested?

47.* Tom is a 75.0 kg bicyclist who recently averaged 42.0 km/h during a 240. km race with an 8.40 kg racing bike. If he consumes 4100. L of oxygen and has a muscle energy conversion efficiency of 25.0%, determine
   a. His rate of conversion of oxygen in L/min.
   b. His locomotion transport number.

48. Sharla, weighing 140. lbf, absorbs 0.500 hp in her muscles while pedaling a 7.00 lbf bicycle at 25.0 mph into a 5.00 mph head wind in air at 70.0°F. Her frontal cross-section is 4.00 ft by 2.00 ft, and her drag coefficient is 0.500.
   a. Compute her locomotion transport number.
   b. Determine her most efficient velocity on the bicycle.

49.* Jim has a frontal cross-section of 2.00 m high by 0.500 m wide and can run at 4.00 m/s and swim at 1.00 m/s. His drag coefficients in air and water are 1.30 and 1.10, respectively. What speeds should Jim run and swim at to be most efficient? Assume Jim's BMR is 0.300 MJ/h.

50.* Determine which of the following expends the most energy over a 20.0 mi course:
   a. A 126 kg pedal-powered aircraft flying at a speed of 15.0 mph with a locomotion transport number of 0.134.
   b. A fast walking 75.0 kg person walking at a speed of 2.50 mph,
   c. Determine the energy expenditure *rates* of the person powering the aircraft and the person walking and comment on the feasibility of maintaining these rates over the 20.0 mi course.

51. If King Kong was ten times bigger than a normal human being,
   a. Find his locomotion transport number while riding a bicycle (assume his bicycle locomotion transport number is 25.0% of the minimum value shown on Figure 17.12).
   b. What was his most efficient walking velocity (assume the same drag coefficient as for a human).

52. Convert the k_d/α information associated with Eq. (17.33) from metric units into Engineering English units.

53.* If a cold-blooded animal has an activation entropy of death of 3088 kJ/(kgmole · K) at 25.0°C, what is the change in k_d/α for the animal if it moves to an environment at 20.0°C?

54.* If the specific molar activation enthalpy of death is 800. MJ/kgmole, the compensation temperature is 330. K, and the constant $\beta = -276\,\text{kJ/(kgmole·K)}$, determine the specific molar activation entropy of death.

Design Problems

The following are open-ended design problems. The objective is to carry out a preliminary design as indicated. A detailed design with working drawings is not required unless otherwise specified by your instructor. These problems have no specific answers, so each student's design is unique.

55. Design an inexpensive apparatus that measures the energy conversion efficiency of an in vivo human arm or leg muscle. Measure the oxygen consumption rate of the test subject to determine the energy input rate. Include proper transducer instrumentation for the necessary input data, and specify adequate output electronics. Provide assembly and detailed drawings sufficient to allow a technician to fabricate, assemble, and test your design.

56. One of the problems with the commercially available bomb calorimeters is that they can test only small samples on the order of a few grams. Design a bomb calorimeter large enough to burn a sample as large as 1 pound. Pay special attention to safety considerations in your design. Do not attempt to construct or test your design.

57. Design a system that measures the rate of metabolic heat production of a small warm-blooded animal. You may use either a direct or an indirect calorimetry technique. Provide engineering drawings and instrumentation specifications.

58. Design a whole body calorimeter that measures the instantaneous metabolic heat loss rate from an entire human body. Your system must be large enough or else sufficiently mobile that measurements can be made while the test subject is doing physical labor without restraint from your system.

59. Design a variable resistance rowing exercise machine that has a direct digital readout of the instantaneous energy expenditure rate of the user. This means you have to specify or design transducers that measure the instantaneous work rate (i.e., power) done on the machine. This power can be absorbed by the machine either electrically or mechanically. Provide assembly and detail drawings of your design plus specify all the electronics necessary to process the transducer signals and provide the proper digital output.

60. Design an apparatus that measures the metabolic heat loss rate and surface temperature of a yeast culture or a small insect at various environmental temperatures. Construct and calibrate this apparatus if possible, and make enough measurements to plot $\dot{Q}/T_b$ for some living system vs. time at various environmental temperatures. Does $\dot{Q}/T_b$ increase or decrease as the environmental temperature decreases?

Computer Problems

The following open-ended computer problems are designed to be done on a personal computer using a spreadsheet, equation solver, or programming language.

61. Develop an interactive computer program that returns the user's basal metabolic rate, oxygen uptake rate, carbon dioxide production rate, pulse rate, and breathing rate when the user inputs his or her mass or weight.

62. Develop an interactive computer program that outputs the energy conversion efficiency of a person or an animal. The user must be prompted for input data regarding work performed, energy output resulting in increases in potential or kinetic energies, and changes in body total internal energy. You may assume that the specific internal energy of the body is constant for activities that occur over short time periods.

63. Develop an interactive computer program that provides the user with the metabolizable energy content of foods chosen from a menu. Allow the user to specify the desired energy units (Calories, Btu, MJ) of the output.

64. Develop an interactive computer program that returns the user's daily caloric food intake needs when the user selects his or her activities from a screen menu and inputs the time devoted to each activity.
65. Combine the programs of Problems 63 and 64 to produce an interactive computer program that returns a series of three possible exercise programs to achieve a weight-loss goal input by the user. The user must also input his or her current caloric consumption and physical activities. (Note: This is a computer exercise only. You are not qualified to give medical advice to anyone regarding eating or exercise habits, so do not allow *anyone*, including yourself, to use your program to develop an actual weight-loss schedule. Anyone seeking this advice must consult a qualified physician.)

CHAPTER 18

Introduction to Statistical Thermodynamics

CONTENTS

18.1 INTRODUCTION

In this chapter, we explore some of the basic concepts of statistical thermodynamics that lead to useful engineering results. In Chapter 2, we discuss the difference between *microscopic* and *macroscopic* systems and note that classical thermodynamics is based on a continuum macroscopic system approach. Recognition of the existence of atoms and molecules was not necessary for the development of classical thermodynamics, the results of which are valid for all processes in which the continuum hypothesis holds. Statistical thermodynamics, on the other hand, is based on the use of standard statistical methods in the analysis of molecular behavior and, therefore, corresponds to a microscopic system approach.

There are four basic attributes of statistical thermodynamics. First, it can be used to explain certain apparent discontinuities in physical behavior, such as superconductivity. Second, it can be used to extend classical thermodynamic results into regions where the continuum hypothesis is no longer valid, as in the case of rarefied gases. Third, it can often provide a molecular interpretation of physical phenomena that are observed at the macroscopic level but originate at the molecular level (such as fluid viscosity). Fourth and perhaps most important, it can function as a tool to provide accurate equations of state that describe the behavior of nonmeasurable thermodynamic properties, such as internal energy, enthalpy, and entropy, as a function of measurable properties, such as pressure, temperature, and density, without resorting to experimental measurements. These

HOW DID IT ALL BEGIN?

The development of statistical thermodynamics began in the late 19th century, shortly after William Thomson (1824–1907) and Rudolf Clausius (1822–1888) unified classical thermodynamics in the 1860s. Starting from basic mechanics principles, James Clerk Maxwell (1831–1879) developed a simple molecular interpretation of ideal gas behavior called the *kinetic theory* of gases, which led many physicists to conclude that all thermodynamic phenomena could be fully explained from mechanics principles. However, the mechanical approach was never able to predict the classical thermodynamic laws of the conservation of energy and positive entropy production; consequently, thermodynamics has held its own as an independent science.

In the 1870s, Ludwig Boltzmann (1844–1906) made great progress in the understanding of entropy, when he postulated that a mathematical relationship existed between entropy and mathematical probability by arguing that equilibrium states are not simply inevitable but merely highly probable states of molecular order.

Between 1900 and 1930. quantum mechanics blossomed under Max Planck (1858–1947), Albert Einstein (1879–1955), Peter Debye (1884–1966), Niels Bohr (1885–1962), Enrico Fermi (1901–1954), Erwin Schrodinger (1887–1961), and many others. It was only natural that their results be extended into the thermodynamic area whenever possible; thus evolved the new area of "quantum statistical thermodynamics," which is still an important research area.

equations of state are very useful when dealing with a substance for which empirically derived thermodynamic tables and charts do not yet exist but the basic molecular structure of the substance is known.

In this chapter, we survey the main engineering results of statistical thermodynamics by treating its two main components, kinetic theory and quantum statistical thermodynamics, as separate topics. The goal is the development of thermodynamic property relationships and equations of state of engineering value.

18.2 WHY USE A STATISTICAL APPROACH?

To begin with, we should explain why we resort to a statistical approach rather than simply a molecular approach. Suppose we have a large number of particles N in a box. To find out what happens inside the box without using a statistical analysis, we would have to follow the motion of each of the N individual particles. The motion of each particle must satisfy Newton's second law, and because of collisions and long-range forces between particles, each particle could conceivably influence the motion of every other particle in the box. Let F_{ij} be the force exerted on particle i by particle j. Then, the sum of all the forces on particle i due to all the other j particles must equal the mass m_i of particle i times its acceleration a_i, or

$$\sum_{j=1}^{N-1}\mathrm{F}_{ij} = m_i a_i = m_i\left(\frac{dV_i}{dt}\right) = m_i\left(\frac{d^2\mathrm{x}_i}{dt^2}\right) \tag{18.1}$$

where the terms $\mathbf{V}_i$ and $\mathbf{x}_i$ are the time-dependent velocity and position vectors of particle i. Since each of the N particles must obey Eq. (18.1), and particle-particle interactions couple all N Eqs. (18.1) together, they must all be solved simultaneously. Also, since Eq. (18.1) is a vector equation, there are really $3N$ scalar second-order coupled differential equations to be solved.

For a typical gas at standard temperature and pressure, $N \approx 10^{20}$ molecules/cm^3. Therefore, if we were to try to follow the molecules of the gas contained in one cubic centimeter at STP using the methods of classical mechanics, we would need to solve about 3×10^{20} scalar second-order coupled differential equations, each containing 10^{20} terms. This is impossible today, even with the fastest digital computers. Hence, we must abandon the approach of applying the equations of classical mechanics to each particle in the system. Instead of formulating a theory based on knowing the exact position of each particle in time and space, we develop a theory based on knowing only the *average* behavior of the particle.

18.3 KINETIC THEORY OF GASES

The elements of kinetic theory were developed by Maxwell, Boltzmann, Clausius, and others between 1860 and 1880. Though kinetic theory results are currently available for solids, liquids, and gases, we are concerned only with the behavior of gases. The following eight assumptions underlie the kinetic theory of gases:

1. The gas is composed of N identical molecules moving in random directions.
2. There is always a large number of molecules ($N \gg 1$) in the system.

3. All the molecules behave like rigid elastic spheres.
4. The molecules exert no forces on each other except when they collide (i.e., there are no long-range forces).
5. All molecular collisions are perfectly elastic.
6. The molecules are always distributed uniformly in their container.
7. The molecular velocities range continuously between zero and infinity.[1]
8. The laws of classical mechanics govern the behavior of all molecules in the system.

Each of the N molecules has its own unique velocity V_i. Using this velocity, we can define the following concepts for the system of molecules.

$$\text{The average molecular velocity: } V_{\text{avg}} = \sum_{i=1}^{N} \frac{V_i}{N} \tag{18.2}$$

$$\text{The root mean square molecular velocity: } V_{\text{rms}} = \left(\sum_{i=1}^{N} \frac{V_i^2}{N} \right)^{1/2} \tag{18.3}$$

and in the limit as $N \to \infty$, we can extend the summations in Eqs. (18.2) and (18.3) into integrals as follows

$$V_{\text{avg}} = \int_0^{\infty} \left(\frac{V}{N}\right) dN_V \tag{18.4}$$

$$V_{\text{rms}} = \left(\int_0^{\infty} \left(\frac{V^2}{N}\right) dN_V \right)^{1/2} \tag{18.5}$$

where dN_V is the number of molecules with velocities between V and $V + dV$. We also define the *translational total internal energy* U_{trans} as the sum of the kinetic energies of all the molecules in the system, or[2]

$$U_{\text{trans}} = \sum_{i=1}^{N} (m_i V_i^2/2) \tag{18.6}$$

Since assumption 1 requires that all the molecules have identical mass, we can set $m_i = m$, and Eq. (18.6) becomes

$$U_{\text{trans}} = \frac{m}{2} \left(\sum_{i=1}^{N} V_i^2 \right) = \frac{1}{2} Nm$$

Consider now a spherical shell of radius R containing $N \gg 1$ molecules. The radial force F_r on the shell due to a single molecular collision is (see Figure 18.1)

$$(F_r)_{\substack{\text{per}\\ \text{molecule}}} = ma_r = m\frac{dV_r}{dt} \approx m\frac{\Delta V_r}{\Delta t}$$

where a_r and V_r are the radial components of the acceleration and velocity. Using the geometry shown in Figure 18.1, this equation becomes

$$(F_r)_{\substack{\text{per}\\ \text{molecule}}} \approx m\frac{V_i \cos\theta - (-V_i \cos\theta)}{\Delta t} = \frac{2mV_i \cos\theta}{\Delta t}$$

and the total radial force on the shell due to collisions by all N molecules is

$$(F_r)_{\text{total}} = \sum_{i=1}^{N} (F_r)_{\substack{\text{per}\\ \text{molecule}}} = \sum_{i=1}^{N} \frac{2mV_i \cos\theta}{\Delta t}$$

The internal pressure inside the shell can now be computed from

$$p = \frac{(F_r)_{\text{total}}}{\text{Area}} = \frac{1}{4\pi R^2} \sum_{i=1}^{N} \frac{2mV_i \cos\theta}{\Delta t} \tag{18.7}$$

---

[1] Clearly, no molecule can have a velocity greater than the speed of light, but allowing the velocities to range to infinity is of tremendous mathematical value in the development of this theory. Though fundamentally wrong, we find that this assumption adds little error to the results.

[2] Since the formulae presented in this chapter were developed by physicists using the SI units system wherein $g_c = 1$, I elected to set $g_c = 1$ in all the relevant equations in this chapter to simplify them somewhat. Thus, we write $mV^2/2$ instead of $mV^2/(2g_c)$ for kinetic energy and so forth.

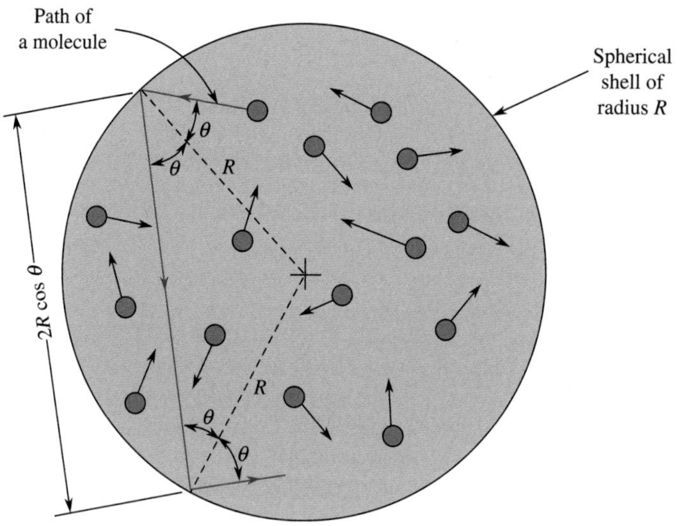

FIGURE 18.1
Motion of molecules inside a spherical shell of radius R.

In these equations, Δt is the time increment between successive molecular collisions. This can be calculated by dividing the distance that a molecule travels between successive collisions by its velocity, or

$$\Delta t = \frac{2R\cos\theta}{V_i}$$

Then, Eq. (18.7) becomes

$$p = \frac{1}{4\pi R^2}\sum_{i=1}^{N}\frac{2mV_i^2\cos\theta}{2R\cos\theta} = \frac{m}{4\pi R^3}\sum_{i=1}^{N}V_i^2 \tag{18.8}$$

Since the volume of the spherical shell is $\forall = \frac{4}{3}\pi R^3$, we can then write Eq. (18.8) as

$$p\forall = p\frac{4\pi R^3}{3} = \frac{m}{3}\sum_{i=1}^{N}V_i^2 = \frac{1}{3}NmV_{\text{rms}}^2 \tag{18.9}$$

In this equation, the product Nm is equal to the total mass of gas in the shell, m_T, and therefore, Eq. (18.9) can be written as

$$p\forall = \frac{1}{3}m_T V_{\text{rms}}^2 \tag{18.10}$$

If we now limit our attention to gases that obey the ideal gas equation of state, then Eq. (18.10) becomes

$$p\forall = \frac{1}{3}m_T V_{\text{rms}}^2 = m_T RT \tag{18.11}$$

where R is the specific gas constant given by

$$R = \frac{\Re}{M} = \frac{N_o k}{M} = \frac{k}{m} \tag{18.12}$$

where

$\Re$ = universal gas constant, 8314.3 J/(kgmole·K) or 1545.35 ft·lbf/(lbmole·R)
M = molecular mass (kg/kgmole or lbm/lbmole) of the gas
N_o = Avogadro's number[3] (or constant), 6.022×10^{26} molecules/kgmole

[3] In 1909, the French physicist Jean Perrin proposed naming this in honor of the Italian scientist Amedeo Avogadro, who, in 1811, proposed that the volume of a gas at a given pressure and temperature is proportional to the number of atoms or molecules in the gas regardless of the type of gas. Perrin won the 1926 Nobel Prize in Physics, mainly for his work in determining the value of the Avogadro constant by several different methods. In 1971, the name was officially changed from *Avogadro's number* to *Avogadro constant* (N_A) when the "mole" was introduced as a new fundamental unit in the International System of Units (SI). The change in name from the possessive form *Avogadro's* to the nominative form *Avogadro* is common usage today for all physical constants.

k = Boltzmann's constant, 1.380×10^{-23} J/(molecule·K)
$m = M/N_o$, which is the mass of one molecule of the gas

Combining Eqs. (18.11) and (18.12), we find that $p\forall = NkT$, where $N = m_T/m$ and N is the total number of molecules. Also, combining the second half of Eq. (18.11) and the first and last parts of Eq. (18.12), we get

$$V_{rms} = \sqrt{3RT} = \sqrt{\frac{3kT}{m}} \tag{18.13}$$

therefore, from Eq. (18.6), the kinetic theory interpretation of the translational total internal energy of an ideal gas is

$$U_{trans} = \frac{1}{2}NmV_{rms}^2 = \frac{3}{2}NkT \tag{18.14}$$

We see from Eq. (18.14) that U_{trans} depends only on temperature, which is in agreement with our definition of an ideal gas given in Chapter 3.

EXAMPLE 18.1

The qualifying examination for a very exclusive preschool in the future to which you want to send your child requires that your child answer questions on the kinetic theory of gases. For example, every five-year-old child must (Figure 18.2) be able to compute

a. The root mean square molecular velocity.
b. The total translational internal energy.

for 1.00 kg of nitrogen gas at 20.0°C. How is this done?

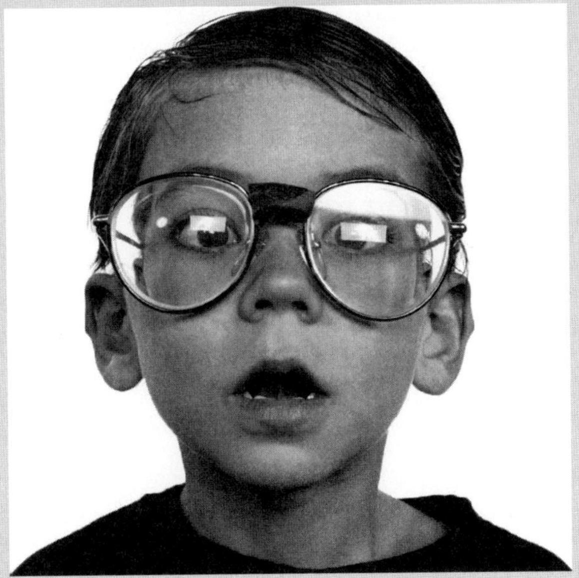

FIGURE 18.2
Example 18.1.

Solution

a. The kinetic theory root mean square molecular velocity is given by Eq. (18.13) as

$$V_{rms} = \sqrt{3RT}$$

For nitrogen, $R = 296$ J/kg·K = 296 N·m/kg·K = 296 m^2/s^2·K (from Table C.15b in *Thermodynamic Tables to accompany Modern Engineering Thermodynamics*), then Eq. (18.13) gives

$$V_{rms} = \sqrt{3[296\text{m}^2/(\text{s}^2\cdot\text{K})](20.0 + 273.15\text{K})} = 510.\text{m/s}$$

b. The kinetic theory total translational internal energy of a gas is given by Eq. (18.14) as

$$U_{trans} = \frac{3}{2}NkT$$

(Continued)

EXAMPLE 18.1 (*Continued*)

Since the total mass of gas present is $m_{total} = m_{molecule}N$, where N is the number of molecules present, and $m_{molecule}$ = molecular mass/Avogadro's number = M/N_o; for nitrogen,

$$m_{molecule} = \frac{M}{N_o} = \frac{28.0\text{ kg/kgmole}}{6.022\times10^{26}\text{ molecules/kgmole}} = 4.65\times10^{-26}\text{ kg/molecule}$$

then,

$$N = \frac{m}{m_{molecule}} = \frac{1.00\text{ kg}}{4.65\times10^{-26}\text{ kg/molecule}} = 2.15\times10^{25}\text{ molecules}$$

The total translational internal energy in 1.00 kg of nitrogen gas is

$$U_{trans} = \frac{3}{2}NkT = \frac{3}{2}\left[(2.15\times10^{25}\text{ molecules})\left(1.380\times10^{-23}\frac{\text{J}}{\text{molecule}\cdot\text{K}}\right)(20.0+273.15\text{ K})\right]$$
$$= 131{,}000\text{ J} = 131\text{ kJ}$$

Note that the kinetic theory internal energy is independent of gas pressure and depends only on gas temperature, which is a characteristic of ideal gas behavior.

Exercises

1. Determine the root mean square velocity of the nitrogen gas in Example 18.1 when the temperature is increased from 20.0°C to 200.°C. **Answer:** V_{rms} = 648 m/s.
2. Suppose the nitrogen gas in Example 18.1 is replaced with water vapor. Determine the root mean square velocity and mass of a water vapor molecule at 20.0°C. **Answer:** $(m_{molecule})_{water} = 2.99 \times 10^{-26}$ kg and $(V_{rms})_{water}$ = 637 m/s.
3. Determine the number of water molecules in 1.00 m^3 of saturated water vapor at 100.°C. **Answer:** $N_{water} = 2.00 \times 10^{25}$ water molecules.

18.4 INTERMOLECULAR COLLISIONS

To better understand molecule-molecule collisions, imagine that all the molecules except one are frozen in space. Then, the moving molecule travels through this stationary forest of molecules colliding with them at random. This model can be further simplified if the moving molecule is enlarged to twice its normal diameter while all the stationary molecules are reduced to points of zero diameter (see Figure 18.3). The area swept out by the motion of the enlarged molecule is called its molecular *collision cross-section*, given by

$$\sigma = \pi(2r)^2 = 4\pi r^2.$$

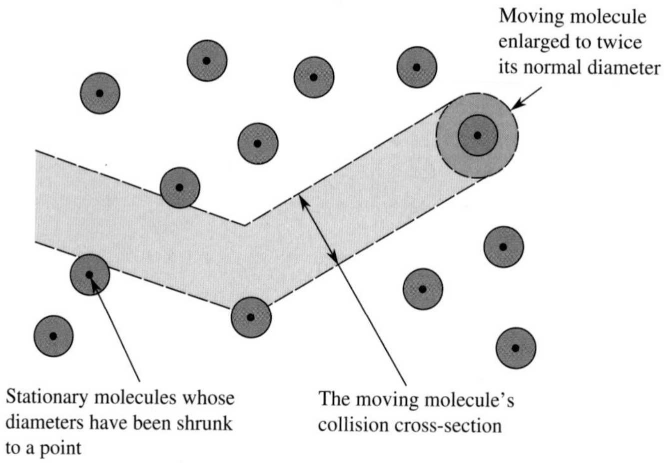

FIGURE 18.3
A simplified model illustrating the molecular collision cross-section.

where r is the effective radius of the kinetic theory spherical molecule. Typical values for the effective radius of simple molecules are given in Table 18.1.

The *collision frequency* $\mathcal{F}$ is the number of collisions per unit time made by the moving molecule, determined from

$$\mathcal{F} = \sigma V_{rms} \frac{N}{\forall}\left(\frac{8}{3\pi}\right)^{1/2}$$

where $N/\forall = N_o/\bar{v} = pN_o/\Re T = p/kT$ is the number of molecules per unit volume, σ is the molecular collision moving cross section, and V_{rms} is the root mean square velocity of the average molecule. The molecular *mean free path* λ is defined to be the distance traveled between molecular collisions, calculated from

$$\lambda = \frac{1}{(N/\forall)\sigma} \tag{18.15}$$

Table 18.1 Typical Values of the Effective Molecular Radius

Molecule	Effective Radius, *r* (m)
He	1.37×10^{-10}
Ne	1.30×10^{-10}
Ar	1.82×10^{-10}
H_2	0.74×10^{-10}
N_2	1.10×10^{-10}
O_2	1.21×10^{-10}
Br_2	2.28×10^{-10}
Cl_2	1.99×10^{-10}
F_2	1.41×10^{-10}
I_2	2.67×10^{-10}
HBr	1.41×10^{-10}
HCl	1.27×10^{-10}
HF	0.92×10^{-10}
HI	1.60×10^{-10}
H_2O	1.50×10^{-10}
CO	1.13×10^{-10}
NO	1.15×10^{-10}
CO_2	2.30×10^{-10}
NH_3	2.22×10^{-10}
CH_4	2.07×10^{-10}

Source: Material drawn from the JANAF Thermochemical Tables, first ed., 1961, Thermal Research Laboratory, Dow Chemical Corporation, Midland, MI. Reprinted by permission.

WHAT ARE ATOMIC AND MOLECULAR RADII?

The distance from the center of the nucleus to the outer electron shell of an atom is called the *atomic radius* of that atom. The distance from the nucleus to the outer shell depends on the electrostatic attraction that the nucleus exerts on the electrons of the outer shell. The atomic radii increase as you move down in the periodic table, as electrons fill outer electron shells. However, the atomic radii decrease as you move from left to right, across the periodic table, even though more electrons are added to atoms. This is because the increasing nuclear charge "pulls" the electron clouds inward, making the atomic radii smaller.

Molecules are much more complex and much less spherical. There are various techniques for measuring molecular geometry such as the *effective* molecular size a molecule displays in a solution. However, a molecule's diameter can be estimated as the cube root of the volume it sweeps out as it moves through space.

EXAMPLE 18.2

Determine the collision frequency and mean free path for neon at 273 K and 0.113 MPa. The molecular mass of neon is 20.183 kg/kgmole.

Solution

For neon,

$$m = \frac{M}{N_o} = \frac{20.183\ \text{kg/kgmole}}{6.022 \times 10^{26}\ \text{molecules/kgmole}} = 3.35 \times 10^{-26}\text{kg/molecule}$$

Then, the root mean square velocity of the neon molecules is given by Eq. (18.13) as

$$V_{rms} = \left(\frac{3kT}{m}\right)^{1/2} = \left[\frac{3[1.38 \times 10^{-23}\text{J/(molecule}\cdot\text{K)}](273\ \text{K})}{3.35 \times 10^{-26}\text{kg/molecule}}\right]^{1/2} = 581\ \text{m/s}$$

From Table 18.1, we find that the radius of the neon molecule is 1.30×10^{-10} m,

so the collision cross-section is

$$\sigma = 4\pi r^2 = 4\pi(1.30 \times 10^{-10}\text{m})^2 = 2.12 \times 10^{-19}\text{m}^2$$

and the collision frequency is

$$\mathcal{F} = \sigma V_{rms}\frac{N}{\forall}\left(\frac{8}{3\pi}\right)^{1/2}$$

where

$$\frac{N}{\forall} = \frac{p}{kT} = \frac{0.113 \times 10^6\ \text{N/m}^2}{[1.380 \times 10^{-23}\text{N}\cdot\text{m/(molecule}\cdot\text{K)}](273\ \text{K})} = 3.00 \times 10^{25}\text{molecules/m}^3$$

then,

$$\mathcal{F} = \sigma V_{avg}\frac{N}{V}\left(\frac{8}{3\pi}\right)^{1/2} = (3.00 \times 10^{25}\ \text{molecules/m}^3)\left(\frac{8}{3\pi}\right)^{1/2}(2.12 \times 10^{-19}\text{m}^2)(581\ \text{m/s})$$

$$= 3.40 \times 10^9\text{collisions/s}$$

so that the molecular mean free path is given by Eq. (18.15) as

$$\lambda = \frac{1}{(N/\forall)\sigma} = \frac{1}{(3.00 \times 10^{25}\text{molecules/m}^3)(2.12 \times 10^{-19}\text{m}^2)} = 1.57 \times 10^{-7} m$$

Exercises

4. If the temperature of the neon in Example 18.2 is increased from 273 K to 1000. K, determine the new root mean square velocity and collision frequency of the neon molecules. **Answer**: V_{rms} = 112 m/s and $\mathcal{F} = 6.5 \times 10^9$collisions/s.
5. Determine the collision cross-sectional area of methane molecules. **Answer**: $\sigma = 5.38 \times 10^{-19}$ m^2.
6. Compute the molecular mean free path for carbon dioxide molecules at 300. K and 1.50 kPa. **Answer**: $\lambda = 4.15 \times 10^{-6}$ m.

18.5 MOLECULAR VELOCITY DISTRIBUTIONS

Theories attempting to explain population behavior in living systems often begin with the following simple differential equation for the time rate of change of the population N:

$$\frac{dN}{dt} = \pm\alpha N \tag{18.16}$$

which says that the rate of change of the population N depends directly on the instantaneous value of the population. If α is a constant, Eq. (18.16) can be integrated to give

$$N = N_0 e^{\pm\alpha t} \tag{18.17}$$

where N_0 is the initial population at time zero. This equation predicts an exponential growth or decay in population depending on the sign of α (see Figure 18.4). In biological systems, Eq. (18.17) is usually inaccurate over long time intervals, because α is not constant but instead depends on a number of variables and is often dependent upon N itself.

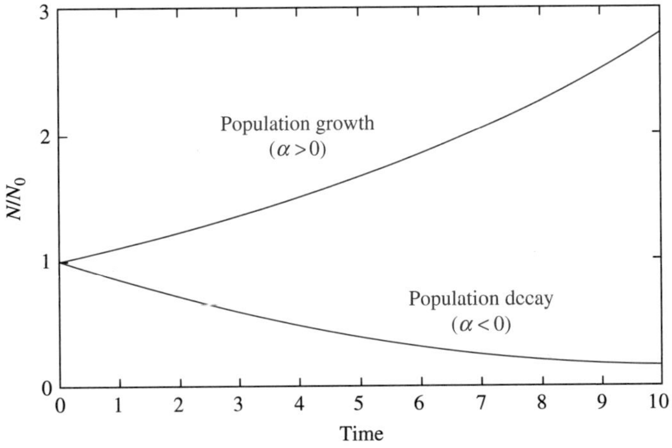

FIGURE 18.4
Population growth and decay as predicted by Eq. (18.17).

The problem of determining the distribution of velocities among the molecules of a gas can be thought of as a population problem, except that we are no longer interested in how the population size N varies with time, but rather how it varies with molecular velocity. Using the general form of Eq. (18.16), we can postulate a velocity distribution population model as follows:

$$\frac{dN_V}{dV} = f(V)N \tag{18.18}$$

where α has been replaced by a more general function $f(V)$, called the *velocity distribution function*. The problem now is to find the mathematical form of $f(V)$. For example, if we assume a Gaussian velocity distribution, we have

$$f(V) = \frac{1}{\sqrt{2\pi}\,\delta}\exp\left(-\frac{V^2}{2\delta^2}\right)$$

where δ is the standard deviation.

By utilizing the assumptions stated at the beginning of this section, Maxwell was able to show that $f(V)$ is not Gaussian, but instead has the following form:

$$f(V) = \frac{4}{\sqrt{\pi}}\left(\frac{m}{2kT}\right)^{3/2} V^2 \exp\left(-\frac{mV^2}{2kT}\right) \tag{18.19}$$

Substituting Eq. (18.19) into Eqs. (18.4) and (18.5), one finds that the average and root mean square velocities have the following simple formulae:

$$V_{avg} = \sqrt{\frac{8kT}{\pi m}} \tag{18.20}$$

$$V_{rms} = \sqrt{\frac{3kT}{m}} \tag{18.21}$$

Figure 18.5 shows the shape of the distribution function $f(V)$ for oxygen at 300 K as described by Eq. (18.19).

We call the velocity at which $f(V)$ has a maximum the *most probable velocity* V_{mp}. It is determined by setting $df(V)/dV = 0$ and solving for $V = V_{mp}$, which, using Eq. (18.19), gives

$$V_{mp} = \sqrt{\frac{2kT}{m}} \tag{18.22}$$

By comparing Eqs. (18.20), (18.21), and (18.22), it is clear that $V_{mp} < V_{avg} < V_{rms}$, as shown in Figure 18.5.

Let $\int_{V_1}^{V_2} dN_V = N(V_1 \rightarrow V_2)$ be the number of molecules with velocities between V_1 and V_2. Then, it follows from Eq. (18.18) that

$$\frac{N(V_1 \rightarrow V_2)}{N} = \int_{V_1}^{V_2} f(V)\,dV \tag{18.23}$$

Substituting Eq. (18.19) into Eq. (18.23) and carrying out the integration gives

$$\frac{N(V_1 \to V_2)}{N} = \text{erf}(x_2) - \text{erf}(x_1) - \frac{2}{\sqrt{\pi}}\left(x_2 e^{-x_2^2} - x_1 e^{-x_1^2}\right) \tag{18.24}$$

where

$$x_1 = V_1/V_{mp},\ x_2 = V_2/V_{mp}, \text{ and } \text{erf}(x) = \text{error function of } x = 2/\sqrt{\pi}\int_0^x e^{-x^2}dx$$

Representative values for the error function can be found in Table 18.2. Note that erf(0) = 0 and erf(∞) = 1.

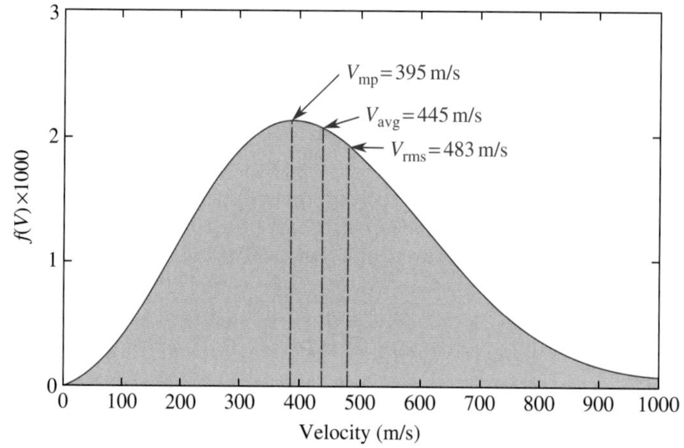

FIGURE 18.5
The Maxwell velocity distribution function $f(V)$ for oxygen (O_2) at 300 K as defined by Eq. (18.19).

Table 18.2 Values of the Error Function

x	erf(x)
0.0	0.0
0.1	0.1125
0.2	0.2227
0.3	0.3286
0.4	0.4284
0.5	0.5205
0.6	0.6039
0.7	0.6778
0.8	0.7421
0.9	0.7969
1.0	0.8427
1.2	0.9103
1.4	0.9523
1.6	0.9764
1.8	0.9891
2.0	0.9953
2.2	0.9981
2.4	0.9993
2.6	0.9998
2.8	0.9999
∞	1.0

Note: For all x, $\text{erf}(x) = \frac{2}{\sqrt{\pi}}\left(x - \frac{x^3}{3(1!)} + \frac{x^5}{5(2!)} - \frac{x^7}{7(3!)} + \ldots\right)$, *and* $\exp(-x^2) = 1 - x^2/1! + x^4/2! - x^6/3! + x^8/4! - \ldots$

Equation (18.24) can be evaluated to find the fraction of molecules whose velocities lie in the range from 0 to V as

$$\frac{N(0 \to V)}{N} = \text{erf}(x) - \frac{2}{\sqrt{\pi}} x e^{-x^2} \tag{18.25}$$

and to find the fraction of molecules whose velocities lie in the range from V to ∞ as

$$\frac{N(V \to \infty)}{N} = 1 - \frac{N(0 \to V)}{N} = 1 - \text{erf}(x) + \frac{2}{\sqrt{\pi}} x e^{-x^2} \tag{18.26}$$

where $x = V/V_{mp}$ in each case.

EXAMPLE 18.3

Test assumption 7 at the beginning of this section by computing the fraction of neon molecules at 273 K whose velocities are faster than (a) V_{mp}, (b) V_{avg}, (c) V_{rms}, and (d) c (the speed of light). Use the molecular data for neon given in Example 18.2.

Solution

a. The fraction having velocities greater than V_{mp} is given by Eq. (18.26) with $x = V_{mp}/V_{mp} = 1.0$ as

$$\frac{N(V_{mp} \to \infty)}{N} = 1 - \text{erf}(1.0) + \frac{2}{\sqrt{\pi}}(1.0)\, e^{-1.0}$$

or

$$\frac{N(V_{mp} \to \infty)}{N} = 1 - 0.8427 + 0.4151 = 0.5724$$

Thus, 57.24% of the molecules have velocities faster than V_{mp}.

b. Here, Eq. (18.20) gives

$$V_{\text{avg}} = \sqrt{\frac{8kT}{\pi m}}, \quad V_{mp} = \sqrt{\frac{2kT}{m}},$$

and

$$x = \frac{V_{\text{avg}}}{V_{mp}} = \sqrt{\frac{8}{2\pi}} = 1.128$$

Therefore, the fraction of molecules having velocities greater than V_{avg} is given by interpolating in Table 18.2 to find

$$\frac{N(V_{\text{avg}} \to \infty)}{N} = 1 - \text{erf}(1.128) + \frac{2}{\sqrt{\pi}}(1.128)e^{-1.272}$$

$$= 1 - 0.8893 + 0.3566 = 0.4673$$

Consequently, 46.73% of the molecules have velocities faster than V_{avg}.

c. Here, Eq. (18.21) gives $V_{\text{rms}} = \sqrt{3kT/m}$, so

$$x = V_{\text{rms}}/V_{mp} = \sqrt{\tfrac{3}{2}} = 1.225$$

and

$$\frac{N(V_{\text{rms}} \to \infty)}{N} = 1 - \text{erf}(1.225) + \frac{2}{\sqrt{\pi}}(1.225)e^{-1.501}$$

$$= 1 - 0.9168 + 0.3081 = 0.3913$$

or 39.13% of the molecules have velocities greater than V_{rms}.

d. It can be shown (see Problem 10 at the end of this chapter) that, when $V/V_{mp} \gg 1$, the fraction of molecules with velocities in the range from V to ∞ is approximately given by

$$\frac{N(V \to \infty)}{N} \approx \frac{2}{\sqrt{\pi}}\left(x + \frac{1}{2x}\right)e^{-x^2} \text{ (for } x \gg 1 \text{ only)}$$

(Continued)

EXAMPLE 18.3 *(Continued)*

Consider $x = V/V_{mp} = 10.00$ then,

$$\frac{N(V \to \infty)}{N} \approx \frac{2}{\sqrt{\pi}}(10.05)e^{-100} = 4.22 \times 10^{-43}$$

Thus, only one molecule in about 10^{20} moles of a gas has a velocity ten times greater than V_{mp}. Now, the velocity of light c is 3.00×10^8 m/s, and for neon at 273 K. we have $m = 3.35 \times 10^{-26}$ kg (see Example 18.2), and $V_{mp} = \sqrt{2kT/m} = 474$ m/s. Thus,

$$x = \frac{V}{V_{mp}} = \frac{c}{V_{mp}} = \frac{3.00 \times 10^8\,\text{m/s}}{474\,\text{m/s}} = 6.33 \times 10^5$$

so that

$$\frac{N(c \to \infty)}{N} \approx \frac{2}{\sqrt{\pi}}(6.33 \times 10^5)e^{-4.0\times10^{11}} \approx 0$$

Even though we allow molecules to move faster than the speed of light in our mathematical model, we find that, for all practical purposes, this model predicts that virtually no molecules have velocities this fast at ordinary temperatures.

Exercises

7. Plot the actual value and the four-term series expansion of the error function given in Table 18.2 for $0.1 \le x \le 2.0$ and compare the results. **Partial answer**: The series expansion is valid only for $x < 1.0$, see Figure 18.6.
8. Determine the percentage of molecules in Example 18.3 that have velocities greater than twice the most probable velocity. **Answer**: $N(2V_{mp} \to \infty)/N = 0.046 = 4.6\%$.
9. Determine the *mean* velocity in Example 18.3 (i.e., find the velocity V for which half the molecules have a velocity greater than V and half have a velocity less than V) as defined by $N(V \to \infty)/N = 0.5 = 50\%$. **Answer**: $V = 1.09 \times V_{mp}$.

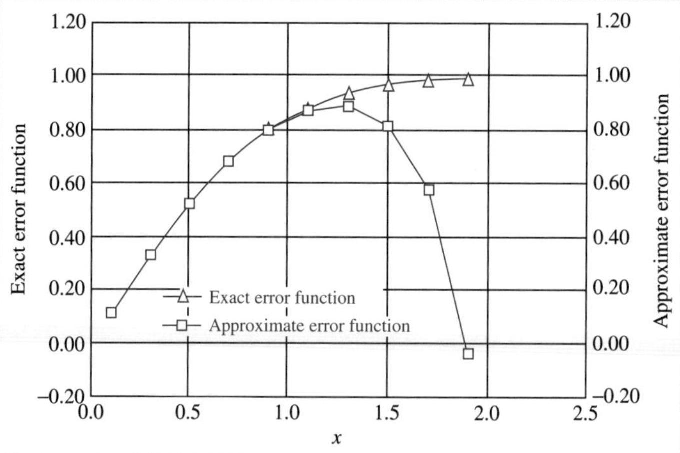

FIGURE 18.6
Example 18.3, Exercise 7.

18.6 EQUIPARTITION OF ENERGY

Equation (18.14) gives the total translational kinetic energy of a system of N molecules as $\frac{3}{2}NkT$. The principle of *equipartition of energy* requires that the translational kinetic energy of an unrestricted molecule be equally divided among the three translational degrees of freedom (one for each independent coordinate direction). Therefore, the translational total internal energy in each of the x, y, and z coordinate directions must be one third of that given in Eq. (18.14), or

$$(U_{\text{trans}})_x = (U_{\text{trans}})_y = (U_{\text{trans}})_z = U_{\text{trans}}/3 = \frac{1}{2}NkT$$

We can therefore conclude that the total energy of a system of N molecules with F degrees of freedom per molecule is given by

$$U = \frac{F}{2}NkT = \frac{F}{2}m_T RT$$

and that its specific internal energy is $u = U/m_T = FRT/2$. From the definition of constant volume specific heat introduced in Chapter 3, we have

$$c_v = \left(\frac{\partial u}{\partial T}\right)_v = \frac{\partial}{\partial T}\left(\frac{F}{2}RT\right)_v = \frac{FR}{2} \quad (18.27)$$

and, since $c_p - c_v = R$ for an ideal gas,

$$c_p = R + c_v = \left(1 + \frac{F}{2}\right)R \quad (18.28)$$

finally the specific heat ratio k becomes

$$k = \frac{c_p}{c_v} = \frac{F+2}{F} \quad (18.29)$$

For a molecule containing b atoms, there are $F = 3b$ degrees of freedom. If $b = 1$, then $F = 3$ and the three degrees of freedom are all translational. If $b = 2$, then $F = 6$ and there are three degrees of freedom in translation, two in rotation and one in vibration. If $b > 2$, then $F = 3b$ and there are three degrees of freedom in translation, three in rotation and $3b - 6$ in vibration. In the case of monatomic (single-atom) molecules like He, Ne, Ar, Kr, or Xe, there are only three degrees of freedom (all translational). Then Eqs. (18.27), (18.28), and (18.29) give

$$\left.\begin{aligned} c_v &= 1.5R \\ c_p &= 2.5R \\ k &= 1.67 \end{aligned}\right\}\text{Monatomic gases}$$

For diatomic (two-atom) molecules, such as H_2, O_2, CO, or NO, we have $b = 2$ and consequently $F = 3(2) = 6$. Then, we have

$$\left.\begin{aligned} c_v &= 3R \\ c_p &= 4R \\ k &= 1.33 \end{aligned}\right\}\text{Diatomic gases}$$

Similarly for triatomic gases, such as CO_2, H_2O, NO_2, or SO_2, we have $b = 3$ and $F = 9$. Then,

$$\left.\begin{aligned} c_v &= 4.5R \\ c_p &= 5.5R \\ k &= 1.22 \end{aligned}\right\}\text{Triatomic gases}$$

A comparison of these values with the measured specific heats of some real gases, given in Table 18.3, reveals that, for simple molecules (e.g., monatomic gases), the kinetic theory works quite well. For complex molecules, however, kinetic theory predictions are much less accurate.

In summary, the thermodynamic properties of an ideal gas as predicted by Maxwell's kinetic theory are

$$\begin{aligned} p\forall &= mRT \\ u_2 - u_1 &= c_v(T_2 - T_1) \\ h_2 - h_1 &= c_p(T_2 - T_1) \\ s_2 - s_1 &= c_p \ln(T_2/T_1) - R\ln(p_2/p_1) = c_v \ln(T_2/T_1) + R\ln(v_2/v_1) \\ c_p - c_v &= R \end{aligned}$$

where

$$\begin{aligned} c_v &= FR/2 \\ c_p &= (1 + F/2)R \\ F &= 3b \\ b &= \text{number of atoms in the molecule} \end{aligned}$$

Table 18.3 Measured Values of the Specific Heats of Various Gases at 20.0°C

Gas	c_v/R	c_p/R	$k = c_p/c_v$
Monatomic			
He	1.50	2.50	1.67
Ne	1.50	2.50	1.67
Ar	1.51	2.52	1.67
Kr	1.00	1.68	1.68
Xe	1.52	2.51	1.65
Diatomic			
CO	2.51	3.51	1.40
NO	2.51	3.51	1.40
H_2	2.44	3.42	1.40
O_2	2.53	3.53	1.40
N_2	2.50	3.50	1.40
Triatomic			
CO_2	3.48	4.48	1.29
SO_2	3.97	4.97	1.25
H_2O	3.05	4.05	1.33

EXAMPLE 18.4

Estimate the heat transfer rate required to heat low-pressure gaseous carbon tetrachloride (CCl_4) from 500. to 1200. K in a steady state, steady flow, single-inlet, single-outlet, aergonic (i.e., zero work) process at a flow rate of 1.00 kg/min.

Solution

The system here is just the gas in the heating zone. Neglecting the flow stream kinetic and potential energies, the energy rate balance for this system reduces to

$$\dot{Q} + \dot{m}(h_{\text{in}} - h_{\text{out}}) = 0$$

so that

$$\dot{Q} = \dot{m}(h_{\text{out}} - h_{\text{in}}) = \dot{m}c_p(T_{\text{out}} - T_{\text{in}})$$

For $CC1_4$, $b = 5$; consequently, $F = 3b = 15$. Then, Eq. (18.28) gives

$$c_p = (1 + 15/2)R = 8.5 \times R$$

Now, the molecular mass of carbon tetrachloride is

$$M = 12.0 + 4(35.5) = 154 \text{ kg/kgmole}$$

and its gas constant is

$$R = \frac{\Re}{M} = \frac{8.3143 \text{ kJ/(kgmole} \cdot \text{K)}}{154 \text{ kg/kgmole}} = 0.0540 \text{ kJ/(kg} \cdot \text{K)}$$

so

$$c_p = 8.5[0.0540 \text{ kJ/(kg} \cdot \text{K)}] = 0.459 \text{ kJ/(kg} \cdot \text{K)}$$

Therefore,

$$\dot{Q} = (1.00 \text{ kg/min})[0.459 \text{ kJ/(kg} \cdot \text{K)}](1200. - 500.\text{K}) = 321 \text{ kJ/min}$$

Exercises

10. If the mass flow rate of gaseous carbon tetrachloride in Example 18.4 is suddenly increased from 1.00 to 3.50 kg/min, determine the new heat transfer rate, assuming all the other variables remain unchanged. **Answer:** $\dot{Q} = 1120$ kJ/min.

11. The exit temperature of the gaseous carbon tetrachloride in Example 18.4 is increased from 1200. K to 2000. K. Determine the heat transfer rate, assuming all the other variables remain unchanged. **Answer:** $\dot{Q} = 688$ kJ/min.

12. The gas in Example 18.4 is changed from carbon tetrachloride to gaseous dichlorodifluoromethane (CCl_2F_2). Estimate the heat transfer rate for this gas, assuming all the other variables in Example 18.4 remain unchanged. **Answer:** $\dot{Q} = 409$ kJ/min.

18.7 INTRODUCTION TO MATHEMATICAL PROBABILITY

If there are N mutually exclusive[4] equally likely outcomes of an experiment, M of which results in event A, then we write the *probability of the occurrence* of event A, P_A, as

$$P_A = \frac{M}{N} \tag{18.30}$$

Now, if $P_A = 0$, then event A is *impossible* and if $P_A = 1$, then event A is a *certainty*. Generally, probabilities do not take on these extreme values but instead lie somewhere in the region

$$0 \leq P_A \leq 1$$

Let $\sim P_A$ be the probability of event A *not* occurring. Then, because it is a certainty that either event A will occur or it will not occur, we can write

$$P_A + \sim P_A = 1$$

EXAMPLE 18.5

a. Consider the toss of a single evenly weighted die. What are the six possible mutually exclusive results that can occur?
b. Now consider the toss of a pair of evenly weighted dice and add their individual results. What is the probability of each of the resulting sums?
c. What is the most probable sum in item b?

Solution

a. Since the die is an evenly weighted cube, all six sides have the same probability of landing face up, so we have $N = 6$. Consequently the six possible mutually exclusive results are

$$P_1 = P_2 = P_3 = P_4 = P_5 = P_6 = 1/6$$

Also, we can write

$$\sim P_1 = \sim P_2 = \sim P_3 = \sim P_4 = \sim P_5 = \sim P_6 = 5/6$$

b. The total number of combinations of results is shown in Table 18.4. From this table, it is seen that $N = 6 \times 6 = 36$. Using this table, we can construct an event vs. frequency table, as shown in Table 18.5. Thus, we can compute the following probabilities for the sum of the results of the individual die:

$$P_0 = P_1 = 0 \qquad P_5 = P_9 = 4/36 = 1/9$$
$$P_2 = P_{12} = 1/36 \qquad P_6 = P_8 = 5/36$$
$$P_3 = P_{11} = 2/36 = 1/18 \qquad P_7 = 6/36 = 1/6$$
$$P_4 = P_{10} = 3/36 = 1/12$$

c. From Table 18.5, it is clear that in the toss of two evenly weighted dice, the number 7 is the most probable outcome, appearing on the average of once every six tosses. It has a probability given by Eq. (18.30) of

$$P_7 = 1/6 = 0.1667 = 16.67\%$$

Exercises

13. If you toss an evenly weighted die, what is the probability that it will come up with an even number? **Answer**: *P*(even number) = 1/2.
14. Suppose you have a four-sided instead of a six-sided die. How many outcomes for the sum when tossing two such dice are possible, and what is the most probable outcome for the sum and what is its probability? **Answer**: $M = 16$, and most probable sum = 5 (from 1 + 4, 4 + 1, 2 + 3, and 3 + 2), then *P*(most probable) = $P_5 = 4/16 = 0.25$.
15. The probability of X not occurring is simply $P(\text{not } X) = 1 - P(X)$. Using this concept, determine the probability of not throwing a 7 as the sum of two six-sided dice. **Answer**: $P(\text{not } 7) = 1 - P(7) = 1 - 1/6 = 5/6$.

(Continued)

[4] *Mutually exclusive* means that no two of the N outcomes can occur simultaneously.

EXAMPLE 18.5 *(Continued)*

Table 18.4 The Total Number of Combinations of Tossing Two Dice

Die 1	Die 2	Die 1	Die 2
1	1	4	1
1	2	4	2
1	3	4	3
1	4	4	4
1	5	4	5
1	6	4	6
2	1	5	1
2	2	5	2
2	3	5	3
2	4	5	4
2	5	5	5
2	6	5	6
3	1	6	1
3	2	6	2
3	3	6	3
3	4	6	4
3	5	6	5
3	6	6	6

Table 18.5 An Event-Frequency Table for Tossing Two Dice

Sum M of Die Values	Number of Results Producing Sum M	Ways of Obtaining Sum M
0	0	
1	0	
2	1	1 + 1
3	2	1 + 2, 2 + 1
4	3	2 + 2, 1 + 3, 3 + 1
5	4	2 + 3, 3 + 2, 1 + 5, 4 + 1
6	5	3 + 3, 4 + 2, 2 + 5, 5 + 1, 1 + 5
7	6	6 + 1, 1 + 6, 5 + 2, 2 + 5, 4 + 3, 3 + 4
8	5	4 + 4, 5 + 3, 3 + 6, 6 + 2, 2 + 6
9	4	4 + 5, 5 + 4, 3 + 7, 6 + 3
10	3	6 + 4, 4 + 6, 5 + 6
11	2	6 + 5, 5 + 6
12	1	6 + 6

This probability concept can be used to further investigate the collision frequency characteristics of the kinetic theory model of ideal gases. Let the number of molecular collisions occurring in an ideal gas during some time interval $\Delta t = t_2 - t_1$ be δN, and let $N(t)$ be the number of molecules of the gas at time t that have not yet had a collision. Then, $\delta N = N(t_1) - N(t_2)$. Since we define the Δ symbol to be evaluated at time t_2 minus time t_1, ΔN becomes

$$\Delta N = N(t_2) - N(t_1) = -\delta N$$

We postulate that δN will be proportional to the product of $N(t)$, Δt, and the average molecular velocity V_{avg} as follows:

$$\delta N = -\Delta N \propto -N V_{avg}(\Delta t)$$

WHAT ARE THE GAMES CRAPS AND DICE?

Craps is a simplified version of the old game of *hazard* introduced in Europe during the Crusades. Craps is played between two players. If the first throw is a 7 or 11 (a natural), the thrower wins immediately. If it a 2, 3, or 12 (craps, which was called *crabs* in the old game of hazard), the thrower loses immediately. If any other number is thrown the thrower goes on throwing until either the same number or a 7 occurs. If a 7 comes up before the first number thrown, the thrower loses; otherwise, the thrower wins.

Most casino dice are custom-made of celluloid with a tolerance of one 10,000th of an inch (even the heat of a shootor's hands may slightly alter the size of a pair of dice). The celluloid is "cured" over a period of time to dry it out and give it stability. Then, it goes through a series of milling operations to form it into a perfect cube. The shallow indentations that form the spots are made on all six sides. These indentations are filled with a polyester resin, and the dice are then subjected to a final grinding, which leaves them with smooth surfaces. Cheap plastic dice are polished in a mechanical tumbler similar to those used by lapidarists, but such mechanical polishing also gives them rounded edges which can affect the outcome of a throw. Some casinos have dice made in odd sizes to prevent a dishonest player from switching loaded dice for honest dice. Dice in most casinos are changed every 30 days to maintain precise sizing, and pairs of used dice are often given to customers as souvenirs.

This proportionality can be reduced to an equality by the introduction of a proportionality constant P_c, which we call the *collision probability*. Then, the number of collisions that occur in time interval Δt becomes

$$\delta N = -P_c N V_{avg}(\Delta t)$$

In the limit as $\Delta t \to 0$, this equation becomes

$$dN = -P_c N V_{avg} dt$$

Let $V_{avg} dt = dX$, where X is the distance between collisions. This equation can be written as

$$\frac{dN}{N} = -P_c\, dX$$

which can be integrated to give

$$N = N(t) = N_0 \exp(-P_c X) \tag{18.31}$$

where $N_0 = N(t = 0)$. In Eq. (18.15), we define the mean free path λ as the average distance between collisions, which can be determined from X as

$$\lambda = X_{avg} = \frac{1}{N_0} \int_0^\infty X(-dN)$$

where the minus sign has been introduced because $N(t)$ decreases with time, and therefore, $dN < 0$. From Eq. (18.31), we find that

$$dN = -N_0 P_c \exp(-P_c X) dX$$

and the mean free path is given by

$$\lambda = \frac{1}{N_0} \int_0^\infty X N_0 P_c \exp(-P_c X) dX = \frac{1}{P_c}$$

Consequently, we have the result that the collision probability is exactly equal to the inverse of the mean free path:

$$P_c = \frac{1}{\lambda} = \sigma N \forall$$

where σ is the collision cross-section discussed earlier. Finally, Eq. (18.31) takes the form

$$N(t) = N_0 \exp(-X/\lambda)$$

so that, by the time all the molecules in the gas have traveled a distance of only one mean free path ($X = \lambda$),

$$\frac{N}{N_0} = e^{-1} = 0.368 = 36.8\%$$

of them have not yet had a collision.

Probabilities of related events that are not mutually exclusive have the following interpretation. Assume that event B depends in some way on another event A. Then, the *conditional* probability of event B occurring given the fact that event A has already occurred is written $P_{B/A}$. The *compound* probability that both events A and B will occur is written as P_{AB}. The relation between these two probabilities is the basic **and** logic probability statement, written as

$$P(A\,\text{and}\,B) = P_{AB} = P_A P_{B/A} = P_B P_{A/B} \tag{18.32}$$

If events A and B are totally *independent*, their probabilities are uncoupled and $P_{A/B} = P_A$ and $P_{B/A} = P_B$. Then,

$$P(A\,\text{and}\,B) = P_{AB} = P_A P_B \tag{18.33}$$

Similarly, the basic **or** logic probability statement can be written as

$$P(A\,\text{or}\,B) = P_A + P_B - P_{AB} \tag{18.34}$$

If events A and B are mutually exclusive (i.e., they cannot occur simultaneously), then $P_{AB} = 0$ and $P\,(A \text{ or } B) = P_A + P_B$.

EXAMPLE 18.6

In the draw of a single card from a full deck of playing cards, what is the probability that it will be an ace or a spade?

Solution

Equation (18.34) gives P (ace or spade) $= P_{\text{ace}} + P_{\text{spade}} - P_{\text{ace of spades}}$. Now,

$$P_{\text{ace}} = \frac{4}{52} = \frac{1}{13},\quad P_{\text{spade}} = \frac{13}{52} = \frac{1}{4},\quad P_{\text{ace of spades}} = \frac{1}{52}$$

so

$$P\,(\text{ace or spade}) = \frac{4}{52} + \frac{13}{52} - \frac{1}{52} = \frac{16}{52} = 0.308 = 30.8\%$$

Exercises

16. Instead of drawing either an ace or a spade in Example 18.6, determine the probability of drawing either a 2 or a 3 of any suit. **Answer**: P(2 or 3) = 0.154 = 15.4% of the time.
17. What is the probability of drawing an ace of spades in Example 18.6? **Answer**: P(ace and spade) = 0.0192 = 1.92% of the time.
18. Find the probability of drawing four aces in a row out of a standard 52 card deck. Hint: Initially, there are four aces scattered among the 52 cards, but after the first ace is drawn, only three aces remain among 51 cards, and so forth. **Answer**: P(ace and ace and ace and ace) = 3.7×10^{-6}.

Table 18.6 lists the probability of the different five-card poker hands.

Table 18.6 Probabilities of Five-Card Poker Hands

Name of Hand	Number of Card Combinations Giving This Hand	Probability of the Hand
Royal flush	4	1.54×10^{-6}
Straight flush	36	1.38×10^{-5}
Four of a kind	624	0.00024
Full house	3,744	0.00144
Flush	5,108	0.00197
Straight	10,200	0.00392
Three of a kind	54,912	0.02113
Two pair	123,552	0.0475
One pair	1,098,240	0.42257
All other hands	1,302,540	0.50118
Total	2,598,960	1.00000

The last mathematical concepts needed in our study of probability are those of permutations and combinations. A specific ordered arrangement of N distinguishable objects is called a *permutation*. The total number of ways of making different ordered arrangements of the N objects taken R at a time using no object more than once is given by

$$P_R^N = \frac{N!}{(N-R)!} \tag{18.35a}$$

where $N! = N(N-1)(N-2)(N-3)\ldots(3)(2)(1)$ is N *factorial*. Note that we define $0! = 1$. Thus, the total number of permutations of N distinct objects taken N at a time is

$$P_R^N = N!/0! = N! \tag{18.35b}$$

However, if the objects are allowed to be repeated within an arrangement, then the total number of arrangements becomes

$$P_N^N = N^R \tag{18.35c}$$

For example, the total number of permutations (arrangements) of the ten digits 0, 1, 2, 3, 4, 5, 6, 7, 8, and 9 taken three at a time, using no digit more than once, is given by Eq. (18.35a) with $N = 10$ and $R = 3$ as

$$(P_3^{10})_{\substack{\text{using each}\\ \text{digit only once}}} = \frac{10!}{(10-3)!} = \frac{10!}{7!} = \frac{10\times 9\times 8\times 7!}{7!} = 720 \text{ arrangements}$$

But, if we allow each digit to be used more than once, then Eq. (18.35c) tells us that the number of permutations increases to

$$(P_3^{10})_{\substack{\text{using each}\\ \text{digit more than once}}} = 10^3 = 1000 \text{ arrangements}$$

However, the total number of permutations (arrangements) of just the three digits 1, 2, and 3, using none of these digits more than once is given by Eq. (18.35b) as

$$(P_3^3)_{\substack{\text{using each}\\ \text{digit only once}}} = 3! = 3\times 2\times 1 = 6 \text{ arrangements}$$

(they are 1, 2, 3; 1, 3, 2; 3, 1, 2; 2, 3, 1; 3, 2, 1; and 2, 1, 3), but if we allow these digits to be repeated, the Eq. (18.35c) shows that the number of permutations increases dramatically to

$$(P_3^3)_{\substack{\text{using each}\\ \text{digit more than once}}} = 3^3 = 27 \text{ arrangements}$$

We define *combinations* as the ways of choosing a sample of R objects from a group of N objects without regard to order within the sample (e.g., the groupings AB and BA are different permutations of A and B, but they are the same combination of A and B). The total number of combinations of N unique objects taken R at a time without using any object more than once is given by

$$C_R^N = \frac{P_R^N}{R!} = \frac{N!}{(N-R)!R!} \tag{18.36a}$$

But, if the objects are allowed to be repeated within the sample, then the number of combinations becomes

$$C_R^N = \frac{P_R^N}{R!} = \frac{(N+R-1)!}{(N-1)!R!} \tag{18.36b}$$

For example, we can determine the number of five-card hands that can be dealt out of a standard deck of 52 distinct cards (Table 18.6). Since the order in which the cards are received is not important, the number of combinations of 52 cards taken 5 at a time is given by Eq. (18.36a) as

$$(C_5^{52})_{\substack{\text{using each}\\ \text{card only once}}} = \frac{52!}{(52-5)!5!} = \frac{52\times 51\times 50\times 49\times 48\times 47!}{47!\times 5!} = 2{,}598{,}960 \text{ different hands}$$

and since there are 13 possible four-of-a-kind hands in the deck with each hand having any of the remaining 48 cards as the fifth card in the hand, the probability of getting a five-card hand with four-of-a-kind is

$$P_{\text{four-of-a-kind hand}} = \frac{13 \times 48}{2{,}598{,}960} = 0.00024$$

or only 0.024% of the time. However, suppose each card is returned to the deck before the next card is drawn so that a card can be drawn more than once (i.e., repeated), then the number of combinations of five-card hands is given by Eq. (18.36b) as

$$(C_5^{52})_{\substack{\text{using each card}\\ \text{more than once}}} = \frac{(52+5-1)!}{(52-1)!5!} = \frac{56 \times 55 \times 54 \times 53 \times 52 \times 51!}{51! \times 5!} = 3{,}819{,}816 \text{ different hands}$$

Suppose that not all objects in the group of N are different from each other. The number of permutations of N objects, R_1 of one kind, R_2 of a second kind, ..., R_k of a kth kind is given by

$$P^N_{R_1,R_2,\ldots,R_k} = \frac{N!}{R_1!R_2!R_3!\ldots R_k!} \tag{18.37}$$

since the objects within the k groups are no longer unique, the total number of combinations and permutations are equal, or

$$C^N_{R_1,R_2,\ldots,R_k} = P^N_{R_1,R_2,\ldots,R_k}$$

EXAMPLE 18.7

Suppose you have to form a team of five students from a group of ten available students.

a. How many different five-person groups of officers could you form from the ten students if each student filled one of the positions of president, vice president, secretary, treasurer, and events coordinator (i.e., the group is *ordered*), without using any student more than once?
b. How many ordered officer groups could you form if you allowed students to be in more than one group?
c. How many officer groups could you form if the students were not assigned a position (i.e., the groups were not ordered) but a student could not be in more than one group?
d. How many teams could you form if the students were not assigned a position (i.e., the groups were not ordered) and students could be in more than one group?
e. If there are four men and six women in the group, how many different unordered ten-person groups could be formed?

Solution

a. Equation (18.35a) gives the number of ordered groups of $R = 5$ things chosen from a group of $N = 10$ as

$$(P_5^{10})_{\substack{\text{using each}\\ \text{student only once}}} = \frac{10!}{(10-5)!} = \frac{10!}{5!} = \frac{10 \times 9 \times 8 \times 7 \times 6 \times 5!}{5!}$$
$$= 10 \times 9 \times 8 \times 7 \times 6 = 30{,}240 \text{ groups}$$

b. Equation (18.35c) gives the result when the students are allowed to be in more than one group:

$$(P_3^{10})_{\substack{\text{using each student}\\ \text{more than once}}} = 10^5 = 100{,}000 \text{ groups}$$

c. If the students are not assigned a position within the group, but they are allowed to belong to only one group, then Eq. (18.36a) gives the possible number of groups as

$$(C_5^{10})_{\substack{\text{using each}\\ \text{student only once}}} = \frac{10!}{(10-5)!5!} = \frac{10 \times 9 \times 8 \times 7 \times 6 \times 5!}{5! \times 5!} = 252 \text{ groups}$$

d. If the students are not assigned a position within the group, but they are allowed to belong to more than one group, then Eq. (18.36b) gives the possible number of groups as

$$(C_5^{10})_{\substack{\text{using each student}\\ \text{more than once}}} = \frac{(10+5-1)!}{(10-1)!5!} = \frac{14 \times 13 \times 12 \times 11 \times 10 \times 9!}{9! \times 5!} = 2002 \text{ groups}$$

e. If there are four men and six women in the group, then the number of different unordered ten-person groups that can be formed is given by Eq. (18.37) as

$$P_{4,6}^{10} = \frac{10!}{4! \times 6!} = \frac{10 \times 9 \times 8 \times 7 \times 6!}{4 \times 3 \times 2 \times 1 \times 6!} = 210 \text{ groups}$$

Exercises

19. The available group in Example 18.7 suddenly drops from ten to six students from which to form the five-person officer groups. How many different officer groups could you form without using any student more than once?
Answer: $P_5^6 5 = 720$ groups.

20. Suppose you only need four officers instead of five in Example 18.7. How many different officer groups could you form if you allowed students to be in more than one group? Assume ten students are still available for the officer positions.
Answer: $P_4^{10} 4 = 10{,}000$ groups.

21. How many different arrangements are there of three black, seven red, and four green marbles?
Answer: $P_{3,7,4}^{14} = 120{,}120$ arrangements.

WHAT ARE PERMUTATIONS AND COMBINATIONS?

A *permutation*[5] is an arrangement of a group of items in specific order. Consider the group of three items denoted by *X*, *Y*, and *Z*. The number of ordered arrangements of these three items taking two at a time without allowing repetition (this is called *permutations without repetition*) is

$$P_2^3 = \frac{3!}{(3!2)!} = 6$$

These arrangements are *XY*, *YX*, *YZ*, *ZY*, *XZ*, and *ZX*. However, the number of ordered arrangements when repetition is allowed (called *permutations with repetition*) is

$$P_2^3 = 3^2 = 9$$

and these arrangements are *XY*, *YX*, *YZ*, *ZY*, *XZ*, and *ZX* plus the repeats *XX*, *YY*, and *ZZ*.

A *combination* is an arrangement of a group of items where the order does not matter. If order is not important within the group, then the number of arrangements of the three items *X*, *Y*, and *Z* taking two at a time but not allowing items to be repeated (called *combinations without repetition*) is

$$C_2^3 = \frac{3!}{(3!2)!2!} = 3$$

and these arrangements are *XY*, *YZ*, and *XZ* (note that *XY* is the same as *YX* when the order within the group is not important). But, if we allow repetition of the items within the groups, then the number of arrangements (called combinations with repetition) becomes

$$C_2^3 = \frac{(3+2!1)!}{(3!2)!2!} = 6$$

and they are *XY*, *YZ*, and *XZ* plus the repeats *XX*, *YY*, and *ZZ*.

In summary, if the order does not matter, it is a *combination*. If the order does matter, it is a *permutation*.

[5] *The word* permutation *is from the Latin "per" (thoroughly) + "mutare" (to change).*

18.8 QUANTUM STATISTICAL THERMODYNAMICS

We begin by defining the *microstate* of a group of molecules as the state produced by specifying the instantaneous energy state of each molecule of the group. We define the *macrostate* as the instantaneous average state of the collection of molecules, and the thermodynamic *equilibrium state* as being the most probable macrostate. The mathematical probability of macrostate *A* is defined as

$$p_A = \frac{W(A)}{\sum_i W(i)} \tag{18.38}$$

where $W(A)$ is the number of ways that macrostate A can occur (i.e., the number of microstates per macrostate A), and $\sum_i W(i)$ is the total number of macrostates possible.

Then, the condition we call *thermodynamic equilibrium* is simply the macrostate that has the largest value of P.

A macrostate is an overview of a complex situation, whereas a microstate describes the details of how each element of the system functions. As an analogy, consider a national presidential election. The macrostates are the various possible winners of the election, and the microstates are the various combinations of ways in which the voters may cast their ballots.

Late in the 19th century, electrical discharge experiments in various gases produced light emission spectra that were very unusual. Instead of being an emission with a continuous color frequency (like *white* light), the emissions consisted of discrete spectral lines located at fixed wavelengths. Figure 18.7 shows the emission spectrum of atomic hydrogen in the visible region of the electromagnetic spectrum. From these emission spectra, it was clear that, if the emission phenomenon is attributed to photon ejection by electrons as they move from an atom's outer orbit to an inner orbit, then the electrons must occupy discrete orbits and consequently are not simply clustered around the nucleus in a random manner.

In 1913, Niels Bohr hypothesized that the electron orbits of an atom were *quantitized* (i.e., made discrete) according to the value of the electron's angular momentum as

$$m\omega = mVr = n\frac{\hbar}{2\pi}$$

where ω is the angular velocity of the electron, V is the electron's orbital velocity, r is the radius of the orbit, $\hbar$ is Planck's constant, and $n = 1, 2, 3, \ldots$ is the (primary) quantum number. Therefore, the radius of an electron's orbit is given by

$$r = \frac{\hbar}{2\pi mV}n$$

As the years passed, other quantum numbers had to be introduced to account for such things as the elliptical shape of the orbit (this accounted for the finite width of the emission lines and was called the *azimuthal* quantum number), the splitting of the spectral lines in a strong magnetic field (the *magnetic* quantum number), the magnetic moment associated with the direction of electron spin (the *electron spin* quantum number), and so forth.

The continual modification of the original Bohr model required to make it conform to experimental observations started physicists looking for a new model. In 1924, Louis Victor Pierre Raymond de Broglie (1892–1987) used an analogy between classical mechanics and geometric optics to formulate a dual particle-wave model for matter. He argued that since the energy ε of a photon is given by

$$\varepsilon = \hbar v$$

where $\hbar$ is Planck's constant and v is the photon's frequency, and since Einstein's mass-energy relation for the photon is

$$\varepsilon = mc^2$$

where m is its mass and c is the velocity of light, then the linear momentum p of the photon can be written as

$$p = mc = \frac{\hbar v}{c} = \frac{\hbar}{\lambda}$$

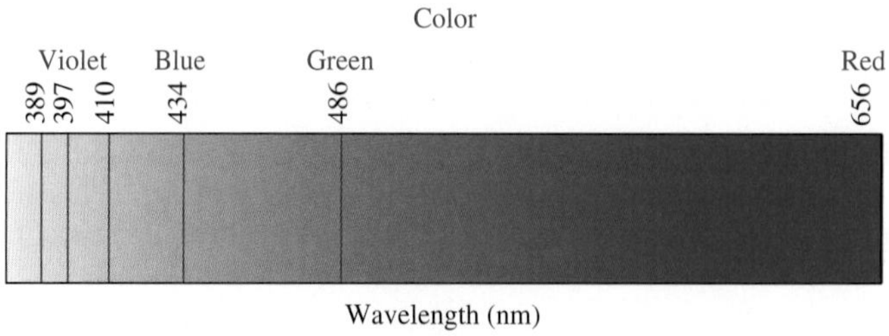

FIGURE 18.7
The emission spectrum of atomic hydrogen.

where $\lambda = c/v$ is the wavelength of the photon. De Broglie then extended the argument to mass particles (like electrons) by postulating that, for them,

$$p = mV = \frac{\hbar}{\lambda}$$

where λ is the particle's *wavelength*. This postulation was experimentally verified in 1927, when it was demonstrated that electrons could be diffracted in a wavelike manner from a ruled diffraction surface. Thus, electrons appeared to have both particlelike and wavelike behavior, and the *duality* principle of matter was established.

Once the wavelike character of matter was recognized, it became clear that the kinetic behavior of atomic particles ought to be governed by the same equations that govern the propagation of waves in a continuum. In 1926, Erwin Schrödinger developed an unsteady wave equation appropriate to matter waves of the form

$$\nabla^2\psi = \frac{2m(\varepsilon - \varepsilon_p)}{\varepsilon^2}\left(\frac{\partial^2\psi}{\partial t^2}\right) \tag{18.39}$$

where ψ is the *wave function* (the wave amplitude), ε is the total energy of the particle, ε_p is the potential energy of the particle, m is the particle's mass, and ∇^2 is the differential operator defined by

$$\nabla^2(\,) = \frac{\partial^2}{\partial x^2}(\,) + \frac{\partial^2}{\partial y^2}(\,) + \frac{\partial^2}{\partial z^2}(\,)$$

Remarkably, the solutions to (18.39) are inherently quantitized (i.e., solutions exist for only discrete values of ε); thus, it has become a fundamental equation in quantum mechanics.

18.9 THREE CLASSICAL QUANTUM STATISTICAL MODELS

Consider a system composed of $N = \sum N_i$ particles that are distributed in some manner among ε_i energy levels. Then, the total internal energy of the system is $U = \sum N_i\varepsilon_i$. The most probable distribution $(N_i)_{mp}$ of the N particles is the one that corresponds to the macrostate with the maximum probability P, and the total internal energy of that macrostate is $U_{mp} = \sum (N_i)_{mp}\varepsilon_i$. Once an equation for W is found, the distribution N_i that maximizes it can easily be found by setting $d(W) = 0$ subject to the constraint that the total energy and total number of particles in the system are constant (i.e., $dN = dU = 0$), and solving for $N_i = (N_i)_{mp}$.

A particle N_i has a total energy ε_i, which, in general, is made up of a number of energy modes. For example, we could partition the total energy of the particle into kinetic energy, rotational energy, vibrational energy, and so forth; and the particle's total energy can be divided among these modes in many ways. The total number of arrangements of a particle's different energy modes that add up to a given energy level ε_i is called the *degeneracy* of that energy level and is given the symbol g_i.

The following three classical statistical models have been developed to describe the basic particle-wave nature of certain material particles, and their corresponding W and $(N_i)_{mp}$ equations can be found in Table 18.7.

1. **The Maxwell-Boltzmann model.**[6] Here, all the N_i particles are assumed to be indistinguishable from each other and distributed among various degenerate energy levels. This model accurately represents the behavior of most simple gases at low pressures.
2. **The Fermi-Dirac model**. Here, the particles are assumed to be indistinguishable and are distributed among various degenerate energy levels with only one particle per degeneracy (g_i) value. This model accurately represents the behavior of electron and proton gases.
3. **The Bose-Einstein model**. Here, the particles are assumed to be indistinguishable and are distributed among various degenerate energy levels with no limit on the number of particles per degeneracy. This model accurately represents the behavior of photon and phonon gases.

The second law of thermodynamics states that S_p or $\dot{S}_p \geq 0$, which implies that, at equilibrium, the entropy of a closed system is a maximum. Also, since thermodynamic equilibrium corresponds to the system's being in its most probable macrostate, it is logical to assume that a functional relation exists between the entropy S of the system and the statistical probability of the most probable macrostate, W_{mp}. We postulate that this relation has the form:

$$S = f(W_{mp})$$

[6] This is called the "corrected" Maxwell-Boltzmann model in most statistical thermodynamics texts.

Table 18.7 Formula for Computing the Number of Microstates in the *i*th Macrostate for Various Statistical Models

Model	Number of Microstates per Macrostate, *W*	Most Probable Distribution $(N_i)_{mp}$
Maxwell-Boltzmann	$\prod_i \frac{g_i^{N!}}{N_i!}$	$\left(\frac{N}{Z}\right) g_i \exp\left(-\frac{\varepsilon_i}{kT}\right)$
Fermi-Dirac	$\prod_i \frac{g_i!}{N_i!(g_i - N_i)!}$	$g_i\left[B\exp\left(\frac{\varepsilon_i}{kT}\right)+1\right]^{-1}$
Bose-Einstein	$\prod_i \frac{(g_i + N_i - 1)!}{N_i!(g_i - 1)!}$	$g_i\left[B\exp\left(\frac{\varepsilon_i}{kT}\right)-1\right]^{-1}$

Note: Here, $Z = \sum g_i \exp(-\varepsilon_i/kT)$ *= partition function, and* $B = \exp(-\bar{\mu}/kT)$*, where* $\bar{\mu}$ *is the molar chemical potential.*

The problem we now face is that the total entropy S of a system is an additive property whereas W_{mp} is not. In the probability mathematics presented earlier, we found that the probability that independent events A and B simultaneously occur is $P\,(A \text{ and } B) = P_{AB} = P_A P_B$, and since W_{mp} is related to mathematical probability through Eq. (18.38), we can write

$$W(A\,\text{and}\,B) = W_A W_B$$

but

$$S(A\,\text{and}\,B) = S_A + S_B$$

Therefore, we must find a function f such that

$$S(A\,\text{and}\,B) = f(W_A) + f(W_B) = f[W(A\,\text{and}\,B)] = f(W_A W_B)$$

The only general function that satisfies this relation is the logarithm, since

$$\ln W_A + \ln W_B = \ln W_A W_B$$

Therefore, we choose to set S proportional to $\ln W_{mp}$. It can be shown that the constant of proportionality in this relation is just Boltzmann's constant k, so we end up with the following entropy-probability relation:

$$S = k \ln W_{mp} \tag{18.40}$$

Thus, we see that entropy is a measure of the molecular order within a system.

18.10 MAXWELL-BOLTZMANN GASES

To limit the algebraic complexity of the resulting property formula, we restrict our attention to the Maxwell-Boltzmann model. It can be shown that, for Maxwell-Boltzmann gases with $N \gg 1$,

$$u = RT^2\left(\frac{\partial \ln Z}{\partial T}\right) \tag{18.41}$$

and

$$h = u + RT \tag{18.42}$$

where

$$Z = \sum g_i \exp(-\varepsilon_i/kT)$$

Z is called the *partition function* of the system and g_i is the degeneracy of the ith energy level ε_i. At high temperatures, the number of quantum states (or degeneracy levels g_i) available at any energy level is much larger than the number of particles N_i in that energy level, or

$$\frac{g_i}{N_i} \gg 1$$

Then the number of microstates per macrostate for the three statistical models shown in Table 18.7 are approximately the same; that is,

$$W_{BE} \approx W_{FD} \approx W_{MB} = \prod_i \frac{g_i^{N_i}}{N_i!}$$

Also, under this condition, the most probable particle distribution of these three models are approximately the same

$$\left(N_i\right)_{\substack{BE\\mp}} \approx \left(N_i\right)_{\substack{FD\\mp}} \approx \left(N_i\right)_{\substack{MB\\mp}} = \left(\frac{N}{Z}\right) g_i \exp(-\varepsilon_i/kT) \tag{18.43}$$

These two results can be inserted into Eq. (18.4) to produce an equation for entropy as follows. First, we calculate $\ln W_{mp}$ from Table 18.7 as

$$\ln W_{mp} = \sum_i (N_i \ln g_i - \ln N_i!)_{mp}$$

We then use Stirling's approximation:

$$\ln N! \approx N \ln N - N$$

for the factorial term to obtain

$$\ln W_{mp} \approx \sum_i (N_i)_{mp} \left[\ln (g_i/N_i)_{mp} + 1\right]$$

Then, we use Eq. (18.43) to evaluate the term

$$(g_i/N_i)_{mp} = \frac{Z}{N} \exp(\varepsilon_i/kT)$$

which produces the result

$$\ln W_{mp} \approx \sum_i (N_i)_{mp} [\ln (Z/N) + \varepsilon_i/kT + 1]$$

which simplifies to

$$\ln W_{mp} \approx N[\ln (Z/N) + 1] + U/kT$$

The total entropy is now given by Eq. (18.40) as

$$S = Nk[\ln (Z/N) + 1] + U/T$$

and since $Nk = N\Re/N_o = n\Re = (m/M)\Re = mR$, the specific entropy can be written as

$$s = R[\ln (Z/N) + 1] + u/T \tag{18.44}$$

The molecular model we consider here is that of a relatively simple molecule, in which the total molecular energy can be separated into only three modes: translational, rotational, and vibrational. Then, the partition function Z is made up of translational, rotational, and vibrational molecular energy storage mechanisms and can be written as

$$Z = (Z_{\text{trans}})(Z_{\text{rot}})(Z_{\text{vib}})$$

Consequently, we can determine the molecular translational, rotational, and vibrational contribution to each of the properties, u, h, and s. Because these partition functions depend on the geometry of the molecule, we begin their study with the simplest possible structure, a monatomic gas.

18.11 MONATOMIC MAXWELL-BOLTZMANN GASES

For a Maxwell-Boltzmann monatomic gas, it can be shown that

$$Z_{\text{trans}} = \forall(2\pi mkT/\hbar^2)^{3/2} \tag{18.45}$$

where $\forall$ is the total volume of the system, $\hbar$ and k are Planck's constant and Boltzmann's constant, and $Z_{rot} = Z_{vib} = 0$. Then, Eqs. (18.41), (18.42), and (18.44) give

$$u = \frac{3}{2}RT \tag{18.46a}$$

$$h = u + RT = \frac{5}{2}RT \tag{18.46b}$$

$$c_v = \frac{3}{2}R \tag{18.46c}$$

$$c_p = \frac{5}{2}R \tag{18.46d}$$

$$s = R\left\{ \ln\left[(2\pi m/\hbar^2)^{3/2}(kT)^{5/2}/p\right] + \frac{5}{2}\right\} \tag{18.46e}$$

For monatomic gases only

These formulae for u, h, c_v, and c_p are the same as those obtained from the kinetic theory of gases discussed earlier in the chapter. The equation for entropy, on the other hand, is more complex and therefore one would expect it to be more accurate. We now progress to the next most complex geometric molecular structure, diatomic gases.

EXAMPLE 18.8

In a fiendish plan to incapacitate Superman, the arch villain Dorkmann proposes to compress 3.50 kg of krypton gas from 1.00 atmosphere, 20.0°C to 10.0 MPa, producing a concentrated and possibly toxic concentration of kryptonite. The foolish, unschooled fiend intends to try to carry out the compression process adiabatically using only 100. kJ of work. But you, as the ever-present hero Thermoperson, seeker of truth and wisdom, hold the power to foil the plan by computing

a. The final temperature of the krypton gas after compression.
b. The entropy production of the compression process.

Solution

a. The final temperature of the gas can be found from an energy balance and Eq. (18.46a) as

$$_1Q_2 - {}_1W_2 = m(u_2 - u_1) = m\left(\tfrac{3}{2}R\right)(T_2 - T_1)$$

where $R_{krypton} = \Re/M_{krypton} = 8.3143/83.80 = 0.0992$ kJ/kg·K. Since $_1Q_2 = 0$ for an adiabatic compression, the energy balance equation can then be solved for T_2 as

$$T_2 = T_1 - \frac{_1W_2}{3mR/2} = (20.0 + 273.15\text{ K}) - \frac{-100.\text{ kJ}}{3(3.50\text{ kg})(0.0992\text{ kJ/kg}\cdot\text{K})/2} = 458\text{ K}$$

b. An entropy balance using Eq. (18.46e) produces

$$\frac{_1Q_2}{T_b} + {}_1(S_P)_2 = m(s_2 - s_1)$$

$$= m\left(R\left\{\ln\left[\frac{(2\pi m/\hbar^2)^{3/2}(kT_2)^{5/2}}{p_2}\right] + \frac{5}{2}\right\} - R\left\{\ln\left[\frac{(2\pi m/\hbar^2)^{3/2}(kT_1)^{5/2}}{p_1}\right] + \frac{5}{2}\right\}\right)$$

Solving this equation for $_1(S_P)_2$ gives

$$_1(S_P)_2 = mR\ln\left[\left(\frac{T_2}{T_1}\right)^{5/2}\left(\frac{p_1}{p_2}\right)\right]$$

$$= (3.50\text{ kg})(0.0992\text{ kJ/kg}\cdot\text{K})\ln\left[\left(\frac{485\text{ K}}{293.15\text{ K}}\right)^{5/2}\left(\frac{0.101325\text{ MPa}}{10.0\text{ MPa}}\right)\right] = -1.16\frac{\text{kJ}}{\text{kg}\cdot\text{K}}$$

Since $_1(S_P)_2$ is less than zero here, it violates the second law of thermodynamics and this process can not possibly occur. Therefore, Superman has nothing to worry about.

Exercises

22. If Dorkmann in Example 18.8 increases the work input to the process from 100. kJ to 1000. kJ with all the other variables unchanged, would the process work then? **Answer**: Possibly, because now $_1(S_P)_2 = 0.160$ kJ/kg·K, which, being positive, does not violate the second law of thermodynamics.

23. Alternatively, Dorkmann in Example 18.8 might choose to lower the final compression pressure to 1.00 MPa. Would the process work under this condition, if all the other variables remain unchanged? **Answer**: No, because now ${}_1(S_P)_2 = -0.358$ kJ/kg·K, which, being negative, violates the second law of thermodynamics.

24. However, instead of choosing either of the alternatives proposed in Exercises 22 and 23, Dorkmann in Example 18.8, for some unknown reason, chooses to change the gas from krypton to helium. Will his original process work with this new gas? **Answer**: No, since now ${}_1(S_P)_2 = -32.8$ kJ/k·K, which, being negative, violates the second law of thermodynamics.

18.12 DIATOMIC MAXWELL-BOLTZMANN GASES

For a diatomic Maxwell-Boltzmann gas, it can be shown that the translational partition function is the same as that for a monatomic gas, Eq. (18.45). However, now, 2 rotational and 1 vibrational degrees of freedom are present. In this case, the rotational and vibrational partition functions are

$$Z_{\text{rot}} = \frac{T}{\sigma\,\Theta_r} \tag{18.47}$$

and

$$Z_{\text{vib}} = [1 - \exp(-\Theta_v/T)]^{-1} \tag{18.48}$$

where σ is the rotational symmetry number (the number of axes about which the molecule can be rotated 180° and be indistinguishable from the original configuration), Θ_r is called the *characteristic rotational temperature*, and Θ_v is called the *characteristic vibrational temperature*. Tables 18.8 and 18.9 give values for Θ_r, Θ_v, and σ for various substances. The various components to the resulting specific property equations then become

$$u_{\text{trans}} = \frac{3}{2}RT \tag{18.49a}$$

$$u_{\text{rot}} = RT \tag{18.49b}$$

$$u_{\text{vib}} = (u_o)_{\text{vib}} + \frac{R\Theta_v}{[\exp(\Theta_v/T) - 1]} \tag{18.49c}$$

Table 18.8 Characteristic Vibrational and Rotational Temperatures of Some Common Diatomic Materials

Material	Θ_v(K)	Θ_r(K)
H_2	6140	85.5
HF	5954	30.3
OH	5360	27.5
HCl	4300	15.3
CH	4100	20.7
N_2	3340	2.86
HBr	3700	12.1
HI	3200	9.0
Co	3120	2.77
NO	2740	2.47
O_2	2260	2.09
Cl_2	810	0.35
Br_2	470	0.12
I_2	309	0.05
Na_2	230	0.22
K_2	140	0.08

Source: From Lee, John F., Sears, Francis W., Turcotte, Donald L. Statistical Thermodynamics, © 1963. Addison-Wesley Publishing Co., Inc, Reading, MA. Adapted from Table 10-1 on page 204. Reprinted with permission.

Table 18.9 Rotational Symmetry Number for Some Simple Materials

Material	σ
Any diatomic molecule with two different atoms (e.g., HCl, HI, or NO)	1
Any diatomic molecule with two identical atoms (e.g., H_2, O_2, or N_2)	2
Any triatomic molecule with two different atoms forming an isosceles triangle (such as H_2O) or any linear triatomic molecule (e.g., CO_2 or NO_2,)	2
Any quatratomic molecule with two different atoms forming an equilateral triangular pyramid (e.g., NH_3)	3
Any molecule forming a plane rectangle (e.g., C_2H_4)	4
Any pentatomic molecule with two different atoms forming a regular tetrahedron with the carbon atom at the center of mass (e.g., CCl_4 or CH_4)	12

Thus,

$$u = u_{\text{trans}} + u_{\text{rot}} + u_{\text{vib}} = (u_o)_{\text{vib}} + \frac{5}{2}RT + \frac{R\Theta_v}{[\exp(\Theta_v/T) - 1]} \tag{18.49d}$$

where

$$(u_o)_{\text{vib}} = R\Theta_v/2 \tag{18.50}$$

is the vibrational energy at absolute zero temperature. Similarly,

$$h_{\text{trans}} = \frac{5}{2}RT \tag{18.51a}$$

$$h_{\text{rot}} = u_{\text{rot}} = RT \tag{18.51b}$$

$$h_{\text{vib}} = u_{\text{vib}} = (u_o)_{\text{vib}} + \frac{R\Theta_v}{[\exp(\Theta_v/T) - 1]} \tag{18.51c}$$

Thus,

$$h = h_{\text{trans}} + h_{\text{rot}} + h_{\text{vib}} = (u_o)_{vib} + \frac{7}{2}RT + \frac{R\Theta_v}{[\exp(\Theta_v/T) - 1]} \tag{18.51d}$$

Then, we can find

$$(c_v)_{\text{trans}} = \frac{3}{2}R \tag{18.52a}$$

$$(c_v)_{\text{rot}} = R \tag{18.52b}$$

$$(c_v)_{\text{vib}} = \frac{R(\Theta_v/T)^2[\exp(\Theta_v/T)]}{[\exp(\Theta_v/T) - 1]^2} \tag{18.52c}$$

Thus,

$$c_v = (c_v)_{\text{trans}} + (c_v)_{\text{rot}} + (c_v)_{\text{vib}} = \frac{5}{2}R + \frac{R(\Theta_v/T)^2[\exp(\Theta_v/T)]}{[\exp(\Theta_v/T) - 1]^2} \tag{18.52d}$$

WHAT IS ABSOLUTE ZERO TEMPERATURE?

In the diatomic Maxwell-Boltzmann model, the concept of absolute zero temperature corresponds to the cessation of all translational and rotational molecular motion, but vibrational motion is still allowed to occur. Thus, the internal energy does not vanish at absolute zero temperature in this model.

and

$$(c_p)_{\text{trans}} = \frac{5}{2}R \tag{18.53a}$$

$$(c_p)_{\text{rot}} = (c_v)_{\text{rot}} = R \tag{18.53b}$$

$$(c_p)_{\text{vib}} = (c_v)_{\text{vib}} = \frac{R(\Theta_v/T)^2[\exp(\Theta_v/T)]}{[\exp(\Theta_v/T) - 1]^2} \tag{18.53c}$$

Thus,

$$c_p = (c_p)_{\text{trans}} + (c_p)_{\text{rot}} + (c_p)_{\text{vib}} = \frac{7}{2}R + \frac{R(\Theta_v/T)^2[\exp(\Theta_v/T)]}{[\exp(\Theta_v/T) - 1]^2} \tag{18.53d}$$

Finally,

$$s_{\text{trans}} = R\left\{\ln\left[\left(2\pi m/\hbar^2\right)^{3/2}(kT)^{5/2}/p\right] + \frac{5}{2}\right\} \tag{18.54a}$$

$$s_{\text{rot}} = R\{\ln[T/(\sigma\Theta_r)] + 1\} \tag{18.54b}$$

$$s_{\text{vib}} = R\left\{\ln\left[1 - \exp(-\Theta_v/T)\right]^{-1} + (\Theta_v/T)/[\exp(\Theta_v/T) - 1]\right\} \tag{18.54c}$$

Thus,

$$\begin{aligned} s &= s_{\text{trans}} + s_{\text{rot}} + s_{\text{vib}} \\ &= R\left\{\ln\left[\left(2\pi m/\hbar^2\right)^{3/2}(kT)^{5/2}/p\right] + \frac{5}{2}\right\} + R\{\ln[T/(\sigma\Theta_r)] + 1\} \\ &\quad + R\left\{\ln\left[1 - \exp(-\Theta_v/T)\right]^{-1} + (\Theta_v/T)/[\exp(\Theta_v/T) - 1]\right\} \end{aligned} \tag{18.54d}$$

These equations produce reasonably accurate results for diatomic gases at moderate to high temperatures, as illustrated by the following example.

EXAMPLE 18.9

To test the diatomic Maxwell-Boltzmann gas equations just developed, compute the value of c_v/R for nitrous oxide (NO) at 20.0°C and compare it with the measured result given in Table 18.3.

Solution

For a diatomic Maxwell-Boltzmann gas, the constant volume specific heat is given by Eq. (18.52d) as

$$c_v = \frac{5}{2}R + \frac{R(\Theta_v/T)^2\exp(\Theta_v/T)}{[\exp(\Theta_v/T) - 1]^2}$$

then,

$$\frac{c_v}{R} = \frac{5}{2} + \frac{(\Theta_v/T)^2\exp(\Theta_v/T)}{[\exp(\Theta_v/T) - 1]^2}$$

From Table 18.8, we find that $\Theta_v = 2740$ K for NO, so,

$$\left(\frac{c_v}{R}\right)_{\text{NO}} = \frac{5}{2} + \frac{\left(\dfrac{2740}{20.0 + 273.15}\right)^2 \exp\left(\dfrac{2740}{20.0 + 273.15}\right)}{\left[\exp\left(\dfrac{2740}{20.0 + 273.15}\right) - 1\right]^2} = 2.51$$

as in Table 18.3. Note that, since $R_{\text{NO}} = \mathcal{R}/M_{\text{NO}} = 8.3143/30.01 = 0.2771$ kJ/kg·K, then $(c_v)_{\text{NO}} = 0.2771 \times 2.51 = 0.695$ kJ/kg·K at 20.0°C.

(Continued)

EXAMPLE 18.9 (*Continued*)

Exercises

25. Determine the values of c_v and the ratio of c_v/R for the nitrous oxide in Example 18.9 at 2000. K. **Answer:** $(c_v)_{NO} =$ 0.930 kJ/kg·K and $(c_v/R)_{NO} = 3.36$.

26. Using Eqs. (18.51a–d), compute the values for the specific enthalpy and the ratio of h/R for the nitrous oxide in Example 18.9 at 20.0°C. **Answer:** $h_{NO} = 948$ kJ/kg ($h_{trans} = 203$ kJ/kg, $h_{rot} = 81.2$ kJ/kg, $h_{vib} = 664$ kJ/kg), and $(h/R)_{NO} = 3420$ K.

27. Using Eqs. (18.54a–d), compute the values of the specific entropy and the ratio s/R for the nitrous oxide in Example 18.9 at 20.0°C and a pressure of 1 atm. **Answer:** $s_{NO} = 6.63$ kJ/kg·K ($s_{trans} = 5.03$ kJ/kg·K, $s_{rot} = 1.60$ kJ/kg·K, $s_{vib} = 0.0003$ kJ/kg·K) and $(s/R)_{NO} = 23.9$.

18.13 POLYATOMIC MAXWELL-BOLTZMANN GASES

Polyatomic gases are divided into two categories of molecular geometry: linear and nonlinear. Linear polyatomic molecules have only 2 degrees of rotational freedom, as in the case of diatomic molecules. However, they have $3b-5$ degrees of vibrational freedom, where b is the number of atoms in the molecule. Therefore, the equations used to calculate the translational and rotational contributions to the molecular energy are the same as those used for the diatomic molecule (i.e., parts a and b of Eqs. (18.49), and (18.51) through (18.54)). The equations for the vibrational contribution to the molecular energy of a linear polyatomic Maxwell-Boltzmann gas are

$$u_{vib} = (u_o)_{vib} + R\sum_{i=1}^{3b-5}\frac{\Theta_{vi}}{[\exp(\Theta_{vi}/T) - 1]} \tag{18.55a}$$

$$h_{vib} = u_{vib} = (u_o)_{vib} + R\sum_{i=1}^{3b-5}\frac{\Theta_{vi}}{[\exp(\Theta_{vi}/T) - 1]} \tag{18.55b}$$

$$(c_v)_{vib} = R\sum_{i=1}^{3b-5}\frac{(\Theta_{vi}/T)^2[\exp(\Theta_{vi}/T)]}{[\exp(\Theta_{vi}/T) - 1]^2} \tag{18.55c}$$

$$(c_p)_{vib} = (c_v)_{vib} = R\sum_{i=1}^{3b-5}\frac{(\Theta_{vi}/T)^2[\exp(\Theta_{vi}/T)]}{[\exp(\Theta_{vi}/T) - 1]^2} \tag{18.55d}$$

and

$$s_{vib} = R\sum_{i=1}^{3b-5}\left\{\ln[1 - \exp(-\Theta_{vi}/T)]^{-1} + (\Theta_{vi}/T)/[\exp(\Theta_{vi}/T) - 1]\right\} \tag{18.55e}$$

where the vibrational internal energy at absolute zero temperature is now found from

$$(u_o)_{vib} = \sum_{i=1}^{3b-5} R\Theta_{vi}/2 \tag{18.56}$$

since there are now $3b-5$ characteristic vibrational temperatures Θ_{vi}.

EXAMPLE 18.10

Carbon dioxide is a linear triatomic molecule that has the following characteristic temperatures

$$\Theta_r = 0.562\ \text{K}$$
$$\Theta_{v1} = 1932\ \text{K}$$
$$\Theta_{v2} = \Theta_{v3} = 960.\ \text{K}$$
$$\Theta_{v4} = 3380\ \text{K}$$

Determine the specific internal energy, specific enthalpy, and specific entropy of CO_2 at a temperature of 1000. K and a pressure of 1.00 atm.

Solution

The mass of the CO_2 molecule is

$$m = M/N_o = 44.01/6.023\times 10^{26} = 7.31\times 10^{-26}\ \text{kg/molecule}$$

and the gas constant for CO_2 is

$$R = \Re/M = 8.3143/44.01 = 0.1889\ \text{KJ/(kg}\cdot\text{K)}$$

Equation (18.56) gives the vibrational specific internal energy at absolute zero temperature as

$$(u_o)_{vib} = (0.1889)(1932+960.+960.+3380)/2 = 683\ \text{kJ/kg}$$

and Eq. (18.55a) gives the vibrational component of the specific internal energy as

$$\begin{aligned} u_{vib} = 683 + (0.1889)\{&(1932)[\exp(1.932)-1]^{-1} \\ &+2(960.)[\exp(0.960.)-1]^{-1} \\ &+(3380)[\exp(3.380)-1]^{-1}\} = 992\ \text{kJ/kg} \end{aligned}$$

The translational and rotational components are given by Eqs. (18.49a) and (18.49b) as

$$u_{trans} = \frac{3}{2}RT = \frac{3}{2}(0.1889)(1000.) = 283.4\ \text{kJ/kg}$$

$$u_{rot} = RT = (0.1889)(1000.) = 188.9\ \text{kJ/kg}$$

Then,

$$u = u_{trans} + u_{rot} + u_{vib} = 283.4 + 188.9 + 992 = 1465\ \text{kJ/kg}$$

The specific enthalpy is now given simply by

$$h = u + RT = 1465 + (0.1889)(1000.) = 1654\ \text{kJ/kg}$$

The translational and rotational specific entropy values are calculated from Eqs. (18.54a) and (18.54b). First, we calculate

$$\begin{aligned} (2\pi m/\hbar^2)^{3/2}(kT)^{5/2}/p &= \left[2\pi(7.31\times 10^{-26})/(6.626\times 10^{-34})^2\right]^{3/2} \\ &\quad\times[(1.38\times 10^{-23})(1000)]^{5/2}/101{,}325 \\ &= 2.36\times 10^8\ \text{per molecule} \end{aligned}$$

and then Eq. (18.54a) gives

$$s_{trans} = (0.1889)\left[\ln(2.36\times 10^8) + \frac{5}{2}\right] = 4.11\ \text{kJ/(kg}\cdot\text{K)}$$

and Eq. (18.54b) with $\sigma = 2$ from Table 18.9 gives

$$s_{rot} = (0.1889)\{\ln[1000/(2)(0.562)] + 1\} = 1.47\ \text{kJ/(kg}\cdot\text{K)}$$

Equation (18.55e) is then used to find the vibrational component of the specific entropy as

$$\begin{aligned} s_{vib} = (0.1889)\{&\ln[1-\exp(-1.932)]^{-1} + (1.932)[\exp(1.932)-1]^{-1} \\ &+\ln[1-\exp(-0.960)]^{-1} + (0.960)[\exp(0.960)-1]^{-1} \\ &+\ln[1-\exp(-0.960)]^{-1} + (0.960)[\exp(0.960)-1]^{-1} \\ &+\ln[1-\exp(-3.380)]^{-1} + (3.380)[\exp(3.380)-1]^{-1} \\ = 0.527\ &\text{kJ/(kg}\cdot\text{K)} \end{aligned}$$

Then, the specific entropy is

$$s = s_{trans} + s_{rot} + s_{vib} = 4.11 + 1.47 + 0.527 = 6.11\ \text{kJ/(kg}\cdot\text{K)}$$

(Continued)

EXAMPLE 18.10 *(Continued)*

Exercises

28. Note that the only property of a polytropic Maxwell-Boltzmann gas that depends on pressure is entropy. Compute the entropy for the carbon dioxide in Example 18.10 if the pressure is increased from 1.00 to 10.0 atm. Assume all other variables remain unchanged. **Answer:** $s = 5.68$ kJ/kg·K ($s_{trans} = 3.68$ kJ/kg·K, $s_{rot} = 1.47$ kJ/kg·K, $s_{vib} = 0.528$ kJ/kg·K).
29. Recompute the specific internal energy, enthalpy, and entropy of the carbon dioxide in Example 18.10 when the temperature is lowered from 1000. K to 300. K but the pressure remains constant at 1.00 atm. **Answer:** $u = 841$ kJ/kg ($u_{trans} = 85.0$ kJ/kg, $u_{rot} = 56.7$ kJ/kg, $u_{vib} = 699$ kJ/kg); $h = 898$ kJ/kg; $s = 4.86$ kJ/kg·K ($s_{trans} = 3.55$ kJ/kg·K, $s_{rot} = 1.24$ kJ/kg·K, $s_{vib} = 0.0694$ kJ/kg·K).
30. Determine the specific internal energy, enthalpy, and entropy of the carbon dioxide in Example 18.10 for a temperature of 5000. K and a pressure of 0.100 atmosphere. **Answer:** $u = 6190$ kJ/kg ($u_{trans} = 1420$ kJ/kg, $u_{rot} = 945$ kJ/kg, $u_{vib} = 3832$ kJ/kg); $h = 7138$ kJ/kg; $s = 8.72$ kJ/kg·K ($s_{trans} = 5.31$ kJ/kg·K, $s_{rot} = 1.78$ kJ/kg·K, $s_{vib} = 1.64$ kJ/kg·K).

The nonlinear polyatomic molecule has 3 translational degrees of freedom, 3 rotational degrees of freedom, and $3b-6$ degrees of vibrational freedom. Thus, it has the same equations for the translational molecular energy as in the linear polyatomic case, but the rotational and vibrational contribution equations are different. In this case,

$$u_{rot} = h_{rot} = \frac{3}{2}RT \tag{18.57a}$$

$$(c_v)_{rot} = (c_p)_{rot} = \frac{3}{2}R \tag{18.57b}$$

$$s_{rot} = R\left\{\ln[T/(\sigma\Theta_r)] + \frac{3}{2}\right\} \tag{18.57c}$$

and

$$u_{vib} = (u_o)_{vib} + R\sum_{i=1}^{3b-6}\Theta_{vi}/[\exp(\Theta_{vi}/T)-1] \tag{18.58a}$$

$$h_{vib} = u_{vib} = (u_o)_{vib} + R\sum_{i=1}^{3b-6}\Theta_{vi}/[\exp(\Theta_{vi}/T)-1] \tag{18.58b}$$

$$(c_v)_{vib} = R\sum_{i=1}^{3b-6}(\Theta_{vi}/T)^2[\exp(\Theta_{vi}/T)]/[\exp(\Theta_{vi}/T)-1]^2 \tag{18.58c}$$

$$(c_p)_{vib} = (c_v)_{vib} + R\sum_{i=1}^{3b-6}(\Theta_{vi}/T)^2[\exp(\Theta_{vi}/T)]/[\exp(\Theta_{vi}/T)-1]^2 \tag{18.58d}$$

and

$$s_{vib} = R\sum_{i=1}^{3b-6}\{\ln[1-\exp(-\Theta_{vi}/T)]^{-1} + (\Theta_{vi}/T)[\exp(\Theta_{vi}/T)-1]^{-1}\} \tag{18.58e}$$

where the vibrational internal energy at absolute temperature is now found from

$$(u_o)_{vib} = \sum_{i=1}^{3b-6} R\Theta_{vi}/2 \tag{18.59}$$

since the nonlinear polyatomic molecule has $3b-6$ characteristic vibrational temperatures Θ_{vi}.

SUMMARY

The subject of statistical thermodynamics is inherently mathematically complex and conceptually difficult. It is often the subject of an entire advanced engineering course, usually at the graduate level. The material presented in this chapter is intended only as an introduction to this subject.

In this chapter, we summarize the essential concepts of statistical thermodynamics and present them in a simple enough manner to make the results useful. This subject is very effective when experimental data are not available and values of various thermodynamic properties are needed to carry out an engineering analysis. The interested reader is encouraged to fill in the various theoretical gaps in the material presented in this chapter by additional reading in this area.

Some of the more important equations introduced in this chapter follow. Do not attempt to use them blindly without understanding their limitations. Please refer to the text material where they were introduced to gain an understanding of their use and limitations.

1. The collision frequency $\mathcal{F}$ and mean free path λ of a moving gas molecule in a volume $\forall$ containing N molecules are given by

$$\mathcal{F} = \sigma V_{\text{rms}} \frac{N}{\forall}\left(\frac{8}{3\pi}\right)^{1/2} \qquad \lambda = \frac{1}{(N/\forall)\sigma}$$

where $\sigma = 4\pi r^2_{\text{molecule}}$ is the collision cross-section.

2. The average, root mean square and most probable molecular velocities are given by

$$V_{\text{avg}} = \sqrt{\frac{8kT}{\pi m}},\ V_{\text{rms}} = \sqrt{\frac{3kT}{m}},\ V_{mp} = \sqrt{\frac{2kT}{m}}$$

3. The fraction of molecules with velocities between V_1 and V_2 is given by

$$\frac{N(V_1 \to V_2)}{N} = \text{erf}(x_2) - \text{erf}(x_1) - \frac{2}{\sqrt{\pi}}\left(x_2 e^{-x_2^2} - x_1 e^{-x_1^2}\right)$$

$$x_1 = V_1/V_{mp},\ x_2 = V_2/V_{mp}$$

4. The compound probability that both independent events A and B will occur is given by

$$P(A \text{ and } B) = P_{AB} = P_A P_B$$

5. The probability that either event A or B will occur is given by

$$P(A \text{ or } B) = P_A + P_B - P_{AB}$$

6. The number of different ordered arrangements (permutations) of N things taken R at a time without repetition is given by

$$P_R^N = \frac{N!}{(N-R)!}$$

7. The number of different ordered arrangements (permutations) of N things taken R at a time allowing repetition is given by

$$P_N^N = N^R$$

8. The number of different unordered arrangements (combinations) of N things taken R at a time without repetition is given by

$$C_R^N = \frac{P_R^N}{R!} = \frac{N!}{(N-R)!R!}$$

9. The number of unordered arrangements (combinations) of N things taken R at a time allowing repetition is given by

$$C_R^N = \frac{P_R^N}{R!} = \frac{(N+R-1)!}{(N-1)!R!}$$

10. The number of permutations or combinations of N things, R_1 of one kind, R_2 of a second kind, ..., and R_k of a kth kind is given by

$$P^N_{R_1,R_2,\ldots,R_k} = C^N_{R_1,R_2,\ldots,R_k} = \frac{N!}{R_1!R_2!R_3!,\ldots,R_k!}$$

11. The statistical thermodynamic equations for a monatomic Maxwell-Boltzmann gas are

$$\left.\begin{aligned} u &= \frac{3}{2}RT \\ h &= u + RT = \frac{5}{2}RT \\ c_v &= \frac{3}{2}R \\ c_p &= \frac{5}{2}R \\ s &= R\left\{\ln\left[(2\pi m/\hbar^2)^{3/2}(kT)^{5/2}/p\right] + \frac{5}{2}\right\} \end{aligned}\right\} \text{For monatomic gases only}$$

12. The statistical thermodynamic equations for a diatomic Maxwell-Boltzmann gas are

$$u = \frac{R\Theta_v}{2} + \frac{5}{2}RT + \frac{R\Theta_v}{[\exp(\Theta_v/T) - 1]}$$

$$h = \frac{R\Theta_v}{2} + \frac{7}{2}RT + \frac{R\Theta_v}{[\exp(\Theta_v/T) - 1]}$$

$$s = R\left\{\ln\left[(2\pi m/\hbar^2)^{3/2}(kT)^{5/2}/p\right] + \frac{5}{2}\right\} + R\{\ln[T/\sigma\Theta_r)] + 1\}$$
$$+ R\{\ln[1 - \exp(-\Theta_v/T)]^{-1} + (\Theta_v/T)/[\exp(\Theta_v/T) - 1]\}$$

$$c_v = \frac{5}{2}R + \frac{R(\Theta_v/T)^2[\exp(\Theta_v/T)]}{[\exp(\Theta_v/T) - 1]^2}$$

$$c_p = \frac{7}{2}R + \frac{R(\Theta_v/T)^2[\exp(\Theta_v/T)]}{[\exp(\Theta_v/T) - 1]^2}$$

13. The statistical thermodynamic equations for a linear polyatomic Maxwell-Boltzmann gas molecule are

$$u = \frac{5}{2}RT + \sum_{i=1}^{3b-5}\frac{R\Theta_{vi}}{2} + R\sum_{i=1}^{3b-5}\frac{\Theta_{vi}}{[\exp(\Theta_{vi}/T) - 1]}$$

$$h = \frac{7}{2}RT + \sum_{i=1}^{3b-5}\frac{R\Theta_{vi}}{2} + R\sum_{i=1}^{3b-5}\frac{\Theta_{vi}}{[\exp(\Theta_{vi}/T) - 1]}$$

$$s = R\left\{\ln[(2\pi m/\hbar^2)^{3/2}(kT)^{5/2}/p] + \frac{5}{2}\right\} + R\{\ln[T/(\sigma\Theta_r)] + 1\}$$
$$+ R\sum_{i=1}^{3b-5}\{\ln[1 - \exp(-\Theta_{vi}/T)]^{-1} + (\Theta_{vi}/T)/[\exp(\Theta_{vi}/T) - 1]\}$$

$$s_{\text{vib}} = R\sum_{i=1}^{3b-5}\{\ln[1 - \exp(-\Theta_{vi}/T)]^{-1} + (\Theta_{vi}/T)/[\exp(\Theta_{vi}/T) - 1]\}$$

14. The statistical thermodynamic equations for a nonlinear polyatomic Maxwell-Boltzmann gas molecule are the same as those for a linear molecule except the upper limit of $3b-5$ in the summations is replaced by $3b-6$.

Problems (* indicates problems in SI units)

1. Determine the number of diatomic nitrogen (N_2) molecules in 1 in^3 at 70.0°F and a pressure of 1.00×10^{-10} mm of mercury absolute (a very high vacuum).
2.* Determine the mean free path and the collision frequency of diatomic nitrogen at 1000. K and 10.0 MPa. The effective radius of the nitrogen molecule is 1.10×10^{-10} m.
3.* For 10^{10} bromine (Br_2) molecules confined in a volume of 1.00 m^3 at a pressure of 1.00 Pascal, assume ideal gas kinetic theory behavior and determine
   a. The temperature in the container.
   b. The mean free path between collisions.
   c. The collision frequency $\mathcal{F}$.

4. Find the temperature T at which $V_{mp} = c$ (the velocity of light) for the neon atoms discussed in Examples 18.2 and 18.3.

5.* Hydrogen (H_2) at 3.00 MPa and 1000. K is confined in a volume of 1.00 m^3. Determine V_{avg}, V_{rms}, V_{mp}, and the collision frequency $\mathcal{F}$.

6.* Oxygen (O_2) at 300. K and 1.00 atm pressure is confined in a volume of 1.00×10^{-4} m^3. Determine

a. The number of oxygen molecules present.
b. The average molecular velocity.
c. The rms molecular velocity.
d. The most probable molecular velocity.
e. The number of molecules with a velocity in the range of 0 to 10^{-4} m/s.
f. The number of molecules with velocities greater than 10^6 m/s.

7. Using the following infinite series expansion

$$\exp(-x^2) = 1 - x^2/1! + x^4/2! - x^6/3! + x^8/4! - \cdots$$

show that the equivalent expansion for the error function is

$$\operatorname{erf}(x) = \frac{2}{\sqrt{\pi}}\left(x - \frac{x^3}{3(1!)} + \frac{x^5}{5(2!)} - \frac{x^7}{7(3!)} + \cdots\right)$$

8. Show that Eq. (18.26) can be written as

$$\frac{N(V \to \infty)}{N} = \frac{2}{\sqrt{\pi}}\left(\int_x^{\infty} e^{-x^2}\,dx + xe^{-x^2}\right)$$

where $x = V/V_{mp}$.

9. Show that if $a \gg 1$, then

$$\int_a^{\infty} e^{-x^2}\,dx \approx \frac{1}{2a}e^{-a^2}$$

(*Hint*: Set $x = a + y$, then $dx = dy$ and the integral over dy have limits from 0 to ∞.)

10. Using the results from Problems 8 and 9, show that for $x \gg 1$,

$$\frac{N(V \to \infty)}{N} = \frac{2}{\sqrt{\pi}}\left(x + \frac{1}{2x}\right)e^{-x^2}$$

11. Using the equations of kinetic theory, it can be shown that the number of molecules per unit time that leak out of an isothermal pressurized container of volume $\forall$ through a small hole of area A is

$$\dot{N}_{leak} = (N/\forall)(A/4)V_{avg}$$

Show that the pressure in the container then decays according to

$$p = p_0 \exp(-AV_{avg}t/4\forall)$$

where $V_{avg} = (8kT/\pi m)^{1/2}$, and p_0 is the pressure in the container at time $t = 0$.

12.* Using the information given in Problem 11 and the equations of kinetic theory, calculate the mass rate of separation of the isotope U^{235} from a gaseous mixture of U^{238} and U^{235} by a molecular sieve. The sieve is a porous pipe 0.0300 m in diameter and 2000. m long. The area of each pore is 2.60×10^{-19} m^2, and there are 10^9 pores per meter of pipe length. The internal sieve temperature is 1000. K, and the partial pressure difference of the U^{235} across the sieve is 10.0 Pa.

13. The probability of failure of a space shuttle primary computer system is 1.50×10^{-3}. This computer system has a secondary backup computer system with the same failure probability. What is the probability of simultaneous failure of both the primary and secondary computer systems?

14. An eight-cylinder engine has one bad spark plug. If the mechanic removes two spark plugs at random, what is the probability that the defective spark plug is found on the first try?

15. A single die is tossed. What is the probability that it will come up with a value greater than 4?

16. Two dice are tossed. Determine the probability that their sum will be greater than (a) 2, (b) 4, (c) 6, (d) 8, and (e) 10.

17. Two cards are to be drawn from a standard deck of 52 cards. Calculate the probability that these two cards are an ace and a 10, drawn in any order.

18. A coin is flipped twice. Determine the probability that only one head results.

19. Two coins are flipped simultaneously. Determine the probability that at least one is a head.

20. If three coins are simultaneously flipped, what is the probability of getting (a) at least two heads and (b) exactly two heads.

21. Show that $C_R^N \equiv C_{N-R}^N$ for any N and R.

22. An electronic component is available from five suppliers. How many different ways can two suppliers be chosen from the five available?

23. How many different ten-digit phone numbers can be made from the digits 0 through 9 if the first three digits must be 414?

24. How many different six-digit automobile license plates can be made using only the digits 0 through 9 if the digits may be repeated?

25. How many different nine-digit social security numbers can be made if

a. No digit is allowed to be repeated.
b. The digits can be repeated.

26. How many different three-letter "words" can be made from the 26 letters of the English alphabet without regard to vowels if the letters can be used more than once per word?

27. A component subassembly consists of five pieces which can be assembled in any order. A production test is to be designed to determine the minimum time required to assemble the subassembly. If each sequence of assembly is to be tested once, how many tests need to be conducted?

28. A person is shopping for a new car. One dealer offers a choice of five body styles, four engine types, ten color combinations, three transmissions, and three accessory packages. How many different cars are there to choose from?

29. Determine the collision probability of the bromine molecules in Problem 4.

30.* A typical galaxy occupies a volume of about 1.00×10^{61} m^3 and contains 1.00×10^{11} stars, each with an effective radius of 1.00×10^9 m. Determine

a. The collision probability of two stars within the galaxy.
b. The number of stars that will have experienced a collision by the time they have traveled a distance of 0.0100 mean free paths.

31. Show that, for a Maxwell-Boltzmann gas, the total enthalpy H is given by

$$H = -N\partial(\ln Z)/\partial\beta + NkT\forall[\partial(\ln Z)/\partial\forall]$$

where $\beta = 1/kT$ and the pressure is $p = NkT(\partial \ln Z/\partial\forall)$.

32. Compute the percent error in Stirling's approximation, $\ln N! \approx N \ln N - N$, for the following values of N: (a) $N = 5$, (b) $N = 10$, and (c) $N = 50$.
33.* Use the Maxwell-Boltzmann formulae for diatomic gases and calculate the value of c_p for molecular iodine I_2 at 0.00°C.
34. Use the Maxwell-Boltzmann formulae for diatomic gases and calculate the value of c_p/R at 2000.°C for (a) hydrogen, (b) carbon monoxide, and (c) oxygen. Use the characteristic vibrational temperatures found in Table 18.8. Compare your results with the values given in Table 18.3.
35. An experimentally determined relation for the temperature dependence of the constant pressure specific heat of molecular oxygen O_2 over a wide temperature range is

$$c_p = 0.3598 + 47.81/T - 5.406/\sqrt{T}$$

where c_p is in Btu/(lbm·R) and T is in R. Use the Maxwell-Boltzmann equations for a diatomic molecule and attempt to predict the coefficients of the first two terms (i.e., 0.3598 and 47.81) of the equation. (Hint: At very high temperatures, $\exp(\Theta_v/T) \approx 1 + \Theta_v/T$). Explain why your results may not be very accurate.
36. Use the Maxwell-Boltzmann formulae for diatomic gases and calculate the value of c_p for water vapor at 400.°F and 1.00 psia. The characteristic temperatures of H_2O are $\Theta_r = 0.337$ R, $\Theta_{v1} = 4131$ R, $\Theta_{v2} = 9459$ R, and $\Theta_{v3} = 9720$ R.
37.* The experimentally measured value for c_v/R for ammonia, NH_3, at 15.0°C is 3.42. Use the Maxwell-Boltzmann formulae for nonlinear polyatomic gases to calculate the value of c_v/R for ammonia, then compute the percent error in your result. The characteristic vibrational temperatures of ammonia are $\Theta_{v1} = 1367$ K, $\Theta_{v2} = \Theta_{v3} = 2341$ K, $\Theta_{v4} = 4801$ K, and $\Theta_{v5} = \Theta_{v6} = 4955$ K.
38.* The experimentally measured value for c_v/R for methane, CH_4, at 300. K is 3.2553. Use the Maxwell-Boltzmann formulae for nonlinear polyatomic gases to calculate the value of c_v/R for methane, and then compute the percent error in your result. The characteristic vibrational temperatures of methane are $\Theta_{v1} = \Theta_{v2} = \Theta_{v3} = 1879$ K, $\Theta_{v4} = \Theta_{v5} = 2207$ K, $\Theta_{v6} = 4197$ K, and $\Theta_{v7} = \Theta_{v8} = \Theta_{v9} = 4344$ K.
39.* Calculate the specific entropy of HF at 0.00°C and atmospheric pressure.
40.* Calculate the change in specific entropy as HCl gas is heated at a constant pressure of 2.00 atm from 300. to 3000. K.
41. Determine the heat transfer required to heat 11.3 lbm of HBr gas from 100. to 1000.°F in a closed, rigid, 1.50 ft^3 container.
42.* Determine the power produced as 0.300 kg/s of HI passes through a steady state, adiabatic turbine from 2000. K, at 50.0 atm, to 1000. K at 1.00 atm pressure.
43.* Determine the entropy production rate for the turbine in Problem 42.

Design Problems

The following are open-ended design problems. The objective is to carry out a preliminary design as indicated. A detailed design with working drawings is not required unless otherwise specified by the instructor. These problems have no specific answers, so each student's design is unique.

44. Design a heater that raises the temperature of diatomic chlorine gas from 70°F to 2500°F without changing the pressure significantly. Do not assume ideal gas behavior. Use Eqs. (18.49a) through (18.54d) to calculate the necessary thermodynamic properties of chlorine. Provide an assembly drawing of your design along with all the relevant thermodynamic and design calculations.
45. Design a flow meter that uses a measurement of one or more of the thermodynamic u, h, or s to calculate the mass flow rate $\dot{m}$. For example, we can construct an open system such that $\dot{m}$ can be calculated from $\dot{Q}$, h_1, and h_2 as $\dot{m} = \dot{Q}/(h_2 - h_1)$. Provide an assembly drawing of your design along with all the relevant thermodynamic and design calculations.
46. Using the equation given in Problem 11, design an instrument that determines the porosity of a small, square, flat test sample of material. Be sure to explain how to calculate the porosity from the measurements taken. If possible, set up an experiment and test your technique with samples of known porosity. Provide an assembly drawing of your design along with all the relevant thermodynamic and design calculations.
47. Design a spring-loaded throttling valve that isothermally throttles 8.30 lbm/s of molecular bromine gas from 1000. psia, 70.0°F, to atmospheric pressure using choked flow conditions (see Chapter 16). Provide an assembly drawing of your design along with all the relevant thermodynamic and design calculations.
48.* Design a system that increases the temperature of 0.350 kg of diatomic sodium gas Na_2 from 1500. K at 20.0 kPa to 3000. K simply by compressing it by some mechanism in a closed system. No auxiliary heaters or coolers may be used. Provide an assembly drawing of your design along with all the relevant thermodynamic and design calculations.

Computer Problems

The following open-ended computer problems are designed to be done on a personal computer using a spreadsheet, equation solver, or programming language.

49. Using the equations from the kinetic theory of gases, write an interactive computer program that returns values for u, h, and s when p, T, M, and b are input by the user from the keyboard. Allow the user to choose either the English or the SI units system for the input values, and output values in the same units system.
50.* Using the data on NO given in Example 18.9, plot curves of c_p and c_v for NO vs. temperature for $0 \leq T \leq 5000$ K. Compute at least 100 points for each curve.
51. Using the equations for the thermodynamic properties of diatomic gases given in the text, write an interactive computer program that returns values for u, h, and s when p and T are input by the user from the keyboard for one or more (instructor's choice) of the gases listed in Table 18.8. Assume $u_0 = 0$, and allow the user to choose to work in either the English or the SI units system. Output values in the same units system that was chosen for input values.
52.* Using the appropriate equations from the text, plot $k = c_p/c_v$ vs. T for molecular oxygen over the range $0 \leq T \leq 5000$ K.

CHAPTER 19

Introduction to Coupled Phenomena

CONTENTS

19.1 INTRODUCTION

Physical phenomena are generally considered independent of each other. However, phenomenon can become coupled whenever the presence of one physical phenomenon induces one or more other physical phenomena to occur simultaneously. For example, the heat transport of energy through a system can induce a flow of electrical energy under certain circumstances; similarly, the flow of electrical energy can induce the flow of heat. This particular coupling is called the *thermoelectric effect.* Since energy is always conserved, in this case, some of the thermal energy is converted directly into electrical energy by an internal mechanism. The energy conversion efficiency from thermal energy directly into electrical energy is normally quite low; however, the reverse energy conversion process (from electrical to thermal energy) can be 100% efficient. Some of the more common types of direct energy conversion coupled phenomena are shown later in Tables 19.1 and 19.2. Notice that most of these couplings have been known for a very long time.

19.2 COUPLED PHENOMENA

Coupled phenomena have the potential for future technological utilization as important direct energy conversion processes. Today, they are used mainly in sensors to produce low-level electrical signals that are proportional to the magnitude of the other phenomenon present. For example, the thermoelectric effect is commonly used to produce a voltage proportional to the local temperature that can be read by an instrument or a computer. Such a device is called a *thermocouple,* and it is used as a temperature sensor.

Table 19.1 Some Examples of Known Coupled Phenomena

Electrical	Magnetic	Mechanical
Thermoelectric	Thermomagnetic	Thermomechanical
Photoelectric	Photomagnetic	Mechanochemical
Electrokinetic	Galvanomagnetic	Thermoelastic
Electrostriction	Magnetostriction	Piezoelectric
Electroluminescence	Gyromagnetic	Mechanocaloric
Electro-optical	Magneto-optical	Piezo-optical
Electrorheologic	Electromagnetic	Triboelectric

Table 19.2 Some Common Types of Direct Energy Conversion Coupled Phenomena

Energy Conversion Type	Common Name	Discoverer	Approximate Maximum Efficiency
Thermal to mechanical (thermomechanical)	Heat engine	Hero of Alexandria (~150 AD)	~60% (Carnot)
Chemical to electrical (electrochemical)	Battery and fuel cell	Volta (1800) and Davy (1802)	80–100%
Thermal to electric (thermoelectric)	Thermocouple	Seebeck (1821), Peltier (1834), Kelvin (1854)	~10%
	Photoelectric	Becquerel (1839)	~25%
	Thermopile	Weston (1884)	35–50%
	Thermionics	Edison (1883)	~40%
	Vacuum tube	Richardson (1912)	~40%
Mechanical to electrical (mechanoelectric)	MHD and EHD Piezoelectric	Faraday (1831) Curie (1883)	85–95% 10–30%
Thermal to kinetic energy	Thermal ionization	Saha (1920)	~10%
Chemical to mechanical (mechanochemical)	Animal muscles	Appears in nature	~25%
Potential energy to electrical energy	Kelvin water dropper	Kelvin (1860)	~50%
Chemical to thermal (thermochemical)	Combustion	Appears in nature	~100%

The reverse thermoelectric process is also possible. The application of a voltage to the leads of a thermocouple produces a heating or a cooling of the thermocouple junction. Though not very energy efficient, thermoelectric cooling is an important source of localized cooling in industries where space is a premium.

Some of the many known types of coupled phenomena and their uses are illustrated in Tables 19.1 and 19.2.

Nearly every energy conversion process has an efficiency of less than 100%, due to energy dissipation within the system, resulting from the inherent irreversibilities of the energy conversion process. The overall energy conversion efficiency is the product of all the individual energy conversion efficiencies within the system. For example, the technology in use today for providing electrical energy in your home consists of the following series of energy conversion processes.

Chemical Energy →	Thermal Energy → (Combustion)	Mechanical Energy → (Turbine)	Electrical Energy → (Generator)	Home Use Energy → (Appliances)	Environment

so that $\eta_{\text{overall}} = \eta_T \eta_m \eta_g \eta_E \eta_H$, and since each η is less than 1.0 (100%), the overall η can be quite small. Consequently, there is considerable interest today in developing direct energy conversion technologies whose efficiencies are competitive with those of the conventional chemical and nuclear multilevel indirect energy conversion technologies.

To understand the analysis of coupled phenomena, we introduce the concept of entropy production per unit time per unit volume. Because these terms have the form of an entropy production rate per unit volume, they are more conveniently called *entropy production rate densities*, abbreviated EPRD for convenience.

$$\text{EPRD} = \text{Entropy production rate density}$$

Table 19.3 Entropy Production Rate Density One-Dimensional and Three-Dimensional Formulae

Entropy Production Rate Density Due To	EPRD Formula
Heat transport of energy	$\sigma_Q = -\left(\frac{\dot{q}}{T^2}\right)\left(\frac{dT}{dx}\right) = \vec{q}\cdot\vec{\nabla}\left(\frac{1}{T}\right)$
Viscous dissipation	$(\sigma_W)_{vis} = \frac{\mu}{T}\left(\frac{dV}{dx}\right)^2 = \underset{\sim}{\tau}:\vec{\nabla}\vec{V}$
Electrical energy dissipation	$(\sigma_W)_{elect} = \rho_e J_E^2/T = -\vec{E}\cdot\vec{J_E}/T$
Diffusion of n dissimilar chemical species	$(\sigma_m)_{diff} = \sum_{i=1}^{n} j_{ix}\cdot\frac{d}{dx}\left(\frac{\hat{\mu}_i}{T}\right) = \sum_{i=1}^{n} \vec{J_i}\cdot\vec{\nabla}\left(\frac{\mu_i}{T}\right)$

Table 19.4 Generalized EPRD One-Dimensional Flux and Force Formulae

Entropy Production Rate Density (σ)	Generalized Energy Transport Flux (J)	Generalized Energy Transport Force (X)
σ_Q	$\dot{q} = \vec{q}/A$ = heat flux vector	$-\frac{1}{T^2}\left(\frac{dT}{dx}\right)$
$(\sigma_W)_{vis}$	$\mu\left(\frac{dV}{dx}\right)$	$\frac{1}{T}\left(\frac{dV}{dx}\right)$
$(\sigma_W)_{elect}$	$J_E = I/A$ = electron flux	$\frac{\rho_e J_e}{T} = -\frac{1}{T}\left(\frac{d\phi}{dx}\right)$
$(\sigma_m)_{diff}$	j_{ix} = mass flux of chemical species i in the x direction	$\frac{d}{dx}\left(\frac{\hat{\mu}_i}{T}\right)$

In Table 19.3 $J_E = I/A$ = the current (electron) flux. Note that each σ listed in Table 19.3 has the form

$$\sigma = JX = \vec{J}\cdot\vec{X} = \underset{\sim}{J} : \underset{\sim}{X} \tag{19.1}$$

where J, $\vec{J}$, $\underset{\sim}{J}$ = generalized (scalar, vector, or tensor) energy transport flux, and X, $\vec{X}$, $\underset{\sim}{X}$ = generalized (scalar, vector, or tensor) energy transport driving force producing the flux.

Table 19.4 identifies the various flux and force terms associated with the entropy production rate density formulae given in Table 19.3.

For any system containing a number of entropy production rate densities resulting from the simultaneous operation of a number of irreversible processes, the total EPRD is simply the sum of all the individual EPRDs present within the system. Also, the second law of thermodynamics requires that the total EPRD be positive, so

$$\sigma_{total} = \sigma_Q + (\sigma_W)_{vis} + (\sigma_W)_{elect} + (\sigma_m)_{diff} + \cdots > 0 \tag{19.2}$$

Using the generalized flux-force formulation, this can be written for m such densities as

$$\sigma_{total} = J_Q X_Q + (J_W X_W)_{vis} + \cdots = \sum_{i=1}^{m} J_i X_i > 0 \tag{19.3}$$

19.3 LINEAR PHENOMENOLOGICAL EQUATIONS

We begin by postulating that all simultaneously occurring physical phenomena that can be coupled are coupled to each other in some way. This is succinctly stated in the "coupling postulate."

For example, consider a system that has two energy transport fluxes, J_1 and J_2, which result from two generalized forces, X_1 and X_2. The coupling postulate states that

$$J_1 = J_1(X_1, X_2)$$

and

$$J_2 = J_2(X_1, X_2)$$

Expanding J_1 about the "equilibrium" state $X_1 = X_2 = 0$ using a Taylor's series gives

$$J_1(X_1, X_2)\, J_1(0, 0) + \frac{\partial J_1}{\partial X_1}dX_1 + \frac{\partial J_2}{\partial X_2}dX_2 + \text{higher order terms}$$

THE COUPLING POSTULATE

All the generalized flows (J_i) are dependent on (or coupled to) all the generalized forces (X_i) present within the system.[1] That is, a functional "coupling" relationship exists between all generalized fluxes and all generalized forces within a system of the form

$$J_i = J_i(X_1, X_2, X_3, X_4, \ldots) \tag{19.4}$$

[1] *This is limited by the Curie principle, which states that, in an isotropic system, the forces and fluxes must be of the same tensor rank before they can be coupled.*

Now $J_1(0, 0) = 0$ (if there are no forces, there are no flows, or fluxes) and very near the equilibrium state of the system, we can write

$$dX_1 \approx \Delta X_1 = X_1 - 0 = X_1$$

and

$$dX_2 \approx \Delta X_2 = X_2 - 0 = X_2$$

so that near equilibrium, we have

$$J_1 = \left(\frac{\partial J_1}{\partial X_1}\right)X_1 + \left(\frac{\partial J_1}{\partial X_2}\right)X_2 \tag{19.5}$$

and, similarly,

$$J_2 = \left(\frac{\partial J_2}{\partial X_1}\right)X_1 + \left(\frac{\partial J_2}{\partial X_2}\right)X_2 \tag{19.6}$$

Now, we define the "primary" coefficients, L_{11} and L_{22}, as

$$L_{11} = \frac{\partial J_1}{\partial X_1}$$

$$L_{22} = \frac{\partial J_2}{\partial X_2}$$

and the "secondary" or "coupling" coefficients, L_{12} and L_{21}, as

$$L_{12} = \frac{\partial J_1}{\partial X_2}$$

$$L_{21} = \frac{\partial J_2}{\partial X_1}$$

Then, Eqs. (19.5) and (19.6) become

$$J_1 = L_{11}X_1 + L_{12}X_2 \tag{19.7}$$

$$J_2 = L_{21}X_1 + L_{22}X_2 \tag{19.8}$$

or, in general,

$$J_i = \sum_{j=1}^{m} L_{ij}X_j \tag{19.9}$$

where the summation notation implies that the summation is to take place over all m fluxes and forces present in the system.

Equations (19.9) are called the *linear phenomenological equations*, because they resulted from a linearization of the general coupling postulate relationship (Eq. (19.4)) via a Taylor's series expansion around the equilibrium state $X_1 = X_2 = 0$. These equations therefore are valid for systems that are only slightly nonequilibrium. Consequently, the linear phenomenological equations have a limited range of usage, but they are sufficiently accurate to explain many of the known coupled effects shown in Table 19.1.

In 1931, Lars Onsager (1903–1976) proved that the coupling coefficients, L_{ij}, formed a symmetrical matrix and established what is now called the *reciprocity relationship* for these coefficients as

$$L_{ij} = L_{ji} \tag{19.10}$$

The total EPRD for a system near equilibrium is found by combining Eqs. (19.3), (19.9), and (19.10) to yield

$$\sigma_{\text{total}} = \sum_{i=1}^{m} J_i X_i = \sum_{i=1}^{m}\sum_{j=1}^{m} L_{ij} X_j X_i > 0 \tag{19.11}$$

where $L_{ij} = L_{ji}$.

19.4 THERMOELECTRIC COUPLING

Consider the simultaneous near equilibrium flow of thermal and electrical energy in a system. Then, Eq. (19.3) for the total EPRD is

$$\sigma_{\text{total}} = J_Q X_Q + J_E X_E$$

where the fluxes J_Q, J_E and the forces X_Q, X_E are defined in Table 19.4. Also, Eq. (19.9) gives

$$J_Q = L_{QQ} X_Q + L_{QE} X_E \tag{19.12}$$

and

$$J_E = L_{EQ} X_Q + L_{EE} X_E \tag{19.13}$$

Equation (19.11) then becomes

$$\sigma_{\text{total}} = L_{QQ} X_Q + L_{QE} X_Q X_E + L_{EQ} X_Q X_E + L_{EE} X_E$$

and combining this with Onsager's reciprocity relationship, $L_{EQ} = L_{QE}$, gives

$$\sigma_{\text{total}} = L_{QQ} X_Q + 2L_{QE} X_E X_Q + L_{EE} X_E > 0 \tag{19.14}$$

Equation (19.12) explains how a heat flow J_Q can occur even without the presence of a temperature difference ($X_Q = 0$), if the coupling coefficient L_{QE} is nonzero and a voltage source X_E exists within the system. Similarly, Eq. (19.13) illustrates how a flow of electrical energy J_E can exist with no apparent voltage source ($X_E = 0$), when heat transport occurs J_Q with a nonzero coupling coefficient L_{EQ}. Considerable thermoelectric technology has developed around this simple coupled effect.

Thermoelectric heaters, coolers, and temperature measurement instruments (thermocouples) are common devices today. To fully understand and utilize thermoelectric coupling, we must evaluate the primary (L_{QQ}, L_{EE}) and secondary or coupling ($L_{EQ} = L_{QE}$) coefficients for any given system. Historically, many of the thermoelectric coupling effects were discovered empirically by individuals long before an accurate thermodynamic understanding of this phenomenon was known. These effects were originally thought to be independent phenomena and had no adequate explanation. The important empirical thermoelectric discoveries occurred as follows.

19.4.1 The Seebeck Effect[2]

Heating one junction of a bimetallic (or semiconductor) closed circuit while simultaneously cooling the other junction produces a flow of current in the circuit without an apparent voltage source being present. If two such materials are joined at a single junction to produce an open circuit, then heating or cooling that junction produces an electrical potential (voltage) difference across the circuit that is directly proportional to the temperature of the junction (see Figure 19.1).

[2] Discovered in 1821 by the German physicist Thomas Johann Seebeck (1770–1831). It occurs because the free electron densities of certain materials (metals and semiconductors) differ from one another at a given temperature. When two such materials are joined, their junction appears as a voltage source due to electrons diffusing down the electron concentration gradient at the junction.

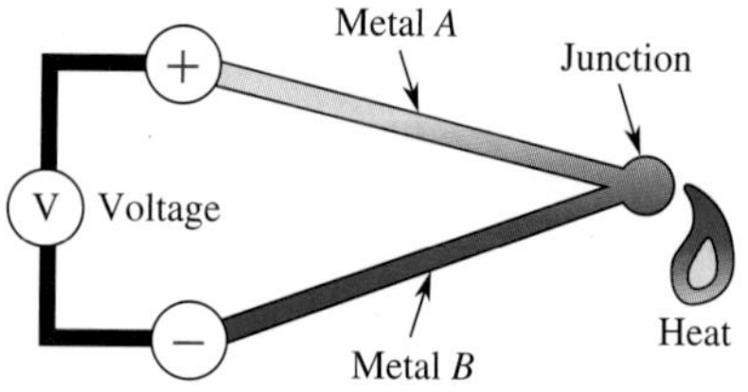

FIGURE 19.1
A schematic of the Seebeck effect.

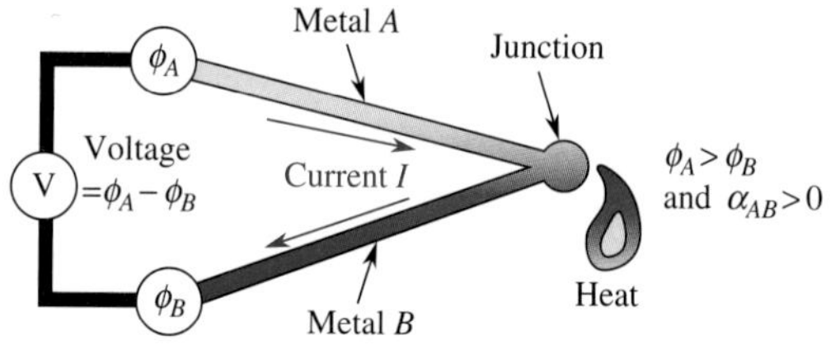

FIGURE 19.2
The sign convention for the Seebeck coefficient, α_{AB}.

This early discovery gave rise to the relative Seebeck coefficient, α_{AB}, defined as (see Figure 19.2).

$$\sigma_{AB} = -\lim_{\Delta T \to 0} \left(\frac{\phi_A - \phi_B}{\Delta T} \right) \Bigg|_{I=0} = -\frac{d\phi_{AB}}{dT} \Bigg|_{I=0} \tag{19.15}$$

where ϕ_{AB} is the potential difference (known as the Seebeck voltage), $\phi_A - \phi_B$. Thus, α_{AB} is simply the negative value of the slope of the open circuit voltage-temperature relationship for the pair of conductors. The sign convention usually adopted for α_{AB} is as follows. If the current flow in conductor A is from the cooler junction (T_C) to the hotter junction (T_H), then α_{AB} is positive (see Figure 19.2). Both the terminals across which ϕ_{AB} is measured must be at the same temperature, as shown in Figure 19.1b, or else additional Seebeck voltages are generated at these terminals.

A positive value for α_{AB} corresponds to current flowing from T_C to T_H in conductor A, and from T_H to T_C in conductor B. Also, it is possible to assign "absolute" Seebeck coefficients to the pure conductors, α_A and α_B, defined from

$$\alpha_{AB} = \alpha_A - \alpha_B$$

since a superconducting material below its transition temperature (i.e., where it becomes superconducting) has no measurable thermoelectric effect ($\alpha_S = 0$). We can determine the absolute Seebeck coefficient of any material, say A, by joining it to a superconductor and measuring the resultant open circuit voltage-temperature slope.[3] Then,

$$-\frac{d\phi_{AB}}{dT} \Bigg|_{I=0} = \alpha_{AS} = \alpha_A - \alpha_S = \alpha_A - 0 = \alpha_A$$

and once we know the absolute Seebeck coefficient for any one material, that material can be used as a reference material (α_R) to determine the absolute Seebeck coefficients of other materials as

$$\alpha_B = \alpha_{BR} + \alpha_R$$

Finally, we can easily determine a relationship between relative Seebeck coefficients as

$$\alpha_{AB} = \alpha_A - \alpha_B = (\alpha_A - \alpha_C) - (\alpha_B - \alpha_C) = \alpha_{AC} - \alpha_{BC}$$

19.4.2 The Peltier Effect[4]

Passing a current I through an isothermal bimetallic (or semiconductor) closed circuit causes heat absorption at one junction and heat release at the other junction (see Figure 19.3).

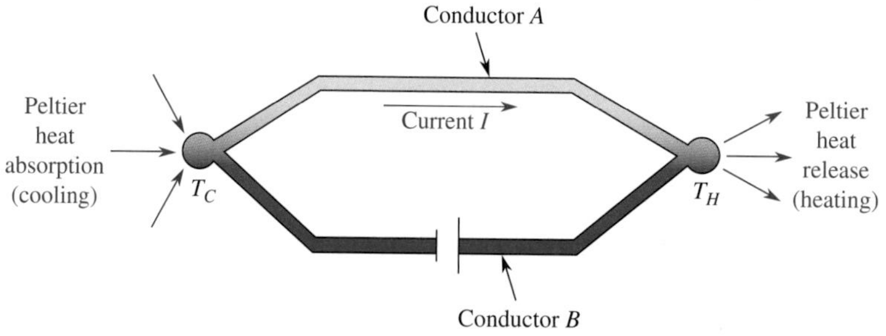

FIGURE 19.3
A schematic of the Peltier effect.

[3] This technique works only at very low temperatures (below 18 K). Other techniques are used at higher temperatures.
[4] Discovered in 1834 by the French watchmaker turned physicist Jean Charles Athanase Peltier (1785–1845).

This discovery was originally thought to be independent of the Seebeck effect and gave rise to a new parameter called the *relative Peltier coefficient*, π_{AB}, defined as

$$\pi_{AB} = \pi_A - \pi_B = \frac{\dot{Q}_P}{I} \tag{19.16}$$

where $\dot{Q}_P$ is the Peltier heating or cooling rate, and π_A and π_B are the absolute Peltier coefficients for the pure conductors. Later (see Eq. (19.33)), it was discovered that the Peltier and Seebeck coefficients are related by $\pi_{AB} = T\alpha_{AB}$ (or $\pi_A = T\alpha_A$), where T is the absolute temperature of the junction.

19.4.3 The Kelvin Effect[5]

When an electric current passes through a single homogeneous conductor along which a temperature difference exists, heating or cooling of the conductor occurs, depending on the direction of the current flow relative to the temperature difference (see Figure 19.4).

The Kelvin effect can be demonstrated by heating the center of a uniform wire while cooling its ends and passing a current through it. If we measure the temperatures at two points, A and B, equidistant from the center, we find that $T_A \neq T_B$. The electrical current has disturbed the temperature profile in the wire. This effect gave rise to the Kelvin coefficient, τ, defined as

$$\tau = \lim_{\Delta T \to 0} \frac{\dot{Q}_K}{I(\Delta T)} = \frac{\dot{q}'_K}{J_E(dT/dX)} \tag{19.17}$$

where $\dot{q}'_K$ is the Kelvin heating or cooling rate per unit volume, and $J_E = I/A$ is the electrical current density (or electron flux). It can be shown that the Kelvin coefficient is related to the absolute Seebeck coefficient by $\tau = -T(d\alpha/dT)$, where T is the absolute temperature of the junction. For a thermocouple with conductors A and B, the difference in the Kelvin coefficients for the two conductors is

$$\tau_A - \tau_B = \tau_{AB} = -T\frac{d}{dT}(\alpha_A - \alpha_B) = -T\frac{d\alpha_{AB}}{dT}$$

19.4.4 The Fourier Effect[6]

A temperature difference in a homogeneous conductor produces a heat flow in the direction of decreasing temperature (see Figure 19.5).

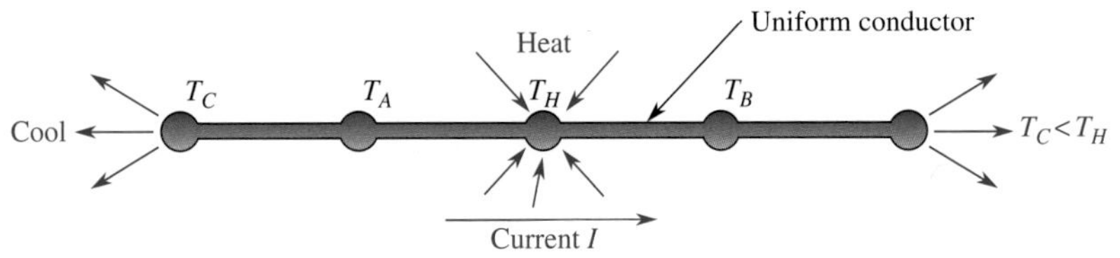

FIGURE 19.4
A schematic of the Kelvin effect.

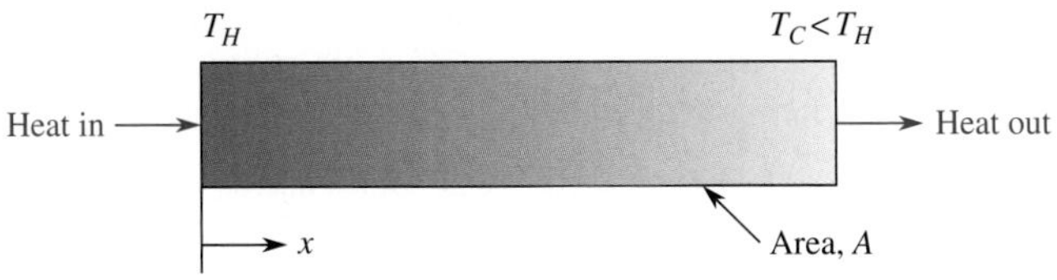

FIGURE 19.5
A schematic of the Fourier effect.

---

[5] Discovered in 1854 by the Irish mathematician, physicist, and engineer William Thomson (Lord Kelvin) (1824–1907).
[6] Discovered in 1822 by the French mathematician Jean Baptiste Joseph Fourier (1768–1830).

This effect gave rise to Fourier's law of heat conduction,

$$\dot{Q}_F = -k_t A\left(\frac{dT}{dx}\right) \quad (19.18)$$

or

$$\dot{q}_F = \frac{\dot{Q}_F}{A} = -k_t\left(\frac{dT}{dx}\right)$$

where $\dot{Q}_F$ is the Fourier heat transfer rate, $\dot{q}_F$ is the Fourier heat flux, and k_t is the Fourier coefficient (known today as the *thermal conductivity*).

19.4.5 The Joule Effect[7]

An electrical current passing through a homogeneous isothermal conductor produces an internal heating of the conductor which is independent of the direction of current flow (see Figure 19.6).

This effect gave rise to the Joule heating formula

$$\dot{Q}_J = k_J I^2 R_e \quad (19.19)$$

or

$$\dot{q}_J = k_J I^2 R_e / A$$

where $\dot{Q}_J$ is the Joule heating rate, $\dot{q}_F$ is the Joule heat flux, R_e is the electrical resistivity of the conductor, and k_J is the Joule coefficient,

$$k_J = 778.17\,\text{ft}\cdot\text{lbf/Btu} = 0.293\,\text{W}\cdot\text{h/Btu} = 1\,\text{WLs/joule}$$

Today k_J is recognized as being merely a units conversion factor and it is not normally written in the equation. Therefore, from now on we write Eq. (19.19) simply as

$$\dot{Q}_J = I^2 R_e \quad (19.20)$$

19.4.6 The Ohm Effect[8]

When an electrical current is passed through a homogeneous isothermal conductor, a voltage drop occurs in the direction of current flow (see Figure 19.7).

This effect gave rise to Ohm's law,

$$\phi_1 - \phi_2 = R_e I \quad (19.21)$$

where R_e is Ohm's coefficient, today called the *electrical resistance* of the conductor.

All six of these effects occur simultaneously when thermal and electrical energy simultaneously flow through a system. It should be apparent by now that it is easier to comprehend thermoelectric effects through the coupling Eqs. (19.12) and (19.13) than to master the six interrelated effects just discussed. However, these six effects are now an integral part of our engineering jargon and technical literature, and they cannot be dispensed with so easily. Therefore, we develop formulae for the thermoelectric primary and secondary coefficients in terms of the coefficients of the six empirically discovered thermoelectric effects discussed. This provides continuity between the old and the new interpretations of thermoelectricity.

Since the primary electrical coefficient L_{EE} is the easiest to deal with, we begin with it. Primary coefficients result from noncoupled phenomena, and pure electrical effects are described by Ohm's law. Ohm's law can be cast into a number of forms; for example,

$$\phi_1 - \phi_2 = R_e I = I(\rho_e L/A) = \rho_e (I/A) L = \rho_e J_E L$$

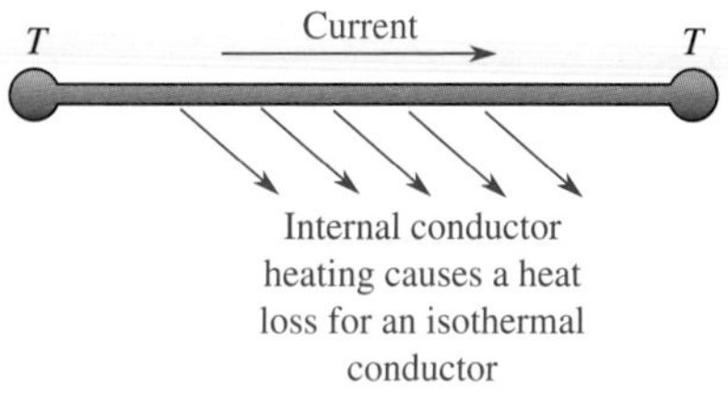

FIGURE 19.6
A schematic of the Joule effect.

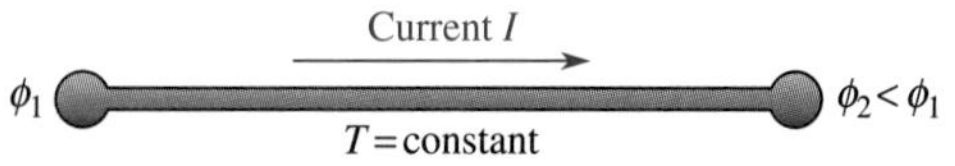

FIGURE 19.7
A schematic of the Ohm effect.

---

[7] Discovered in 1841 by the English physicist James Prescott Joule (1818–1889).

[8] Discovered in 1827 by the German physicist Georg Simon Ohm (1787–1854).

where $R_e = \rho_e L/A$ is the electrical resistance, ρ_e is the electrical resistivity, L is the length of the conductor, A is the cross-sectional area of the conductor, and $J_E = I/A$ is the electron flux in the conductor. Now, let

$$\frac{\phi_1 - \phi_2}{L} = -\left(\frac{\phi_2 - \phi_1}{L}\right) = -\frac{d\phi}{dx} = \rho_e J_E$$

and let the electrical conductivity k_e be defined as

$$k_e = 1/\rho_e$$

then we can rearrange Ohm's law as

$$J_E = I/A = -k_e\left(\frac{d\phi}{dx}\right) \tag{19.22}$$

For purely electrical effects (no thermal effects, i.e., no temperature differences), Eq. (19.13) reduces to

$$J_E\big|_{T=\text{constant}} = L_{EE}X_E$$

and Table 19.4 gives

$$X_E = -\frac{1}{T}\left(\frac{d\phi}{dx}\right)$$

Combining these equations produces

$$J_E\big|_{T=\text{constant}} = -\frac{L_{EE}}{T}\left(\frac{d\phi}{dx}\right) = -k_e\left(\frac{d\phi}{dx}\right)$$

consequently, we find that

$$L_{EE} = Tk_e = T/\rho_e \tag{19.23}$$

The next easiest term to deal with is the coupling coefficient ($L_{EQ} = L_{QE}$), because it is simply related to the Seebeck coefficient.[9] The absolute Seebeck coefficient α is defined as

$$\alpha = -\frac{d\phi}{dT}\bigg|_{I=0} = -\frac{d\phi}{dT}\bigg|_{J_E=0}$$

Equation (19.13) then gives, for zero current flow,

$$L_E = L_{EQ}X_Q + L_{EE}X_E = 0$$

and introducing the formula for X_Q and X_E from Table 19.4 provides

$$L_{EQ}\left(-\frac{1}{T^2}\frac{dT}{dx}\right)\bigg|_{I=0} + L_{EE}\left(-\frac{1}{T}\frac{d\phi}{dx}\right)\bigg|_{I=0} = 0$$

or

$$\left(\frac{d\phi/dx}{dTdx}\right)\bigg|_{I=0} = \frac{d\phi}{dT}\bigg|_{I=0} = -\frac{L_{EQ}}{TL_{EE}} = -\alpha \tag{19.24}$$

So that

$$L_{EQ} = L_{QE} = \alpha TL_{EE}$$

and introducing L_{EE} from Eq. (19.23) gives the coupling coefficient as

$$L_{EQ} = L_{QE} = \alpha T^2 k_e = \alpha T^2/\rho_e \tag{19.25}$$

Combining Eqs. (19.12) and (19.13) with $X_Q = dT/dx = 0$ (i.e., isothermal conditions) gives

$$\frac{J_Q}{J_E}\bigg|_{T=\text{constant}} = \frac{L_{QE}}{L_{EE}} = \frac{\dot{Q}_i}{T} \tag{19.26}$$

---

[9] The phenomenological coefficients have been developed in terms of the Seebeck coefficient, because it is the easiest thermoelectric coefficient to measure.

where $\dot{Q}_i$ is the isothermal heat transport rate "induced" by the presence of the electrical current I. From Eqs. (19.24) and (19.26), we see that

$$\alpha = \frac{\dot{Q}_i}{IT} = \frac{\dot{S}_i}{I} \tag{19.27}$$

where $\dot{S}_i = \dot{Q}_i/T$ is the isothermal entropy transport rate "induced" by the presence of the electrical current I. This is the entropy transport rate due to the flow of electrons in the conductor.[10]

Fourier's law represents pure thermal effects, but it is somewhat more difficult to interpret since it corresponds to a condition of zero electron flow ($J_E = 0$) but not necessarily a zero voltage difference. From Eq. (19.13) with $J_E = 0$, we find that

$$X_E = -\frac{L_{EQ}}{L_{EE}} X_Q$$

then, Eq. (19.12) gives

$$J_Q\big|_{J_E=0} = L_{QQ}X_Q + L_{QE}\left(-\frac{L_{EQ}}{L_{EE}}\right)X_Q = \left(\frac{L_{QQ}L_{EE} - L_{QE}{}^2}{L_{EE}}\right)X_Q$$

where we use Onsager's reciprocity relationship, $L_{EQ} = L_{QE}$. Now, by definition,

$$J_Q\big|_{J_E=0} = \frac{\dot{Q}_F}{A} = \dot{q}_F$$

and Table 19.4 gives

$$X_Q = \frac{1}{T^2}\left(\frac{dT}{dx}\right)$$

so

$$J_Q\big|_{J_E=0} = \dot{q}_F = -k_t\left(\frac{dT}{dx}\right) = -\left(\frac{L_{QQ}L_{EE} - L_{QE}{}^2}{L_{EE}T^2}\right)\left(\frac{dT}{dx}\right)$$

consequently,

$$k_t = \frac{L_{QQ}L_{EE} - L_{QE}^2}{L_{EE}T^2} \tag{19.28}$$

Therefore, the thermal conductivity of a substance is not a simple quantity. It is composed of all three phenomenological coefficients combined with the inverse of the absolute temperature squared. Substituting the formulae for L_{EE} and L_{QE} from Eqs. (19.23) and (19.25) into Eq. (19.28) and solving for the remaining coefficient, L_{QQ}, gives

$$L_{QQ} = T^2(k_t + \alpha^2 Tk_e) \tag{19.29}$$

Substituting Eqs. (19.23), (19.25), and (19.29) along with the formulae for X_Q and X_E from Table 19.4 into Eqs. (19.12) and (19.13) produces the linear phenomenological equations (in terms of the pure conductor Seebeck coefficient) as

$$J_Q = -(k_t + \alpha^2 Tk_e)\left(\frac{dT}{dx}\right) - \alpha Tk_e\left(\frac{d\phi}{dx}\right) \tag{19.30}$$

and

$$J_E = -k_e\left(\frac{d\phi}{dx}\right) - \alpha k_e\left(\frac{dT}{dx}\right) \tag{19.31}$$

Also, substitution into the thermoelectric total EPRD formula of Eq. (19.14) gives

$$\sigma_{\text{thermoelectric}} = \frac{1}{T^2}\left[(k_t + \alpha^2 Tk_e)\left(\frac{dT}{dx}\right) + \alpha Tk_e\left(\frac{d\phi}{dx}\right)\right]\left(\frac{dT}{dx}\right) + \frac{1}{T}\left[k_e\left(\frac{d\phi}{dx}\right) + \alpha k_e\left(\frac{dT}{dx}\right)\right]\left(\frac{d\phi}{dx}\right) > 0 \tag{19.32}$$

---

[10] Curiously, though work mode transport of energy cannot produce a direct entropy transport, it can induce a secondary entropy transport.

The Peltier and Kelvin coefficients are not independent from the Seebeck coefficient, so they need not be introduced into the phenomenological coefficients. For example, the Peltier heat $\dot{Q}_P$ is the heat transfer rate that occurs when there is no temperature difference, $dT/dx = 0 = X_Q$. Therefore, from Eq. (19.12),

$$J_Q\big|_{X_Q=0} = \frac{\dot{Q}_P}{A} = \dot{q}_P = L_{QE}X_E$$

or

$$\dot{Q}_P = A(\alpha T^2 k_e)\left(-\frac{1}{T}\frac{d\phi}{dx}\right) = -\alpha A T k_e\left(\frac{d\phi}{dx}\right)$$

Also, from Eq. (19.13),

$$J_E\big|_{X_Q=0} = I/A = L_{EE}X_E$$

or

$$I = ATk_e\left(-\frac{1}{T}\frac{d\phi}{dx}\right) = -Ak_e\left(\frac{d\phi}{dx}\right)$$

Then, the Peltier coefficient π given in Eq. (19.16) becomes

$$\pi = \frac{\dot{Q}_P}{I} = \alpha T \tag{19.33}$$

Thus, the Seebeck and Peltier coefficients for a pure conductor are directly related by Eq. (19.33).

EXAMPLE 19.1

A 0.0100 m diameter copper bus bar is maintained at a constant temperature of 20.0°C and is subjected to a voltage gradient of 1.00 V/m. Determine the Peltier heat flow, $\dot{Q}_P$. The Seebeck coefficient for the copper is $\alpha_{cu} = 3.50 \times 10^{-6}$ V/K, and its resistivity is $\rho_e = 5.00 \times 10^{-9}$ ohm meters.

Solution

The Peltier heat flow is given by Eq. (19.33) as

$$\dot{Q}_P = \pi I = (\alpha T)I$$

where

$$I = AJ_E = \frac{A}{\rho_e}\left(-\frac{d\phi}{dx}\right)$$

Here, $-d\phi/dx$ = voltage gradient = 1.00 V/m, so

$$I = \frac{(\pi/4)(0.0100\,\text{m})^2(1.00\,\text{V/m})}{5.00\times10^{-9}\,\Omega\,\text{gm}} = 1.57\times10^4\,\text{V}/\Omega = 1.57\times10^4\,\text{A}$$

Then

$$\dot{Q}_P = (3.50\times10^{-6}\,\text{V/K})(20,0+273.16\,\text{K})(1.57\times10^4\,\text{A}) = 16.2\,\text{V}\cdot\text{A} = 16.2\,\text{W}$$

EXAMPLE 19.2

The open circuit voltage of an iron-copper thermocouple is approximately given by

$$\phi_{\text{fe-cu}} = (-13.4\,T + 0.014\,T^2 + 0.00013\,T^3)\times10^{-6}\,\text{V}$$

where T is in °C, not K. At 100.°C, determine

a. The relative Seebeck coefficient $\alpha_{\text{fe-cu}}$.
b. The relative Peltier coefficient $\pi_{\text{fe-cu}}$.

(Continued)

EXAMPLE 19.2 *(Continued)*

Solution

a. From Eq. (19.15), we have that

$$\alpha_{\text{fe-cu}} = -\left.\frac{d\phi_{\text{fe-cu}}}{dT}\right|_{I=0}$$

Thus,

$$\alpha_{\text{fe-cu}} = -(-13.4 + 0.028\,T + 0.00039\,T^2) \times 10^{-6}\ \text{V/K}$$

where T is in °C not K. At $T = 100.$°C, this becomes

$$\alpha_{\text{fe-cu}}(100.°\text{C}) = 6.70 \times 10^{-6}\ \text{V/K}$$

b. From Eq. (19.33) we have[11]

$$\pi_{\text{fe-cu}} = T\alpha_{\text{fe-cu}} = (100. + 273.15\ \text{K})(6.70 \times 10^{-6}\ \text{V/K}) = 2.50 \times 10^{-3}\ \text{V}$$

[11] *Note that, even though most of the temperature to voltage correlations published for thermocouples are written in terms of relative temperature units (°C or °F), the multiplying temperature factor T in the Peltier and Kelvin coefficient equations is always in absolute units (K or R).*

The following example illustrates the thermoelectric effect used in a temperature measurement circuit. Note particularly the result obtained in part c, where the effect of connecting the thermocouple wires to a remote instrument using standard copper lead wires is investigated.

EXAMPLE 19.3

The chromel-alumel thermocouple circuit shown below has its cold junction at 0°C and its hot junction at 100.°C. Assuming that the absolute Seebeck coefficients $\alpha_{\text{ch}} = 23.0 \times 10^{-6}$ V/K and $\alpha_{\text{al}} = -18.0 \times 10^{-6}$ V/K are constant over this temperature range, determine

a. The open circuit thermoelectric (Seebeck) voltage for the chromel-alumel portion of the circuit.
b. The absolute and relative Peltier coefficients for each chromel-alumel junction.
c. The influence of the copper lead wires on the potentiometer voltage reading for the lead wire junction temperatures shown in Figure 19.8.

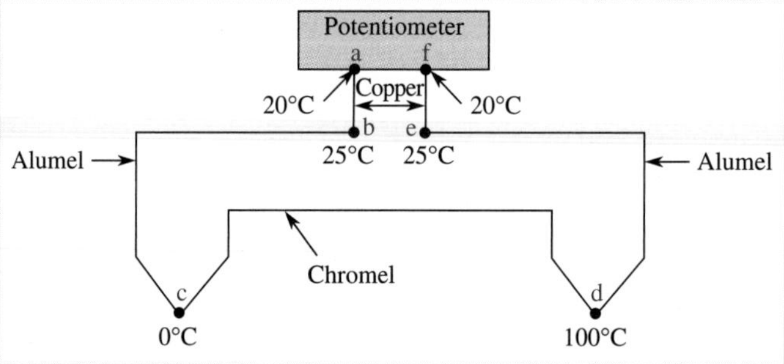

FIGURE 19.8
Example 19.3.

Solution

a. The relative Seebeck coefficient for the chromel-alumel circuit is

$$\alpha_{\text{ch-al}} = \alpha_{\text{ch}} - \alpha_{\text{al}} = [23.0 - (-18.0)] \times 10^{-6}\ \text{V/K} = 41.0 \times 10^{-6}\ \text{V/K}$$

and for a constant $\alpha_{\text{ch-al}}$, Eq. (19.15) can be integrated to give

$$-\phi_{\text{ch-al}} = \alpha_{\text{ch-al}}(T_{\text{H}} - T_{\text{C}}) = (41.0 \times 10^{-6}\ \text{V/K})(100. - 0\ \text{K}) = 4.1 \times 10^{-3}\ \text{V} = \phi_{\text{al-ch}}$$

(note that $\phi_{\text{al-ch}} = -\phi_{\text{ch-al}}$).

b. The absolute Peltier coefficients at each junction are given by Eq. (19.33) as $\pi_{ch} = \alpha_{ch}T$ and $\pi_{al} = \alpha_{al}T$, and we can compute $\pi_{ch\text{-}al} = \pi_{ch} - \pi_{al}$. At the 0.00°C = 273.15 K junction,

$$\pi_{ch} = (23.0\times10^{-6}\,\text{V/K})(273.15\,\text{K}) = 6.28\times10^{-3}\,\text{V}$$

$$\pi_{al} = (-18.0\times10^{-6}\,\text{V/K})(273.15\,\text{K}) = -4.91\times10^{-3}\,\text{V}$$

and

$$\pi_{ch\text{-}al} = [6.28-(-4.91)]\times10^{-3}\,\text{V} = 11.2\times10^{-2}\,\text{V}$$

At the 100.°C = 373.15 K junction,

$$\pi_{ch} = (23.0\times10^{-6}\,\text{V/K})(373.15\,\text{K}) = 8.58\times10^{-3}\,\text{V}$$

$$\pi_{al} = (-18.0\times10^{-6}\,\text{V/K})(373.15\,\text{K}) = -6.71\times10^{-3}\,\text{V}$$

and

$$\pi_{ch\text{-}al} = [8.58-(-6.71)]\times10^{-3}\,\text{V} = 15.3\times10^{-3}\,\text{V}$$

c. To determine the influence of the copper lead wires, we use Eq. (19.15) for the complete cu-al-ch-al-cu circuit connected to the potentiometer. Using the junction notation shown in Figure 19.8, the potentiometer reading is $\phi_{af} = \phi_a - \phi_f$. We can move around the circuit to evaluate this reading as

$$\phi_a - \phi_f = (\phi_a - \phi_b) + (\phi_b - \phi_c) + (\phi_c - \phi_d) + (\phi_d - \phi_e) + (\phi_e - \phi_f)$$

where

$$\phi_a - \phi_b = -\int_{T_b}^{T_a} \alpha_{cu}\,dT = \int_{T_a}^{T_b} \alpha_{cu}\,dT$$

$$\phi_b - \phi_c = -\int_{T_b}^{T_c} \alpha_{al}\,dT$$

$$\phi_c - \phi_d = -\int_{T_c}^{T_d} \alpha_{ch}\,dT$$

$$\phi_d - \phi_e = -\int_{T_e}^{T_e} \alpha_{al}\,dT$$

and

$$\phi_e - \phi_f = -\int_{T_e}^{T_f} \alpha_{cu}\,dT$$

Now, the contribution of the copper lead wires is

$$\Delta\phi_{cu} = \phi_a - \phi_b + \phi_e - \phi_f = \int_{T_a}^{T_b} \alpha_{cu}\,dT + \int_{T_e}^{T_f} \alpha_{cu}\,dT$$

and if, as the circuit diagram shows, $T_a = T_f$ and $T_b = T_e$, then the influence of the copper leads is

$$\Delta\phi_{cu} = \int_{T_a}^{T_b} \alpha_{cu}\,dT + \int_{T_b}^{T_a} \alpha_{cu}\,dT = \int_{T_a}^{T_b} \alpha_{cu}\,dT + \left(-\int_{T_a}^{T_b} \alpha_{cu}\,dT\right) = 0$$

(Continued)

EXAMPLE 19.3 (*Continued*)

Thus, if the copper lead wires have the same end point junction temperatures, then they do not contribute any net Seebeck voltage to the circuit. Adding up all the potential differences around the circuit and using $T_a = T_f$ and $T_b = T_e$ gives

$$\Delta\phi_{af} = \int_{T_a}^{T_b} \alpha_{\text{cu}} dT + \int_{T_b}^{T_a} \alpha_{\text{cu}} dT + \int_{T_b}^{T_c} \alpha_{\text{al}} dT + \int_{T_c}^{T_d} \alpha_{\text{ch}} dT + \int_{T_d}^{T_b} \alpha_{\text{al}} dT$$

Now,

$$\int_{T_b}^{T_c} \alpha_{\text{al}} dT + \int_{T_d}^{T_b} \alpha_{\text{al}} dT = \int_{T_d}^{T_c} \alpha_E dT = -\int_{T_c}^{T_d} \alpha_{\text{al}} dT$$

so

$$\phi_{af} = \int_{T_c}^{T_d} (\alpha_{\text{ch}} - \alpha_E) dT = \int_{T_c}^{T_d} (\alpha_{\text{ch-al}}) dT = \alpha_{\text{ch-al}}(T_d - T_c)$$

Consequently, the potentiometer measures the chromal-alumel thermoelectric effects only so long as all the lead wires have equal junction temperatures.

19.5 THERMOMECHANICAL COUPLING

In this section, we investigate the open system coupling phenomena resulting from the coupling of simultaneous heat transfer and mass flow. This type of coupling is commonly called the *thermomechanical effect.*

In 1873, W. Feddersen reported having observed a flow of air through a porous plug brought about by only a temperature difference on the opposite sides of the plug. There was no pressure drop across the plug, yet there was a flow of air. We call the phenomenon of fluid flow induced by the presence of a temperature difference rather than a pressure difference thermal osmosis.[12] This effect has been observed in both gases and liquids.

In 1939, J. G. Daunt and K. Mendelssohn reported observing isothermal heat transfer caused by only a pressure difference in a fluid. This is the reciprocal of the thermal osmosis effect and is often called the *mechanocaloric effect.*

The thermal osmosis and mechanocaloric effects are secondary or coupled effects induced by the primary effects of Fourier conduction heat transfer and Darcy-Weisbach pressure-driven mass flow. In thermomechanical systems, the fluxes are

1. Heat flux: $J_Q = \dot{Q}/A = \dot{q}$
2. Mass flux: $J_M = \dot{m}/A = \rho V$, where ρ is the local density and V is the local velocity of the moving fluid. The generalized forces are

$$\text{Temperature gradient: } X_Q = -\frac{1}{T^2}\left(\frac{dT}{dx}\right)$$

$$\text{Pressure gradient: } X_M = -\frac{v}{T}\left(\frac{dp}{dx}\right)$$

where v is the local specific volume of the moving fluid. Then, for near equilibrium conditions, Eq. (19.9) gives the coupled heat and mass flux equations as

$$J_Q = -\frac{L_{QQ}}{T^2}\left(\frac{dT}{dx}\right) - \frac{vL_{QM}}{T}\left(\frac{dp}{dx}\right) \tag{19.34}$$

and

$$J_M = -\frac{L_{QQ}}{T^2}\left(\frac{dT}{dx}\right) - \frac{vL_{MM}}{T}\left(\frac{dp}{dx}\right) \tag{19.35}$$

[12] This phenomenon is also known as *Knudsen effect* and the *fountain effect* in the literature.

Under isothermal conditions ($dT = 0$), Eq. (19.34) gives

$$J_Q\big|_{T=\text{constant}} = -\frac{vL_{QM}}{T}\left(\frac{dp}{dx}\right) \tag{19.36}$$

This equation describes the mechanocaloric effect, which is usually modeled after Fourier's law as

$$J_Q\big|_{T=\text{constant}} = -k_o\left(\frac{dp}{dx}\right) \tag{19.37}$$

where k_o is an empirical material constant called the *osmotic heat conductivity* coefficient. Comparing Eqs. (19.36) and (19.37), using Onsager's reciprocity relation and $\rho = 1/v$, we see that

$$L_{QM} = Tk_o/v = \rho Tk_o = L_{MQ} \tag{19.38}$$

Similarly, the mass flux under isothermal conditions ($dT = 0$) is given by Eq. (19.35) as

$$J_M\big|_{T=\text{constant}} = \frac{vL_{MM}}{T}\left(\frac{dp}{dx}\right) \tag{19.39}$$

Now, two flow models can be used to interpret this equation. The first was formulated empirically in 1856 by the French hydraulic engineer Henri Philibert Gaspard Darcy (1803–1858) for flow through porous media. Appropriately called *Darcy's law*, it states that the bulk fluid velocity in a porous material is given by

$$V = -\frac{k_p}{\mu}\left(\frac{dp}{dx}\right) \tag{19.40}$$

where μ is the fluid viscosity and k_p is called the *permeability* of the porous material.[13] Comparing Eqs. (19.39) and (19.40) and introducing the mass flux definition along with the relation $\rho = 1/v$ produces

$$J_M\big|_{T=\text{constant}} = \rho V = -\frac{\rho k_p}{\mu}\left(\frac{dp}{dx}\right) = -\frac{vL_{MM}}{T}\left(\frac{dp}{dx}\right) \tag{19.41}$$

from which it is clear that the mass flow primary coefficient L_{MM} is

$$L_{MM} = \rho^2 Tk_p/\mu \tag{19.42}$$

The other flow model that can be used in Eq. (19.39) was developed for flow through a circular tube and is called the *Darcy-Weisbach* equation after Henri Darcy and the German engineer Julius Ludwig Weisbach (1806–1871). In this model, the isothermal mass flux is given by

$$J_M\big|_{T=\text{constant}} = \rho V = -\frac{2Dg_c}{fV}\left(\frac{dp}{dx}\right) \tag{19.43}$$

where D is the tube diameter and f is an empirically determined "friction factor." The value of f can be found in most fluid mechanics textbooks from a generalized curve of f versus the dimensionless Reynolds number, $\rho VD/\mu$. It can be shown that, if the flow is laminar ($\rho VD/\mu < 2000$), then

$$J_M\big|_{T=\text{constant}} = \rho V = -\frac{\rho D^2}{32\mu}\left(\frac{dp}{dx}\right) \tag{19.44}$$

By comparing Eqs. (19.41) and (19.44), we see that these equations are similar and they become identical when we define the effective permeability of a circular tube as

$$k_p = D^2/32 \tag{19.45}$$

Thus, for very slow flow through a circular tube, we can write

$$L_{MM} = \frac{T(\rho D)^2}{32\mu} \tag{19.46}$$

Finally, combining Eqs. (19.34) and (19.35) to eliminate the term $(v/T)(dp/dx)$ gives

$$J_Q = -\left(\frac{L_{QQ}L_{MM} - L_{QM}{}^2}{L_{MM}T^2}\right)\left(\frac{dp}{dx}\right) - \frac{L_{QM}}{T}J_M \tag{19.47}$$

[13] The common unit of measure of permeability is the darcy, which has the rather awkward definition of 1 darcy being the permeability that allows the flow of 1 cm^3/s of fluid with a pressure gradient of 1 atm/cm. The darcy is not an SI unit. The SI unit for permeability is m^2, where 1 darcy = 10^{-12} m^2.

and, when $J_M = 0$, we get a situation of pure thermal conduction, which is described by Fourier's law, Eq. (19.18), as

$$J_Q = \dot{q}_F = \dot{Q}_F/A = -k_t\left(\frac{dT}{dx}\right)$$

where k_t is the thermal conductivity of the fluid. Setting $J_M = 0$ in Eq. (19.47) and using Eq. (19.18) to solve for L_{QQ} gives

$$L_{QQ} = k_t T^2 + L_{QM}^2/L_{MM}$$

Using the results from Eqs. (19.38) and (19.42) for L_{QM} and L_{MM} in this equation produces

$$L_{QQ} = k_t T^2 + T\mu k_o^2/k_p \tag{19.48}$$

Substituting the formula for L_{QQ}, $L_{QM} = L_{MQ}$, and L_{MM} back into Eqs. (19.34) and (19.35) gives the final formula for the coupled fluxes as

$$J_Q = -\left[k_t + \frac{\mu k_o^2}{Tk_p}\right]\left(\frac{dT}{dx}\right) - k_o\left(\frac{dp}{dx}\right) \tag{19.49}$$

and

$$J_M = -\frac{\rho k_o}{T}\left(\frac{dT}{dx}\right) - \frac{\rho k_p}{\mu}\left(\frac{dp}{dx}\right) \tag{19.50}$$

Substituting these results into the total EPRD formula of Eq. (19.14) gives

$$\sigma_{\text{thermomechanical}} = \left[k_t + \frac{\mu k_o}{Tk_p}\right]\left(\frac{1}{T}\frac{dT}{dx}\right)^2 + \frac{2k_o}{T^2}\left(\frac{dT}{dx}\right)\left(\frac{dp}{dx}\right) + \left(\frac{k_p}{\mu T}\right)\left(\frac{dp}{dx}\right)^2 \tag{19.51}$$

When the system has reached a steady state condition with $J_M = 0$, then Eq. (19.50) shows us that the pressures and temperatures are coupled, so that

$$\left(\frac{dp/dx}{dT/dx}\right)\Bigg|_{J_M=0} = \left(\frac{dp}{dT}\right)\Bigg|_{J_M=0} = -\frac{\mu k_o}{Tk_p} \tag{19.52}$$

Here, dp is the pressure change "induced" by the temperature change dT under the condition of zero mass flow rate. It is called the *thermomolecular pressure difference*.

Similarly, when the system has reached a steady state condition with $J_Q = 0$, Eq. (19.49) gives

$$\left(\frac{dp}{dT}\right)\Bigg|_{J_Q=0} = -\frac{k_t}{k_o} - \frac{\mu k_o}{Tk_p} = -\frac{k_t}{k_o} + \left(\frac{dp}{dT}\right)_{J_M=0}\Bigg| \tag{19.53}$$

It is conceivable that these equations may some day be found to provide a basic understanding of various phenomena of industrial value, such as the vortex tube temperature separation effect discussed earlier.

EXAMPLE 19.4

A membrane with a permeability of 1.00×10^{-6} m^2 separates two chambers filled with carbon dioxide gas. The gas has a temperature of 300. K on one side of the membrane and 305 K on the other side. The osmotic heat conductivity (k_o) of the membrane with CO_2 is 2.00×10^4 m^2/s and the viscosity of the CO_2 is 1.50×10^{-5} kg/(m·s). Determine the steady state thermomolecular pressure difference across the membrane.

Solution

Since this is a steady state problem, we can use the result given in Eq. (19.52),

$$dp = -\frac{\mu k_o}{k_p\left(\frac{dT}{T}\right)}$$

and, assuming μ, k_o, and k_p are all constants over the small temperature range of 300. to 305 K, we can integrate this equation to find

$$p_2 - p_1 = -\left(\frac{\mu k_o}{k_p}\right)\ln\left(\frac{T_2}{T_1}\right) = -\left[\frac{(1.50\times10^{-5}\,\text{kg/(m·s)})(2.00\times10^4\,\text{m}^2/\text{s})}{1.00\times10^{-6}\,\text{m}^2}\right]\ln\left(\frac{305}{300.}\right) = -4960\,\text{N/m}^2$$

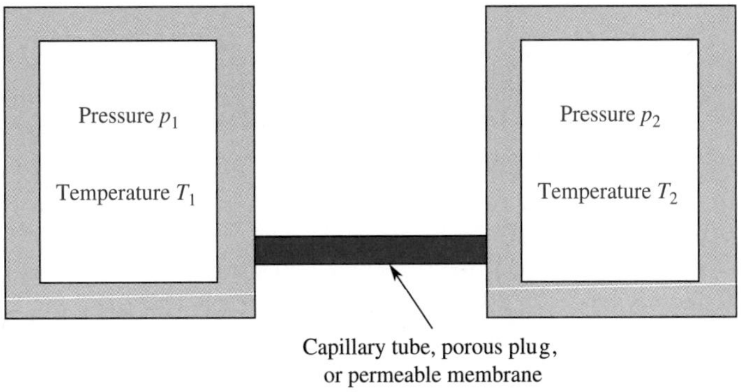

FIGURE 19.9
The thermomechanical effect.

In Example 19.4, the pressure difference and the temperature difference have opposite signs. However, the osmotic heat conductivity (k_o) can be negative for certain substances, and this produces pressure and temperature differences with the same algebraic sign.

Figure 19.9 shows two rigid insulated vessels connected by a capillary tube (or a tube filled with a porous substance, or one having a permeable membrane). If we construct this system so that there is either (1) zero heat transfer between the vessels, (2) zero temperature difference between the vessels, or (3) zero pressure difference between the vessels, then the following phenomena occur.

19.5.1 Zero Heat Transfer

When the vessels are at different pressures and temperatures but no heat transfer occurs between the vessels ($J_Q = 0$), Eq. (19.49) requires that a temperature gradient exists along the connecting tube of the form

$$\frac{dT}{dx} = -\left[\frac{k_o}{k_t + \mu k_o^2/(Tk_p)}\right]\left(\frac{dp}{dx}\right)$$

Then, from Eq. (19.50), the following mass flux occurs through the connecting tube:

$$J_M\big|_{J_Q=0} = \left[\frac{\rho k_o}{Tk_t + \mu k_o^2/k_p} - \frac{\rho k_p}{\mu}\right]\left(\frac{dp}{dx}\right) \tag{19.54}$$

which vanishes only under the improbable circumstance of $Tk_t = 0$.

19.5.2 Zero Temperature Gradient

When both vessels are at different pressures but equal temperatures, $T =$ constant and $dT/dx = 0$. Then, Eqs. (19.49) and (19.50) give

$$J_Q\big|_{T=\text{constant}} = -k_o\left(\frac{dp}{dx}\right)$$

and

$$J_M\big|_{T=\text{constant}} = -\frac{\rho k_p}{\mu}\left(\frac{dp}{dx}\right)$$

So that, using Eq. (19.50), we can write

$$\left(\frac{J_Q}{J_M}\right)\bigg|_{T=\text{constant}} = \frac{\mu k_o}{\rho k_p} = -\left(\frac{T}{\rho}\right)\left(\frac{dp}{dx}\right)\bigg|_{J_M=0}$$

At this point, we define the isothermal energy transport rate $\dot{Q}_i$ that is "induced" by the thermomechanical mass flow rate $\dot{m}$ with[14]

$$\left(\frac{J_Q}{J_M}\right)\bigg|_{T=\text{constant}} = \frac{\dot{Q}_i}{\dot{m}} = \frac{\mu k_o}{\rho k_p} \tag{19.55}$$

---

[14] Notice that we appear to have a heat transfer here that is *not* due to a temperature difference. It is because such thermal energy transports exist that we chose earlier to define heat transfer in general terms as a nonwork, nonmass flow energy transport.

Then, these equations give

$$\left(\frac{dp}{dT}\right)\bigg|_{J_M=0} = -\frac{\rho \dot{Q}_i}{T\dot{m}} = -\frac{\rho \dot{S}_i}{\dot{m}} \tag{19.56}$$

where $\dot{S}_i = \dot{Q}_i/T$ is the isothermal entropy transport rate "induced" by the thermomechanical mass flow rate $\dot{m}$.

19.5.3 Zero Pressure Gradient

When both vessels are at different temperatures but at equal pressures, $p = \text{constant}$ and $dp/dx = 0$. Then, Eqs. (19.49) and (19.50) give

$$J_Q\big|_{p=\text{constant}} = -\left[k_t + \frac{\mu k_o^2}{Tk_p}\right]\left(\frac{dT}{dx}\right)$$

and

$$J_M\big|_{p=\text{constant}} = -\left(\frac{\rho k_o}{T}\right)\left(\frac{dT}{dx}\right)$$

Then,

$$\left(\frac{J_Q}{J_M}\right)\bigg|_{p=\text{constant}} = \frac{Tk_t}{\rho k_o} + \frac{\mu k_o}{\rho k_p} \tag{19.57}$$

This time, we define the isobaric mass flow rate $\dot{m}$ "induced" by the thermomechanical heat transfer rate $\dot{Q}_i$ as

$$\left(\frac{J_Q}{J_M}\right)\bigg|_{p=\text{constant}} = \frac{\dot{Q}}{\dot{m}} = \frac{Tk_t}{\rho k_o} + \frac{\mu k_o}{\rho k_p} \tag{19.58}$$

Combining Eqs. (19.57) and (19.58) gives the isobaric mass flow rate induced by the thermomechanical heat transfer in the thermomechanical system as

$$\dot{m}_i = \frac{\rho \dot{Q}}{Tk_t/k_o + \mu k_o/k_p} \tag{19.59}$$

EXAMPLE 19.5

Both of the large vessels shown in Figure 19.10 are filled with saturated liquid water at 30.0°C. They are maintained isothermal but have a pressure difference of 10.0 kPa. The interconnecting tube has an inside diameter of 0.0100 m and is 0.100 m in length. It is filled with a porous material having a permeability of 1.00×10^{-12} m^2. Careful measurements reveal that the isothermal energy transport rate in this system is 15.0 J/s. Determine

a. The thermomechanical mass flow rate between the vessels.
b. The osmotic heat conductivity coefficient.
c. The isothermal entropy transport rate induced by the thermomechanical mass flow rate.

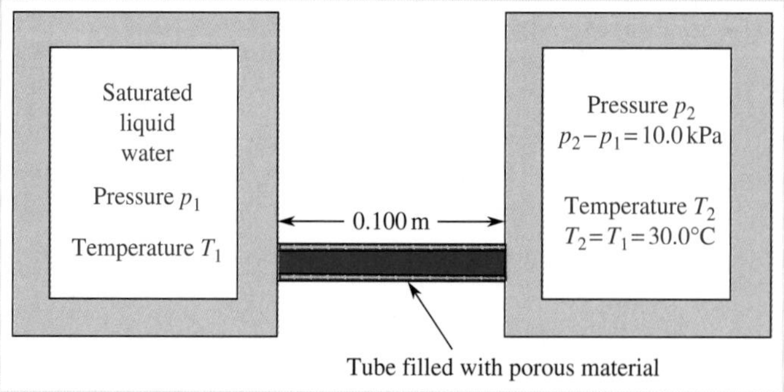

FIGURE 19.10
Example 19.5.

Solution

a. For an isothermal system, $dT = 0$ and Eq. (19.50) gives

$$J_M = \frac{\dot{m}}{A} = -\frac{\rho k_p}{\mu}\left(\frac{dp}{dx}\right)$$

so

$$\dot{m} = -\frac{\rho A k_p}{\mu}\left(\frac{dp}{dx}\right)$$

For saturated liquid water at 30.0°C, $\rho = 996$ kg/m^3 and $\mu = 891 \times 10^{-6}$ kg/(s·m). Then,

$$\dot{m} = \frac{(996\,\text{kg/m}^3)(\pi/4)(0.0100\,\text{m})^2(10^{-12})}{891\times 10^{-6}\,\text{kg/(s·m)}}\left(-\frac{1.00\times 10^4\,\text{N/m}^2}{0.100\,\text{m}}\right) = -8.78\times 10^{-6}\,\text{kg/s}$$

(Note that the mass flow rate is in the same direction as the pressure drop.)

b. We also have

$$J_Q\big|_{T = \text{constant}} = \frac{\dot{Q}_i}{A} = -k_o\left(\frac{dp}{dx}\right)$$

so that

$$k_o = -\frac{\dot{Q}/A}{(dp/dx)} = -\frac{(15.0\,\text{J/s})/[(\pi/4)(0.0100\,\text{m})^2]}{(-1.00\times 10^4\,\text{N/m}^2)/(0.100\,\text{m})} = 1.91\,\text{m}^2/\text{s}$$

c. From Eq. (19.56), we have

$$\dot{S}_i = \frac{\dot{Q}_i}{T} = \frac{15.0\,\text{J/s}}{(273.15 + 30.0\,\text{K})} = 0.0500\,\text{J/(s·K)}$$

EXAMPLE 19.6

If the vessels in Example 19.5 were maintained isobaric at a mean temperature of 30.0°C and the measured thermomechanical heat transfer rate was 8.70 J/s, then find the induced isobaric mass flow rate and the resulting temperature difference between the vessels.

Solution

From Eq. (19.59), we have

$$\dot{m}_i = \frac{\rho \dot{Q}}{T k_t/k_o + \mu k_o/k_p}$$

where the values of μ, k_o, and k_p are the same as in Example 19.5; and for saturated liquid water at 30.0°C, we have $k_t = 0.610$ W/(K·m). Then, Eq. (19.59) gives

$$\dot{m} = \frac{(996\,\text{kg/m}^3)(8.70\,\text{J/s})}{\dfrac{(303\,\text{K})[0.610\,\text{J/(s·K·m)}]}{1.91\,\text{m}^2/\text{s}} + \dfrac{[891\times 10^{-6}\,\text{kg/(s·m)}](1.91\,\text{m}^2/\text{s})}{1.00\times 10^{-12}\,\text{m}^2}} = 5.10\times 10^{-6}\,\text{kg/s}$$

Then,

$$\frac{dT}{dx} = -\frac{T}{\rho k_o} J_M\big|_{p = \text{constant}} = -\frac{T\dot{m}}{\rho k_o} = -\frac{(303\,\text{K})(5.10\times 10^{-6}\,\text{kg/s})}{(996\,\text{kg/m}^3)(1.91\,\text{m}^2/\text{s})} = -8.11\times 10^{-7}\,\text{K/m}$$

and so

$$dT = \Delta T = (-8.11\times 10^{-7}\,\text{K/m})(0.100\,\text{m}) = -8.11\times 10^{-8}\,\text{K}$$

CASE STUDY: ELECTROHYDRODYNAMIC COUPLING

The term *electrohydrodynamics* is used to describe the phenomena associated with the conversion of electrical energy into kinetic energy and vice versa. For example, electrostatic fields can create hydrostatic pressure (or motion) in dielectric fluids, and conversely, a flow of dielectric fluid in an electrostatic field can produce a voltage difference. This is now called this the *viscoelectric* effect.

Recently, various researchers attempted to formulate and solve the combined conservation of mass, momentum, and charge equations for these flows. The drawback to this approach is that it requires the solution to extremely complex partial differential equations. While using analytical or numerical methods may be very effective for simple cases, it is still unrealistic for general flow-induced electrostatic charging in complex geometries.

In 1985, the Renewable Energy Management Laboratory (REMLAB) at the University of Wisconsin–Milwaukee began studying the curious effect of electrostatic generation in flowing dielectric fluids and solid granules. Electrostatic generation has been known since ancient times. It can easily be observed when certain materials (e.g., fur and amber) are rubbed together to produce a noticeable electrostatic charge. This phenomenon, known as the *triboelectric* effect, has a large literature. It has a significant group of industrial applications (painting, printing, air purification, etc.) but it is most commonly encountered today as an electrical hazard causing unwanted shock, electrical interference, and occasional explosions when electrostatic discharges occur. Since the middle of the 18th century, it has been known that electrostatic effects can occur in certain (dielectric) moving fluids as well as solids. Today, this effect creates large-scale electrostatic hazards in the petroleum, shipping, aircraft, and agricultural industries.

The electrohydrodynamic, or viscoelectric, effect is similar to the well-known electrokinetic effect, except that the electric field is perpendicular rather than parallel to the flow. It is also similar to the electrorheological effect, except that a macroscopic conductive second phase is not required. The viscoelectric effect is a transverse field phenomenon similar to the Hall effect.

While hydrocarbons do not normally ionize appreciably, it only takes one singly ionized impurity particle in $2 \leftrightarrow 10^{12}$ molecules to produce large electrostatic charging in a moving dielectric fluid. Such low-impurity trace concentrations are not yet easily detectable, but they can have a devastating results on the low conductivities of these fluids. In the area of hydrocarbons, very large electrostatic charges have been known to develop in fuel-transport vehicles, refueling of aircraft, filling of fuel storage tanks, filtering, and washing of fuel shipping tanker compartments.

The production of excessive electrostatic charging in the petroleum, aircraft, and combustible dust areas causes serious explosion and shock hazards. There are numerous reports of fuel storage tanks exploding while being filled and fuel tankers exploding while being cleaned due to an apparent discharge arc forming between the fuel and an unbonded conductor within the container. Also, a helicopter may carry an electrostatic potential of as much as 100,000 V, and anyone touching it before it has been grounded during a landing operation is seriously injured by the electrical discharge through his or her body.

The most frequent form of electrostatic charging, called *contact charging*, occurs at the molecular level at an interface of dissimilar materials. The development of a large electrostatic potential requires the physical separation of the materials, one of which must be dielectric. Typical examples are a hydrocarbon fluid flowing out of a metal pipe or into a metal vessel, film or paper moving across a conductive web or roller, synthetic fabric rubbing on a human, adhesive tape being applied or removed from a conductor, plastic pellets filling a metal hopper, and so forth. While much less of an electrostatic hazard, "inductive charging" can also occur, as in the electrostatic generator used to test the continuity of the transatlantic cable as it was being laid in the late 19th century.

In recent times, explosive electrostatic conditions have been produced by the flow of dielectric petroleum products during the filling of marine oil tankers and the refueling of aircraft. Dry or wet air is also a good dielectric and, consequently, is the source of numerous motion-generated electrostatic hazards. The generation of atmospheric lightning due to mesoscale circulation of air containing water droplets, snow, sand, and volcanic dust is well known. Moving aircraft and helicopter rotors and spacecraft also produce well-known electrostatic hazards resulting from atmospheric air moving over solid objects. Since 1979, there has been increasing interest in dust explosions (in grain products, plastics, metals, etc.) caused by electrostatic generation and discharge. In addition, the washing of cargo tanks on marine chemical tankers (especially petroleum) with water sprays has been identified as the source of several tanker explosions due to electrostatic generation in the tank by the motion of the water spray. Also, in the late 1970s, detrimental static electrification produced by the flow of oil and other dielectric fluids used for cooling and insulation in transformers began to be a source of concern in the power system equipment industry.

The nonequilibrium thermodynamics theory for fluid electrodynamics is based on a two-flow model: (1) fluid (mass) flow and (2) electron flow. The corresponding linearly coupled nonequilibrium thermodynamics flux equations are:

$$J_{\text{current}} = i/A_i = L_{\phi\phi}\nabla\phi + L_{\phi m}\nabla(p/\rho) \tag{19.60}$$

and

$$J_{\text{mass}} = \dot{m}/A_m = L_{mp}\nabla\phi + L_{mm}\nabla(p/\rho) \tag{19.61}$$

where i is the electrostatic electrical current, $\dot{m}$ is the mass flow rate of the fluid, A_i and A_m are the current and mass flux cross-sectional areas, $\nabla\phi$ is the electrical energy gradient normal to the direction of flow, and $\nabla(p/\rho)$ is the pressure energy gradient in the direction of flow. $L_{\phi\phi}$ and L_{mm} are the primary coefficients obtained from Ohm's and Bernoulli's laws and (assuming reciprocity holds) $L_{\phi m} = L_{m\phi}$ is the secondary, or coupling, coefficient. In the SI system, these quantities have the units shown in Table 19.5.

Lightning as a Renewable Energy Source

The idea of harnessing lightning to supplement our electrical power needs has been considered numerous times in the past. But, knowing when and where lightning will strike, capturing the lightning bolt, and finding the right materials that could withstand and store the sudden surge of electricity are still substantial engineering challenges.

Table 19.5 Case Study SI Units

Quantity	Units
$J_{current}$	A/m^2
J_{mass}	$kg/(s \cdot m^2)$
$\nabla\phi$	$V/m = kg \cdot m/(s^3 \cdot A)$
$\nabla(p/\rho)$	$N/kg = m/s^2$
$L_{\phi\phi}$	$A^2 \cdot s^3/(m^3 \cdot kg)$
L_{mm}	$kg \cdot s/m^3$
$L_{\phi m}$	$A \cdot s^2/m^3$
$L_{m\phi}$	$A \cdot s^2/m^3$

Recent satellite data suggest that there are more than 3 million lightning flashes worldwide per day, or more than 30 flashes per second on average. An average bolt of lightning can carry a current of 300,000 A, transfers a charge of up to 300 coulombs, has a potential difference up to 10 GV (10,000 million volts), and lasts for tens or hundreds of milliseconds. A moderate thunderstorm generates several hundred megawatts of electrical power, which is enough energy to supply the entire United States with electricity for 20 min.

The theoretical basis behind the existance of lightning has been a source of debate for many years. However, since air is a dielectric fluid that contains small charged particles (dust, ice, and liquid water), it seems reasonable that lightning is simply a large-scale manifestation of the viscoelectric effect. The conditions and parameters necessary to understand its formation can be found from the viscoelectric coupling coefficients. The development of strategically placed large lightning capture power plants, which would rapidly convert the electrical energy of lightning into another form of energy, would be a practical method of dealing with the highly transient nature of lightning. For example, a lightning strike could be used to rapidly dissociate water into hydrogen and oxygen. The oxygen could be released into the atmosphere and the hydrogen could be used to supply domestic energy through fuel cells or direct combustion.

Figure 19.11 illustrates where lightning is abundant. The area on earth with the highest lightning activity is located over the Democratic Republic of the Congo in Central Africa. This area has thunderstorms all year round as a result of moisture-laden air masses from the Atlantic Ocean encountering mountains.

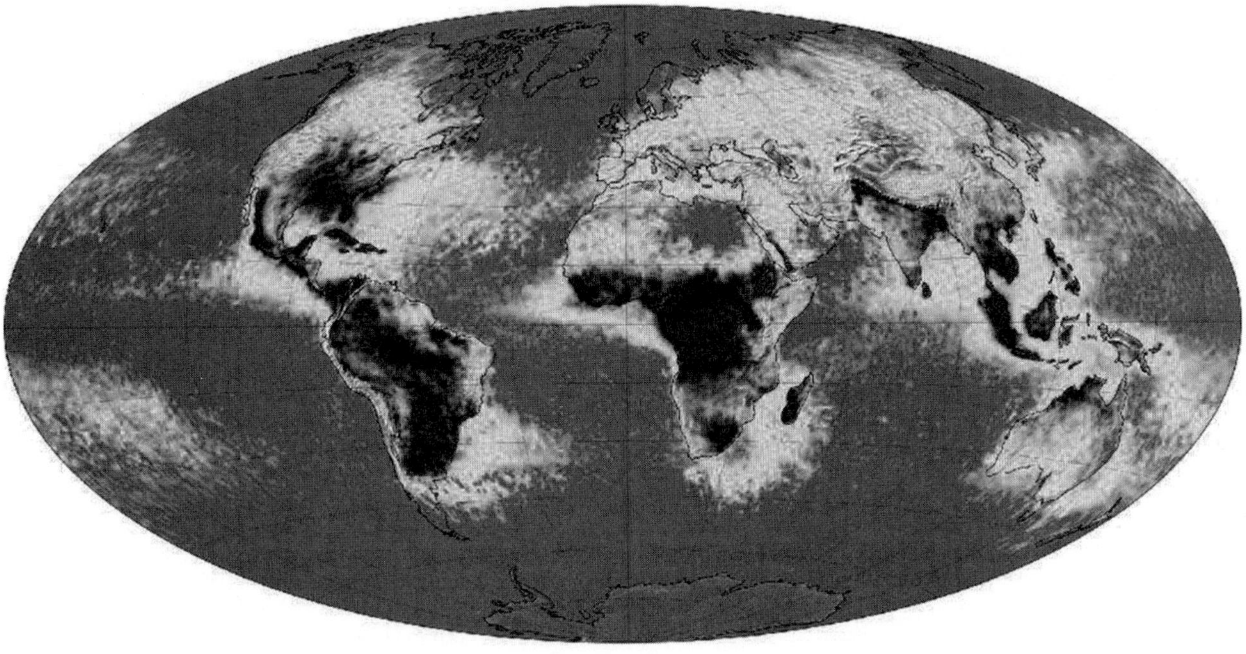

FIGURE 19.11

The average yearly lightning flashes per square kilometer based on data collected by NASA satellites between 1995 and 2002. Places where less than one flash occurred each year are light gray. The places with the largest number of lightning strikes are black. Much more lightning occurs over land than ocean because daily sunshine heats up the land surface faster than the ocean. The map also shows that more lightning occurs near the equator than near the poles.

SUMMARY

In this chapter, we discuss the basic concepts of coupled phenomena. The linear phenomenological equations are shown to accurately model near equilibrium thermoelectric and thermomechanical coupling. Thermoelectric coupling is shown to consist of the Seebeck, Peltier, Kelvin, Fourier, Joule, and Ohm effects. Thermomechanical coupling deals mainly with thermal osmosis and mechanocaloric effects. Equations are developed to describe the entropy production rates of both the thermoelectric and the thermomechanical coupling processes. A modern case study of electrohydrodynamic coupling is discussed and the concept of the viscoelectric effect is introduced to explain a number of known electrostatic phenomena in moving dielectric fluids and solids. Modeling lightning as a viscoelectric effect may lead to its use as an important new renewable energy source.

Problems (* indicates problems in SI units)

1. Equation (19.11) gives the EPRD for a general binary system as $\sigma = L_{11}X_{12} + 2L_{12}X_1X_2 + L_{22}X_{22} > 0$. Show that, if X_1 is held constant, a minimum value in σ occurs when $J_2 = L_{12}X_1 + L_{22}X_2 = 0$. Assume all the L_{ij} are constant here.
2. Equation (19.11) gives the EPRD for a general binary system as $\sigma = L_{11}X_{12} + 2L_{12}X_1X_2 + L_{22}X_{22} > 0$. This quadratic form has to be positive for all positive or negative values of X_1 and X_2. Show that this requires that
   a. $L_{11} > 0$.
   b. $L_{22} > 0$.
   c. $L_{11}L_{22} > 2L_{12}$.
3. Show that the difference in Kelvin coefficients for a thermocouple made of conductors *A* and *B* can be written in terms of the open circuit voltage as $\tau_{AB} = T(d^2\phi_{AB}/dT^2)$.
4. Show that, if the Kelvin thermoelectric effect did not exist (i.e., if τ_{AB} were zero), then the voltage produced by a thermocouple would always depend linearly on temperature.
5.* Table 19.6 gives the temperature-voltage conversion for an iron-constantan thermocouple. Estimate its Seebeck coefficient at 2.00°C and at 200.°C, and compute the percent difference based on the 20.0°C value.

Table 19.6 Problem 5

Temp. (°C)	Seebeck voltage (mV)
18	0.916
19	0.967
20	1.019
21	1.070
22	1.122
–	–
198	10.666
199	10.721
200	10.777
201	10.832
202	10.888

6.* The absolute Seebeck coefficient for a material is given by $\alpha = (300. + 150.\ T - T^2) \times 10^{-6}$ V/K, where *T* is in °C. Determine the formula for the (a) Peltier and (b) Kelvin coefficients.
7.* Using the temperature to Seebeck voltage relationship for the iron-copper thermocouple given in Example 19.2, determine formulae for the Seebeck, Peltier, and difference in Kelvin coefficients and evaluate them at 0.00°C and 200.°C.
8. Seebeck voltage (ϕ) to temperature (*T*) conversion for thermocouples is often written in a power series of the form $T = a_0 + a_1\phi + a_2\phi^2 + a_3\phi^3 + \ldots + a_n\phi^n$. Using this representation, determine the formula for the (a) Seebeck, (b) Peltier, and (c) Kelvin coefficients.
9.* Determine the EPRD in a semiconductor thermoelectric junction that has the following physical properties:
   Electrical resistivity = 1.10×10^{-5} ohm meters
   thermal conductivity = 1.30 W(m · K)
   Seebeck coefficient = $(200. + 10.0\ T - 0.0100\ T^2) \times 10^{-6}$ V/K
   Junction temperature = 400. K
   The temperature gradient at the junction is 1000. K/m and the voltage gradient is 0.0300 V/m.
10.* A thermocouple is connected to a battery. The cold junction is maintained at 0.00°C while the temperature of the hot junction varies. When the hot junction is at 100.°C, the relative Peltier coefficient is measured to be 2.75×10^{-3} W/A, and when it is at 200.°C, it is 4.51×10^{-3} W/A. If the thermocouple potential is given by $\phi = aT + b^2$, where *T* is in K, determine
   a. The constants *a* and *b*.
   b. The value of ϕ when the hot junction is at 100.°C and 200.°C.
11. Show that the entropy production rate per unit volume (σ) is always positive for a thermocouple regardless of the values of α, dT/dx, and $d\phi/dx$.
12. Show that the entropy production rate per unit volume (σ) for a thermocouple must always be greater than $(k_t/T)(dT/dx)^2$.
13.* Suppose the vessels used in Example 19.5 and 19.6 are arranged so that no heat transfer occurs between them along the interconnecting tube. If the pressure difference between the vessels is 10.0 kPa, determine the required temperature difference and mass flow rate in the interconnecting tube.
14.* Determine the numerical values of the thermomechanical coupling coefficients (L_{11}, L_{12}, L_{21}, and L_{22}) for a liquid undergoing very slow flow in a 1.00 mm diameter circular tube at 0.00°C. The fluid data are
   Viscosity = 9.20×10^{-5} kg/(m·s)
   Thermal conductivity = 0.105 W/(m · K)
   Density = 927 kg/m^3
   Osmotic heat conductivity = 3.72×10^{-3} m^2/s.
15.* Determine the numerical values of the thermomechanical coupling coefficients (L_{11}, L_{12}, L_{21}, and L_{22}) for a liquid that has the same physical properties as saturated liquid water. The isothermal heat flux is 2.76 W/m^2 and the isothermal mass flux is 0.0100 kg/(m^2 · s) both at 20.0°C and a pressure gradient of –14.0 MPa/m.
16. Starting with Eq. (19.2), use Eqs. (19.38), (19.41), (19.42), and (19.43) to derive the formula for σ for
   a. Viscous flow through porous media.
   b. Turbulent flow through circular pipes.
   c. Laminar flow through circular pipes.
17. a. Show that, for an ideal gas in a thermomechanical system, Eq. (19.54) can be written as

$$\left(\frac{T}{p}\right)\left(\frac{dp}{dT}\right)_{J_M=0} = \left[\frac{d(\ln(p))}{d(\ln(T))}\right]_{J_M=0} = -\frac{\dot{S}_i}{\dot{m}R}$$

where *R* is the specific gas constant.

b. For a Knudson gas, we know that $\dot{S}_i/\dot{m}R = -1/2$. Integrate the result of part a to show that, for a Knudson gas in a thermomechanical system, $p_2/p_1 = (T_2/T_1)^{1/2}$.

18. Figure 19.12 illustrates the basic operation of an electrohydrodynamic (EHD) generator. This device exploits the coupling of mass flow and electric current such that an electric current is induced (i.e., generated) simply by the mass flow of the liquid as shown. The system contains no moving parts except for the liquid and easily produces a 20.0×10^3 V potential difference. This device was used by William Thomson (Lord Kelvin) to test the continuity of the transatlantic cable as it was being laid at sea in 1858. The mass and electrical current fluxes are related by

$$J_{\text{mass}} = \rho V = -(L_{ME}/T)(d\phi/dx) - (\nu L_{MM}/T)(dp/dx)$$

and

$$J_{\text{current}} = I/A = -(L_{EE}/T)(d\phi/dx) - (\nu L_{EM}/T)(dp/dx)$$

subject to the following special "effects":

$$\text{Ohm's law,}\ (J_{\text{current}})_{p=\text{constant}} = -k_e(d\phi/dx)$$

$$\text{Darcy-Weisbach law,}\ (J_{\text{mass}})_{\phi=\text{constant}} = -(\rho k_p/\mu)(dp/dx) = -(\rho D^2/32\mu)(dp/dx)$$

The mechanoelectric effect, $(d\phi/dp)_{I=0} = k_m/p^{1/2}$, where k_m is the mechanoelectric coefficient.

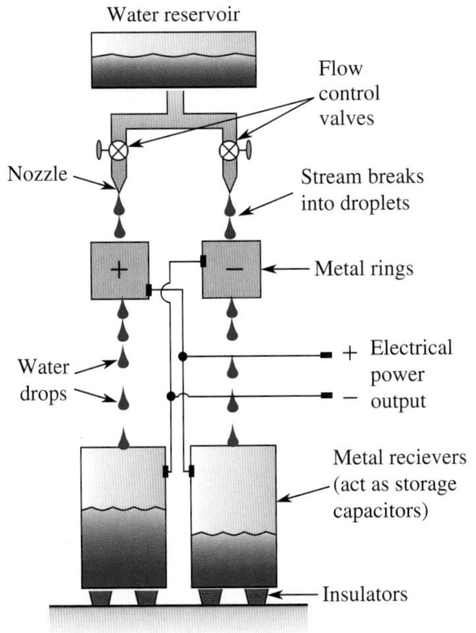

FIGURE 19.12
Problem 18.

a. Find formulae for L_{EE}, L_{MM}, L_{EM}, and L_{ME} in terms of measurable quantities (i.e., T, k_e, D, μ, etc.).

b. Predict the existence of and find a formula for the mass flow induced by the flow of an electric current when there is no pressure drop. This is the reverse of the EHD generator shown previously and is called *EHD pumping*.

c. Find a formula for the short-circuit current induced by the mass flow of the EHD generator. This is called the *streaming current* of the device.

Computer Problems

19.* The Seebeck voltage (ϕ) to relative temperature (T) conversion for an iron-constantan thermocouple with an ice point (0.00°C) reference is given over the range 0.00°C to 760.°C by $T = a_0 + a_1\phi + a_2\phi^2 + a_3\phi^3 + a_4\phi^4 + a_5\phi^5$, where T is in °C and ϕ is the Seebeck voltage in microvolts (i.e., 10^{-6} V). The polynomial coefficients given by the National Bureau of Standards are

$a_0 = -0.048868252$
$a_1 = 19873.14503$
$a_2 = -218614.5353$
$a_3 = 11569199.78$
$a_4 = -264917531.4$
$a_5 = 2018441314.$

a. Write an interactive computer program that asks the user for the thermocouple voltage (in the proper units) and returns the temperature, Seebeck, Peltier, and difference in Kelvin coefficients (in proper units) to the screen.

b. Plot the Seebeck voltage and the Seebeck and Peltier coefficients vs. temperature over the temperature range of 0.00°C to 700.°C using at least 100 points per curve.

20.* The Seebeck voltage (ϕ) to relative temperature (T) conversion for a copper-constantan thermocouple with an ice point (0.00°C) reference is given over the range −160.°C to 400.°C by $T = a_0\ a_1\phi + a_2\phi^2 + a_3\phi^3 + \cdots + a_7\phi^7$, where T is in °C and ϕ is the Seebeck voltage in microvolts (i.e., 10^{-6} V). The polynomial coefficients given by the National Bureau of Standards are

$a_0 = 0.100860910$
$a_1 = 25727.94369$
$a_2 = -767345.8295$
$a_3 = 78025595.81$
$a_4 = -9247486589$
$a_5 = 6.97688 \times 10^{11}$
$a_6 = -2.66192 \times 10^{13}$
$a_7 = 3.94078 \times 10^{14}$

a. Write an interactive computer program that asks the user for the thermocouple voltage (in the proper units) and returns the temperature, Seebeck, Peltier, and difference in Kelvin coefficients (in proper units) to the screen.

b. Plot the Seebeck voltage and the Seebeck and Peltier coefficients vs. temperature over the temperature range of 100.°C to 400.°C using at least 100 points per curve.

Appendix A: Physical Constants and Conversion Factors

PHYSICAL CONSTANTS

Avogadro's number, $N_A = 6.023 \times 10^{26}$ molecules/kgmole
Boltzmann's constant, $k = 1.381 \times 10^{-23}$ J/(molecule·K)
Electron charge, $e = 1.602 \times 10^{-19}$C
Electron mass, $m_e = 9.110 \times 10^{-31}$ kg
Faraday's constant, F = 96,487 kC/kgmole electrons = 96,487 kJ/(V·kgmole electrons)
Gravitational acceleration (standard), $g = 32.174\ \text{ft/s}^2 = 9.807\ \text{m/s}^2$
Gravitational constant, $k_G = 6.67 \times 10^{-11}\text{m}^3/(\text{kg}\cdot\text{s}^2)$
Newton's second law constant, $g_c = 32.174\ \text{lbm}\cdot\text{ft}/(\text{lbf}\cdot\text{s}^2) = 1.0\ \text{kg}\cdot\text{m}/(\text{N}\cdot\text{s}^2)$
Planck's constant, $\hbar = 6.626 \times 10^{-34}$ J·s/molecule
Stefan-Boltzmann constant, $\sigma = 0.1714 \times 10^{-8}\ \text{Btu}/(\text{h}\cdot\text{ft}^2\cdot\text{R}^4) = 5.670 \times 10^{-8}\ \text{W}/(\text{m}^2\cdot\text{k}^4)$
Universal gas constant $\Re$ = 1545.35 ft·lbf/(lbmole·R) = 8314.3 J/(kgmole·K)
= 8.3143 kJ/(kgmole·K) = 1.9858 Btu/(lbmole·R)
= 1.9858 kcal/(kgmole·K) = 1.9858 cal/(gmole·K)
= 0.08314 bar·m^3/(kgmole·K) = 82.05 L·atm/(kgmole·K)
Velocity of light in a vacuum, $c = 9.836 \times 10^8$ ft/s $= 2.998 \times 10^8$ m/s

UNIT DEFINITIONS

1 coulomb (C) = 1 A·s	1 ohm (Ω) = 1 V/A
1 dyne = 1 g·cm/s^2	1 pascal (Pa) = 1 N/m^2
1 erg = 1 dyne·cm	1 poundal = 1 lbm·ft/s^2
1 farad (F) = 1 C/V	1 siemens (S) = 1 A/V
1 henry (H) = 1 Wb/A	1 slug = 1 lbf·s^2/ft
1 hertz (Hz) = 1 cycle/s	1 tesla (T) = 1 Wb/m^2
1 joule (J) = 1 N·m	1 volt (V) = 1 W/A
1 lumen = 1 candela·steradian	1 watt (W) = 1 J/s
1 lux = 1 lumen/m^2	1 weber (Wb) = 1 V·s
1 newton (N) = 1 kg·m/s^2	

CONVERSION FACTORS

Length	Energy
1 m = 3.2808 ft = 39.37 in = 10^2 cm = 10^{10} Å	1 J = 1 N·m = 1 kg·m^2/s^2 = 9.479×10^{-4} Btu
1 cm = 0.0328 ft = 0.394 in = 10^{-2} m = 10^8 Å	1 kJ = 1000 J = 0.9479 Btu = 238.9 cal
1 mm = 10^{-3} m = 10^{-1} cm	1 Btu = 1055.0 J = 1.055 kJ = 778.16 ft·lbf = 252 cal
1 km = 1000 m = 0.6215 miles = 3281 ft	1 cal = 4.186 J = 3.968×10^{-3} Btu
1 in = 2.540 cm = 0.0254 m	1 Cal (in food value) = 1 kcal = 4186 J = 3.968 Btu
1 ft = 12 in = 0.3048 m	1 erg = 1 dyne·cm = 1 g·cm^2/s^2 = 10^{-7} J
1 mile = 5280 ft = 1609.36 m = 1.609 km	1 eV = 1.602×10^{-19} J

(Continued)

CONVERSION FACTORS *(Continued)*

Area

1 m^2 = 10^4 cm^2 = 10.76 ft^2 = 1550 in^2

1 ft^2 = 144 in^2 = 0.0929 m^2 = 929.05 cm^2

1 cm^2 = 10^{-4} m^2 = 1.0764 × 10^{-3} ft^2 = 0.155 in^2

1 in^2 = 6.944 × 10^{-3} ft^2 = 6.4516 × 10^{-4} m^2 = 6.4516 cm^2

Volume

1 m^3 = 35.313 ft^3 = 6.1023 × 10^4 in^3 = 1000 L = 264.171 gal

1 L = 10^{-3} m^3 = 0.0353 ft^3 = 61.03 in^3 = 0.2642 gal

1 gal = 231 in^3 = 0.13368 ft^3 = 3.785 × 10^{-3} m^3

1 ft^3 = 1728 in^3 = 28.3168 L = 0.02832 m^3 = 7.4805 gal

1 in^3 = 16.387 cm^3 = 1.6387 × 10^{-5} m^3 = 4.329 × 10^{-3} gal

Mass

1 kg = 1000 g = 2.2046 lbm = 0.0685 slug

1 lbm = 453.6 g = 0.4536 kg = 3.108 × 10^{-2} slug

1 slug = 32.174 lbm = 1.459 × 10^4 g = 14.594 kg

Force

1 N = 10^5 dyne = 1 kg·m/s^2 = 0.225 lbf

1 lbf = 4.448 N = 32.174 poundals

1 poundal = 0.138 N = 3.108 × 10^{-2} lbf

Power

1 W = 1 J/s = 1 kg·m^2/s^3 = 3.412 Btu/h = 1.3405 × 10^{-3} hp

1 kW = 1000 W = 3412 Btu/h = 737.3 ft·lbf/s = 1.3405 hp

1 Btu/h = 0.293 W = 0.2161 ft·lbf/s = 3.9293 × 10^{-4} hp

1 hp = 550 ft·lbf/s = 33000 ft·lbf/min = 2545 Btu/h = 746 W

Pressure

1 Pa = 1 N/m^2 = 1 kg/(m·s^2) = 1.4504 × 10^{-4} lbf/in^2

1 lbf/in^2 = 6894.76 Pa = 0.068 atm = 2.036 in Hg

1 atm = 14.696 lbf/in^2 = 1.01325 × 10^5 Pa

= 101.325 kPa = 760 mm Hg

1 bar = 10^5 Pa = 0.987 atm = 14.504 lbf/in^2

1 dyne/cm^2 = 0.1 Pa = 10^{-6} bar = 145.04 × 10^{-7} lbf/in^2

1 in Hg = 3376.8 Pa = 0.491 lbf/in^2

1 in H_2O = 248.8 Pa = 0.0361 lbf/in^2

MISCELLANEOUS UNIT CONVERSIONS

Specific Heat Units

1 Btu/(lbm·°F) = 1 Btu/(lbm·R)

1 kJ/(kg·K) = 0.23884 Btu/(lbm·R) = 185.8 ft·lbf/(lbm·R)

1 Btu/(lbm·R) = 778.16 ft·lbf/(lbm·R) = 4.186 kJ/(kg·K)

Energy Density Units

1 kJ/kg = 1000 m^2/s^2 = 0.4299 Btu/lbm

1 Btu/lbm = 2.326 kJ/kg = 2326 m^2/s^2

Energy Flux

1 W/m^2 = 0.317 Btu/(h·ft^2)

1 Btu/(h·ft^2) = 3.154 W/m^2

Heat Transfer Coefficient

1 W/(m^2·K) = 0.1761 Btu/(h·ft^2·R)

1 Btu/(h·ft^2·R) = 5.679 W/(m^2·K)

Thermal Conductivity

1 W/(m·K) = 0.5778 Btu/(h·ft·R)

1 Btu/(h·ft·R) = 1.731 W/(m·K)

Temperature

$T(°F) = \frac{9}{5}T(°C) + 32 = T(R) - 459.67$

$T(°C) = \frac{5}{9}[T(°F) - 32] = T(K) - 273.15$

$T(R) = \frac{9}{5}T(K) = (1.8)T(K) = T(°F) + 459.67$

$T(K) = \frac{5}{9}T(R) = T(R)/1.8 = T(°C) + 273.15$

Density

1 lbm/ft^3 = 16.0187 kg/m^3

1 kg/m^3 = 0.062427 lbm/ft^3 = 10^{-3} g/cm^3

1 g/cm^3 = 1 kg/L = 62.4 lbm/ft^3 = 10^3 kg/m^3

Viscosity

1 Pa·s = 1 N·s/m^2 = 1 kg/(m·s) = 10 poise

1 poise = 1 dyne·s/cm^2 = 1 g/(cm·s) = 0.1 Pa·s

1 poise = 2.09 × 10^{-3} lbf·s/ft^2 = 6.72 × 10^{-2} lbm/(ft·s)

1 centipoise = 0.01 poise = 10^{-3} Pa·s

1 lbf·s/ft^2 = 1 slug/(ft·s) = 47.9 Pa·s = 479 poise

1 stoke = 1 cm^2/s = 10^{-4} m^2/s = 1.076 × 10^{-3} ft^2/s

1 centistoke = 0.01 stoke = 10^{-6} m^2/s = 1.076 × 10^{-5} ft^2/s

1 m^2/s = 10^4 stoke = 10^6 centistoke = 10.76 ft^2/s

Appendix B: Greek and Latin Origins of Engineering Terms

English is a complex combination of numerous languages and is thought by some to be the most colorful and expressive language we have. To understand why English is such a complicated language, you need to be aware of several milestones in the history of the English culture. Around 1000 BC, Celtic-speaking armies from central Europe conquered the British Isles, and after several centuries of occupation, the language of the prehistoric Britains was completely replaced by Celtic. Gaelic, Welsh, Irish, and Scotch are modern remnants of Celtic.

Britain was again conquered in the first century AD by the Roman empire, and for the next several centuries, Latin words began to be absorbed into the British Celtic tongue. Also, some Greek words were introduced during this period by Roman Christian missionaries.

In the fifth century, Britain was conquered by the Teutonic Angles, Saxons, and Jutes from Germanic Europe. Their combined language, called *Anglo-Saxon,* was the basis of modern English (in fact the word *English* is really *Anglish*).

In the eighth century, the Danish invaded parts of Britain and fragments of their language were assimilated into the Celtic-Anglo-Saxon English.

In the 11th century, Britain was again conquered, this time by the French. Norman French then became the official language of the country, but the masses continued to speak English. By 1500, the French had been driven out, and English, now ripe with French words, was reestablished as the official language of the land. It is clear that the English language has had a long, colorful history and the assimilation of numerous words from different languages is one of the things that gives the English language versatility and complexity.

New words normally enter a language not as replacements for existing words but to describe new concepts or ideas. The words most often taken from the occupying forces in Britain were of this type. The new concepts were often, as they are today, from science, technology, or religion. During the Middle Ages, Arabian science and technology were quite advanced. Consequently, technical words, such as algebra, alcohol, and alkali, entered the English language (they can be recognized by the fact that they begin with the Arabic definite article, *al*). The Greek language has had a much smaller impact than Latin on the evolution of English. Greek words have generally entered the English language indirectly, having been absorbed into Latin first, borrowed directly from Greek authors, or through the coining of new scientific or technical terms.

Long ago, it became customary for professionals to carry out their business in a dead language from an earlier culture. Thus, the Romans used Greek, and the English used Latin (with a smattering of Greek). Whether this was to keep the masses from understanding professional dialogue or merely to exercise scholarly activity is not known. Even today, we go to either Latin or Greek to name a new scientific phenomenon or technology. Xerox, for example, is a trade name that comes from the word xerography, which is Greek for "dry" (*xero*) "copying process" (*graphy*).

Many long technical words are compound words formed from the following basic elements: (a) the root word, (b) the combining vowel, and (c) the suffix or prefix. The root word is the core of any technical term. It can be either Latin or Greek, and it is often linked to the prefix or suffix by a combining vowel. For example, *therm* is a root word meaning "heat," *o* is a connecting vowel, and *dynamics* is a suffix meaning "able to produce power." Thus, the term *thermodynamics* loosely translates as "the process of converting heat into power."

A prefix is a syllable or syllables placed in front of a root word to alter its meaning. Table B.1 is a list of common Greek and Latin technical prefixes.

A suffix is a syllable or syllables added to the end of a word to change the meaning of the root word or to form a new word. Table B.2 is a list of common Greek and Latin technical suffixes.

Finally, Table B.3 is a short list of common technical Greek and Latin root words and their normal connecting vowels.

There are also many hybrid technical words, such as *automobile*, which is composed of the Greek word *aut* ("self"), a connecting vowel *o*, and the Latin word *mobilis* ("movable").

Table B.1 Common Greek and Latin Technical Prefixes

Prefix	Meaning	Example
a-	without or not	asymmetric
dia-	through	diameter
hemi-	half	hemisphere
hyper-	above	hypersonic
hypo-	under	hypodermic
infra-	below	infrared
is-	equal or constant	isothermal
para-	beside	paramagnetic
peri-	around	perimeter
semi-	half	semicircle
syn-	with or together	synthetic
trans-	across	transport

Table B.2 Common Greek and Latin Technical Suffixes

Suffix	Meaning	Example
-e	noun-forming, often means *instrument*	metronome
-er	one who	worker
-ic, -al, -ar	pertaining to	electric, electrical
-ist	one who specializes	scientist
-graph	instrument to record	polygraph
-graphy	process of recording	xerography
-gram	record	electrocardiogram
-meter	measuring instrument	psychrometer
-ology	the study of	technology
-ologist	one who studies	technologist

Table B.3 Common Technical Greek and Latin Root Words

Root Word	Meaning	Example
aer/o	air	aeroplane
bar/o	air weight	barometer
electr/o	electricity	electromagnetic
therm/o	heat	thermodynamics
phon/o	sound or voice	phonograph
lith/o	stone	lithographer
flu/+i	flow	fluid, flux
hydr/o	water	hydrodynamic
hygr/o	moist	hygrocyst
psychr/o	cold	psychrometer
cry/o	extreme cold	cryogenic
xer/o	dry	xerography
meteor/o	weather	meteorology

Table B.4 Plural Endings

Singular Ending	Plural Ending	Example
-um	-a	datum, data
-a	-ae	formula, formulae
-on	-a	phenomenon, phenomena
-ix	-ices	matrix, matrices

Table B.5 Numbers

Greek Number	Latin Number	English Meaning	Example
mon/o	uni	one or single	monoplane, unicycle
di	bi	two or double	disulfate, bicycle
tri, tripl/i	tri	three or triple	triplicate, tripod
tetr/a	quadr/a	four or fourfold	tetrahedron, quadrangle
pent/a	quinqu/e	five	pentagon, quinquevalent
hex/a	sex/a	six, sixth	hexadecimal, sexagenarian
hept/a	sept/a	seven, seventh	heptagon, septet
oct/a	oct/a	eight, eighth	octane
enne/a	non/i	nine or ninth	ennead, nonillion
dec/i	dec/i	ten or tenth	decimal
kil/o	mille	one thousand	kilometer
	mill/i	one thousandth	millimeter
poly	mult/i	many	polytropic, multistage
hemi	semi	half	hemisphere, semicircle

In English, the plural of a word is usually formed by adding *s* to the end of the word. However, in many Greek and Latin technical terms, we retain all of their original spelling, including their plural ending. Table B.4 list illustrates the proper plural ending for these words.

The first ending in the list is particularly important because it occurs in so many common technical terms. Other examples are continuum, continua; medium, media; symposium, symposia; colloquium, colloquia; quantum, quanta; and so forth. Thus, in technical writing, we should not refer to a singular *data point,* but rather to a singular *datum point,* or to a set of *data points.*

Occasionally, different Greek and Latin words mean the same thing, and this can be confusing when both are used in the literature. This occurs, for example, in the names of numbers. Table B.5 illustrates this problem.

Thus, we begin to see the structure of thermodynamic terms such as energy (*en* meaning "in" plus *ergon* meaning "work"), adiabatic (*a* meaning "without" plus *dia* meaning "through" plus *bainein* meaning "to go"), aergonic (*a* meaning "without" plus *ergon* meaning "work"), entropy (*en* meaning "in" plus *trope* meaning "turning"), enthalpy (*en* meaning "in" plus *thalpos* meaning "warmth"), isochoric, isothermal, isenthalpic, polytropic, and so forth. If you understand the etymology of a word, it will cease to be a mysterious sound without connotation or meaning.

Index

Page numbers followed by *f* indicates a figure, *t* indicates a table and *n* indicates a footnote.

D

E

F

Y

Z